Bringing Earth to Life

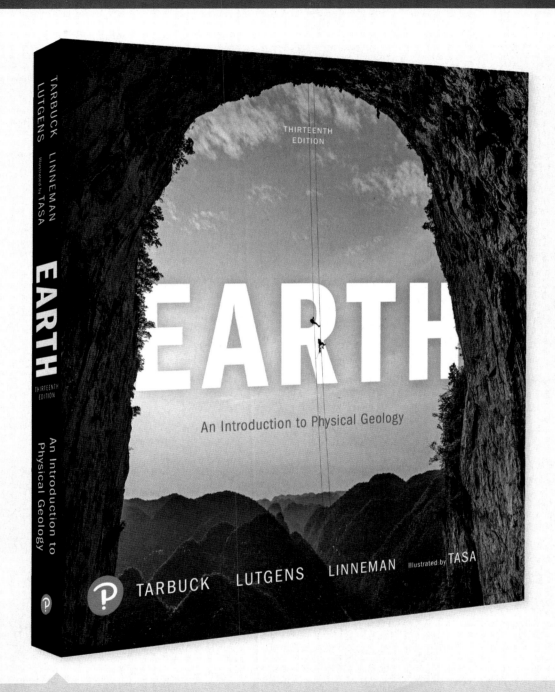

THIRTEENTH EDITION

EARTH

An Introduction to Physical Geology

TARBUCK   LUTGENS   LINNEMAN   Illustrated by TASA

Earth: An Introduction to Physical Geology is a leading text in the field, characterized by no-nonsense, student-friendly writing, excellent illustrations, and a modular learning path driven by learning objectives. The new edition is the first to integrate 3D technology that brings geology to life. This edition has also been updated with significant content and digital updates—read on to see what's new!

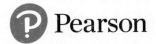
Pearson

Bring Earth to Life with 3D and . . .

NEW! 3D models allow students to get "virtually hands on" with rocks, minerals, and outcrops through guided exploration.

NEW! 3D models are embedded in the eText, available in the Study Area, and assignable with assessment in Mastering Geology.

. . . Other Dynamic Media

SmartFigure Project Condor Videos were created using a quadcopter-mounted GoPro camera. Ten key geological locations and processes were filmed. These process-oriented videos are designed to bring the field to the classroom and improve the learning experience within the text. Available for assignment in Mastering Geology with assessment.

SmartFigure Mobile Field Trips take students to iconic locations with geologist–pilot–photographer Michael Collier in the air and on the ground to learn about iconic landscapes in North America and beyond that relate to discussions in the chapter.

Mobile Field Trips include Formation of a Water Gap, Ice Sculpts Yosemite, Fire and Ice Land, Dendrochronology, and Desert Geomorphology. In Mastering Geology, these videos are accompanied by auto-gradable assessments.

Engage Students with Real-World Connections and Embedded Videos

In The NEWS

Geothermal Energy: Clean Energy to Power the Future?

Iceland, a small island country that straddles the Mid-Atlantic Ridge, is a major producer of *geothermal energy*, where heat captured from Earth, in the form of hot water or steam, is used to produce electricity. The source of geothermal energy is hot igneous rocks below the surface of Earth's crust. Unlike fossil fuel-derived electricity, geothermal power is renewable, and it is less polluting.

Iceland's abundant volcanic fields allow it easy access to geothermal energy. This surplus of relatively cheap electricity has attracted new industries, including aluminum smelting and processing. New geothermal tapping methods may further expand how much geothermal power generation is possible.

A typical geothermal well in Iceland is now about 2.5 kilometers (1.5 miles) deep and can produce enough electricity to supply roughly 4000 homes. A new government-sponsored study is researching whether drilling twice as deep could tap into a water reservoir hotter than 450°C (about 840°F). This super-heated steam might produce 10 times the electricity of today's typical borehole. If ultimately successful in Iceland, this form of geothermal electricity generation may also be implemented on a wider scale throughout other parts of the world, including the South Pacific, eastern Africa, and the western United States.

▶ Bathers at the Blue Lagoon bath near Reykjanes, Iceland, relax in view of thermal wells used to produce geothermal energy. The Iceland Deep Drilling Project (IDDP) will use a large drilling rig to drill several boreholes to depths of about 5 kilometers (3 miles) in a field of volcanic rock located a short drive from Iceland's capital, Reykjavik.

NEW! In the News features begin each chapter illustrating to students the importance and relevance of the chapter's core topics by providing real-world connections.

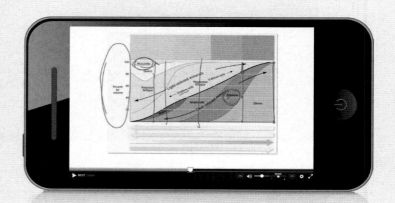

NEW! Quick Response (QR) codes link to SmartFigures giving readers immediate access to over 200 videos and animations, inlcuding *Project Condor* Videos, Mobile Field Trips, Tutorials, Animations, and Videos to help visualize physical processes and concepts. SmartFigures will be embedded directly in the eText.

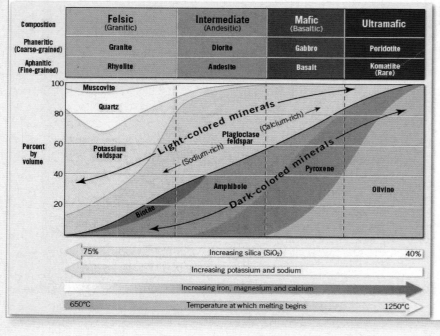

◀ SmartFigure 4.5
Mineralogy of common igneous rocks

Tutorial
https://goo.gl/8Z4g2Y

Composition	Felsic (Granitic)	Intermediate (Andesitic)	Mafic (Basaltic)	Ultramafic
Phaneritic (Coarse-grained)	Granite	Diorite	Gabbro	Peridotite
Aphanitic (Fine-grained)	Rhyolite	Andesite	Basalt	Komatiite (Rare)

75%   Increasing silica (SiO₂)   40%

Increasing potassium and sodium

Increasing iron, magnesium and calcium

650°C   Temperature at which melting begins   1250°C

Clear Learning Path in Each Chapter

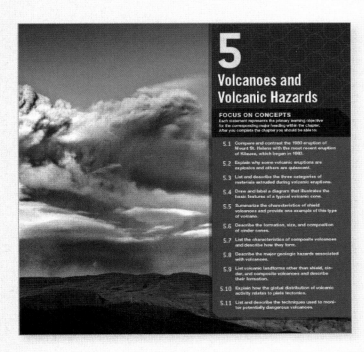

5
Volcanoes and Volcanic Hazards

FOCUS ON CONCEPTS

Each statement represents the primary learning objective for the corresponding major heading within the chapter. After you complete the chapter you should be able to:

5.1 Compare and contrast the 1980 eruption of Mount St. Helens with the most recent eruption of Kilauea, which began in 1983.

5.2 Explain why some volcanic eruptions are explosive and others are quiescent.

5.3 List and describe the three categories of materials extruded during volcanic eruptions.

5.4 Draw and label a diagram that illustrates the basic features of a typical volcanic cone.

5.5 Summarize the characteristics of shield volcanoes and provide one example of this type of volcano.

5.6 Describe the formation, size, and composition of cinder cones.

5.7 List the characteristics of composite volcanoes and describe how they form.

5.8 Describe the major geologic hazards associated with volcanoes.

5.9 List volcanic landforms other than shield, cinder, and composite volcanoes and describe their formation.

5.10 Explain how the global distribution of volcanic activity relates to plate tectonics.

5.11 List and describe the techniques used to monitor potentially dangerous volcanoes.

The chapter-opening **Focus on Concepts** lists the learning objectives for the chapter. Each section of the chapter is tied to a specific learning objective, providing students with a clear learning path to the chapter content.

CONCEPT CHECKS 5.2

1. List these magmas in order, from the highest to lowest silica content: mafic (basaltic) magma, felsic (granitic/rhyolitic) magma, intermediate (andesitic) magma.

2. List the two primary factors that determine the manner in which magma erupts.

3. Define *viscosity*.

4. Are volcanoes fed by highly viscous magma *more* or *less* likely to be a greater threat to life and property than volcanoes supplied with very fluid magma?

Concept Checker
https://goo.gl/NhoRXH

Each chapter section concludes with **Concept Checks**, a feature that lists questions tied to the section's learning objective, allowing students to monitor their grasp of significant facts and ideas.

NEW! Podcast-style Concept Checkers offer an audio review of the Concept Checks to aid in student self-study.

5
CONCEPTS IN REVIEW
Volcanoes and Volcanic Hazards

5.1 Mount St. Helens Versus Kilauea

Compare and contrast the 1980 eruption of Mount St. Helens with the most recent eruption of Kilauea, which began in 1983.

• Volcanic eruptions cover a broad spectrum from explosive eruptions, like that of Mount St. Helens in 1980, to the comparatively quiet eruptions of Kilauea.

silica-rich intermediate (andesitic) and felsic (rhyolitic) magmas, which are the most viscous and contain the greatest quantity of gases, are the most explosive.

Q Although Kilauea mostly erupts in a gentle manner, what risks might you encounter if you chose to live nearby?

5.2 The Nature of Volcanic Eruptions

Explain why some volcanic eruptions are explosive and others are quiescent.

Key Terms:
magma
lava
effusive eruption
viscosity
eruption column

• The two primary factors determining the nature of a volcanic eruption are a magma's viscosity (a fluid's resistance to flow) and its gas content. In general, magmas containing more silica are more viscous. Temperature also influences viscosity; hot lavas are more fluid than relatively cool lavas.

• Mafic (basaltic) magmas, which are fluid and have low gas content, tend to generate effusive (nonexplosive) eruptions. In contrast,

5.3 Materials Extruded During an Eruption

List and describe the three categories of materials extruded during volcanic eruptions.

Key Terms:
aa flow
pahoehoe flow
lava tube
block lava
pillow lava
volatile
pyroclastic material
tephra
scoria
pumice

• Volcanoes erupt lava, gases, and solid pyroclastic materials.

• Low-viscosity basaltic lava can flow great distances. On the surface, they travel as pahoehoe or aa flows. Fluid lavas congeal and harden at the surface, while the lava below the surface continues to flow in tunnels called *lava tubes*. When lava erupts underwater, the outer surface is instantly chilled, while the inside continues to flow, producing pillow lavas.

• The gases most commonly emitted by volcanoes are water vapor and carbon dioxide. Upon reaching the surface, gases rapidly expand, leading to explosive eruptions that can generate a mass of lava fragments called *pyroclastic materials*.

• Pyroclastic materials come in several sizes. From smallest to largest, they are ash, lapilli, and blocks or bombs. Blocks exit the

volcano as solid fragments, whereas bombs are ejected as liquid blobs.

• Bubbles of gas in lava can be preserved as voids in rocks called *vesicles*. Especially frothy, silica-rich lava can cool to form lightweight *pumice*, while basaltic lava with lots of bubbles cools to form *scoria*.

Q This photo shows layers of volcanic material ejected by a violent eruption and deposited roughly horizontally. What term describes this type of volcanic material?

Concepts in Review provides students with a structured review of the chapter. Consistent with the Focus on Concepts and Concept Checks, Concepts in Review is structured around the learning objective for each section.

GIVE IT SOME THOUGHT

1. Examine the accompanying photo and complete the following:

   a. What type of volcano is shown? What features helped you classify it as such?

   b. What is the eruptive style of such volcanoes? Describe the likely composition and viscosity of its magma.

   c. Which type of plate boundary is the likely setting for this volcano?

   d. Name a city that is vulnerable to the effects of a volcano of this type.

2. Answer the following questions about divergent plate boundaries, such as the Mid-Atlantic Ridge, and their associated lavas:

   a. Divergent boundaries are characterized by eruptions of what type of lava: andesitic, basaltic, or rhyolitic?

4. Explain why an eruption of Mount Rainier similar to the 1980 eruption of Mount St. Helens could be considerably more destructive.

5. For each of the volcanoes or volcanic regions listed below, identify whether it is associated with a *convergent* or *divergent* plate boundary or with *intraplate volcanism*.

   a. Crater Lake
   b. Hawaii's Kilauea
   c. Mount St. Helens
   d. East African Rift
   e. Yellowstone
   f. Mount Pelée
   g. Deccan Traps
   h. Fujiyama

6. The accompanying image shows a geologist at the end of an unconsolidated flow consisting of lightweight lava blocks that rapidly descended the flank of Mount St. Helens.

   a. What term best describes this type of flow: an aa flow, a pahoehoe flow, or a pyroclastic flow?

   b. What lightweight (vesicular) igneous rock type is likely the main constituent of this flow?

Give It Some Thought activities challenge students to analyze, synthesize, and apply chapter material.

Mastering Geology Delivers Personalized Learning . . .

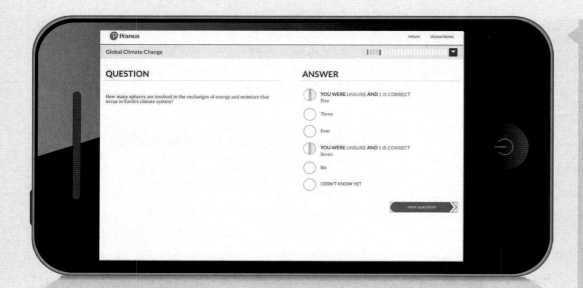

Dynamic Study Modules help students study effectively—and at their own pace. How? By keeping them motivated and engaged. The assignable modules rely on the latest research in cognitive science, using methods—such as adaptivity, gamification, and intermittent rewards—to stimulate learning and improve retention. Each module poses a series of questions about a course topic. These question sets adapt to each student's performance and offer personalized, targeted feedback.

Learning Catalytics™ helps you generate class discussion, customize your lecture, and promote peer-to-peer learning with real-time analytics. As a student response tool, Learning Catalytics uses students' smartphones, tablets, or laptops to engage them in more interactive tasks and thinking.

- **NEW!** Upload a full PowerPoint® deck for easy creation of slide questions.
- Help your students develop critical thinking skills.
- Monitor responses to find out where your students are struggling.
- Rely on real-time data to adjust your teaching strategy.
- Automatically group students for discussion, teamwork, and peer-to-peer learning.

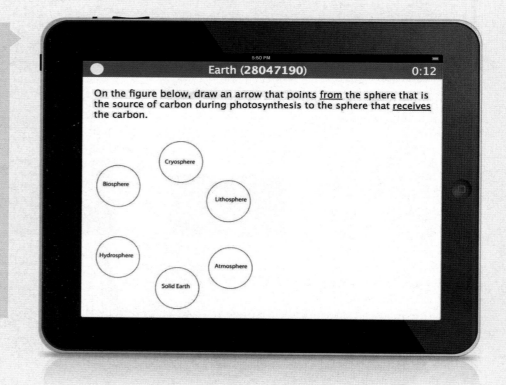

. . . and Interactive Media

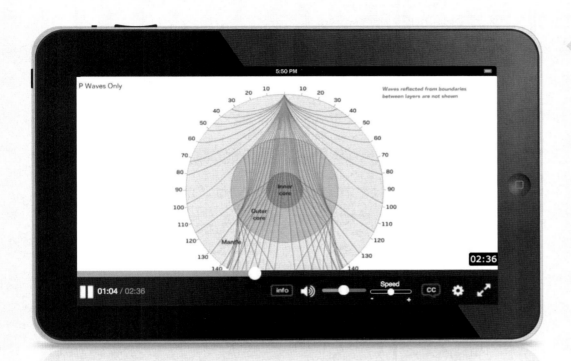

SmartFigure Animations are brief videos created by Dennis Tasa that animate a process or concept depicted in the textbook's figures. With QR codes, students are given a view of moving figures rather than static art to depict how geologic processes move throughout time.

GigaPan Activities allow students to take advantage of a virtual field experience with high-resolution picture technology that has been developed by Carnegie Mellon University in conjunction with NASA.

Dig into Data Analysis

NEW! **Data Analysis end-of-chapter features** send students outside of the book to online science tools and data sets and help develop their data analysis and critical thinking skills.

DATA ANALYSIS

Recent Volcanic Activity

The Smithsonian Institution Global Volcanism Program and the USGS work together to compile a list of new and changing volcanic activity worldwide. NOAA also uses this information to issue Volcanic Ash Advisories to alert aircraft of volcanic ash in the air.

https://goo.gl/eD226M

ACTIVITIES

Go to the Weekly Volcanic Activity Report page at http://volcano.si.edu.

1. What information is displayed on this page?

2. Look at the "New Activity/Unrest" section of the homepage. In what areas is most of the volcanic activity concentrated?

3. Under the "Reports tab," click on the "Weekly Volcanic Activity Report." List three ongoing volcanic activity locations shown on the map.

Click on the name of a volcano under "New Activity Highlights."

4. Where is this volcano located? Be sure to include the city, country, volcanic region name, latitude and longitude, and type of volcano.

5. Do some investigating online and in your text book. What are the key characteristics for this type of volcano?

6. Briefly describe the dates of the most recent activity. How was this activity observed?

Go to the Volcanic Ash Advisory Center (VAAC) page at https://www.ospo.noaa.gov/Products/atmosphere/vaac/.

7. This page shows the Washington VAAC area. Are any active volcanoes listed? Name them.

8. Are any advisories listed for eruptions? If so, list the volcano, the date, and the eruption height.

Give It Some Thought Activities ask students to analyze, synthesize, and think critically about geology to encourage active learning.

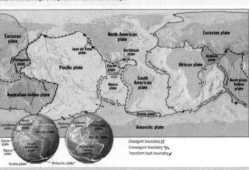

Get Hands-On with Maps in MapMaster 2.0

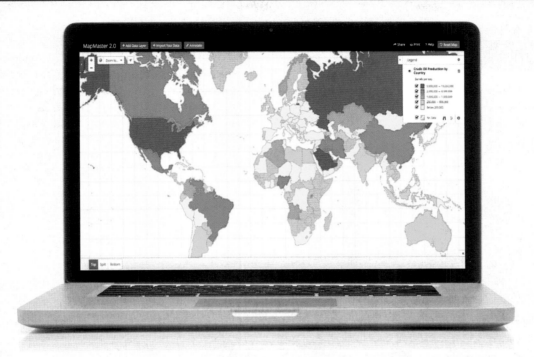

NEW! MapMaster 2.0 Interactive Map Activities Inspired by GIS, these activities allow students to layer various thematic maps to analyze spatial patterns and data at regional and global scales. Now fully mobile, with enhanced analysis tools, the ability for students to geolocate themselves in the data, and the ability for students to upload their own data for advanced map making, this tool includes zoom and annotation functionality with hundreds of map layers leveraging recent data from sources such as the PRB, the World Bank, NOAA, NASA, USGS, United Nations, the CIA, and more.

A Whole New Reading Experience

NEW! Pearson eText is a simple-to-use, mobile-optimized, personalized reading experience available within Mastering. It allows students to easily highlight, take notes, and review key vocabulary all in one place—even when offline. Seamlessly integrated videos, 3D models, and other rich media engage students and give them access to the help they need, when they need it.

"The most interactive, customizable eText I have seen to date."
—Instructor, Ruben Murcia

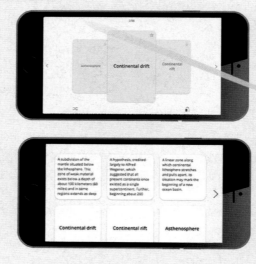

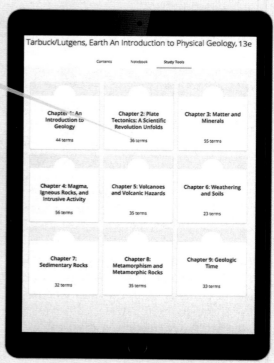

Customizable for Your Course Instructors can add their own notes, embed videos and links, share highlights and notes with students, and rearrange chapters and sections to ensure the Pearson eText fits every unique course.

THIRTEENTH EDITION

EARTH

An Introduction to Physical Geology

EDWARD J. TARBUCK

FREDERICK K. LUTGENS

SCOTT R. LINNEMAN

Illustrated by
DENNIS TASA

 Pearson

Courseware Portfolio Management, Analyst: Cady Owens
Courseware Portfolio Management, Specialist: Christian Botting
Director of Portfolio Management: Beth Wilbur
Content Producer: Heidi Allgair
Managing Producer: Mike Early
Courseware Director, Content Development: Ginnie Simione Jutson
Courseware Sr. Analyst: Erin Strathmann
Courseware Sr. Analyst: Margot Otway
Courseware Editorial Assistant: Sherry Wang
Rich Media Content Producer: Ziki Dekel
Full-Service Vendor: SPi Global
Compositor: SPi Global

Design Manager: Mark Ong
Interior Designer: Jeff Puda
Cover Designer: Jeff Puda
Photo and Illustration Support: International Mapping
Rights & Permissions Project Manager: Matthew Perry/ SPi Global
Rights & Permissions Management: Ben Ferrini
Photo Researcher: Kristin Piljay
Manufacturing Buyer: Stacey Weinberger
Director of Field Marketing: Tim Galligan
Director of Product Marketing: Allison Rona
Executive Field Marketing Manager: Mary Salzman
Product Marketing Manager: Alysun Estes

Cover Photo Credit: Jimmy Chin/Jimmy Chin Productions

1 2019

Library of Congress Cataloging-in-Publication Data

Names: Tarbuck, Edward J., author. | Lutgens, Frederick K., author. | Linneman, R. Scott, author. | Tasa, Dennis, illustrator.
Title: Earth : an introduction to physical geology / Edward J. Tarbuck, Frederick K. Lutgens, Scott R. Linneman ; illustrated by Dennis Tasa.
Description: Thirteenth edition. | Hoboken : Pearson Education, Inc., [2020] | Includes index.
Identifiers: LCCN 2018057366 | ISBN 9780135188316
Subjects: LCSH: Physical geology--Textbooks. | Geology--Textbooks.
Classification: LCC QE28.2 .T37 2020 | DDC 550--dc23 LC record available at https://lccn.loc.gov/2018057366

ISBN 10: 0-135-18831-8; ISBN 13: 978-0-135-18831-6 (Student Edition)
ISBN 10: 0-135-20389-9; ISBN 13: 978-0-135-20389-7 (Looseleaf Edition)

www.pearson.com

TABLE OF CONTENTS

Page 36

Page 70

4
Igneous Rocks and Intrusive Activity 102

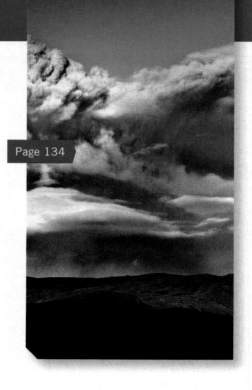

Page 134

5
Volcanoes and Volcanic Hazards 134

Page 202

Page 234

9
Geologic Time 262

10
Crustal Deformation 290

Page 310

11
Earthquakes and
Earthquake Hazards 310

Page 362

Page 472

Page 438

Page 536

Page 592

Page 624

Page 660

Page 690

SmartFigures

Use your mobile device and a free Quick Response (QR) code reader app to scan a SmartFigure identified by a QR code, and a video or animation illustrating the SmartFigure's concept launches immediately. No slow websites or hard-to-remember logins required. These mobile media transform textbooks into convenient digital platforms, breathe life into your learning experience, and help you grasp challenging Physical Geology concepts.

PREFACE

Earth: An Introduction to Physical Geology, 13th edition, is a college-level text that is intended to be a meaningful, nontechnical survey for students taking their first course in geology. In addition to being informative and up-to-date, *Earth* is a readable and user-friendly text that is a valuable tool for learning the basic principles and concepts of geology.

We were fortunate to have Dr. Scott Linneman join the *Earth* team as we prepared the 13th edition. A professor of geology and science education at Western Washington University in Bellingham, Washington, Scott brought an experienced and fresh perspective to this revision. In addition to providing many thoughtful suggestions, he was responsible for revising Chapter 6, *Weathering and Soils*; Chapter 7, *Sedimentary Rocks*; Chapter 9, *Geologic Time*; Chapter 15, *Mass Movement: The Work of Gravity*; and Chapter 16, *Running Water.* Having earned a B.A. from Carleton College in Minnesota and a Ph.D. from the University of Wyoming, Scott's research spans the fields of geomorphology as well as igneous petrology and volcanology. In 2011, he was named the Higher Education Science Teacher of the Year for Washington by the Washington Science Teachers Association, and in 2013 he was chosen the Carnegie Professor of the Year for Washington State. He currently serves as the Director of the Western Washington University Honors Program.

New and Important Features

Extensive revisions and new additions took place in the 13th edition. In addition, integrated textbook and digital resources enhance the learning experience.

- **New:** *In the News* is a chapter-opening current-events story and related two-page photo spread that illustrates real-world connections to the concepts at hand. Topics are fun and engaging; for example, students learn how volcanic ash wreaks havoc on airplanes, why "sand pirates" are stealing whole beaches and islands to manufacture concrete, and that the Sierra Nevada grew taller during a recent California drought.
- **New:** *Data Analysis* end-of-chapter question sets give students practice with data interpretation and problem solving.
- **New: 3D models** are embedded in the eText and can be assigned in Mastering Geology. These activities allow students to get (virtually) hands on with rocks, minerals, and outcrops in guided explorations.
- **New:** *Concept Checkers* are podcast-style audio reviews that can be reached via quick response (QR) codes found in the Concept Check sections that conclude learning objective sections.
- **Significant updating and revision** of book content has occurred in every chapter. In addition to updating data

throughout the chapters, the authors have streamlined discussions for easier reading. New photos of recent geologic events—including mass movements, earthquakes, and volcanic eruptions—keep the discussion fresh. The authors added more than two dozen brand new figures and substantially revised many more. (For a chapter-by-chapter summary of important changes, see the "Major Changes to this Edition" section on page xviii.)

- ***SmartFigures*** enhance the student learning experience. QR codes allow you to use your mobile device to link to 200 unique and innovative learning opportunities beyond the printed text. Each *SmartFigure* also displays a short URL to type in instead of using the QR code. *SmartFigures* are truly media that teach! Types include:
  - ***SmartFigure Tutorials:*** These 2- to 4-minute mini-lessons bring key figure topics to life. The Tutorials are prepared and narrated by Professor Callan Bentley.
  - ***SmartFigure Mobile Field Trips:*** More than two dozen video field trips explore classic geologic sites, from Hawaii to Iceland. Each Mobile Field Trip follows geologist/photographer/pilot Michael Collier into the air and onto the ground to learn about landscapes featured in various chapters. New mobile field trips in this edition include formation of a water gap, how ice sculpts Yosemite, and desert geomorphology.
  - ***SmartFigure Condor:*** Ten Project Condor videos take you to mountain sites in America's West. These videos combine footage taken from a quadcopter-mounted GoPro camera with ground-level views, narration, and animation.
  - ***SmartFigure Animations:*** Certain book topics include animations that use static book art as the starting point to clarify difficult-to-visualize processes.
  - ***SmartFigure Videos:*** Brief video clips cement student understanding of topics such as mineral properties and the structure of ice sheets.
- **An objective-driven learning path** guides students through each chapter. The *Focus on Concepts* section lists all chapter learning objectives at the start of each chapter. The main chapter is organized by learning objective, providing an active, modular path through content. At the end of each chapter, a *Concepts in Review* section concisely summarizes the key content once again. *Give It Some Thought, Eye on Earth,* and *Data Analysis* questions allow students to apply chapter information and stretch their understanding.
- **Unparalleled visual program**, including maps, satellite images, photos, and diagrams that are frequently paired to enhance understanding. The visual program was prepared by the gifted and highly respected geoscience illustrator Dennis Tasa.

Digital and Print Resources

Mastering Geology™ with Pearson eText

Used by more than 3 million science students, the Mastering platform is the most effective and widely used online tutorial, homework, and assessment system for the sciences. Now available with *Earth*, 13th edition, **Mastering Geology™** offers tools for use before, during, and after class:

- **Before class:** Assign adaptive Dynamic Study Modules and reading assignments from the eText with Reading Quizzes to ensure that students come to class prepared, having done the reading.
- **During class:** Learning Catalytics, a "bring your own device" student engagement, assessment, and classroom intelligence system, allows students to use a smartphone, tablet, or laptop to respond to questions in class. With Learning Catalytics, you can assess students in real-time, using open-ended question formats to uncover student misconceptions and adjust lectures accordingly.
- **After class:** Assign an array of assessment resources such as Mobile Field Trips, Project Condor videos, GigaPan activities, Google Earth Encounter activities, geoscience animations, and much more. Students receive wrong-answer feedback personalized to their answers, to help them get back on track.

The Mastering Geology Student Study Area also provides students with self-study materials including videos, geoscience animations, *In the News* articles, Self-Study Quizzes, Web Links, a Glossary, and Flashcards.

Pearson eText gives students access to the text whenever and wherever they can access the Internet. Features of Pearson eText include:

- Now available on smartphones and tablets using the Pearson eText app
- Seamlessly integrated videos and other rich media, including new 3D models
- Fully accessible (screen-reader ready)
- Configurable reading settings, including resizable type and night reading mode
- Instructor and student note-taking, highlighting, bookmarking, and search

For more information or access to Mastering Geology, please visit www.masteringgeology.com.

For Instructors

All of your resources are now easily available in the Mastering Instructor Resource Area.

Instructor's Resource Materials (Download Only)

- Download all the line art, tables, and photos from the text in JPEG files.
- Three PowerPoint files are available for each chapter. Cut down on your preparation time, no matter what your lecture needs, by taking advantage of these components of the PowerPoint files:
  - **Exclusive art:** All the photos, art, and tables from the text have been loaded into PowerPoint slides, in book order.
  - **Lecture outline:** This set averages 50 slides per chapter and includes customizable lecture outlines with supporting art.
  - **Classroom Response System (CRS) questions:** Authored for use in conjunction with classroom response systems, these PowerPoint files allow you to electronically poll your class for responses to questions, pop quizzes, attendance, and more.

Instructor Resource Manual (Download Only)

The *Instructor Resource Manual* has been designed to help seasoned and new instructors alike, offering the following sections in each chapter: an introduction to the chapter, outline, learning objectives/Focus on Concepts; teaching strategies; teacher resources; and answers to *Concept Checks*, *Eye on Earth*, *Give It Some Thought*, and *Data Analysis* questions from the textbook.

TestGen Computerized Test Bank (Download Only)

TestGen is a computerized test generator that lets instructors view and edit Test Bank questions, transfer questions to tests, and print tests in a variety of customized formats. The Test Bank includes more than 2,000 multiple-choice, matching, and essay questions. Questions are correlated to Bloom's Taxonomy, each chapter's learning objectives, the Earth Science Learning Objectives, and the Pearson Science Global Outcomes to help instructors better map the assessments against both broad and specific teaching and learning objectives. The Test Bank is also available in Microsoft Word and can be imported into Blackboard.

For Students

Laboratory Manual in Physical Geology, 11th edition, by the American Geological Institute and the National Association of Geoscience Teachers, edited by Vincent Cronin, illustrated by Dennis G. Tasa (0134986962)
This user-friendly bestselling lab manual examines the basic processes of geology and their applications to everyday life. Featuring contributions from more than 170 highly regarded geologists and geoscience educators, along with an exceptional illustration program by Dennis Tasa, *Laboratory Manual in Physical Geology*, 11th edition, offers an inquiry- and activities-based approach that builds skills and gives students a more complete learning experience in the lab. Pre-lab videos linked from the print labs introduce students to the content, materials, and techniques they will use in each lab. These teaching videos help TAs prepare for lab setup and learn new teaching skills. The lab manual is available in Mastering Geology with Pearson eText, allowing teachers to use activity-based exercises to build students' lab skills.

***Dire Predictions: Understanding Global Climate Change*, 2nd edition, by Michael Mann and Lee R. Kump (0133909778)**

Periodic reports from the Intergovernmental Panel on Climate Change (IPCC) evaluate the risk of climate change brought on by humans. However, the sheer volume of scientific data remains inscrutable to the general public, particularly to those who may still question the validity of climate change. In just over 200 pages, this practical text presents and expands upon the latest climate change data and scientific consensus of the IPCC's *Fifth Assessment Report* in a visually stunning and undeniably powerful way to the lay reader. Scientific findings that provide validity to the implications of climate change are presented in clear-cut graphic elements, striking images, and understandable analogies. The second edition integrates mobile media links to online media. The text is also available in various eText formats, including an eText upgrade option from Mastering Geology courses.

Acknowledgments

Writing and revising a college textbook requires the talents and cooperation of many people. It is truly a team effort, and we authors are fortunate to be part of an extraordinary team at Pearson Education. In addition to being great people to work with, they are all committed to producing the best textbooks possible. Special thanks to our editors, Cady Owens and Christian Botting: We appreciate their enthusiasm, hard work, and quest for excellence. The 13th edition of *Earth* was greatly improved by the talents and input of our senior content analysts, Erin Strathmann and Margot Otway. Many thanks. We also thank our content producer, Heidi Allgair, who did a terrific job of keeping this project on track. It was the job of the production team, led by Heidi, to turn our manuscript into a finished product. The team included copyeditor Kitty Wilson, proofreader Heather Mann, project manager Francesca Monaco, and photo researcher Kristin Piljay. As always, our marketing managers, Mary Salzman and Alysun Estes, who engage with faculty daily, provided us with helpful advice and many valuable ideas. We thank these talented people, all true professionals, with whom we are very fortunate to be associated.

The authors owe special thanks to three people who were very important contributors to this project:

- **Dennis Tasa:** Working with Dennis Tasa, who is responsible for all of the text's outstanding illustrations and several of its animations, is always special for us. He has been part of our team for more than 30 years. We not only value his artistic talents, hard work, patience, and imagination but his friendship as well.

- **Michael Collier:** As you read this text, you will see dozens of extraordinary photographs by Michael Collier. Most are aerial shots taken from his nearly 60-year-old Cessna 180. Michael was also responsible for preparing the remarkable Mobile Field Trips that are scattered throughout the text. Among the many awards he has received is the American Geological Institute Award for Outstanding Contribution to the Public Understanding of Geosciences. We think that Michael's photographs and field trips are the next best thing to being there. We were very fortunate to have had Michael's assistance on *Earth*, 13th edition. Thanks, Michael.

- **Callan Bentley:** Callan is an assistant professor of geology at Northern Virginia Community College in Annandale, where he has been honored many times as an outstanding teacher. He is a frequent contributor to *EARTH* magazine and is author of the popular geology blog *Mountain Beltway*. Callan was responsible for preparing the SmartFigure Tutorials that appear throughout the text. As you take advantage of these outstanding learning aids, you can hear Callen's passion for geology and engaging with students as he explains the concepts illustrated in these features. In addition, Callan was responsible for preparing *Concept Checkers,* the podcast-style audio reviews now available with the Concept Check sections that conclude each learning objective section. We appreciate Callan's contributions to this edition of *Earth*.

Thanks go to Professor Redina Herman at Western Illinois University for her work on the new end-of-chapter *Data Analysis* feature. Thanks also go to our colleagues who prepared in-depth reviews. Their critical comments and thoughtful input helped guide our work and clearly strengthened the text. Special thanks to:

Sulaiman Abushagur, El Paso Community College
Evan Bagley, University of Southern Mississippi
Martin Balinsky, Tallahassee Community College
Holly Brunkal, Western State Colorado University
Alvin Coleman, Cape Fear Community College
Ellen Cowan, Appalachian State University
Can Denizman, Valdosta State University
Gail Holloway, University of Oklahoma
Rebecca Jirón, College of William & Mary
Beth Johnson, University of Wisconsin, Fox Valley
Steve Kadel, Glendale Community College
James Kaste, College of William & Mary
Dan Kelley, Bowling Green State University
Pam Nelson, Glendale Community College
Bill Richards, North Idaho College
Jeffrey Ryan, University of South Florida
Jinny Sisson, University of Houston
Christiane Stidham, Stony Brook University
Donald Thieme, Valdosta State University

Last, but certainly not least, we gratefully acknowledge the support and encouragement of our wives, Joanne Bannon, Nancy Lutgens, and Rebecca Craven. Preparation of this edition of *Earth* would have been far more difficult without their patience and understanding.

Ed Tarbuck
Fred Lutgens
Scott Linneman

Chapter 1: An Introduction to Geology

- Added new *In the News* on why flash floods and mudflows often follow wildfires.
- Added a global climate change head and added updated material to Section 1.1.
- Updated current events references in Section 1.4 so that the California wildfires/mass flows of 2017 are now referenced.
- Reworded solar system formation discussion in Section 1.5.
- Added new Figure 1.3 on world population growth.
- Added new Figure 1.4 on natural hazards, featuring 2016 earthquake in Ecuador.
- Revised Figures 1.7 (geologic time scale) and 1.8 (geologic time) for easier reader comprehension.
- Added new Figure 1.16 on deadly debris flows, featuring Montecito, California, in 2018.
- Added new *Data Analysis* end-of-chapter section on the Swift Creek landslide.

Chapter 2: Plate Tectonics

- Added new *In the News* on the search for plate tectonics and alien life on distant planets.
- Streamlined the introductory section on what drives plate tectonics. Per newer research, this edition also omits mention of the "layer cake model."
- Combined old Sections 2.2 and 2.3 and reworked the section *The Great Debate*.
- Revised Figure 2.4 (fossil evidence of continental drift) to be easier to read and more colorblind accessible.
- Figure 2.10 now combines both spherical and flat views of Earth's lithospheric plates.
- Figures 2.29, 2.30, and 2.31 (all covering magnetic reversals) now have color palettes with more contrasting colors.
- Added new *Data Analysis* end-of-chapter section on tectonic plate movement.

Chapter 3: Matter and Minerals

- Added new *In the News* on the Cave of Crystals in Chihuahua, Mexico.
- Revised Figure 3.17 (hardness scales) to make comparing panels easier.
- Added new *Data Analysis* end-of-chapter section on global mineral resources.

Chapter 4: Igneous Rocks and Intrusive Activity

- Added new *In the News* on research that could greatly leverage how much power can be tapped from geothermal energy.
- Heavily revised Section 4.2 (igneous compositions).
- Revised Figure 4.3 (intrusive and extrusive igneous rocks) so that the magma chamber and eruption column are more realistic.
- Added new Figure 4.11, which shows a photo of an explosive volcanic eruption.
- Revised Figure 4.18 (why the mantle is mainly solid) to include a color key.
- Redrew Figure 4.21 on Bowen's Reaction series to be easier to understand.
- Redrew Figure 4.24 (assimilation of host rock by a magma body) to look more realistic.
- Figure 4.27 (Formation of felsic magma) now includes numbered segments to make it easier to follow.
- Added new *Data Analysis* end-of-chapter section on generating magma from solid rock.

Chapter 5: Volcanoes and Volcanic Hazards

- Added new *In the News* on the perils volcanic eruptions pose to air travel.
- Added new Learning Objective Section 5.1 comparing Mount St. Helens and Kilauea eruptions.
- Redrew Figure 5.11 (anatomy of a volcano) to have a more realistic magma chamber.
- Added information on the 2018 Kilauea eruption.
- Added new Figures 5.33 (InSAR image of ground deformation at Mount Etna), 5.34 (satellite image of an ash plume), and 5.35 (volcano hazard map for Mount Rainier region).
- Added new *Data Analysis* end-of-chapter section on recent volcanic activity.

Chapter 6: Weathering and Soils

- Added new *In the News* on soil erosion as a threat to civilization.
- Added new Figure 6.3 showing a rock crushing plant and quarry.
- Added a soil productivity index rating column to Table 6.2.
- Added new *Data Analysis* end-of-chapter section on soil types.

Chapter 7: Sedimentary Rocks

- Added new *In the News* on how changes in our understanding of when terrestrial plants evolved could change models that explain climate change over time.
- The main terminology in Section 7.2 now refers to "clastic" rather than "detrital" sedimentary rocks.
- Added mention of the simple rock cycle model to Section 7.1.
- Added new separate figures (Figures 7.25 and 7.26) on continental sedimentary environments and marine/transitional sedimentary environments.
- Added new *Data Analysis* end-of-chapter section on sedimentary rocks near you.

Chapter 8: Metamorphism and Metamorphic Rocks

- Added new *In the News* on lapis lazuli mining that funds terrorism groups in Afghanistan.
- Added new *Data Analysis* end-of-chapter section on the formation of metamorphic rocks.

Chapter 9: Geologic Time

- Added new *In the News* on zircons, tiny crystals that give evidence of a lost continent.
- Reordered Sections 9.4 and 9.5 for better concept building and reading flow.
- Figure 9.25 (geologic time scale) has a new color design for easier reading.
- Added new *Data Analysis* end-of-chapter section on fossils and geologic time.

Chapter 10: Crustal Deformation

- Added new *In the News* on the New Madrid seismic zone and its major earthquakes in the past.
- Heavily revised and updated Section 10.1 on how rocks deform.
- Simplified Figure 10.8 (plunging anticline) for easier learning.
- Heavily revised Section 10.4 on mapping geologic structures.
- Added new *Data Analysis* end-of-chapter section on measuring movement of the land.

Chapter 11: Earthquakes and Earthquake Hazards

- Added new *In the News* on earthquake-resistant building and retrofitting.

- Combined old Sections 11.1 and 11.2 into a single Section (11.1).
- Revised Figure 11.7 (fault propagation) to reduce extraneous cognitive load in third panel.
- Added new passage in Section 11.5 on 2015 Nepal earthquake, landslides, and ground subsidence.
- In Section 11.7, included mention of Shanghai Tower and modern skyscraper earthquake-safety building techniques.
- Added new *Data Analysis* end-of-chapter section on earthquakes around the world.

Chapter 12: Earth's Interior

- Added new *In the News* on unusual data sources for studying the historical geomagnetic record.
- Figure 12.1 (Earth's layered structure) has a new layout for easier reading.
- Added new Figure 12.4 on seismic waves providing a way to "see" into Earth.
- Reordered chapter for better concept building and learning progression.
- Added new *Data Analysis* end-of-chapter section on seismic tomography.

Chapter 13: Origin and Evolution of the Ocean Floor

- Added new *In the News* on searching for a missing jet and finding shipwrecks through mapping the uncharted ocean floor.
- In Section 13.1, added information on the latest bathymetric technologies used to create maps.
- In Section 13.2, expanded information on turbidity currents and submarine canyons.
- Added new Figure 13.10 on Challenger Deep.
- Added new *Data Analysis* end-of-chapter section on exploring the ocean surface.

Chapter 14: Mountain Building

- Added new *In the News* on how drought made the Sierra Nevada mountains grow taller.
- Revised Figure 14.12 (collision and accretion of small crustal fragments to a continental margin) so asthenosphere is easier to see.
- Added new *Data Analysis* end-of-chapter section on isostasy at work.

Chapter 15: Mass Movement: The Work of Gravity

- Added new *In the News* on air pollution possibly triggering landslides.
- Changed the term *mass wasting* to *mass movement* throughout the chapter to reflect more current terminology.
- Clarified the terms *slump* and *rockfall* to reflect more current terminology.
- Added new section on detecting, monitoring, and mitigating landslides.
- Added new Figures 15.23 (landslide monitoring) and 15.24 (slide displacement map).
- Added new *Data Analysis* end-of chapter section on landslides in Oregon.

Chapter 16: Running Water

- Added new *In the News* on predicting and monitoring floods with social media.
- Added new SmartFigure about water gaps in Section 16.2.
- Updated examples of floods throughout chapter.
- Completely rewrote content on the hydrologic cycle for Section 16.1.
- Added new *Data Analysis* end-of-chapter section on streamflow rates.

Chapter 17: Groundwater

- Added new *In the News* on a giant sinkhole that swallowed three homes in Florida.
- In Section 17.1, under "Groundwater: A Basic Resource," updated statistics to the most recently available, and added the new section "Trends in Water Use."
- In Section 17.6, updated the text on environmental problems and added a new passage on the impact of prolonged drought and groundwater pumping in California.
- In Section 17.7, added a paragraph and illustration on flowstone and augmented Figure 17.33 with diagrams showing how stalactites and stalagmites grow and may join to form a column.
- In *Give It Some Thought*, removed four questions and added two new ones.
- Added new *Data Analysis* end-of-chapter section on sinkholes in Tennessee.

Chapter 18: Glaciers and Glaciation

- Added new *In the News* on the impending loss of glaciers in Glacier National Park.
- Updated the discussion of ice shelves and the Larsen C shelf in Section 18.1.

- In Section 18.2, revised and improved the description of glacial ice formation.
- Updated the Section 18.2 information on glacier retreat.
- Expanded the Section 18.5 passage on crustal subsidence and rebound to include interactions between crustal movements and sea-level rise.
- Substantially revised or updated Figures 18.3, 18.4, 18.6, 18.8, 18.14, 18.23, and 18.25.
- Added new SmartFigure 18.15 on declining ice mass in Greenland.
- Added new *Data Analysis* end-of-chapter section on glacial flow patterns.

Chapter 19: Deserts and Wind

- Added new *In the News* on the world's sand shortage.
- Revised the Section 19.2 passage on desert varnish to reflect new science, with a new illustration.
- Added new Figure 19.5 on petroglyphs.
- Rewrote Section 19.3 to contrast the desert landforms of the Basin and Range with those of the Colorado Plateau. Three new figures (Figures 19.10, 19.11, and 19.12) now illustrate the Colorado Plateau, buttes and mesas, and Monument Valley.
- Combined old Sections 19.4 and 19.5 into a single Section 19.4, on wind erosion.
- Revised the section "Armoring the Desert Surface" to discuss lag deposits and desert pavement as phenomena with distinct mechanistic explanations.
- Substantively altered or updated Figures 19.2, 19.3, 19.9, 19.13, 19.18, and 19.22.
- Added two new *Give It Some Thought* questions and removed one old question.
- Added new *Data Analysis* end-of-chapter section on the Aral Sea.

Chapter 20: Shorelines

- Added new *In the News* on erosion threatening the Alaska coast now that protective sea ice has melted.
- Reorganized Sections 20.1–20.3. After introducing the shoreline as a dynamic interface, the text now covers the generation and characteristics of ocean waves, then beaches and shoreline processes, and finally the features found on shorelines.
- Added information on lagoon filling to Section 20.3.
- Section 20.5 now covers the 2017 hurricanes Harvey, Maria, and Irma, and the subsection "Heavy Rain and Inland Flooding" has been rewritten.
- Thoroughly updated and revised the section "Monitoring Hurricanes."

- Added four new figures on bluff failure (Figure 20.1), hurricane categories (Figure 20.25), Hurricane Harvey (Figure 20.28), and hurricane track forecasts (Figure 20.31).
- Substantially updated or revised Figures 20.9, 20.18, 20.23, 20.27, and 20.34.
- Added one new *Give It Some Thought* question, deleted two old questions, and modified one question.
- Added new *Data Analysis* end-of-chapter section on tides and sea level rise.

Chapter 21: Global Climate Change

- Added new *In the News* on the effect on roads and buildings when sea ice melts.
- Updated numbers and statistics throughout the chapter to the most recent available.
- Broadly revised and updated Section 21.2, including an improved treatment of oxygen isotope analysis.
- Broadly revised the Section 21.5 passage on volcanic activity and climate change.
- Generally revised the Section 21.6 passage on the role of trace gases.
- Retitled Section 21.7 "Predicting Future Climate Change" and revised to give more thorough discussion on computer models and their use in predicting future changes.
- Provided more detail on Arctic sea ice in Section 21.8.
- Added new figures on ocean floor sediment core studies (Figure 21.5), concentration of oxygen isotopes in seawater by climate (Figure 21.6), sulfur dioxide from Kilauea Volcano (21.18), methane (21.24), projected global temperature changes based on two emission scenarios (Figure 21.30), and a graph on sea level rise from 1993 to 2017 (Figure 21.32).
- Substantively altered Figures 21.4, 21.9, 21.13, 21.17, 21.19, 21.25, 21.26, and 21.34.
- Replaced one *Give It Some Thought* question.
- Added new *Data Analysis* end-of-chapter section on Arctic sea ice.

Chapter 22: Earth's Evolution Through Geologic Time

- Added new *In the News* on Earth's first known animal.
- Expanded Section 22.1 to discuss exoplanets, the Kepler mission, and what is meant by a habitable zone.
- Updated Section 22.2 to reflect additional modes of nucleosynthesis, including neutron-star mergers.

- Updated the Section 22.3 passage on oxygen in the atmosphere.
- Reorganized the Section 22.7 passage on mid-Paleozoic life passage and made it into a separate subsection.
- Expanded the Section 22.7 passage on reptiles to include a discussion of amniotic eggs and their importance from an evolutionary point of view.
- Updated the Section 22.9 information on mammal groups and the demise of dinosaurs and expansion of mammals.
- Added four new figures on the Laramide Rockies (Figure 22.18), placoderms (Figure 22.24), anatomy of a reptile egg (Figure 22.28), and giant sequoia trees (Figure 22.29).
- Substantively altered Figures 22.2, 22.3, 22.12, 22.26, and 22.33.
- Added new *Data Analysis* end-of-chapter section on fossils in your area.

Chapter 23: Energy and Mineral Resources

- Added new *In the News* on sand pirates getting rich by stealing beaches, riverbeds, and islands.
- Added new statistics on coal energy usage and oil sands in Section 23.1.
- Expanded information on solar energy and updated information on geothermal energy.
- Added new *Data Analysis* end-of-chapter section on global mineral resources.

Chapter 24: Touring Our Solar System

- Added new *In the News* on potentially using a giant cavern as a future moonbase.
- Expanded information on the formation of gas giants in Section 24.1.
- Added discussion of issues the nebular theory of solar system formation does not address in Section 24.1.
- Added new Figure 24.10 on the lobate scarps of Mercury.
- Updated and expanded discussion of possible existence of liquid water on Mars in Section 24.3.
- Updated Section 24.4 on Jupiter, including revised Figure 24.18.
- Section 24.5 now mentions *Hayabusa-2*, an unstaffed spacecraft currently studying the asteroid Ryugu.
- Added to Section 24.5 mention of carbonaceous chondrite, a subcategory of stony meteorite.
- Added new *Data Analysis* end-of-chapter section exploring the topography of Venus.

Why Do Flash Floods and Mud Flows Often Follow Wildfires?

In December 2017, wildfires burned large areas in the rugged hills of southern California. The fires were especially fierce because the weather had been unusually dry, and strong, hot seasonal Santa Ana winds fanned the flames.

▲ Raging wildfire in southern California.

Then, in early January 2018, this same area experienced extraordinarily heavy rains. One might think rain would be welcomed by fire-weary locals, but people immediately went on alert because they knew from experience that in California, flash floods and mudflows often follow fires. As much as 10 centimeters (4 inches) of rain fell in 2 days in the areas around Santa Barbara and Montecito, California. With vegetation that normally anchors steep hillsides burned away, the rain-soaked slopes became unstable. This paved the way for large debris flows and flash floods to destroy property and take lives.

As this example illustrates, atmospheric conditions such as drought and the processes that move water from the hydrosphere to the atmosphere and then to the solid Earth can have a profound impact on plants and animals (including humans). As you will learn in this chapter, Earth is a complex system, and geology gives us a way to study how parts of the system influence and interact with each other.

► Wildfires swept through the Montecito, California, area in December 2017, burning away much of the hillside vegetation. When heavy rains fell in January 2018, massive mudflows inundated the town.

1

An Introduction to Geology

FOCUS ON CONCEPTS

Each statement represents the primary learning objective for the corresponding major heading within the chapter. After you complete the chapter, you should be able to:

1.1 Distinguish between physical and historical geology and describe the connections between people and geology.

1.2 Summarize early and modern views on how change occurs on Earth and relate them to the prevailing ideas about the age of Earth.

1.3 Discuss the nature of scientific inquiry, including the construction of hypotheses and the development of theories.

1.4 List and describe Earth's four major spheres. Define *system* and explain why Earth is considered a system.

1.5 Outline the stages in the formation of our solar system.

1.6 Sketch Earth's internal structure and label and describe the main subdivisions.

1.7 Sketch, label, and explain the rock cycle.

1.8 List and describe the major features of the ocean basins and continents.

The spectacular eruption of a volcano, the terror brought by an earthquake, the magnificent scenery of a mountain range, and the destruction created by a landslide or flood are all subjects for a geologist. The study of geology deals with many fascinating and practical questions about our physical environment. What forces produce mountains? When will the next major earthquake occur in California? What was the Ice Age like, and will there be another? How are ore deposits formed? Where should we search for water? Will we find plentiful oil if we drill a well in a particular location? Geologists seek to answer these and many other questions about Earth, its history, and its resources.

1.1 Geology: The Science of Earth

Distinguish between physical and historical geology and describe the connections between people and geology.

The subject of this text is **geology**, from the Greek *geo* (Earth) and *logos* (discourse). Geology is the science that pursues an understanding of planet Earth. Understanding Earth is challenging because our planet is a dynamic body with many interacting parts and a complex history. Throughout its long existence, Earth has been changing. In fact, it is changing as you read this page, and it will continue to do so. Sometimes the changes are rapid and violent, as when landslides or volcanic eruptions occur. Just as often, change takes place so slowly that it goes unnoticed during a lifetime. Scales of size and space also vary greatly among the phenomena that geologists study. Sometimes geologists must focus on phenomena that are microscopic, such as the crystalline structure of minerals, and at other times they must deal with processes that are continental or global in scale, such as the formation of major mountain ranges.

▼ Figure 1.1
Internal and external processes The processes that operate upon and beneath Earth's surface are an important focus of physical geology.

Physical and Historical Geology

Geology is traditionally divided into two broad areas: physical and historical. **Physical geology**, which is the primary focus of this book, examines the materials composing Earth and seeks to understand the many processes that operate beneath and upon its surface (**Figure 1.1**). The aim of **historical geology**, on the other hand, is to understand Earth's origins and its development through time. Thus, it strives to establish an orderly chronological arrangement of the multitude of

A.

External processes erode and sculpt surface features. Alaska's Buckskin Glacier is covered by landslide debris.

B.

Internal processes are those that occur beneath Earth's surface. Sometimes they lead to major features forming at the surface, such as Hawaii's Kilauea Volanco, seen here erupting in June 2018

A.

B.

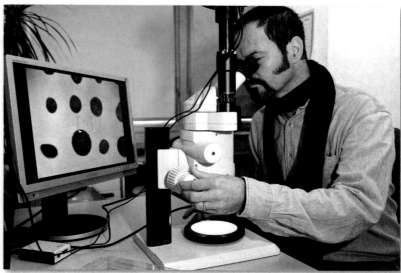

▲ **Figure 1.2**
In the field and in the lab Geology involves not only outdoor fieldwork but work in the laboratory as well. **A.** This geologist from the U.S. Geological Survey's Hawaiian Volcano Observatory is taking measurements near Kilauea Volcano during activity in May 2018. **B.** This researcher is studying micrometeorites, microscopic particles of rocks that fell to Earth from space.

physical and biological changes that have occurred in the geologic past. The study of physical geology logically precedes the study of Earth history because we must first understand how Earth works before we attempt to unravel its past. It should also be pointed out that physical and historical geology are divided into many areas of specialization. Every chapter of this book represents one or more areas of specialization in geology.

Geology is perceived as a science that is done outdoors—and rightly so. A great deal of geology is based on observations, measurements, and experiments conducted in the field. But geologists also work in the laboratory, where, for example, their analysis of minerals and rocks provides insights into many basic processes and the microscopic study of fossils unlocks clues to past

environments (**Figure 1.2**). Geologists must also understand and apply knowledge and principles from physics, chemistry, and biology. Geology is a science that seeks to expand our knowledge of the natural world and our place in it.

Geology, People, and the Environment

Geology is a science that explores many important relationships between people and the natural environment. Complicating all environmental issues is the rapid growth of world population and everyone's aspiration to a better standard of living (**Figure 1.3**).

Natural Hazards Natural hazards are a part of living on Earth. Every day they adversely affect millions of people worldwide and are responsible for staggering

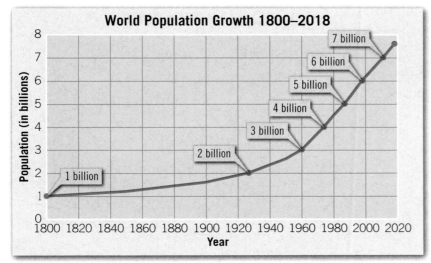

◄ **Figure 1.3**
World population growth Our planet is currently adding an estimated 83 million people per year. The demand for resources is ballooning, and there is growing pressure for people to live in environments facing significant geologic hazards.
(Data from *World Population Prospects, 2017 Revision*, from the United Nations Department of Economics and Social Affairs)

▶ Figure 1.4
Earthquake in Ecuador On April 16, 2016, a magnitude 7.8 earthquake struck coastal Ecuador. It was the strongest quake in that region in 40 years. There were nearly 700 fatalities and more than 7000 people injured. Natural hazards are *natural processes*. They become hazards only when people try to live where the processes occur.

damages. Geologists study volcanoes, floods, tsunamis, earthquakes, and landslides. Of course, these geologic hazards are *natural* processes. They become hazards only when people try to live where these processes occur (**Figure 1.4**).

According to the United Nations, more people live in cities than in rural areas. This global trend toward urbanization concentrates millions of people into megacities, many of which are susceptible to natural hazards. Coastal sites are especially vulnerable because development often destroys natural storm defenses such as wetlands and sand dunes. Some megacities are exposed to seismic (earthquake) and volcanic hazards, and inappropriate land use and poor construction practices, coupled with rapid population growth, increase the risk of death and damage.

Global Climate Change Among the most important environmental issues linked to human activities is global climate change. As you will learn in Chapter 21, climate is naturally variable. However, when recent and future climate changes are considered,

▶ Figure 1.5
Mineral resources represent an important link between people and geology Each year an average American requires huge quantities of Earth materials. According to the USGS, each person in the country uses more than 6 tons of stone, 4.5 tons of sand and gravel, nearly a half ton of cement, almost 400 pounds of salt, 360 pounds of phosphate, and about a half ton of other nonmetals. In addition, per capita consumption of metals such as iron, aluminum, and copper exceeds 700 pounds. This open pit copper mine is in southern Arizona.

natural variability is overshadowed by the influence of humans. Human activities, especially those associated with burning fossil fuels (coal, oil, and natural gas), cause changes in the composition of the atmosphere, which in turn cause global temperatures to rise. Among the many potential impacts of global warming are a rise in sea level, more extreme weather events, and the extinction of many plant and animal species (see Figure 21.26, page 611).

Resources   Resources are an important focus of geology that is of great practical value to people. Resources include water and soil, a great variety of metallic and nonmetallic minerals, and energy (**Figure 1.5**). Together resources form the very foundation of modern civilization. Geology deals not only with how and where these vital resources form but also with maintaining supplies and with the environmental impacts of their extraction and use. Chapter 23 focuses on this aspect of geology.

People Influence Geologic Processes   At the same time that geologic processes have impacts on people, we humans dramatically influence geologic processes as well. For example, landslides and river flooding occur naturally, but the magnitude and frequency of these

events can be affected significantly by human activities such as clearing forests, building cities, and constructing roads and dams. Unfortunately, natural systems do not always adjust to artificial changes in ways that we can anticipate. Thus, an alteration to the environment that was intended to benefit society sometimes has the opposite effect.

At many places throughout this book, you will examine different aspects of our relationship with the physical environment. Nearly every chapter addresses some aspect of natural hazards, resources, and the environmental issues associated with each. Significant parts of some chapters provide the basic geologic knowledge and principles needed to understand environmental problems.

CONCEPT CHECKS 1.1

1. Name and distinguish between the two broad subdivisions of geology.

2. List at least three different geologic hazards.

3. Aside from geologic hazards, describe another important connection between people and geology.

1.2 The Development of Geology

Summarize early and modern views on how change occurs on Earth and relate them to the prevailing ideas about the age of Earth.

The nature of Earth—its materials and processes—has been a focus of study for centuries. Writings about such topics as fossils, gems, earthquakes, and volcanoes date back to the early Greeks, more than 2300 years ago.

Certainly the most influential Greek philosopher was Aristotle. Unfortunately, Aristotle's explanations about the natural world were not based on keen observations and experiments. He arbitrarily stated that rocks were created under the "influence" of the stars and that earthquakes occurred when air crowded into the ground, was heated by central fires, and escaped explosively. When confronted with a fossil fish, he explained that "a great many fishes live in the earth motionless and are found when excavations are made." Although Aristotle's explanations may have been adequate for his day, they continued to be viewed as authoritative for many centuries, thus inhibiting the acceptance of more up-to-date ideas. After the Renaissance of the 1500s, however, more people became interested in finding answers to questions about Earth.

Catastrophism

In the mid-1600s James Ussher, Archbishop of Armagh, Primate of all Ireland, published a major work that had

immediate and profound influences. A respected biblical scholar, Ussher constructed a chronology of human and Earth history in which he calculated that Earth was only a few thousand years old, having been created in 4004 B.C.E. Ussher's treatise was widely accepted by Europe's scientific and religious leaders, and his chronology was soon printed in the margins of the Bible itself.

During the seventeenth and eighteenth centuries, Western thought about Earth's features and processes was strongly influenced by Ussher's calculation. The result was a guiding doctrine called **catastrophism**. Catastrophists believed that Earth's landscapes were shaped primarily by great catastrophes. Features such as mountains and canyons, which today we know form only over great spans of time, were explained as resulting from sudden, often worldwide disasters produced by unknowable causes that no longer operate. This philosophy was an attempt to fit the rates of Earth processes to then-current ideas about the age of Earth.

The Birth of Modern Geology

Against the backdrop of Aristotle's views and a conception of an Earth created in 4004 B.C.E., a Scottish physician and gentleman farmer, James Hutton, published *Theory of the Earth* in 1795. In this work, Hutton put forth a fundamental principle that is a pillar of geology today: **uniformitarianism**. It states that the *physical, chemical, and biological processes that operate today have also operated in the geologic past*. This means that the forces that we observe presently shaping our planet have been at work for a very long time. Thus, to understand ancient rocks, we must first understand present-day processes and their results. This idea is commonly stated as *the present is the key to the past*.

Prior to Hutton's *Theory of the Earth*, no one had effectively demonstrated that many geologic processes must occur over extremely long periods of time. However, Hutton persuasively argued that seemingly small forces can, over long spans of time, produce effects that are just as great as those resulting from sudden catastrophic events. Unlike his predecessors, Hutton carefully cited verifiable observations to support his ideas.

For example, Hutton argued that mountains are sculpted and ultimately destroyed by weathering and the work of running water, and that their waste materials are carried to the oceans by observable processes. He said, "We have a chain of facts which clearly demonstrate . . . that the materials of the wasted mountains have traveled through the rivers"; and further, "there is not one step in all this progress . . . that is not to be actually perceived." He summarized this thought by asking a question and immediately providing the answer: "What more can we require? Nothing but time."

Geology Today

Today the basic tenets of uniformitarianism are just as viable as in Hutton's day. Indeed, we realize more strongly than ever before that the present gives us insight into the past and that the physical, chemical, and biological laws that govern geologic processes remain unchanging through time. However, we also understand that the doctrine should not be taken too literally. To say that geologic processes in the past were the same as those occurring today is not to suggest that they have always had the same relative importance or that they have operated at precisely the same rate. Moreover, some important geologic processes are not currently observable, but there is well-established evidence that they occur. For example, we know that Earth has experienced impacts from large meteorites even though we have no human witnesses to those impacts. Nevertheless, such events have altered Earth's crust, modified its climate, and strongly influenced life on the planet.

Acceptance of uniformitarianism meant the acceptance of a very long history for Earth. Although Earth processes vary in intensity, they still take a very long time to create or destroy major landscape features. The Grand Canyon provides a good example (**Figure 1.6**).

► SmartFigure 1.6
Earth history—Written in the rocks The erosional work of the Colorado River along with other external processes created this natural wonder. For someone studying historical geology, hiking down the South Kaibab Trail in Grand Canyon National Park is a trip through time. The canyon's rock layers hold clues to more than 1.5 billion years of Earth history.

Mobile Field Trip
https://goo.gl/e225m

The rock record contains evidence showing that Earth has experienced many cycles of mountain building and erosion (**Figure 1.7**). Concerning the ever-changing nature of Earth through great expanses of geologic time, Hutton famously stated in 1788, "The results, therefore, of our present enquiry is, that we find no vestige of a beginning—no prospect of an end."

In the chapters that follow, we will examine the materials that compose our planet and the processes that modify it. It is important to remember that, although many features of our physical landscape may seem to be unchanging over the decades we observe them, they are nevertheless changing—but on time scales of hundreds, thousands, or even many millions of years.

The Magnitude of Geologic Time

Among geology's important contributions to human knowledge is the discovery that Earth has a very long and complex history. Although Hutton and others recognized that geologic time is exceedingly long, they had no methods to accurately determine the age of Earth. Early time scales simply placed the events of Earth history in the proper sequence or order, without indicating how long ago in years they occurred.

Today our understanding of radioactivity—and the fact that rocks and minerals contain certain radioactive isotopes having decay rates ranging from decades to billions of years—allows us to accurately determine numerical dates for rocks that represent important events in Earth's distant past (see Figure 1.7). For example, we know that the dinosaurs died out about 65 million years ago. Today geologists put the age of Earth at about 4.6 billion years. Chapter 9 is devoted to a much more complete discussion of geologic time and the geologic time scale.

The concept of geologic time is new to many nongeologists. People are accustomed to dealing with increments of time measured in hours, days, weeks, and years. History books often examine events over spans of centuries, but even a century is difficult to appreciate fully. For most of us, someone or something that is 90 years old is *very old*, and a 1000-year-old artifact is *ancient*.

By contrast, geologists must routinely deal with vast time periods—millions or billions (thousands of millions) of years. When viewed in the context of Earth's 4.6-billion-year history, a geologic event that occurred 50 million years ago may be characterized as "recent" by a geologist, and a rock sample that has been dated at 5 million years may be called "young." An appreciation for the magnitude of geologic time is important in the study of geology because many processes are so gradual that vast spans of time are needed before significant changes occur. How long is 4.6 billion years? If you were to begin counting at the rate of one number per second and continued 24 hours a day, 7 days a week and never stopped,

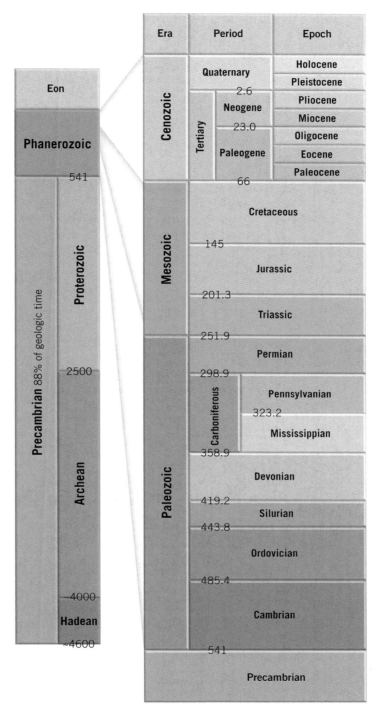

▲ **Figure 1.7**
Geologic time scale: A basic reference The time scale divides the vast 4.6-billion-year history of Earth into eons, eras, periods, and epochs. Numbers on the time scale represent time in millions of years before the present. The Precambrian accounts for more than 88 percent of geologic time. The geologic time scale is a dynamic tool that is periodically updated. Numerical ages appearing on this time scale are those that were currently accepted by the International Commission on Stratigraphy (ICS) in 2018. The ICS is responsible for establishing global standards for the time scale.

◀ **SmartFigure 1.8**
Magnitude of geologic time

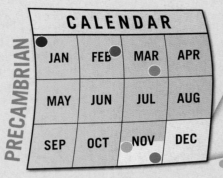

What if we compress the 4.6 billion years of Earth history into a single year?

PRECAMBRIAN

CALENDAR

JAN	FEB	MAR	APR
MAY	JUN	JUL	AUG
SEP	OCT	NOV	DEC

● 6. December 15 to 26
Dinosaurs dominate

DECEMBER

1	2	3	4	5	6	7
8	9	10	11	12	13	14
15	16	17	18	19	20	21
22	23	24	25	26	27	28
29	30	31				

● 1. January 1: Origin of Earth

● 2. February 12: Oldest known rocks

● 3. Late March: Earliest evidence for life (bacteria)

● 4. Mid-November: Beginning of the Phanerozoic eon. Animals having hard parts become abundant

● 5. Late November: Plants and animals move to the land

● 7. December 31: the last day of the year (all times are P.M.)

● 8. Dec. 31 (11:49) Humans (*Homo sapiens*) appear

● 9. Dec. 31 (11:58:45) Ice Age glaciers recede from the Great Lakes

● 10. Dec. 31 (11:59:45 to 11:59:50) Rome rules the Western world

● 11. Dec. 31 (11:59:57) Columbus arrives in the New World

● 12. Dec. 31 (11:59:59.999) Turn of the millennium

it would take about two lifetimes (150 years) to reach 4.6 billion! **Figure 1.8** provides another interesting way of viewing the expanse of geologic time. Although helpful in conveying the magnitude of geologic time, this figure and other analogies, no matter how clever, only begin to help us comprehend the vast expanse of Earth history. Nevertheless, they help us shift from thinking a million years is impossibly long ("never in a million years") to thinking that a million years is a "blink of an eye" in the history of Earth.

CONCEPT CHECKS 1.2

1. Describe Aristotle's influence on geology.

2. Contrast catastrophism and uniformitarianism. How did each view the age of Earth?

3. How old is Earth?

4. Refer to Figure 1.7 and list the eon, era, period, and epoch in which we live.

1.3 The Nature of Scientific Inquiry

Discuss the nature of scientific inquiry, including the construction of hypotheses and the development of theories.

Developing an understanding of how science is done and how scientists work is an important theme in this book. As members of a modern society, we are constantly reminded of the benefits derived from science. But what exactly is the nature of scientific inquiry? Science is a process of producing knowledge. The process depends both on making careful observations and on creating explanations that make sense of the observations. The types of data collected often help to answer well-defined questions about the natural world. In this book you will explore the difficulties in gathering data and some of the ingenious methods that have been developed to overcome these difficulties (**Figure 1.9**). You will also see many examples of how hypotheses are formulated and tested, and you will learn about the evolution and development of some major scientific theories.

All science is based on the assumption that the natural world behaves in a consistent and predictable manner that is comprehensible through careful, systematic study. The overall goal of science is to discover the underlying patterns in nature and then to use the knowledge to make predictions about what should or should not be expected, given certain facts or circumstances. For example, by knowing how oil deposits form, geologists can predict the most favorable sites for exploration and,

perhaps as importantly, avoid regions that have little or no potential.

Hypothesis

A scientific **hypothesis** is a proposed explanation for a certain phenomenon that occurs in the natural world. Before a hypothesis can become an accepted part of scientific knowledge, it must pass objective testing and analysis. Therefore, a hypothesis must be *testable*, and it must be possible to make *predictions* based on the hypothesis being considered. Put another way, hypotheses must fit observations other than those used to formulate them in the first place. Hypotheses that fail rigorous testing are ultimately discarded. The history of science is littered with discarded hypotheses. One of the best known is the Earth-centered model of the universe—a proposal that was supported by the apparent daily motion of the Sun, Moon, and stars around Earth. More detailed astronomical observations disproved this hypothesis.

◀ Figure 1.9
Observation and measurement Scientific data are gathered in many ways. This piece of equipment produces artificial earthquake waves, which are used to probe the underlying rock structures. Vibrations from the 26-ton truck are transmitted to the ground by way of a baseplate (visible between the wheels). The waves reflect from rock layers and structures and are recorded by a network of seismographs.

Theory

When a hypothesis has survived extensive scrutiny and when competing hypotheses have been eliminated, it may be elevated to the status of a scientific **theory**. In everyday language, we may say that something is "only a theory." But among the scientific community, a theory is a well-tested and widely accepted view that the scientific community agrees best explains certain observable facts.

Some theories that are extensively documented and extremely well supported are comprehensive in scope. For example, the theory of plate tectonics provides a framework for understanding the origins of mountains, earthquakes, and volcanic activity. It also explains the evolution of continents and ocean basins through time— ideas that are explored in detail in Chapters 2, 13, and 14. As you will see in Chapter 21, this theory also helps us understand some important aspects of climate change through long spans of geologic time.

Scientific Methods

The process just described, in which scientists gather data through observations and formulate scientific hypotheses and theories, is called the **scientific method**. Contrary to popular belief, the scientific method is not a standard recipe that scientists apply in a routine manner to unravel the secrets of our natural world. Rather, it is an endeavor that involves creativity and insight. Rutherford and Ahlgren put it this way: "Inventing hypotheses or theories to imagine how the world works and then figuring out how they can be put to the test of reality is as creative as writing poetry, composing music, or designing skyscrapers."[*]

There is no fixed path that scientists always follow that leads unerringly to scientific knowledge. However, many scientific investigations involve the steps outlined in

Figure 1.10. In addition, some scientific discoveries result from purely theoretical ideas that stand up to extensive examination. Some researchers use high-speed computers to create models that simulate what is happening in the real world. These models are useful when dealing with natural processes that occur on very long time scales or take place in extreme or inaccessible locations. Still other scientific advancements are made when a totally

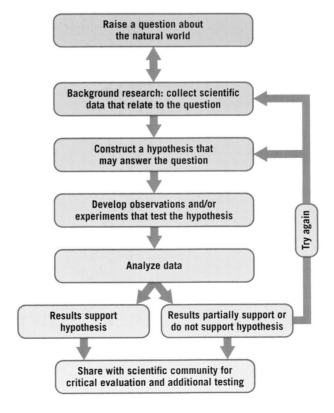

◀ Figure 1.10
Steps frequently followed in scientific investigations The diagram depicts the steps involved in the process many refer to as the *scientific method*.

Raise a question about the natural world

Background research: collect scientific data that relate to the question

Construct a hypothesis that may answer the question

Develop observations and/or experiments that test the hypothesis

Analyze data

Results support hypothesis

Results partially support or do not support hypothesis

Try again

Share with scientific community for critical evaluation and additional testing

---

*F. James Rutherford and Andrew Ahlgren, *Science for All Americans* (New York: Oxford University Press, 1990), p. 7.

unexpected happening occurs during an experiment. These serendipitous discoveries are more than pure luck.

Scientific knowledge is acquired through several avenues, so it might be best to describe the nature of scientific inquiry as the *methods* of science rather than as *the* scientific method. In addition, we should always remember that even the most compelling scientific theories are still simplified explanations of the natural world.

Plate Tectonics and Scientific Inquiry

This book offers many opportunities to develop and reinforce understanding of how science works. In particular, you will learn about data-gathering methods and the observational techniques and reasoning processes used by geologists.

During the past several decades, we have learned a great deal about the workings of our dynamic planet. The revolution in our understanding began in the early part of the twentieth century, with the radical proposal of *continental drift*—the idea that the continents move about the face of the planet. This hypothesis contradicted the established view that the continents and ocean basins are permanent and stationary features on the face of Earth. For that reason, the notion of drifting continents was received with great skepticism and even ridicule. More than 50 years passed before enough data were gathered to transform this controversial hypothesis into a sound theory that wove together the basic processes known to operate on Earth. What finally emerged—called the *theory of plate tectonics*—provided geologists with the first comprehensive model of Earth's internal workings.

In Chapter 2, which covers plate tectonics in detail, you will not only gain insights into the workings of our planet but also see an excellent example of the way geologic "truths" are uncovered and reworked.

CONCEPT CHECKS 1.3

1. How is a scientific hypothesis different from a scientific theory?

2. Summarize the basic steps followed in many scientific investigations.

Concept Checker
https://goo.gl/S2StjP

1.4 Earth as a System

List and describe Earth's four major spheres. Define *system* and explain why Earth is considered a system.

Anyone who studies Earth soon learns that our planet is a dynamic body with many separate but interacting parts, or *spheres*. The hydrosphere, atmosphere, biosphere, and geosphere and all of their components can be studied separately. However, the parts are *not* isolated. Each is related in many ways to the others, producing a complex and continuously interacting whole that we call the *Earth system*.

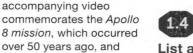
▼ **SmartFigure 1.11**
Two classic views of Earth from space The accompanying video commemorates the *Apollo 8 mission*, which occurred over 50 years ago, and was the first time a spacecraft carrying humans reached the Moon's orbit. The video re-creates the moment when the crew first saw and photographed Earth rising from behind the Moon.

Video
https://goo.gl/uiYNAe

A.
View called "Earthrise" that greeted *Apollo 8* astronauts as their spacecraft emerged from behind the Moon in December 1968. This classic image let people see Earth differently than ever before.

B.
This image taken from *Apollo 17* in December 1972 is perhaps the first to be called "The Blue Marble." The dark blue ocean and swirling cloud patterns remind us of the importance of the oceans and atmosphere.

Earth's Spheres

The images in **Figure 1.11** are important because they let humanity see Earth differently than ever before. These early views from the NASA Moon missions of the late 1960s and early 1970s profoundly altered our conceptualizations of Earth. Seen from space, Earth is breathtaking in its beauty and startling in its solitude. The photos remind us that our home is, after all, a planet—small, self-contained, and in some ways even fragile. Bill Anders, the *Apollo 8* astronaut who took the "Earthrise" photo, expressed it this way: "We came all this way to explore the Moon, and the most important thing is that we discovered the Earth."

The closer view of Earth from space shown in Figure 1.11 helps us appreciate why the physical environment is traditionally divided into three major parts: the water portion of our planet, the *hydrosphere*; Earth's gaseous envelope, the *atmosphere*; and, of course, the solid Earth, or *geosphere*. It needs to be emphasized that our environment is highly integrated and not dominated by rock, water, or air alone. Rather, it is characterized by continuous interactions as air comes in contact with rock, rock with water, and water with air. Moreover, the *biosphere*, which is the totality of all plant and animal life on our planet, interacts with each of the three physical realms and is an equally integral part of the planet.

The interactions among Earth's spheres are incalculable. **Figure 1.12** provides us with one easy-to-visualize example. The shoreline is an obvious meeting place for rock, water, and air, and these spheres in turn support life-forms in and near the water. In this scene,

> *"We came all this way to explore the Moon, and the most important thing is that we discovered the Earth."—Bill Anders, Apollo 8 astronaut*

ocean waves created by the drag of air moving across the water are breaking against the rocky shore. The force of water, in turn, erodes the shoreline.

Hydrosphere   Earth is sometimes called the *blue planet*. Water, more than anything else, makes Earth unique. The **hydrosphere** is a dynamic mass of water that is continually on the move, evaporating from the oceans to the atmosphere, precipitating to the land, and running back to the ocean again. The global ocean is certainly the most prominent feature of the hydrosphere, blanketing nearly 71 percent of Earth's surface to an average depth of about 3800 meters (12,500 feet). It accounts for more than 96 percent of Earth's water (**Figure 1.13**). The hydrosphere also includes the freshwater found underground and in streams, lakes, glaciers, and clouds. Moreover, water is an important component of all living things.

Although freshwater constitutes just a tiny fraction of the total, it is much more important than its meager percentage indicates. Clouds play a vital role in many weather and climate processes, and streams, glaciers, and groundwater are responsible for sculpting and creating many of our planet's varied landforms.

◄ **Figure 1.12**
Interactions among Earth's spheres The shoreline is one obvious interface—a common boundary where different parts of a system interact. In this scene, ocean waves (hydrosphere) that were created by the force of moving air (atmosphere) break against a rocky shore (geosphere). The force of the water can be powerful, and the erosional work that is accomplished can be great.

Hydrosphere

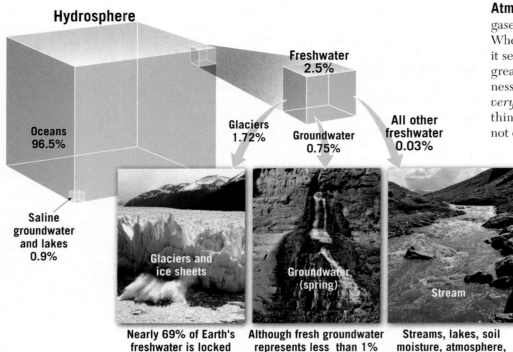

▲ **Figure 1.13**
The water planet Distribution of water in the hydrosphere.

Nearly 69% of Earth's freshwater is locked up in glaciers.

Although fresh groundwater represents less than 1% of the hydrosphere, it accounts for 30% of all freshwater and about 96% of all liquid freshwater.

Streams, lakes, soil moisture, atmosphere, etc. account for 0.03% (3/100 of 1%)

▶ **Figure 1.14**
The atmosphere Earth's shallow envelope of air is an integral part of the planet. Average sea-level pressure is slightly more than 1000 millibars (about 14.7 lb/in²). There is no sharp upper boundary, but the atmosphere thins rapidly as you travel to greater heights.

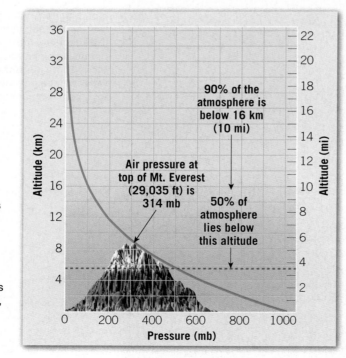

90% of the atmosphere is below 16 km (10 mi)

Air pressure at top of Mt. Everest (29,035 ft) is 314 mb

50% of atmosphere lies below this altitude

Atmosphere Earth is surrounded by a life-giving gaseous envelope called the **atmosphere** (**Figure 1.14**). When we watch a high-flying jet plane cross the sky, it seems that the atmosphere extends upward for a great distance. However, when compared to the thickness (radius) of the solid Earth, the atmosphere is a *very* shallow layer. Despite its modest dimensions, this thin blanket of air is an integral part of the planet. It not only provides the air we breathe but also protects us from the Sun's intense heat and dangerous ultraviolet radiation. The energy exchanges that continually occur between the atmosphere and Earth's surface and between the atmosphere and space produce the effects we call *weather* and *climate*. Climate has a strong influence on the nature and intensity of Earth's external processes. When climate changes, these processes respond.

If, like the Moon, Earth had no atmosphere, our planet would be lifeless, and many of the processes and interactions that make the surface such a dynamic place could not operate. Without weathering and erosion, the face of our planet might more closely resemble the lunar surface, which has not changed appreciably in nearly 3 billion years.

Biosphere The **biosphere** includes all life on Earth (**Figure 1.15**). Ocean life is concentrated in the sunlit surface waters of the sea. Most life on land is also concentrated near the surface, with tree roots and burrowing animals tunneling a few meters underground and flying insects and birds reaching a kilometer or so into the atmosphere. A surprising variety of life-forms are also adapted to extreme environments. For example, on the ocean floor, where pressures are extreme and no light penetrates, there are places where vents spew hot, mineral-rich fluids that support communities of exotic life-forms. On land, some bacteria thrive in rocks as deep as 4 kilometers (2.5 miles) and in boiling hot springs. Moreover, air currents can carry microorganisms many kilometers into the atmosphere. But even when we consider these extremes, life still must be thought of as being confined to a narrow band very near Earth's surface.

Plants and animals depend on the physical environment for the basics of life. However, organisms do not just respond to their physical environment. Through countless interactions, life-forms help maintain and alter the physical environment. Without life, the makeup and nature of the geosphere, hydrosphere, and atmosphere would be very different.

Geosphere Beneath the atmosphere and the oceans is the solid Earth, or **geosphere**. The geosphere extends from the surface to the center of the planet, a depth of nearly 6400 kilometers (nearly 4000 miles), making it by far the largest of Earth's four spheres. Much of our study of the solid Earth focuses on the more accessible surface features. Fortunately, many of these features represent the outward expressions of the dynamic behavior of Earth's interior. Examining the most prominent surface features and their global extent gives us clues to the dynamic processes that have shaped our planet. A first look at the structure of Earth's interior and at the major surface features of the geosphere comes later in the chapter.

Soil, the thin veneer of material at Earth's surface that supports the growth of plants, may be thought of as part of all four spheres. The solid portion is a mixture of weathered rock debris (geosphere) and organic matter from decayed plant and animal life (biosphere). The decomposed and disintegrated rock debris is the product of weathering processes that require air (atmosphere) and water (hydrosphere). Air and water also occupy the open spaces between the solid particles.

◄ **Figure 1.15**
The biosphere The biosphere, one of Earth's four spheres, includes all life.

The ocean contains a significant portion of Earth's biosphere. Modern coral reefs are unique and complex examples and are home to about 25% of all marine species. Because of this diversity, they are sometimes referred to as the ocean equivalent of a rain forest.

Tropical rain forests are characterized by hundreds of different species per square kilometer.

Earth System Science

Scientists have recognized that in order to more fully understand our planet, they must learn how its individual components (land, water, air, and life-forms) are interconnected. This endeavor, called **Earth system science**, aims to study Earth as a *system* composed of numerous interacting parts, or *subsystems*. Rather than look through the limited lens of only one of the traditional sciences—geology, atmospheric science, chemistry, biology, and so on—Earth system science attempts to integrate the knowledge of several academic fields. Using an interdisciplinary approach, those engaged in Earth system science attempt to achieve the level of understanding necessary to comprehend and solve many of our global environmental problems.

A **system** is a group of interacting, or interdependent, parts that form a complex whole. Most of us hear and use the term *system* frequently. We may service our car's cooling *system*, make use of the city's transportation *system*, and participate in our political *system*. A news report might inform us of an approaching

weather *system*. Further, we know that Earth is just a small part of a larger system known as the *solar system*, which in turn is a subsystem of an even larger system, the Milky Way Galaxy. The *In the News* feature on the opening page of the chapter and **Figure 1.16** provide examples of the interactions among different parts of the Earth system.

The Earth System

The Earth system has a nearly endless array of subsystems in which matter is recycled over and over. One familiar loop, or subsystem, is the *hydrologic cycle* (see Figure 16.2, page 441). It represents the unending circulation of Earth's water among the hydrosphere, atmosphere, biosphere, and geosphere. Water enters the atmosphere through evaporation from Earth's surface and transpiration from plants. Water vapor condenses in the atmosphere to form clouds, which in turn produce precipitation that falls back to Earth's surface. Some of the rain that falls onto the land infiltrates (soaks in) to be taken up by plants or become groundwater, and some flows across the surface toward the ocean.

▲ Figure 1.16
Deadly debris flow This image provides an example of interactions among different parts of the Earth system. Extraordinary rains triggered the debris flow (popularly called a mudslide) that buried this house in Montecito, California, in January 2018.

Viewed over long time spans, the rocks of the geosphere are constantly forming, changing, and re-forming. The loop that involves the processes by which one rock changes to another is called the *rock cycle* and is discussed at some length later in the chapter. The cycles of the Earth system are not independent of one another. To the contrary, there are many places where the cycles come in contact and interact.

The Parts Are Linked The parts of the Earth system are linked so that a change in one part can produce changes in any or all of the other parts. For example, when a volcano erupts, lava from Earth's interior may flow out at the surface and block a nearby valley. This new obstruction influences the region's drainage system by creating a lake or causing streams to change course. The large quantities of volcanic ash and gases that can be emitted during an eruption might be blown high into the atmosphere and influence the amount of solar energy that can reach Earth's surface. The result could be a drop in air temperatures over the entire hemisphere.

Where the surface is covered by lava flows or a thick layer of volcanic ash, existing soils are buried. This causes soil-forming processes to begin anew to transform the new surface material into soil (**Figure 1.17**). The soil that eventually forms will reflect the interactions among many parts of the Earth system—the volcanic parent material, the climate, and the impact of biological activity. Of course, there would also be significant changes in the biosphere. Some organisms and their habitats would be eliminated by the lava and ash, whereas new settings for life, such as a lake formed by a lava dam, would be created. The potential climate change could also impact sensitive life-forms.

Time and Space Scales The Earth system is characterized by processes that vary on spatial scales from fractions of millimeters to thousands of kilometers. Time scales for Earth's processes range from seconds to billions of years. As we learn about Earth, it becomes increasingly clear that despite significant separations in distance or time, many processes are connected, and a change in one component can influence the entire system.

Energy for the Earth System The Earth system is powered by energy from two sources. The Sun drives external processes that occur in the atmosphere, in

▲ Figure 1.17
Change is a geologic constant Mount St. Helens erupted in May 1980. **A.** The eruption. **B.** The immediate aftermath. **C.** The recovery that followed.

the hydrosphere, and at Earth's surface. Weather and climate, ocean circulation, and erosional processes are also driven by energy from the Sun. Earth's interior is the second source of energy. Heat remaining from when our planet formed and heat that is continuously generated by radioactive decay power the internal processes that produce volcanoes, earthquakes, and mountains.

People and the Earth System Humans are *part of* the Earth system, a system in which the living and non-living components are entwined and interconnected. Therefore, our actions produce changes in all the other parts. When we burn gasoline and coal, dispose of our wastes, and clear the land, we cause other parts of the system to respond, often in unforeseen ways. Throughout this book you will learn about many of Earth's subsystems, including the hydrologic system, the

tectonic (mountain-building) system, the rock cycle, and the climate system. Remember that these components and we humans are all part of the complex interacting whole we call the Earth system.

CONCEPT CHECKS 1.4

1. List and briefly contrast the four spheres that constitute the Earth system.

2. How much of Earth's surface do oceans cover? What percentage of Earth's water supply do oceans represent?
   Concept Checker
   https://goo.gl/KL6CSh

3. What is a system? List three examples.

4. What are the two sources of energy for the Earth system?

1.5 Origin and Early Evolution of Earth

Outline the stages in the formation of our solar system.

Recent earthquakes caused by displacements of Earth's crust and lavas spewed from active volcanoes are only the latest in a long line of events by which our planet has attained its present form and structure. The geologic processes operating in Earth's interior can be best understood when viewed in the context of much earlier events in Earth history.

Origin of Planet Earth

This section describes the most widely accepted views on the origin of our solar system. The theory described here represents the most consistent set of ideas we have to explain what we know about our solar system today. **GEOgraphics 1.1** provides a useful perspective on size and scale in our solar system. In addition, the origins of Earth and other bodies of our solar system are discussed in more detail in Chapters 22 and 24.

The Universe Begins Our scenario begins about 13.7 billion years ago with the *Big Bang*, an incomprehensibly large explosion that sent all matter of the universe flying outward at incredible speeds. In time, the debris from this explosion, which was almost entirely hydrogen and helium, began to cool and condense into the first stars and galaxies. It was in one of these galaxies, the Milky Way, that our solar system and planet Earth took form.

The Solar System Forms Earth is one of eight planets that, along with several dozen moons and numerous smaller bodies, revolve around the Sun. The orderly nature of our solar system leads researchers to conclude that Earth and the other planets formed at essentially the same time and from the same primordial material as

the Sun. The **nebular theory** proposes that the bodies of our solar system evolved from an enormous rotating cloud called the **solar nebula** (**Figure 1.18**). Besides the hydrogen and helium atoms generated during the Big Bang, the solar nebula consisted of microscopic dust grains and the ejected matter of long-dead stars. (Nuclear fusion in stars converts hydrogen and helium into the other elements found in the universe.)

Nearly 5 billion years ago, something—perhaps a shock wave from an exploding star (*supernova*)—caused this nebula to start collapsing in response to its own gravitation. As it collapsed, it evolved from a huge, vaguely rotating cloud to a much smaller, faster-spinning disk. Gradually, through collisions and other interactions, gases and particles began to orbit in one plane. The disk spun faster as it shrank, and most of the cloud's matter ended up in the center of the disk, where it formed the *protosun* (pre-Sun). Astronomers have observed many such disks around newborn stars in neighboring regions of our galaxy.

The protosun and inner disk were heated by the gravitational energy of infalling matter. In the inner disk, temperatures became high enough to cause the dust grains to evaporate. However, at distances beyond the orbit of Mars, temperatures probably remained quite low. At −200°C (−328°F), the tiny particles in the

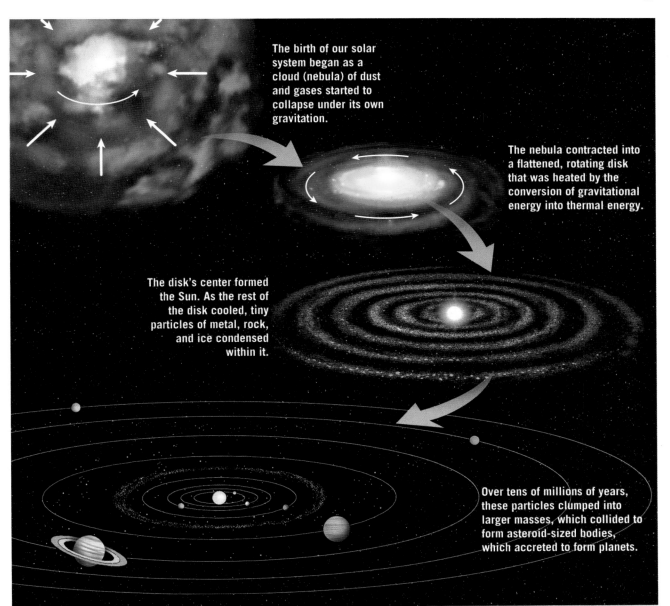

The birth of our solar system began as a cloud (nebula) of dust and gases started to collapse under its own gravitation.

The nebula contracted into a flattened, rotating disk that was heated by the conversion of gravitational energy into thermal energy.

The disk's center formed the Sun. As the rest of the disk cooled, tiny particles of metal, rock, and ice condensed within it.

Over tens of millions of years, these particles clumped into larger masses, which collided to form asteroid-sized bodies, which accreted to form planets.

◄ SmartFigure 1.18
Nebular theory The nebular theory explains the formation of the solar system.

Tutorial
https://goo.gl/DQZRDb

outer portion of the nebula were likely covered with a thick layer of frozen water, carbon dioxide, ammonia, and methane. The disk-shaped cloud also contained appreciable amounts of the lighter gases hydrogen and helium.

The Inner Planets Form The formation of the Sun marked the end of the period of contraction and thus the end of gravitational heating. Temperatures in the region where the inner planets now reside began to decline. This decrease in temperature caused those substances with high melting points to condense into tiny particles that began to coalesce (join together). Materials such as iron and nickel and the elements of which the rock-forming minerals are composed—silicon, calcium, sodium, and so forth—formed metallic and rocky clumps that orbited the Sun (see Figure 1.18). Repeated collisions

caused these masses to coalesce into larger asteroid-size bodies, called *planetesimals*, which in a few tens of millions of years accreted into the four inner planets we call Mercury, Venus, Earth, and Mars (Figure 1.19). Not all of these clumps of matter were incorporated into the planetesimals. The rocky and metallic pieces that remained in orbit are called *meteorites* when they survive an impact with Earth.

As more and more material was swept up by these growing planetary bodies, the high-velocity impact of nebular debris caused their temperatures to rise. Because of their relatively high temperatures and weak gravitational fields, the inner planets were unable to accumulate much of the lighter components of the nebular cloud. The lightest of these, hydrogen and helium, were eventually whisked from the inner solar system by the solar wind.

► **Figure 1.19**
A remnant planetesimal This image of Asteroid 21 Lutetia was obtained by special cameras aboard the *Rosetta* spacecraft on July 10, 2010. Spacecraft instruments showed that Lutetia is a primitive body (planetesimal) left over from when the solar system formed.

surface and solidified to produce a primitive crust. These rocky materials were enriched in oxygen and "oxygen-seeking" elements, particularly silicon and aluminum, along with lesser amounts of calcium, sodium, potassium, iron, and magnesium. In addition, some heavy metals such as gold, lead, and uranium, which have low melting points or were highly soluble in the ascending molten masses, were scavenged from Earth's interior and concentrated in the developing crust. This early period of chemical differentiation established the three basic divisions of Earth's interior: the iron-rich *core*; the thin *primitive crust*; and Earth's largest layer, called the *mantle*, which is located between the core and crust.

An Atmosphere Develops An important consequence of the early period of chemical differentiation is that large quantities of gaseous materials were allowed to escape from Earth's interior, as happens today during volcanic eruptions. By this process, a primitive atmosphere gradually evolved. It is on this planet, with this atmosphere, that life as we know it came into existence.

The Outer Planets Develop At the same time that the inner planets were forming, the larger outer planets (Jupiter, Saturn, Uranus, and Neptune), along with their extensive satellite systems, were also developing. Because of low temperatures far from the Sun, the material from which these planets formed contained a high percentage of ices—frozen water, carbon dioxide, ammonia, and methane—as well as rocky and metallic debris. The accumulation of ices partly accounts for the large size and low density of the outer planets. The two most massive planets, Jupiter and Saturn, had a surface gravity sufficient to attract and hold large quantities of even the lightest elements—hydrogen and helium.

Formation of Earth's Layered Structure

As material accumulated to form Earth (and for a short period afterward), the high-velocity impact of nebular debris and the decay of radioactive elements caused the temperature of our planet to steadily increase. During this time of intense heating, Earth became hot enough that iron and nickel began to melt. Melting produced liquid blobs of dense metal that sank toward the center of the planet. This process occurred rapidly on the scale of geologic time and produced Earth's dense, iron-rich core.

Chemical Differentiation and Earth's Layers

The early period of heating resulted in another process of chemical differentiation, whereby melting formed buoyant masses of molten rock that rose toward the

Continents and Ocean Basins Evolve Following the events that established Earth's basic structure, the primitive crust was lost to erosion and other geologic processes, so we have no direct record of its makeup. When and exactly how the continental crust—and thus Earth's first landmasses—came into existence is a matter of ongoing research. Nevertheless, there is general agreement that the continental crust formed gradually over the past 4 billion years. (The oldest rocks yet discovered are isolated fragments found in the Northwest Territories of Canada that have radiometric dates of about 4 billion years.) In addition, as you will see in subsequent chapters, Earth is an evolving planet whose continents and ocean basins have continually changed shape and even location.

CONCEPT CHECKS 1.5

1. Name and briefly outline the theory that describes the formation of our solar system and then describe the steps in the formation of Earth's layered structure.

2. List the inner planets and outer planets. Describe basic differences in size and composition.  Concept Checker https://goo.gl/jwf1Eb

3. Explain why density and buoyancy were important in the development of Earth's layered structure.

Solar System: Size and Scale

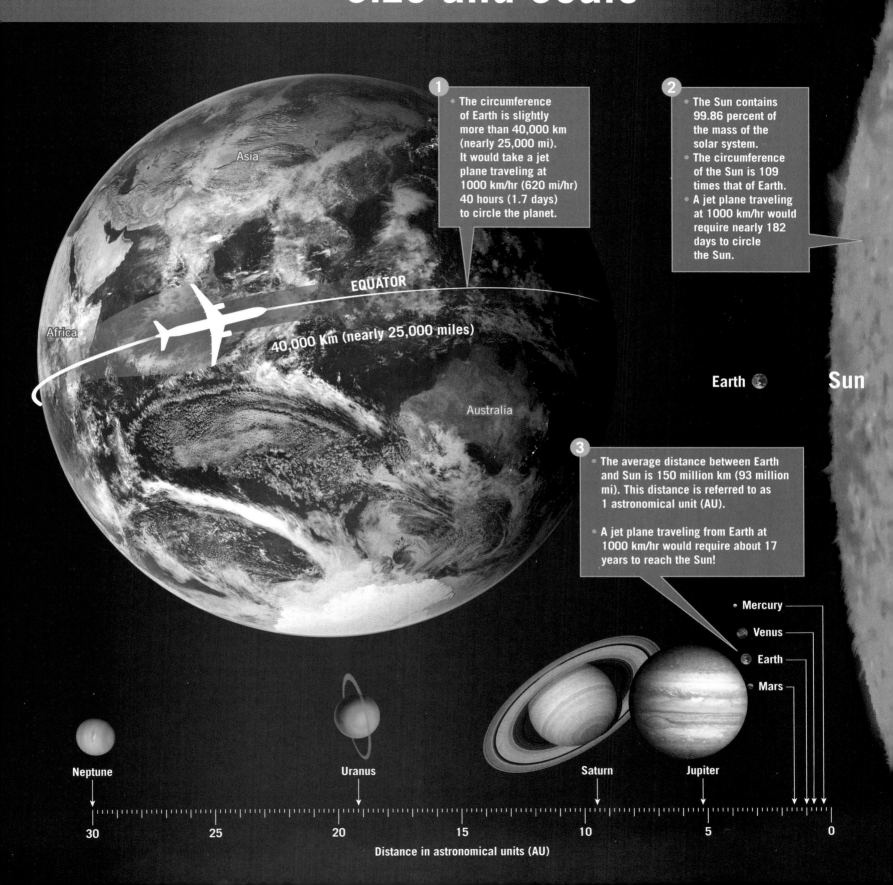

1
- The circumference of Earth is slightly more than 40,000 km (nearly 25,000 mi). It would take a jet plane traveling at 1000 km/hr (620 mi/hr) 40 hours (1.7 days) to circle the planet.

2
- The Sun contains 99.86 percent of the mass of the solar system.
- The circumference of the Sun is 109 times that of Earth.
- A jet plane traveling at 1000 km/hr would require nearly 182 days to circle the Sun.

3
- The average distance between Earth and Sun is 150 million km (93 million mi). This distance is referred to as 1 astronomical unit (AU).
- A jet plane traveling from Earth at 1000 km/hr would require about 17 years to reach the Sun!

Asia

Africa

EQUATOR

40,000 km (nearly 25,000 miles)

Australia

Earth

Sun

Mercury
Venus
Earth
Mars

Neptune

Uranus

Saturn

Jupiter

30   25   20   15   10   5   0

Distance in astronomical ynits (AU)

1.6 Earth's Internal Structure

Sketch Earth's internal structure and label and describe the main subdivisions.

The preceding section outlined how differentiation of material early in Earth's history resulted in the formation of three major layers defined by their chemical composition: the crust, mantle, and core. In addition to these compositionally distinct layers, Earth is divided into layers based on physical properties. The physical properties used to define such zones include whether the layer is solid or liquid and how weak or strong it is. Important examples include the lithosphere, asthenosphere, outer core, and inner core. Knowledge of both chemical and physical layers is important to our understanding of many geologic processes, including volcanism, earthquakes, and mountain building. **Figure 1.20** shows different views of Earth's layered structure.

How did we learn about the composition and structure of Earth's interior? The nature of Earth's interior is primarily determined by analyzing seismic waves from earthquakes. As these waves of energy penetrate the planet, they change speed and are bent and reflected as they move through zones that have different properties. Monitoring stations around the world detect and record this energy. With the aid of computers, these data are analyzed and used to build a detailed picture of Earth's interior. There is much more about this in Chapter 12.

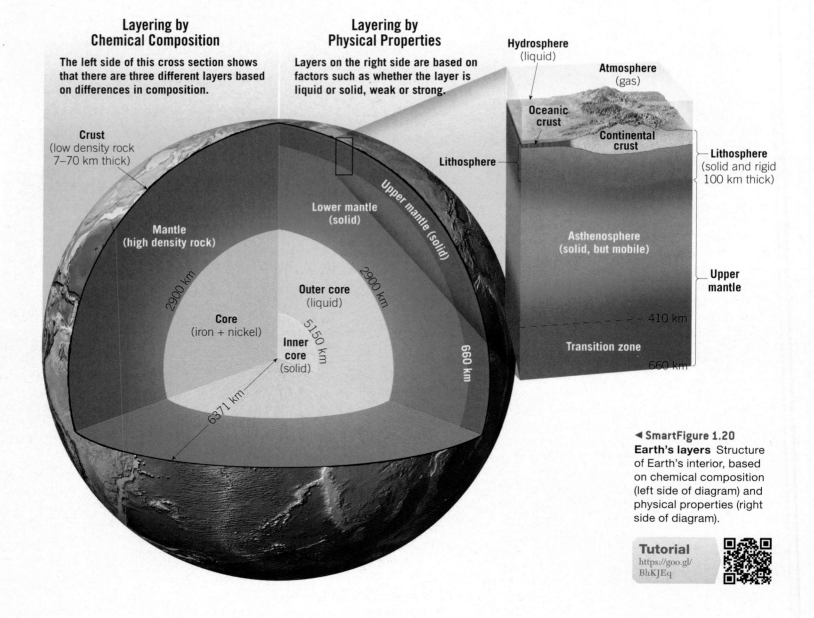

◄ SmartFigure 1.20
Earth's layers Structure of Earth's interior, based on chemical composition (left side of diagram) and physical properties (right side of diagram).

Tutorial
https://goo.gl/BhKJEq

Earth's Crust

The **crust**, Earth's relatively thin, rocky outer skin, consists of continental crust and oceanic crust. The oceanic crust is roughly 7 kilometers (4.5 miles) thick and composed of the dark igneous rock *basalt*. By contrast, the continental crust averages about 35 kilometers (22 miles) thick but may exceed about 70 kilometers (45 miles) in some mountainous regions such as the Rockies and Himalayas. Unlike the oceanic crust, which has a relatively homogeneous chemical composition, the continental crust consists of many rock types. Although the upper crust has an average composition of a *granitic rock* called *granodiorite*, it varies considerably from place to place.

Continental rocks have an average density of about 2.7 g/cm^3, and some have been found to be more than 4 billion years old. The rocks of the oceanic crust are younger (180 million years or less) and denser (about 3.0 g/cm^3) than continental rocks. For comparison, liquid water has a density of 1 g/cm^3; therefore, the density of basalt, the primary rock composing oceanic crust, is three times that of water.

Earth's Mantle

More than 82 percent of Earth's volume is contained in the **mantle**, a solid, rocky shell that extends to a depth of about 2900 kilometers (1800 miles). The boundary between the crust and mantle represents a marked change in chemical composition. The dominant rock type in the uppermost mantle is *peridotite*, which contains minerals richer in the metals magnesium and iron compared to the minerals found in either the continental or oceanic crust.

The Upper Mantle   The upper mantle extends from the crust–mantle boundary down to a depth of about 660 kilometers (410 miles). The upper mantle can be divided into three different parts. The top portion is part of the stronger *lithosphere*, and beneath it is the weaker *asthenosphere*. The bottom part of the upper mantle is called the *transition zone*.

The **lithosphere** ("rock sphere") consists of the entire crust plus the uppermost mantle and forms Earth's relatively cool, rigid outer shell (see Figure 1.20). Averaging about 100 kilometers (60 miles) thick, the lithosphere can be as much as two and a half times this thick below the oldest portions of the continents. Beneath this stiff layer to a depth of about 410 kilometers (255 miles) lies a soft, comparatively weak layer known as the **asthenosphere** ("weak sphere"). The top portion of the asthenosphere has a temperature/pressure regime that results in a small amount of melting. Within this very weak zone, the lithosphere is mechanically detached from the layer below. The lithosphere thus is able to move independently of the asthenosphere, a fact we will consider in Chapter 2, when we discuss plate tectonics.

It is important to emphasize here that the strength of various Earth materials is a function of both their composition and the temperature and pressure of their environment. You should not get the idea that the entire lithosphere behaves like a rigid or brittle solid similar to rocks found on the surface. Rather, the rocks of the lithosphere get progressively hotter and weaker (more easily deformed) with increasing depth. At the depth of the uppermost asthenosphere, the rocks are close enough to their melting temperature (some melting may actually occur) that they are very easily deformed. Thus, the uppermost asthenosphere is weak because it is near its melting point, just as hot wax is weaker than cold wax.

From about 410 kilometers (255 miles) to about 660 kilometers (410 miles) in depth is the part of the upper mantle called the **transition zone** (see Figure 1.20). The top of the transition zone is identified by a sudden increase in density from about 3.5 to 3.7 g/cm^3. This change occurs because minerals in the rock peridotite respond to the increase in pressure by forming new minerals with closely packed atomic structures.

The Lower Mantle   The **lower mantle** is a zone that exists from a depth of 660 kilometers (410 miles) to 2900 kilometers (1800 miles). The lower mantle ends at the top of the core. Because of an increase in pressure (caused by the weight of the rock above), the mantle gradually strengthens with depth. Despite their strength, however, the rocks in the lower mantle are very hot and capable of extremely gradual flow.

In the bottom few hundred kilometers of the mantle is a highly variable and unusual layer called the D″ (pronounced "dee double-prime") layer. The nature of this boundary layer between the rocky mantle and the hot liquid iron outer core will be examined in Chapter 12.

Earth's Core

The **core** is, as the name implies, the region at the center of Earth's interior. It is composed of an iron–nickel alloy with minor amounts of oxygen, silicon, and sulfur—elements that readily form compounds with iron. At the extreme pressure found in the core, this iron-rich material has an average density of nearly 11 g/cm^3 and approaches 14 times the density of water at Earth's center.

The core is divided into two regions that exhibit very different mechanical strengths. The **outer core** is a *liquid layer* 2250 kilometers (1395 miles) thick. The movement of metallic iron within this zone generates Earth's magnetic field. The **inner core** is a sphere that has a radius of 1221 kilometers (757 miles). Despite the higher temperature, the iron in the inner core is *solid* due to the immense pressures that exist in the center of the planet.

CONCEPT CHECKS 1.6

1. Name and describe the three major layers defined by their chemical composition.

2. Contrast the characteristics of the lithosphere and the asthenosphere.

   Concept Checker
   https://goo.gl/ZCjUrJ

3. Why is the inner core solid?

1.7 Rocks and the Rock Cycle

Sketch, label, and explain the rock cycle.

Rock is the most common and abundant material on Earth. To a curious traveler, the variety seems nearly endless. When a rock is examined closely, we find that it usually consists of smaller crystals called minerals. *Minerals* are chemical compounds (or sometimes single elements), each with its own composition and physical properties. The grains or crystals may be microscopically small or easily seen with the unaided eye.

The minerals that compose a rock strongly influence its nature and appearance. In addition, a rock's *texture*—the size, shape, and/or arrangement of its constituent minerals—also has a significant effect on its appearance. A rock's mineral composition and texture, in turn, reflect the geologic processes that created it (**Figure 1.21**). Such analyses are critical to understanding our planet. This understanding also has many practical applications, including finding energy and mineral resources and solving environmental problems.

Geologists divide rocks into three major groups: igneous, sedimentary, and metamorphic. **Figure 1.22** provides some examples. As you will learn, each group is linked to the others by the processes that act upon and within the planet.

Earlier in this chapter, you learned that Earth is a system. This means that our planet consists of many interacting parts that form a complex whole. Nowhere is this idea better illustrated than when we examine the rock cycle (**Figure 1.23**, page 26). The **rock cycle** allows us to view many of the interrelationships among different parts of the Earth system. It helps us understand the origin of igneous, sedimentary, and metamorphic rocks and to see that each type is linked to the others by processes that act upon and within the planet. Consider the rock cycle to be a simplified but useful overview of physical geology. *Learn the rock cycle well*; you will be examining its interrelationships in greater detail throughout this book.

The Basic Rock Cycle

Magma is molten rock that forms deep beneath Earth's surface. Over time, magma cools and solidifies. This process, called *crystallization*, may occur either beneath the surface or, following a volcanic eruption, at the surface. In either situation, the resulting rocks are called **igneous rocks**.

If igneous rocks are exposed at the surface, they undergo *weathering*, in which the day-in and day-out influences of the atmosphere, hydrosphere, and biosphere slowly disintegrate and decompose rocks. The materials that result are often moved downslope by gravity before being picked up and transported by any of a number of erosional agents, such as running water, glaciers, wind, or waves. Eventually these particles and dissolved substances, called **sediment**, are deposited. Although most sediment ultimately comes to rest in the ocean, other sites of deposition include river floodplains, desert basins, swamps, and sand dunes.

Next, sediments undergo *lithification*, a term meaning "conversion into rock." Sediment is usually lithified into **sedimentary rock** when compacted by the weight of overlying layers and/or cemented as percolating groundwater fills the pores with mineral matter.

If the resulting sedimentary rock is buried deep within Earth and involved in the dynamics of mountain building or intruded by a mass of magma, it is subjected to great pressures and/or intense heat. The sedimentary rock reacts to the changing environment and turns into the third rock type, **metamorphic rock**.

► Figure 1.21
Two basic rock characteristics Texture and mineral composition are basic rock features. These two samples are the common igneous rocks granite (**A**) and basalt (**B**).

A. The large crystals of light-colored minerals in granite result from the slow cooling of molten rock deep beneath the surface. Granite is abundant in the continental crust.

B. Basalt is rich in dark minerals. Rapid cooling of molten rock at Earth's surface is responsible for the rock's microscopically small crystals. The oceanic crust is a basalt-rich layer.

Igneous rocks form when molten rock solidifies at the surface (extrusive) or beneath the surface (intrusive). The lava flow in the foreground is the fine-grained rock basalt and came from SP Crater in northern Arizona.

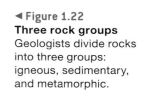

◄ Figure 1.22
Three rock groups
Geologists divide rocks into three groups: igneous, sedimentary, and metamorphic.

Sedimentary rocks consist of particles derived from the weathering of other rocks. This layer consists of durable sand-size grains of the mineral quartz that are cemented into a solid rock. The grains were once a part of extensive dunes. This rock layer, called the Navajo Sandstone, is prominent in southern Utah.

The metamorphic rock pictured here, known as the Vishnu Schist, is exposed in the inner gorge of the Grand Canyon. It formed deep below Earth's surface where temperatures and pressures are high and in association with mountain-building episodes in Precambrian time.

When metamorphic rock is subjected to additional pressure changes or to still higher temperatures, it melts, creating magma, which eventually crystallizes into igneous rock, starting the cycle all over again.

Where does the energy that drives Earth's rock cycle come from? Processes driven by heat from Earth's interior are responsible for creating igneous and metamorphic rocks. Weathering and erosion, external processes powered by energy from the Sun, produce the sediment from which sedimentary rocks form.

Alternative Paths

The paths shown in the basic rock cycle are not the only ones possible. The alternatives indicated by the light blue arrows in Figure 1.23 are just as likely to be followed as those described in the preceding section.

Rather than being exposed to weathering and erosion at Earth's surface, igneous rocks may remain deeply buried. Eventually these masses may be subjected to the strong compressional forces and high temperatures

ROCK CYCLE

Viewed over long time spans,
rocks are constantly forming,
changing, and re-forming.

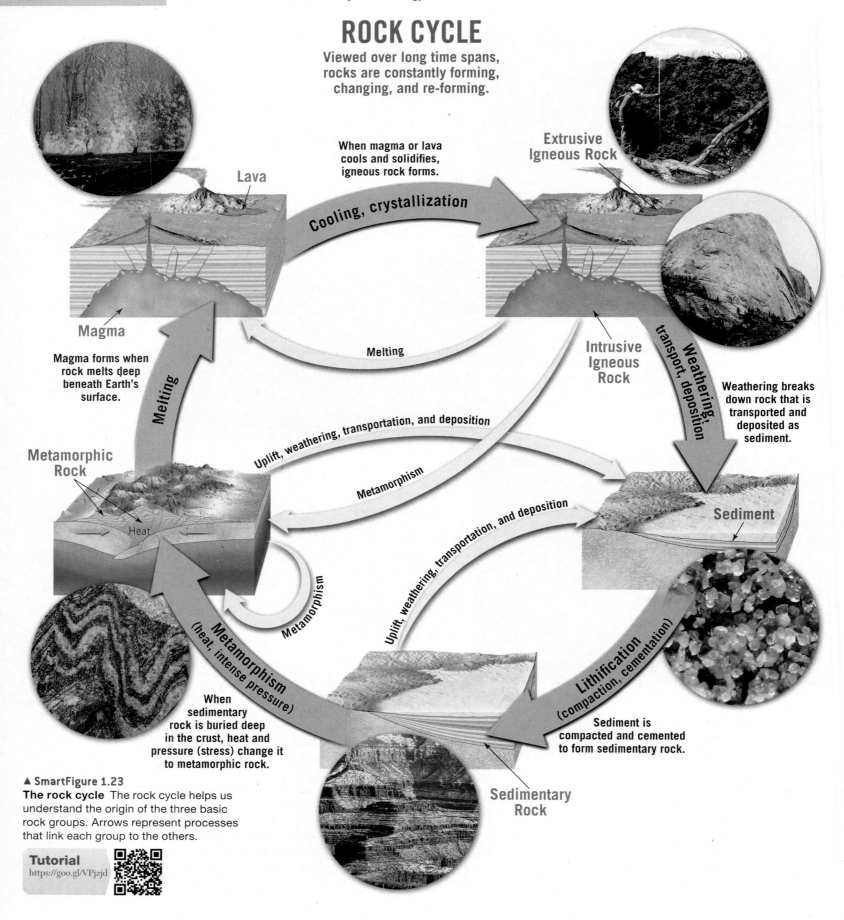

When magma or lava
cools and solidifies,
igneous rock forms.

**Extrusive
Igneous Rock**

Lava

Cooling, crystallization

Magma

Magma forms when
rock melts deep
beneath Earth's
surface.

Melting

**Intrusive
Igneous
Rock**

Melting

**Weathering,
transport, deposition**

Weathering breaks
down rock that is
transported and
deposited as
sediment.

Uplift, weathering, transportation, and deposition

**Metamorphic
Rock**

Metamorphism

Heat

Metamorphism

Uplift, weathering, transportation, and deposition

Sediment

**Metamorphism
(heat, intense pressure)**

When
sedimentary
rock is buried deep
in the crust, heat and
pressure (stress) change it
to metamorphic rock.

**Lithification
(compaction, cementation)**

Sediment is
compacted and cemented
to form sedimentary rock.

**Sedimentary
Rock**

▲ SmartFigure 1.23
The rock cycle The rock cycle helps us
understand the origin of the three basic
rock groups. Arrows represent processes
that link each group to the others.

Tutorial
https://goo.gl/VPjzjd

associated with mountain building. When this occurs, they are transformed directly into metamorphic rocks.

Metamorphic and sedimentary rocks, as well as sediment, do not always remain buried. Rather, overlying layers may be stripped away, exposing the once-buried rock. This exposed material is attacked by weathering processes and turned into new raw materials for sedimentary rocks.

Although rocks may seem to be unchanging masses, the rock cycle shows that they are often in a state of transformation. The changes, however, take time—vast amounts of time. We can observe different parts of the cycle operating all over the world. Today new magma is forming beneath the island of Hawaii. When it erupts at the surface, the lava flows add to the size of the island.

Meanwhile, the Colorado Rockies are gradually being worn down by weathering and erosion. Some of this weathered debris will eventually be carried to the Gulf of Mexico, where it will add to the already substantial mass of sediment that has accumulated there.

CONCEPT CHECKS 1.7

1. List two rock characteristics that are used to determine the processes that created a rock.

2. Sketch and label a basic rock cycle. Make sure to include alternate paths.

 **Concept Checker** https://goo.gl/pkQJNA

1.8 The Face of Earth

List and describe the major features of the ocean basins and continents.

The two principal divisions of Earth's surface are the **ocean basins** and the **continents** (**Figure 1.24**). A significant difference between these two areas is their relative levels. The elevation difference between the ocean basins and the continents is primarily due to differences in their respective densities and thicknesses:

- **Ocean basins.** The average depth of the ocean floor is about 3.8 kilometers (2.4 miles) below sea level, or about 4.5 kilometers (2.8 miles) lower than the average elevation of the continents. The basaltic rocks that comprise the oceanic crust average only 7 kilometers (about 4.5 miles) thick and have an average density of about 3.0 g/cm^3.
- **Continents.** The continents are remarkably flat features that have the appearance of plateaus protruding above sea level. With an average elevation of about 0.8 kilometer (0.5 mile), continental blocks lie close to sea level, except for limited areas of mountainous terrain. Recall that the continents average about 35 kilometers (22 miles) thick and are composed of granitic rocks that have a density of about 2.7 g/cm^3.

The thicker, less dense continental crust is more buoyant than the oceanic crust. As a result, continental crust floats on top of the deformable rocks of the mantle at a higher level than oceanic crust for the same reason that a large, empty (less dense) cargo ship rides higher than a small, loaded (denser) one.

Major Features of the Ocean Floor

If all water were drained from the ocean basins, a great variety of features would be seen, including chains of volcanoes, deep canyons, plateaus, and large expanses of monotonously flat plains. In fact, the scenery would be nearly as diverse as that on the continents (see Figure 1.24). These features and the processes that form them are covered in detail in Chapter 13.

During the past 75 years, oceanographers have used modern depth-sounding equipment and satellite technology to map significant portions of the ocean floor. These studies have led them to identify three major regions: *continental margins*, *deep-ocean basins*, and *oceanic (mid-ocean) ridges*.

Continental Margins   The **continental margin** is the portion of the seafloor adjacent to major landmasses. It may include the *continental shelf*, the *continental slope*, and the *continental rise*.

Although land and sea meet at the shoreline, this is *not* the boundary between the continents and the ocean basins. Rather, along most coasts, a gently sloping platform of material, called the **continental shelf**, extends seaward from the shore. Because it is underlain by continental crust, it is clearly a flooded extension of the continents. A glance at Figure 1.24 shows that the width of the continental shelf varies. For example, it is broad along the East and Gulf coasts of the United States but relatively narrow along the Pacific margin of the continent.

The boundary between the continents and the deep-ocean basins lies along the **continental slope**, a relatively steep dropoff that extends from the outer edge of the continental shelf to the floor of the deep ocean

► Figure 1.24
The face of Earth Major surface features of the geosphere.

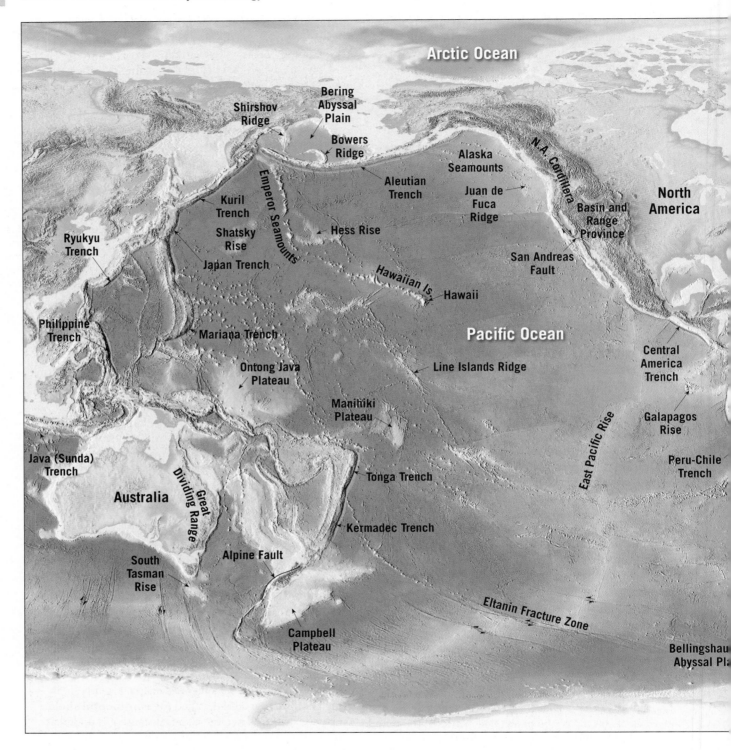

(see Figure 1.24). Using this as the dividing line, we find that about 60 percent of Earth's surface is represented by ocean basins and the remaining 40 percent by continents.

In regions where trenches do not exist, the steep continental slope merges into a more gradual incline known as the **continental rise**, a thick wedge of sediment that moved downslope from the continental shelf and accumulated on the deep-ocean floor.

Deep-Ocean Basins   Situated between the continental margins and oceanic ridges are **deep-ocean basins**. Parts of these regions consist of incredibly flat features called **abyssal plains**. The ocean floor also contains extremely deep depressions, some more than 11,000 meters (36,000 feet) deep. Although these **deep-ocean trenches** are relatively narrow and represent only a small fraction of the ocean floor, they are nevertheless very significant features. Some trenches are located adjacent to

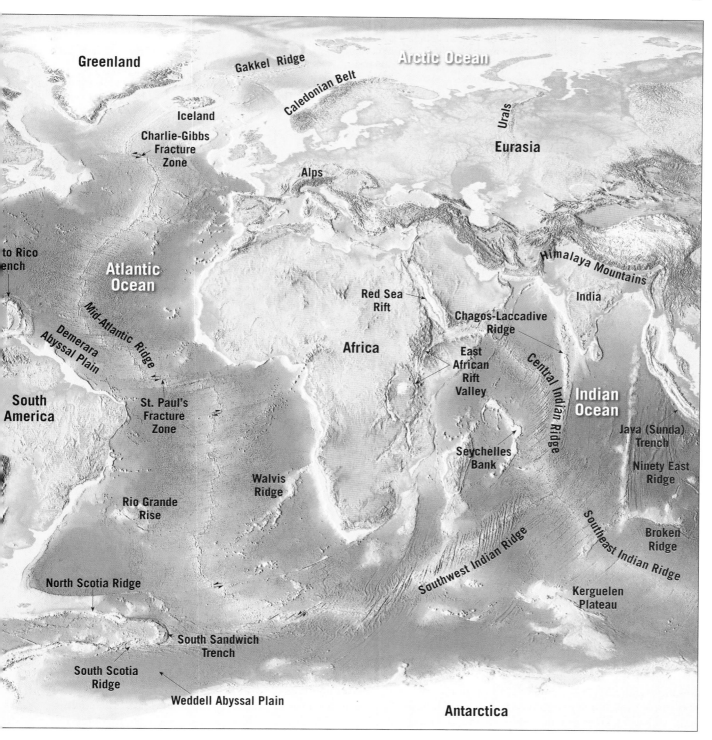

Greenland

Gakkel Ridge

Arctic Ocean

Caledonian Belt

Urals

Iceland

Eurasia

Charlie-Gibbs
Fracture
Zone

Alps

to Rico
ench

Atlantic
Ocean

Red Sea
Rift

India

Himalaya Mountains

Mid-Atlantic Ridge

Chagos-Laccadive
Ridge

Demerara
Abyssal Plain

Africa

East
African
Rift
Valley

Central Indian Ridge

Indian
Ocean

South
America

St. Paul's
Fracture
Zone

Seychelles
Bank

Java (Sunda)
Trench

Ninety East
Ridge

Walvis
Ridge

Rio Grande
Rise

Southwest Indian Ridge

Southeast Indian Ridge

Broken
Ridge

North Scotia Ridge

Kerguelen
Plateau

South Sandwich
Trench

South Scotia
Ridge

Weddell Abyssal Plain

Antarctica

young mountains that flank the continents. For example, in Figure 1.24 the Peru–Chile trench off the west coast of South America parallels the Andes Mountains. Other trenches parallel island chains called *volcanic island arcs*.

Dotting the ocean floor are submerged volcanic structures called **seamounts**, which sometimes form long, narrow chains. Volcanic activity has also produced several large *lava plateaus*, such as the Ontong Java Plateau located northeast of New Guinea. In addition, some submerged plateaus are composed of continental-type crust. Examples include the Campbell Plateau southeast of New Zealand and the Seychelles Bank northeast of Madagascar.

Oceanic Ridges   The most prominent feature on the ocean floor is the **oceanic ridge**, or **mid-ocean ridge**. As shown in Figure 1.24, the Mid-Atlantic Ridge and the East Pacific Rise are parts of this system. This

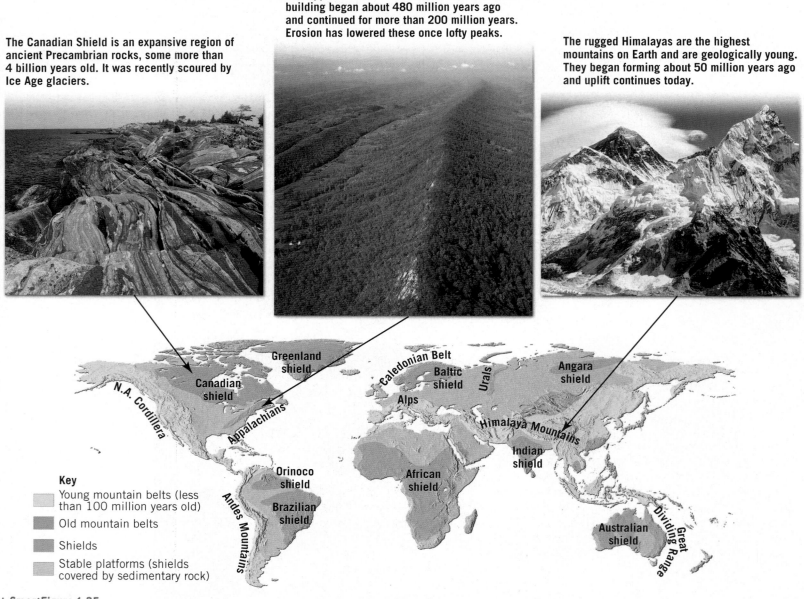

The Canadian Shield is an expansive region of ancient Precambrian rocks, some more than 4 billion years old. It was recently scoured by Ice Age glaciers.

The Appalachians are old mountains. Mountain building began about 480 million years ago and continued for more than 200 million years. Erosion has lowered these once lofty peaks.

The rugged Himalayas are the highest mountains on Earth and are geologically young. They began forming about 50 million years ago and uplift continues today.

Key
- Young mountain belts (less than 100 million years old)
- Old mountain belts
- Shields
- Stable platforms (shields covered by sedimentary rock)

Greenland shield
Canadian shield
N.A. Cordillera
Caledonian Belt
Baltic shield
Urals
Angara shield
Alps
Appalachians
Himalaya Mountains
Indian shield
Orinoco shield
African shield
Andes Mountains
Brazilian shield
Australian shield
Great Dividing Range

▲ SmartFigure 1.25
The continents
Distribution of mountain belts, stable platforms, and shields.

Tutorial
https://goo.gl/
hMT8aK

broad elevated feature forms a continuous belt winding more than 70,000 kilometers (43,500 miles) around the globe, in a manner similar to the seam of a baseball. Unlike most continental mountains that consist of highly deformed rock, the oceanic ridge system consists of layer upon layer of igneous rock that has been fractured and uplifted.

Major Features of the Continents

The major features of the continents can be grouped into two distinct categories: uplifted regions of deformed rocks that make up present-day mountain belts, and extensive flat, stable areas that have been eroded nearly to sea level. Notice in **Figure 1.25** that the young mountain belts tend to be long, narrow features at the margins of continents and that the flat, stable areas are typically located in the interiors of the continents. Mountain building is discussed in more detail in Chapter 14.

Mountains The most prominent continental features are **mountain belts**. Although their distribution appears to be random, this is not the case. The youngest mountains (those less than 100 million years old) are located principally in two major zones. The circum-Pacific belt (the region surrounding the Pacific Ocean) includes the mountains of the western Americas and continues into the western Pacific in the form of volcanic island arcs (see Figure 1.24). Island arcs are active mountainous regions composed largely of volcanic rocks and deformed sedimentary rocks. Examples include the Aleutian Islands, Japan, the Philippines, and New Guinea.

The other major mountain belt extends eastward from the Alps through Iran and the Himalayas and then dips southward into Indonesia. Careful examination of mountainous terrains reveals that most are places where thick sequences of rocks have been squeezed and highly deformed, as if placed in a gigantic vise. Older mountains are also found on the continents. Examples include the Appalachians in the eastern United States and the Urals in Russia. Their once lofty peaks are now worn low as a result of millions of years of weathering and erosion.

The Stable Interior   Unlike the young mountain belts that have formed within the past 100 million years, the interiors of the continents, called **cratons**, have been relatively stable (undisturbed) for the past 600 million years or even longer. Typically, these regions were involved in mountain-building episodes much earlier in Earth's history.

Within the stable interiors are areas known as **shields**—expansive, flat regions composed largely of deformed igneous and metamorphic rocks. Notice in Figure 1.25 that the Canadian Shield is exposed in much of the northeastern part of North America. Radiometric dating of shields indicates that they are truly ancient regions. All contain Precambrian-age rocks more than 1 billion years old, with some samples approaching 4 billion years in age. Even these oldest-known rocks exhibit evidence of enormous forces that have folded, faulted, and metamorphosed them. Thus, we conclude that these rocks were once part of an ancient mountain system that has since been eroded away to produce these expansive, flat regions.

Other flat areas of the craton exist, where highly deformed rocks, like those found in the shields, are covered by a relatively thin veneer of sedimentary rocks. These areas are called **stable platforms**. The sedimentary rocks in stable platforms are nearly horizontal, except where they have been warped to form large basins or domes. In North America a major portion of the stable platform is located between the Canadian Shield and the Rocky Mountains.

Being familiar with the topographic features that comprise the face of Earth is essential to understanding the mechanisms that have shaped our planet. What is the significance of the enormous ridge system that extends through all the world's oceans? What is the connection, if any, between young, active mountain belts and oceanic trenches? What forces crumple rocks to produce majestic mountain ranges? These are a few of the questions that will be addressed beginning in the next chapter, as we investigate the dynamic processes that shaped our planet in the geologic past and will continue to shape it in the future.

CONCEPT CHECKS 1.8

1. Contrast ocean basins and continents.

2. Name the three major regions of the ocean floor. What are some features associated with each?

   Concept Checker
   https://goo.gl/t9zcsQ

3. Describe the general distribution of Earth's youngest mountains.

4. What is the difference between shields and stable platforms?

CONCEPTS IN REVIEW

An Introduction to Geology

1.1 Geology: The Science of Earth

Distinguish between physical and historical geology and describe the connections between people and geology.

Key Terms:
geology

physical geology
historical geology

- Geologists study Earth. Physical geologists focus on the processes by which Earth operates and the materials that result from those processes. Historical geologists apply an understanding of Earth materials and processes to reconstruct the history of our planet.

- Humans' relationship with planet Earth can be positive and negative. Earth processes and products sustain us every day, but they can also harm us. Similarly, people have the ability to alter or harm natural systems, including those that sustain civilization.

1.2   The Development of Geology

Summarize early and modern views on how change occurs on Earth and relate them to the prevailing ideas about the age of Earth.

Key Terms: catastrophism   uniformitarianism

- Early ideas about the nature of Earth were based on religious traditions and notions of great catastrophes. In 1795, James Hutton emphasized that the same slow processes have acted over great spans of time and are responsible for Earth's rocks, mountains, and landforms. This similarity of processes over vast spans of time led to this principle being dubbed *uniformitarianism*.
- Based on the rate of radioactive decay of certain elements, the age of Earth has been calculated to be about 4.6 billion years.

Q In what eon, era, period, and epoch do we live?

1.3   The Nature of Scientific Inquiry

Discuss the nature of scientific inquiry, including the construction of hypotheses and the development of theories.

Key Terms: theory
hypothesis   scientific method

- Geologists make observations, construct tentative explanations for those observations (*hypotheses*), and then test those hypotheses with field investigations and laboratory work. In science, a *theory* is a well-tested and widely accepted view that the scientific community agrees best explains certain observable facts.
- As we discard flawed hypotheses, scientific knowledge moves closer to a correct understanding, but we can never be fully confident that we know all the answers. Scientists must always be open to new information that forces changes in our model of the world.

1.4   Earth as a System

List and describe Earth's four major spheres. Define *system* and explain why Earth is considered a system.

Key Terms: biosphere   system
hydrosphere   geosphere
atmosphere   Earth system science

- Earth's physical environment is traditionally divided into three major parts: the solid Earth, called the *geosphere*; the water portion of our planet, called the *hydrosphere*; and Earth's gaseous envelope, called the *atmosphere*.
- A fourth Earth sphere is the *biosphere*, the totality of life on Earth. It is concentrated in a relatively narrow zone that extends a few kilometers into the hydrosphere and geosphere and a few kilometers up into the atmosphere.
- Of all the water on Earth, more than 96 percent is in the oceans, which cover nearly 71 percent of the planet's surface.
- Although each of Earth's four spheres can be studied separately, they are all related in a complex and continuously interacting whole that is called the *Earth system*.
- *Earth system science* uses an interdisciplinary approach to integrate the knowledge of several academic fields in the study of our planet and its global environmental problems.
- The two sources of energy that power the Earth system are (1) the Sun, which drives the external processes that occur in the atmosphere, in the hydrosphere, and at Earth's surface, and (2) heat from Earth's interior, which powers the internal processes that produce volcanoes, earthquakes, and mountains.

Q Is glacial ice part of the geosphere, or does it belong to the hydrosphere? Explain your answer.

1.5   Origin and Early Evolution of Earth

Outline the stages in the formation of our solar system.

Key Terms: nebular theory   solar nebula

- The *nebular theory* describes the formation of the solar system. The planets and Sun began forming about 4.6 billion years ago from a large cloud of dust and gases.
- As the cloud contracted, it began to rotate and assume a disk shape. Material that was gravitationally pulled toward the center became the protosun. Within the rotating disk, small centers, called planetesimals, swept up more and more of the cloud's debris.
- Because of their high temperatures and weak gravitational fields, the inner planets were unable to accumulate and retain many of the lighter components. Because of the very cold temperatures existing far from the Sun, the large outer planets consist of huge amounts of lighter materials. These gaseous substances account for the comparatively large sizes and low densities of the outer planets.

1.6 Earth's Internal Structure

Sketch Earth's internal structure and label and describe the main subdivisions.

Key Terms:	asthenosphere	outer core
crust	transition zone	inner core
mantle	lower mantle	
lithosphere	core	

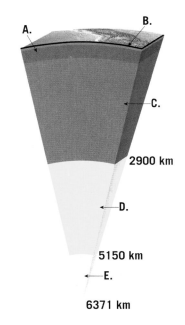

- Compositionally, the solid Earth has three layers: *core*, *mantle*, and *crust*. The core is most dense, and the crust is least dense.
- Earth's interior can also be divided into layers based on physical properties. The crust and upper mantle make a two-part layer called the *lithosphere*, which is broken into the plates of plate tectonics. Beneath that is the "weak" *asthenosphere*. The *lower mantle* is stronger than the asthenosphere and overlies the molten *outer core*. This liquid is made of the same iron–nickel alloy as the *inner core*, but the extremely high pressure of Earth's center compacts the inner core into a solid form.

Q The diagram represents Earth's layered structure. Does it show layering based on physical properties or layering based on composition? Identify the lettered layers.

A.

B.

C.

2900 km

D.

5150 km

E.

6371 km

1.7 Rocks and the Rock Cycle

Sketch, label, and explain the rock cycle.

Key Terms:	igneous rock	sedimentary rock
rock cycle	sediment	metamorphic rock

- The *rock cycle* is a model for thinking about the transformation of one rock to another due to Earth processes. All *igneous rocks* are made from molten rock. All *sedimentary rocks* are made from weathered products of other rocks. All *metamorphic rocks* are the products of preexisting rocks that are transformed at high temperatures or pressures. Given the right conditions, any kind of rock can be transformed into any other kind of rock.

Q Name the processes that are represented by each of the letters in this simplified rock cycle diagram.

A.

B.

C.

D.

E.

Rock cycle

1.8 The Face of Earth

List and describe the major features of the ocean basins and continents.

Key Terms:	continental rise	mountain belt
ocean basin	deep-ocean basin	craton
continent	abyssal plain	shield
continental margin	deep-ocean trench	stable platform
continental shelf	seamount	
continental slope	oceanic ridge	

- Two principal divisions of Earth's surface are *ocean basins* and *continents*. A significant difference is their relative levels. The elevation differences between ocean basins and continents is primarily due to differences in their respective densities and thicknesses.

- There are shallow portions of the oceans that are essentially flooded margins of the continents, and there are deeper portions that include vast *abyssal plains* and *deep-ocean trenches*. Seamounts and lava plateaus interrupt the abyssal plain in some places.
- Continents consist of relatively flat, stable areas called *cratons*. Where a craton is blanketed by a relatively thin layer of sediment or sedimentary rock, it is called a *stable platform*. Where a craton is exposed at the surface, it is known as a *shield*. Wrapping around the edges of some cratons are *mountain belts*, linear zones of intense deformation and metamorphism.

Q Put these features of the ocean floor in order from shallowest to deepest: continental slope, deep-ocean trench, continental shelf, abyssal plain, continental rise.

GIVE IT SOME THOUGHT

1. The length of recorded history for humankind is about 5000 years. How does this compare to the length of geologic time? Calculate the percentage or fraction of geologic time that is represented by recorded history. To make calculations easier, round the age of Earth to the nearest billion.

2. Refer to the graph in Figure 1.14 to answer the following questions.

    a. If you were to climb to the top of Mount Everest, how many breaths of air would you have to take at that altitude to equal one breath at sea level?

    b. If you are flying in a commercial jet at an altitude of 12 kilometers (about 39,000 feet), about what percentage of the atmosphere's mass is below you?

3. The accompanying photo shows a mudflow that was triggered by extraordinary rains in March 2014. Describe how each of Earth's four spheres was influenced by and/or involved in this natural disaster that buried a 1-square-mile rural neighborhood near Oso, Washington, and caused more than 40 fatalities.

4. Refer to Figure 1.23. How does the rock cycle diagram—particularly the information pointed out by the process arrows—support the fact that sedimentary rock is the most abundant rock type on the surface of Earth?

5. This is a shoreline scene along the coast of Maine in Acadia National Park.

    a. This area, like other shorelines, is described as an *interface*. What does this mean? Describe another interface in the Earth system.

    b. Does the shoreline, the line where the water meets the land, mark the outer edge of the North American continent? Explain.

6. After entering a dark room, you turn on a wall switch, but the light does not come on. Suggest at least three hypotheses that might explain this observation. Once you have formulated your hypotheses, what is the next logical step?

EYE ON EARTH

1. These rock layers consist of sediments such as sand, mud, and gravel that were deposited by rivers, waves, wind, and glaciers. The material was buried and eventually compacted and cemented into solid rock. Later, erosion uncovered the layers.

   a. Can you establish a relative time scale for these rocks? That is, can you determine which one of the layers shown here is likely oldest and which is probably youngest?

   b. Explain the logic you used.

DATA ANALYSIS

Swift Creek Landslide

The Swift Creek landslide is located in northwestern Washington State, on the west side of Sumas Mountain. The landslide moves on average 3 to 4 meters per year, which is about 1 centimeter per day.

https://goo.gl/eD226M

ACTIVITIES

Go to the Western Washington University Swift Creek Landslide Observatory (SCLO) page at http://landslide.geol.wwu.edu and click About SCLO.

1. Why are geologists, government agencies, and local residents interested in the Swift Creek landslide?

Click TimeLapse to open the video generator. Choose a start date of January 1, 2007, and an end date of January 1, 2014. For time of day select between 12pm–3pm for the best lighting. Set the framerate to Medium, for 20 frames per second. Select the camera position and click Generate. It may take about 30 seconds to generate the video.

2. Compare the Upper camera to the Lower camera. Which section of the landslide has a more constant flow rate? How can you tell?

Use the TimeLapse video generator to answer the following question. You may need to watch multiple years and examine each season separately.

3. Is the flowrate of the upper part of the landslide faster during the wet season (October–March) or the dry season (April–September)?

Mastering Geology

Looking for additional review and test prep materials? Visit pearson.com/mastering/geology to enhance your understanding of this chapter's content by accessing a variety of resources, including **Self-Study Quizzes, Geoscience Animations, SmartFigure Tutorials,** *Project Condor* **Videos, Mobile Field Trips, Dynamic Study Modules,** and an optional **Pearson eText.**

The Search for Plate Tectonics— And Alien Life

Earth, the only planet in our solar system known to sustain life, is also the only planet known to exhibit active plate tectonics. The link is so compelling that astronomers on the hunt for extraterrestrial life now focus on studying larger, rocky planets—those that are most likely to have plate tectonics. But why are these two things thought to go hand in hand?

▲ Development of life on Earth may be tied to plate tectonics.

Plate tectonics, where large pieces of a planet's outer shell slide around and interact, gives us mountains, earthquakes, and volcanic eruptions. It also provides a mechanism by which carbon dioxide, an important greenhouse gas, moves through planet layers, from atmosphere to the interior and back again.

This cycle acts as a thermostat, keeping temperatures within a *not-too-hot* and *not-too-cold* range that preserves liquid water at the planet's surface. The presence of liquid water for billions of years on Earth has been essential to the development of life as we know it.

How common are rocky exoplanets that might also have plate tectonics? A recently published study indicates that about one-third of the Earthlike planets in the galaxy have the potential to support plate tectonics. This means there are *billions* of planets with these essential characteristics— making them good candidates for where life may exist.

▶ Our home planet Earth as seen from space.

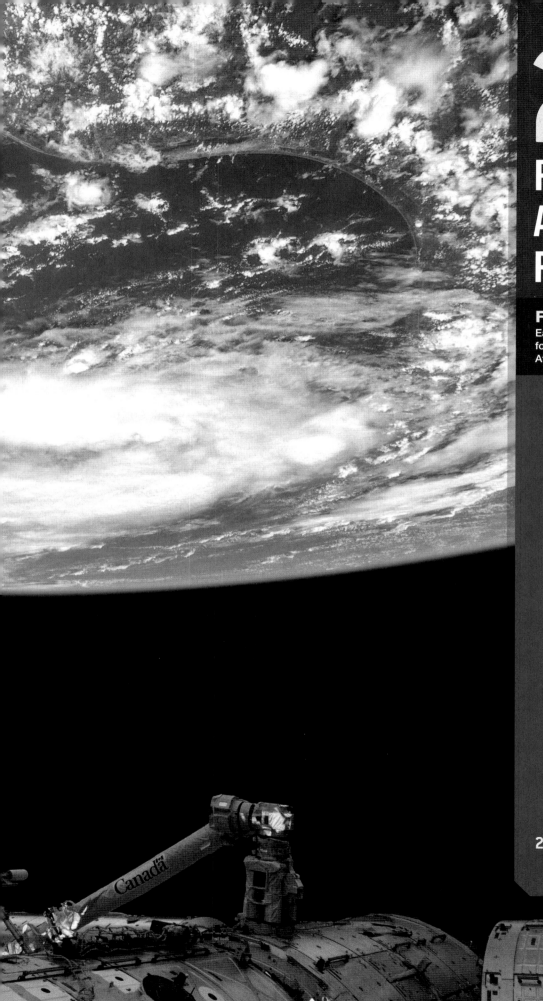

2
Plate Tectonics: A Scientific Revolution Unfolds

FOCUS ON CONCEPTS

Each statement represents the primary learning objective for the corresponding major heading within the chapter. After you complete the chapter, you should be able to:

2.1 Summarize the view that most geologists held prior to the 1960s regarding the geographic positions of the ocean basins and continents.

2.2 List and explain the evidence Wegener presented to support his continental drift hypothesis.

2.3 List the major differences between Earth's lithosphere and asthenosphere and explain the importance of each in the plate tectonics theory.

2.4 Sketch and describe the movement along a divergent plate boundary that results in the formation of new oceanic lithosphere.

2.5 Compare and contrast the three types of convergent plate boundaries and name a location where each type can be found.

2.6 Describe the relative motion along a transform fault boundary and locate several examples of transform faults on a plate boundary map.

2.7 Explain why plates such as the African and Antarctic plates are increasing in size, while the Pacific plate is decreasing in size.

2.8 List and explain the evidence used to support the plate tectonics theory.

2.9 Describe two methods researchers use to measure relative plate motion.

2.10 Describe plate–mantle convection and explain two of the primary driving forces of plate motion.

Plate tectonics is the first theory to provide a comprehensive view of the processes that produced Earth's major surface features, including the continents and ocean basins. Within the framework of this model, geologists have found explanations for the basic causes and distribution of earthquakes, volcanoes, and mountain belts. Further, the theory of plate tectonics helps explain the formation and distribution of igneous and metamorphic rocks and their relationship with the rock cycle.

2.1 From Continental Drift to Plate Tectonics

Summarize the view that most geologists held prior to the 1960s regarding the geographic positions of the ocean basins and continents.

Plate tectonics refers to the movement of lithospheric plates that shifts continents and causes volcanism, earthquakes, and mountain building. *Tectonic processes* (*tekto* = to build) are processes that deform Earth's crust to create major structural features, such as mountains, continents, and ocean basins.

▼ Figure 2.1
The Himalayan mountains were created when the subcontinent of India collided with southeastern Asia.

Until the late 1960s, most geologists held the view that the ocean basins and continents were all ancient and had fixed geographic positions. However, Earth's continents are not static; instead, they gradually migrate across the globe. These movements cause blocks of continental material to collide, deforming the intervening crust and thereby creating Earth's great mountain chains (**Figure 2.1**). Occasionally landmasses split apart, and as

continental blocks separate, a new ocean basin emerges between them. Meanwhile, other portions of the sea-floor plunge into the mantle.

This profound reversal in scientific thought has been appropriately called a *scientific revolution*. The revolution began early in the twentieth century as a proposal termed *continental drift*. For more than 50 years, the scientific community rejected the idea that continents are capable of movement. North American geologists in particular had difficulty accepting continental drift, per-haps because much of the supporting evidence had been gathered from Africa, South America, and Australia, continents with which most North American geologists were unfamiliar.

After World War II, modern instruments replaced rock hammers as the tools of choice for many Earth scientists. Researchers, including *geophysicists* and *geochemists*, made several surprising discoveries that rekindled interest in the drift hypothesis. By 1968, these developments led to the far more encompassing explana-tion known as the *theory of plate tectonics.*

In this chapter, we will examine the events that led to this dramatic reversal of scientific opinion. We will also briefly trace the development of the *continental drift hypothesis*, examine why it was initially rejected, and consider the evidence that finally led to the acceptance of its direct descendant—the theory of plate tectonics.

CONCEPT CHECKS 2.1

1. Briefly describe the view held by most geologists prior to the 1960s regarding the ocean basins and continents.

2. Name the early-twentieth-century hypothesis that was at first rejected by geologists and the more comprehensive theory that later replaced it.

Concept Checker
https://goo.gl/EkF6BN

2.2 Continental Drift: An Idea Before Its Time

List and explain the evidence Wegener presented to support his continental drift hypothesis.

During the 1600s, as better world maps became available, people noticed that continents, particularly South America and Africa, could be fit together like pieces of a jigsaw puzzle. How-ever, little significance was given to this observation until 1915, when German meteorologist and geophysicist Alfred Wegener wrote *The Origin of Continents and Oceans*. This book outlined Wegener's **continental drift hypothesis**, which challenged the long-held assumption that the continents and ocean basins have fixed geographic positions.

Wegener suggested that a single **supercontinent** con-sisting of all Earth's landmasses once existed.[*] He named this giant landmass **Pangaea** (pronounced "Pan-jee-ah," meaning "all lands") **(Figure 2.2)**. Wegener further hypothesized that about 200 million years ago, during a time period called the *Mesozoic era* (see Figure 1.7, page 9), this supercontinent began to fragment into smaller land-masses. These continental blocks then "drifted" to their present positions over a span of millions of years.

Wegener and others collected substantial evidence to support the continental drift hypothesis. The fit of South America and Africa and the geographic distribution of fossils and ancient climates all seemed to buttress the idea that these now separate landmasses had once been joined.

Evidence: The Continental Jigsaw Puzzle

Wegener and others pointed to the remarkable similarity between the coastlines on opposite sides of the Atlantic

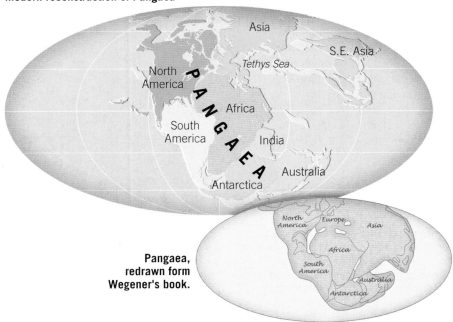

Modern reconstruction of Pangaea

Asia
S.E. Asia
Tethys Sea
North America
P A N G A E A
Africa
South America
India
Australia
Antarctica

Pangaea, redrawn form Wegener's book.

North America
Europe
Asia
Africa
South America
Australia
Antarctica

◀ **SmartFigure 2.2**
Reconstructions of Pangaea The supercon-tinent, as it is thought to have formed in the late Paleozoic and early Mesozoic eras, more than 200 million years ago.

Tutorial
https://goo.gl/vB94e1

*Wegener was not the first to conceive of a long-vanished super-continent. Geologist Eduard Suess pieced together evidence for a giant landmass comprising South America, Africa, India, and Australia before him.

Ocean as evidence that the continents may have once been joined. However, other Earth scientists challenged this evidence. These opponents (correctly) argued that wave erosion and depositional processes continually modify shorelines. Even if continental displacement had taken place, a good fit today would be unlikely. Because Wegener's original jigsaw fit of the continents was crude, it is assumed that he was aware of this problem (see Figure 2.2).

Scientists later determined that a much better approximation of the outer boundary of a continent is the seaward edge of its continental shelf, which lies submerged a few hundred meters below sea level. In the early 1960s, Sir Edward Bullard and two associates constructed a map that pieced together the edges of the continental shelves of South America and Africa at a depth of about 900 meters (3000 feet) (**Figure 2.3**). The fit obtained through these measurements was remarkably precise.

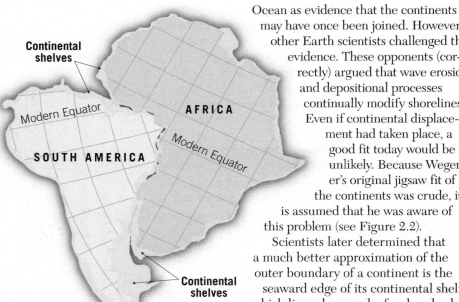

▲ Figure 2.3
Two of the puzzle pieces The best fit of South America and Africa occurs along the continental slope at a depth of 500 fathoms (about 900 meters [3000 feet]).

Evidence: Fossils Matching Across the Seas

Wegener learned that identical fossil organisms had been discovered in rocks from both South America and Africa. In fact, most *paleontologists* (scientists who study the fossilized remains of ancient organisms) of the time agreed that some type of land connection was needed to explain the existence of similar Paleozoic-age life-forms on widely separated landmasses. Just as modern life-forms native to North America are not the same as those of Africa and Australia, Paleozoic-age organisms on widely separated continents should have been distinctly different.

Mesosaurus and Glossopteris Wegener documented several cases in which the same fossil organism is found on landmasses that are now widely separated, even though it is unlikely that the living organism could have crossed the barrier of a broad ocean (**Figure 2.4**).

A classic example is *Mesosaurus*, a small aquatic freshwater reptile whose fossil remains are mainly found in shales of Permian age (about 260 million years ago) in eastern South America and southwestern Africa. If *Mesosaurus* had been able to make the long journey across the South Atlantic, its remains should be more widely distributed. Because this is not the case, Wegener asserted that South America and Africa must have been joined during that period of Earth history.

Wegener also cited the distribution of the fossil "seed fern" *Glossopteris* as evidence for Pangaea's existence. With tongue-shaped leaves and seeds too large to be carried by the wind, this plant was known to be widely dispersed throughout Africa, Australia, India, and South America. Later, fossil remains of *Glossopteris* were also discovered in Antarctica.† Wegener also learned that these seed ferns and associated flora grew only in cool climates—similar to central Canada. Therefore, he concluded that when these landmasses were joined, they were located much closer to the South Pole.

How did opponents of continental drift explain the existence of identical fossil organisms in places separated by thousands of kilometers of open ocean? Rafting, transoceanic land bridges (isthmian links), and island stepping stones were the most widely invoked explanations for these migrations (**Figure 2.5**). We know, for example, that during the Ice Age that ended about 8000 years ago, the lowering of sea level allowed mammals (including humans) to cross the narrow Bering Strait that separates Russia and Alaska. Was it

---

†In 1912 Captain Robert Scott and two companions froze to death lying beside 16 kilograms (35 pounds) of rock on their return from a failed attempt to be the first to reach the South Pole. These samples, collected on Beardmore Glacier, contained fossil remains of *Glossopteris*.

▶ Figure 2.4
Fossil evidence supporting continental drift Fossils of identical organisms have been discovered in rocks of similar age in Australia, Africa, South America, Antarctica, and India—continents that are currently widely separated by ocean barriers. Wegener accounted for these occurrences by placing these continents in their pre-drift locations.

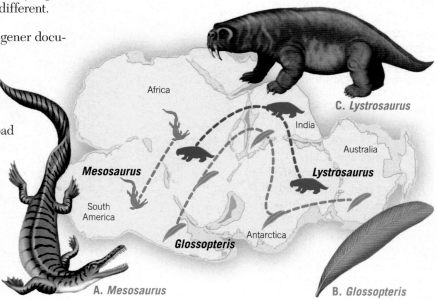

Africa

India

Australia

C. *Lystrosaurus*

Mesosaurus

Lystrosaurus

South America

Glossopteris

Antarctica

A. *Mesosaurus*

B. *Glossopteris*

possible that land bridges once connected Africa and South America but later subsided below sea level? Modern maps of the seafloor show no sunken land bridges though, which substantiates Wegner's views.

Evidence: Rock Types and Geologic Features

Successfully completing a jigsaw puzzle requires maintaining the continuity of the picture while fitting the pieces together. In the case of continental drift, this means that the rocks on either side of the Atlantic that predate the proposed Mesozoic split should match up to form a continuous "picture" when the continents are fitted together as Wegener proposed.

Indeed, Wegener found such "matches" across the Atlantic. For instance, highly deformed igneous rocks in eastern South America closely resemble similar rocks of the same age in Africa. Also, the mountain belt that includes the Appalachians trends northeastward through the eastern United States and disappears off the coast of Newfoundland (**Figure 2.6A**). Mountains of comparable age and structure are found in the British Isles and Scandinavia. When these landmasses are positioned as Wegener proposed (**Figure 2.6B**), the mountain chains form a nearly continuous belt.

Evidence: Ancient Climates

Because Alfred Wegener was a student of world climates, he suspected that paleoclimatic (*paleo* = ancient, *climatic* = climate) data might also support the idea of mobile continents. His assertion was bolstered by the discovery of evidence for a glacial period dating to the late *Paleozoic era* (see Figure 1.7, page 9) in southern

Africa, South America, Australia, and India. During that period, about 300 million years ago, vast ice sheets covered extensive portions of the Southern Hemisphere as well as India (**Figure 2.7A**). Much of the land that contains evidence of this Paleozoic glaciation presently lies within 30 degrees latitude of the equator in subtropical or tropical climates.

How could extensive ice sheets form near the equator? One proposal suggested that our planet experienced a period of extreme global cooling. Wegener rejected this because during the same geologic time span, large tropical swamps existed in several locations in the Northern Hemisphere. The lush vegetation in those swamps was eventually buried and converted to coal (**Figure 2.7B**).

▲ Figure 2.5
How do land animals cross vast oceans?
These sketches illustrate various early proposals to explain the occurrence of similar species on landmasses now separated by vast oceans.

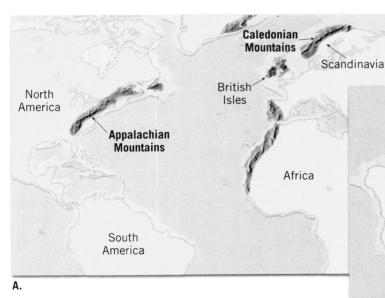

◄ Figure 2.6
Matching mountain ranges across the North Atlantic
A. The current locations of the continents surrounding the Atlantic.
B. The configuration of the continents about 200 million years ago.

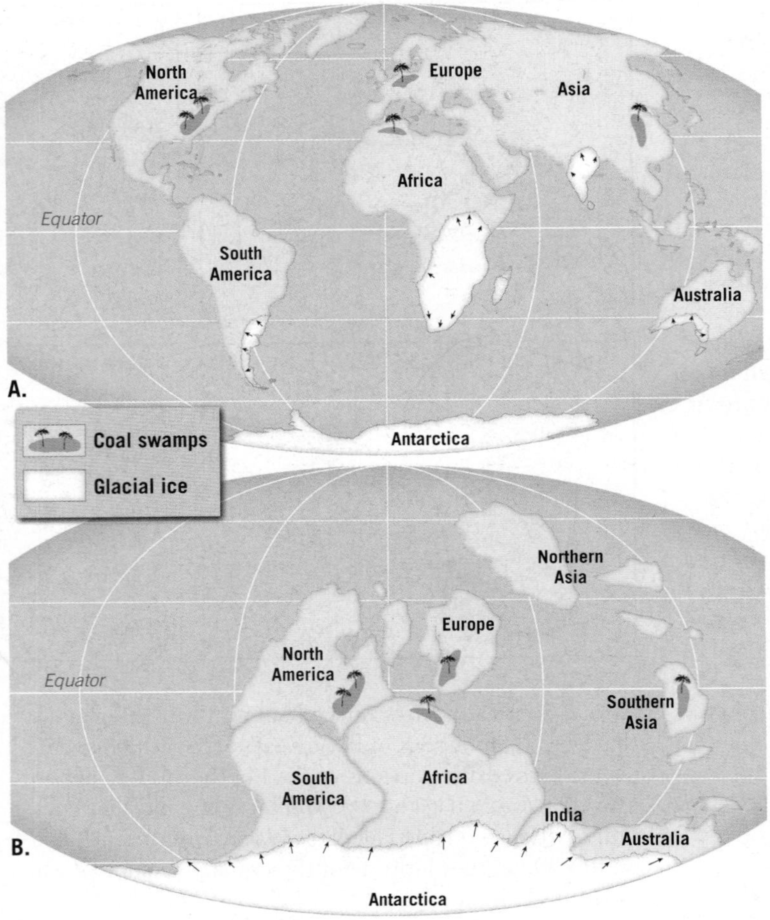

▶ Figure 2.7
Paleoclimatic evidence for continental drift
A. Continents in their current locations provide evidence that about 300 million years ago, ice sheets covered extensive areas of the Southern Hemisphere and India. Arrows show the direction of ice movement that can be inferred from the pattern of glacial scratches and grooves found in the bedrock. Tropical coal swamps also existed in areas that are now temperate.
B. Restoring the continents to their pre-drift positions creates a single glaciation centered on the South Pole and puts the coal swamps near the equator.

‡Note that coal can form in a variety of climates, provided that large quantities of plant life are buried.

continents nearer the equator and accounts for the tropical swamps that generated the vast coal deposits.

The Great Debate

From publication of his book until well past his death, Wegener's drift hypothesis encountered a great deal of hostile criticism. One of the main objections stemmed from his inability to identify a credible mechanism for continental drift. Wegener proposed that gravitational forces of the Moon and Sun that produce Earth's tides were also capable of gradually moving the continents across the globe. However, the prominent physicist Harold Jeffreys correctly argued that tidal forces strong enough to move Earth's continents would have resulted in halting our planet's rotation—which, of course, has not happened.

Wegener also incorrectly suggested that the larger and sturdier continents broke through thinner oceanic crust, much as icebreakers cut through ice. However, no evidence existed to suggest that the ocean floor was weak enough to permit passage of the continents without the continents being appreciably deformed in the process.

In 1930, Wegener made his fourth and final trip to the Greenland Ice Sheet (**Figure 2.8**). Although the primary focus of this expedition was to study this great ice cap and its climate, Wegener continued to test his continental drift hypothesis. While returning from Eismitte, an experimental station located in the center of Greenland, Wegener perished along with his Greenland companion. His intriguing idea, however, did not die.

Why was Wegener unable to overturn the established scientific views of his day? Foremost was the fact that, although the central theme of Wegener's drift hypothesis was correct, some details were incorrect. For example, continents do not break through the ocean floor, and tidal energy is much too weak to move continents. For any comprehensive scientific hypothesis to gain wide acceptance, it must withstand critical testing from all areas of science. Despite Wegener's great contribution to our understanding of Earth, not *all* of the evidence supported the continental drift hypothesis as he had

Today these deposits comprise major coal fields in the eastern United States and Northern Europe. Many of the fossils found in these coal-bearing rocks were produced by tree ferns with large fronds—ferns that would have grown in warm, moist climates.‡ The existence of these large tropical swamps, Wegener argued, was inconsistent with the proposal that extreme global cooling caused glaciers to form in areas that are currently tropical.

Wegener suggested a more plausible explanation for the late Paleozoic glaciation: The southern continents were joined together in the supercontinent of Pangaea and located near the South Pole (see Figure 2.7B). This would account for the polar conditions required to generate extensive expanses of glacial ice over much of these landmasses. At the same time, this geography places today's northern

"It is just as if we were to refit the torn pieces of a newspaper by matching their edges and then check whether the lines of print run smoothly across. If they do, there is nothing left but to conclude that the pieces were in fact joined in this way."
—Alfred Wegener

proposed it. As a result, most of the scientific community (particularly in North America) rejected or treated continental drift with considerable skepticism. However, some scientists recognized the strength of the evidence Wegner had accumulated and continued to pursue the idea.

CONCEPT CHECKS 2.2

1. What was the first line of evidence that led early investigators to suspect that the continents were once connected?

2. Explain why the discovery of the fossil remains of *Mesosaurus* in both South America and Africa, but nowhere else, supports the continental drift hypothesis.

Concept Checker
https://goo.gl/ARX1mS

3. Early in the twentieth century, what was the prevailing view of how land animals apparently migrated across vast expanses of open ocean?

4. Describe two aspects of Wegener's continental drift hypothesis that were objectionable to most Earth scientists.

Alfred Wegener, shown waiting out the 1912–1913 Arctic winter during an expedition to Greenland, where he made a 1200-kilometer traverse across the widest part of the island's ice sheet.

▲ Figure 2.8
Alfred Wegener during an expedition to Greenland

2.3 The Theory of Plate Tectonics

List the major differences between Earth's lithosphere and asthenosphere and explain the importance of each in the plate tectonics theory.

Following World War II, oceanographers equipped with new marine tools and ample funding embarked on an unprecedented period of oceanographic exploration. Over the next two decades, a much better picture of large expanses of the seafloor began to emerge. From this work came the discovery of a global oceanic ridge system that winds through all the major oceans.

Studies conducted in the western Pacific also demonstrated that earthquakes were occurring at great depths beneath deep-ocean trenches. Of equal importance was the fact that dredging of the seafloor did not bring up any oceanic crust that was older than 180 million years. Further, sediment accumulations in the deep-ocean basins were found to be thin, not the thousands of meters that had been predicted. By 1968 these developments, among others, led to the unfolding of a far more encompassing theory than continental drift, known as the **theory of plate tectonics**.

Rigid Lithosphere Overlies Weak Asthenosphere

According to the plate tectonics model, the crust and the uppermost, and therefore coolest, part of the mantle constitute Earth's strong outer layer, the **lithosphere** (*lithos* = stone). The lithosphere varies in both thickness and density, depending on whether it is oceanic or continental (**Figure 2.9**). Oceanic lithosphere is about 100 kilometers (60 miles) thick in the deep-ocean basins but is considerably thinner along the crest of the oceanic ridge system—a topic we will consider later. In contrast, continental lithosphere averages about 150 kilometers (90 miles) thick but may extend to depths of 200 kilometers (125 miles) or more beneath the stable interiors of the continents. Further, oceanic and continental crust differ in density. Oceanic crust is composed of *basalt*, a rock rich in dense iron and magnesium, whereas continental crust is composed largely of less dense *granitic* rocks. Because of these differences, the overall density of oceanic lithosphere (crust and upper mantle) is greater than the overall density of continental lithosphere. This important difference will be considered in greater detail later in this chapter.

The **asthenosphere** (*asthenos* = weak) is a hot, comparatively weak region in the mantle that lies directly below the lithosphere (see Figure 2.9). In the upper asthenosphere (located between 100 and 200 kilometers [60 to 125 miles] depth), the high pressure and

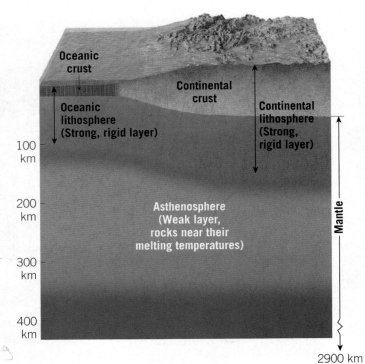

The asthenosphere is heated but still stay solid. However, the react by pushing upwards towards the lithosphere bending or breaking it forming many plates

temperature bring rock very near to melting. Consequently, although the rock remains largely solid, it responds to forces by *flowing*, similarly to the way clay may deform if you compress it slowly. By contrast, the relatively cool and rigid lithosphere tends to respond to forces acting on it by *bending or breaking but not flowing*. Because of these differences, Earth's rigid outer shell is effectively detached from the asthenosphere, which allows these layers to move independently.

Earth's Major Plates

The lithosphere is broken into numerous segments of irregular size and shape called **lithospheric plates**, or simply **plates**, which are in constant motion with respect to one another (**Figure 2.10**). Seven major lithospheric plates are recognized and account for 94 percent of Earth's surface area: the *North American*, *South American*, *Pacific*, *African*, *Eurasian*, *Australian-Indian*, and *Antarctic plates*. The largest is the Pacific plate, which encompasses a significant portion of the Pacific basin. Each of the six other large plates consists of an entire continent, as well as a significant amount of oceanic crust. Notice in Figure 2.10 that the South American plate encompasses almost all of South America and about one-half of the floor of the South Atlantic. Note also that none of the plates are defined entirely by the margins of a single continent. This is a major departure from Wegener's continental drift hypothesis, which proposed that the continents move through the ocean floor, not with it.

Intermediate-sized plates include the *Caribbean, Nazca, Philippine, Arabian, Cocos, Scotia,* and *Juan de Fuca plates*. These plates, with the exception of the Arabian plate, are composed mostly of oceanic lithosphere. In addition, many smaller plates, called *microplates*, have been identified but are not shown in Figure 2.10.

Plate Movement

One of the main tenets of the plate tectonics theory is that as plates move, the distance between two locations on different plates, such as New York and London, gradually changes, whereas the distance between sites on the same plate—New York and Denver, for example—remains relatively constant. However, parts of some plates are comparatively "weak." For example, southern China is literally being squeezed as the Indian subcontinent rams into Asia.

Because plates are in constant motion relative to each other, most major interactions occur along their *boundaries*, and this is therefore where most crustal deformation occurs. In fact, plate boundaries were first established by plotting the locations of earthquakes and volcanoes. Plates are delimited by three distinct types of boundaries that exhibit different types of movement:

- Divergent plate boundaries—where two plates move apart, resulting in upwelling and partial melting of hot material from the mantle to create new seafloor (see Figure 2.11, page 46)
- Convergent plate boundaries—where two plates move toward each other, resulting either in oceanic lithosphere descending beneath an overriding plate, eventually to be reabsorbed into the mantle, or possibly in the collision of two continental blocks to create a mountain belt (see Figure 2.15, page 49)
- Transform plate boundaries—where two plates grind past each other without the production or destruction of lithosphere (see Figure 2.20, page 52)

Divergent and convergent plate boundaries each account for about 40 percent of all plate boundaries. Transform boundaries account for the remaining 20 percent. In the following sections we will discuss the three types of plate boundaries.

CONCEPT CHECKS 2.3

1. What new findings about the ocean floor did oceanographers discover after World War II?

2. Compare and contrast Earth's lithosphere and asthenosphere. **Concept Checker** https://goo.gl/i5xLo8

3. List the three types of plate boundaries and describe the relative motion along each.

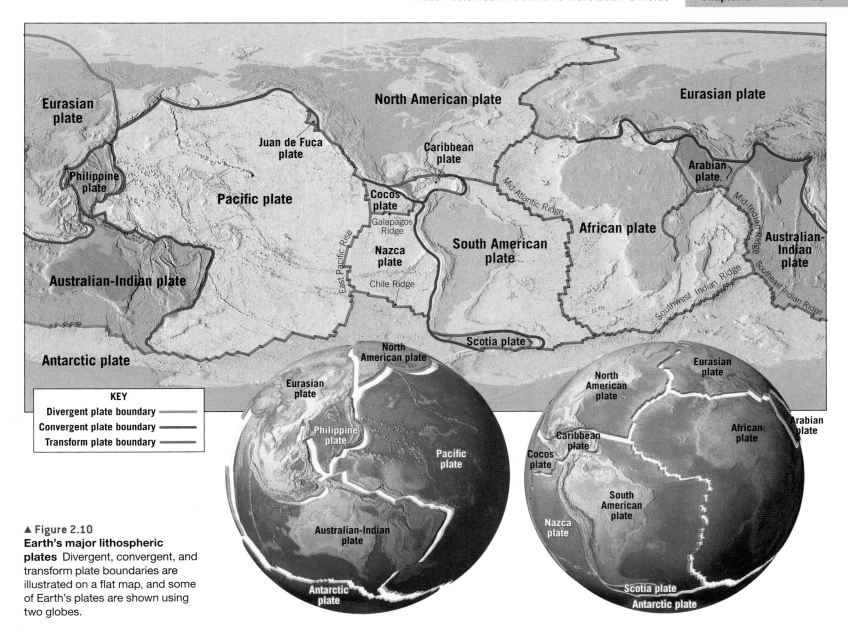

KEY
Divergent plate boundary
Convergent plate boundary
Transform plate boundary

▲ **Figure 2.10**
Earth's major lithospheric plates Divergent, convergent, and transform plate boundaries are illustrated on a flat map, and some of Earth's plates are shown using two globes.

2.4 Divergent Plate Boundaries and Seafloor Spreading

Sketch and describe the movement along a divergent plate boundary that results in the formation of new oceanic lithosphere.

Most **divergent plate boundaries** (*di* = apart, *vergere* = to move) are located along the crests of oceanic ridges and can be thought of as *constructive plate margins*, because this is where new ocean floor is generated (**Figure 2.11**). Oceanic ridges are elevated areas of the seafloor characterized by high heat flow and volcanism. The global **oceanic ridge system** is the longest topographic feature on Earth's surface, exceeding 70,000 kilometers (43,000 miles) in length. As shown in Figure 2.10, various segments of the global ridge system have been named, including the Mid-Atlantic Ridge, East Pacific Rise, and Mid-Indian Ridge.

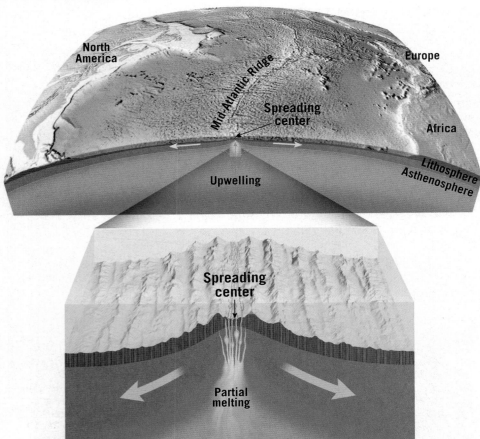

▲ Figure 2.11
Seafloor spreading Most divergent plate boundaries are situated along the crests of oceanic ridges—the sites of seafloor spreading.

Representing 20 percent of Earth's surface, the oceanic ridge system winds through all major ocean basins—much like the seams on a baseball. Although the crest of the oceanic ridge is commonly 2 to 3 kilometers (1 to 2 miles) higher than the adjacent ocean basins, the term *ridge* may be misleading because it implies "narrow" when, in fact, ridges vary in width from 1000 kilometers (600 miles) to more than 4000 kilometers (2500 miles). Further, along the crest of some ridge segments

▶ SmartFigure 2.12
Rift valley in Iceland Thingvellir National Park, Iceland, is located on the western margin of a rift valley roughly 30 kilometers (20 miles) wide. This rift valley is connected to a similar feature that extends along the crest of the Mid-Atlantic Ridge. The cliff in the left half of the image approximates the eastern edge of the North American plate.

Mobile Field Trip
https://goo.gl/RsbHWM

is a deep canyonlike structure called a **rift valley** (**Figure 2.12**). These structures are evidence that tensional (pulling apart) forces are actively pulling apart the ocean crust at the ridge crest.

Seafloor Spreading

The mechanism that operates along the oceanic ridge system to create new seafloor is called **seafloor spreading**. Spreading typically averages around 5 centimeters (2 inches) per year, roughly the same rate at which human fingernails grow. Comparatively slow spreading rates of 2 centimeters (about 1 inch) per year are found along the Mid-Atlantic Ridge, whereas spreading rates exceeding 15 centimeters (6 inches) per year have been measured along sections of the East Pacific Rise. Although these rates of seafloor production are slow on a human time scale, they are rapid enough to have generated all of Earth's current oceanic lithosphere within the past 200 million years.

Seafloor spreading occurs where two adjacent plates move away from each other, producing long, narrow fractures in the ocean crust. The resulting upwelling from below causes a small percentage of the rising hot mantle rock of the asthenosphere to melt. This process

generates magma that rises and builds the 7-kilometer-thick oceanic crust along the ridge axis—at the edges of the two diverging plates. Thus, in a slow yet unending manner, adjacent plates spread apart, and new oceanic crust forms between them. For this reason, divergent plate boundaries are also called **spreading centers**.

How Does Oceanic Lithosphere Change as It Ages?

Beneath the newly formed oceanic crust, the lithospheric mantle is quite thin because the temperatures in the zone of upwelling are unusually hot. The hot, and therefore low-density, rock of the asthenosphere supports the elevated ridge above, which can be more than 2.5 kilometers (about 1.5 miles) higher than the adjacent ocean basin.

As soon as the oceanic crust forms, it is slowly and continually displaced away from the zone of hot mantle upwelling. The newly formed crust, which is in contact with relatively cold ocean water, cools quickly and maintains its thickness of about 7 kilometers (about 4 miles). Simultaneously, the underlying asthenosphere cools at a slower pace by radiating heat to the crust above. Cooling causes the upper asthenosphere to become increasingly rigid, and eventually it becomes part of the oceanic lithosphere. Stated another way, cooling of the warm, weak asthenosphere as it moves away from the zone of upwelling produces cooler, stronger lithospheric rock. As the lithosphere grows thicker, cooling also causes it to contract and increase in density.

It takes about 80 million years for the temperature of oceanic lithosphere to stabilize. At this point, it will have reached its maximum thickness of about 100 kilometers (60 miles). Therefore, to a certain degree, the thickness of oceanic lithosphere is age dependent: The older (cooler) it is, the greater its thickness.

The contraction of the oceanic lithosphere and its increase in density account for the increase in ocean depth away from the ridge crest. Thus, rock that was once part of the elevated oceanic ridge system will be located in the deep-ocean basin, roughly 5 kilometers (3 miles) below sea level. One way to think about the change in elevation of newly formed oceanic lithosphere is that the plate is young, hot, and high at the ridge crest and becomes old, cold, and low during the aging and cooling process.

Continental Rifting

Divergent boundaries can develop within a continent and may cause the landmass to split into two or more smaller segments separated by an ocean basin. Continental rifting begins when plate motions produce tensional forces that pull

and stretch the lithosphere. This stretching, in turn, promotes mantle upwelling and broad upwarping of the overlying lithosphere (**Figure 2.13A**). This process thins the lithosphere and breaks the brittle crustal rocks into large blocks. As the tectonic forces continue to pull apart the crust, the broken crustal fragments sink, generating an elongated depression called a **continental rift**, which can widen to form a narrow sea (**Figure 2.13B,C**) and eventually a new ocean basin (**Figure 2.13D**).

An example of an active continental rift is the East African Rift (**Figure 2.14**). Whether this rift will eventually result in the breakup of Africa is a topic of ongoing research. Nevertheless, the East African Rift is an excellent model of the initial stage in the breakup of a continent. Here, tensional forces have stretched and thinned the lithosphere, allowing molten rock to ascend from the upper mantle. Evidence for this upwelling includes several large volcanic mountains, including

▼ SmartFigure 2.13
Continental rifting: Formation of new ocean basins

Tutorial
https://goo.gl/F7EEjD

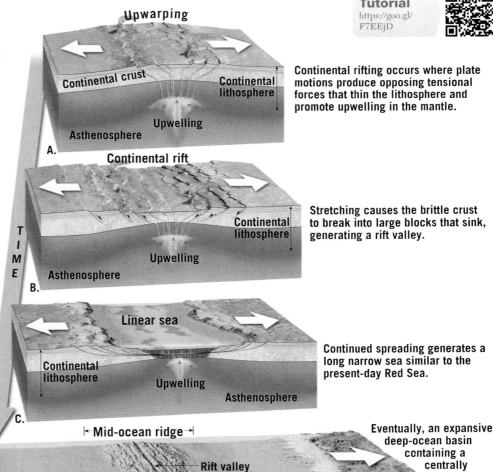

Continental rifting occurs where plate motions produce opposing tensional forces that thin the lithosphere and promote upwelling in the mantle.

Stretching causes the brittle crust to break into large blocks that sink, generating a rift valley.

Continued spreading generates a long narrow sea similar to the present-day Red Sea.

Eventually, an expansive deep-ocean basin containing a centrally located oceanic ridge is formed by continued seafloor spreading.

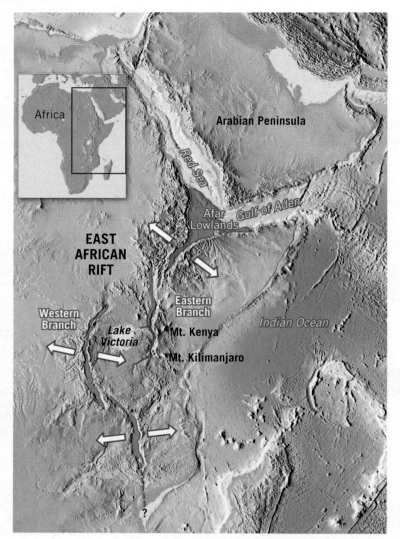

◄ SmartFigure 2.14
East African Rift Valley The East African Rift Valley represents the early stage in the breakup of a continent. Areas shown in red consist of lithosphere that has been stretched and thinned, allowing magma to well up from the mantle.

Condor Video
https://goo.gl/beYtRz

Mount Kilimanjaro and Mount Kenya, Africa's tallest peaks. Research suggests that if rifting continues, the rift valley will lengthen and deepen (see Figure 2.13C). At some point, the rift valley will become a narrow sea with an outlet to the ocean. The Red Sea, formed when the Arabian Peninsula split from Africa, is a modern example of such a feature and provides us with a view of how the Atlantic Ocean may have looked in its infancy (see Figure 2.13D).

CONCEPT CHECKS 2.4

1. Sketch or describe how two plates move in relation to each other along divergent plate boundaries.

2. What is the average rate of seafloor spreading in modern oceans?

Concept Checker
https://goo.gl/8WjCDw

3. List four features that characterize the oceanic ridge system.

2.5 Convergent Plate Boundaries and Subduction

Compare and contrast the three types of convergent plate boundaries and name a location where each type can be found.

New lithosphere is constantly being produced at the oceanic ridges. However, our planet is not growing larger; its total surface area remains constant. A balance is maintained because older, denser portions of oceanic lithosphere descend into the mantle at a rate equal to seafloor production. This activity occurs along **convergent plate boundaries**, where two plates move toward each other, and the leading edge of one is bent downward as it slides beneath the other.

Convergent boundaries are also called **subduction zones** because they are sites where lithosphere is descending (being subducted) into the mantle. A lithospheric plate subducts (sinks) when its density is greater than the density of the underlying asthenosphere. In general, old oceanic lithosphere, which is largely composed of dense ferromagnesian-rich minerals, is about 2 percent more dense than the underlying asthenosphere, and it sinks much like a ship's anchor. Continental lithosphere, in contrast, is less dense than both oceanic lithosphere and the underlying asthenosphere and tends to resist subduction. However, there are a few locations where continental lithosphere is thought to have been forced below an overriding plate, albeit to relatively shallow depths.

Geologists agree that the denser lithospheric plates subduct into the underlying asthenosphere. However, the Earth's strong and rigid outer shell must break before subduction can begin. But just exactly what causes the lithosphere to break remains strongly debated by researchers.

Deep-ocean trenches are long, linear depressions in the seafloor, generally located a few hundred kilometers off-shore from either a continent or a chain of volcanic islands such as the Aleutian chain (see Figure 13.8, page 369). These underwater surface features are produced where oceanic lithosphere bends as it descends into the mantle along subduction zones (see Figure 2.15A). An example is the Peru–Chile trench located along the west coast of South America. It is more than 4500 kilometers (3000 miles) long, and its floor is as much as 8 kilometers (5 miles) below sea level.

Slabs of oceanic lithosphere descend into the mantle at angles that vary from a few degrees to nearly vertical (90 degrees). The angle at which oceanic lithosphere subducts depends largely on its age and, therefore, its density. For example, when seafloor spreading occurs relatively near a subduction zone, as is the case along the coast of Chile (see Figure 2.10B), the subducting litho-sphere is young and buoyant, which results in a low angle of descent. As the two plates converge, the overriding plate scrapes over the top of the subducting plate below—a type of forced subduction. Consequently, the region around the Peru–Chile trench experiences great earthquakes, including the 2010 Chilean earthquake—one of the 10 largest quakes on record.

As oceanic lithosphere ages (moves farther from the spreading center), it gradually cools, which causes it to thicken and increase in density. In parts of the western Pacific, some oceanic lithosphere is 180 million years old—the thickest and densest in today's oceans. The very dense slabs in this region typically plunge into the mantle at angles approaching 90 degrees. This largely explains why most trenches in the western Pacific, including the Mariana and Tonga trenches, are deeper than trenches in the eastern Pacific.

Although all convergent zones have the same basic characteristics, they may vary considerably depending on the type of crustal material involved and the tectonic setting. Convergent boundaries can form *between one oce-anic plate and one continental plate, between two oceanic plates,* or *between two continental plates* (**Figure 2.15**).

Oceanic–Continental Convergence

When the leading edge of a plate capped with continen-tal crust converges with a slab of oceanic lithosphere, the buoyant continental block remains "floating," while the denser oceanic slab sinks into the mantle (see Figure 2.15A). This process triggers melting in the wedge of mantle rock directly above the descending plate. How does the subduction of a cool slab of oceanic lithosphere cause mantle rock to melt?

The oceanic crust contains large amounts of water, which is carried to great depths by a descending plate. As the plate plunges downward, heat and pressure drive out water from the hydrated (water-rich) minerals in the

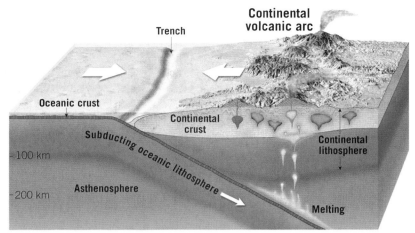

A. Convergent plate boundary where oceanic lithosphere is subducting beneath continental lithosphere.

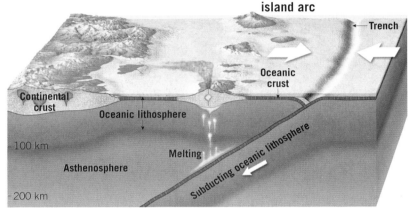

B. Convergent plate boundary involving two slabs of oceanic lithosphere.

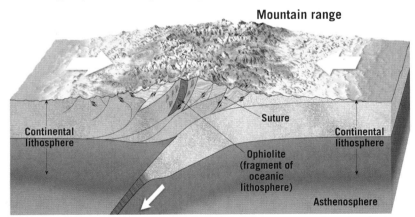

C. Continental collisions occur along convergent plate boundaries when both plates are capped with continental crust.

▲ **SmartFigure 2.15**
Three types of convergent plate boundaries

Tutorial
https://goo.gl/ ywGx8h

subducting slab. At a depth of roughly 100 kilometers (60 miles), the wedge of mantle rock is sufficiently hot that the introduction of water from the slab below leads to some melting (**Figure 2.16A**). Water released by the descending plates acts the way salt does to melt ice. That is, "wet" rock in a high-pressure environment melts at substantially lower temperatures than does "dry" rock of the same composition.

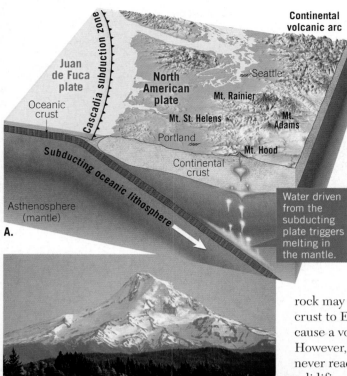

▲ **Figure 2.16**
Example of an oceanic–continental convergent plate boundary
A. The Cascade Range is a *continental volcanic arc* formed by the subduction of the Juan de Fuca plate beneath the North American plate.
B. Mount Hood, Oregon, is one of more than a dozen large composite volcanoes in the Cascade Range.

This process, called **partial melting**, is thought to generate some magma, which is mixed with unmelted mantle rock. Being less dense than the surrounding mantle, this hot mobile material gradually rises toward the surface. Depending on the environment, these mantle-derived masses of molten rock may ascend through the crust to Earth's surface and cause a volcanic eruption. However, much of this material never reaches the surface but solidifies at depth—a process that thickens the crust.

The volcanoes of the towering Andes were produced by molten rock generated by the subduction of the Nazca plate beneath the South American continent (see Figure 2.10). Mountain systems like the Andes, which are produced in part by volcanic activity associated with the subduction of oceanic lithosphere, are called **continental volcanic arcs**. The Cascade Range in Washington, Oregon, and California is another mountain system consisting of several well-known volcanoes, including Mount Rainier, Mount Shasta, Mount St. Helens, and Mount Hood (**Figure 2.16B**). This active volcanic arc also extends into Canada, where it includes Mount Garibaldi and Mount Meager.

Oceanic–Oceanic Convergence

An *oceanic–oceanic convergent boundary* has many features in common with oceanic–continental plate margins (see Figure 2.15A,B). Where two oceanic slabs converge, one descends beneath the other, initiating volcanic activity by the same mechanism that operates at all subduction zones (see Figure 2.10). Water released from the subducting slab of oceanic lithosphere triggers melting in the hot wedge of mantle rock above. In this setting, volcanoes grow up from the ocean floor rather than upon a continental platform. Sustained subduction eventually results in a chain of volcanic structures large enough to emerge as islands. The newly formed land, consisting of an arc-shaped chain of volcanic islands,

is called a **volcanic island arc** or simply an **island arc** (Figure 2.17).

Island arcs are generally located 120 to 360 kilometers (75 to 225 miles) from a deep-ocean trench. The Aleutian, Mariana, and Tonga Islands are examples of relatively young volcanic island arcs; they are each located adjacent to trenches of the same names.

Most volcanic island arcs are located in the western Pacific. Only two are located in the Atlantic—the Lesser Antilles arc, on the eastern margin of the Caribbean Sea, and the Sandwich Islands, located off the tip of South America. The Lesser Antilles are a product of the subduction of the Atlantic seafloor beneath the Caribbean plate. Located within this volcanic arc are the Virgin Islands of the United States and Britain as well as Martinique, where Mount Pelée erupted in 1902, destroying the town of St. Pierre and killing an estimated 28,000 people. This chain of islands also includes Montserrat, where volcanic activity has occurred as recently as 2010.

Island arcs are typically simple structures made of numerous volcanic cones underlain by oceanic crust that is generally less than 20 kilometers (12 miles) thick. Some island arcs, however, are more complex and are underlain by highly deformed crust that may reach 35 kilometers (22 miles) in thickness. Examples include Japan, Indonesia, and the Alaskan Peninsula. These island arcs are built on material generated by earlier episodes of subduction, or on small slivers of continental crust that have rafted away from the mainland.

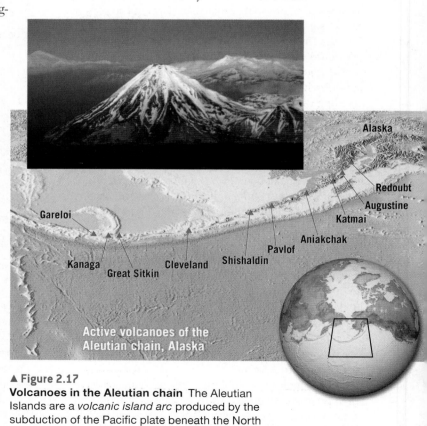

▲ **Figure 2.17**
Volcanoes in the Aleutian chain The Aleutian Islands are a *volcanic island arc* produced by the subduction of the Pacific plate beneath the North American plate. Notice that the volcanoes of the Aleutian chain extend into Alaska proper.

Continental–Continental Convergence

The third type of convergent boundary results when one landmass collides with the margin of another because of subduction of the intervening seafloor (**Figure 2.18A**). Whereas oceanic lithosphere tends to be dense and readily sinks into the mantle, the buoyancy of continental material generally inhibits it from being subducted, at least to any great depth. Consequently, a collision between two converging continental fragments ensues (**Figure 2.18B**). This process folds and deforms the accumulation of sediments and sedimentary rocks along the continental margins as if they had been placed in a gigantic vise. The result is the formation of a new mountain belt composed of deformed sedimentary and metamorphic rocks that often contain slivers of oceanic lithosphere.

Such a collision began about 50 million years ago, when the subcontinent of India "rammed" into Asia, producing the Himalayas—the most spectacular mountain range on Earth (**Figure 2.18C**). During this collision, the continental crust buckled and fractured and was generally shortened horizontally and thickened vertically. In addition to the Himalayas, several other major mountain systems, including the Alps, Appalachians, and Urals, formed as continental fragments collided. This topic will be considered further in Chapter 14.

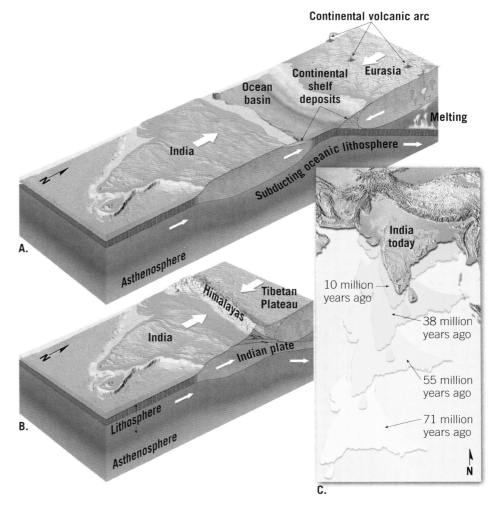

CONCEPT CHECKS 2.5

1. Why does oceanic lithosphere subduct, while continental lithosphere does not?

2. What characteristic of a slab of oceanic lithosphere explains the formation of a deep oceanic trench as opposed to one that is less deep?

Concept Checker
https://goo.gl/fJJCF5

3. What distinguishes a continental volcanic arc from a volcanic island arc?

▲ SmartFigure 2.18
The collision of India and Eurasia formed the Himalayas The ongoing collision of the subcontinent of India with Eurasia began about 50 million years ago and produced the majestic Himalayas. Although the map in part C illustrates only the movement of India, it should be noted that both India and Eurasia were moving as these landmasses collided.

Animation
https://goo.gl/Wy8mSX

2.6 Transform Plate Boundaries

Describe the relative motion along a transform fault boundary and locate several examples of transform faults on a plate boundary map.

Along a **transform plate boundary**, also called a **transform fault**, plates slide horizontally past one another without the production or destruction of lithosphere. The nature of transform faults was discovered in 1965 by Canadian geologist J. Tuzo Wilson, who proposed that these large faults connect two spreading centers (divergent boundaries) or, less commonly, two trenches (convergent boundaries). Most transform faults are found on the ocean floor, where they offset segments of the oceanic ridge system, producing a steplike plate margin (**Figure 2.19A**). Notice that the zigzag shape of the Mid-Atlantic Ridge in Figure 2.10 (see page 45) roughly reflects the shape of the original rifting that caused the breakup of the supercontinent Pangaea. (Compare the shapes of the continental margins of the landmasses on both sides of the Atlantic with the shape of the Mid-Atlantic Ridge.)

A. The Mid-Atlantic Ridge, with its zigzag pattern, roughly reflects the shape of the zone of rifting that resulted in the breakup of Pangaea.

B. Fracture zones are long, narrow scar-like features in the seafloor that are roughly perpendicular to the offset ridge segments. They include both the active transform fault and its preserved trace.

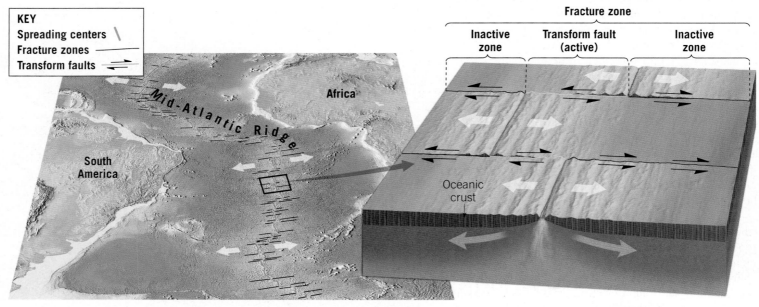

KEY
Spreading centers ╲
Fracture zones ──
Transform faults ⇄

▲ SmartFigure 2.19
Transform plate boundaries Most transform faults offset segments of a spreading center, producing a plate margin that exhibits a zigzag pattern.

Tutorial
https://goo.gl/
N76csp

Typically, transform faults are part of prominent linear breaks in the seafloor known as **fracture zones**, which include both active transform faults and their inactive extensions into the plate interior (**Figure 2.19B**). In a fracture zone, the active transform fault lies *only between* the two offset ridge segments; it is generally defined by weak, shallow earthquakes. On each side of the fault, the seafloor moves away from the corresponding ridge

segment. Thus, between the ridge segments, these adjacent slabs of oceanic crust are grinding past each other along a transform fault. Beyond the ridge crests, these faults are inactive because the rock on either side moves in the same direction. However, these inactive faults are preserved as linear topographic depressions, which are evidence of past transform fault activity. The trend (orientation) of these fracture zones roughly parallels the

▶ **Figure 2.20**
Transform faults facilitate plate motion
Seafloor generated along the Juan de Fuca Ridge moves southeastward, past the Pacific plate. Eventually it subducts beneath the North American plate. Thus, this transform fault connects a spreading center (divergent boundary) to a subduction zone (convergent boundary). Also shown is the San Andreas Fault, a transform fault connecting a spreading center located in the Gulf of California with the Mendocino Fault.

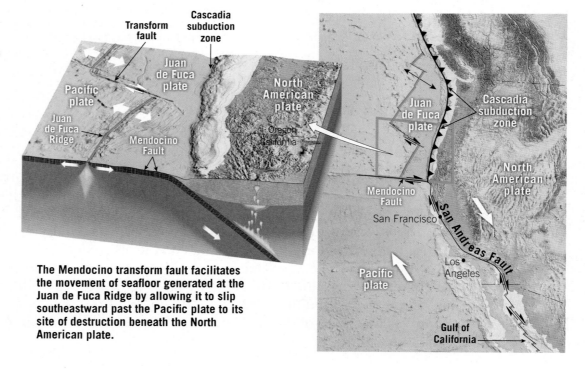

The Mendocino transform fault facilitates the movement of seafloor generated at the Juan de Fuca Ridge by allowing it to slip southeastward past the Pacific plate to its site of destruction beneath the North American plate.

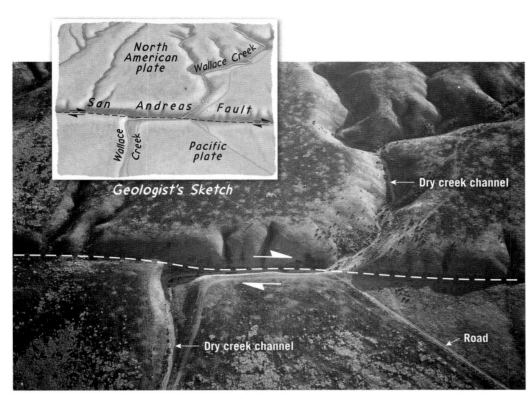

Geologist's Sketch

← Dry creek channel

← Dry creek channel

Road →

◄ **SmartFigure 2.21**
Movement along the San Andreas Fault This aerial view shows the offset in the dry channel of Wallace Creek near Taft, California.

Mobile Field Trip
https://goo.gl/3Zj4o4

direction of plate motion at the time of their formation. Thus, these structures help geologists map the direction of plate motion in the geologic past.

Transform faults also provide the means by which the oceanic crust created at ridge crests can be transported to a site of destruction—the deep-ocean trenches. **Figure 2.20** illustrates this scenario. Notice that the Juan de Fuca plate moves in a southeasterly direction, eventually being subducted under the West Coast of the United States and Canada. The southern end of this plate is bounded by a transform fault called the Mendocino Fault. This transform boundary connects the Juan de Fuca Ridge to the Cascadia subduction zone. Therefore, this fault facilitates the movement of the crustal material created at the Juan de Fuca Ridge to its destination beneath the North American continent.

Most transform fault boundaries (such as the Mendocino Fault) are located within the ocean basins; however, a few cut through continental crust. Two examples found on continental crust are the San Andreas Fault of California and New Zealand's Alpine Fault. Notice in Figure 2.20 that the San Andreas Fault connects a spreading center located in the Gulf of California to the Cascadia subduction zone and the Mendocino Fault. Along the San Andreas Fault, the Pacific plate is moving toward the northwest, past the North American plate (**Figure 2.21**). If this movement continues, the part of California west of the fault zone, including Mexico's Baja Peninsula, will become an island off the West Coast of the United States and Canada. However, a more immediate concern is the earthquake activity triggered by movements along this fault system.

CONCEPT CHECKS 2.6

1. Sketch or describe how two plates move in relationship to each other along a transform plate boundary.

Concept Checker
https://goo.gl/3SvmYN

2. List two characteristics that differentiate transform faults from the two other types of plate boundaries.

2.7 How Do Plates and Plate Boundaries Change?

Explain why plates such as the African and Antarctic plates are increasing in size, while the Pacific plate is decreasing in size.

Although Earth's total surface area does not vary, the size and shape of individual plates constantly changes. For example, the African and Antarctic plates, which are mainly bounded by divergent boundaries—sites of seafloor production—continually grow as new lithosphere is added to their margins. By contrast, the Pacific plate is getting smaller because it is being consumed into the mantle along much of its outer margin faster than it is being generated along the East Pacific Rise.

Another result of plate motion is that boundaries migrate. For example, the position of the Peru–Chile trench, which is the result of the Nazca plate being bent downward as it descends beneath the South American plate, has changed over time (see Figure 2.10). Because of the westward drift of the South American plate relative to the Nazca plate, the Peru–Chile trench has migrated in a westerly direction as well.

Plate boundaries can also be created or destroyed in response to changes in the forces acting on the lithosphere. For example, some plates carrying continental crust are presently moving toward one another. In the South Pacific, Australia is moving northward. If this continues, the boundary separating Australia from southern Asia will eventually disappear as these plates become one. Other plates are moving apart. Recall that the Red Sea is the site of a relatively new spreading center that came into existence less than 20 million years ago, when the Arabian Peninsula began to break apart from Africa.

The Breakup of Pangaea

The breakup of Pangaea is a classic example of how plate boundaries change through geologic time. By employing modern tools not available to Wegener, geologists re-created the steps in the breakup of this supercontinent, an event that began about 180 million years ago (**Figure 2.22A,B**). This work established the dates and relative motions of when individual crustal fragments separated from one another (**Figure 2.22**).

An important consequence of Pangaea's breakup was the creation of a new ocean basin: the Atlantic. As you can see in Figure 2.22, splitting of the supercontinent did not occur simultaneously along the margins of the Atlantic. The first split developed between North America and Africa. Here, the continental crust was highly fractured, providing pathways for huge quantities of fluid lavas to reach the surface. Today, these solidified lavas are represented by weathered igneous rocks found along the eastern seaboard of the United States—primarily buried beneath the sedimentary rocks that form the continental shelf. Radiometric dating indicates rifting began about 180 million years ago. This time span represents the "birth date" for this section of the North Atlantic.

By 130 million years ago, the South Atlantic began to open near the tip of what is now South Africa. As this zone of rifting migrated northward, it gradually opened the South Atlantic (**Figures 2.22B,C**). Continued breakup of the southern landmass led to

► Figure 2.22
The breakup of Pangaea

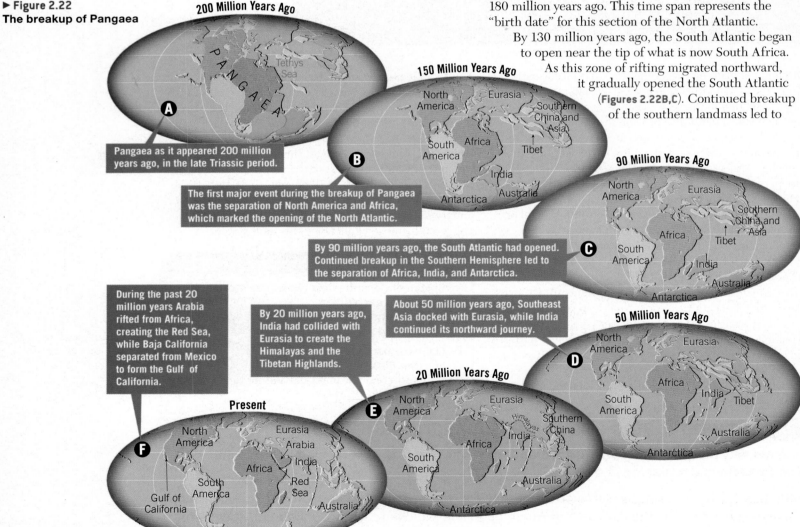

Pangaea as it appeared 200 million years ago, in the late Triassic period.

The first major event during the breakup of Pangaea was the separation of North America and Africa, which marked the opening of the North Atlantic.

By 90 million years ago, the South Atlantic had opened. Continued breakup in the Southern Hemisphere led to the separation of Africa, India, and Antarctica.

During the past 20 million years Arabia rifted from Africa, creating the Red Sea, while Baja California separated from Mexico to form the Gulf of California.

By 20 million years ago, India had collided with Eurasia to create the Himalayas and the Tibetan Highlands.

About 50 million years ago, Southeast Asia docked with Eurasia, while India continued its northward journey.

the separation of Africa and Antarctica and sent India on a northward journey. By about 50 million years ago (early Cenozoic era), Australia had separated from Antarctica, and the South Atlantic had become a full-fledged ocean (**Figure 2.22D**).

India eventually collided with Asia (**Figure 2.22E**), an event that began about 50 million years ago and created the Himalayas and the Tibetan Highlands. About the same time, the separation of Greenland from Eurasia completed the breakup of the northern landmass. During the past 20 million years or so of Earth's history, Arabia has rifted from Africa to form the Red Sea, and Baja California has separated from Mexico to form the Gulf of California (**Figure 2.22F**). Meanwhile, the Panama Arc joined North America and South America to produce our globe's familiar modern appearance.

Plate Tectonics in the Future

Geologists have extrapolated present-day plate movements into the future. **Figure 2.23** illustrates where Earth's landmasses may be 50 million years from now if today's plate movements persist during this time span.

In North America, we see that the Baja Peninsula and the portion of southern California that lies west of the San Andreas Fault will have slid past the North American plate. If this northward migration continues, Los Angeles and San Francisco will pass each other in about 10 million years, and in about 60 million years the Baja Peninsula will begin to collide with the Aleutian Islands.

If Africa maintains its northward path, it will continue to collide with Eurasia. The result will be the closing of the Mediterranean Sea, the last remnant of a once-vast ocean called the Tethys Ocean, and the initiation of another major mountain-building episode (see Figure 2.23). Australia will be astride the equator and, along with New Guinea, will be on a collision course with Asia. Meanwhile, North and South America will begin to separate, while the Atlantic and Indian Oceans will continue to grow at the expense of the Pacific Ocean, which will shrink.

If movement along the San Andreas fault continues as projected, then about 10 million years in the future, Los Angeles and San Francisco will become neighboring cities.

A few geologists have even speculated on the nature of the globe 250 million years in the future. In this scenario, the Atlantic seafloor will eventually become old and dense enough to form subduction zones around much of its margins, not unlike the present-day Pacific basin. Continued subduction of the Atlantic Ocean floor will result in the closing of the Atlantic basin and the collision of the Americas with the Eurasian–African landmass to form the next supercontinent, shown in Figure 2.24. Support for the possible closing of the Atlantic comes from evidence for a similar event, when an ocean predating the Atlantic closed during Pangaea's formation. Australia is also projected to collide with Southeast Asia by that time. If this scenario is accurate, the dispersal of Pangaea will end when the continents reorganize into the next supercontinent.

Such projections, although interesting, must be viewed with considerable skepticism because many assumptions must be correct for these events to unfold as just described. Nevertheless,

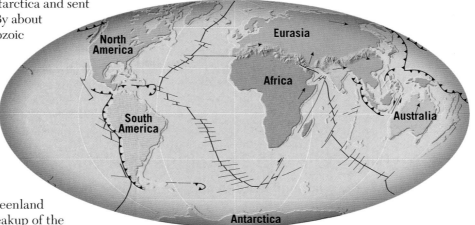

◄ Figure 2.23
The world as it may look 50 million years from now This reconstruction is highly idealized and based on the assumption that the processes that caused the breakup and dispersal of the supercontinent of Pangaea will continue to operate.

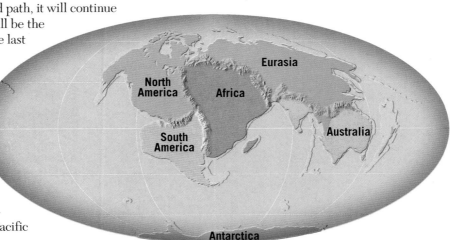

◄ Figure 2.24
Earth as it may appear 250 million years from now

changes in the shapes and positions of continents that are equally profound will undoubtedly occur for many hundreds of millions of years to come. Only after much more of Earth's internal heat has been lost will the engine that drives plate motions cease. (The prevailing theory on what causes plate tectonics is discussed in Section 2.10.)

CONCEPT CHECKS 2.7

1. Name two plates that are growing in size. Name a plate that is shrinking in size.

2. What new ocean basin was created by the breakup of Pangaea?

Concept Checker
https://goo.gl/b23gBK

2.8 Testing the Plate Tectonics Model

List and explain the evidence used to support the plate tectonics theory.

Some of the evidence supporting continental drift was presented earlier in this chapter. With the development of plate tectonics theory, researchers began testing this model of how Earth works. In addition to new supporting data, new interpretations of already existing data often swayed the tide of opinion.

Evidence of Seafloor Spreading: Ocean Drilling

Some of the most convincing evidence for seafloor spreading came from the Deep Sea Drilling Project, which operated from 1966 until 1983. One of the early goals of the project was to gather samples of the ocean floor in order to establish its age. A drilling ship capable of working in water thousands of meters deep drilled hundreds of holes through the layers of sediments that blanket the oceanic crust, as well as into the basaltic rocks below. Rather than use radiometric dating, which can be unreliable on oceanic rocks because of the alteration of basalt by seawater, researchers dated the seafloor by examining the fossil remains of microorganisms found in the sediments resting directly on the crust at each drill site.

Researchers recorded the age of the sediment samples from each site and noted each sample's distance from the ridge crest. They found that the sediments increased in age with increasing distance from the ridge. This finding supported the seafloor-spreading hypothesis, which predicted that the youngest oceanic crust would be found at the ridge crest—the site of seafloor production—and the oldest oceanic crust would be located adjacent to the continents.

Distribution and thickness of ocean-floor sediments provided additional verification of seafloor spreading. Drill cores from the *Glomar Challenger* revealed that sediments are almost entirely absent on the ridge crest and that sediment thickness increases with increasing distance from the ridge (**Figure 2.25A**). This pattern of sediment distribution should be expected if the seafloor-spreading hypothesis is correct.

The data collected by the Deep Sea Drilling Project also reinforced the idea that the ocean basins are geologically young because no seafloor older than 180 million years was found. By comparison, most continental crust

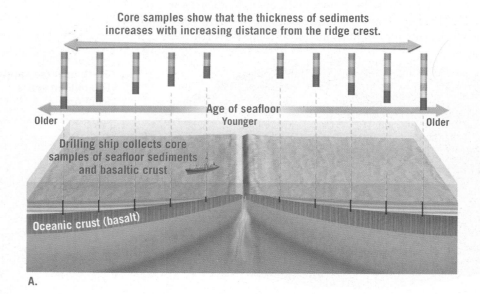

Core samples show that the thickness of sediments increases with increasing distance from the ridge crest.

Older   Age of seafloor   Older
               Younger

Drilling ship collects core samples of seafloor sediments and basaltic crust

Oceanic crust (basalt)

A.

B.

▲ Figure 2.25
Deep-sea drilling A. Data collected through deep-sea drilling have shown that the ocean floor is indeed youngest at the ridge axis.
B. The Japanese deep-sea drilling ship *Chikyu*, designed to drill up to 7000 meters (more than 4 miles) below the seafloor, became operational in 2007.

exceeds several hundred million years in age, and some samples are more than 4 billion years old.

In 1983, a new ocean-drilling program was launched by the Joint Oceanographic Institutions for Deep Earth Sampling (JOIDES). Now the International Ocean Discovery Program (IODP), this ongoing international effort uses multiple vessels for exploration, including the massive 210-meter-long (nearly 690-foot-long) *Chikyu* ("planet Earth" in Japanese), which began operations in 2007 (**Figure 2.25B**). One of the goals of the IODP is to recover a complete section of the oceanic crust, from top to bottom.

Evidence: Mantle Plumes, Hot Spots, and Island Chains

Mapping volcanic islands and *seamounts* (submarine volcanoes) in the Pacific Ocean revealed several linear chains of volcanic structures. One of these chains consists of at least 129 volcanic structures extending from the Hawaiian Islands to Midway Island and continuing northwestward toward the Aleutian trench (**Figure 2.26**). Radiometric dating of this linear feature, called the Hawaiian Island–Emperor Seamount chain, showed that the farther the volcano is from the Big Island of Hawaii, the older it is.

One widely accepted hypothesis[§] proposes that a roughly cylindrical upwelling of hot rock that originates deep in the mantle, called a **mantle plume**, is located beneath the island of Hawaii. As the hot, rocky plume

ascends through the mantle, the confining pressure drops, which triggers partial melting. (This process, called *decompression melting*, is discussed in Chapter 4.) The surface manifestation of this activity is a **hot spot**, an area of volcanism, high heat flow, and crustal uplifting measuring a few hundred kilometers across.

A hot spot is thought to maintain a relatively fixed position within the mantle. So, as the Pacific plate moved over it, a chain of volcanic structures known as a **hot-spot track** was built. As shown in Figure 2.26, the age of each volcano indicates how much time has elapsed since it was situated over the mantle plume. The youngest volcanic island in the chain, Hawaii, rose from the ocean floor less than 1 million years ago, whereas Midway Island is 27 million years old, and Detroit Seamount, near the Aleutian trench, is about 80 million years old.[**] A closer look at the five largest islands of Hawaii reveals a similar pattern of ages, from the volcanically active island of Hawaii to the inactive volcanoes that make up the oldest island, Kauai (see Figure 2.26). Five million years ago, when Kauai was positioned over the hot spot, it was the *only* modern Hawaiian island in existence. Kauai's age is evident in the island's inactive

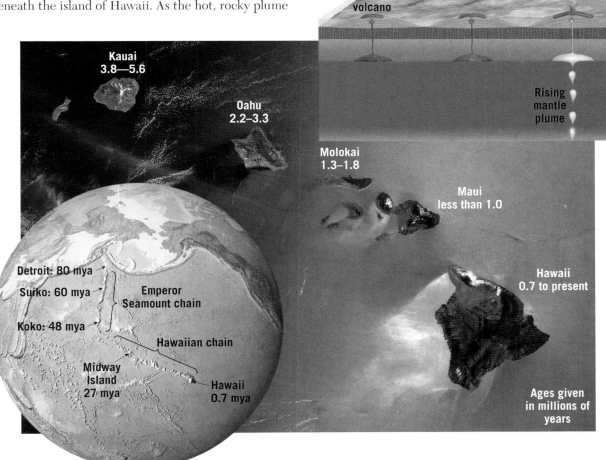

Plate motion

Extinct volcano

Hot-spot volcanism

Rising mantle plume

Kauai
3.8—5.6

Oahu
2.2–3.3

Molokai
1.3–1.8

Maui
less than 1.0

Hawaii
0.7 to present

Detroit: 80 mya
Suiko: 60 mya
Emperor Seamount chain
Koko: 48 mya
Hawaiian chain
Midway Island
27 mya
Hawaii
0.7 mya

Ages given in millions of years

◄ **Figure 2.26**
Hot-spot volcanism and the formation of the Hawaiian chain
Radiometric dating of the Hawaiian Islands shows that volcanic activity increases in age moving away from the Big Island of Hawaii.

[§]Although widely accepted, the validity of the mantle plume hypothesis, unlike the theory of plate tectonics, remains unresolved. The existence of slim mantle plumes that originate near Earth's core–mantle boundary has not been verified by seismic studies. As a result, some geologists propose that the source of magma that generated the Hawaiian chain originated from localized melting in the upper mantle.

[**]Of approximately 40 hot spots thought to have formed because of upwelling of hot mantle plumes, most, but not all, have hot-spot tracks.

volcanoes, which have been eroded into jagged peaks and vast canyons. By contrast, the relatively young island of Hawaii exhibits many fresh lava flows, and one of its five major volcanoes, Kilauea, is active today.

Evidence: Paleomagnetism

You are probably aware that Earth has a magnetic field, with invisible lines of force that pass through the planet and extend from one magnetic pole to the other out into space (**Figure 2.27**). Today the magnetic north and south poles roughly align with the geographic poles that are located where Earth's rotational axis intersects the surface. Earth's magnetic field is less obvious to us than the pull of gravity because we cannot feel it. However, movement of a compass needle confirms its presence. A traditional compass needle, a small magnet free to rotate on an axis, becomes aligned with the magnetic lines of force. When laid flat, one end of a compass needle points to magnetic north, while the other points to magnetic south.

In addition, some naturally occurring minerals are magnetic and influenced by Earth's magnetic field. One of the most common is the iron-rich mineral *magnetite*, which is abundant in lava flows of basaltic composition.†† Basaltic lavas erupt at the surface at temperatures greater than 1000°C (1800°F), exceeding a threshold temperature for magnetism known as the **Curie point** (about 585°C [1085°F]). The magnetite grains in molten lava are nonmagnetic, but as the lava cools, these iron-rich grains become magnetized and align themselves in the direction of the existing magnetic lines of force. Once the minerals solidify, the magnetism they possess usually remains "frozen" in this position. Thus, they act like a compass needle because they "point" toward the position of the magnetic poles at the time of their formation. Rocks that formed

thousands or millions of years ago and contain a "record" of the direction of the magnetic poles at the time of their formation are said to possess **paleomagnetism**, or **preserved magnetism**.

Apparent Polar Wandering   A study of paleomagnetism in ancient lava flows throughout Europe led to an interesting discovery. A plot of the location of the magnetic north pole, as measured from Europe, indicated that during the past 500 million years, the pole had gradually "wandered" from a location near Hawaii northeastward to its present location over the Arctic Ocean (**Figure 2.28**). This was strong evidence that either the magnetic north pole had migrated, an idea known as *polar wandering*, or that the poles had remained in place and the continents

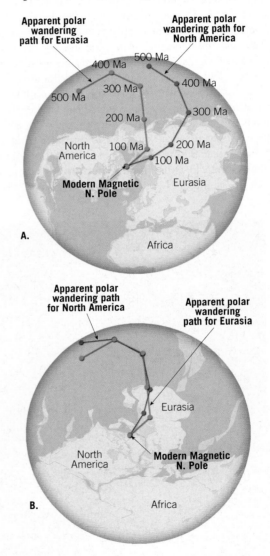

▲ Figure 2.28
Apparent polar-wandering paths A. Scientists believe that the more westerly path determined from North American data was caused by the westward drift of North America by about 24 degrees from Eurasia.
B. The positions of the wandering paths when the landmasses are reassembled in their pre-drift locations.

▶ Figure 2.27
Earth's magnetic field Earth's magnetic field consists of lines of force much like those a giant bar magnet would produce if placed at the center of Earth.

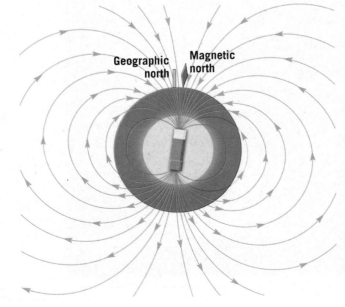

††Some sediments and sedimentary rocks also contain enough iron-bearing mineral grains to acquire a measurable amount of magnetization.

had drifted beneath them—in other words, Europe had drifted relative to the magnetic north pole.

Taken at face value, the magnetic alignment of iron-rich minerals in lava flows of different ages would indicate that the position of the paleomagnetic poles changed through time. However, although the magnetic poles are known to move in a somewhat erratic path, studies of paleomagnetism from numerous locations show that the positions of the magnetic poles, averaged over thousands of years, correspond closely to the positions of the geographic poles. Therefore, a more acceptable explanation was provided by Wegener's hypothesis: If the magnetic poles remain stationary, their *apparent movement* is produced by the drift of the seemingly fixed continents.

Further evidence for continental drift came a few years later, when a polar-wandering path was constructed for North America (see Figure 2.28A). For the first 200 million years or so, the paths for North America and Europe were found to be similar in direction—but separated by about 5000 kilometers (3000 miles). Then, during the middle of the Mesozoic era (180 million years ago), they began to converge on the present North Pole. The explanation for these curves is that North America and Europe were joined until the Mesozoic, when the Atlantic began to open. From this time forward, these continents continuously moved apart. When North America and Europe are moved back to their pre-drift positions, as shown in Figure 2.28B, these paths of apparent polar wandering coincide. This is evidence that North America and Europe were once joined and moved relative to the poles as part of the same continent.

Magnetic Reversals and Seafloor Spreading   More evidence emerged when geophysicists discovered that over periods of hundreds of thousands of years, Earth's magnetic field periodically reverses polarity. During a **magnetic reversal**, the magnetic north pole becomes the magnetic south pole and vice versa. Lava that solidified during a period of reverse polarity has been magnetized with the polarity opposite that of volcanic rocks being formed today. When rocks exhibit the same magnetism as the present magnetic field, they are said to possess **normal polarity**, whereas rocks exhibiting the opposite magnetism are said to have **reverse polarity**.

Once the concept of magnetic reversals was confirmed, researchers set out to establish a time scale for these occurrences. The task was to measure the magnetic polarity of hundreds of lava flows and use radiometric dating techniques to establish the age of each flow. **Figure 2.29** shows the **magnetic time scale** established using this technique for the past few million years. The major divisions of the magnetic time scale, termed *chrons*, last roughly 1 million years each. As more measurements became available, researchers realized that several short-lived reversals (less than 200,000 years long) sometimes occurred during a single chron.

Meanwhile, oceanographers had begun magnetic surveys of the ocean floor in conjunction with their efforts to construct detailed maps of seafloor topography. These magnetic surveys were accomplished by towing very sensitive instruments, called **magnetometers**, behind research vessels (**Figure 2.30A**). The goal of these geophysical surveys was to map variations in the strength of Earth's magnetic field that arise from differences in the magnetic properties of the underlying crustal rocks.

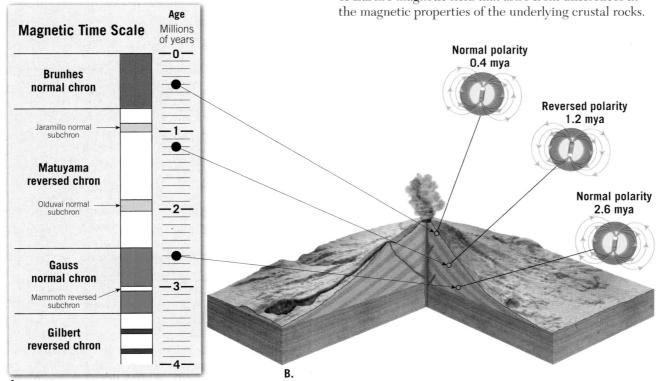

A.

B.

◄ **SmartFigure 2.29**
Time scale of magnetic reversals A. Time scale of Earth's magnetic reversals for the past 4 million years.
B. This time scale was developed by establishing the magnetic polarity for lava flows of known age. (Data from Allen Cox and G. B. Dalrymple)

Tutorial
https://goo.gl/qa2tE5

► Figure 2.30

Ocean floor as a magnetic recorder
A. A magnetometer towed across a segment of the oceanic floor records magnetic intensities.
B. Notice the symmetrical stripes of low- and high-intensity magnetism that parallel the axis of the Juan de Fuca Ridge. The colored stripes of high-intensity magnetism occur where normally magnetized oceanic rocks enhance the existing magnetic field. Conversely, the white low-intensity stripes are regions where the crust is polarized in the reverse direction, which weakens the existing magnetic field.

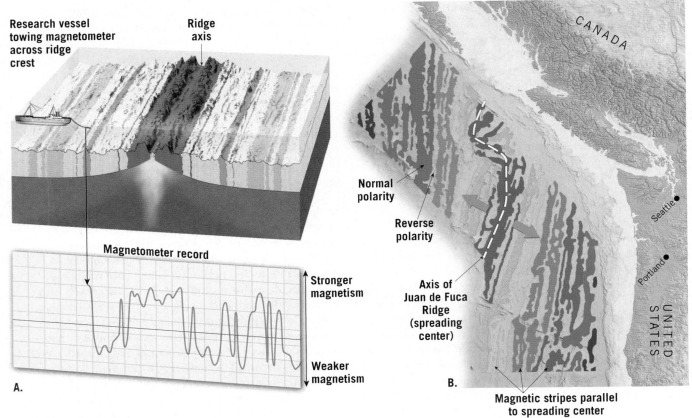

Research vessel towing magnetometer across ridge crest

Ridge axis

Magnetometer record

Stronger magnetism

Weaker magnetism

A.

Normal polarity

Reverse polarity

Axis of Juan de Fuca Ridge (spreading center)

CANADA

Seattle

Portland

UNITED STATES

B.

Magnetic stripes parallel to spreading center

▼ SmartFigure 2.31

Magnetic reversals and seafloor spreading
When new basaltic rocks form at mid-ocean ridges, they magnetize according to Earth's existing magnetic field. Hence, oceanic crust provides a permanent record of each reversal of our planet's magnetic field over the past 200 million years.

Animation
https://goo.gl/jAQaSt

Normal magnetic polarity

Reversed magnetic polarity

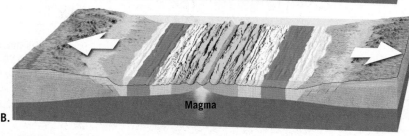

Axis of mid-ocean ridge

Magma

A.

Magma

B.

Magma

C.

The first comprehensive study of this type was performed off the Pacific coast of North America and had an unexpected outcome. Researchers discovered alternating stripes of high- and low-intensity magnetism, as shown in **Figure 2.30B**. This relatively simple pattern of magnetic variation defied explanation until 1963, when Fred Vine and D. H. Matthews demonstrated that the high- and low-intensity stripes supported the concept of seafloor spreading. Vine and Matthews suggested that the stripes of high-intensity magnetism are regions where the paleomagnetism of the ocean crust exhibits normal polarity (see Figure 2.29A). Consequently, these rocks *enhance* (reinforce) Earth's magnetic field. Conversely, the low-intensity stripes are regions where the ocean crust is polarized in the reverse direction and, therefore, *weaken* the existing magnetic field.

But how do parallel stripes of normally and reversely magnetized rock become distributed across the ocean floor? Vine and Matthews reasoned that as magma solidifies at the crest of an oceanic ridge, it is magnetized with the polarity of Earth's magnetic field at that time (**Figure 2.31**).

Because of seafloor spreading, this strip of magnetized crust would gradually increase in width. When a reversal in the polarity of Earth's magnetic field occurred, newly generated seafloor with the opposite polarity would form in the middle of the old strip. Gradually, the two halves of the old strip would be carried in opposite directions, away from the ridge crest. Subsequent reversals would build a pattern of normal and reverse magnetic stripes, as shown in Figure 2.31. Because new rock is added in equal amounts to both trailing edges of the spreading ocean floor, we should expect the pattern of stripes (width and polarity) found on one side of an oceanic ridge to be a mirror image of those on the other side. In fact, a survey across the Mid-Atlantic Ridge just south of Iceland reveals a pattern of magnetic stripes exhibiting a remarkable degree of symmetry in relationship to the ridge axis.

CONCEPT CHECKS 2.8

1. What is the age of the oldest sediments recovered using deep-ocean drilling? How do the ages of these sediments compare to the ages of the oldest continental rocks?

   Concept Checker
   https://goo.gl/nYy2xA

2. What did the study of preserved magnetism in ancient lava flows tell researchers about the geographic locations of North America and Europe about 180 million years ago?

3. Assuming that hot spots remain fixed, in what direction was the Pacific plate moving while the Hawaiian Islands were forming?

4. Describe how magnetic reversals provide evidence of the seafloor-spreading hypothesis.

2.9 How Is Plate Motion Measured?

Describe two methods researchers use to measure relative plate motion.

A number of methods are used to establish the direction and rate of plate motion. Some of these techniques not only confirm that lithospheric plates move but also allow us to trace those movements back in geologic time.

Geologic Measurement of Plate Motion

Using ocean-drilling ships, researchers obtain dates for hundreds of locations on the ocean floor. By knowing the age of a rock sample and its distance from the ridge axis where it was generated, an average rate of plate motion can be calculated.

Scientists used these data, combined with their knowledge of paleomagnetism stored in hardened lavas on the ocean floor and seafloor topography, to create maps that show the age of the ocean floor. The reddish-orange bands shown in **Figure 2.32** range in age from the present to about 40 million years ago. The width of

▼ Figure 2.32
Age of the ocean floor

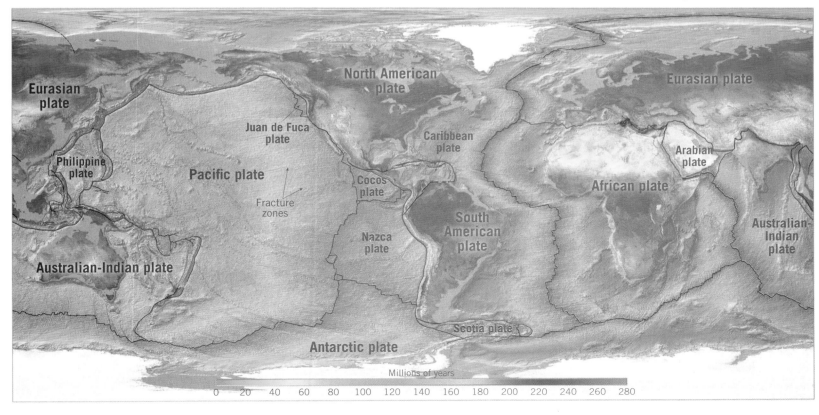

Millions of years

0   20   40   60   80   100   120   140   160   180   200   220   240   260   280

► Figure 2.33
Rates of plate motion
Each red arrow shows
plate motion at the
location of the arrow's
base, determined from
GPS data. Longer arrows
represent faster spread-
ing rates. The small
black arrows and labels
show seafloor spreading
velocities based mainly
on paleomagnetic data.
(Seafloor data from DeMets
and others; GPS data from Jet
Propulsion Laboratory)

Directions and rates of plate motions measured in centimeters per year

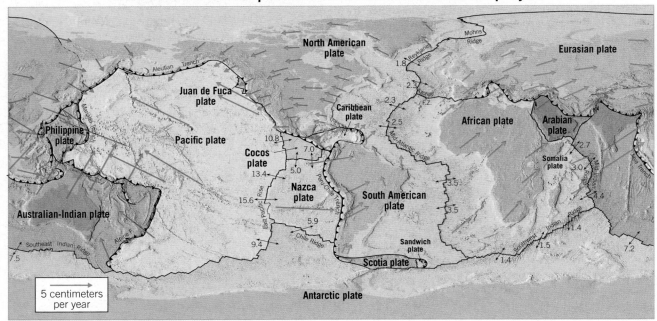

a band indicates how much crust formed during that
time period. For example, the reddish-orange band
along the East Pacific Rise is more than three times
wider than the same-color band along the Mid-Atlantic
Ridge. Therefore, the rate of seafloor spreading has
been approximately three times faster in the Pacific
basin than in the Atlantic.

Maps of this type also provide clues to the current direction of plate movement. Notice the offsets
in the ridges; these are transform faults that connect
the spreading centers. Recall that transform faults are
aligned parallel to the direction of spreading. Careful
measurement of transform faults reveals the direction
of plate movement.

To establish the direction of plate motion in
the past, geologists can examine the long *fracture
zones* that extend for hundreds or even thousands of
kilometers from ridge crests. Fracture zones include
inactive extensions of transform faults and therefore
provide a record of past directions of plate motion.
Unfortunately, most of the ocean floor is less than
180 million years old, so to look deeper into the past,
researchers must rely on paleomagnetic evidence
provided by continental rocks.

Measuring Plate Motion from Space

If you use a map app on your phone to navigate
while driving, then you've experienced the technology called Global Positioning System (GPS). Using
satellites to send radio signals that are intercepted by
receivers located at Earth's surface, GPS determines

the exact position of a site by simultaneously
establishing the distance from the receiver to four or
more satellites. Researchers use specially designed
equipment to locate a point on Earth to within a few
millimeters. To establish plate motions, GPS data are
collected at numerous sites repeatedly over a number
of years.

Data obtained from GPS and other techniques are
shown in **Figure 2.33**. Calculations show that Hawaii
is moving in a northwesterly direction toward Japan
at a rate of 8.3 centimeters per year. A location in
Maryland is retreating from a location in England at a
speed of 1.7 centimeters per year—a value close to the
2.0-centimeters-per-year spreading rate established
from paleomagnetic evidence obtained for the North
Atlantic. Techniques involving GPS devices have also
been useful in confirming small-scale crustal movements, like those occurring along faults in regions
known to be tectonically active (for example, the San
Andreas Fault).

CONCEPT CHECKS 2.9

1. What does the orientation of transform faults
   indicate about plate
   motion?

2. Based on what you
   see in Figure 2.33,
   which three plates
   appear to exhibit the highest rates of motion?

Concept Checker
https://goo.gl/oGrjfY

2.10 What Drives Plate Motions?

Describe plate–mantle convection and explain two of the primary driving forces of plate motion.

When the theory of plate tectonics was first proposed, geologists thought that convective flow within Earth's mantle actively pulled along the lithospheric plates. Using mathematical modeling, we have discovered that mantle convection alone is not strong enough to propel thick tectonic plates.

Forces That Drive Plate Motion

Geologists have discovered that the lithospheric plates are part of the larger convection system that produces plate motion. **Convection** is the way heat transfers through liquids and gases. To understand, imagine heating a beaker of water using a Bunsen burner. Water near the flame becomes warmer and therefore less dense, rising in relatively thin sheets or blobs that spread out at the surface. As the surface layer of the water cools, its density increases, causing it to sink toward the bottom of the kettle. This cycle of the hot liquid rising and cool liquid sinking creates a convection current. The convection system that drives plate tectonics is similar to, but considerably more complex than, the model just described.

Slab Pull   Geologists agree that subduction of cold, dense slabs of oceanic lithosphere is the major driving force of plate motion. This phenomenon, called **slab pull**, occurs because cold slabs of oceanic lithosphere are denser than the underlying warm asthenosphere, and they sink much like a ship's anchor—meaning the slabs are pulled down into the mantle by gravity (**Figure 2.34**). Although a subduction slab begins to warm as it descends, it is continually being replaced, in a conveyor belt fashion, by cooler, denser lithosphere from above. As a result, as a slab of lithosphere is subducted, it remains hundreds of degrees *cooler* than the surrounding asthenosphere. The cooler temperatures result in the subducting slab being more dense than the surrounding asthenosphere, and it is therefore pulled down harder by the force of gravity than are the warmer rocks of the asthenosphere—hence the name *slab pull*.

Ridge Push   Another important driving force occurs along the ocean ridges, where newly formed oceanic lithosphere is being pushed away from the ridge axis. This gravity-driven mechanism, called **ridge push**, results from the elevated position of oceanic ridges, which cause slabs of lithosphere to "slide" down both flanks of the ridge crest (see Figure 2.34). This is analogous to how gravity "pulls" snow down a mountain slope during an avalanche. Despite the fact that oceanic ridges have gentle slopes, the amount of material involved is enormous, so the *forces* that drive ridge push are also great.

Studies have shown that ridge push contributes less to plate motion than slab pull. The primary evidence for this is that the *fastest-moving* plates—the Pacific, Nazca, and Cocos plates—have extensive subduction zones along their margins. By contrast, the average spreading rate in the Atlantic basin, which is nearly devoid of subduction zones, is one of the slowest, at about 2.5 centimeters (1 inch) per year.

A Model of Plate–Mantle Convection

Although the large-scale convection system that drives plate tectonics is not yet fully understood, researchers generally agree on the following:

- Plate tectonics and convective flow in the mantle are part of the same system. Subducting oceanic plates drive the cold downward-moving portion of convective flow, while shallow upwelling of hot rock along the oceanic ridge and buoyant mantle plumes are the upward-flowing arms of the convective mechanism.
- The energy source for plate tectonics is Earth's internal heat. This heat continually moves to the surface, where it is eventually radiated into space.

What remains uncertain is the exact structure of this convective flow. Several models have been proposed for plate–mantle convection, and we will examine one that is considered viable.

Some researchers favor a type of *whole-mantle convection* model, also called the *plume model*, in which cold oceanic lithosphere sinks to great depths and stirs

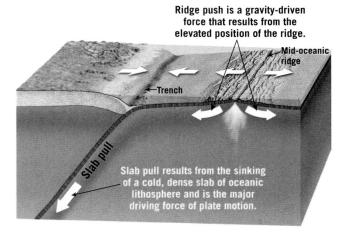

Ridge push is a gravity-driven force that results from the elevated position of the ridge.

Mid-oceanic ridge

Trench

Slab pull

Slab pull results from the sinking of a cold, dense slab of oceanic lithosphere and is the major driving force of plate motion.

◄ **Figure 2.34**
Major forces that act on lithospheric plates

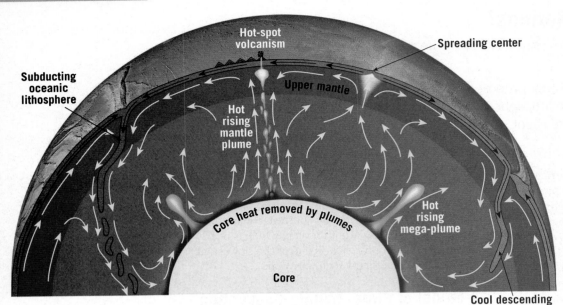

▲ Figure 2.35
One model of mantle convection In the whole-mantle model, sinking slabs of cold oceanic lithosphere are the downward limbs of convection cells, while rising mantle plumes carry hot material from the core–mantle boundary toward the surface.

the entire mantle (**Figure 2.35**). This model proposes that the ultimate burial ground for the subducting lithospheric slabs is Earth's core–mantle boundary. The downward flow of these subducting slabs is balanced by buoyantly rising mantle plumes that transport hot mantle rock toward the surface.

Two kinds of plumes are thought to exist—narrow tube-like plumes, and giant upwellings, often referred to as *mega-plumes*. The long, narrow plumes originate from the core–mantle boundary and produce hot-spot

volcanism of the type associated with the Hawaiian Islands, Iceland, and Yellowstone. Scientists believe that areas of large mega-plumes, as shown in Figure 2.35, occur beneath the Pacific basin and southern Africa. These mega-plumes are thought to explain why southern Africa has an elevation much higher than would be predicted for a stable continental landmass. In this whole-mantle convection model, heat for both the narrow plumes and the mega-plumes arises mainly from Earth's core. However, some researchers have questioned that idea and instead propose that the source of magma for most hot-spot volcanism is in the upper mantle (asthenosphere).

It should be noted that geologists continue to debate the nature of plate–mantle flow. As they continue to investigate the possibilities, a widely accepted hypothesis will emerge.

CONCEPT CHECKS 2.10

https://goo.gl/NyE9YB

1. Which of these forces—slab pull or ridge push—contributes more to plate motion?

2. Briefly describe the *whole-mantle convection (plume) model.*

---

2 CONCEPTS IN REVIEW

Plate Tectonics: A Scientific Revolution Unfolds

2.1 From Continental Drift to Plate Tectonics

Summarize the view that most geologists held prior to the 1960s regarding the geographic positions of the ocean basins and continents.

Key Term: plate tectonics

- Geologists once thought that ocean basins were very old and continents were fixed in place. The theory of *plate tectonics* changed that. Supported by multiple kinds of evidence, plate tectonics is the foundation of modern Earth science.

2.2 Continental Drift: An Idea Before Its Time

List and explain the evidence Wegener presented to support his continental drift hypothesis.

Key Terms: continental drift hypothesis supercontinent Pangaea

- German meteorologist Alfred Wegener formulated the *continental drift hypothesis* in 1915, suggesting that Earth's continents are not fixed but move slowly over geologic time.

- Wegener proposed that *Pangaea*, a *supercontinent*, existed about 200 million years ago during the late Paleozoic and early Mesozoic eras.

- Evidence that Pangaea existed and later broke into pieces that drifted apart includes (1) the shape of the continents, (2) continental fossil organisms that matched across oceans, (3) matching rock types and modern mountain belts on separate continents, and (4) sedimentary rocks that recorded ancient climates, including glaciers on the southern portion of Pangaea.

- Continental drift hypothesis suffered from two flaws: It proposed tidal forces as the mechanism for motion of continents, and it implied that continents plowed through weaker oceanic crust, like boats cutting through a thin layer of sea ice. Most geologists rejected continental drift when Wegener proposed it.

Q Why did Wegener choose organisms such as *Glossopteris* and *Mesosaurus* as evidence for continental drift, as opposed to other fossil organisms such as sharks or jellyfish?

2.3 The Theory of Plate Tectonics

List the major differences between Earth's lithosphere and asthenosphere and explain the importance of each in the plate tectonics theory.

Key Terms:

theory of plate tectonics	lithosphere	lithospheric plate
	asthenosphere	

- Research conducted after World War II led to new insights that revived interest in continental drift hypothesis. Seafloor exploration uncovered an extremely long mid-ocean ridge system. Sampling of the oceanic crust revealed that it was quite young relative to the continents.
- The *lithosphere*, Earth's outermost rocky layer, is relatively stiff and deforms by bending or breaking. The lithosphere consists both of crust (either oceanic or continental) and underlying upper mantle. Beneath the lithosphere is the *asthenosphere*, a relatively weak layer of solid rock that deforms by flowing.
- The lithosphere consists of numerous segments of irregular size and shape called *lithospheric plates*. There are seven large plates, another seven intermediate-size plates, and many relatively small microplates. Plates meet along boundaries that may be divergent (moving apart from each other), convergent (moving toward each other), or transform (moving laterally past each other).

Q Compare and contrast Earth's lithosphere and asthenosphere.

2.4 Divergent Plate Boundaries and Seafloor Spreading

Sketch and describe the movement along a divergent plate boundary that results in the formation of new oceanic lithosphere.

Key Terms:

divergent plate boundary	oceanic ridge system	seafloor spreading
	rift valley	continental rift

- *Seafloor spreading* leads to formation of new oceanic lithosphere at *oceanic ridge systems*. As two plates move apart from one another, tensional forces open cracks in the plates, allowing magma to well up and generate new slivers of seafloor. This process generates new oceanic lithosphere at a rate of 2 to 15 centimeters (1 to 6 inches) each year.
- As it ages, oceanic lithosphere cools and becomes denser. It therefore subsides as it is transported away from the mid-ocean ridge. At the same time, the underlying asthenosphere cools, adding new material to the underside of the plate, which consequently thickens.
- *Divergent plate boundaries* (spreading centers) are not limited to the seafloor. Continents can break apart, too, starting with a *continental rift* (as in modern-day east Africa) that produced a *rift valley*, potentially producing a new ocean basin between the two sides of the rift.

2.5 Convergent Plate Boundaries and Subduction

Compare and contrast the three types of convergent plate boundaries and name a location where each type can be found.

Key Terms:

convergent plate boundary	deep-ocean trench	continental volcanic arc
	partial melting	volcanic island arc

- Plates move toward one another at *convergent plate boundaries* (subduction zones). Here oceanic lithosphere subducts into the mantle, where it is recycled. Subduction manifests as a *deep-ocean trench*. A subducting slab of oceanic lithosphere can descend at a variety of angles, from nearly horizontal to nearly vertical.
- Aided by the presence of water, subducted oceanic lithosphere triggers *partial melting* in the mantle, which produces magma. The magma

is less dense than the surrounding rock and will rise. It may cool at depth, thickening the crust, or it may make it all the way to Earth's surface, where it erupts as a volcano.
- A line of volcanoes emerging through continental crust is a *continental volcanic arc*. A line of volcanoes arising in an overriding plate of oceanic lithosphere is a *volcanic island arc*.
- Continental crust resists subduction due to its relatively low density, and so when an intervening ocean basin is completely destroyed through subduction, the continents on either side collide, generating a new mountain range.

Q Sketch a typical continental volcanic arc and label the key parts. Then repeat the drawing with an overriding plate made of oceanic lithosphere.

2.6 Transform Plate Boundaries

Describe the relative motion along a transform fault boundary and locate several examples of transform faults on a plate boundary map.

Key Terms: transform plate boundary   fracture zone

- At a *transform plate boundary* (transform fault), lithospheric plates slide horizontally past one another. No new lithosphere is generated, and no old lithosphere is consumed. Shallow earthquakes signal the movement of these slabs of rock as they grind past their neighbors.
- The San Andreas Fault in California is an example of a transform boundary in continental crust, while the *fracture zones* between segments of the Mid-Atlantic Ridge are transform faults in oceanic crust.

Q On the accompanying tectonic map of the Caribbean, find the Enriquillo Fault. (The location of the 2010 Haiti earthquake is shown as a yellow star.) What kind of plate boundary is shown here? Are there any other faults in the area that show the same type of motion?

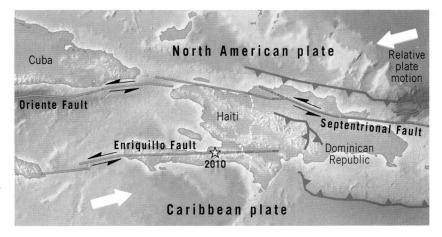

2.7 How Do Plates and Plate Boundaries Change?

Explain why plates such as the African and Antarctic plates are increasing in size, while the Pacific plate is decreasing in size.

- Although the total surface area of Earth does not change, the shape and size of individual plates constantly change as a result of subduction and seafloor spreading. Plate boundaries can also be created or destroyed in response to changes in the forces acting on the lithosphere.
- The breakup of Pangaea and the collision of India with Eurasia are two examples of how plates change through geologic time.

2.8 Testing the Plate Tectonics Model

List and explain the evidence used to support the plate tectonics theory.

Key Terms:

mantle plume	Curie point	reverse polarity
hot spot	paleomagnetism	magnetic time scale
hot-spot track	magnetic reversal	magnetometer
	normal polarity	

- Multiple lines of evidence verify the plate tectonics model. For instance, deep-sea drilling found that seafloor age increases with distance from a mid-ocean ridge. Thickness of sediment atop this seafloor is also proportional to distance from the ridge: Older lithosphere has had more time to accumulate sediment.
- A *hot spot* is an area of volcanic activity where a *mantle plume* reaches Earth's surface. Volcanic rocks generated by *hot-spot tracks* provide evidence of both the direction and rate of plate movement over time.
- Magnetic minerals such as magnetite align themselves with Earth's magnetic field as rock forms. This *paleomagnetism (preserved magnetism)* is a record of the ancient orientation of Earth's magnetic field. This allows a given stack of rock layers to be interpreted in terms of their orientation relative to the magnetic poles through time. *Magnetic reversals* in the orientation of the Earth's magnetic field are preserved as "stripes" of *normal polarity* and *reversed polarity* in the oceanic crust. *Magnetometers* reveal this signature of seafloor spreading as a symmetrical pattern of magnetic stripes parallel to the axis of the mid-ocean ridge.

2.9 How Is Plate Motion Measured?

Describe two methods researchers use to measure relative plate motion.

- Ocean floor data established direction and rate of motion of lithospheric plates. Transform faults point in the direction the plate is moving. Dating seafloor rocks helps calibrate the rate of motion.
- GPS can be used to accurately measure the motion of receivers on Earth's surface to within a few millimeters. The "real-time" data collected support the inferences made from seafloor observations. On average, plates move at about the same rate human fingernails grow: about 5 centimeters (2 inches) per year.

2.10 What Drives Plate Motions?

Describe plate–mantle convection and explain two of the primary driving forces of plate motion.

Key Terms: convection slab pull ridge push

- In general, *convection* (upward movement of less dense material and downward movement of more dense material) appears to drive the motion of plates.
- Slabs of oceanic lithosphere sink at subduction zones because a subducted slab is denser than the underlying asthenosphere. In this process, called *slab pull*, Earth's gravity tugs at the slab, drawing the rest of the plate toward the subduction zone. As oceanic lithosphere slides down the mid-ocean ridge, it exerts an additional but smaller force, called *ridge push*.
- Plate tectonics and convective flow in the mantle are part of the same system. Subducting oceanic plates drive the cold downward-moving portion of convective flow, while shallow upwelling of hot rock along the oceanic ridge and buoyant mantle plumes are the upward-flowing arms of the convective mechanism.

Q Compare and contrast mantle convection with the operation of a lava lamp.

GIVE IT SOME THOUGHT

1. **Refer to Section 1.3, "The Nature of Scientific Inquiry," to help answer the following:**

   a. What observations led to the continental drift hypothesis?

   b. Why did most of the scientific community reject the continental drift hypothesis?

   c. Do you think Wegener followed the basic principles of scientific inquiry? Support your answer.

A.   B.   C.

2. Some people think that California will sink into the ocean. Is this idea consistent with the theory of plate tectonics? Explain.

3. Australian marsupials (kangaroos, koalas, etc.) have direct fossil links to marsupial opossums found in the Americas. Yet the modern marsupials in Australia are markedly different from their American relatives. How does the breakup of Pangaea help to explain these differences? (*Hint:* See Figure 2.22.)

4. Explain how the processes that create hot-spot volcanic chains differ from the processes that generate volcanic island arcs.

5. Refer to the accompanying diagrams (to the right) illustrating the three types of convergent plate boundaries to answer the following:

   a. Identify each type of convergent boundary.

   b. On what type of crust do volcanic island arcs develop?

   c. Describe two ways in which oceanic–oceanic convergent boundaries differ from oceanic–continental boundaries. How are they similar?

6. Refer to the accompanying hypothetical plate map to answer the following questions:

   a. This diagram shows portions of how many plates?

   b. Explain why active volcanoes are more likely to be found on continents A and B than on continent C.

   c. Provide one scenario in which volcanic activity might be triggered on continent C.

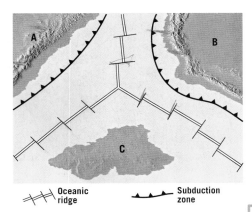

Oceanic ridge   Subduction zone

7. Volcanic islands that form over mantle plumes, such as the Hawaiian chain, are home to some of Earth's largest volcanoes. However, several volcanoes on Mars are gigantic compared to any on Earth. What does this difference tell us about the role of plate motion in shaping the Martian surface?

8. Imagine that you are studying seafloor spreading along two different oceanic ridges. Using data from a magnetometer, you produced the two accompanying maps.

   a. From these maps, what can you determine about the relative rates of seafloor spreading along these two ridges? Explain.

   b. Using the scale provided, measure the distance in kilometers from ridge axis to the beginning (right edge) of the yellowish normal polarity stripe for each ridge. How many kilometers has each side of these ocean basins spread during the past 1 million years?

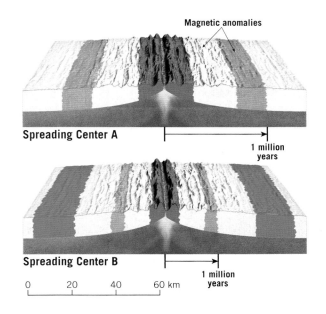

Magnetic anomalies

Spreading Center A

1 million years

Spreading Center B

1 million years

0   20   40   60 km

9. Refer to the accompanying map and the pairs of cities to complete the question. (Boston/Denver, London/Boston, and Honolulu/Beijing)

   a. Which pair of cities is moving apart as a result of plate motion?

   b. Which pair of cities is moving closer together as a result of plate motion?

   c. Which pair of cities is not presently moving relative to each other?

Plate motion measured in centimeters per year

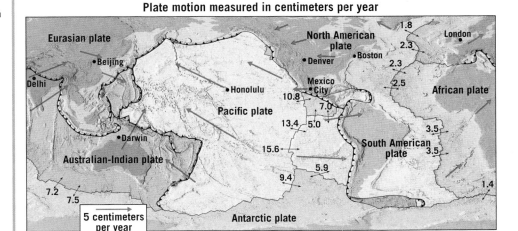

EYE ON EARTH

1. Baja California is separated from mainland Mexico by a long narrow sea called the Gulf of California (also known to local residents as the Sea of Cortez). The Gulf of California contains many islands created by volcanic activity.

   a. What type of plate boundary is responsible for opening the Gulf of California?

   b. What major U.S. river originates in the Colorado Rockies and created a large delta at the northern end of the Gulf of California?

   c. If the material carried by the river in Question b had *not* been deposited, the Gulf of California would extend northward to include the inland sea shown in this satellite image. What is the name of this inland sea?

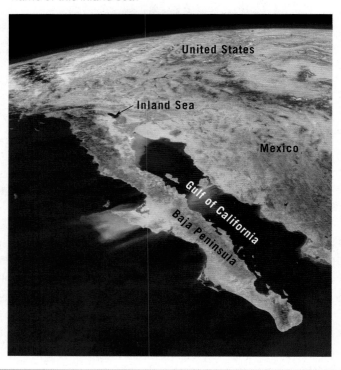

2. In December 2011 a new volcanic island formed near the southern end of the Red Sea. People fishing in the area witnessed lava fountains reaching up to 30 meters (100 feet). This volcanic activity occurred off the west coast of Yemen, along the Red Sea Rift, among a collection of small islands in the Zubair Group.

   a. What two plates border the Red Sea Rift?

   b. Are these two plates moving *toward* or *away from* each other?

   c. What type of plate boundary produced this new volcanic island?

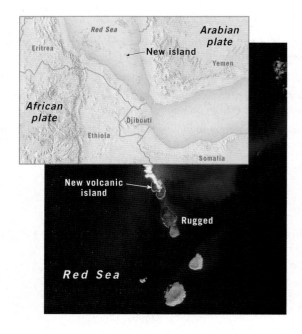

DATA ANALYSIS

Tectonic Plate Movement

The Global Positioning System (GPS), a set of 30 satellites used for navigation, is also used to track plate movement.

https://goo.gl/eD226M

ACTIVITIES

Go to the Jet Propulsion Lab's GPS Time Series map, at http://sideshow. jpl.nasa.gov/post/series.html. Click the plus sign in the bottom-right corner of the map twice to enlarge the view and then center the map over Portugal. The green dots are locations whose positions are being continually tracked. The yellow lines extending from each dot indicates the relative velocities (speed and direction) that each of these locations is moving—mostly due to motion of Earth's tectonic plates. Longer lines indicate faster rates of plate movement.

1. **From what source do the data for this map originate? (*Hint*: Read the chapter introduction.)**

2. **In what general direction is the station located in Portugal (Lisbon) moving?**

3. **Are the other European cities shown on this map moving in the *same direction as* or a *different direction than* Lisbon, Portugal?**

4. **Based on your answer to Question 3, are these cities most likely located on the same plate or on different plates?**

5. **Examine the station located in the North Atlantic and describe its motion as compared to the European cities shown on this map.**

Next, drag the map so the United States is centered on your screen and examine the station in the Atlantic Ocean off the East Coast of the United States. This station is located on the island of Bermuda.

6. **In what general direction is the station on Bermuda moving?**

7. **Describe how the direction of plate movement changes as we move from the East Coast of the United States westward into the Rocky Mountains. (The stations in the Rockies are located directly west of the "United States" label on the map.)**

8. **What is the direction of plate motion for the stations located in southwestern California?**

9. **Compare the rate and direction of plate motion detected for stations in southern California with those located in the Rockies.**

10. **Explain why the differences you described in Question 9 occur.**

Drag the map so that the upper part of South America is visible and the word "Brazil" is centered on your screen.

11. **Examine the station in the Pacific Ocean off the coast of Ecuador. In what general direction is that station moving?**

12. **Compare the speed and direction of the station you examined in Question 11 with that of the station located in Colombia.**

13. **On what plate is the station you examined in Question 11 located? (Hint: See Figure 2.10.)**

Mastering Geology

Looking for additional review and test prep materials? Visit pearson.com/mastering/geology to enhance your understanding of this chapter's content by accessing a variety of resources, including **Self-Study Quizzes, Geoscience Animations, SmartFigure Tutorials,** *Project Condor* **Videos, Mobile Field Trips, Dynamic Study Modules,** and an optional **Pearson eText.**

Discovered and Then Resubmerged: The Cave of Crystals

When miners in Naica, Mexico, drilled into a submerged cavern 1000 feet down and pumped out the groundwater, their discovery made news around the world. They had found a cave filled with enormous gypsum crystals, the largest of which was as tall as a three-story building and weighed 55 tons.

Nearly as noteworthy as the crystals were the lengths to which people had to go to view them. A magma body lying just a few miles below heated the cave's air to 136°F, and the relative humidity exceeded 90 percent. As adventurer and television host George Kourounis put it, "As soon as you walk in, you start to die."

To survive for more than 10 minutes in this extreme environment, each researcher needed to don a special refrigerated suit with an ice-cooled breathing system. After years of study, mining operations in the area ended, and so did the constant pumping out of water. In 2017, the cave reverted to its original, submerged state.

The miners who discovered the Naica Crystal Cave hunted for lead, silver, zinc, and copper. These metals are just a few of the myriad of minerals that exist. Every process that geologists study in some way depends upon the properties of these basic materials of Earth.

▶ The Cave of Crystals in Naica, Mexico, contains giant gypsum crystals, some of the largest natural crystals ever found.

3
Matter and Minerals

EARTH'S CRUST AND OCEANS are home to a wide variety of useful and essential minerals. Most people are familiar with the common uses of many basic metals, including aluminum in beverage cans, copper in electrical wiring, and gold and silver in jewelry. However, some people are not aware that pencil "lead" contains the greasy-feeling mineral graphite and that bath powders and many cosmetics contain the mineral talc. Moreover, many do not know that dentists use drill bits impregnated with diamonds to drill through tooth enamel. In fact, practically every manufactured product contains materials obtained from minerals.

In addition to the economic uses of rocks and minerals, every geologic process in some way depends on the properties of these basic Earth materials. Events such as volcanic eruptions, mountain building, weathering and erosion, and even earthquakes involve rocks and minerals. Consequently, a basic knowledge of Earth materials is essential to understanding all geologic phenomena.

3.1 Minerals: Building Blocks of Rocks

List and describe the main characteristics that an Earth material must possess to be considered a mineral.

Minerals are the building blocks of rocks. Human cultural development has been closely tied to mineral use, which is why we mark eras of human civilization with the terms *Stone Age*, *Bronze Age*, and *Iron Age*. During the Stone Age (starting in prehistoric times), people fashioned weapons and cutting tools from flint and chert. As early as 3700 B.C.E. Egyptians mined gold, silver, and copper for jewelry and art. By 2200 B.C.E. the Bronze Age commenced as humans discovered how to combine copper with tin to make bronze—a strong, hard alloy used for weapons, tools, and other goods. The Iron Age began later, when people learned to extract iron from minerals such as hematite. By the Middle Ages, mining of a variety of minerals became common, and the impetus for the formal study of minerals (known as **mineralogy**) was in place. Today mining and use of minerals continues to expand and evolve. For instance, the common mineral quartz is now the source of silicon for computer chips (**Figure 3.1**).

Those concerned with health extol the benefits of consuming vitamins and minerals, and the guessing game *Twenty Questions* usually begins with the question "Is it animal, vegetable, or mineral?" The mining industry typically uses the word *mineral* to refer to anything extracted from Earth, such as coal, iron ore, sand, and gravel. But what criteria do geologists use to determine whether something is a mineral?

► Figure 3.1
Quartz crystals A collection of well-developed quartz crystals found near Hot Springs, Arkansas.

Defining a Mineral

Geologists define a **mineral** as any Earth material that exhibits the following characteristics:

1. **Naturally occurring.** Minerals are naturally occurring substances produced by geologic processes. By contrast, synthetic materials that are manufactured by humans (such as cubic zirconia) are not considered minerals.

2. **Generally inorganic.** Minerals are usually inorganic compounds, meaning they are not derived from living things and do not contain hydrocarbons. This explains why ordinary table salt (halite, NaCl) is a mineral, while similar-looking table sugar ($C_6H_{12}O_6$) is not. An exception to this are the material secreted by marine animals, such as calcium carbonate (calcite) to form shells and coral reefs. If these materials are buried and become part of the rock record, then geologists consider them minerals.

3. **Orderly crystalline structure.** Minerals contain orderly arrangements of atoms that result in regularly shaped units called *crystals* (**Figure 3.2**). Some naturally occurring solids, such as volcanic glass (obsidian), lack a repetitive atomic structure and, hence, are *not* considered minerals.

4. **Solid.** All minerals are solids. Interestingly, ice (frozen water) is classified as a mineral because it meets these criteria. However, liquid water and water vapor are not considered minerals.

5. **Definite chemical composition that allows for some variation.** Most minerals have compositions that can be expressed by chemical formulas. For example, the mineral quartz has the formula SiO_2, which indicates that quartz consists of silicon (Si) and oxygen (O) atoms, in a 1:2 ratio. This proportion of silicon to oxygen is true for any sample of pure quartz, regardless of its origin. However, the compositions of some minerals can vary *within specific, well-defined limits*. This occurs because certain elements can substitute for others of similar size without changing the mineral's internal structure.

What Is a Rock?

Rocks are more loosely defined than minerals. A **rock** is any solid mass of mineral, or mineral-like, matter that occurs naturally as part of our planet. Most rocks, like the sample of granite shown in **Figure 3.3**, are aggregates of several different minerals. The term *aggregate* implies that the minerals joined in such a way that their individual properties are retained. Note that the different minerals that make up granite can be easily identified. However, some rocks

◀ **Figure 3.2**
Atomic arrangements
A. Minerals have orderly, repetitive arrangements of atoms.
B. Substances like glass have unordered or irregular atomic structures.

A.

B.

are composed almost entirely of one mineral. A common example is the sedimentary rock *limestone*, which is an impure mass of the mineral calcite. In addition, some rocks are composed of nonmineral matter. These include the volcanic rocks *obsidian* and *pumice*, which are noncrystalline glassy substances, and *coal*, which consists of solid organic debris.

Before we discuss rocks and minerals further, it is helpful to review some details on atoms, the building blocks of all matter.

CONCEPT CHECKS 3.1

1. List five characteristics of a mineral.

2. Based on the definition of a mineral, which of the following—gold, liquid water, synthetic diamonds, ice, and wood—are not classified as minerals?

 Concept Checker
https://goo.gl/Xh8z3t

3. Define the term *rock*. How do rocks differ from minerals?

Granite
(Rock)

◀ **SmartFigure 3.3**
Most rocks are aggregates of minerals Shown here are a hand sample of the igneous rock granite and three of its major constituent minerals.

Tutorial
https://goo.gl/TQ6K3j

Quartz
(Mineral)

Hornblende
(Mineral)

Feldspar
(Mineral)

3.2 Atoms: Building Blocks of Minerals

Compare and contrast the three primary particles contained in atoms.

All matter, including minerals, is composed of minute building blocks called **atoms**—the smallest particles that constitute specific elements and cannot be split by chemical means. Atoms, in turn, contain even smaller particles—*protons* and *neutrons* located in a central **nucleus** that is surrounded by *electrons* (**Figure 3.4**).

Properties of Protons, Neutrons, and Electrons

Protons and **neutrons** are very dense particles with almost identical masses. By contrast, **electrons** have a negligible mass, about 1/2000 that of a proton. To visualize this difference, imagine a scale on which a proton or neutron has the mass of a baseball, whereas an electron has the mass of a single grain of rice.

Both protons and electrons share a fundamental property called *electrical charge*. Protons have an electrical charge of +1, and electrons have a charge of −1. Neutrons, as the name suggests, have no charge. The charges of protons and electrons are equal in magnitude but opposite in polarity, so when a proton and an electron are paired, the charges cancel each other out. Since matter typically contains equal numbers of positively charged protons and negatively charged electrons, most substances are electrically neutral.

Illustrations sometimes show electrons orbiting the nucleus in a manner that resembles the planets of our solar system orbiting the Sun (see **Figure 3.4A**). However, electrons do not actually behave this way. A more realistic depiction is to show electrons as a cloud of negative charges surrounding the nucleus (see **Figure 3.4B**). Studies of the arrangements of electrons show that they move about the nucleus in regions called *principal shells*, each with an associated energy level. Each shell can hold a specific number of electrons, with the outermost shell generally containing **valence electrons**. These outer-shell electrons can be transferred to or shared with other atoms to form chemical bonds.

Most of the atoms in the universe (except hydrogen and helium) were created inside massive stars by nuclear fusion and then released into interstellar space during hot, fiery supernova explosions. As this ejected material cooled, the newly formed nuclei attracted electrons to complete their atomic structure. At the temperatures found at Earth's surface, free atoms (those not bonded to other atoms) generally have a full complement of electrons—one for each proton in the nucleus.

Elements: Defined by the Number of Protons

The simplest atoms have only 1 proton in their nuclei, whereas others have more than 100. The number of protons in the nucleus of an atom, called the **atomic number**, determines the atom's chemical nature. All atoms with the same number of protons have the same chemical and physical properties; collectively they constitute an **element**. There are about 90 naturally occurring elements, and several more have been synthesized in the laboratory. You are probably familiar with the names of many elements, including carbon, nitrogen, and oxygen. Every carbon atom has 6 protons, whereas every nitrogen atom has 7 protons, and every oxygen atom has 8.

▶ **Figure 3.4**
Two models of an atom
A. Simplified view of an atom's central nucleus, composed of protons and neutrons, encircled by high-speed electrons.
B. An atom model showing spherically shaped electron clouds (shells) surrounding the central nucleus. (Not to scale)

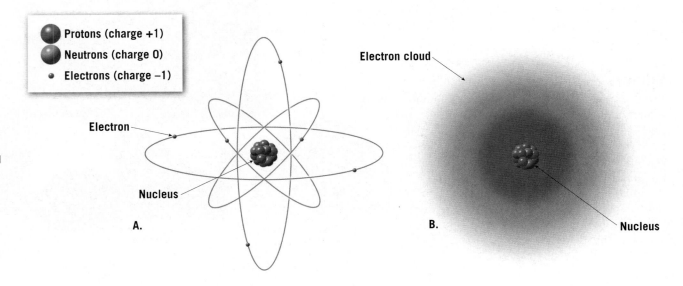

Protons (charge +1)
Neutrons (charge 0)
Electrons (charge −1)

Electron
Nucleus
A.

Electron cloud
Nucleus
B.

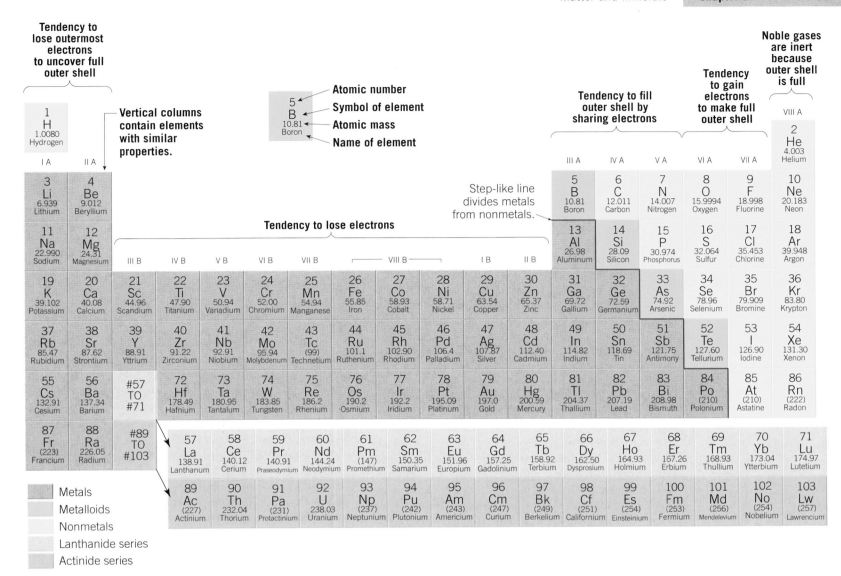

Tendency to lose outermost electrons to uncover full outer shell

Vertical columns contain elements with similar properties.

Atomic number
Symbol of element
Atomic mass
Name of element

Tendency to fill outer shell by sharing electrons

Tendency to gain electrons to make full outer shell

Noble gases are inert because outer shell is full

Step-like line divides metals from nonmetals.

Tendency to lose electrons

Metals
Metalloids
Nonmetals
Lanthanide series
Actinide series

▲ Figure 3.5
Periodic table of the elements

Scientists use the **periodic table**, shown in Figure 3.5, to organize the known elements. Within this important reference tool, elements with similar properties are aligned in columns, referred to as *groups*. Each element is assigned a one- or two-letter symbol. The atomic number and atomic weight for each element are also included in the periodic table.

Atoms are the basic building blocks of Earth's minerals. Most join to form **chemical compounds**—substances composed of atoms of two or more elements held together by chemical bonds. Examples include the common minerals quartz (SiO_2), halite (NaCl), and calcite ($CaCO_3$). However, a few minerals, such as diamonds, sulfur, and native gold and copper, are made entirely of atoms of only one element (Figure 3.6). (The word "native" is used as a prefix to describe a metal that is found in nature in its pure form.)

A. Gold on quartz

B. Sulfur

C. Copper

▲ Figure 3.6
Examples of minerals each composed of a single element

CONCEPT CHECKS 3.2

1. Make a simple sketch of an atom and label its three main particles. Explain how these particles differ from one another. Concept Checker https://goo.gl/mjXEMG

2. What is the significance of valence electrons?

3.3 How Atoms Bond to Form Minerals

Distinguish among ionic bonds, covalent bonds, and metallic bonds and explain how they form minerals.

Under the conditions found on Earth, most elements form bonds with atoms of other elements. (A group of elements known as the noble gases are an exception; they do not readily bond with other elements.) Some atoms bond to form *ionic compounds*, some form *molecules*, and still others form *metallic substances*. Why does this happen? Experiments show that electrical forces hold atoms together and bond them to each other. These electrical attractions lower the total energy of the bonded atoms, which, in turn, makes them more stable.

The Octet Rule and Chemical Bonds

A **chemical bond** is a transfer or sharing of electrons that allows each atom to attain a full valence shell of electrons. As noted earlier, valence (outer-shell) electrons are generally involved in this process. **Figure 3.7** shows a shorthand representation of the number of valence electrons for selected elements. Notice that the elements in Group I have one valence electron each, those in Group II have two valence electrons each, and so on, up to eight valence electrons each in Group VIII.

A chemical guideline known as the **octet rule** states that an *atom tends to gain, lose, or share electrons until it is surrounded by eight valence electrons.* Although there are exceptions to the octet rule, it is a useful *rule of thumb* for understanding chemical bonding. When an atom's outer shell does not contain eight electrons, it is likely to chemically bond to other atoms to achieve an octet in its outer shell. The noble gases have very stable electron arrangements with eight valence electrons (except helium, which has two), and this explains their lack of chemical reactivity. There are three primary types of bonds: *ionic, covalent,* and *metallic*.

Ionic Bonds: Electrons Transferred

The type of bond that is perhaps easiest to visualize is the **ionic bond**, in which one atom gives up one or more valence electrons to another atom to form **ions**—*positively and negatively charged atoms*. The atom that loses electrons becomes a positive ion, and the atom that gains electrons becomes a negative ion. Oppositely charged ions are strongly attracted to one another and join to form *ionic compounds*.

Consider the ionic bonding that occurs between sodium (Na) and chlorine (Cl) to produce the solid ionic compound sodium chloride—the mineral halite (common table salt). Notice in **Figure 3.8A** that a sodium atom gives up its single valence electron to chlorine and, as a result, becomes a positively charged sodium ion (Na^+). Chlorine, on the other hand, gains one electron and becomes a negatively charged chloride ion (Cl^-). Because ions with unlike charges attract, an ionic bond is an attraction of oppositely charged ions to produce an electrically neutral ionic compound.

Figure 3.8B illustrates the arrangement of sodium and chlorine ions in ordinary table salt. Notice that salt consists of alternating sodium and chlorine ions, positioned so that each positive ion is attracted to and surrounded on all sides by negative ions and vice versa. This arrangement maximizes the attraction between ions with opposite charges while minimizing the repulsion between ions with identical charges. Thus, ionic compounds consist of an orderly arrangement of oppositely charged ions assembled in a definite ratio that provides overall electrical neutrality.

The properties of a chemical compound are dramatically different from the properties of the various elements comprising it. For example, sodium is a soft silvery metal that is extremely reactive and poisonous. If you were to consume even a small amount of elemental sodium, you would need immediate medical attention. Chlorine, a green poisonous gas, is so toxic that it was used as a chemical weapon during World War I. Together, however, these elements produce sodium chloride, the harmless flavor enhancer that we call table salt. Thus, when elements combine to form compounds, their properties change significantly.

Covalent Bonds: Electron Sharing

With **covalent bonds**, a pair of atoms share one or more valance electrons between them. One example of this is the hydrogen molecule (H_2). Hydrogen is one of

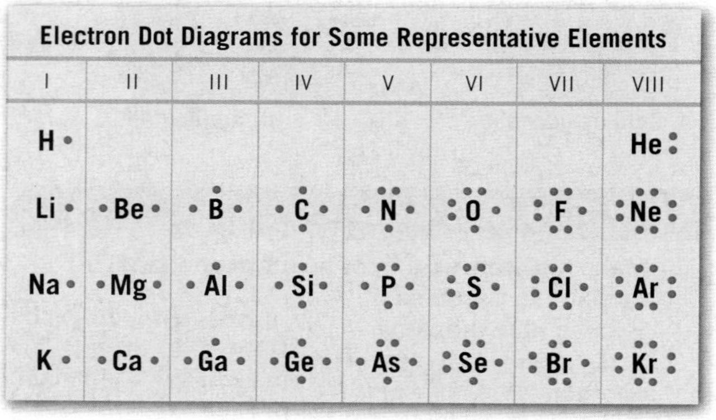

▶ **Figure 3.7**
Dot diagrams for certain elements Each dot represents a valence electron found in the outermost principal shell.

Electron Dot Diagrams for Some Representative Elements							
I	II	III	IV	V	VI	VII	VIII
H •							He :
Li •	• Be •	• B •	• C •	• N •	: O •	: F •	: Ne :
Na •	• Mg •	• Al •	• Si •	• P •	: S •	: Cl •	: Ar :
K •	• Ca •	• Ga •	• Ge •	• As •	: Se •	: Br •	: Kr :

◀ **Figure 3.8**
Formation of the ionic compound sodium chloride

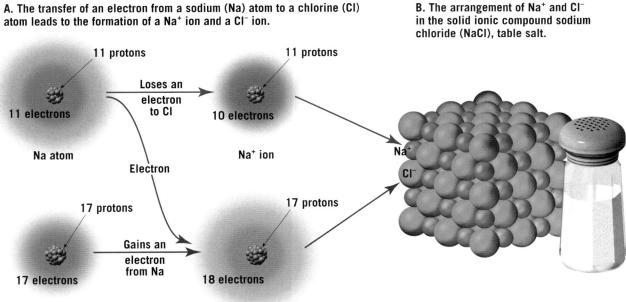

A. The transfer of an electron from a sodium (Na) atom to a chlorine (Cl) atom leads to the formation of a Na⁺ ion and a Cl⁻ ion.

11 protons

11 electrons

Na atom

Loses an electron to Cl

11 protons

10 electrons

Na⁺ ion

Electron

17 protons

17 electrons

Cl atom

Gains an electron from Na

17 protons

18 electrons

Cl⁻ ion

B. The arrangement of Na⁺ and Cl⁻ in the solid ionic compound sodium chloride (NaCl), table salt.

Na⁺

Cl⁻

the exceptions to the octet rule: Its single shell is full with just two electrons. Imagine two hydrogen atoms (each with one proton and one electron) approaching one another, as shown in **Figure 3.9**. Once they meet, the electron configuration changes so that both electrons primarily occupy the space between the atoms. In other words, the two electrons are shared by both hydrogen atoms and are attracted simultaneously by the positive charge of the proton in the nucleus of each atom. In this situation, the hydrogen atoms do not form ions. Instead, the force that holds these atoms together arises from the attraction of oppositely charged particles—positively charged protons in the nuclei and negatively charged electrons that surround these nuclei.

Metallic Bonds: Electrons Free to Move

A few minerals, such as native gold, silver, and copper, are made entirely of metal atoms packed tightly together in an orderly way. The bonding that holds these atoms together results from each atom contributing its valence electrons to a common pool of electrons, which freely move throughout the entire metallic structure. The contribution of one or more valence electrons leaves an array of positive ions immersed in a "sea" of valence electrons, as shown in **Figure 3.10**.

The attraction between this sea of negatively charged electrons and the positive ions produces the **metallic bonds** that give metals unique properties. Metals are good conductors of electricity because the valence electrons are free to move from one atom to another. Metals are also *malleable*, which means they

can be hammered into thin sheets, and *ductile*, which means they can be drawn into thin wires. By contrast, ionic and covalent solids tend to be *brittle* and fracture when stress is applied. To visualize the difference between metallic, ionic, and covalent bonds, compare what happens when a metal fork is dropped to the floor compared to what happens when a ceramic dinner plate is dropped.

Two hydrogen atoms combine to form a hydrogen molecule, held together by the attraction of oppositely charged particles—positively charged protons in each nucleus and negatively charged electrons that surround these nuclei.

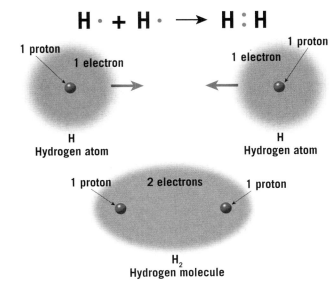

$$H \cdot + H \cdot \longrightarrow H : H$$

1 proton

1 electron

H
Hydrogen atom

1 proton

1 electron

H
Hydrogen atom

1 proton

2 electrons

1 proton

H_2
Hydrogen molecule

◀ **Figure 3.9**
Formation of a covalent bond When hydrogen atoms bond, the negatively charged electrons are shared by both hydrogen atoms and attracted simultaneously by the positive charge of the proton in the nucleus of each atom.

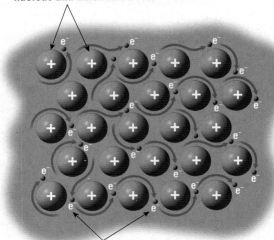

The central core of each metallic atom, which has an overall positive charge, consists of the nucleus and inner electrons.

A "sea" of negatively charged outer electrons, that are free to move throughout the structure, surround the positive ions.

How Minerals Form

Minerals form through a wide variety of chemical reactions in many different environments, and their properties are determined in part by how their atoms bonded. Three general processes that lead to mineral formation are *precipitation*, *crystallization*, and *deposition*.

Mineral Precipitation   Recall from chemistry that when a solid forms and then falls out of a solution, we call the process *precipitation*. The precipitation of mineral matter (ions) occurs when a saltwater body—typically inland bodies of water like the Great Salt Lake in Utah—evaporates. When this mineral-rich water reaches saturation, the ions within it begin to bond, forming crystalline solids called *salts*, that settle out of the solution. Some common *evaporate deposits* include *halite* (NaCl), *sylvite* (KCl), and *gypsum* ($CaSO_4 \cdot 2H_2O$). Worldwide, extensive evaporite deposits, some exceeding hundreds of meters in thickness, provide evidence of ancient seas that have long since evaporated (see Figure 3.40, page 96).

The Great Salt Lake is one of many locations worldwide where minerals are mined using the process of solar evaporation. Brine from the lake is pumped into a series of shallow ponds, where the Sun evaporates the water over months or years. The first mineral to precipitate is halite (common salt), which is used mainly in water softeners or to melt ice on highways. The well-known Morton Salt Company is one of several mining operations that extract a variety of minerals from the Great Salt Lake.

Minerals can also precipitate from slowly moving groundwater that fills fractures and voids in rocks and sediments. One interesting example, called a *geode*,

is a somewhat spherically shaped object with inward-projecting crystals that were gradually deposited by groundwater (**Figure 3.11**). Geodes often contain spectacular crystals of quartz, calcite, or other less common minerals.

Crystallization of Molten Rock   The crystallization of minerals from molten rock is a process similar to (though more complex than) the process of water freezing. (See Figure 4.3, page 105.) The atoms of magma are very mobile, but as the molten material cools, the atoms slow and begin to chemically combine. Crystallization of a molten mass generates igneous rocks that consist of a mosaic of intergrown crystals that tend to lack well-developed planar surfaces, or faces (see Figure 3.3). This process is discussed in greater detail in Chapter 4.

Deposition as a Result of Biological Processes

As mentioned previously, water-dwelling organisms transform substantial quantities of dissolved material into mineral matter. For example, corals create large quantities of marine limestones—rocks composed of the mineral calcite. These relatively simple invertebrate animals absorb calcium (Ca) ions from seawater and use it to secrete external skeletons composed of calcium carbonate ($CaCO_3$). Over time, these small organisms are capable of creating massive limestone structures called coral reefs.

Geodes like this one form when silica dissolved in groundwater precipitates to form quartz crystals that grow within cavities in rocks.

Groundwater containing dissolved silica

Silica (SiO_2)

Purple quartz crystals

Geologist's Sketch

▲ Figure 3.11
Geode partially filled with amethyst Geodes form in cavities in rocks, such as limestone and volcanic rocks. Slowly moving groundwater deposits dissolved mineral matter in these voids.

Some marine invertebrates, such as clams and oysters, also secrete shells composed of the carbonate minerals calcite and aragonite. When the remains of these shells are buried, they become the major component of the sedimentary rock limestone. Other organisms, such as diatoms and radiolarians, produce glasslike silica skeletons. During burial, this material forms microscopic silicon dioxide (quartz) crystals that are the main constituents of rocks such as chert and flint.

CONCEPT CHECKS 3.3

1. Explain the difference between an atom and an ion.
2. How does an atom become a positive ion? A negative ion?
3. Briefly distinguish between ionic, covalent, and metallic bonding and discuss the role that electrons play in each.
4. Describe three ways minerals can form.

Concept Checker
https://goo.gl/bvFXR3

3.4 Properties of Minerals

List and describe the properties used in mineral identification.

Minerals have definite crystalline structures and chemical compositions that give them unique sets of physical and chemical properties shared by all specimens of that mineral, regardless of when or where they formed. For example, two samples of the mineral quartz will be equally hard and equally dense, and they will break in a similar manner. However, the physical properties of individual samples may vary within specific limits due to ionic substitutions, inclusions of foreign elements (impurities), and defects in the crystalline structure. Certain aspects, called **diagnostic properties**, are particularly useful in identifying an unknown mineral. For example, the mineral halite has a salty taste that very few others do, making the taste a diagnostic property of halite. Other properties, particularly color, may vary among different specimens of the same mineral; these are referred to as **ambiguous properties**.

Optical Properties

Optical characteristics such as luster, color, streak, and ability to transmit light are frequently used for mineral identification.

Luster The appearance or quality of light reflected from the surface of a mineral is known as **luster**. Minerals that are shiny like a metal, regardless of color, are said to have a *metallic luster* (**Figure 3.12A**). Some metallic minerals, such as native copper and galena, develop a dull coating or tarnish when exposed to the atmosphere. Because they are not as shiny as samples with freshly

◀ **SmartFigure 3.13**
Color variations in minerals Some minerals, such as fluorite, exhibit a variety of colors.

Tutorial
https://goo.gl/rr7dsx

A. This freshly broken sample of galena displays a metallic luster.

B. This sample of galena is tarnished and has a submetallic luster.

Metallic

Submetallic

▲ **Figure 3.12**
Metallic versus submetallic luster

broken surfaces, these samples are often said to exhibit a *submetallic luster* (**Figure 3.12B**).

Most minerals have a *nonmetallic luster* and are described using various adjectives. For example, some minerals are described as being *vitreous*, or *glassy*. Other nonmetallic minerals are described as having a *dull*, or *earthy* luster, or a *pearly luster* (like a pearl or the inside of a clamshell). Still others exhibit a *silky luster* (like satin cloth) or a *greasy luster* (as though coated in oil).

Color Although **color** is generally the most conspicuous characteristic of any mineral, it is considered a diagnostic property of only a few minerals. Slight impurities in fluorite, for example, give this common mineral a variety of tints, including pink, purple, yellow, white, and green (**Figure 3.13**). Other minerals, such as quartz, also

▶ SmartFigure 3.14
Streak

Video
https://goo.gl/
GsbNKY

Mineral
(Pyrite)

Streak
(Black)

Color
(Brass yellow)

Although the color of a mineral is not always helpful in identification, the streak, which is the color of the powdered mineral, can be very useful.

exhibit a variety of hues, with multiple colors sometimes occurring in the same sample. Thus, the use of color as a means of identification is often ambiguous or even misleading.

Streak   The color of a mineral in powdered form, called **streak**, is often useful in identification. A mineral's streak is obtained by rubbing it across a *streak plate* (a piece of unglazed porcelain) and observing the color of the mark it leaves (**Figure 3.14**). Although a mineral's color may vary from sample to sample, its streak is usually consistent in color. (Note that not all minerals produce a streak when rubbed across a streak plate. Quartz, for example, is harder than a porcelain streak plate and therefore leaves no streak.)

Streak can also help distinguish between minerals with metallic luster and those with nonmetallic luster. Metallic minerals generally have a dense, dark streak, whereas minerals with nonmetallic luster typically produce a light-colored streak.

Ability to Transmit Light   Another optical property used to identify minerals is the ability to transmit light. When no light is transmitted through a mineral sample, that mineral is described as *opaque*; when light, but not an image, is transmitted, the mineral is said to be *translucent*. When both light and an image are visible through the sample, the mineral is described as *transparent*.

Crystal Shape, or Habit

Mineralogists use the term **crystal shape**, or **habit**, to refer to the common or characteristic shape of individual crystals or aggregates of crystals. Some minerals tend to grow equally in all three dimensions, whereas others tend to be elongated in one direction or flattened if growth in one dimension

is suppressed. The crystals of a few minerals can have a regular polygonal shape that is helpful in identification. For example, magnetite crystals sometimes occur as octahedrons, garnets often form dodecahedrons, and halite and fluorite crystals tend to grow as cubes or near-cubes. Most minerals have just one common crystal shape, but a few, such as the pyrite samples shown in **Figure 3.15**, have two or more characteristic crystal shapes.

In addition, some mineral samples consist of numerous intergrown crystals exhibiting characteristic shapes that are useful for identification. Terms commonly used to describe these and other crystal habits include *equant* (equidimensional), *bladed*, *fibrous*, *tabular*, *cubic*, *prismatic*, *platy*, *blocky*, and *banded*. Some of these habits are pictured in **Figure 3.16**.

Mineral Strength

Mineralogists use the terms *hardness*, *cleavage*, *fracture*, and *tenacity* to describe mineral strength and how minerals break when stress is applied. How easily minerals break or deform under stress is determined by the type and strength of the chemical bonds holding the crystals together.

Hardness   One of the most useful diagnostic properties is **hardness**, a measure of the resistance of a mineral to abrasion or scratching. It is determined by rubbing a mineral of unknown hardness against one of known hardness or vice versa. A numerical value of hardness can be obtained by using the **Mohs scale**, which consists of 10 minerals arranged in order from 1 (softest) to 10 (hardest), as shown in **Figure 3.17A**. It should be noted that the Mohs scale is a relative ranking and does not imply that a mineral with a hardness of 2, such as gypsum, is twice as hard as mineral with a hardness of 1, like talc. In fact, gypsum is only slightly harder than talc, as **Figure 3.17B** indicates.

▼ Figure 3.15
Common crystal shapes of pyrite

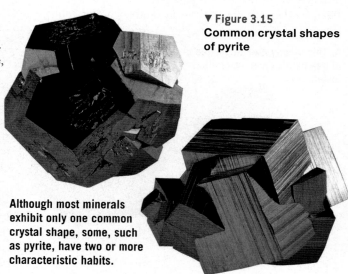

Although most minerals exhibit only one common crystal shape, some, such as pyrite, have two or more characteristic habits.

◄ **SmartFigure 3.16**
Some common crystal habits **A.** Thin, rounded crystals that break into fibers. **B.** Elongated crystals that are flattened in one direction. **C.** Minerals that have stripes or bands of different color or texture. **D.** Groups of crystals that are cube shaped.

Tutorial
https://goo.gl/Ei3Kp7

A. Fibrous (Sample: asbestos)

B. Bladed (Sample: selenite gypsum)

C. Banded (Sample: agate)

D. Cubic crystals (Sample: halite)

In the laboratory, common objects used to determine the hardness of a mineral can include a human fingernail, which has a hardness of about 2.5, a copper penny (3.5), and a piece of glass (5.5). The mineral gypsum, which has a hardness of 2, can be easily scratched with a fingernail. On the other hand, the mineral calcite, which has a hardness of 3, will scratch a fingernail but will not scratch glass. Quartz, one of the hardest common minerals, will easily scratch glass. Diamonds, hardest of all, scratch anything, including other diamonds.

Cleavage   In the crystal structure of many minerals, some atomic bonds are weaker than others. It is along these weak bonds that minerals tend to break when they are stressed. **Cleavage** (*kleiben* = carve) is the tendency of a mineral to break along planes of weak bonding. Not all minerals have cleavage, but those that do can be identified by the relatively smooth, flat surfaces that are produced when the mineral is broken.

The simplest type of cleavage is exhibited by the micas. Because these minerals have very weak bonds in one direction, they cleave to form thin, flat sheets

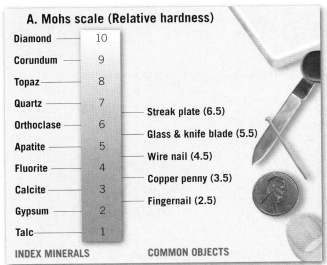

A. Mohs scale (Relative hardness)

INDEX MINERALS		COMMON OBJECTS
Diamond	10	
Corundum	9	
Topaz	8	
Quartz	7	
Orthoclase	6	Streak plate (6.5)
Apatite	5	Glass & knife blade (5.5)
Fluorite	4	Wire nail (4.5)
Calcite	3	Copper penny (3.5)
Gypsum	2	Fingernail (2.5)
Talc	1	

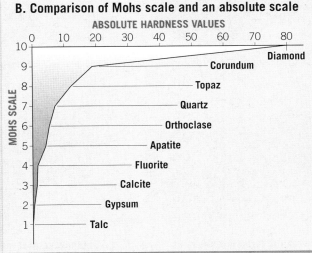

B. Comparison of Mohs scale and an absolute scale
ABSOLUTE HARDNESS VALUES
MOHS SCALE

Diamond
Corundum
Topaz
Quartz
Orthoclase
Apatite
Fluorite
Calcite
Gypsum
Talc

◄ **SmartFigure 3.17**
Hardness scales **A.** The Mohs scale of hardness ranks the hardness of some common objects. **B.** Relationship between the Mohs relative hardness scale and an absolute hardness scale.

Tutorial
https://goo.gl/HrPWG1

▶ SmartFigure 3.18
Micas exhibit perfect cleavage The thin sheets shown here exhibit one plane of cleavage.

Animation
https://goo.gl/ahFP6X

Knife blade   **Weak bonds**

Strong bonds

(**Figure 3.18**). Some minerals have excellent cleavage in one, two, three, or more directions, whereas others exhibit fair or poor cleavage, and still others have no cleavage at all. When minerals break evenly in more than one direction, cleavage is described by *the number of cleavage directions and the angle(s) at which they meet* (**Figure 3.19**).

Each cleavage surface that has a different orientation is counted as a different direction of cleavage. For example, some minerals, such as halite, cleave to form six-sided cubes. Because a cube is defined by three different sets of parallel planes that intersect at 90-degree

angles, cleavage for the mineral halite is described as *three directions of cleavage that meet at 90 degrees*.

Do not confuse cleavage with crystal shape. When a mineral exhibits cleavage, it breaks into pieces that all have the same geometry. By contrast, the smooth-sided quartz crystals shown in Figure 3.1 do not have cleavage. If broken, they fracture into shapes that do not resemble one another or the original crystals.

Fracture   Minerals having chemical bonds that are equally, or nearly equally, strong in all directions exhibit a property called **fracture**. When minerals fracture, most produce uneven surfaces and are described as exhibiting *irregular fracture* (**Figure 3.20A**). However, some minerals, including quartz, sometimes break into smooth, curved surfaces resembling broken glass; such breaks are called *conchoidal fractures* (**Figure 3.20B**). Still other minerals exhibit fractures that produce splinters or fibers; these types of fracture are referred to as *splintery fracture* and *fibrous fracture*, respectively.

Tenacity   The term **tenacity** describes how a mineral responds to stress—for instance, whether it tends to break in a brittle fashion or bend elastically. As mentioned earlier, nonmetallic minerals such as quartz and minerals that are ionically bonded, such as fluorite and halite, tend to be *brittle* and fracture or exhibit cleavage when struck. By contrast, native metals, such as copper and gold, are *malleable*, which means they can be hammered without breaking. In addition, minerals that can be cut into thin shavings, including gypsum and talc, are described as *sectile*. Still others, notably the micas, are *elastic* and bend and snap back to their original shape after stress is released.

▶ SmartFigure 3.19
Cleavage directions exhibited by minerals

Tutorial
https://goo.gl/3TQwNF

  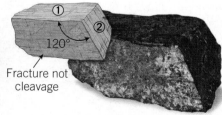

A. Cleavage in one direction. (Sample: muscovite)

B. Cleavage in two directions at 90° angles. (Sample: feldspar)
Fracture not cleavage

C. Cleavage in two directions not at 90° angles. (Sample: hornblende)
Fracture not cleavage

D. Cleavage in three directions at 90° angles. (Sample: halite)

E. Cleavage in three directions not at 90° angles. (Sample: calcite)

F. Cleavage in four directions. (Sample: fluorite)

Density and Specific Gravity

Density is defined as mass per unit volume. Mineralogists often use **specific gravity**, the ratio of a mineral's weight to the weight of an equal volume of water. Most common minerals have a specific gravity between 2 and 3. Quartz has a specific gravity of 2.65. By contrast, some metallic minerals, such as pyrite, native copper, and magnetite, are more than twice as dense and thus have more than twice the specific gravity of quartz. 24-karat gold has a specific gravity of approximately 20. You can estimate the specific gravity of a mineral by hefting it in your hand. Does this mineral feel about as "heavy" as similarly sized rocks you have handled? If "yes," the specific gravity of the sample will likely be between 2 and 3.

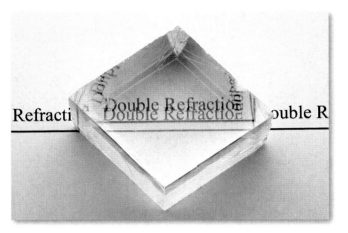

◄ Figure 3.21
Double refraction This calcite sample exhibits double refraction.

Other Properties of Minerals

Some minerals can be recognized by other distinctive properties. Halite tastes salty, talc feels soapy, and graphite feels greasy. Further, the streaks of many sulfur-bearing minerals smell like rotten eggs. A few minerals, such as magnetite, have high iron content and can be picked up with a magnet, while some varieties (such as lodestone) are themselves natural magnets and will pick up small iron-based objects such as pins and paper clips (see Figure 3.39F, page 94).

Moreover, some minerals exhibit special optical properties. For example, when a transparent piece of calcite is placed over printed text, the letters appear twice. This optical property is known as *double refraction* (**Figure 3.21**).

One very simple chemical test to detect carbonate minerals involves placing a drop of dilute hydrochloric acid from a dropper bottle onto a freshly broken mineral surface. Samples containing carbonate minerals will effervesce (fizz) as carbon dioxide gas is released (**Figure 3.22**). This test is especially useful in identifying calcite, a common carbonate mineral.

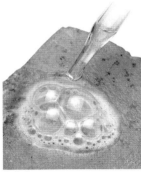

Calcite

▲ SmartFigure 3.22
Calcite reacting with a weak acid

Video
https://goo.gl/
G7BPzm

A. Irregular fracture (Quartz)

B. Conchoidal fracture (Quartz)

▲ Figure 3.20
Irregular versus conchoidal fracture

CONCEPT CHECKS 3.4

1. Why is color not always a useful property in mineral identification? Give an example of a mineral that supports your answer.

2. What differentiates cleavage from fracture?

3. What is meant by a mineral's *tenacity*? List three terms that describe tenacity.

 Concept Checker
https://goo.gl/nUwzB4

4. Describe a simple chemical test that is useful in identifying the mineral calcite.

3.5 Mineral Structures and Compositions

Distinguish between compositional and structural variations in minerals and provide one example of each.

Many people associate the word *crystal* with delicate wine goblets or glassy objects with smooth sides and gem-like shapes. In geology, the term **crystal**, or **crystalline**, refers to *any natural solid with an orderly, repeating internal structure*. Therefore, all mineral samples are crystals or crystalline solids, even if they lack smooth-sided faces. The specimen shown in Figure 3.1, for example, exhibits the characteristic crystal form associated with quartz—a six-sided prismatic shape with pyramidal ends. However, the quartz crystals in the sample of granite shown in Figure 3.3 do not display well-defined faces. Both quartz samples are nonetheless crystalline.

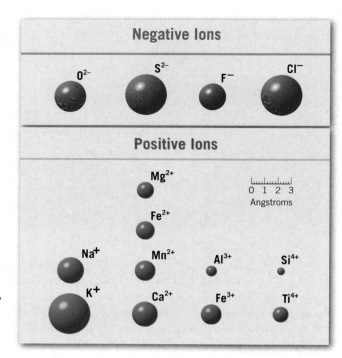

Negative Ions

O^{2-} S^{2-} F^- Cl^-

Positive Ions

Mg^{2+}

0 1 2 3
Angstroms

Fe^{2+}

Na^+ Mn^{2+} Al^{3+} Si^{4+}

K^+ Ca^{2+} Fe^{3+} Ti^{4+}

▶ Figure 3.23
Relative sizes and charges of selected ions Ionic radii are usually expressed in angstroms (1 angstrom equals 10^{-8} cm).

Mineral Structures

The smooth faces and symmetry possessed by well-developed crystals are surface manifestations of the orderly packing of the atoms or ions that constitute a mineral's internal structure. This highly ordered atomic arrangement within minerals can be illustrated by using spherically shaped atoms held together by ionic, covalent, or metallic bonds.

The simplest crystal structures are those of native metals, such as gold and silver, which are composed of only one element. These materials consist of atoms packed together in a rather simple three-dimensional network that minimizes voids. Imagine a group of cannon balls stacked in layers such that the spheres in one layer nestle in the hollows between spheres in the adjacent layers.

The atomic structure of most minerals consists of at least two different ions (often of very different sizes). **Figure 3.23** illustrates the relative sizes of some of the most common ions found in minerals. Notice that the negative ions, which are atoms that gained electrons, tend to be larger than the positive ions, which lost electrons.

Crystal structures can be considered three-dimensional stacks of larger spheres (negative ions) with smaller spheres (positive ions) located in the spaces between them, so that the positive and negative charges cancel each other out. Consider the mineral halite (NaCl), which has a relatively simple framework composed of an equal number of positively charged sodium ions and negatively charged chlorine ions. Because ions of similar charge repel, they are spaced as far apart from each other as possible. Consequently, in halite, each sodium ion (Na^+) is surrounded on all sides by chlorine ions and vice versa (**Figure 3.24**). This particular arrangement forms basic building blocks, called **unit cells**, that have cubic shapes. As shown in **Figure 3.24C**, these cubic unit cells combine to form cube-shaped halite crystals, including those that come out of salt shakers.

The shape and symmetry of these building blocks relate to the shape and symmetry of the entire crystal. It is important to note, however, that two minerals can be constructed of geometrically similar building blocks yet exhibit different external forms. For example, fluorite, magnetite, and garnet are minerals constructed of cubic unit cells, but these unit cells can join to produce crystals of many shapes. Typically, fluorite crystals are cubes, whereas magnetite crystals are octahedrons, and garnets form dodecahedrons built up of many small cubes, as

A. Sodium and chlorine ions.

B. Basic building block of the mineral halite.

C. Collection of basic building blocks (crystal).

D. Crystals of the mineral halite.

▶ Figure 3.24
Arrangement of sodium and chloride ions in the mineral halite The arrangement of atoms into basic building blocks that have a cubic shape results in regularly shaped cubic crystals.

A. Cube (fluorite)

B. Octahedron (magnetite)

C. Dodecahedron (garnet)

◄ **Figure 3.25**
Cubic unit cells These cells stack together in different ways to produce crystals that exhibit different shapes. **A.** Fluorite tends to display cubic crystals, whereas **B.** magnetite crystals are typically octahedrons, and **C.** garnets usually occur as dodecahedrons.

shown in **Figure 3.25**. Because the building blocks are so small, the resulting crystal faces are smooth and flat.

Despite the fact that natural crystals are rarely perfect, the angles between equivalent crystal faces of the same mineral are remarkably consistent. This observation was first made by Nicolas Steno in 1669. Steno found that the angles between adjacent prism faces of quartz crystals are 120 degrees, regardless of the size of the sample, the size of the crystal faces, or where the crystals were collected (**Figure 3.26**). This observation is commonly called **Steno's law**, or the **law of constancy of interfacial angles**, because it applies to all minerals. Because Steno's law holds for all minerals, crystal shape is frequently a valuable tool in mineral identification.

Compositional Variations in Minerals

Mineralogists have determined that the chemical composition of some minerals varies substantially from sample to sample. These compositional variations in minerals are possible because ions of similar size can readily substitute for one another without disrupting a mineral's internal framework. This is analogous to a wall made of bricks of different colors and materials. As long as the bricks are roughly the same size, the shape of the wall is unaffected; only its composition changes.

For example, olivine's chemical formula— $(Mg,Fe)_2SiO_4$—has the variable components magnesium and iron in parentheses. Magnesium (Mg^{2+}) and iron (Fe^{2+}) readily substitute for one another because they are nearly the same size and have the same electrical charge. At one extreme, olivine may contain magnesium without iron—a variety called forsterite (Mg_2SiO_4). At the other extreme, olivine may contain only iron, resulting in the favalite variety (Fe_2SiO_4). However, most samples of olivine have some of both of these ions in their structure, and all olivine types have the same internal structure and exhibit very similar, but not identical, properties. For example, iron-rich olivines have a higher density than magnesium-rich specimens, a reflection of the greater atomic weight of iron as compared to magnesium.

In contrast to olivine, minerals such as quartz (SiO_2) and fluorite (CaF_2) tend to have chemical compositions that differ very little from their chemical formulas. However, even these minerals often contain tiny amounts of other, less common elements, referred to as *trace elements*. Although trace elements have little effect on most mineral properties, they can significantly influence a mineral's color.

Structural Variations in Minerals

It is possible for two minerals with exactly the same chemical composition to have different internal structures and, hence, different external forms. Minerals of this type are called **polymorphs** (*poly* = many, *morph* = form). Graphite and diamond are particularly good examples of polymorphism because, when pure, they are both made up exclusively of carbon atoms. Graphite is the soft gray mineral from which pencil "lead" is made, whereas diamond is the hardest-known mineral. The differences between these minerals can be attributed to the conditions

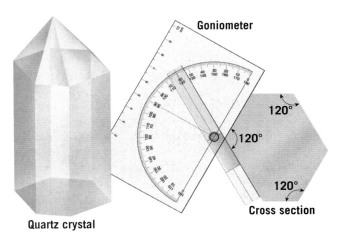

Goniometer

120°

120°

120°

Cross section

Quartz crystal

◄ **Figure 3.26**
Steno's law Because some faces of a crystal may grow larger than others, two crystals of the same mineral may *not* have identical shapes. Nevertheless, the angles between equivalent faces are remarkably consistent.

▶ Figure 3.27
Diamond versus graphite Both diamond and graphite are natural substances with the same chemical composition: carbon atoms. Nevertheless, their internal structures and physical properties reflect the fact that each formed in a very different environment.

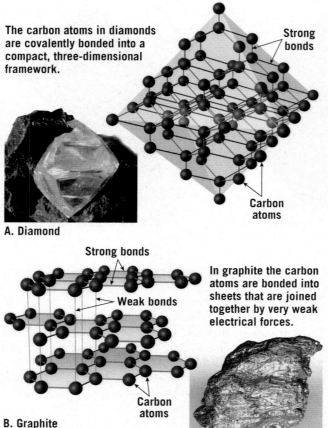

The carbon atoms in diamonds are covalently bonded into a compact, three-dimensional framework.

Strong bonds

Carbon atoms

A. Diamond

Strong bonds

Weak bonds

In graphite the carbon atoms are bonded into sheets that are joined together by very weak electrical forces.

Carbon atoms

B. Graphite

under which they form. Diamonds form at depths that may exceed 200 kilometers (nearly 125 miles), where extreme pressures and temperatures produce the compact structure shown in **Figure 3.27A**. Graphite, on the other hand, forms under comparatively low pressures and consists of sheets of carbon atoms that are widely spaced and weakly bonded (**Figure 3.27B**). Because the carbon sheets in graphite easily slide past one another, graphite has a greasy feel and makes an excellent lubricant.

Scientists have learned that by heating graphite under high confining pressures, they can generate synthetic diamonds. Because diamonds form in environments of extreme pressure and temperature, they are somewhat unstable at Earth's surface. Fortunately for jewelers, "diamonds are forever" because the rate at which diamonds change to their more stable form, graphite, is infinitesimally slow.

The transformation of one polymorph to another is an example of a *phase change*. In nature, certain minerals go through phase changes as they move from one environment to another. For example, when a slab of ocean crust composed of olivine-rich basalt is carried to great depths by a subducting plate, the olivine changes to a more compact, denser polymorph with the same structure as the mineral *spinel*.

Recall that oceanic lithosphere sinks because it is colder and denser than the underlying mantle. It follows, therefore, that during subduction, the transformation of olivine from its low-density form to its high-density form would contribute to plate subduction. Stated another way, this phase change causes an increase in the overall density of the slab, thereby enhancing its rate of descent.

CONCEPT CHECKS 3.5

1. Explain Steno's law in your own words.

2. Define *polymorph* and give an example.

 Concept Checker
https://goo.gl/ab1dHd

3.6 Mineral Groups

Explain how minerals are classified and name the most abundant mineral group in Earth's crust.

More than 4000 minerals have been named, and several new ones are identified each year. Fortunately for students who are beginning to study minerals, no more than a few dozen are abundant. Collectively, these few make up most of the rocks of Earth's crust and, as such, they are often referred to as the **rock-forming minerals**.

Although less abundant, many other minerals are used extensively in the manufacture of products; these minerals are called **economic minerals**. However, rock-forming minerals and economic minerals are not mutually exclusive groups. When found in large deposits, some rock-forming minerals are economically significant. One example is calcite, the primary component of the sedimentary rock limestone. Calcite has many uses, including the production of concrete.

Classifying Minerals

In much the same way that plants and animals are classified, mineralogists use the term *mineral species* for a collection of specimens that exhibit similar internal structures and chemical compositions. Some common mineral species are quartz, calcite, galena, and pyrite. However, just as individual plants and animals within a species differ somewhat from one another, so do most specimens of the same mineral.

Some mineral species are further subdivided into *mineral varieties*. For example, pure quartz (SiO_2) is colorless and transparent. However, when small amounts of aluminum are incorporated into its atomic structure, in a variety called *smoky quartz*, quartz appears quite dark. *Amethyst*, another variety of quartz, owes its violet color to the presence of trace amounts of iron.

Mineral species are assigned to *mineral groups*. Some important mineral groups are the silicates, carbonates, halides, and sulfates. Minerals within each group tend to have similar internal structures and, hence, similar properties. For example, minerals of the carbonate group react chemically with acid—albeit to varying degrees—and many exhibit similar cleavage. Furthermore, minerals within the same group are often found together in the same rock. For example, halite (NaCl) and silvite (KCl) belong to the halide class and commonly occur together in evaporite deposits.

Silicate Versus Nonsilicate Minerals

Only *eight elements* make up the vast majority of the rock-forming minerals and represent more than 98 percent (by weight) of Earth's continental crust (**Figure 3.28**). These elements, in order of most to least abundant, are oxygen (O), silicon (Si), aluminum (Al), iron (Fe), calcium (Ca), sodium (Na), potassium (K), and magnesium (Mg). As shown in Figure 3.28, silicon and oxygen are by far the most common elements in Earth's crust. They readily combine to form the basic "building block" for the most common mineral group, the **silicates**. More than 800 silicate minerals are known, and they account for about 92 percent of Earth's crust.

Because other mineral groups are far less abundant in Earth's crust than the silicates, they are often grouped together under the heading **nonsilicates**. Although the nonsilicate minerals are not as common

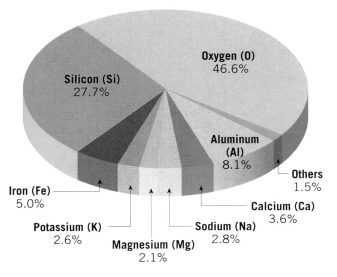

as silicates, many of them are economically important. They provide us with iron and aluminum to build automobiles, gypsum for plaster and drywall for home construction, and copper wire that carries electricity and connects us to the Internet. Common nonsilicate mineral groups include the carbonates, sulfates, and halides. In addition to their economic importance, these groups include minerals that are major constituents in sediments and sedimentary rocks.

CONCEPT CHECKS 3.6

1. Distinguish between *rock-forming minerals* and *economic minerals*.

2. List the eight most common elements in Earth's crust.

 **Concept Checker**
https://goo.gl/h2rhxk

3.7 The Silicates

Sketch the silicon–oxygen tetrahedron and explain how this fundamental building block joins together to form various silicate structures.

Every silicate mineral contains oxygen and silicon, and most of them also contain one or more of the other common elements. These elements give rise to hundreds of silicate minerals, including hard quartz, soft talc, sheet-like mica, fibrous asbestos, green olivine, and blood-red garnet.

Silicate Structures

The structure of every silicate mineral includes the **silicon–oxygen tetrahedron** (SiO_4^{4-}). In this structure, four oxygen ions covalently bond to one comparatively small silicon ion, forming a *tetrahedron*—a pyramid shape with four identical planar surfaces, or faces (**Figure 3.29**). The tetrahedra are complex ions (SiO_4^{4-}) having a net charge of −4. To become electrically balanced, these complex ions bond to positively charged metal ions. Specifically, each O^{2-} has one of its valence

electrons bonding with the Si^{4+} located at the center of the tetrahedron. The remaining negative charge on each oxygen ion is available to bond with another positive ion or with the silicon ion in an adjacent tetrahedron.

Minerals with Independent Tetrahedra One of the simplest silicate structures consists of independent tetrahedra that have their four oxygen ions bonded to positive ions, such as Mg^{2+}, Fe^{2+}, and Ca^{2+}. The mineral olivine, with the formula $(Mg, Fe)_2 SiO_4$, is a good

◀ Figure 3.29
Two representations of the silicon–oxygen tetrahedron

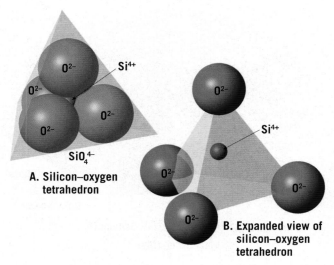

A. Silicon–oxygen tetrahedron

B. Expanded view of silicon–oxygen tetrahedron

example. In olivine, magnesium (Mg^{2+}) and/or iron (Fe^{2+}) ions pack between comparatively large independent SiO_4 tetrahedra, forming a dense, three-dimensional structure. Garnet, another common silicate, is also composed of independent tetrahedra ionically bonded to positive ions. Both olivine and garnet form dense, hard, equidimensional crystals that lack cleavage.

Minerals with Chain or Sheet Structures   One reason for the great variety of silicate minerals is the ability of silicon–oxygen tetrahedra to link to one another in a variety of configurations. This important phenomenon, called **polymerization**, is achieved by the sharing of one, two, three, or all four of the oxygen atoms with adjacent tetrahedra. Vast numbers of tetrahedra join together to form single chains, double chains, sheet structures, or three-dimensional frameworks, as shown in **Figure 3.30**.

To see how oxygen atoms are shared between adjacent tetrahedra, select one of the silicon ions (small blue spheres) near the middle of the single chain shown in **Figure 3.30B**. Notice that this silicon ion is completely surrounded by four larger oxygen ions (red spheres). Also notice that two of the four oxygen atoms are bonded to two silicon atoms, whereas the other two are not shared in this manner. It is the linkage across the shared oxygen ions that joins the tetrahedra into a chain structure. Now examine a silicon ion near the middle of the sheet structure (see Figure 3.30D) and count the number of shared and unshared oxygen ions surrounding it. As you have likely observed, the sheet structure is the result of three of the four oxygen atoms being shared by adjacent tetrahedra.

Minerals with Three-Dimensional Frameworks
In the most common silicate structure, all four oxygen ions are shared, producing a complex three-dimensional framework (see Figure 3.30E). Quartz and the most common mineral group, the feldspars, exhibit this type of structure.

The ratio of oxygen ions to silicon ions differs in each type of silicate structure. In independent tetrahedra, there are four oxygen ions for every silicon ion (SiO_4). In single chains, the oxygen-to-silicon ratio is 3:1 (SiO_3), and in three-dimensional frameworks, as found in quartz, the ratio is 2:1 (SiO_2). As more oxygen ions are shared, the percentage of silicon in the structure increases. Silicate minerals are therefore described as having a low or high silicon content, based on their ratio of oxygen to silicon. Silicate minerals with three-dimensional structures have the highest silicon content, while those composed of independent tetrahedra have the lowest.

◀ SmartFigure 3.30
Five basic silicate structures A. Independent tetrahedra. **B**. Single chains. **C**. Double chains. **D**. Sheet structures. **E**. Three-dimensional framework.

Tutorial
https://goo.gl/ypDNQS

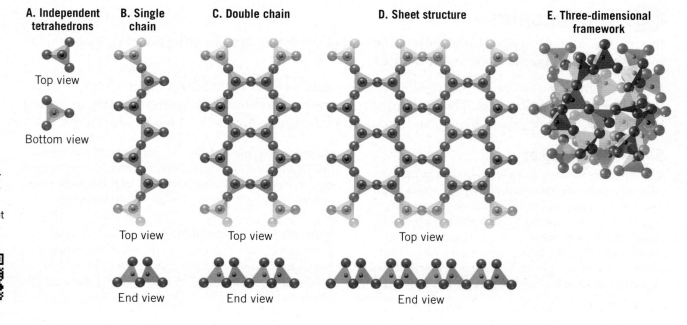

A. Independent tetrahedrons   B. Single chain   C. Double chain   D. Sheet structure   E. Three-dimensional framework

Top view

Bottom view

Top view   Top view   Top view

End view   End view   End view

Joining Silicate Structures

Except for quartz (SiO_2), the basic structure (chains, sheets, or three-dimensional frameworks) of most silicate minerals has a net negative charge. Therefore, metal ions are required to bring the overall charge into balance and to serve as the "mortar" that holds these structures together. The positive ions that most often link silicate structures are iron (Fe^{2+}), magnesium (Mg^{2+}), potassium (K^+), sodium (Na^+), aluminum (Al^{3+}), and calcium (Ca^{2+}). These positively charged ions bond with the unshared oxygen ions that occupy the corners of the silicate tetrahedra.

As a general rule, the covalent bonds between silicon and oxygen are stronger than the ionic bonds that hold one silicate structure to the next. Consequently, properties such as cleavage and, to some extent, hardness are controlled by the nature of the silicate framework. Quartz (SiO_2), which has only silicon–oxygen bonds, has great hardness and lacks cleavage, mainly because it has equally strong bonds in all directions. By contrast, the mineral talc (the source of talcum powder) has a sheet structure because magnesium ions are found between the sheets and weakly join them together. The slippery feel of talcum powder is due to the silicate sheets sliding relative to one another, in much the same way that sheets of carbon atoms in graphite slide, giving graphite its lubricating properties.

Recall that atoms of similar size can substitute freely for one another without altering a mineral's structure. For example, in olivine, iron and magnesium substitute for each other. This also holds true for the third most common element in Earth's crust, aluminum (Al^{3+}), which often substitutes for silicon (Si^{4+}) in the center of silicon–oxygen tetrahedra.

Because most silicate structures will readily accommodate two or more different positive ions at a given bonding site, individual specimens of a particular mineral may contain varying amounts of certain elements. As a result, many silicate minerals form mineral groups that exhibit a range of compositions between two end members. Examples include the olivines, pyroxenes, amphiboles, micas, and feldspars.

CONCEPT CHECKS 3.7

1. Sketch the silicon–oxygen tetrahedron and label its parts.

2. What is the ratio of oxygen to silicon found in single tetrahedral? How about in framework structures? Which has the highest silicon content?

Concept Checker
https://goo.gl/SWoVxe

3.8 Common Silicate Minerals

Compare and contrast the light (nonferromagnesian) silicates with the dark (ferromagnesian) silicates and list three common minerals in each group.

The major groups of silicate minerals and common examples are given in **Figure 3.31**. Most silicate minerals form when molten rock cools and crystallizes. Cooling can occur at or near Earth's surface (low temperature and pressure) or at great depths (high temperature and pressure). The environment during crystallization and the chemical composition of the molten rock determine, to a large degree, which minerals are produced. For example, olivine crystallizes early, whereas quartz forms much later in the crystallization process.

In addition to silicate minerals that crystallize from molten rock, some form at Earth's surface from other silicate minerals through the process of weathering. Still others are formed under the extreme pressures associated with mountain building. Each silicate mineral, therefore, has a structure and a chemical composition that *indicate the conditions under which it formed.* By carefully examining the mineral constituents of rocks, geologists can usually determine the circumstances under which the rocks formed.

We will now examine some of the most common silicate minerals, which we divide into two major groups on the basis of their chemical makeup: the light silicates and the dark silicates.

The Light Silicates

The **light** (or **nonferromagnesian**) **silicates** are generally light in color and have a specific gravity of about 2.7, lower than that of the dark (ferromagnesian) silicates. These differences are mainly attributable to the presence or absence of iron and magnesium, which are "heavy" elements. The light silicates contain varying amounts of aluminum, potassium, calcium, and sodium rather than iron and magnesium.

Feldspar Group *Feldspar minerals* are by far the most plentiful silicate group in Earth's crust, comprising about 51 percent of the crust (**Figure 3.32**, page 91). Their abundance can be partially explained by the fact

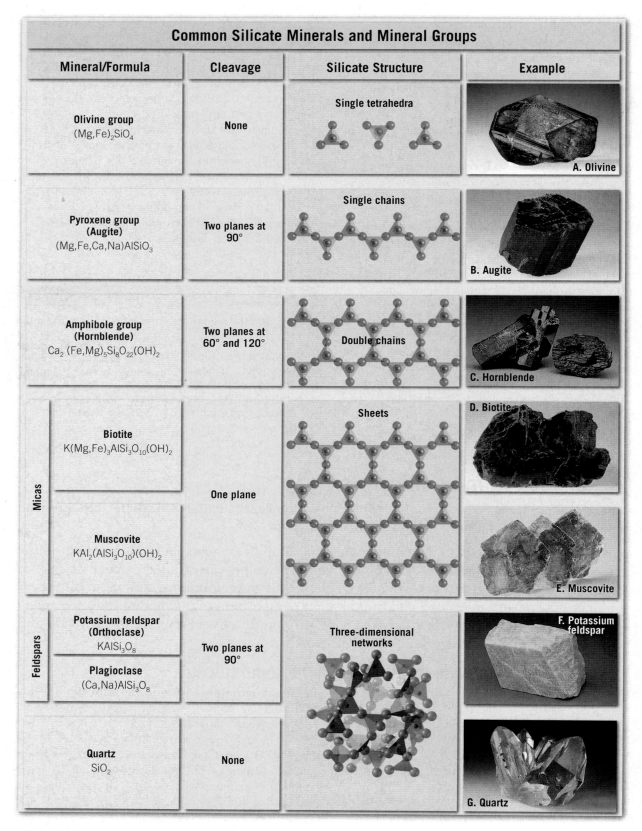

Common Silicate Minerals and Mineral Groups			
Mineral/Formula	**Cleavage**	**Silicate Structure**	**Example**
Olivine group $(Mg,Fe)_2SiO_4$	None	Single tetrahedra	A. Olivine
Pyroxene group (Augite) $(Mg,Fe,Ca,Na)AlSiO_3$	Two planes at 90°	Single chains	B. Augite
Amphibole group (Hornblende) $Ca_2(Fe,Mg)_5Si_8O_{22}(OH)_2$	Two planes at 60° and 120°	Double chains	C. Hornblende
Micas — Biotite $K(Mg,Fe)_3AlSi_3O_{10}(OH)_2$	One plane	Sheets	D. Biotite
Micas — Muscovite $KAl_2(AlSi_3O_{10})(OH)_2$	One plane	Sheets	E. Muscovite
Feldspars — Potassium feldspar (Orthoclase) $KAlSi_3O_8$	Two planes at 90°	Three-dimensional networks	F. Potassium feldspar
Feldspars — Plagioclase $(Ca,Na)AlSi_3O_8$	Two planes at 90°	Three-dimensional networks	
Quartz SiO_2	None	Three-dimensional networks	G. Quartz

▲ **Figure 3.31**

Common silicate minerals Note that the complexity of the silicate structure increases from the top of the chart to the bottom.

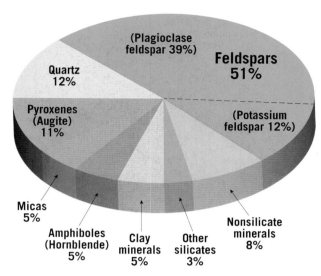

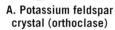
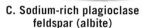
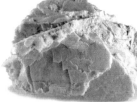

▲ Figure 3.32
Feldspar minerals make up about 51 percent of Earth's crust Also, note from this chart that the silicate minerals make up about 92 percent of Earth's crust.

Potassium Feldspar

A. Potassium feldspar crystal (orthoclase)

B. Potassium feldspar showing cleavage (orthoclase)

Plagioclase Feldspar

C. Sodium-rich plagioclase feldspar (albite)

D. Plagioclase feldspar showing striations (labradorite)

◀ Figure 3.33
Some common feldspar minerals A. Characteristic crystal form of potassium feldspar. **B.** Most salmon-colored feldspar belongs to the potassium feldspar subgroup. **C.** Sodium-rich plagioclase feldspar tends to be light in color and have a pearly luster. **D.** Calcium-rich plagioclase feldspar tends to be gray, blue-gray, or black in color. Labradorite, the sample shown here, exhibits striations on one of its crystal faces.

that they can form under a wide range of temperatures and pressures. Two different feldspar structures exist. One of the feldspar structures, which contains potassium ions, is termed **potassium feldspar** (**Figure 3.33A,B**). (*Orthoclase* and *microcline* are common members of the potassium feldspar group.) The other structure, called **plagioclase feldspar**, contains both sodium and calcium ions that freely substitute for one another, depending on the environment during crystallization (**Figure 3.33C,D**). Despite these differences, all feldspar minerals have similar physical properties. They have two planes of cleavage meeting at or near 90-degree angles, are relatively hard (6 on the Mohs scale), and have a luster ranging from glassy to pearly. As a component in igneous rocks, feldspar crystals can be identified by their rectangular shape and rather smooth, shiny faces.

Potassium feldspar is usually light cream, salmon pink, or occasionally blue-green in color. The plagioclase feldspars, on the other hand, range from gray to blue-gray, or sometimes black. However, color should not be used to distinguish these groups, as the only way to distinguish the feldspars by looking at them is through the presence of a multitude of fine parallel lines, called *striations*. Striations are found on some cleavage planes of plagioclase feldspar (see Figure 3.33D) but are not present on potassium feldspar (see Figure 3.33B).

Quartz The second most abundant mineral in the continental crust, **quartz** (SiO_2) is the only common silicate mineral that consists entirely of silicon and oxygen. As such, the term *silica* is commonly applied to quartz. Because quartz contains a ratio of two oxygen ions (O^{2-}) to every silicon ion (Si^{4+}), no other positive ions are needed to attain neutrality.

In quartz, a three-dimensional framework develops through the complete sharing of oxygen by adjacent silicon atoms (see Figure 3.31). Thus, all the bonds in quartz are of the strong silicon–oxygen type. Consequently, quartz is hard, resists weathering, and does not have cleavage. When broken, quartz generally exhibits conchoidal fracture. When pure, quartz is clear and, if allowed to grow without interference, will develop hexagonal crystals that develop pyramid-shaped ends. However, like most other clear minerals, quartz is often colored by inclusions of various ions (impurities) and often forms without developing good crystal faces. The most common varieties of quartz are milky quartz (white), smoky quartz (gray), rose quartz (pink), amethyst (purple), citrine (yellow to brown), and rock crystal (clear) (**Figure 3.34**).

A. Smoky quartz

B. Rose quartz

C. Milky quartz

D. Amethyst

◀ Figure 3.34
Quartz, the second most common mineral in Earth's crust, has many varieties
A. *Smoky quartz* is commonly found in coarse-grained igneous rocks. **B.** *Rose quartz* owes its color to small amounts of titanium. **C.** *Milky quartz* often occurs in veins, which occasionally contain gold. **D.** *Amethyst*, a purple variety of quartz often used in jewelry, is the birthstone for February.

Muscovite   A common member of the mica family, **muscovite** is light in color and has a pearly luster (see Figure 3.18). Like other micas, muscovite has excellent cleavage in one direction. In thin sheets, muscovite is clear, a property that accounts for its use as window "glass" during the Middle Ages. It can often be identified by the shiny sparkle it gives to a rock. If you look closely at beach sand, you may see the glimmering brilliance of the mica flakes scattered among the other sand grains.

Clay Minerals   **Clay** is a term used to describe a category of complex minerals that, like the micas, have a sheet structure. Unlike other common silicates, most clay minerals originate as products of the chemical breakdown (chemical weathering) of other silicate minerals. Thus, clay minerals make up a large percentage of the surface material we call soil. (Weathering and soils are discussed in detail in Chapter 6.) Because of soil's importance to agriculture and because of its role as a supporting material for buildings, clay minerals are extremely important to humans. In addition, clays account for nearly half the volume of sedimentary rocks. Clay minerals are generally very fine grained, making them difficult to identify unless they are studied microscopically. Their layered structure and the weak bonding between layers give them a characteristic slippery feel when wet. Clays are common in shales, mudstones, and other sedimentary rocks.

One of the most common clay minerals is *kaolinite* (**Figure 3.35**), which is used in the manufacture of fine china and as a coating for high-gloss paper, such as that used in this textbook. Further, some clay minerals absorb large amounts of water, which allows them to swell to several times their normal size. These clays have been used commercially in a variety of ingenious ways, including as an additive to thicken milkshakes in fast-food restaurants.

▲ **Figure 3.35**
Kaolinite Kaolinite is a common clay mineral formed by weathering of feldspar minerals.

Olivine-rich peridotite (variety dunite)

▲ **Figure 3.36**
Olivine Commonly black to olive green in color, olivine has a glassy luster and is often granular in appearance. Olivine is commonly found in the igneous rock basalt.

The Dark Silicates

As their name implies, the **dark** (or **ferromagnesian**) **silicates** are minerals containing ions of iron and/or magnesium in their structure. Because of their iron content, ferromagnesian silicates are dark in color and have a greater specific gravity, between 3.2 and 3.6, than nonferromagnesian silicates. The most common dark silicate minerals are olivine, the pyroxenes, the amphiboles, dark mica (biotite), and garnet.

Olivine Group   **Olivine**, a family of high-temperature silicate minerals, are black to olive green in color and have a glassy luster and a conchoidal fracture (see Figure 3.31A). Transparent olivine is occasionally used as a gemstone called peridot. Olivine commonly forms small, rounded crystals that give olivine-rich rocks a granular appearance (**Figure 3.36**). Olivine and related forms are typically found in basalt (a common igneous rock of the oceanic crust and volcanic areas on the continents) and are thought to constitute up to 50 percent of Earth's upper mantle.

Pyroxene Group   The *pyroxenes* are a group of diverse minerals that are important components of many dark-colored igneous rocks. The most common member, **augite**, is black, opaque, and one of the dominant minerals in basalt (see Figure 3.31B).

Amphibole Group   **Hornblende** is the most common member of a chemically complex group of minerals called *amphiboles.* Hornblende, which is usually dark green to black in color, forms elongated crystals *(see Figure 3.31C).* When found in igneous rocks, hornblende often makes up

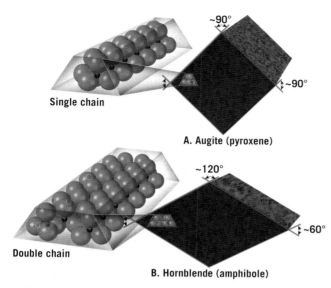

~90°

Single chain

~90°

A. Augite (pyroxene)

~120°

Double chain

~60°

B. Hornblende (amphibole)

▲ **Figure 3.37**
Comparing the cleavage of augite and horn-blende Because the silicate structures in hornblende are more weakly bonded than in augite, hornblende exhibits better cleavage.

Biotite **Biotite** is a dark, iron-rich member of the mica family (see Figure 3.31D). Like other micas, biotite possesses a sheet structure that gives it excellent cleavage in one direction. Its shininess helps distinguish it from the other dark ferromagnesian minerals. Like hornblende, biotite is a common constituent of igneous rocks, including granite.

Garnet **Garnet** is similar to olivine in that its structure is composed of individual tetrahedral linked by metallic ions. Also like olivine, garnet has a glassy luster, lacks cleavage, and exhibits conchoidal fracture. Although the colors of garnet are varied, this mineral is most often brown to deep red. Well-developed garnet crystals have 12 diamond-shaped faces and are most commonly found in metamorphic rocks (**Figure 3.38**). When transparent, garnets are prized as semiprecious gemstones.

◄— 2 cm —►

▲ **Figure 3.38**
Well-formed garnet crystal Garnets come in a variety of colors and are commonly found in mica-rich metamorphic rocks.

the dark portion of an otherwise light-colored rock (see Figure 3.3). Hornblende and augite are often mistaken for one another due to their similar appearance and dark color. Their most distinguishing characteristic is cleavage. The two planes of cleavage of hornblende meet at angles that are about 60 degrees and 120 degrees, whereas the cleavage exhibited by augite is roughly 90 degrees (**Figure 3.37**).

CONCEPT CHECKS 3.8

1. Apart from their difference in color, what is one main distinction between light and dark silicates? What accounts for this difference?

   ◄)) Concept Checker
   https://goo.gl/JXkPVX

2. Based on the chart in Figure 3.31, what do muscovite and biotite have in common? How do they differ?

3.9 Important Nonsilicate Minerals

List the common nonsilicate minerals and explain why each is important.

Although the nonsilicates make up only about 8 percent of Earth's crust, some nonsilicate minerals, such as gypsum, calcite, and halite, occur as constituents in sedimentary rocks in significant amounts. Many nonsilicates are also economically important.

Nonsilicate minerals are typically divided into groups based on the negatively charged ion or complex ion that the members have in common. For example, the *oxides* contain negative oxygen ions (O^{2-}), which bond to one or more kinds of positive ions. Thus, within each mineral group, the basic structure and type of bonding is similar. As a result, the minerals in each group have similar physical properties that are useful in mineral identification. **Figure 3.39** lists some of the major nonsilicate mineral groups and includes a few examples of each.

Some of the most common nonsilicate minerals belong to one of three groups of minerals: the carbonates (CO_3^{2-}), the sulfates (SO_4^{2-}), and the halides (Cl^{1-}, F^{1-}, Br^{1-}). The carbonate minerals have much simpler structures than the silicates. This mineral group is composed of the carbonate ion (CO_3^{2-}) and one or more kinds of positive ions. The two most common carbonate minerals are **calcite**, $CaCO_3$ (calcium carbonate), and **dolomite**, $CaMg(CO_3)_2$ (calcium/magnesium carbonate) (see Figure 3.39A,B). Calcite and dolomite are usually found together as the primary constituents in

Common Nonsilicate Mineral Groups

Mineral Group (key ion(s) or element(s))	Mineral Name	Chemical Formula	Economic Use	Examples
Carbonates (CO_3^{2-})	Calcite Dolomite	$CaCO_3$ $CaMg(CO_3)_2$	Portland cement, lime Portland cement, lime	  B. Dolomite A. Calcite
Halides $(Cl^{1-}, F^{1-}, Br^{1-})$	Halite Fluorite Sylvite	$NaCl$ CaF_2 KCl	Common salt Used in steel making Used as fertilizer	 C. Halite  D. Fluorite
Oxides (O^{2-})	Hematite Magnetite Corundum Ice	Fe_2O_3 Fe_3O_4 Al_2O_3 H_2O	Ore of iron, pigment Ore of iron Gemstone, abrasive Solid form of water	 E. Hematite  F. Magnetite
Sulfides (S^{2-})	Galena Sphalerite Pyrite Chalcopyrite Cinnabar	PbS ZnS FeS_2 $CuFeS_2$ HgS	Ore of lead Ore of zinc Sulfuric acid production Ore of copper Ore of mercury	  H. Chalcopyrite G. Galena
Sulfates (SO_4^{2-})	Gypsum Anhydrite Barite	$CaSO_4 \cdot 2H_2O$ $CaSO_4$ $BaSO_4$	Plaster Plaster Drilling mud	  J. Anhydrite I. Gypsum
Native elements (single elements)	Gold Copper Diamond Graphite Sulfur Silver	Au Cu C C S Ag	Trade, jewelry Electrical conductor Gemstone, abrasive Pencil lead Sulfa drugs, chemicals Jewelry, photography	 K. Copper  L. Sulfur

▲ Figure 3.39
Important nonsilicate mineral groups

Gemstones

Important Gemstones

Gemstones are classified in one of two categories: precious or semiprecious. Precious gems are rare and generally have hardnesses that exceed 9 on the Mohs scale. Therefore, they are more valuable and thus more expensive than semiprecious gems.

GEM	MINERAL NAME	PRIZED HUES
PRECIOUS		
Diamond	Diamond	Colorless, pinks, blues
Emerald	Beryl	Greens
Ruby	Corundum	Reds
Sapphire	Corundum	Blues
Opal	Opal	Brilliant hues
SEMIPRECIOUS		
Alexandrite	Chrysoberyl	Variable
Amethyst	Quartz	Purples
Cat's-eye	Chrysoberyl	Yellows
Chalcedony	Quartz (agate)	Banded
Citrine	Quartz	Yellows
Garnet	Garnet	Red, greens
Jade	Jadeite or nephrite	Greens
Moonstone	Feldspar	Transparent blues
Peridot	Olivine	Olive greens
Smoky quartz	Quartz	Browns
Spinel	Spinel	Reds
Topaz	Topaz	Purples, reds
Tourmaline	Tourmaline	Reds, blue-greens
Turquoise	Turquoise	Blues
Zircon	Zircon	Reds

Precious stones have been prized since antiquity. Although most gemstones are varieties of a particular mineral, misinformation abounds regarding gems and their mineral makeup.

The Famous Hope Diamond

The deep-blue Hope Diamond is a 45.52-carat gem that is thought to have been cut from a much larger 115-carat stone discovered in India in the mid-1600s. The original 115-carat stone was cut into a smaller gem that became part of the crown jewels of France and was in the possession of King Louis XVI and Marie Antoinette before they attempted to escape France. Stolen during the French Revolution in 1792, the gem is thought to have been recut to its present size and shape. In the 1800s, it became part of the collection of Henry Hope (hence its name) and is on display at the Smithsonian in Washington, DC.

What Constitutes a Gemstone?

When found in their natural state, most gemstones are dull and would be passed over by most people as "just another rock." Gems must be cut and polished by experienced professionals before their true beauty is displayed. Cutting and polishing is accomplished using abrasive material, most often tiny fragments of diamonds that are embedded in a metal disk.

Naming Gemstones

Most precious stones are given names that differ from their parent mineral. For example, sapphire is one of two gems that are varieties of the same mineral, corundum. Trace elements can produce vivid sapphires of nearly every color. Tiny amounts of titanium and iron in corundum produce the most prized blue sapphires. When the mineral corundum contains a sufficient quantity of chromium, it exhibits a brilliant red color. This variety of corundum is called ruby.

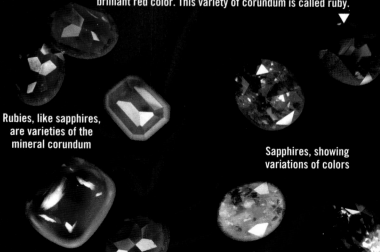

Rubies, like sapphires, are varieties of the mineral corundum

Sapphires, showing variations of colors

the sedimentary rocks limestone and dolostone. When calcite is the dominant mineral, the rock is called *limestone*, whereas *dolostone* results from a predominance of dolomite. Limestone, used in road aggregate and as a building stone, is the main ingredient in Portland cement.

Two other nonsilicate minerals frequently found in sedimentary rocks are **halite** and **gypsum** (see Figure 3.39C,I). Both minerals are commonly found in thick layers that are the last vestiges of ancient seas that have

While nonsilicates make up only about 8 percent of Earth's crust, many of these minerals are economically important.

long since evaporated (**Figure 3.40**). Like limestone, both halite and gypsum are important nonmetallic resources. Halite is the mineral name for common table salt (NaCl). Gypsum ($CaSO_4 \cdot 2H_2O$), which is calcium sulfate with water bound into the structure, is the mineral from which plaster and other similar building materials are composed.

Most nonsilicate mineral classes contain minerals that are prized for their economic value. This include the oxides, whose members *hematite* and *magnetite* are important ores of iron (see Figure 3.39E,F). Also significant are the sulfides, which are basically compounds of sulfur (S) and one or more metals. Important sulfide minerals include galena (lead sulfide), sphalerite (zinc sulfide), and chalcopyrite (copper sulfide). In addition, native elements—including gold, silver, and carbon (diamonds)—are economically important, as are a host of other nonsilicate minerals—such as fluorite (used as a flux in making steel), corundum (gemstone, abrasive), and uraninite (a uranium source).

▶ **Figure 3.40**
Thick bed of halite exposed in an underground mine Halite (salt) mine in Grand Saline, Texas. (Note the person for scale.)

CONCEPT CHECKS 3.9

1. List eight common nonsilicate minerals and their economic uses.

2. What is the most common carbonate mineral?

Concept Checker
https://goo.gl/CcSfZi

---

3 CONCEPTS IN REVIEW
Matter and Minerals

3.1 Minerals: Building Blocks of Rocks

List and describe the main characteristics that an Earth material must possess to be considered a mineral.

Key Terms: mineralogy mineral rock

- Geologists use the word *mineral* to refer to naturally occurring inorganic solids that possess an orderly crystalline structure and a characteristic chemical composition.
- Minerals are the building blocks of rocks. *Rocks* are naturally occurring masses of minerals or mineral-like matter such as natural glass or organic material.

3.2 Atoms: Building Blocks of Minerals

Compare and contrast the three primary particles contained in atoms.

Key Terms:
neutron
atom
nucleus
proton
electron
valence electron
atomic number
element
periodic table
chemical compound

- Minerals are composed of *atoms* of one or more elements. All atoms consist of the same three basic components: *protons*, *neutrons*, and *electrons*.

- The *atomic number* represents the number of protons found in the nucleus of an atom of a particular element.
- Protons and neutrons have approximately the same size and mass, but protons are positively charged, whereas neutrons have no charge.
- Electrons weigh only about 1/2000 as much as protons or neutrons. They occupy the space around the nucleus, where they form a structure consisting of several distinct energy levels called principal shells. The electrons in the outermost principal shell, called *valence*

electrons, are responsible for the bonds that hold atoms together to form chemical compounds.
- The *periodic table* is organized so elements with the same number of valence electrons form a column or group consisting of elements that tend to behave similarly.

Q Use the periodic table (see Figure 3.5) to identify these geologically important elements by their number of protons: (A) 14, (B) 6, (C) 13, (D) 17, and (E) 26.

(3.3) How Atoms Bond to Form Minerals

Distinguish among ionic bonds, covalent bonds, and metallic bonds and explain how they form minerals.

Key Terms:
- ionic bond
- chemical bond
- octet rule
- ion
- covalent bond
- metallic bond

- *Chemical bonds* form when atoms are attracted to other atoms, leading to the transfer or sharing of valence electrons. The most stable arrangement for most atoms is to have eight electrons in the outermost principal shell, according to the *octet rule*.
- To form *ionic bonds*, atoms of one element give up one or more valence electrons to atoms of another element, forming positively and negatively charged atoms called *ions*. The ionic bond results from the attraction between oppositely charged ions.
- *Covalent bonds* form when adjacent atoms share valence electrons.
- *Metallic bonds* involve extensive sharing of the valence electrons, such that the electrons move freely through the substance.

- Minerals form in various ways. They may precipitate out of a solution, forming bonds as the water in which they are suspended evaporates. As molten rock cools, the atoms form bonds with other atoms to produce a crystalline material. And some marine organisms extract ions from the surrounding seawater and secrete exoskelatal material usually made of calcium carbonate or silica.

Q Match the diagrams labeled A–C with the type of bonding each illustrates: ionic, covalent, or metallic. What are the distinguishing characteristics of each type of bonding?

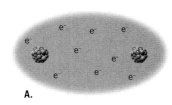
A.

B.

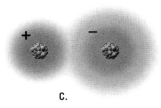
C.

(3.4) Properties of Minerals

List and describe the properties used in mineral identification.

Key Terms:
- diagnostic property
- ambiguous property
- luster
- color
- streak
- crystal shape
- hardness
- Mohs scale
- cleavage
- fracture
- tenacity
- density
- specific gravity

- *Luster* is a mineral's ability to reflect light. *Transparent, translucent,* and *opaque* describe the degree to which a mineral can transmit light.
- *Color* can be unreliable for mineral identification. A more reliable identifier is *streak*, the color of the powder generated by scraping a mineral against a porcelain streak plate.
- *Crystal shape*, also called crystal *habit*, is often useful for mineral identification.
- Variations in the strength of chemical bonds give minerals properties such as resistance to being scratched, called *hardness* (measured by the *Mohs scale*) and resistance to deforming stresses, called *tenacity* (which could be brittle breakage, bending, or malleable deformation).
- *Cleavage*, the preferential breakage of a mineral along planes of weakly bonded atoms, is very useful in identifying minerals.
- The amount of matter packed into a given volume determines a mineral's *density*. To compare the densities of minerals,

mineralogists use a related quantity, known as *specific gravity,* which is the ratio between a mineral's density and the density of water.
- *Diagnostic properties* are characteristics used for identifying minerals. Certain properties, such as smell, taste, feel, reaction to hydrochloric acid, magnetism, and double refraction, are diagnostic for certain minerals because they are rarely seen attributes.

Q Research the minerals *quartz* and *calcite*. List five physical characteristics that may be used to distinguish one from the other.

Quartz   **Calcite**

3.5 Mineral Structures and Compositions

Distinguish between compositional and structural variations in minerals and provide one example of each.

Key Terms: Steno's law   polymorph
crystal
unit cell

- *Crystals* are naturally occurring solids with orderly, repeating internal structures. The atoms forming a mineral are packed together in some geometrically regular fashion. The smallest expression of that arrangement is called a *unit cell*; many unit cells repeating in three dimensions yield a mineral crystal.
- *Steno's law*, or the *law of constancy of interfacial angles*, says that no matter how big a crystal of a given mineral may be, the angles between its faces will always be the same.

- The composition of minerals may vary within certain parameters. Generally, atoms of similar size and charge can "swap out" for one another.
- The same chemical compound can grow into different mineral crystals, with different arrangements of atoms. Calcite and aragonite are examples of this *polymorph* relationship: Though they are different minerals, they both have the same chemical composition—$CaCO_3$.

Q A crystal of the mineral corundum measuring 1 millimeter long has an angle between two crystal faces that measures 60 degrees. Assume that the crystal continues to grow until it has tripled in size. What is the angle between the crystal faces now?

3.6 Mineral Groups

Explain how minerals are classified and name the most abundant mineral group in Earth's crust.

Key Terms: economic mineral   nonsilicate
rock-forming mineral   silicate

- More than 4000 different minerals have been identified, but only a few dozen are common in Earth's crust: These are the *rock-forming minerals*.
- Minerals are placed into groups on the basis of similar crystal structures and compositions and are found in similar geologic settings.
- Silicon and oxygen are the most common elements in Earth's crust, and so the most common minerals in the crust are *silicate* minerals. *Nonsilicate* minerals are less common but include many *economic minerals* used in manufacturing.

3.7 The Silicates

Sketch the silicon–oxygen tetrahedron and explain how this fundamental building block joins together to form various silicate structures.

Key Terms: silicon–oxygen   polymerization
tetrahedron

- Silicate minerals all have a four-sided pyramid-like structure, called the *silicon–oxygen tetrahedron*, consisting of one silicon atom surrounded by four oxygen atoms. Individual tetrahedral can bond to other elements, including aluminum, iron, and potassium.
- Neighboring tetrahedral can share some of their oxygen atoms, causing them to develop long chains in a process called *polymerization*. This process can produce single chains, double chains, sheets, or even complicated three-dimensional framework of tetrahedral that share all the oxygen atoms in the mineral.

3.8 Common Silicate Minerals

Compare and contrast the light (nonferromagnesian) silicates with the dark (ferromagnesian) silicates and list three common minerals in each group.

Key Terms: muscovite   augite
light silicate   clay   hornblende
potassium feldspar   dark silicate   biotite
plagioclase feldspar   olivine   garnet
quartz

- Silicate minerals are the most common mineral group on Earth. They are subdivided into minerals that contain iron and/or magnesium (*dark*, or *ferromagnesian*, *silicates*) and those that do not (*light*, or *nonferromagnesian*, *silicates*).
- Nonferromagnesian silicates are generally light in color and of relatively low specific gravity. *Feldspar*, *quartz*, *muscovite*, and *clays* are all examples.

- Ferromagnesian silicates are generally dark in color and relatively dense. *Olivine, pyroxene, amphibole, biotite,* and *garnet* are all examples.

Q In general, nonferromagnesian minerals are light in color—shades of peach, tan, clear, or white. What could account for the fact that some nonferromagnesian minerals are dark colored, like the smoky quartz in this photo?

Smoky quartz

3.9 Important Nonsilicate Minerals

List the common nonsilicate minerals and explain why each is important.

Key Terms: dolomite   gypsum
calcite   halite

- Nonsilicate mineral groups, including the oxides, carbonates, sulfates, and halides, consist of negatively charged ions or complex ions.
- Many nonsilicate minerals have economic value. Hematite is an important source of industrial iron, *calcite* is a critical component of cement, *halite* is table salt, *dolomite* is an important soil amender for agriculture, and *gypsum* is used to make drywall for buildings.

GIVE IT SOME THOUGHT

1. Using the geologic definition of *mineral* as your guide, determine which of the items in this list are minerals and which are not. If something in this list is not a mineral, explain.

   a. Gold nugget

   b. Seawater

   c. Quartz

   d. Cubic zirconia

   e. Obsidian

   f. Ruby

   g. Glacial ice

   h. Amber

2. Assume that the number of protons in a neutral atom is 92 and the atomic mass is 238.03. (*Hint:* Refer to the periodic table in Figure 3.5 to answer this question.)

   a. What is the name of the element?

   b. How many electrons does it have?

3. Gold has a specific gravity of almost 20. A 5-gallon bucket of water weighs 40 pounds. How much would a 5-gallon bucket of gold weigh?

4. Examine the accompanying photo of a mineral that has several smooth, flat surfaces that resulted when the specimen was broken.

   a. How many flat surfaces are present on this specimen?

   b. How many different directions of cleavage does this specimen have?

   c. Do the cleavage directions meet at 90-degree angles?

Cleaved sample

5. Each of the following statements describes a silicate mineral or mineral group. In each case, provide the appropriate name:

   a. The most common member of the amphibole group

   b. The most common light-colored member of the mica family

   c. The only common silicate mineral made entirely of silicon and oxygen

   d. A silicate mineral with a name based on its color

   e. A silicate mineral characterized by striations

   f. Silicate minerals that originate as a product of chemical weathering

6. What mineral property is illustrated in the accompanying photo?

7. Do an Internet search to determine which minerals are used to manufacture the following products:

   a. Stainless steel utensils

   b. Cat litter

   c. Tums brand antacid tablets

   d. Lithium batteries

   e. Aluminum beverage cans

8. The accompanying diagram shows one of several possible arrangements for bonding silicon–oxygen tetrahedral. Describe the structure shown and name a mineral group that displays this type of structure.

EYE ON EARTH

1. The accompanying image shows one of the world's largest open pit gold mines, located near Kalgoorlie, Australia. Known as the Super Pit, it originally consisted of a number of small underground mines that were consolidated into a single, open-pit mine. Each year, about 28 metric tons of gold are extracted from the 15 million tons of rock shattered by blasting and then transported to the surface.

   a. What is one environmental advantage of underground mining over open pit mining?

   b. For those employed at this mine, what change in working conditions would have occurred as it evolved from an underground mine to an open pit mine?

2. Glass bottles, like most other manufactured products, contain substances obtained from minerals extracted from Earth's crust and oceans. The primary ingredient in commercially produced glass bottles is the mineral quartz. Glass also contains lesser amounts of the mineral calcite.

   a. In what mineral group does quartz belong?

   b. Glass beer bottles are usually clear, green, or brown. Based on what you know about how the mineral quartz is colored, what do glass manufacturers do to make bottles green and brown?

DATA ANALYSIS

Global Mineral Resources

Mineral resources are important for the global economy. Some minerals are found in only a few locations, while others are more evenly distributed around the globe. The U.S. Geological Survey (USGS) maintains a mineral resource database that will help us explore the distribution of major mineral deposits worldwide.

https://goo.gl/eD226M

ACTIVITIES

Go to the Mineral Resources Online Spatial Data interactive map at https://mrdata.usgs.gov/#mineral-resources and scroll down to "Major mineral deposits of the world." Click on the map to reveal a close-up map of the contiguous United States. (Zoom in further if necessary to see the 48 states clearly.) Click on the "Select map layers" icon in the top-left corner (above the question mark) to reveal the legend and answer the following questions.

1. Look at "Deposits by commodity." PGE stands for *platinum-group elements*. Where are most of the major gold deposits located in the lower 48 states? Consult the Internet to determine if the United States is an importer or an exporter of gold.

2. Where are most of the major iron deposits located in the lower 48 states? What is one economic use for iron?

3. What region of the United States has the greatest concentration of clay deposits? Consult the Internet and give one economic use for clay.

4. What region of the United States has the greatest concentration of copper deposits?

5. Name two materials that appear to lack economically important deposits. Name one economic use for each.

6. Consult the Internet and name the only mineral that is an economic source for aluminum.

7. Is the United States a net importer or a net exporter of aluminum?

Go back to https://mrdata.usgs.gov/#mineral-resources and click on the map next to "Mineral Resources Data System (MRDS)."

8. What information is displayed on this map?

9. Notice that Missouri and Iowa have many mineral resources, but Kansas and Nebraska have few. Why might the mining of mineral deposits change so abruptly at state boundaries?

Mastering Geology

Looking for additional review and test prep materials? Visit pearson.com/mastering/geology to enhance your understanding of this chapter's content by accessing a variety of resources, including **Self-Study Quizzes, Geoscience Animations, SmartFigure Tutorials,** *Project Condor* **Videos, Mobile Field Trips, Dynamic Study Modules,** and an optional **Pearson eText.**

Geothermal Energy: Clean Energy to Power the Future?

Iceland, a small island country that straddles the Mid-Atlantic Ridge, is a major producer of *geothermal energy*, where heat captured from Earth, in the form of hot water or steam, is used to produce electricity. The source of geothermal energy is hot igneous rocks below the surface of Earth's crust. Unlike fossil fuel–derived electricity, geothermal power is renewable, and it is less polluting.

Iceland's abundant volcanic fields allow it easy access to geothermal energy. This surplus of relatively cheap electricity has attracted new industries, including aluminum smelting and processing. New geothermal tapping methods may further expand how much geothermal power generation is possible.

A typical geothermal well in Iceland is now about 2.5 kilometers (1.5 miles) deep and can produce enough electricity to supply roughly 4000 homes. A new government-sponsored study is researching whether drilling twice as deep could tap into a water reservoir hotter than 450°C (about 840°F). This super-heated steam might produce 10 times the electricity of today's typical borehole. If ultimately successful in Iceland, this form of geothermal electricity generation may also be implemented on a wider scale throughout other parts of the world, including the South Pacific, eastern Africa, and the western United States.

► Bathers at the Blue Lagoon bath near Reykjanes, Iceland, relax in view of thermal wells used to produce geothermal energy. The Iceland Deep Drilling Project (IDDP) will use a large drilling rig to drill several boreholes to depths of about 5 kilometers (3 miles) in a field of volcanic rock located a short drive from Iceland's capital, Reykjavik.

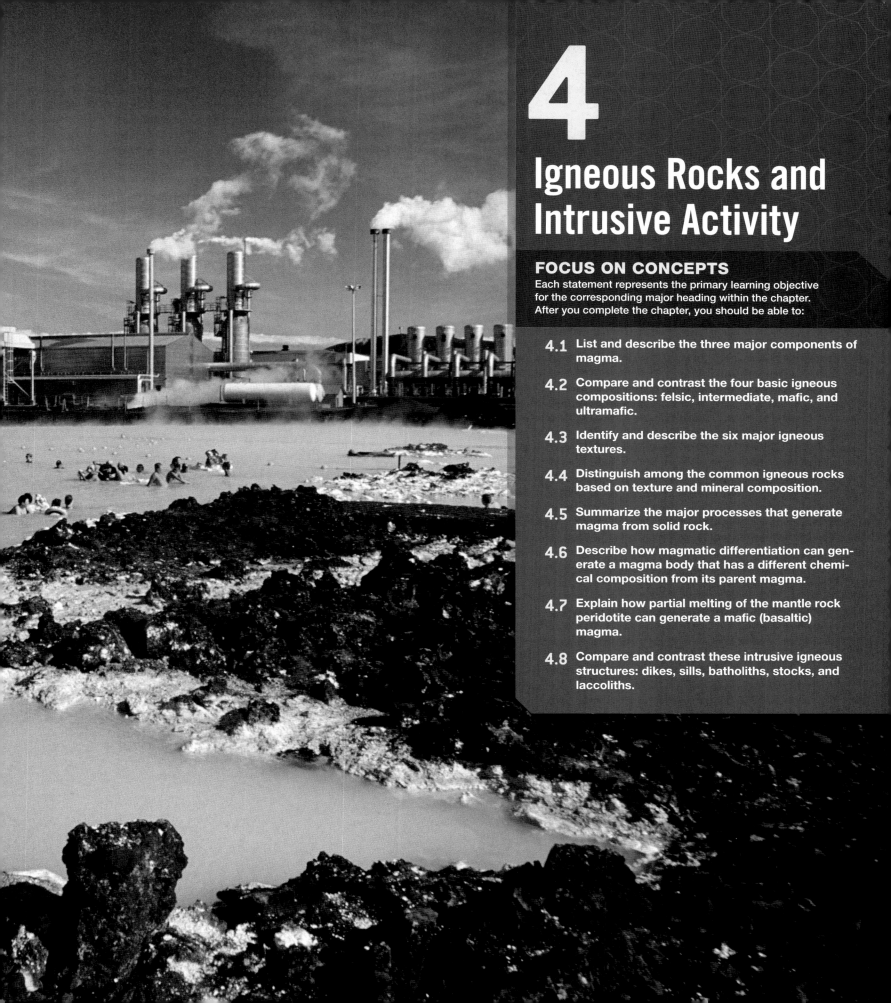

4

Igneous Rocks and Intrusive Activity

FOCUS ON CONCEPTS

Each statement represents the primary learning objective for the corresponding major heading within the chapter. After you complete the chapter, you should be able to:

4.1 List and describe the three major components of magma.

4.2 Compare and contrast the four basic igneous compositions: felsic, intermediate, mafic, and ultramafic.

4.3 Identify and describe the six major igneous textures.

4.4 Distinguish among the common igneous rocks based on texture and mineral composition.

4.5 Summarize the major processes that generate magma from solid rock.

4.6 Describe how magmatic differentiation can generate a magma body that has a different chemical composition from its parent magma.

4.7 Explain how partial melting of the mantle rock peridotite can generate a mafic (basaltic) magma.

4.8 Compare and contrast these intrusive igneous structures: dikes, sills, batholiths, stocks, and laccoliths.

Understanding the structure, composition, and internal workings of our planet requires a basic knowledge of igneous rocks, which are a major constituent of Earth's crust and mantle. Volcanoes such as Mount Rainier are composed of igneous rocks, as are large portions of the Sierra Nevada, Black Hills, and high peaks of the Adirondacks. Igneous rocks also make excellent building materials for homes and stone monuments.

4.1 Magma: Parent Material of Igneous Rock

List and describe the three major components of magma.

Magma is completely or partly molten rock, which, when cooled, solidifies into **igneous rocks** composed mainly of silicate minerals. Considerable evidence indicates that parent material for igneous rocks is formed by partial melting of solid rock that occurs at various levels within Earth's crust and upper mantle to depths of about 250 kilometers (about 150 miles). Once formed, a magma body buoyantly rises toward the surface because it is less dense than the surrounding rocks. When rock melts, it takes up more space and, hence, becomes less dense than the surrounding solid rock. Occasionally, molten rock reaches Earth's surface, where it is called **lava**. Sometimes escaping gases propel lava in a fountain-like spray from a magma chamber (**Figure 4.1**). On other occasions, lava is explosively ejected, generating dramatic steam and ash eruptions. However, not all eruptions are violent; many volcanoes emit quieter outpourings of fluid lava.

▼ Figure 4.1
Fluid lava emitted from Bardarbunga Volcano, Iceland Iceland is home to some of the world's most active volcanoes.

The Nature of Magma

Most magmas consist of three materials: a *liquid* component, a *solid* component, and a *gaseous* component. The liquid portion, called **melt**, is composed mainly of mobile ions of the eight most common elements found in Earth's crust—silicon and oxygen, along with lesser amounts of aluminum, potassium, calcium, sodium, iron, and magnesium.

The solid components (if any) in magma are crystals of silicate minerals. As a magma body cools, the size and number of crystals increase. During the last stage of cooling, a magma body is like a "crystalline mush," resembling a bowl of very thick oatmeal and containing only small amounts of melt.

The gaseous components of magma, called **volatiles**, are materials that vaporize (form a gas) at surface pressures. The volatiles most commonly present in magma are water vapor (H_2O), carbon dioxide (CO_2), and sulfur dioxide (SO_2), and they are confined in the magma by the immense pressure exerted by the overlying rocks. These gases tend to separate from the melt as it moves toward the surface (from a high- to a low-pressure environment). As the gases build up, they may eventually propel magma from the vent. When deeply buried magma bodies crystallize, the remaining volatiles collect as hot, water-rich fluids that migrate through the surrounding rocks. These hot fluids play an important role in metamorphism and will be considered in Chapter 8.

From Magma to Crystalline Rock

To better understand how magma crystallizes, let us first consider how a simple crystalline solid melts.

Recall that, in any crystalline solid, the ions are arranged in a closely packed regular pattern. However, they are not without some motion; they exhibit a restricted vibration about fixed points. As the temperature rises, ions vibrate more rapidly and, consequently, collide with their neighbors with ever-increasing vigor. Thus, heating causes the ions to occupy more space, which in turn causes the solid to expand. When the ions are vibrating rapidly enough to overcome the force of their chemical bonds, melting occurs. At this stage, the ions are able to slide past one another, and the orderly crystalline structure disintegrates. Thus, melting converts a solid consisting of tight, uniformly packed ions into a liquid composed of unordered ions moving randomly about.

In the process called **crystallization**, cooling reverses the events of melting. As the temperature of the liquid drops, ions pack more closely together as their rate of movement slows. When they are cooled sufficiently, the forces of the chemical bonds again confine the ions to an orderly crystalline arrangement.

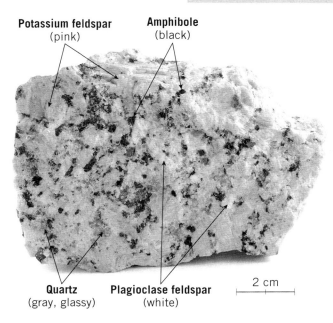

Potassium feldspar (pink)   Amphibole (black)

Quartz (gray, glassy)   Plagioclase feldspar (white)

2 cm

▲ Figure 4.2
Igneous rock composed of interlocking crystals The largest crystals measure about 1 centimeter in length.

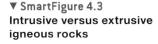

▼ SmartFigure 4.3
Intrusive versus extrusive igneous rocks

Tutorial
https://goo.gl/6qwRRP

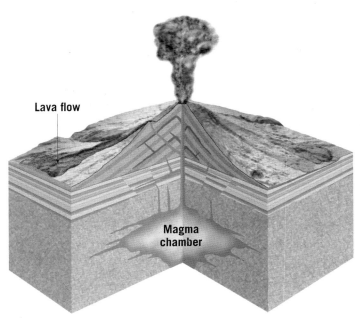

Lava flow

Magma chamber

A. Molten rock may crystallize at depth or at Earth's surface.

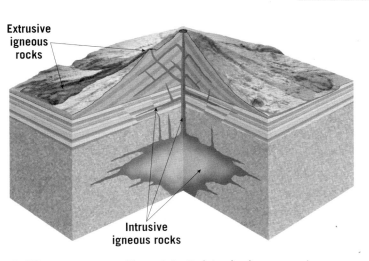

Extrusive igneous rocks

Intrusive igneous rocks

B. When magma crystallizes at depth, intrusive igneous rocks form. When lava solidifies on Earth's surface, extrusive igneous rocks form.

▶ Figure 4.4
Mount Rushmore National Memorial This memorial, located in the Black Hills of South Dakota, is carved from intrusive igneous rocks.

When a magma body cools, the silicon and oxygen atoms link together first to form silicon–oxygen tetrahedra, the basic building blocks of the silicate minerals (see Figure 3.29, page 88). As magma continues to lose heat to its surroundings, the tetrahedra join with each other and with other ions to form embryonic crystal nuclei. Each nucleus slowly grows as ions lose their mobility and join the crystalline network.

The minerals that form the earliest have space to grow and tend to have better-developed crystal faces than do the minerals that form later and occupy the remaining spaces. Eventually all of the melt is transformed into a solid mass of interlocking silicate minerals that we call an *igneous rock* (**Figure 4.2**).

Igneous Processes

Igneous rocks form in two basic settings. Molten rock may crystallize within Earth's crust over a range of depths, or it may solidify at Earth's surface (**Figure 4.3**).

When magma crystallizes *at depth*, it forms **intrusive igneous rocks**, also known as **plutonic rocks**—after Pluto, the god of the underworld in classical mythology. These rocks are observed at Earth's surface in locations where uplifting and erosion have stripped away the overlying rocks. Exposures of intrusive igneous rocks occur in many places, including the White Mountains, New Hampshire; Stone Mountain, Georgia; Mount Rushmore in the Black Hills of South Dakota; and Yosemite National Park, California (**Figure 4.4**).

Igneous rocks that form when molten rock solidifies *at the surface* are classified as **extrusive igneous rocks**. They are also called **volcanic rocks**—after Vulcan, the Roman fire god. Extrusive igneous rocks form when lava solidifies or when erupted volcanic debris falls to Earth's surface. Extrusive igneous rocks are abundant in western portions of the Americas, where they make up the volcanic peaks of the Cascade Range and the Andes Mountains. In addition, many oceanic islands, including the Hawaiian chain and Alaska's Aleutian Islands, are composed almost entirely of extrusive igneous rocks. The nature of volcanic activity will be addressed in more detail in Chapter 5.

CONCEPT CHECKS 4.1

1. What is magma? How does magma differ from lava?

2. List and describe the three components of magma.

Concept Checker
https://goo.gl/jzWuNn

3. Compare and contrast extrusive and intrusive igneous rocks.

4.2 Igneous Compositions

Compare and contrast the four basic igneous compositions: felsic, intermediate, mafic, and ultramafic.

Igneous rocks are composed mainly of silicate minerals, meaning that silicon (Si) and oxygen (O) are by far the most abundant constituents of igneous rocks. These two elements, plus ions of aluminum (Al), calcium (Ca), sodium (Na), potassium (K), magnesium (Mg), and iron (Fe), make up roughly 98 percent, by weight, of most magmas. In addition, magma contains small amounts of many other elements, including titanium and manganese, and trace amounts of rare elements, such as gold, silver, and uranium.

As magma cools and solidifies, these elements combine to form two major groups of silicate minerals. The *light* (or *nonferromagnesian*) *silicates* contain greater amounts of potassium, sodium, and calcium. Light silicate minerals are richer in silica than dark silicates; they include *quartz, muscovite mica*, and the most abundant mineral group, the *feldspars*. By contrast, the *dark* (or *ferromagnesian*) *silicates* are rich in iron and/or magnesium and comparatively low in silica. *Olivine, pyroxene, amphibole*, and *biotite mica* are the common dark silicate minerals of Earth's crust.

Compositional Categories

Despite the great compositional diversity of igneous rocks, geologists classify these rocks (and the magmas from which they form) into four broad groups, according to their proportions of light and dark minerals. As shown in **Figure 4.5**, these compositional groups are *felsic, intermediate, mafic*, and *ultramafic*.

Felsic Versus Mafic   Near one end of the continuum are rocks composed almost entirely of light-colored silicates—quartz and potassium feldspar. The composition

◄ **SmartFigure 4.5**
Mineralogy of common igneous rocks

Tutorial
https://goo.gl/
8Z4g2Y

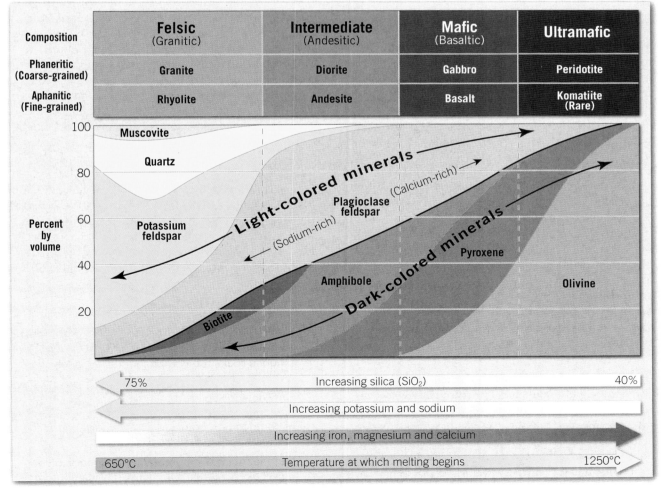

Composition	**Felsic** (Granitic)	**Intermediate** (Andesitic)	**Mafic** (Basaltic)	**Ultramafic**
Phaneritic (Coarse-grained)	Granite	Diorite	Gabbro	Peridotite
Aphanitic (Fine-grained)	Rhyolite	Andesite	Basalt	Komatiite (Rare)

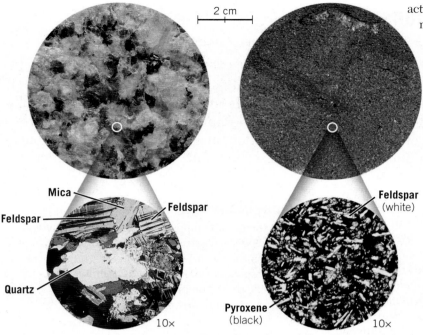

2 cm

Mica

Feldspar

Feldspar

Quartz

10×

Feldspar (white)

Pyroxene (black)

10×

A. Granite is a felsic, coarse-grained igneous rock composed of light-colored silicates—quartz and potassium feldspar.

B. Basalt is a fine-grained mafic igneous rock containing substantial amounts of dark-colored silicates and plagioclase feldspar.

of igneous rocks dominated by these minerals is classified as **felsic**, a term derived from *fel*dspar and *si*lica (quartz). Because felsic magmas most commonly solidify to form *granite*, geologists also refer to this type of magma as having **granitic composition**. In addition to quartz and feldspar, most granitic rocks contain about 10 percent dark silicate minerals, usually biotite mica and amphibole (**Figure 4.6A**). Felsic rocks and the magma from which they were derived are also notable for being rich in silica (about 70 percent or more) and for being major constituents of Earth's continental crust.

Rocks that contain at least 45 percent dark silicates (ferromagnesian minerals) are classified as **mafic** (from *ma*gnesium and *f*errum, the Latin name for iron). Mafic rocks are typically darker and denser than felsic rocks due to their iron content. Because mafic magmas most often solidify to form the igneous rock *basalt*, geologists also refer to this type of magma as having a **basaltic composition** (**Figure 4.6B**). Basaltic rocks make up the ocean floor as well as many of the volcanic islands located within the ocean basins. Basalt also forms extensive lava flows on the continents.

Other Compositional Groups   Rocks with a composition between felsic and mafic rocks are said to have an **intermediate**, or **andesitic composition**, after the common volcanic rock *andesite*. Intermediate rocks contain at least 25 percent dark silicate minerals—mainly amphibole, pyroxene, and biotite mica—with the other dominant mineral being plagioclase feldspar. This important category of igneous rocks is often associated with volcanic

activity on the seaward margins of continents and on volcanic island arcs such as the Aleutian chain.

Another important igneous rock, **peridotite**, contains mostly olivine and pyroxene and thus falls on the opposite side of the compositional spectrum from felsic rocks (see Figure 4.5). Because peridotite is composed almost entirely of ferromagnesian minerals, its chemical composition is referred to as **ultramafic**. Although ultramafic rocks are rare at Earth's surface, peridotite is the main constituent of the upper mantle.

Silica Content as an Indicator of Composition

An important aspect of the chemical composition of igneous rocks is silica (SiO_2) content. Typically, the silica content of crustal rocks ranges from as low as about 40 percent in ultramafic rocks to a high of more than 70 percent in felsic rocks. The percentage of silica in igneous rocks varies in a systematic manner that corresponds to the abundance of other elements. For example, rocks that are relatively low in silica contain large amounts of iron, magnesium, and calcium. By contrast, rocks that are high in silica contain comparatively less iron, magnesium, and calcium but are enriched with sodium and potassium. Consequently, the chemical makeup of an igneous rock can be inferred directly from its silica content.

Further, the amount of silica present in magma strongly influences the magma's behavior. Felsic magma, which has a high silica content, is quite viscous ("thick") and may erupt at temperatures as low as 650°C (1200°F), whereas mafic (basaltic) magmas, which are low in silica, are generally more fluid. Mafic magmas also erupt at higher temperatures than felsic magmas—usually between 1050° and 1250°C (1920° and 2280°F).

CONCEPT CHECKS 4.2

1. Igneous rocks are composed mainly of which group of minerals?

2. How do light-colored igneous rocks differ in composition from dark-colored igneous rocks?

Concept Checker
https://goo.gl/znhQ7N

3. List the four basic compositional groups of igneous rocks, in order from highest silica content to lowest silica content.

4.3 Igneous Textures: What Can They Tell Us?

Identify and describe the six major igneous textures.

Mineralogists use the term **texture** to describe the overall appearance of a rock based on the size, shape, and arrangement of its mineral grains—not how it feels to touch. Texture is an important property because it reveals a great deal about the environment in which the rock formed (**Figure 4.7**). Geologists can make inferences about a rock's origin based on careful observations of grain size and other characteristics of the rock.

Three factors influence the textures of igneous rocks:

- The rate at which molten rock cools
- The amount of silica present in the magma
- The amount of dissolved gases in the magma

Among these, the rate of cooling tends to be the dominant factor. A very large magma body located many kilometers beneath Earth's surface remains insulated from lower surface temperatures by the surrounding rock and thus will cool very slowly over a period of tens

Tutorial
https://goo.gl/Va155T

▼ **SmartFigure 4.7**
Igneous rock textures

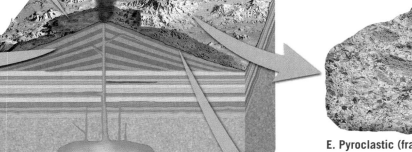

D. Vesicular texture
Extrusive rock containing voids left by gas bubbles that escape as lava solidifies. (Pumice is a frothy volcanic glass that displays a vesicular texture.)

A. Glassy texture
Composed of unordered atoms and resembles dark manufactured glass. (Obsidian is a natural glass that usually forms when highly silica-rich magmas solidify.)

B. Porphyritic texture
Composed of two distinctly different crystal sizes.

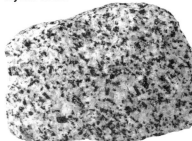
C. Phaneritic (coarse-grained) texture
Composed of mineral grains that are large enough to be identified without a microscope.

E. Pyroclastic (fragmental) texture
Produced by the consolidation of fragments that may include ash, once molten blobs, or large angular blocks that were ejected during an explosive volcanic eruption.

F. Aphanitic (fine-grained) texture
Composed of crystals that are too small for the individual minerals to be identified without a microscope.

Groundmass

Phenocryst

1 cm

▲ Figure 4.8
Porphyritic texture The large crystals in porphyritic rocks are called *phenocrysts*, and the matrix of smaller crystals is called *groundmass*.

of thousands to millions of years. Initially, relatively few crystal nuclei form. Slow cooling permits ions to migrate freely until they eventually join one of the existing crystals. Consequently, slow cooling promotes the growth of fewer but larger crystals.

On the other hand, when cooling occurs rapidly—for example, in a thin lava flow—the ions quickly lose their mobility and readily combine to form crystals. This results in the development of numerous embryonic nuclei, all of which compete for the available ions. The result is a solid mass of many tiny intergrown crystals.

Types of Igneous Textures

A magma body may migrate to a new location or erupt at the surface before it completely solidifies. As a result, several types of igneous textures exist, including aphanitic (fine-grained), phaneritic (coarse-grained), porphyritic, vesicular, glassy, and pyroclastic (fragmented).

Aphanitic (Fine-Grained) Texture

Igneous rocks that form at the surface or as small intrusive masses within the upper crust where cooling is relatively rapid exhibit a **fine-grained texture**, termed **aphanitic** (*a* = not, *phaner* = visible). By definition, the crystals that make up aphanitic rocks are so small that individual minerals can be distinguished only with the aid of a polarizing microscope or other sophisticated techniques (see Figure 4.6B and 4.7F). Therefore, we commonly characterize fine-grained rocks as being light, intermediate, or dark in color. Using this system of grouping, light-colored aphanitic rocks are those containing primarily light-colored nonferromagnesian silicate minerals.

Phaneritic (Coarse-Grained) Texture

When large masses of magma slowly crystallize at great depth, they form igneous rocks that exhibit a **coarse-grained texture**, described as **phaneritic** (*phaner* = visible). Coarse-grained rocks consist of a mass of intergrown crystals that are roughly equal in size and large enough to distinguish the individual minerals without the aid of a microscope (see Figure 4.6A and Figure 4.7C). Geologists often use a small magnifying lens to aid in identifying minerals in a phaneritic rock.

Porphyritic Texture

A large mass of magma may require thousands or millions of years to solidify. Because different minerals crystallize under different environmental conditions (temperatures and pressure), it is possible for crystals of one mineral to become quite large before others even begin to form. If molten rock containing some large crystals moves to a different environment—for example, by erupting at the surface—the remaining liquid portion of the lava cools more quickly. The resulting rock, which has large crystals embedded in a matrix of smaller crystals, is said to have a **porphyritic texture** (**Figure 4.8**). The large crystals in porphyritic rocks are referred to as **phenocrysts** (*pheno* = show, *cryst* = crystal), and the matrix of smaller crystals is called **groundmass**. A rock with a *porphyritic* texture is termed a **porphyry**.

Vesicular Texture

Common features of many extrusive rocks are the voids left by gas bubbles that escape as lava solidifies. These nearly spherical openings are called *vesicles*, and the rocks that contain them are said to have a **vesicular texture**. Rocks that exhibit a vesicular texture often form in the upper zone of a lava flow, where cooling occurs rapidly enough to preserve the openings produced by the expanding gas bubbles (**Figure 4.9**).

Lava flow

Vesicular texture

▲ Figure 4.9
Vesicular texture The larger image shows a lava flow on Hawaii's Kilauea Volcano. The inset photo is a close-up showing the vesicular texture of hardened lava. Vesicles are small holes left by escaping gas bubbles.

Another common vesicular rock, called *pumice*, forms when silica-rich lava is ejected during an explosive eruption (see Figure 4.7D).

Glassy Texture   During some volcanic eruptions, molten rock is ejected into the atmosphere, where it is quenched (very quickly cooled) and becomes a solid. Rapid cooling of this type may generate rocks having a **glassy texture**. Glass results when unordered ions are "frozen in place" before they are able to unite into an orderly crystalline structure.

Obsidian, a common type of natural glass, is similar in appearance to dark chunks of manufactured glass. Because of its excellent conchoidal fracture and ability to hold a sharp, hard edge, obsidian was a prized material from which Native Americans chipped arrowheads and cutting tools (**Figure 4.10**).

Obsidian flows, typically a few hundred feet thick, provide evidence that rapid cooling is not the only mechanism that produces a glassy texture. Magmas with high silica content tend to form long, chainlike structures (polymers) before crystallization is complete. These structures, in turn, impede the movement of ions and increase the magma's viscosity. (*Viscosity* is a measure of a fluid's resistance to flow.) So, granitic magma, which is rich in silica, may be extruded as an extremely viscous mass that eventually solidifies to form obsidian.

By contrast, basaltic magma, which is low in silica, forms very fluid lavas that usually generate fine-grained crystalline rocks as they cool. However, when a basaltic lava flow enters the ocean, its surface is quenched rapidly enough to form a thin, glassy skin.

Pyroclastic (Fragmental) Texture   Another group of igneous rocks is formed from the consolidation of individual rock fragments ejected during explosive volcanic eruptions. The ejected particles might be very fine ash, molten blobs, or large angular blocks torn from the walls of a vent during an eruption (**Figure 4.11**). Igneous rocks composed of these rock fragments are said to have a **pyroclastic texture**, or **fragmental texture** (see Figure 4.7E).

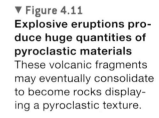

◄ **Figure 4.10**
Obsidian arrowhead Native Americans made arrowheads and cutting tools from obsidian, a natural glass.

▼ **Figure 4.11**
Explosive eruptions produce huge quantities of pyroclastic materials These volcanic fragments may eventually consolidate to become rocks displaying a pyroclastic texture.

► Figure 4.12
Pegmatitic texture This granite pegmatite, found in the inner gorge of the Grand Canyon, is composed mainly of quartz and feldspar.

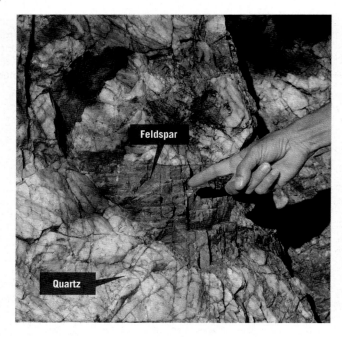

Feldspar

Quartz

Pegmatitic Texture Under special conditions, exceptionally coarse-grained igneous rocks, called **pegmatites**, may form. Rocks of this type, in which most of the crystals are larger than 1 centimeter in diameter, are described as having a **pegmatitic texture** (Figure 4.12). Most pegmatites occur as small masses or thin veins within or around the margins of large intrusive igneous bodies.

Pegmatites form late in the crystallization of a magma, when water and other elements, such as carbon dioxide, chlorine, and fluorine, make up an unusually high percentage of the melt. Because ions can readily move in these hot fluid-rich environments, the crystals that form are abnormally large. Thus, the large crystals in pegmatites do not result from inordinately long cooling histories; rather, they are the consequence of a fluid-rich environment that enhances crystallization.

The composition of most pegmatites is similar to that of granite. Thus, pegmatites contain large crystals of quartz, feldspar, and muscovite. However, some contain significant quantities of relatively rare and hence valuable elements—including gold, tungsten, beryllium, and other rare-earth elements used in modern high-technology devices, including cell phones and hybrid vehicles.

A common type of pyroclastic rock, called *welded tuff*, is composed of fine fragments of glass that remained hot enough to fuse together. Other pyroclastic rocks are composed of fragments that solidified before impact and became cemented together at some later time. Because pyroclastic rocks are made of individual particles or fragments rather than interlocking crystals, their textures often resemble those exhibited by sedimentary rocks rather than those associated with igneous rocks.

CONCEPT CHECKS 4.3

1. How does the rate of cooling influence crystal size? What other factors influence the texture of igneous rocks?

2. List the six major igneous rock textures.

3. What does a porphyritic texture indicate about the cooling history of an igneous rock?

🔊 **Concept Checker**
https://goo.gl/6rWXxa

4.4 Naming Igneous Rocks

Distinguish among the common igneous rocks based on texture and mineral composition.

The various igneous textures described in the previous section result mainly from different cooling histories, whereas the mineral composition of an igneous rock depends on the chemical makeup of its parent magma. Because geologists classify igneous rocks on the basis of both texture and mineral composition, some rocks with similar mineral constituents but differing textures are given different names. Figure 4.13 shows examples of all the igneous rock types discussed in the next section.

Felsic Igneous Rocks

Granite Of all the igneous rocks, **granite** is perhaps the best known (see GEOgraphics 4.1, page 114). This is because of its natural beauty, which can be enhanced with polishing, along with its abundance in the continental crust. Slabs of polished granite are commonly used for tombstones and monuments and as building stones. Well-known U.S. locations where granite is quarried include Barre, Vermont; Mount Airy, North Carolina; and St. Cloud, Minnesota.

Granite is a coarse-grained rock composed of about 10 to 20 percent quartz and roughly 50 percent feldspar. When examined close up, the quartz grains appear somewhat rounded in shape, glassy, and clear to gray in color. By contrast, feldspar crystals are generally white, gray, or salmon pink in color, and they are blocky or rectangular in shape. Other minor constituents of granite include small amounts of dark silicates, particularly biotite and amphibole, and sometimes muscovite. Although the dark components generally make up less than 10 percent of most granites, dark minerals appear to be more prominent than their percentage indicates.

IGNEOUS ROCK CLASSIFICATION CHART

◄ SmartFigure 4.13
Classification of igneous rocks Igneous rocks are classified based on mineral composition and texture.

Tutorial
https://goo.gl/JqC1V5

	MINERAL COMPOSITION			
	Felsic	**Intermediate**	**Mafic**	**Ultramafic**
Dominant Minerals	Quartz Potassium feldspar	Amphibole Plagioclase feldspar	Pyroxene Plagioclase feldspar	Olivine Pyroxene
Accessory Minerals	Plagioclase feldspar Amphibole Muscovite Biotite	Pyroxene Biotite	Amphibole Olivine	Plagioclase feldspar

TEXTURE

Texture	Felsic	Intermediate	Mafic	Ultramafic
Phaneritic (coarse-grained)	Granite	Diorite	Gabbro	Peridotite
Aphanitic (fine-grained)	Rhyolite	Andesite	Basalt	Komatiite (rare)
Porphyritic (two distinct grain sizes)	Granite porphyry	Andesite porphyry	Basalt porphyry	Uncommon
Glassy	Obsidian	Less common	Less common	Uncommon
Vesicular (contains voids)	Pumice (also glassy)	Scoria		Uncommon
Pyroclastic (fragmental)	Most fragments < 4mm Tuff or welded tuff	Most fragments > 4mm Volcanic breccia		Uncommon

Rock Color (based on % of dark minerals)	0% to 25%	25% to 45%	45% to 85%	85% to 100%

Granite:
An Intrusive Igneous Rock

Where is granite found?

This map shows the distribution of exposed granitic rocks in the United States, Canada, and Mexico. Granitic rocks labeled batholiths along the west coast are mainly Mesozoic in age. Those in the north central United States and the Canadian Shield are intermixed with metamorphic rocks and are Precambrian, while the granitic rocks of the Appalachian Mountains are mostly of Paleozoic age.

Locations of granite outcrops

1. Coast Range batholith, British Columbia
2. Canadian Shield, Ontario, Canada
3. Cadillac Mountain, Acadia National Park, Maine
4. Cathedral Ledge State Park, White Mountains, New Hampshire
5. Stone Mountain, Georgia
6. Mount Lemmon from Catalina Highway, Tuscon, Arizona
7. Joshua Tree National Park, California
8. Canyon Creek Lakes, Trinity Alps, California

Granite

At a distance, most granitic rocks appear gray in color (**Figure 4.14**). However, when composed of dark pink feldspar grains, granite can exhibit a reddish color. In addition, some granites have a porphyritic texture. These specimens contain elongated feldspar crystals a few centimeters in length that are scattered among smaller crystals of quartz and amphibole.

Rhyolite   The fine-grained equivalent of granite is **rhyolite**, which, like granite, is composed essentially of the light-colored silicates. This fact accounts for its color, which is usually buff to pink, but rhyolite may also appear as light gray. Rhyolite frequently contains glass fragments and voids, indicating that it cooled rapidly at or near Earth's surface. In contrast to granite, which is widely distributed as large intrusive masses, rhyolite deposits are less common and generally less voluminous. The thick rhyolite lava flows and extensive ash deposits in and around Yellowstone National Park are well-known exceptions to this generalization.

Obsidian   As mentioned previously, the dark, glassy rock called **obsidian** cools quickly at the surface, and so its silica-rich arrangement of ions is not orderly. Consequently, glassy rocks such as obsidian are not composed of minerals in the same sense as most other rocks.

Although usually black or reddish-brown in color, obsidian has a chemical composition that is roughly equivalent to that of the light-colored igneous rock granite rather than dark rocks such as basalt. Obsidian's dark color results from small amounts of metallic ions in an otherwise relatively clear, glassy substance. If you examine a thin edge, obsidian appears nearly transparent (see Figure 4.7A).

Pumice   **Pumice** is a glassy volcanic rock with a vesicular texture that forms when large amounts of gas escape through silica-rich lava to generate a gray, frothy rock. In some samples, the voids are quite noticeable, whereas in others the pumice resembles fine shards of intertwined glass. Because of the large percentage of voids, many samples of pumice float in water (**Figure 4.15**). Often, flow lines are visible in pumice, indicating that some movement occurred before solidification was complete. Moreover, pumice and obsidian are often found in the same rock mass, where they exist in alternating layers.

▲ **SmartFigure 4.14**
Rocks contain information about the processes that produced them This massive granitic monolith (El Capitan) located in Yosemite National Park, California, was once a molten mass deep within Earth.

Mobile Field Trip
https://goo.gl/z8XJ1i

▶ Figure 4.15
Pumice, a vesicular (and also glassy) igneous rock Most samples of pumice will float in water because they contain numerous vesicles.

←2 cm→

Intermediate Igneous Rocks

Andesite   **Andesite** is a medium-gray, fine-grained rock typically of volcanic origin. Its name comes from South America's Andes Mountains, where it is found in the form of numerous volcanoes. In addition, the volcanoes of the Cascade Range and many of the volcanic structures occupying the continental margins that surround the Pacific Ocean are composed mostly of andesite. It commonly exhibits a porphyritic texture. When this is the case, the phenocrysts are often light, rectangular crystals of plagioclase feldspar or black, elongated amphibole crystals. Andesite may also resemble rhyolite, so its identification usually requires microscopic examination to verify mineral makeup.

Diorite   The intrusive equivalent of andesite, **diorite**, is a coarse-grained rock that looks somewhat like gray granite. However, it can be distinguished from granite because it contains little or no visible quartz crystals and has a higher percentage of dark silicate minerals. The mineral makeup of diorite is primarily plagioclase feldspar and amphibole. Because the presence of light-colored feldspar grains appears to be roughly equal to the amount of dark amphibole crystals, diorite has a salt-and-pepper appearance.

Mafic Igneous Rocks

Basalt   **Basalt** is a very dark green to black, fine-grained rock composed primarily of pyroxene and calcium-rich plagioclase feldspar, with lesser amounts of olivine and amphibole (see Figure 4.13). When it is porphyritic, basalt commonly contains small, light-colored feldspar phenocrysts or green, glassy-appearing olivine grains embedded in a dark groundmass.

Basalt is the most common extrusive igneous rock. Many volcanic islands, such as the Hawaiian Islands and Iceland, are composed mainly of basalt (**Figure 4.16**). Further, the upper layers of the oceanic crust consist of basalt. In the United States, large portions of central Oregon and Washington were the sites of

extensive basaltic outpourings (discussed in detail in Chapter 5). At some locations, these once-fluid basaltic flows have accumulated to a combined thickness exceeding 3 kilometers (nearly 2 miles).

Gabbro   The intrusive equivalent of basalt is **gabbro**, which, like basalt, tends to be dark green to black in color and composed primarily of pyroxene and calcium-rich plagioclase feldspar. Although gabbro is uncommon in the continental crust, it makes up a significant percentage of oceanic crust.

Pyroclastic Rocks

Pyroclastic rocks are composed of fragments ejected during a volcanic eruption. One of the most common pyroclastic rocks, called **tuff**, is composed mainly of tiny, ash-size fragments cemented together. In situations where the ash particles remained hot enough to fuse, the rock is called **welded tuff**. Although welded tuff consists mostly of tiny glass shards, it may contain walnut-size pieces of pumice and other rock fragments.

▼ Figure 4.16
Basaltic lava flowing from Kilauea Volcano, Hawaii

◀ **Figure 4.17**
Welded tuff, a pyroclastic igneous rock Outcrop of welded tuff from Valles Caldera near Los Alamos, New Mexico. Tuff is composed mainly of ash-sized particles and may contain larger fragments of pumice or other volcanic rocks.

Close-up

per hour. Early investigators of these deposits incorrectly classified them as rhyolite lava flows. Today, we know that silica-rich lava is too viscous (thick) to flow more than a few miles from a vent.

Pyroclastic rocks composed mainly of particles larger than ash are called *volcanic breccia*. The particles in volcanic breccia may consist of streamlined lava blobs that solidified in air, blocks broken from the walls of the vent, ash, and glass fragments.

Unlike most igneous rock names, such as granite and basalt, the terms *tuff* and *volcanic breccia* do not imply mineral composition. Instead, they are frequently identified with a modifier; for example, *rhyolite tuff* indicates a rock composed of ash-size particles having a felsic composition.

Welded tuff deposits cover vast portions of previously volcanically active areas of the western United States (**Figure 4.17**). Some of these tuff deposits are hundreds of meters thick and extend for more than 100 kilometers (60 miles) from their source. Most formed millions of years ago as volcanic ash spewed from large volcanic structures (calderas), sometimes spreading laterally at speeds approaching 100 kilometers

---

CONCEPT CHECKS 4.4

1. List the two criteria by which igneous rocks are classified.

2. How are granite and rhyolite different? In what way are they similar?

 Concept Checker
https://goo.gl/SpwEWu

3. Describe each of the following in terms of composition and texture: diorite, rhyolite, and basalt porphyry.

4.5 Origin of Magma

Summarize the major processes that generate magma from solid rock.

Workers in underground mines know that temperatures increase as they descend deeper below Earth's surface. The rate of temperature change varies considerably from place to place, but it averages about 25°C per kilometer in the *upper* crust; this increase in temperature with depth is known as the **geothermal gradient**. However, Figure 4.18 shows that when we compare a typical geothermal gradient to the melting point curve for the mantle rock peridotite, the temperature at which peridotite melts is higher than the geothermal gradient. Thus, under normal conditions, the mantle is mostly solid rock. The study of earthquake waves also confirms that the crust and mantle are mostly solid rock.

Generating Magma from Solid Rock

So if the crust and mantle are mostly solid, how is magma produced? The answer is that tectonic processes trigger melting by various means.

Decrease in Pressure: Decompression Melting

If temperature alone determined whether rock melts, our planet would be a molten ball covered with a thin, solid outer shell. However, *pressure* increases

► SmartFigure 4.18
Why the mantle is mainly solid This diagram shows the geothermal gradient (the increase in temperature with depth) for the crust and upper mantle. Also illustrated is the melting point curve for the ultramafic mantle rock peridotite.

Animation
https://goo.gl/
qotVYV

Temperature (°C) ➡

The geothermal gradient (red curve) shows how the temperature rises with increasing depth.

At the depth of the upper asthenosphere, the red curve is close to the orange zone—so that a slight change in conditions can cause mantle rock to partially melt.

The fact that the geothermal gradient lies completely in the green area means that mantle rock is typically solid.

Key   Solid rock   Partial melting   Complete melting

► Figure 4.19
Decompression melting As hot mantle rock ascends, it continually moves into zones of lower and lower pressure. This drop in confining pressure initiates *decompression melting* in the upper mantle.

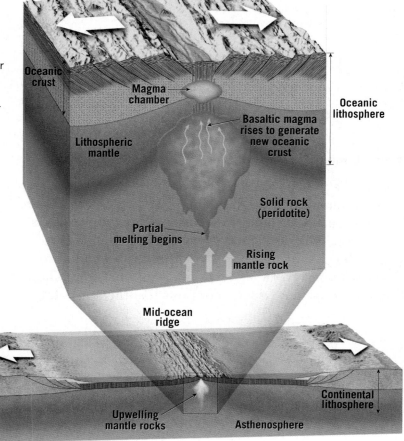

with depth, too, and also affects the melting temperatures of rocks.

When rocks melt, they increase in volume. Because the confining pressure exerted by the weight of overlying rocks steadily increases with depth, the melting temperature of a rock also increases with increasing depth. The opposite is also true: Reducing confining pressure *lowers* a rock's melting temperature. When confining pressure drops sufficiently, **decompression melting** is triggered. Decompression melting occurs where hot, solid mantle rock ascends, thereby moving into regions of lower pressure, which reduces the rock's melting temperature.

Most decompression melting occurs along spreading centers (divergent plate boundaries), where two plates are moving away from each other, creating fractures in the oceanic crust. As a result, hot mantle rock rises and melts, generating basaltic magma that solidifies to form new oceanic crust between the two diverging plates (**Figure 4.19**).

Decompression melting also occurs when ascending mantle plumes reach the uppermost mantle. If this rising magma reaches the surface, it will trigger an episode of hot-spot volcanism.

Addition of Water to Trigger Melting

Another factor that affects the melting temperature of rock is its water content. Water and other volatiles act on rock much the way salt acts on ice. That is, water causes rock to melt at lower temperatures, just as rock salt on an icy sidewalk enhances melting.

The introduction of water to generate magma occurs mainly at convergent plate boundaries, where cold slabs of oceanic lithosphere descend into the mantle (**Figure 4.20**). Recall that after the oceanic crust forms along an oceanic ridge, seafloor spreading continually moves this crust away from the ridge. Gradually, the crust cools as it interacts with the cold ocean water that percolates through fractures to a depth of several kilometers. As seawater circulates through the young, hot crust, the water becomes hot enough to chemically react with the layers of basalt to form *hydrated* (water-bearing) minerals. Mineral hydration is a chemical reaction in which water is added to the crystal structure of a mineral, usually creating a new mineral. For example, the mineral olivine chemically reacts with hot seawater to produce the hydrated mineral serpentine [$Mg_3Si_2O_5(OH)_4$].

When a slab of hydrated mineral-rich oceanic crust reaches a subduction zone, it begins to descend into the hot mantle below. As the oceanic plate sinks, it warms and causes the hydrated minerals to release water—a process called *dehydration*. Because the newly released fluids are buoyant (less dense), they migrate into the wedge of hot mantle that lies directly above the subducting slab. At a depth of about 100 kilometers (60 miles), the wedge of mantle rock is sufficiently hot that this addition of water leads to some melting.

Partial melting of the mantle rock peridotite generates a basaltic magma with temperatures that may exceed 1250°C (nearly 2300°F). The magma generated by this process is enriched in volatiles (mainly water and carbon dioxide) compared to the basaltic magmas that form along the oceanic ridges by the process of decompression melting. This difference contributes to the explosive volcanic eruptions that we associate with subduction zone volcanism—a topic we will cover in Chapter 5.

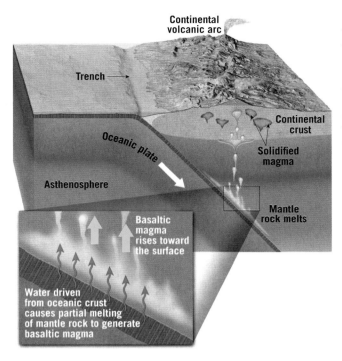

▲ **Figure 4.20**
Water lowers the melting temperature of hot mantle rock to trigger partial melting As an oceanic plate descends into the mantle, water and other volatiles are driven from the subducting crustal rocks into the mantle above.

Magma can be generated three ways: (1) A decrease in pressure (without an increase in temperature) can result in decompression melting; (2) the introduction of water can lower the melting temperature of hot mantle rock sufficiently to generate magma; and (3) heating of crustal rocks above their melting temperature produces magma.

low-density felsic magmas reach the surface, they tend to produce explosive volcanic eruptions.

Crustal rocks can also melt during continental collisions that result in the formation of a large mountain belt. During these events, the crust is greatly thickened, and some crustal rocks are buried to depths where the temperatures are elevated sufficiently to cause some melting. The felsic (granitic) magmas produced in this manner usually solidify before reaching the surface, so volcanism is not typically associated with these collision-type mountain belts.

Temperature Increase: Melting Crustal Rocks
Mantle-derived basaltic magma tends to be less dense than the surrounding rocks, which causes the magma to buoyantly rise toward the surface. In oceanic settings, these basaltic magmas often erupt on the ocean floor to produce seamounts, which may grow to form volcanic islands, as exemplified by the Hawaiian Islands. However, in continental settings, basaltic magma often "ponds" beneath low-density crustal rocks. These overlying rocks have lower melting temperatures than basaltic magmas, and the hot basaltic magma may heat them sufficiently to generate a secondary melt of silica-rich, felsic magma. If these

---

CONCEPT CHECKS 4.5

1. Explain the process of decompression melting.

2. What role does water play in the formation of magma?   **Concept Checker**
   https://goo.gl/4HAPSr

3. Briefly explain one way that basaltic magma can generate felsic magma.

4.6 How Magmas Evolve

Describe how magmatic differentiation can generate a magma body that has a different chemical composition from its parent magma.

Geologists have observed that, over time, a volcano may extrude lavas that vary in composition. Based on these observations, scientists have reasoned that magma might change, and thus one magma body could become the parent to a variety of igneous rocks. To explore this idea, N. L. Bowen carried out a pioneering investigation into the crystallization of magma early in the twentieth century.

Bowen's Reaction Series and the Composition of Igneous Rocks

Whereas ice freezes at a specific temperature, mafic magma crystallizes over a range of at least 200°C of cooling (from about 1200° to 1000°C). In a laboratory setting, Bowen and his coworkers demonstrated that as a mafic magma cools through various temperatures, minerals tend to crystallize in a systematic and ordered fashion, based on each mineral's melting temperature. This order of mineral formation became known as **Bowen's reaction series** (**Figure 4.21**). The first mineral to crystallize is the ferromagnesian mineral olivine. Further cooling generates calcium-rich plagioclase feldspar as well as pyroxene, then amphibole, and so forth across the diagram.

During the crystallization process, the composition of the remaining liquid portion of the magma, called *melt*, continually changes. For example, at the stage when about one-third of the magma has solidified, the remaining molten material will only contain small amounts of iron, magnesium, and calcium because these elements are major constituents of the minerals that form earliest in the crystallization process. The absence of these elements causes the melt to become enriched in sodium and potassium. Further, because the original mafic magma contained about 50 percent silica (SiO_2), the crystallization of the earliest-formed mineral, olivine (which is only about 40 percent silica), results in more SiO_2 being left in the remaining melt. Thus, the silica component of the remaining melt becomes enriched as the magma evolves.

Bowen also demonstrated that when the outer regions of crystals forming in a magma remain in contact with the remaining melt, they continue to exchange ions (react chemically) with the melt. As a result, the periphery of these mineral grains have a different, more evolved composition than the interiors. Stated another way, minerals that remain in contact with a melt gradually change composition to become the next mineral in the crystallization series Bowen identified.

▶ **Figure 4.21**
Bowen's reaction series This diagram shows the sequence in which minerals crystallize from a mafic magma. Compare this figure to the mineral composition of the rock groups in Figure 4.13. Note that each rock group consists of minerals that crystallize in the same temperature range.

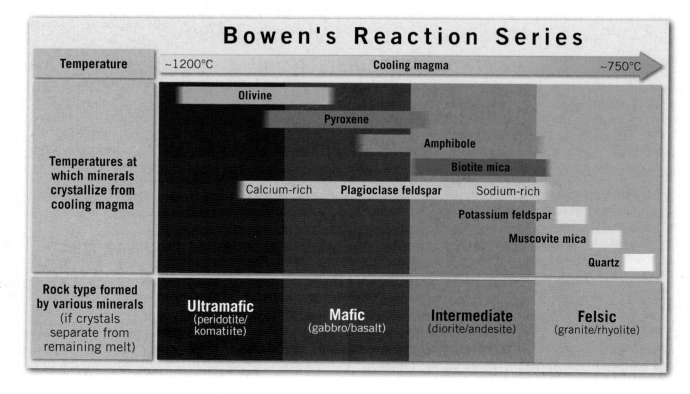

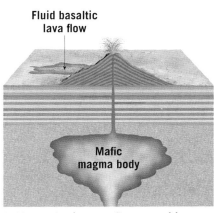

Fluid basaltic lava flow

Mafic magma body

A. Magma having a mafic composition erupts fluid basaltic lavas.

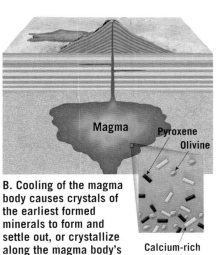

Magma

Pyroxene

Olivine

Calcium-rich plagioclase

B. Cooling of the magma body causes crystals of the earliest formed minerals to form and settle out, or crystallize along the magma body's cool margins.

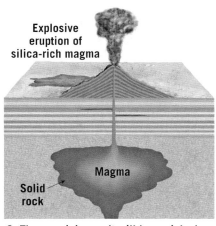

Explosive eruption of silica-rich magma

Magma

Solid rock

C. The remaining melt will be enriched with silica, and should a subsequent eruption occur, the rocks generated will be more silica-rich and closer to the felsic end of the compositional range than the initial magma.

◄ **Figure 4.22**
Crystal settling results in a change in the composition of the remaining melt A magma evolves as the earliest-formed minerals (those richer in iron, magnesium, and calcium) crystallize and settle to the bottom of the magma chamber, leaving the remaining melt richer in sodium, potassium, and silica (SiO_2).

Bowen's reaction series is highly idealized, showing crystallization order and composition of rock under precise laboratory conditions. In nature, the process is almost never so pure and tidy; for instance, the earliest-formed minerals can separate from the melt, thus halting further chemical reactions. However, evidence that Bowen's crystallization model approximates what can happen in nature comes from analysis of igneous rocks. In particular, geologists note that minerals forming in the same general temperature regime depicted in Bowen's reaction series are found together in the same igneous rocks. For example, notice in Figure 4.21 that the minerals quartz, potassium feldspar, and muscovite, which are located in the same temperature region of Bowen's diagram, are typically found together as major constituents of the intrusive igneous rock granite.

Magmatic Differentiation and Crystal Settling

Bowen demonstrated that minerals crystallize from magma in a systematic fashion. But how do Bowen's findings account for the great diversity of igneous rocks? It has been shown that, at one or more stages during the crystallization of magma, a separation of various components can occur. One mechanism that causes this to happen is called **crystal settling**. This process occurs when the earlier-formed minerals are denser (heavier) than the liquid portion and sink toward the bottom of the magma chamber, as shown in **Figure 4.22**. When the remaining melt solidifies—either in place or in another location, if it migrates into fractures in the surrounding rocks—it

will form a rock with a mineral composition that is different from that of the parent magma. The formation of a magma body having a mineralogy (chemical composition) that is different from that of the parent magma is called **magmatic differentiation**.

A classic example of magmatic differentiation is found in the Palisades Sill, which is a 300-meter-thick (1000-foot-thick) slab of dark igneous rock exposed along the west bank of the lower Hudson River across from New York City (**Figure 4.23**). Because of its great thickness and subsequent slow rate of solidification, crystals of olivine (the first mineral to form) sank and make up about 25 percent of the lower portion of the Palisades Sill. By contrast, near the top of this igneous body, where the last melt crystallized, olivine represents only 1 percent of the rock mass.*

*Recent studies indicate that the Palisades Sill was produced by multiple injections of magma and does not represent a simple case of crystal settling. However, it is nonetheless an instructional example of that process.

◄ **Figure 4.23**
The Palisades Sill, as seen from New York City The Palisades form impressive cliffs along the west side of the Hudson River for more than 80 kilometers (50 miles). This structure, which is visible from Manhattan, formed when magma was injected between layers of sandstone and shale.

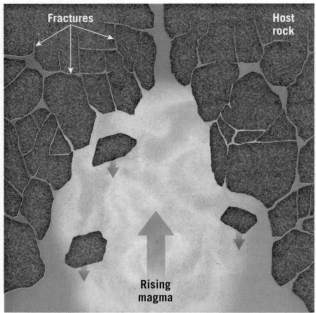

Fractures

Host rock

Rising magma

As magma rises through Earth's brittle upper crust, it may dislodge and incorporate the surrounding host rocks. Melting of these blocks, a process called *assimilation*, changes the overall composition of the rising magma body.

▲ Figure 4.24
Assimilation of the host rock by a magma body The composition of magma changes when the molten mass incorporates pieces of surrounding host rock, in a process called assimilation.

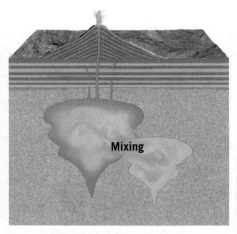

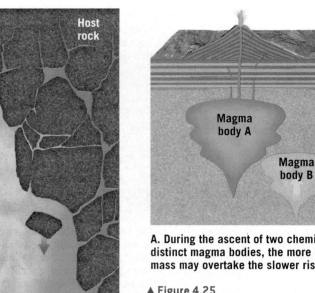

A. During the ascent of two chemically distinct magma bodies, the more buoyant mass may overtake the slower rising body.

Magma body A

Magma body B

Mixing

B. Once joined, convective flow mixes the two magmas, generating a mass that is a blend of the two magma bodies.

▲ Figure 4.25
Magma mixing

Assimilation and Magma Mixing

Bowen successfully demonstrated that through magmatic differentiation, a single parent magma can generate several mineralogically different igneous rocks. However, more recent work indicates that magmatic differentiation involving crystal settling cannot, by itself, account for the entire compositional spectrum of igneous rocks. Once a magma body forms, the incorporation of foreign material can also change its composition.

For example, in near-surface environments where rocks are brittle, the magma pushing upward can cause numerous fractures in the overlying rock. The force of the injected magma is often sufficient to dislodge and incorporate some of the surrounding host rock (**Figure 4.24**). Melting of these blocks, a process called **assimilation**, changes the overall chemical composition of the magma body.

Another means by which the composition of magma can be altered is called **magma mixing**. Magma mixing may occur during the ascent of two chemically distinct magma bodies as the more buoyant mass overtakes the more slowly rising body (**Figure 4.25**). Once they are joined, convective flow stirs the two magmas, generating a single mass that has an intermediate composition.

CONCEPT CHECKS 4.6

1. Define *magmatic differentiation*.

2. How does the crystallization and settling of the earliest formed minerals affect the composition of the remaining magma?

    **Concept Checker** https://goo.gl/fCE3dS

3. Describe the processes of assimilation.

4.7 Partial Melting and Magma Composition

Explain how partial melting of the mantle rock peridotite can generate a mafic (basaltic) magma.

Recall that igneous rocks are composed of a mixture of minerals and, therefore, tend to melt over a temperature range of at least 200°C (nearly 400°F). As rock begins to melt, the minerals with the lowest melting temperatures are the first to melt. If melting continues, minerals with higher melting points begin to melt, and the composition of the melt steadily approaches the overall composition of the rock from which it was derived. Most often, however, only a small percentage of mantle rock melts in most tectonic settings, a process known as **partial melting**.

Recall from Bowen's reaction series (see Figure 4.21) that rocks with a felsic (granitic) composition are composed of minerals with the lowest melting (crystallization) temperatures—namely, quartz and potassium feldspar. Also note that as we move up Bowen's reaction series, the minerals have progressively higher melting temperatures and that olivine, which is found at the top, has the highest melting point. When a rock undergoes partial melting, it forms a melt that is enriched in ions from minerals with the lowest melting temperatures, while the unmelted portion is composed of minerals with higher melting temperatures (**Figure 4.26**). Separation of these two fractions yields a melt with a chemical composition that is richer in silica and nearer the felsic end of the spectrum than the rock from which it formed. In general, partial melting of *ultramafic* rocks tends to yield *mafic (basaltic) magmas*, partial melting of *mafic* rocks generally yields *intermediate (andesitic) magmas*, and partial melting of *intermediate* rocks can generate *felsic (granitic) magmas*.

Formation of Mafic Magma

Most magma that erupts at Earth's surface is mafic in composition, with a temperature range of 1000° to 1250°C. Experiments show that under the high-pressure conditions calculated for the upper mantle, *partial melting of the ultramafic rock peridotite generates a mafic magma*.

Mafic (basaltic) magmas that originate from partial melting of mantle rocks are called *primary*, or *primitive*, magmas because they have not yet evolved. Recall that partial melting, which produces mantle-derived magmas, may be triggered by a reduction in confining pressure during the process of decompression melting. This can occur, for example, where hot mantle rock ascends as part of slow-moving convective flow at mid-ocean ridges (see Figure 4.19). Basaltic magmas are also generated at subduction zones, where water driven from the descending slab of oceanic crust promotes partial melting of the mantle rocks that lie above the slab (see Figure 4.20).

Formation of Intermediate and Felsic Magmas

Recall that silica-rich magmas erupt mainly along the continental margins. This provides strong evidence that continental crust, which is thicker and has a lower density than oceanic crust, must play a role in generating the more highly evolved intermediate and felsic magmas.

One way that andesitic magma can form is when a rising mantle-derived basaltic magma undergoes magmatic differentiation as it slowly makes its way through the continental crust. Recall from our discussion of Bowen's reaction series that as basaltic magma solidifies, the silica-poor ferromagnesian minerals crystallize first. If these iron-rich components are separated from the

liquid by crystal settling, the remaining melt will have an intermediate or andesitic composition (see Figure 4.22).

Intermediate (andesitic) magmas can also form when rising mafic magmas assimilate crustal rocks that tend to be silica rich. Partial melting of basaltic rocks is yet another way in which at least some andesitic magmas are thought to be produced.

Although felsic (granitic) magmas can be formed through magmatic differentiation of andesitic magmas, most granitic magmas are thought to form when hot basaltic magma becomes trapped, because of its greater density, below continental crust (**Figure 4.27**). The process of

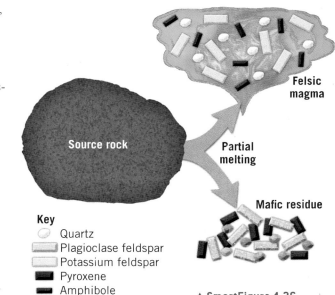

Key
- Quartz
- Plagioclase feldspar
- Potassium feldspar
- Pyroxene
- Amphibole

▲ **SmartFigure 4.26**
Partial melting Partial melting generates a magma that is nearer the felsic (granitic) end of the compositional spectrum than the parent rock from which it was derived.

Tutorial
https://goo.gl/3KzHtp

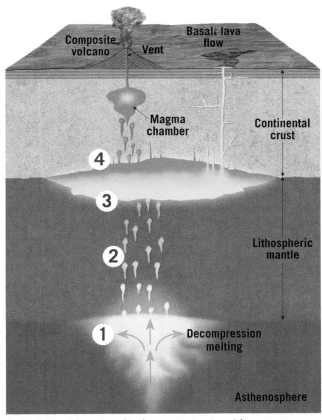

1. **Partial melting of peridotite generates basaltic magma.**
2. **Basaltic magma rises through lithospheric mantle.**
3. **Basaltic magma ponds beneath less dense crustal rocks.**
4. **Partial melting of continental crust generates magma with a felsic composition.**

◄ **SmartFigure 4.27**
Formation of felsic magma Felsic (granitic) magmas are generated by the partial melting of continental crust.

Animation
https://goo.gl/ihWfp3

becoming trapped is called *ponding*. When the heat from this hot basaltic magma partially melts the silica-rich overlying crustal rocks having a much lower melting temperature, the result can be the production of large quantities of granitic magmas. This process is thought to have been responsible for the volcanic activity in and around Yellowstone National Park in the distant past.

CONCEPT CHECKS 4.7

1. Briefly describe why partial melting results in a magma having a composition different from the rock from which it was derived.

Concept Checker
https://goo.gl/Entx6q

2. What is the process thought to generate most basaltic magmas? Most granitic magmas?

4.8 Intrusive Igneous Activity

Compare and contrast these intrusive igneous structures: dikes, sills, batholiths, stocks, and laccoliths.

Although volcanic eruptions can be violent and spectacular events, most magma crystallizes at depth, without fanfare. Therefore, understanding the igneous processes that occur deep underground is as important to geologists as studying volcanic events.

Nature of Intrusive Bodies

When magma rises through the crust, it forcefully displaces preexisting crustal rocks, termed **host rock**, or **country rock**. The structures that result from the emplacement of magma into preexisting rocks are called **intrusions**, or **plutons**. Because all intrusions form far below Earth's surface, they are studied primarily after uplifting and erosion (covered in later chapters) expose them. The challenge lies in reconstructing the events that generated these structures in vastly different conditions deep underground, millions of years ago.

Intrusions are known to occur in a great variety of sizes and shapes. Some of the most common types are illustrated in **Figure 4.28**. Notice that some plutons have

▼ **SmartFigure 4.28**
Intrusive igneous structures

Animation
https://goo.gl/erhgH7

A. Relationship between volcanism and intrusive igneous activity.

B. Basic intrusive structures, some of which have been exposed by erosion.

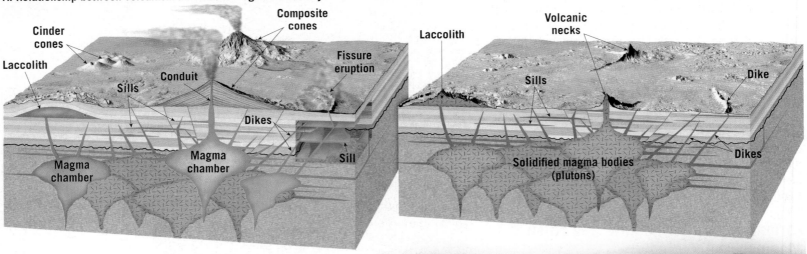

C. Extensive uplift and erosion exposed a batholith composed of several smaller intrusive bodies (plutons).

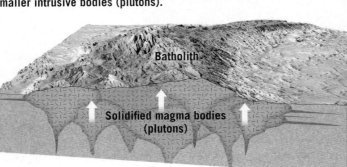

Exposed portion of the Sierra Nevada Batholith

a **tabular** (*tabula* = table) shape, whereas others are best described as **massive** (blob-shaped plutons). Also, observe that some of these bodies cut *across* existing structures, such as sedimentary strata, whereas others form when magma is injected *between* sedimentary layers. Because of these differences, intrusive igneous bodies are generally classified both according to their shape and based on their orientation with respect to the host rock. Igneous bodies are said to be **discordant** (*discordare* = to disagree) if they cut across existing structures and **concordant** (*concordare* = to agree) if they inject parallel to features such as sedimentary strata.

Tabular Intrusive Bodies: Dikes and Sills

Dikes and Sills   Tabular intrusive bodies are produced when magma is forcibly injected into a rock fracture or zone of weakness, such as a bedding surface (see Figure 4.28). **Dikes** are discordant bodies that form when magma is forcibly injected into fractures and cut across bedding surfaces and other structures in the host rock. By contrast, **sills** tend to be horizontal, concordant bodies that form when magma exploits weaknesses between sedimentary beds or other rock structures (**Figure 4.29**). In general,

◄ SmartFigure 4.30
Dike exposed in the Spanish Peaks, Colorado This wall-like dike is composed of igneous rock that is more resistant to weathering than the surrounding material.

Condor Video
https://goo.gl/cr1MUE

dikes serve as tabular conduits that transport magma upward, whereas sills accumulate magma and increase in thickness.

Dikes and sills are typically shallow features, occurring where the country rocks are sufficiently brittle to fracture. They can range in thickness from about a centimeter to more than 1 kilometer.

While dikes and sills can occur as solitary bodies, dikes tend to form in roughly parallel groups called *dike swarms*. These multiple structures reflect the tendency for fractures to form in sets when tensional forces pull apart brittle country rock. Dikes can also radiate from an eroded volcanic neck, like spokes on a wheel. Where such formations are found, the active ascent of magma generated fissures in the volcanic cone out of which lava flowed and later solidified. Dikes frequently are more resistant and thus weather more slowly than the surrounding rock. Consequently, when exposed by erosion, dikes tend to have a wall-like appearance, as shown in **Figure 4.30**.

Because dikes and sills are relatively uniform in thickness and can extend for many kilometers, they are assumed to be the product of very fluid, and therefore mobile, magmas. One of the largest and most studied of all sills in the United States is the Palisades Sill (see Figure 4.23). Exposed for 80 kilometers (50 miles) along the west bank of the Hudson River in southeastern

▲ SmartFigure 4.29
Sill exposed in Sinbad Country, Utah Each dark, essentially horizontal band is a sill of basaltic composition that intruded horizontal layers of sedimentary rock.

Mobile Field Trip
https://goo.gl/KxGBxP

New York and northeastern New Jersey, this sill is about 300 meters (1000 feet) thick. Because it is resistant to erosion, the Palisades Sill forms an imposing cliff that can be easily seen from the opposite side of the Hudson.

Columnar Jointing   In many respects, sills closely resemble buried lava flows. Both are tabular and can extend over a wide area, and both may exhibit columnar jointing. **Columnar jointing** occurs when igneous rocks cool and develop shrinkage fractures that produce elongated, pillar-like columns that most often have six sides (**Figure 4.31**). Further, because sills and dikes generally form in near-surface environments and may be only a few meters thick, the emplaced magma often cools quickly enough to generate a fine-grained texture. (Recall that most intrusive igneous bodies have a coarse-grained texture.)

Massive Plutons: Batholiths, Stocks, and Laccoliths

Batholiths and Stocks   By far the largest intrusive igneous plutons are **batholiths** (*bathos* = depth, *lithos* = stone). Batholiths occur as mammoth linear structures several hundred kilometers long and up to 100 kilometers (60 miles) wide (**Figure 4.32**). The Sierra Nevada batholith, for example, is a continuous granitic structure that forms much of the "backbone" of the Sierra Nevada in California. An even larger batholith extends for over 1800 kilometers (1100 miles) along the Coast Mountains of western Canada and into southern Alaska. Although batholiths can cover a large area, recent geophysical studies indicate that most are less than 10 kilometers (6 miles) thick. Some are even thinner; the coastal batholith of Peru, for example, is essentially a flat slab with an average thickness of only 2 to 3 kilometers (1 to 2 miles). Batholiths are typically composed of felsic (granitic) and intermediate rock types and are often called "granite batholiths."

Early investigators thought the Sierra Nevada batholith was a huge single body of intrusive igneous rock. Today we know that large batholiths are produced by hundreds of discrete injections of magma that form smaller plutons that intimately crowd against or penetrate one another. These bulbous masses are emplaced over spans of millions of years. The intrusive activity that created the Sierra Nevada batholith, for example, occurred nearly continuously over a 130-million-year period that ended about 80 million years ago (see Figure 4.32).

A batholith is generally defined as a pluton having a surface exposure greater than 100 square kilometers (40 square miles). Smaller plutons are termed **stocks**. However, many stocks appear to be portions of much larger intrusive bodies that would be classified as batholiths if they were fully exposed at the surface.

Laccoliths   A nineteenth-century study by G. K. Gilbert of the U.S. Geological Survey in the Henry Mountains of Utah produced the first clear evidence that igneous intrusions can lift the sedimentary strata they penetrate. Gilbert named the igneous intrusions he observed **laccoliths**, which he envisioned as igneous rock forcibly injected between sedimentary strata, so as to arch the beds above while leaving those below relatively flat. It is now known that the five major peaks of the Henry Mountains are not laccoliths but stocks. However, these central magma bodies are the source

Geologist's Sketch

Columnar jointing

Rapid cooling from the outside causes shrinkage cracks

Columnar jointing tends to produce 6-sided columns

▶ **Figure 4.31**
Columnar jointing Columnar jointing on Akun Island located in the Aleutian Islands, Alaska.

material for branching offshoots that are true laccoliths, as Gilbert defined them (**Figure 4.33**).

Numerous other granitic laccoliths have since been identified in Utah. The largest is a part of the Pine Valley Mountains located north of St. George, Utah. Others are found in the La Sal Mountains near Arches National Park and in the Abajo Mountains directly to the south.

Emplacement of Large Plutons   How do magma bodies make their way through several kilometers of solid rock? What happened to the rock that was displaced by these huge igneous masses? (Geologists call this the "room problem.")

We know that magma rises because it is less dense than the surrounding rock, much as a cork held at the bottom of a container of water will rise when it is released. In the upper mantle and lower crust, where temperatures and pressures are high, rock is ductile (able to flow). In this setting, buoyant magma bodies are assumed to rise in the form of *diapirs*, inverted-teardrop-shaped masses with rounded heads and tapered tails. However, in the upper crust, which is more brittle, large fractures provide conduits for the ascent of magma.

Depending on the tectonic environment, geologists have proposed several mechanisms to solve the room problem. At great depths, where rock is ductile, a mass of buoyant, rising magma can forcibly make room for itself by pushing aside the overlying rock—a process called *shouldering*. As the magma continues to move upward, some of the host rock that was displaced will fill in the space left by the magma body as it passes.[†]

As a magma body nears the surface, it encounters relatively cool, brittle country rock that is not

[†]An analogous situation occurs when a can of oil-based paint is left in storage. The oily component of the paint is less dense than the pigments used for coloration; thus, oil collects into drops that slowly migrate upward, while the heavier pigments settle to the bottom.

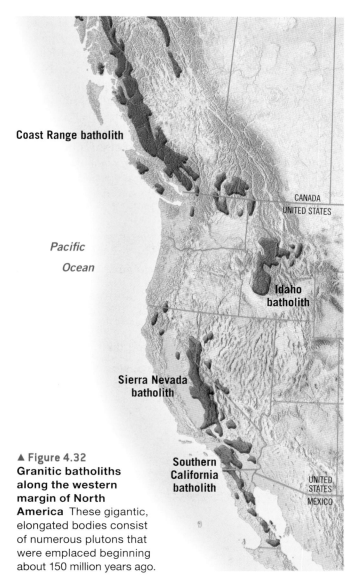

▲ Figure 4.32
Granitic batholiths along the western margin of North America These gigantic, elongated bodies consist of numerous plutons that were emplaced beginning about 150 million years ago.

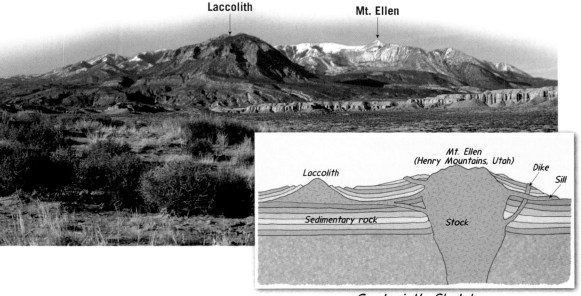

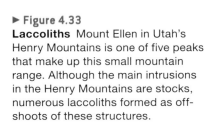

▶ Figure 4.33
Laccoliths Mount Ellen in Utah's Henry Mountains is one of five peaks that make up this small mountain range. Although the main intrusions in the Henry Mountains are stocks, numerous laccoliths formed as offshoots of these structures.

Geologist's Sketch

▲ Figure 4.34
Xenolith Xenoliths are inclusions of host rock contained within igneous bodies. This unmelted chunk of dark (mafic) rock was incorporated into a felsic magma in Rock Creek Canyon the eastern Sierra Nevada of California.

easily pushed aside. Further upward movement may be accomplished by a process called *stoping*, in which blocks of the roof overlying a hot, rising mass become dislodged and sink through the magma (see Figure 4.24). Evidence supporting stoping is found in plutons that contain suspended blocks of country rock called **xenoliths** (*xenos* = a stranger, *lithos* = stone) (**Figure 4.34**).

Magma may also *melt* and *assimilate* some of the overlying host rock. However, this process is greatly limited by the available thermal energy contained in the magma body. When plutons are emplaced near the surface, the room problem may be solved by "lifting the roof" that overlies the intrusive body.

CONCEPT CHECKS 4.8

1. What is meant by the term *country rock*?

2. Describe *dikes* and *sills*, using the appropriate terms from the following list: massive, discordant, tabular, and concordant.
   Concept Checker
   https://goo.gl/2GjzPs

3. Distinguish among batholiths, stocks, and laccoliths in terms of size and shape.

---

4 CONCEPTS IN REVIEW

Igneous Rocks and Intrusive Activity

4.1 Magma: Parent Material of Igneous Rock

List and describe the three major components of magma.

Key Terms:	lava	crystallization
magma	melt	intrusive igneous rock
igneous rock	volatile	extrusive igneous rock

- Completely or partly molten rock is called *magma* if it is below Earth's surface and *lava* if it has erupted. It consists of a liquid melt plus solids (mineral crystals) and gases (*volatiles*), including water vapor or carbon dioxide.

- As magma and lava cools, silicate minerals begin to form through the addition of ions to their outer surface. *Crystallization* gradually transforms magma into a solid mass of interlocking crystals—an *igneous rock*.

- Magmas that cool below the surface produce *intrusive igneous rocks*, while lava that erupt onto Earth's surface generate *extrusive igneous rocks*.

4.2 Igneous Compositions

Compare and contrast the four basic igneous compositions: felsic, intermediate, mafic, and ultramafic.

Key Terms:	mafic composition	peridotite
felsic composition	intermediate composition	ultramafic

- Igneous rocks are primarily composed of silicate minerals. Igneous rocks of *felsic composition* mostly contain nonferromagnesian minerals. Igneous rocks of *mafic composition* have a greater

proportion of ferromagnesian minerals. Mafic rocks are generally darker in color and of greater density than their felsic counterparts. Broadly, continental crust is felsic in composition, and oceanic crust is mafic.

- Rocks of *intermediate composition*, in which plagioclase feldspar predominates, are compositionally between felsic and mafic. They are typical of continental volcanic arcs. *Ultramafic* rocks, which are rich in the minerals olivine and pyroxene, dominate in the upper mantle.

- The amount of silica (SiO_2) in an igneous rock is an indication of its overall composition. Rocks rich in silica (up to 70 percent, or more) are felsic, while rocks that are poor in silica (as low as 40 percent) are ultramafic.

Q Describe igneous rocks having the compositions of samples A and D, using terms such as mafic, felsic, etc. Would you ever expect to find quartz and olivine in the same igneous rock? Why or why not?

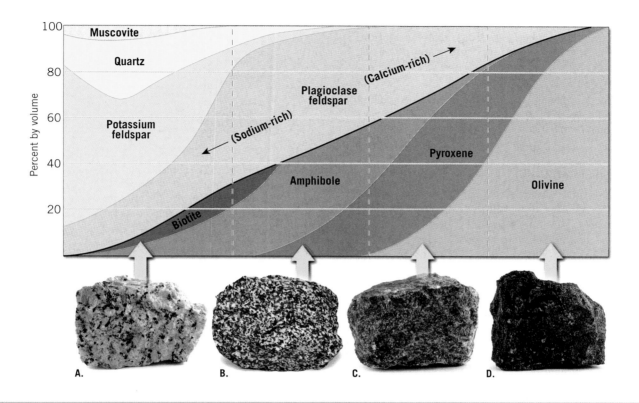

4.3 Igneous Textures: What Can They Tell Us?

Identify and describe the six major igneous textures.

Key Terms:

texture	phenocryst	pyroclastic texture
aphanitic	groundmass	pegmatite
phaneritic	porphyry	pegmatitic texture
porphyritic texture	vesicular texture	
	glassy texture	

- *Texture* describes the size, shape, and arrangement of mineral grains in a rock. The rate at which magma or lava cools largely determines a rock's texture and tells us about the conditions under which it formed.

- The rate of cooling is fast for lava (magma) at or close to the surface, which results in a large number of very small crystals. The result is an *aphanitic*, or *fine-grained*, *texture*. Magma cooling at depth loses heat more slowly. This allows sufficient time for the magma's ions to be organized into larger crystals, resulting in a rock with a *phaneritic*, or *coarse-grained*, *texture*. If crystals begin to form at depth and then the magma rises to a shallow depth or erupts at the surface, it will have a two-stage cooling history, resulting in a rock with a *porphyritic texture*.

- Volcanic rocks may exhibit additional textures: *vesicular* if the lava had a high gas content, *glassy* if the lava was high in silica, or *pyroclastic* if it erupted explosively. The large crystals that characterize *pegmatitic* textures result from the crystallization of magmas with high water content.

4.4 Naming Igneous Rocks

Distinguish among the common igneous rocks based on texture and mineral composition.

Key Terms:

granite	pumice	gabbro
rhyolite	andesite	pyroclastic rocks
obsidian	diorite	tuff
	basalt	welded tuff

- Igneous rocks are classified on the basis of their textures and compositions. Figure 4.13 summarizes the naming system based on these two criteria. Two magmas with the same composition can cool at different rates, resulting in different textures.

4.5 Origin of Magma

Summarize the major processes that generate magma from solid rock.

Key Terms: geothermal gradient   decompression melting

- Solid rock can melt through three processes: (1) raising its temperature by adding heat; (2) dropping the pressure on already hot rock, which causes *decompression melting* (as occurs at mid-ocean ridges); and (3) adding water to hot rock, which decreases its melting point (as occurs at subduction zones).

Q Different processes produce magma in different tectonic settings. Consider situations A, B, and C in the diagram and describe the processes that would be most likely to trigger melting in each one.

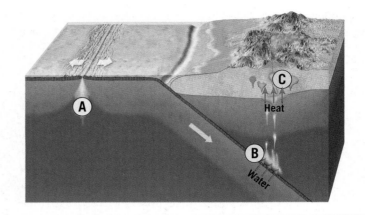

4.6 How Magmas Evolve

Describe how magmatic differentiation can generate a magma body that has a different chemical composition from its parent magma.

Key Terms:	crystal settling	assimilation
Bowen's reaction series	magmatic differentiation	magma mixing

Q Consider the accompanying diagram, which shows a cross-sectional view of a hypothetical magma chamber. Using your understanding of Bowen's reaction series and magma evolution, interpret the layered structure by explaining how crystallization occurred.

- Experimentation by N. L. Bowen revealed that when magma cools, minerals crystallize in a specific order. Ferromagnesian silicates such as olivine crystallize first, at the highest temperatures (1250°C), and nonferromagnesian silicates such as quartz crystallize last, at lowest temperatures (650°C). Bowen found that in between these temperatures, chemical reactions take place between the crystallized minerals and the remaining melt, resulting in the formation of new minerals.
- Various physical processes can cause changes in the composition of magma. For instance, when the minerals that crystallize are denser than the remaining magma, they sink to the bottom of the magma chamber. Because the early-formed minerals are rich in iron and magnesium (mafic), the remaining magma will become more felsic in composition.
- *Assimilation* of host rock and *magma mixing* alter the composition of magma.

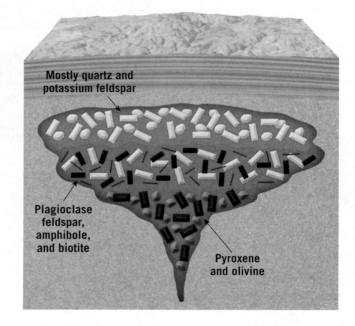

4.7 Partial Melting and Magma Composition

Explain how partial melting of the mantle rock peridotite can generate a mafic (basaltic) magma.

Key Term: partial melting

- Rocks do not always fully melt. Different minerals melt at different temperatures, and minerals with the lowest melting temperatures melt first.

- *Partial melting* of the mantle, which is ultramafic in composition, yields mafic magmas. Partial melting of the lower continental crust at subduction zones produces magmas that have intermediate or felsic compositions.

4.8 Intrusive Igneous Activity

Compare and contrast these intrusive igneous structures: dikes, sills, batholiths, stocks, and laccoliths.

Key Terms:	tabular	concordant
host rock	massive	dike
intrusion	discordant	sill

columnar jointing	stock	xenolith
batholith	laccolith	

- When magma intrudes other rocks, it may cool and crystallize before reaching the surface, producing *intrusions* called *plutons*. Plutons may cut across the host rocks without regard for preexisting structures, or the magma may flow along weak zones in the host rock, such as between the horizontal layers of sedimentary bedding.

- *Tabular* intrusions may be *concordant* (*sills*) or *discordant* (*dikes*). Massive plutons may be small (*stocks*) or large (*batholiths*). Blister-like intrusions are called *laccoliths*. As solid igneous rock cools, its volume decreases. Contraction can produce a distinctive fracture pattern called *columnar jointing*.

- Several processes contribute to magma's intrusion into host rocks. Rising diapirs are one possibility, and another is shouldering aside of host rocks. Stoping of xenoliths from the host rock can open up more room, or the magma can melt and assimilate some of the host rock.

GIVE IT SOME THOUGHT

1. **Apply your understanding of igneous rock textures to describe the cooling history of each of the igneous rocks pictured here.**

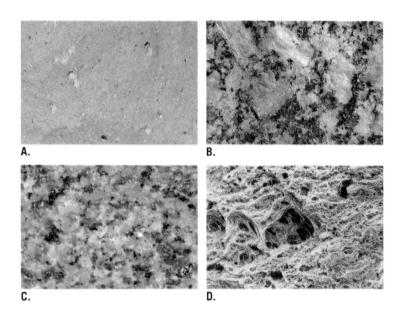

A.

B.

C.

D.

2. **Use Figure 4.5 to classify the following igneous rocks.**

   a. An aphanitic rock containing about 30 percent calcium-rich plagioclase feldspar, 55 percent pyroxene, and 15 percent olivine

   b. A phaneritic rock containing about 20 percent quartz, 40 percent potassium feldspar, 20 percent sodium-rich plagioclase feldspar, a few percent muscovite, and the remainder dark-colored silicate

   c. An aphanitic rock containing about 50 percent plagioclase feldspar, 35 percent amphibole, 10 percent pyroxene, and minor amounts of other light-colored silicates

   d. A phaneritic rock made mainly of olivine and pyroxene, with lesser amounts of calcium-rich plagioclase feldspar

3. **Identify the igneous rock textures described by each of the following statements.**

   a. Openings produced by escaping gases

   b. A matrix of fine crystals surrounding phenocrysts

   c. Consists of crystals that are too small to be seen without a microscope

   d. A texture characterized by rock fragments welded together

   e. Coarse grained, with crystals of roughly equal size

   f. Exceptionally large crystals, most exceeding 1 centimeter in diameter

4. **During a hike, you pick up the igneous rock shown in the accompanying photo.**

   a. What is the mineral name of the small, rounded, glassy green crystals?

   b. Did the magma from which this rock formed likely originate in Earth's mantle or crust? Explain.

   c. Was the magma likely a high-temperature magma or a low-temperature magma? Explain.

   d. Describe the texture of this rock, using terms introduced in this chapter.

5. **A common misconception about Earth's upper mantle is that it is a thick shell of molten rock. Explain why Earth's mantle is actually solid under most conditions.**

6. **Describe two mechanisms by which mantle rock can melt without an increase in temperature. Name the plate tectonic settings in which these magma-generating mechanisms are found.**

7. **Use your understanding of Bowen's reaction series (see Figure 4.21) to explain how partial melting can generate magmas that have different compositions.**

8. During a field trip with your geology class, you visit an exposure of rock layers similar to the one sketched here. A fellow student suggests that the layer of basalt is a sill. You disagree. Why do you think the other student is incorrect? What is a more likely explanation for the basalt layer?

Shale

Vesicles

Basalt

Sandstone

Shale

Limestone

9. Mount Whitney, the highest summit (4421 meters [14,505 feet]) in the contiguous United States, is located in the Sierra Nevada batholith. Based on its location, is Mount Whitney likely composed mainly of felsic (granitic), intermediate (andesitic), or mafic (basaltic) rocks?

Mount Whitney ⟶

EYE ON EARTH

1. Shiprock, New Mexico, is an igneous structure that rises more than 510 meters (1700 feet) above the surrounding desert in northwestern New Mexico. It consists of rock that accumulated in the vent of a volcano that has since been eliminated by erosion.

   a. What type of landform is Shiprock?

   b. What type of structure is the long, narrow ridge extending away from Shiprock?

DATA ANALYSIS

Generating Magma from Solid Rock

Temperature and pressure increase as you go deeper below Earth's surface. Both of these variables are important in determining the physical state—solid or melted—of a given rock.

https://goo.gl/eD226M

ACTIVITIES

1. **Is the rock labeled in this diagram in a solid or liquid state?**

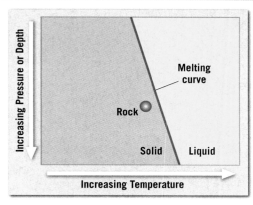

2. **This diagram shows how the melting curve shifts when water is added to the system. Would the rock shown be remain solid, or melt?**

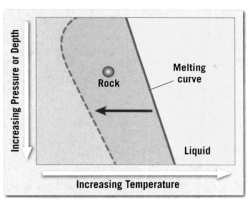

For questions 3–6, match the diagram below with the appropriate scenario description.

3. **An area of solid rock in the subsurface is heated quickly in place and starts to melt.**

4. **An area of solid rock undergoes increases in pressure and temperature as it is buried.**

5. **An area of solid rock deep in the subsurface quickly rises to a lower depth/pressure and melts.**

6. **Magma erupts at the surface and then rapidly cools to a solid. Note: There are two motions here.**

7. **Which of the scenarios (2–5) are associated with divergent plate boundaries?**

8. **Which of the scenarios (2–5) are associated with continental volcanic arcs near convergent plate boundaries?**

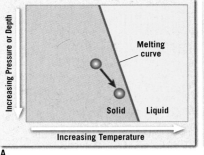

A.

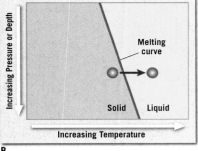

B.

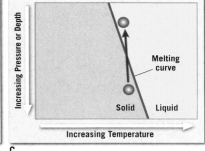

C.

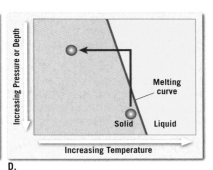

D.

Mastering Geology

Looking for additional review and test prep materials? Visit pearson.com/mastering/geology to enhance your understanding of this chapter's content by accessing a variety of resources, including **Self-Study Quizzes, Geoscience Animations, SmartFigure Tutorials,** *Project Condor* **Videos, Mobile Field Trips, Dynamic Study Modules,** and an optional **Pearson eText.**

Volcanic Ash Adds Peril to Air Travel

Eruption dangers aren't confined to land. If volcanic ash reaches altitudes where airplanes cruise, it poses serious safety risks. According to the U.S. Geological Survey (USGS), between 1953 and 2009, at least 26 planes sustained significant damage due to flying through volcanic ash clouds, including 9 events of engine failure.

▲ These black deposits inside a jet engine resulted from flying through a volcanic ash cloud from Mount Redoubt in 1989.

Unlike ash from wood fires, volcanic ash is abrasive. Some planes emerge from ash clouds with windows badly pitted. A damaged windscreen impedes a pilot's vision, and worse, if it breaks, sudden cabin decompression may result. The tiny rock and mineral particles in ash also penetrate critical engine filters, damage jet turbines, clog fuel nozzles, and interfere with temperature sensors. Further, silicate rock particles melt at lower temperatures than jet fuel burns. That means volcanic ash may liquefy as it passes through an engine, then cool and solidify into glassy residue that coats other parts of the plane. This, in turn, can cause problems with radio transmissions, navigation, and airspeed monitoring.

Due to these risks, airlines cancel or reroute flights to avoid ash clouds in Earth's troposphere. A 2010 volcanic eruption in Iceland resulted in widespread flight delays and cancellations throughout Europe. And a 2017 eruption of Mount Agung in Indonesia caused the Bali airport to close for over 2 days, impacting travel for 57,000 people, according to airport officials.

▶ Iceland's 2010 Eyjafjallajökull Volcano eruption caused airlines across Europe to cancel thousands of flights. Several weeks passed before air travel resumed a normal schedule.

5
Volcanoes and Volcanic Hazards

FOCUS ON CONCEPTS

Each statement represents the primary learning objective for the corresponding major heading within the chapter. After you complete the chapter, you should be able to:

5.1 Compare and contrast the 1980 eruption of Mount St. Helens with the most recent eruption of Kilauea, which began in 1983.

5.2 Explain why some volcanic eruptions are explosive and others are quiescent.

5.3 List and describe the three categories of materials extruded during volcanic eruptions.

5.4 Draw and label a diagram that illustrates the basic features of a typical volcanic cone.

5.5 Summarize the characteristics of shield volcanoes and provide one example of this type of volcano.

5.6 Describe the formation, size, and composition of cinder cones.

5.7 List the characteristics of composite volcanoes and describe how they form.

5.8 Describe the major geologic hazards associated with volcanoes.

5.9 List volcanic landforms other than shield, cinder, and composite volcanoes and describe their formation.

5.10 Explain how the global distribution of volcanic activity relates to plate tectonics.

5.11 List and describe the techniques used to monitor potentially dangerous volcanoes.

because volcanoes extrude molten rock that formed at great depth, they provide our only means of directly observing processes that occur many kilometers below Earth's surface. Furthermore, Earth's atmosphere and oceans have evolved from gases emitted during volcanic eruptions. Either of these facts is reason enough for igneous activity to warrant our attention.

5.1 Mount St. Helens Versus Kilauea

Compare and contrast the 1980 eruption of Mount St. Helens with the most recent eruption of Kilauea, which began in 1983.

On May 18, 1980, the largest volcanic eruption to occur in North America in historic times transformed a picturesque volcano into a decapitated remnant (**Figure 5.1**). On that date in southwestern Washington State, Mount St. Helens erupted with tremendous force. The blast blew out the entire north flank of the volcano, leaving a gaping hole. In one brief moment, a prominent volcano whose summit had been more than 2900 meters (9500 feet) above sea level was lowered by more than 400 meters (1350 feet).

▼ Figure 5.1
Before-and-after photographs show the transformation of Mount St. Helens The May 18, 1980, eruption of Mount St. Helens occurred in southwestern Washington.

The event devastated a wide swath of timber-rich land on the north side of the mountain (**Figure 5.2**). Trees were flattened, stripped of branches and strewn about like giant toothpicks. The accompanying mudflows carried ash, trees, and water-saturated rock debris 29 kilometers (18 miles) down the Toutle River. The eruption claimed 59 lives; some died from the intense heat and the suffocating cloud of ash and gases, others from the impact of the blast, and still others from being trapped in mudflows.

The initial blast ejected nearly a cubic kilometer of ash and rock debris horizontally, and it blanketed a 400-square-kilometer (160-square-mile) area. Following the initial explosion, ash and hot gases were propelled skyward more than 18 kilometers (11 miles) into the

Spirit Lake

1350 feet

The blast blew out the entire north flank of Mount St. Helens, leaving a gaping hole. In a brief moment, a prominent volcano was lowered by 1350 feet.

Spirit Lake, largely covered by fallen trees.

stratosphere. During the next few days, this very fine-grained material was carried around Earth by strong upper-level winds. Crops were damaged in central Montana, and measurable deposits were reported as far away as Oklahoma and Minnesota. Meanwhile, ash fallout in the immediate vicinity exceeded 2 meters (6 feet) in depth. The air over Yakima, Washington (130 kilometers [80 miles] to the east), was so filled with ash that residents experienced midnight-like darkness at noon.

Not all volcanic eruptions are this violent. Some, like the Hawaii's Kilauea Volcano, generate relatively calm outpourings of fluid lavas that pour from the vent and flow downslope. These quiet (nonexplosive) eruptions are not without some fiery displays; occasionally fountains of incandescent lava spray hundreds of meters into the air (see Figure 5.4). A testimony to the quiescent nature of Kilauea's eruptions is the fact that the Hawaiian Volcano Observatory has operated on its summit since 1912, despite the fact that Kilauea has had more than 50 eruptive phases since record keeping began in 1823. Nevertheless, as we will consider later in the chapter, the incandescent lava emitted from Kilauea over this time period has been quite destructive to property.

◀ Figure 5.2
Douglas fir trees snapped off or uprooted by the lateral blast of Mount St. Helens

CONCEPT CHECK 5.1

1. Briefly compare the 1980 eruption of Mount St. Helens to a typical eruption of Hawaii's Kilauea Volcano.

Concept Checker
https://goo.gl/9YfSm4

5.2 The Nature of Volcanic Eruptions

Explain why some volcanic eruptions are explosive and others are quiescent.

Volcanic activity is commonly perceived as a process that produces a picturesque, cone-shaped structure that periodically erupts in a violent manner. However, many eruptions are not explosive. What determines the manner in which volcanoes erupt?

Magma: Source Material for Volcanic Eruptions

Recall that **magma**, molten rock that may contain some solid crystalline material and also contains varying amounts of dissolved gas (mainly water vapor and carbon dioxide), is the parent material of igneous rocks. Erupted magma is called **lava**.

Composition of Magma   As we discussed in Chapter 4, *mafic* (*basaltic*) igneous rocks contain a high percentage of dark, iron- and magnesium-rich silicate minerals and calcium-rich plagioclase feldspar and, therefore, tend to be dark in color. By contrast, *felsic* rocks (*granite* and its extrusive equivalent *rhyolite*) contain mainly light-colored silicate minerals—quartz and potassium feldspar. *Intermediate* (*andesitic*) igneous rocks have a composition between that of basaltic rocks and that of granitic rocks. Correspondingly, mafic magmas contain a much *lower percentage of silica* (SiO_2) than felsic magmas.

The compositional differences between magmas also affect several other properties, as summarized in **Figure 5.3**. For example, mafic (basaltic) magmas have the lowest silica and gas content, and they erupt at the highest temperatures. By contrast, felsic (granitic and rhyolitic) magmas have the highest silica content and the highest gas content, and they can erupt at relatively low temperatures. Intermediate (andesitic) magmas have characteristics between those of mafic and felsic magmas.

Properties of Magma Bodies with Differing Compositions

Composition	Silica Content (SiO₂)	Gas Content (% by weight)	Eruptive Temperature	Viscosity	Tendency to Form Pyroclastics	Volcanic Landform
MAFIC (Basaltic) High in Fe, Mg, Ca, low in K, Na	**Least** (~50%)	**Least** (0.5–2%)	**Highest** 1000–1250°C	**Least**	**Least**	Shield volcanoes, basalt plateaus, cinder cones
INTERMEDIATE (Andesitic) Varying amounts of Fe, Mg, Ca, K, Na	Intermediate (~60%)	Intermediate (3–4%)	Intermediate 800–1050°C	Intermediate	Intermediate	Composite cones
FELSIC (Rhyolitic/ Granitic) High in K, Na, low in Fe, Mg, Ca	Most (~70%)	Most (5–8%)	Lowest 650–900°C	Greatest	Greatest	Pyroclastic flow deposits, lava domes

Effusive Versus Explosive Eruptions

Geologists refer to quiescent (nonviolent) eruptions that produce outpourings of fluid lava as **effusive eruptions** (*effus* = pour forth). At the other end of the spectrum are explosive eruptions that send gases and solid materials high into the air.

The two primary factors that determine how magma erupts are its *viscosity* and *gas content*. **Viscosity** (*viscos* = sticky) is a measure of a fluid's mobility. The more viscous a material, the greater its resistance to flow. For example, syrup is more *viscous*, and thus more resistant to flow, than water.

Factors Affecting Viscosity
Magma's viscosity depends primarily on its temperature and silica content: *The more silica in magma, the greater its viscosity.* Early in the crystallization process, silicon–oxygen tetrahedra begin to link together into long chains, which makes the magma more rigid and impedes its flow. Consequently, silica-rich rhyolitic lavas are the most viscous and tend to travel at imperceptibly slow speeds to form comparatively short, thick flows. By contrast, basaltic lavas, which contain much less silica, are relatively fluid and have been known to travel 150 kilometers (90 miles) or more before solidifying. Andesitic magmas, which are intermediate in composition, have flow rates between these extremes.

Temperature affects the viscosity of magma in much the same way it affects the viscosity of pancake syrup: The hotter the temperature, the more fluid (less viscous) it will be. As lava cools and begins to congeal, its viscosity increases, and the flow eventually halts.

The more silica present in magma, the greater its viscosity.

Role of Gases The nature of volcanic eruptions also depends on the amount of dissolved gases held within the magma body by the pressure exerted by the overlying rock (confining pressure). Typically, the most abundant gases in magma are water vapor and carbon dioxide. These dissolved gases tend to come out of solution when the confining pressure is reduced. This is analogous to how carbon dioxide behaves in cans and bottles of soft drinks: When you reduce the pressure on a soft drink by opening the cap, the dissolved carbon dioxide quickly separates from the solution to form bubbles that rise and escape.

Like viscosity, a magma's gas content is directly related to its composition, as shown in Figure 5.3. At one end of the spectrum are mafic (basaltic) magmas, which are very fluid and have a low gas content, sometimes as little as 0.5 percent by weight. At the other extreme are felsic (rhyolitic) magmas, which are highly viscous (sticky) and contain a lot of gas—as much as 8 percent by weight.

Effusive Eruptions

All magmas contain some water vapor and other gases that are kept in solution by the immense pressure of the overlying rock. As magma rises toward Earth's surface, or as the rocks confining the magma fail, the pressure drops, causing the dissolved gases to separate from the melt and form large numbers of tiny bubbles. When fluid mafic magmas erupt, these pressurized gases readily escape. At temperatures often exceeding 1100°C (2000°F), these gases can quickly expand to occupy hundreds of times their original volumes. This expansion can propel incandescent (glowing) lava hundreds of meters into the air, producing lava fountains (**Figure 5.4**). Although spectacular, these fountains are usually harmless and generally not associated with major explosive events that cause great loss of life.

Effusive eruptions of very fluid mafic lavas, such as the recent eruptions of Kilauea on Hawaii's Big Island, are often triggered by the arrival of a new batch of molten rock, which accumulates in a near-surface magma chamber. Geologists can usually detect such an impending event because the summit of the volcano begins to inflate and rise months or even years before an eruption. The injection of a fresh supply of hot molten rock heats and remobilizes the semi-liquid magma in the chamber. Swelling of the magma chamber fractures the rock above, allowing the fluid magma to move upward along the newly formed fissures, often generating effusions of fluid lava for weeks, months, or possibly years. The most recent eruptive phase of Kilauea began in 1983 and continues today.

How Explosive Eruptions Are Triggered

Recall that silica-rich rhyolitic magmas have a relatively high gas content and are quite viscous (sticky) compared to basaltic magmas. As rhyolitic magma rises, the gases remain dissolved until the confining pressure drops sufficiently, at which time tiny bubbles begin to form and increase in size. The highly viscous rhyolitic magma tends to trap these gas bubbles, forming a sticky froth.

When the pressure exerted by the expanding magma exceeds the strength of the overlying rock, fracturing occurs. As the frothy magma moves up through the fractures, the resulting drop in confining pressure creates even more gas bubbles. This chain reaction generates an explosive event in which magma is literally blown into fragments (ash and pumice) that are carried to great heights by the escaping hot gases. (The collapse of a volcano's flank can also greatly reduce the pressure on the magma below, causing an explosive eruption, as exemplified by the 1980 eruption of Mount St. Helens.)

When molten rock in the uppermost portion of the magma chamber is forcefully ejected by the escaping gases, the confining pressure on the magma directly below also drops suddenly. Thus, rather than being a single "bang," an explosive eruption is really a series of violent explosions that can last for a few days.

Because highly gaseous magmas expel fragmented lava at nearly supersonic speeds, they are associated with hot, buoyant **eruption columns** consisting mainly of volcanic ash and gases (**Figure 5.5**). Eruption

▲ Figure 5.4
Lava fountain produced by gases escaping fluid mafic lava

Eruptions of highly viscous lavas may produce explosive clouds of hot ash and gases called eruption columns.

◀ SmartFigure 5.5
Eruption column generated by viscous, silica-rich magma Steam and ash eruption column from Mount Tavurvur in eastern Papua New Guinea, 2014.

Video
https://goo.gl/QHcrMR

columns can rise 40 kilometers (25 miles) into the atmosphere. It is not uncommon for a portion of an eruption column to collapse, sending hot ash rushing down the volcanic slope at speeds exceeding 100 kilometers (60 miles) per hour. As a result, volcanoes that erupt highly viscous magmas having a high gas content are the most destructive to property and human life. Following explosive eruptions, partially degassed lava tends to slowly ooze out of the vent to form thick lava flows or dome-shaped lava bodies that grow over the vent.

CONCEPT CHECKS 5.2

1. List these magmas in order, from the highest to lowest silica content: mafic (basaltic) magma, felsic (granitic/rhyolitic) magma, intermediate (andesitic) magma.

2. List the two primary factors that determine the manner in which magma erupts.

3. Define *viscosity*.

Concept Checker
https://goo.gl/NhoRXH

4. Are volcanoes fed by highly viscous magma *more* or *less* likely to be a greater threat to life and property than volcanoes supplied with very fluid magma?

5.3 Materials Extruded During an Eruption

List and describe the three categories of materials extruded during volcanic eruptions.

Volcanoes erupt lava, large volumes of gas, and pyroclastic materials (broken rock, lava "bombs," and ash). In this section, we will examine each of these materials.

Lava Flows

More than 90 percent of Earth's lava is estimated to be mafic (basaltic) in composition. Most of this type of lava erupts along oceanic ridges (divergent plate boundaries), generating new oceanic crust. Lavas having an intermediate (andesitic) composition account for most of the rest and are common components of volcanic island arcs and volcanic chains that form along the margins of landmasses. Rhyolitic (felsic) flows, which make up as little as 1 percent of the total, are found mostly in continental settings. Rhyolitic magmas tend to extrude mostly hot gases and volcanic ash rather than lava.

When hot mafic lavas erupt on land, they generally flow in thin, broad sheets or streamlike ribbons. These fluid lavas have been clocked at speeds exceeding 30 kilometers (19 miles) per hour down steep slopes, although slower flow rates are more common. Silica-rich rhyolitic lava, by contrast, often moves too slowly to be observed. Furthermore, rhyolitic lavas seldom travel more than a few kilometers from their vents. As you might expect, andesitic lavas that are intermediate in composition exhibit flow characteristics between these extremes.

Aa and Pahoehoe Flows   Fluid basaltic magmas tend to generate two types of lava flows, which are known by their Hawaiian names. The first, called **aa** (pronounced "ah-ah") **flows**, have surfaces of rough jagged blocks with dangerously sharp edges and spiny projections (**Figure 5.6A**). Crossing a hardened aa flow can be a trying and miserable experience. The second type,

▼ **Figure 5.6**
Lava flows A. A slow-moving, basaltic aa flow advancing over hardened pahoehoe lava.
B. A typical fluid pahoehoe (ropy) lava. Both of these lava flows erupted from a rift on the flank of Hawaii's Kilauea Volcano.

A. Active aa flow overriding an older pahoehoe flow.

B. Pahoehoe flow displaying the characteristic ropy appearance.

A. Lava tubes are cave-like tunnels that once served as conduits carrying lava from an active vent to the flow's leading edge.

Valentine Cave, a lava tube at Lava Beds National Monument, California.

B. Skylights develop where the roofs of lava tubes collapse and reveal the hot lava flowing through the tube.

▲ Figure 5.7
Lava tubes A. Some lava tubes exhibit extraordinary dimensions. Kazumura Cave, located on the southeastern slope of Hawaii's Mauna Loa Volcano, is a lava tube extending more than 60 kilometers (40 miles).
B. The collapsed section of the roof of a lava tube results in a skylight.

pahoehoe (pronounced "pah-hoy-hoy") **flows**, exhibit smooth surfaces that sometimes resemble twisted braids of ropes (**Figure 5.6B**).

Although both lava types can erupt from the same volcano, pahoehoe lavas are hotter and more fluid than aa flows. In addition, pahoehoe lavas can change into aa lava flows, although the reverse (aa to pahoehoe) does not occur. Cooling that occurs as the flow moves away from the vent is one factor that facilitates the change from pahoehoe to aa. The lower temperature increases viscosity and promotes bubble formation. Escaping gas bubbles produce numerous voids (vesicles) and sharp spines in the surface of the congealing lava. The interior part of the lava flow remains molten, and as it advances, the outer crust breaks, transforming the relatively smooth surface of a pahoehoe flow into an aa flow made up of an advancing mass of rough, sharp, broken lava blocks.

Pahoehoe flows often contain cave-like tunnels, called **lava tubes**, that form because the lava in the interior of the flow remains fluid, while the exposed surface cools and hardens (**Figure 5.7**). These insulated pathways serve as conduits for carrying lava from an active vent to the flow's leading edge. As a result, lava tubes facilitate the flow of fluid lava great distances from its source.

Block Lavas In contrast to fluid basaltic magmas that can travel many kilometers, viscous andesitic and rhyolitic magmas tend to generate relatively short but prominent flows—a few hundred meters to a few kilometers long. Their upper surface consists largely of massive, detached blocks—hence the name **block lava**. Although similar to aa flows, these lavas consist of blocks with slightly curved, smooth surfaces rather than the rough, spiny surfaces typical of aa flows.

Pillow Lavas When outpourings of lava occur on the ocean floor, the flow's outer skin quickly "freezes" (solidifies) to form volcanic glass. However, the interior lava is able to move forward by breaking through the hardened surface. This process occurs over and over, as molten basalt is extruded like toothpaste from a tightly squeezed tube. The result is a lava flow composed of numerous tube-like structures called **pillow lavas**, stacked one atop the other (**Figure 5.8**). Pillow lavas are useful when

▼ Figure 5.8
Pillow lava formation The pillows vary in shape but tend to be elongated tube-like structures. The photo shows an undersea pillow lava flow off the coast of Hawaii.

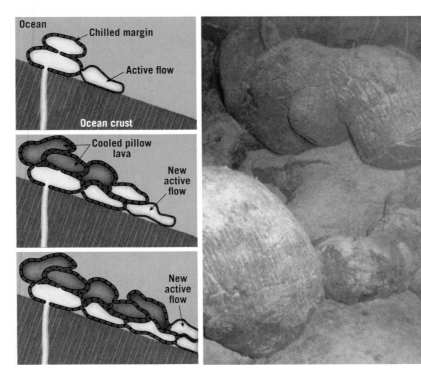

Ocean
Chilled margin
Active flow
Ocean crust
Cooled pillow lava
New active flow
New active flow

► Figure 5.9
Types of pyroclastic materials Pyroclastic materials are also commonly referred to as tephra.

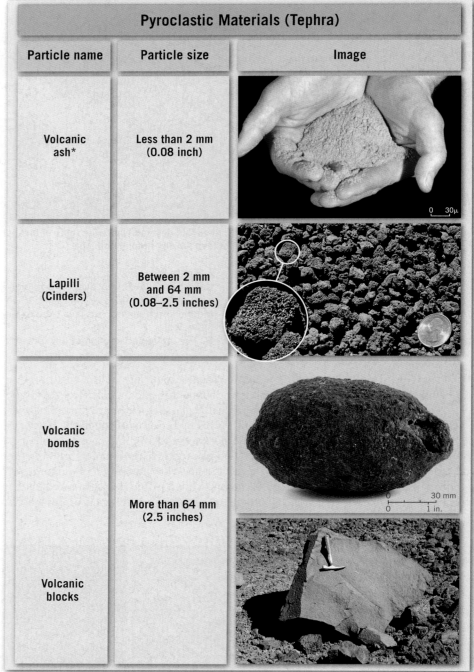

Pyroclastic Materials (Tephra)		
Particle name	Particle size	Image
Volcanic ash*	Less than 2 mm (0.08 inch)	
Lapilli (Cinders)	Between 2 mm and 64 mm (0.08–2.5 inches)	
Volcanic bombs	More than 64 mm (2.5 inches)	
Volcanic blocks		

*The term volcanic dust is used for fine volcanic ash less than 0.063 mm (0.0025 inch).

► Figure 5.9
Types of pyroclastic materials Pyroclastic materials are also commonly referred to as tephra.

reconstructing geologic history because their presence indicates that the lava flow formed below the surface of a water body.

Gases

The dissolved gases in magmas are referred to as **volatiles**. As mentioned earlier, they remain in molten rock because of confining pressure, just as carbon dioxide is held in a soft drink can. As with soft drinks, as soon as the pressure is reduced, the gases begin to escape. Obtaining gas samples from an erupting volcano is difficult and dangerous, so geologists usually must estimate the amount of gas originally contained in the magma.

The gaseous portion of most magma bodies ranges from less than 1 percent to about 8 percent of the total weight, with most of this in the form of water vapor (H_2O), followed by carbon dioxide (CO_2) and sulfur dioxide (SO_2), along with lesser amounts of hydrogen sulfide (H_2S), carbon monoxide (CO), and nitrogen (N_2). The relative proportion of each gas varies significantly by region. Although the percentage may be small, the actual quantity of emitted gases can exceed thousands of tons per day. These gases contribute significantly to our planet's atmosphere. Volcanoes are also natural sources of air pollution; some emit large quantities of sulfur dioxide (SO_2), which readily combines with atmospheric gases to form toxic sulfuric acid and other sulfate compounds.

Pyroclastic Materials

When volcanoes erupt energetically, they eject pulverized rock and fragments of lava and glass from the vent. The particles ejected, **pyroclastic materials** (*pyro* = fire, *clast* = fragment), are also called **tephra**. These fragments range in size from very fine dust- and sand-sized particles (less than 2 millimeters) to pieces that weigh several tons (**Figure 5.9**).

Fine particles, called *volcanic ash* and *dust*, result when gas-rich viscous magma erupts explosively. As magma moves up in the vent, the gases rapidly expand, generating a melt that resembles the froth that flows from a bottle of champagne. As the hot gases expand explosively, the froth is blown into fine glassy fragments. When the hot ash falls, the glassy shards often fuse to form a rock called *welded tuff.* Sheets of this material, as well as ash deposits that later consolidate, cover vast portions of the western United States.

Somewhat larger pyroclasts ranging from the size of small beads to the size of walnuts (2–64 millimeters [0.08–2.5 inches] in diameter) are known as *lapilli* ("little stones"), or *cinders.* Particles larger than 64 millimeters (2.5 inches) in diameter are called *blocks* when they are made of hardened lava and *bombs* when they are ejected as incandescent lava (see Figure 5.9). Because bombs are semi-molten when ejected, they often take on a streamlined shape as they hurl through the air. Bombs and blocks usually fall near the vent due to their weight and size, but they are occasionally propelled great distances. For instance, bombs 6 meters (20 feet) long and weighing about 200 tons were blown 600 meters (2000 feet) from the vent during an eruption of the Japanese volcano Asama.

Pyroclastic materials can be classified by texture and composition as well as by size. For instance, **scoria** is the term for vesicular ejecta produced most often during the eruption of basaltic magmas (**Figure 5.10A**). These black to reddish-brown fragments are generally found in the size range of lapilli and resemble cinders and clinkers produced by furnaces used to smelt iron.

By contrast, when magmas with andesitic (intermediate) or rhyolitic (felsic) compositions erupt explosively, they emit ash and the vesicular rock **pumice** (**Figure 5.10B**). Pumice is usually lighter in color and less dense than scoria, and many pumice fragments have so many vesicles that they are light enough to float (see Figure 4.15, page 116).

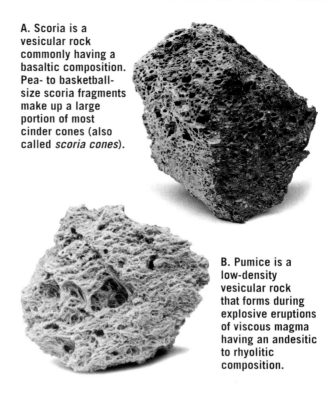

A. Scoria is a vesicular rock commonly having a basaltic composition. Pea- to basketball-size scoria fragments make up a large portion of most cinder cones (also called *scoria cones*).

B. Pumice is a low-density vesicular rock that forms during explosive eruptions of viscous magma having an andesitic to rhyolitic composition.

◄ **Figure 5.10**
Common vesicular rocks Scoria and pumice are volcanic rocks that exhibit a vesicular texture. Vesicles are small holes left by escaping gas bubbles.

CONCEPT CHECKS 5.3

1. Contrast pahoehoe and aa lava flows.

2. How do lava tubes form?

3. List the main gases released during a volcanic eruption.

 Concept Checker
https://goo.gl/mvwjbL

4. How do volcanic bombs differ from blocks of pyroclastic debris?

5. What is scoria? How is it different from pumice?

5.4 Anatomy of a Volcano

Draw and label a diagram that illustrates the basic features of a typical volcanic cone.

A popular image of a volcano is a solitary, graceful, snowcapped cone, such as Mount Hood in Oregon or Japan's Fujiyama. These picturesque conical mountains are produced by volcanic activity that occurred intermittently over thousands, or even hundreds of thousands, of years. However, many volcanoes do not fit this image. Cinder cones are quite small and form during a single eruptive phase that lasts a few days to a few years. Alaska's Valley of Ten Thousand Smokes is a flat-topped ash deposit that blanketed a river valley to a depth of 200 meters (600 feet). The eruption that produced it lasted less than 60 hours yet emitted more than 20 times more volcanic material than the 1980 Mount St. Helens eruption.

Volcanic landforms come in a wide variety of shapes and sizes, and each volcano has a unique eruptive history. Nevertheless, volcanologists have been able to classify volcanic landforms and determine their eruptive patterns. In this section, we will consider the general anatomy of an idealized volcanic cone.

Volcanic activity frequently begins when a **fissure** (crack) develops in Earth's crust as magma moves forcefully toward the surface. As the gas-rich magma moves up through a fissure, its path is usually localized into a somewhat pipe-shaped **conduit** that terminates at a surface opening called a **vent** (**Figure 5.11**). The cone-shaped

▶ SmartFigure 5.11
Anatomy of a volcano Compare the structure of the "typical" composite cone shown here to that of a shield volcano (Figure 5.12) and a cinder cone (Figure 5.14).

Tutorial
https://goo.gl/N61SH5

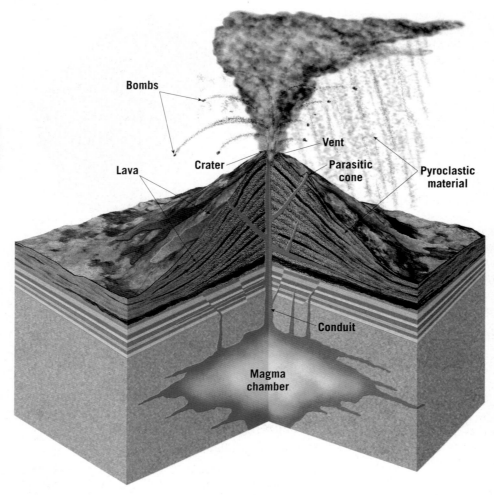

Bombs

Lava   Crater   Vent

Parasitic cone   Pyroclastic material

Conduit

Magma chamber

called **calderas**, which have diameters that are greater than 1 kilometer (0.6 mile) and that in rare cases exceed 50 kilometers (30 miles). Calderas usually form when the summit area of a volcano collapses following an eruption.

In an idealized volcanic cone, most volcanic discharges come from the vent within the central summit crater. However, material can also be emitted from fissures that develop along the flanks (sides) or at the base of the volcano. Continued activity from a flank eruption may produce one or more smaller **parasitic cones**. Italy's Mount Etna, for example, has more than 200 secondary vents, some of which have built parasitic cones. Many of these vents, however, emit only hot gases and are more appropriately called **fumaroles** (*fumus* = smoke).

Next, we will explore the three major types of volcanic cones—shield volcanoes, cinder cones, and composite volcanoes.

structure we call a **volcanic cone** is often created by successive eruptions of lava, pyroclastic material, or frequently a combination of both, often separated by long periods of inactivity.

Located at the summit of most volcanic cones is a somewhat funnel-shaped depression called a **crater** (*crater* = bowl). Volcanoes built primarily of pyroclastic materials typically have craters that form by gradual accumulation of volcanic debris on the surrounding rim. Other craters form during explosive eruptions as the rapidly ejected particles erode the crater walls. Some volcanoes have very large circular depressions,

---

CONCEPT CHECKS 5.4

1. Distinguish among a conduit, a vent, and a crater.

2. How is a crater different from a caldera?

3. What is a parasitic cone, and where does it form?

Concept Checker
https://goo.gl/gCGYPh

5.5 Shield Volcanoes

Summarize the characteristics of shield volcanoes and provide one example of this type of volcano.

A **shield volcano** is produced by the accumulation of fluid basaltic lavas and exhibits the shape of a broad, slightly domed structure that resembles a warrior's shield (**Figure 5.12**). Most shield volcanoes begin on the ocean floor as **seamounts** (submarine volcanoes), and a few of them grow large enough to form volcanic islands. In fact, many oceanic islands are either a single shield volcano or, more often, the coalescence of two or more shields built upon massive amounts of pillow lavas.

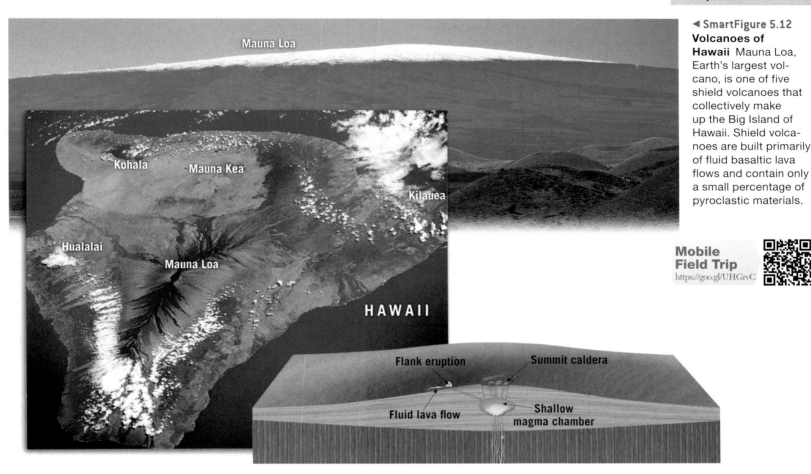

◄ SmartFigure 5.12
Volcanoes of Hawaii Mauna Loa, Earth's largest volcano, is one of five shield volcanoes that collectively make up the Big Island of Hawaii. Shield volcanoes are built primarily of fluid basaltic lava flows and contain only a small percentage of pyroclastic materials.

Mobile Field Trip
https://goo.gl/UHGrvC

Examples include the Hawaiian Islands, the Canary Islands, Iceland, the Galapagos Islands, and Easter Island. Although less common, some shield volcanoes form on continental crust, including Nyamuragira, Africa's most active volcano, and Newberry Volcano in Oregon.

Mauna Loa: Earth's Largest Shield Volcano

Extensive study of the Hawaiian Islands has revealed that they are constructed of a myriad of thin basaltic lava flows, each averaging a few meters thick, intermixed with relatively minor amounts of pyroclastic material. Mauna Loa is the largest of five overlapping shield volcanoes that comprise the Big Island of Hawaii (see Figure 5.12). From its base on the floor of the Pacific Ocean to its summit, Mauna Loa is over 9 kilometers (6 miles) high, exceeding the height of Mount Everest. The volume of material composing Mauna Loa is roughly 200 times greater than that of the large composite cone Mount Rainier, located in Washington (**Figure 5.13**).

Like Hawaii's other shield volcanoes, Mauna Loa has flanks with gentle slopes of only a few degrees. This low angle is due to the very hot, fluid lava that traveled fast and far from the vent. In addition, most of the lava (perhaps 80 percent) flowed through a well-developed system of lava tubes. Another feature common to active shield volcanoes is having one or more large, steep-walled calderas that occupy the summit (see Figure 5.12). Calderas on shield volcanoes usually form when the roof above the magma chamber collapses. This occurs after the magma reservoir empties, either following a large eruption or as magma migrates to the flank of a volcano to feed a fissure eruption.

In their final stage of growth, shield volcanoes tend to erupt more sporadically, with pyroclastic ejections being more common. The lava is also more viscous, resulting in thicker, shorter flows. These eruptions steepen the slope of the summit area, which often becomes capped with clusters of cinder cones. This explains why Mauna Kea, a more mature volcano that has not erupted in historic times, has a steeper summit than Mauna Loa, which has erupted continuously since 1984. Scientists are so certain that Mauna Kea is "over the hill" that astronomers built an elaborate observatory on its summit to house some of the world's most advanced telescopes.

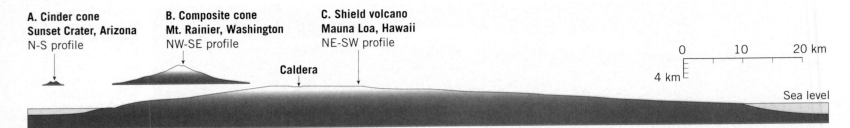

A. Cinder cone
Sunset Crater, Arizona
N-S profile

B. Composite cone
Mt. Rainier, Washington
NW-SE profile

C. Shield volcano
Mauna Loa, Hawaii
NE-SW profile

Caldera

0   10   20 km

4 km

Sea level

▲ **SmartFigure 5.13**
Comparing scales of different volcanoes
A. Profile of Sunset Crater, Arizona, a typical steep-sided cinder cone.
B. Profile of Mount Rainier, Washington. Note how it dwarfs a typical cinder cone.
C. Profile of Mauna Loa, the largest shield volcano in the Hawaiian chain.

Animation
https://goo.gl/M6aeGE

Kilauea: Hawaii's Most Active Volcano

Volcanic activity on the Big Island of Hawaii began on what is now the northwestern flank of the island and has gradually migrated southeastward. It is currently centered on Kilauea Volcano, one of the most active and intensely studied shield volcanoes in the world. Located in the shadow of Mauna Loa, Kilauea has experienced more than 60 eruptions since record keeping began in 1823.

Several months before each eruptive phase, Kilauea inflates as magma gradually migrates upward and accumulates in a central reservoir, located a few kilometers below the summit. For up to 24 hours before an eruption, swarms of small earthquakes warn of the impending activity. Most of the recent activity on Kilauea has occurred along the flanks of the volcano, in a region called the *East Rift Zone*. The longest and largest rift eruption ever recorded on Kilauea began in January 1983 and continues today, with no signs of abating (see **GEOgraphics 5.1**, page 147). Since this eruptive phase began, lava flows have added over 570 acres of new land and paved over 40 square miles of existing land that included many historical sites and several communities.

One of the most destructive phases of this eruption began on May 3, 2018, when multiple fissure eruptions sent lava flowing into the streets of a nearby neighborhood. Throughout the following month, this eruption produced four large channelized flows that marched toward the sea. The first of these flows heavily damaged the community of Leilani Estates. In June, a river of lava engulfed two seaside subdivisions—Vacationland and Kapoho Beach Lots. In total about 600 houses and other structures were destroyed. Lava from this flow eventually filled Kapoho Bay, a popular tourist spot with tidal pools and black sand beaches.

CONCEPT CHECKS 5.5

1. Describe the composition and viscosity of the lava associated with shield volcanoes.

2. Are pyroclastic materials a significant component of shield volcanoes?

Concept Checker
https://goo.gl/C55w6e

3. Where do most shield volcanoes form—on the ocean floor or on the continents?

5.6 Cinder Cones

Describe the formation, size, and composition of cinder cones.

Cinder cones (also called **scoria cones**) are built from ejected basaltic lava fragments that begin to harden in flight, producing the vesicular rock *scoria* (**Figure 5.14**). These pyroclastic fragments range in size from fine ash to bombs that may exceed 1 meter (3 feet) in diameter. However, most of the volume of a cinder cone consists of pea- to walnut-sized fragments that are markedly vesicular and have a black to reddish-brown color (see Figure 5.10A).

Although cinder cones are composed mostly of loose scoria fragments, some produce extensive lava fields. These lava flows generally form in the final stages of the volcano's life span, when the magma body has lost most of its gas content. Because cinder cones are composed of loose fragments rather than solid rock, the lava usually flows out from the unconsolidated base of the cone rather than from the crater. Due to

their unconsolidated nature, cinder cones more readily weather and erode than do other types of volcanoes.

Cinder cones have very simple, distinct shapes (see Figure 5.14). Because cinder cones have a high angle of repose (the steepest angle at which a pile of loose material remains stable), these volcanoes are steep-sided, having slopes between 30 and 40 degrees. In addition, a

Kilauea's East Rift Eruption

Kilauea, one of **the most active volcanoes in the world,** is located on the island of **Hawaii** in the shadow of **Mauna Loa. Most** of the recent activity on Kilauea **has occurred along the flanks of the** volcano in a region called the **East Rift Zone. The longest and largest** eruption ever recorded on **Kilauea began in 1983, and continues with** no signs of abating.

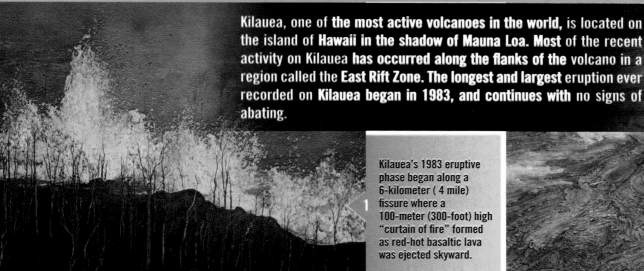

1 Kilauea's 1983 eruptive phase began along a 6-kilometer (4 mile) fissure where a 100-meter (300-foot) high "curtain of fire" formed as red-hot basaltic lava was ejected skyward.

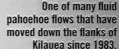

2 One of many fluid pahoehoe flows that have moved down the flanks of Kilauea since 1983.

3 The activity became localized at a single vent and a series of 44 short-lived episodes of lava fountaining built a cinder and spatter cone—given the Hawaiian name *Puu Oo*.

4 One of the most destructive phases of this eruption began on May 3, 2018 when multiple fissure eruptions sent lava flowing into the streets of a nearby neighborhood. Throughout the following month this eruption produced four large channelized flows that marched toward the sea, engulfing three subdivisions.

▶ SmartFigure 5.14
Cinder cones Cinder cones are built from ejected lava fragments (mostly cinders and bombs) and are relatively small—usually less than 300 meters (1000 feet) in height.

Mobile Field Trip
https://goo.gl/an3tkG

Lava flow

Crater

Pyroclastic material

SP Crater is a classic cinder cone located north of Flagstaff, Arizona.

Central vent filled with rock fragments

cinder cone has a large, deep crater relative to the over-all size of the structure. Although they tend to be rela-tively symmetrical, some cinder cones are elongated and higher on the side that was downwind during the final eruptive phase.

Most cinder cones are produced by a single, short-lived eruptive event. One study found that half of all cinder cones examined were constructed in less than 1 month, and 95 percent of them formed in less than 1 year. Once the event ceases, the magma in the con-duits—or "plumbing" connecting the vent to the magma source—solidifies, and the volcano usually does not

erupt again. One exception is Cerro Negro, a cinder cone in Nicaragua, which has erupted more than 20 times since it formed in 1850. Cinder cones tend to be small, usually between 30 and 300 meters (100 and 1000 feet) tall, due to their typically short life span. A few, like Cerro Negro, exceed 700 meters (2300 feet) in height.

Cinder cones number in the thousands around the globe. Some occur in groups, such as the volcanic field near Flagstaff, Arizona, which consists of about 600 cones. Others are parasitic cones that are found on the flanks or within the calderas of larger volcanic structures.

Parícutin, a cinder cone located in Mexico, erupted for 9 years.

An aa flow emanating from the base of the cone buried much of the village of San Juan Parangaricutiro, leaving only remnants of the village's church exposed.

▶ SmartFigure 5.15
Parícutin, a well-known cinder cone The village of San Juan Parangaricutiro was engulfed by aa lava from Parícutin. Only por-tions of the church remain.

Condor Video
https://goo.gl/ioX6gL

Parícutin: Life of a Garden-Variety Cinder Cone

One of the very few volcanoes studied by geologists from its very beginning is the cinder cone called Parícutin, located about 320 kilometers (200 miles) west of Mexico City. In 1943, its eruptive phase began in a cornfield owned by Dionisio Pulido, who witnessed the event.

Numerous tremors caused apprehension in the nearby village of Parícutin for weeks prior to the first eruption. Then, on February 20, sulfurous gases billowed from a small depression that had been in the cornfield for as long as local residents could remember. During the night, hot, glowing rock fragments were ejected from the vent, producing a spectacular fireworks display. Explosive discharges continued, throwing hot fragments and ash occasionally as high as 6000 meters (20,000 feet) into the air. Larger fragments fell near the crater, some remaining incandescent as they rolled down the slope. These materials built an aesthetically pleasing cone, while finer ash fell over a much larger area, burning and eventually covering the village of Parícutin. In the first day, the cone grew to 40 meters (130 feet), and by the fifth day it was more than 100 meters (330 feet) high.

The first lava flow came from a fissure that opened just north of the cone, but after a few months, flows emerged from the base of the cone. In June 1944, a clinkery aa flow 10 meters (30 feet) thick moved over much of the village of San Juan Parangaricutiro, leaving only remnants of the church exposed (**Figure 5.15**). After 9 years of intermittent pyroclastic explosions and nearly continuous discharge of lava from vents at its base, the activity ceased almost as quickly as it began. Today, Parícutin is just another one of the scores of cinder cones dotting the landscape in this region of Mexico. Like the others, it will not erupt again.

CONCEPT CHECKS 5.6

1. Describe the composition of a cinder cone.

2. How do cinder cones compare with shield volcanoes in terms of size and the steepness of their flanks?

Concept Checker
https://goo.gl/MbeRTS

5.7 Composite Volcanoes

List the characteristics of composite volcanoes and describe how they form.

Earth's most picturesque yet potentially dangerous volcanoes are **composite volcanoes**, also known as **stratovolcanoes**. Most are located in a relatively narrow zone that rims the Pacific Ocean, appropriately called the *Ring of Fire* (see Figure 5.28). This active zone includes a chain of continental volcanoes distributed along the west coast of the Americas, including the large cones of the Andes in South America and the Cascade Range of the western United States and Canada.

Classic composite cones are large, nearly symmetrical structures consisting of layers of explosively erupted cinders and ash interbedded with lava flows. Just as shield volcanoes owe their shape to fluid mafic (basaltic) lavas, composite cones reflect the viscous nature of the material from which they are made. In general, composite cones are the product of silica-rich magma having an andesitic (intermediate) composition. However, many composite cones also emit various amounts of fluid basaltic lava and, occasionally, pyroclastic material having a felsic (rhyolitic) composition. The andesitic magmas typical of composite cones generate thick, viscous lavas that travel less than a few kilometers. Composite cones are also noted for generating explosive eruptions that eject huge quantities of pyroclastic material.

A conical shape, with a steep summit area and gradually sloping flanks, is typical of most large composite cones. This classic profile, which adorns calendars and postcards, is partially a result of the way viscous lavas and pyroclastic ejected materials contribute to the cone's growth. Coarse fragments ejected from the summit crater tend to accumulate near their source and contribute to the steep slopes around the summit. Finer ejected materials, on the other hand, are deposited as a thin layer over a large area and hence tend to flatten the flank of the cone. In addition, during the early stages of growth, lavas tend to be more abundant and flow greater distances from the vent, which contributes to the cone's broad base. As a composite volcano matures, the shorter flows that come from the central vent serve to armor and strengthen the summit area. Consequently, steep slopes exceeding 40 degrees are possible. Two of the most perfect cones—Mount Mayon in the Philippines and Fujiyama in Japan—exhibit the classic form we expect of composite cones, with steep summits and gently sloping flanks (**Figure 5.16**).

Despite the symmetrical forms of many composite cones, most have complex histories. Many composite volcanoes have secondary vents on their flanks that have produced cinder cones or even much larger volcanic structures. Huge mounds of volcanic debris surrounding these structures provide evidence that in the past, large sections of these volcanoes slid downslope as massive landslides. Some develop

▲ Figure 5.16
Fujiyama, a classic composite volcano Japan's Fujiyama exhibits the classic form of a composite cone, with a steep summit and gently sloping flanks.

amphitheater-shaped depressions at their summits as a result of explosive lateral eruptions—as occurred during the 1980 eruption of Mount St. Helens. Often, so much rebuilding occurs after these eruptions that eventually no trace of these amphitheater-shaped scars remain. Other stratovolcanoes, such as Crater Lake in Oregon, have been truncated by the collapse of their summit (see Figure 5.22).

CONCEPT CHECKS 5.7

1. What name is given to the region having the greatest concentration of composite volcanoes?

2. Describe the materials that compose composite volcanoes.
https://goo.gl/jxKAFa

3. How do the composition and viscosity of lava flows differ between composite volcanoes and shield volcanoes?

5.8 Volcanic Hazards

Describe the major geologic hazards associated with volcanoes.

Roughly 1500 of Earth's known volcanoes have erupted at least once, and some several times, in the past 10,000 years. Based on historical records and studies of active volcanoes, 70 volcanic eruptions can be expected each year. In addition, 1 large-volume eruption can be expected every decade. These large eruptions account for the vast majority of volcano-related human fatalities.

Pyroclastic flows

A.

B.

◀ **Figure 5.17**
Pyroclastic flows, some of the most destructive volcanic forces A. These pyroclastic flows occurred on Mount Mayon, Philippines. Pyroclastic flows are composed of hot ash and pumice and/or blocky lava fragments that race down the slopes of volcanoes.
B. Residents running away from a pyroclastic flow that reached the base of Mount Sinabung, Indonesia, in 2014.

Today, an estimated 500 million people in places such as Japan, Indonesia, Italy, and Oregon live near active volcanoes. They face a number of volcanic hazards, such as destructive pyroclastic flows, molten lava flows, mudflows called lahars, and falling ash and volcanic bombs.

Pyroclastic Flows: A Deadly Force

One of the most destructive volcanic hazards is **pyroclastic flow**, consisting of hot gases infused with incandescent ash and larger lava fragments. Also known as **nuée ardentes** ("glowing avalanches"), these fiery flows can race down steep volcanic slopes at speeds exceeding 100 kilometers (60 miles) per hour (**Figure 5.17**). Pyroclastic flows have two components: a low-density cloud of hot expanding gases containing fine ash particles and a ground-hugging portion composed of pumice and other vesicular pyroclastic material.

Driven by Gravity   Pyroclastic flows are propelled by the force of gravity and tend to move in a manner similar to snow avalanches. They are mobilized by expanding volcanic gases released from the lava fragments and by the expansion of heated air that is overtaken and trapped in the moving front. These gases reduce friction between ash and pumice fragments, which gravity propels downslope in a nearly frictionless environment. This is why some pyroclastic flow deposits are found many miles from their source.

Occasionally, powerful hot blasts that carry small amounts of ash separate from the main body of a pyroclastic flow. These low-density clouds, called *surges*, can be deadly but seldom have sufficient force to destroy buildings in their paths. Nevertheless, in 2014, a hot ash cloud from Japan's Mount Ontake killed 47 hikers and injured 69 more.

Pyroclastic flows may originate in a variety of volcanic settings. Some occur when a powerful eruption blasts pyroclastic material out of the side of a volcano.

More frequently, however, pyroclastic flows are generated by the collapse of tall eruption columns during an explosive event. When gravity eventually overcomes the initial upward thrust provided by the escaping gases, the ejected materials begin to fall, sending massive amounts of incandescent blocks, ash, and pumice cascading downslope.

The Destruction of St. Pierre   In 1902, an infamous pyroclastic flow and associated surge from Mount Pelée, a small volcano on the Caribbean island of Martinique, destroyed the port town of St. Pierre. Although the main pyroclastic flow was largely confined to the valley of Rivière Blanche, a low-density fiery surge spread south of the river and quickly engulfed the entire city. The destruction happened in moments and was so devastating that nearly all of St. Pierre's 28,000 inhabitants were killed. At least 1 person on the outskirts of town—a prisoner whose dungeon provided protection—and a few people on ships in the harbor were spared (**Figure 5.18**).

Scientists who arrived on the scene within days of the disaster reported that although St. Pierre was mantled by only a thin layer of volcanic debris, masonry walls nearly 1 meter (3 feet) thick had been knocked over like dominoes, large trees had been uprooted, and cannons had been torn from their mounts.

The Destruction of Pompeii   Another well-documented event of historic proportions was the C.E. 79 eruption of the Italian volcano we now call Mount Vesuvius. For centuries prior to this eruption, Vesuvius had been dormant, with vineyards adorning its sunny slopes.

► Figure 5.18
Destruction of St. Pierre
A. St. Pierre as it appeared shortly after the eruption of Mount Pelée in 1902.
B. St. Pierre before the eruption. Many vessels were anchored offshore when this photo was taken, as was the case on the day of the eruption.

B. **St. Pierre before the 1902 eruption.**

A. **St. Pierre following the eruption of Mount Pelée.**

Yet in less than 24 hours, the entire city of Pompeii (near Naples) and a few thousand of its residents were entombed beneath a layer of volcanic ash and pumice. The city and the victims of the eruption remained buried and forgotten for nearly 17 centuries. The excavation of Pompeii gave archaeologists a superb picture of ancient Roman life (**Figure 5.19A**).

By reconciling historical records with detailed scientific studies of the region, volcanologists reconstructed the sequence of events. During the first day of the eruption, a rain of ash and pumice accumulated at a rate of 12 to 15 centimeters (5 to 6 inches) per hour, causing most of the roofs in Pompeii to eventually give way. Then, suddenly, a surge of searing hot ash and gas swept rapidly down the flanks of Vesuvius. This deadly pyroclastic flow killed those who had somehow managed to survive the initial ash and pumice fall. Their remains were quickly buried by falling ash, and subsequent rainfall caused the ash to harden. Over the centuries, the remains decomposed, creating cavities that were discovered by nineteenth-century excavators. Casts were then produced by pouring plaster of Paris into the voids (**Figure 5.19B**). Mount Vesuvius has had more than two dozen explosive eruptions since C.E. 79, the most recent occurring in 1944. Today, Vesuvius towers over the Naples skyline, a region occupied by roughly 3 million people. Such an image should prompt us to consider how volcanic crises might be managed in the future.

Lahars: Mudflows on Active and Inactive Cones

In addition to producing violent eruptions, large composite cones may generate a type of fluid mudflow known by its Indonesian name, **lahar**. These destructive flows occur when volcanic

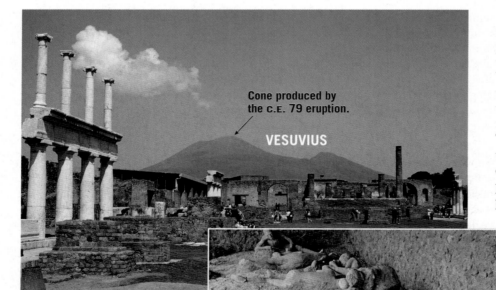

Cone produced by the C.E. 79 eruption.

VESUVIUS

A.

B.

► Figure 5.19
Pompeii's destruction from the C.E. 79 eruption of Mount Vesuvius A. The ruins of the Roman city of Pompeii as they appear today. In less than 24 hours, Pompeii and all its residents were buried under a layer of volcanic ash and pumice that fell like rain.
B. Plaster casts of Pompeii eruption victims are on display at the archeological site.

debris becomes saturated with water and rapidly moves down steep volcanic slopes, generally following stream valleys. Some lahars are triggered when rising magma nears the surface of a glacially clad volcano, causing large volumes of ice and snow to melt. Others happen when heavy rains saturate weathered volcanic deposits. Thus, lahars may occur even when a volcano is *not* erupting.

When Mount St. Helens erupted in 1980, several lahars were generated. These flows and accompanying floodwaters raced down nearby river valleys at speeds exceeding 30 kilometers (20 miles) per hour. These raging rivers of mud destroyed or severely damaged nearly all the homes and bridges along their paths (**Figure 5.20**). Fortunately, the area was not densely populated.

In 1985, another deadly lahar was triggered during a small eruption of Nevado del Ruiz, a 5300-meter (17,400-foot) volcano in the Andes Mountains of Colombia. Hot pyroclastic material melted ice and snow that capped the mountain (*nevado* means "snowy" in Spanish) and sent torrents of ash and debris down three major river valleys that flank the volcano. Reaching speeds of 100 kilometers (60 miles) per hour, these mudflows tragically claimed 25,000 lives.

Many consider Mount Rainier, Washington, to be America's most dangerous volcano because, like Nevado del Ruiz, it has a thick, year-round mantle of snow and glacial ice. Adding to the risk is the fact that more than 100,000 people live in the valleys around Rainier, and many homes are built on deposits left by lahars that flowed down the volcano hundreds or thousands of years ago. A future eruption, or perhaps just a period of heavier-than-average rainfall, may produce lahars that could be similarly destructive.

Other Volcanic Hazards

Volcanoes can be hazardous to human health and property in other ways. Ash and other pyroclastic material can collapse the roofs of buildings or may be drawn into the lungs of humans and other animals or into aircraft engines (**Figure 5.21**). Volcanic gases, most notably sulfur dioxide, pollute the air and, when mixed with rainwater, can destroy vegetation and reduce the quality of groundwater. Despite the known risks, millions of people live in close proximity to active volcanoes.

Volcano-Related Tsunamis

Although **tsunamis** are most often associated with displacement along a fault located on the seafloor (see Chapter 11), some result from the collapse of a volcanic cone. This was dramatically demonstrated during the 1883 eruption on the Indonesian island of Krakatau, when the northern half of a volcano plunged into the Sunda Strait, creating a tsunami that exceeded 30 meters (100 feet) in height. Although Krakatau was uninhabited, an estimated 36,000 people were killed along the coastline of the islands of Java and Sumatra.

Volcanic Ash and Aviation

As described in this chapter's *In the News*, commercial jets have been damaged by inadvertently flying into clouds of volcanic ash. The 2010 eruption of Iceland's Eyjafjallajökull Volcano sent ash high into the atmosphere. This thick plume of ash drifted over Europe, causing airlines all across Europe to cancel thousands of flights and leaving hundreds of thousands of travelers stranded. Several weeks passed before air travel resumed its normal schedule.

A.

B.

◀ Figure 5.20
Lahars, mudflows that originate on volcanic slopes
A. This lahar raced down the snow-covered slopes of Mount St. Helens following an eruption on March 19, 1982.
B. The aftermath of a lahar that formed following the 1982 eruption of Galunggung Volcano in Indonesia.

▶ **Figure 5.21**
Volcanic hazards In addition to generating destructive pyroclastic flows and lahars, volcanoes can be hazardous to human health and property in many other ways.

Ash and other pyroclastic materials can collapse roofs, or completely cover buildings.

Lava flows can destroy homes, roads, and other structures in their paths.

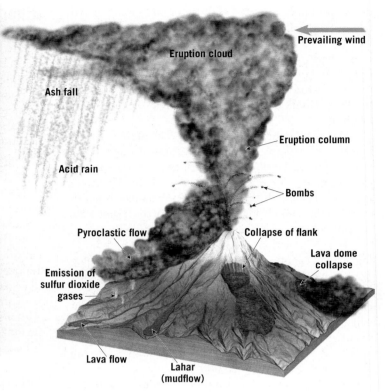

Prevailing wind
Eruption cloud
Ash fall
Acid rain
Eruption column
Bombs
Pyroclastic flow
Collapse of flank
Lava dome collapse
Emission of sulfur dioxide gases
Lava flow
Lahar (mudflow)

Volcanic Gases and Respiratory Health One of the most destructive volcanic events, called the Laki eruptions, began in 1783 along a large fissure in southern Iceland. An estimated 14 cubic kilometers (3.4 cubic miles) of fluid basaltic lavas were released, along with 130 million tons of sulfur dioxide and other poisonous gases. When sulfur dioxide is inhaled, it reacts with moisture in the lungs to produce sulfuric acid, a deadly toxin. More than half of Iceland's livestock died, and the ensuing famine killed 25 percent of the island's human population.

This huge eruption also endangered people and property all across Europe. Crop failure occurred in parts of Western Europe, and thousands of residents perished from lung-related diseases. One report estimated that a similar eruption today would cause more than 140,000 cardiopulmonary fatalities in Europe alone.

Effects of Volcanic Ash and Gases on Weather and Climate Volcanic eruptions can eject dust-sized particles of volcanic ash and sulfur dioxide gas high into the atmosphere. The ash particles reflect sunlight back to space, producing temporary atmospheric cooling. The 1783 Laki eruptions in Iceland appear to have affected atmospheric circulation around the globe. Drought

conditions prevailed in the Nile River valley, and the winter of 1784 saw the longest period of below-zero temperatures in New England's history.

Other eruptions that have produced significant effects on climate worldwide include the eruption of Indonesia's Mount Tambora in 1815, which produced the "year without a summer" (1816), and the eruption of El Chichón in Mexico in 1982. El Chichón's eruption, although small, emitted an unusually large quantity of sulfur dioxide that reacted with water vapor in the atmosphere to produce a dense cloud of tiny sulfuric acid droplets. Such particles, called *aerosols*, take several years to settle out of the atmosphere. Like fine ash, these aerosols lower the mean temperature of the atmosphere by reflecting solar radiation back to space.

CONCEPT CHECKS 5.8

1. Describe pyroclastic flows and explain why they are capable of traveling great distances.

2. What is a lahar?

3. List at least three volcanic hazards besides pyroclastic flows and lahars.

 Concept Checker
https://goo.gl/wwGZRU

5.9 Other Volcanic Landforms

List volcanic landforms other than shield, cinder, and composite volcanoes and describe their formation.

The most widely recognized volcanic structures are the cone-shaped edifices of composite volcanoes that dot Earth's surface. However, volcanic activity also produces other distinctive landforms.

Calderas

Recall that *calderas* are large steep-sided depressions that have diameters exceeding 1 kilometer (0.6 miles) and have a somewhat circular form. Those less than 1 kilometer across are called *collapse pits* or *craters*. Most calderas are formed by one of the following processes: (1) the collapse of the summit of a large composite volcano following an explosive eruption of silica-rich pumice and ash fragments (*Crater Lake–type calderas*); (2) the collapse of the top of a shield volcano caused by subterranean drainage from a central magma chamber (*Hawaiian-type calderas*); or (3) the collapse of a large area caused by the discharge of colossal volumes of silica-rich pumice and ash along ring fractures (*Yellowstone-type calderas*).

Crater Lake–Type Calderas
Crater Lake, Oregon, is situated in a caldera approximately 10 kilometers (6 miles) wide and 600 meters (more than 1970 feet) deep. This caldera formed about 7000 years ago, when a composite cone named Mount Mazama violently extruded 50 to 70 cubic kilometers of pyroclastic material (**Figure 5.22**). The eruption compromised the volcano's structure, and 1500 meters (nearly 1 mile) of the summit of this once-prominent cone collapsed, producing a caldera that eventually filled with water. Later, volcanic activity built a small cinder cone in the caldera. Today this cinder cone, called Wizard Island, provides a mute reminder of past activity.

Hawaiian-Type Calderas
Many calderas form gradually because of the loss of lava from a shallow magma chamber underlying a volcano's summit. For example, Hawaii's active shield volcanoes, Mauna Loa and Kilauea, both have large calderas at their summits. Kilauea's measures 3.3 by 4.4 kilometers (about 2 by 3 miles) and is 150 meters (500 feet) deep. The walls are almost vertical, and as a result, the caldera looks like a vast, nearly flat-bottomed pit. Kilauea's caldera formed through gradual subsidence as magma slowly drained laterally from the underlying magma chamber, leaving the summit unsupported.

Yellowstone-Type Calderas
All historic volcanic eruptions pale in comparison to what happened 630,000 years ago in the region now occupied by Yellowstone National Park. Back then, a catastrophic eruption moved 1000 cubic kilometers of material, sending showers of ash as far as the Gulf of Mexico. The eruption formed a caldera 70 kilometers (43 miles) across (**Figure 5.23A**). Vestiges of this event are the many hot springs and geysers in the Yellowstone region.

Yellowstone-type eruptions eject huge volumes of pyroclastic materials, mainly in the form of ash and pumice fragments. Upon coming to rest, the hot fragments of ash and pumice fuse together, forming a welded tuff that closely resembles a solidified lava flow. Despite the immense size of these calderas, the eruptions that produce them are brief, lasting hours to perhaps a few days.

▼ **SmartFigure 5.22**
Formation of Crater Lake–type calderas About 7000 years ago, a violent eruption partly emptied the magma chamber of former Mount Mazama, causing its summit to collapse. Precipitation and groundwater contributed to forming Crater Lake, the deepest lake in the United States—594 meters (1949 feet) deep—and the ninth-deepest lake in the world.

Animation
https://goo.gl/jr7kzw

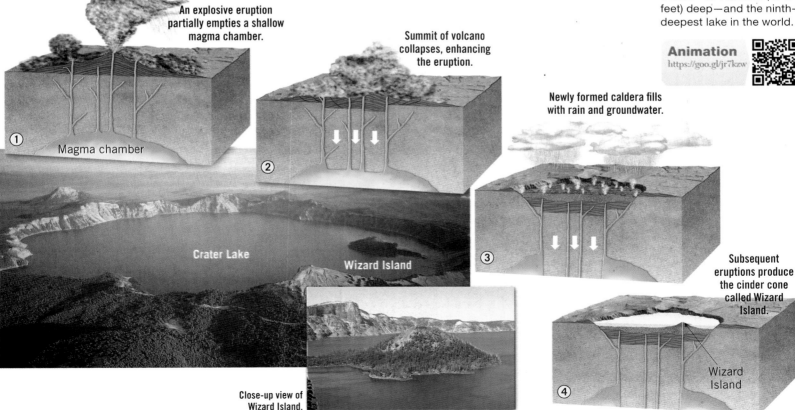

An explosive eruption partially empties a shallow magma chamber.

① Magma chamber

Summit of volcano collapses, enhancing the eruption.

②

Newly formed caldera fills with rain and groundwater.

③

Crater Lake

Wizard Island

Subsequent eruptions produce the cinder cone called Wizard Island.

④ Wizard Island

Close-up view of Wizard Island.

Mountains of southern Colorado; and the Valles Caldera, west of Los Alamos, New Mexico. These and similar calderas found around the globe are among the largest volcanic structures on Earth and hence are called *supervolcanoes*. Volcanologists compare their destructive force to that of the impact of a small asteroid. Fortunately, no Yellowstone-type eruption has occurred in historic times.

Fissure Eruptions and Basalt Plateaus

The greatest volume of volcanic material is extruded from fractures in Earth's crust, called *fissures*. Rather than building cones, **fissure eruptions** usually emit fluid basaltic lavas that blanket wide areas (**Figure 5.24**). In some locations, extraordinary amounts of lava have been extruded along fissures in a relatively short time, geologically speaking. These voluminous accumulations are commonly called **basalt plateaus** because most have a basaltic composition, and they tend to be rather flat

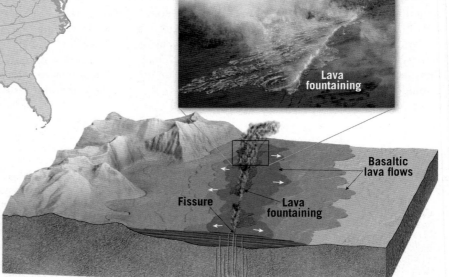

These large calderas tend to exhibit a complex eruptive history. In the Yellowstone region, for example, three caldera-forming episodes are known to have occurred over the past 2.1 million years (**Figure 5.23B**). The most recent eruption (630,000 years ago) was followed by episodic outpourings of degassed rhyolitic and basaltic lavas. In the intervening years, a slow upheaval of the floor of the caldera produced two elevated regions called *resurgent domes* (see Figure 5.23A). A recent study has determined that a huge magma reservoir still exists beneath Yellowstone; thus, another caldera-forming eruption is likely—but not necessarily imminent.

Yellowstone-type calderas are so vast and poorly defined that many were undetected until high-quality aerial and satellite images became available. Other examples of Yellowstone-type calderas are California's Long Valley Caldera; LaGarita Caldera in the San Juan

▲ Figure 5.24
Basaltic fissure eruptions Lava fountaining from a fissure and formation of fluid lava flows called *flood basalts*. The lower photo shows flood basalt flows near Idaho Falls.

and broad. The Columbia Plateau in the northwestern United States, which consists of the Columbia River Basalts, is a product of this type of activity (**Figure 5.25**). Numerous fissure eruptions have buried the landscape, creating a lava plateau nearly 1500 meters (1 mile) thick. Some of the lava remained molten long enough to flow 150 kilometers (90 miles) from its source. The term **flood basalts** appropriately describes these extrusions.

Massive accumulations of basaltic lava have also occurred elsewhere in the world. One of the largest is known as the Deccan Plateau or Deccan Traps (*traps* = stairs), a thick sequence of flat-lying basalt flows covering nearly 500,000 square kilometers (195,000 square miles) of west-central India. When the Deccan Traps formed about 66 million years ago, nearly 2 million cubic kilometers of lava were extruded over a period of approximately 1 million years. Several other massive accumulations of flood basalts, including the Ontong Java Plateau, have been discovered in the deep-ocean basins (see Figure 5.31).

KEY

Columbia River Basalts

Other basaltic rocks

▲ Large Cascade volcanoes

WASHINGTON

Cascade Range

Columbia River Basalts

Yellowstone National Park

Snake River Plain

OREGON   IDAHO

A.

The Palouse River in Washington State has cut a canyon about 300 meters (1000 feet) deep into the flood basalts of the Columbia Plateau.

B.

◀ **Figure 5.25**
Columbia River Basalts
A. The Columbia River Basalts cover an area of nearly 164,000 square kilometers (63,000 square miles) that is commonly called the Columbia Plateau. Activity here began about 17 million years ago, as lava began to pour out of large fissures, eventually producing a basalt plateau with an average thickness of more than 1 kilometer.
B. Basalt flows exposed in the Palouse River Canyon in southwestern Washington State.

eruption of silica-rich (andesitic or rhyolitic) magma. A recent example is the dome that began to grow in the crater of Mount St. Helens immediately following the 1980 eruption (**Figure 5.26A**). If this activity were to continue over hundreds of years, the dome could fill the amphitheater-shaped structure on the north slope of the volcano.

The collapse of lava domes, particularly those that form on the summits or along the steep flanks of composite cones, often produce powerful pyroclastic flows

Lava Domes

In contrast to hot basaltic lavas, relatively cool silica-rich lavas are so viscous that they hardly flow at all. As the thick lava is "squeezed" out of a vent, it often produces a dome-shaped mass called a **lava dome**. Lava domes are usually only a few tens of meters high, and they come in a variety of shapes that range from pancake-like flows to steep-sided plugs that were pushed upward like pistons. Most lava domes grow over a period of several years, following an explosive

Lava dome

A.

Lava domes are produced when highly viscous magma slowly rises over a period of months or years.

When a growing lava dome becomes over-steepened, it may collapse producing a blocky pyroclastic flow.

Blocky pyroclastic flow

Decompression of the interior magma may produce an explosive eruption and pyroclastic flow.

Fiery pyroclastic flow

B.

◀ **Figure 5.26**
Lava domes can generate pyroclastic flows
A. This lava dome began to develop in the vent of Mount St. Helens following the May 1980 eruption.
B. The collapse of a lava dome often results in a powerful pyroclastic flow.

▶ **SmartFigure 5.27**
Volcanic neck Shiprock, New Mexico, is a volcanic neck that stands about 520 meters (1700 feet) high. It consists of igneous rock that crystallized in the vent of a volcano that has long since eroded.

Tutorial
https://goo.gl/SBvY8j

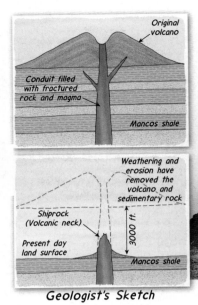

Geologist's Sketch

victims were journalists and filmmakers who ventured too close to the volcano in order to obtain photographs and document the event.

Volcanic Necks

Most of the lava and materials erupted from a volcano travel through conduits that connect shallow magma chambers to vents located at the surface. When a volcano becomes inactive, congealed magma is often preserved in the feeding conduit of the volcano as a crudely cylindrical mass. As the volcano succumbs to weathering and erosion, the rock occupying the volcanic conduit, which is highly resistant to weathering, may remain standing above the surrounding terrain long after the cone wears away. Shiprock, New Mexico, is a widely recognized and spectacular example of these structures, which geologists call **volcanic necks**, or **plugs** (Figure 5.27). More than 510 meters (1700 feet) high, Shiprock is taller than most skyscrapers and is one of many such landforms that protrude conspicuously from the red desert landscapes of the American Southwest.

(Figure 5.26B). These flows result from highly viscous magma slowly entering the dome, causing it to expand and steepen its flanks. Over time, the cooler outer layer of the dome may start to crumble, producing relatively small pyroclastic flows consisting of dense blocks of lava. Occasionally, rapid removal of the outer layer causes a significant decrease in pressure on the dome's interior, leading to explosive degassing of the interior magma that, in turn, triggers a fiery pyroclastic flow.

Since 1995, pyroclastic flows generated by the collapse of several lava domes on the Soufriére Hills Volcano have rendered more than half of the Caribbean island of Montserrat uninhabitable. The capital city, Plymouth, was destroyed, and two-thirds of the population has evacuated. In the early 1990s, a collapsed lava dome at the summit of Japan's Mount Unzen produced a pyroclastic flow that claimed 42 lives. Many of the

CONCEPT CHECKS 5.9

1. Describe the formation of Crater Lake.

2. How do the eruptions that created the Columbia Plateau differ from the eruptions that create large composite volcanoes?

   Concept Checker
   https://goo.gl/jgJuHR

3. What type of volcanic structure is Shiprock, New Mexico, and how did it form?

5.10 Plate Tectonics and Volcanism

Explain how the global distribution of volcanic activity relates to plate tectonics.

Geologists have known for decades that the global distribution of most of Earth's volcanoes is not random. Most active volcanoes on land are located along the margins of the ocean basins—notably within the circum-Pacific belt known as the **Ring of Fire** (Figure 5.28), where denser oceanic lithosphere subducts under continental lithosphere. Another group of volcanoes includes the innumerable seamounts that form along the crest of the mid-ocean ridges. Some volcanoes, however, appear to be randomly distributed around the globe. These volcanic structures comprise most of the islands of the deep-ocean basins, including the Hawaiian Islands, the Galapagos Islands, and Easter Island.

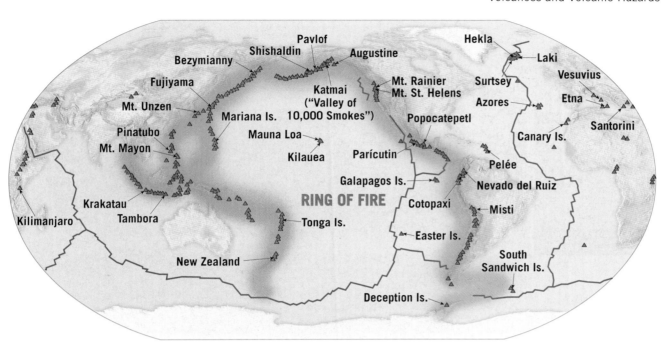

◀ **Figure 5.28**
Ring of Fire Most of Earth's major volcanoes are located in a zone around the Pacific called the Ring of Fire. Another large group of active volcanoes lies unseen along the mid-ocean ridge system.

Setting aside for the moment the volcanoes that form in the deep-ocean basins, the development of the theory of plate tectonics provided geologists with a plausible explanation for the distribution of Earth's volcanoes and established the basic connection between plate tectonics and volcanism.

Volcanism at Divergent Plate Boundaries

The greatest volume of magma erupts along divergent plate boundaries associated with seafloor spreading—out of human sight (**Figure 5.29B**). Below the ridge axis where lithospheric plates are continually being pulled apart, the solid yet mobile mantle rises to fill the rift. Recall from Chapter 4 that as hot rock rises, it experiences a decrease in confining pressure and may undergo *decompression melting*. This activity continuously adds new basaltic rock to plate margins, temporarily welding them together, although they break again as spreading continues. Along some ridge segments, extrusions of pillow lavas build numerous volcanic structures, the largest of which is Iceland.

Although most spreading centers are located along the axis of an oceanic ridge, some are not. The East African Rift is a notable example of a site where continental lithosphere is being pulled apart (see **Figure 5.29F**). This region of the globe is home to vast outpourings of fluid basaltic lavas as well as several active volcanoes.

Plate motion provides the mechanism by which mantle rocks undergo partial melting to generate magma.

Volcanism at Convergent Plate Boundaries

Recall that along convergent plate boundaries, two plates move toward each other, and a slab of dense oceanic lithosphere descends into the mantle. In these settings, water released from hydrated (water-rich) minerals found in the subducting oceanic crust is driven upward and triggers partial melting in the hot mantle above (**Figure 5.29A**).

Volcanism at a convergent plate margin results in the development of a slightly curved chain of volcanoes that develop roughly parallel to the associated trench—at distances of 200 to 300 kilometers (100 to 200 miles). Volcanic arcs that develop within the ocean and grow large enough for their tops to rise above the surface are labeled *archipelagos* in most atlases. Geologists prefer the more descriptive term **volcanic island arcs**, or simply **island arcs** (see Figure 5.29A). Several young volcanic island arcs border the western Pacific basin, including the Aleutians, the Tongas, and the Marianas.

Volcanism associated with convergent plate boundaries may also take place where slabs of oceanic lithosphere are subducted under continental lithosphere to produce a **continental volcanic arc** (**Figure 5.29E**). The mechanisms that generate these mantle-derived magmas are essentially the same as those that create volcanic island arcs. The most significant difference is that continental

A. Convergent Plate Volcanism When an oceanic plate subducts, melting in the mantle produces magma that gives rise to a volcanic island arc on the overlying oceanic crust.

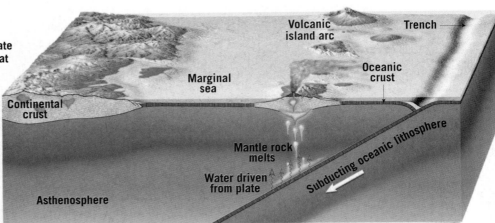

C. Intraplate Volcanism When an oceanic plate moves over a hot spot, a chain of volcanic structures such as the Hawaiian Islands is created.

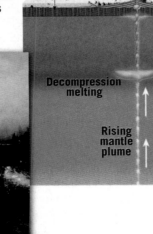

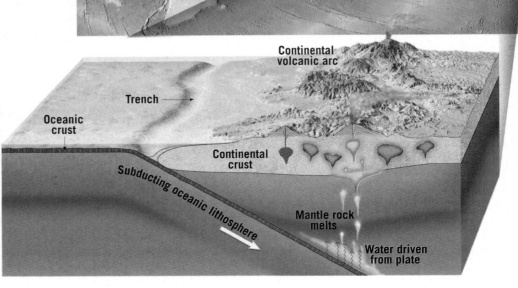

E. Convergent Plate Volcanism When oceanic lithosphere descends beneath a continent, magma generated in the mantle rises to form a continental volcanic arc.

▶ SmartFigure 5.29
Earth's zones of volcanism

Tutorial
https://goo.gl/G51ynb

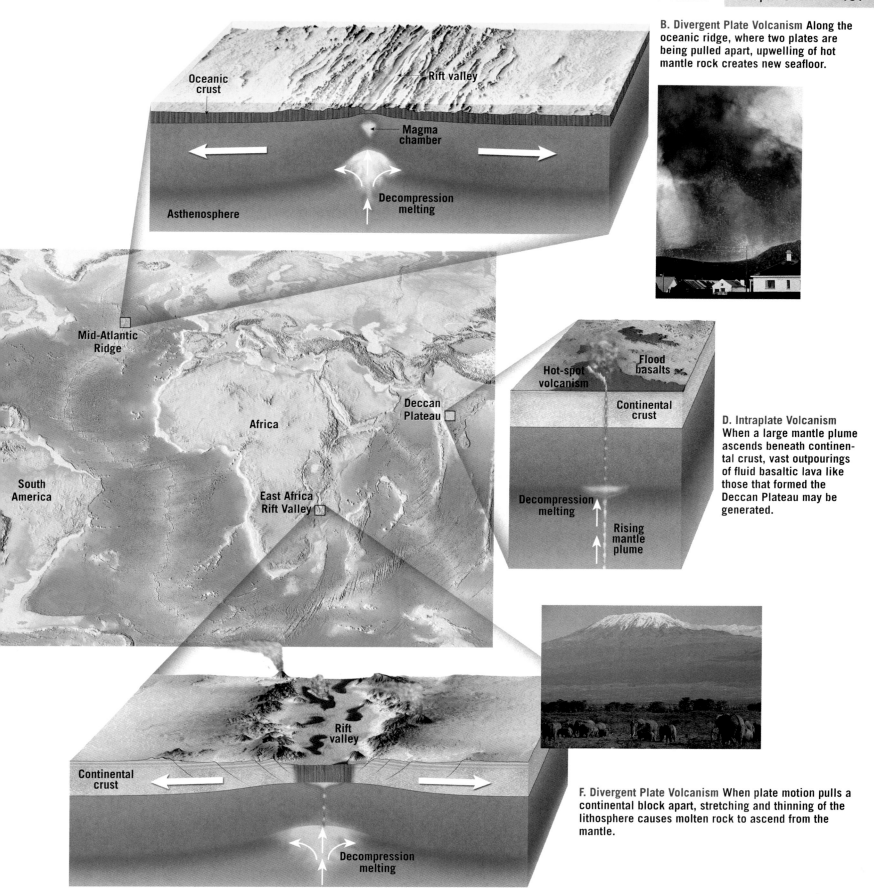

B. Divergent Plate Volcanism Along the oceanic ridge, where two plates are being pulled apart, upwelling of hot mantle rock creates new seafloor.

Oceanic crust

Rift valley

Magma chamber

Decompression melting

Asthenosphere

Mid-Atlantic Ridge

South America

Africa

Deccan Plateau

East Africa Rift Valley

D. Intraplate Volcanism When a large mantle plume ascends beneath continental crust, vast outpourings of fluid basaltic lava like those that formed the Deccan Plateau may be generated.

Hot-spot volcanism

Flood basalts

Continental crust

Decompression melting

Rising mantle plume

Rift valley

Continental crust

Decompression melting

F. Divergent Plate Volcanism When plate motion pulls a continental block apart, stretching and thinning of the lithosphere causes molten rock to ascend from the mantle.

► SmartFigure 5.30
Cascade Range volcanoes Subduction of the Juan de Fuca plate along the Cascadia subduction zone produced the volcanoes of the Cascade Range.

Tutorial
https://goo.gl/
SCYXSq

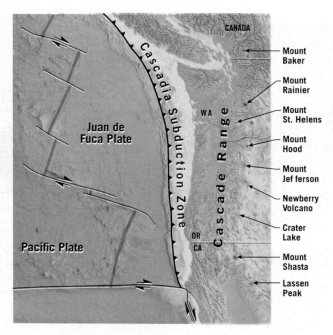

Intraplate Volcanism

We know why igneous activity is initiated along plate boundaries, but why do eruptions occur in the interiors of plates? Hawaii's Kilauea, considered one of the world's most active volcanoes, is situated thousands of kilometers from the nearest plate boundary, in the middle of the vast Pacific plate (**Figure 5.29C**). Sites of **intraplate** (meaning "within the plate") **volcanism** include the large outpourings of fluid basaltic lavas such as those that compose the Columbia Plateau, the Siberian Traps in Russia, and several submerged oceanic plateaus, including the Ontong Java Plateau in the western Pacific (**Figure 5.31**).

One widely accepted hypothesis proposes that most intraplate volcanism occurs when a relatively narrow mass of hot material, called a **mantle plume**, ascends toward the surface (**Figure 5.32A**).° Although the depth at which mantle plumes originate is a topic of debate, some are thought to form deep within Earth, at the core–mantle boundary. These plumes of solid yet mobile rock rise toward the surface in a manner similar to the blobs that form within a lava lamp. A lava lamp contains two immiscible liquids in a glass container; as the base of the lamp is heated, the liquid at the bottom becomes buoyant and forms blobs that rise to the top. Like the blobs in a lava lamp, a mantle plume has a bulbous head that draws out a narrow stalk or tail beneath it as it rises. The

crust is much thicker and composed of rocks having higher silica content than oceanic crust. Hence, by melting the surrounding silica-rich crustal rocks, mantle-derived magma changes composition as it rises through the crust. The volcanoes of the Cascade Range in the northwestern United States, including Mount Hood, Mount Rainier, Mount Shasta, and Mount St. Helens, are examples of volcanoes generated at a convergent plate boundary along a continental margin (**Figure 5.30**).

°Recall from Section 1.2 that a *hypothesis* is a tentative scientific explanation for a given set of observations. Although widely accepted, the validity of the mantle plume hypothesis, unlike the theory of plate tectonics, remains unresolved.

► SmartFigure 5.31
Global distribution of large basalt plateaus The basalt plateaus (shown in red) are thought to have been produced by bursts of volcanism generated by partial melting of the bulbous heads of hot mantle plumes. The orange dashed lines represent the chain of volcanic structures produced by partial melting of the plume tail. The orange dots indicate the current surface locations of the hot mantle plumes that likely generated the associated basalt plateaus.

Tutorial
https://goo.gl/nXystm

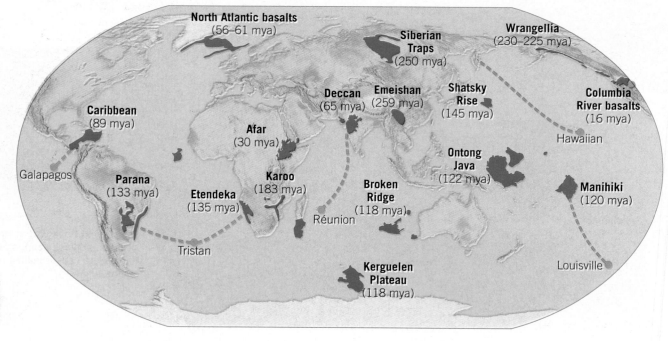

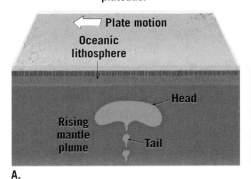

A rising mantle plume with a large bulbous head is thought to generate Earth's large basalt plateaus.

Plate motion

Oceanic lithosphere

Head

Rising mantle plume

Tail

A.

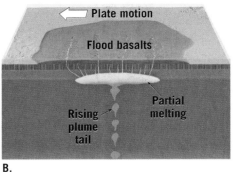

Rapid decompression melting of the plume head produces extensive outpourings of flood basalts over a relatively short time span.

Plate motion

Flood basalts

Partial melting

Rising plume tail

B.

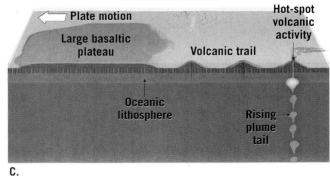

Because of plate movement, volcanic activity from the rising tail of the plume generates a linear chain of smaller volcanic structures.

Plate motion

Hot-spot volcanic activity

Large basaltic plateau

Volcanic trail

Oceanic lithosphere

Rising plume tail

C.

▲ Figure 5.32
Mantle plumes and large basalt plateaus Hot-spot volcanism is thought to explain the formation of large basalt plateaus and the chains of volcanic islands associated with these features.

surface manifestation of this activity is called a **hot spot**, an area of volcanism, high heat flow, and crustal uplifting a few hundred kilometers wide.

Large mantle plumes, dubbed **superplumes**, are thought to be responsible for the vast outpourings of basaltic lava that created the large basalt plateaus. When the head of the plume reaches the base of the lithosphere, decompression melting progresses rapidly. This causes a burst of volcanism that emits voluminous flows of lava over a period of 1 million or so years (**Figure 5.32B**). Extreme eruptions of this type would have affected Earth's climate, causing (or at least contributing to) the extinction events recorded in the fossil record.

The comparatively short initial eruptive phase is often followed by millions of years of less voluminous activity, as the plume tail slowly rises to the surface. Extending away from some large flood basalt plateaus is a chain of volcanic structures similar to the Hawaiian chain (**Figure 5.32C**).

Intraplate volcanism associated with mantle plumes is also thought to be responsible for the

massive eruptions of silica-rich pyroclastic material that occurred in continental settings. Perhaps the best known of these hot-spot eruptions are the three caldera-forming eruptions that occurred in the Yellowstone region over the past 2.1 million years (see Figure 5.23).

CONCEPT CHECKS 5.10

1. Are volcanoes in the Ring of Fire generally described as effusive or explosive? Provide an example that supports your answer.

2. How is magma generated along convergent plate boundaries?  **Concept Checker** https://goo.gl/yr3v1h

3. Volcanism at divergent plate boundaries is most often associated with which magma type?

4. What is thought to be the source of magma for most intraplate volcanism?

5.11 Monitoring Volcanic Activity

List and describe the techniques used to monitor potentially dangerous volcanoes.

Volcano monitoring is critical because it provides hazard assessment for the millions of people who live and work on or near volcanoes. Monitoring addresses questions such as "How likely is it that an eruption is imminent?" Volcanologists utilize several volcano monitoring techniques, most of which detect movement of magma from a subterranean reservoir (typically several kilometers deep) toward Earth's surface. The three most noticeable changes in a volcano caused by the migration of magma are *changes in earthquake patterns, inflation or deflation of a volcanic cone*, and *changes in the amount and/or composition of gases released from a volcano*.

► **Figure 5.33**
InSAR image of ground deformation at Mount Etna, Italy The red and yellow areas indicate that the volcano was inflated. When this image was acquired, the volcano had gone through a 2-year cycle of inflation (expansion) that was followed by a summit eruption.

Displacement (cm)

-10   0   +15

Monitoring Earthquake Patterns

Almost one-third of all volcanoes that have erupted in historic times are now monitored using *seismographs* (see Figure 11.9, page 316), instruments that detect earthquake tremors. In general, a sharp increase in seismic unrest followed by a period of relative quiet has been shown to be a precursor of many volcanic eruptions. Sometimes, it is possible to track the upward movement of magma by analyzing earthquake data.

However, some volcanoes have exhibited lengthy periods of seismic unrest. For example, in 1981, a strong increase in seismicity was recorded for Rabaul

Caldera in New Guinea. The activity continued for the next 13 years, and the ensuing eruption finally occurred in 1994.

Occasionally, a large earthquake triggers a volcanic eruption or at least disturbs the volcano's "plumbing." Hawaii's Kilauea, for example, produced a strong earthquake (magnitude 6.9) in 2018 near Leilani Estates as one of the most destructive eruptions on record began.

Remote Sensing of Volcanoes

Because many active volcanoes are located in remote areas, remote-sensing devices are invaluable monitoring tools. The Global Positioning System (GPS) and InSAR—a radar system used to map surface deformation—are important tools for detecting inflation (swelling) and/or deflation of a volcano. Inflation occurs when the roof of a volcano rises as new magma accumulates in its interior—a phenomenon that precedes many volcanic eruptions. For example, when Mount Etna was being monitored using InSAR, a large flank eruption caused the volcano to deflate. When the flank eruption ceased, a 2-year cycle of inflation followed. This swelling, shown in **Figure 5.33**, was caused by an increase in the volume of a shallow magma chamber centered about 5 kilometers (3 miles) beneath the cone. As expected, this period of inflation preceded an energetic summit eruption.

Modern monitoring technology also makes it possible to remotely detect changes in the quantity and/or composition of volcanic gases being released. This is an important advancement because direct sampling of gases such as sulfur dioxide can be time-consuming and hazardous to humans. Monitoring has revealed that some volcanoes show an increase in sulfur dioxide emissions prior to an eruption.

The magma body emits gases as it moves toward the surface, and remote monitoring can also determine its likely composition. This information helps volcanologists determine the nature of a future eruption and the associated hazards to life and property.

Some remote-sensing techniques are useful for monitoring eruptions in progress. Satellite images, for example, can detect lava flows and ash plumes rising from volcanoes anywhere on Earth—information critical to emergency-management personnel and air traffic controllers around the globe (**Figure 5.34**).

Despite the many technological advances in monitoring, accurate prediction of the timing and likelihood of volcanic eruptions still eludes scientists. This is perhaps best demonstrated by the February 2015 eruption of Sinabung volcano in Sumatra, Indonesia, which claimed at least 16 lives and occurred just days after authorities gave the "all-clear" for residents to return to their homes on Sinabung's slopes.

► **Figure 5.34**
A plume of ash and steam emanating for Eyjafjallajökull Volcano, Iceland This image was acquired by NASA's *Aqua* satellite on May 10, 2010. On this day, airports in Ireland and Portugal were closed because of the clouds of volcanic ash from previous days.

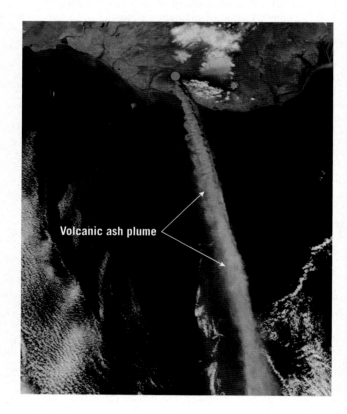

Volcanic ash plume

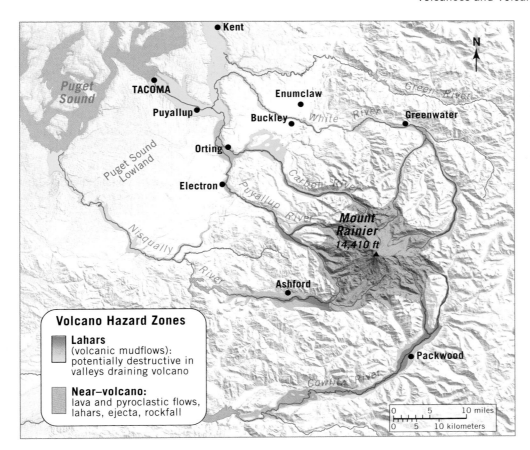

◀ **Figure 5.35**
Hazard map for the region around Mount Rainier This map shows areas that could be affected by lahars (volcanic mudflows), lava flows, and pyroclastic flows from Mount Rainier. Mount Rainier is considered the most threatening volcano in the Cascades. The greatest risk comes from its potential for generating huge lahars.

Volcanic Hazard Maps

Monitoring technology can be used to produce volcanic hazard maps like the one for the areas around Mount Rainier shown in **Figure 5.35**. This simplified volcanic hazard map shows the zones that could potentially be affected by volcanic activity, including lava flows, volcanic bombs, and ash falls, as well as more distant areas likely to be affected by lahars (volcanic mudflows).

Mount Rainier produces a major lahar every 500 to 1000 years and smaller ones more frequently. The most recent major lahar, the Electron Lahar, occurred about 600 years ago and was more than 30 meters (100 feet) thick where the community of Electron, Washington, is located today. A much more massive lahar, called the Osceola Lahar, occurred about 5600 years ago. Osceola's deposits cover an area of about 550 square kilometers (212 square miles), including the area that is now the Port of Tacoma. In response to lahar hazard

concerns, state emergency-management agencies, in coordination with the USGS, developed the Mount Rainier lahar warning system in the Carbon and Puyallup River valleys. This warning system is designed to give local residents between 40 minutes and 3 hours to move to higher ground to avoid these destructive phenomena.

CONCEPT CHECKS 5.11

1. What three factors do volcanologists monitor in order to determine whether magma is migrating toward Earth's surface?

  **Concept Checker** https://goo.gl/H2nKHP

2. What volcanic hazard does the warning system installed around Mount Rainier aim to identify?

CONCEPTS IN REVIEW

Volcanoes and Volcanic Hazards

5.1 Mount St. Helens Versus Kilauea

Compare and contrast the 1980 eruption of Mount St. Helens with the most recent eruption of Kilauea, which began in 1983.

- Volcanic eruptions cover a broad spectrum from explosive eruptions, like that of Mount St. Helens in 1980, to the comparatively quiet eruptions of Kilauea.

5.2 The Nature of Volcanic Eruptions

Explain why some volcanic eruptions are explosive and others are quiescent.

Key Terms:

magma	lava	viscosity
	effusive eruption	eruption column

- The two primary factors determining the nature of a volcanic eruption are a magma's viscosity (a fluid's resistance to flow) and its gas content. In general, magmas containing more silica are more viscous. Temperature also influences viscosity; hot lavas are more fluid than relatively cool lavas.
- Mafic (basaltic) magmas, which are fluid and have low gas content, tend to generate effusive (nonexplosive) eruptions. In contrast, silica-rich intermediate (andesitic) and felsic (rhyolitic) magmas, which are the most viscous and contain the greatest quantity of gases, are the most explosive.

Q Although Kilauea mostly erupts in a gentle manner, what risks might you encounter if you chose to live nearby?

5.3 Materials Extruded During an Eruption

List and describe the three categories of materials extruded during volcanic eruptions.

Key Terms:

aa flow	block lava	tephra
pahoehoe flow	pillow lava	scoria
lava tube	volatile	pumice
	pyroclastic material	

- Volcanoes erupt lava, gases, and solid pyroclastic materials.
- Low-viscosity basaltic lava can flow great distances. On the surface, they travel as pahoehoe or aa flows. Fluid lavas congeal and harden at the surface, while the lava below the surface continues to flow in tunnels called *lava tubes*. When lava erupts underwater, the outer surface is instantly chilled, while the inside continues to flow, producing pillow lavas.
- The gases most commonly emitted by volcanoes are water vapor and carbon dioxide. Upon reaching the surface, gases rapidly expand, leading to explosive eruptions that can generate a mass of lava fragments called *pyroclastic materials*.
- Pyroclastic materials come in several sizes. From smallest to largest, they are ash, lapilli, and blocks or bombs. Blocks exit the volcano as solid fragments, whereas bombs are ejected as liquid blobs.
- Bubbles of gas in lava can be preserved as voids in rocks called *vesicles*. Especially frothy, silica-rich lava can cool to form lightweight pumice, while basaltic lava with lots of bubbles cools to form scoria.

Q This photo shows layers of volcanic material ejected by a violent eruption and deposited roughly horizontally. What term describes this type of volcanic material?

5.4 Anatomy of a Volcano

Draw and label a diagram that illustrates the basic features of a typical volcanic cone.

Key Terms:	vent	caldera
fissure	volcanic cone	parasitic cone
conduit	crater	fumarole

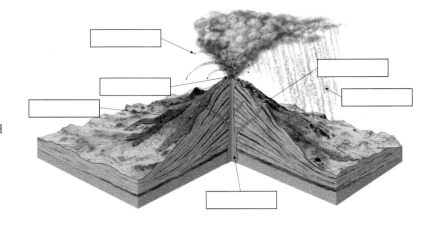

- Volcanoes vary in size and form but share a few common features. Most are roughly conical piles of extruded material that collect around a central *vent*. The vent is usually within a summit crater or caldera. On the flanks of the volcano, there may be smaller vents marked by small parasitic cones, or there may be *fumaroles*, spots where only gas is expelled.

Q Label the diagram using the following terms: conduit, vent, lava, parasitic cone, bombs, pyroclastic material.

5.5 Shield Volcanoes

Summarize the characteristics of shield volcanoes and provide one example of this type of volcano.

Key Terms:	shield volcano	seamount

- Shield volcanoes consist of many successive flows of low-viscosity basaltic lava but lack significant amounts of pyroclastic debris. Lava tubes help transport lava far from the main vent, resulting in very gentle, shield-like profiles.
- Most shield volcanoes begin as seamounts that grow from Earth's seafloor. Mauna Loa, Mauna Kea, and Kilauea in Hawaii are classic examples of the low, wide form characteristic of shield volcanoes.

5.6 Cinder Cones

Describe the formation, size, and composition of cinder cones.

Key Term: cinder cone

- Cinder cones are steep-sided structures composed mainly of pyroclastic debris, typically having a basaltic composition. Lava flows sometimes emerge from the base of a cinder cone.
- Cinder cones are small relative to the other kinds of volcanoes, reflecting the fact that most form quickly, as single eruptive events. Because they are unconsolidated, cinder cones easily succumb to weathering and erosion.

5.7 Composite Volcanoes

List the characteristics of composite volcanoes and describe how they form.

Key Term: composite volcano

- *Composite volcanoes* have a classic symmetrical cone shape and are so named because they consist of both pyroclastic material and lava flows. They typically erupt silica-rich magmas of andesitic or rhyolitic composition. They are much larger than cinder cones and form from multiple eruptions over millions of years.
- Because andesitic and rhyolitic lavas are more viscous than basaltic lava, they accumulate at a steeper angle than does the lava from shield volcanoes.
- Mount Rainier and the other volcanoes of the Cascade Range in the northwestern United States are good examples of composite volcanoes.

5.8 Volcanic Hazards

Describe the major geologic hazards associated with volcanoes.

Key Terms:	lahar	tsunami
pyroclastic flow		

- The greatest volcanic hazard to human life is the *pyroclastic flow*. This dense mix of hot gas and pyroclastic materials races downhill at great speed, incinerating everything in its path. A pyroclastic flow can travel many kilometers from its source. Because pyroclastic flows are hot, their deposits frequently "weld" together into a solid rock called *welded tuff*.
- Lahars are mudflows that form on volcanoes. These rapidly moving slurries of ash and debris suspended in water tend to follow stream valleys and can result in loss of life and/or significant damage to structures.
- Volcanic ash in the atmosphere can be a risk to air travel when it is sucked into airplane engines. Volcanoes at sea level can generate tsunamis when they erupt or when their flanks collapse into the ocean. Those that spew large amounts of gas such as sulfur dioxide can cause respiratory problems. If volcanic gases reach the stratosphere, they screen out a portion of incoming solar radiation and can trigger short-term cooling at Earth's surface.

5.9 Other Volcanic Landforms

List volcanic landforms other than shield, cinder, and composite volcanoes and describe their formation.

Key Terms:

basalt plateau   lava dome
fissure eruption   flood basalt   volcanic neck

- Calderas, which can be among the largest volcanic structures, form when the rigid, cold rock above a magma chamber cannot be supported and collapses, creating a broad, roughly circular depression. On a shield volcano, a caldera forms slowly as lava drains from the magma chamber beneath the volcano. On a composite volcano, caldera collapse often follows an explosive eruption that can result in significant loss of life and destruction of property.

- Fissure eruptions occasionally produce massive floods of fluid basaltic lava from large cracks, called fissures, in the crust. Layer upon layer of these flood basalts may accumulate to significant thicknesses and blanket a wide area. The Columbia Plateau in the northwestern United States is an example.

- Lava domes are thick masses of high-viscosity, silica-rich lava that accumulate in the summit crater or caldera of a composite volcano. When they collapse, lava domes can produce extensive pyroclastic flows.

- Shiprock, New Mexico, is an example of a volcanic neck where the lava in the "throat" of an ancient volcano congealed to form a plug of solid rock that weathered more slowly than the surrounding volcanic rocks. The surrounding pyroclastic debris eroded, and the resistant neck remains as a distinctive landform.

5.10 Plate Tectonics and Volcanism

Explain how the global distribution of volcanic activity relates to plate tectonics.

Key Terms:

Ring of Fire   continental volcanic arc   mantle plume
volcanic island arc   intraplate volcanism   hot spot   superplume

- Volcanoes form at both convergent and divergent plate boundaries, as well as in intraplate settings.

- At divergent plate boundaries, where lithosphere is being rifted apart, decompression melting is the dominant generator of magma. As warm rock rises, it can begin to melt without the addition of heat.

- Convergent plate boundaries that involve the subduction of oceanic crust are the most common site for explosive volcanoes—most prominently in the Pacific Ring of Fire. The release of water from the subducting plate triggers partial melting in the overlying mantle. The ascending magma interacts with the lower crust of the overlying plate and can form a volcanic arc at the surface.

- In intraplate settings, the source of magma is a mantle plume—a column of mantle rock that is warmer and more buoyant than the surrounding mantle.

Q The accompanying diagram shows one of the tectonic settings where volcanism is a dominant process. Name the setting and briefly explain how magma is generated here.

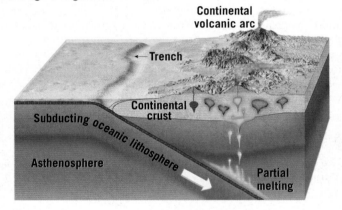

5.11 Monitoring Volcanic Activity

List and describe the techniques used to monitor potentially dangerous volcanoes.

- Volcanoes provide a variety of signals, which volcanologists can directly or remotely monitor to determine whether a volcano has the potential to erupt.

- Volcano monitoring involves observing changes in the pattern of earthquakes beneath a volcano that could signal magma movement, changes in the shape of a volcano, and changes in the composition and quantity of gas output.

GIVE IT SOME THOUGHT

1. **Examine the accompanying photo and complete the following:**

   a. What type of volcano is shown? What features helped you classify it as such?

   b. What is the eruptive style of such volcanoes? Describe the likely composition and viscosity of its magma.

   c. Which type of plate boundary is the likely setting for this volcano?

   d. Name a city that is vulnerable to the effects of a volcano of this type.

2. **Answer the following questions about divergent plate boundaries, such as the Mid-Atlantic Ridge, and their associated lavas:**

   a. Divergent boundaries are characterized by eruptions of what type of lava: andesitic, basaltic, or rhyolitic?

   b. What is the main source of the lavas that erupt at divergent plate boundaries?

   c. What process causes the source rocks to melt?

3. **For each of the accompanying four sketches, identify the geologic setting (zone of volcanism). Which of these settings will most likely generate explosive eruptions? Which will produce outpourings of fluid basaltic lavas?**

A.

B.

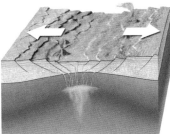

C.

D.

4. **Explain why an eruption of Mount Rainier similar to the 1980 eruption of Mount St. Helens could be considerably more destructive.**

5. **For each of the volcanoes or volcanic regions listed below, identify whether it is associated with a *convergent* or *divergent* plate boundary or with *intraplate volcanism*.**

   a. Crater Lake

   b. Hawaii's Kilauea

   c. Mount St. Helens

   d. East African Rift

   e. Yellowstone

   f. Mount Pelée

   g. Deccan Traps

   h. Fujiyama

6. **The accompanying image shows a geologist at the end of an unconsolidated flow consisting of lightweight lava blocks that rapidly descended the flank of Mount St. Helens.**

   a. What term best describes this type of flow: an aa flow, a pahoehoe flow, or a pyroclastic flow?

   b. What lightweight (vesicular) igneous rock type is likely the main constituent of this flow?

7. **Assume that you are monitoring a volcano that has erupted several times in the recent past but appears to be quiet now. How might you determine whether magma is actually moving through the crust beneath the volcano? Suggest at least two phenomena you would observe or measure.**

8. **The formula for the volume of a cone is $V = 1/3\pi r^2 h$ (where V = volume, π = 3.14, r = radius, and h = height). If Mauna Loa is 9 kilometers high and has a radius of roughly 85 kilometers, what is its approximate total volume?**

EYE ON EARTH

1. This photo shows the February 3, 2015, eruption of Mount Sinabung in North Sumatra, Indonesia. Before awaking in 2010, Mount Sinabung had been dormant since 1600. In recent years it has erupted several times, resulting in at least 23 deaths and the evacuation of more than 30,000 people.

   a. What name is given to the ash- and pumice-laden cloud that is racing down this volcano?

   b. What type of volcano is associated with such destructive eruptions?

2. This image shows the Buddhist monastery Taung Kalat, located in central Myanmar (Burma). The monastery sits high on a sheer-sided rock made mainly of magmas that solidified in the conduit of an ancient volcano. The volcano has since been worn away.

   a. Based on this information, what volcanic structure do you think is shown in this photo?

   b. Would this volcanic structure most likely have been associated with a composite volcano or cinder cone? Explain how you arrived at your answer.

DATA ANALYSIS

Recent Volcanic Activity

The Smithsonian Institution Global Volcanism Program and the USGS work together to compile a list of new and changing volcanic activity worldwide. NOAA also uses this information to issue Volcanic Ash Advisories to alert aircraft of volcanic ash in the air.

https://goo.gl/eD226M

ACTIVITIES

Go to the Weekly Volcanic Activity Report page at http://volcano.si.edu.

1. **What information is displayed on this page?**

2. **Look at the "New Activity/Unrest" section of the homepage. In what areas is most of the volcanic activity concentrated?**

3. **Under the "Reports tab," click on the "Weekly Volcanic Activity Report." List three ongoing volcanic activity locations shown on the map.**

Click on the name of a volcano under "New Activity Highlights."

4. **Where is this volcano located? Be sure to include the city, country, volcanic region name, latitude and longitude, and type of volcano.**

5. **Do some investigating online and in your text book. What are the key characteristics for this type of volcano?**

6. **Briefly describe the dates of the most recent activity. How was this activity observed?**

Go to the Volcanic Ash Advisory Center (VAAC) page at https://www.ospo.noaa.gov/Products/atmosphere/vaac/.

7. **This page shows the Washington VAAC area. Are any active volcanoes listed? Name them.**

8. **Are any advisories listed for eruptions? If so, list the volcano, the date, and the eruption height.**

Mastering Geology

Looking for additional review and test prep materials? Visit pearson.com/mastering/geology to enhance your understanding of this chapter's content by accessing a variety of resources, including **Self-Study Quizzes, Geoscience Animations, SmartFigure Tutorials,** *Project Condor* **Videos, Mobile Field Trips, Dynamic Study Modules,** and an optional **Pearson eText.**

Does Soil Loss Pose the Greatest Threat to Civilization?

When we imagine a civilization-killing apocalypse, we usually see a new disease, nuclear conflict, runaway artificial intelligence, or meteor strike being the cause. Soil loss is as real a threat as these, but it tends to be much subtler: It's quietly happening right now, mostly unnoticed.

▲ Not all soil loss is as dramatic as the 1930s Dust Bowl. In this scene, Iowa topsoil is lost after hard rains form gullies that wash it away.

Dramatic events such as the Great Dust Bowl of the 1930s (see **GEOgraphics 6.1**) have provided alarming visuals that help illuminate their damage. Although most soil loss doesn't look particularly remarkable, it occurs nearly everywhere we grow food, resulting from modern agricultural practices and other human activities. A 2018 Intergovernmental Science-Policy Platform on Biodiversity and Ecosystem Services (IPBES) report stated, "Degradation of the Earth's land surface through human activities is negatively impacting the wellbeing of at least 3.2 billion people, pushing the planet toward a sixth mass species extinction."

We lose more than 1 percent of productive agricultural soils per year, on average, and the rate of loss is even more dramatic in regions projected to have the greatest population growth, such as sub-Saharan Africa.

This chapter describes weathering and the other slow, steady processes that lead to the formation of soil. It also discusses the impact of human activities on soil. Today, humans move more Earth materials, mostly through agriculture, than *all* natural processes combined, including rivers, landslides, and glaciers. What took thousands of years to develop on Earth, humans have substantially altered in just the few centuries since the dawn of the Industrial Revolution.

► A large 1935 dust storm looms over Stratford, Texas.

6
Weathering and Soils

FOCUS ON CONCEPTS

Each statement represents the primary learning objective for the corresponding major heading within the chapter. After you complete the chapter, you should be able to:

6.1 Define *weathering* and distinguish between the two main types.

6.2 List and describe four examples of mechanical weathering.

6.3 Discuss the role of water in each of three chemical weathering processes.

6.4 Summarize the factors that influence the type and rate of rock weathering.

6.5 Define *soil* and explain why soil is referred to as an *interface*.

6.6 List and briefly discuss five controls of soil formation.

6.7 Sketch, label, and describe an idealized soil profile. Explain the need for classifying soils.

6.8 Explain the detrimental impact of human activities on soil and list several ways to combat soil erosion.

E arth's surface constantly changes. Rock disintegrates and decomposes. It moves to lower elevations by gravity, or gets carried away by water, wind, or ice. In this manner, Earth's physical landscape is sculpted. This chapter focuses on weathering, the first step of this never-ending process. It looks at what causes solid rock to crumble, and why the type and rate of weathering vary from place to place. Soil, a vital resource and product of weathering, is also examined.

▼ SmartFigure 6.1
Arches National Park Mechanical and chemical weathering contributed greatly to the creation of North Window Arch and all of the other arches and other rock formations in Utah's Arches National Park.

Animation
https://goo.gl/VrJztQ

6.1 Weathering

Define *weathering* and distinguish between the two main types.

Weathering involves the physical breakdown (disintegration) and chemical alteration (decomposition) of rock at or near Earth's surface. Weathering goes on all around us, but it seems like such a slow and subtle process that it is easy to underestimate its importance. It is a basic part of the rock cycle and thus a key process in the Earth system. Weathering of Earth's crustal rocks plays an important role in the cycling of carbon as it moves from the solid Earth, to the atmosphere and ocean, to the shells of marine organisms, and ultimately back to sediment on the ocean floor.

Many of the life-sustaining minerals and elements found in soil, and ultimately in the food we eat, were freed from solid rock by weathering processes. As **Figure 6.1** and many other images throughout this book illustrate, weathering also contributes to the formation of some of Earth's most spectacular scenery, including most of our national parks and monuments. These same processes are also responsible for causing the deterioration of many of the structures we build.

There are two basic categories of weathering. **Mechanical weathering** (or disintegration) is accomplished by physical forces that break rock into smaller and smaller pieces without changing the rock's mineral composition. **Chemical weathering** (or decomposition) involves a chemical transformation of rock into one or more new compounds. These two concepts can be illustrated by a piece of firewood. The original log disintegrates when it is split into smaller and smaller pieces, whereas decomposition occurs when the wood is burned.

Why does rock weather? Simply, weathering is the response of Earth materials to a changing environment. For instance, after millions of years of uplift and erosion (the removal and transport of weathered rock material by gravity-driven movement of water, wind, or ice), the rocks overlying a large, intrusive igneous body may be removed, exposing it at the surface. This mass of crystalline rock—formed deep below ground, where temperatures and pressures are high—is now subjected to a very different surface environment. In response, this rock mass will gradually change. In the following sections we will examine the various types of mechanical and chemical weathering. Although we will consider these two categories separately, keep in mind that they usually work simultaneously in nature and reinforce each other.

CONCEPT CHECKS 6.1

1. What are the two basic categories of weathering?

2. How do the products of each category of weathering differ?

 Concept Checker https://goo.gl/GAhtRz

6.2 Mechanical Weathering

List and describe four examples of mechanical weathering.

When a rock undergoes mechanical weathering, it breaks into smaller and smaller pieces, each retaining the characteristics of the original material. The end result is many small pieces from a single large one. **Figure 6.2** shows that breaking a rock into smaller pieces increases the surface area available for chemical reactions. An analogous situation occurs when sugar is added to a liquid. A sugar cube dissolves much more slowly than an equal volume of sugar granules because the cube has much less surface area available for dissolution. Hence, by breaking rocks into smaller pieces, mechanical weathering increases the amount of surface area available for chemical weathering.

In nature, four important physical processes lead to the fragmentation of rock: frost wedging, salt crystal growth, sheeting, and biological activity. In addition, although the work of erosional agents such as wind, waves, glacial ice, and running water is usually considered separately from mechanical weathering, this work is nevertheless related. As these mobile agents (discussed in detail in later chapters) transport rock debris, particles continue to break and abrade. In addition to these natural processes, humans are prone to making big pieces of rock into smaller pieces. We do this in most mining operations in order to obtain gravel for roads, cement

As mechanical weathering breaks rock into smaller pieces, more surface area is exposed to chemical weathering.

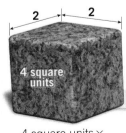  

Increase in surface area

4 square units ×
6 sides ×
1 cube =
_____
24 square units of surface

1 square unit ×
6 sides ×
8 cubes =
_____
48 square units of surface

.25 square unit ×
6 sides ×
64 cubes =
_____
96 square units of surface

◄ SmartFigure 6.2
Mechanical weathering (breaking) increases surface area Mechanical weathering adds to the effectiveness of chemical weathering because chemical weathering can occur only on exposed surfaces.

Tutorial https://goo.gl/2r6vjG

▲ **Figure 6.3**
Humans crush a lot of rock Each American requires about 25,000 pounds of minerals (not including petroleum or natural gas) each year! Most of the operations to produce these require breaking rocks.

for concrete structures, and molybdenum for steel (**Figure 6.3**).

Frost Wedging

If you leave a glass bottle filled with water in the freezer too long, the bottle will fracture (**Figure 6.4**). The bottle breaks because liquid water has the unique property of expanding about 9 percent upon freezing. This is why poorly insulated or exposed water pipes rupture during frigid weather.

This is also the basis for the traditional explanation of **frost wedging**: After water works its way into rock cracks, it freezes, expands, and enlarges the cracks,

*Bernard Hallet, "Why Do Freezing Rocks Break?" *Science* 314(17): 1092–1093, November 2006.

which causes angular fragments to eventually break off (**Figure 6.5**). According to conventional wisdom, most frost wedging occurs this way. However, newer research indicates that frost wedging can also occur due to a process related to *ice lenses.**

It has long been known that when moist soils freeze, they expand, or *frost heave*, due to the growth of ice lenses. These masses of ice grow larger because they are supplied with water migrating from unfrozen areas as thin liquid films. As more water accumulates and freezes, the soil is pushed upward. A similar process occurs within the cracks and pore spaces of rocks. Lenses of ice grow larger as they attract liquid water from surrounding pores. The growth of these ice masses gradually weakens the rock, causing it to fracture.

Salt Crystal Growth

Growth of salt crystals creates another expansive force that can split rocks. Rocky shorelines and arid regions are common settings for this process. It begins when sea spray from breaking waves or salty groundwater penetrates crevices and pore spaces in rock. As this water evaporates, salt crystals form. As these crystals gradually grow larger, they weaken the rock by pushing apart the surrounding grains or by enlarging tiny cracks.

This same process contributes to the crumbling of roadways where salt is spread to melt snow and ice during winter. The salt dissolves in water and seeps into cracks that quite likely originated from frost action. When the water evaporates, the growth of salt crystals further breaks the pavement.

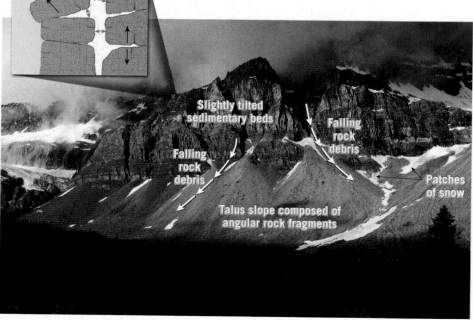

Frost wedging

Slightly tilted sedimentary beds

Falling rock debris

Falling rock debris

Talus slope composed of angular rock fragments

Patches of snow

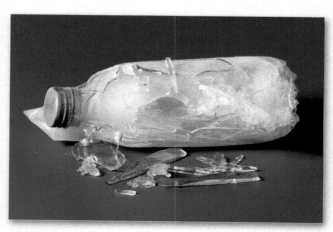

▲ **Figure 6.4**
Ice breaks bottle The bottle broke because water expands about 9 percent when it freezes.

▲ **SmartFigure 6.5**
Ice breaks rock In mountainous areas, frost wedging creates angular rock fragments that accumulate to form piles of debris called talus slopes.

 Tutorial https://goo.gl/aRoP18

Fractures: Jointing and Sheeting

Many fractures are created by expansion of rock masses. Contraction during the crystallization of magma forms other fractures, as do tectonic forces involved in mountain building. Fractures produced by these activities generally form a definite pattern and are called **joints** (**Figure 6.6**).

Joints are important rock structures that allow water to penetrate to depth and start the process of weathering long before the rock is exposed. Rock-breaking thermal expansion can occur during daily temperature changes, wildfires, and even lightning strikes. When large masses of igneous rock, particularly granite, are exposed at Earth's surface by erosion, concentric slabs begin to break loose. The process that generates these onion-like layers is called **sheeting**. It takes place, at least in part, due to the great reduction in pressure that occurs as the overlying rock is eroded away, a process called *unloading*.

Figure 6.7 illustrates what happens: As the overburden is removed, the outer parts of the granitic mass expand more than the rock below and separate from the rock body. Continued weathering eventually causes the slabs to separate and peel off, creating an **exfoliation dome** (*ex* = off, *folium* = leaf). Excellent examples of exfoliation domes are Stone Mountain, Georgia, and Half Dome and Liberty Cap in Yosemite National Park.

A process analogous to sheeting can also occur when human activities reduce the confining pressure, similar to what occurs during unloading. For example, in deep mines, large rock slabs have been known to explode off

◄ Figure 6.6
Joints aid weathering
Aerial view of nearly parallel joints near Moab, Utah.

Moab Vally, Utah

Parallel joints produced by bending of sandstone layer

Erosion along fractures carved sandstone into fins

Entrada sandstone

▼ SmartFigure 6.7
Unloading leads to sheeting Sheeting leads to the formation of an exfoliation dome.

Tutorial
https://goo.gl/FgTZ2g

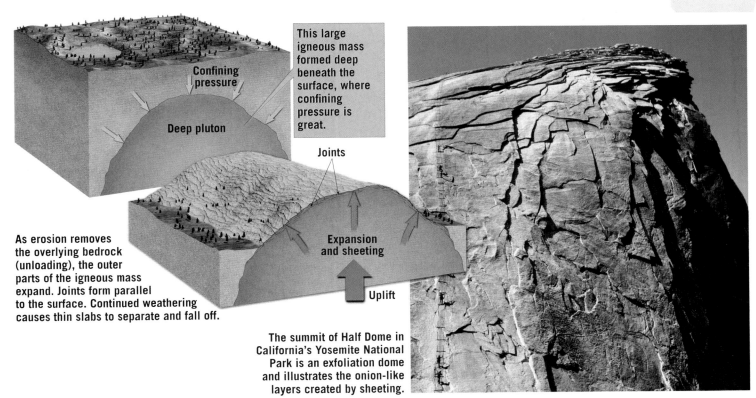

Confining pressure

Deep pluton

This large igneous mass formed deep beneath the surface, where confining pressure is great.

Joints

As erosion removes the overlying bedrock (unloading), the outer parts of the igneous mass expand. Joints form parallel to the surface. Continued weathering causes thin slabs to separate and fall off.

Expansion and sheeting

Uplift

The summit of Half Dome in California's Yosemite National Park is an exfoliation dome and illustrates the onion-like layers created by sheeting.

► Figure 6.8
Plants can break rock
Root wedging.

the walls of newly cut tunnels. In quarries, fractures occur parallel to the floor when large blocks of rock are removed.

Biological Activity

Weathering can be accomplished by the activities of organisms, including plants, burrowing animals, and humans. Plant roots in search of nutrients and water grow into fractures, and as the roots grow, they wedge apart the rock (**Figure 6.8**). Burrowing animals further break down rock by moving fresh material to the surface, where physical and chemical processes can more effectively attack it. Decaying organisms also produce acids that contribute to chemical weathering. Where rock has been blasted in search of minerals or for road construction, the impact of humans is particularly noticeable (see Figure 6.3).

CONCEPT CHECKS 6.2

1. When a rock is mechanically weathered, how does its surface area change? How does this influence chemical weathering?

2. How do joints promote weathering?

Concept Checker
https://goo.gl/e8Us3b

3. Explain how water can cause mechanical weathering.

4. How does biological activity contribute to weathering?

6.3 Chemical Weathering

Discuss the role of water in each of three chemical weathering processes.

In the preceding discussion, you learned that mechanical weathering breaks rock into smaller pieces, which increases the total surface area of rock fragments available for chemical attack. It should also be pointed out that chemical weathering contributes to mechanical weathering by weakening the outer portions of some rocks, which, in turn, makes them more susceptible to being broken by mechanical weathering processes. This breaking then exposes more surface area that is then susceptible to chemical weathering. These two types of weathering are therefore interrelated and encourage each other.

Chemical weathering involves the complex processes that break down rock components and internal structures of minerals. Such processes convert the constituents to new minerals or release them to the surrounding environment. During this transformation, the original rock decomposes into substances that are stable in the surface environment. Consequently, the products of chemical weathering will remain essentially unchanged as long as they remain in an environment similar to the one in which they formed.

Water is by far the most important agent of chemical weathering. Pure water alone is a good solvent, and small amounts of dissolved materials result in increased chemical activity for weathering solutions. The major processes of chemical weathering are dissolution, oxidation, and hydrolysis. Water plays a leading role in each of them.

Dissolution

Perhaps the easiest type of chemical weathering process to envision is **dissolution**, in which certain minerals dissolve in water. One of the most water-soluble minerals is halite (common salt), which, as you may recall, is composed of sodium and chloride ions. Halite readily

Water molecules are polar because both hydrogen atoms bond to the same side of an oxygen atom. Thus, the hydrogen side of the molecule is slightly positive, and the oxygen side is slightly negative.

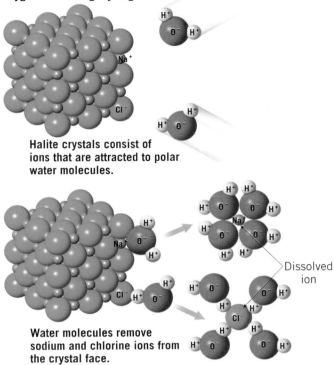

Halite crystals consist of ions that are attracted to polar water molecules.

Water molecules remove sodium and chlorine ions from the crystal face.

Dissolved ion

▲ Figure 6.9
Water can dissolve some rocks Dissolution is chemical weathering in which minerals dissolve in water. Halite readily dissolves in pure water. Other minerals, such as calcite, readily dissolve only in water that is acidic.

dissolves in water because, although this compound maintains overall electrical neutrality, the individual ions retain their respective charges. Moreover, the surrounding water molecules are polar—that is, the oxygen end of the molecule has a small residual negative charge, and the hydrogen end has a small positive charge. As the water molecules come into contact with halite, their negative ends approach sodium ions, and their positive ends cluster about chloride ions. This disrupts the attractive forces in the halite crystal and releases the ions to the water solution (**Figure 6.9**).

If most water on Earth were pure, there would be no reaction when it came into contact with the majority of minerals. However, water on the planet usually mixes with other molecules and ions that can lead to chemical reactions. The presence of even a small amount of acid in water dramatically increases the corrosive force, leading to dissolution of rock. (An acidic solution contains the reactive hydrogen ion, H^+.) In nature, acids are produced by a number of processes. For example, carbon dioxide in the atmosphere dissolves in raindrops, forming carbonic acid. As acidic rainwater soaks into the ground, carbon dioxide in the soil may increase the acidity of the weathering solution. Various organic acids are also released into the soil as organisms decay, and sulfuric acid is produced by the weathering of pyrite and other sulfide minerals. Most rocks readily decompose in these acidic waters, producing certain products that are water soluble along the way.

For example, calcite ($CaCO_3$), the mineral that composes marble and limestone, is easily attacked by even a weakly acidic solution. The overall reaction by which calcite dissolves in water containing carbon dioxide is:

$$CaCO_3 + (H^+ + HCO_3^-) \longrightarrow Ca^{2+} + 2HCO_3^-$$

calcite   carbonic acid   calcium ion   bicarbonate ion

During this process, the insoluble calcium carbonate is transformed into soluble products. In nature, large quantities of limestone dissolve and get carried away by underground water. This activity, occurring over thousands of years, is largely responsible for the formation of limestone caverns (**Figure 6.10**). Monuments and buildings made of limestone or marble are also subjected to the corrosive work of acids, particularly in urban and industrial areas that have smoggy, polluted air.

The soluble ions from these reactions are retained in our underground water supply, resulting in "hard water" for many cities or towns. Hard water is considered

▼ Figure 6.10
Acidic waters create caves The dissolving power of carbonic acid plays an important role in creating limestone caverns, such as Baredine Cave in Croatia.

► Figure 6.11
Iron oxides add color Many sedimentary rocks are very colorful. The most important "pigments" are small amounts of iron oxide. Just as iron oxide colors the rusty barrels in **A**, this product of chemical weathering is also responsible for the reds and oranges seen in the rocks composing the Supai Formation in the Grand Canyon in **B**.

A.

these minerals to form the reddish-brown iron oxide called *hematite* (Fe_2O_3) or, in other cases, a yellowish-colored rust called *limonite* [$FeO(OH)$]. These products are responsible for the rusty color on the surfaces of dark igneous rocks, such as basalt, as they begin to weather. Hematite and limonite are also important cementing and coloring agents in many sedimentary rocks (**Figure 6.11**). However, oxidation can occur only after iron is freed from the silicate structure by another process, called *hydrolysis*.

Another important oxidation reaction occurs when sulfide minerals such as pyrite decompose. Sulfide minerals are major constituents of many metallic ores, and pyrite is frequently associated with coal deposits as well. In a moist environment, chemical weathering of pyrite (FeS_2) yields sulfuric acid (H_2SO_4) and iron oxide [$FeO(OH)$]. In many mining locales, this weathering process creates a serious environmental hazard, particularly in humid areas where abundant rainfall infiltrates spoil banks (waste material left after coal or other minerals are removed). This so-called acid mine drainage eventually makes its way to streams, killing aquatic organisms and degrading aquatic habitats (**Figure 6.12**).

Hydrolysis

The most common mineral group, the silicates, decomposes primarily by the process of **hydrolysis** (*hydro* = water, *lysis* = loosening). Basically, hydrolysis is the reaction of any substance with water. Ideally, the hydrolysis of a mineral could take place in pure water as some of the water molecules dissociate to form the very reactive hydrogen (H^+) and hydroxyl (OH^-) ions. The hydrogen ion attacks and replaces other positive ions in the crystal lattice. With the introduction of hydrogen ions into the crystalline structure, the original orderly arrangement of atoms is destroyed, and the mineral decomposes.

Hydrolysis in the Presence of Acids
In nature, water usually contains other substances that contribute additional hydrogen ions, thereby greatly accelerating hydrolysis. As mentioned previously, when we discussed dissolution, carbon dioxide (CO_2) in the atmosphere and in decaying organic matter in the soil dissolves in water to form carbonic acid (H_2CO_3). In water, carbonic acid dissociates to form hydrogen ions

undesirable because the active ions in it react with soap, producing an insoluble material that can leave a residue instead of removing dirt. To solve this problem, a water softener can be used to remove these ions (Ca^{2+}), generally by replacing them with others that do not chemically react with soap, such as sodium ions ($Na+$).

Oxidation

Everyone has seen rusty iron and steel objects, such as gardening tools left out in the rain. The rusting process can also happen with iron-rich minerals. The process occurs when oxygen combines with iron to form iron oxide, as follows:

$$4Fe + 3O_2 \longrightarrow 2Fe_2O_3$$
iron   oxygen   iron oxide (hematite)

This type of **oxidation**[†] occurs when electrons are lost from one element during the chemical reaction. In this case, we say that iron was oxidized because it lost electrons to oxygen. This kind of reaction happens in the presence of water because the lost electrons make (OH)$^-$, which then reacts with the liberated iron ions. Although the oxidation of iron occurs quickly in the presence of water, the reaction still happens, though very slowly, in a dry environment.

Oxidation is important in decomposing such ferromagnesian minerals as olivine, pyroxene, hornblende, and biotite. Oxygen readily combines with the iron in

[†]Note that *oxidation* is a general term, and this reaction can occur with or without oxygen being involved.

◀ **Figure 6.12**
Acid mine drainage In 2015 the Gold King Mine in Colorado accidentally released millions of gallons of mine waste water into the Animas River. The river turned yellow in the immediate aftermath, and suffered widespread environmental damage from pollutants in the long term.

(H^+) and bicarbonate ions (HCO_3^-). To illustrate how a rock undergoes hydrolysis in the presence of carbonic acid, let's examine the chemical weathering of granite, a common continental rock. Recall that granite consists mainly of quartz and feldspars. The weathering of the potassium feldspar component of granite is as follows:

$$2KAlSi_3O_8 + 2(H^+ + HCO_3^-) + H_2O \rightarrow$$

potassium feldspar   carbonic acid   water

$$Al_2Si_2O_5(OH)_4 + 2K^+ + 2HCO_3^- + 4SiO_2$$

kaolinite   potassium   bicarbonate   silica
(residual clay)   ion   ion

in solution

In this reaction, the hydrogen ions (H^+) attack and replace potassium ions (K^+) in the feldspar structure, thereby disrupting the crystalline network. Once removed, the potassium is available as a nutrient for plants or becomes the soluble salt potassium bicarbonate $(KHCO_3)$, which may be incorporated into other minerals or carried to the ocean.

Products of Silicate-Mineral Weathering

Even in complex dynamic systems like Earth, which are inherently chaotic and unstable, there are relatively stable steady states that chaos theory calls *attractors*. In the context of sedimentary rocks, these are rock types that other rocks naturally tend to develop into.[‡] The attractors of this part of the Earth system are shale, sandstone, and limestone (see the rock cycle model in Figure 1.23 on page 26). A large continuum of possible combinations of grain size and mineral content exists, yet these three rock types form most frequently due to the processes of weathering and

transport. It's a bit like how when you roll two dice, the most common result is seven because that happens to be the number that the greatest number of dice combinations sum to.

To illustrate the idea of attractors and sedimentary rocks, let's look at granite, the most common igneous rock (an igneous attractor). During chemical weathering, the most abundant breakdown product of the feldspars within granite is the clay mineral *kaolinite*. Clay minerals are the end products of weathering and are very stable under surface conditions. Consequently, clay minerals make up a high percentage of the inorganic material in soils. But most clay is eventually washed to the ocean as suspended sediment. Eventually it settles out onto the seafloor, where it becomes a high proportion of the first sedimentary rock attractor, shale.

During the weathering of feldspar minerals, in addition to the formation of clay minerals, some silica is removed from the feldspar structure and carried away by groundwater during chemical weathering. This dissolved silica will eventually precipitate, producing nodules of chert or flint. Other times, the dissolved silica will fill in the pore spaces between grains of sediment, or it will be carried to the ocean, where microscopic animals will remove it from the water to build hard silica shells. To summarize, the weathering of potassium feldspar generates a residual clay mineral, a soluble salt (potassium bicarbonate), and some silica, which enters into solution.

Quartz, the other main component of granite, is *very* resistant to chemical weathering and remains substantially unaltered when attacked by weak acidic solutions (most commonly carbonic acid in rain). As a result, as the feldspar in granite dulls and turns to clay, it releases the once-interlocked quartz grains, which still retain their fresh, glassy appearance. Although some of the quartz remains in the soil, much is eventually transported as *bed load* (weathered particles moved along the bed of a stream) to the ocean, where it becomes the main constituent of such features as sandy beaches and sand dunes. In time these quartz grains may be lithified to form the second sedimentary rock attractor, sandstone.

[‡]Dr. Lynn Fichter (James Madison University) came up with the idea of sedimentary rock types as attractors.

TABLE 6.1	Products of Chemical Weathering	
Mineral	**Residual Products**	**Material in Solution**
Quartz	Quartz grains	Silica (SiO_2)
Feldspars	Clay minerals	SiO_2 potassium ions (K^+) sodium ions (Na^+) calcium ions (Ca^{2+})
Amphibole (hornblende)	Clay minerals Limonite Hematite	SiO_2 Ca^{2+} magnesium ions (Mg^{2+})
Olivine	Limonite Hematite	SiO_2 Mg^{2+}

Table 6.1 lists the weathered products of some of the most common silicate minerals. Remember that silicate minerals make up most of Earth's crust and that these minerals are essentially composed of only eight elements. When chemically weathered, silicate minerals yield sodium, calcium, potassium, and magnesium ions that form soluble products, which may be removed from groundwater. The element iron combines with oxygen, producing relatively insoluble iron oxides, most notably hematite and limonite, which give soil a reddish-brown or yellowish color. When rivers and groundwater carry dissolved ions to the ocean, they are available for tropical organisms to build their shells. The accumulation of these shells, particularly those originating from calcium ions, contributes to the formation of the third sedimentary rock attractor, limestone.

Spheroidal Weathering

Many rock outcrops have a rounded appearance. This occurs because chemical weathering works inward from exposed surfaces. **Figure 6.13** illustrates how angular masses of jointed rock change through time. The process is aptly called **spheroidal weathering**. Because weathering attacks edges from two sides and corners from three sides, these areas wear down faster than a single flat surface. Gradually, sharp edges and corners become smooth and rounded. Eventually an angular block may evolve into a nearly spherical boulder. Once this occurs, the boulder's shape does not change, but the spherical mass continues to get smaller.

CONCEPT CHECKS 6.3

Concept Checker
https://goo.gl/9vJz9a

1. How is carbonic acid formed in nature?

2. What occurs when carbonic acid reacts with calcite-rich rocks such as limestone?

3. What products result when carbonic acid reacts with potassium feldspar?

4. List several minerals that are especially susceptible to oxidation and list two common products of oxidation.

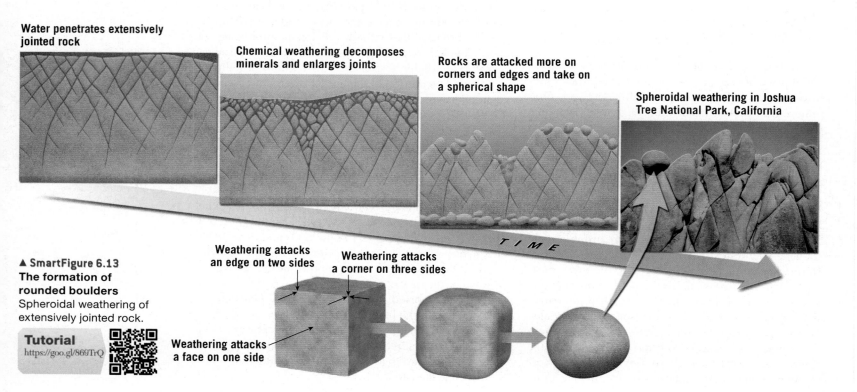

Water penetrates extensively jointed rock

Chemical weathering decomposes minerals and enlarges joints

Rocks are attacked more on corners and edges and take on a spherical shape

Spheroidal weathering in Joshua Tree National Park, California

TIME

▲ SmartFigure 6.13
The formation of rounded boulders
Spheroidal weathering of extensively jointed rock.

Tutorial
https://goo.gl/869TrQ

Weathering attacks an edge on two sides

Weathering attacks a corner on three sides

Weathering attacks a face on one side

6.4 Rates of Weathering

Summarize the factors that influence the type and rate of rock weathering.

As discussed earlier in this chapter, when rock is mechanically weathered, the amount of surface area exposed to chemical weathering increases. Other important factors that influence the type and rate of rock weathering include rock characteristics and climate.

Rock Characteristics

Rock characteristics encompass all of the chemical traits of rocks, including mineral composition and solubility. In addition, physical features, such as the spacing of joints and the size of minerals grains, can be important because they influence water's ability to penetrate rock.

The variations in weathering rates due to the mineral constituents can be demonstrated by comparing old headstones in a cemetery. Gravestones made from granite, which is composed of silicate minerals, are relatively resistant to chemical weathering. By contrast, marble headstones show signs of extensive chemical alteration over a relatively short period. We can see this by examining the inscriptions on the headstones shown in **Figure 6.14**. Marble is composed of calcite (calcium carbonate), which slowly dissolves even in a weakly acidic solution.

The silicates, the most abundant mineral group, chemically weather in essentially the same order in which they crystallize. By examining Bowen's reaction series (see Figure 4.21 on page 120), you can see that olivine crystallizes first and is therefore least resistant to chemical weathering, whereas quartz, which crystallizes last, is the most resistant.

Climate

Climatic factors, particularly temperature and precipitation, are key determinants in the rate of rock weathering. For example, areas with more freeze–thaw cycles will have more frost wedging, an important type of mechanical weathering. Temperature and moisture also exert a strong influence on rates of chemical weathering and determine the kind and amount of vegetation present. Regions with lush vegetation often have a thick mantle of soil, rich in decayed organic matter, from which chemically active fluids such as carbonic acid and humic acids are derived. The optimum environment for chemical weathering is a combination of warm temperatures and abundant moisture. Higher temperatures increase the number of high-speed collisions in a solution, which speeds up the reaction. In polar regions, frigid temperatures keep the available moisture locked up as ice, preventing most chemical weathering. Likewise, in arid regions, there is insufficient moisture to promote rapid chemical weathering.

Human activities often produce pollutants that alter the composition of the atmosphere. Such changes can, in turn, influence the rate of chemical weathering. One well-known example is acid rain (**Figure 6.15**).

▼ Figure 6.15
Acid rain accelerates the chemical weathering of stone monuments and structures As a result of burning large quantities of coal and petroleum, tens of millions of tons of sulfur and nitrogen oxides are released into the atmosphere each year worldwide. Through a series of complex chemical reactions, some of these pollutants are converted into acids that then fall to Earth's surface as rain or snow.

This granite headstone was erected in 1868. The inscription still looks fresh.

This headstone of calcite-rich marble dates from 1874, six years after the granite stone. The inscription is barely legible.

▲ SmartFigure 6.14
Rock type influences weathering An examination of headstones in the same cemetery shows that the rate of chemical weathering depends on rock type.

Tutorial
https://goo.gl/WvGdK5

► SmartFigure 6.16
Monuments to weathering This example of differential weathering is in New Mexico's Bisti Badlands. When weathering accentuates differences in rocks, spectacular landforms are sometimes created.

Mobile Field Trip
https://goo.gl/wiDZ3Q

Differential Weathering

Rock masses do not weather uniformly; this is known as **differential weathering**. The results vary in scale from the rough, uneven surface of the marble headstone in Figure 6.14 to the boldly sculpted exposures of bedrock in New Mexico's Bisti Badlands (**Figure 6.16**).

Differential weathering and subsequent erosion are responsible for creating many unusual, often spectacular, rock formations and landforms. Many factors influence the different rates of rock weathering. Among the most important are variations in rock composition. More resistant rock protrudes as ridges or pinnacles or as steeper cliffs on irregular hillsides (see Figure 7.5, page 208). The number and spacing of joints can also be a significant factor (see Figure 6.13).

CONCEPT CHECKS 6.4

1. Explain why the headstones in Figure 6.14 weathered so differently.

2. How does climate influence weathering?

Concept Checker
https://goo.gl/iEEhkD

6.5 Soil

Define *soil* and explain why it is referred to as an *interface*.

Weathering is important to the formation of soil. Along with air and water, soil is one of our most indispensable resources. Also, like air and water, soil is often taken for granted.

Soil has accurately been called "the bridge between life and the inanimate world." All life—the entire biosphere—owes its existence to a dozen or so elements that must ultimately come from Earth's crust. Once weathering and other processes create soil, plants carry out the intermediary role of assimilating the necessary elements and making them available to animals, including humans.

An Interface in the Earth System

Soil forms where the geosphere, atmosphere, hydrosphere, and biosphere meet. This is why soil is considered an *interface*: It is a common boundary where different parts of a system interact (see Chapter 1). Soil is a material that develops in response to complex environmental interactions among different parts of the Earth system. Over time, soil gradually evolves to a state of equilibrium, or balance, with the environment. Soil is dynamic and sensitive to almost every aspect of its surroundings. Thus, when environmental changes occur, such as changes in climate, vegetative cover, and animal (including human) activity, the soil responds. Any such change gradually alters soil characteristics until a new balance is reached. Although thinly distributed over the land surface, soil functions as a fundamental interface, providing an excellent example of the integration among many parts of the Earth system.

What Is Soil?

With few exceptions, Earth's land surface is covered by **regolith**, a layer of rock and mineral fragments produced by weathering. **Soil** is the portion of the regolith that supports growth of plants. It is a combination of minerals, organic matter, water, and air. Although the proportions of these major components vary, soil contains all of them (**Figure 6.17**). About one-half of the total volume of good-quality surface soil is a mixture of disintegrated and decomposed rock (mineral matter) and **humus**, the decayed remains of animal and plant life (organic matter). The remaining half consists of pore spaces among the solid particles where air and water circulate.

Although the mineral portion of soil is usually much greater than the organic portion, humus is an essential component that supplies plants with nutrients and enhances the soil's ability to retain water. Soil water is a complex solution that contains many dissolved nutrients plants can utilize, in addition to the moisture necessary for life-sustaining chemical reactions. Finally, the pore spaces that are not filled with water contain air. This air is the source of necessary oxygen (for certain microbes) and carbon dioxide for plants that live in the soil.

Soil Texture and Structure

Most soils contain particles in a range of different sizes. **Soil texture**, like *rock texture*, refers to the proportions of different particle sizes. Texture strongly influences the ability of soil to retain and transmit water and air, which therefore also determines how well plants can grow. Sandy soils may drain too rapidly and dry out quickly to facilitate much plant growth. At the opposite extreme, the pore spaces of dense, clay-rich soils may be so small that they inhibit drainage, and long-lasting

puddles result. Moreover, when the clay and silt content is very high, plant roots may have difficulty penetrating the soil.

Because soils rarely consist of particles of only one size, *textural categories* have been established based on the varying proportions of clay (<0.002 mm), silt (0.002–0.06 mm), and sand (0.06–2.0 mm). The standard system of classes used by the U.S. Department of Agriculture is shown in **Figure 6.18**. For example, point A on this triangular diagram (left center) represents a soil composed of 10 percent silt, 40 percent clay, and 50 percent sand. Such a soil is called a *sandy clay*. Soils called *loam*, which occupy the central portion of the diagram, are those in which no single particle size predominates over the other two. Loam soils are best suited to support plant life because they generally hold moisture and

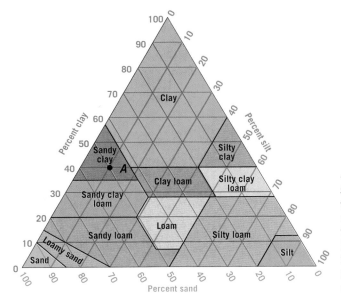

◄ Figure 6.18
Soil texture diagram The texture of any soil can be represented by a point on this diagram. Soil texture is one of the factors used to estimate agricultural potential and engineering characteristics.

nutrients better than do soils composed predominantly of clay or coarse sand.

Soil particles are seldom completely independent of one another. Rather, they usually form clumps called *peds* that give soils a particular structure. The four basic structures recognized in well-developed soils are platy, prismatic, blocky, and spheroidal. Soil structure is important because it influences how easily a soil can be cultivated as well as how susceptible the soil is to erosion. Soil structure also affects a soil's porosity and permeability (the ease with which water can penetrate). This in turn influences the movement of nutrients to plant roots. Prismatic and blocky peds usually allow for moderate water infiltration, whereas platy and spheroidal structures are characterized by slower infiltration rates.

CONCEPT CHECKS 6.5

1. Explain why soil is considered an interface in the Earth system.

2. How is regolith different from soil?

3. Why is texture an important soil property for agriculture?

Concept Checker
https://goo.gl/Fx51Aj

4. Using the soil texture diagram in Figure 6.18, name the soil that consists of 60 percent sand, 30 percent silt, and 10 percent clay.

6.6 Controls of Soil Formation

List and briefly discuss five controls of soil formation.

Soil is the product of the complex interplay of several factors, including parent material, climate, plants and animals, time, and topography. Although all these factors are interdependent, their roles will be examined separately.

Parent Material

The source of the weathered mineral matter from which soils develop, called the **parent material**, is a major factor influencing newly forming soil. Gradually this weathered matter undergoes physical and chemical changes as soil formation progresses. Parent material can either be the underlying bedrock or a layer of unconsolidated deposits. When the parent material is bedrock, the soils are termed *residual soils*, while soils developed on unconsolidated sediment are called *transported soils* (**Figure 6.19**). It should be pointed out that transported soils form *in place* on parent materials that have been carried from elsewhere and deposited by gravity, water, wind, or ice.

Parent material influences soils in two ways. First, the type of parent material influences the rate of weathering and thus the rate of soil formation. Also, because unconsolidated deposits are already partly weathered, soil development on such material will likely progress more rapidly than when bedrock is the parent material. Second, the chemical makeup of the parent material will affect the soil's fertility. This influences the character of the natural vegetation the soil can support.

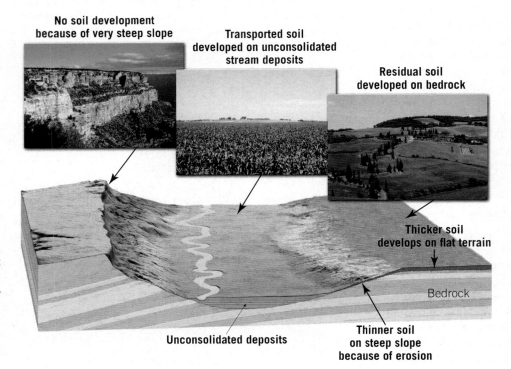

▶ Figure 6.19
Slopes and soil development The parent material for residual soils is the underlying bedrock. Transported soils form on unconsolidated deposits. Also note that as slopes become steeper, soil becomes thinner.

No soil development because of very steep slope

Transported soil developed on unconsolidated stream deposits

Residual soil developed on bedrock

Thicker soil develops on flat terrain

Bedrock

Unconsolidated deposits

Thinner soil on steep slope because of erosion

At one time, the parent material was thought to be the primary factor causing differences among soils. However, soil scientists have come to understand that other factors, especially climate, are more important. In fact, similar soils often develop from different parent materials, while dissimilar soils may develop from the same parent material. Such discoveries reinforce the importance of other soil-forming factors.

Climate

Climate is considered to be the most influential control of soil formation. As noted earlier in this chapter, variations in temperature and precipitation determine whether chemical or mechanical weathering will predominate and also greatly influence the rate and depth of weathering. For instance, a hot, wet climate may produce a thick layer of chemically weathered soil in the same amount of time that a cold, dry climate produces only a thin mantle of mechanically weathered debris. Also, the amount of precipitation influences the degree to which various materials are removed from the soil by surface erosion or by percolating water (a process called *leaching*), thereby affecting soil fertility. Finally, climatic conditions are an important control on the types and numbers of plant and animal life present.

Plants, Animals, and Microbes

Plants, animals, and microbes play vital roles in soil formation. The types and abundance of organisms strongly influence the physical and chemical properties of a soil (**Figure 6.20**). In fact, for well-developed soils

in many regions, the significance of natural vegetation on soil type is implied in the names commonly used by soil scientists, such as *prairie soil*, *forest soil*, and *tundra soil*.

Plants and animals furnish organic matter to the soil. Certain bog soils are composed almost entirely of organic matter, whereas desert soils might contain as little as a small fraction of 1 percent. Although the quantity of organic matter varies substantially among soils, it is a rare soil that completely lacks it.

The primary source of organic matter in soil is plants, although animals and an infinite number of microorganisms also contribute. Decomposed organic matter supplies important nutrients to plants, as well as to animals and microorganisms living in the soil. Consequently, soil fertility is in part related to the amount of organic matter present. Furthermore, the decay of plant and animal remains causes the formation of various organic acids. These complex acids hasten the weathering process. Organic matter also has a high water-holding ability and thus aids water retention in a soil.

Microorganisms, including fungi, bacteria, and single-celled protozoa, play an active role in the decay of plant and animal remains. The end product is humus, a material that no longer resembles the plants and animals from which it is formed. In addition, certain microorganisms aid soil fertility by converting atmospheric nitrogen (N_2) into types of nitrogen that plants can use, such as ammonium (NH_4^-).

Earthworms and other burrowing animals mix the mineral and organic portions of a soil. Earthworms, for example, feed on organic matter and thoroughly

Meager desert rainfall means reduced rates of weathering and relatively meager vegetation. Desert soils are typically thin and lack much organic matter.

In the northern coniferous forest, the organic litter is high in acid resin, which contributes to an accumulation of acid in the soil. As a result, acid leaching is an important soil-forming process.

Soils that develop in well-drained prairie regions typically have a humus-rich surface horizon that is rich in calcium and magnesium. Fertility is usually excellent.

◄ Figure 6.20
Plants influence soil The nature of the vegetation in an area can have a significant influence on soil formation.

mix soils in which they live, often moving and enriching many tons per acre each year. We call this mixing *bioturbation*. Burrows and holes also aid the passage of water and air through the soil.

Time

Time is an important component of *every* geologic process, including soil formation. The nature of soil is strongly influenced by the length of time processes have been operating. If weathering has been going on for a comparatively short time, the character of the parent material strongly influences soil characteristics. As weathering processes continue, the influence of parent material on soil is overshadowed by other soil-forming factors, especially climate. The time required for various soils to evolve cannot be listed because the soil-forming processes act at varying rates under different circumstances. However, as a rule, the longer a soil has been forming, the thicker it becomes and the less it resembles the parent material.

Topography Within a Climate Zone

Variations in land topography can be extreme over short distances, and this can lead to the development of a variety of localized soil types. Many of the differences exist because the length and steepness of slopes significantly affect the amount of erosion and the water content of soil.

On steep slopes, soils are often poorly developed. Due to rapid runoff, the quantity of water soaking in is slight; as a result, the soil's moisture content may not be sufficient for vigorous plant growth. Further, because of accelerated erosion on steep slopes, the soils are thin or in some cases nonexistent (see Figure 6.19).

In contrast, poorly drained and waterlogged soils in bottomlands have a much different character. Such soils are usually thick and dark. The dark color results from the large quantity of organic matter that accumulates as saturated conditions retard the decay of vegetation. The optimum terrain for soil development is a flat-to-undulating upland surface. Here we find good drainage, minimal erosion, and sufficient infiltration of water into the soil.

Slope orientation, or the direction a slope is facing, is another consideration. In the midlatitudes of the Northern Hemisphere, a south-facing slope receives a great deal more sunlight than a north-facing slope. In fact, a steep north-facing slope may receive no direct sunlight at all. The difference in the amount of solar radiation received causes differences in soil temperature and moisture, which in turn influence the nature of the vegetation and the character of the soil.

Although this section deals separately with each of the soil-forming factors, remember that all of them work together to form soil. No single factor is responsible for a soil's character; rather, it is the combined influence of parent material, climate, plants and animals, time, and topography that determines this character.

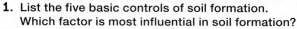

CONCEPT CHECKS 6.6

1. List the five basic controls of soil formation. Which factor is most influential in soil formation?

2. How might the direction a slope is facing influence soil formation?

 **Concept Checker**
https://goo.gl/LXkWPx

6.7 Describing and Classifying Soils

Sketch, label, and describe an idealized soil profile. Explain the need for classifying soils.

The factors controlling soil formation vary greatly from place to place and from time to time, leading to an amazing variety of soil types.

The Soil Profile

Because soil-forming processes operate from the surface downward, soil composition, texture, structure, and color gradually evolve differently at varying depths. These vertical differences, which usually become more pronounced as time passes, divide the soil into zones or layers known as **horizons**. If you were to dig a pit in soil, you would see that its walls are layered. Such a vertical section through all of the soil horizons constitutes the **soil profile**.

Figure 6.21 presents an idealized view of a well-developed soil profile in which five horizons are

identified. From the surface downward, they are designated *O*, *A*, *E*, *B*, and *C*. These five horizons are common to soils in temperate regions; not all soils have these five layers. The characteristics and extent of horizon development vary in different environments. Thus, different localities exhibit soil profiles that can contrast greatly with one another:

- The *O soil horizon* consists largely of organic material, in contrast to the layers beneath it, which consist mainly of mineral matter. The upper portion of the *O* horizon is primarily plant litter, such as loose leaves and other still-recognizable organic debris.

"E" horizon). Water percolating downward also dissolves soluble inorganic soil components and carries them to deeper zones. This depletion of soluble materials from the upper soil is termed **leaching**. The E horizon is also known as the *zone of leaching.*

- The *B horizon*, or *subsoil*, is where much of the material removed by eluviation from the E horizon is deposited. Thus, the B horizon is often referred to as the *zone of accumulation.* The accumulation of the fine clay particles enhances this horizon's ability to hold water. In extreme cases, clay accumulation can form a very compact, impermeable layer called *hardpan.*

- The *C horizon* is characterized by partially altered parent material. Whereas the parent material is difficult to see in the O, A, E, and B horizons, it is easily identifiable in the C horizon. Although this material is undergoing changes that will eventually transform it into soil, it has not yet crossed the threshold that separates regolith from soil.

The characteristics and extent of development can vary greatly among soils in different environments (**Figure 6.22**). The boundaries between soil horizons may be sharp, or the horizons may blend gradually from one to another. Consequently, a well-developed soil profile indicates that environmental conditions have been relatively stable over an extended time span and that the soil is *mature.* By contrast, some soils lack horizons altogether. Such soils are called *immature* because soil building has been going on for only a short

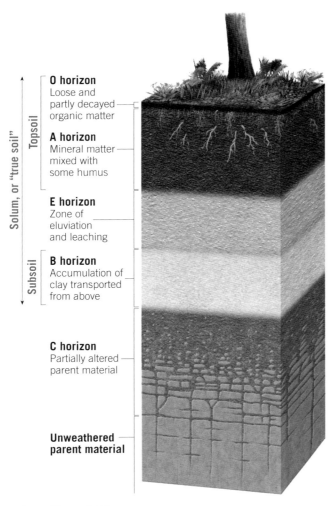

O horizon
Loose and partly decayed organic matter

A horizon
Mineral matter mixed with some humus

E horizon
Zone of eluviation and leaching

B horizon
Accumulation of clay transported from above

C horizon
Partially altered parent material

Unweathered parent material

Solum, or "true soil"
Topsoil
Subsoil

▲ SmartFigure 6.21
Soil horizons Idealized soil profile from a humid climate in the middle latitudes.

Tutorial
https://goo.gl/9G8gc5

By contrast, the lower portion of the O horizon is made up of partly decomposed organic matter (humus) in which plant structures can no longer be identified. The O horizon is also teeming with microscopic life, including bacteria, fungi, algae, and insects. All these organisms contribute oxygen, carbon dioxide, and organic acids to the developing soil.

- The *A horizon* is largely mineral matter, yet biological activity is high, and humus is generally present—up to 30 percent in some instances. Together, O and A horizons make up what is commonly called the *topsoil.*

- The *E horizon* is a light-colored layer that contains little organic material. As water percolates downward through this zone, finer particles are carried away. This washing out of fine soil components is termed **eluviation** (thus the out-of-order

Horizons are indistinct in this soil in Puerto Rico, giving it a relatively uniform appearance.

This profile shows a soil in southeastern South Dakota with well-developed horizons.

◄ Figure 6.22
Contrasting soil profiles
Soil characteristics and development vary greatly in different environments.

TABLE 6.2   Basic Soil Orders

Soil Order	Description	Productivity Index Rating	World's Surface Percentage*
Alfisol	Moderately weathered soils formed under boreal forests or broadleaf deciduous forests; rich in iron and aluminum. Clay particles accumulate in a subsurface layer due to leaching in moist environments. Fertile and productive because they are neither too wet nor too dry.	10	9.65
Andisol	Young soils in which the parent material is volcanic ash and cinders, deposited by recent volcanic activity.	11	0.7
Aridosol	Soils that develop in dry places with insufficient water to remove soluble minerals; may have calcium carbonate, gypsum, or salt accumulation in subsoil; low organic content.	5	12.02
Entisol	Young soils with limited development and exhibiting properties of the parent material. Productivity ranges from very high for some forming on recent river deposits to very low for those forming on shifting sand or rocky slopes.	6	16.16
Gelisol	Young soils with little profile development, found in regions with permafrost. Low temperatures and frozen conditions for much of the year; slow soil-forming processes.	8	8.61
Histosol	Organic soils found in any climate where organic debris accumulates to form a bog soil. Dark, partially decomposed organic material commonly referred to as *peat*.	14	1.17
Inceptisol	Weakly developed young soils showing the beginning (inception) of profile development. Most common in humid climates but found from the arctic to the tropics. Native vegetation is most often forest.	9	9.81
Mollisol	Dark, soft soils developed under grass vegetation, generally found in prairie areas. Humus-rich surface horizon that is rich in calcium and magnesium; excellent fertility. Also found in hardwood forests with significant earthworm activity. Climatic range is boreal or alpine to tropical. Dry seasons are normal.	13	6.89
Oxisol	Soils formed on old land surfaces unless parent materials were strongly weathered before they were deposited. Generally found in the tropics and subtropical regions. Rich in iron and aluminum oxides, oxisols are heavily leached and hence are poor soils for cultivation.	3	7.5
Spodosol	Soils found only in humid regions on sandy material. Common in northern coniferous forests and cool humid forests. Beneath the dark upper horizon of weathered organic material lies a light-colored leached horizon, the distinctive property of this soil.	7	2.56
Ultisol	Soils representing the products of long periods of weathering. Percolating water concentrates clay particles in the lower horizons. Restricted to humid climates in the temperate regions and the tropics, where the growing season is long. Abundant water and a long frost-free period contribute to extensive leaching and poor fertility.	4	8.45
Vertisol	Soils containing large amounts of clay, which shrink when dry and swell with the addition of water. Found in subhumid to arid climates if sufficient water is available to saturate the soil after periods of drought. Soil expansion and contraction exert stresses on human structures.	12	2.24

*Percentages refer to the world's ice-free surface.

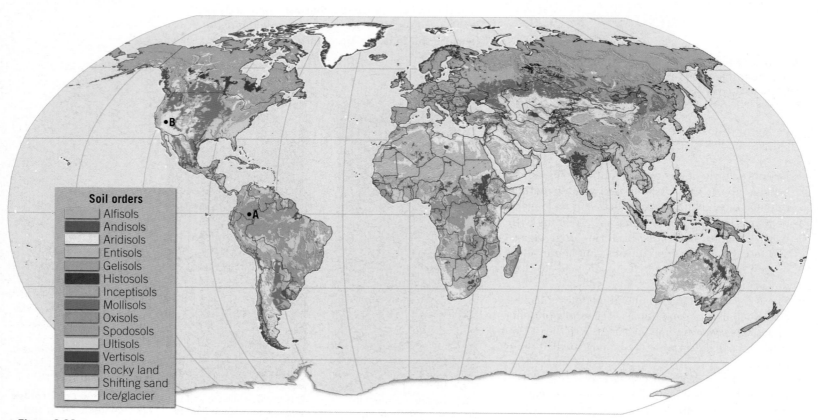

Soil orders
- Alfisols
- Andisols
- Aridisols
- Entisols
- Gelisols
- Histosols
- Inceptisols
- Mollisols
- Oxisols
- Spodosols
- Ultisols
- Vertisols
- Rocky land
- Shifting sand
- Ice/glacier

▲ Figure 6.23
Global soil regions Worldwide distribution of the Soil Taxonomy's 12 soil orders. Points A and B are references for a *Give It Some Thought* item at the end of the chapter.

time, such as in recently deglaciated landscapes. Immature soils are also characteristic of steep slopes, where erosion continually strips away the soil, preventing full development.

Classifying Soils

The great variety of soils on Earth makes it essential to devise some means of classifying the vast array of soil data. Soil scientists in the United States have devised a system for classifying soils known as the **soil taxonomy**. It emphasizes the physical and chemical properties of the soil profile and is organized on the basis of observable soil characteristics. There are 6 hierarchical categories of classification, ranging from *order*, the broadest category, to *series*, the most specific category. The system recognizes 12 soil orders and more than 19,000 soil series.

The names of the classification units are mostly combinations of Latin or Greek descriptive terms. For example, soils of the order aridosol (from the Latin *aridus* = dry and *solum* = soil) are characteristically dry soils in arid regions. Soils in the order inceptisol (from Latin *inceptum* = beginning and *solum* = soil) are soils with only the beginning, or inception, of profile development.

How productive, or able to grow specific crops, a soil is can be described using the *productivity index*.

This scale goes from 0 (least productive) to 19 (most productive). Brief descriptions of the 12 basic soil orders are provided in **Table 6.2**. The complex worldwide distribution pattern of the soil taxonomy's 12 soil orders is shown in **Figure 6.23**. Like many other classification systems, the soil taxonomy is not suitable for every purpose. It is especially useful for agricultural and related land-use purposes, but it is not a useful system for engineers who are preparing evaluations of potential construction sites.

CONCEPT CHECKS 6.7

1. Sketch and label the main soil horizons in a well-developed soil profile.

2. Describe the following features or processes: eluviation, leaching, zone of accumulation, and hardpan.

Concept Checker
https://goo.gl/5p4Rt4

3. Why are soils classified?

4. Examine Figure 6.23 and identify three particularly extensive soil orders that occur in the contiguous 48 United States. Describe two soil orders in Alaska.

6.8 The Impact of Human Activities on Soil

Explain the detrimental impact of human activities on soil and list several ways to combat soil erosion.

Soils are just a tiny fraction of all Earth materials, yet they are vital for the growth of rooted plants and thus are a basic foundation of the human life-support system. Because soil forms very slowly, it must be considered a nonrenewable resource. Just as human ingenuity can increase the agricultural productivity of soils through fertilization and irrigation, soils can be damaged or destroyed by careless activities. Despite their role in providing food, fiber, and other basic materials, soils are among our most abused resources.

Clearing the Tropical Rain Forest: A Case Study of Human Impact on Soil

Over the past few decades, the destruction of tropical forests has become a serious environmental issue. Each year millions of acres are cleared for agriculture and logging (**Figure 6.24**). This clearing results in soil degradation, loss of biodiversity, and climate change.

Thick red-orange soils (oxisols) are common in the wet tropics and subtropics (see Figure 6.23). They are the end product of extreme chemical weathering. Because lush tropical rain forests are associated with these soils, many people assume that they are fertile and have great potential for agriculture. However, just the

◀ Figure 6.24
Tropical deforestation Clearing the Amazon rain forest in Suriname. The thick soils (oxisols) are highly leached. Clearing of the tropical rain forest is a serious environmental problem.

opposite is true: Oxisols are among the poorest soils for farming. How can this be?

Rain forest soils develop under conditions of high temperature and heavy rainfall and are therefore severely leached. Not only does leaching remove the soluble materials such as calcium carbonate, but the great quantities of percolating rainwater also remove much of the silica, and as a result, insoluble oxides of iron and aluminum become concentrated in the soil. Iron oxides give the soil its distinctive color. Because bacterial activity is high in the wet tropics, organic matter quickly breaks down, and rain forest soils contain very little humus. Moreover, leaching destroys fertility because most plant nutrients in the soil are removed by the large volume of downward-percolating water. Despite the dense and luxuriant rain forest vegetation, the soil itself contains few available nutrients.

Most nutrients that support the rain forest are locked up in the trees themselves. As vegetation dies and decomposes, the roots of the rain forest trees quickly absorb the nutrients before they are leached from the soil. The nutrients are continuously recycled as trees die and decompose. Therefore, when forests are cleared to provide land for farming or to harvest timber, most of the nutrients are removed as well. What remains is a soil that contains little to nourish planted crops.

Rain forest clearing not only removes plant nutrients but also accelerates soil erosion. The roots of rain forest vegetation anchor the soil, and leaves and branches provide a canopy that protects the ground by deflecting the full force of the frequent heavy rains. When the protective vegetation is gone, soil erosion increases.

"Homo sapiens, wise man indeed. There's still time to live up to our name—if only we stop treating our soil like dirt."—Dr. David Montgomery, author of Dirt: The Erosion of Civilizations

The removal of vegetation also exposes the ground to strong direct sunlight. When baked by the Sun, these tropical soils can harden to a bricklike consistency and become practically impenetrable to water and crop roots. In just a few years, a freshly cleared area may no longer be cultivable.

Soil Erosion: Losing a Vital Resource

Many people do not realize that soil erosion—the removal of topsoil—is a serious environmental problem. Perhaps this is the case because a substantial amount of soil seems to remain even where soil erosion is serious. Nevertheless, although the loss of fertile topsoil may not be obvious to the untrained eye, it is a significant and growing problem as human activities expand and disturb more and more of Earth's surface.

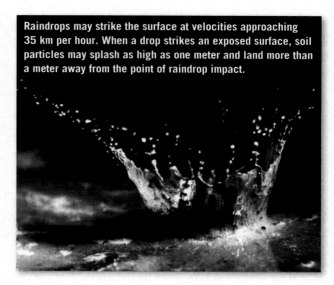

Raindrops may strike the surface at velocities approaching 35 km per hour. When a drop strikes an exposed surface, soil particles may splash as high as one meter and land more than a meter away from the point of raindrop impact.

▲ **Figure 6.25**
Raindrop impact Soil dislodged by raindrop impact is more easily moved by sheet erosion.

Soil erosion is a natural process; it is part of the constant recycling of Earth materials that we call the *rock cycle*. Once soil forms, erosional forces, especially water and wind, move soil components from one place to another. Every time it rains, raindrops strike the land with surprising force (**Figure 6.25**). Each drop acts like a tiny bomb, blasting movable soil particles out of their positions in the soil mass. Then, water flowing across the surface carries away the dislodged soil particles. Because the soil is moved by thin sheets of water, this process is termed *sheet erosion*.

After the water flows as a thin, unconfined sheet for a relatively short distance, threads of current typically develop, and tiny channels called *rills* begin to form. Still deeper cuts in the soil, known as *gullies*, are created as rills enlarge (**Figure 6.26**). When normal farm cultivation cannot eliminate the channels, we know the rills have grown large enough to be called gullies. Although most dislodged soil particles move only a short distance during each rainfall, substantial quantities eventually leave the fields and make their way downslope to a stream. Once in the stream channel, these soil particles, which can now be called *sediment*, are transported downstream and eventually deposited.

Rates of Erosion   We know that soil erosion is the ultimate fate of practically all soils. In the past, erosion occurred at slower rates than it does today because more of the land surface was covered with and protected by

Severe sheet and rill erosion on an Iowa farm following heavy rains. Just one millimeter of soil from a single acre amounts to about 5 tons.

A.

Gully erosion on unprotected soil on a Wisconsin farm.

B.

◀ **Figure 6.26**
Erosion of unprotected soil
A. Sheetflow and rills.
B. Rills can grow into deep gullies.

trees, shrubs, grasses, and other plants. However, human activities such as farming, logging, and construction, which remove or disrupt the natural vegetation, have greatly accelerated the rate of soil erosion. Without the stabilizing effect of plants, the soil is more easily swept away by the wind or carried downslope by sheet wash.

Natural rates of soil erosion vary greatly from one place to another and depend on soil characteristics as well as factors such as climate, slope, and type of vegetation. Over a broad area, erosion caused by surface runoff may be estimated by determining how much sediment is carried by the streams that drain the region. Studies of this kind made on a global scale indicate that prior to the appearance of humans, sediment transport by rivers to the ocean amounted to just over 9 billion metric tons per year. In contrast, the amount of material currently transported to the sea by rivers is about 24 billion metric tons per year—or more than 2.5 times the pre-human rate.

It is estimated that flowing water is responsible for about two-thirds of the soil erosion in the United States. Much of the remainder is caused by wind. When dry conditions prevail, strong winds can remove large quantities of soil from unprotected fields (**Figure 6.27**). At present, it is estimated that topsoil is

eroding faster than it forms on more than one-third of the world's croplands. The results—lower productivity, poorer crop quality, and reduced agricultural income— add up to an ominous future.

Controlling Soil Erosion   On every continent, perilous soil loss is occurring because well-known conservation measures are not being taken. Although we recognize that soil erosion can never be completely eliminated, soil conservation programs can substantially reduce the loss of this basic resource.

Steepness of slope is an important factor in soil erosion. The steeper the slope, the faster the water runs off and the greater the erosion. It is best to leave steep slopes undisturbed, but when such slopes are farmed, terraces

The man is pointing to where the ground surface was when the grasses began to grow. Wind erosion lowered the land surface to the level of his feet.

Clumps of anchored soil

Unanchored soil

Sand dune

1.2 meters

◀ **Figure 6.27**
Wind erosion When the land is dry and largely unprotected by anchoring vegetation, soil erosion by wind can be significant.

The 1930s Dust Bowl
An Environmental Disaster

During a span of dry years in the 1930s, large dust storms plagued the Great Plains. Topsoil was stripped from millions of acres. Because of the size and severity of these storms, the region came to be called the Dust Bowl and the time period the Dirty Thirties.

In places, dust drifted like snow, covering farm buildings, fences, and fields. Crop failure and economic hardship resulted in many farms being abandoned.

OTHER AREAS SEVERELY AFFECTED BY DUST STORMS

HEART OF THE DUST BOWL

The southern Great Plains were most severely affected.

WA
ND
MT
SD
MN
WY
IA
OR
ID
UT
NE
CA
NV
CO
KS
OK
AZ
NM
TX

Dust blackens the sky near Elkhart, Kansas, on May 21, 1937. The transformation of semiarid grasslands into farms during an unusually wet period set the stage for this disastrous period of soil erosion. When drought struck, the unprotected soils were vulnerable to the wind.

◄**Figure 6.28**
Soil conservation Crops on a farm in northeastern Iowa are planted to decrease water erosion.

Crops planted in strips along contours of hillside

Corn

Grass planted along drainage

Hay

Corn and hay have been planted in strips that follow the contours of the hillside. This pattern reduces soil loss because it slows the rate of water runoff.

can be constructed. These nearly flat, steplike surfaces slow runoff and thus decrease soil loss while allowing more water to soak into the ground.

Soil erosion by water also occurs on gentle slopes. **Figure 6.28** illustrates one conservation method in which crops are planted parallel to the contours of the slope. This pattern reduces soil loss by slowing runoff. Strips of grass or cover crops such as hay slow runoff even more and act to promote water infiltration and trap sediment.

Probably the most effective soil preservation measure is "no-till" farming, where crop residues are left on fields, and the soil is not disturbed by tilling. Creating grassed waterways is another common practice (**Figure 6.29**). Natural drainageways are shaped to form smooth, shallow channels and then planted with grass. The grass prevents

the formation of gullies and traps soil washed from cropland. To protect fields from excessive wind erosion, rows of trees and shrubs are planted as windbreaks to slow the wind and deflect it upward (**Figure 6.30**).

These flat expanses are susceptible to wind erosion, especially when the fields are bare. The rows of trees slow and deflect the wind, which decreases the loss of top soil.

◄**Figure 6.30**
Reducing wind erosion Windbreaks protect wheat fields in North Dakota.

The grassed waterway prevents the formation of gullies and traps soil washed from cropland.

▲**Figure 6.29**
Reducing erosion by water Grassed waterway on a Pennsylvania farm.

CONCEPT CHECKS 6.8

1. Why are soils in tropical rain forests not well suited for intensive farming?

2. Place these phenomena related to soil erosion in the proper sequence: sheet erosion, gullies, raindrop impact, rills, stream.

 **Concept Checker**
https://goo.gl/GLcVAd

3. Explain how human activities have affected the rate of soil erosion.

4. Briefly describe three ways to control soil erosion.

6 CONCEPTS IN REVIEW

Weathering and Soils

6.1 Weathering

Define *weathering* and distinguish between the two main types.

Key Terms:
weathering

mechanical
weathering

chemical weathering
erosion

- *Weathering* is the disintegration and decomposition of rocks on the surface of Earth. Rocks may break into many smaller pieces through physical processes called *mechanical weathering*. Rocks also decompose through *chemical weathering*, where minerals react with environmental agents, such as oxygen and water, to produce new substances that are stable at Earth's surface.

Q Does the bicycle shown in the photo provide an example of mechanical or chemical weathering?

6.2 Mechanical Weathering

List and describe four examples of mechanical weathering.

Key Terms:
frost wedging

joint
sheeting

exfoliation dome

- Mechanical weathering forces include the expansion of ice, the crystallization of salt, and the growth of plant roots. All work to pry apart grains and enlarge fractures in rock.

- Rocks that form under lots of pressure deep in Earth expand when exposed at the surface. Sometimes this expansion is great enough to cause the rock to break into onion-like layers. This *sheeting* can generate broad dome-shaped exposures of rock called *exfoliation domes*.

6.3 Chemical Weathering

Discuss the role of water in each of three chemical weathering processes.

Key Terms:
dissolution

oxidation
hydrolysis

spheroidal weathering

- Water plays an important role in the chemical reactions that take place at the surface of Earth. It frees and transports ions from some minerals through *dissolution*. Water can also facilitate reactions such as rusting, an example of *oxidation*. Acid mine drainage is an environmental consequence of the oxidation of pyrite in old coal mines.

- Water also directly reacts with exposed minerals, producing new minerals that are stable at Earth's surface. The *hydrolysis* of feldspar that forms kaolinite clay is an example. Clays are stable minerals at Earth's surface conditions, and they are profusely generated by the hydrolysis of silicate minerals. As a result, clay is a common constituent of soil and sedimentary rocks.

- *Spheroidal weathering* results when sharp edges and corners of rocks are chemically weathered more rapidly than flat rock faces. The higher proportion of surface area for a given volume of rock at the edges and corners means there is more mineral material exposed to chemical attack. Faster weathering at the corners produces weathered rocks that become increasingly sphere shaped over time.

6.4 Rates of Weathering

Summarize the factors that influence the type and rate of rock weathering.

Key Term:
differential weathering

- Some rocks are more stable at Earth's surface than others, due to the minerals they contain. Different minerals break down at different rates under the same conditions. Quartz is the most stable silicate mineral, while minerals that crystallize early in Bowen's reaction series, such as olivine, tend to decompose more rapidly.

- Rock weathers most rapidly in an environment with lots of heat to drive reactions and water to facilitate those reactions. Consequently, rocks decompose relatively quickly in hot, wet climates and slowly in cold, dry conditions.

- Frequently, rocks exposed at Earth's surface do not weather at the same rate. This *differential weathering* of rocks is influenced by factors such as mineral composition and degree of jointing. In addition, if a rock mass is protected from weathering by another, more resistant rock, then it will weather at a slower rate than a fully exposed equivalent rock. Differential weathering produces many of our most spectacular landforms.

Q Imagine that we broke an unweathered sample of granite into two equal pieces. We put one piece in the Dry Valleys of Antarctica and the the other in the Amazon rain forest. Which one will weather more rapidly, and why? How would the products of weathering differ between the two places?

6.5 Soil

Define *soil* and explain why soil is referred to as an *interface*.

Key Terms:	soil	soil texture
regolith	humus	

- *Soils* are vital combinations of organic and inorganic components found at the interface where the geosphere, atmosphere, hydrosphere, and biosphere meet. This dynamic zone is the overlap between different parts of the Earth system. It includes the *regolith*'s rocky debris, mixed with *humus*, water, and air.

- *Soil texture* refers to the proportions of different particle sizes (clay, silt, and sand) found in soil. Soil particles often form clumps called peds that give soils a particular structure.

Q Label the four components of a soil on this pie chart.

6.6 Controls of Soil Formation

List and briefly discuss five controls of soil formation.

Key Term: parent material

- Residual soils form in place due to the weathering of bedrock, whereas transported soils develop on unconsolidated sediment.
- Soils formed in different climates are different in part due to temperature and moisture differences but also due to the organisms that live in those different environments. These organisms can add organic matter or chemical compounds to the developing soil or can help mix the soil through their growth and movement.
- It takes time for soil to form. Soils that have developed for a longer period of time will have different characteristics than young soils. In addition, some minerals break down more readily than others. Soils produced from the weathering of different parent rocks are produced at different rates.
- The steepness of the slope on which a soil is forming is a key variable, with shallow slopes retaining their soils and steeper slopes losing them to accumulate elsewhere.

6.7 Describing and Classifying Soils

Sketch, label, and describe an idealized soil profile. Explain the need for classifying soils.

Key Terms:	soil profile	leaching
horizon	eluviation	Soil Taxonomy

- Despite the great diversity of soils around the world, there are some broad patterns to the vertical anatomy of soil layers. Organic material, called humus, is added at the top (*O* horizon), mainly from plant sources. There, it mixes with mineral matter (*A* horizon). At the bottom, bedrock breaks down and contributes mineral matter (*C* horizon). In between, some materials are leached out or *eluviated* from higher levels (*E* horizon) and transported to lower levels (*B* horizon), where they may form an impermeable layer called hardpan.
- The need to bring order to huge quantities of data motivated the establishment of a classification scheme for the world's soils. This *Soil Taxonomy* features 12 broad orders.

6.8 The Impact of Human Activities on Soil

Explain the detrimental impact of human activities on soil and list several ways to combat soil erosion.

- The clearing of tropical rain forests is an issue of concern. Most of the nutrients in the tropical rain forest ecosystem are not in the soil but in the trees themselves. When the trees are removed, most of the nutrients are removed. The loss of vegetation also makes the soils highly susceptible to erosion. Once cleared of vegetation, soils may also be baked by the Sun into a bricklike consistency.
- Soil erosion is a natural process, part of the constant recycling of Earth materials that we call the rock cycle. But human activities have increased soil erosion rates over the past several hundred years. Because natural soil production rates are constant, there is a net loss of soil at a time when a record-breaking number of people live on the planet.
- Using no-till farming, plowing the land along horizontal contour lines, and installing windbreaks, terraces, and grassed waterways are all practices that have been shown to reduce soil erosion.

Q Why was this row of evergreens planted on an Indiana farm?

GIVE IT SOME THOUGHT

1. How are the two main categories of weathering represented in this image that shows human-made objects?

2. Describe how plants contribute to mechanical and chemical weathering but inhibit erosion.

3. Granite and basalt are exposed at Earth's surface in a hot, wet region. Which rock will weather more rapidly? Why?

4. The accompanying photo shows Shiprock, a well-known landmark in the northwestern corner of New Mexico. It is a mass of igneous rock that represents the "plumbing" of a now-vanished volcanic feature. Extending toward the upper left is a related wall-like igneous structure known as a dike. The igneous features are surrounded by sedimentary rocks. Explain why these once deeply buried igneous features now stand high above the surrounding terrain. What term in Section 6.4, "Rates of Weathering," applies to this situation?

5. Due to burning of fossil fuels such as coal and petroleum, the level of carbon dioxide in the atmosphere has been increasing for more than 150 years. Will this increase tend to accelerate or slow down the rate of chemical weathering of Earth's surface rocks? Explain how you arrived at your conclusion.

6. In Chapter 4, you learned that feldspars are very common minerals in igneous rocks. When you learn about the common minerals that compose sedimentary rocks in Chapter 7, you will find that feldspars are relatively rare. Applying what you have learned about chemical weathering, explain why this is true. Based on this explanation, what mineral might you expect to be common in sedimentary rocks that is not found in igneous rocks?

7. What might cause different soils to develop from the same kind of parent material or similar soils to form from different parent materials?

8. Using the map of global soil regions in Figure 6.23, identify the main soil order in the region adjacent to South America's Amazon River (point A on the map) and the predominant soil order in the American Southwest (point B). Briefly contrast these soils. Do they have anything in common? Referring to Table 6.2. might be helpful.

9. This soil sample is from a farm in the Midwest. From which horizon was the sample most likely taken— *A, E, B,* or *C*? Explain.

EYE ON EARTH

1. This is a close-up view of a massive granite feature in the Sierra Nevada of California.

   a. Relatively thin slabs of granite are separating from this rock mass. Describe the process that caused this to occur.

   b. What term is applied to this process? What term describes the dome-like feature that results?

2. This sample of granite is rich in potassium feldspar and quartz. There are minor quantities of biotite and hornblende.

   a. If this rock were to undergo chemical weathering, how would its minerals change? Describe the products you would expect to result from each of the minerals in the sample.

   b. Would all of the minerals decompose? If not, which mineral would likely be most resistant and remain relatively intact?

3. The rounded boulders in this image gradually formed in place from a rock mass that had many fractures. Initially the rocks had sharp corners and edges.

    a. Explain the process that transformed angular blocks of bedrock into rounded boulders.

    b. What term is applied to this process?

4. This thick red soil is exposed somewhere in the United States and is either a gelisol, a mollisol, or an oxisol.

    a. Refer to the descriptions in Table 6.2. and determine the likely soil order shown in this image. Explain your choice.

    b. Which state is the most likely location of the soil: Alaska, Illinois, or Hawaii?

DATA ANALYSIS

Soil Types

Soil types have been mapped across the United States. This information is used by farmers and planners to determine how the land can be used.

https://goo.gl/eD226M

ACTIVITIES

Go to the USDA's Web Soil Survey, at http://websoilsurvey.sc.egov.usda.gov. Click on the "Start WSS" button. Select "State and County" under "Quick Navigation"; enter your state and county and then click "View." Click on the "i" (information) tool and then click on your location to display location information.

1. What are the latitude and longitude for the location you chose?

2. When was this aerial photograph taken?

Create an area of interest (AOI) around your location by clicking the "AOI" rectangle tool and dragging the mouse to create a space around the location you chose. This area should be larger than just a few streets in order to capture some variation. When a striped area appears, click the "Soil Map" tab near the top of the page. Click the "Map Unit Legend" to see soil types in your area and then click on items in the "Map Unit Name" column for details.

3. What type(s) of soil is(are) present at your location?

4. Based on Figure 6.18 (page 185), what is the percentage range in your soil for sand? Silt? Clay?

Click the "Soil Data Explorer" tab near the top of the page and be sure "Suitabilities and Limitations for Use" is selected. Select "Land Classifications" and choose "Farmland Classification"; then click "View Rating."

5. Click "View Description" and determine what information is displayed.

6. What is the predominant rating for this area? What other ratings are available in your area?

7. What percentage is prime farmland? Does this make sense, based on what you know about your area?

Select "Land Management" and choose "Erosion Hazard (Off-Road, Off-Trail)"; then click "View Rating."

8. Click "View Description" and determine what information is displayed.

9. What is the predominant rating for this area? What other ratings are available in your area? If your area is not rated, find a nearby place that has ratings.

10. Based on what you know about your area's physical features (flat or hilly, urban or rural, and so on), what can you say about the relationship between erosion hazards and the characteristics of the area?

Mastering Geology

Looking for additional review and test prep materials? Visit pearson.com/mastering/geology to enhance your understanding of this chapter's content by accessing a variety of resources, including **Self-Study Quizzes, Geoscience Animations, SmartFigure Tutorials,** *Project Condor* **Videos, Mobile Field Trips, Dynamic Study Modules,** and an optional **Pearson eText.**

When Did Plants Begin to Live on Dry Land— And How Did That Impact Earth's Geology?

For years teachers have told students that the fossil record in sedimentary rocks tells us that Earth's continents were devoid of plant life until about 420 million years ago, when various species began evolving in ways that allowed them to leave water environments and venture onto dryer land. However, recent studies are turning that "fact" on its head.

▲ 1.6-billion-year-old fossils of red algae from rocks in India.

Application of newer molecular biology techniques to the incomplete fossil record estimate that ancestors of land plants were likely in place about *100 million years* earlier than originally thought. This research uses a "molecular clocks" method to compare the genetic differences among the four main kinds of land plants to estimate their evolutionary history. The studies concluded that plants likely began colonizing land around 520 million years ago, about the same time that animals did.

Why does this matter? Recall from earlier chapters that weathering of silicate minerals and deposition of sedimentary rocks pulls carbon out of the atmosphere, a process that ultimately leads to climate cooling. Terrestrial plant life-forms interact with sedimentary rocks and increase their weathering. Therefore, understanding the timing of land plant colonization is critical for understanding Earth's climate history.

► Weathering of (mostly) sedimentary rocks on the continents greatly increased after plants evolved to live on land. This is one example of the important interactions of Earth's biosphere, geosphere, hydrosphere, and atmosphere in our planet's history (Cretaceous age Elbe sandstone in the Saxony region of Germany).

7
Sedimentary Rocks

FOCUS ON CONCEPTS

Each statement represents the primary learning objective
for the corresponding major heading within the chapter.
After you complete the chapter, you should be able to:

7.1 Explain the importance of sedimentary rocks
as indicators of past environments and sources
of resources necessary for modern society.
Summarize the part of the rock cycle that pertains to sediments and sedimentary rocks. List
the three categories of sedimentary rocks.

7.2 Describe the primary basis for distinguishing
among clastic rocks and describe how the origin
and history of such rocks might be determined.

7.3 Explain the processes involved in the formation
of chemical sedimentary rocks and describe
several examples.

7.4 Outline the successive stages in the formation
of coal.

7.5 Describe the processes that convert sediment
into sedimentary rock and other changes associated with burial.

7.6 Summarize the criteria used to classify sedimentary rocks.

7.7 Distinguish among three broad categories
of sedimentary environments and provide an
example of each. List several sedimentary structures and explain why these features are useful
to geologists.

7.8 Relate weathering processes and sedimentary
rocks to the carbon cycle.

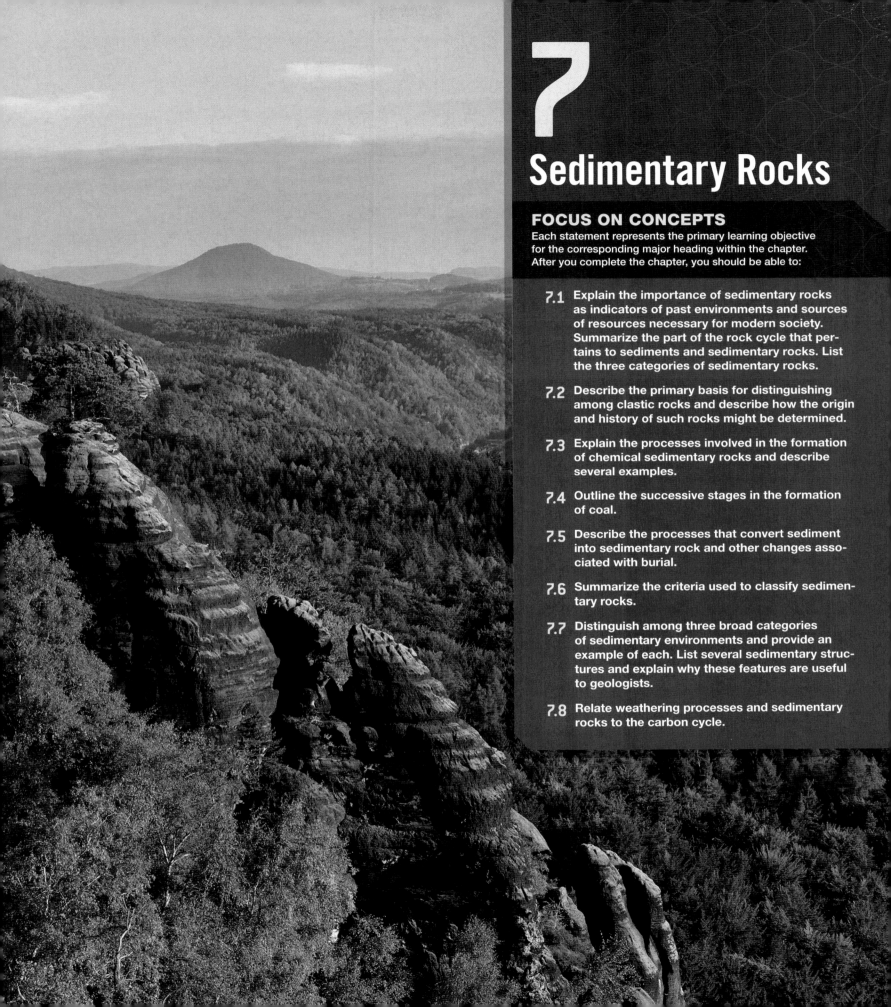

Chapter 6 provided the background you need to understand the origin of sedimentary rocks. Recall that weathering of existing rocks begins the process. Next, gravity and agents of erosion such as running water, wind, and glacial ice remove the products of weathering and carry them to a new location, where they are deposited. Usually the particles are broken down further during this transport phase. Following deposition, this material, which is now called *sediment*, becomes *lithified* (turned to rock). It is from sedimentary rocks that geologists reconstruct many details of Earth's history. Because sediments are deposited in a variety of settings at the surface, the rock layers that they eventually form hold many clues about past surface environments. A layer may represent a desert sand dune, the muddy floor of a swamp, or a tropical coral reef. There are many possibilities. Many sedimentary rocks are associated with important energy and mineral resources and are therefore important economically as well.

7.1 An Introduction to Sedimentary Rocks

Explain the importance of sedimentary rocks as indicators of past environments and sources of resources necessary for modern society. Summarize the part of the rock cycle that pertains to sediments and sedimentary rocks. List the three categories of sedimentary rocks.

Sediments or sedimentary rock comprise about three-quarters of the Earth's solid materials at the surface. On the ocean floor, which represents about 70 percent of Earth's surface, virtually everything is covered by sediment, with igneous rock exposed only at the crests of mid-ocean ridges and in some volcanic areas. Interestingly, looking at a fuller cross section of the total volume of crust, going from surface to 16 kilometers (10 miles) below, we find that sedimentary rock is only 5 percent of the total rock volume, with the vast majority being igneous or metamorphic rock.

▶ Figure 7.1
Sedimentary rocks record change Because they contain fossils and other clues about the geologic past, sedimentary rocks are important in the study of Earth history. Vertical changes in rock types represent environmental changes through time. These strata are exposed at Karijini National Park, Western Australia.

Importance

Being concentrated at or near the surface, sedimentary rock is an interface among the geosphere, hydrosphere, atmosphere, and biosphere. Sediments and the rock layers they eventually form therefore contain evidence of past conditions and events at the surface. Based on the compositions, textures, structures, and fossils in sedimentary rock, experienced geologists can decipher clues that provide insights into past climates, ecosystems, and ocean environments. Furthermore, by studying the many kinds of sedimentary rock, geologists can reconstruct the configuration of ancient landmasses and the locations and compositions of long-vanished mountain systems. In short, sedimentary rock provides geologists with much of the basic information needed to reconstruct the details of Earth history (**Figure 7.1**).

The study of sedimentary rock has economic significance as well. Coal, which still provides a significant portion of our electrical energy, is classified as a type of sedimentary rock. Other major energy sources—including oil and natural gas—occur in pores with sedimentary rock. Sedimentary rock is also the major source of iron, aluminum, manganese, and phosphate for manufacturing goods and fertilizers, plus numerous materials that are essential to the construction industry, such as cement, aggregate, and gypsum. Sediments and sedimentary rock are also the primary reservoir of groundwater. Thus, having an understanding of these rocks and the processes that form and modify them is basic to locating and maintaining supplies of many important resources.

Origins

Like other rocks, **sedimentary rocks** have their origin in the rock cycle. **Figure 7.2** illustrates the portion of the rock cycle that occurs near Earth's surface and pertains to sediments and sedimentary rocks. A brief overview of these processes provides a useful perspective:

- **Weathering:** Weathering preexisting igneous, metamorphic, and sedimentary rocks generates a variety of products subject to erosion, including various solid particles and ions in solution. These are the raw materials for sedimentary rocks.

- **Transport:** Sediment usually moves from sites of origination to places of accumulation. Soluble constituents dissolve and are carried away by runoff and groundwater. Solid particles move downslope by gravity, a process termed *mass wasting*, and then running water, groundwater, wave activity, wind, and glacial ice remove them. Sediment transport is usually intermittent. For example, during a flood, a rapidly moving river moves large quantities of sand and gravel. As the floodwaters recede, particles are temporarily deposited, only to be moved again by a subsequent flood.

- **Deposition:** Transported solid particles eventually get deposited when wind and water currents slow down and as glacial ice melts. The word *sedimentary* actually refers to this process. It is derived from the Latin *sedimentum*, which means "to settle," a reference to solid material settling out of a fluid (water or air). The mud on the floor of a lake, a delta at the mouth of a river, a gravel bar in a streambed, the particles in a desert sand dune, and even household dust are examples of sediment produced by this never-ending process. The deposition of material dissolved in water is not related to the strength of water currents. Rather, ions in solution are removed when chemical or temperature changes cause material to crystallize and precipitate (solidify out of a liquid

▼ **SmartFigure 7.2**
From sediment to sedimentary rock This diagram shows the portion of the rock cycle that pertains to the formation of sedimentary rocks.

Tutorial
https://goo.gl/
nn9CNo

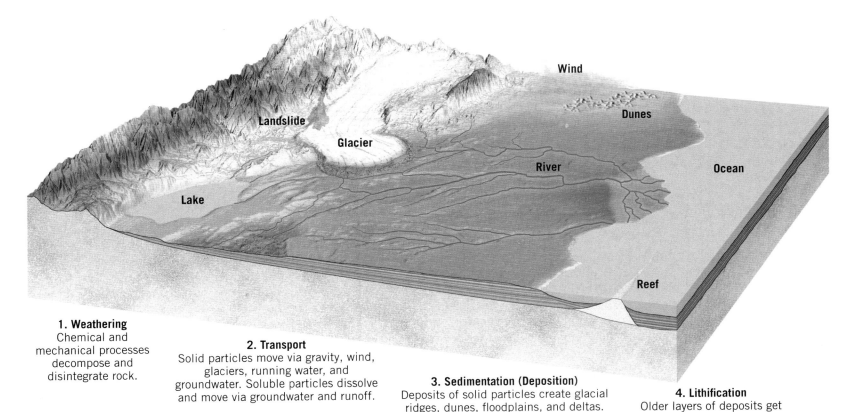

1. Weathering
Chemical and mechanical processes decompose and disintegrate rock.

2. Transport
Solid particles move via gravity, wind, glaciers, running water, and groundwater. Soluble particles dissolve and move via groundwater and runoff.

3. Sedimentation (Deposition)
Deposits of solid particles create glacial ridges, dunes, floodplains, and deltas. Dissolved material precipitates to form reefs. Eventually, much sediment reaches the ocean floor.

4. Lithification
Older layers of deposits get covered by new layers and gradually lithify into rock through compaction and cementation.

solution) or when organisms remove dissolved material to build hard parts such as shells.

- **Lithification:** As deposition continues, older sediments get buried beneath younger layers and gradually convert to sedimentary rock (lithify) by compaction and cementation. This and other changes are referred to as *diagenesis* (*dia* = change; *genesis* = origin). Diagenesis is a collective term for all the changes (short of metamorphism, discussed in Chapter 8) that take place in texture, composition, and other physical properties after sediments are deposited.

Categories of Sedimentary Rock   Geologists recognize three categories of sedimentary rocks. **Clastic sedimentary rocks** are transported accumulations of rocks weathered by both mechanical and chemical processes. Another term for these rocks is *detrital* (like "detritus"). **Chemical sedimentary rock**, as the name implies, is soluble material produced largely by chemical weathering. Ions in solution are precipitated by either inorganic or biological processes.

The third category is **organic sedimentary rock**. These types of rocks form from carbon-rich remains of organisms. The primary example is coal, a black combustible rock that consists of organic carbon from the remains of plants that died and accumulated on the floor of a swamp. The bits and pieces of undecayed plant material that constitute the "sediments" in coal are quite unlike the weathering products that make up detrital and chemical sedimentary rocks.

Finally, it's worth recalling that, according to the *simple rock cycle model* (Figure 1.23, page 26), the diverse processes involved in creating sedimentary rocks tend to keep transforming materials until they become one of three most common categories of end products, called *attractors* (see Chapter 6). These attractors are shale, quartz sandstone, and limestone. All of them will be described in more detail in upcoming sections of the chapter.

CONCEPT CHECKS 7.1

1. How does the volume of sedimentary rocks in Earth's crust compare to the volume of igneous and metamorphic rocks?

2. List two ways in which sedimentary rocks are important. **Concept Checker** https://goo.gl/3ueGVr

3. Outline the steps that would transform an exposure of granite in the mountains into various sedimentary rocks.

4. List and briefly describe the differences among the three basic sedimentary rock categories.

7.2 Clastic Sedimentary Rocks

Describe the primary basis for distinguishing among clastic rocks and describe how the origin and history of such rocks might be determined.

Though a wide variety of minerals and rock fragments (*clasts*) may be found in clastic rocks, clay minerals and quartz are the chief constituents of most sedimentary rocks in this category. Recall from Chapter 6 that clay minerals are the most abundant product of the chemical weathering of silicate minerals, especially the feldspars. Clays are fine-grained minerals with sheet-like crystalline structures similar to the micas. The other common mineral, quartz, is abundant because it is extremely durable and very resistant to chemical weathering. Thus, when igneous rocks such as granite are attacked by weathering processes, individual quartz grains are freed.

Other common minerals in clastic rocks are feldspars and micas. Because chemical weathering rapidly transforms these minerals into new substances, their presence in sedimentary rocks indicates that erosion and deposition occurred fast enough to preserve some of the primary minerals from the source rock before they could be decomposed.

Particle size is the primary basis for distinguishing among various clastic sedimentary rocks. **Figure 7.3** presents the size categories for particles making up clastic rocks. Particle size allows us to distinguish kinds of clastic rocks, and the sizes of the component grains also provide useful information about environments of deposition. Currents of water or air sort the particles by size; the stronger the current, the larger the particle size that can be carried. Gravel, for example, is moved by swiftly flowing rivers as well as by landslides and glaciers. Less energy is required to transport sand; thus, sand is found in such features as windblown dunes and some river deposits and beaches. Very little energy is needed to transport clay, so it settles very slowly. Accumulation of these tiny particles is generally associated with the quiet water of a lake, lagoon, swamp, or certain marine environments.

In order of increasing particle size, common clastic sedimentary rocks include shale, sandstone, and conglomerate and breccia. We will now look at each type and how it forms.

Shale

Shale is a sedimentary rock consisting of silt- and clay-size particles (**Figure 7.4**). This fine-grained clastic rock type accounts for well over half of all sedimentary rocks, making it the number-one attractor among sedimentary rock types. The particles in shale are so small that they cannot be readily identified without great magnification, and this makes shale more difficult to study and analyze than most other sedimentary rocks.

How Does Shale Form? Much of what can be learned about the process that forms shale is related to particle size. The tiny grains in shale indicate that deposition occurs as a result of gradual settling from relatively quiet, nonturbulent currents. Such environments include lakes, river floodplains, lagoons, and portions of the deep-ocean basins. Even in these "quiet" environments, there is usually enough turbulence to keep clay-size particles suspended almost indefinitely. Consequently, much of the clay is deposited only after the individual particles coalesce to form larger aggregates in a process called flocculation.

Sometimes the chemical composition of the rock provides additional information. For example, black shale is dark because it contains abundant organic matter (carbon). When such a rock is found, it strongly implies that deposition occurred in an oxygen-poor environment such as a swamp, where organic materials do not readily oxidize and decay.

Thin Layers As silt and clay accumulate, they tend to form thin layers, which are commonly referred to as *laminae* (*lamin* = thin sheet). Initially the particles in the laminae are oriented randomly. This disordered arrangement leaves a high percentage of open space (called *pore space*) that is filled with water. However, as additional layers of sediment pile up and compact the sediment below, the clay and silt particles take on a more nearly parallel alignment and become tightly packed. This rearrangement of grains reduces the size

Clastic Sedimentary Rocks

Size Range (millimeters)	Particle Name	Common Name	Clastic Rock
>256	Boulder	Gravel	Conglomerate / Breccia
64–256	Cobble		
4–64	Pebble		
2–4	Granule		
1/16–2	Sand	Sand	Sandstone / Arkose
1/256–1/16	Silt	Mud	Shale or Mudstone / Siltstone
<1/256	Clay		

0 10 20 30 40 50 60 70 mm

◀ **Figure 7.3**
Clastic rocks by particle size Particle size is the primary basis for distinguishing among various clastic sedimentary rocks.

◀ **Figure 7.4**
Shale—the most abundant sedimentary rock Dark shale containing fossilized plant remains is relatively common.

▶ **Figure 7.5**
Shale crumbles easily Beds of resistant sandstone and limestone produce bold cliffs. Weaker, poorly cemented shale produces gentler slopes of weathered debris.

Beds of resistant sandstone and limestone

Beds of weak shale

of the pore spaces and forces out much of the water. Once the grains are pressed closely together, the tiny spaces between particles do not readily permit solutions containing cementing material to circulate. Therefore, geologists often describe shales as being weak because they are poorly cemented and therefore not well lithified.

The inability of water to penetrate shale's microscopic pore spaces explains why this rock often forms barriers to the subsurface movement of water and petroleum. Indeed, rock layers that contain groundwater are commonly underlain by shale beds that block further downward movement. The opposite is true for underground reservoirs of petroleum. They are often capped by shale beds that effectively prevent oil and gas from escaping to the surface.°

* The relationship between impermeable beds and the occurrence and movement of groundwater is examined in Chapter 17. Shale beds can be cap rocks in oil traps and are discussed in Chapter 23.

▶ **Figure 7.6**
Quartz sandstone After shale, sandstone is the next most abundant sedimentary rock.

0   2 cm

Close up

Shale, Mudstone, or Siltstone? It is common to apply the term *shale* to all fine-grained sedimentary rocks, especially in a nontechnical context. However, be aware that geologists have a more restricted use of the term. In this narrower usage, shale must exhibit the ability to split into thin layers along well-developed, closely spaced planes. This property is termed **fissility** (*fissilis* = that which can be cleft or split). If the rock breaks into chunks or blocks, then geologists call it *mudstone*. Another fine-grained sedimentary rock that, like mudstone, is often grouped with shale but lacks fissility is *siltstone* (see Figure 7.3). As its name implies, siltstone is composed largely of silt-size particles and contains less clay-size material than shale and mudstone.

Gentle Slopes Although shale is far more common than other sedimentary rocks, it does not usually attract as much notice as other, less abundant, members of this group. The reason is that shale does not form prominent outcrops, as sandstone and limestone often do. Rather, shale crumbles easily and usually forms a cover of soil that hides the unweathered rock below. This is illustrated nicely in the Grand Canyon, where the gentler slopes of weathered shale are quite inconspicuous with sparse vegetation, in sharp contrast with the bold cliffs produced by more durable rocks (**Figure 7.5**).

Although shale beds may not form striking cliffs and prominent outcrops, some deposits have economic value. Certain shales are quarried to obtain raw material for pottery, brick, tile, and china. Moreover, when mixed with limestone, shale is used to make Portland cement. In the future, one type of shale, called oil shale, may become a valuable energy resource.

Oil Shale Versus Shale Oil These two terms can be confusing. *Oil shale* is a kind of sedimentary rock that contains kerogen, a solid organic compound derived from the remains of algae that was deposited at the same time as the clay minerals. Kerogen is not a mature hydrocarbon, so getting useful energy from oil shale requires substantial processing.

Contrast this with *shale oil*, where the kerogen has been converted to mature oil and natural gas by deep burial and is trapped in the tiny pores of the fine-grained clastic rock. Technological advances such as horizontal drilling and hydraulic fracturing have provided the means to extract the oil and gas from their fine-grained host shale. Oil shale and shale oil potential are discussed in Chapter 23.

Sandstone

Sandstone is the name given to rocks in which sand-size grains predominate (**Figure 7.6**). After shale, sandstone is the next most abundant sedimentary rock, accounting for approximately 20 percent of the entire group, making it another attractor among sedimentary rock types. Sandstone forms in a variety of environments, and observations such as the composition, shape, and sorting of the grains may provide information about the environment in which the sediment was deposited.

Sorting All the particles in sandstone are not necessarily identical in size. **Sorting** refers to the degree of similarity in particle size in a sedimentary rock. For example, if all the grains in a sample of sandstone are about the same size, the sand is considered *well sorted*. Conversely, if the rock contains mixed large and small particles, the sand is said to be *poorly sorted* (**Figure 7.7**).

By studying the degree of sorting, we can learn much about the depositing current. Deposits of wind-blown sand are usually better sorted than deposits sorted by wave activity (**Figure 7.8**). Particles washed by waves are commonly better sorted than materials deposited by streams. Sediment accumulations that exhibit poor sorting usually result when particles are transported for only a relatively short time and then rapidly deposited. For example, when a turbulent stream reaches the gentler slopes at the base of a steep mountain, its velocity is quickly reduced, and poorly sorted sands and gravels are deposited.

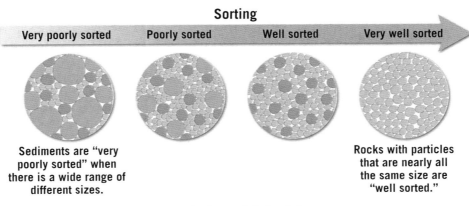

Sorting

Very poorly sorted — Poorly sorted — Well sorted — Very well sorted

Sediments are "very poorly sorted" when there is a wide range of different sizes.

Rocks with particles that are nearly all the same size are "well sorted."

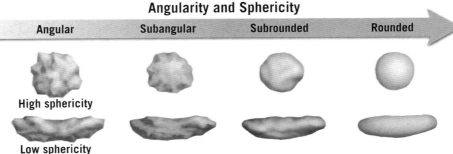

Angularity and Sphericity

Angular — Subangular — Subrounded — Rounded

High sphericity

Low sphericity

Transportation reduces the size and angularity of particles but does not change their general shape.

◄ SmartFigure 7.7
Sorting and particle shape *Sorting* refers to the range of particle sizes present in a rock. Geologists describe a particle's shape in terms of its *angularity* (the degree to which edges and corners are rounded) and *sphericity* (how close the shape is to a sphere).

Tutorial
https://goo.gl/cYspGL

Particle Shape The shapes of sand grains can also help decipher the history of a sandstone (see Figure 7.7). When streams, winds, or waves move sand and other larger sedimentary particles, the grains lose their sharp edges and corners and become more rounded as they collide with other particles during transport. Thus, rounded grains likely have been airborne or waterborne. Further, the degree of rounding indicates the distance or time involved in the transportation of sediment by currents of air or water. Highly rounded grains indicate that a great deal of abrasion—and hence a great deal of transport—has occurred.

▼ **Figure 7.8**
Sand dunes consist of well-sorted sediment
A. The Navajo Sandstone of southern Utah and Arizona represents a vast area of ancient sand dunes that once covered an area the size of California.
B. These modern dunes are among the highest in North America.

A. The orange and yellow cliffs of Utah's Zion National Park expose thousands of feet of Jurassic-age Navajo Sandstone.

B. The quartz grains composing the Navajo Sandstone were deposited by wind as dunes similar to these in Colorado's Great Sand Dunes National Park. The sand is well sorted because all of the particles are practically the same size.

▶ Figure 7.9
Conglomerate The gravel-size particles in this rock are rounded.

```
0                    5 cm
```

Very angular grains, on the other hand, imply two things: that the rock materials were transported only a short distance before they were deposited or that some other medium may have transported them. For example, when glaciers move sediment, the particles are usually made more irregular by the crushing and grinding action of the ice.

Transport Affects Mineral Composition

In addition to affecting the degree of rounding and the amount of sorting that particles undergo, the length of transport by turbulent air and water currents also influences the mineral composition of a sedimentary deposit. Substantial weathering and long transport lead to the gradual destruction of weaker and less stable minerals, including the feldspars and ferromagnesians. Because quartz is very durable, it is usually the mineral that survives a long trip in a turbulent environment.

To summarize, the origin and history of sandstone can often be deduced by examining the sorting, roundness, and mineral composition of its constituent grains. Knowing this information allows us to infer that a well-sorted, quartz-rich sandstone consisting of highly rounded grains must be the result of a great deal of

transport. Quartz grains are so durable that the extensive transport necessary to produce highly sorted, very rounded, all-quartz sand grains may have happened over generations of rock cycle processes: Earlier-formed sandstone was uplifted, weathered, and eroded, and the grains were transported and deposited to another sedimentary basin. Conversely, we may also conclude that a sandstone containing significant amounts of feldspar and angular grains of ferromagnesian minerals underwent little chemical weathering and transport and was probably deposited close to the source area of the rock particles.

Varieties of Sandstone Due to its durability, quartz is the predominant mineral in most sandstones. Such rock is often simply called *quartz sandstone* (see Figure 7.6). When a sandstone contains appreciable quantities of feldspar (25 percent or more), the rock is called *arkose*. In addition to feldspar, arkose usually contains quartz and sparkling bits of mica. The mineral composition of arkose indicates that the grains were derived from granitic source rocks. The particles are generally poorly sorted and angular, which suggests short-distance transport, minimal chemical weathering in a relatively dry climate, and rapid deposition and burial.

A third variety of sandstone is known as *graywacke*. Along with quartz and feldspar, this dark-colored rock contains abundant rock fragments and matrix—finer-grained material in which the fragments are embedded. More than 15 percent of graywacke's volume is matrix. The poor sorting and angular grains characteristic of graywacke suggest that the particles were transported only a relatively short distance from their source area and were then rapidly deposited. Before the sediment could be reworked and sorted further, it was buried by additional layers of material. Graywacke is frequently associated with submarine deposits made by dense sediment-choked torrents called *turbidity currents*.

Conglomerate and Breccia

Conglomerate consists largely of gravel (**Figure 7.9**). As Figure 7.3 indicates, these rounded particles can range in size from large boulders to particles as small as peas. The particles are often large enough to be identified as distinctive rock types; thus, they can be valuable in identifying the source areas of sediments. More often than not, conglomerates are poorly sorted because the openings between the large gravel particles contain sand or mud (**Figure 7.10**).

Gravels accumulate in a variety of environments and usually indicate the existence of steep slopes or very turbulent currents. The coarse particles in a conglomerate may reflect the action of energetic mountain streams or may result from strong wave activity along a rapidly eroding coast. Some glacial and landslide deposits also contain plentiful gravel.

If the large particles are angular rather than rounded, the rock is called **breccia** (**Figure 7.11**). Because

▶ Figure 7.10
Poorly sorted sediments Gravel deposits along Carbon Creek in Grand Canyon National Park are poorly sorted.

large particles abrade and become rounded very rapidly during transport, the pebbles and cobbles in a breccia indicate that they did not travel far from their source area before they were deposited.

▲ Figure 7.11
Breccia The gravel-size particles in this rock are sharp and angular.

0   2 cm

7.3 Chemical Sedimentary Rocks

Explain the processes involved in the formation of chemical sedimentary rocks and describe several examples.

In contrast to clastic rocks, which form from the solid products of weathering, chemical sediments derive from ions that are carried *in solution* to lakes and seas. This material does not remain dissolved in the water indefinitely, however. Some of it precipitates to form chemical sediments. These become rocks such as limestone, chert, and rock salt.

This precipitation of material occurs in two ways. *Inorganic* (*in* = not, *organicus* = life) processes such as evaporation and chemical activity can produce chemical sediments. *Organic* (life) processes of water-dwelling organisms also form chemical sediments, said to be of **biochemical** origin.

One example of a deposit resulting from inorganic chemical processes is the dripstone that decorates many caves (**Figure 7.12**). Another is the salt left behind as a body of seawater evaporates. In contrast, many water-dwelling animals and plants extract dissolved mineral matter to form shells and other hard parts. After the organisms die, their skeletons collect by the millions on the floor of a lake or an ocean as biochemical sediment (**Figure 7.13**).

Limestone

Representing about 10 percent of the total volume of all sedimentary rocks, **limestone** is the most abundant chemical sedimentary rock (and one of the three main attractors among sedimentary rock types), and it has economic significance as well (see **GEOgraphics 7.1**). Composed chiefly of the mineral calcite ($CaCO_3$), it forms either through inorganic or biochemical processes. Although the mineral composition of all limestone is similar, many different types exist, reflecting

Delicate calcite crystals forming in a drop of water at the tip of soda straw stalactite. The formation of crystals is triggered when some carbon dioxide escapes from the water drop.

◀ Figure 7.12
Cave deposits These precipitated crystals are an example of a chemical sedimentary rock with an inorganic origin.

▶ **Figure 7.13**
Coquina This variety of limestone consists of shell fragments; therefore, it has a biochemical origin.

Close up

the varied conditions under which limestone originates. Forms that have a marine biochemical origin are by far the most common.

Carbonate Reefs   Corals are one important example of organisms that are capable of creating large quantities of marine limestone. These relatively simple invertebrate animals secrete a calcareous (calcium carbonate) external skeleton. Although they are small, corals are capable of creating massive structures called *reefs* (**Figure 7.14**). Reefs consist of coral colonies made up of great numbers of individuals that live side by side on a calcite structure secreted by the animals. In addition, calcium carbonate–secreting algae live with the corals and help cement the entire structure into a solid mass. A wide variety of other organisms also live in and near the reefs.

The best-known modern reef is Australia's 2600-kilometer-long (1600-mile-long) Great Barrier Reef, but many lesser reefs also exist. They develop in the shallow, warm waters of the tropics and subtropics equatorward of about 30 degrees latitude. Striking examples exist in The Bahamas, Hawaii, and the Florida Keys.

Modern corals were not the first reef builders. Earth's first reef-building organisms were photosynthesizing bacteria that lived during Precambrian time, more than 2 billion years ago. From fossil remains, it is known that a variety of organisms have constructed reefs, including bivalves (clams and oysters), bryozoans

▶ **Figure 7.14**
Carbonate reefs Large quantities of biochemical limestone are created by reef-building organisms.

Great Barrier Reef

Australia

This image in Guadalupe Mountains National Park, Texas, shows a small portion of an ancient reef complex that once formed a 600-kilometer loop around the margin of a Permian sedimentary basin.

Aerial view showing a small portion of Australia's Great Barrier Reef. Located off the coast of Queensland, it extends for 2600 kilometers and consists of more than 2900 individual reefs.

Limestone
An Important and Versatile Commodity

Limestone, as defined by the minerals industry, refers to any rock composed mostly of the carbonate minerals calcite and dolomite. The U.S. Geological Survey characterizes limestone as an *"essential mineral commodity of national importance."*

Why is limestone important?

Limestone that has undergone metamorphism is called marble. It is used as floor tile, table and countertops, and as a building stone.

Purified limestone is added to bread and cereal as a source of calcium and is an ingredient in antacid tablets and calcium supplements. It is even used to neutralize acids in wine and beer making!

Limestone is used as a filler and white pigment in many products including paper, plastics, paint, and even toothpaste.

Limestone is a key ingredient in making Portland cement, an essential product to the building industry.

White roofing granules.

Huge quantities of limestone are crushed and used as aggregate—the solid base of many roads and an ingredient in concrete. It is the raw material for making lime (CaO), which is used to treat soils, purify water, and smelt copper among many uses.

Limestone has been used as a building stone for centuries—from Egypt's ancient pyramids and Europe's medieval castles, to modern buildings such as this.

UNITED STATES

FEDERAL TRADE COMMISSION BUILDING

The massive White Chalk Cliffs. Chalk is a biochemical limestone made up almost entirely of the tiny hard parts of microscopic marine organisms, mainly plankton.

View of a group of plankton called *coccolithophores* from a scanning electron microscope. Individual plates shaped like hubcaps are only three one-thousandths of a millimeter in diameter—so tiny they could pass through the eye of a needle.

▲ **Figure 7.15**
The White Cliffs of Dover This prominent deposit underlies large portions of southern England as well as parts of northern France.

(coral-like animals), and sponges. Corals have been found in fossil reefs as ancient as 500 million years old, but corals similar to the modern colonial varieties have constructed reefs only during the past 60 million years.

In the United States, reefs of Silurian age (416 to 444 million years ago) are prominent features in Wisconsin, Illinois, and Indiana. In western Texas and adjacent southeastern New Mexico, a massive reef complex that formed during the Permian period (251 to 299 million years ago) is strikingly exposed in Guadalupe Mountains National Park (see Figure 7.14).

Coquina and Chalk   Although a great deal of limestone is produced by biological processes, this origin is not always evident because shells and skeletons may undergo considerable change before lithifying into rock. However, one easily identified biochemical limestone is *coquina*, a coarse-grained rock composed of poorly cemented shells and shell fragments (see Figure 7.13). Another less obvious but

nevertheless familiar example is *chalk*, a soft, porous rock made up almost entirely of the hard parts of microscopic marine organisms. Among the most famous chalk deposits are those exposed along the southeastern coast of England (**Figure 7.15**).

Inorganic Limestones   When chemical changes or high water temperatures increase the water's concentration of calcium carbonate to the point that it precipitates, inorganic limestones occur. *Travertine*, the type of limestone commonly seen in caves, is an example (see Figure 7.12). When travertine is deposited in caves, groundwater is the source of the calcium carbonate. As water droplets become exposed to the air in a cavern, some of the carbon dioxide dissolved in the water escapes, causing calcium carbonate to precipitate.

Another variety of inorganic limestone is *oolitic limestone*, a rock composed of small spherical grains called *ooids*. Ooids form in shallow marine waters as tiny "seed" particles (commonly small shell fragments) are moved back and forth with currents. As the grains are rolled about in the warm water, which is supersaturated with calcium carbonate, they become coated with layer upon layer of the chemical precipitate (**Figure 7.16**).

Small spherical grains called *ooids* are formed by chemical precipitation of calcium carbonate around a tiny nucleus and are the raw material for oolitic limestone.

▲ **Figure 7.16**
Oolitic limestone Oolitic limestone is an inorganic limestone composed of ooids. Note the dime coin for scale.

Flint

Jasper

Chert arrowhead

Petrified wood

◀ **Figure 7.17**
Colorful chert Chert is the name applied to a number of dense, hard chemical sedimentary rocks made of microcrystalline quartz.

Dolostone

Closely related to limestone is **dolostone**, a rock composed of the calcium-magnesium carbonate mineral dolomite [$CaMg(CO_3)_2$]. Although dolostone and limestone sometimes closely resemble one another, they can be easily distinguished by observing their reaction to dilute hydrochloric acid. When a drop of acid is placed on limestone, the reaction (fizzing) is obvious. However, unless dolostone is powdered, it does not visibly react to the acid.

Dolostone's origins remain a subject of discussion among geologists. No marine organisms produce hard parts of dolomite, and the chemical precipitation of dolomite from seawater occurs only under conditions of unusual water chemistry in certain near-shore sites. Yet dolostone is abundant in many ancient sedimentary rock successions. It appears that significant quantities of dolostone are produced when magnesium-rich waters circulate through limestone and convert calcite to dolomite when some calcium ions are replaced by magnesium ions (a process called *dolomitization*). However, not all dolostones appear to be formed by such a process, and their origin remains uncertain.

Chert

Chert is a name used for a number of very compact and hard rocks made of microcrystalline quartz (SiO_2). It has a wide range of coloration due to trace elements present in the rock, with nearly white, black, gray, brown, red, or greenish all being possible hues. **Figure 7.17** shows some varieties. One well-known form is *flint*, whose dark color results from the organic matter it contains. *Jasper*, a red variety, gets its bright color from iron oxide. *Petrified wood* is chert that is made when silica-rich material such as volcanic ash buries trees. Groundwater rich in dissolved silica from the ash penetrates the wood. As the dissolved silica precipitates, it gradually replaces the wood. The shape and structures such as growth rings are often preserved. Like glass, most chert has a conchoidal fracture. Its hardness, ease of chipping, and ability to hold a sharp edge made chert a favorite of Native Americans for fashioning points for spears and arrows. Because of chert's durability and extensive use, arrowheads are found in many parts of North America.

Chert most commonly occurs as layered deposits called *bedded cherts* or as somewhat spherical masses called *nodules*, which form within other rock types (commonly limestone) and range in diameter from pea-sized to several centimeters. Most bedded cherts are believed to originate from tiny marine organisms (diatoms and radiolarians) that produce hard parts from silica rather than calcium carbonate. These organisms extract silica from seawater (where it is present in only tiny quantities); when they die, their hard parts accumulate on the seafloor. Some bedded cherts occur in association with lava flows and layers of volcanic ash. For these occurrences, it is probable that the silica was derived from the decomposition of the volcanic ash and not from biochemical sources. Chert nodules are sometimes referred to as *secondary cherts* or *replacement cherts*, and they most often occur within beds of limestone. They form when silica, originally deposited in one place, dissolves, migrates, and then chemically precipitates elsewhere, replacing older material.

Evaporites

In the geologic past, many areas that are now dry land were shallow marine bays with only narrow connections to the open ocean. Seawater continually moved into the bays to replace water lost by evaporation. Eventually the waters of the bay became saturated, and salt deposition began. Such deposits are called **evaporites**.

Minerals commonly precipitated in this fashion include halite (sodium chloride, $NaCl$), which is the chief component of *rock salt*, and gypsum (hydrous calcium sulfate, $CaSO_4 \cdot 2H_2O$), which is the main ingredient in *rock gypsum* (**Figure 7.18**). Both have economic importance. In addition to its familiar role as table salt, halite has many other uses, from melting ice on roads to making hydrochloric acid. People have sought, traded, and fought over it for much of

▶ Figure 7.18
Solar salt Seawater is one important source of common salt. In this process, seawater is held in shallow ponds, while solar energy evaporates the water. The nearly pure salt deposits that eventually form are essentially artificial evaporite deposits.

process, potassium and magnesium salts precipitate. One of these last-formed salts, the mineral *sylvite*, is mined as a significant source of potassium ("potash") for fertilizer.

On a smaller scale, evaporite deposits can be seen in such places as Death Valley, California, and Utah's Bonneville Salt Flats. Following rains or periods of snowmelt in the mountains, streams flow from the surrounding mountains into an enclosed basin. As the water evaporates, **salt flats** form when dissolved materials are precipitated as a white crust on the ground (**Figure 7.19**).

human history. Gypsum is the basic ingredient in plaster of Paris and is used for making wallboard (drywall) and interior plaster.

When a body of seawater evaporates, the minerals that precipitate do so in a sequence that is determined by their solubility. Less soluble minerals precipitate first, and more soluble minerals precipitate later, as salinity increases. For example, gypsum precipitates when about 80 percent of the seawater has evaporated, and halite crystals precipitate when 90 percent of the water has been removed. During the last stages of this

CONCEPT CHECKS 7.3

1. Explain how the formation of biochemical sediments differs from the formation of sediments by inorganic processes. Use examples as part of your explanation.

2. Distinguish among limestone, dolostone, and chert. Describe several varieties of each.

◀)) **Concept Checker**
https://goo.gl/ZFeXDc

3. How do evaporites form? What are some examples?

▶ SmartFigure 7.19
Bonneville Salt Flats This well-known Utah site was once a large salt lake.

Tutorial
https://goo.gl/
3sdVVF

This extensive evaporite deposit is a 30,000-acre expanse of hard white salt that in places is nearly 2 meters thick.

7.4 Coal: An Organic Sedimentary Rock

Outline the successive stages in the formation of coal.

Unlike limestone and chert, which are rich in calcite and silica, **coal** is made of organic matter. Close examination of coal under a magnifying glass often reveals plant structures such as leaves, bark, and wood that have been chemically altered but are still identifiable. This supports the conclusion that coal is the end product of large amounts of plant material being buried for millions of years (**Figure 7.20**). The formation of coal involves these stages:

1. **Accumulation of large amounts of plant remains.** Dead plants readily decompose when exposed to the atmosphere or other oxygen-rich environments, so the plants must die and accumulate in an oxygen-poor environment, such as a swamp, in order to transform into coal. Stagnant swamp water is oxygen deficient, so complete decay (oxidation) of the plant material is not possible. Instead, the plants are attacked by certain bacteria that partially decompose the organic material and liberate oxygen and hydrogen. As these elements escape, the percentage of carbon in the plant matter gradually increases. The bacteria are not able to finish the job of decomposition because their growth is impeded by acids liberated from the plants.

2. **Formation of peat and lignite.** The partial decomposition of plant remains in an oxygen-poor swamp creates a layer of *peat*, a soft brown material in which plant structures are still easily recognized. With shallow burial, peat slowly changes to *lignite*, a soft brown coal. Burial increases the temperature of sediments as well as the pressure on them.

3. **Formation of bituminous coal.** Higher temperatures bring about chemical reactions within the plant materials and yield water and organic gases (volatiles). As the load increases from more sediment on top of the developing coal, the water and volatiles are pressed out, and the proportion of *fixed carbon* (the remaining solid combustible material) increases. The greater the carbon content, the greater the coal's energy ranking as a fuel. During burial, the coal also becomes increasingly compact. For example, deeper burial transforms lignite into a harder, more compacted black rock called *bituminous* coal. A bed

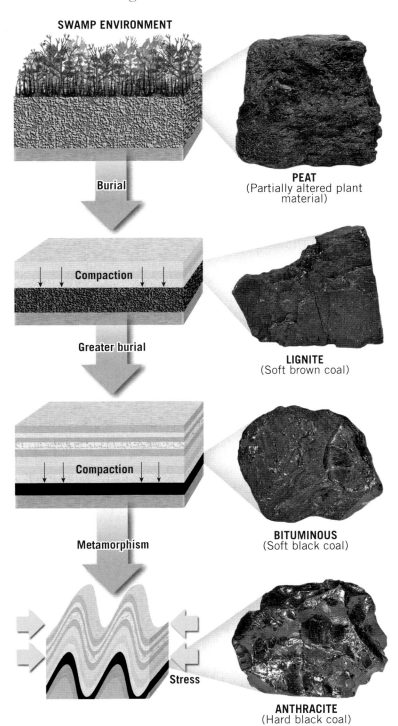

SWAMP ENVIRONMENT

Burial

PEAT
(Partially altered plant material)

Compaction

LIGNITE
(Soft brown coal)

Greater burial

Compaction

BITUMINOUS
(Soft black coal)

Metamorphism

Stress

ANTHRACITE
(Hard black coal)

◀ SmartFigure 7.20
From plants to coal
Successive stages in the formation of coal.

Tutorial
https://goo.gl/2Dp44X

of bituminous coal may be only one-tenth as thick as the peat bed from which it formed.

4. **Formation of anthracite coal.** Lignite and bituminous coals are sedimentary rocks. However, when sedimentary layers are subjected to the folding and deformation associated with mountain building, the heat and pressure cause a further loss of volatiles and water, thus increasing the concentration of fixed carbon. This metamorphoses bituminous coal into *anthracite*, a very hard, shiny, black *metamorphic* rock (see Chapter 8). Although anthracite produces more energy per mass and is cleaner-burning than its bituminous cousins, only a relatively small amount is mined. Anthracite is not widespread and is more difficult and expensive to extract than the relatively flat-lying layers of bituminous coal.

Coal is a major energy resource. Its role as a fuel and some of the problems associated with burning coal are discussed in Chapter 23.

CONCEPT CHECKS 7.4

1. What is the "raw material" for coal? Under what circumstances does it accumulate?

2. Outline the successive stages in the formation of coal.

Concept Checker
https://goo.gl/YHNebC

7.5 Turning Sediment into Sedimentary Rock: Diagenesis and Lithification

Describe the processes that convert sediment into sedimentary rock and other changes associated with burial.

We've seen that sediment changes as it is converted into sedimentary rock. The term **diagenesis** (*dia* = change, *genesis* = origin) is a collective term for all the chemical, physical, and biological changes that take place after sediments are deposited and then transformed into sedimentary rock. Diagenesis includes lithification, where the sediments turn to rock, as well as certain changes that occur after lithification. However, it's important to understand that diagenesis is distinct from the metamorphic processes that may later convert sedimentary rock into metamorphic rock.

Diagenesis

As sediments are buried, they are subjected to increasingly higher temperatures and pressures. Diagenesis occurs within the upper few kilometers of Earth's crust at temperatures that are generally less than 150° to 200°C (300° to 400°F). Beyond this somewhat arbitrary temperature threshold, metamorphism is said to occur.

One example of diagenetic change is *recrystallization*, the development of more stable minerals from less stable ones. For example, the mineral aragonite is the less stable form of calcium carbonate ($CaCO_3$). Aragonite is secreted by many marine organisms to form shells and other hard parts, such as the skeletal structures produced by corals. In some environments, large quantities of these solid materials accumulate as sediment. As burial takes place, aragonite recrystallizes to form calcite, the more stable form of calcium carbonate. Calcite, as mentioned previously, is the main constituent in the sedimentary rock limestone.

Another example of diagenesis is the chemical alteration of organic matter in an oxygen-poor environment evolving into peat and then into coal. Instead of completely decaying, as would occur in the presence of oxygen, the organic matter is slowly transformed into solid carbon.

Lithification

Diagenesis includes **lithification**, the processes by which unconsolidated sediments are transformed into solid sedimentary rocks (*lithos* = stone, *fic* = making). Basic lithification processes include compaction and cementation (**Figure 7.21**).

Compaction As described earlier, as sediment accumulates, the weight of overlying material compresses and compacts the deeper sediments. The deeper a sediment is buried, the greater its **compaction** and the firmer it becomes. As the grains are pressed closer and closer, there is considerable reduction in pore space (the open spaces between particles). For example, when clays are buried beneath several thousand meters of material, the volume of the clay layer may be reduced by as much as 40 percent. As pore space decreases, much of the water that was trapped in the sediments is squeezed out.

COMPACTION

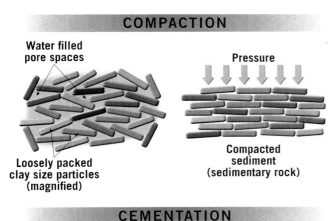

Water filled pore spaces

Loosely packed clay size particles (magnified)

Pressure

Compacted sediment (sedimentary rock)

CEMENTATION

Circulation of mineral-bearing groundwater

Loosely packed sand or gravel size particles (magnified)

Cement

Gradually the cementing material fills much of the pore space and "glues" the grains together

▲ **Figure 7.21**
Lithification processes Compaction and cementation both reduce porosity as sediment becomes sedimentary rock.

Because sands and other coarse sediments are less compressible, compaction is most significant as a lithification process in fine-grained sedimentary rocks like shale.

Cementation   Ions in solution get transported by groundwater and then crystallize in spaces between sediment grains to form minerals that glue the grains together. This process, called **cementation**, is the most important process by which sediments change into sedimentary rock. Like compaction, cementation reduces the porosity of the rock.

Calcite, silica, and iron oxide are the most common cements. It is often a relatively simple matter to identify the cementing material. Calcite cement effervesces (bubbles) in contact with dilute hydrochloric acid. Silica is the hardest cement and thus produces the hardest sedimentary rocks. An orange or dark-red color in a sedimentary rock means that iron oxide is present.

While compaction and cementation are the main lithification processes for most types of sediment, in some cases the original sedimentary particles themselves undergo dissolution and recrystallization. For example, with time and burial, loose sediment consisting of delicate calcium carbonate–rich skeletal debris may recrystallize into a relatively dense crystalline limestone. Because crystals grow until they fill all the available space, crystalline sedimentary rocks often lack pore spaces. Unless the rocks later develop joints and fractures, they will be relatively impermeable to fluids such as water and oil. Evaporites form initially as solid masses of intergrown crystals and hence can be similarly impermeable.

CONCEPT CHECKS 7.5

1. If *lithos* means "stone," then what is *lithification*?

2. Compaction is most important as a lithification process with which clastic sediment size?

Concept Checker
https://goo.gl/btvPcf

3. List three common cements of sedimentary clasts. How might each be identified?

7.6 Classification of Sedimentary Rocks

Summarize the criteria used to classify sedimentary rocks.

The classification scheme in **Figure 7.22** divides sedimentary rocks into major groups: clastic on the left side and chemical/organic on the right. Further, we can see that the main criterion for subdividing the clastic rocks is particle size, whereas the primary basis for distinguishing among different rocks in the chemical group is their mineral composition.

As is the case with many (perhaps most) classifications of natural phenomena, the categories presented in Figure 7.22 are more rigid than the actual state of nature. Many of the sedimentary rocks classified into the chemical group also contain at least small quantities of detrital sediment. Many limestones, for example, contain varying amounts of mud or sand, giving them a "sandy" or "shaly" quality. Conversely, because practically all clastic rocks are cemented with material that was originally dissolved in water, they too are far from being "pure."

Texture, one part of sedimentary rock classification, describes the relationship of mineral grains in rock. There are two major textures used in the classification

► Figure 7.22
**Identification of
sedimentary rocks** The
main criterion for naming
clastic rocks is particle
size. The primary basis
for naming chemical and
organic sedimentary rocks
is their composition.

Clastic Sedimentary Rocks

Clastic Texture (particle size)		Sediment Name	Rock Name
Coarse (over 2 mm)		Gravel (Rounded particles)	Conglomerate
Coarse (over 2 mm)		Gravel (Angular particles)	Breccia
Medium (1/16 to 2 mm)		Sand	Sandstone (Arkose)*
Fine (1/16 to 1/256 mm)		Silt (mud)	Siltstone
Very fine (less than 1/256 mm)		Clay (mud)	Shale or Mudstone

*If abundant feldspar is present the rock is called Arkose.

Chemical and Organic Sedimentary Rocks

Composition	Texture	Rock Name	
Calcite, CaCO₃	Nonclastic: Fine to coarse crystalline	Crystalline Limestone	
Calcite, CaCO₃	Nonclastic: Fine to coarse crystalline	Travertine	
Calcite, CaCO₃	Clastic: Visible shells and shell fragments loosely cemented	Coquina	Biochemical Limestone
Calcite, CaCO₃	Clastic: Various size shells and shell fragments cemented with calcite cement	Fossiliferous Limestone	Biochemical Limestone
Calcite, CaCO₃	Clastic: sand-sized spherical grains of concentric layers of calcite	Oolitic Limestone	Biochemical Limestone
Calcite, CaCO₃	Clastic: Microscopic shells and clay	Chalk	Biochemical Limestone
Quartz, SiO₂	Nonclastic: Very fine crystalline	Chert (light colored) Flint (dark colored) Jasper (red) Agate (banded)	
Gypsum CaSO₄•2H₂O	Nonclastic: Fine to coarse crystalline	Rock Gypsum	
Halite, NaCl	Nonclastic: Fine to coarse crystalline	Rock Salt	
Altered plant fragments	Nonclastic: Fine-grained organic matter	Bituminous Coal	

of sedimentary rocks. The term **clastic** texture (Greek for "broken") consists of discrete fragments and particles that are cemented and compacted together. Although cement is present in the spaces between particles, these openings are rarely filled completely. In addition, some chemical sedimentary rocks exhibit this texture. For example, coquina, the limestone composed of shells and shell fragments, is obviously as clastic as a conglomerate or sandstone. The same applies for some varieties of oolitic limestone.

Some chemical sedimentary rocks have a **nonclastic,** or **crystalline**, texture, in which the minerals form a pattern of interlocking crystals. The crystals

► Figure 7.23
Rock salt Like other
evaporites, rock salt
has a nonclastic texture
because it is composed of
intergrown crystals.

may be microscopically small or large enough to be visible without magnification. Evaporites are this type (**Figure 7.23**). As described earlier, some limestones originate as clastic deposits of shell fragments and subsequently undergo recrystallization. Chert can have a similar history.

Because they consist of intergrown crystals, nonclastic sedimentary rocks may resemble igneous rocks. The two rock types are usually easy to distinguish because the minerals contained in nonclastic sedimentary rocks (such as halite, gypsum, and calcite) are quite different from those found in most igneous rocks.

CONCEPT CHECKS 7.6

1. What is the primary basis for distinguishing (naming) different chemical sedimentary rocks? How is the naming of clastic rocks different?

2. Distinguish between the texture of clastic and nonclastic sedimentary rock.

 Concept Checker

https://goo.gl/RXVAh5

7.7 Sedimentary Rocks Represent Past Environments

Distinguish among three broad categories of sedimentary environments and provide an example of each. List several sedimentary structures and explain why these features are useful to geologists.

Sedimentary rocks are important to interpreting Earth's history (**Figure 7.24**). By understanding the conditions under which sedimentary rocks form, geologists can often deduce the history of a rock, including information about the origin of its component particles, the method of sediment transport, and the nature of the place where the grains eventually came to rest—that is, the environment of deposition.

An **environment of deposition,** or **sedimentary environment**, is a geographic setting where sediment is accumulating. Each site is characterized by a particular combination of geologic processes and environmental conditions. Some sediments, such as the chemical sediments that precipitate in water bodies, are solely products of their sedimentary environment. That is, their component minerals originated and were deposited in the same place. Other sediments originate far from the site where they accumulate and are transported from their source by some combination of gravity, water, wind, and ice.

At any given time, the geographic setting and environmental conditions of a sedimentary environment determine the nature of the sediments that accumulate. Geologists carefully study the sediments in present-day depositional environments because the features they find can also be observed in ancient sedimentary rocks.

Geologists apply a thorough knowledge of present-day conditions to reconstruct the ancient environments and geographic relationships of an area at the time a particular set of sedimentary layers was deposited. This process is an excellent example of applying a fundamental principle of modern geology: "The present is the key to the past."[†] Such analyses often lead to the creation of maps depicting the past geographic distribution of land and sea, mountains and river valleys, deserts and glaciers, and other environments of deposition.

"The present is the key to the past." —nineteenth-century geologist Charles Lyell

◄ **SmartFigure 7.24 Utah's Capitol Reef National Park** These tilted sedimentary strata are part of the Waterpocket Fold at Halls Creek. The strata here record changing environments during the Mesozoic era.

Mobile Field Trip https://goo.gl/dVAVGi

environments. Chapters 16 through 20 examine many of these environments in greater detail. Each of these three categories is an area where sediment accumulates and where organisms live and die. Each produces a characteristic sedimentary rock or assemblage that reflects prevailing conditions.

Types of Sedimentary Environments

Sedimentary environments fall into three broad categories: continental, marine, or transitional (shoreline). Each category includes many specific subenvironments. **Figure 7.25** and **Figure 7.26** are idealized diagrams illustrating a number of important sedimentary environments associated with each category. Realize that this is just a sampling of the great diversity of depositional

Continental Environments

Continental environments are dominated by the erosion and deposition associated with streams. In some cold regions, movement of glacial ice replaces running water as the dominant process. In arid regions (as well as some coastal settings), wind takes on greater importance. Clearly, the nature of the sediments deposited in continental environments is strongly influenced by climate.

Streams erode more land and then transport and deposit more sediment than any other process; as such, they are the biggest modifier of landscapes. In addition to channel deposits, large quantities of sediment are dropped when floodwaters periodically inundate broad, flat valley floors (called *floodplains*). Where rapid

[†] For more on this idea, see Section 1.2 in Chapter 1.

Caves that develop in limestone are sites where calcium carbonate is deposited as dripstone.

Sand dunes consist of well-sorted sand grains deposited by the wind.

Alluvial fans consist of coarse sediments that are deposited when mountain streams reach flat lowlands.

Inland seas and **lakes** in arid environments where evaporation exceeds precipitation, produce evaporite deposits such as rock salt and gypsum.

Dunes

Salt flat

Alluvial fans

Inland lake

Lake

Swamp

Landslides produce an unsorted jumble of many sediment sizes.

Swamps and **bogs** are quiet-water environments where mud and decayed plant material accumulate.

Streams in mountainous areas erode and deposit a wide variety of sediment, while those in lowlands transport and deposit mostly mud (silt and clay) and sand.

Glacial deposits often consist of a poorly-sorted mixture of many different sediment sizes ranging from clay to boulders.

▲ Figure 7.25
Continental sedimentary environments Note that in reality, there is some overlap between continental, transitional, and marine environments. Each is characterized by certain physical, chemical, and biological conditions. This illustration shows an idealized view highlighting various terrestrial environments.

Beaches that form where wave activity is strong, consist mainly of pebbles and cobbles.

Beaches, **bars**, and **spits** along low-lying coasts and in sheltered coves are typically composed of well-sorted sand and/or shell fragments.

Tidal flats and **lagoons** are areas where fine clay particles or carbonate-rich muds accumulate.

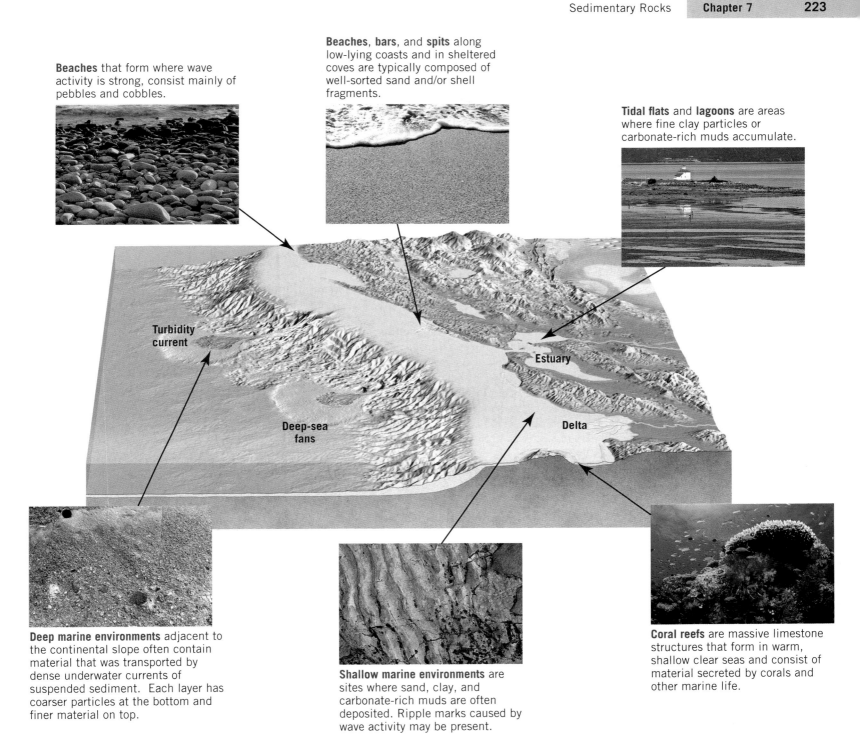

Deep marine environments adjacent to the continental slope often contain material that was transported by dense underwater currents of suspended sediment. Each layer has coarser particles at the bottom and finer material on top.

Shallow marine environments are sites where sand, clay, and carbonate-rich muds are often deposited. Ripple marks caused by wave activity may be present.

Coral reefs are massive limestone structures that form in warm, shallow clear seas and consist of material secreted by corals and other marine life.

▲ Figure 7.26
Marine and transitional sedimentary environments Note that in reality, there is some overlap between continental, transitional, and marine environments. This illustration shows an idealized view highlighting various marine and near-shore environments.

streams emerge from a mountainous area onto a flatter surface, a distinctive cone-shaped accumulation of sediment known as an *alluvial fan* forms.

In frigid, high-latitude or high-altitude settings, glaciers pick up and transport huge volumes of sediment. Materials deposited directly from ice are typically unsorted mixtures of particles that range in size from fine clay to huge boulders. Water from melting glaciers transports and redeposits some of this glacial sediment, creating stratified, sorted accumulations.

The work of wind and its resulting deposits are referred to as *eolian*, after Aeolus, the Greek god of wind. Unlike glacial deposits, eolian sediments are well sorted. Wind can lift fine dust high into the atmosphere and transport it great distances. Where winds are strong and the surface is not anchored by vegetation, sand is transported closer to the ground, where it accumulates in *dunes*. Deserts and coasts are common sites for this type of deposition.

In addition to being areas where dunes sometimes develop, desert basins are sites where shallow *playa lakes* occasionally form following heavy rains or periods of snowmelt in adjacent mountains. They rapidly dry up, sometimes leaving behind evaporites and other characteristic deposits. Figures 19.9 and 19.10 on page 545 illustrate such an environment. In humid regions, lakes are more enduring features, and their quiet waters are excellent sediment traps. Small deltas, beaches, and bars form along the lakeshore, with finer sediments coming to rest on the lake floor.

Marine Environments   Marine depositional environments are divided according to depth. The *shallow marine environment* reaches to depths of about 200 meters (nearly 700 feet) and extends from the shore to the outer edge of the continental shelf. The *deep marine environment* lies seaward of the continental shelf in waters deeper than 200 meters.

The shallow marine environment borders all of the world's continents. Its width varies greatly, from practically nonexistent in some places to broad expanses extending as far as 1500 kilometers (more than 900 miles) in other locations. On average this zone is about 80 kilometers (50 miles) wide. The kind of sediment deposited here depends on distance from shore, elevation of the adjacent land area, water depth, water temperature, and climate.

Due to the ongoing erosion of the adjacent continent, the shallow marine environment receives huge quantities of land-derived sediment. Where the influx of sediment is small and the seas are relatively warm, carbonate-rich muds may be the predominant sediment. Most of this material consists of the skeletal debris of carbonate-secreting organisms mixed with inorganic precipitates. Coral reefs are also associated with warm, shallow marine environments. In hot regions where the sea occupies a basin with restricted circulation, evaporation triggers the precipitation of soluble materials and the formation of marine evaporite deposits.

Deep marine environments include all the floors of the deep ocean. Far from landmasses, tiny particles from many sources remain adrift for long spans. Gradually these small grains "rain" down on the ocean floor, where they accumulate very slowly. Significant exceptions are thick deposits of relatively coarse sediment that occur at the base of the continental slope. These materials move down from the continental shelf as turbidity currents—dense gravity-driven masses of sediment and water (see Figure 7.26).

Transitional Environments   The shoreline is the transition zone between marine and continental environments. Here we find the familiar deposits of sand or gravel called *beaches*. Mud-covered *tidal flats* are alternately covered with shallow sheets of water and then exposed to air as tides rise and fall. Along and near the shore, the work of waves and currents distributes sand, creating *spits*, *bars*, and *barrier islands*. Offshore bars and reefs create *lagoons*. The quieter waters in these sheltered areas are another site of deposition in the transition zone.

Deltas are among the most significant deposits associated with transitional environments. The complex accumulations of sediment build outward into the sea when rivers experience an abrupt loss of velocity and deposit their load of detrital material.

Sedimentary Facies

When we study a series of sedimentary layers, we can see the successive changes in environmental conditions that occurred at a particular place over time. Changes in past environments may also be seen when we laterally trace a single layer of sedimentary rock. This is true because at any one time, many different depositional environments can exist over a broad area. For example, when sand is accumulating in a beach environment, finer muds are often being deposited in quieter offshore waters. Still farther out, perhaps in a zone where biological activity is high and land-derived sediments are scarce, the deposits consist largely of the calcite-rich remains of small organisms. In this example, different sediments are accumulating adjacent to one another at the same time. Different parts of each layer possess a distinctive set of characteristics that reflect the conditions in a particular environment. The term **facies** is used to describe such sets of sediments. When a sedimentary layer is examined in cross section from one end to the other, each facies grades

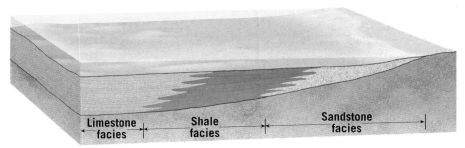

◄ SmartFigure 7.27
Lateral change When a sedimentary layer is traced laterally, we may find that it is made up of several different rock types. This occurs because many sedimentary environments can exist at the same time over a broad area. The term *facies* is used to describe such sets of sedimentary rocks. Each facies grades laterally into another that formed at the same time but in a different environment.

Tutorial
https://goo.gl/
46f765

laterally into another that formed at the same time but that exhibits different characteristics (**Figure 7.27**). The merging of adjacent facies tends to be a gradual transition rather than a sharp boundary, but abrupt changes do sometimes occur.

Sedimentary Structures

In addition to variations in grain size, mineral composition, and texture, sediments exhibit a variety of structures that are often preserved when they change to sedimentary rock. Some, such as graded beds, build as sediments accumulate and are a reflection of the transporting medium. Others, such as *mud cracks*, form after the materials have been deposited and result from processes occurring in the environment. When present, sedimentary structures provide additional information that can be useful in interpreting Earth's history.

Sedimentary rocks form as layers of sediment accumulate in various depositional environments. These layers, called **strata** (singular: stratum), or **beds**, are probably *the single most common and characteristic feature of sedimentary rocks*. Each stratum is unique. It may be a coarse sandstone, a fossil-rich limestone, a black shale, and so on. When you look at **Figure 7.28** or look back at the chapter-opening photo and at Figures 7.1 and 7.5, you see many such layers, each different from the others. The variations in texture, composition, and thickness reflect the different conditions under which each layer was deposited.

Beds range from microscopically thin to tens of meters thick. Separating these strata are **bedding planes**, relatively flat surfaces along which rocks tend to separate or break. Changes in the grain size or in the composition of the sediment being deposited can create bedding planes. Pauses in deposition can also lead to layering because chances are slight that newly deposited material will be exactly the same as previously deposited sediment. Generally, each bedding plane marks the end of one episode of sedimentation and the beginning of another.

◄ Figure 7.28
Layers are called strata This outcrop of sedimentary strata illustrates the characteristic layering of this group of rocks. This exposure of sedimentary strata is in New York's Shawangunk Mountains.

▶ Figure 7.29
Cross-bedding Sand dunes typically exhibit thin layers inclined at an angle to the main bedding.

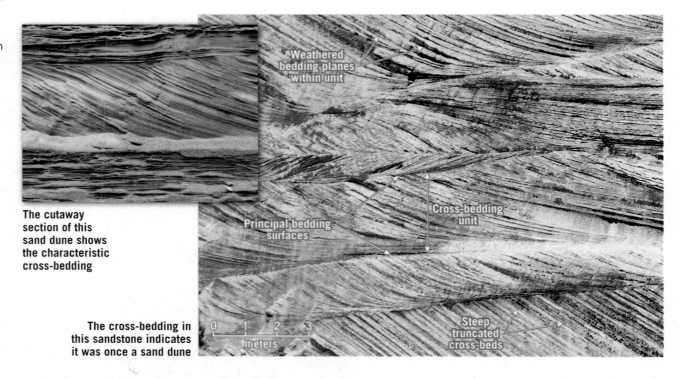

The cutaway section of this sand dune shows the characteristic cross-bedding

The cross-bedding in this sandstone indicates it was once a sand dune

Because sediments usually accumulate as particles that settle from a fluid, most strata are originally deposited as horizontal layers. There are circumstances, however, in which sediments do not accumulate in horizontal beds. Sometimes when a bed of sedimentary rock is examined, we see layers within it that are inclined to the horizontal. This is called **cross-bedding**, and it is most characteristic of sand dunes, river deltas, and certain stream channel deposits (see Figure 7.8 and **Figure 7.29**).

Graded beds are another special type of bedding (**Figure 7.30**). In this case, the particles within a single

▶ Figure 7.30
Graded bedding A graded bed is characterized by a decrease in sediment size from bottom to top. Graded beds are associated with submarine currents known as *turbidity currents*.

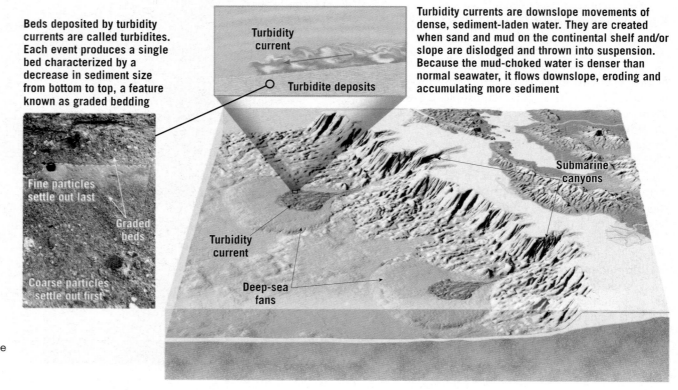

Beds deposited by turbidity currents are called turbidites. Each event produces a single bed characterized by a decrease in sediment size from bottom to top, a feature known as graded bedding

Turbidity currents are downslope movements of dense, sediment-laden water. They are created when sand and mud on the continental shelf and/or slope are dislodged and thrown into suspension. Because the mud-choked water is denser than normal seawater, it flows downslope, eroding and accumulating more sediment

▶ **Figure 7.31**
Frozen in stone
A. These current ripple marks formed in sandy sediment and are now preserved in rock.
B. When muddy sediments dry out, they shrink and create cracks. When the sediment is turned into rock, the mud cracks are preserved.

A.

B.

sedimentary layer gradually change from coarse at the bottom to fine at the top. Graded beds are most characteristic of rapid deposition from water containing sediments of varying sizes. When a current experiences a rapid energy loss, the largest particles settle first, followed by successively smaller grains. The deposition of a graded bed is most often associated with a turbidity current, a mass of sediment-choked water that is denser than clear water and that moves downslope along the bottom of a lake or an ocean.

Geologists can deduce quite a bit about past sedimentary environments by examining sedimentary rocks. A conglomerate, for example, may indicate a high-energy environment such as a surf zone or rushing stream, where only coarse materials remain and finer particles are kept suspended. If the rock is arkose, it may signify a dry climate, where little chemical alteration of feldspar occurred. Carbonaceous shale is a sign of a low-energy, organic-rich environment, such as a swamp or lagoon.

Other features found in some sedimentary rocks also give clues to past environments. **Ripple marks** are small waves of sand that develop on the surface of a sediment layer through the action of moving water or air (**Figure 7.31A**). The ridges form at right angles to the direction of motion. If the ripple marks were formed by air or water moving in essentially one direction, their form will be asymmetrical. These *current ripple marks* will have steeper sides in the downcurrent direction and more gradual slopes on the upcurrent side. Ripple marks produced by a stream flowing across a sandy channel and by wind blowing over a sand dune are two common examples of current ripples. When present in solid rock, they may be used to determine the direction of movement of ancient wind or water currents. Other ripple marks have a symmetrical form. These features, called *oscillation ripple marks*, result from the back-and-forth movement of surface waves in a shallow near-shore environment.

Mud cracks (**Figure 7.31B**) indicate that the sediment in which they were formed was alternately wet and dry. When exposed to air, wet mud dries out and shrinks, producing cracks. Mud cracks are associated with environments such as tidal flats, shallow lakes, and desert basins.

Fossils, the remains or traces of prehistoric life, are important inclusions in sediment and sedimentary rocks. They are important tools for interpreting the geologic past. Knowing the nature of the life-forms that existed at a particular time helps researchers understand past environmental conditions. In addition, fossils are important time indicators and play a key role in correlating rocks of similar ages that are from different places. Fossils are discussed further in Chapter 9.

CONCEPT CHECKS 7.7

1. What are the three broad categories of sedimentary environments? List a specific example associated with each category (see Figures 7.25 and 7.26).

 **Concept Checker**

https://goo.gl/q3S9L4

2. Why might a single sedimentary layer be made up of different types of sedimentary rock? What term applies to these different parts of such a layer?

3. What is the single most characteristic feature of sedimentary rocks?

4. How might mud cracks and ripple marks be useful clues about the geologic past?

7.8 The Carbon Cycle and Sedimentary Rocks

Relate weathering processes and sedimentary rocks to the carbon cycle.

To illustrate the movement of material and energy among the spheres of the Earth system, let us take a brief look at the **carbon cycle** (Figure 7.32). The carbon cycle refers to the movement of carbon (in various forms) throughout the atmosphere, hydrosphere, geosphere, and biosphere. Most carbon is bonded chemically to other elements to form compounds such as carbon dioxide (in the atmosphere and hydrosphere), calcium carbonate (in the geosphere), and the hydrocarbons found in coal and petroleum. Carbon is also the basic building block of life, as it readily combines with hydrogen and oxygen to form the fundamental organic compounds that compose living things (the biosphere).

Movement of Carbon Between the Atmosphere and Biosphere

Certainly one of the most active parts of the carbon cycle is the movement of carbon from the atmosphere to the biosphere and back again. In the atmosphere, carbon exists mainly as carbon dioxide (CO_2). Atmospheric carbon dioxide is significant because it is a greenhouse gas, which means it is an efficient absorber of energy emitted by Earth and thus influences the heating of the atmosphere, a process described in more detail in Chapter 21. Many of the processes that operate on Earth involve carbon dioxide, so the gas constantly moves into and out of the atmosphere. Through photosynthesis, plants absorb carbon dioxide from the atmosphere to produce the essential organic compounds needed for growth. Animals that consume these plants (or consume other animals that eat plants) use these organic compounds as a source of energy and, through the process of respiration, return carbon

dioxide to the atmosphere. (Plants also return some CO_2 to the atmosphere via respiration.) Further, when plants die and decay or are burned, this biomass is oxidized, and carbon dioxide is returned to the atmosphere.

Movement of Carbon Between the Geosphere, Hydrosphere, Atmosphere, and Biosphere

Carbon also moves from the geosphere and hydrosphere to the atmosphere (and back again). For example, volcanic activity early in Earth's history is thought to have been the source of much of our atmospheric carbon dioxide. Some of that carbon dioxide dissolves into water vapor in air, forming carbonic acid (H_2CO_3), which then attacks the rocks that compose Earth's crust when it rains. One product of this chemical weathering of solid rock is the soluble bicarbonate ion (HCO_3^-), which is carried by groundwater and streams to the ocean. Water-dwelling organisms extract this dissolved material to produce hard parts of calcium carbonate ($CaCO_3$). When the organisms die, these skeletal remains settle to the ocean floor as biochemical sediment and become sedimentary rock. In fact, the crust is by far Earth's largest depository of carbon, where it is a constituent of a variety of rocks, the most abundant being limestone. Eventually the limestone may be exposed at Earth's surface, where chemical weathering will cause the carbon stored in the rock to be released to the atmosphere as carbon dioxide.

Movement of Carbon from Biosphere to Geosphere (Fossil Fuels)

Not all dead plant material decays immediately back to carbon dioxide. A small percentage is deposited as sediment at any given moment, and over long spans of geologic time, considerable biomass is buried with sediment. Under the right conditions, some of these carbon-rich deposits are converted to fossil fuels—coal, petroleum, or natural gas. Some of these fuels have been recovered (mined or pumped from a well) and burned to generate electricity and fuel vehicles. One result of fossil-fuel combustion is the release of huge quantities of carbon dioxide back into the atmosphere.

▼ Figure 7.32
Carbon cycle This simplified diagram emphasizes the flow of carbon between the atmosphere and the hydrosphere, geosphere, and biosphere. The arrows show whether the flow of carbon is into or out of the atmosphere.

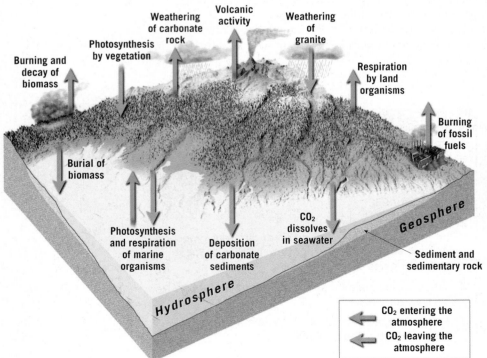

In summary, carbon moves among all four of Earth's major spheres. It is essential to every living thing in the biosphere. In the atmosphere, carbon dioxide is an important greenhouse gas. In the hydrosphere, carbon dioxide is dissolved in lakes, rivers, and the ocean. In the geosphere, carbon is contained in carbonate sediments and sedimentary rocks and is stored as organic matter dispersed through sedimentary rocks and as deposits of coal and petroleum.

CONCEPT CHECKS 7.8

1. Describe how chemical weathering and the formation of biochemical sediment remove carbon from the atmosphere and store it in the geosphere.

2. Provide an example by which carbon moves from the geosphere to the atmosphere.

 Concept Checker https://goo.gl/KcoCzG

CONCEPTS IN REVIEW

Sedimentary Rocks

7.1 An Introduction to Sedimentary Rocks

Explain the importance of sedimentary rocks as indicators of past environments and sources of resources necessary for modern society. Summarize the part of the rock cycle that pertains to sediments and sedimentary rocks. List the three categories of sedimentary rocks.

Key Terms: sedimentary rock · clastic sedimentary rock · chemical sedimentary rock · organic sedimentary rock

- Though igneous and metamorphic rocks make up most of Earth's crust by volume, sediment and sedimentary rocks are concentrated near the surface. There, at the interface between Earth's four spheres, sediments and the rock layers they eventually form make a record of past conditions and events at the surface. *Sedimentary rocks* contain fossils that show the evolution of life over time.

- Numerous geologic resources are restricted to sedimentary rocks, such as coal, oil, uranium, and several major ores of metals.

- Sediments are the raw materials from which sedimentary rocks are made. Sediments are produced from the weathering of preexisting rocks. Both solid particles of many sizes and chemical residues, including ions in solution, qualify as sediments.

- Once produced, sediments are transported from their source area by water currents, wind, glacial ice, or simply downhill movement under the influence of gravity. Eventually, these will be deposited at a new site, where they will be compacted and/or cemented so that the individual sediments become bonded together into a sedimentary rock.

- There are three main varieties of sedimentary rocks: *clastic, chemical,* and *organic.*

7.2 Clastic Sedimentary Rocks

Discuss the primary basis for distinguishing among clastic rocks and describe how the origin and history of such rocks might be determined.

Key Terms: fissility · shale · sorting · sandstone · breccia · conglomerate

- Clastic sedimentary rocks are made of solid particles, mostly quartz grains and microscopic clay minerals. Quartz and clay dominate because unlike most other minerals, they are stable at Earth's surface. Feldspars and micas are conspicuous additions to certain detrital sediments, indicating a relatively short time in the chemical weathering environment; these rocks were mechanically weathered, transported a relatively short distance, and deposited with minimal decomposition.

- Clastic sedimentary rocks are classified mainly based on the size of the sedimentary grains of which they are composed. Grain size is a clue about how energetic the environment of deposition was. Bigger particles indicate more powerful transporting currents; finer grains can be deposited only where current energy is relatively low.

- *Shale* is made mostly of small grains of clay minerals that accumulate in low-energy depositional environments such as the deep sea, lake bottoms, and floodplains adjacent to rivers. Shale is *fissile* due to the alignment of microscopic clay flakes parallel to bedding. Shale containing a lot of organic material forms in low-oxygen environments and is characterized by a black color.

- *Sandstone* is dominated by sand-sized grains and may exhibit various degrees of *sorting.* Sorting is a reflection of how abruptly or gradually the sand was deposited. Rounding of the individual sand grains is another important aspect of a sandstone's texture: Rounder grains signify further transport, while angular grains imply a shorter transport distance. Sandstones also vary in their composition: The presence of minerals that are relatively unstable at the surface (like feldspar) implies that the sand did not undergo much chemical weathering prior to deposition. The greater the proportion of quartz, the more the source sediment was chemically weathered before deposition. Three main varieties of sandstone worth knowing are quartz sandstone, arkose, and graywacke.

- *Conglomerate* and *breccia* are characterized by a high proportion of gravel-sized grains. If deposited by water, a conglomerate implies a very energetic current. Grains in a breccia are angular, indicating that the material was deposited closer to its source area. Conglomerate is made of rounded grains, implying a significant amount of transport before deposition.

7.3 Chemical Sedimentary Rocks

Explain the processes involved in the formation of chemical sedimentary rocks and describe several examples.

Key Terms: limestone   chert   salt flat
biochemical   dolostone   evaporite

- Chemical sedimentary rocks are formed when ions dissolved in solution link together to form mineral crystals. Sometimes this happens inorganically (without life being involved), and other times living organisms biochemically extract the ions to precipitate mineral matter as bone or shell.
- *Limestone* is the most common chemical sedimentary rock. It forms mainly in shallow, warm ocean settings. Limestone is dominated by calcium carbonate. This is the material from which corals construct reefs. Coquina and chalk are also examples of *biochemical* limestone. Travertine and oolitic limestone are examples of inorganic limestone.
- *Dolostone* is a chemical sedimentary rock that is dominated by the mineral dolomite. Like calcite, dolomite is a carbonate mineral, but about half of its calcium ions have been replaced by magnesium ions.
- *Chert* is the general term for rocks made of microcrystalline silica. If the chert is red, it is called jasper. Black chert is flint. Agate is multicolored. When silica replaces plant matter to make petrified wood, often a variety of colors are present.
- *Evaporite* deposits form when minerals precipitate from an ever-more-concentrated solution of dissolved ions. This is how the vast *salt flats* of the American West formed. Digging into these deposits reveals rock salt, rock gypsum, and the potassium salt sylvite.

7.4 Coal: An Organic Sedimentary Rock

Outline the successive stages in the formation of coal.

Key Term: coal

- *Coal* forms from large amounts of plant matter buried in low-oxygen depositional environments such as swamps and bogs. Through compression, the peat formed becomes compressed into a low-grade form of coal called lignite. Lignite can be compressed further, driving out volatile components and concentrating carbon to make higher-grade bituminous coal. Metamorphism accompanying mountain building can take this concentration process even further, producing the highest-grade coal, anthracite.

7.5 Turning Sediment into Sedimentary Rock: Diagenesis and Lithification

Describe the processes that convert sediment into sedimentary rock and other changes associated with burial.

Key Terms: diagenesis   lithification   compaction   cementation

- When sediments are buried to relatively shallow depths (the upper few kilometers of the crust), changes in temperature and pressure trigger a variety of processes called *diagenesis*, a collective term for all the chemical, physical, and biological changes that occur after sediments are deposited and during and after lithification.
- The transformation of sediment into sedimentary rock is called *lithification*. The two main processes contributing to lithification are *compaction* (a reduction in pore space that involves packing grains more tightly together) and *cementation* (a reduction in pore space that involves adding new mineral material that acts as a "glue" to bind the grains to each other).

7.6 Classification of Sedimentary Rocks

Summarize the criteria used to classify sedimentary rocks.

Key Terms: clastic   nonclastic

- Sedimentary rocks are classified primarily based on whether they are *clastic*, chemical, or organic. Clastic rocks are subdivided by grain size, while among the chemical rocks, mineral composition is the key distinguishing characteristic. An additional characteristic is whether the rocks exhibit a clastic or *nonclastic (crystalline)* texture.
- Crystalline texture is common to nonclastic sedimentary rocks and igneous rocks and is particularly apparent under magnification. However, the minerals involved are totally different and allow the two types of rock to be distinguished.

7.7 Sedimentary Rocks Represent Past Environments

Distinguish among three broad categories of sedimentary environments and provide an example of each. List several sedimentary structures and explain why these features are useful to geologists.

Key Terms:

environment   bedding plane   mud crack   facies
  of deposition   cross-bedding   graded bed   fossils
strata   ripple mark

- Different combinations of tectonic, climatic, and biological conditions result in different types of sediment accumulating. The principle of uniformity suggests that the sedimentary record can be interpreted in light of modern depositional environments. Continental, marine, and transitional (shoreline) environments all have distinctive characteristics that allow geologists to identify sedimentary rocks formed in those environments.
- Sedimentary *facies* are lateral equivalents that represent different depositional conditions operating in adjacent areas at the same time. For instance, a beach today may be depositing sand, while a kilometer or two offshore only mud is being deposited, and further beyond that, carbonate minerals may be precipitating. All are the same age but represent neighboring areas governed by different conditions.
- Sedimentary structures are patterns that form in sedimentary rock at the time of deposition (or shortly thereafter), before the sediments become lithified. They can provide powerful clues to the conditions under which the sediment accumulated.
- *Beds* (or *strata*) are sheets of sediment deposited in a more-or-less continuous layer. Sometimes *cross-beds* are preserved within bedding, allowing geologists to deduce the direction of the depositional current. *Ripple marks*, small waves created by wind or water on the surface of a sediment layer, also provide useful clues. *Graded beds* indicate

depositional currents that quickly lost their energy. Larger grains settled out first, and the smallest grains settled out last. *Mud cracks* form when mud contracts upon drying out, and they indicate that the sediment was exposed to air. *Fossils* serve as useful tools that allow geologists to date and interpret past environmental conditions.

Q Moving toward deeper water, do the lateral equivalent rock layers become finer or coarser grained?

7.8 The Carbon Cycle and Sedimentary Rocks

Relate weathering processes and sedimentary rocks to the carbon cycle.

Key Term: carbon cycle

- Carbon is a vital component of the atmosphere, biosphere, geosphere, and hydrosphere. The same atom of carbon that is currently part of

your nose may have entered your body in a piece of bread. Prior to that, it may once have been part of a plant, and before that it may have been a carbon dioxide molecule in the atmosphere. How did it get to the atmosphere? Perhaps it escaped into the air after originally being dissolved in the ocean, and perhaps it got to the ocean by being weathered from limestone and then washed down a stream. While individual details vary, the key point is that carbon is a reactive element that is equally at home in rocks, water, the air, and living tissue.

GIVE IT SOME THOUGHT

1. Develop a geologic "life history" of a sedimentary rock. Begin with a mass of igneous bedrock in a mountain area and end with your sedimentary rock being collected by a future geology student. Be as complete as possible.

2. How is the use of the term *clay* in Chapter 3 different from the use of the term in Figure 7.3? How are these two uses of the term related?

3. If you hiked to a mountain peak and found limestone at the top, what would that indicate about the likely geologic history of the rock atop the mountain?

4. Why is it *not* necessary to indicate the texture of clastic rocks on the identification chart for sedimentary rocks (see Figure 7.22)?

5. During a hike in Utah's Zion National Park, you pick up a sedimentary rock sample. When you examine the sample with your hand lens, you see that the rock consists mainly of rounded glassy particles that appear to be quartz. To be sure, you conduct two basic tests. When you check for hardness, the rock easily scratches glass, which is what quartz would do. However, when you place a drop of acid on the sample, it fizzes. Explain how a rock that appears to be rich in quartz could effervesce with acid.

6. Examine the accompanying sketch, which shows three sediment layers on the ocean floor. What term is applied to such layers? What process was responsible for creating these layers? Are these layers more likely part of an offshore lagoon or a deep-sea fan?

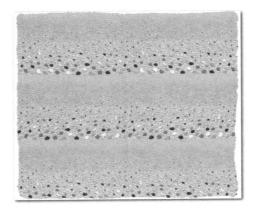

7. In which of the environments illustrated in Figure 7.25 would you expect to find:

   a. An evaporite deposit?

   b. A well-sorted sand deposit?

   c. A deposit that includes a high percentage of partially decomposed plant material?

   d. A jumbled mix of many sediment sizes?

8. This image shows the surface of a sand dune. What term is applied to the wave-like ridges on the surface? Be as specific as possible. Is the prevailing wind direction most likely from the left or from the right? Explain.

9. While on a field trip with your geology class, you stop at an outcrop of sandstone. An examination with a hand lens shows that the sandstone is poorly sorted and rich in feldspar and quartz. Your instructor tells you that the sediment was derived from one of two sites in the area:

Site 1: A nearby exposure of weathered basaltic lava flows

Site 2: An outcrop of granite at the previous field trip stop up the road

Select the most likely site and explain your choice. What name is given to this type of sandstone?

10. This rock sample consists of intergrown crystals. How would you determine whether the rock is sedimentary or igneous? If it is sedimentary, what term describes its texture?

EYE ON EARTH

1. This clastic rock consists of angular grains and is rich in potassium feldspar and quartz.

a. What do the angular grains indicate about the distance the sediment was transported?

b. The source of the sediment in this rock was an igneous mass. Name the likely rock type.

c. Did the sediment in this sample undergo a great deal of chemical weathering? Explain.

2. This is a mass of chemical sedimentary rock in Yellowstone National Park. It was formed by the following process: Rainwater became acidic when it absorbed carbon dioxide in the air. As the water seeped beneath the surface, it dissolved calcite in the limestone bedrock. Eventually, Yellowstone's underground plumbing returned the water, now saturated with calcium carbonate, to the surface as a hot spring. When the water emerged, some carbon dioxide escaped into the air, triggering the deposition of the rock seen here.

a. Did this rock have a biochemical origin or an inorganic origin?

b. Is the rock most likely chert or limestone? Explain.

c. Name the particular variety of this rock. What figure in this chapter provides another example?

3. This is an aerial view of North Carolina's Hatteras Island (looking toward the south). To the left (east) of the island is the Atlantic Ocean. The sheltered waters of Pamlico Sound are on the right, the side of the island facing the mainland. Assume that the sediments accumulating in the area are primarily detrital.

a. If you were to sample the sediments at points A and B, which would likely have the coarser particles?

b. Explain why the sediments at these two sites would probably be different.

c. Which one of the three broad categories of sedimentary environments is represented in this image?

DATA ANALYSIS

Sedimentary Rock Near You

Sedimentary rocks cover much of the surface of the United States and provide many valuable resources for the country. What kind of sedimentary rock is near your location?

https://goo.gl/eD226M

ACTIVITIES

Go to the U.S. Geological Survey's Mineral Resources site, found at https://mrdata.usgs.gov/geology/state/map.html. The interactive map can show you the geologic surface units as mapped over the years. Use double-clicks (to zoom) and drag to find a location. You can open the "Map Layers" feature (top left menu button) to turn on/off layers, such as county names or OpenStreetMap, that might help you navigate. You can also use the "Find Geographic Areas by Name" feature (globe icon at the left) to find a particular named location.

The different areas of color correspond to different geologic units. To discover what kind of rock makes up a particular unit, single-click it to reveal a legend on the right. Then you can click on the name of any geologic unit (formation) to get more information, in a new tab, about the kind of rocks, their age, and what resources they might provide.

1. **Find three different kinds of sedimentary rock in your area. If you live in an area that's mostly igneous or metamorphic rock, you might need to explore a bit farther from home.**

   a. Are they clastic, chemical, or organic sedimentary rock?

   b. What age range does each unit (formation) span?

   c. Does a single unit (formation) have more than one kind of sedimentary rock?

   d. What resources, if any, can be derived from each unit (formation)?

2. **Can you determine, based on the age in the legend or from the topography, which unit is youngest? Which is the oldest?**

3. **Are there any fossils described for any of your units (formations)?**

4. **For more detailed geologic maps of each state, browse to https://www.nps.gov/subjects/geology/state-geologic-maps.htm and select your state of interest. Each state has a different system for presenting its geologic information, so we cannot provide instructions that work for all. Does your state map provide any different information compared to the national map used above?**

Mastering Geology

Looking for additional review and test prep materials? Visit pearson.com/mastering/geology to enhance your understanding of this chapter's content by accessing a variety of resources, including **Self-Study Quizzes, Geoscience Animations, SmartFigure Tutorials,** *Project Condor* **Videos, Mobile Field Trips, Dynamic Study Modules,** and an optional **Pearson eText.**

"Blood Lapis" Funding Conflict in Afghanistan

While lapis lazuli (commonly known as *lapis*) is found in small deposits throughout the world, Afghanistan's mines are ancient and famed as a high-quality source of this vivid blue gemstone.

▲ Raw lapis lazuli

Although lapis is commonly referred to as a mineral, it is a metamorphic rock composed of a variety of minerals—most commonly lazurite, calcite, and pyrite. Unfortunately, in recent decades lapis mining and trade have funded conflict in this war-torn area.

The Global Witness organization estimates that the Taliban and other armed groups have earned up to $20 million a year in Afghanistan from trading lapis. Numerous advocacy groups call for declaring lapis from this region a conflict mineral, akin to the "blood diamonds" that finance dictatorships in Africa. Assigning lapis the designation of conflict mineral could lead to better regulation of its trade.

Although people may not recognize lapis by name, they likely know the famous works made with it. For instance, the prominent eyebrows in King Tut's famous gold funerary mask were created with lapis. In powdered form mixed in paint, it is known as *ultramarine*, a highly sought-after pigment. In Renaissance Europe, lapis was more expensive than gold, and many artists used ultramarine in paintings and statuary for garments of Mary, the mother of Jesus. Today, synthetic pigments have mostly replaced lapis in paint, but lapis endures as a popular gemstone in jewelry.

► The Hindu Kush Mountains in Afghanistan are home to a lapis lazuli mine that's over 6000 years old.

8
Metamorphism and Metamorphic Rocks

FOCUS ON CONCEPTS

Each statement represents the primary learning objective for the corresponding major heading within the chapter. After you complete the chapter, you should be able to:

8.1 Compare and contrast the environments that produce metamorphic, sedimentary, and igneous rocks.

8.2 List and distinguish among the four agents that drive metamorphism.

8.3 Explain how foliated and nonfoliated textures develop.

8.4 List and describe the most common metamorphic rocks.

8.5 Write a description for each of the following environments: contact metamorphism, hydrothermal metamorphism, subduction zone metamorphism, and regional metamorphism.

8.6 Explain how index minerals are used to establish the metamorphic grade of a rock body.

Subject any existing rock to great stresses in the form of heat and pressure, and over time, it will transform—or be metamorphosed—into a new type of rock. This chapter looks at the tectonic forces that forge metamorphic rocks and how these rocks change in appearance, mineralogy, and (sometimes) overall chemical composition.

8.1 What Is Metamorphism?

Compare and contrast the environments that produce metamorphic, sedimentary, and igneous rocks.

Recall from the discussion of the rock cycle in Chapter 1 that metamorphism is the transformation of one rock type into another rock type. Metamorphic rocks come from preexisting sedimentary and igneous rocks, as well as from other metamorphic rocks. The rock that is transformed is called the **parent rock**. **Metamorphism**, which means "a change in form," is a process that transforms the **mineralogy** (a rock's mineral constituents), texture, and sometimes chemical composition of the parent rock. Usually these changes involve elevated temperatures and pressures, which are significantly different from those of the environment in which the parent rock formed.

Figure 8.1 illustrates the relationships among metamorphic, sedimentary, and igneous environments. Metamorphism occurs over a range of temperatures that lie between those experienced during sedimentary rock formation (up to about 200°C [400°F]) and temperatures approaching those at which rocks begin to melt (about 700°C [1300°F]). However, during metamorphism, *the rock remains essentially solid*. If significant melting occurs, then the rock has entered the realm of igneous activity, as discussed in Chapter 4.

Clay minerals, the most common minerals found in sedimentary rocks, provide an example of how metamorphic rocks form. When buried to a depth where temperatures exceed 200°C (nearly 400°F), the clay minerals in sedimentary rock transform into the minerals chlorite and/or muscovite mica. (Kaolinite is one example of a clay mineral; see Figure 3.35, page 92.) Chlorite is a

mica-like mineral formed by the metamorphism of dark iron- and magnesium-rich silicate minerals. Under even more extreme conditions, chlorite becomes biotite mica.

Metamorphic Grade

The degree to which a parent rock changes during metamorphism is called its **metamorphic grade**. *Low-grade metamorphic environments* have low temperatures and pressures. An example of a low-grade metamorphic rock is slate, which transforms from the common sedimentary rock shale. During this transformation, shale's clay minerals turn into tiny chlorite and muscovite mica flakes. Hand samples of slate and shale are sometimes difficult to distinguish from one another, illustrating that the transition from sedimentary to metamorphic rock is often gradual and subtle (**Figure 8.2A**).

▶ Figure 8.1
Metamorphic versus sedimentary and igneous environments Metamorphism occurs over a range of temperatures that lie between those experienced in sedimentary environments and temperatures that approach those at which rocks melt. Pressure, which includes confining pressure and differential stress, also plays a major role in metamorphism.

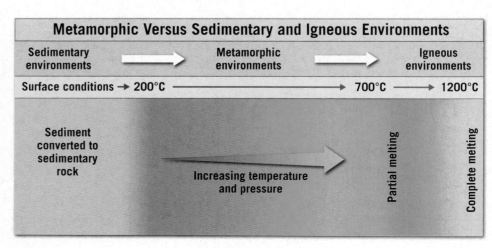

Metamorphic Versus Sedimentary and Igneous Environments

Sedimentary environments → Metamorphic environments → Igneous environments

Surface conditions → 200°C —————→ 700°C ——→ 1200°C

Sediment converted to sedimentary rock

Increasing temperature and pressure

Partial melting

Complete melting

By contrast, *high grade metamorphic environments* have more extreme temperatures and pressures. This environment causes a transformation so complete that the identity of the parent rock cannot be easily determined. In high-grade metamorphism, features such as bedding planes, fossils, and vesicles that existed in the parent rock are obliterated. Further, when rocks deep in the crust are subjected to *compressional stress* (like being placed in a giant vise), the entire rock body may be deformed, usually by folding (**Figure 8.2B**).

Pressure also plays an important role in metamorphism. Although confining pressure acts to compact sediment to form sedimentary rocks, the pressures involved in metamorphism are even greater—sufficient to convert mineral matter into denser forms having more compact crystalline structures. Thus, metamorphism involves the formation of new minerals from preexisting ones.

Unlike some igneous and sedimentary processes that occur in surface or near-surface environments, metamorphism most often occurs deep within Earth, beyond direct observation. Despite this significant obstacle, metamorphic rocks contain clues to the environment in which they formed. These clues include their texture and mineral composition.

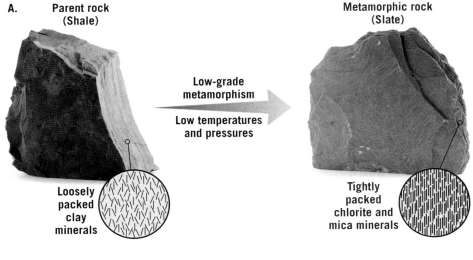

A. Parent rock (Shale)

Low-grade metamorphism — Low temperatures and pressures

Metamorphic rock (Slate)

Loosely packed clay minerals

Tightly packed chlorite and mica minerals

B. Parent rock (Granodiorite)

High-grade metamorphism — Strong compressional forces, high temperatures and pressures

Metamorphic rock (Folded gneiss)

Randomly oriented minerals

Deformed layers of segregated minerals

◄ **Figure 8.2**
Metamorphic grade
A. Low-grade metamorphism, illustrated with the transformation of the sedimentary rock shale to the more compact metamorphic rock slate.
B. High-grade metamorphism, which occurs at temperatures approaching rock melting points, obliterates existing texture.

CONCEPT CHECKS 8.1

1. *Metamorphism* means "a change in form." Describe how a rock may change during metamorphism.

2. What is meant by the statement "Every metamorphic rock has a *parent rock*"?

3. Define *metamorphic grade*.

Concept Checker
https://goo.gl/gbAvha

8.2 What Drives Metamorphism?

List and distinguish among the four agents that drive metamorphism.

The agents of metamorphism include *heat, pressure, differential stress*, and *chemically active fluids*. During metamorphism, rocks may be subjected to all four metamorphic agents simultaneously. However, the degree of metamorphism and the contribution of each agent vary greatly from one environment to another.

Heat as a Metamorphic Agent

The most important factor driving metamorphism is *heat*, which provides the energy needed to produce the chemical reactions that result in the recrystallization of existing minerals. Recall from the discussion of igneous rocks in Chapter 4 that an increase in temperature causes the atoms within a mineral to vibrate more rapidly. Even in a crystalline solid, where atoms are strongly

▶ SmartFigure 8.3
Sources of heat for heat metamorphism The main sources for metamorphism are the increasing temperature that occurs as we travel deeper into Earth's interior and heat released to surrounding rocks when a magma body cools.

Tutorial
https://goo.gl/mNNzTL

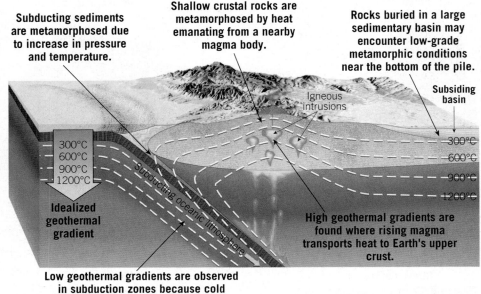

Subducting sediments are metamorphosed due to increase in pressure and temperature.

Shallow crustal rocks are metamorphosed by heat emanating from a nearby magma body.

Rocks buried in a large sedimentary basin may encounter low-grade metamorphic conditions near the bottom of the pile.

Igneous intrusions

Subsiding basin

300°C
600°C
900°C
1200°C

Idealized geothermal gradient

Subducting oceanic lithosphere

300°C
600°C
900°C
1200°C

High geothermal gradients are found where rising magma transports heat to Earth's upper crust.

Low geothermal gradients are observed in subduction zones because cold oceanic crust and overlying sediments are descending into the mantle.

bonded, this elevated level of activity allows individual atoms to *migrate* more freely between sites in the crystalline structure.

Changes Caused by Heat

The formation of new or enlarged mineral grains at the expense of original grains is called **recrystallization**. During this process, regardless of whether the mineralogy of the rock changes, its grains tend to become larger. For example, when quartz sandstone metamorphoses to form quartzite, the mineralogy of both parent and new rock remains quartz grains. But the new quartzite rock has fewer and larger quartz grains than its parent sandstone does.

By contrast, when shale metamorphoses to slate, the clay minerals recrystallize and become new minerals—usually chlorite and muscovite. Although the mineralogy changes in the transition from shale to slate, the overall chemical composition remains essentially unchanged. Instead, the existing atoms are rearranged into new crystalline structures that are more stable in the new environment. However, in some environments, ions may actually migrate into or out of a rock, thereby changing its overall chemical composition.

Heat Sources

There are two primary ways in which rocks may become hotter. First, rocks become hotter when *heat is transported upward* from the mantle into the shallowest layers of the crust. Rising mantle plumes, upwelling at mid-ocean ridges, and magma generated by partial melting of mantle rock at subduction zones are three such examples (see **Figure 8.3**). When magma

intrudes rocks at shallow depths, the magma cools and releases heat, which "bakes" and transforms the surrounding host rock.

The second way rocks become hotter is that they may *become deeply buried* in the crust, where temperature rises with depth. The rate of temperature increase with depth is known as the **geothermal gradient**. In the upper crust, this increase in temperature averages about 25°C (45°[kw1]F) per kilometer (see Figure 8.3). Thus, rocks that formed at Earth's surface experience a gradual increase in temperature if they are transported to greater depths. As described earlier, clay minerals tend to become unstable when buried to a depth of about 8 kilometers (5 miles), where temperatures average about 200°C (400°F). The clay minerals begin to recrystallize into new minerals, such as chlorite and muscovite, both of which are stable in this new environment. However, many silicate minerals, particularly those found in crystalline igneous rocks—such as quartz and feldspar—remain stable at these temperatures. Thus, metamorphic changes in quartz and feldspar occur only at higher temperatures.

Environments where rocks may be carried to great depths and heated include convergent plate boundaries, where slabs of sediment-laden oceanic crust are being subducted. Rocks may also become deeply buried in large basins, where gradual subsidence results in thick accumulations of sediment. These basins, exemplified by the Gulf of Mexico, are known to develop low-grade metamorphic conditions near the base of the pile. In addition, continental collisions, which result in mountain building, cause some rocks to be uplifted while others are thrust downward, where elevated temperatures and pressures trigger metamorphism.

Confining Pressure

Pressure, like temperature, increases with depth because the thickness of the overlying rock increases. Buried rocks are subjected to **confining pressure**, which is analogous to water pressure, in which the forces are applied equally in all directions. Think about how scuba divers experience greater confining pressure the deeper they dive.

Confining pressure causes the spaces between mineral grains to close, producing more compact rocks that have greater densities. If the pressure becomes extreme enough, it can cause the atoms in a mineral to pack more closely together to produce a new, denser mineral (**Figure 8.4A**). Recall from Chapter 3 that the transformation of one mineral (polymorph) to another is called a *phase change*.

Differential Stress

In addition to experiencing confining pressure, rocks may be subjected to directed pressure. This occurs, for example, at convergent plate boundaries where slabs of lithosphere collide. Here the forces that deform rock have different magnitudes in different directions; this is known as **differential stress**. (A discussion of various types of *differential stress* is provided in Chapter 10.) Unlike confining pressure, which "squeezes" rock equally in all directions, differential stresses are greater in one direction than in other directions.

Differential stress that squeezes a rock mass as if it were placed in a vise is termed **compressional stress**. As shown in **Figure 8.4B**, rocks subjected to compressional stress are shortened in the direction of greatest stress and elongated, or lengthened, in the direction perpendicular to that stress. Along convergent plate boundaries, the greatest differential stress is directed horizontally in the direction of plate motion. Consequently, in these settings, the crust is greatly shortened (horizontally) and thickened (vertically), resulting in mountainous topography.

In high-temperature, high-pressure environments, rocks are *ductile*, which allows their mineral grains to flatten (like what happens when you step on a piece of clay) when subjected to differential stress. The *metaconglomerate*, also called a *stretched pebble conglomerate*, shown in **Figure 8.5**, illustrates this tendency. The parent rock, a conglomerate, consisted of nearly spherical pebbles that have been flattened into elongated structures by differential stress. On a larger scale, rocks that are ductile deform by flowing rather than breaking or fracturing. As a result, deeply buried rocks subjected to extremely high temperature and pressure can develop intricate folds when deformed by differential stress (**Figure 8.6**).

By contrast, in near-surface environments where temperature and pressure are comparatively low, rocks are *brittle* and tend to fracture when subjected to differential stress. Continued deformation grinds and pulverizes the mineral grains into smaller and smaller fragments.

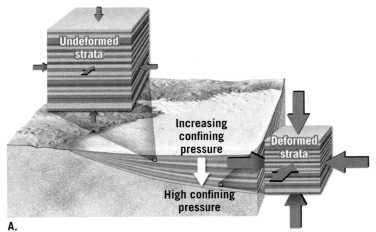

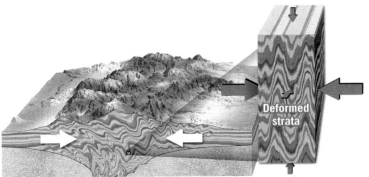

A.

B.

Chemically Active Fluids

Water is abundant in Earth's crust. In the upper crust, it occurs as groundwater. At depth, water at extremely high temperatures is released into the surrounding rocks when a magma body cools and solidifies. In addition,

◀ SmartFigure 8.4
Confining pressure and differential stress
A. In a depositional environment, as confining pressure increases, rocks deform by decreasing in volume.
B. Rocks subjected to differential stress during mountain building are shortened in the direction of maximum stress and lengthened in the direction of minimum stress.

Tutorial
https://goo.gl/
MSyTrK

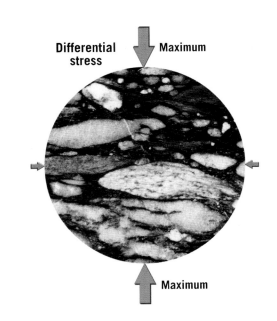

◀ Figure 8.5
Metaconglomerate, also called stretched pebble conglomerate This metaconglomerate is made of once nearly spherical pebbles that have been heated and flattened into elongated structures by differential stress.

▶ **Figure 8.6**
Deformed and folded gneiss, Anza-Borrego Desert State Park, California

many minerals, including clays, micas, and amphiboles, are hydrated—which means they contain water in their crystalline structures. Elevated temperatures and pressures cause these minerals to dehydrate, expelling hot, mineral-laden water.

Hot, chemically active fluids enhance metamorphism by dissolving and transporting ions from one site in the crystal structure to another, thereby facilitating the process of recrystallization. In increasingly hot environments, these fluids become correspondingly more reactive.

In some metamorphic environments, hot fluids transport mineral matter over considerable distances. This occurs, for example, when hot, ion-rich fluids, called a *hydrothermal solution*, are expelled from a magma body as it cools and solidifies. If the rocks that surround the pluton differ markedly in chemical composition from the invading fluids, there may be an exchange of ions between these fluids and host rocks. In other words, these fluids bring in new atoms or take out atoms rather than simply reorganizing what is already present. When this occurs, the overall chemical composition of the surrounding rock changes, in a process called *metasomatism*. One example of metasomatism is the formation of the mineral wollastonite ($CaSiO_3$) from calcite ($CaCO_2$), which is the primary ingredient in limestone. When a silica-rich hydrothermal solution invades limestone, calcite reacts

with silica (SiO_2) to generate wollastonite, and the gas carbon dioxide (CO_2) is driven off.

The Importance of Parent Rock

Most metamorphic rocks have the same overall chemical composition as the parent rocks from which they formed, except for the possible loss or acquisition of volatiles such as water (H_2O) and carbon dioxide (CO_2).

Therefore, when geologists try to establish from what parent material a metamorphic rock derived, the most important clue is the rock's overall chemical composition.

Consider the large exposures of the metamorphic rock marble found high in the Alps of Southern Europe. Because marble and the common sedimentary rock limestone have the same mineralogy (calcite), it seems reasonable to conclude that limestone is the parent rock of marble. Furthermore, because limestone usually forms in warm, shallow marine environments, we can surmise that considerable deformation must have occurred to convert limy deposits in a shallow sea into marble crags in the lofty Alps.

The mineral makeup of the parent rock also largely determines the degree to which each metamorphic agent will cause change. For example, when

The mineral makeup of the parent rock also largely determines the degree to which each metamorphic agent will cause change.

magma forces its way into an existing body of rock, high temperatures and hot fluids may alter the host rock. If the host rock is composed of minerals that are comparatively nonreactive, such as quartz, any alterations that may occur will be confined to a narrow zone next to the igneous intrusion. However, when the host rock is limestone, which is highly reactive, the zone of metamorphism may extend far from the intrusion.

8.3 Metamorphic Textures

Explain how foliated and nonfoliated textures develop.

The term **texture** is used to describe the size, shape, and arrangement of the mineral grains within a rock.

Foliation

Most igneous and many sedimentary rocks consist of mineral grains or crystals that have a random orientation and thus appear uniform when viewed from any direction. By contrast, metamorphic rocks that contain platy minerals (such as micas) and/or elongated minerals (such as amphiboles) typically display some kind of *preferred orientation*, in which the mineral grains exhibit a parallel to subparallel alignment. Like a fistful of pencils, rocks containing elongated mineral grains that are oriented parallel to each other appear different when viewed from the side than when viewed head-on. A rock that exhibits a planar (nearly flat) preferred orientation of its mineral grains or crystals is said to possess **foliation**.

In metamorphic environments, foliation is usually driven by compressional stress that shortens rock units, causing mineral grains in preexisting rocks to develop parallel, or nearly parallel, alignments. Examples of foliation include the parallel alignment of platy minerals through *rotation, recrystallization*, and *flattening of mineral grains or pebbles*.

Rotation of Platy Mineral Grains   The rotation of existing mineral grains is the easiest of the foliation mechanisms to envision. **Figure 8.7** illustrates the mechanics by which platy or elongated mineral grains are rotated. Note that the new alignment of the grains is roughly perpendicular to the direction of maximum stress. Although physical rotation of platy minerals contributes to the development of foliation in low-grade metamorphism, other mechanisms dominate in more extreme environments.

Recrystallization That Produces New Minerals

Recall that *recrystallization* is the creation of new mineral grains from preexisting ones. When recrystallization occurs as rock is being subjected to differential stress, any elongated minerals (such as amphiboles) and platy minerals (such as micas) that form tend to recrystallize perpendicular to the direction of maximum stress. Thus, the newly formed mineral grains exhibit distinct layering.

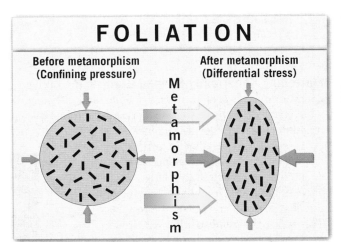

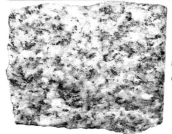
Platy and elongated mineral grains having random orientation.

When differential stress causes rocks to flatten, the mineral grains rotate and align roughly perpendicular to the direction of maximum differential stress.

◀ **SmartFigure 8.7**
Mechanical rotation of platy mineral grains to produce foliation

Animation
https://goo.gl/9Q51mS

▶ Figure 8.8
Solid-state flow of mineral grains Mineral grains can be flattened by solid-state flow when units of a mineral's crystalline structure slide relative to each other. This mechanism involves breaking existing chemical bonds and forming new ones.

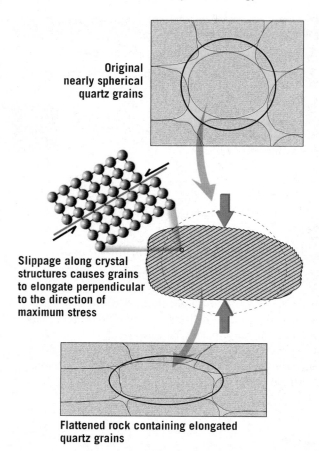

Original nearly spherical quartz grains

Slippage along crystal structures causes grains to elongate perpendicular to the direction of maximum stress

Flattened rock containing elongated quartz grains

Flattening Spherically Shaped Grains

Rocks that contain minerals such as quartz, calcite, and olivine, which develop roughly spherical crystals, can be flattened during metamorphism by two processes. First, a change in grain shape can occur as distinct units of a mineral's crystalline structure slide relative to one another along discrete planes, thereby distorting the grain, as shown in **Figure 8.8**. This type of gradual **solid-state flow** involves slippage that disrupts the crystalline structure as atoms shift positions by breaking existing chemical bonds and forming new ones.

Second, the shape of a mineral may be altered by a process in which individual atoms move from a location along the margin of a mineral grain that is highly stressed to a less-stressed position on the same grain. This mechanism, called **pressure solution**, is significantly aided by hot, ion-rich water. Mineral matter (ions) dissolves where grains are in contact with each other (areas of high stress) and is deposited in pore spaces (areas of low stress). As a result, the mineral grains

tend to become shortened in the direction of maximum stress and elongated in the direction of minimum stress. While both of these mechanisms flatten mineral grains, the mineralogy of the rock does not change.

Foliated Textures

The type of foliation depends largely upon the grade of metamorphism and the mineralogy of the parent rock. We will look at three: *rock (slaty) cleavage*, *schistosity*, and *gneissic texture (banding)*.

Rock (Slaty) Cleavage Rocks that split into thin slabs when hit with a hammer exhibit **rock cleavage**. Excellent rock cleavage often occurs in slate, which is why it is also called **slaty cleavage** (**Figure 8.9**). Because it splits easily, slate is used for building materials such as roof and floor tiles as well as billiard table surfaces.

Slaty cleavage commonly develops where beds of shale (and related sedimentary rocks) are metamorphosed to form slate (**Figure 8.10**). The process begins when compressional stress begins to deform rock units, producing broad folds. Further deformation causes the clay minerals in shale, which initially aligned roughly parallel to the bedding surfaces, to begin to recrystallize into tiny flakes of chlorite and mica. However, these new platy mineral grains grow so they are aligned roughly perpendicular to the maximum differential stress, as shown in **Figure 8.10B**.

Because slate typically forms during the low-grade metamorphism of shale, evidence of the original sedimentary bedding surfaces is often preserved. However, as **Figure 8.10C** illustrates, the orientation of slate's cleavage usually develops at an angle to the

▲ Figure 8.9
Excellent slaty cleavage Slaty cleavage is exhibited by the rock in this slate quarry. Because slate breaks into flat slabs, it has many uses. The inset photo shows slate used for the roof of a house in Switzerland.

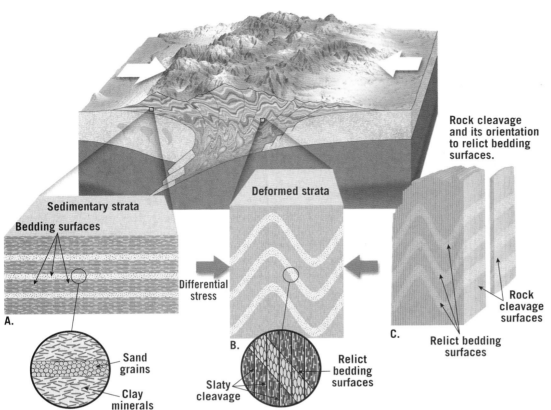

◄ SmartFigure 8.10
Development of slaty cleavage When shale that is interbedded with sandstone is strongly folded and metamorphosed, the clay minerals begin to recrystallize into tiny flakes of chlorite and mica. These new platy minerals grow so they are aligned roughly perpendicular to the directed stress, which gives slate its foliation.

Tutorial
https://goo.gl/
yGXXwD

sedimentary beds. Thus, unlike shale, which splits *along* bedding planes, slate usually splits *across* bedding surfaces.

Other metamorphic rocks, such as schists and gneisses, sometimes split along planar surfaces and thus also exhibit rock cleavage.

Schistosity At higher temperatures and pressures, the minute mica and chlorite flakes in slate begin to recrystallize into larger muscovite and biotite crystals. When these platy crystals are large enough to be discernible with the unaided eye, they exhibit planar or layered structures called **schistosity**. Rocks that have this type of foliation are termed *schist*. In addition to containing platy minerals, schist may contain deformed quartz and feldspar crystals that appear flattened or shaped like a lens embedded among the mica grains.

Gneissic Texture (Banding) During high-grade metamorphism, ion migration can result in the segregation of minerals, as shown in **Figure 8.11**. Notice that the dark biotite and amphibole crystals and light silicate minerals (quartz and feldspar) have separated, giving the rock a banded appearance called **gneissic texture,** or **gneissic banding**. Metamorphic rocks with this texture are called *gneiss* (pronounced "nice"). Although they are foliated, gneisses do not usually split as easily as slates and some schists.

Other Metamorphic Textures

Metamorphic rocks that *do not* develop a layered or banded appearance as a result of metamorphism are referred to as **nonfoliated**. Nonfoliated metamorphic rocks typically form in metamorphic environments where compressional stress is minimal and when the parent rock is composed of minerals that develop equidimensional crystals rather than flat or tabular-shaped crystals. For example, when a fine-grained limestone (made of calcite) is metamorphosed in an environment where differential stress is absent, the small calcite grains in limestone recrystallize to form larger

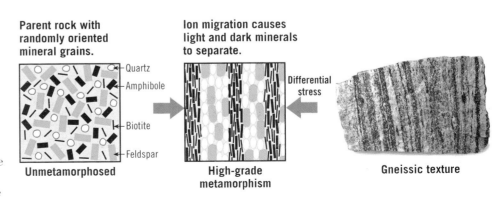

▲ **Figure 8.11**
Development of gneissic banding Migration of ions causes felsic and mafic minerals to grow in separate layers, resulting in gneissic banding.

▶ **Figure 8.12**
Garnet–mica schist The dark red garnet crystals (porphyroblasts) are embedded in a matrix of fine-grained micas.

Porphyroblasts

Close-up of porphyroblast

(picture clay pottery in a kiln) to produce a tough rock that lacks alignment of its platy minerals.

Some metamorphic rocks contain unusually large grains, called *porphyroblasts*, that are surrounded by a fine-grained matrix of other minerals. **Porphyroblastic textures** develop in a wide range of rock types when minerals in the parent rock recrystallize to form new minerals. During recrystallization, some metamorphic minerals, such as garnet, tend to develop *a small number of very large crystals*. By contrast, minerals such as muscovite and biotite typically form *a large number of smaller grains*. Thus, recrystallization produces relatively common metamorphic rocks that contain large crystals (porphyroblasts) of garnet embedded in a finer-grained matrix of biotite and muscovite (**Figure 8.12**).

uniform-shaped crystals. The resulting metamorphic rock, *marble*, consists of intergrown calcite crystals that lack banding and are similar in appearance to the crystals in coarse-grained igneous rocks.

Another common nonfoliated metamorphic rock, called *hornfels*, is generated when clay-rich rocks like shale and mudstone are intruded by a hot magma body. In this environment, the clay minerals are baked

---

CONCEPT CHECKS 8.3

1. Define *foliation*.

2. Distinguish among *slaty cleavage*, *schistosity*, and *gneissic* textures.

3. What is meant by *nonfoliated texture*? Name one rock that exhibits this texture.

Concept Checker
https://goo.gl/x7TSuT

8.4 Common Metamorphic Rocks

List and describe the most common metamorphic rocks.

Most metamorphic rocks that we observe at Earth's surface derive from the three most common sedimentary rocks: shale (or mudstone), limestone, and quartz sandstone. Shale is the most likely parent of most slate, phyllite, schist, and gneiss. This sequence of metamorphic rocks exhibits an increase in grain size, a change in rock texture, and a change in mineralogy.

Limestone, which is composed of the mineral calcite ($CaCO_3$), is the parent rock of marble, while quartz (SiO_2) sandstone is the parent of quartzite. Calcite and quartz are simple chemical compounds compared to clay minerals, and their mineralogy does not typically change during metamorphism. Rather, these minerals tend to recrystallize to produce larger fused grains of calcite and quartz, which are the main constituents of marble and quartzite, respectively.

The major characteristics of the most common metamorphic rocks are summarized in **Figure 8.13**. Notice that metamorphic rocks are broadly classified by the type of foliation exhibited and, to a lesser extent, the chemical composition of the parent rock. It is worth noting that certain *rock names* (slate, schist, and gneiss) are also used to describe *texture*.

Foliated Metamorphic Rocks

Slate A very fine-grained (less than 0.5 millimeter) foliated rock composed mainly of minute chlorite and mica flakes (too small to be visible to the human eye) is termed **slate**. Slate may also contain tiny quartz and feldspar crystals. Because slate is fine grained, it tends to appear dull and closely resembles shale. A noteworthy characteristic of slate is its excellent rock cleavage, or tendency to break into flat slabs (see Figure 8.9).

Slate is most often generated by the low-grade metamorphism of shale, mudstone, or siltstone but is sometimes produced when volcanic ash is metamorphosed. Slate's color depends on its mineral constituents: Black (carbonaceous) slate contains organic material, red slate

► **Figure 8.13**
Classification of common metamorphic rocks

gets its color from iron oxide, and green slate is composed mainly of chlorite.

Phyllite   **Phyllite** consists of platy minerals that are larger than those in slate but not large enough to be readily identifiable with the unaided eye. Although similar in appearance to slate, phyllite can be easily distinguished from slate by its glossy sheen and wavy surface (see Figure 8.13). Phyllite exhibits rock cleavage and is composed mainly of very fine crystals of muscovite, chlorite, or both.

Schist   Medium- to coarse-grained metamorphic rocks in which platy minerals dominate are called **schists**. These flat components commonly include muscovite and biotite with parallel alignments that give the rock its foliated texture (**Figure 8.14**). In addition, schists contain smaller amounts of other minerals, termed *accessory minerals*. As with slate, the parent rock of most schists is shale or mudstone that has undergone medium- to high-grade metamorphism.

   Recall that the term *schist* is also used to describe the texture of a rock, so rocks called schists may have a wide variety of chemical compositions. To indicate composition, mineral names are added. For example, schists composed primarily of muscovite and biotite are called *mica schist* (see Figure 8.14). Some mica schists contain accessory minerals that are unique to metamorphic rocks. Common accessory minerals that occur as porphyroblasts include *garnet, staurolite*, and *andalusite*, in which case the rock is called *garnet-mica schist, staurolite-mica schist*, or *andalusite-mica schist*, respectively.

   In addition, schists may be composed largely of the minerals chlorite, talc, or hornblende, in which case they are called *chlorite schist, talc schist*, and *hornblende schist*, respectively. Both chlorite and hornblende schists form when rocks having a mafic (basaltic) composition undergo metamorphism.

Gneiss   **Gneiss** is the term applied to medium- to coarse-grained banded metamorphic rocks in which granular and elongated (as opposed to platy) minerals predominate. The most common minerals in gneiss are quartz, potassium feldspar, and plagioclase feldspar. Most gneisses also contain lesser amounts of biotite, muscovite, and amphibole. Some gneisses split along the layers of platy minerals, but most break in an irregular fashion.

   Recall that during high-grade metamorphism, the light and dark components separate, giving gneisses their characteristic banded or layered appearance. Thus,

Metamorphic Rock	Texture	Comments	Parent Rock
Slate	Foliated	Composed of tiny chlorite and mica flakes, breaks in flat slabs via slaty cleavage, smooth dull surfaces	Shale, mudstone, or siltstone
Phyllite	Foliated	Fine-grained, glossy sheen, breaks along wavy surfaces	Shale, mudstone, or siltstone
Schist	Foliated	Medium- to coarse-grained, scaly foliation, micas dominate	Shale, mudstone, or siltstone
Gneiss	Foliated	Coarse-grained, compositional banding due to segregation of light- and dark-colored minerals	Shale, granite, or volcanic rocks
Marble	Nonfoliated	Medium- to coarse-grained, relatively soft (3 on the Mohs scale), interlocking calcite or dolomite grains	Limestone, dolostone
Quartzite	Nonfoliated	Medium- to coarse-grained, very hard, massive, fused quartz grains	Quartz sandstone
Hornfels	Nonfoliated	Very fine-grained, often exceedingly tough and durable, usually dark colored	Often shale, but can have any composition

Mica schist

Parallel alignment of mineral grains

◄ **Figure 8.14**
Mica schist This sample of schist, composed mostly of muscovite and biotite, exhibits foliation.

▲ Figure 8.15
Banded gneiss found in the Adirondacks, New York

▲ Figure 8.16
Marble is a widely used building stone The exterior of India's Taj Mahal is constructed primarily of the metamorphic rock marble.

most gneisses consist of alternating bands of white or reddish feldspar-rich zones and layers of dark ferromagnesian minerals (**Figure 8.15**). These banded gneisses often exhibit evidence of deformation, including folds and sometimes faults.

Gneisses having a felsic composition may be derived from granite or its fine-grained equivalent, rhyolite. However, most gneisses are generated through high-grade metamorphism of shale. Therefore, gneiss represents the highest-grade metamorphic rock in the sequence of shale, slate, phyllite, schist, and gneiss. Like schists, gneisses may also include large crystals of accessory minerals such as garnet. Gneisses made up primarily of dark minerals also occur. For example, an amphibole-rich rock that exhibits a gneissic texture is called *amphibolite*.

Nonfoliated Metamorphic Rocks

Marble   The metamorphism of limestone or dolostone produces the crystalline metamorphic rock called **marble** (see Figure 8.13). Pure marble is white and composed essentially of the mineral calcite. Because of its relative softness (3 on the Mohs scale), marble is easy to cut and shape. White marble is particularly prized as a stone from which monuments and statues are carved, such as the Lincoln Memorial in Washington, DC, and the Taj Mahal in India (**Figure 8.16**). Unfortunately, when exposed to acid rain, marble's composition (calcium carbonate) makes it susceptible to chemical weathering (see Chapter 6).

The parent rocks of most marbles contain impurities that color the stone. Thus, marble can be pink, gray, green, or even black and may contain a variety of accessory minerals (such as chlorite, mica, garnet, and wollastonite). When marble forms from limestone interbedded with shales, it appears banded and exhibits visible foliation. When deformed, these banded marbles

may develop highly contorted mica-rich folds that enhance the rocks' artistic appearance. These decorative marbles have been used as building stones since prehistoric times.

Quartzite   **Quartzite** is a very hard metamorphic rock formed from quartz sandstone (**Figure 8.17**). Under moderate- to high-grade metamorphism, the quartz grains in sandstone fuse together (inset in Figure 8.17). Recrystallization is often so complete that, when broken, quartzite splits across the original quartz grains rather than along their boundaries. In some instances, sedimentary features such as cross-bedding are preserved and give the rock a banded appearance. Pure quartzite is white, but iron oxide may produce reddish or pinkish stains, while dark mineral grains may impart shades of green or gray.

Hornfels   A fine-grained nonfoliated metamorphic rock, **hornfels** is unlike marble and quartzite in that it has a variable mineral composition. The parent rock of most hornfels is shale or another clay-rich rock that has been "baked" by a hot intruding magma body. Hornfels tends to be gray to black in color and quite hard (see Figure 8.13).

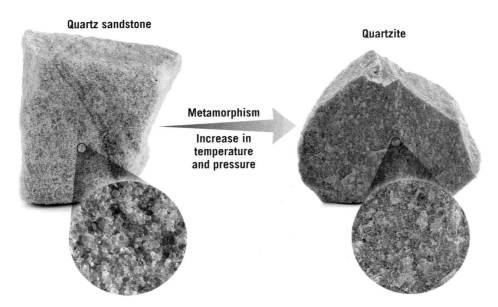

Quartz sandstone

Quartzite

Metamorphism

Increase in temperature and pressure

◄ **Figure 8.17**
Quartzite Quartzite is a nonfoliated metamorphic rock formed from quartz sandstone. The close-up images show the loosely bound grains in a quartz sandstone compared to the interlocking quartz grains typical of quartzite.

8.5 Metamorphic Environments

Write a description for each of the following environments: contact metamorphism, hydrothermal metamorphism, subduction zone metamorphism, and regional metamorphism.

Most metamorphism occurs along plate margins. Here we consider four common metamorphic environments—*contact (thermal) metamorphism, hydrothermal metamorphism, burial* and *subduction zone metamorphism,* and *regional metamorphism*—as well as a few types of metamorphism that generate relatively small quantities of metamorphic rock. As **Figure 8.18** illustrates, each type of metamorphism occurs over a specific range of temperatures and pressures.

Contact, or Thermal, Metamorphism

Contact (thermal) metamorphism occurs in Earth's upper crust when rocks immediately surrounding a molten igneous body are "baked" by the high temperatures. Because contact metamorphism occurs under low pressure and does not involve differential stress, the resulting metamorphic rocks are not foliated.

Contact metamorphism alters rocks in a discrete zone adjacent to the heat source, called an **aureole** (**Figure 8.19**). The emplacement of small intrusions such as dikes and sills typically form aureoles only a few centimeters thick. By contrast, large molten bodies that eventually cool to form batholiths can produce aureoles that extend outward for several kilometers. These large aureoles

often consist of distinct *zones of metamorphism*. Close to the magma body, high-temperature minerals such as garnet may form, whereas farther away, low-grade metamorphism produces minerals such as chlorite.

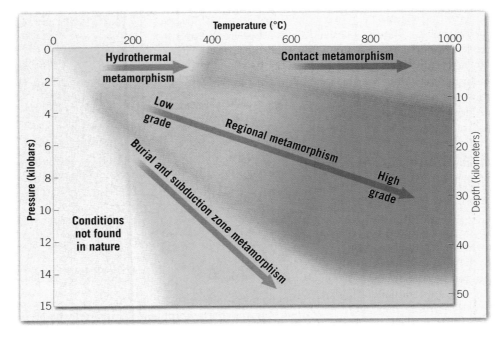

◄ **Figure 8.18**
Metamorphic environments The graph illustrates temperatures and pressures typically associated with the major types of metamorphic environments.

▶ **SmartFigure 8.19**
Contact metamorphism
In the photo (Sierra Nevada Mountains, near Bishop, California) the dark layer is a type of metamorphic aureole called a *roof pendant*, which consists of metamorphosed host rocks that are in contact with the upper part of the light-colored igneous pluton. *Roof pendant* rocks were once the top of a magma chamber.

Tutorial
https://goo.gl/
Fk1ZeT

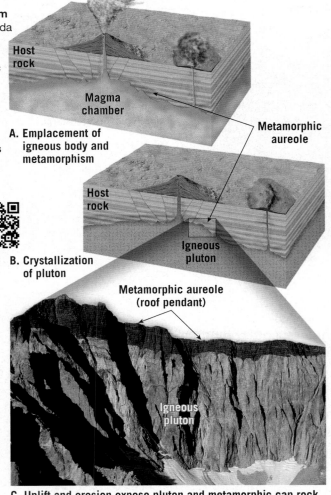

A. Emplacement of igneous body and metamorphism

B. Crystallization of pluton

C. Uplift and erosion expose pluton and metamorphic cap rock

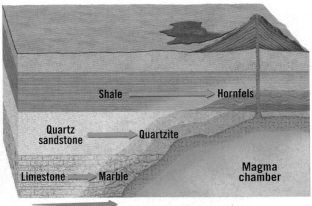

▲ **Figure 8.20**
Rocks produced by contact metamorphism Contact metamorphism of shale yields hornfels, while contact metamorphism of quartz sandstone and limestone produces quartzite and marble, respectively.

by enhancing the recrystallization of existing minerals. Sometimes, hot ion-rich fluids facilitate the movement of mineral matter into and out of rock bodies, thereby changing their overall chemical composition.

The water for hydrothermal metamorphism can be groundwater that has percolated down from the surface,

Hydrothermal metamorphism can occur at shallow crustal depths in regions where geysers and hot springs are active.

Depending mainly on the composition of the parent rock, a variety of metamorphic rocks can form in the same setting (**Figure 8.20**). For example, during contact metamorphism of mudstones and shales, the clay minerals are baked into very hard, fine-grained hornfels. Hornfels can also form from a variety of other materials, including volcanic ash and basaltic rocks. Other metamorphic rocks that are produced by contact metamorphism are marble and quartzite.

Hydrothermal Metamorphism

When hot, ion-rich water circulates through pore spaces or fractures in rock, a chemical alteration called **hydrothermal metamorphism** may occur (**Figure 8.21**). The hot, mineral-laden fluids, called *hydrothermal solutions*, contribute to metamorphism

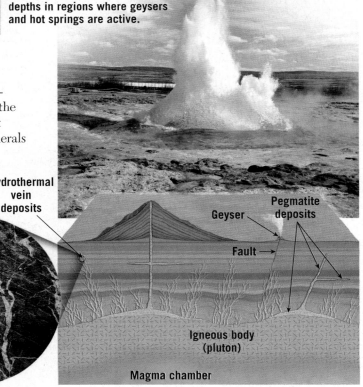

▲ **Figure 8.21**
Hydrothermal metamorphism associated with an intrusive igneous body Pegmatites and hydrothermal mineral deposits form adjacent to an igneous intrusion (pluton).

where it is heated and circulates upward. This type of metamorphism tends to occur at shallow depths with low pressures and relatively low to moderate temperatures.

Water that drives hydrothermal metamorphism may also arise from igneous activity. As large magma bodies cool and solidify, ion-rich water is released into the surrounding host rocks. When the host rock is porous or highly fractured, mineral matter contained in these fluids may precipitate to form deposits of copper, silver, and gold. These ion-rich fluids can also generate pegmatites—very coarse-grained granitic (felsic) igneous rocks (see Chapter 4).

The most widespread occurrence of hydrothermal metamorphism is along the axis of the mid-ocean ridge system (**Figure 8.22**). As plates move apart, upwelling magma from the mantle generates new seafloor. Seawater percolating through the young, hot oceanic crust is heated and chemically reacts with the newly formed basaltic rocks. The result is the conversion of ultramafic rocks of the oceanic crust and uppermost mantle into the hydrated rocks *serpentinite* and *soapstone* (**Figure 8.23**).

Hydrothermal solutions circulating through the seafloor also remove large amounts of metals, such as iron, cobalt, nickel, silver, gold, and copper, from the newly formed crust. These hot, metal-rich fluids eventually rise along fractures and gush from the seafloor, generating particle-filled clouds called *black smokers*. Upon mixing with the cold seawater, sulfides and carbonate minerals containing these heavy metals precipitate to form metallic deposits. Geologists credit this process with the formation of the copper ores mined today on the Mediterranean island of Cyprus.

Burial and Subduction Zone Metamorphism

Burial metamorphism occurs where massive amounts of sedimentary or volcanic material accumulate in a subsiding sedimentary basin (see Figure 8.3). Here, low-grade metamorphic conditions may be produced within the deepest layers. Confining pressure and heat drive the recrystallization of the constituent minerals, changing the texture and/or mineralogy of the rock without appreciable deformation. Metaconglomerate is an example of a metamorphic rock formed by burial metamorphism (see Figure 8.5).

Rocks and sediments can also be carried to great depths along convergent boundaries where oceanic lithosphere is being subducted (see Figure 8.3). In this setting, cold, dense oceanic crust and sediments are subducting rapidly enough that pressure increases faster than temperature. This phenomenon, called **subduction zone metamorphism**, differs from burial metamorphism in that differential stress plays a major role in deforming rock as it is being metamorphosed.

Regional Metamorphism

Regional metamorphism is a widespread type of metamorphism associated with mountain building, where large

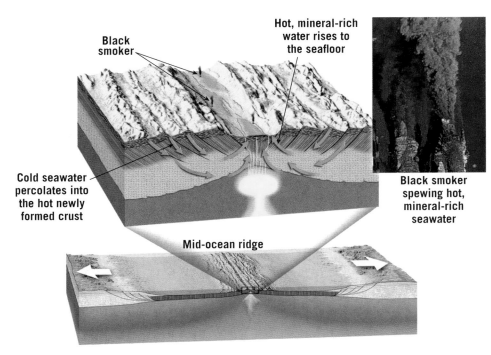

▲ Figure 8.22
Hydrothermal metamorphism along a mid-ocean ridge

segments of Earth's crust are intensely deformed by the collision of two continental blocks (**Figure 8.24**). Recall that denser oceanic crust subducts under more buoyant continental crust but that two landmasses will instead collide and deform. Sediments and crustal rocks that form the margins of the colliding continents are folded and faulted and, as a result, shorten and thicken like a rumpled carpet. Continental collisions may also cause crystalline basement rocks lying under sedimentary layers, as well as slices of oceanic crust that once floored the intervening ocean basin, to be uplifted and deformed.

The general thickening of the crust that occurs during mountain building results in buoyant lifting, in which deformed rocks are elevated high above sea level. Crustal thickening also results in the deep burial of large quantities of rock as one crustal block is thrust beneath another. Deep in the roots of mountains, elevated temperatures caused by deep burial are responsible for the most intense metamorphic activity within a mountain belt.

In some settings, deeply buried rocks become heated beyond their melting points, producing magma. When these magma bodies grow large enough to buoyantly rise, they intrude the overlying metamorphic and sedimentary rocks (see Figure 8.24). Consequently, the cores of many mountain belts consist of folded and faulted metamorphic

Serpentinite   Soapstone

◀ Figure 8.23
Serpentinite and soapstone These metamorphic rocks are produced by hydrothermal alteration of ultramafic rocks along the mid-ocean ridge system.

► SmartFigure 8.24
Regional metamorphism Regional metamorphism occurs during continental collisions where rocks are squeezed between two converging crustal blocks, resulting in mountain building.

Animation
https://goo.gl/
hjMshR

Sediments deposited on
continental margins

Ocean
basin

Continental
lithosphere

Subducting oceanic lithosphere

Asthenosphere

TIME

Regions of low-grade
metamorphism

Continental
lithosphere

Partial melting
of deeply buried
crustal rocks

Region of high-grade
metamorphism

Asthenosphere

rocks, often intertwined with igneous bodies. Over time, these deformed rock masses are uplifted, and erosion removes the overlying material to expose the igneous and metamorphic core of the mountain range.

Regional metamorphism produces some of the most common metamorphic rocks. In these settings, shale is metamorphosed to produce the sequence slate, phyllite, schist, and gneiss. In addition, quartz sandstone and limestone are metamorphosed into quartzite and marble.

Other Metamorphic Environments

Other, less-common types of metamorphism generate relatively small amounts of metamorphic rock that tends to be geographically localized.

Metamorphism Along Fault Zones

Near Earth's surface, rock behaves like a brittle solid. Consequently, movement along a fault zone fractures and pulverizes rock (**Figure 8.25A**). The result is a loosely coherent rock called *fault breccia*, composed of broken and crushed rock fragments. Displacements along California's San Andreas Fault have created a zone of fault breccia and related rock types more than 1000 kilometers (600 miles) long and up to 3 kilometers (2 miles) wide.

Much of the deformation associated with fault zones, however, occurs at great depth and thus at high temperatures. In this environment, preexisting minerals deform by ductile flow (**Figure 8.25B**). As large slabs of rock move in opposite directions, the minerals in the fault zone between them tend to form elongated grains that give the rock a foliated appearance. Rocks formed in these zones of intense ductile deformation are termed *mylonites*.

Impact Metamorphism

Impact, or **shock, metamorphism** occurs when meteorites (fragments of comets or asteroids) strike Earth's surface at high speeds (see **GEOgraphics 8.1**). Upon impact, the energy of the once rapidly moving meteorite is transformed into heat energy and

► SmartFigure 8.25
Metamorphism along a fault zone

Tutorial
https://goo.gl/
zPMdwG

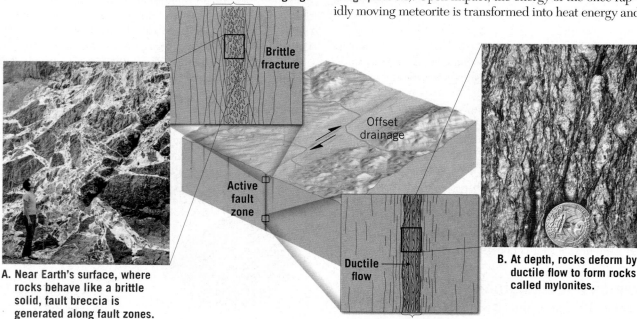

Zone of fault breccia and gouge

Brittle
fracture

Active
fault
zone

Offset
drainage

Ductile
flow

Zone of mylonite

A. Near Earth's surface, where rocks behave like a brittle solid, fault breccia is generated along fault zones.

B. At depth, rocks deform by ductile flow to form rocks called mylonites.

Impact Metamorphism

Recently it became clear that comets and asteroids have collided with Earth far more frequently than was once assumed. The evidence: More than 100 giant impact structures have been identified to date.

Impact breccia
Courtesy of Aerolite Meteorites Inc.

One signature of impact craters is shock metamorphism

When high-velocity projectiles (comets, asteroids) impact Earth's surface, pressures are extreme and temperatures momentarily exceed 2000°C. The result is pulverized and shattered rock that may form impact breccia.

Tektites: Products of impact metamorphism

Tektites are silica-rich glass beads no more than a few centimeters across that are jet black, dark green, or yellowish in color. Most researchers agree that tektites are the result of impacts of large projectiles that are capable of melting crustal rock. In Australia, millions of tektites are strewn over an area seven times the size of Texas. Several other tektite groupings (called strewnfields) have been identified worldwide.

Tektites recovered from Nullarbor Plain, Australia.

Most meteorite impacts occurred millions of years ago.

Meteor Crater, Arizona

Comparatively young impact craters, such as Meteor Crater located west of Winslow, Arizona, appear fresh with rock fragments (ejecta) that ring the impact site.

shock waves that pass through the surrounding rocks. The result is pulverized, shattered, and sometimes melted rock.

The products of these impacts, called *impactiles*, include mixtures of fused fragmented rock plus glass-rich ejecta that resemble volcanic bombs. In some cases, a very dense form of quartz (*coesite*) and minute diamonds can be found. The existence of these high-pressure minerals provides convincing evidence that pressures and temperatures involved in impact metamorphism can be as great as those found in the upper mantle.

8.6 Determining Metamorphic Environments

Explain how index minerals are used to establish the metamorphic grade of a rock body.

Metamorphic rocks offers clues that help geologists determine the metamorphic environments in which the rocks formed. Some of these clues come from *textural differences* found in rocks that have the same parent; other rocks contain *index minerals* that are good indicators of the environment in which they formed. Further clues are gleaned from assemblages of minerals called *metamorphic facies*.

Textural Variations

Across areas where regional metamorphism has occurred, rock textures vary based on the intensity of metamorphism. If we begin with a parent rock composed of clay-rich minerals such as shale or mudstone, a gradual increase from low- to high-grade metamorphism is accompanied by a general coarsening of the grain size. **Figure 8.26** illustrates this principle: As metamorphic intensity increases, shale changes to a fine-grained slate, which then forms phyllite, which, through continued recrystallization, generates a medium-grained schist. Under more intense conditions, a gneissic texture that exhibits layers of dark and light minerals may develop.

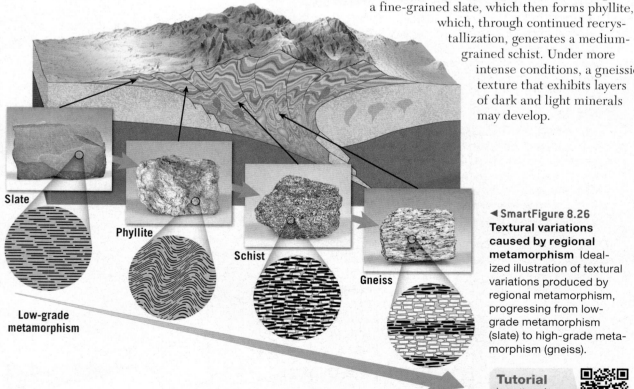

Slate

Phyllite

Schist

Gneiss

Low-grade metamorphism

High-grade metamorphism

◀ **SmartFigure 8.26 Textural variations caused by regional metamorphism** Idealized illustration of textural variations produced by regional metamorphism, progressing from low-grade metamorphism (slate) to high-grade metamorphism (gneiss).

Tutorial https://goo.gl/ 5YTvS6

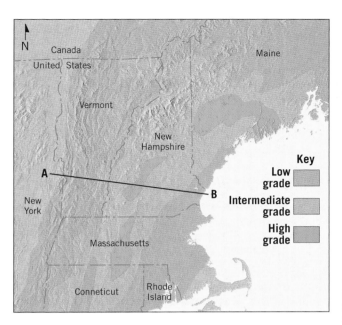

▲ Figure 8.27
Zones of metamorphic intensities in New England This highly generalized map shows areas of low- to high-grade metamorphism in New England.

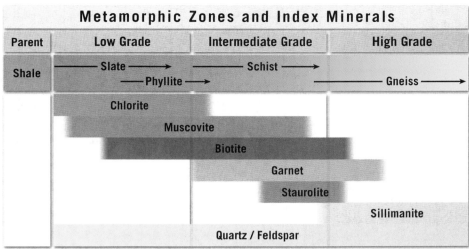

▲ Figure 8.28
Metamorphic zones and index minerals A typical transition of various index minerals associated with the progression from low-grade to high-grade metamorphism of the rock shale.

This systematic transition in metamorphic textures can be observed as we approach the Appalachian Mountains from the west. Beds of shale, which once extended over large areas of the eastern United States, still occur as nearly flat-lying strata in Ohio. However, in the broadly folded Appalachians of central Pennsylvania, the rocks that once formed flat-lying beds are folded and display a preferred orientation of platy mineral grains, as exhibited by well-developed slaty cleavage. As we move farther east, toward the intensely deformed crystalline Appalachians, we find large exposures of schists. Some of the most intense zones of metamorphism are found in Vermont and New Hampshire, where gneissic rocks are exposed at Earth's surface (**Figure 8.27**).

Using Index Minerals to Determine Metamorphic Grade

In addition to textural changes, metamorphic rocks undergo corresponding changes in mineralogy that are evident as we look at specimens from areas of low- to high-grade metamorphism. An idealized transition in mineralogy that results from the regional metamorphism of shale is shown in **Figure 8.28**. The first new mineral to form as shale changes to slate is chlorite. At higher temperatures, flakes of muscovite and biotite begin to dominate. Under more extreme conditions, metamorphic rocks may contain garnet and/or staurolite crystals (**Figure 8.29**). At temperatures approaching the melting point of rock, sillimanite forms. Sillimanite

is a high-temperature metamorphic mineral used to produce porcelain, a material that can withstand extreme environments and that is used in products such as spark plugs for engines.

Through the study of metamorphic rocks in their natural settings (called *field studies*) and through experimentation, geologists have learned that certain **index minerals** are good indicators of the metamorphic environment in which they formed (see Figure 8.28). Using index minerals, geologists distinguish among different zones of regional metamorphism. For example, the mineral chlorite begins to form when temperatures are relatively low—less than 200°C (400°F). Thus, rocks containing chlorite (usually slates) are categorized as *low grade*. By contrast, the mineral sillimanite forms only in extreme environments where temperatures exceed 600°C (1100°F), and rocks containing it are considered

◄ SmartFigure 8.29
Garnet, an index mineral, provides evidence of medium- to high-grade metamorphism These garnet porphyroblasts are found in a gneiss in the Adirondacks, New York.

Mobile Field Trip
https://goo.gl/J3XMH3

► Figure 8.30
Migmatite Under high-grade metamorphism, light-colored (felsic) minerals in a gneiss may begin to melt, while the dark-colored (mafic) minerals remain solid. If this melt solidifies in place, the rock—called a migmatite—will contain light-colored igneous rock intermixed with metamorphic rock composed of dark-colored (mafic) minerals.

high grade. By mapping the occurrences of index minerals, geologists can identify zones of varying metamorphic grades (see Figure 8.27).

Migmatites In the most extreme environments, even the highest-grade metamorphic rocks undergo change. For example, gneissic rocks may be heated sufficiently to trigger partial melting. The light-colored (felsic) minerals, usually quartz and potassium feldspar, have the lowest melting temperatures and begin to melt first, while the dark-colored (mafic) minerals, such as amphibole and biotite, remain solid. This results in rocks of intermixed metamorphic and magmatic (made from magma) components called **migmatites** (mixed rocks). Migmatites often are often intricately folded, as shown in **Figure 8.30**, and represent the highest grade of metamorphism.

Each tectonic setting has temperature and pressure conditions that produce a unique assemblage of minerals.

Migmatites also serve to illustrate that some rocks are transitional because they do not fit neatly into the description of either igneous or metamorphic rocks. In addition, some migmatites originate when a gneiss is injected with magma. Because all migmatites form in very high-temperature regimes, it is usually not possible to discern what role an outside source of magma played in their formation, if any.

Metamorphic Facies as an Indicator of Metamorphic Environments

Geologists discovered that groups of associated minerals could be used to determine the pressure and temperature regimes at which rocks undergo metamorphism. Simply, metamorphic rocks containing the same assemblage of minerals belong to the same **metamorphic facies**—indicating that they formed in very similar metamorphic environments. Using metamorphic facies to determine a metamorphic environment is analogous to using a group of plants to define a climatic zone—a region characterized by similar precipitation and temperature conditions.

For instance, sparsely vegetated regions dominated by cacti identify the desert climate zone characterized by low precipitation and high temperatures.

The common metamorphic facies are shown in **Figure 8.31**. These include the *hornfels, zeolite, greenschist, amphibolite, granulite, blueschist,* and *eclogite facies.* Facies names are based on the minerals that define them. For example, rocks of the amphibolite facies are characterized by hornblende (a common amphibole); the greenschist facies consists of schists in which the green minerals chlorite, epidote, and serpentine are prominent. Similar groups of minerals are found in rocks of all ages and in all parts of the world. Thus, the concept of metamorphic facies is useful in interpreting Earth's history. Rocks belonging to the same metamorphic facies all formed under the same conditions of temperature and pressure, and therefore in similar *plate tectonic settings*, regardless of their location or age.

► Figure 8.31
Metamorphic facies and corresponding temperature and pressure conditions Note the metamorphic rocks produced from regional metamorphism of basalt versus shale under similar conditions of temperature and pressure.

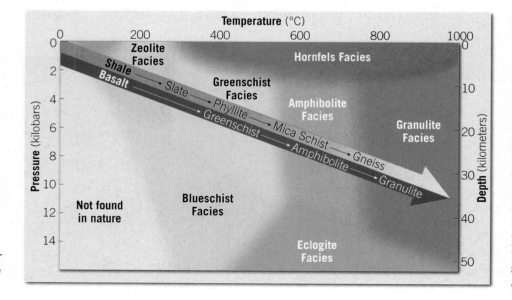

Plate Tectonics and Metamorphic Facies

Figure 8.32 shows how the concept of facies fits into the context of plate tectonics. Each tectonic setting has its characteristic temperature and pressure conditions and therefore produces a unique assemblage of minerals. For example, near deep-ocean trenches, slabs of relatively cool oceanic lithosphere and the overlying crust

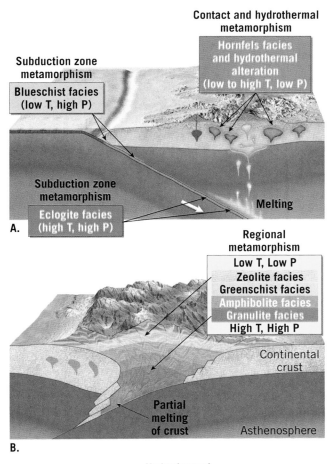

A.

B.

C.

▲ Figure 8.32
Metamorphic facies and plate tectonics These block diagrams show various metamorphic facies and the tectonic environments that generate them.

◄ Figure 8.33
Blueschist is produced by subduction zone metamorphism Blue-colored amphibole called glaucophane gives blueschist its hue.

Blueschist forms in low-temperature, high-pressure environments

are subducted. As the lithosphere descends, crustal rocks are subjected to steadily increasing temperatures and pressures (see Figure 8.32A).

However, pressure increases much more rapidly than temperature because the rocks in the cool subducting slab are poor conductors of heat. The metamorphic facies associated with this type of high-pressure but relatively low-temperature environment is called *blueschist facies* because of the presence of the blue-colored variety of amphibole called *glaucophane* (**Figure 8.33**). The rocks of California's Coast Range belong to the blueschist facies; these highly deformed rocks were once deeply buried but have been uplifted because of a change in the plate boundary. In some areas, subduction carries rocks to even greater depths, producing the *eclogite facies* that indicates both extreme temperatures and pressures (**Figure 8.34**).

◄ Figure 8.34
The eclogite facies forms at great depth in a subduction zone This sample of eclogite contains reddish grains of garnet and green grains of pyroxene.

Eclogite forms in high-temperature and extreme high-pressure environments

Along some convergent zones, continental plates collide to form extensive mountain belts (see Figure 8.32B,C). This activity results in large areas of regional metamorphism that often include zones of contact and hydrothermal metamorphism. Evidence of the gradually increasing temperatures and pressures associated with regional metamorphism are recorded by the *zeolite–greenschist–amphibolite–granulite facies* sequence shown in Figure 8.31.

CONCEPT CHECKS 8.6

1. Briefly describe the different grades of metamorphism that might be encountered moving west to east from Ohio to the crystalline core of the Appalachians.

2. How are index materials used to determine *metamorphic grade*?

Concept Checker
https://goo.gl/sn6bDk

3. What is a metamorphic facies? What two physical conditions vary within Earth to produce different metamorphic mineral groupings?

8 CONCEPTS IN REVIEW

Metamorphism and Metamorphic Rocks

8.1 What Is Metamorphism?

Compare and contrast the environments that produce metamorphic, sedimentary, and igneous rocks.

Key Terms: metamorphism · metamorphic grade
parent rock · mineralogy

- Rocks subjected to elevated temperatures and pressures can change form to produce metamorphic rocks. Every metamorphic rock has a *parent rock*, the original rock before it was transformed by metamorphosis.
- *Metamorphic grade* describes the intensity of metamorphism. Low-grade metamorphic rocks resemble their parent rock. High-grade metamorphism destroys the textures of parent rocks and features such as fossils.
- *Metamorphism* takes place in the solid state and in most cases does not involve any melting.

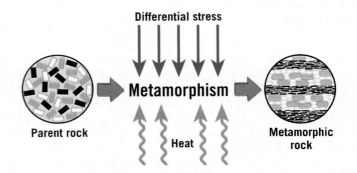

Q Compare the processes that form metamorphic rocks with those that generate igneous or sedimentary rocks.

8.2 What Drives Metamorphism?

List and distinguish among the four agents that drive metamorphism.

Key Terms: geothermal gradient · differential stress
recrystallization · confining pressure · compressional stress

- Heat, pressure, differential stress, and chemically active fluids are four agents that drive metamorphic reactions. All of these, alone or together, trigger metamorphism.
- Deep burial or the intrusion of a nearby magma body will raise a rock's temperature. Heat provides energy that drives chemical reactions and results in the *recrystallization* of the existing minerals. Different minerals have different levels of susceptibility to recrystallization: Some crystals just grow larger, while others react to form new minerals.

- *Confining pressure*, exerted on buried rocks, is force exerted equally in all directions. Increased confining pressure causes rocks and minerals to compact into denser configurations.
- Tectonic forces that cause greater stress in one direction than in other directions are called *differential stress*. Deep in Earth's crust, rocks tend to respond in a ductile manner, shortening in the direction of greatest stress and elongating in the direction(s) of least stress, producing flattened or stretched grains. At shallow depths, differential stress deforms rock in a brittle fashion, causing it to break.
- Water found deep within Earth's crust is extremely hot, making it a chemically active fluid that can cause minerals within a rock to recrystallize and form new minerals. Hot water can also transport dissolved mineral matter over great distances, which may alter rock composition by introducing or removing certain elements.

8.3 Metamorphic Textures

Explain how foliated and nonfoliated textures develop.

Key Terms:

texture	pressure solution	nonfoliated
foliation	rock cleavage	porphyroblastic
solid-state flow	schistosity	texture
	gneissic texture	

- *Foliation*, the planar arrangement of mineral grains, is a common metamorphic rock *texture*. Varieties of foliation include *slaty (rock) cleavage*, *schistosity*, and *gneissic banding*. Foliation forms perpendicular to the direction of maximum differential stress through a combination of processes: rotation of mineral grains, recrystallization and growth of new mineral grains, and flattening of grains by *solid-state flow* or *pressure solution*.
- Metamorphic rocks that do not exhibit foliation are described as *nonfoliated*. For example, marble, a nonfoliated metamorphic rock, forms when small calcite grains in a parent limestone grow into a mass of large, uniformly shaped crystals.
- *Porphyroblasts* are unusually large crystals of certain minerals, such as garnet, that form in an otherwise finer-grained metamorphic rock. They represent the tendency of some minerals to form a small number of large crystals during recrystallization, while other minerals form a large number of smaller crystals.

Q Examine the accompanying photograph. Is this rock foliated or nonfoliated? Which pair of arrows shows the direction of maximum stress?

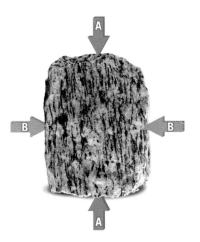

8.4 Common Metamorphic Rocks

List and describe the most common metamorphic rocks.

Key Terms:

slate	schist	quartzite
phyllite	gneiss	hornfels
	marble	

- Common foliated metamorphic rocks include (in order of increasing metamorphic grade) *slate, phyllite, schist*, and *gneiss*.
- Common nonfoliated metamorphic rocks include *quartzite, marble*, and *hornfels*, which form from quartz sandstone, limestone, and shale, respectively.

Q If you were on a class field trip and found an outcrop of nonfoliated light-colored rock, how would you determine whether it is quartzite or marble?

8.5 Metamorphic Environments

Write a description for each of the following environments: contact metamorphism, hydrothermal metamorphism, subduction zone metamorphism, and regional metamorphism.

Key Terms:

contact metamorphism	hydrothermal metamorphism	regional metamorphism
aureole	burial metamorphism	impact metamorphism
	subduction zone metamorphism	

- *Contact metamorphism* occurs in Earth's upper crust when heat from a molten igneous body "bakes" the surrounding rocks in a zone called the *aureole*. Differential stress is typically absent, so the rocks are not foliated.
- *Hydrothermal metamorphism* occurs when hot, ion-rich water circulates through pore spaces or fractures in rock. These fluids enhance recrystallization and can also import or export elements, changing the rock's chemical composition. Hydrothermal metamorphism produces economically important deposits of metal ore.
- *Burial metamorphism* occurs when rocks are buried beneath kilometers of overlying rock and are subjected to confining pressure and elevated temperatures. *Subduction zone metamorphism* is similar but with the added influence of differential stress.
- Continental collisions produce *regional metamorphism*. Resulting belts of metamorphic rock (and associated igneous intrusions) mark zones where continental blocks collided to produce mountains. Long after the mountains have worn away, the collision site is marked by a deformed zone of regional metamorphic rocks that may include slate, schist, marble, and gneiss.
- Other distinctive varieties of metamorphism are associated with faulting and meteorite impacts.

8.6 Determining Metamorphic Environments

Explain how index minerals are used to establish the metamorphic grade of a rock body.

Key Terms:
index mineral

migmatite
metamorphic facies

- Metamorphic rocks reveal the degree of metamorphism through their textures and the minerals they contain.
- Grain size increases with higher levels of metamorphism. Under progressively higher temperatures and pressures, shale may transform first to slate, then to phyllite, to schist, and to gneiss.
- Certain minerals can act as indicators of the environment (temperature/pressure) in which a metamorphic rock formed. The mineral chlorite is associated with low-grade metamorphism, whereas garnet and staurolite are *index minerals* for intermediate metamorphic grades. Sillimanite indicates a high metamorphic grade.
- In extreme metamorphic environments, the minerals with the lowest melting points (usually quartz and potassium feldspar) may melt while the mafic minerals with higher melting points remain solid. The resulting rocks, called *migmatites*, consist of light and dark bands that may be intricately folded.

- Blueschist and eclogite facies are typical of subduction zone metamorphism, where pressure is the dominant driver of change. The zeolite–greenschist–amphibolite–granulite sequence of *metamorphic facies* is typical of regional metamorphism at convergent plate boundaries, where both temperature and pressure conditions are important.

Q When exposed to metamorphic conditions over sufficient time, shale becomes schist. This sample of schist contains porphyroblasts of sillimanite. Do these porphyroblasts indicate low-, intermediate-, or high-grade metamorphism?

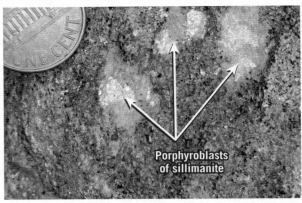

Porphyroblasts of sillimanite

GIVE IT SOME THOUGHT

1. Each of the following statements describes one or more characteristics of a particular metamorphic rock. For each statement, identify the metamorphic rock or rocks being described:

   a. calcite-rich and nonfoliated

   b. composed of alternating bands of light and dark silicate minerals

   c. represents a grade of metamorphism between slate and schist

   d. composed of tiny chlorite and mica grains and displaying excellent rock cleavage

   e. foliated and composed predominantly of platy materials

   f. hard and nonfoliated, often produced by contact metamorphism

   g. loosely coherent and composed of broken fragments that formed along a fault zone

2. Examine the accompanying close-up images of a conglomerate and a metaconglomerate.

   a. Describe how the conglomerate is different from the metaconglomerate.

   b. Does this metaconglomerate appear to have been exposed to differential stress? Explain.

3. Examine the accompanying photos that show the geology of the Grand Canyon. Notice that most of the canyon consists of layers of sedimentary rocks, but if you were to hike into the inner gorge, you would encounter the Vishnu Schist, a metamorphic rock.

   a. What process might have been responsible for the formation of the Vishnu Schist?

   b. What does the Vishnu Schist tell you about the history of the Grand Canyon prior to the formation of the canyon itself?

   c. Why is the Vishnu Schist visible at Earth's surface?

   d. Is it likely that rocks similar to the Vishnu Schist exist elsewhere but are not exposed at Earth's surface? Explain.

Inner Gorge

A. Inner Gorge of the Grand Canyon

B. Close-up of Vishnu Schist (dark color)

A. Conglomerate

B. Metaconglomerate

4. **Which of the accompanying photos of rock outcrops consists of metamorphic rock? Explain how you determined your answer.**

A.

B.

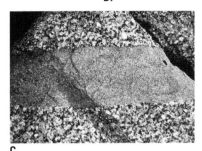

C.

5. **Refer to the accompanying diagram and match each labeled area with the appropriate environment listed below:**

   a. contact metamorphism

   b. subduction metamorphism

   c. regional metamorphism

   d. burial metamorphism

   e. hydrothermal metamorphism

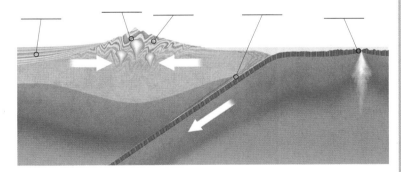

6. **Based on the information provided in Figure 8.27, complete the following:**

   a. Describe how the metamorphic grade changes from west to east across New England along line A–B.

   b. How might these metamorphic rocks have formed?

   c. Are these zones of metamorphism consistent with the current tectonic setting of New England? Explain.

7. **Examine the accompanying close-up images of six different rocks labeled A–F. Classify them as igneous, sedimentary, or metamorphic, based on texture. (*Hint*: There are two of each rock type.)**

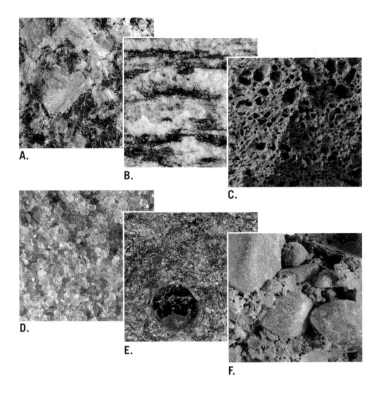

A.

B.

C.

D.

E.

F.

8. **The accompanying image shows a metamorphic rock located in Purgatory Chasm in Newport, Rhode Island, that is made of elongated cobbles composed mainly of quartz.**

   a. What name would you give to this metamorphic rock?

   b. Which set of arrows (red or blue) best represents the direction of maximum differential stress?

   c. Is this type of deformation best described as ductile or brittle?

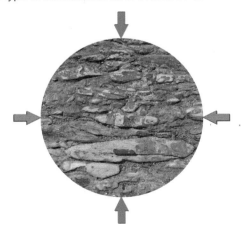

EYE ON EARTH

1. This metamorphic rock outcrop (bedrock exposed at the surface) is found in the Southern Alps, located on the South Island of New Zealand. The continued growth of the Southern Alps is somewhat unique in that these mountains lie where the Pacific and Australian plates collide and simultaneously slide past one another along a large transform fault called the Alpine Fault.

   a. Do the rocks in this outcrop display foliation? Explain.

   b. Do these rocks appear to have experienced high-grade or low-grade metamorphism? Explain.

2. This rock outcrop, located in Joshua Tree National Park, California, consists of dark-colored metamorphic rocks that overlie light-colored igneous rocks.

   a. Name the type of metamorphism—contact, hydrothermal, burial, subduction zone, or regional metamorphism—that likely produced these metamorphic rocks.

   b. Write a brief statement that outlines the geologic history of this area, based on what you observe in this image.

DATA ANALYSIS

The Story of Metamorphic Rocks

Pressure, temperature, and time are the tools of metamorphism that transform a parent rock of one type into a new rock type with different appearance, mineralogy, and (sometimes) chemical composition. If you can examine the appearance of a metamorphic rock and determine its mineral composition, then you can also "work backward" to figure out the temperature and pressure conditions that existed during formation.

https://goo.gl/eD226M

ACTIVITIES

Examine Figure 8.28.

1. **Is it possible for a rock to contain chlorite and:**

   a. Muscovite?

   b. Biotite?

   c. Garnet?

   d. Staurolite?

   e. Sillimanite?

2. **Say that you have a rock that contains garnet but not muscovite or sillimanite.**

   a. What is the rock's metamorphic grade?

   b. Is the rock foliated?

   c. Describe the rock's appearance.

   d. What is the rock's texture?

Refer to Figure 8.31 while considering the rock from Question 2

3. **Which two metamorphic facies could the rock belong to?**

4. **What is the temperature range where this rock formed? (Include units in your answer.)**

5. **What is the pressure range where this rock formed? (Include units in your answer.)**

6. **What is the depth range where this rock formed? (Include units in your answer.)**

Examine Figure 8.18.

7. **What metamorphic environment was responsible for the formation of the rock in Question 2?**

Mastering Geology

Looking for additional review and test prep materials? Visit pearson.com/mastering/geology to enhance your understanding of this chapter's content by accessing a variety of resources, including **Self-Study Quizzes, Geoscience Animations, SmartFigure Tutorials,** *Project Condor* **Videos, Mobile Field Trips, Dynamic Study Modules,** and an optional **Pearson eText.**

Zircons: Tiny Minerals Tell Big Tales of Earth's Story

For geologists studying Earth's past, zircons—simple rock crystals—are the star when it comes to deciphering what happened millions of years ago.

Zircons are composed of zirconium silicate ($ZrSiO_4$). Typically they are tiny (0.1–0.3 mm), very common in igneous rocks, and extremely durable to rock weathering processes. Zircons make their way through the rock cycle as sediment

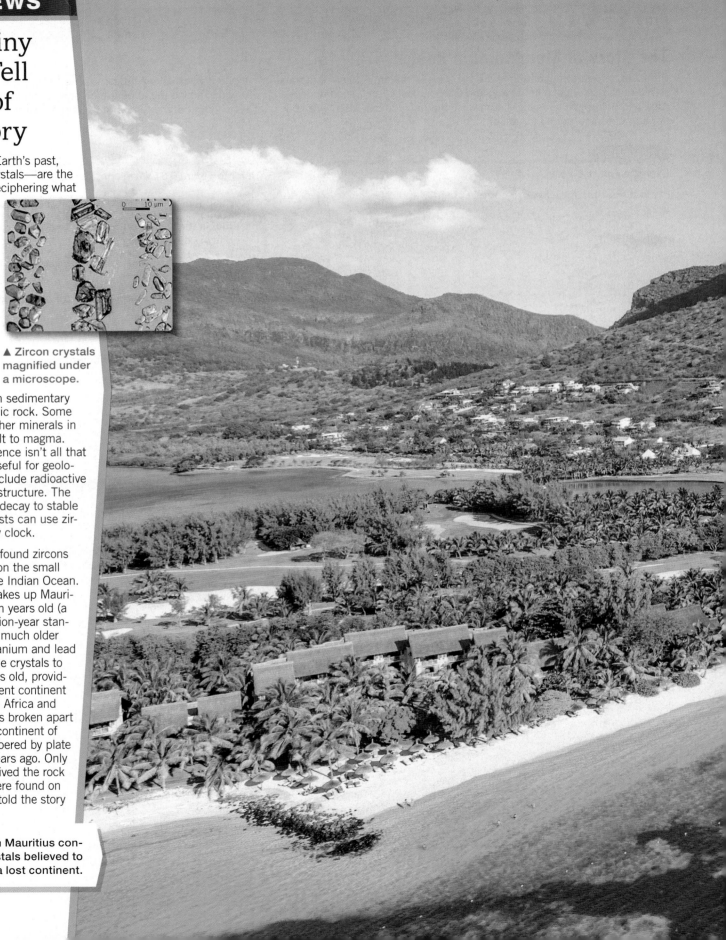

▲ Zircon crystals magnified under a microscope.

in rivers and beaches, in sedimentary rock, and in metamorphic rock. Some even survive when all other minerals in a metamorphic rock melt to magma. This remarkable persistence isn't all that makes zircon crystals useful for geologists. Zircon can also include radioactive uranium in the mineral structure. The uranium atoms steadily decay to stable lead atoms, and geologists can use zircons as an Earth history clock.

Geologists recently found zircons among the beach sand on the small island of Mauritius in the Indian Ocean. Though the rock that makes up Mauritius is less than 8 million years old (a baby, by Earth's 4.5-billion-year standard), the zircons told a much older story. Analysis of the uranium and lead in the zircons showed the crystals to be at least 3 billion years old, providing evidence for an ancient continent that used to lie between Africa and India. This continent was broken apart when the ancient supercontinent of Gondwana was dismembered by plate tectonics 200 million years ago. Only the durable zircons survived the rock cycle, and when they were found on a beautiful beach, they told the story of a lost continent.

▶ This beach on Mauritius contains zircon crystals believed to be from a lost continent.

9
Geologic Time

FOCUS ON CONCEPTS

Each statement represents the primary learning objective for the corresponding major heading within the chapter. After you complete the chapter, you should be able to:

9.1 Distinguish between numerical and relative dating and apply relative dating principles to determine a time sequence of geologic events.

9.2 Define *fossil* and discuss the conditions that favor the preservation of organisms as fossils. List and describe various types of fossils.

9.3 Explain how rocks of similar age that are in different places can be matched up.

9.4 Discuss three types of radioactive decay and explain how radioactive isotopes are used to determine numerical dates.

9.5 Explain how reliable numerical dates are determined for layers of sedimentary rock.

9.6 Distinguish among the four basic time units that make up the geologic time scale and explain why the time scale is considered to be a dynamic tool.

must be very old and that mountains and some other surface features might
develop over very long periods of time. However, pioneering geologists had no
way to accurately determine whether the time scale involved was thousands, millions,
or billions of years. Relative dating techniques allowed for the assembly of a useful and
accurate geologic time scale. The development of absolute dating techniques in the
twentieth century made it possible to add ages to the story. This chapter explores the
geologic time scale and various methods used for determining geologic dates and ages.

9.1 Creating a Time Scale: Relative Dating Principles

Distinguish between numerical and relative dating and apply relative dating principles to determine a time sequence of geologic events.

In the late eighteenth century, James Hutton recognized the immensity of Earth history and the importance of time as a component in all geologic processes. In the nineteenth century, Sir Charles Lyell and others effectively demonstrated that Earth had experienced many episodes of mountain building and erosion, which must have required great spans of geologic time. Although these pioneering scientists understood that Earth was very old, they had no way of determining its age in years. Was it tens of millions, hundreds of millions, or even billions of years old? Long before geologists could establish a geologic time scale that included numerical dates in years, they gradually assembled a time scale using relative dating principles. With the discovery of radioactivity and the development of radiometric dating techniques, geologists now can assign quite accurate dates to many of the events in Earth history.

► Figure 9.1
Contemplating geologic time This hiker is resting atop the Kaibab Formation, the uppermost layer in the Grand Canyon.

Figure 9.1 shows a hiker resting atop the Permian-age Kaibab Formation on the Grand Canyon's North Rim. Beneath him are thousands of meters of sedimentary strata that were deposited as far back as Cambrian time, more than 540 million years ago. These strata rest atop even older sedimentary, metamorphic, and igneous rocks from a span known as the Precambrian. Some of these rocks are 2 billion years old. Although the Grand Canyon's rock record has numerous interruptions, the rocks beneath the hiker contain clues to great spans of Earth history.

The Importance of a Time Scale

Geologic history is like a record from a number of different "books" in different geographic locales. We get a complete geologic record by combining all of the books together. The books, however, are not complete. Many pages, especially in the early chapters, are missing. Others are tattered, torn, or smudged. Yet enough of the books remain to allow much of the story to be deciphered.

Interpreting Earth history is an important goal of the science of geology. Like a modern-day sleuth, a geologist must interpret the clues found preserved in the rocks. By studying rocks, especially sedimentary rocks, and the features they contain, geologists can unravel the complexities of the past.

Geologic events by themselves, however, have little meaning until they are put into a time perspective. Studying history, whether it is the Civil War or the age of dinosaurs, requires a calendar. Among geology's major contributions to human knowledge are the *geologic time scale* and the discovery that Earth history is exceedingly long.

Numerical and Relative Dates

The geologists who developed the geologic time scale revolutionized the way people perceive our planet. They learned that Earth is much older than anyone had previously imagined and its surface and interior have been changed over and over again by the same geologic processes that operate today.

Numerical Dates During the late 1800s and early 1900s, scientists attempted to determine Earth's age. Although some of the methods appeared promising at the time, none proved to be reliable. What those scientists were seeking was a **numerical date**, which specifies the actual number of years that have passed since an event occurred. Today, our understanding of radioactivity allows us to accurately determine numerical dates for rocks that represent important events in Earth's distant past. (We will study radioactivity later in this chapter.) Prior to the discovery of radioactivity, geologists had no reliable method of carrying out numerical dating.

Relative Dates When we place rocks in their proper *sequence of formation*—indicating which formed first, second, third, and so on—we are establishing **relative dates**. Because accurate numerical dating still remained elusive in the eighteenth and nineteenth centuries, geologists developed principles of relative dating. Such dates cannot tell us how long ago something took place, but they can indicate whether it followed one event and preceded another. Relative dating techniques are valuable and still widely used today, often in conjunction with numerical dating techniques.

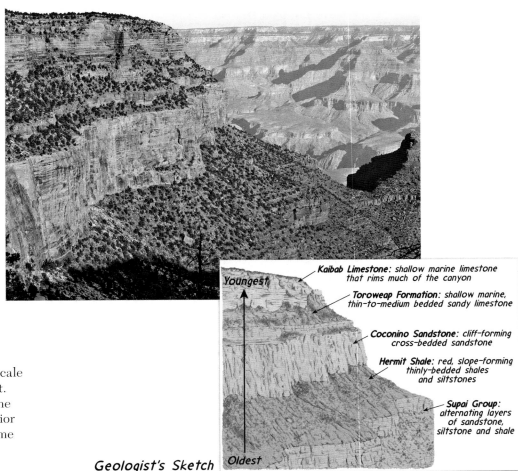

Geologist's Sketch   Youngest   Oldest

Kaibab Limestone: shallow marine limestone that rims much of the canyon

Toroweap Formation: shallow marine, thin-to-medium bedded sandy limestone

Coconino Sandstone: cliff-forming cross-bedded sandstone

Hermit Shale: red, slope-forming thinly-bedded shales and siltstones

Supai Group: alternating layers of sandstone, siltstone and shale

▲ **Figure 9.2**
Superposition Applying the principle of superposition to these layers in the upper portion of the Grand Canyon, the Supai Group is oldest and the Kaibab Limestone is youngest.

Principle of Superposition

Nicolas Steno (1638–1686), a Danish anatomist, geologist, and priest, was the first to recognize a sequence of historical events in an outcrop of sedimentary rock layers. Working in the mountains of western Italy, Steno applied a very simple rule that has become the most basic principle of relative dating—the **principle of superposition** (*super* = above; *positum* = to place). This principle states that in an undeformed sequence of sedimentary rocks, each bed is older than the one above and younger than the one below. Although it may seem obvious that a rock layer could not be deposited with nothing beneath it for support, it was not until 1669 that Steno clearly stated this principle.

This rule also applies to other surface-deposited materials, such as lava flows and beds of ash from volcanic eruptions. Applying the law of superposition to the beds exposed in the upper portion of the Grand Canyon, we can easily place the layers in their proper order. Among those pictured in **Figure 9.2**, the sedimentary rocks in the Supai Group are the oldest, followed in order by the Hermit Shale, Coconino Sandstone, Toroweap Formation, and Kaibab Limestone.

▶ Figure 9.3
Original horizontality Most layers of sediment are deposited in a nearly horizontal position. When we see strata that are folded or tilted, we can assume that they were moved into that position by crustal disturbances *after* their deposition.

observe rock layers that are flat, it means they have not been disturbed and still have their *original* horizontality. The layers in the Grand Canyon illustrate this in Figures 9.1 and 9.2. But if layers are folded or inclined at a steep angle (discussed in detail in Chapter 10), they must have been moved into that position by crustal disturbances sometime *after* their deposition (**Figure 9.3**).

Principle of Original Horizontality

Steno is also credited with recognizing the importance of another basic rule, the **principle of original horizontality**, which says that layers of sediment are generally deposited in a horizontal position. Thus, if we

▶ Figure 9.4
Lateral continuity
A. Sediments are deposited over a large area in a continuous sheet. Sedimentary strata extend continuously in all directions until they thin out at the edge of a depositional basin or grade into a different type of sediment.
B. Although rock exposures are separated by many miles, we can infer that they were once continuous.
C. The idea depicted in B is illustrated in this image of the Grand Canyon.

Layer ends by thinning at margin of sedimentary basin

Layer ends by grading into a different kind of sediment

A.

Lateral continuity allows us to infer that the layers were originally continuous across the canyon

B.

Coconino Sandstone

Redwall Limestone

C.

Principle of Lateral Continuity

The **principle of lateral continuity** refers to the fact that sedimentary beds originate as continuous layers that extend in all directions until they grade into a different type of sediment or until they thin out at the edge of the basin of deposition (**Figure 9.4A**). For example, when a river creates a canyon, we can assume that identical or similar strata on opposite sides once spanned the canyon (**Figure 9.4B**). Although rock outcrops may be separated by a considerable distance, the principle of lateral continuity tells us those outcrops once formed a continuous layer (**Figure 9.4C**). This principle allows geologists to relate rocks in isolated outcrops to one another. Combining the principles of lateral continuity and superposition, we can extend relative age relationships over broad areas. This process, called *correlation*, is examined in Section 9.3.

Principle of Cross-Cutting Relationships

The **principle of cross-cutting relationships** states that geologic features that cut across rocks must form *after* the rocks they cut through. One easy-to-understand example relates to faults, fractures in rock along which displacement occurs. Clearly, the layers of rock offset by a fault must be older than the fault that then breaks them. (See **Figure 9.5**). Igneous intrusions (see Chapter 4) provide another example. The dikes shown in **Figure 9.6** are tabular masses of igneous rock that cut through the surrounding rock. The magmatic heat from igneous intrusions often creates a narrow "baked" zone of contact metamorphism on the adjacent rock, also indicating that the intrusion occurred after the surrounding rocks were in place.

▲ SmartFigure 9.5
Cross-cutting fault The rocks are older than the fault that displaced them.

Principle of Inclusions

Inclusions are fragments of one rock unit that have been enclosed within another. The **principle of inclusions** is that the rock mass adjacent to the one containing the inclusions must have been there first, in order to provide the rock fragments. Therefore, the rock mass that contains inclusions is the younger of the two. For example, when magma intrudes into surrounding rock, blocks of the surrounding rock may become dislodged and incorporated into the magma. If these pieces do not melt, they remain as inclusions known as *xenoliths* (see Chapter 4). In another example, when sediment is deposited atop a weathered mass of bedrock, pieces of the weathered rock become incorporated into the younger sedimentary layer (**Figure 9.7**).

Unconformities

When we observe layers of rock that were deposited essentially without interruption, we call them **conformable**. Particular sites exhibit conformable beds representing certain spans of geologic time. However, no place on Earth has a complete set of conformable strata.

Throughout Earth history, the deposition of sediment has been interrupted over and over again. Such breaks in the rock record are termed *unconformities*. An **unconformity** represents a long period during which deposition ceased, erosion removed previously formed rocks, and then deposition resumed. In each case, uplift and erosion were followed by subsidence and renewed sedimentation. Unconformities are important features because they represent significant geologic events in

▲ Figure 9.6
Cross-cutting dikes The dark igneous intrusion is younger than the light rock that is intruded.

▼ SmartFigure 9.7
Inclusions Rock containing inclusions is younger than the inclusions themselves.

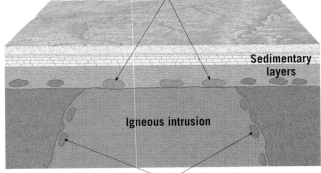

These inclusions of igneous rock contained in the adjacent sedimentary layer indicate that the sediments were deposited atop the weathered igneous mass and thus are younger.

Xenoliths are inclusions in an igneous intrusion that form when pieces of surrounding rock are incorporated into magma.

▶ SmartFigure 9.8
Formation of an angular unconformity An angular unconformity represents an extended period during which deformation and erosion occurred.

Tutorial
https://goo.gl/GaMXhn

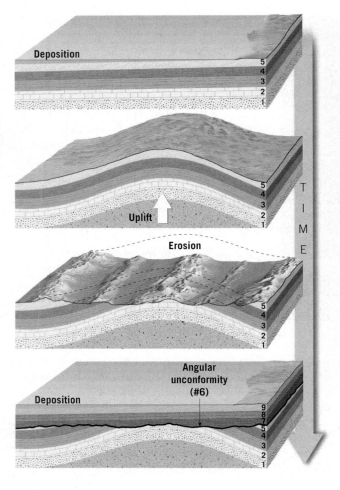

▲ Figure 9.9
Siccar Point, Scotland James Hutton studied this famous unconformity in the late 1700s.

Earth history. Three basic types of unconformities (discussed next) help geologists identify what intervals of time are not represented by strata and thus are missing from the geologic record.

Angular Unconformity

Perhaps the most easily recognized unconformity is an **angular unconformity**. It consists of tilted or folded sedimentary rocks that are overlain by younger, more flat-lying strata. It indicates that during the pause in deposition, a period of deformation (folding or tilting) and erosion occurred (**Figure 9.8**).

When geologist James Hutton studied an angular unconformity in Scotland more than 200 years ago,* he understood that it represented a major episode of geologic activity (**Figure 9.9**). He and his colleagues also appreciated the immense time span implied by such relationships; a companion later wrote of their visit to this site, "the mind seemed to grow giddy by looking so far into the abyss of time."

Disconformity

A **disconformity** is a gap in the rock record that represents a period during which erosion rather than deposition occurred. Imagine that a series of sedimentary layers is deposited in a shallow marine setting. Then sea level falls or the land rises, exposing

some of the sedimentary layers. During this above-sea-level time period, no new sediment accumulates, and some of the existing layers erode away. Later, sea level rises or the land subsides, resulting in the landscape becoming submerged once more. This causes a new series of sedimentary beds to be deposited. The boundary separating the two sets of beds is a disconformity—a span for which there is no rock record (**Figure 9.10**).

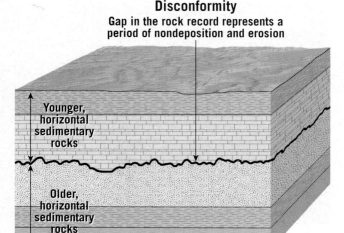

Disconformity
Gap in the rock record represents a period of nondeposition and erosion

▲ Figure 9.10
Disconformity The layers on both sides of this gap in the rock record are essentially parallel.

* Hutton is discussed in the section on the birth of modern geology in Chapter 1.

Because the layers above and below a disconformity are parallel, these features are sometimes difficult to identify unless you notice evidence of erosion such as a buried stream channel.

Nonconformity   The third basic type of unconformity is a **nonconformity**, in which younger sedimentary strata overlie older metamorphic or intrusive igneous rocks (**Figure 9.11**). Just as angular unconformities and some disconformities imply crustal movements, so too do nonconformities. Intrusive igneous masses and metamorphic rocks originate far below the surface. Thus, for a nonconformity to develop, there must be a period of uplift and erosion of overlying rocks. Once exposed at the surface, the igneous or metamorphic rocks are subjected to weathering and erosion, and then they undergo subsidence and renewed sedimentation.

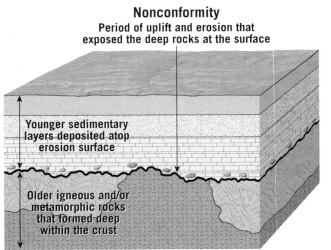

Nonconformity
Period of uplift and erosion that exposed the deep rocks at the surface

Younger sedimentary layers deposited atop erosion surface

Older igneous and/or metamorphic rocks that formed deep within the crust

◄ **Figure 9.11**
Nonconformity Younger sedimentary rocks rest atop older metamorphic or igneous rocks.

Unconformities in the Grand Canyon   The rocks exposed in the Grand Canyon of the Colorado River represent a tremendous span of geologic history. It is a wonderful place to take a trip through time. The canyon's colorful strata record a long history of sedimentation in a variety of environments—advancing seas, rivers and deltas, tidal flats and sand dunes. But the record is not continuous. Unconformities represent amounts of time that have not been recorded in the canyon's layers. The amount of time missing from the record can vary from a few millions years to billions of years. **Figure 9.12** provides a geologic cross section of the Grand Canyon. All three types of unconformities can be seen in the canyon walls.

▼ **Figure 9.12**
Cross section of the Grand Canyon All three types of unconformities are present.

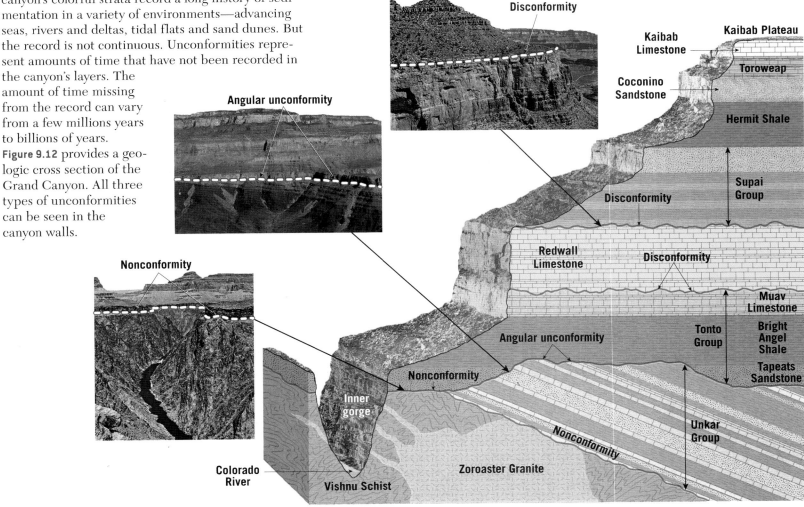

Disconformity

Angular unconformity

Kaibab Plateau

Kaibab Limestone

Toroweap

Coconino Sandstone

Hermit Shale

Disconformity

Supai Group

Nonconformity

Redwall Limestone

Disconformity

Muav Limestone

Tonto Group

Bright Angel Shale

Tapeats Sandstone

Angular unconformity

Nonconformity

Inner gorge

Unkar Group

Nonconformity

Zoroaster Granite

Colorado River

Vishnu Schist

Working out the geologic history of a hypothetical region

Interpretation:

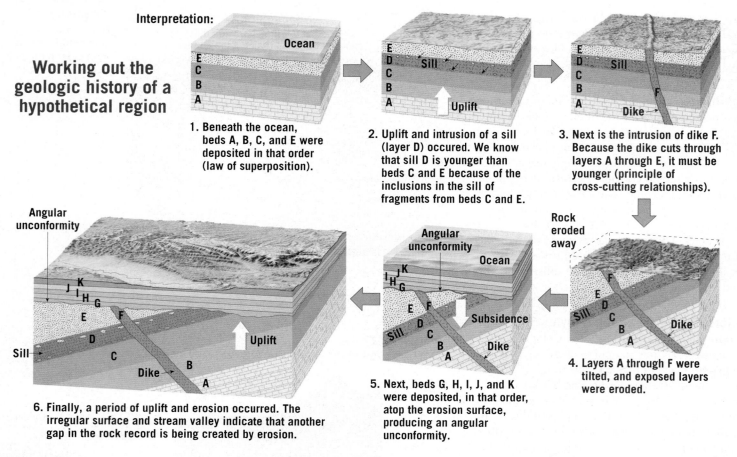

1. Beneath the ocean, beds A, B, C, and E were deposited in that order (law of superposition).

2. Uplift and intrusion of a sill (layer D) occured. We know that sill D is younger than beds C and E because of the inclusions in the sill of fragments from beds C and E.

3. Next is the intrusion of dike F. Because the dike cuts through layers A through E, it must be younger (principle of cross-cutting relationships).

4. Layers A through F were tilted, and exposed layers were eroded.

5. Next, beds G, H, I, J, and K were deposited, in that order, atop the erosion surface, producing an angular unconformity.

6. Finally, a period of uplift and erosion occurred. The irregular surface and stream valley indicate that another gap in the rock record is being created by erosion.

▲ SmartFigure 9.13
Applying principles of relative dating

Tutorial
https://goo.gl/RpFf54

Applying Relative Dating Principles

Scientists can apply the principles of relative dating to place in proper sequence rocks and the events they represent. The hypothetical geologic cross section in **Figure 9.13** provides an example, summarizing the logic used to interpret the cross section.

Keep in mind that although this example establishes a relative time scale for the rocks and events shown, this method gives us no idea how many years of Earth history are represented: We have no numerical dates, and we do not know how this area compares to any other. See **GEOgraphics 9.1** for an extraterrestrial example of applying relative dating principles.

CONCEPT CHECKS 9.1

1. Distinguish between numerical and relative dates.

2. Sketch and label four simple diagrams that illustrate each of the following: superposition, original horizontality, lateral continuity, and cross-cutting relationships.

 **Concept Checker**
https://goo.gl/w9iosG

3. What is the significance of an unconformity?

4. Sketch and explain the difference between an angular unconformity and a nonconformity.

9.2 Fossils: Evidence of Past Life

Define *fossil* and discuss the conditions that favor the preservation of organisms as fossils. List and describe various types of fossils.

Fossils, the remains or traces of prehistoric life, are inclusions in sediment and sedimentary rocks. They are basic and important tools for interpreting the geologic past. The scientific study of fossils and prehistoric life is called **paleontology**. It is an interdisciplinary science that blends geology and biology in an attempt to understand all aspects of the succession of life over the vast expanse of geologic time (see **GEOgraphics 9.2**). Knowing the nature of the life-forms that existed at a particular time helps researchers understand past environmental conditions. Further, fossils are important time indicators and play a key role in correlating rocks of similar ages that are from different places.

Dating the Lunar Surface

Just as we use relative dating principles to determine the sequence of geologic events on Earth, so too can we apply such principles to the surface of the Moon.

Numerical Dates

Radiometric dating of Moon rocks brought back from *Apollo* missions showed that the age of the highlands is greater than 4 billion years, whereas the maria have ages ranging from 3.2 to 3.9 billion years.

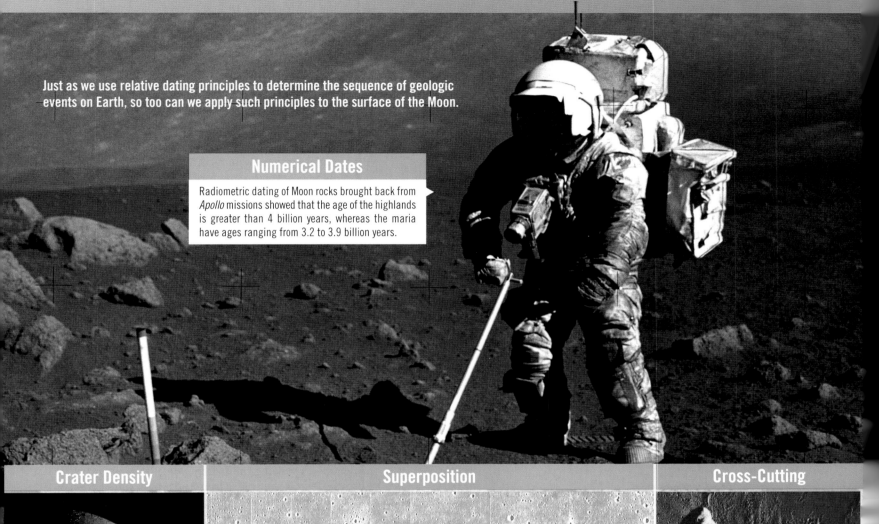

Crater Density

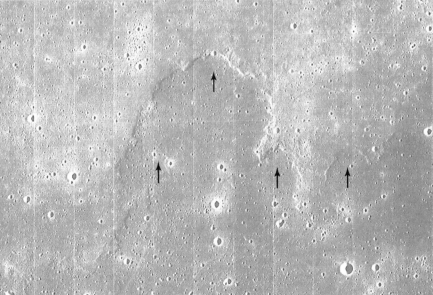

Maria

Highlands

Older regions have been exposed to meteorite impact longer and therefore have more craters. Using this technique, we can infer that the highly cratered highlands are older than the dark areas called maria. The number of craters per unit area (called crater density) is obviously much greater in the highlands.

Superposition

This map view image shows the forward margin of a lava flow "frozen" in place. By applying the law of superposition, we know that this flow is younger than the adjacent layer that disappears beneath it.

Cross-Cutting

Older

Younger

The most obvious features on the lunar surface are craters. Most were produced by the impact of rapidly moving objects called meteorites. When we observe one impact crater that overlaps another, we know the continuous unbroken crater came after the one it cuts across.

Types of Fossils

Certain parts of organisms may not have been structurally altered very much when preserved as fossils in the landscape. Such objects as teeth, bones, and shells are common examples (**Figure 9.14**). Rarely entirely Ice-Age animals, flesh included, are preserved because of rather unusual circumstances. Remains of prehistoric elephants called mammoths that were frozen in the Arctic tundra of Siberia and Alaska are examples, as are the mummified remains of sloths preserved in a dry cave in Nevada. In this section, we consider some of the processes responsible for preserving evidence of life from the distant past.

Permineralization   When mineral-rich groundwater permeates porous tissue such as bone or wood, minerals precipitate out of solution and fill pores and empty spaces, a process called *permineralization*. The formation of *petrified wood* involves permineralization with silica, often from a volcanic source such as a surrounding layer of volcanic ash. The wood is gradually transformed into chert, sometimes with colorful bands from impurities such as iron or carbon (**Figure 9.15A**). The word *petrified* literally means "turned into stone." Sometimes the microscopic details of the petrified structure are faithfully retained.

The fossil record of organisms with hard parts that lived in areas of sedimentation is quite abundant.

Molds and Casts   Other common classes of fossils are *molds* and *casts*. When a shell or other structure is buried in sediment and then dissolved by underground water, a *mold* is created. The mold faithfully reflects only the shape and surface marking of the organism; however, it does not reveal any information concerning the organism's internal structure. If these hollow spaces are subsequently filled with mineral matter, a mineral or rock replica of the organism, called a *cast*, is created (**Figure 9.15B**).

Carbonization and Impressions   *Carbonization* is a process that is particularly effective at preserving leaves and delicate animal forms. It occurs when fine sediment encases the remains of an organism. As time passes, pressure squeezes out the liquid and gaseous components and leaves behind a thin residue of carbon (**Figure 9.15C**). Black shale deposited as organic-rich mud in oxygen-poor environments often contains abundant carbonized remains. If the film of carbon is lost from a fossil preserved in fine-grained sediment, a replica of the surface, called an *impression*, may still show considerable detail (**Figure 9.15D**).

Amber   Delicate organisms, such as insects, are not easily preserved, and consequently they are relatively rare in the fossil record. Not only must they be protected from decay, but they must not be subjected to any pressure that would crush them. One way in which some insects have been preserved is in *amber*, the hardened resin of ancient trees. The spider in **Figure 9.15E** was preserved after being trapped in a drop of sticky resin. Resin sealed off the spider from the atmosphere and protected the remains from damage by water and air. As the resin hardened, a protective pressure-resistant case was formed.

Trace Fossils   In addition to the fossils already mentioned, there are numerous other types, many of them providing indirect evidence of prehistoric life. Examples include:

- Tracks—animal footprints or trails made in soft sediment that later turned into sedimentary rock.
- Burrows—tubes in sediment, wood, or rock made by an animal. These holes may later become filled with mineral matter and preserved. Some of the oldest-known fossils are believed to be worm burrows.
- Coprolites—fossil dung and stomach contents that can provide useful information about the size and food habits of organisms (**Figure 9.15F**).
- Gastroliths—highly polished stomach stones that were used in the grinding of food by some extinct reptiles.

▼ **Figure 9.14**
La Brea tar pits The fossils here are actual (unaltered) remains.

Skeleton of a mammoth, a prehistoric relative of the modern elephant, from the La Brea tar pits.

Excavating bones from pit 91. It is a site rich in unaltered Ice Age organisms. Scientists have been excavating here since 1915.

Conditions Favoring Preservation

Only a tiny fraction of the organisms that have lived throughout history are preserved as fossils. Normally, when an organism perishes, its soft parts are quickly eaten by scavengers or decomposed by bacteria. Two specific conditions increase the probability that an organism will be fossilized: a rapid burial in sediment after death, and the possession of hard parts. Burial protects remains from the surface environment, where destructive processes operate. The hard parts of organisms—for example, bones, shells, teeth, woody stems, or tough seeds—have a much better chance of being preserved as part of the fossil record than mostly soft organisms. The reason for this is straightforward: Flesh and other soft parts decay quickly, before they can be preserved. Although traces and imprints of soft-bodied animals such as jellyfish, worms, and insects exist, they are not common.

Because preservation is affected by several on special conditions, the record of life in the geologic past is biased. There are more fossils of the hard parts of organisms than of fossils of soft tissues. Due to this bias, and the ease at which biological material can be destroyed before having a chance to fossilize, we will at best only be able to study less than 10% of all organism types that have ever existed on Earth.

Permineralization
A. Petrified wood

Mold and cast
B. A trilobite

Carbonization

C. A fossil bee

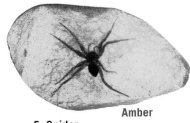
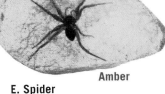

Impressions

D. Fish fossils show considerable detail

Amber

E. Spider

Trace fossils

F. Fossil dung is coprolite

▲ Figure 9.15
Types of fossils

> **CONCEPT CHECKS 9.2**
>
> 1. Describe several ways that an animal or a plant can be preserved as a fossil.
> 2. List two examples of trace fossils.
> 3. What conditions favor the preservation of an organism as a fossil?
>
> **Concept Checker**
> https://goo.gl/LJnUSC

9.3 Correlation of Rock Layers

Explain how rocks of similar age that are in different places can be matched up.

To develop a geologic time scale that is applicable to the entire Earth, rocks of similar age in different regions must be matched up, in a process called **correlation**. Correlating the rocks from one place to another makes possible a more comprehensive view of the geologic history of a region. **Figure 9.16**, for example, shows the correlation of strata at three sites on the Colorado Plateau in southern Utah and northern Arizona. No single locale exhibits the entire sequence, but correlation reveals a more complete picture of the sedimentary rock record.

Correlation Within Limited Areas

Within a limited area, geologists can correlate rocks of one locality with those of another simply by walking along the outcropping edges. However, this may not be possible when the rocks are mostly concealed by soil and vegetation. Correlation over short distances is often achieved by noting the position of a bed in a sequence of strata. Alternatively, a layer may be identified in another location if it is composed of distinctive or uncommon minerals. For correlation between widely separated areas or between continents, geologists must rely on both rock strata and fossils.

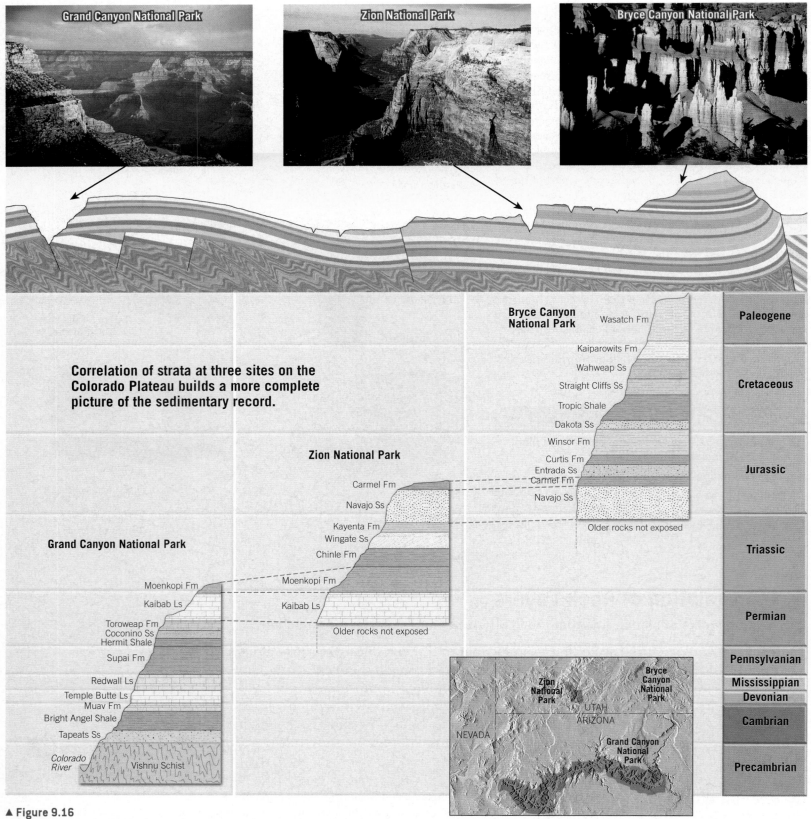

Grand Canyon National Park

Zion National Park

Bryce Canyon National Park

Correlation of strata at three sites on the Colorado Plateau builds a more complete picture of the sedimentary record.

Bryce Canyon National Park

Wasatch Fm	Paleogene
Kaiparowits Fm	
Wahweap Ss	Cretaceous
Straight Cliffs Ss	
Tropic Shale	
Dakota Ss	
Winsor Fm	
Curtis Fm	Jurassic
Entrada Ss	
Carmel Fm	
Navajo Ss	
Older rocks not exposed	

Zion National Park

Carmel Fm
Navajo Ss
Kayenta Fm
Wingate Ss
Chinle Fm
Moenkopi Fm
Kaibab Ls
Older rocks not exposed

Grand Canyon National Park

Moenkopi Fm
Kaibab Ls
Toroweap Fm
Coconino Ss
Hermit Shale
Supai Fm
Redwall Ls
Temple Butte Ls
Muav Fm
Bright Angel Shale
Tapeats Ss
Colorado River   Vishnu Schist

Triassic

Permian

Pennsylvanian

Mississippian

Devonian

Cambrian

Precambrian

NEVADA   UTAH   ARIZONA
Zion National Park
Bryce Canyon National Park
Grand Canyon National Park

▲ Figure 9.16
Correlation Matching strata at three locations on the Colorado Plateau.

Fossils and Correlation

The existence of fossils had been known for centuries, yet it was not until the late 1700s and early 1800s that their significance as geologic tools became evident. During this period, English engineer and canal builder William Smith discovered that each rock formation in the canals he worked on contained fossils unlike those in the beds either above or below. Further, he noted that sedimentary strata in widely separated areas could be identified—and correlated—based on their distinctive fossil content.

Principle of Fossil Succession Based on Smith's observations and the findings of many later geologists, the **principle of fossil succession** was formulated: *Fossil organisms succeed one another in a definite and determinable order, and therefore any time period can be recognized by its fossil content.* In other words, when fossils are arranged according to their age, they do not present a random or haphazard picture. To the contrary, fossils document the evolution of life through time.

For example, an Age of Trilobites is recognized quite early in the fossil record. Then, in succession, paleontologists recognize an Age of Fishes, Age of Coal Swamps, Age of Reptiles, and Age of Mammals. These "ages" pertain to groups that were especially plentiful and characteristic during particular time periods. Within each of the "ages" are many subdivisions, based on predominating species. We find this same succession of dominant organisms, in the same order, on every continent.

Index Fossils and Fossil Assemblages Fossils are the most useful means of correlating rocks of similar age in different regions. Geologists pay particular attention to certain **index fossils** (Figure 9.17). Index fossils are widespread geographically but limited to a short span of geologic time, so their presence provides an important method of matching rocks of the same age. Rock formations do not always contain a specific index fossil. In such situations, however, a group of fossils, called a **fossil assemblage**, may be used to establish the age of the bed. Figure 9.18 illustrates how an assemblage of fossils may be used to date rocks more precisely than could be accomplished by the use of any single fossil.

Environmental Indicators In addition to often being essential tools for correlation, fossils are important environmental indicators. Although we can deduce much about past environments by studying the nature and characteristics of sedimentary rocks, a close examination of the fossils present can usually provide a great deal more information. For example, when the remains of certain clam shells are found in limestone, a geologist quite reasonably assumes that the region was once covered by a shallow sea.

Fossils can also at times be used to identify the approximate position of an ancient shoreline. Given what we know of living organisms, we can conclude that fossil animals with thick shells, capable of withstanding pounding and surging waves, inhabited shorelines. On the other hand, animals with thin, delicate shells probably indicate deep, calm offshore waters.

Fossils also can be used to indicate the former temperature of the water. Certain kinds of present-day corals must live in warm and shallow tropical seas like those around Florida and The Bahamas. When similar types of coral are found in ancient limestones, they indicate the marine environment that must have existed when they were alive. These examples illustrate how fossils can help unravel the complex story of Earth history.

▲ **Figure 9.17**
Index fossils Because microfossils are often very abundant, widespread, and quick to appear and become extinct, they constitute ideal index fossils. This scanning electron micrograph shows marine microfossils from the Miocene epoch.

CONCEPT CHECKS 9.3

1. What is the goal of correlation?

2. State the principle of fossil succession in your own words.

Concept Checker
https://goo.gl/PNCXCS

3. Along with being important in correlation, how else are fossils useful to geologists?

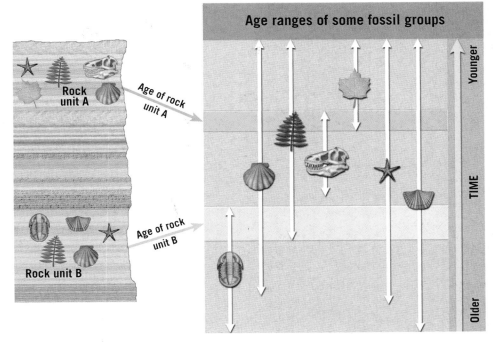

Age ranges of some fossil groups

▲ **SmartFigure 9.18**
Fossil assemblage Overlapping ranges of fossils help date rocks more exactly than is possible using a single fossil.

Tutorial
https://goo.gl/fk2464

How is paleontology different from archaeology?

People frequently confuse these two areas of study because a common perception of both paleontologists and archaeologists is of scientists carefully extracting important clues about the past from layers of rock or sediment. While it is true that scientists in both disciplines "dig" a lot, the focus of each is different.

Archaeology

Archaeologists help us learn about how our human ancestors met the challenges of life in the past.

Archaeologists focus on the material remains of past human life. These remains include both the objects used by people long ago, called *artifacts*, and the buildings and other structures associated with where people lived, called *sites*.

Paleontology

Paleontologists study fossils and are concerned with all life forms in the geologic past. These scientists are excavating the fossil remains of *Albertasaurus*, a carnivore similar to *Tyrannosaurus rex*.

9.4 Numerical Dating with Nuclear Decay

Discuss three types of radioactive decay and explain how radioactive isotopes are used to determine numerical dates.

In this section you will learn about radioactivity and its application in radiometric dating, which allows us to attach numerical dates to important events, such as setting Earth's age at 4.6 billion years. Our understanding of changes in the nuclei of atoms has allowed us to determine that geologic time is vast. This immense span is often referred to as *deep time*. Radiometric dating allows us to measure it quantitatively.

Reviewing Basic Atomic Structure

Recall from Chapter 3 that each atom has a *nucleus* containing protons and neutrons that is orbited by electrons. *Electrons* have a negative electrical charge, and *protons* have a positive charge. A *neutron* has no charge (it is neutral), but it can be converted to a positively charged proton and a negatively charged electron.

Atomic number refers to the number of protons in the nucleus. Every element has a different number of protons and thus a different identifying atomic number (carbon = 6, oxygen = 8, uranium = 92, and so on). Atoms of the same element always have the same number of protons, so the atomic number stays constant.

Practically all of an atom's mass (99.9 percent) is in the nucleus, and electrons have virtually no mass. By adding the protons and neutrons in an atom's nucleus, we derive the atom's *mass number.* The number of neutrons can vary, and these variants, called *isotopes*, have different mass numbers. For example, uranium's nucleus always has 92 protons, so its atomic number is always 92. But its neutron population varies, and uranium has three isotopes: uranium-234 (protons + neutrons= 234),

uranium-235, and uranium-238. All three isotopes are mixed in nature. They look the same and behave the same in chemical reactions.

Changes to Atomic Nuclei

Usually, the forces that hold together atomic nuclei are strong. However, in some isotopes, the forces that bind protons and neutrons are not strong enough to keep them together forever. Such nuclei are unstable, spontaneously breaking apart in a process called **nuclear decay** (also called **radioactive decay**). As time goes by, more and more of the unstable atoms decay, producing an ever-growing number of stable isotopes. Not all isotopes are unstable—there are stable isotopes, too—but here we focus on the unstable isotopes and the stable isotopes they produce.

What happens when unstable nuclei break apart? There are several possibilities. Three common types of radioactive decay that are important for dating are illustrated in **Figure 9.19**:

- In the process of *alpha decay, alpha particles* (α particles) are emitted from the nucleus. An alpha

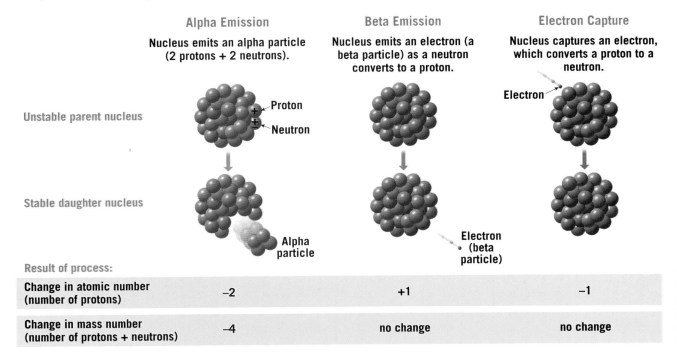

	Alpha Emission	Beta Emission	Electron Capture
	Nucleus emits an alpha particle (2 protons + 2 neutrons).	Nucleus emits an electron (a beta particle) as a neutron converts to a proton.	Nucleus captures an electron, which converts a proton to a neutron.
Unstable parent nucleus	Proton / Neutron		Electron
Stable daughter nucleus	Alpha particle	Electron (beta particle)	
Result of process:			
Change in atomic number (number of protons)	–2	+1	–1
Change in mass number (number of protons + neutrons)	–4	no change	no change

◀ **Figure 9.19**
Common types of radioactive decay Notice that in each example, the number of protons (atomic number) in the nucleus changes, thus producing a different element.

particle is composed of 2 protons and 2 neutrons. Thus, emission of an alpha particle means that the mass number of the isotope is reduced by 4, and the atomic number is lowered by 2.

- *Beta decay* occurs when an electron (often confusingly referred to as a "beta particle," or β particle) is emitted from a nucleus. In this case, the mass number remains unchanged because electrons have practically no mass. However, the electron is produced when a neutron (which has no charge) decays to produce the electron plus a proton. Because the nucleus now contains one more proton than before, the atomic number increases by 1—and it's no longer the same element!

- *Electron capture* happens when an electron is captured by the nucleus. The electron combines with a proton and forms an additional neutron. As with beta decay, the mass number remains unchanged. However, because the nucleus now contains one fewer proton, the atomic number decreases by 1.

An unstable radioactive isotope is called the *parent*, and isotopes resulting from the decay of the parent

Our understanding of changes in the nuclei of atoms has allowed us to determine that geologic time is vast. This immense span is often referred to as deep time.

are termed the *daughter products*. But the path from parent to daughter isn't always direct. Uranium-238, one of the most important isotopes for geologic dating, provides an example of the complexity (**Figure 9.20**). When the radioactive parent, uranium-238 (atomic number 92, mass number 238) decays, it follows a number of steps, emitting a total of 8 alpha particles and 6 electrons before finally becoming the stable daughter product lead-206 (atomic number 82, mass number 206). One of the unstable daughter products produced during this decay series is the radioactive gas radon.

Radiometric Dating

Nuclear decay occurs at a steady rate in Earth's outer layers, and the rates of decay for many unstable isotopes have been precisely measured. **Radiometric dating** is a process in which we use our knowledge of decay to calculate precise dates for certain rock or mineral samples.

Half-Life   The time required for half of the nuclei in a given unstable isotope to decay is called the **half-life** of that isotope. Half-life is a common way of expressing the rate of radioactive decay. When the quantities of parents and daughters are equal (ratio 1:1), we know that one half-life has transpired. When one-quarter of the original parent atoms remain and three-quarters have decayed to the daughter product, the parent/daughter ratio is 1:3, and we know that two half-lives have passed. After three half-lives, the ratio of parent atoms to daughter atoms is 1:7 (one parent atom for every seven daughter atoms).

If the half-life of a radioactive isotope is known and the parent/daughter ratio can be determined, the age of the sample can be calculated. For example, assume that the

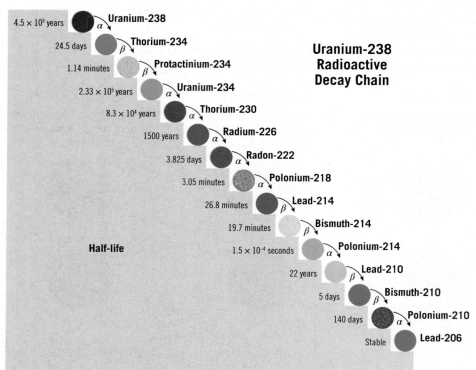

Uranium-238 Radioactive Decay Chain

Half-life	Isotope
4.5×10^9 years	Uranium-238 α
24.5 days	Thorium-234 β
1.14 minutes	Protactinium-234 β
2.33×10^5 years	Uranium-234 α
8.3×10^4 years	Thorium-230 α
1500 years	Radium-226 α
3.825 days	Radon-222 α
3.05 minutes	Polonium-218 α
26.8 minutes	Lead-214 β
19.7 minutes	Bismuth-214 β
1.5×10^{-4} seconds	Polonium-214 α
22 years	Lead-210 β
5 days	Bismuth-210 β
140 days	Polonium-210 α
Stable	Lead-206

▶ Figure 9.20
Decay of U-238
Before the stable end product (Pb-206) is reached, many isotopes are produced as intermediate steps.

half-life of a hypothetical unstable isotope is 1 million years, and the parent/daughter ratio in a sample is 1:15. This ratio indicates that four half-lives have passed and that the sample must be 4 million years old.

Notice that the *percentage* of radioactive atoms that decay during one half-life is always the same: 50 percent (see **Figure 9.21**). However, the *actual number* of atoms that decay with the passing of each half-life continually decreases. Thus, as the percentage of radioactive parent atoms declines, the proportion of stable daughter atoms rises, and the increase in daughter atoms just matches the drop in parent atoms. This fact is the key to radiometric dating.

Calculating Numerical Dates for Rock Samples

Understanding decay rates and the progression through various parent and daughter products gives us a reliable method for numerically dating rock samples. The key is to look at the proportions of certain isotopes in a sample.

For example, some minerals contain uranium atoms in their crystal lattice. When such a mineral crystallizes from magma, it contains no lead (the stable daughter product) from previous decay. The radiometric "clock" starts at this point, when the newly formed sample contains only uranium. As the uranium decays, atoms of the daughter product accumulate in the crystal, eventually building up to measurable levels. Similarly, when a crystal of feldspar forms, some of the potassium atoms incorporated into its lattice will be the unstable isotope potassium-40. These atoms will decay at a steady rate by electron capture to produce the daughter argon-40. Over time, there is less and less of the parent potassium and more and more of the daughter argon.

Unstable Isotopes that Are Useful for Radiometric Dating

Of the many radioactive isotopes that exist in nature, five are particularly useful in providing radiometric ages for ancient rocks (**Table 9.1**). Rubidium-87, thorium-232, and the two isotopes of uranium are used only for dating rocks that are millions of years old, but potassium-40 is more versatile. Although the half-life of potassium-40 is 1.3 billion years, analytical techniques make it possible to detect tiny amounts of its stable daughter product, argon-40, in some rocks that are younger than 100,000 years. Another important reason for its frequent use is that potassium is an abundant constituent of many common minerals, particularly micas and feldspars.

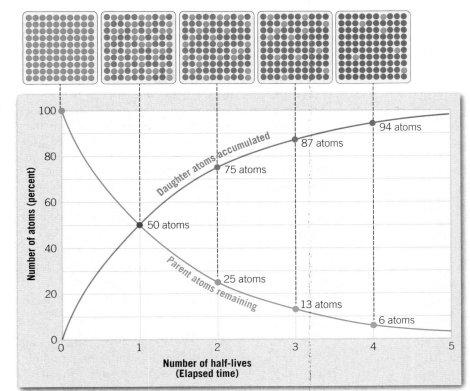

▲ SmartFigure 9.21
Radioactive decay curve Change is exponential. Half of the radioactive parent remains after one half-life. After a second half-life, one-quarter of the parent remains, and so forth.

Tutorial
https://goo.gl/hkbqK2

A Complex Process Although the basic principle of radiometric dating is simple, the actual procedure is quite complex. The chemical analysis that determines the quantities of parent and daughter must be painstakingly precise. In addition, some radioactive materials do not decay directly into the stable daughter product, and this fact may further complicate the analysis. In the case of uranium-238, there are 13 intermediate unstable daughter products formed before the fourteenth and last daughter product, the stable isotope lead-206 (see Figure 9.20).

TABLE 9.1	Isotopes Frequently Used in Radiometric Dating	
Radioactive Parent	**Stable Daughter Product**	**Currently Accepted Half-Life Values**
Uranium-238	Lead-206	4.5 billion years
Uranium-235	Lead-207	704 million years
Thorium-232	Lead-208	14.1 billion years
Rubidium-87	Strontium-87	47.0 billion years
Potassium-40	Argon-40	1.3 billion years

Sources of Error   It is important to understand that an accurate radiometric date can be obtained only if there has been no leakage of parent or daughter isotopes between the mineral crystal and its surroundings in the time since the mineral formed. This is not always the case. In fact, a limitation of the potassium-argon method arises from the fact that argon is a gas, and it may leak from minerals, resulting in a radiometric age that is lower than the actual age. Indeed, losses can be significant if the rock is subjected to high temperatures. If rock is heated to the point where all of the argon in its minerals escapes, then its radiometric clock will be reset, and radiometric dating will give the time of thermal resetting, not the true age of the rock

For other radiometric clocks, a loss of daughter atoms can occur if the rock has been subjected to weathering or leaching. To avoid such a problem, one simple safeguard is to use only fresh, unweathered material and not samples that exhibit signs of chemical alteration.

To guard against errors in radiometric dating, scientists often use cross-checks, such as subjecting a sample to two different methods. If the results agree, the likelihood is high that the date is reliable. If the results are appreciably different, other cross-checks must be employed to determine whether either of them is correct and, if so, which one.

Earth's Oldest Rocks   Radiometric dating methods have produced literally thousands of dates for events in Earth history. Rocks exceeding 3.5 billion years in age are found on all of the continents. Earth's oldest rocks (so far) may be as old as 4.28 Ga (Ga is a gigaannum, ie gigyear, or one billion years). Discovered in northern Quebec, Canada, on the shores of Hudson Bay, these rocks may be remnants of Earth's earliest crust. Rocks from western Greenland have been dated at 3.7 to 3.8 Ga, and rocks nearly as old are found in the Minnesota River Valley and northern Michigan (3.5 to 3.7 Ga.), in southern Africa (3.4 to 3.5 Ga), and

in western Australia (3.4 to 3.6 Ga). Tiny crystals of the mineral zircon having radiometric ages as old as 4.3 Ga have been found in younger sedimentary rocks in western Australia. The source rocks for these tiny durable grains either no longer exist or have not yet been found.

Radiometric dating vindicated the ideas of Hutton, Darwin, and others, who more than 150 years ago inferred that geologic time must be immense. Indeed, modern dating methods prove that there has been enough time for the processes we observe to have accomplished tremendous tasks.

Dating with Carbon-14

To date relatively recent events, scientists can use carbon-14, the radioactive isotope of carbon (**Figure 9.22**). The process is often called **radiocarbon dating**. It is important to emphasize that carbon-14 can only be used to date organic materials, such as wood, charcoal, bones, flesh, and cloth. Because the half-life of carbon-14 is only 5730 years, radiocarbon dating can only be used to date events from the historic past as well as those from very recent geologic history. In some cases carbon-14 can be used to date events as far back as 70,000 years.

Carbon-14 (also written as ^{14}C) is continuously produced in the upper atmosphere as a result of cosmic-ray bombardment. Cosmic rays (high-energy nuclear particles) shatter the nuclei of gas atoms, releasing neutrons. Some of the neutrons are absorbed by nitrogen atoms (atomic number 7, mass number 14), causing each nucleus to emit a proton. As a result, the atomic number decreases by 1 (to 6), and a different element, carbon-14, is created (see **Figure 9.22**). This isotope of carbon quickly becomes incorporated into carbon dioxide, which circulates in the atmosphere and is absorbed by living matter. As a result, all organisms—including you—contain a small amount of carbon-14. You "top off" your ^{14}C levels every time you eat something.

As long as an organism is alive, the decaying radiocarbon is continually replaced, and the proportions of carbon-14 and carbon-12 remain constant. Carbon-12 is the stable and most common isotope of carbon. However, when any plant or animal dies, the amount of carbon-14 gradually decreases as it decays to nitrogen-14 by beta emission (see Figure 9.22). By comparing the proportions of

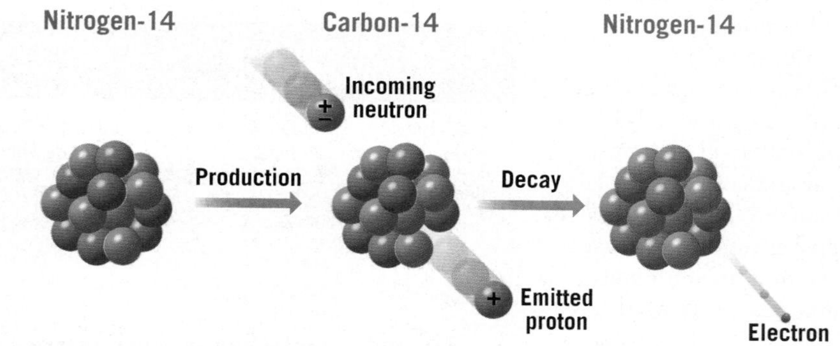

► Figure 9.22
Carbon-14 Production and decay of radiocarbon. These sketches represent the nuclei of the respective atoms.

	Nitrogen-14	Carbon-14	Nitrogen-14
Atomic number	7	6	7
Mass number	14	14	14

carbon-14 and carbon-12 in a sample, radiocarbon dates can be determined.

Although carbon-14 is only useful in dating the last small fraction of geologic time, it is a valuable tool for anthropologists, archaeologists, and historians, as well as for geologists who study very recent Earth history (**Figure 9.23**). In fact, the development of radiocarbon dating was considered so important that the chemist who discovered this application, Willard F. Libby, received a Nobel Prize in 1960.

▲ **Figure 9.23**
Cave art Chauvet Cave in southern France, discovered in 1994, contains some of the earliest-known cave paintings. Radiocarbon dating indicates that most of the images were drawn between 30,000 and 32,000 years ago.

CONCEPT CHECKS 9.4

1. List three types of radioactive decay. For each type, describe how the atomic number and atomic mass change.

2. Sketch a simple diagram to explain the idea of half-life.    **Concept Checker** https://goo.gl/nNqhDB

3. For what time span does radiocarbon dating apply?

9.5 Determining Numerical Dates for Sedimentary Strata

Explain how reliable numerical dates are determined for layers of sedimentary rock.

Although reasonably accurate numerical dates have been worked out for the periods of the geologic time scale, the task is not without difficulty because not all rocks can be dated by using radiometric methods. For a radiometric date to be useful, all the minerals in the rock must have formed at about the same time. Unstable isotopes can therefore be used to determine when minerals in an igneous rock crystallized and when pressure and heat created new minerals in a metamorphic rock.

However, samples of sedimentary rock can only rarely be dated directly by radiometric means. Although a detrital sedimentary rock may include particles that contain radioactive isotopes, the rock's age cannot be accurately determined because the grains composing the rock are not the same age as the rock in which they occur. Rather, the sediments have been weathered and eroded from rocks of diverse ages.

Radiometric dates obtained from metamorphic rocks may also be difficult to interpret because the age of a particular mineral in a metamorphic rock does not necessarily represent the time when the rock initially formed. Instead, the date might indicate any one of a number of subsequent metamorphic phases.

If samples of sedimentary rocks rarely yield reliable radiometric ages, how can numerical dates be assigned to sedimentary layers? Usually a geologist must relate the strata to datable igneous masses, as in **Figure 9.24**.

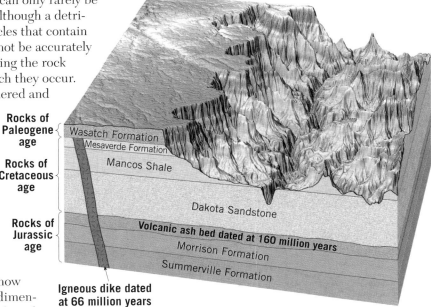

Rocks of Paleogene age
Wasatch Formation
Mesaverde Formation

Rocks of Cretaceous age
Mancos Shale
Dakota Sandstone

Rocks of Jurassic age
Volcanic ash bed dated at 160 million years
Morrison Formation
Summerville Formation

Igneous dike dated at 66 million years

◄ **Figure 9.24**
Dating sedimentary strata Numerical dates for sedimentary layers are usually determined by examining their relationship to igneous rocks.

In this example, radiometric dating has been used to determine the ages of the volcanic ash bed in the Morrison Formation and the dike cutting the Mancos Shale and Mesaverde Formation. The sedimentary beds below the ash are obviously older than the ash, and all the layers above the ash are younger. The dike is younger than the Mancos Shale and the Mesaverde Formation but older than the Wasatch Formation because the dike does not intrude these Paleogene-age rocks.

From this kind of evidence, geologists estimate that the last part of the Morrison Formation was deposited about 160 million years ago, as indicated by the ash bed. Further, they conclude that the deposition of the Wasatch Formation began after the intrusion of the dike, 66 million years ago. This is one example of literally thousands that illustrate how datable materials are used to bracket the various episodes in Earth history within specific time periods. It shows the necessity of combining laboratory dating methods with relative dating principles applied to field observations of rocks.

CONCEPT CHECKS 9.5

1. Briefly explain why it is often difficult to assign a reliable numerical date to a sample of sedimentary rock.

2. How might a numerical date for a layer of sedimentary rock be determined?

 Concept Checker
https://goo.gl/FGtFDr

9.6 The Geologic Time Scale

Distinguish among the four basic time units that make up the geologic time scale and explain why the time scale is considered to be a dynamic tool.

Geologists have divided the whole of geologic history into units of varying length that compose the **geologic time scale** of Earth history. The major units of the time scale were set during the nineteenth century, principally by scientists in Western Europe. Because radiometric dating was unavailable at that time, the entire time scale was originally based only on relative dating techniques. In the twentieth century radiometric methods permitted numerical dates to be added.

Structure of the Time Scale

The geologic time scale subdivides the 4.6-billion-year history of Earth into many different units of time. As shown in **Figure 9.25**, **eons** represent the broadest expanses of time. The eon that began about 541 million years ago is the **Phanerozoic**, a term derived from Greek words meaning "visible life." It is an appropriate description because the rocks and deposits of the Phanerozoic eon contain abundant fossils that document major evolutionary trends.

Eons are divided into **eras**. The Phanerozoic eon consists of the **Paleozoic era** (*paleo* = ancient, *zoe* = life), the **Mesozoic era** (*meso* = middle, *zoe* = life), and the **Cenozoic era** (*ceno* = recent, *zoe* = life). As the names imply, these eras are bounded by profound worldwide changes in life-forms.*

Each era of the Phanerozoic eon is further divided into time units known as **periods**. The Paleozoic has seven, and the Mesozoic and Cenozoic each have three periods. A period is characterized by more specific changes in the evolution of life-forms compared to the broader changes represented by eras.

Finally, each period is divided into still smaller units called **epochs**. Seven epochs have been named for the periods of the Cenozoic. The epochs of other periods usually are simply termed *early*, *middle*, and *late*.

Precambrian Time

The time scale is divided into more detail in the Paleozoic era than in the prior era. The nearly 4 billion years prior to the Cambrian are divided into two eons, the **Archean** (*archaios* = ancient) and the **Proterozoic** (*proteros* = before, *zoe* = life). It is also common for this vast expanse of time to simply be referred to as the **Precambrian**.

Why is the huge expanse of Precambrian time, which represents about 88 percent of Earth history, not divided into epochs? The reason is that Precambrian history is not known in great enough detail. In geology, as in human history, the farther back we go, the less we know because records and clues become fragmented and incomplete. The first abundant fossil evidence also does not appear in the geologic record until the beginning of the Cambrian period. Prior to the Cambrian, simple life-forms such as algae, bacteria, fungi, and worms predominated. All these organisms lack hard parts, an important condition favoring preservation. For this reason, there is only a meager Precambrian fossil record. Many exposures of Precambrian rocks have been studied in some detail, but correlation can be difficult when fossils are lacking.

Another reason the Precambrian can't easily be subdivided is that the rocks dating to this time are very

---

* Major changes in life-forms are discussed in Chapter 22.

old, so most have been subjected to a great many changes. Much of the Precambrian rock record is composed of highly distorted metamorphic rocks. This makes it difficult to interpret past environments because many of the clues present in the original sedimentary rocks have been destroyed.

Radiometric dating has provided a partial solution to the troublesome task of dating and correlating Precambrian rocks. But untangling the complex Precambrian record still remains a daunting task.

Terminology and the Geologic Time Scale

Some terms associated with the geologic time scale are not officially recognized as being a part of it. The best known, and most common, example is *Precambrian*—the informal name for the eons that came before the current Phanerozoic eon. Although the term *Precambrian* has no formal status on the geologic time scale, it has been traditionally used as though it does.

Hadean is another informal term that is used by many geologists and some versions of the geologic time scale. It refers to the earliest interval (eon) of Earth history—before the oldest-known rocks. When the term was coined in 1972, the age of Earth's oldest rocks was hypothesized to be about 3.8 billion years. Today that number stands at slightly greater than 4 billion, and, of course, is subject to revision. The name *Hadean* derives from *Hades*, Greek for *underworld*—a reference to the "hellish" conditions that prevailed on Earth early in its history.

Effective communication in the geosciences requires that the geologic time scale consist of standardized divisions and dates. The organization that is largely responsible for maintaining and updating this important document is the International Commission on Stratigraphy (ICS), a committee of the International Union of Geological Sciences. Advances in the geosciences require that the scale be periodically updated to include changes in unit names and boundary age estimates.

For example, the geologic time scale shown in Figure 9.25 was updated in 2017. After considerable dialogue among geologists who focus on very recent Earth history, the ICS changed the date for the start of the Quaternary period and the Pleistocene epoch from 1.8 million to 2.6 million years ago. If you were to examine a geologic time scale from just a few years ago, it is quite possible that you would see the Cenozoic era divided into the Tertiary and Quaternary periods. However, on

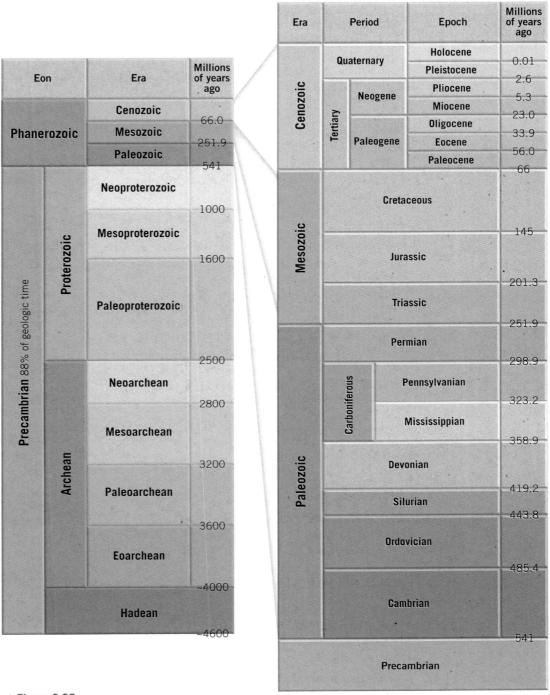

▲ **Figure 9.25**

Geologic time scale Numbers on the time scale represent time in millions of years before present. Numerical dates were added long after the time scale was established using relative dating techniques. The dates on this time scale are those accepted by the International Commission on Stratigraphy (ICS) in 2017. The color scheme used on this chart was selected because it is similar to that used by the ICS.

more recent versions, the space formerly designated as Tertiary is divided into the Paleogene and Neogene periods. Today, the Tertiary period is also considered a "historic" name and is given no official status on the ICS version of the time scale. Many time scales still contain references to the Tertiary period, though, including Figure 9.25. One reason for this is that a great deal of past (and some current) geologic literature uses this name.

> *The geologic time scale is a dynamic tool that continues to be refined as our knowledge and understanding of Earth history evolves.*

this designation is currently used as an informal metaphor for human-caused global environmental change, a number of scientists feel that there is merit in recognizing the Anthropocene as a new official geologic epoch. For those studying historical geology, it is important to realize that the geologic time scale is a dynamic tool that continues to be refined as our knowledge and understanding of Earth history progresses.

Anthropocene Some scientists have suggested that the Holocene epoch has ended and that we have entered a new epoch called the *Anthropocene*. It is considered to be the span in which the global environmental effects of increased human population and economic development have dramatically transformed Earth's surface. There is much debate about when to set the beginning of the Anthropocene—such as when we started altering the landscape by farming or when we started altering the atmosphere by burning lots of fossil fuels or when we first exposed Earth's surface to certain isotopes through nuclear bombs and tests. Although

---

CONCEPT CHECKS 9.6

1. List the four basic units that make up the geologic time scale.

2. What term applies to *all* of geologic time prior to the Phanerozoic eon? Why is this span *not* divided into epochs as is the Phanerozoic eon?

3. Explain why scientists occasionally change the geologic time scale.

 Concept Checker
https://goo.gl/RnF4hb

---

9 CONCEPTS IN REVIEW

Geologic Time

9.1 Creating a Time Scale: Relative Dating Principles

Distinguish between numerical and relative dating and apply relative dating principles to determine a time sequence of geologic events.

Key Terms:

numerical date	principle of lateral continuity	unconformity
relative date	principle of cross-cutting relationships	angular unconformity
principle of superposition	principle of inclusions	disconformity
principle of original horizontality	conformable	nonconformity

- *Relative dates* put events in their proper sequence of formation, while *numerical dates* pinpoint the time, in years, since an event took place.
- Relative dates can be established using the *principles of superposition, original horizontality, cross-cutting relationships,* and *inclusions. Unconformities*, gaps in the geologic record, may be identified during the relative dating process.

Q The accompanying diagram is a cross section of a hypothetical area. Place the lettered features in the proper sequence, from oldest to youngest. Where in the sequence can you identify an unconformity? Which principles did you use to establish the sequence?

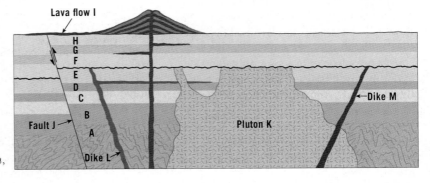

Oldest ___ ___ — — — — — — — — ___ ___ Youngest

9.2 Fossils: Evidence of Past Life

Define *fossil* and discuss the conditions that favor the preservation of organisms as fossils. List and describe various types of fossils.

Key Terms: fossil   paleontology

- *Fossils* are physical remains or indirect traces of ancient life (such as animal paw prints). *Paleontology* is the branch of science that studies fossils.
- Fossils form through many processes. Preservation as a fossil occurs more frequently when there is rapid burial to prevent decomposition. Also, an organism's hard parts are most likely to be preserved because soft tissue decomposes rapidly in most circumstances.
- Types of fossils include permineralized hard parts (shells, bones, teeth, wood), molds, casts, and impressions of organisms, as well as their tracks and burrows.

Q What term is used to describe the type of fossil that is shown here? Briefly explain how it formed.

9.3 Correlation of Rock Layers

Explain how rocks of similar age that are in different places can be matched up.

Key Terms: principle of fossil   index fossil
correlation   succession   fossil assemblage

- Matching up exposures of rock that are the same age but are in different places is called *correlation*. By correlating rocks from around the world, geologists developed the geologic time scale and obtained a fuller perspective on Earth history.
- The *principle of fossil succession* states that fossil organisms succeed one another in a definite and determinable order, and, therefore, a geologic time period can be recognized by examining its fossil content. Distinctive fossil content in rocks can be used to correlate sedimentary rocks in widely separated places.
- *Index fossils* are particularly useful in correlation because they are widespread and associated with a relatively narrow time span. The overlapping ranges of *fossils in an assemblage* may be used to establish an age for a rock layer that contains multiple fossils.
- Fossils may be used to establish ancient environmental conditions that existed when sediment was deposited.

9.4 Numerical Dating with Nuclear Decay

Discuss three types of radioactive decay and explain how radioactive isotopes are used to determine numerical dates.

Key Terms: radiometric dating   radiocarbon
nuclear decay   half-life   dating

- *Nuclear decay* is the spontaneous breaking apart (decay) of certain unstable atomic nuclei. Three common forms of nuclear decay are (1) emission of an alpha particle from the nucleus, (2) emission of a beta particle (electron) from the nucleus, and (3) capture of an electron by the nucleus.
- An unstable radioactive isotope, called a parent, will decay and form daughter products. The length of time for one-half of the nuclei of a radioactive isotope to decay is called the half-life of the isotope. If the *half-life* of an isotope is known, and the parent/daughter ratio can be measured, the age of the sample can be calculated in a procedure called *radiometric dating*.

Q Measurements of zircon crystals containing trace amounts of uranium from a specimen of granite yield parent/daughter ratios of 25 percent parent (uranium-235) and 75 percent daughter (lead-206). The half-life of uranium-235 is 704 million years. How old is the granite?

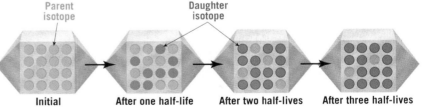

Parent isotope   Daughter isotope

Initial   After one half-life   After two half-lives   After three half-lives

9.5 Determining Numerical Dates for Sedimentary Strata

Explain how reliable numerical dates are determined for layers of sedimentary rock.

- Sedimentary strata are usually not directly datable using radiometric techniques because they consist of the material produced by the weathering of other rocks. A particle in a sedimentary rock comes from some older source rock. If you were to date the particle using isotopes, you would get the age of the source rock, not the age of the sedimentary deposit.
- One way geologists assign numerical dates to sedimentary rocks is to use relative dating principles to relate them to datable igneous masses, such as dikes and volcanic ash beds. A layer may be older than one igneous feature and younger than another.

Q Express the numerical age of the sandstone layer in the diagram as accurately as possible.

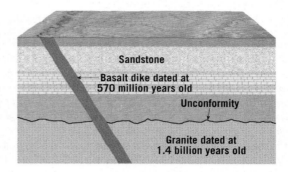

Sandstone

Basalt dike dated at 570 million years old

Unconformity

Granite dated at 1.4 billion years old

9.6 The Geologic Time Scale

Distinguish among the four basic time units that make up the geologic time scale and explain why the time scale is considered to be a dynamic tool.

Key Terms:
geologic time scale	Paleozoic era	Archean
eon	Mesozoic era	Proterozoic
Phanerozoic	Cenozoic era	Precambrian
era	period	epoch

- Earth history is divided into units of time on the *geologic time scale*. *Eons* are divided into *eras*, which each contain multiple *periods*. Periods are divided into *epochs*.
- *Precambrian* time includes the *Archean* and *Proterozoic* eons. It is followed by the *Phanerozoic* eon, which is well documented by abundant fossil evidence, resulting in many subdivisions.
- The geologic time scale is a work in progress, continually being refined as new information becomes available.

GIVE IT SOME THOUGHT

1. The accompanying image shows the metamorphic rock gneiss, a basaltic dike, and a fault. Place these three features in their proper sequence in terms of which came first, second, and third and explain your logic.

Gneiss

Dike

Fault

Dike

Gneiss

2. A mass of granite is in contact with a layer of sandstone. Using a principle described in this chapter, explain how you might determine whether the sandstone was deposited on top of the granite or whether the magma that formed the granite was intruded after the sandstone was deposited.

3. This scenic image is from Monument Valley in the northeastern corner of Arizona. The bedrock in this region consists of layers of sedimentary rocks. Although the prominent rock exposures ("monuments") in this photo are widely separated, we can infer that they represent a once-continuous layer. Discuss the principle that allows us to make this inference.

4. The accompanying photo shows two layers of sedimentary rock. The lower layer is shale from the late Mesozoic era. Note the old river channel that was carved into the shale after it was deposited. Above is a younger layer of boulder-rich breccia. Are these layers conformable? Explain why or why not. What term from relative dating applies to the line separating the two layers?

5. These polished stones are called gastroliths. Explain how such objects can be considered fossils. What category of fossil are they? Name another example of a fossil in this category.

0 1 2
Centimeters

6. If a radioactive isotope of thorium (atomic number 90, mass number 232) emits 6 alpha particles and 4 beta particles during the course of radioactive decay, what are the atomic number and mass number of the stable daughter product?

7. A hypothetical radioactive isotope has a half-life of 10,000 years. If the ratio of radioactive parent to stable daughter product is 1:3, how old is the rock that contains the radioactive material?

8. This scene in Montana's Glacier National Park shows layers of Precambrian sedimentary rocks. The darker layer contained within the sedimentary layers is igneous. The narrow, light-colored areas adjacent to the igneous rock were created when molten material that formed the igneous rock baked the adjacent rock.

   a. Is the igneous layer more likely a lava flow that was laid down at the surface prior to the deposition of the layers above it or a sill that was intruded after all the sedimentary layers were deposited? Explain.

   b. Is it likely that the igneous layer will exhibit a vesicular texture? Explain.

   c. To which group (igneous, sedimentary, or metamorphic) does the light-colored rock belong? Relate your explanation to the rock cycle.

9. Solve the problems related to the magnitude of Earth history below. To make calculations easier, round Earth's age to 5 billion years.

   a. What percentage of geologic time is represented by recorded history? (Assume 5000 years for the length of recorded history.)

   b. Humanlike ancestors (hominids) have been around for roughly 5 million years. What percentage of geologic time is represented by these ancestors?

   c. The first abundant fossil evidence does not appear until the beginning of the Cambrian period, about 540 million years ago. What percentage of geologic time is represented by abundant fossil evidence?

10. A portion of a popular college text in historical geology includes 10 chapters (281 pages) in a unit titled "The Story of Earth." Two chapters (49 pages) are devoted to Precambrian time. By contrast, the last two chapters (67 pages) focus on the most recent 23 million years, with 25 of those pages devoted to the Holocene Epoch, which began 10,000 years ago.

   a. Compare the percentage of pages devoted to the Precambrian to the percentage of geologic time that this span represents.

   b. How does the number of pages about the Holocene compare to its percentage of geologic time?

   c. Suggest some reasons the text seems to have such an unequal treatment of Earth history.

EYE ON EARTH

1. This image shows **West Cedar Mountain in southern Utah**. The gray rocks are shale that originated as muddy river delta deposits. The sediments composing the orange sandstone were deposited by a river.

   a. Place the events related to the geologic history of this area in proper sequence. Explain your logic. Include the following: uplift, sandstone, erosion, and shale.

   b. What term is applied to the type of dates you established for this site?

2. This image is from the bottom of the Grand Canyon, in a zone called the Inner Gorge. The dark rock is the Vishnu Schist. The light-colored rock near the bottom of the photograph, called the Zoroaster Granite, is a series of dikes. Both rocks date from Precambrian time. Suppose you are on a raft trip through the Grand Canyon. Your companions are bright and curious but are not trained in geology, as you are. Sitting around the campfire the night before you reached the site pictured here, you and your fellow rafters discussed the geologic time scale and the magnitude of geologic time.

   a. When you arrive at this site, someone asks why Precambrian history seems so sketchy and why this time span is not divided into nearly as many subdivisions as Phanerozoic time. Use this setting as you answer these questions.

   b. Which is older, the Vishnu Schist or the Zoroaster Granite? Explain.

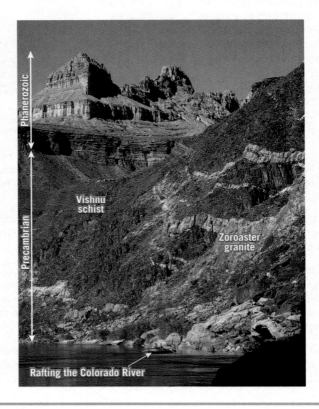

DATA ANALYSIS

Fossils and Geologic Time

Fossils are extremely useful in stratigraphy. In the Phanerozoic, time unit boundaries typically represent widespread changes in fossil assemblages. Often, these changes represent mass extinctions.

https://goo.gl/eD226M

ACTIVITIES

Go to the Paleobiology Database at https://paleobiodb.org and click "Explore" Then "Go to application." Click on the "taxa browser" (image of an insect) and type *Life* in the search box. Click "Life (unranked)" to see the locations of all fossils.

1. **What type of organism is most prevalent in this record? What percentage of all listed fossils does it represent?**

2. **What percentage of the listed fossils are mammals (Mammalia)?**

3. **What is the total number of occurrences displayed? What is the total number of collections (dig sites)?**

4. **One explanation for the prevalence of the organism you named for Question 1 is that it was the dominant species at the time. What might be another explanation?**

Foraminifera (forams) are small marine organisms that are used widely for stratigraphy. In the taxa browser search box, type *Foraminifera* and click "Foraminifera (phylum)." Click on the "Stats" button (image of a line graph) to display the diversity of foram fossils over time. Mouse over the colored blocks to see the names of the time periods.

5. **Identify points on this graph that could represent a mass extinction among forams, followed by the appearance of new types. Would such a feature make a good stratigraphic marker?**

6. **Identify three places where mass extinctions occur at boundaries between geologic time units.**

Double-click on the black "Geologic Time" band at the bottom to see the full range of geologic time. Double-click on a time unit to zoom in and to see numerical ages. Mouse over a time unit to see its full name.

7. **Single-click successively on the Mesoproterozoic, Neoproterozoic, and Paleozoic eras. Judging from these data, when do fossils first become abundant?**

8. **Why is the geologic time scale much more finely divided in the Phanerozoic than in earlier eons?**

9. **Within the Phanerozoic, roughly how long does a typical period last? What are the longest and shortest periods?**

Mastering Geology

Looking for additional review and test prep materials? Visit pearson.com/mastering/geology to enhance your understanding of this chapter's content by accessing a variety of resources, including **Self-Study Quizzes, Geoscience Animations, SmartFigure Tutorials,** *Project Condor* **Videos, Mobile Field Trips, Dynamic Study Modules,** and an optional **Pearson eText.**

When an Earthquake Made the Mississippi Flow Backward: The New Madrid Seismic Zone

The New Madrid seismic zone spans a large area where Missouri, Kentucky, Arkansas, and Tennessee meet. During a 3-month period spanning 1811 and 1812, the New Madrid seismic zone experienced three major earthquakes of magnitude 7.0 or higher.

▲ New Madrid seismic zone map

As a result some land was uplifted, while large areas bordering the Mississippi River subsided. One area where subsidence ranged from 5 to 20 feet became Tennessee's Reelfoot Lake. People in the area reported having seen the Mississippi River temporarily flow *backward* for a number of hours during the lake's formation.

In the early 1800s the population in the New Madrid seismic zone numbered in the low thousands, and there were very few buildings. Today, about 12 million people call the area between Memphis and Saint Louis home. Many local buildings are made of bricks and other materials that don't withstand earthquakes well. So another "big one" occurring in the seismic zone could produce catastrophic loss of life and property. The USGS estimates a 7–10 percent chance of another magnitude 7.0 or higher quake occurring in the region sometime in the next half-century.

► Reelfoot Lake, Tennessee, is a swampy, shallow lake that formed after a major earthquake in 1812.

10
Crustal Deformation

FOCUS ON CONCEPTS

Each statement represents the primary learning objective for the corresponding major heading within the chapter. After you complete the chapter, you should be able to:

10.1 Describe the three types of differential stress and name the type of plate boundary most commonly associated with each.

10.2 List and describe five common folded structures.

10.3 Sketch and briefly describe the relative motion of rock bodies located on opposite sides of normal, reverse, and thrust faults as well as both types of strike-slip faults.

10.4 Explain how strike and dip are measured and what these measurements tell geologists about the orientations of rock structures located below Earth's surface.

Earth is dynamic. Shifting lithospheric plates gradually change the face of our planet by moving continents across the globe. The results of this tectonic activity are perhaps most strikingly apparent in Earth's major mountain belts. Rocks containing fossils of marine organisms are found thousands of meters above sea level, and massive rock units are bent, contorted, overturned, and sometimes riddled with fractures.

This chapter explores the forces that deform rock and the rock structures that result. Foliation and rock cleavage were examined in Chapter 8; this chapter is devoted to the other major structural features of Earth's crust and the tectonic forces that produce them.

10.1 How Rocks Deform

Describe the three types of differential stress and name the type of plate boundary most commonly associated with each.

Tectonic forces can cause rocks to *move, tilt*, and/or *change shape*. For instance, colliding plates can uplift flat-lying beds of marine limestone so they are exposed at the surface, rotate them so they lie at a steep angle, or crumple the rock layers into folds. Collectively, all of these types of change are called **deformation**. Rock deformation is caused by tectonic forces, and it occurs mostly along *plate boundaries*—the places where lithospheric plates push together, pull apart, or scrape past each other.

Rocks deform in characteristic ways that can be observed at an **outcrop**—places where rocks are exposed on Earth's surface. For example, the folds in **Figure 10.1** represent a typical response to compressional forces at depth. Geologists use the term **geologic structures**, or **rock structures**, for the structural features that reflect a rock's tectonic history when observed in outcrops. This chapter will examine three types of tectonic structures: folds (the bending of rock layers without breakage), faults (fractures along which one rock sides past another), and joints (cracks in the rock).

To understand rock deformation, we must first look more closely at the concepts of *stress* and *strain*.

Stress: The Force That Deforms Rocks

So far, we've said that tectonic *forces* cause deformation. More precisely, rocks respond to **stress**, which takes into

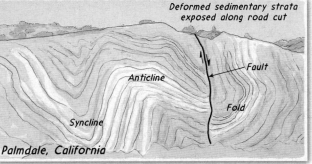

Deformed sedimentary strata exposed along road cut

Anticline

Fault

Fold

Syncline

Palmdale, California

Geologist's Sketch

▲ SmartFigure 10.1
Deformed sedimentary strata Deformed strata are exposed in a road cut near Palmdale, California. In addition to the obvious folding, light-colored beds are offset along a fault located on the right side of the photograph.

Tutorial
https://goo.gl/e9GWrD

consideration the area over which a force acts (Stress = Force/Area). Recall that a force applied equally in all directions is called **confining pressure**; this type of force compacts mineral grains and reduces the volume of a rock body. However, confining pressure does not cause deformation; instead, deformation is caused by **differential stress**, in which the force is stronger in one direction and weaker in another.

We will consider three types of differential stress: *compressional, tensional,* and *shear* (**Figure 10.2**). On a large scale, each type of stress tends to be associated with one type of plate boundary:

1. **Compression.** Differential stress that squeezes a rock mass as if it were in a vise is called **compressional stress**. Compressional stresses are most often associated with *convergent plate boundaries*, where plates collide. Earth's crust generally becomes laterally shortened and vertically thickened. Over millions of years, this deformation produces mountain belts.

2. **Tension.** Differential stress that pulls apart rock bodies is known as **tensional stress**. Tensional stress occurs most often along *divergent plate boundaries*, where plates move apart. For example, in the region around the East African Rift, tensional forces have fractured, stretched, and thinned the crust sufficiently to produce deep rift valleys.

3. **Shear.** Differential stress can also cause a rock body to **shear**, which involves the movement of one part of a rock body relative to another. An everyday example of shear is the slippage that occurs between individual playing cards when the top of the deck is moved relative to the bottom (**Figure 10.3**). Shear is the dominant force at *transform plate boundaries*, such as the San Andreas Fault, where large segments of Earth's crust slip horizontally past one another.

Strain: A Change in Shape Caused by Stress

Recall that differential stress can deform a rock body by causing it to *move, tilt*, and/or *change shape*. When differential stress changes a rock's shape, the resulting deformation (distortion) is called **strain**. By observing and measuring the strain imprinted on a rock body, we can infer the type of stress that deformed the rock.

How does a rigid object like a rock change shape? One way is by undergoing slippage along parallel surfaces of weakness, such as microscopic fractures or foliation surfaces. As with the deck of cards in Figure 10.3, many tiny slips can add up to a significant change in shape.

Mineral grains can also change shape in response to differential stress that does not involve slippage along zones of weakness. Instead, the movement of atoms from a location that is highly stressed to a less-stressed position on the same grain triggers a change in shape—a process called *recrystallization* (see Section 8.3, page 241).

Types of Deformation

Rocks experience three types of deformation that lead to shape changes (strain): *elastic, brittle,* and *ductile*.

Elastic Deformation   If you open a spring-loaded door, the spring stretches and then returns to its original shape as it closes the door. We say the spring undergoes **elastic deformation** because it changes shape temporarily in response to a stress (the force used to open the door) and then returns to its original shape when the stress is removed. Rocks can also undergo elastic deformation. Like a spring, chemical bonds in a mineral grain stretch while under stress and then snap back to their original length.

Brittle Deformation   When rock is deformed beyond its ability to respond elastically, it either breaks or is permanently bent. Rocks that break into smaller pieces exhibit **brittle deformation**. From our everyday experience, we know that glass objects, wooden pencils, ceramic plates, and even our bones exhibit brittle deformation when their strength is surpassed. Brittle deformation occurs when stress breaks the chemical bonds that hold a material together.

Ductile Deformation   When an object changes shape without breaking, we say that it has undergone **ductile deformation**. When you knead clay or taffy, you are deforming it in a ductile way. Ductile deformation in rocks takes place largely through the mechanisms described earlier—slippage along surfaces of weakness within the rock and the gradual reshaping of mineral grains. These processes enable rock to flow very slowly, even though it remains in a solid state. Folds are an example of ductile deformation (see Figure 10.1).

Factors That Affect How Rocks Deform

As shown in **Figure 10.4**, rock deformation tends to be brittle at shallow depths and ductile at greater depths. Four factors influence how a rock deforms: *temperature, confining pressure*, the *type of rock*, and *time*. The first two factors, temperature and confining pressure, are influenced mainly by the depth of the rock's burial. As depth increases, so do the temperatures and pressures. Deeper depths exhibit more extreme environments.

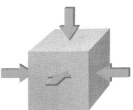

A. Confining pressure

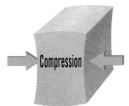

B. Compressional stress (shortening)

C. Tensional stress (stretching)

D. Shear stress (sliding and tearing)

▲ Figure 10.2
Confining pressure and three types of differential stress: Compression, tension, and shear

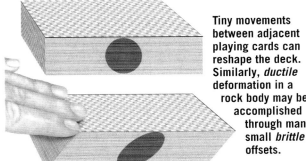

Tiny movements between adjacent playing cards can reshape the deck. Similarly, *ductile* deformation in a rock body may be accomplished through many small *brittle* offsets.

◄ Figure 10.3
Shearing and the resulting deformation (strain) A deck of playing cards with a circle embossed on its side illustrates shearing and the resulting strain.

▶ Figure 10.4
Deformation caused by three types of stress Brittle deformation (fracturing and faulting) dominates in the upper crust, where the temperatures are comparatively cool. By contrast, at depths greater than about 10 kilometers (6 miles), where temperatures are high, rock deforms through ductile flow and folding.

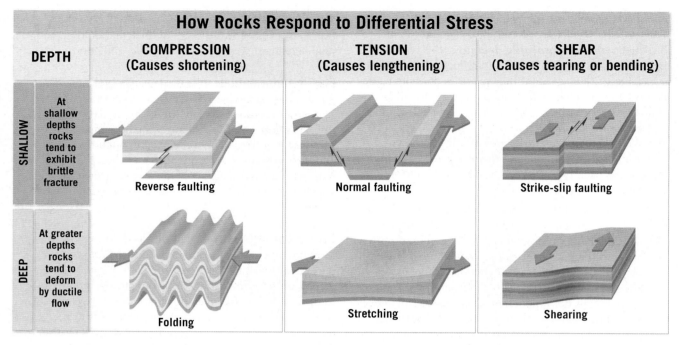

How Rocks Respond to Differential Stress

DEPTH	COMPRESSION (Causes shortening)	TENSION (Causes lengthening)	SHEAR (Causes tearing or bending)
SHALLOW — At shallow depths rocks tend to exhibit brittle fracture	Reverse faulting	Normal faulting	Strike-slip faulting
DEEP — At greater depths rocks tend to deform by ductile flow	Folding	Stretching	Shearing

The Role of Temperature Where temperatures are high (deep in Earth's crust or adjacent to a heat source such as a magma chamber), rocks are nearer their melting points and are therefore weaker and more capable of ductile deformation. Near the surface or in a comparatively cool environment such as a subduction zone, rocks are more brittle and prone to fracture. This behavior should be familiar: You can snap a cold chocolate bar in two with your hands, while the same bar will bend in your hand if it is warm.

When stressed beyond their strength, rocks near Earth's surface generally undergo brittle deformation and break, while rocks deep in the Earth's crust undergo ductile deformation and bend.

The Role of Confining Pressure Recall that the confining pressure on rocks increases with depth as the thickness of the overlying rock increases. Because confining pressure squeezes rocks equally from all directions, it tends to make them harder to break and hence less brittle. Thus, at depth, the increase in temperature and the increase in pressure have complementary effects: Higher temperatures enhance ductile behavior, and greater pressures tend to keep the rock intact—and thus more likely to *bend* rather than fracture.

The Influence of Rock Type The manner in which a particular rock type responds to stress is greatly influenced by its mineral composition and texture. Granite, basalt, and well-cemented quartz sandstones are examples of strong, brittle rocks that tend to fail by breaking (brittle deformation) when subjected to stresses that exceed their strength. By contrast, clay-rich or weakly cemented sedimentary rocks and foliated metamorphic rocks more readily exhibit ductile deformation. Weak rocks that are likely to behave in a ductile manner (bend or flow) when subjected to differential stress include rock salt, shale, limestone, and schist. Also, glacial ice, which is technically a rock, deforms easily, accommodating internal flow.

Figure 10.5 shows an example of this type of behavior in a metamorphic rock. The central nonfoliated layer behaved in a brittle

▶ Figure 10.5
How rock type influences the type of deformation This structure is an example of *boudinage*, which forms when some portions of a rock body deform in a ductile fashion and others act as brittle units. The central greenish brittle layer has broken into a series of chunks, and the surrounding gray clay material has flowed into the gaps between the chunks.

Brittle layer breaks into chunks

Ductile layer flows into gaps

manner, breaking into chunks, while the flanking foliated layers responded to the same differential stress in a ductile manner, flowing into the gaps between the chunks. You can create the same effect by biting into a s'more: The graham crackers break, while the warmed marshmallow and chocolate layers ooze.

Time as a Factor   The folded rocks of mountain belts show that tens of kilometers of compressional strain (shortening) can be accommodated by ductile deformation. For this to happen, stress has to be applied slowly enough that the sluggish processes of ductile deformation can keep up. If stress is applied to a rock unit too quickly, the rock will deform elastically until its strength is exceeded, and then it will fracture. Taffy exhibits the

same behavior on a more familiar time scale: If you hit a bar of taffy against the edge of a table, it will break, but if you put a weight on it and leave it overnight, it will gradually spread and flatten.

CONCEPT CHECKS 10.1

1. List the three types of differential stress and briefly describe the changes they can impart to rock bodies.
2. Explain how strain differs from stress.
3. How is brittle deformation different from ductile deformation? **Concept Checker** https://goo.gl/yUbjYq
4. List and describe the four factors that affect whether a rock deforms in a brittle or ductile manner.

10.2 Folds: Rock Structures Formed by Ductile Deformation

List and describe five common folded structures.

Along convergent plate boundaries, rock strata are often bent into a series of wave-like undulations called **folds** (Figure 10.6). Folds come in a wide variety of sizes and configurations. Some are broad structures in which strata hundreds of meters thick have been slightly warped, while others are very tight, even microscopic structures found in metamorphic rocks. *Compressional stresses* that result in a lateral shortening and vertical thickening of the crust create most folds.

Folds are geologic structures consisting of stacks of originally horizontal surfaces, such as sedimentary strata, that have been bent as a result of permanent deformation. Each layer is bent around an imaginary axis called a *hinge line*, or simply a *hinge*.

Folds are also described by their *axial plane*, which is a surface that connects all the hinge lines of the folded strata. In simple folds, the axial plane is vertical and divides the fold into two roughly symmetrical *limbs*. However, the axial plane often leans so that one limb is steeper than the other.

Anticlines and Synclines

The two most common types of folds are anticlines and synclines (Figure 10.7). **Anticlines** usually form by the upfolding, or arching, of sedimentary layers.[*] Typically found in association with anticlines are downfolds, or troughs, called **synclines**. Notice in Figure 10.7 that the limb of an anticline is also a limb of an adjacent syncline.

Depending on their orientation, these basic folds are described as *symmetrical* when the limbs are mirror images of each other and *asymmetrical* when they are not. The limbs of a symmetrical fold dip at the same angle, resulting in a vertical axial plane. By contrast, the limbs of an asymmetrical fold dip at different angles, which results in an inclined axial plane. An asymmetrical fold is said to be *overturned* if both limbs dip in the same

direction, with one limb tilted beyond the vertical (see Figure 10.7). An overturned fold can also "lie on its side" so that the axial plane is horizontal. These *recumbent* folds are common in highly deformed mountain belts such as the Alps.

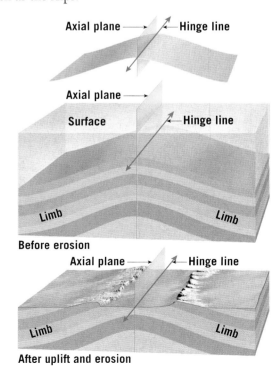

Before erosion

After uplift and erosion

◄ **SmartFigure 10.6**
Features associated with symmetrical folds The axial plane divides a fold as symmetrically as possible, while the hinge line traces the points of maximum curvature of any layer.

Condor Video https://goo.gl/zqfzK3

---

[*]By strict definition, an anticline is a structure in which the oldest strata are found in the center, and a syncline is a structure where the youngest strata are found in the center.

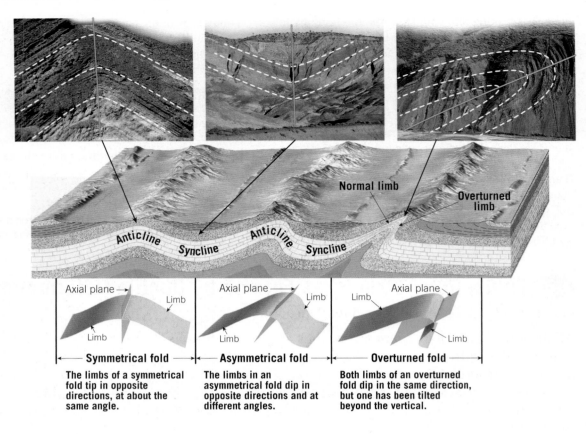

Symmetrical fold
The limbs of a symmetrical fold tip in opposite directions, at about the same angle.

Asymmetrical fold
The limbs in an asymmetrical fold dip in opposite directions and at different angles.

Overturned fold
Both limbs of an overturned fold dip in the same direction, but one has been tilted beyond the vertical.

Folds can also be tilted by tectonic forces so that their hinge lines slope downward (**Figure 10.8A**). Folds of this type are said to *plunge* because the hinge lines penetrate Earth's surface. Sheep Mountain, Wyoming, is an example of a plunging anticline. Notice in **Figure 10.8B** that a V-shaped pattern is produced when erosion removes the upper layers of a plunging fold and exposes its interior. In the case of an anticline, such as Sheep Mountain, the tip of the V points in the direction of plunge (**Figure 10.8C**); the opposite is true for a syncline.

A good example of the topography that can result when erosional forces attack folded sedimentary strata is found in the Valley and Ridge Province of the Appalachians (see Figure 14.11, page 402). It is important to realize that anticlines typically do not

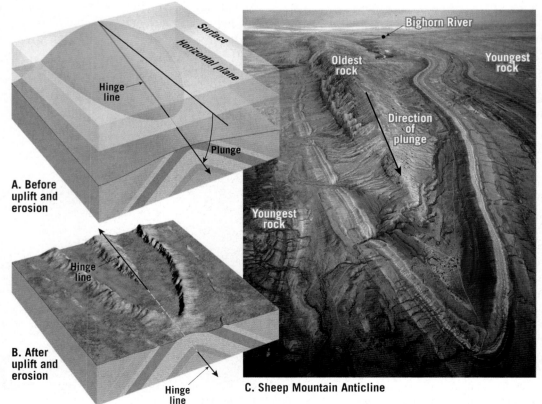

A. Before uplift and erosion

B. After uplift and erosion

C. Sheep Mountain Anticline

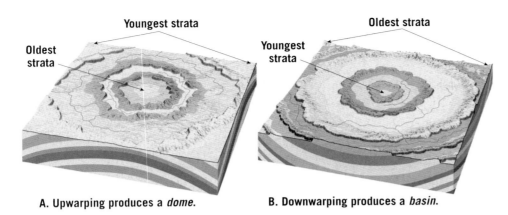

A. Upwarping produces a *dome*.   **B. Downwarping produces a *basin*.**

◄ **SmartFigure 10.9**
Domes versus basins Gentle upwarping and downwarping of crustal rocks produce (**A**) domes and (**B**) basins. Erosion of these structures results in an outcrop pattern that ranges from roughly circular to more elongated (elliptical).

Tutorial
https://goo.gl/
SVJCLP

show up as ridges, nor synclines as valleys. Rather, ridges and valleys result from differential weathering and erosion. For example, in the Valley and Ridge Province, resistant sandstone beds remain as imposing ridges separated by valleys that are cut into more easily eroded shale or limestone beds.

Domes and Basins

A **dome** is a structure that occurs when a broad upwarping of basement rock deforms the overlying cover of sedimentary strata to produce a circular or slightly elongated bulge (**Figure 10.9A**). The Black Hills of western South Dakota represent a large structural dome generated by upwarping. Here erosion has stripped away the highest portions of the overlying sedimentary beds, exposing older igneous and metamorphic rocks in the center (**Figure 10.10**).

Structural domes can also form through the intrusion of magma (laccoliths), as shown in Figure 4.34, page 128. In addition, the upward migration of buried salt deposits can produce salt domes like those beneath the Gulf of Mexico. Salt domes are economically important rock structures because when salt migrates upward, the surrounding oil-bearing sedimentary strata may deform to form oil reservoirs (see Figure 23.6, page 666).

The inverse of a dome is a downwarped structure termed a **basin** (**Figure 10.9B**). Several large structural basins exist in the United States (**Figure 10.11**). The basins of Michigan and

▲ **Figure 10.10**
A large structural dome The central core of the Black Hills, North Dakota, is composed of resistant Precambrian-age igneous and metamorphic rocks. The surrounding rocks are mainly younger limestones and sandstones.

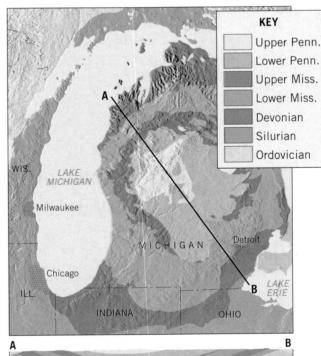

KEY
- Upper Penn.
- Lower Penn.
- Upper Miss.
- Lower Miss.
- Devonian
- Silurian
- Ordovician

▲ **Figure 10.11**
The bedrock geology of the Michigan basin The youngest rocks are centrally located, and the oldest beds flank this structure.

► SmartFigure 10.12
The East Kaibab Monocline, Arizona This monocline consists of bent sedimentary beds that were deformed by faulting in the bedrock below. The thrust fault does not reach the surface. The inclined strata once extended over the sedimentary layers now exposed at the surface—evidence that a tremendous volume of rock has been eroded from this area.

Condor Video
https://goo.gl/6RMZHN

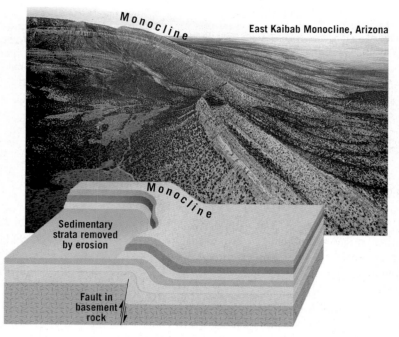

East Kaibab Monocline, Arizona

Monocline

Monocline

Sedimentary strata removed by erosion

Fault in basement rock

Monoclines

Although folds and faults are discussed separately, they may occur together as a result of the same tectonic stresses. Particularly prominent features of the Colorado Plateau region, **monoclines** (*mono* = one, *kleinen* = incline) are large, steplike folds in otherwise horizontal sedimentary strata (**Figure 10.12**). These folds appear to have resulted from the reactivation of ancient, steep-dipping reverse faults located in basement rocks beneath the plateau. As large blocks of basement rock were displaced upward, the comparatively ductile sedimentary strata above responded by draping over the fault like clothes hanging over a bench. Displacement along these reactivated faults can exceed 1 kilometer (0.6 miles).

Illinois have gently sloping beds similar to saucers and are thought to have resulted from large accumulations of sediment, whose weight caused the crust to subside. (See the section "The Principle of Isostasy," in Chapter 14, page 406.) A few structural basins were produced by giant meteorite impacts.

Because the sedimentary beds in basins usually slope at low angles, basins are identified mainly by the age sequence of their strata. In a structural basin, the youngest rocks are at the center and the oldest on the flanks. In a dome, such as the Black Hills, the reverse is true: The oldest rocks form the core.

CONCEPT CHECKS 10.2

1. Sketch and distinguish between anticlines and synclines and between domes and basins.
2. The Black Hills of South Dakota is a good example of what type of geologic structure?  **Concept Checker** https://goo.gl/tS1wov
3. Where are the youngest rocks in a structural basin found: near its center or near its margin?
4. Describe how a monocline forms.

10.3 Faults and Joints: Rock Structures Formed by Brittle Deformation

Sketch and briefly describe the relative motion of rock bodies located on opposite sides of normal, reverse, and thrust faults as well as both types of strike-slip faults.

Faults and joints are both structures that form where brittle deformation leads to fracturing of Earth's crust. A *joint* is a fracture, whereas a **fault** is a fracture along which motion has occurred, so that the rocks on either side are offset from each other. **Figure 10.13** shows a small fault revealed in a road cut, where the sedimentary beds have been offset by a few meters. Faults of this scale often occur as single discrete breaks.

By contrast, large faults, like the San Andreas Fault in California, have displacements of hundreds of kilometers and consist of many interconnecting fault surfaces. These structures, described as *fault zones*, can be several kilometers wide and are often easier to identify from aerial photographs than at ground level. Sudden movements along faults cause most earthquakes. However, the vast majority of faults are remnants of past deformation and are inactive.

▲ SmartFigure 10.13
Faults are fractures where slip has occurred

Condor Video
https://goo.gl/MPPfzi

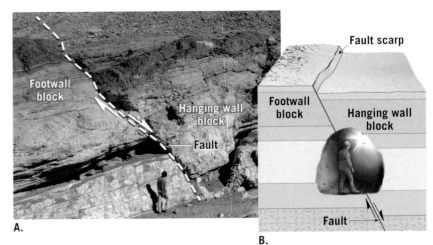

A.

B.

▲ SmartFigure 10.14
Hanging wall block and footwall block The rock immediately above a fault surface is the *hanging wall block*, and the one below is called the *footwall block*. These terms were coined by miners who excavated ore deposits along fault zones. The miners hung their lanterns on the rocks above the fault trace (hanging wall block) and walked on the rocks below the fault trace (footwall block).

Animation
https://goo.gl/zyRHvV

Describing the Orientation of Geologic Structures: Strike and Dip

Because faults have nearly planar surfaces, geologists describe their orientations in terms of their strike and dip. **Strike** is the *compass direction* of the line produced by the intersection of a fault (or inclined sedimentary layer) with a horizontal plane. For example, the strike (or *trend*) of the San Andreas Fault is northwesterly. **Dip** is the *angle of inclination* of a fault from a horizontal plane. The San Andreas is nearly vertical, so it has a dip, or *inclination*, of about 90 degrees. By contrast, some faults are nearly horizontal, with dips of 10 degrees or less.

Strike and dip can be used to describe the orientation of inclined sedimentary layers, dikes, joints, and other planar structures, in addition to faults. For example, the hinge and limbs of an anticline can be described by their strike and dip. Strike and dip are also used to map geologic structures, topics considered later in this chapter.

Dip-Slip Faults

Faults where movement is primarily parallel to the slope of the fault surface are called **dip-slip faults**. Geologists identify the rock body that is above the fault's surface as the **hanging wall block** and the rock body below the fault as the **footwall block** (Figure 10.14). These names were first used by prospectors and miners who excavated metallic ore deposits such as gold that had precipitated from hydrothermal solutions along inactive fault zones. The miners would walk on the rocks below the mineralized fault zone (the *footwall block*) and hang their lanterns

on the rocks above (the *hanging wall block*). The two main types of dip-slip faults are *normal faults* and *reverse faults*.

Normal Faults When the hanging wall block moves down relative to the footwall block, dip-slip faults are classified as **normal faults** (Figure 10.15). These faults are called

A. Fault Motion

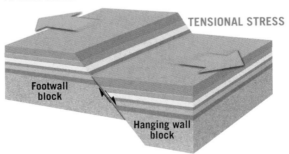

TENSIONAL STRESS

Footwall block

Hanging wall block

B. After Erosion

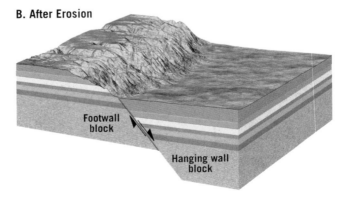

Footwall block

Hanging wall block

◀ SmartFigure 10.15
Normal dip-slip fault
A. illustrates the relative displacement that occurs between the blocks on either side of a fault, while B. shows how erosion may alter the up-faulted blocks.

Tutorial
https://goo.gl/YXMyyv

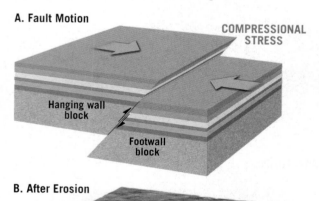

Normal faults range significantly in size. Some are small, having displacements of only a meter or so, like the one shown in the road cut in Figure 10.13. Others, however, extend for tens of kilometers. Most large normal faults have relatively steep dips at the surface but tend to flatten out with depth. Large normal faults are mainly associated with *divergent plate boundaries*. Examples include normal faults that bound the oceanic rift valleys along the global ridge system as well as the rift valleys that form on land. The latter is represented by the East African Rift—a location where continental crust is thinning as it is being stretched and pulled apart.

Fault-Block Mountains and Horsts and Grabens

In the western United States, large normal faults are associated with structures called **fault-block mountains**. Excellent examples of fault-block mountains are found in the Basin and Range Province, a region that encompasses much of the inland western United States as well as part of Mexico (**Figure 10.16**). Here the crust has been uplifted and elongated, generating more than 200 comparatively small mountain ranges. Averaging about 80 kilometers (50 miles) in length, the ranges rise 900 to 1500 meters (3000 to 5000 feet) above the adjacent down-faulted basins.

The topography of the Basin and Range Province evolved in association with a system of normal faults trending roughly north–south. Movements along these faults produced alternating elevated fault blocks called **horsts** (*horst* = hill) and down-dropped blocks called **grabens** (*graben* = ditch). Horsts form the mountain ranges and are the source of sediments that have accumulated in the basins created by the grabens. As Figure 10.16 illustrates, structures called *half-grabens*, which are tilted fault blocks, also contribute to the alternating topographic highs and lows in the Basin and Range Province.

Also notice in Figure 10.16 that the slopes of many of the large normal faults in the Basin and Range Province decrease with depth and eventually join to form a low-angle, nearly horizontal fault called a **detachment fault**. These faults represent a major boundary between the rocks below, which exhibit ductile deformation, and the rocks above, which exhibit mainly brittle deformation.

"normal" because one would expect gravity to pull a block of rock down an inclined plane (the fault surface). Normal faults are associated with tensional stresses that pull rock units apart, thereby lengthening the crust horizontally and thinning it vertically. This "pulling apart" can be accomplished either by uplift that causes the crust to stretch and break or by tensional forces that elongate the crust.

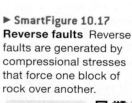

A. Fault Motion

COMPRESSIONAL STRESS

Hanging wall block

Footwall block

B. After Erosion

Hanging wall block

Footwall block

Reverse and Thrust Faults

Dip-slip faults in which the hanging wall block moves up relative to the footwall block are called **reverse faults** (Figure 10.17). A **thrust fault** is a type of reverse fault in which the fault's angle is less than 45 degrees. Both reverse and thrust faults

A. Fault Motion

COMPRESSIONAL STRESS

Hanging wall block

Footwall block

B. After Erosion

Hanging wall block

Footwall block

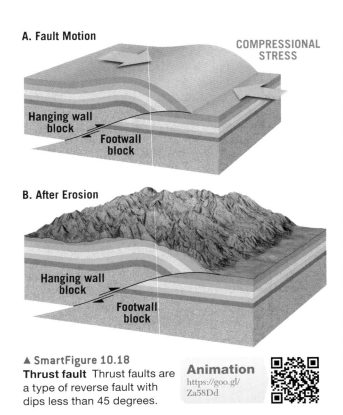

▲ **SmartFigure 10.18**
Thrust fault Thrust faults are a type of reverse fault with dips less than 45 degrees.

Animation
https://goo.gl/Za58Dd

▲ **Figure 10.19**
Montana's Glacier National Park is a classic site of thrust faulting

result from compressional stresses that produce horizontal shortening of the crust.

While reverse faults tend to be small, thrust faults exist at all scales, with some large thrust faults having displacements ranging from tens to hundreds of kilometers. Movement along a thrust fault may cause the hanging wall block to be thrust nearly horizontally over the footwall block, as shown in **Figure 10.18**.

Thrust faulting is common along *convergent plate boundaries* where two landmasses collide. In these settings, compressional forces generate folds as well as thrust faults that thicken and shorten the crust to produce mountainous topography. The Northern Rockies, which include Montana's Glacier National Park, is a classic site of thrust faulting (**Figure 10.19**). Mountain peaks that provide the park's majestic scenery have been carved mainly from a thick sheet of ancient limestones that were displaced, as essentially one unit, over much younger shale deposits. This displaced block of limestone was 6 kilometers (4 miles) thick and slid a distance of about 100 kilometers (60 miles) along the Lewis Thrust Fault.

Even larger thrust faults form the convergent plate boundaries between subducting slabs of oceanic lithosphere and the overlying lithospheric plates. This type of thrust fault is so massive, it is termed a **megathrust fault**. Megathrust faults have produced the majority of Earth's most powerful earthquakes. Because megathrust faults lie beneath the ocean floor, rupture of these faults can displace the overlying water, generating destructive *tsunamis*, including those associated with the 2011 Japan quake and the 2004 Indian Ocean quake near Sumatra (see Section 11.5, page 326).

Strike-Slip Faults

A fault in which the dominant displacement is horizontal and parallel to the *strike* (direction) of the fault surface is called a **strike-slip fault** (**Figure 10.20**). The earliest scientific records of strike-slip faulting were made following surface ruptures that produced large earthquakes. One of the most noteworthy was the great San Francisco

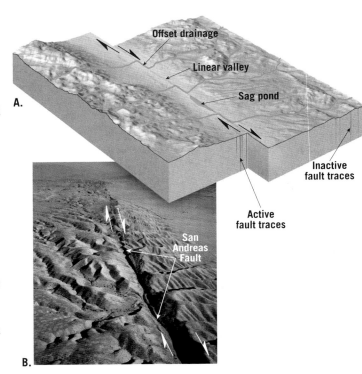

A.
Offset drainage
Linear valley
Sag pond
Inactive fault traces
Active fault traces
San Andreas Fault
B.

◄ **Figure 10.20**
Strike-slip faults
A. The block diagram illustrates the features associated with large strike-slip faults. Notice how the stream channels have been offset by fault movement.
B. Aerial view of the San Andreas Fault.

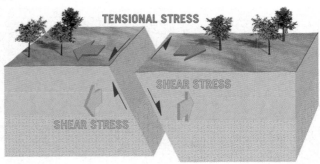

▲ SmartFigure 10.21
Oblique-slip faults Oblique-slip faults exhibit a combination of dip-slip and strike-slip movement.

Animation
https://goo.gl/CrWRGt

earthquake of 1906 (see **GEOgraphics 10.1**). During this strong earthquake, structures such as fences and roads that were built across the San Andreas Fault were displaced as much as 4.7 meters (15 feet).

Right- and Left-Lateral Strike-Slip Faults

Because movement along the San Andreas Fault causes the crustal block on the opposite side of the fault to move to the right as you face the fault, it is called a *right-lateral* strike-slip fault (see Figure 10.20). The Great Glen Fault in Scotland, which exhibits the opposite sense of displacement, is a well-known example of a *left-lateral* strike-slip fault. The total displacement along the Great Glen Fault is estimated to exceed 100 kilometers (60 miles). Also associated with this fault trace are numerous lakes, including Loch Ness, in Scottish folklore, the home of the Loch Ness monster.

Transform Plate Boundaries As discussed in Chapter 2, **transform faults** are strike-slip faults that accommodate motion between two tectonic plates. However, only those strike-slip faults that form *transform plate boundaries* are named *transform faults*. Numerous transform faults cut the oceanic lithosphere and link spreading oceanic ridges (see Figure 2.19, page 52). Others accommodate displacement between continental blocks that slip horizontally past each other. Some of the best-known transform faults include California's San Andreas Fault, New Zealand's Alpine Fault, the Middle East's Dead Sea Fault, and Turkey's North Anatolian Fault. Large transform faults like these accommodate relative displacements of up to several hundred kilometers.

Rather than being a single fracture, most continental transform faults consist of a zone of roughly parallel fractures. While this zone may be up to several kilometers wide, the most recent movement is often along a strand only a few meters wide, which may offset features such as stream channels. Crushed and broken rocks produced during faulting are more easily weathered and eroded, so linear valleys and shallow lakes often identify the existence of a transform fault.

Oblique-Slip Faults

Faults that exhibit both dip-slip and strike-slip movement, called **oblique-slip faults**, are caused by a combination of shearing and tensional or compressional stress (**Figure 10.21**). Nearly all faults have minor components of both dip-slip and strike-slip movement, so defining a fault as oblique requires that both types of slip be significant enough to be observed and measured.

Structures Associated with Faulting

Although all large faults are unique, they often exhibit similar characteristics.

Fault Scarps Vertical displacements along faults can produce long, low cliffs called **fault scarps** (*scarpe* = slope). Fault scarps, such as the one shown in **Figure 10.22**, are usually generated by rapid vertical displacements associated with earthquakes. Occasionally, horizontal movement along a strike-slip fault produces a fault scarp when an area of higher ground is displaced next to lower terrain.

Slickensides The majority of fault surfaces that have been uplifted and exposed by erosion are remnants of past deformation. On some of these fault surfaces, the rocks became highly polished and striated (scratched), or grooved, as the crustal blocks slid past one another. These polished and striated surfaces, called **slickensides** (*sliken* = smooth), provide geologists with evidence for the direction of the most recent displacement along the fault (**Figure 10.23**). By swiping your hand back and forth

▼ Figure 10.22
Fault scarp produced by a 2016 earthquake in New Zealand At this site the amount of vertical uplift exceeded 3 meters (about 20 feet).

▲ Figure 10.23
Slickensides As two rock bodies slide past one another, their fault surfaces often become polished and grooved into *slickensides*.

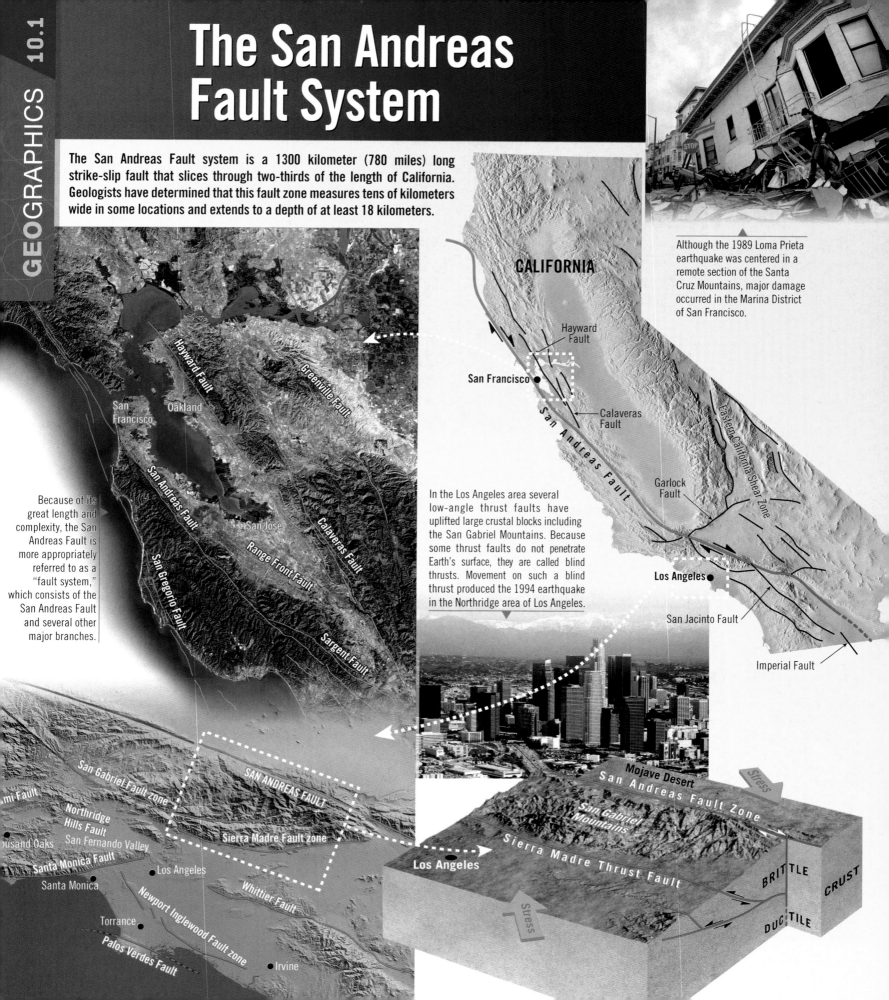

The San Andreas Fault System

The San Andreas Fault system is a 1300 kilometer (780 miles) long strike-slip fault that slices through two-thirds of the length of California. Geologists have determined that this fault zone measures tens of kilometers wide in some locations and extends to a depth of at least 18 kilometers.

Although the 1989 Loma Prieta earthquake was centered in a remote section of the Santa Cruz Mountains, major damage occurred in the Marina District of San Francisco.

CALIFORNIA

Hayward Fault

San Francisco

Calaveras Fault

San Andreas Fault

Garlock Fault

Eastern California Shear Zone

Because of its great length and complexity, the San Andreas Fault is more appropriately referred to as a "fault system," which consists of the San Andreas Fault and several other major branches.

Hayward Fault

San Francisco

Oakland

San Andreas Fault

Greenville Fault

San Jose

Calaveras Fault

Range Front Fault

San Gregorio Fault

Sargent Fault

In the Los Angeles area several low-angle thrust faults have uplifted large crustal blocks including the San Gabriel Mountains. Because some thrust faults do not penetrate Earth's surface, they are called blind thrusts. Movement on such a blind thrust produced the 1994 earthquake in the Northridge area of Los Angeles.

Los Angeles

San Jacinto Fault

Imperial Fault

mi Fault

Northridge Hills Fault

San Gabriel Fault zone

SAN ANDREAS FAULT

ousand Oaks

San Fernando Valley

Sierra Madre Fault zone

Santa Monica Fault

Santa Monica

Los Angeles

Newport Inglewood Fault zone

Whittier Fault

Torrance

Palos Verdes Fault

Irvine

Mojave Desert

San Andreas Fault Zone

Stress

San Gabriel Mountains

Los Angeles

Sierra Madre Thrust Fault

Stress

BRITTLE

DUCTILE

CRUST

► Figure 10.24
Nearly parallel joints in Entrada sandstone, Arches National Park, Utah These two sets of joints formed through regional upwarping, which caused the rigid Entrada sandstone to fracture. The orientation of the dominant joint set is shown with arrows. The other joint set is oriented perpendicular (90 degrees) to the dominant set.

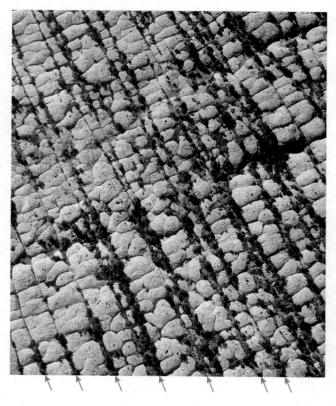

► Figure 10.25
How joints influence the development of landforms Weathering along a prominent set of joints in what is now Arches National Park produced a topography called *fins*.

Geologist's Sketch

Dominant joint set   Stretching

Upwarping

Development of fins at Arches N.P.

Upwarping creates parallel joints in hard Entrada S.S

Weathering widens joints

With time these joints are enlarged to generate fins

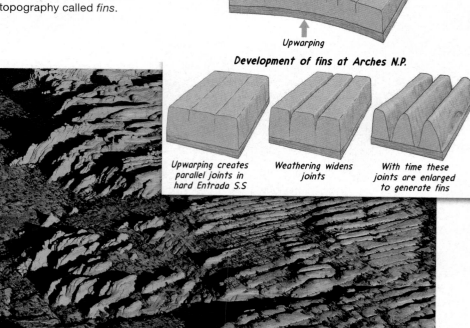

across these surfaces, you should be able to identify the direction of relative movement.

Fault Breccia   Recall that near Earth's surface, rock behaves like a brittle solid. Consequently, movement along a large fault zone can fracture and pulverize the intervening rock (see Figure 8.25A, page 250). The result is a loosely coherent rock called **fault breccia**, composed of broken and crushed rock fragments. Displacement along California's San Andreas Fault has created a massive zone of fault breccia that is up to 3 kilometers (2 miles) wide in some places.

Joints

Joints are among the most common geologic structures and can be found in nearly all rock outcrops. As mentioned earlier, **joints** differ from faults in that no appreciable displacement has occurred along the fracture. Although some joints have random orientations, most occur in roughly parallel groups (**Figure 10.24**).

Most joints are produced when rocks in Earth's outermost crust are deformed by tensional stresses that stretch the rock layer and cause it to fail by brittle fracture. One way in which rock layers are stretched is in response to relatively subtle regional upwarping and downwarping of Earth's crust. This is illustrated in **Figure 10.25**, which shows how regional upwarping of the Entrada sandstone in Arches National Park generated a set of nearly parallel joints. Joints profoundly affect the weathering of bedrock by allowing ion-rich water to penetrate to depth and start the weathering process long before the rock is exposed at Earth's surface. As a result, joints strongly influence how landforms develop. The iconic landscape in Figure 10.25 formed as weathering enlarged joints, creating long, narrow walls called *fins*. These narrow rock walls, in turn, are the setting in which differential weathering created Arches National Park's famed arches (see Figure 6.1, page 174).

Not all joints are produced by regional tensional stresses, however. Recall that *columnar joints* form when igneous rocks cool and develop shrinkage fractures that produce elongated, pillar-like columns (see Figure 4.31, page 126). Another example of jointing can be seen in Figure 6.6, page 177, which illustrates *sheeting*, a process that produces joints that form parallel to the surface of large igneous masses.

Because jointing weakens rocks, highly jointed rocks present a risk to construction projects, including bridges, highways, and dams. On June 5, 1976, the Teton Dam in Idaho failed, taking 14 lives and causing nearly $1 billion in property damage. This earthen dam, constructed of easily eroded clays and silts, was situated on highly fractured volcanic rocks. Although attempts were made to fill the voids in the jointed rock, water gradually penetrated the subsurface fractures and undermined the dam's foundation. Eventually the moving water cut a tunnel into the easily erodible clays and silts. Within minutes the dam failed, sending a 20-meter- (65-foot-) high wall of water down the Teton and Snake Rivers.

However, in some settings, jointed rocks provide economic benefits. For example, highly jointed rocks are often a significant source of groundwater. In addition, some of the world's largest and most important mineral deposits are located along joint systems. Hydrothermal solutions (mineralized fluids) can migrate into fractured host rocks and precipitate economically significant amounts of copper, silver, gold, zinc, lead, and uranium.

CONCEPT CHECKS 10.3

1. Contrast the movements that occur along normal and reverse faults. What type of stress is responsible for each kind of fault?

2. How are reverse faults different from thrust faults? In what way are they similar?

Concept Checker
https://goo.gl/h6QvU5

3. Describe the relative movement along a strike-slip fault.

4. How are joints different from faults?

10.4 Mapping Geologic Structures

Explain how strike and dip are measured and what these measurements tell geologists about the orientations of rock structures located below Earth's surface.

By studying the orientations of faults, folds, and tilted sedimentary strata, geologists can reconstruct the geologic structures that are hidden beneath the surface and determine the nature of the forces that generated these structures. In this way, the complex events of Earth's geologic history are unraveled.

Frequently, geologic structures are so large that only a small portion of a structure is visible from any particular vantage point. In many situations, the bedrock is concealed by vegetation or buried by recent sedimentation. Consequently, reconstruction must be done using data gathered from a limited number of *outcrops*—sites where bedrock is exposed at the surface.

Despite these challenges, a number of mapping techniques enable geologists to infer the shape and orientation of rock structures below the surface. In recent years, this work has been aided by advances in aerial photography and satellite imagery. In addition, seismic reflection profiling (see Chapter 13) and drill holes provide data on the composition and orientation of rock structures that lie at depth.

Measuring Strike and Dip

Because they are usually deposited in horizontal layers, sedimentary rock units are most useful when studying rock deformation. If sedimentary strata are horizontal, geologists infer that the area is undisturbed structurally. Inclined, bent, or broken rock layers indicate that a period of deformation occurred following deposition.

In the previous section, you learned that geologists use *strike* (trend) and *dip* (inclination) to describe the orientation of planar rock features such as sedimentary strata and fault surfaces (**Figure 10.26**).

Recall that *strike* is the compass bearing (direction) of the line produced by the intersection of an inclined rock layer (or fault) with a horizontal surface. Strike is expressed as an angle relative to north—for example, N10°E (north 10° east) means the line of strike is 10 degrees

to the east of north. The strike of the rock units illustrated in Figure 10.26 is north 42° east (N42°E).

Dip is the angle of inclination of the surface of a rock unit or fault, measured from a horizontal surface. Dip includes both an angle of inclination and a direction toward which the rock is inclined. In Figure 10.26 the dip angle of the rock layer is 30 degrees. The direction of dip is always at a 90-degree angle to the strike. To illustrate, hold a closed book at an angle to a tabletop. The upper edge of your book represents the strike. Regardless of the direction you point the book, the direction of dip of the book is always at 90 degrees, or at a right angle, to the strike.

Geologic Maps and Block Diagrams

When doing field research, geologists measure the strike and dip of sedimentary strata at as many outcrops as is practical. These data are then plotted on

▼ Figure 10.26
Strike and dip of rock layers

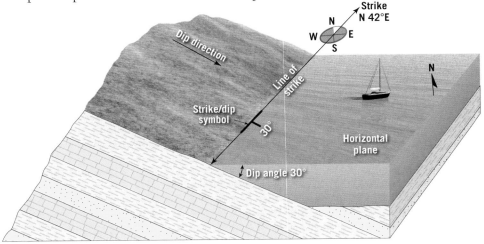

▶ **Figure 10.27**
Mapping geologic structures By establishing the strike and dip of outcropping sedimentary beds and placing symbols on a map, geologists can infer the orientation of the rock structures below the surface.

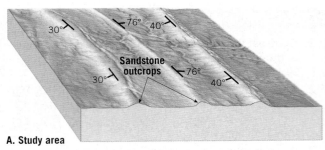

A. Study area

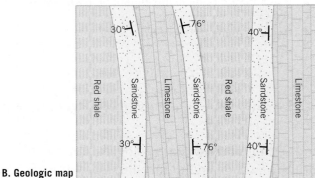

B. Geologic map

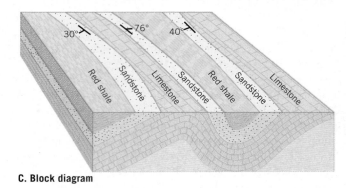

C. Block diagram

a topographic map or an aerial photograph, using T-shaped symbols, as shown in **Figure 10.27A**. The long line shows the strike direction, the short line indicates the direction of the dip, and the dip angle is noted—for example, 30°.

This information, as well as notes about the composition of each rock unit, is used to prepare a graphical depiction of the study area, called a *geologic map*. A **geologic map** is a representation of Earth's surface, as viewed from above, that shows the locations and orientations of the rock units that outcrop at the surface. **Figure 10.27B** shows a simplified geologic map of a hypothetical study area.

Geologists also construct *block diagrams* to illustrate both the *map view* and usually two *cross-sectional views* of the area. Hence, a **block diagram** is a three-dimensional view of a portion of Earth's crust that allows you to visualize rock layers at the surface as well as underground (**Figure 10.27C**).

Geologic maps and block diagrams are valuable tools used to infer the orientation and shape of buried rock structures. Using this information, geologists can reconstruct the pre-erosional structures and begin to interpret the region's geologic history. Geologic maps and block diagrams are also used in the exploration of mineral resources and for the assessment of hazards, such as the potential for groundwater contamination from a proposed waste disposal site.

CONCEPT CHECKS 10.4

1. Describe how the dip of an inclined rock structure is oriented relative to its strike.

2. Why are geologic maps useful?

 **Concept Checker**
https://goo.gl/GxGNQS

10 CONCEPTS IN REVIEW

Crustal Deformation

10.1 How Rocks Deform

Describe the three types of differential stress and name the type of plate boundary most commonly associated with each.

Key Terms:

deformation	confining pressure	strain
outcrop	differential stress	elastic deformation
geologic structure	compressional stress	brittle deformation
stress	tensional stress	ductile deformation
	shear	

- *Geologic (rock) structures* include folds, faults, and joints. They develop when rocks deform by bending or breaking.
- *Stress* is the force that drives rock deformation. When stress acts equally from all directions, we call it *confining pressure*. When the stress is greatest in one direction, we call it *differential stress*. There

are three main types of differential stress: *compressional, tensional,* and *shear*.

- A rock's strength is its ability to resist permanent *deformation*. When the stresses on a rock exceed its strength, the rock deforms, usually by folding or faulting.
- *Elastic deformation* is caused by a temporary stretching of the chemical bonds in a rock. When the stress releases, the rock returns to its original shape. When the rock's strength is exceeded, bonds break, and the rock deforms in either a brittle manner (by fracturing) or in ductile fashion (changing shape).
- Whether a rock deforms in a brittle or ductile manner depends on temperature and confining pressure. The hotter a rock, the more likely it is to experience *ductile deformation*. Greater confining pressure makes a rock stronger and less likely to break. Thus, rock deformation tends to be brittle in the shallow crust and ductile at deeper levels.

- Whether deformation is brittle or ductile also depends on the type of rock. For example, shale is weaker than granite, so it is more prone to ductile deformation. If a rock is forced to deform more quickly than can be accommodated by ductile deformation, it will break.

Q Did the quarter on the right in the accompanying image experience brittle or ductile deformation?

10.2 Folds: Rock Structures Formed by Ductile Deformation

List and describe five common folded structures.

Key Terms:
fold
anticline

syncline
dome
basin

monocline

- *Folds* are wave-like undulations in layered rocks that develop through ductile deformation caused by compressional stresses—usually along convergent plate boundaries.
- *Anticlines* usually arise by upfolding, or arching, of sedimentary layers, whereas *synclines* are downfolds, or troughs. Anticlines and synclines may be symmetrical, asymmetrical, overturned, or recumbent.
- A fold is said to plunge when its axis penetrates the ground at an angle. This results in a V-shaped outcrop pattern.
- *Domes* and *basins* are large bowl or saucer-shaped folds that produce roughly circular outcrop patterns. When eroded, a dome's oldest beds are in the middle, and a basin's oldest beds are around the margin.

- *Monoclines* are large steplike folds in otherwise horizontal strata that develop when beds drape over a vertical offset produced by subsurface faulting.

Q What name is given to the geologic structure shown in the accompanying image?

10.3 Faults and Joints: Rock Structures Formed by Brittle Deformation

Sketch and briefly describe the relative motion of rock bodies located on opposite sides of normal, reverse, and thrust faults as well as both types of strike-slip faults.

Key Terms:
fault
strike
dip
dip-slip fault
hanging wall block
footwall block
normal fault

fault-block mountain
horst
graben
detachment fault
reverse fault
thrust fault
megathrust fault
strike-slip fault

transform fault
oblique-slip fault
fault scarp
slickenside
fault breccia
joint

- A *fault* is a fracture along which one rock body slides past another.
- Faults in which movement is primarily parallel to the *dip* of the fault surface are called *dip-slip faults*. Dip-slip faults are classified as *normal faults* if the *hanging wall* moves down relative to the *footwall* and *reverse faults* if the hanging wall moves up relative to the footwall. Reverse faults with low dip angles are called *thrust faults*.
- Divergent plate boundaries, such as the East African Rift, are regions where tensional stresses and normal faulting are dominant and cause the crust to be stretched and thinned.
- Convergent plate boundaries, areas where compressional stresses are dominant, are associated with reverse and thrust faults that shorten the crust horizontally while thickening it vertically.

- The movement along a *strike-slip fault* is along the *strike* (trend) of the fault trace. *Transform faults* are large strike-slip faults that serve as tectonic boundaries between lithospheric plates that are sliding past one another.
- *Oblique-slip faults* display characteristics of both dip-slip and strike-slip faults.
- *Joints* are fractures in rocks along which no appreciable movement has occurred.

Q What type of rock structure is shown in each of the accompanying photos: faults or joints? Explain how you arrived at your answer.

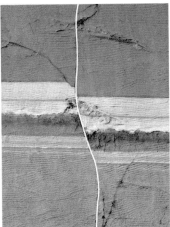

A.

B.

10.4 Mapping Geologic Structures

Explain how strike and dip are measured and what these measurements tell geologists about the orientations of rock structures located below Earth's surface.

Key Terms: geologic map   block diagram

- Strike and dip are measurements of the orientation of planar rock features such as sedimentary bedding and fault surfaces. Strike is the compass direction of the line at the intersection between a horizontal plane and a geologic surface. Dip is the angle of inclination between the horizontal plane and a line perpendicular to the strike.

- *Geologic maps* and *block diagrams* are valuable tools for establishing the shape and orientation of subterranean rock structures—an important step for resource extraction and hazard assessment. Geologic maps and block diagrams are also used to reconstruct the geologic history of a region.

GIVE IT SOME THOUGHT

1. Which rock—granite or mica schist—is more likely to fold or flow rather than fracture when subjected to differential stress? Explain.

2. Refer to the accompanying diagrams to answer the following questions:

   a. What type of dip-slip fault is shown in Diagram 1? Were the dominant forces during faulting tensional, compressional, or shear?

   b. What type of dip-slip fault is shown in Diagram 2? Were the dominant forces during faulting tensional, compressional, or shear?

   c. Match the correct pair of arrows in Diagram 3 to the faults in Diagrams 1 and 2.

| Diagram 1 (cross section) | Diagram 2 (cross section) | Diagram 3 |

3. Refer to the accompanying photo to answer the following:

   a. The white line shows the approximate location of a fault that displaced these furrows created by a plow. What type of fault caused the offset shown?

   b. Is this a right-lateral or left-lateral fault? Explain.

4. With which type of plate boundary does normal faulting predominate? Thrust faulting? Strike-slip faulting?

5. Write a brief statement describing each of the accompanying photos, using terms from the following list: strike-slip, dip-slip, normal, reverse, right-lateral, left-lateral.

A.

B.

   a. What type of deformation is exhibited—ductile or brittle?

   b. Did this deformation most likely occur near Earth's surface or at great depth?

6. The accompanying simplified geologic map shows the strike and dip of outcropping sedimentary beds. From these measurements, describe the rock structure that is most likely beneath Earth's surface, using terms from the following list: anticline, syncline, dome, basin, symmetrical, asymmetrical.

Map view

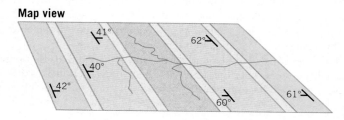

7. Examine the diagrams depicting the idealized geologic structure of the East African Rift and Montana's Glacier National Park.

   a. Characterize the type of faulting found at each location and identify the type of differential stress that resulted in these landforms.

   b. Along which type of plate boundary did each of these structures form?

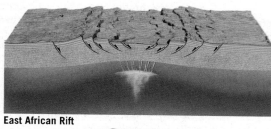

East African Rift

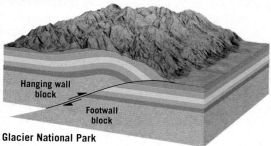

Glacier National Park

EYE ON EARTH

1. **This image features a large geologic structure that outcrops in Death Valley National Park, California.**

   a. What name would you give to this geologic structure?

   b. Based on this image, would you describe this fold as symmetrical or asymmetrical?

   c. Do these rock units mainly display ductile deformation or brittle deformation?

2. **This satellite image shows an area just south of the Tien Shan Mountains, China. The distinctive red, green, and cream-colored bands are sedimentary rock layers that were tilted by compressional forces as various landmasses collided with southern Asia. Also visible in this image is the Piqiang Fault, which runs roughly perpendicular to the colored sedimentary layer for about 70 kilometers (40 miles).**

   a. What type of fault is the Piqiang Fault?

   b. Is it a right-lateral or left-lateral fault?

Piqiang Fault

DATA ANALYSIS

Measuring the Movement of Land

As tectonic plates shift, Earth's crust deforms along faults. Scientists measure these movements to investigate the physical processes responsible for fault movement and to develop early warnings for potential earthquakes.

https://goo.gl/eD226M

ACTIVITIES

Go to the USGS Earthquake Hazards Program, at http://earthquake.usgs.gov. Click on Monitoring, choose Crustal Deformation Monitoring, and select Fault Creep, Borehole Strain, and Tiltmeter Monitoring Measurements.

1. **Click on Monitoring Instruments. Which instruments are used to measure land movement? Describe what each instrument measures.**

Go back to the Crustal Deformation Monitoring page and click on Real-Time GPS Data. Click on GPS Real-Time PPP Displacements to display the data regions. Hover over each regions to see its name. Click on the Northern California region. Each triangle represents a data collection location. Click on a location with recent data (less than 30 seconds old) and select View Processed GPS Data.

2. **What are the latitude and longitude for this location? Over what time period are data available?**

3. **Using the North American Fixed graphs, determine the total northward movement from the beginning to the end of the data record. Also determine the total eastward movement. (It may help to use the trend line showing the average motion.) What direction is this location moving toward?**

4. **What is the range (upward plus downward) of upward motions experienced by this location? What is the average (using the trend line) upward motion?**

5. **Based on the table below the graph, in which direction is the station you selected moving?**

6. **Repeat 2–5 for a station in the Southern California region.**

Go back to Real-Time GPS Data and click on the GPS Velocities option above the map.

7. **Examine the velocities in Northern California, and Southern California. Do most or all of the stations move in the way you determined above? What are the overall directions of movement for these regions?**

8. **Examine the Northern Rockies region. How would you describe the motion of this region?**

Mastering Geology

Looking for additional review and test prep materials? Visit pearson.com/mastering/geology to enhance your understanding of this chapter's content by accessing a variety of resources, including **Self-Study Quizzes, Geoscience Animations, SmartFigure Tutorials, *Project Condor* Videos, Mobile Field Trips, Dynamic Study Modules,** and an optional **Pearson eText.**

Constructed on Shaky Ground: Keeping the Built Landscape Safe in Earthquake Zones

A 2018 U.S. Geological Survey report noted that dozens of high-rise buildings in San Francisco are at greater-than-expected risk of damage from a major earthquake. Based on the report, the *New York Times* ran a story about the buildings and their design flaw. These structures were erected between the 1960s and the early 1990s—fairly modern, by most standards. People expected the buildings to be very safe in the event of a major earthquake. What went wrong?

▲ An earthquake-damaged building in Kathmandu, Nepal.

News about the flawed buildings wasn't actually new—or a secret. The 1994 Northridge, California, earthquake revealed vulnerabilities in steel-frame buildings that were constructed using similar techniques. Building codes in most earthquake-prone areas—San Francisco included—have been updated multiple times since the Northridge quake to correct the issue. But the codes only apply to new construction. Retrofitting older buildings in earthquake-prone areas has long been neglected, mostly due to cost.

Bolting foundations and reinforcing walls in a single-family home costs about $5000, while retrofitting high-rises can cost hundreds of thousands of dollars.

▶ San Francisco has numerous buildings deemed by the U.S. Geological Survey to be vulnerable in a major earthquake, including the pyramid-shaped Transamerica Building shown near the center of this image.

11

Earthquakes and Earthquake Hazards

On April 25, 2015, Nepal experienced its worst natural disaster in over 80 years, when a magnitude 7.8 earthquake struck this mountainous region of South Asia. Geologists had anticipated a significant earthquake for decades because of Nepal's location high in the Himalayas, above the collisional boundary where India is being thrust into Asia. The quake killed nearly 9000 people and resulted in injuries to more than 22,000 others. The shallow nature of the 50-second quake resulted in widespread destruction and triggered avalanches on Mt. Everest; it claimed the lives of 19 people, including international hikers and Sherpa guides. Numerous landslides throughout the region blocked roads and delayed relief efforts. Nepal's capital, Kathmandu, reportedly shifted 3 meters (10 feet) to the south during the course of the event.

11.1 What Is an Earthquake?

Sketch and describe the mechanism that generates most earthquakes.

An **earthquake** is ground shaking caused by the sudden and rapid movement of one block of rock slipping past another along fractures in Earth's crust, called **faults**. Most of the time, a fault is locked because the confining pressure exerted by the overlying crust is enormous, causing these fractures in the crust to be "squeezed shut." An earthquake triggers when differential stress builds to such a level that it overcomes the frictional forces holding the rock bodies together. The location where the rock slippage begins is called the **hypocenter**, or **focus**. The point on Earth's surface directly above the hypocenter is called the **epicenter** (Figure 11.1).

Large earthquakes release huge amounts of stored-up energy as **seismic waves** that travel through the lithosphere and Earth's interior. The waves spread out in all directions from the site of the earthquake, in a manner similar to how waves move outward from a stone thrown into a calm pond. The energy carried by these waves causes the material that transmits them to vibrate, which causes the "shaking" that occurs during the event. Although seismic energy dissipates rapidly as it moves away from the source (or hypocenter) of an earthquake, sensitive instruments can detect earthquakes even when they occur on the opposite side of Earth.

While thousands of earthquakes occur around the world every day, only about 15 strong earthquakes (magnitude 7 or greater) are recorded each year, and many of them occur in remote regions. The rest are small events that don't cause damage or much notice. The occasional large earthquakes that are triggered near major population centers are among the most destructive natural events on Earth. The shaking of the ground, coupled with the liquefaction of soils, wreaks havoc on buildings, roadways, and other structures. In addition, a quake occurring in a populated area can rupture power and gas lines, causing numerous fires. In the famous 1906 San Francisco earthquake, much of the damage was caused by fires that became uncontrollable when broken water mains left firefighters with only trickles of water (Figure 11.2).

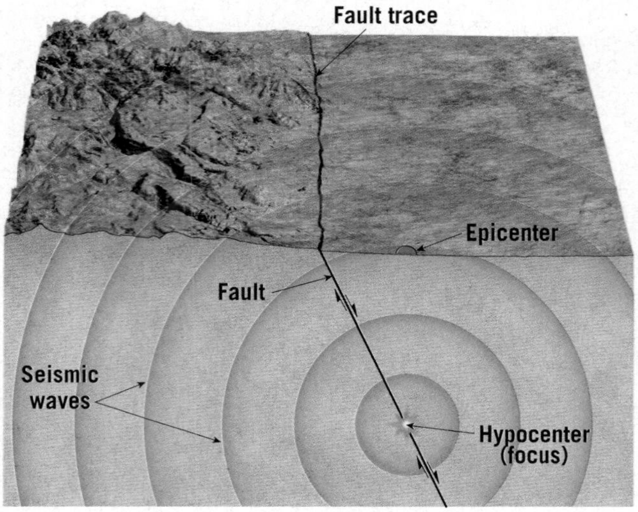

Discovering the Causes of Earthquakes

The energy released by volcanic eruptions, massive landslides, and meteorite impacts can generate earthquake-like waves, but these events are usually weak. What mechanism produces a destructive earthquake?

► Figure 11.1
An earthquake's hypocenter and epicenter The *hypocenter* is the zone at depth where the initial displacement occurs. The *epicenter* is the surface location directly above the hypocenter.

San Francisco in flames following the 1906 quake. Broken water mains left firefighters without water.

Fire triggered when a gas line ruptured during the Northridge earthquake in southern California in 1994.

▲ Figure 11.2
Earthquakes can trigger fires

The mechanism of earthquake generation eluded geologists until H. F. Reid conducted a landmark study following the 1906 San Francisco earthquake. This earthquake was accompanied by horizontal surface displacements of several meters along the northern portion of the San Andreas Fault (Figure 11.3). Field studies determined that during this single earthquake, the Pacific plate lurched as much as 9.7 meters (32 feet) northward, past the adjacent North American plate. To better visualize this, imagine standing on one side of the fault and watching a person on the other side suddenly slide horizontally about 30 feet to your right.

An earthquake is ground shaking caused by the sudden and rapid movement of one block of rock slipping past another along fractures in Earth's crust, called faults.

What Reid concluded from his investigations is illustrated in **Figure 11.4**. Over tens to hundreds of years, differential stress slowly bends the crustal rocks on both sides of a fault. This is much like a person bending a limber wooden stick, as shown in Figure 11.4A,B. Frictional resistance keeps the fault from rupturing and slipping. (Friction inhibits slippage and is enhanced by irregularities that occur along the fault surface.) At some point, the stress along the fault overcomes this frictional resistance, and slip initiates. Slippage allows the deformed (bent) rock to "snap back" to its original, stress-free, shape; a series of earthquake waves radiate outward as the rock slides (see Figure 11.4C,D). Reid termed this "springing back" **elastic rebound** because the rock behaves elastically, much as a stretched rubber band does when it is released.

Foreshocks and Aftershocks

Small earthquakes called **foreshocks** often, but not always, precede major earthquakes by days or, in some cases, several years. Monitoring of foreshocks to predict forthcoming earthquakes has been attempted with only limited success. Strong earth quakes are followed by numerous earthquakes of lesser magnitude, called **aftershocks**. When a fault ruptures and produces a large earthquake, the movement along the fault surface stresses and deforms the surrounding rock, which cause the aftershocks. In other words, aftershocks are mainly caused by the stress added to the rocks on both sides of the fault plane by the initial earthquake. Because most aftershocks occur along the same fault surface as the main quake, they help geologists determine the size of the area where slippage occurred.

▼ Figure 11.3
Displacement of human-made structures along a fault

This fence was offset 2.5 meters (8.5 feet) during the 1906 San Francisco earthquake.

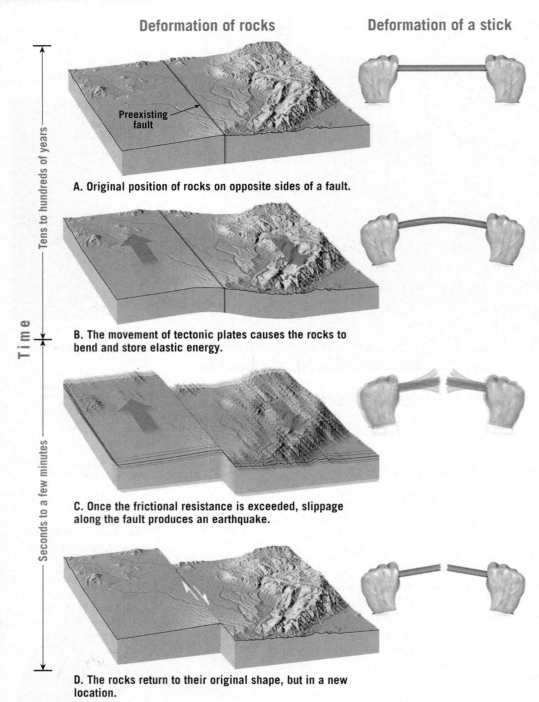

Deformation of rocks   **Deformation of a stick**

Time

Tens to hundreds of years

Preexisting fault

A. Original position of rocks on opposite sides of a fault.

B. The movement of tectonic plates causes the rocks to bend and store elastic energy.

Seconds to a few minutes

C. Once the frictional resistance is exceeded, slippage along the fault produces an earthquake.

D. The rocks return to their original shape, but in a new location.

▲ SmartFigure 11.4
Elastic rebound

Tutorial
https://goo.gl/
7Bx2jN

weakened structures. For example, in north-western Armenia in 1988, where many people lived in large apartment buildings constructed of brick and concrete slabs, a moderate earthquake of magnitude 6.9 weakened many structures, and a strong aftershock of magnitude 5.8 completed the demolition.

Because faults rarely occur in isolation, an earthquake on one fault may trigger an earthquake on a nearby fault that has been accumulating stress for many decades or centuries. The difference between this secondary earthquake and an aftershock is that the stress added by the first earthquake accounts for only a minor component of the stress released when the adjacent fault ruptures.

Plate Tectonics and Large Earthquakes

Recall that large slabs of Earth's lithospheric plates are continually grinding past one another. As these mobile plates interact with neighboring plates, they strain and deform the rocks along their margins. Faults associated with convergent and transform plate boundaries are the source of most large earthquakes.

Convergent Plate Boundaries Along convergent boundaries where one continent is colliding with another, the resulting compressional forces slice Earth's crust along numerous large *thrust faults* (**Figure 11.5**). The 2015 Nepal earthquake is one example of an earthquake generated along a thrust fault. The epicenter of the quake was located about 80 kilometers (50 miles) north of Kathmandu, where the Indian plate is advancing into the Eurasian plate at a rate of 4.5 centimeters (about 2 inches) per year, driving the uplift of the Himalayas.

When convergence entails the subduction of oceanic lithosphere under another plate, the area of contact between the two plates forms an extensive fault zone, called a **megathrust fault**, that can be several thousand kilometers long (see Figure 11.5). Along subduction zones these megathrust faults remain locked for decades or even centuries. As the subducting plate slowly descends, it drags and bends the leading edge of the overlying plate, sometimes producing a bulge on the ocean floor (see Figure 11.26A, page 326). Once the frictional forces between the two stuck plates are exceeded, the overriding plate snaps back to its original shape. This snapping back generates an earthquake of a magnitude that largely depends on the size of the zone of slippage.

Aftershocks gradually diminish in frequency and intensity over a period of several months following an earthquake. In about a month following the M 7.0 earthquake that devastated the island of Haiti in 2010, the U.S. Geological Survey detected nearly 60 aftershocks with magnitudes of 4.5 or greater. The two largest aftershocks had magnitudes of 6.0 and 5.9, both large enough to inflict further damage. Hundreds of minor tremors were also felt.

Although aftershocks are weaker than the main earthquake, they often trigger the destruction of already

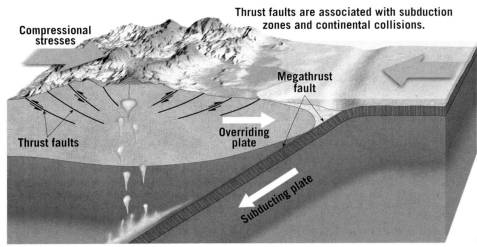

Compressional stresses

Thrust faults are associated with subduction zones and continental collisions.

Megathrust fault

Thrust faults

Overriding plate

Subducting plate

▲ **Figure 11.5**
Large thrust and megathrust faults form at convergent plate boundaries A convergent plate boundary is a site where one plate is subducting beneath another. Megathrust faults that separate these plates generate most of Earth's largest earthquakes.

Megathrust faults have produced the majority of Earth's most powerful and destructive earthquakes, including the 2011 Japan quake (M 9.0), the 2004 Indian Ocean (Sumatra) quake (M 9.1), the 1964 Alaska quake (M 9.2), and the largest earthquake yet recorded, the 1960 Chile quake (M 9.5).

Transform Plate Boundaries   Faults in which the dominant displacement is horizontal and parallel to the direction of the fault trace (the line where the fault intersects Earth's surface) are called *strike-slip faults*.

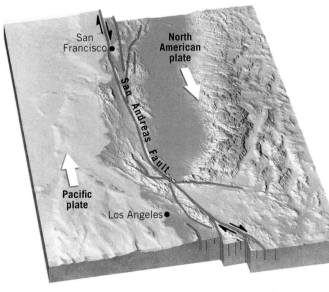

San Francisco

North American plate

San Andreas Fault

Pacific plate

Los Angeles

▲ **Figure 11.6**
The San Andreas Fault is a transform plate boundary The San Andreas Fault is a large fault system separating the Pacific plate from the North American plate. This type of large strike-slip fault, called a transform fault, can generate destructive earthquakes.

Recall from Chapter 2 that *transform plate boundaries*, or simply *transform faults*, accommodate this type of motion between two tectonic plates. For example, the San Andreas Fault is a large transform fault that lies between the North American plate and the Pacific plate (**Figure 11.6**). Most large transform faults are not perfectly straight or continuous; instead, they consist of numerous branches and smaller fractures that display kinks and offsets (see Figure 11.6). Earthquakes can occur along any of these branches.

Fault Rupture and Propagation

By studying earthquakes around the globe, geologists learned that with large faults, displacement occurs along discrete fault segments that often behave differently from one another. Some sections of the San Andreas, for example, exhibit slow, gradual displacement known as **fault creep** and produce little seismic shaking. Other segments slip at relatively closely spaced intervals, producing numerous small to moderate earthquakes. Still other segments remain locked and store elastic energy for up to a few hundred years before they break loose. Ruptures on segments that have been locked for a hundred years or longer usually result in major earthquakes.

Geologists also discovered that slippage along large faults, such as the San Andreas, does not occur instantaneously (**Figure 11.7**). The initial slip occurs at the hypocenter and propagates (travels) along the fault surface. As each section slips, it puts strain on the next section, causing it to slip as well. Slippage

Rupture begins at the earthquake hypocenter.

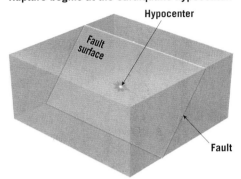

Hypocenter

Fault surface

Fault

Rupture propagates across fault surface at 2 to 3 kilometers per second.

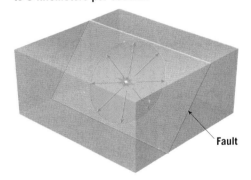

Fault

Rupture halts when it reaches parts of the fault surface that are not strained enough to rupture.

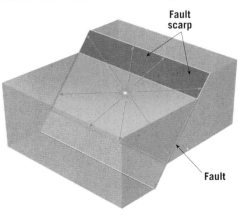

Fault scarp

Fault

▲ **Figure 11.7**
Fault propagation

propagates at 2 to 4 kilometers per second—faster than a rifle shot. Rupture of a 100-kilometer (60-mile) fault segment takes about 30 seconds, and rupture of a 300-kilometer (200-mile) segment takes about 90 seconds. As rupturing progresses, it can slow down, speed up, or even jump to a nearby fault segment. Earthquake waves are generated at every point along the fault as that portion of the fault begins to slip.

CONCEPT CHECKS 11.1

1. What is an earthquake?
2. How are hypocenters and epicenters related?
3. Explain what is meant by *elastic rebound*.

Concept Checker
https://goo.gl/2Qnz28

11.2 Seismology: The Study of Earthquake Waves

Compare and contrast the types of seismic waves and describe how a seismograph works.

The study of earthquake waves called **seismology**, dates back to attempts made in China almost 2000 years ago to determine the direction from which these waves originated. The earliest known instrument, invented by Zhang Heng, was a large hollow jar containing a weight suspended from the top (**Figure 11.8**). The suspended weight (similar to a clock pendulum) was connected to the jaws of several large dragon figurines that encircled the container. The jaws of each dragon held a metal ball. When earthquake waves reached the instrument, the relative motion between the suspended mass and the jar would dislodge some of the metal balls into the waiting mouths of frog figurines directly below.

Ball dropping

▲ Figure 11.8
Ancient Chinese seismograph During an Earth tremor, a dragon located in the direction of the main vibrations would drop a ball into the mouth of a frog below.

Instruments That Record Earthquakes

In principle, modern **seismographs**, or **seismometers**, are similar to the instruments used in ancient China. A seismograph has a weight freely suspended from a support that is securely attached to bedrock (**Figure 11.9**). When vibrations from an earthquake reach the instrument, the **inertia** of the weight keeps it relatively stationary, while Earth and the support move. Inertia can be simply described by this statement: *Objects at rest tend to stay at rest, and objects in motion tend to remain in motion, unless acted upon by an outside force.* You have experienced inertia when you have tried to stop your automobile quickly and your body has continued to move forward.

Most seismographs are designed to amplify ground motion in order to detect very weak earthquakes or a great earthquake that has occurred in another part of the world. Instruments used in earthquake-prone areas are designed to withstand the violent shaking that can occur near a quake's epicenter.

Seismic Waves

The records obtained from seismographs, called **seismograms**, provide useful information about the nature of seismic waves. Seismograms reveal that two main types

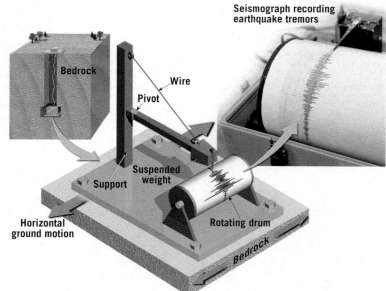

▲ SmartFigure 11.9
Principle of the seismograph The inertia of the suspended weight tends to keep it motionless, while the recording drum, which is anchored to bedrock, vibrates in response to seismic waves. The stationary weight provides a reference point from which to measure the amount of displacement occurring as a seismic wave passes through the ground.

Animation
https://goo.gl/cgR39h

of seismic waves are generated by the slippage of a rock mass. One of these wave types, **body waves**, travel through Earth's interior. The other type, **surface waves**, travel in the rock layers just below Earth's surface (**Figure 11.10**).

Body Waves Body waves are further divided into two types—**primary waves**, or **P waves**, and **secondary waves**, or **S waves**—and are identified by their mode of travel through intervening materials. P waves are "push/pull" waves; they momentarily push (compress) and pull (stretch) rocks in the direction the waves are traveling (**Figure 11.11A**). This wave motion is similar to that generated by striking a drum, which moves air back and forth to create sound. Solids, liquids, and gases resist stresses that change their volume when compressed and, therefore, elastically spring back once the stress is removed. Therefore, P waves can travel through all these materials.

By contrast, S waves "shake" material at right angles to their direction of travel. This can be illustrated by fastening one end of a rope and shaking the other end, as shown in **Figure 11.11B**. Unlike P waves, which temporarily change the *volume* of intervening material by alternately squeezing and stretching it, S waves change the *shape* of the material that transmits them. Because fluids (gases and liquids) do not resist stresses that cause changes in shape—meaning fluids do not return to their original shape once the stress is removed—liquids and gases do not transmit S waves.

Surface Waves There are two types of surface waves. One type causes Earth's surface and anything resting on it to move up and down, much as ocean swells toss a ship (**Figure 11.12A**). The second type of surface wave causes Earth's surface to move from side to side. This motion is

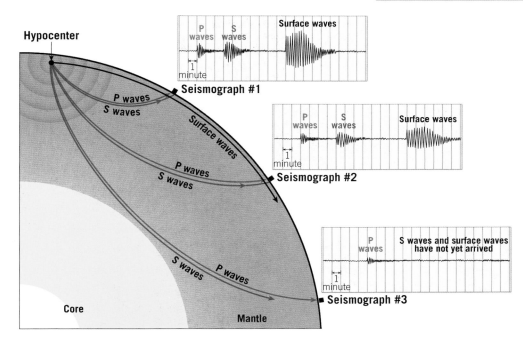

Hypocenter

Seismograph #1

P waves S waves Surface waves

P waves
S waves

Surface waves

P waves
S waves

Seismograph #2

P waves S waves Surface waves

S waves P waves

P waves S waves and surface waves have not yet arrived

Seismograph #3

Core   Mantle

particularly damaging to the foundations of structures (**Figure 11.12B**).

Comparing the Speed and Size of Seismic Waves

By examining the seismogram shown in **Figure 11.13**, you can see that another major difference among the types of seismic waves is their speed of travel. P waves are the first to arrive at a recording station, then S waves, and finally surface waves. Generally, in any solid Earth material, P waves travel about 70 percent faster than S waves, and S waves are roughly 10 percent faster than surface waves.

In addition to the velocity differences in the waves, notice in Figure 11.13 that their seismograph recordings differ in height, or *amplitude*, which reflects the amount of shaking they cause. S waves have slightly greater amplitudes than P waves, and surface waves exhibit even

▲ **SmartFigure 11.10**
Body waves (P and S waves) versus surface waves P and S waves travel through Earth's interior, while surface waves travel in the layer directly below the surface. P waves are the first to arrive at a seismic station, followed by S waves and then surface waves.

Tutorial
https://goo.gl/1ZzWQK

A. As illustrated by a toy Slinky, P waves alternately compress and expand the material through which they pass.

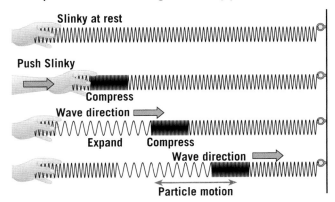

Slinky at rest

Push Slinky
Compress

Wave direction
Expand Compress
Wave direction
Particle motion

B. S waves cause material to oscillate at right angles to the direction of wave motion.

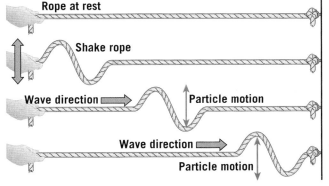

Rope at rest

Shake rope

Wave direction Particle motion

Wave direction
Particle motion

◄ **Figure 11.11**
The characteristic motion of P and S waves During a strong earthquake, ground shaking consists of a combination of various kinds of seismic waves.

▶ SmartFigure 11.12
Two types of surface waves

Animation
https://goo.gl/
cV7eTx

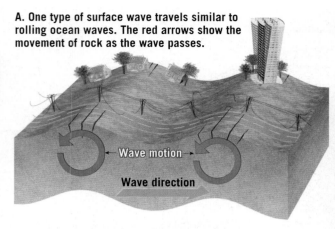

A. One type of surface wave travels similar to rolling ocean waves. The red arrows show the movement of rock as the wave passes.

Wave motion

Wave direction

B. A second type of surface wave moves the ground from side to side and can be particularly damaging to building foundations.

Side-to-side wave motion

Wave direction

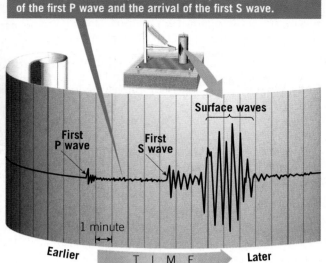

Note the time interval (about 5 minutes) between the arrival of the first P wave and the arrival of the first S wave.

Surface waves

First P wave   First S wave

1 minute

Earlier   T I M E   Later

▲ Figure 11.13
Idealized seismogram

greater amplitudes. Surface waves also retain their maximum amplitude longer than P and S waves. As a result, surface waves tend to cause greater ground shaking and, hence, greater property damage, than either P or S waves.

CONCEPT CHECKS 11.2

1. Briefly describe how a seismograph works.

2. List the major differences between P, S, and surface waves.

3. Which of the three basic types of seismic wave is likely to cause the greatest destruction to buildings?

Concept Checker
https://goo.gl/P3YuNb

11.3 Locating the Source of an Earthquake

Explain how seismographs locate the epicenter of an earthquake.

When seismologists analyze an earthquake, they first determine its *epicenter*, the point on Earth's surface directly above the hypocenter, or focus (see Figure 11.1). One method used for locating an earthquake's epicenter relies on the fact that P waves travel faster than S waves.

The traveling waves are analogous to two racing automobiles, one faster than the other. The first P wave, like the faster automobile, always wins the race, arriving ahead of the first S wave. The greater the length of the race, the greater the difference in their arrival times at the finish line (the seismic station). Therefore, the longer the interval between the arrival of the first P wave and the arrival of the first S wave, the greater the distance to the epicenter. **Figure 11.14** shows three simplified seismograms for the same earthquake. Based on the P–S interval, which city—New York, Nome, or Mexico City—is farthest from the epicenter?

The system for locating earthquake epicenters was developed by using seismograms from earthquakes whose epicenters could be easily pinpointed based on

physical evidence. From these seismograms, travel–time graphs were constructed (**Figure 11.15**). Using the sample seismogram for New York in Figure 11.14 and the travel–time curve in Figure 11.15, we can determine the distance separating the recording station from the earthquake in three steps:

1. Using the seismogram for New York, we determine that the time interval between the arrival of the first P wave and the arrival of the first S wave is 5 minutes.

2. Using the travel–time graph, we find the location where the vertical separation between P and S curves is equal to the P–S time interval (5 minutes in this example).

THREE SEISMOGRAMS

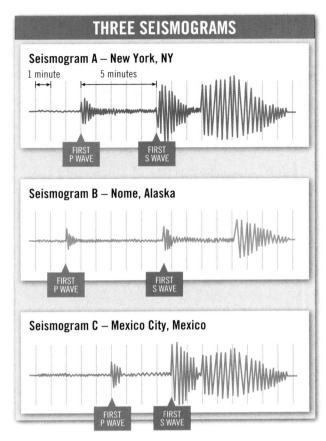

Seismogram A – New York, NY

1 minute | 5 minutes

FIRST P WAVE   FIRST S WAVE

Seismogram B – Nome, Alaska

FIRST P WAVE   FIRST S WAVE

Seismogram C – Mexico City, Mexico

FIRST P WAVE   FIRST S WAVE

▲ Figure 11.14
Seismograms of the same earthquake recorded at three different locations

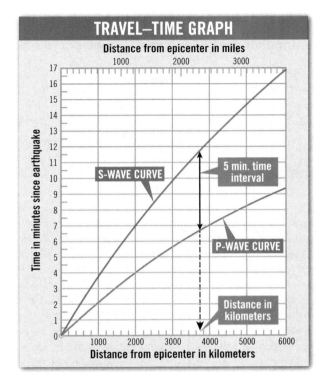

TRAVEL–TIME GRAPH

Distance from epicenter in miles

Time in minutes since earthquake

S-WAVE CURVE

5 min. time interval

P-WAVE CURVE

Distance in kilometers

Distance from epicenter in kilometers

▲ Figure 11.15
Travel–time graph A travel–time graph is used to determine the distance to an earthquake's epicenter. The difference in arrival times of the first P and S waves in the example shown is 5 minutes.

3. From the position in step 2, we draw a vertical line to the horizontal axes and read the distance to the epicenter.

Using these steps, we determine that the earthquake occurred 3700 kilometers (2300 miles) from the recording instrument in New York City.

Now we know the *distance*, but what about *direction*? The epicenter could be in any direction from the seismic station. Using a method called *triangulation*, we can determine the location of an epicenter if we know the distance to it from two or more additional seismic stations (**Figure 11.16**). On a map or globe, we draw a circle around each seismic station with a radius equal to the distance from that station to the epicenter. The point where the three circles intersect is the approximate epicenter of the quake.

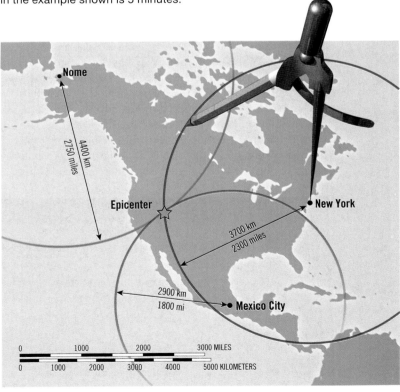

Nome

4400 km
2750 miles

Epicenter

New York

3700 km
2300 miles

2900 km
1800 mi

Mexico City

0   1000   2000   3000 MILES
0   1000   2000   3000   4000   5000 KILOMETERS

▲ Figure 11.16
Triangulation to locate an earthquake This method involves using the distance obtained from three or more seismic stations to establish the location of an earthquake.

CONCEPT CHECKS 11.3

1. What information does a travel–time graph provide?

2. Briefly describe the *triangulation* method used to locate the epicenter of an earthquake.

Concept Checker
https://goo.gl/qj3FLa

11.4 Determining the Size of an Earthquake

Distinguish between intensity scales and magnitude scales.

Seismologists use a variety of methods to determine two fundamentally different measures that describe the size of an earthquake: *intensity* and *magnitude*. An **intensity** scale uses observed property damage to estimate the amount of ground shaking at a particular location. **Magnitude** scales, which were developed more recently, use data from seismographs to estimate the amount of energy released at an earthquake's source.

Intensity Scales

Until the mid-1800s, historical records provided the only accounts of the severity of earthquake shaking and destruction. Perhaps the first attempt to scientifically describe the aftermath of an earthquake came following the great Italian earthquake of 1857. By systematically mapping effects of the earthquake, a measure of the intensity of ground shaking was established. The map generated by this study used lines to connect places of equal damage and hence equal ground shaking. Using this technique, zones of intensity were identified, with the zone of highest intensity representing the location of maximum ground shaking, which often (but not always) surrounds the earthquake epicenter.

In 1902, Giuseppe Mercalli developed a more reliable intensity scale, which is still used today in a modified form. The **Modified Mercalli Intensity scale**, shown in **Table 11.1**, was developed using California buildings as its standard. Based on the 12-point Mercalli Intensity scale, an area in which some well-built wood structures and most masonry buildings are destroyed by an earthquake would be assigned a Roman numeral X.

The U.S. Geological Survey has developed a website called "Did You Feel It," where Internet users experiencing a quake can enter their zip code and answer questions such as "Did objects fall off shelves?" Within a few hours, a Community Internet Intensity Map, like the one in **Figure 11.17** for the 2011 central Virginia earthquake (M 5.8), is generated. As shown in Figure 11.17, shaking strong enough to be felt was reported from Maine to Florida, an area occupied by one-third of the U.S. population. Several national landmarks were damaged, including the Washington Monument and the National Cathedral, located about 130 kilometers (80 miles) away from the epicenter.

Magnitude Scales

To more accurately compare earthquakes around the globe, scientists searched for a way to describe the energy released by earthquakes that did not rely on factors such as building practices, which vary considerably from one part of the world to another.

Richter-Like Magnitude Scales   In 1935 Charles Richter of the California Institute of Technology developed the first magnitude scale to use seismic recordings. Seismologists have since modified Richter's work and developed other *Richter-like magnitude scales*, which we will consider next.

As shown in **Figure 11.18** (top), these magnitude scales are calculated by measuring the amplitude of the largest seismic wave (usually an S wave) recorded on a seismogram. Because seismic waves weaken as the distance between the hypocenter and the seismograph increases, methods were developed to compensate for the decrease in wave amplitude with increasing distance. Theoretically, as long as equivalent instruments are used, monitoring stations at different locations will obtain the same magnitude for each recorded

TABLE 11.1   Modified Mercalli Intensity Scale

I	Not felt except by a very few under especially favorable circumstances.
II	Felt only by a few persons at rest, especially on upper floors of buildings.
III	Felt quite noticeably indoors, especially on upper floors of buildings, but many people do not recognize it as an earthquake.
IV	During the day felt indoors by many, outdoors by few. Sensation like heavy truck striking building.
V	Felt by nearly everyone, many awakened from sleep. Disturbances of trees, poles, and other tall objects sometimes noticed.
VI	Felt by all; many frightened and run outdoors. Some heavy furniture moved; few instances of fallen plaster or damaged chimneys. Damage slight.
VII	Everybody runs outdoors. Damage negligible in buildings of good design and construction; slight to moderate in well-built ordinary structures; considerable in poorly built or badly designed structures.
VIII	Damage slight in specially designed structures; considerable, with partial collapse, in ordinary buildings; great in poorly built structures (falling chimneys, factory stacks, columns, monuments, walls).
IX	Damage considerable in specially designed structures. Buildings shifted off foundations. Ground cracked conspicuously.
X	Most masonry and frame structures destroyed. Some well-built wooden structures destroyed. Ground badly cracked.
XI	Few, if any, masonry structures remain standing. Bridges destroyed. Broad fissures in ground.
XII	Damage total. Waves seen on ground surfaces. Objects thrown upward into air.

The green dots on the map show locations of people who reported feeling earthquakes of similar magnitude. The difference is attributable to the rigidity of the bedrock.

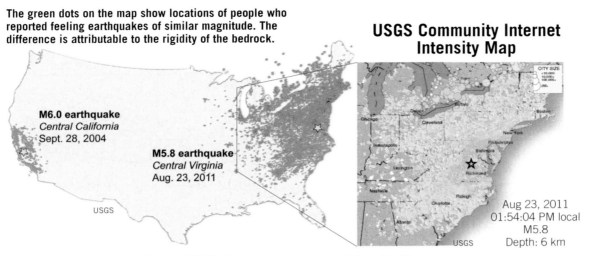

USGS Community Internet Intensity Map

M6.0 earthquake
Central California
Sept. 28, 2004

M5.8 earthquake
Central Virginia
Aug. 23, 2011

USGS

Aug 23, 2011
01:54:04 PM local
M5.8
Depth: 6 km
USGS

◄ SmartFigure 11.17
USGS Community Internet Intensity Map Maps like this one are prepared using data collected on the Internet from people responding to questions such as "Did objects fall off shelves?"

Tutorial
https://goo.gl/
2JvUvh

Key for USGS Community Internet Intensity Map

INTENSITY	I	II-III	IV	V	VI	VII	VIII	IX	X
SHAKING	Not felt	Weak	Light	Moderate	Strong	Very strong	Severe	Violent	Extreme
DAMAGE	None	None	None	Very light	Light	Moderate	Moderate/Heavy	Heavy	Very heavy

◄ Figure 11.18
Determining the magnitude of an earthquake using seismic waves

1. Measure the height (amplitude) of the largest wave on the seismogram (23 mm) and plot it on the amplitude scale (right).

2. Determine the distance to the earthquake using the time interval separating the arrival of the first P wave and the arrival of the first S wave (24 seconds) and plot it on the distance scale (left).

24 sec.   Amplitude
           23 mm
Seismograph record   P   S
Time sec. 0 10 20

30 mm
20
10

S-P, sec.	Distance, km	Magnitude, M_L	Amplitude, mm
50	500		100
40	400	6	50
30	300		20
	200	5	10
20		4	5
10	100	3	2
8	60		1
6	40	2	0.5
4			
	20	1	1.2
2			0.1
	0.5	0	

3. Draw a line connecting the two plots and read the Richter magnitude (M_L 5) from the magnitude scale (center).

earthquake. In practice, however, different recording stations often obtain slightly different magnitudes for the same earthquake—a result of the variations in the rock types through which the waves travel.

Earthquakes vary enormously in strength, and great earthquakes produce wave amplitudes thousands of times larger than those generated by weak tremors. To accommodate this wide variation, magnitude scales use a *logarithmic scale* to express magnitude, in which a *10-fold* increase in wave amplitude corresponds to an increase of 1 on the magnitude scale. Thus, the intensity of ground shaking for a magnitude 5 earthquake is 10 times greater than that produced by an earthquake having a magnitude of 4 (**Figure 11.19**).

In addition, each unit on these magnitude scales equates to roughly a *32-fold increase in the energy released*. Thus, an earthquake with magnitude 6.5 releases 32 times more energy than one with magnitude 5.5 and roughly 1000 times (32×32) more energy than a magnitude 4.5 quake. A major earthquake with a magnitude of

Magnitude vs. Ground Motion and Energy

Difference in Magnitude	Difference in Ground Motion (amplitude)	Difference in Energy Release (approximate)
4.0	10,000 times	1,000,000 times
3.0	1000 times	32,000 times
2.0	100 times	1000 times
1.0	10 times	32 times
0.5	3.2 times	5.5 times
0.1	1.3 times	1.4 times

◄ Figure 11.19
Magnitude versus ground motion and energy released An earthquake that is 1 unit of magnitude stronger than another (such as M 6 versus M 5) produces seismic waves that have a maximum amplitude 10 times greater and releases about 32 times more energy than the weaker quake.

Frequency and Energy Released by Earthquakes of Different Magnitudes

Magnitude (Mw)	Average Per Year	Description	Examples	Energy Release (equivalent kilograms of explosive)
<1	**Largest recorded earthquakes–** destruction over vast area, massive loss of life possible	Chile, 1960 (M 9.5); Alaska, 1964 (M 9.2); Sumatra, 2004 (M 9.1); Japan, 2011 (M 9.0)	56,000,000,000,000	
1	**Great earthquakes–** severe economic impact, large loss of life	Chile, 2010 (M 8.8); Chiapas, Mexico, 2017 (M 8.2); Mexico City, 1980 (M 8.1)	1,800,000,000,000	
15	**Major earthquakes–** damage ($ billions), loss of life	San Francisco, California, 1906 (M 7.9); Haiti, 2012 (M 7.0); Nepal, 2015 (M 7.8); New Zealand, 2016 (M 7.8)	56,000,000,000	
134	**Strong earthquakes–** can be destructive in populated areas, loss of life	Kobe, Japan, 1995 (M 6.9); Loma Prieta, California, 1989 (M 6.9); Northridge, California, 1994 (M 6.7)	1,800,000,000	
1319	**Moderate earthquakes–** property damage to poorly constructed buildings	Mineral, Virginia, 2011 (M 5.8); Northern New York, 1994 (M 5.8); East of Oklahoma City, Oklahoma, 2011 (M 5.6)	56,000,000	
13,000	**Light earthquakes–** noticeable shaking of items indoors, some property damage	Western Minnesota, 1975 (M 4.6); Arkansas, 2011 (M 4.7)	1,800,000	
130,000	**Minor earthquakes–** felt by humans, very light property damage, if any	New Jersey, 2009 (M 3.0); Maine, 2006 (M 3.8); Texas, 2015 (M 3.6)	56,000	
1,300,000	**Very minor earthquakes–** felt by humans, no property damage		1,800	
Unknown	**Very minor earthquakes–** generally not felt by humans, but may be recorded		56	

Data from USGS

▲ **Figure 11.20**
Frequency of earthquakes with various moment magnitudes

8.5 releases millions of times more energy than the smallest earthquakes felt by humans (**Figure 11.20**).

The convenience of describing the size of an earthquake by a single number that can be calculated quickly from seismograms makes magnitude scales a powerful tool. Despite their usefulness, Richter-like magnitude scales are not adequate for describing very large earthquakes. For example, the 1906 San Francisco earthquake and the 1964 Alaska earthquake have roughly the same magnitudes on these scales. However, based on the relative size of the affected areas and the associated tectonic changes, the Alaska earthquake released considerably more energy than the San Francisco quake. Thus, these magnitude scales are considered *saturated* for major earthquakes because the scales cannot distinguish among them. Despite this shortcoming, Richter-like scales are still used because they allow for quick calculations.

Moment Magnitude   For measuring medium and large earthquakes, seismologists now favor a newer scale called **moment magnitude** (M_W) which estimates the total energy released during an earthquake. Moment magnitude is calculated by determining the average amount of slip on the fault plane, the area of the fault surface that slipped, and the strength of the failed rocks.

Moment magnitude can also be calculated by modeling data obtained from seismograms. The results are converted to a magnitude number, similar to other magnitude scales. Thus, each unit increase on the moment magnitude scale equates to roughly a 32-fold increase in the energy released.

Because the moment magnitude scale is better than other magnitude scales at estimating the relative sizes of very large earthquakes, seismologists have used it to recalculate the magnitudes of older strong earthquakes (see Figure 11.20). For example, the 1964 Alaska earthquake, given a Richter magnitude of 8.3, has since been recalculated using the moment magnitude scale, resulting in an upgrade to M_W 9.2. Conversely, the 1906 San Francisco earthquake's Richter magnitude of 8.3 was downgraded to M_W 7.9. The strongest earthquake on record is the 1960 Chilean megathrust earthquake, at a moment magnitude of 9.5.

CONCEPT CHECKS 11.4

1. What does the Modified Mercalli Intensity scale tell us about an earthquake?

2. How much more energy does a magnitude 7.0 earthquake release than a magnitude 6.0 earthquake?

3. Why is the moment magnitude scale favored over the Richter-like magnitude scales for large earthquakes?

Concept Checker
https://goo.gl/v9LgRz

11.5 Earthquake Destruction

List and describe the major destructive forces that earthquake vibrations can trigger.

On January 12, 2010, an estimated 100,000 to 316,000 people lost their lives when a magnitude 7.0 earthquake struck the small Caribbean nation of Haiti, the poorest country in the Western Hemisphere. In addition to the staggering death toll, more than more than 280,000 houses and commercial buildings were destroyed or damaged, mainly in and around the capital city of Port-au-Prince. Because of the quake's shallow depth, ground shaking was extreme for an event of this magnitude.

Other factors that contributed to the Port-au-Prince disaster included the city being built on unconsolidated sediments, which is quite susceptible to ground shaking during an earthquake. More importantly, inadequate or nonexistent building codes meant that numerous buildings, including the Presidential Palace, collapsed or were severely damaged.

Destruction from Seismic Vibrations

During an earthquake, the region within 20 to 50 kilometers (12 to 30 miles) of the epicenter tends to experience roughly the same degree of ground shaking, and beyond that limit, vibrations usually diminish rapidly. However, earthquakes that occur in the stable continental interior, such as the New Madrid, Missouri, earthquakes of 1811–1812, are generally felt over a much larger area than those in earthquake-prone areas such as California.

As the energy released by an earthquake travels along Earth's surface, it causes the ground to vibrate in a complex manner involving up-and-down as well as side-to-side motion. The amount of damage to human-made structures attributable to the vibrations depends on several factors, including (1) the *intensity* and *duration of the vibrations*, (2) the *construction practices of the region*, and (3) the *nature of the material on which structures rest*.

Intensity and Duration Strong earthquakes (those with the greatest intensity) exhibit longer durations than those of modest magnitudes. Recall that the initial slip begins at the hypocenter and propagates (travels) along the fault surface at 2 to 4 kilometers per second. Thus, the rupture of a 100-kilometer (60-mile) fault segment takes about 30 seconds, while the rupture of a 200-kilometer (120-mile) segment lasts about a minute. Therefore, the shaking produced by a large earthquake lasts longer and has the potential to produce more structural damage than that of a smaller quake.

For example, the 1964 Alaska earthquake—the most violent earthquake ever recorded in North America—had a magnitude of 9.2 and was felt for about 3 to 4 minutes. By comparison, the strong vibrations of the 1989 Loma Prieta earthquake, which had a modest magnitude of 6.9, lasted less than 15 seconds.

Construction Practices Structural engineers have discovered that buildings constructed of blocks and bricks that are *not reinforced* with steel rods are the most serious safety threats to property and life during an earthquake. By contrast, the Transamerica Pyramid, San Francisco's tallest building, which is buttressed with reinforcing steel rods at each level, survived the 1989 Loma Prieta earthquake without any structural damage—despite the fact that it swayed back and forth by at least 1 foot. A recent U.S. Geological Survey report indicates that the Transamerica Pyramid will be able to withstand even stronger quakes.

Wood-frame residential buildings common in the United States are relatively quake resistant because they flex but are not prone to collapse during an earthquake. They can, however, sustain structural damage, particularly if they have brick facades or brick chimneys. Unfortunately, most of the structures in the developing world are constructed of concrete blocks, unreinforced concrete slabs, and bricks made of dried mud—a primary reason the death toll in poor and developing countries such as Haiti, Nepal, and Mexico is usually higher compared to earthquake events of similar size in the United States (**Figure 11.21**). For example, nearly two-thirds of the buildings that collapsed in Mexico City's 2017

▼ Figure 11.21
Destruction caused by 2017 Chiapas, Mexico, earthquake

► Figure 11.22
Effects of liquefaction on buildings

These tilted buildings rested on unconsolidated sediment that behaved like quicksand during the 1964 Niigata, Japan, earthquake.

extensive damage caused by moderate earthquakes. The most recent Italian event occurred in 2016 and displaced about 25,000 people. The destruction was most evident in buildings constructed of nonreinforced blocks and bricks.

A discussion of how buildings can be designed to withstand a destructive earthquake is found in Section 11.7.

Amplification of Seismic Waves and Liquefaction

The impact of earthquake shaking may vary considerably in an area, depending upon the nature of the ground underlying built structures. Water-saturated, loose sediments, for example, *amplify the up-and-down seismic vibrations more than solid bedrock*. This was exemplified by the 1989 Loma Prieta earthquake. Although the earthquake was centered 100 kilometers (60 miles) to the south in a remote section of California's Santa Cruz Mountains, major damage occurred in San Francisco's Marina District, which was built on landfill consisting of water-saturated sand and rubble. The 1989 earthquake caused the unconsolidated, waterlogged ground in this upscale neighborhood to be transformed into a substance that acts like a fluid—a phenomenon called **liquefaction**.

When liquefaction occurs, the ground becomes mobile and unable to support buildings, and underground storage tanks and sewer lines may literally float toward the surface (**Figure 11.22**). In the Marina District, foundations failed, and geysers of sand and water shot from the ground, evidence that liquefaction had occurred (**Figure 11.23**). More than 70 structures in the Marina District either collapsed or were severely damaged.

The most tragic result of the 1989 earthquake in the San Francisco Bay Area was the collapse of a stretch of a double-decked highway in Oakland known as the Cypress Viaduct. The collapsed section of the roadway sat on soft, muddy marshland that caused the earthquake's up-and-down shock waves to ripple the ground more severely than a nearby stretch of road built on compacted sand and gravel (**Figure 11.24**). The intense ground motion shattered concrete support columns and sent the upper deck crashing onto the lower roadway. In an instant, 42 people traveling in cars perished, and many more were injured.

▼ Figure 11.23
Liquefaction These sand volcanoes, produced by the Christchurch, New Zealand, earthquake of 2011, formed when "geysers" of sand and water shot from the ground, an indication that liquefaction occurred.

earthquake were built using a construction method that is now prohibited in the United States, Chile, and New Zealand, according to a team of structural engineers at Stanford University. Similarly, mountainous regions in central Italy with small historic towns have suffered

Sand volcanoes

Landslides and Ground Subsidence

The greatest earthquake-related damage is often caused by landslides and ground subsidence triggered by vibrations. This was the case when the magnitude 7.8 earthquake struck the steep Himalaya Mountains of central Nepal in 2015. The earthquake unleashed more than 10,000 landsides that blocked rivers and damaged roads, houses, and other vital infrastructure across the country. (Some estimates place the landside total much higher.) The earthquake also triggered the Mount Everest slide

Well-packed soil

Loose, saturated, sandy material

A. Before earthquake

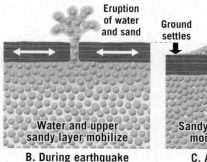

Eruption of water and sand

Water and upper sandy layer mobilize

B. During earthquake

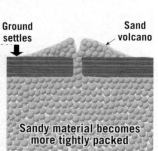

Ground settles

Sand volcano

Sandy material becomes more tightly packed

C. After earthquake

that claimed the lives of 19 hikers and Sherpas. The landslides did not end when the shaking stopped. The mountainous terrain, which was severely weakened by the quake, continued to slide following periods of heavy rains and during aftershocks that followed the main quake.

The largest and most devastating landslide submerged the Nepalese village of Langtang under a huge mass of ice and rock (**Figure 11.25**). What began as an ice and snow avalanche gathered rock debris as a 500-meter-high (over 1600-foot-high) section of the mountain slope plummeted onto the valley floor. This slide displaced 2 million cubic meters of debris and essentially buried or destroyed nearly all the houses in the village, killing nearly half of the villagers and at least a few foreign trekkers who were staying in this popular hiking destination.

Fire

More than a century ago, San Francisco was the economic center of the western United States, largely because of gold and silver mining. Then, at dawn on April 18, 1906, a violent earthquake struck, triggering an enormous firestorm (see Figure 11.2). Much of the city was reduced to ashes and ruins. It is estimated that 3000 people died and more than half of the city's 400,000 residents were left homeless.

The historic San Francisco earthquake reminds us of the formidable threat of fire, which started when the quake severed gas and electrical lines. The initial ground shaking broke the city's water lines into hundreds of disconnected pieces, which made controlling the fires virtually impossible. The fires, which raged out of control for 3 days, were finally contained when expensive houses along Van Ness Avenue were dynamited to provide a fire break, similar to the strategy used to fight forest fires.

While few deaths were attributed to the San Francisco fires, other earthquake-initiated fires have been more destructive, claiming many more lives. For example, the 1923 earthquake in Japan triggered an estimated 250 fires, devastating the city of Yokohama and destroying more than half the homes in Tokyo. More than 100,000 deaths were attributed to the fires, which were driven by unusually high winds.

Tsunamis

Major undersea earthquakes may set in motion a series of large ocean waves that are known by the Japanese name **tsunami** ("harbor wave"). Most tsunamis are generated by displacement along a megathrust fault that suddenly lifts a large slab of seafloor (**Figure 11.26**). Once generated, a tsunami resembles a series of ripples formed when a pebble is dropped into a pond. In contrast to ripples, however, tsunamis advance across the ocean at amazing speeds, about 800 kilometers (500 miles) per hour—equivalent to the cruising speed of a commercial airliner. Despite this striking characteristic, a tsunami in the open ocean can pass undetected because its height (amplitude) is usually less than 1 meter (3 feet), and the distance

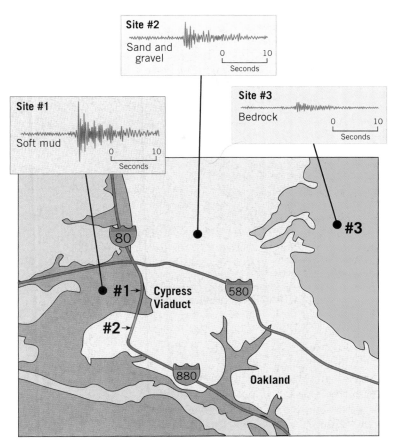

◄ **Figure 11.24**
Collapse of the double-decked section of I-880 The collapsed section of this roadway was built on soft mud where the earthquake's up-and-down shock waves rippled the ground more severely than at a nearby stretch built on compacted sand and gravel.

▶ **Figure 11.25**
Landslide triggered by a quake buried the tiny village of Langtang, Nepal

A. Ground view

B. Aerial view

▼ **SmartFigure 11.26**
How a tsunami is generated by displacement of the ocean floor during an earthquake The speed of a tsunami wave correlates with ocean depth. In deep water, these waves can advance at speeds exceeding 800 kilometers (500 miles) per hour. When they enter coastal waters and begin to "feel bottom," they slow down and grow in height. They are still very fast moving, with a speed of 50 kilometers (30 miles) per hour at a depth of 20 meters (65 feet). The size and spacing of the swells in this figure are not to scale.

Tutorial
https://goo.gl/
T4ApaZ

separating wave crests ranges from 100 to 700 kilometers (60 to 425 miles). However, upon entering shallow coastal waters, these destructive waves "feel bottom" and slow due to friction, causing the water to pile up (see Figure 11.26). A few exceptional tsunamis have approached 20 meters (65 feet) in height. As the crest of a tsunami approaches the shore, it appears as a rapid rise in sea level with a turbulent and chaotic surface; it usually does not resemble a breaking wave (**Figure 11.27**).

The first warning of an approaching tsunami is often the rapid withdrawal of water from beaches, which is the result of the trough of the first large wave preceding the crest. Some inhabitants of the Pacific basin have learned to heed this warning and quickly move to higher ground. Approximately 5 to 30 minutes after the retreat of water, a surge capable of extending several kilometers inland occurs. In a successive fashion, each surge is followed by a rapid oceanward retreat of the sea. Therefore, people experiencing a tsunami should not return to the shore when the first surge of water retreats.

Tsunami Damage from the 2004 Indonesia Earthquake A massive undersea earthquake of M_W 9.1 occurred near the island of Sumatra on December 26, 2004, sending waves of water racing across the Indian Ocean and Bay of Bengal. It was one of the deadliest natural disasters of any kind in modern times, claiming more than 230,000 lives. As water surged several kilometers inland, cars and trucks were flung around like toys in a bathtub, and fishing boats were rammed into homes. In some locations, the backwash of water dragged bodies and huge amounts of debris out to sea.

The destruction was indiscriminate, afflicting luxury resorts as well as poor fishing hamlets along the Indian Ocean. Damage was reported as far away as the coast of Somalia in Africa, 4100 kilometers (2500 miles) west of the earthquake epicenter.

Japan Tsunami Because of Japan's location along the circum-Pacific belt and its extensive coastline, it is especially vulnerable to tsunami destruction. The most powerful earthquake to strike Japan in the age of modern seismology was the 2011 Tohoku earthquake (M_W 9.0). This historic earthquake and devastating tsunami resulted in at least 15,890 deaths, more than 3000 people missing, and 6107 injured. Nearly 400,000 buildings, 56 bridges, and 26 railways were destroyed or severely damaged.

A. Before Earthquake: When a megathrust fault is locked, the subducting plate gradually drags and bends the leading edge of overlying plate, sometimes producing a bulge on the ocean floor.

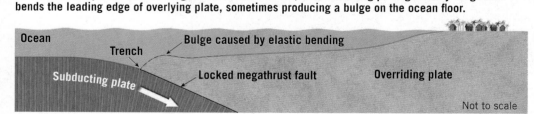

B. During Earthquake: During a megathrust earthquake, the overlying plate kicks seaward and upward, while the seafloor landward of this zone stretches and subsides. These rapid vertical motions trigger a tsunami, which consists of a train of waves.

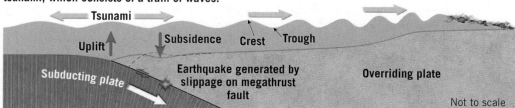

The majority of human casualties and damage after the Tohoku earthquake were caused by a Pacific-wide tsunami that reached a maximum height of about 8.5 meters (28 feet) and traveled inland 10 kilometers (6 miles) in the region of Sendai, Japan (**Figure 11.28**). Based on physical evidence, the *run-up height*, which is the maximum elevation on land that is inundated by water from a tsunami, was almost twice as high—15 meters (50 feet) above sea level. In addition, the tsunami disabled the power supply and cooling mechanisms, which caused the meltdown of three inundated nuclear reactors in Japan's Fukushima Daiichi Nuclear Complex. Across the Pacific in California, Oregon, Peru, and Chile, some loss of life occurred, and several houses, boats, and docks were destroyed.

◀ **SmartFigure 11.27**
Tsunami generated off the coast of Sumatra, 2004

Animation
https://goo.gl/x4nV2d

Tsunami Warning System   In 1946, a large tsunami struck the Hawaiian Islands without warning. A wave more than 15 meters (50 feet) high left several coastal villages in shambles. This destruction motivated the U.S. Coast and Geodetic Survey to establish a tsunami warning system for coastal areas of the Pacific that today includes 26 countries. Seismic observatories throughout the region report large earthquakes to the Tsunami Warning Center in Honolulu. Scientists at the center use deep-sea buoys equipped with pressure sensors to detect energy released by an earthquake. In addition, tidal gauges measure the rise and fall in sea level that accompany tsunamis, and warnings are issued within an hour. Although tsunamis travel very rapidly, there is sufficient time to warn all except those in the areas nearest the epicenter.

◀ **Figure 11.28**
Japan tsunami, March 2011 This tsunami breached a seawall and devastated the city of Miyako, Japan, shortly after a magnitude 9.0 earthquake hit northern Japan in 2011.

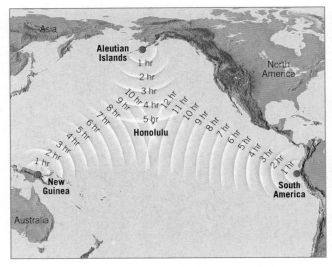

▲ **Figure 11.29**
Tsunami travel times Travel times to Honolulu, Hawaii, from selected locations throughout the Pacific.

For example, a tsunami generated near the Aleutian Islands would take 5 hours to reach Hawaii, and one generated near the coast of Chile would travel 15 hours before reaching the shores of Hawaii (**Figure 11.29**).

CONCEPT CHECKS 11.5

1. List three factors that influence the amount of destruction that seismic vibrations cause to human-made structures.

2. In addition to the destruction created directly by seismic vibrations, list three other types of destruction associated with earthquakes.

Concept Checker
https://goo.gl/uiQ58p

3. What is a tsunami? How are tsunamis generated?

11.6 Where Do Most Destructive Earthquakes Occur?

Locate Earth's major earthquake belts on a world map.

About 95 percent of the energy released by earthquakes originates in a few relatively narrow zones, shown in **Figure 11.30**. These earthquake zones are found mainly along the three types of plate boundaries—convergent, divergent, and transform plate boundaries.

Earthquakes Associated with Plate Boundaries

The zone of greatest seismic activity, called the **circum-Pacific belt**, encompasses the coastal regions of Chile, Central America, Indonesia, Japan, and Alaska, including the Aleutian Islands (see Figure 11.30). Most of the large earthquakes in the circum-Pacific belt occur along convergent plate boundaries, where one plate subducts beneath another at a comparatively low angle. Recall that the contacts between the subducting and overlying plates are called *megathrust faults* (see Figure 11.5). There are more than 40,000 kilometers (25,000 miles) of subduction boundaries where displacement is dominated by thrust faulting. Ruptures occasionally occur along fault segments that are 1000 kilometers (600 miles) long, generating catastrophic earthquakes.

Another major concentration of strong seismic activity, referred to as the *Alpine–Himalayan belt*, runs through the mountainous regions that flank the Mediterranean Sea and extends past the Himalaya Mountains (see Figure 11.30). Tectonic activity in this region is mainly attributed to collisions of the African plate and the Indian subcontinent with the vast Eurasian plate. These plate interactions created both thrust and strike-slip faults.

Transform faults, another type of plate boundary, are also a source of strong earthquakes. Examples include California's San Andreas Fault, New Zealand's Alpine Fault, and Turkey's North Anatolian Fault, all of which have produced deadly earthquakes.

Figure 11.30 shows another continuous earthquake belt that extends thousands of

▼ **Figure 11.30**
Global earthquake belts Distribution of nearly 15,000 earthquakes with magnitudes equal to or greater than 5 for a 10-year period, indicated by red dots.

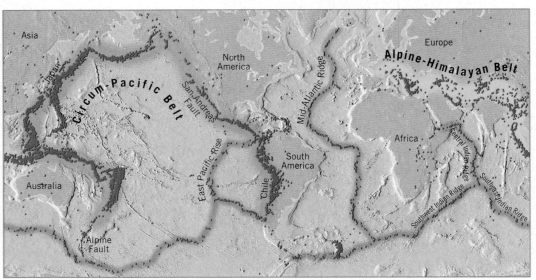

Historic Earthquakes East of the Rockies, 1755–2018				
LOCATION	DATE	INTENSITY	MAGNITUDE*	COMMENTS
1 East of Oklahoma City	2011	VII	5.6	Destroyed 14 homes
2 Mineral, Virginia	2011	VII	5.8	Felt by many
3 Southeastern Illinois	2008	VII	5.4	Occurred along the Wabash Valley Seismic Zone
4 Northeast Kentucky	1980	VII	5.2	Largest earthquake ever recorded in Kentucky
5 Merriman, Nebraska	1964	VII	5.1	Largest earthquake ever recorded in Nebraska
6 Northern New York	1944	VIII	5.8	Left several structures unsafe for occupancy
7 Ossipee Lake, New Hampshire	1947	VII	5.5	Two earthquakes occurred 4 days apart
8 Western Ohio	1937	VIII	5.4	Extensive damage to plaster walls
9 Valentine, Texas	1931	VIII	5.8	Brick buildings were severely damaged
10 Giles County, Virginia	1897	VIII	5.9	Changed the flow of natural springs
11 Charleston, Missouri	1895	VIII	6.6	Structural damage and liquefaction reported
12 Charleston, South Carolina	1886	X	7.3	Caused 60 deaths, destroyed many buildings
13 New Madrid, Missouri	1811-1812	X	7.0-7.7	Three strong earthquakes occurred
14 Cape Ann, Massachusetts	1755	VIII	?	Buildings damaged in Boston

Source: U.S. Geological Survey
*Intensity and magnitudes have been estimated for many of these events.

▲ Figure 11.31
Historical earthquakes east of the Rockies Large earthquakes are uncommon in the middle of continents, far from the places where plates collide or grind past one another or where one plate slides beneath another. Nevertheless, several damaging earthquakes have occurred in the central and eastern United States since colonial times.

kilometers through the world's oceans. This zone coincides with the oceanic ridge system—a divergent plate boundary—which is an area of frequent but weak seismic activity.

Damaging Earthquakes East of the Rockies

When you think "earthquake," you probably imagine the western United States or Japan. However, six major earthquakes and several others that inflicted considerable damage have occurred in the central and eastern United States since colonial times (**Figure 11.31**).

Three of these quakes, which occurred as a cluster, had estimated magnitudes of about 7.0 and destroyed what was then the frontier town of New Madrid, Missouri, located in the Mississippi River valley. It has been estimated that if an earthquake the size of the 1811–1812 New Madrid event were to strike in the same location in the next decade, it would result in casualties in the thousands and damages in the tens of billions of dollars. (See the Chapter 10 opening page for more information on the famous New Madrid earthquakes of 1811–1812.)

The greatest historical earthquake in the eastern states occurred on August 31, 1886, in Charleston, South Carolina. This earthquake resulted in 60 deaths, numerous injuries, and great economic loss. Within 8 minutes, the quake was felt as far away as Chicago and St. Louis, where strong vibrations shook the

upper floors of buildings and caused people to rush outdoors. In Charleston alone, more than 100 buildings were destroyed, and 90 percent of the remaining structures were damaged (**Figure 11.32**).

Earthquakes in the central and eastern United States occur far less frequently than in California, yet history indicates that the East is vulnerable. Further, these shocks east of the Rockies have generally produced

▲ Figure 11.32
Damage to Charleston, South Carolina, caused by the August 31, 1886, earthquake

structural damage over a larger area than earthquakes of similar magnitude in California. This is because the underlying bedrock in the central and eastern United States is older and more rigid. As a result, seismic waves can travel greater distances with less attenuation (loss of strength) than in the western United States.

Earthquakes that occur away from plate boundaries are called *intraplate earthquakes*. Intraplate earthquakes can be caused by a variety of factors. For example, stress can rejuvenate ancient fault systems that formed when crustal fragments collided to generate the continent billions of years ago. In addition, the process called *fracking*, in which a solution is injected into the ground under high pressure to enhance oil and

gas production, has contributed to the recent increase in earthquake activity east of the Rockies. Most of these intraplate earthquakes, fortunately, are weak.

CONCEPT CHECKS 11.6

1. What zone on Earth has the greatest amount of seismic activity?

2. Which type of plate boundary is associated with Earth's most destructive earthquakes?

Concept Checker
https://goo.gl/FmN1vN

11.7 Earthquakes: Predictions, Forecasts, and Mitigation

Compare and contrast the goals of short-range earthquake predictions and long-range forecasts.

Strong earthquakes cause loss of life primarily because they strike without warning. For example, the vibration that unexpectedly shook the San Francisco area in 1989 caused 63 deaths, heavily damaged the Marina District, and caused the collapse of a double-decked section of I-880 in Oakland, California (see Figure 11.24). This level of destruction was the result of an earthquake of moderate magnitude (M_W 6.9). Seismologists warn that other earthquakes of comparable or greater strength can be expected along the San Andreas system, which cuts a nearly 1300-kilometer (800-mile) path through one-third of the western portion of the state (see GEOgraphics 11.1). Can earthquakes be predicted?

Short-Range Predictions

It would be extremely desirable if earthquakes could be predicted in the way in which volcanic eruptions sometimes can. Unfortunately, despite substantial efforts in earthquake-prone countries such as Japan, the United States, China, and Russia, no reliable method exists at present for making short-range earthquake predictions.

To be useful, such a method would have to be both accurate and reliable. That is, *it must have a small range of uncertainty with regard to location and timing, and it must produce few failures or false alarms.* Can you imagine the debate that would precede an order to evacuate a large U.S. city, such as Los Angeles or San Francisco, and the expense of doing so?

Research on short-range prediction has concentrated on monitoring possible **precursors**—events or changes that precede an earthquake and thus might provide a warning. In California, for example, seismologists monitor changes in ground elevation and variations in strain levels near active faults. Other researchers measure changes in groundwater levels, while still others try to predict earthquakes based on an increase in the frequency of the foreshocks that precede some, but not all, earthquakes. To date, neither these nor other avenues of research have yielded a reliable method for short-range earthquake prediction.

Long-Range Forecasts

In contrast to short-range predictions, which aim to predict earthquakes within a time frame of hours or days, long-range forecasts are estimates of the likelihood that an earthquake of a certain magnitude will occur in a given place on a time scale of 30 to 100 years or more. Although long-range forecasts are not as informative as we might like, these data provide important guides for building codes so that buildings, dams, and roadways are constructed to withstand expected levels of ground shaking.

Most long-range forecasting strategies are based on the observation that large faults often break in a cyclical manner, producing similar quakes at roughly similar intervals. In other words, as soon as a section of a fault ruptures, the continuing motions of Earth's plates begin to deform (bend) the rocks until they fail (rupture) once more. To determine the characteristic interval for a given fault, seismologists study the historical and geologic record of earthquakes generated by that fault.

Paleoseismology   Paleoseismology (*paleo* = ancient, *seismos* = shake) is the study of the timing, location, and size of prehistoric earthquakes. Paleoseismology studies are often conducted by digging a trench across a fault zone and looking for evidence of ancient faulting, such as offset sedimentary strata or mud volcanoes. A large

Seismic Risks on the San Andreas Fault System

California's San Andreas Fault runs diagonally from southeast to northwest for nearly 1300 kilometers (800 miles) through much of the western part of the state. For years researchers have been trying to predict the location of the next "Big One"—an earthquake with a magnitude of 8 or greater—along this fault system.

CALIFORNIA

The 1906 San Francisco earthquake caused displacement on the northernmost section of the fault. This event, likely relieved much of the strain that had been building during the previous 200 years or so.

1906 epicenter
San Francisco ●

Located just south of the 1906 rupture is a section of the San Andreas Fault that exhibits fault creep. When plates gradually slide past each other, less strain accumulates than when the fault is locked.

1857 epicenter

● Los Angeles

The 1906 San Francisco earthquake was the most devastating in California's history. The quake and resulting fires caused an estimated 3000 deaths and extensively damaged buildings throughout the city.

This 300-kilometer-long section of the San Andreas Fault System produced the Fort Tejon earthquake of 1857. Because a portion of the fault has likely accumulated considerable strain since the Fort Tejon quake, the U.S. Geological Survey gives it a 60 percent probability of producing a major earthquake in the next 30 years.

The next major quake on the San Andreas may well be on its southernmost 200 kilometers—an area that has not produced a large event in about 300 years.

On October 17, 1989, millions of television viewers around the world were settling in to watch the third game of the World Series. Instead, they saw their TVs go to black as tremors hit San Francisco's Candlestick Park. Although the Loma Prieta earthquake was centered in a remote section of the Santa Cruz Mountains (100 miles to the south) major damage occurred in the Marina District of San Francisco.

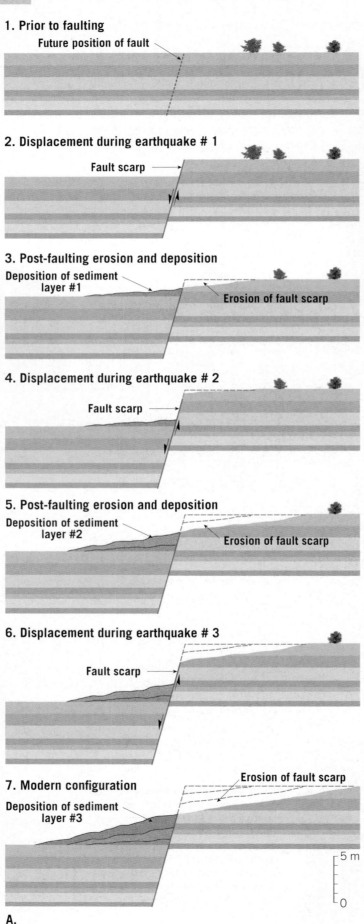

▶ Figure 11.33
Paleoseismology: The study of prehistoric earthquakes
This simplified diagram **(A)** shows that vertical displacement occurred on this fault three different times, with each event producing an earthquake. Based on the size of the vertical displacements, these ancient earthquakes had estimated magnitudes between 6.8 and 7.4. This study was conducted in the Pallet Creek area by digging a trench **(B)** across a branch of the San Andreas Fault and then looking for evidence of ancient displacements, such as offset sedimentary strata.

1. Prior to faulting
Future position of fault

2. Displacement during earthquake # 1
Fault scarp

3. Post-faulting erosion and deposition
Deposition of sediment layer #1
Erosion of fault scarp

4. Displacement during earthquake # 2
Fault scarp

5. Post-faulting erosion and deposition
Deposition of sediment layer #2
Erosion of fault scarp

6. Displacement during earthquake # 3
Fault scarp

7. Modern configuration
Erosion of fault scarp
Deposition of sediment layer #3

5 m

0

A.

B. The events depicted in the accompanying diagrams were deciphered by digging a trench (shown here) across the fault zone and studying the displaced sedimentary beds.

vertical offset of the layers of accumulated sediments indicates a large earthquake. Sometimes buried plant debris can be carbon dated, making it possible to establish the timing of recurrence.

One investigation that employed paleoseismology focused on a segment of the San Andreas Fault that lies north and east of Los Angeles. At this site, the drainage of Pallet Creek has been repeatedly disturbed by successive ruptures along the fault zone (**Figure 11.33**). Trenches excavated across the creek bed exposed sediments that were displaced by several large earthquakes over a span of 1500 years. From these data, researchers determined that strong earthquakes occur an average of once every 135 years. The last major event, the Fort Tejon earthquake, occurred on this segment of the San Andreas Fault in 1857, roughly 160 years ago. Because earthquakes occur on a cyclical basis, a major event in southern California may be imminent.

Paleoseismology has also revealed that powerful earthquakes (magnitude 8 or larger) and associated tsunamis have repeatedly struck the coastal Pacific Northwest over the past several thousand years. Each of these events is attributed to slippage along a section of the megathrust fault associated with the Cascadia subduction zone located off the west coast from southern British Columbia to northern California. The most recent event, which occurred in January 1700, generated a destructive tsunami that caused massive flooding in the coastal lowlands of western North America and was recorded as far away as Japan.

As a result of these findings, public officials have updated building codes and retrofitted some of the

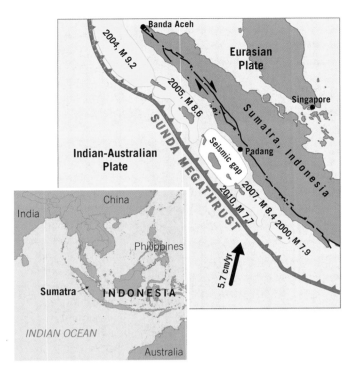

▲ **Figure 11.34**
Seismic gaps: Tools for forecasting earthquakes Seismic gaps are "quiet zones" that are thought to be storing elastic strain and will eventually produce major earthquakes. This seismic gap is located near Padang, a low-lying coastal city with a population of 800,000 people.

region's existing buildings, dams, bridges, and water systems to be more earthquake resistant. One notable example is the recent action taken by the Ocosta School District of southwestern Washington State.° This coastal community, located in an area inundated by the 1700 tsunami, took the initiative to build North America's first engineered tsunami refuge atop a new elementary school. The structure consists of a flat roof about 10 meters (30 feet) above ground, accessible from four heavily reinforced stairways. This initiative took cues from the 2011 Japan tsunami, but more importantly, was founded on scientific discoveries about Cascadia's earthquake and tsunami risks.

Seismic Gaps When seismologists mapped the rupture zones associated with great earthquakes around the globe, they discovered that these zones tend to lie next to each other, without appreciable overlap. A segment that has not experienced an earthquake within the past one to several centuries is called a **seismic gap**. Typically, these gaps are accumulating seismic strain and thus represent places where future large earthquakes are likely to occur. **Figure 11.34** shows a seismic gap on the megathrust fault that lies offshore of Padang, a low-lying Sumatran city of 800,000 people, which has not ruptured since

1797. Scientists are particularly concerned about this seismic gap because rupture of an adjacent segment of the fault caused the 2004 Indian Ocean (Sumatra) earthquake and tsunami that claimed 230,000 lives.

Minimizing Earthquake Hazards

The statement "earthquakes don't kill people; buildings kill people" succinctly expresses the fact that falling structures are by far the greatest cause of casualties during an earthquake. Thus, in regions where there is a known earthquake hazard, houses, roads, bridges, dams, and other structures should be constructed to withstand at least moderate shaking. In addition, communities need to adopt building codes that require inspection and retrofitting of existing structures to withstand seismic shaking (**Figure 11.35**). But due to prohibitive costs, the important step of retrofitting existing structures is often not a high priority. For example, a recent study indicated that more than half of the hospitals in Los Angeles County would likely collapse in a strong earthquake.

Earthquake-Resistant Structures The importance of designing new buildings to resist earthquakes and retrofitting older structures is perhaps best illustrated by comparing two earthquakes of similar magnitude—the 1988 Armenian earthquake (M 6.8) and the 1989 San Francisco earthquake (M 6.9).

▲ **Figure 11.35**
Retrofitted building Webb Tower, located on the campus of the University of Southern California, was retrofitted to help the structure withstand shaking in the event of a strong earthquake. Notice the cross braces on the building's exterior.

---

*Based on an article by Marcia McNutt while she was editor-in-chief of *Science*.

▶ Figure 11.36
Poorly constructed buildings destroyed during the 1988 Armenian earthquake

▼ Figure 11.37
Shanghai Tower, China The Shanghai Tower, which stands more than 2000 feet high, was made more quake resistant with a reinforced concrete pad 6 meters (20 feet) thick supported by 947 pilings—some of which are nearly 200 feet deep.

proactive measures to minimize casualties and property damage in the event of a strong earthquake.

As metropolitan areas around the world become more densely populated, architects and engineers are designing increasingly tall skyscrapers. However, some of the most populated areas on the globe are also prone to earthquakes. Consequently, earthquake-resistant buildings are becoming more common.

Most of the buildings leveled by the Armenian quake were constructed of unreinforced concrete, which collapsed into rubble—killing an estimated 25,000 people (**Figure 11.36**). Although the 1989 San Francisco earthquake was very destructive, the death toll (63 lives) was 400 times lower than that of the Armenian quake.

The difference was due mainly to building practices; in California, most buildings are either wood framed or built with reinforced concrete that resists collapse, whereas in Armenia the buildings that collapsed were constructed of unreinforced concrete. In areas that are not seismically active but have the potential for experiencing a strong earthquake, the challenge is to establish appropriate building codes and educate residents on

One striking example is China's Shanghai Tower, which stands more than 2000 feet high and is the second tallest building on Earth (**Figure 11.37**). Commissioned to be built on a site composed primarily of soft, clay-rich material, the massive structure was made more quake-resistant with a reinforced concrete pad 6 meters (20 feet) thick supported by 947 pilings—some of them nearly 200 feet deep.

Like other modern skyscrapers, the Shanghai Tower also incorporates a *mass damper*, which consists of a large pendulum—1000 tons of steel suspended from cables located near the top of the building. As the building sways, either during an earthquake or in high winds, the inertia of the pendulum acts as a counterweight, pulling in the opposite direction. Studies indicated that, because of its height, a traditional damper would not prevent the Shanghai Tower from swaying too fast or too far. With that critical information in mind, engineers incorporated a magnetic system that "pulls" on the massive pendulum to further counter the sway of this unusually tall structure during a quake.

Earthquake Preparedness Although the collapse of buildings is the largest cause of earthquake destruction, other hazards such as ground subsidence caused by liquefaction, landslides, and tsunamis can also be devastating. The catastrophic 2011 Japan (Tohoku) earthquake (M 9.0) is a sobering reminder of this fact. Because of Japan's strict building codes, the buildings, bridges, and other structures built or retrofitted since 1995 withstood the ground shaking of this powerful earthquake extremely well; only a few buildings collapsed as a direct result of shaking. It was the massive tsunami triggered by the quake that claimed the lives of 93 percent of the estimated 16,000 people who perished (**Table 11.2**). Ground subsidence caused by liquefaction and landslides also caused significant damage to buildings, roads, and utilities, such as water and gas pipes.

Japan's earthquake preparedness is arguably the best in the world, in large part because it has been struck by 19 major quakes (M 7–7.9) in the past 2 decades. Had the 2011 Japan earthquake occurred in a region where structures are less well engineered, the casualties attributable to collapse caused by shaking would have been considerably higher.

Earthquake Warning Systems   Because P waves are less destructive and travel faster than both S waves and surface waves, they can be used to provide a type of earthquake warning system. Japan has operated such a system for over 25 years. One use of this system is to trigger automated shutdown of power-generating facilities and braking of high-speed bullet trains.

In addition, during the 2011 Tohoku earthquake, this system was used to send an automated alert to the nation's television stations as well as to more than 50 million telephones. In the area closest to the epicenter, this alert gave residents about a 10-second chance to take cover. In Tokyo, a city of around 9 million people located about 370 kilometers (230 miles) south of the

TABLE 11.2 Some Notable Earthquakes

Year	Location	Deaths (est.)	Magnitude*	Comments
856	Iran	200,000		
893	Iran	150,000		
1138	Syria	230,000		
1268	Asia Minor	60,000		
1290	China	100,000		
1556	Shensi, China	830,000		Possibly the greatest natural disaster
1667	Caucasia	80,000		
1727	Iran	77,000		
1755	Lisbon, Portugal	70,000		Tsunami damage extensive
1783	Italy	50,000		
1908	Messina, Italy	120,000		
1920	China	200,000	7.5	Landslide buried a village
1923	Tokyo, Japan	143,000	7.9	Fire caused extensive destruction
1948	Turkmenistan	110,000	7.3	Almost all brick buildings near epicenter collapsed
1960	Southern Chile	5700	9.5	The largest-magnitude earthquake ever recorded
1964	Alaska	131	9.2	Greatest-magnitude North American earthquake
1970	Peru	70,000	7.9	Great rockslide
1976	Tangshan, China	242,000	7.5	Estimates for the death toll are as high as 655,000
1985	Mexico City	9500	8.1	Major damage occurred 400 km from epicenter
1988	Armenia	25,000	6.9	Poor construction practices increased the death toll
1990	Iran	50,000	7.4	Landslides and poor construction practices led to great damage
1993	Latur, India	10,000	6.4	Located in stable continental interior
1995	Kobe, Japan	5472	6.9	Damages estimated to exceed $100 billion
1999	Izmit, Turkey	17,127	7.4	Nearly 44,000 injured and more than 250,000 displaced
2001	Gujarat, India	20,000	7.9	Millions homeless
2003	Bam, Iran	31,000	6.6	Ancient city with poor construction
2004	Indian Ocean (Sumatra)	230,000	9.1	Devastating tsunami damage
2005	Pakistan/Kashmir	86,000	7.6	Many landslides; 4 million homeless
2008	Sichuan, China	87,000	7.9	Millions homeless, some towns will not be rebuilt
2010	Port-au-Prince, Haiti	100,000–316,000	7	More than 300,000 injured and 1.3 million homeless
2011	Japan	16,000	9	Majority of the casualties due to a tsunami
2015	Nepal	9000	7.8	About 21,000 injured
2017	Chiapas, Mexico	98	8.2	Largest earthquake of 2017; 41,000 structures damaged

*Widely differing magnitudes and death tolls have been estimated for some of these earthquakes, particularly those events that occurred prior to the eighteenth century. When available, moment magnitudes are used.
Source: U.S. Geological Survey.

epicenter, citizens had more than a minute's warning before the strong shaking began.

Several governmental agencies in California have been working to produce an early warning system. Unfortunately, the lack of funding required to install or upgrade an additional 283 seismic sensors has delayed this effort. It could take several years before a fully operational early warning system that covers the entire state will be operational.

11 CONCEPTS IN REVIEW

Earthquakes and Earthquake Hazards

11.1 What Is an Earthquake?

Sketch and describe the mechanism that generates most earthquakes.

Key Terms:

earthquake	epicenter	aftershock
fault	seismic wave	megathrust fault
hypocenter	elastic rebound	fault creep
	foreshock	

- The sudden movements of large blocks of rock on opposite sides of faults cause most *earthquakes*. The location where the rock begins to slip is called the *hypocenter*, or *focus*. During an earthquake, *seismic waves* radiate outward from the hypocenter in all directions. The point on Earth's surface directly above the hypocenter is the *epicenter*.

- Earthquakes are caused by differential stress that gradually bends Earth's crust over tens to hundreds of years. Up to a point, frictional resistance along the *fault* keeps the rock from rupturing and slipping. Once that point is reached, the fault slips, allowing the bent rock to "spring back" to its original shape, generating an earthquake. The springing back is called *elastic rebound*.

- Convergent plate boundaries and associated subduction zones are marked by *megathrust faults*, which are responsible for most of the largest earthquakes in recorded history.

- The San Andreas Fault in California is an example of a large strike-slip fault that forms a transform plate boundary capable of generating destructive earthquakes.

Q Label the blanks on the diagram to show the relationship between earthquakes and faults using the following terms: epicenter, seismic waves, fault, fault trace, and hypocenter.

11.2 Seismology: The Study of Earthquake Waves

Compare and contrast the types of seismic waves and describe how a seismograph works.

Key Terms:

seismology	inertia	surface waves
seismograph	seismogram	primary waves
	body waves	secondary waves

- *Seismology* is the study of seismic waves. A *seismograph* measures these waves, using the principle of inertia. While the body of the instrument moves with the waves, the *inertia* of a suspended weight keeps a sensor stationary to record the displacement between the two.

- A *seismogram*, a record of seismic waves, reveals two main categories of earthquake waves: *body waves* (*primary* [P] *waves* and

secondary [S] *waves*), which are capable of moving through Earth's interior, and *surface waves*, which travel only along the upper layers of the crust. P waves are the fastest, S waves are intermediate in speed, and surface waves are the slowest. However, surface waves tend to have the greatest amplitude and produce the strongest shaking, so surface waves usually account for most damage during earthquakes.

- P waves momentarily push (compress) and pull (stretch) rocks as they travel through a rock body, thereby changing the volume of the rock. S waves impart a shaking motion as they pass through rock, changing the rock's shape but not its volume. Because fluids do not resist forces that change their shape, S waves cannot travel through fluids, whereas P waves can.

11.3 Locating the Source of an Earthquake

Explain how seismographs locate the epicenter of an earthquake.

- The distance separating a recording station from an earthquake's epicenter can be determined by using the difference in arrival times between P and S waves. When the distances are known from three or more seismic stations, the epicenter can be located using a method called triangulation.

11.4 Determining the Size of an Earthquake

Distinguish between intensity scales and magnitude scales.

Key Terms: magnitude   moment magnitude
intensity   Modified Mercalli
               Intensity scale

- *Intensity* is a measurement of the amount of ground shaking at a location due to an earthquake, and *magnitude* is an estimate of the amount of energy released during an earthquake.
- The *Modified Mercalli Intensity scale* is a 12-point scale that uses structural damage to quantify earthquake intensity.
- Richter-like scales take into account both the maximum amplitude of the seismic waves measured at a given seismograph and that seismograph's distance from the earthquake. These scales are logarithmic, meaning that a number on the scale represents seismic amplitudes that are 10 times greater than those represented by the next lower number. Furthermore, each larger number on a Richter-like scale represents the release of about 32 times more energy than the number below it.
- Because the Richter scale does not effectively differentiate between very large earthquakes, the *moment magnitude* scale was devised. This scale measures the total energy released from an earthquake by considering the strength of the faulted rock, the amount of slippage, and the area of the fault surface that slipped.

11.5 Earthquake Destruction

List and describe the major destructive forces that earthquake vibrations can trigger.

Key Terms: liquefaction   tsunami

- Factors influencing how much destruction an earthquake might inflict on human-made structures include (1) intensity and length of time the shaking persists, (2) building construction, and (3) the nature of the ground that underlies the structure. Buildings constructed of unreinforced bricks and blocks are more likely than other types of structures to be severely damaged in a quake.
- In general, unconsolidated sediments amplify seismic shaking, while bedrock-supported buildings fare best in an earthquake.
- *Liquefaction* may occur when waterlogged sediment or soil is severely shaken during an earthquake. Liquefaction can reduce the strength of the ground to the point at which it may not support buildings.
- Earthquakes may trigger landslides or ground subsidence, and they can break gas lines, which may initiate devastating fires.
- *Tsunamis* are large ocean waves that form when water is displaced, usually by a megathrust fault rupturing on the seafloor. Traveling at the speed of a jet aircraft, a tsunami is hardly noticeable in deep water. However, upon arrival in shallower coastal waters, the tsunami slows down and piles up, producing a wall of water sometimes more than 30 meters (100 feet) in height. Tsunami warning systems have been established in most of the large ocean basins.

11.6 Where Do Most Destructive Earthquakes Occur?

Locate Earth's major earthquake belts on a world map.

Key Term: circum-Pacific belt

- Most earthquake energy is released in the *circum-Pacific belt*, the ring of megathrust faults rimming the Pacific Ocean. Another earthquake belt is the Alpine–Himalayan belt, which runs along the zone where the Eurasian plate collides with the Indian subcontinent and African plate.
- Earth's oceanic ridge system produces another belt of frequent but small-magnitude quakes. Transform faults in the continental crust, including the San Andreas Fault, can produce large earthquakes.
- Some earthquakes occur at considerable distances from plate boundaries. Examples include the 1811–1812 New Madrid, Missouri, earthquakes and the 1886 Charleston, South Carolina, earthquake.

Q Outline the circum-Pacific earthquake belt on the accompanying map that has plate boundaries drawn in red. Do the same for the Alpine–Himalayan belt.

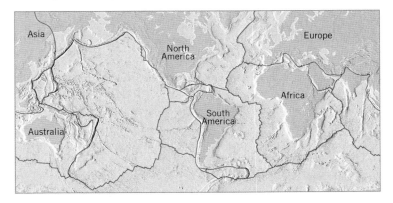

11.7 Earthquakes: Predictions, Forecasts, and Mitigation

Compare and contrast the goals of short-range earthquake predictions and long-range forecasts.

Key Terms: paleoseismology
precursor   seismic gap

- Successful earthquake prediction has been an elusive goal in seismology for many years. Attempts at short-range predictions (for hours or days) focus on *precursor* events such as changes in ground elevation or in strain levels near a fault; they have not been reliable.
- Long-range forecasts (for time scales of 30 to 100 years) are statistical estimates of the likelihood that an earthquake of a given magnitude will occur. Long-range forecasts are useful because they can guide development of building codes and infrastructure.
- *Paleoseismology* is a tool used to make long-range forecasts. Because earthquakes occur on a cyclical basis, determining how frequently they have occurred in the past can give some insight into when they are most likely to occur again.
- Scientists have identified *seismic gaps*, which are portions of faults that have been storing strain for a long time, as areas with higher-than-average potential for experiencing an earthquake in the not-too-distant future.

GIVE IT SOME THOUGHT

1. The accompanying map shows the locations of many of the largest earthquakes in the world since 1900. Refer to the map of Earth's plate boundaries in Figure 2.10, page 45, and determine which type of plate boundary is most often associated with these destructive events.

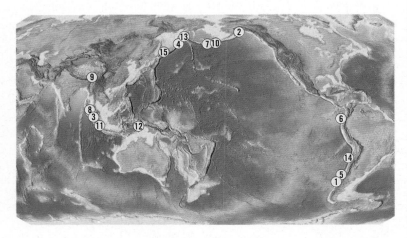

2. Use the accompanying seismogram to answer the following questions:

   a. Which of the three types of seismic waves reached the seismograph first?

   b. What is the time interval between the arrival of the first P wave and the arrival of the first S wave?

   c. Use your answer from Question b and the travel–time graph in Figure 11.15 to determine the distance from the seismic station to the earthquake.

   d. Which of the three types of seismic waves had the highest amplitude when it reached the seismic station?

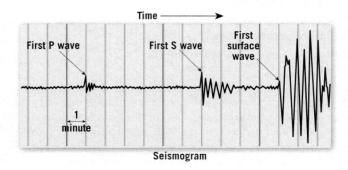

Seismogram

3. You go for a jog on a beach and choose to run near the water, where the sand is saturated. With each step, you notice that your footprint quickly fills with water, but that water is not coming in from the ocean. What is this water's source? For what earthquake-related hazard is this phenomenon a good analogy?

4. On the accompanying diagram, first determine the magnitude for an earthquake at a distance of 400 kilometers, with a maximum amplitude of 0.5 millimeter. Second, for this same earthquake (same magnitude), determine the amplitude of the biggest waves for a seismograph 40 kilometers from the hypocenter.

S-P, sec.	Distance, km	Magnitude, M_L	Amplitude, mm
50	500		100
40	400	6	50
30	300		20
20	200	5	10
		4	5
10	100		
8	60	3	2
6			1
4	40	2	0.5
		1	1.2
2	20		0.1
	0.5	0	

5. Using the accompanying map of the San Andreas Fault, answer the following questions:

   a. Which of the four segments (1–4) of the San Andreas Fault do you think is experiencing fault creep?

   b. Paleoseismology studies have found that the section of the San Andreas Fault that failed during the Fort Tejon quake (segment 3) produces a major earthquake every 135 years, on average. Based on this information, how would you rate the chances of a major earthquake occurring along this section in the next 30 years? Explain.

   c. Do you think San Francisco or Los Angeles has the greater risk of experiencing a major earthquake in the near future? Defend your selection.

EYE ON EARTH

Calaveras Fault

1. The Calaveras Fault, a branch of the San Andreas Fault system, cuts directly through the town of Hollister, California. Rather than being "locked," this section of the fault is slowly slipping, producing noticeable offsets and damage to curbs, sidewalks, roads, and buildings. The concrete wall and sidewalk shown here were straight when they were originally constructed.

   a. What term is used to describe the type of slippage observed along the Calaveras Fault in Hollister?

   b. Are faults that exhibit this type of slippage likely to generate major earthquakes? Explain.

DATA ANALYSIS

Earthquakes Around the World

Earthquakes happen around the world every day. Many of them are minor and go unnoticed, but some are devastating. The United States Geological Survey (USGS) monitors these earthquakes and the damages they cause.

https://goo.gl/eD226M

ACTIVITIES

Go to the USGS Earthquakes page, at http://earthquake.usgs.gov/earthquakes.

1. **Where was the most recent significant earthquake in the past 30 days in the United States? What was its magnitude?** (*Note:* USGS generally uses moment magnitude [M_W] to indicate magnitude.)

Click on the earthquake's link to bring up more information about the earthquake and then click Pager.

2. **What are the estimated fatalities for this earthquake? What are the estimated economic losses?**

3. **Describe the structures located in the vicinity of this earthquake. Are there any secondary effects?**

4. **How many people felt moderate and strong shaking? Very strong and severe shaking? Violent and extreme shaking?**

Go back to the previous page and click Origin.

5. **What was the depth of this earthquake's focus? Be sure to include units and uncertainty.**

Go to the USGS Earthquake Hazards Program's Latest Earthquakes map, at http://earthquake.usgs.gov/earthquakes/map. Use the settings icon (which looks like a gear) to find answers to the following questions. Zoom out to see the entire world.

6. **How many total earthquakes have occurred worldwide in the past 7 days?**

7. **How many earthquakes of magnitude 2.5 or higher have occurred worldwide in the past 7 days?**

8. **How many earthquakes of magnitude 4.5 or higher have occurred worldwide in the past 7 days?**

9. **What fraction of total earthquakes have had a magnitude of 2.5 or higher? 4.5 or higher?**

10. **What conclusion can you draw about the relationship between the frequency and strength of earthquakes?**

Mastering Geology

Looking for additional review and test prep materials? Visit pearson.com/mastering/geology to enhance your understanding of this chapter's content by accessing a variety of resources, including **Self-Study Quizzes, Geoscience Animations, SmartFigure Tutorials,** *Project Condor* **Videos, Mobile Field Trips, Dynamic Study Modules,** and an optional **Pearson eText.**

Ships' Logs and Pottery Shards: Tracking Earth's Shifting Geomagnetic Field

Deep below us, Earth's molten outer core moves around the solid inner core, generating electrical currents responsible for the planet's magnetic field. Paleomagnetism recorded in seabed rocks tells us that the magnetic north and south poles periodically wander and reverse direction. Since 2014, satellites from the European Space Agency have been tracking changes in Earth's surface magnetic field with a high degree of accuracy. Scientists have discovered that since about 1840, Earth's magnetic field has been weakening at a rate of roughly 5 percent per century, possibly signaling that another pole reversal is on the way. How do we obtain information about the state of the magnetic field a few hundred or a few thousand years ago—lengthy by human standards, but just a flash in the geologic time span? We can find the answers in some surprising places.

Sailors have long tracked declination (the difference between the geographic North Pole and magnetic north on a compass) while navigating. Starting around 1700, inclination—the downward dip exerted on one-half of the compass needle—was also noted. By looking at the logbooks of Captain James Cook and others, scientists can compare compass readings from known geographic points in both the past and present to see how the magnetic field has changed over time.

Archaeological sites can also provide geomagnetic data. Clay minerals subjected to high heat lock in the magnetic orientation of the time of their firing. In 2015, reports of remnants of 1000-year-old mud huts in Africa's Limpopo River valley hit the news. Researchers were excited to discover that the clay-rich mud in these ancient structures had burned at high enough temperatures to provide an Iron Age magnetism record for this region.

▲ Painting of *HMS Endeavour*, vessel commanded by Captain James Cook. Records from Cook's voyages helped reconstruct a historical picture of the eighteenth century geomagnetic field in areas where he sailed.

► An archaeological site on the Limpopo River in Africa provides a source of Iron Age geomagnetism data for the region.

12

Earth's Interior*

FOCUS ON CONCEPTS

Each statement represents the primary learning objective for the corresponding major heading within the chapter. After you complete the chapter, you should be able to:

12.1 Explain how Earth acquired its layered structure.

12.2 List and describe each of the three major layers of Earth's interior.

12.3 Explain how seismic waves were used to discover Earth's layered structure.

12.4 Describe the processes of heat transfer that operate in Earth's interior and indicate in which layers these processes are dominant.

12.5 Discuss what seismic tomography has revealed about variations in Earth's layers.

---

*This chapter was originally prepared by Professor Michael Wysession, Washington University.

f you could slice Earth in half, the first thing you would notice is distinct layers. The heaviest materials (metals) appear in the center. Lighter solids (rocks) make up the middle layers, and less dense liquids and gases comprise the outer layer. Within Earth, we know these layers as the *iron core* (center), the *rocky mantle and crust* (middle), the *liquid ocean*, and the *gaseous atmosphere* (outer). More than 95 percent of the variations in composition and temperatures in Earth are due to this seemingly simple layered structure. These layers are, however, complex, dynamic, and part of Earth's geologic history.

12.1 Exploring Earth's Interior

Explain how Earth acquired its layered structure.

Earth's interior consists of three major layers defined by their *chemical composition*—the *core, mantle*, and *crust*. These three compositionally distinct shells can be further subdivided into layers, based on physical properties that include whether the layer is solid or liquid and how weak or strong it is (**Figure 12.1**). Knowledge of both types of layers is essential to our understanding of basic geologic processes such as volcanism, earthquakes, and mountain building.

Formation of Earth's Compositional Layers

If a bottle filled with clay, iron filings, water, and air were shaken, at first it would have a single muddy composition. However, if that bottle were then allowed to sit undisturbed, the different materials would separate and settle into layers. Iron filings are the most dense so would be the first to sink to the bottom of the bottle. Above the iron would be a layer of clay, then water, and, finally, air. Gravity is largely responsible for the layering in the bottle of muddy water, as well as the compositional layering we detect in Earth's interior.

As material accumulated to form Earth, the high-velocity impact of nebular debris and the decay of radioactive elements caused the temperature of our planet to steadily increase. During this time of intense heating, Earth became hot enough that some materials, including iron and nickel, began to melt. Melting produced liquid blobs of these heavy metals that gravitationally sank toward the center of the planet. This process, called *chemical differentiation*, occurred rapidly on the scale of geologic time and produced Earth's dense, iron-rich core.

The period of heating also generated blobs of less dense, molten rock that buoyantly rose toward the surface and solidified to produce a primitive crust. This rocky material consisted mainly of light silicate minerals that were rich in oxygen, silicon, and aluminum, along with lesser amounts of calcium, sodium, potassium, iron,

and magnesium. In addition, some heavy metals such as gold, lead, and uranium, which have low melting points or were highly soluble in the ascending molten masses, were scavenged from Earth's interior and concentrated in the developing crust. The dark silicate minerals that have high melting temperatures—mainly olivine, pyroxene, and garnet—were left behind to become the mantle.

This early period of chemical differentiation established the three basic compositional layers of Earth's interior: (1) an iron-rich *core*, (2) a thin *primitive crust* composed mainly of the light to intermediate silicate minerals, and (3) Earth's largest layer, the *mantle*, composed of dark (ultramafic) silicate minerals.

How Do We Study Earth's Interior?

Before researchers could begin to study Earth's interior, they needed to know the *mass* and *average density* of our planet. Earth's mass, established about 200 years ago, was derived from our planet's gravitational attraction to objects around it. Once the mass was known, its density was easily calculated: The density of an object equals its mass divided by its volume. Earth's average density is about 5.5 gm/cm, about twice that of a typical rock found on Earth's surface. What we can glean from Earth's average density is that material at depth must be denser than rocks found on Earth's surface.

That information, coupled with direct and indirect observations of Earth materials, provided numerous

Layering by Chemical Composition

Earth's three layers based on composition are the crust, mantle, and core.

Layering by Physical Properties

Earth has several layers based on differences in physical properties that include whether the layer is solid or molten, weak or strong.

Crust (low density rock)

Mantle (high density rock)

Outer core (molten)

Core (iron + nickel)

Inner core (solid)

Lower mantle (solid)

Upper mantle (varies)

Lithosphere (solid)

Oceanic crust

Continental crust

Moho

Asthenosphere (solid, but mobile)

410 km

Transition zone

660 km

◄ SmartFigure 12.1
Earth's layers Structure of Earth's interior, based on chemical composition (left) and physical properties (right).

Tutorial
https://goo.gl/
K6EZ2B

clues about the structure and composition of Earth's interior. One of the indirect methods of study, called *mineral physics*, attempts to simulate the high pressures and temperatures found at various depths. By squeezing a tiny sample of a mineral between two diamonds, pressures at various depths in Earth's interior can be simulated. High temperatures are achieved by using a laser to heat the mineral sample (see **GEOgraphics 12.1**, page 344).

In addition, as mantle-derived magmas ascended to the surface, they incorporated other rocks, called **xenoliths**, which are samples of the upper mantle and crust. Based on these data, we know that the upper mantle is composed mainly of the rock **peridotite** (**Figure 12.2**). Peridotite is an ultramafic rock that consists of the mineral *olivine* (about 60%) as well as lesser amounts of *pyroxene* (about 25%), minerals that are rich

Direct Observation of Earth Materials

Sampling of Earth's upper crust has been going on as long as humans have been excavating minerals. This gives us direct knowledge of the mineral makeup of the upper continental crust, which has an average composition of a felsic rock called *granodiorite*. Sampling has also helped us determine that the continental crust becomes more mafic with depth.

Although most oceanic crust forms out of view, far below sea level, geologists have been able to examine the structure of the ocean floor firsthand. In locations such as Newfoundland, Cyprus, Oman, and California, slivers of oceanic crust and underlying mantle have been thrust high above sea level. From these exposures, along with core samples collected by deep-sea drilling ships, researchers concluded that the upper oceanic crust consists mainly of basaltic lavas, while the lower unit is gabbro, the coarse-grained equivalent of basalt.

Peridotite xenolith

Basaltic lava

0   2   4
cm

◄ Figure 12.2
Peridotite: A rock from Earth's mantle This sample of peridotite (olivine-rich rock), carried up from the mantle, provides clues to the composition of Earth's interior. The mantle fragment (xenolith) was found in a basaltic lava flow on the island of Hawaii.

Recreating Deep Earth

Seismology alone cannot determine the nature of the materials deep in Earth's interior. Additional information must be obtained by other techniques. Mineral physics experiments can measure physical properties of rocks and minerals such as stiffness, compressibility, and density while simulating the extreme conditions of the mantle and core.

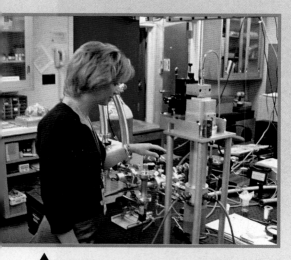

One experiment examines the temperatures and pressures at which one mineral phase will become unstable and convert into a new "high-pressure" phase. These experiments are useful because they help identify where phase changes take place within Earth.

Most mineral physics experiments are conducted using diamond-anvil presses like the one shown here. These take advantage of two important properties of diamonds— hardness and transparency. The tips of two diamonds are cut off, and a small mineral sample is placed between them. By squeezing two diamonds together, pressures as high as those at Earth's center have been simulated. High temperatures are achieved by firing a laser beam through the diamond and into the mineral sample.

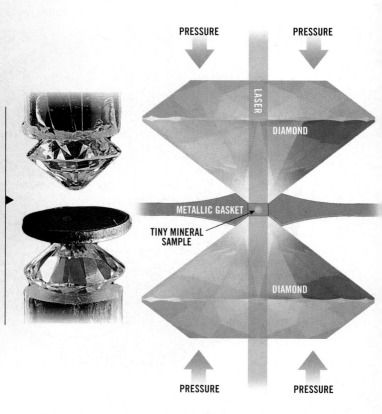

PRESSURE PRESSURE

LASER

DIAMOND

METALLIC GASKET

TINY MINERAL SAMPLE

DIAMOND

PRESSURE PRESSURE

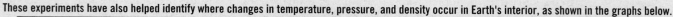

These experiments have also helped identify where changes in temperature, pressure, and density occur in Earth's interior, as shown in the graphs below.

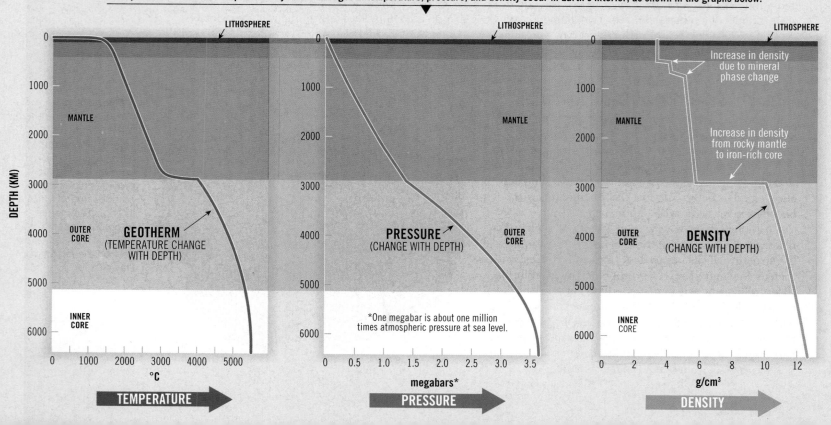

LITHOSPHERE

MANTLE

OUTER CORE

GEOTHERM
(TEMPERATURE CHANGE WITH DEPTH)

INNER CORE

DEPTH (KM)

0 1000 2000 3000 4000 5000
°C

TEMPERATURE

LITHOSPHERE

MANTLE

OUTER CORE

PRESSURE
(CHANGE WITH DEPTH)

*One megabar is about one million times atmospheric pressure at sea level.

0 0.5 1.0 1.5 2.0 2.5 3.0 3.5
megabars*

PRESSURE

LITHOSPHERE

Increase in density due to mineral phase change

MANTLE

Increase in density from rocky mantle to iron-rich core

OUTER CORE

DENSITY
(CHANGE WITH DEPTH)

INNER CORE

0 2 4 6 8 10 12
g/cm³

DENSITY

0
1000
2000
3000
4000
5000
6000

in iron and magnesium. Thus, the mantle is composed of minerals that are denser than those of the continental and oceanic crust that lie above it.

Perhaps the most interesting samples we have of the mantle are igneous rocks called *kimberlites*—a volcanic rock that sometimes contains diamonds. (The volcanic structures that carry kimberlites to the surface, called *kimberlite pipes*, have a narrow pipe-like shape that is wider near the top.) Because diamonds can form only in high-pressure environments that are at least 150 kilometers (about 100 miles) below the surface, these rocks provide samples of the mantle that lies at least that deep. A few recovered diamonds are thought to have formed in the transition zone of the mantle, 410 to 660 kilometers (250 to 400 miles) below the surface. Long after the diamonds formed, magmas containing these high-pressure minerals rapidly carried them up through the lithosphere and produced small but explosive eruptions at the surface.

Many diamonds contain imperfections, called *inclusions*, which are tiny fragments of foreign material that have been trapped in the diamonds for millions of years. Analyses of these materials indicate that diamonds formed in a water-rich environment having a composition similar to that of seawater. This finding has led researchers to conclude that massive amounts of water are stored in the upper mantle.

Meteorites, remnants of small bodies that formed early in the history of the solar system, are compositionally similar to the materials from which Earth accreted.

Fairly recently, a natural diamond was found to contain a tiny piece of a high-pressure form of olivine called *ringwoodite*. (Recall from Chapter 3 that under high pressure, atoms in a mineral can rearrange to form a more compact version of the mineral, in a process called a *phase change*.) The discovery of this rare diamond is further evidence that olivine is a primary component of mantle rocks. It also supports laboratory studies predicting that olivine morphs into various forms that correspond to the depth at which they are found.

In addition, the diamond's fragment of ringwoodite contained about 1 percent water, significantly more than found in other forms of olivine. Although that seems like a miniscule amount, if all the ringwoodite in the upper mantle contains this much water, the mantle would hold as much water as the global oceans combined.

There are two competing hypotheses about where the mantle's water came from. One contends that the source was ocean water carried into the mantle by the subduction of a water-laden seafloor. The other suggests that the mantle contains water from comets that crashed into Earth as it accreted (coalesced) material during its formative period.

Evidence from Meteorites

Meteorites, remnants of small bodies that formed early in the history of the solar system (see Chapter 24), are compositionally similar to the materials from which Earth accreted. The most common meteorites, called *stony meteorites*, are rich in olivine and pyroxene—making them compositionally similar to the peridotite found in the upper mantle.

In addition, *iron meteorites*—primarily alloys of iron with lesser amounts of nickel—are also relatively common (**Figure 12.3**). Because of the abundance of iron in meteorites, we conclude that Earth's composition should be more iron rich than is indicated by the rocky mantle and crust; therefore, the additional iron must be located in the core.

Two other factors point to iron being a major component of the core. First, the density of mantle rock, even if placed under the extreme pressures found in the core, is not high enough to account for Earth's average density, whereas iron's density is. Second, Earth's magnetic field, generated in the core, requires a conductive material like iron to operate. Earth's magnetic field will be considered in more detail in Section 12.5.

Using Seismic Waves to Probe Earth's Interior

The use of **seismic waves** to study Earth's interior has been greatly enhanced during the past few decades due to the growing number of *seismograph networks* that can detect them from distant earthquakes. Each year, about 3000 earthquakes are large enough (about M_w 5.5 or above) to travel through Earth and be recorded by seismographs around the

◄ **Figure 12.3**
Iron meteorite This sample was cut and etched with acid to reveal its crystalline structure. Meteorites accreted to form the young Earth.

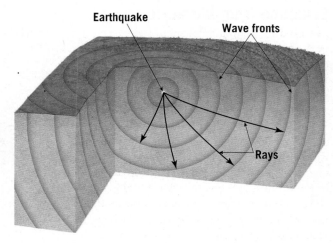

globe (**Figure 12.4**). The P and S waves emanating from strong earthquakes act like medical X-rays, providing the means to "see" into our planet.

What Do Seismic Velocities Tell Us?

The speed at which P waves and S waves travel through Earth's interior depends largely on the properties of the materials that transmit them. In general, seismic waves travel fastest when rock is *stiff* (*rigid*) or *less compressible*. When heated, rock becomes less stiff (imagine warming a frozen chocolate bar), and earthquake waves travel through it more slowly. These properties of stiffness and compressibility are used to interpret temperature—how near a rock is to its melting point at various depths. Specifically, when P waves travel through molten or partially molten rock, they travel much slower than if the rock were solid. Furthermore, recall from Chapter 11 that S waves are not transmitted by liquids. Because S waves do not travel through the outer core, we conclude that the outer core is molten.

The ways in which seismic velocities vary with depth is shown in **Figure 12.5**. Notice that in all layers, P waves

travel at higher velocities than S waves. In addition, in the mantle, both P and S waves exhibit a gradual increase in velocity, indicating that mantle rocks tend to become more rigid and less compressible with depth. One exception occurs in the *asthenosphere*, located directly below the lithosphere, where both the P- and S-wave velocities slow, evidence that the asthenosphere is a weak layer containing rocks near their melting temperature.

The most dramatic drop in P-wave velocities (and the absence of S waves) occurs at the mantle–core boundary—evidence that the outer core is a weak, molten layer. By contrast, an increase in P-wave velocity is observed at the outer core–inner core boundary—indicating that the inner core is more rigid (stronger) than the outer core.

Earthquake waves also travel at different speeds through Earth materials having different compositions. For example, seismic waves travel faster through oceanic crust (composed of basalt) than through continental crust, which has an overall composition similar to that of granite. Thus, the speed at which seismic waves travel has helped researchers determine both the composition of rocks found within Earth and how hot they are.

CONCEPT CHECKS 12.1

1. List the three compositionally distinct layers of Earth's interior.

2. Provide evidence that points to the mantle being composed of the rock peridotite.

Concept Checker
https://goo.gl/njGx9s

3. What do meteorites tell us about the composition of Earth's core?

4. Explain what seismic waves tell us about Earth's interior.

12.2 Earth's Layered Structure

List and describe each of the three major layers of Earth's interior.

Combining data from innumerable studies gives us a layer-by-layer understanding of the structure and composition of Earth's interior (see Figure 12.1).

Crust

Earth's **crust**, a relatively thin, rocky outer skin, is divided into two types —continental crust and oceanic crust (**Figure 12.6**). Continental crust and oceanic crust have very different compositions, histories, and ages.

Oceanic Crust The seafloor is much younger (180 million years old or less), thinner, and denser than the continents. Furthermore, it is compositionally more similar to Earth's mantle than to continental crust. Oceanic crust is about 7 kilometers (4 miles) thick and forms continuously along the mid-ocean ridge system. Having a density of about 3.0 g/cm³, oceanic crust is composed of the dark igneous rock basalt, which forms the upper oceanic crust and gabbro found below. Compared to continental crust, the seafloor has a rather homogeneous chemical composition and structure. The formation and structure of oceanic crust are discussed in greater detail in Chapter 13.

Continental Crust Unlike oceanic crust, continental crust consists of many rock types. Although the upper crust has an average composition of granodiorite, a felsic rock, its composition and structure vary considerably around the globe.

Continental crust averages about 40 kilometers (25 miles) thick but can exceed 70 kilometers (40 miles) thick in mountainous regions such as the Himalayas and the Andes. Further, it has an average density of about 2.7 g/cm³, which is much lower than the density of mantle rock. The low density of continents relative to Earth's mantle explains why continents are buoyant—acting like giant floating rafts that cap tectonic plates—and why continental crust cannot be readily subducted into the mantle. Continental rocks that exceed 4 billion years in age have been discovered using radiometric dating.

Mantle

Beneath Earth's crust lies the mantle. More than 82 percent of Earth's volume is contained within the **mantle**, a nearly 2900-kilometer-thick (1800-mile-thick) shell extending from the base of the crust, called the *Moho*, to the liquid outer core (see Figure 12.6). Because S waves readily travel through the mantle, we know it is a solid rocky layer. However, despite its solid nature, mantle rock is very hot and capable of flow, albeit very slowly.

The Upper Mantle Earth's **upper mantle** extends from the Moho to a depth of about 660 kilometers (410 miles) and can be further divided into three shells (see Figure 12.6):

1. The uppermost mantle, called the **lithospheric mantle**, ranges in thickness from only a few kilometers under the mid-oceanic ridges to perhaps as much as 200 kilometers (125 miles) under the stable continental interiors. This uppermost mantle and the crust make up Earth's rigid outer shell, called the **lithosphere**.
2. Beneath the lithospheric mantle is a weak layer called the **asthenosphere**. The lithospheric mantle and asthenosphere are similar in their compositions; however, the uppermost mantle is strong, whereas the asthenosphere is weak as a result of Earth's temperature structure—a topic we will consider later.
3. The lower portion of the upper mantle, at depths between 410 and 660 kilometers, is called the **transition zone**. The top of the transition zone is identified by a sudden change in seismic velocities, caused by atoms in olivine rearranging to form a denser structure due to the increase in pressure

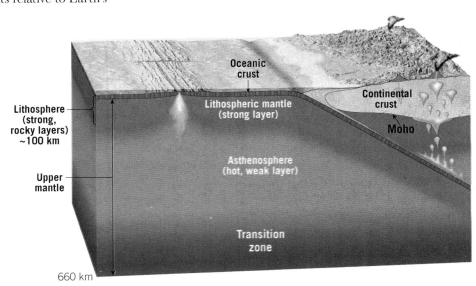

◀ Figure 12.6
Earth's crust and upper mantle

(phase change). A similar phase change is thought to occur at the bottom of the transition zone (see Figure 12.6).

Rocks brought to Earth's surface by volcanism indicate that the upper mantle is composed mainly of peridotite (see Figure 12.2).

The Lower Mantle   Between the transition zone (660 kilometers) and the molten outer core (2900 kilometers) lies the **lower mantle**. Within the lower mantle, both olivine and pyroxene take the chemical structure of a dense mineral called *perovskite* (recently renamed *bridgemanite*) and other related minerals.

The D″ Layer   In the lowest few hundred kilometers of the lower mantle is a highly unusual region called D″ (pronounced "dee double-prime"). The **D″ layer** is thought to have large variations in composition as well as temperature (**Figure 12.7**). Cool areas in the D″ layer are thought to be the graveyard of subducted oceanic lithosphere, whereas hot areas are thought to be the birthplace of deep mantle plumes.

The very base of D″ is where the mantle is in direct contact with the hot liquid iron core. As a result, the base of D″ may be hot enough to be partially molten. Evidence for partial melting comes from S wave velocities, which decrease by 30 percent within D″—an indication that the rocks there are quite weak.

Core

The composition of the **core** is thought to be an iron–nickel alloy, with minor quantities of oxygen, silicon, and sulfur. Because of the extreme pressure found in the core, the iron-rich material has an average density of more than 10 g/cm^3. At Earth's center, the average density is about 13 g/cm^3 (13 times the density of water).

The core accounts for only about one-sixth of Earth's volume but one-third of its mass. This is because the core is composed mainly of iron, which has the greatest density of Earth's most abundant elements.

The Outer Core   The **outer core** is a molten iron-rich layer 2270 kilometers (1410 miles) thick. The nature of the outer core was discovered when researchers noticed that P-wave velocities drop dramatically as they cross the core–mantle boundary, and S waves are not transmitted. The convection of metallic iron within this zone generates Earth's magnetic field, a topic we will consider later.

The Inner Core   At Earth's center lies the **inner core**, a solid dense sphere with a radius of 1216 kilometers (754 miles). Because the inner core is a sphere, whereas Earth's other layers are shells, it appears in illustrations to be much larger than it really is (see Figure 12.1). The inner core is only 1/142 of the volume of Earth (less than 1 percent). Despite its high temperature, the inner core is solid due to the immense pressures in the center of the planet.

The inner core did not exist early in Earth's history, when our planet was hotter. However, as Earth cooled, molten iron began to crystallize and sink to form the solid inner core. Even today, the inner core continues to very gradually increase in size.

Separated from the mantle by the liquid outer core, the solid inner core is free to rotate. It is thought that the inner core rotates faster than the crust and mantle, lapping them every few hundred years.

A recent study found a distinct sphere within the inner core that is about one-half its diameter. The iron crystals that form the outer portion of the inner core roughly align in a north–south direction, while those in the interior of the inner core align in an east–west direction. This finding implies that the innermost part of the core changed its orientation (rotated) at some point in its history.

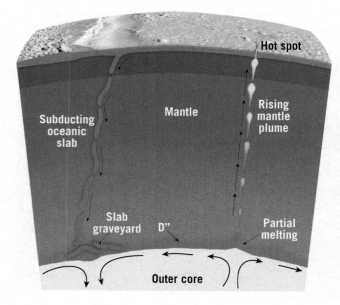

► **Figure 12.7**
The variable and unusual D″ layer lies at the base of the mantle The D″ layer contains large horizontal variations in both temperature and composition. Many geologists think the D″ layer is the graveyard of subducted oceanic lithosphere and the birthplace of some mantle plumes.

CONCEPT CHECKS 12.2

1. How do continental crust and oceanic crust differ?

2. List and briefly describe the composition and physical properties of the layers of the mantle.
Concept Checker
https://goo.gl/xXCMpK

3. Compare and contrast Earth's inner and outer cores.

12.3 Discovering Earth's Layers

Explain how seismic waves were used to discover Earth's layered structure.

Seismic waves do not travel along straight paths; instead, they are *reflected* and/or *refracted* as they pass through our planet (**Figure 12.8**). When seismic waves encounter a boundary between different Earth materials, some of the waves are **reflected**—they "bounce off" the boundary layer. You are familiar with reflected sound waves called echoes. Seismic waves that pass between layers of different density or composition are **refracted** (change direction). This is similar to how light is refracted (bends) as it passes from air to water. These properties of seismic waves were used to discover the major boundaries within our planet.

Discovering the Moho

As mentioned previously, the boundary between the crust and mantle is called the **Moho**. It was the first structure of Earth's interior discovered using seismic waves. In 1909, Croatian seismologist Andrija Mohorovičić identified this boundary, which was then named after him. Mohorovičić made this discovery when he observed that seismograms from shallow earthquakes produce two sets of P waves. One set of waves moved through the ground at about 6 km/s, while the other set of waves traveled about 8 km/s. This difference led Mohorovičić to correctly predict that the waves were traveling through different layers. In particular, he concluded that the slower seismic waves followed a direct path though Earth's crust, whereas the faster, refracted waves traveled through a high-velocity layer below the crust, which today we call the mantle. This finding provided strong evidence that the crust and underlying mantle are composed of different rock types—a hypothesis that has since been verified.

The work done by Mohorovičić also provides a method to calculate the thickness of the crust for any location. *Direct waves* travel along a nearly straight path through the crust, while *refracted waves* follow a path downward through the crust and travel along the top of the mantle (**Figure 12.9**). Seismographs closest to the epicenter record the slower direct waves first. By contrast, seismographs farther from the epicenter record the faster refracted waves first. The point at which both waves arrive at the same time, called the *cross-over*, can be used to determine the depth of the Moho.

The difference between travel times for direct waves and for refracted waves is comparable to the difference between driving to a destination on local roads and on interstate highways. For short distances, you typically arrive sooner if you drive the most direct route—which usually requires going slower on local roads. For long distances, the trip may take less time if you take a less direct route that involves driving on mostly interstate highways. The cross-over point, where both routes take an equal amount of time, is directly related to how far you must drive before reaching the interstate highway.

Applied to determining the depth of the Moho, the cross-over is related to how far seismic waves travel through the crust (slower layer) before they reach the mantle (faster layer): The greater the cross-over distance, the deeper the Moho and, therefore, the thicker the crust. The Moho lies about 25 to 70 kilometers (15 to 45 miles) beneath the continents and about 7 kilometers (4 miles) below the ocean floor.

Discovering the Core

Evidence that Earth has a central core was uncovered in 1906 by British geologist Richard Oldham. At locations beyond approximately 100 degrees (a little more than the distance from the poles to the equator) from the epicenter of an earthquake, Oldham observed that P and S waves were absent or very weak. In other words, Oldham found evidence for a central core that causes P waves to be refracted and thereby produce a seismic

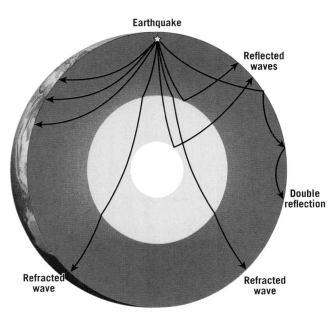

◄ **Figure 12.8**
Possible paths that seismic rays follow through Earth

Earthquake

Reflected waves

Double reflection

Refracted wave

Refracted wave

▶ Figure 12.9
Discovering the Moho The diagram shows how seismic waves were used to discover the Moho and also how seismologists determined its depth.

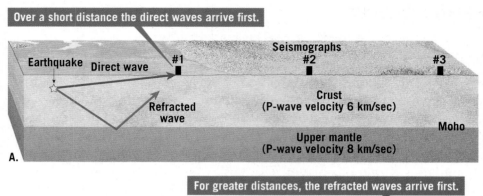

Over a short distance the direct waves arrive first.

Earthquake   Direct wave   #1   Seismographs #2   #3

Refracted wave   Crust (P-wave velocity 6 km/sec)

Moho

Upper mantle (P-wave velocity 8 km/sec)

A.

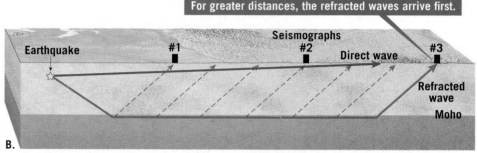

For greater distances, the refracted waves arrive first.

Earthquake   #1   Seismographs #2   #3

Direct wave

Refracted wave

Moho

B.

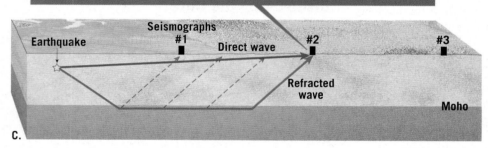

At the cross-over point, both waves arrive at the same time. The distance to the cross-over point increases as the depth of the Moho increases, and therefore can be used to determine the thickness of the crust at different locations.

Earthquake   Seismographs #1   Direct wave   #2   #3

Refracted wave

Moho

C.

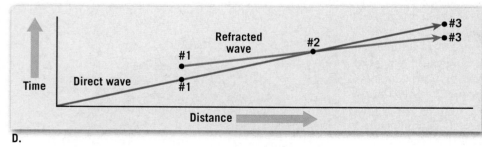

Time

Direct wave
#1
#1

Refracted wave
#2

#3
#3

Distance

D.

Discovering the Inner Core

In 1936, Danish seismologist Inge Lehman discovered that Earth has an inner core that is distinct from its molten outer core. By examining seismograph records of a strong earthquake that occurred in New Zealand, Lehman observed that some weak P waves arrived in the P-wave shadow zone, as shown in Figure 12.10A. From these findings, Lehman concluded that these waves must have been reflected (bounced) off a previously unknown boundary within Earth's core—and, hence, was the first researcher to identify the boundary between the outer core and the inner core.

Shortly after the inner core was discovered, it was hypothesized that it was solid. However, its rigid (solid) nature was not confirmed until 1971. Lehman's work was groundbreaking because seismic waves weaken as they travel through Earth's interior, and the great distances traveled made it difficult to detect these signals. With today's modern seismic networks, much finer details of Earth's interior have been discovered.

shadow zone—much as a tree produces a shadow for light waves. In addition, because the outer core is molten, it blocks the transmission of S waves, which do not travel through liquids.

The locations of the P- and S-wave shadow zones produced by the Earth's core are shown in **Figure 12.10**. Although some P and S waves still arrive in the shadow zone, they are weak because they have either passed though the outer core, a molten region, or have traveled a long distance.

CONCEPT CHECKS 12.3

1. Name the boundary that separates the crust and the mantle and describe how it was discovered.

2. What seismic evidence tells us that Earth's outer core is molten?

 **Concept Checker**
https://goo.gl/2gbKw5

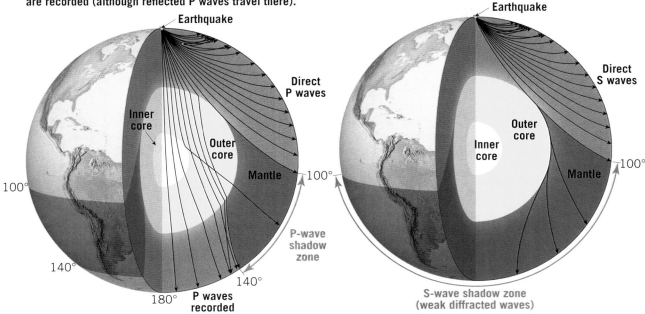

A. The P-wave shadow zone exists because P waves interact with the low-velocity liquid iron of the outer core, which causes their rays to be refracted downward. This creates a shadow zone where no direct P waves are recorded (although reflected P waves travel there).

B. The core is an obstacle to S waves, because they cannot pass through liquids. Therefore, a large shadow zone exists for direct S waves.

◄ SmartFigure 12.10
P- and S-wave shadow zones
A. Some P waves are reflected from the inner core so they arrive within the P-wave shadow zone, shown here as a single ray.
B. Some S waves are bent around the outer core and recorded on the other side of the planet within the S-wave shadow zone.

Tutorial
https://goo.gl/qfwajb

12.4 Earth's Temperature

Describe the processes of heat transfer that operate in Earth's interior and indicate in which layers these processes are dominant.

Like the other planets in our solar system, Earth has experienced two thermal stages (Figure 12.11). The first stage, which occurred during Earth's formation, involved a rapid increase in its internal temperature. The second stage, by contrast, has been very slow cooling over the remainder of Earth's history.

How Did Earth Get So Hot?

Earth formed through a very violent process involving the collisions of countless *planetesimals* (asteroid-size bodies) during the birth of our solar system (see Chapter 22). With each collision, the kinetic energy of motion was converted into thermal energy. As the early Earth grew in size, its temperature rapidly increased. Our young planet also contained many short-lived radioactive isotopes, such as aluminum-26 and calcium-41. As these isotopes decayed to stable forms, they released huge amounts of energy, called *radiogenic heat*.

Another significant event that heated our planet was the collision of a Mars-sized object with Earth that led to the formation of the Moon. At that time, the entire core and most, if not all, of the mantle was molten. From that point—about 4.5 billion years ago—to the present, Earth's interior has gradually cooled.

If Earth's only heat had come from its early formation and the decay of short-lived radioactive isotopes, our planet's interior would have cooled long ago—ending dynamic processes such as plate tectonics, earthquakes, and volcanism. However, the mantle and crust also contain long-lived radioactive isotopes that keep our planet cooking as if on a low burner.

How Does Heat Travel?

As you are probably aware, heat flows from hotter regions toward colder regions. Earth is about 5500°C at its center, about the temperature of the surface of the Sun, and 15°C at its surface. As a result, Earth's internal heat flows toward the surface.

Heat travels from Earth's interior to space via three different mechanisms: *convection*, *conduction*, and *radiation*. Only two of these processes, convection and conduction, operate within Earth's interior. The third, radiation, transports heat away from Earth's surface and eventually to space.

► SmartFigure 12.11
Earth's thermal history through time

Tutorial
https://goo.gl/
S6DWmj

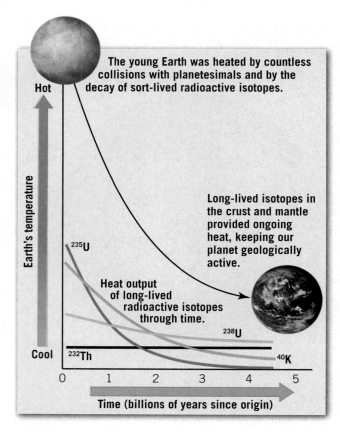

Most rocks are poor conductors of heat, so conduction is *not* an efficient way to move heat through most of Earth. Nevertheless, it is an important heat transfer mechanism in the lithosphere and the D″ layer, and it operates most efficiently in the solid inner core, which is metallic.

How Heat Moves from the Earth's Interior to the Surface The dominant types of heat transfer through Earth's major layers are illustrated in **Figure 12.12**. Notice that conduction is thought to be the most important process in the solid, metallic inner core. When heat conducts from the inner core to the outer core, convection begins to play a more significant role.

The transfer of energy from the outer core to the mantle occurs by conduction because the iron-rich material of the core is too dense to mix with the less dense mantle rocks above. For thermal energy to leave the core, it must conduct across the core–mantle boundary and up through the D″ layer—a relatively slow process. Once it reaches the lower mantle, thermal energy moves upward mainly by convection.

Although the mantle convection process is still not completely understood, an idealized model is shown in **Figure 12.13**. In this model, Earth's rigid outer shell, the *lithosphere*, works in tandem with the mantle to generate convective flow. Notice that after the oceanic lithosphere

Convection Heat transfer by means of hot materials rising to displace cooler materials (or vice versa) is called **convection**, which is the primary means of heat transfer within Earth. You are familiar with convection if you have ever watched boiling water in a pot. The water appears to be rolling: As water at the bottom of a pot is heated, it expands and rises, replacing the cooler, denser water at the top.

Convection can also occur when differences in density are caused by chemical rather than thermal means. *Chemical convection* is an important mechanism operating in the outer core. As iron crystallizes and sinks to form the solid inner core, it leaves behind a molten material that contains a higher percentage of lighter elements. Because this liquid is more buoyant than the surrounding iron-rich material, it rises and contributes to convective flow in the outer core.

Conduction The flow of heat *through a material* is called **conduction**. Heat conducts in two ways: (1) through the collisions of atoms and (2) through the flow of electrons. In rocks, atoms are locked in place but are constantly oscillating. If one side of a rock is heated, the atoms on that side will oscillate more energetically. This increases the intensity of the collisions with neighboring atoms, and in a domino effect, the thermal energy *slowly* propagates through the rock. By contrast, conduction occurs quickly in metals because some of their electrons are free to move and, thus, transport heat quickly from one side of a metallic object to another.

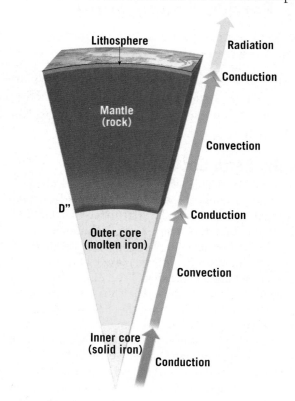

▲ Figure 12.12
Dominant types of heat transfer at various depths Heat travels from Earth's interior to the surface through the processes of convection and conduction. However, Earth ultimately loses its heat to space through radiation.

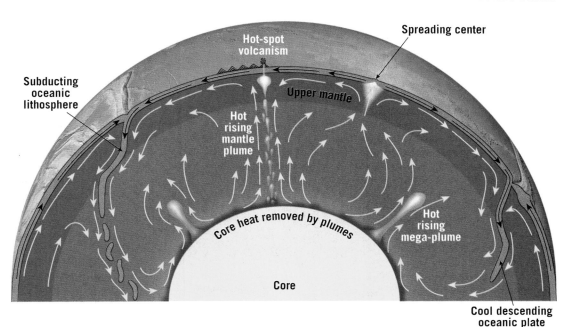

Subducting oceanic lithosphere

Hot-spot volcanism

Spreading center

Upper mantle

Hot rising mantle plume

Core heat removed by plumes

Hot rising mega-plume

Core

Cool descending oceanic plate

◄ **Figure 12.13**
Whole-mantle convection According to this model, the entire mantle is in motion, driven by the sinking of cold oceanic lithosphere back into the deep mantle. The upward flow occurs through a combination of mantle plumes and upwelling of hot mantle rock at oceanic spreading centers. Not all scientists agree that this model is accurate.

forms, it cools rapidly, mainly from the circulation of cool seawater through fractures in the seafloor. This causes oceanic lithosphere to contract, become denser and heavier, and, in time, sink back into the mantle at subduction zones. Cool oceanic lithosphere can be thought of as the top of a mantle–plate convection system. By contrast, the rising hot mantle rock below spreading centers and the warm buoyant rocks called **mantle plumes** are the upward-flowing arms of this convection cell (see Figure 12.13).

Most of the thermal energy that reaches the upper mantle makes its final journey to Earth's surface by slowly conducting across the solid, rigid lithosphere. Heat that reaches Earth's surface does not escape at the same rate in all locations. Heat flow is highest near mid-ocean ridges, where hot magma is found only a few kilometers below the surface. The rate of heat flow is moderate in continental regions, where the rocks are enriched in radioactive isotopes that emit their own source of heat. By contrast, heat flow is lowest in the deep abyssal plains, which are areas of old, cold oceanic seafloor.

Convection is the dominant mode of heat transfer within Earth's mantle and outer core— and possibly within the inner core as well.

Earth's Temperature Profile

Within the crust, temperatures increase rapidly with depth—as much as 30°C per kilometer. Near the top of the mantle, about 100 kilometers below the surface, the temperature is roughly 1400°C. However, for most of the mantle, the temperature increases very slowly—about 0.3°C per kilometer. You would need to descend to nearly the bottom of the mantle before the temperature doubled to 2800°C. The exception to this pattern is within the thin D″ layer, which acts as a thermal boundary. Here temperatures increase by more than 1000°C from top to bottom.

The profile of Earth's average temperature at each depth is called the **geothermal gradient**, or **geotherm**. **Figure 12.14A** illustrates the geotherm, as well as the curve for the melting point of material at each depth. Note that the melting point curves increase gradually with depth as a result of the continual increase in pressure exerted by the overlying material. Most materials are more difficult to melt when they are squeezed (under pressure) because in solid form, materials take up less space than when they are in liquid form. As a result, higher pressures result in higher melting temperatures.

Considered together, the geotherm and the melting point curve are valuable tools for investigating the behavior of Earth's materials. In layers where the geotherm is greater than the melting temperature, the material is molten. As shown in Figure 12.14A, this condition is observed in the outer core.

The relationship between the geotherm and the melting temperature not only determines whether a material is molten but also indicates its **viscosity**— resistance to flow. Notice how viscosity, shown in Figure 12.14B, is directly related to the proximity of the geotherm curves to the melting point curves in Figure 12.14A. When rocks approach their melting point, they begin to weaken and flow more readily. Low-viscosity regions, such as the asthenosphere and D″, are weak. High-viscosity regions, such as the lower mantle and lithosphere, are strong and rigid (see Figure 12.14B).

Research shows that convective flow is several times slower in the lower mantle than in the upper mantle.

▶ SmartFigure 12.14
The geotherm and melting point curve are valuable tools for investigating Earth materials at depth If you compare these two graphs, you can see that rocks are weakest and flow more easily at depths where the temperatures of rocks are close to melting (the asthenosphere and D″ layer). High viscosities, like those of the crust and lithosphere, show rock that is stiffer and flows less easily.

Tutorial
https://goo.gl/
cZGLNt

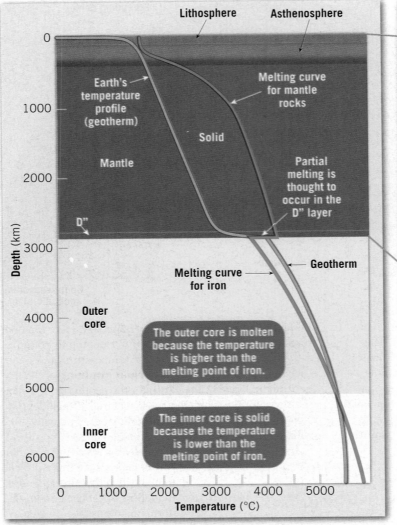

A. Earth's temperature profile with depth and melting point curve for various materials.

B. How viscosity (resistance to flow) changes with depth from Earth's surface to the bottom of the mantle.

Pressure, however, nearly triples, increasing from 1.36 to 3.64 megabars (1 megabar is 1 million times greater than standard atmospheric pressure). Although the temperature is cooler in the outer core than in the inner core, the outer core remains liquid because it is under less pressure. Conversely, iron in the inner core remains solid at these high temperatures because its melting temperature increases dramatically due to the extreme pressures.

However, at the very base of the mantle, temperature increases rapidly with depth. As a result, rock in the D″ layer is relatively weak, flows more easily, and may experience some melting.

In the core, temperature increases at a slower rate than pressure increases. From the core–mantle boundary to Earth's center, temperatures increase by only about 40 percent, or from about 4000° to 5500°C.

CONCEPT CHECKS 12.4

1. What were the two sources of Earth's original internal heat?

2. List the two mechanisms of heat transfer that operate inside Earth.

3. Which mechanism of heat transfer is dominant in the mantle?

Concept
Checker
https://goo.gl/oUEPsr

12.5 Horizontal Variations in Earth's Interior

Discuss what seismic tomography has revealed about variations in Earth's layers.

Earth exhibits not only vertical variations (layering) but horizontal variations as well (differences within a given layer). Most of these horizontal variations, which are directly related to the processes of mantle convection and plate tectonics, have been identified by studying variations in Earth's gravitational and magnetic fields as well as imaging called *seismic tomography*.

Earth's Gravity

Earth's rotation is the most significant cause of the differences in the force of gravity observed at the surface. Because Earth rotates around its axis, the acceleration due to gravity[‡] is less at the equator (9.78 m/s²) than at the poles (9.83 m/s²). Two reasons account for this phenomenon. First, Earth's rotation generates a centrifugal force that is in proportion to the distance from the axis of rotation. Similar to the force that throws you sideways in a vehicle going too quickly around a curve, centrifugal force acts against gravity (outward) at the equator, where the force is greatest.

Second, Earth's rotation also affects its shape—the equator is slightly further from Earth's center (6378 kilometers) than the poles (6357 kilometers) (**Figure 12.15**). Earth, therefore, is not a perfect sphere but instead bulges at the equator—a shape called an *oblate spheroid*. This difference causes the force of gravity to be slightly weaker at the equator than at the poles because gravitational attraction decreases when objects are farther apart. In fact, a person's weight is 0.5 percent less at the equator than at the poles (see Figure 12.15).

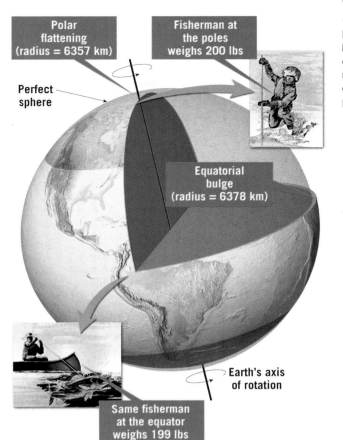

◄ Figure 12.15
Earth is not a sphere but an oblate spheroid Because Earth rotates, it bulges at the equator and flattens at the poles.

What Gravity Tell Us About Earth's Structure

Gravity measurements indicate that some variations cannot be explained by Earth's rotation. For instance, for a large body of unusually dense rock under Earth's surface, the increase in mass will cause a larger-than-average gravitational force, called a **gravity anomaly**, at the surface directly above. Because metals and metallic ores tend to be much denser than silicate rocks, *positive gravity anomalies* (larger-than-average force) have long been used to help prospect for ore deposits.

A map of regional gravity anomalies for the United States is shown in **Figure 12.16**. Notice the narrow positive gravity anomaly that runs down the middle of the country. This is the mid-continent rift (red), where thick, dense volcanic rocks filled a rupture in the crust more than 1 billion years ago. A *negative gravity anomaly* (blue), located in the Basin and Range region of the western United States, is where

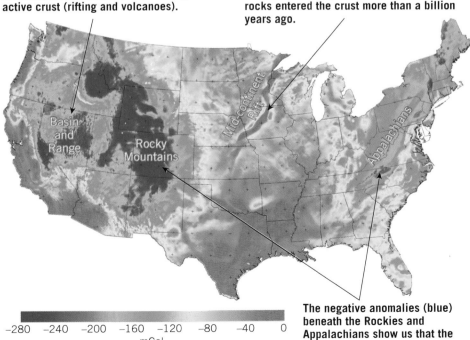

The negative anomaly (blue) in the Basin and Range Province is the result of hotter, less dense, and tectonically active crust (rifting and volcanoes).

The narrow positive anomaly (red) that runs in a line down the middle of the country is the mid-continent rift, where dense volcanic rocks entered the crust more than a billion years ago.

The negative anomalies (blue) beneath the Rockies and Appalachians show us that the crust has deep roots made up of less dense rock beneath the mountains.

−280 −240 −200 −160 −120 −80 −40 0
mGal

◄ Figure 12.16
Gravity anomalies beneath the continental United States

[‡]The force of gravity causes objects, such as apples, to accelerate as they fall to the ground, hence the expression "acceleration due to gravity."

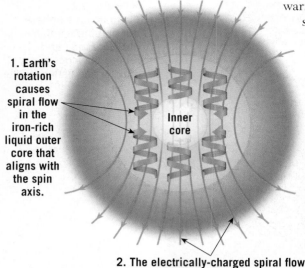

1. Earth's rotation causes spiral flow in the iron-rich liquid outer core that aligns with the spin axis.

Inner core

2. The electrically-charged spiral flow in the outer core generates Earth's magnetic field. This is similar to how an electromagnet works.

▲ Figure 12.17
Earth's magnetic field is generated in the liquid, iron-rich outer core

▼ Figure 12.18
Similarities between Earth's magnetic field and an electromagnet

warm, low-density crust is being stretched and thinned as it is being intruded by hot buoyant magma from below.

Some large-scale differences in density beneath Earth's surface have been detected using satellites. These gravity anomalies result from the large upwellings and downwellings of mantle rocks. Areas of upwelling are associated with hot mantle plumes, whereas downwelling occurs where cold oceanic slabs descend into the mantle.

Earth's Magnetic Field

Convection of molten iron in the outer core is vigorous, which makes the outer core *appear* uniform when viewed with seismic waves. In reality, however, patterns of flow in the outer core create variations in Earth's magnetic field that are measurable at the surface and tell a different story.

The Geodynamo   Earth's **magnetic field** has a complex structure that reflects the processes that generate it. As iron-rich fluid in the outer core rises, its path becomes twisted due to Earth's rotation. As a result, the fluid moves in spiraling columns that align with Earth's axis of rotation (**Figure 12.17**). Because this iron-rich fluid is electrically charged and flowing, it generates a magnetic field—a phenomenon called a **geodynamo**. This process is similar to how an electromagnet generates a magnetic field.

When an electric current is passed through a copper wire wrapped around an iron nail, the nail becomes a simple bar magnet that generates a magnetic field, like the one shown in **Figure 12.18A**. This type of magnetic field is called a *dipolar field* because it has two poles (north and south magnetic poles). As shown in Figure 12.18B, the magnetic field that emanates from Earth's outer core is also dipolar.

However, the magnetic field generated in the outer core is much more complex. More than 90 percent of Earth's magnetic field is dipolar, but the remainder is a result of more complicated patterns of convection that occur in the molten outer core. As a result, the features of Earth's magnetic field change over time.

For centuries, sailors have used compasses to navigate, primarily keeping track of the direction toward which compass needles point. From compass observations, it was determined that the positions of Earth's magnetic poles gradually change. Understanding this phenomenon requires an examination of how the orientation of magnetic field is measured.

Measuring Earth's Magnetic Field   On Earth's surface, the direction of the magnetic field is measured with two angles, called *declination* and *inclination*. The declination measures the direction to the magnetic north pole with respect to the direction to the geographic North Pole (Earth's axis of rotation). The inclination measures the downward tilt of the magnetic lines of force at any location—what a compass would show if tilted on its side. At the magnetic north pole, the field points directly downward, while at the equator, it is horizontal (**Figure 12.19**). In the central United States, it tilts downward at an intermediate angle, depending on the observer's latitude.

From these measurements, it became clear that the locations of both magnetic poles have changed significantly over time. Earth's magnetic north pole was previously located in Canada but moved northward into the Arctic Ocean during the past decade. Currently, it is moving northwestward at over 70 kilometers (40 miles) per year toward the geographic North Pole. Similarly, the magnetic south pole has been moving *away* from the geographic South Pole, going from Antarctica toward the Pacific Ocean.

These changes in location of the magnetic poles are caused by gradual changes in the convection pattern operating in the outer core. Nevertheless, studies have shown that the locations of the magnetic poles averaged over thousands of years align with Earth's axis of rotation (geographic poles).

Magnetic Reversals   One major exception to the averages noted above occurs during periods of **magnetic reversals**. Over relatively short time spans, geologically speaking, Earth's magnetic field reverses polarity such that the *north* needle on a compass instead points *south*. (The importance of these

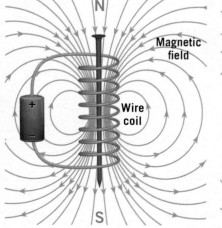

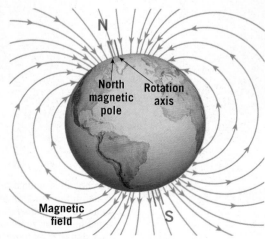

N

Magnetic field

Wire coil

+

−

S

A. Electromagnet (Dipolar field)

N

North magnetic pole

Rotation axis

Magnetic field

S

B. Earth's magnetic field (Dipolar field)

reversals in the study of paleomagnetism is described in Chapter 2.) During a reversal, the strength of the magnetic field decreases to about 10 percent of normal, and the locations of the poles begin to wander, going so far as to cross the equator (**Figure 12.20**). When the strength of the magnetic field returns to normal levels, the field is regenerated with reverse polarity. The entire process takes only a few thousand years.

The rate at which the magnetic field reverses provides evidence that convection patterns in the outer core change over relatively short time spans. This complex process is now being modeled using high-speed computers. In addition, Figure 12.20 illustrates how the magnetic field lines twist in a complex manner before returning to a more uniform, simpler dipolar pattern.

A magnetic reversal could have potentially harmful consequences for Earth's land dwellers. An atmospheric magnetic layer, known as the *magnetosphere*, surrounds our planet and protects Earth's surface from bombardment by ionized particles, called *solar wind*, emitted by the Sun. Prior to a reversal, the strength of the magnetic field would decrease significantly. The result would be an increase in the number of ionized particles reaching Earth's surface—likely becoming a health hazard for humans and other life-forms.

Seismic Tomography

The changes in composition and density detected with gravity measurements can also be viewed using seismology. The technique, called **seismic tomography**, involves collecting signals from many different earthquakes recorded at many seismograph stations in order to "see" all parts of Earth's interior. Seismic tomography is similar to medical tomography, in which doctors use technology such as CT scans to make three-dimensional images of humans' internal organs.

Seismic tomography identifies regions where P or S waves travel faster or slower than average at a particular depth. These *seismic velocity anomalies* are then interpreted as variations in material properties such as temperature, composition, mineral phase, or water content. For instance, increasing the temperature of rock about 100°C can decrease S-wave velocities about 1 percent, so images from seismic tomography are often interpreted in terms of temperature variations.

A seismic tomography cross-section for the mantle is shown in **Figure 12.21**. Regions where seismic waves travel more slowly than average (*negative anomalies*) are red, and regions where waves travel more quickly than average (*positive anomalies*) are blue. Significant patterns can be observed in this diagram. For example, the lithosphere beneath the interiors of North America and Africa exhibits faster seismic velocities than oceanic lithosphere because continental lithosphere is older and stiffer. (Recall that continental lithosphere has been cooling for billions of years.) Seismic imaging also shows

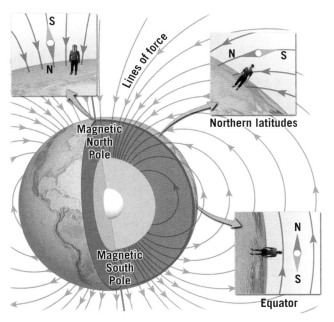

that continental lithosphere (deep-blue areas) can be quite thick, extending more than 300 kilometers (180 miles) into the mantle. Conversely, oceanic ridges such as the Mid-Atlantic Ridge exhibit slow seismic velocities because they are areas of rocks at or above their melting temperatures (see Figure 12.21).

In the mantle beneath North America, note the sloping zone of fast seismic velocities (blue/green), which represent a sheet of descending oceanic lithosphere known as

◀ **Figure 12.19**
Inclination, or dip, of the magnetic field at different locations Although a compass measures only the horizontal direction of the magnetic field (the *declination*), at most locations, the field also dips in or out of the surface at various angles (*inclination*).

A. Normal orientation of magnetic field

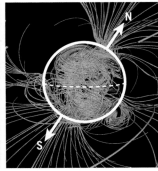

B. Magnetic field weakens and poles begin to wander

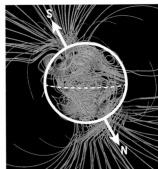

C. Poles wander across the equator

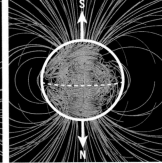

D. Reversal complete with north pole pointing south

◀ **Figure 12.20**
Computer simulations showing how Earth's magnetic field might reverse direction The white circle represents the core–mantle boundary, and the dashed white line represents the projected position of the equator. The arrows point to the north (N) and south (S) magnetic poles. During a reversal, the strength of the magnetic field weakens, and the poles begin to wander greatly, going so far as to cross the equator. When the strength of the field returns to normal levels, the field is regenerated with reverse polarity.

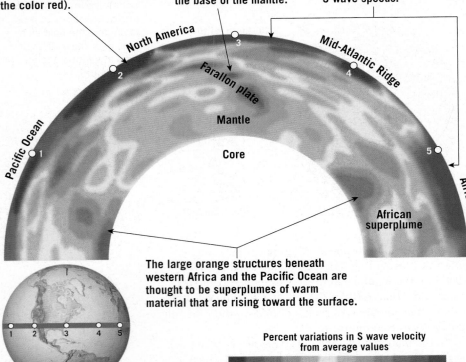

The western United States is tectonically active, making that portion of the continent warmer and weaker (shown by the color red).

The large blue structure extending far beneath North America is likely a slab of cold, dense, ancient Pacific seafloor that is sinking toward the base of the mantle.

Older portions of continents, North America and Africa, are cold and strong—the blue color indicates fast S wave speeds.

North America

Farallon plate

Mid-Atlantic Ridge

Pacific Ocean

Mantle

Core

African superplume

Africa

The large orange structures beneath western Africa and the Pacific Ocean are thought to be superplumes of warm material that are rising toward the surface.

Percent variations in S wave velocity from average values

–1.5% –1.0% –0.5%   0   0.5% 1.0% 1.5%
Slow   Fast

◄ Figure 12.21
A seismic tomographic slice, showing the structure of the mantle Colors show variations in the speed of S waves from their average value. Cool, strong rocks are shown in blue, while warm, weak rocks are shown in red.

the Farallon plate (see Figure 12.21). The cool segment of this former slab of seafloor is currently sinking and warming as it moves toward the core–mantle boundary. Given enough time, material in this slab may become hot and buoyant enough to rise back to the surface.

The large region of slow seismic velocities beneath Africa (the large reddish-orange region at the lower right of Figure 12.21) is called the *African superplume*—a region of upward flow in the mantle. These slow velocities are likely due to unusually high temperatures. The rising rock cannot easily break through the thick African crust, so it is deflected to both sides of the continent, perhaps supplying magma to both the Mid-Atlantic and Indian Ocean spreading centers.

CONCEPT CHECKS 12.5

1. Is Earth's force of gravity equal over its entire surface? Explain why or why not.

2. Describe how the locations of the magnetic poles change through time.

Concept Checker
https://goo.gl/UhYJM9

3. Give an example of a tectonic structure revealed by seismic tomography.

12 CONCEPTS IN REVIEW

Earth's Interior

12.1 Exploring Earth's Interior

Explain how Earth acquired its layered structure.

Key Terms: peridotite   seismic wave
xenolith   meteorite

- The layered structure of Earth developed due to gravitational sorting of Earth materials early in the planet's history. The densest materials sank to the center, and the least dense material rose toward the surface.
- The upper continental crust has an average composition of granodiorite, a felsic rock. This has been determined by direct observation of the crust's mineral makeup.
- Mantle-derived magmas that have ascended to the surface often incorporate other rocks, called *xenoliths*, which are samples of the upper mantle. Based on this and other information, we know that the upper mantle is mainly composed of the rock *peridotite*.

- Iron *meteorites*, remnants of small bodies that formed early in the history of the solar system, provide evidence that Earth's composition should be richer in iron than is indicated by the rocky mantle and crust. Therefore, the additional iron must be located in the core.
- *Seismic waves* generated from large earthquakes allow geoscientists to "look" into Earth's interior. Like the X-rays used to image human bodies, seismic waves reveal details about Earth's layered structure.
- In general, seismic waves travel fastest when rock is stiff (rigid) or less compressible. When heated, rock becomes less stiff, and earthquake waves travel through it more slowly. These properties of stiffness and compressibility are used to interpret temperature—how near a rock is to its melting point at various depths.
- When P waves travel through molten or partially molten rock, they travel much slower than they would if the rock were solid. Furthermore, S waves are not transmitted by liquids. Because S waves do not travel through Earth's outer core, we conclude that it is molten.

12.2 Earth's Layered Structure

List and describe each of the three major layers of Earth's interior.

Key Terms:	lithosphere	D″ layer
crust	asthenosphere	core
mantle	transition zone	outer core
upper mantle	lower mantle	inner core
lithospheric mantle		

- Earth has two distinct kinds of *crust*: oceanic and continental. Oceanic crust is thinner, denser, and younger than continental crust. Oceanic crust also readily subducts, whereas the less dense continental crust does not.
- The uppermost mantle, called the *lithospheric mantle*, makes up the bulk of rigid lithospheric plates, while a relatively weak layer, the *asthenosphere*, lies beneath it.
- Between 410 and 660 kilometers in depth, in the *upper mantle*, lies the *transition zone*. The top of the transition zone is identified by a sudden change in seismic velocities, caused by atoms in olivine rearranging to form a denser structure caused by an increase in pressure (phase change).
- The *lower mantle* lies below the transition zone. At its base, just above the core, is the unusual *D″ layer*. Seismic-wave velocities slow dramatically in the D″ layer, indicating that it may be partially molten.

- The composition of Earth's *core* is a mixture of iron, nickel, and lighter elements. The *outer core* is molten, as S waves cannot pass through it. The *inner core* is solid and very dense—more than 13 times the density of water.

Q Label the accompanying diagram using 6 of the following terms: continental crust, Moho, asthenosphere, lithospheric mantle, oceanic crust, upper mantle, lithosphere.

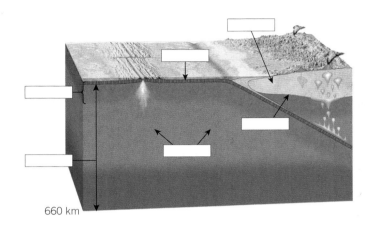

660 km

12.3 Discovering Earth's Layers

Explain how seismic waves were used to discover Earth's layered structure.

Key Terms:	refracted	shadow zone
reflected	Moho	

- When seismic waves encounter a boundary between different Earth materials, they may either reflect or refract. *Reflection* is like a ball bouncing off a wall: The wave "bounces" away from the boundary layer. *Refraction* is the process whereby seismic waves continue into the new layer but bend (change direction).

- The base of the crust is called the Mohorovičić discontinuity, or *Moho*. It was discovered because seismic waves travel faster below the Moho (in the mantle) than above it (in the crust). Beneath the continents, the Moho lies 25 to 70 kilometers below the surface. In oceanic lithosphere, the Moho occurs at a depth of about 7 kilometers.
- The core–mantle boundary causes P waves to be refracted (change direction) and thereby produces a seismic *shadow zone*—much as a tree produces a shadow for light waves.

12.4 Earth's Temperature

Describe the processes of heat transfer that operate in Earth's interior and indicate in which layers these processes are dominant.

Key Terms:	conduction	geothermal gradient
convection	mantle plume	viscosity

- There are two main sources of heat in Earth's interior. Some thermal energy is left over from the collision of countless small bodies during the formation of our solar system. The remainder comes from radioactivity. Early in Earth's history, short-lived isotopes generated heat through radioactive decay. Radioactivity of long-lived isotopes continues to add heat to Earth but at a lower rate.
- Heat flows from Earth's interior outward. *Convection* is the most important mechanism of heat transfer in Earth's interior, occurring in the liquid outer core and in the mantle. *Conduction* is the mechanism by which the core heats the mantle.
- The *geothermal gradient* is the profile of Earth's temperature with depth. Temperature increases rapidly in both the lithosphere and the D″ layer. Other layers experience a relatively gentle *geotherm* of 0.3°C per kilometer.

- The position of the melting point curve relative to the geotherm is a key to the behavior of material in Earth's interior. In layers where the geotherm is greater than the melting point curve, the material is partially or completely molten.
- *Viscosity* describes resistance to flow in a material. Low-viscosity regions of Earth's interior, such as the asthenosphere and D″ layer, are weak. High-viscosity regions, such as the lower mantle and lithosphere, are strong and rigid.

Liquid outer core

Solid inner core

Q Using the accompanying diagram as a reference, explain how the outer core can be molten if the even hotter inner core is solid.

12.5 Horizontal Variations in Earth's Interior

Discuss what seismic tomography has revealed about variations in Earth's layers.

Key Terms:
gravity anomaly
magnetic field
geodynamo
magnetic reversal
seismic tomography

- Earth exhibits not only vertical variations (layering) but horizontal variations (differences within a given layer). The composition and structure of the mantle and core vary from place to place, just as the crust varies.

- *Gravity anomalies* occur where rock is more dense (often because it is cold) or less dense (often because it is warm). Negative gravity anomalies are produced by rising hot mantle rock, whereas positive anomalies are associated with cold subducting lithospheric slabs.

- Convection of the molten, iron-rich outer core generates a *magnetic field* that changes in both polarity and orientations of its poles.

- *Seismic tomography* is a technique for "seeing" Earth's interior that involves collecting signals from many different earthquakes recorded at seismograph stations around the globe. Seismic tomography uses technology similar to medical CT scans to make three-dimensional images of Earth's interior.

GIVE IT SOME THOUGHT

1. Arrange the following Earth layers in order from most dense to least dense: oceanic crust, atmosphere, core, continental crust, mantle, oceans.

2. Explain how the core is only one-sixth of Earth's volume yet comprises one-third of its mass.

3. Earthquakes below the Yellowstone caldera originate at very shallow depths, about 4 kilometers on average. Below this depth, the rocks are at about 400°C, too hot and weak to store elastic energy. Based on this data, answer the following questions:

   a. What is the average geothermal gradient in the first 4 kilometers beneath the Yellowstone caldera, assuming an average surface temperature of 0°C (32°F) and assuming that the temperature at 4 kilometers is 400°C (752°F)?

   b. At about what depth is the groundwater below the Yellowstone caldera hot enough to "boil" and therefore capable of generating a geyser?

4. The accompanying diagram shows the internal structures of Earth, Mars, and Earth's Moon. Based on what you know about the composition of the layers shown and how density increases with depth, list these three bodies in order from most dense to least dense.

5. Describe how Earth's inner core grows in size.

6. Explain why convection is an inefficient means of heat transfer in materials with high viscosity.

7. Use the depth data from this figure to draw a model of Earth's layers. Use a metric measuring tape and a scale of 1 centimeter in the model to represent each kilometer of depth in the actual Earth.

   a. Mark each major boundary.

   b. Using the accompanying diagram, take an imaginary "walk" from the surface to the core and describe the changes you would encounter along the way.

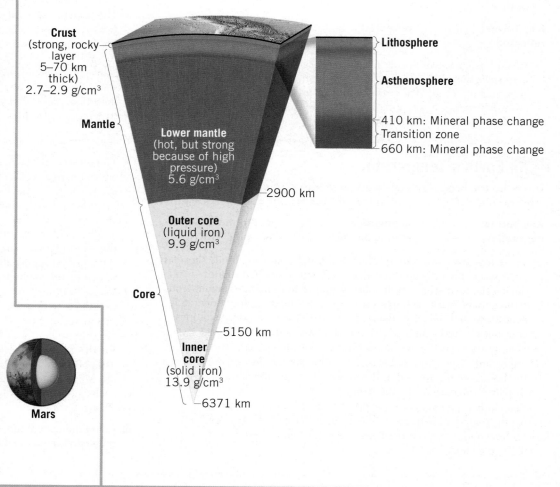

Crust (strong, rocky layer 5–70 km thick) 2.7–2.9 g/cm³

Lithosphere

Asthenosphere

410 km: Mineral phase change
Transition zone
660 km: Mineral phase change

Mantle

Lower mantle (hot, but strong because of high pressure) 5.6 g/cm³

2900 km

Core

Outer core (liquid iron) 9.9 g/cm³

5150 km

Inner core (solid iron) 13.9 g/cm³

6371 km

Moon

Mars

Key
Rocky crust
Rocky mantle
Metallic core

Earth

DATA ANALYSIS

Seismic Tomography

The velocity of earthquake waves is determined by the temperature and density of the medium through which they travel. Seismic tomography allows us to map the 3D structure beneath Earth's surface using the speed of waves produced by earthquakes.

https://goo.gl/eD226M

ACTIVITIES

Go to the USArray webpage at http://www.usarray.org. Click "About" and then click the "What exactly is USArray?" link. Answer the following questions:

1. What exactly is USArray, and what does it do?

2. What are the four interrelated parts of the array? Describe each part in one sentence.

Click "About" and then click the "Transportable Array" link. The map shows the types of instruments and their current locations. TA stands for Transportable Array. Answer the following questions.

3. Which instrument type is the most evenly spread out?

4. Which instrument type is the most clustered? Why do you think the instruments are clustered in this way?

5. Which states are currently being studied by the Flexible Array?

Click "Science" and then click "Earthquake Tomography" and answer the following questions.

6. What is the name of the method used to create images that reveal homogeneities in the geologic structure of the mantle beneath North America?

7. Based on what you see in the text, do regions that are hotter and weaker (tectonically active) have faster or slower velocities?

Now examine the cross sections below the picture of the United States. Blue indicates a region of fast wave velocity, and red indicates a region of slow wave velocity.

8. Which region of the country has cool, dense rock beneath it?

9. Which region of the country has hot, weaker rock beneath it? Are all of the rocks in this region hot and weak?

10. Given what you know about the surface features of these regions, do your answers to Questions 8 and 9 make sense? Why or why not?

Mastering Geology

Looking for additional review and test prep materials? Visit pearson.com/mastering/geology to enhance your understanding of this chapter's content by accessing a variety of resources, including **Self-Study Quizzes, Geoscience Animations, SmartFigure Tutorials,** *Project Condor* **Videos, Mobile Field Trips, Dynamic Study Modules,** and an optional **Pearson eText.**

Jet Remains Lost but Shipwrecks Are Found

The disappearance of a Malaysia Airlines jet in 2014 sparked one of the largest and most expensive searches in aviation history. For years, authorities from multiple nations scoured over 120,000 square miles of remote Indian Ocean seabed. Sadly, the plane's final resting place remains unknown. However, searchers discovered several shipwrecks dating from the 1800s. They lie under more than 3 kilometers (2 miles) of water. Ghostly sonar images of one of the ships reveal that its metal hull and deck remain partially intact.

Today we have maps of Earth's ocean floor to a resolution of about 5 kilometers. These maps were generated using data from satellites and provide twice the resolution of ocean floor maps available prior to 2014. But that's still a vastly less detailed view than what Google Earth provides for dry land, where image resolution ranges from 15 meters to 15 centimeters. Visualizing seafloor topography requires sonar, and we've only collected sonar data on less than 20 percent of deep-ocean regions.

There is still hope that the long-missing Malaysia Airlines jet may someday be found, but more of the world's ocean floor must be charted. Seabed 2030 is an organization whose goal is to create detailed sonar-based maps of the full ocean floor by the year 2030. It is taking on this ambitious task through data sharing—both by signing on ships to be part of the project, and by encouraging other groups to share sonar data they may have accu-mulated for commercial fishing or freight shipping. In fact, some of the sonar data from the missing plane search were donated to Seabed 2030, thereby furthering the public's understanding of this far-flung stretch of ocean.

▲ A multibeam sonar image shows remains of a gunship that sank off North Carolina's coast in 1918.

► Shipwrecks like this one in the Red Sea are popular diving destinations, but many uncharted wrecks lie in deeper, more remote areas.

13

Origin and Evolution of the Ocean Floor

FOCUS ON CONCEPTS

Each statement represents the primary learning objective for the corresponding major heading within the chapter. After you complete the chapter, you should be able to:

13.1 Define *bathymetry* and describe the various bathymetric techniques used to map the ocean floor.

13.2 Compare a passive continental margin with an active continental margin and list the major features of each.

13.3 List and describe the major features of deep-ocean basins.

13.4 Summarize the basic characteristics of oceanic ridges.

13.5 List the four layers of oceanic crust and explain how oceanic crust forms and how it differs from continental crust.

13.6 Outline the steps by which continental rifting results in the formation of new ocean basins.

13.7 Compare and contrast spontaneous subduction and forced subduction.

The ocean is Earth's most prominent feature, covering more than 70 percent of its surface. Yet, prior to the 1950s, information about the ocean floor was extremely limited. With the development of modern instruments, our understanding of the diverse topography of the ocean floor improved dramatically. Particularly significant was the discovery of the global oceanic ridge system, a broad elevated landform that stands 2 to 3 kilometers higher than the adjacent deep-ocean basins and is the longest topographic feature on Earth.

In this chapter, we will examine the topography of the ocean floor and look at the processes that generate its varied features.

13.1 An Emerging Picture of the Ocean Floor

Define *bathymetry* and describe the various bathymetric techniques used to map the ocean floor.

If all water were removed from the ocean basins, a great variety of features would be seen, including broad volcanic peaks, deep trenches, extensive plains, linear elevated landforms called ridges, and large plateaus. The scenery would be strikingly different from what we observe on Earth's continental surface.

Mapping the Seafloor

The complex nature of ocean floor did not begin to become known until the historic voyage of the HMS *Challenger* (**Figure 13.1**). From December 1872 to May 1876, the British research ship conducted the first comprehensive study of the global ocean. During the 127,500-kilometer (79,200-mile) voyage, the ship and its crew of scientists traveled to every ocean except the Arctic. They sampled a multitude of ocean properties, including water depth, which was accomplished by laboriously lowering a long, weighted line overboard and then retrieving it. Using this process, the *Challenger* made the first recording of the deepest-known point on the ocean floor in 1875. This spot, on the floor of the western Pacific, was later named the *Challenger Deep*.

Modern Bathymetric Techniques The measurement of ocean depths and the charting of the shape (topography) of the ocean floor is known as **bathymetry** (*bathos* = depth, *metry* = measurement). Today, sound energy is used to measure water depths. The basic approach employs **sonar**, an acronym for *so*und *na*vigation and *r*anging. The first devices that used sound to measure water depth, called **echo sounders**, were developed early in the twentieth century. Echo sounders work by transmitting a pulse of sound (an acoustic wave called a *ping*) into the water in order to produce an echo when it bounces off an object, such as a large marine organism, or the ocean floor (**Figure 13.2**). A sensitive receiver intercepts the echo reflected from the bottom, and a clock precisely measures the travel time to fractions of a second. By knowing the speed of sound waves in water—about 1500 meters (4900 feet) per second—and the time required for the energy pulse to reach the ocean floor and return, depth can be calculated. Depths determined from continuous monitoring of these echoes are plotted to create a profile of the ocean floor. By carefully combining profiles from several adjacent traverses, a chart of a portion of the seafloor is produced.

Following World War II, the U.S. Navy developed *sidescan sonar* to look for explosive devices that had been deployed in shipping lanes (**Figure 13.3A**). These torpedo-shaped instruments, called *towfish*, can be towed behind a ship, where they send out a fan of sound extending to either side of the ship's track. By combining swaths of sidescan sonar data, researchers produced the first photograph-like images of the seafloor. Although early sidescan sonar

▼ SmartFigure 13.1
HMS *Challenger* The first systematic bathymetric survey of the ocean was made aboard the HMS *Challenger* during its historic 3$\frac{1}{2}$-year voyage.

Tutorial
https://goo.gl/
CCEDdn

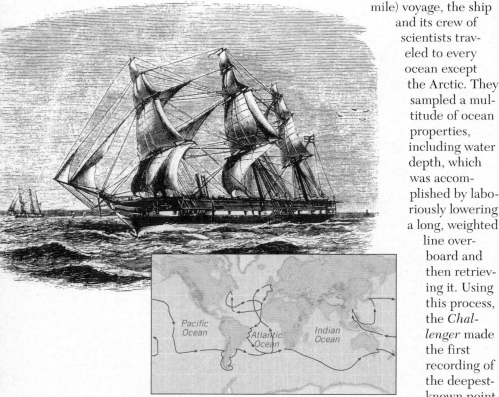

Pacific Ocean

Atlantic Ocean

Indian Ocean

▼ **Figure 13.2**
Echo sounder An echo sounder determines the depth of water by measuring the time interval required for an acoustic wave to travel from a ship to the seafloor and back.

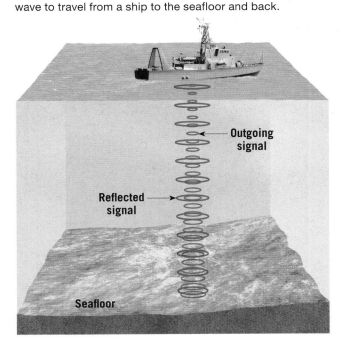

devices provided valuable views of the seafloor, they did not provide accurate bathymetric (water depth) data.

This drawback was resolved in the 1990s, with the development of *high-resolution multibeam sonar* instruments (see Figure 13.3A). These systems use hull-mounted sound sources that send out a fan of sound and then record reflections from the seafloor through a set of narrowly focused receivers aimed at different angles. Rather than obtain the depth of a single point every few seconds, this technique makes it possible for a survey ship to map the features of the ocean floor along a strip tens of kilometers wide. These systems collect bathymetric data of such high resolution that they can distinguish depths that differ by less than a meter (Figure 13.3B). When multibeam sonar is used to map a section of seafloor, the ship travels through the area in a regularly spaced back-and-forth pattern known as "mowing the lawn."

Despite their enhanced efficiencies and detail, research vessels equipped with multibeam sonar travel at a mere 10 to 20 kilometers (6 to 12 miles) per hour. It would take a single ship with modern sonar technology an estimated 350 years to map seafloor that is deeper than 200 meters (650 feet). This explains why less than 15 percent of the ocean depths have been measured directly—and only 50 percent of the world's coastal waters (less than 200 meters deep) have been mapped.

Despite these challenges, the international collaboration Seabed 2030, mentioned in the *In the News* feature at the beginning of this chapter, aims to collect all available bathymetric data into a high-resolution digital map of the ocean floor and to promote international efforts to collect additional data to fill in the gaps. The goal of Seabed 2030 is to map the entire ocean floor by 2030.

Mapping the Ocean Floor from Space

One breakthrough that enhanced our understanding of the seafloor involves measuring the shape of the ocean surface from space. After compensating for waves, tides, currents, and atmospheric effects, it was discovered that the ocean *surface* is not perfectly "flat." Because massive seafloor features exert stronger-than-average gravitational attraction, they produce elevated areas on the ocean surface. Conversely, canyons and trenches create slight depressions.

Satellites equipped with *radar altimeters* are able to measure these subtle differences by bouncing microwaves off the sea surface (**Figure 13.4**). Combined with traditional sonar measurements, these data are used to produce

A. Sidescan sonar and multibeam sonar operating from the same research vessel.

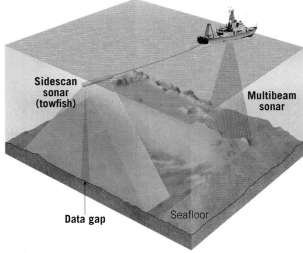

B. Color-enhanced perspective map of the seafloor and coastal landforms in the Los Angeles area of California.

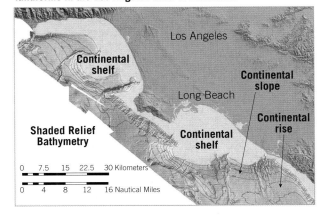

▲ **Figure 13.3**
Sidescan and multibeam sonar

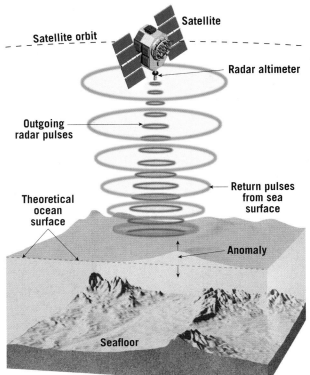

◀ **Figure 13.4**
Satellite altimeter A satellite altimeter measures the variation in sea-surface elevation, which is caused by gravitational attraction and mimics the shape of the seafloor. The sea-surface anomaly is the difference between the measured ocean surface and the theoretical ocean surface.

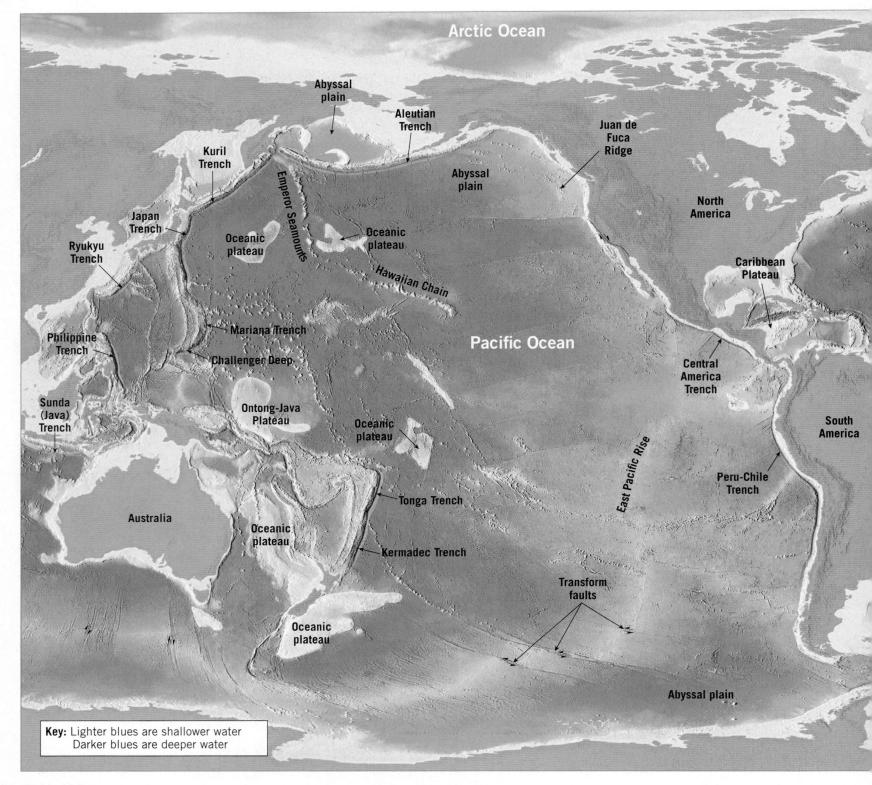

Arctic Ocean

Abyssal
plain

Aleutian
Trench

Juan de
Fuca
Ridge

Kuril
Trench

Emperor Seamounts

Abyssal
plain

Japan
Trench

Oceanic
plateau

Oceanic
plateau

North
America

Ryukyu
Trench

Hawaiian Chain

Caribbean
Plateau

Philippine
Trench

Mariana Trench

Pacific Ocean

Challenger Deep

Central
America
Trench

Sunda
(Java)
Trench

Ontong-Java
Plateau

Oceanic
plateau

South
America

East Pacific Rise

Peru-Chile
Trench

Tonga Trench

Australia

Oceanic
plateau

Kermadec Trench

Transform
faults

Oceanic
plateau

Abyssal plain

Key: Lighter blues are shallower water
Darker blues are deeper water

▲ Figure 13.5
**Major features of the
seafloor**

large-scale ocean-floor maps, such as the one shown in
Figure 13.5. Despite their value in identifying oceanic
topography, the maps generated by these techniques
provide only a low-resolution view of large structural

features of the ocean floor, including the oceanic ridges
and deep-ocean basins. The fine details of the seafloor
can only be obtained from direct measurements obtained
using sonar.

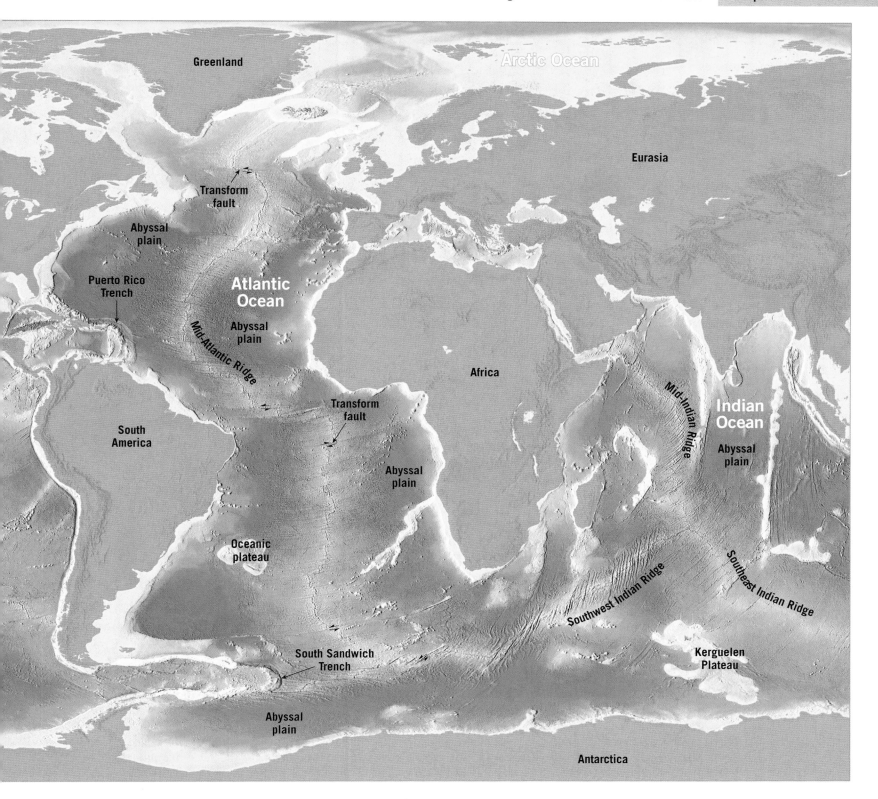

Provinces of the Ocean Floor

Using the maps produced by these various techniques, oceanographers recognize three major units: *continental margins*, *deep-ocean basins*, and *oceanic* *(mid-ocean) ridges*. The map in **Figure 13.6** outlines these provinces for the North Atlantic Ocean, with the profile at the bottom of the illustration showing

► Figure 13.6
Major topographic divisions of the North Atlantic The accompanying profile extends from New England to the coast of North Africa and provides a glimpse of the various structures across the Atlantic basin.

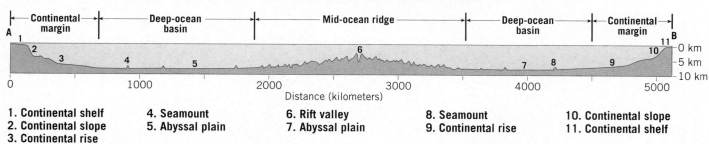

1. Continental shelf
2. Continental slope
3. Continental rise
4. Seamount
5. Abyssal plain
6. Rift valley
7. Abyssal plain
8. Seamount
9. Continental rise
10. Continental slope
11. Continental shelf

topography. Such profiles usually have their vertical dimension exaggerated many times—40 times in this case—to make topographic features more conspicuous. Vertical exaggeration, however, makes slopes shown in seafloor profiles appear to be *much* steeper than they actually are.

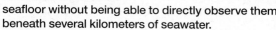

CONCEPT CHECKS 13.1

1. Define *bathymetry*.

2. Describe how satellites orbiting Earth determine features on seafloor without being able to directly observe them beneath several kilometers of seawater.

3. List the three major provinces of the ocean floor.

Concept Checker
https://goo.gl/yVfdGS

13.2 Continental Margins

Compare a passive continental margin with an active continental margin and list the major features of each.

As the name implies, **continental margins** are the outer edges of the continents, where continental crust transitions to oceanic crust. Two types of continental margin have been identified: *passive* and *active*.

Passive Continental Margins

Passive continental margins are geologically inactive regions located a great distance from the closest plate boundary. As a result, they are not associated with strong earthquakes or volcanic activity. Passive continental margins develop when continental blocks rift apart and are separated by continued seafloor spreading. Examples include the continental margins that border the Atlantic Ocean (see Figure 13.5).

Passive margins, most of which are relatively wide, are sites where large quantities of sediments get deposited. The features comprising passive continental margins include the *continental shelf*, the *continental slope*, and the *continental rise* (**Figure 13.7**).

Continental Shelf   The **continental shelf** is a gently sloping, submerged surface that extends from the shoreline toward the ocean basin. It consists mainly of continental crust capped with sedimentary rocks and sediments eroded from the adjacent landmass.

The width of the continental shelf varies greatly: It is very narrow along portions of some continents and extends seaward more than 1500 kilometers (930 miles) in other areas. The average inclination of the continental shelf is only about one-tenth of 1 degree, a slope so slight that it appears to be a horizontal surface when viewed by an observer unaided by measurement tools.

The continental shelf tends to be relatively featureless. However, extensive glacial deposits cover some areas, creating quite a rugged topography. In addition, some continental shelves are dissected by large valleys that run from the coastline into deeper waters. Many of these *shelf valleys* are the seaward extensions of river valleys on the adjacent landmass. They were eroded during the most recent Ice Age, when enormous quantities of water were stored in vast ice sheets on the continents, causing sea level to drop at least 100 meters (330 feet). Because of this sea-level drop, rivers extended their valleys seaward, and land-dwelling plants and animals migrated to the newly exposed portions of the continents. Dredging off the coast of North America has uncovered ancient remains of numerous mammoths, mastodons, and horses, providing further evidence that portions of the continental shelves were once above sea level.

Although continental shelves represent only 7.5 percent of the total ocean area, they have economic and political significance because they contain extensive reservoirs of oil and natural gas, and they support important fishing grounds.

Continental Slope   Marking the seaward edge of the continental shelf is the **continental slope**, a relatively steep zone that marks the boundary between continental crust and oceanic crust (see Figure 13.7). Although the inclination of the continental slope varies greatly from place to place, it averages about 5 degrees and in places exceeds 25 degrees.

Continental Rise   The continental slope merges into a more gradual incline known as the **continental rise** that may extend seaward for hundreds of kilometers. The continental rise consists of a thick accumulation of sediment that has moved down the continental slope and onto deep-ocean floor. Most of the sediments are delivered to the seafloor by *turbidity currents* that periodically flow down *submarine canyons* (discussed below). When these muddy slurries emerge from the mouth of a canyon onto the relatively flat ocean floor, they deposit sediment that forms a **deep-sea fan** (see Figure 13.7). As fans from adjacent submarine canyons grow, they merge to produce a continuous wedge of sediment at the base of the continental slope, forming the continental rise.

Submarine Canyons and Turbidity Currents   Deep, steep-sided valleys known as **submarine canyons** cut into the continental slope and may extend across the entire continental rise to the deep-ocean basin (see Figure 7.30, page 226). Although some of these canyons appear to be the seaward extensions of river valleys, many others do not line up in this manner. In addition, submarine canyons extend to depths far below the maximum lowering of sea level during the Ice Age, so their formation cannot be related to stream erosion.

Instead, it is believed that most submarine canyons have been excavated by **turbidity currents**—episodic

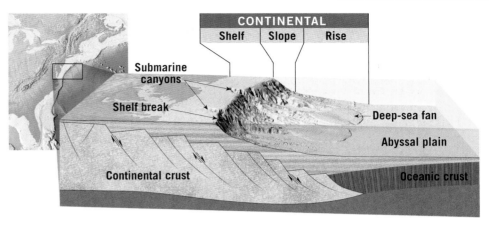

downslope movements of dense, sediment-laden water. They form when rock debris, sand, and mud on the continental shelf and slope are dislodged and form a slurry. The mud-thickened mass is denser than seawater and moves downslope, accumulating more sediment as it goes, eroding along the way. This is similar to how, on land, a flash flood erodes stream channels. Eventually, turbidity currents lose momentum and come to rest along the floor of the deep-ocean basin. As the currents slow, suspended sediments begin to settle out to form the deep-sea fans described above.

Active Continental Margins

Active continental margins are located along convergent plate boundaries where oceanic lithosphere is being subducted beneath the leading edge of a continent. Most of the Pacific Ocean is bordered by subduction zones shown (in red) in **Figure 13.8**. Notice that many of those active subduction zones lie far beyond the margins of the continents.

▲ **SmartFigure 13.7**
Passive continental margins The slopes shown for the continental shelf and continental slope are greatly exaggerated. The continental shelf has an average slope of one-tenth of 1 degree, while the continental slope has an average slope of about 5 degrees.

Tutorial
https://goo.gl/94j3Kj

▼ **Figure 13.8**
Distribution of Earth's subduction zones Notice that most of Earth's active subduction zones surround the Pacific basin.

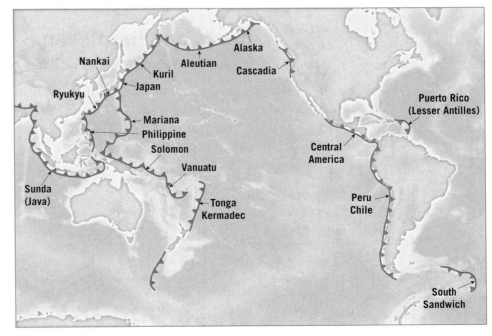

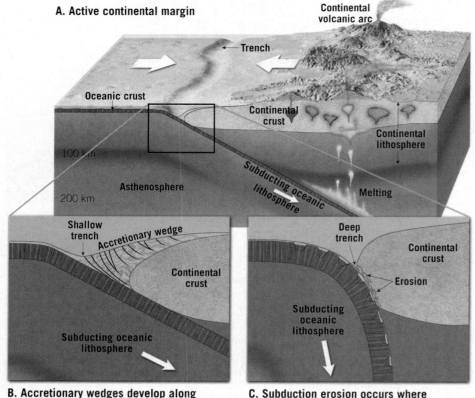

A. Active continental margin

Continental volcanic arc

Trench

Oceanic crust

100 km

Continental crust

Continental lithosphere

Asthenosphere

200 km

Subducting oceanic lithosphere

Melting

Shallow trench

Accretionary wedge

Continental crust

Subducting oceanic lithosphere

Deep trench

Continental crust

Erosion

Subducting oceanic lithosphere

B. Accretionary wedges develop along subduction zones, where sediments from the ocean floor are scraped from the descending oceanic plate and pressed against the edge of the overriding plate.

C. Subduction erosion occurs where sediment and rock scraped off the bottom of the overriding plate are carried into the mantle by the subducting plate.

▲ **Figure 13.9**
Active continental margins

forms the boundary between the Caribbean Sea and the Atlantic Ocean.

Along some subduction zones, sediments from the ocean floor and pieces of oceanic crust are scraped from the descending oceanic plate and plastered against the edge of the overriding plate (**Figure 13.9B**). This chaotic accumulation of deformed sediment and scraps of oceanic crust is called an **accretionary wedge** (*ad* = toward, *crescere* = to grow). Prolonged plate subduction can produce massive accumulations of sediment along active continental margins.

The opposite process, known as **subduction erosion**, characterizes many other active continental margins. Rather than sediment accumulating along the front of the overriding plate, sediment and rock are scraped off the bottom of the overriding plate and transported into the mantle by the subducting plate. Subduction erosion is particularly effective when the angle of descent is steep. Sharp bending of the subducting plate causes faulting in the ocean crust and a rough surface, as shown in **Figure 13.9C**.

CONCEPT CHECKS 13.2

1. List the three major features of a passive continental margin.

2. Describe the differences between active and passive continental margins. Where is each type found?

3. How are active continental margins related to plate tectonics?

Concept Checker

https://goo.gl/kttS2x

Deep-ocean trenches are the major topographic expression at convergent plate boundaries (**Figure 13.9A**). Most of these deep, narrow furrows surround the Pacific basin. One exception is the Puerto Rico trench, which

13.3 Features of Deep-Ocean Basins

List and describe the major features of deep-ocean basins.

Between the continental margin and the oceanic ridge lies the **deep-ocean basin** (see Figure 13.6). It comprises almost 30 percent of Earth's surface. This region includes *deep-ocean trenches*, which are extremely deep linear depressions in the ocean floor; remarkably flat areas known as *abyssal plains*; tall volcanic peaks called *seamounts* and *guyots*; and extensive areas of lava flows piled one atop the other, called *oceanic plateaus*.

Deep-Ocean Trenches

Deep-ocean trenches are long, relatively narrow troughs that are the deepest parts of the ocean. Most trenches sit along the margins of the Pacific Ocean, where many exceed 10 kilometers (6 miles) in depth (see Figure 13.5). A portion of one—the Challenger Deep in the Mariana trench—has been measured at a record 10,994 meters (36,070 feet) below sea level, making it the deepest-known part of the world ocean

(**Figure 13.10**). Only two trenches are located in the Atlantic: the Puerto Rico trench and the South Sandwich trench.

Although deep-ocean trenches represent only a very small portion of the area of the ocean floor, they are nevertheless significant geologic features. Trenches are sites of plate convergence where slabs of oceanic lithosphere subduct and plunge back into the mantle. In addition to generating earthquakes as one plate "scrapes" against another, plate subduction also triggers volcanic activity. As a result, some trenches run parallel to an arc-shaped row of active volcanoes called a **volcanic island arc**. The Mariana Seamounts, shown in Figure 13.10A, are such a feature. Furthermore, **continental volcanic arcs**, such as those making up portions of the Andes and Cascades, are located parallel to trenches that lie adjacent to continental margins. The volcanic activity associated with the trenches that surround the Pacific Ocean explains why the region is called the *Ring of Fire*.

Abyssal Plains

Abyssal plains (*a* = without, *byssus* = bottom) are deep, incredibly flat features. In fact, they are the most level places on Earth. The abyssal plain off the coast of Argentina, for example, has less than 3 meters (10 feet) of relief over a distance exceeding 1300 kilometers (800 miles). The monotonous topography of abyssal plains is occasionally interrupted by the protruding summit of a partially buried volcanic peak (seamount).

Using **seismic reflection profilers**, instruments that generate signals designed to penetrate far below the ocean floor, researchers have determined that abyssal plains owe their relatively featureless topography to thick accumulations of sediment that have buried an otherwise rugged ocean floor (**Figure 13.11**). The nature of the sediment indicates that these plains consist primarily of three types of sediment: (1) fine sediments from land transported far out to sea by winds, ocean currents, and turbidity currents; (2) mineral matter that has precipitated out of seawater; and (3) shells and skeletons of microscopic marine organisms.

Abyssal plains occur in all oceans. However, the floor of the Atlantic has the most extensive abyssal plains because it has few trenches to act as traps for sediment carried down the continental slope.

Volcanic Structures on the Ocean Floor

Dotting the seafloor are numerous volcanic structures of various sizes. Many occur as isolated features that resemble volcanoes on land. Others occur as long, narrow chains that stretch for thousands of kilometers, while still others are massive structures that cover an area the size of Texas.

Seamounts and Volcanic Islands
Submarine volcanoes, called **seamounts**, may rise hundreds of meters above the surrounding seafloor. It is estimated that more than 1 million seamounts exist worldwide. Although some grow large enough to become oceanic islands, most do not have a sufficiently long eruptive history to create a volcano tall enough to emerge above sea level. Seamounts are found on the floors of all oceans but are most common in the Pacific.

If a volcano grows large enough before it is carried from its magma source by plate motion, the structure may emerge as a *volcanic island*. Examples of volcanic islands include Easter Island, Tahiti, Bora Bora, the Galapagos Islands, and the Canary Islands.

Some seamounts and volcanic islands, like the Hawaiian Island–Emperor chain, which stretches from the Hawaiian Islands to the Aleutian trench, form over *volcanic hot spots* (see Figure 2.26, page 57). Others form on oceanic ridges.

◄ Figure 13.10
The Challenger Deep A. Located near the southern end of the Mariana trench, the Challenger Deep is the deepest place in the global ocean, about 10,994 meters (36,070 feet) deep.
B. Film director James Cameron (*Titanic* and *Avatar*) made news when he piloted a deep-diving submersible to the bottom of the Challenger Deep in 2012.

A.

Mariana Seamounts

Saipan

Guam

Mariana Trench

Challenger Deep

B.

▼ Figure 13.11
Seismic profile of the ocean floor This seismic cross section and matching sketch across a portion of the Madeira abyssal plain in the eastern Atlantic Ocean show the irregular oceanic crust buried by sediments.

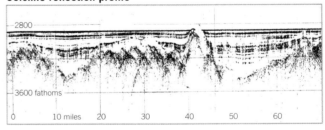

Seismic reflection profile

2800

3600 fathoms

0   10 miles   20   30   40   50   60

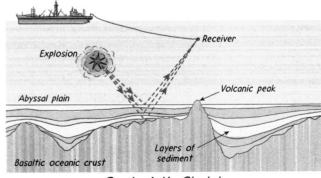

Receiver

Explosion

Volcanic peak

Abyssal plain

Layers of sediment

Basaltic oceanic crust

Geologist's Sketch

▲ Figure 13.12
Darwin's hypothesis Darwin's hypothesis asserts that, in addition to being lowered by erosional forces, many volcanic islands gradually sink. Darwin also suggested that corals respond to the gradual change in water depth caused by the subsiding volcano by building the reef complex upward. Atolls may eventually become guyots, when the rate of subsidence exceeds the ability of corals to build up the reef or when the reef-building organisms die out.

Guyots During their existence, inactive volcanic islands are gradually but inevitably lowered to near sea level by the forces of weathering and erosion. As a moving plate slowly carries inactive volcanic islands away from the elevated oceanic ridge or hot spot over which they formed, they gradually sink and disappear below the water surface. Submerged, flat-topped seamounts that formed in this manner are called **guyots**.[*] (It is pronounced "GEE-oh," with a hard *g*, as in *give*.)

Oceanic Plateaus The ocean floor contains several massive **oceanic plateaus**, which resemble lava plateaus composed of flood basalts found on the continents. Oceanic plateaus, which can be more than 30 kilometers (20 miles) thick, are generated from vast outpourings of fluid basaltic lavas.

Some oceanic plateaus appear to have formed quickly in geologic terms. Examples include the Ontong Java Plateau, which formed in less than 3 million years, and the Kerguelen Plateau, which formed in 4.5 million years (see Figure 13.5). Like basalt plateaus on land, oceanic plateaus are thought to form when the bulbous head of a rising mantle plume melts and produces vast underwater outpourings of basalt (see Figure 5.32, page 163).

Explaining Coral Atolls—Darwin's Hypothesis

Coral **atolls** are ring-shaped structures that extend from slightly above sea level to depths of several thousand meters. *Corals* are tiny animals that generally cluster in large numbers and form colonies when

linked. Most corals create a hard external skeleton made of calcium carbonate. Some build large calcium carbonate structures, called *reefs*, where new colonies grow atop the strong skeletons of previous colonies. Sponges and algae may attach to the reef, enlarging it further.

Reef-building corals grow best in waters with an average annual temperature of about 24°C (75°F). They cannot survive prolonged exposure to temperatures below 18°C (64°F) or above 30°C (86°F). In addition, reef builders require clear, sunlit water. Consequently, the depth of most active reef growth is limited to no more than about 45 meters (150 feet).

Deep-ocean basins comprise about 30 percent of Earth's total surface.

The strict environmental conditions required for coral growth create an interesting paradox: How can corals—which require warm, shallow, sunlit water no deeper than a few dozen meters—create thick structures such as coral atolls, that extend to great depths?

The naturalist Charles Darwin was one of the first to formulate a hypothesis on the origin of ringed-shaped atolls. He did this during a 5-year period (1831–1836) when he sailed aboard the British ship HMS *Beagle* on its famous global circumnavigation. On this journey, Darwin visited many volcanic islands surrounded by coral reefs. A keen observer, Darwin noticed a progression in coral reef development from (1) a *fringing reef* along the margins of a volcano to (2) a *barrier reef* with a volcano in the middle to (3) an *atoll*, consisting of a continuous or broken ring of coral reef surrounding a central lagoon (**Figure 13.12**).

The essence of Darwin's hypothesis is that, in addition to being lowered by erosional forces, many volcanic islands gradually sink. Darwin also

---
[*] The term *guyot* comes from Arnold Guyot, Princeton University's first geology professor.

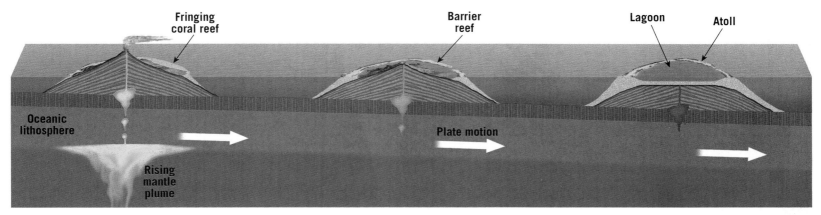

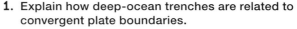

hypothesized that corals responded to the gradual change in water depth caused by the subsiding volcano by building the reef complex upward. Eventually, the volcano would submerge beneath the sea, and its remnant would be covered by a type of reef we call an atoll. During Darwin's time, however, there was no plausible mechanism to account for how or why so many volcanic islands sink.

The plate tectonics theory provides the most current scientific explanation of how volcanic islands become extinct and sink to great depths over long periods of time. Some volcanic islands form over relatively stationary mantle plumes, causing the lithosphere to be warmed and buoyantly uplifted. Over spans of millions of years, these volcanic islands gradually sink as the moving plates carry them away from the region of hot-spot volcanism because the lithosphere cools, becomes

denser, and sinks (**Figure 13.13**). Volcanic islands that form on spreading centers also sink when they move away from these areas of upwelling.

CONCEPT CHECKS 13.3

1. Explain how deep-ocean trenches are related to convergent plate boundaries.

2. Why are abyssal plains more extensive on the floor of the Atlantic than on the floor of the Pacific?

3. How does a flat-topped seamount, called a *guyot*, form? **Concept Checker** https://goo.gl/1jp6LQ

4. Using Darwin's hypothesis, place these coral reefs in order from youngest to oldest: *barrier reef*, *atoll*, and *fringing reef*.

▲ **Figure 13.13**
Plate tectonics and coral atolls The plate tectonics theory provides the most current scientific explanation for how volcanic islands become extinct and sink to great depths over long periods of time. Some volcanic islands form over relatively stationary mantle plumes, causing the lithosphere to be buoyantly uplifted. Over spans of millions of years, these volcanic islands gradually sink as the moving plates carry them away from the region of hot-spot volcanism.

13.4 The Oceanic Ridge System

Summarize the basic characteristics of oceanic ridges.

Along well-developed divergent plate boundaries, the seafloor is elevated, forming a nearly continuous range of underwater volcanic mountains called an **oceanic ridge or rise**, or a **mid-ocean ridge**. Oceanic ridges are characterized by extensive normal and strike-slip faulting, earthquakes, high heat flow, and volcanism. Knowledge of the oceanic ridge system comes from soundings of the ocean floor, core samples from deep-sea drilling, visual inspection using deep-diving submersibles, and firsthand inspection of slices of ocean floor that have been thrust onto dry land along convergent plate boundaries (**Figure 13.14**).

Anatomy of the Oceanic Ridge System

The oceanic ridge system winds through all major oceans in a manner similar to the seam on a baseball. It is the longest topographic feature on Earth, at more than 70,000 kilometers (43,000 miles) in length. The crest of the ridge typically stands 2 to 3 kilometers above the adjacent deep-ocean floor and is associated with a divergent plate boundary where new oceanic crust is created.

Notice in **Figure 13.15** that large sections of the oceanic ridge system have been named based on their locations within the various ocean basins. A ridge that runs through the middle of an ocean basin is appropriately called a *mid-ocean* ridge; the Mid-Atlantic Ridge and the Mid-Indian Ridge are examples. By contrast, the East Pacific Rise is *not* a "mid-ocean" feature but is located in the eastern Pacific, far from the center of the ocean.

▲ **Figure 13.14**
The deep-diving submersible *Alvin* This submersible is 7.6 meters long, weighs 16 tons, has a cruising speed of 1 knot, and can reach depths of 4000 meters. A pilot and two scientific observers are along during a normal 6- to 10-hour dive.

The term *ridge* is somewhat misleading because these features are not narrow and steep, as the term implies, but range in width from 1000 to 4000 kilometers (600 to 2500 miles) and have the appearance of broad, elongated swells that exhibit varying degrees of ruggedness. Furthermore, the ridge system is broken into segments ranging from a few tens to hundreds of kilometers in length. Each segment is offset from the adjacent segment by a transform fault.

Oceanic ridges are as high as some mountains on the continents, but the similarities end there. Whereas most mountain ranges on land form when the compressional forces associated with continental collisions fold and metamorphose thick sequences of sedimentary rocks, oceanic ridges form where upwelling from the mantle generates new oceanic crust. Oceanic ridges consist of layers and piles of newly formed basaltic rocks that are buoyantly uplifted by the hot mantle rocks from which they formed.

Along the axes of some segments of the oceanic ridge system are deep, down-faulted structures that are called **rift valleys** because of their striking similarity to the continental rift valleys in East Africa (**Figure 13.16**). Some rift valleys, including those along the rugged Mid-Atlantic Ridge, are typically 30 to 50 kilometers (20 to 30 miles) wide and have walls that tower 500 to 2500 meters (1640 to 8200 feet) above the valley floor. This makes them comparable to the deepest and widest part of Arizona's Grand Canyon.

Why Is the Oceanic Ridge Elevated?

The greatest volume of magma (more than 60 percent of Earth's total yearly output) is produced along the oceanic ridge system in association with seafloor spreading. As plates diverge, fractures created in the oceanic crust fill with molten rock that gradually wells up from the hot mantle below. This molten material slowly cools and crystallizes, producing new slivers of seafloor. This process repeats in episodic bursts, generating new lithosphere that moves away from the ridge crest in a conveyor belt fashion.

The oceanic ridge system is the longest topographic feature on Earth, at more than 70,000 kilometers (43,000 miles) in length.

The primary reason for the elevated position of the ridge system is that newly created oceanic lithosphere is hot and therefore less dense than cooler rocks of the deep-ocean basin. As the newly formed basaltic crust

▶ **Figure 13.15**
Distribution of the oceanic ridge system The map shows ridge segments that exhibit slow, intermediate, and fast spreading rates.

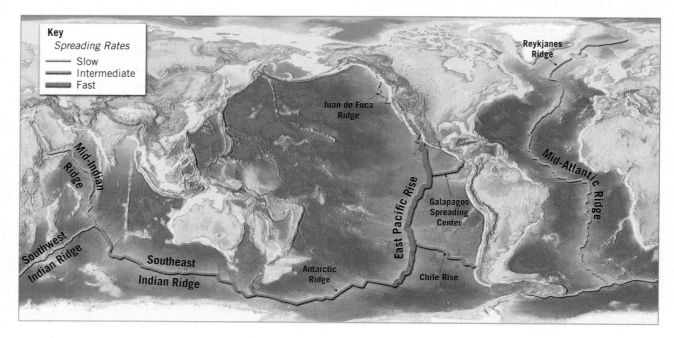

Key
Spreading Rates
— Slow
— Intermediate
— Fast

Reykjanes Ridge

Juan de Fuca Ridge

Mid-Indian Ridge

Southwest Indian Ridge

Southeast Indian Ridge

Antarctic Ridge

East Pacific Rise

Galapagos Spreading Center

Chile Rise

Mid-Atlantic Ridge

travels away from the ridge crest, it is cooled from above as seawater circulates through the pore spaces and fractures in the rock. In addition, it cools because it gets farther and farther from the zone of hot mantle upwelling. As a result, the lithosphere gradually cools and contracts. This thermal contraction accounts for the greater ocean depths that occur away from the ridge. It takes as much as 80 million years of cooling and contraction for rock that formed at the elevated ocean ridge to move away from the ridge crest and become part of the deep-ocean basin.

As lithosphere is displaced away from the ridge crest, cooling also causes a gradual increase in lithospheric thickness. This happens because the boundary between the lithosphere and asthenosphere is a thermal (temperature) boundary. Recall that the lithosphere is Earth's cool, stiff outer layer, whereas the asthenosphere is a comparatively hot and weak layer. As material in the uppermost asthenosphere ages and cools, it becomes stiff and rigid. Thus, the upper portion of the asthenosphere is gradually converted to lithosphere simply by cooling. Oceanic lithosphere continues to thicken until it is about 80 to 100 kilometers (50 to 60 miles) thick. Thereafter, its thickness remains relatively unchanged until it is subducted.

▶ **SmartFigure 13.16**
Rift valleys The axes of some segments of the oceanic ridge system contain deep down-faulted structures called *rift valleys* that may exceed 30 to 50 kilometers in width and 500 to 2500 meters in depth.

Tutorial
https://goo.gl/yD7FMW

Spreading Rates and Ridge Topography

As researchers studied various segments of the oceanic ridge system, it became clear that there are topographic differences resulting from variances in spreading rates—which largely determine the amount of melt generated at a rift zone. More magma wells up from the mantle at fast spreading centers than at slow spreading centers. This difference in output causes differences in the structure and topography of various ridge segments.

Oceanic ridges that exhibit slow spreading rates from 1 to 5 centimeters per year have prominent rift valleys and rugged topography (**Figure 13.17A**). The Mid-Atlantic and Mid-Indian Ridges are examples. The vertical displacement of large slabs of oceanic crust along normal faults is responsible for the steep walls of the rift

valleys. Furthermore, volcanism produces large cones in and around the rift valley, which enhance the rugged topography of the ridge crest.

By contrast, along the Galapagos Ridge, an intermediate spreading rate of 5 to 9 centimeters per year is the norm. The rift valleys found along this ridge segment are relatively shallow—often less than 200 meters (660 feet) deep. In addition, their topography is more subdued compared to ridges that have slower spreading rates.

At fast spreading centers (greater than 9 centimeters per year), such as along much of the East Pacific Rise, rift valleys are generally absent (**Figure 13.17B**). Instead, the ridge axis is elevated. These elevated structures, called *swells*, are built from lava flows up to 10 meters (30 feet) thick that have incrementally paved the ridge crest with

A. At slow spreading rates, a prominent central rift valley develops along the ridge crest, and the topography of the ridge is typically rugged.

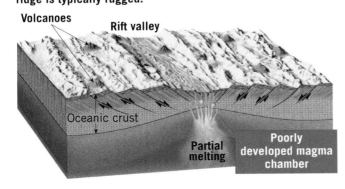

B. Along fast spreading centers, medial rift valleys do not develop, and the topography is comparatively smooth.

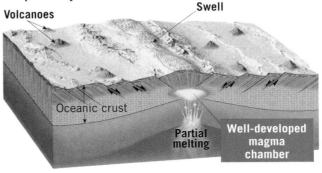

◀ Figure 13.17
Topography of slow and fast spreading centers

Ridges that have fast spreading rates are characterized by smooth topography, gently sloping flanks, and a central swell.

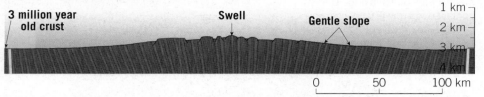

Ridges that have intermediate spreading rates are characterized by small rift valleys (less than 500 meters deep) and moderately sloping flanks.

Ridges that have slow spreading rates are characterized by rugged topography, well-developed rift valleys (500-2500 meters deep), and steeply sloping flanks.

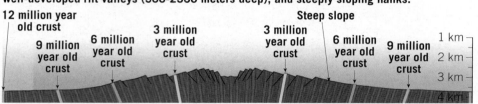

◀ Figure 13.18
Ridge segments that exhibit fast, intermediate, and slow spreading rates

volcanic rocks. In addition, because the depth of the ocean depends largely on the age of the seafloor, ridge segments that exhibit faster spreading rates tend to have more gradual profiles than ridges that have slower spreading rates (**Figure 13.18**). Because of these differences in topography, the gently sloping, less rugged portions of fast spreading ridges are called *rises*.

CONCEPT CHECKS 13.4

1. Briefly describe oceanic ridges.

2. What is the primary reason for the elevated position of the oceanic ridge system?

Concept Checker
https://goo.gl/Vv41k1

3. Compare a slow spreading center such as the Mid-Atlantic Ridge with one that exhibits a faster spreading rate, such as the East Pacific Rise.

13.5 The Nature of Oceanic Crust

List the four layers of oceanic crust and explain how oceanic crust forms and how it differs from continental crust.

An interesting aspect of oceanic crust is that its thickness and structure are remarkably consistent throughout much of the ocean basin. Seismic soundings indicate that its thickness averages only about 7 kilometers (5 miles). Furthermore, it is composed almost entirely of mafic (basaltic) rocks that are underlain by a layer of the ultramafic rock peridotite, which forms the *lithospheric mantle.*

Although most oceanic crust forms out of view, far below sea level, geologists have been able to examine the structure of the ocean floor firsthand. In locations such as Newfoundland, Cyprus, Oman, and California, slivers of oceanic crust and underlying mantle have been thrust high above sea level. From these exposures and from core samples collected by deep-sea drilling ships, researchers have concluded that the ocean crust consists of four distinct layers (**Figure 13.19**):

- **Layer 1.** A sequence of deep-sea sediments or sedimentary rocks compose the upper layer. These sedimentary deposits are very thin near the axis of oceanic ridges but may be several kilometers thick next to continents.
- **Layer 2.** Below the sediments layer lies a rock unit composed mainly of basaltic lavas that contain abundant pillowlike structures called *pillow lavas.*
- **Layer 3.** The middle, rocky layer is made up of numerous interconnected dikes that have a nearly vertical orientation, called the *sheeted dike complex.* These dikes are former pathways where magma rose to feed pillow basalts on the ocean floor.

- **Layer 4.** The lowest unit is mainly *gabbro*, the coarse-grained equivalent of basalt, which crystallized at depth without erupting.

When fragments of oceanic crust and the underlying mantle are discovered on land, they are called an **ophiolite complex**. From studies of various ophiolite complexes around the globe and related data, geologists have pieced together a scenario for the formation of the ocean floor.

How Does Oceanic Crust Form?

The molten rock that forms new oceanic crust originates from partial melting of the ultramafic mantle rock. This process generates basaltic melt that is less dense than the surrounding solid rock from which it formed. The newly formed melt rises through the upper mantle along thousands of tiny conduits that feed into larger channels. These structures, in turn, feed lens-shaped magma chambers located directly beneath the ridge crest. As a result of seafloor spreading, the rocks above these reservoirs are pulled apart and periodically fracture, allowing the melt to ascend along numerous vertical fractures that develop in the ocean crust. Some of the melt cools and solidifies

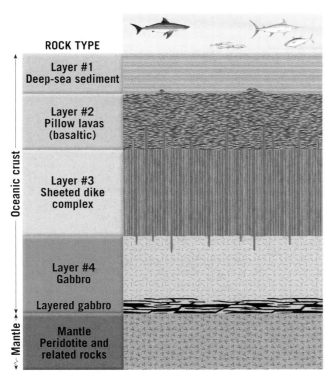
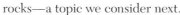

ROCK TYPE

Layer #1
Deep-sea sediment

Layer #2
Pillow lavas
(basaltic)

Layer #3
Sheeted dike
complex

Layer #4
Gabbro

Layered gabbro

Mantle
Peridotite and
related rocks

Oceanic crust

Mantle

▲ **Figure 13.19**
Ophiolite complex: Structure of the oceanic crust This view of the layered structure of oceanic crust is based on data obtained from ophiolite complexes, seismic profiling, and core samples obtained from deep-sea drilling expeditions.

to form new dikes, which intrude older dikes—still warm and weak—to form a **sheeted dike complex**. This portion of the oceanic crust is usually 1 to 2 kilometers thick.

Roughly 10 to 20 percent of the melt reaches the ocean floor and erupts. The surface of these submarine lava flows is chilled quickly by seawater, but as lava accumulates behind the congealed margin of these flows, it breaks through. This process occurs repeatedly as molten basalt is extruded like toothpaste from a tightly squeezed tube (see Figure 5.8, page 141). The result is protuberances resembling large bed pillows stacked one atop the other, hence the name **pillow lavas** (**Figure 13.20**). Pillow lavas form the upper layer of the newly formed oceanic crust. Over time, this crust will be covered by sediments.

In some settings, pillow lavas may build volcano-size mounds that resemble shield volcanoes, whereas in other situations they form elongated ridges tens of kilometers long. These structures eventually separate from their supply of magma as they are carried away from the ridge crest by seafloor spreading.

The lowest unit of the ocean crust develops from crystallization within the central magma chamber itself and accounts for up to 5 of the 7 kilometers of the total crustal thickness. The first minerals to crystallize are olivine and pyroxene, which settle through the magma to form a layered zone near the floor of the reservoir. The remaining melt tends to cool along the walls of magma chambers to form massive amounts of coarse-grained gabbro.

The layered model of the oceanic crust adequately describes the structure of seafloor that formed along fast

spreading centers. However, recent research suggests that other mechanisms may be at work at slower spreading centers, including those of the Atlantic basin. In addition, once the seafloor has been generated, interactions between seawater and the seafloor alter the crustal rocks—a topic we consider next.

▲ **Figure 13.20**
Cross-sectional view of pillow lava This pillow lava is exposed along a sea cliff at Cape Wanbrow, New Zealand. Notice that each "pillow" shows an outer dark glassy layer created by rapid cooling enclosing a dark gray basalt interior.

Interactions Between Seawater and Oceanic Crust

In addition to serving as a mechanism for dissipating Earth's internal heat, the interaction between seawater and the newly formed basaltic crust alters both the seawater and the crust. The permeable and highly fractured lava of the upper oceanic crust allows seawater to penetrate to depths of 2 to 3 kilometers (1 to 2 miles). Seawater circulating through the crust heats up and chemically reacts with the basaltic rock in a process called *hydrothermal* (hot water) *metamorphism* (see Figure 8.21, page 248). This alteration causes the dark silicates (olivine and pyroxene) to form new metamorphic minerals such as chlorite and serpentine. Simultaneously, the hot seawater dissolves ions of silica, iron, copper, and occasionally silver and gold from the hot basalts. When the water temperature reaches a few hundred degrees Celsius, these mineral-rich fluids buoyantly rise along fractures and eventually spew out on the ocean floor. Studies conducted by submersibles along several segments of the oceanic ridge have photographed these metallic-rich solutions gushing from the seafloor to form particle-filled clouds called **black smokers** (see **GEOgraphics 13.1**). As liquid up to 400°C (750°F) mixes with the cold, mineral-laden seawater, the dissolved minerals precipitate to form massive metallic sulfide deposits, some of which are economically important. Occasionally these deposits grow upward to form underwater chimney-like structures equivalent in height to skyscrapers.

CONCEPT CHECKS 13.5

1. Briefly describe the four layers of the ocean crust.

2. How does a sheeted dike complex form?

3. How does hydrothermal metamorphism alter the basaltic rocks that make up the seafloor? How is seawater changed during this process?

4. What is a black smoker?

Concept
Checker
https://goo.gl/FAhRqG

Deep-Sea Hydrothermal Vents

Deep-sea hydrothermal vents are openings in the oceanic crust from which geothermally heated water rises. They are found mainly along the oceanic ridge system where tectonic plates rift apart, resulting in the production of new seafloor by upwelling magma.

When these hot, mineral-rich fluids reach the seafloor their temperatures can exceed 350°C, but because of the extremely high pressures exerted by the water column above, they do not boil. When this hydrothermal fluid comes into contact with the much colder chemical-rich seawater, mineral matter rapidly precipitates to form shimmering smoke-like clouds called "*black smokers*." The particles that compose the black smokers eventually settle out of the seawater. These deposits may contain economically significant amounts of iron, copper, zinc, lead, and occasionally silver and gold.

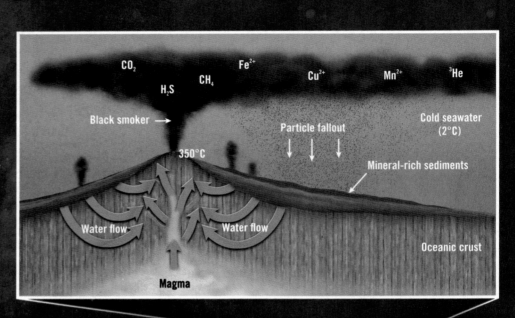

CO_2 Fe^{2+} Cu^{2+} Mn^{2+} ^3He

H_2S CH_4

Black smoker →

350°C

Particle fallout

Cold seawater (2°C)

Mineral-rich sediments

Water flow   Water flow

Oceanic crust

Magma

At oceanic ridges, cold seawater circulates hundreds of meters down into the highly fractured basaltic crust, where it is heated by magmatic sources. Along the way, the hot water strips metals and other elements such as sulfur from surrounding rock. This heated fluid eventually becomes hot and buoyant enough to rise along conduits and fractures toward the surface.

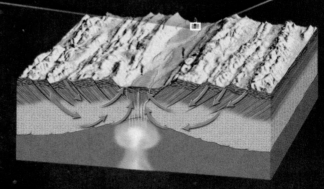

Some minerals immediately solidify and contribute to the formation of spectacular chimney-like structures, which can be as tall as a 15-story building, and are appropriately given names like *Godzilla* and *Inferno*.

13.6 Continental Rifting: The Birth of a New Ocean Basin

Outline the steps by which continental rifting results in the formation of new ocean basins.

The breakup of Pangaea nearly 200 million years ago opened a new ocean basin—the Atlantic. Although geoscientists still debate what initiated this event, Pangaea's breakup illustrates that ocean basins originate when large landmasses break apart.

Evolution of an Ocean Basin

The opening of a new ocean basin begins with the formation of a **continental rift**, an elongated depression along which the lithosphere is stretched and thinned. Where the lithosphere is thick, cool, and strong, rifts tend to be narrow—often less than a few hundred kilometers wide. Modern examples of narrow continental rifts include the East African Rift, the Rio Grande Rift (southwestern United States), the Baikal Rift (south-central Siberia), and the Rhine Valley (northwestern Europe). By contrast, where the crust is thin, hot, and weak, rifts can be more than 1000 kilometers (600 miles) wide, as exemplified by the Basin and Range region in the western United States.

In some settings, the continental rift evolves into a young, narrow ocean basin, such as the present-day Red Sea. Continued seafloor spreading eventually results in the formation of a mature ocean basin bordered by rifted continental margins. The Atlantic Ocean is such a feature.

What follows is an overview of ocean basin evolution, using modern examples to represent the various stages of rifting.

East African Rift   The East African Rift is a continental rift that extends through eastern Africa for approximately 3000 kilometers (2000 miles). It consists of several interconnected rift valleys that split into eastern and western sections around Lake Victoria (**Figure 13.21**). Whether this rift will eventually develop into a spreading center, with the Somali subplate separating from the rest of the continent of Africa, is uncertain.

The most recent period of rifting began about 20 million years ago, as upwelling in the mantle intruded the base of the lithosphere (**Figure 13.22A**). Buoyant uplifting of the heated lithosphere led to doming and stretching of the crust. As a result, the upper crust was broken along high-angle normal faults, producing down-faulted blocks, or *grabens*, while the lower crust deformed by ductile stretching (**Figure 13.22B**).

In the early stages of rifting, magma generated by decompression melting of the rising mantle rocks intruded the crust. Some of the magma migrated upward along fractures and erupted at Earth's surface. This produced extensive basaltic flows within the rift as well as volcanic cones—some forming more

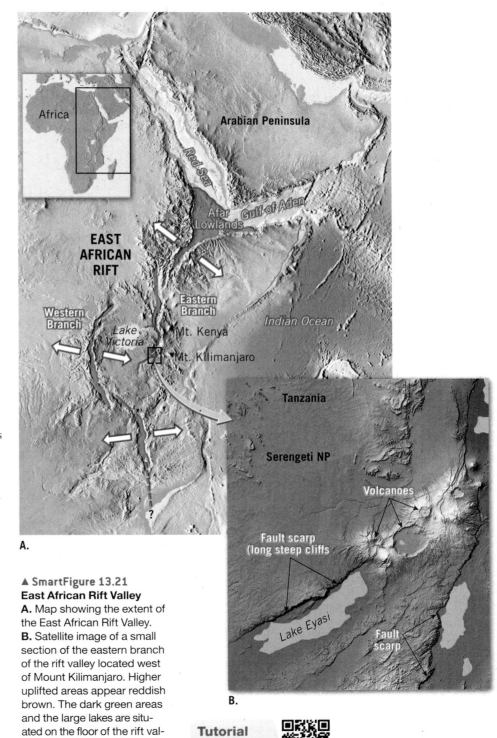

A.

▲ SmartFigure 13.21
East African Rift Valley
A. Map showing the extent of the East African Rift Valley.
B. Satellite image of a small section of the eastern branch of the rift valley located west of Mount Kilimanjaro. Higher uplifted areas appear reddish brown. The dark green areas and the large lakes are situated on the floor of the rift valley. Volcanoes formed during the rifting can also be seen.

Tutorial
https://goo.gl/
jvuykC

▶ SmartFigure 13.22
Formation of an ocean basin

Animation
https://goo.gl/
C7YP8Y

A. Tensional forces and buoyant uplifting of the heated lithosphere cause the upper crust to be broken along normal faults, while the lower crust deforms by ductile stretching.

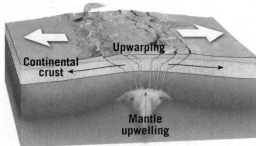

B. As the crust is pulled apart, large slabs of rock sink, generating a rift valley.

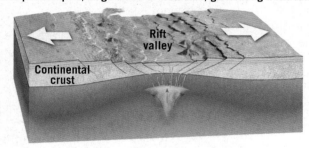

C. Further spreading generates a narrow sea.

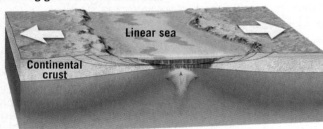

D. Eventually, an expansive ocean basin and ridge system are created.

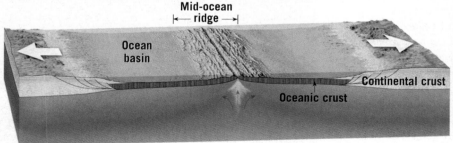

Illustration of the separation of South America and Africa to form the South Atlantic.

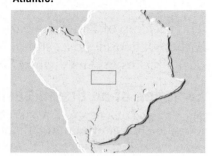

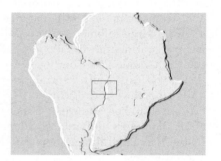

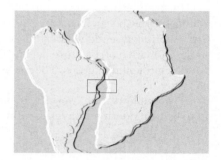

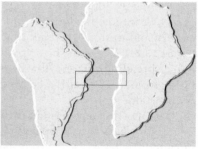

than 100 kilometers (60 miles) from the rift axis. Examples include Mount Kenya and Mount Kilimanjaro, the highest point in Africa, which rises almost 6000 meters (20,000 feet) above the Serengeti Plain.

Red Sea   A rift valley gradually lengthens and deepens, eventually extending to the margin of the continent (**Figure 13.22C**). At this point, the continental rift becomes a narrow linear sea with an outlet to the ocean, similar to the Red Sea.

The Red Sea formed when the Arabian Peninsula rifted from Africa beginning about 30 million years ago (see Figure 13.21A). Steep fault scarps that rise as much as 3 kilometers (2 miles) above sea level flank the margins of this water body. Thus, the escarpments surrounding the Red Sea are similar to the steep cliffs that border the East African Rift. Although the Red Sea reaches oceanic depths of up to 5 kilometers (3 miles) in only a few locations, symmetrical magnetic stripes indicate that typical seafloor spreading has been occurring for at least the past 5 million years.

Atlantic Ocean   If spreading continues, the Red Sea will grow wider and develop an elevated oceanic ridge similar to the Mid-Atlantic Ridge (**Figure 13.22D**). The Atlantic Ocean shows what the Red Sea could eventually become over tens of millions of years. As the Atlantic basin continued to grow in size, the rifted continental margins gradually receded from the region of upwelling. As a result, they cooled, contracted, and sank.

The breakup of Pangaea nearly 200 million years ago opened a new ocean basin—the Atlantic.

Over time, continental margins subsided below sea level, and material that had eroded from the adjacent highlands blanketed this once-rugged topography. The result was a *passive continental margin* on both sides of the Atlantic, consisting of rifted continental crust that has been covered by a thick wedge of relatively undisturbed sediment and sedimentary rock.

Failed Rifting   Not all continental rift valleys develop into full-fledged spreading centers. In the central United States, a failed rift extends from Lake Superior into Kansas (**Figure 13.23**). This once-active rift valley is filled with clastic sedimentary and basaltic rocks that were extruded onto the crust more than 1 billion years ago. Why one rift valley develops into a full-fledged spreading center while others fail to develop is not fully understood.

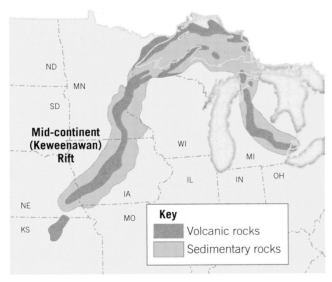

▲ **Figure 13.23**
Mid-continent rift This failed rift extends from the Great Lakes region to Kansas.

Mechanisms for Continental Rifting

At least two supercontinents existed in the geologic past. Pangaea, the most recent, was assembled into a supercontinent between 450 and 230 million years ago, only to break up shortly after it formed. Geologists have concluded that the formation of supercontinents followed by continental splitting is an integral part of plate tectonics. This process, which involves the formation and dispersal of supercontinents, is called the **supercontinent cycle** and is described in detail in Chapter 22.

The supercontinent cycle must involve major changes in the direction and nature of the forces that drive plate motion. In other words, over long periods of geologic time, the forces that drive plate motions tend to organize crustal fragments into a single supercontinent, only to change directions and disperse them again. Mechanisms that are thought to contribute to continental rifting include plumes of hot mobile rock rising from deep in the mantle, upwelling from shallow levels in the asthenosphere, and forces that arise from plate motions.

Mantle Plumes and Hot-Spot Volcanism   A *mantle plume* consists of hotter-than-normal mantle rock that has a large mushroom-shaped head hundreds of kilometers in diameter attached to a long, narrow, trailing tail. As the plume head nears the base of the rigid lithosphere, it spreads laterally. Decompression melting within the plume generates huge volumes of basaltic magma that rises and triggers *hot-spot volcanism* at the surface.

Research suggests that mantle plumes tend to concentrate beneath a supercontinent because once assembled, a large landmass forms an insulating "blanket" that traps heat in the mantle. The resulting temperature increase leads to the formation of mantle plumes that serve to dissipate heat.

Evidence that mantle plumes play a role in the breakup of at least some landmasses can be observed in modern passive continental margins. In several regions on both sides of the Atlantic, continental rifting was preceded by crustal uplift and massive outpourings of basaltic lava. Examples include the Etendeka flood basalts of southwest Africa and the Paraná basalt province of South America.

About 130 million years ago, when South America and Africa were a single landmass, vast outpourings of lava produced a large continental basalt plateau (**Figure 13.24A**). Next, the South Atlantic began to open, splitting the basalt province into two parts—the Etendeka and the Paraná basalt plateaus. As the ocean basin grew, the tail of the plume produced a string of

► Figure 13.24
The role that mantle
plumes might play in
continental rifting

A. Location of these flood basalt plateaus 130 million years ago, just before the South Atlantic began to open.

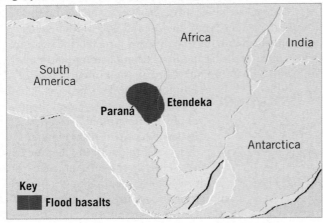

B. Relationship of the Paraná and Etendeka basalt plateaus to the Tristan da Cunha hot spot.

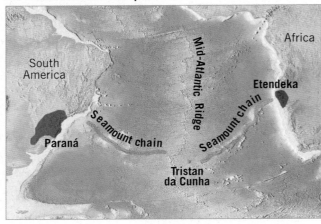

seamounts on each side of the newly formed ridge (**Figure 13.24B**). The modern area of hot-spot activity is centered around the volcanic island of Tristan da Cunha, on the Mid-Atlantic Ridge.

When hot, buoyant mantle plumes reach the base of the lithosphere, they cause the overlying crust to dome and weaken. Doming—of perhaps as much as 1000 meters (3300 feet)—tends to produce three *rifts*, or *arms*, that join in the area above the rising plume, called a **triple junction**. Frequently, continental breakup and the formation of an ocean basin occur along two of the rift arms, whereas the third arm may be less developed and constitute a failed rift that becomes filled with sediments. For example, the Afar plume, associated with the split of the Arabian Peninsula from Africa, is located beneath a region of northeastern Ethiopia called the Afar Lowlands, an area of extensive volcanism (see Figure 13.21). This plume generated a typical rift system consisting of three arms that meet at a triple junction. Two of these

rifts, the Red Sea and the Gulf of Aden, are active spreading centers. The third arm is the East African Rift, which may represent the initial stage in the breakup of a continent, as described earlier, or may be destined to become a failed rift.

Figure 13.25 illustrates the locations of a few mantle plumes that have generated large flood basalt plateaus and were presumably involved in the breakup

Approximate surface location of mantle plumes prior to the breakup of Pangaea.

Timing of the breakup of Pangaea along various rift zones and the plume volcanism that was associated with each period of continental fragmentation. In most cases, volcanism appears to precede breakup by a few million years, or more.

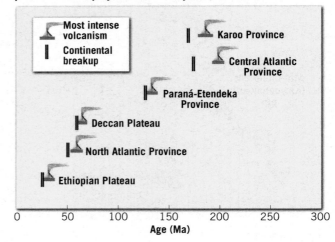

▲ Figure 13.25
The possible role of mantle plumes in the breakup of Pangaea

of Pangaea. One mantle plume is currently located beneath Iceland, near the crest of the Mid-Atlantic Ridge (**Figure 13.26**). Vast outpourings of basaltic lava began about 55 million years ago; evidence of these outpourings can be found in eastern Greenland, as well as across the Atlantic in the Hebrides Islands of northern Scotland. The oldest magnetic stripes between Greenland and Europe are the same age, supporting the connection between the emergence of the Icelandic plume and seafloor spreading in the North Atlantic.

It is important to note that hot-spot volcanism does not necessarily lead to rifting. For example, massive eruptions of basaltic lava that constitute the Columbia River basalts in the Pacific Northwest, as well as Russia's Siberian Traps, are not associated with continental fragmentation. Furthermore, along some rifted continental margins, stretching and thinning of the lithosphere was not accompanied by large-scale volcanism. Consequently, other forces that contribute to continental fragmentation must exist.

Role of Tensional Stress Continental rifting requires tensional stresses sufficiently strong to tear the lithosphere. In the Basin and Range region of the western United States, where the lithosphere is thin, hot, and weak, small stresses are sufficient to cause spreading. During the past 20 million years, a broad zone of upwelling within the asthenosphere is thought to have caused considerable stretching and thinning of the crust in this region (see Figure 14.16, page 405). In such settings, rifting is accompanied by large-scale melting and volcanism.

Tensional stresses resulting from plate motions are also thought to be particularly significant in continental rifting. In settings where a continent is attached to a subducting slab of oceanic lithosphere, the continental crust is pulled along by the descending slab. However, this continental lithosphere is thick and tends to resist being

towed, which creates tensional stresses that may be sufficient to tear the landmass. The zones of rifting in the fragmentation of a supercontinent may be influenced by a preexisting weakness, such as *sutures*—sites where continents once collided to form the supercontinent.

▲ **Figure 13.26**
Eruption of fluid basaltic lava, Iceland, 2010 Iceland is located over a mantle plume that began to build this large volcanic island more than 20 million years ago.

CONCEPT CHECKS 13.6

1. Name a modern example of a continental rift.

2. Briefly describe each of the four stages in the evolution of an ocean basin.

 **Concept Checker** https://goo.gl/pX7Sb3

3. What role do hot spots and mantle plumes play in the breakup of a supercontinent?

13.7 Destruction of Oceanic Lithosphere

Compare and contrast spontaneous subduction and forced subduction.

Although new lithosphere is continually being produced at divergent plate boundaries, Earth's surface area is not growing larger. Therefore, it follows that in order to balance the amount of newly created lithosphere, there must be a process whereby oceanic lithosphere is destroyed.

Why Oceanic Lithosphere Subducts

The process of plate subduction is complex, and the ultimate fate of oceanic lithosphere is still being debated. What is known with some certainty is that oceanic lithosphere resists subduction unless its overall density is greater than that of the underlying asthenosphere. It takes at least 15 million years for a young slab of oceanic

lithosphere to cool sufficiently to become denser than the supporting asthenosphere.

Spontaneous Subduction Subduction zones can be divided into two basic types, based on the nature of the subducting plate. The first type, referred to as a *Mariana-type subduction zone*, is characterized by old, dense lithosphere sinking into the mantle because of its own

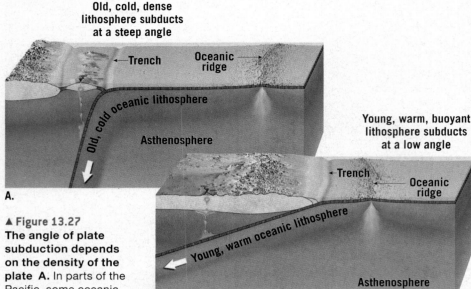

▲ Figure 13.27
The angle of plate subduction depends on the density of the plate A. In parts of the Pacific, some oceanic lithosphere exceeds 180 million years in age and descends into the mantle at angles approaching 90 degrees. **B.** Young oceanic lithosphere is warm and buoyant, and it tends to subduct at a low angle.

weight. The lithosphere entering the Mariana trench is about 185 million years old, some of the oldest and densest in today's oceans. Along this trench, the subducting slab descends into the mantle at a steep angle that approaches 90 degrees (**Figure 13.27A**). Steep subduction angles produce deep trenches, which account in part for the depth of the Challenger Deep, located at the southern limb of the Mariana trench. The Mariana and most of the other subduction zones in the western Pacific involve cold, dense lithosphere and therefore exhibit **spontaneous subduction**.

It is important to note that the *lithospheric mantle*, which makes up about 80 percent of the descending oceanic slab, drives subduction. Even when the overlaying *oceanic crust* is quite old, its density is still less than that of the underlying asthenosphere. Subduction, therefore, depends on lithospheric mantle that is colder and therefore denser than the underlying asthenosphere.

When an oceanic slab descends to about 400 kilometers (250 miles), mineral phase changes (transitions from a low-density to a high-density mineral; see Chapter 12) enhance subduction. At this depth, the transition of olivine to its compact, much denser structure increases the density of the slab, which helps pull the plate into the subduction zone.

Forced Subduction

The second type of subduction zone, called the *Peru–Chile–type subduction zone*, is characterized by younger, hotter, and less-dense lithosphere that dips at shallower angles (**Figure 13.27B**). Along Peru–Chile–type boundaries, the lithosphere is too buoyant to subduct spontaneously; rather, it is *forced* beneath the overlying plate by compressional forces.

In areas where **forced subduction** occurs, a strong coupling develops between the overlying plate and the

subducting plate, which can result in particularly strong and frequent earthquakes. Plate motion here generates horizontal compressional forces that cause the upper plate and underlying plate to grind against each other. The result can be folding and thickening of the upper plate and sometimes the formation of mountainous terrains like those we see today in the Andes. Shallow subduction and strong coupling have also been observed in the past decade along the Sunda subduction zone, off the coast of Sumatra, another region that has experienced several major earthquakes.

Resistance to Subduction Researchers have determined that unusually thick units of oceanic crust, those approaching 30 kilometers (20 miles) in thickness, are likely to resist subduction. The Ontong Java Plateau, for example, is a thick oceanic plateau, about the size of Alaska, located in the western Pacific (see Figure 13.5). About 20 million years ago, this plateau reached the trench that forms the boundary between the subducting Pacific plate and the overriding Australian–Indian plate. Apparently too buoyant to subduct, the Ontong Java Plateau clogged the trench. We will consider the fate of crustal fragments that are too buoyant to subduct in the next chapter.

Subducting Plates: The Demise of Ocean Basins

In the 1970s, geologists began using magnetic stripes and fracture zones on the ocean floor to reconstruct the past 200 million years of plate movement. This research showed that parts of, or even entire, ocean basins have been destroyed along subduction zones. For example, during the breakup of Pangaea (shown in Figure 2.22 on page 54), the African plate moved northward, eventually colliding with Eurasia. During this event, the floor of the intervening Tethys Ocean was almost entirely consumed into the mantle, leaving behind a few small remnants—the Eastern Mediterranean Sea and the Black Sea.

Reconstructions of Pangea's breakup also helped investigators understand the demise of the Farallon plate—a large oceanic plate that once occupied much of the eastern Pacific basin. At the time of the breakup of Pangaea, the Farallon plate was situated on the eastern side of a spreading center, as shown in **Figure 13.28A**. The spreading center, which generated both the Farallon and Pacific plates, is the East Pacific Rise.

Beginning about 180 million years ago, the Americas were propelled westward as Pangaea broke up and the Atlantic Ocean started to open. As a result, the Farallon plate began subducting beneath the Americas faster than it was being generated, causing it to decrease in size (**Figure 13.28B**). The three remaining fragments of the once-extensive Farallon plate include the modern Juan de Fuca, Cocos, and Nazca plates.

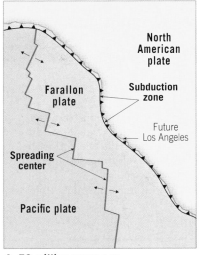

A. 56 million years ago

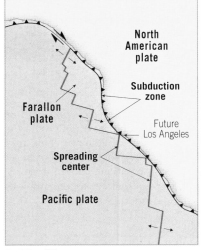

B. 30 million years ago

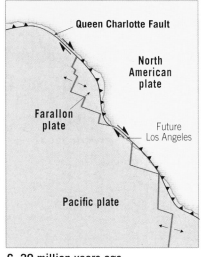

C. 20 million years ago

D. Today

▲ SmartFigure 13.28

The demise of the Farallon plate Because the Farallon plate was subducting faster than it was being generated, it continually got smaller and smaller. The remaining fragments of the once-mighty Farallon plate are the Juan de Fuca, Cocos, and Nazca plates.

Tutorial
https://goo.gl/
DvrfEB

The westward migration of North America also caused a section of the East Pacific Rise to enter the subduction zone that once lay off the coast of California (see Figure 13.28B). As this spreading center subducted, it was destroyed and replaced by a transform fault system that currently accommodates the differential motion between the North American and Pacific plates. Because of this change in plate geometry, the Pacific plate has captured a sliver of North America (the Baja Peninsula and a portion of southern California) and is carrying it northwestward toward Alaska at a rate of about 6 centimeters (2.5 inches) per year.

As more of the ridge subducted, the transform fault system, which we now call the San Andreas Fault, increased in length (**Figure 13.28C**). Today, the southern end of the San Andreas Fault connects to a young spreading center that is generating the Gulf of California (**Figure 13.29**). A similar event generated the Queen Charlotte transform fault, located off the west coast of Canada and southeastern Alaska.

CONCEPT CHECKS 13.7

1. Compare spontaneous subduction and forced subduction. Provide examples of places where each operates.

2. What role do mineral phase changes play in plate subduction?

Concept Checker
https://goo.gl/f2MaiE

3. Explain what happened when the spreading center that generated the Farallon plate collided with the North American plate.

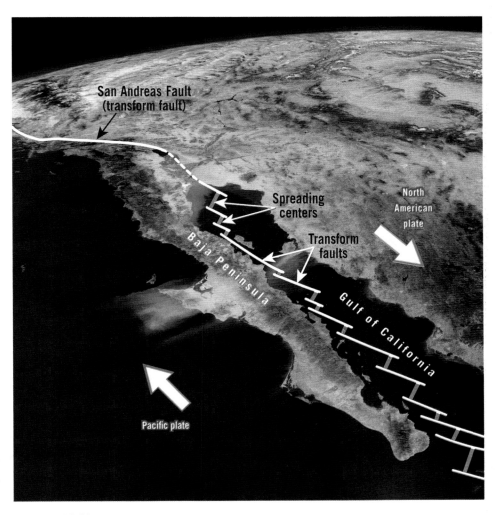

▲ Figure 13.29

The separation between the Baja Peninsula and North America

13 CONCEPTS IN REVIEW

Origin and Evolution of the Ocean Floor

13.1 An Emerging Picture of the Ocean Floor

Define *bathymetry* and describe the various bathymetric techniques used to map the ocean floor.

Key Terms:
sonar
bathymetry
echo sounder

- The measurement of ocean depths and charting of the ocean floor is called *bathymetry*.

- Seafloor mapping is done with *sonar*, using shipboard instruments that emit pulses of sound that "echo" off the bottom. Satellites with gravity-sensing instruments are also used to map the ocean floor by measuring slight variations in sea level that result from differences in the gravitational pull of features on the seafloor.
- Maps of seafloor topography can be made by combining data from these sources.

13.2 Continental Margins

Compare a passive continental margin with an active continental margin and list the major features of each.

Key Terms:
continental margin
passive continental margin
continental shelf
continental slope
continental rise
deep-sea fan
submarine canyon
turbidity current
active continental margin
accretionary wedge
subduction erosion

- *Continental margins* are transition zones between continental and oceanic crust. *Active continental margins* occur where a convergent plate boundary and the edge of a continent collide. *Passive continental margins* are on the trailing edges of continents, far from plate boundaries.
- Heading offshore of a passive margin, a submarine traveler would first encounter the gently sloping *continental shelf* and then the steeper *continental slope*, marking the end of the continental crust and the beginning of the oceanic crust.
- Beyond the continental slope is the gently sloping *continental rise*. Here sediment transported by *turbidity currents* through submarine canyons piles up in *deep-sea fans* atop the oceanic crust.

- At an active continental margin, material may be added to the leading edge of a continent in the form of an *accretionary wedge* (common at shallow-angle subduction zones), or material may be scraped off the edge of a continent by *subduction erosion* (common at steeply dipping subduction zones).

Q What type of continental margin is depicted in this diagram? Be as specific as possible. Name the feature indicated by the question mark.

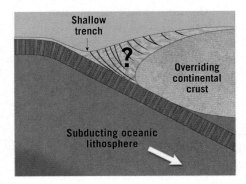

13.3 Features of Deep-Ocean Basins

List and describe the major features of deep-ocean basins.

Key Terms:
deep-ocean basin
deep-ocean trench
volcanic island arc
continental volcanic arc
abyssal plain
seismic reflection profiler
seamount
guyot
oceanic plateau
atoll

- The *deep-ocean basin* makes up about 30 percent of the world's surface area. Much of it is *abyssal plain* (deep, featureless sediment-draped crust).
- Subduction zones and *deep-ocean trenches* also occur in deep-ocean basins. Paralleling trenches are *volcanic island arcs* (if the subduction goes underneath oceanic lithosphere) or *continental volcanic arcs* (if the overriding plate has continental lithosphere on its leading edge).

- There are a variety of volcanic structures on the deep-ocean floor. *Seamounts* are submarine volcanoes; if they pierce the ocean's surface, we call them volcanic islands. *Guyots* are old volcanic islands that had their tops eroded off before they sank below sea level. *Oceanic plateaus* are unusually thick sections of oceanic crust formed by massive underwater lava eruptions.

Q On this cross-sectional view of a deep-ocean basin, label the following features: seamount, guyot, volcanic island, oceanic plateau, and abyssal plain.

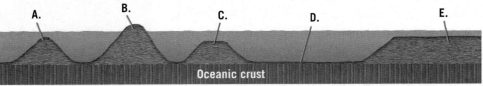

13.4 The Oceanic Ridge System

Summarize the basic characteristics of oceanic ridges.

Key Terms: oceanic ridge   rift valley
or rise

- The oceanic ridge system is the longest topographic feature on Earth, wrapping around the world through all major ocean basins. It is a few kilometers tall, a few thousand kilometers wide, and a few tens of thousands of kilometers long. The crest or axis is the place where new oceanic crust is generated and may be marked by a *rift valley*.

- *Oceanic ridges* are elevated features because they are warm and therefore less dense than older, colder oceanic lithosphere. As oceanic crust moves away from the ridge crest, heat loss causes the oceanic crust to become colder and denser and eventually to subside. After 80 million years, crust that was once part of an oceanic ridge is in the deep-ocean basin, far from the ridge.

- The rate at which seafloor spreading occurs determines the shape of the oceanic ridge. Ridges with slow spreading rates (1 to 5 centimeters per year) have prominent rift valleys and rugged topography. Those with fast spreading rates (greater than 9 centimeters per year) lack rift valleys and show a smoother, more subdued topography.

13.5 The Nature of Oceanic Crust

List the four layers of oceanic crust and explain how oceanic crust forms and how it differs from continental crust.

Key Terms: sheeted dike complex   black smoker
ophiolite complex   pillow lava

- *Ophiolite complexes* contain slices of oceanic crust that have been thrust above sea level. They have four distinct layers: (1) deep-sea sediment, (2) pillow basaltic lava flows, (3) the sheeted dike complex, and (4) the lowermost gabbro layer.

- As two divergent plates move apart, fractures open perpendicular to the stretching direction, and lava moves up through these cracks, toward the seafloor. Once the lava has cooled and sealed these fractures shut, they become a *sheeted dike complex*. Magma that cools at depth crystallizes to become gabbro. Lava that makes it to the seafloor is erupted as *pillow lavas*, which are gradually buried by sedimentation.

- Along mid-ocean ridges, seawater flows through fissures in the oceanic crust and is heated by nearby pockets of magma. The hot water causes hydrothermal metamorphism and dissolves metal ions. These hot, dark solutions may spew out of the crust as *black smokers*.

13.6 Continental Rifting: The Birth of a New Ocean Basin

Outline the steps by which continental rifting results in the formation of new ocean basins.

Key Terms: supercontinent   triple junction
continental rift   cycle

- When continents rift apart, new ocean basins may form. The East African Rift is an example of the initial stages of continental breakup, with rift valleys that are sites of basaltic volcanism. The Red Sea is an example of more advanced rifting, in which seafloor spreading is occurring and the rift is submerged below sea level. Over time, the rift may widen through seafloor spreading, forming an ocean basin flanked by passive continental margins. A modern example of this stage is the Atlantic Ocean.

- *Continental rifting* may be initiated by mantle plumes, or perhaps continents are more likely to break when sites of preexisting mechanical weakness are subjected to tensional stresses. The assembly and breakup of supercontinents is called the supercontinent cycle. At least two supercontinents have formed in Earth's history.

13.7 Destruction of Oceanic Lithosphere

Compare and contrast spontaneous subduction and forced subduction.

Key Terms: spontaneous   forced subduction
subduction

- Most of the oceanic lithosphere produced by seafloor spreading is matched by an equivalent amount that is destroyed though subduction. Subduction carries oceanic lithosphere into the mantle, where its ultimate fate is still uncertain.

- When oceanic lithosphere is sufficiently old (and therefore cold), it may begin to sink because of its increased density. The resulting *spontaneous subduction* zone plunges at an angle near 90 degrees and is marked by a deep trench. The Mariana trench is an example.

- *Forced subduction* is a process in which lithosphere is too buoyant to subduct spontaneously but is forced beneath the overriding plate. This type of subduction results in shallow subduction angles and large earthquakes along megathrust faults. The Peru–Chile trench is an example.

- Regardless of its initial cause, subduction serves to close ocean basins and may eventually bring once widely separated landmasses into contact with one another.

Q Which type of subduction is illustrated in this diagram? Describe the age of the oceanic lithosphere at the deep-ocean trench compared to that at the oceanic ridge.

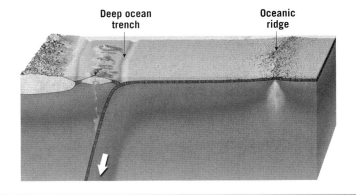
Deep ocean trench   Oceanic ridge

GIVE IT SOME THOUGHT

1. How many seconds would it take an echo sounder's ping to make the trip from a ship to the Challenger Deep (10,994 meters) and back? Recall that depth $= \frac{1}{2}$ (1500 m/sec = echo travel time).

2. Refer to the accompanying map of the eastern seaboard of the United States to complete the following:

   a. Which letter is associated with each of the following terms: *continental shelf*, *continental rise*, and *shelf break*?

   b. How does the size of the continental shelf that surrounds the state of Florida compare with the size of the Florida peninsula?

   c. Why are there no deep-ocean trenches on this map?

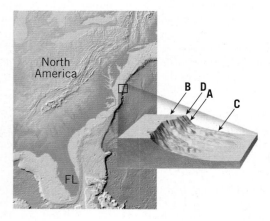

3. Referring to Figure 13.17, compare and contrast the topography of the crest of an oceanic ridge that exhibits a slow spreading rate with one that exhibits a fast spreading rate. Give examples of each.

4. Briefly explain why the ocean floor generally gets deeper the farther one travels from the ridge crest.

5. The accompanying photo is a false-color sonar image that shows the ridge crest (linear whitish-pink area) of a section of a spreading center.

   a. Is the structure along the ridge crest characteristic of a fast or slow spreading center? Explain.

   b. What name is given to the submerged conical-shaped structure in the lower-left portion of this image?

6. Consider the Baja Peninsula of Mexico (see Figure 13.29). Which stage of continental rifting does the Gulf of California most closely match?

7. Refer to Figure 13.28. Predict the fate of the Juan de Fuca plate.

8. This image shows lava at a temperature of about 1200°C erupting on the seafloor west of the Tonga trench. What name is given to lava flows like these that erupt underwater?

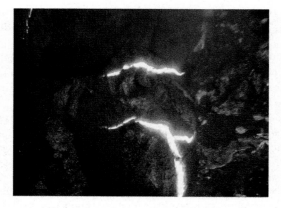

EYE ON EARTH

1. **This image shows a perspective view of the continental margin, looking southwest toward the Palos Verdes Peninsula near Los Angeles.**

   a. Match the features labeled 1 through 4 with the following terms: *continental shelf, continental slope, continental rise,* and *submarine canyon*.

   b. Based on the features in this image, what type of continental margin is shown here?

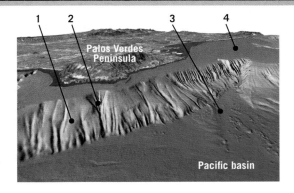

DATA ANALYSIS

Exploring the Ocean Surface

One of the goals of the National Oceanic and Atmospheric Administration (NOAA) is to learn more about Earth's oceans, including ecosystems and surface topography. Research expeditions are conducted to explore various aspects of the ocean world.

https://goo.gl/eD226M

ACTIVITIES

Go to the NOAA's Ocean Exploration and Research Digital Atlas at https://www.ncddc.noaa.gov/website/google_maps/OE/mapsOE.htm.

1. **Where have most of the research cruises been conducted?**

2. **How many missions have occurred during all years displayed on the map?**

3. **How many missions have occurred in the current year?**

4. **Which year had the largest number of missions? How many missions were conducted during that year?**

In the sidebar, click on "Search by Theme" and then select "Canyons." Click on an expedition location on the map to learn more about it. Find an expedition that has the GIS Tools tab. (*Note:* The tabs appear at the top of the pop-up window for each expedition.)

5. **What was the name of the expedition? (See the "Summary" tab for the expedition you selected.) When did it take place?**

6. **What landmass is nearest this research site?**

7. **What are the goals of this expedition? Why is this research mission needed?**

Click the "GIS Tools," tab choose "Ship Track," and click "Plot on Map."

8. **What are the approximate length and width of the area being explored?**

9. **Describe the ship's exploration pattern.**

10. **What named ocean features are near the ship's track?**

Mastering Geology

Looking for additional review and test prep materials? Visit pearson.com/mastering/geology to enhance your understanding of this chapter's content by accessing a variety of resources, including **Self-Study Quizzes, Geoscience Animations, SmartFigure Tutorials,** *Project Condor* **Videos, Mobile Field Trips, Dynamic Study Modules,** and an optional **Pearson eText.**

Sierra Nevada Mountains Grew During California's Extreme Drought

Between 2012 and 2017, California experienced its driest period since officials began recording rainfall levels back in 1840. The drought killed millions of trees, shrank reservoirs, turned green agricultural fields to dust, and led officials to declare a state of emergency for water usage. A recent NASA study found that during this same period, the Sierra Nevada range rose nearly an inch in height, mostly due to the drought.

▲ Cachuma Lake, drought-parched in 2015, is a primary source of drinking water for California's Central Coast region.

The NASA research team looked at data from 1300 Global Positioning System (GPS) monitoring stations located throughout the Sierra Nevada range. The stations, which were originally installed to track tectonic activity along the region's active faults, can measure elevation changes as small as one-tenth of an inch. The team found that the Sierra Nevada range was uplifted 24 millimeters (almost an inch) between 2011 and 2015. How can drought cause a mountain range to rise?

Although some of the uplift was caused by displacements along faults, NASA researchers concluded that the vast majority of the rise was due to loss of water from the mountains themselves. As water was lost from the rocks, they became lighter, and the mountains buoyantly rose—much as a cargo ship rises as it is being unloaded.

This process goes both ways: When abundant rainfall returned to California, researchers found that the Sierra Nevada range had become "heavier" as it soaked up water. This, in turn, led to the mountains losing about half of the uplift generated by the period of extreme drought.

▶ According to GPS data, the Sierra Nevada Mountains grew in response to loss of water during extreme drought in California.

14
Mountain Building

FOCUS ON CONCEPTS

Each statement represents the primary learning objective for the corresponding major heading within the chapter. After you complete the chapter, you should be able to:

14.1 Name and locate Earth's major mountain belts on a world map.

14.2 List and describe the four major features associated with subduction zones.

14.3 Sketch a cross-section of an Andean-type mountain belt and describe how its major features are generated.

14.4 Compare and contrast the formation of an Alpine-type mountain belt with that of a Cordilleran-type mountain belt.

14.5 Summarize the stages in the formation of a fault-block mountain range.

14.6 Explain the principle of isostasy and how it contributes to the elevated topography of mountain belts.

Mountains provide some of the most spectacular scenery on our planet. Poets, painters, and songwriters have captured their splendor. Geologists understand that at some time, all continental regions were mountainous masses and that continents grow through the addition of mountains to their flanks. As geologists unravel the secrets of mountain formation, they gain a deeper understanding of the evolution of Earth's continents. If continents do indeed grow by adding mountains to their flanks, then why do mountains exist in the interior of landmasses? To answer this and related questions, this chapter pieces together the sequence of events that generate these lofty structures.

14.1 Mountain Building

Name and locate Earth's major mountain belts on a world map.

Mountain building has occurred in the recent geologic past at several locations. Young mountain belts include the American Cordillera (*cordillera* means "spine" or "backbone"), which runs along the western margin of the Americas. The American Cordillera extends from southernmost South America to Alaska and includes both the Rockies and the Andes. The Alpine–Himalaya chain, also geologically young, extends along the margin of the Mediterranean through Iran to northern India and into Indochina. Mountainous terrains of the western Pacific, including volcanic island arcs like Japan, the Philippines, and much of Indonesia, are also young. Most of these young mountain belts have formed in the past 100 million years (**Figure 14.1**). Some, including the Himalayas, began their growth as recently as 50 million years ago.

▼ **Figure 14.1**
Earth's major mountain belts Notice the east–west trend of major mountain belts in Eurasia in contrast to the north–south trend of the North and South American Cordillera. The shields and stable platforms shown are composed of old crustal rocks that were highly deformed during ancient mountain-building events. These shields and stable platforms are discussed in Chapter 22.

In addition to these young mountain belts, there are several chains of Paleozoic-age mountains on Earth, including the Appalachians in the eastern United States and the Urals in Russia. Although these older mountain belts are deeply eroded and topographically less prominent, they exhibit the same structural features found in younger mountains.

Key

- **Young mountain belts** (less than 100 million years old)
- **Old mountain belts**
- **Shields**
- **Stable platforms** (shields covered by sedimentary rock)

Greenland shield
Canadian shield
North American Cordillera
Appalachians
Orinoco shield
Brazilian shield
Andes Mountains
Caledonian Belt
Baltic shield
Urals
Alps
Atlas
Caucasus
Zagros
Tien Shan
Altay-Sayan
Himalayas
Angara shield
Indian shield
African shield
Australian shield
Great Dividing Range

The term for the processes that collectively produce a mountain belt is **orogenesis** (*oros* = mountain, *genesis* = to come into being). An episode of mountain building is called an **orogeny**. Most major mountain belts display striking visual evidence of great compressional forces that have shortened the crust horizontally while thickening it vertically. These **collisional mountains** usually result from the collision of one or more small crustal fragments with a continental margin or from the closure of an ocean basin that results in the collision of two major landmasses. As a consequence, collisional mountains contain large quantities of preexisting sediments and sedimentary rocks that at one time lay along the margin of a continent and were subsequently faulted and contorted into a series of folds (**Figure 14.2**). Although folding and thrust faulting are often the most conspicuous signs of orogenesis, varying degrees of metamorphism and igneous activity are always present.

The theory of plate tectonics provides a model for orogenesis that accounts for the origin of virtually all the present mountain belts and most of the ancient ones. According to this model, the tectonic processes that generate Earth's major mountainous terrains occur along convergent plate boundaries. We will next revisit the nature of convergent plate boundaries and then examine how the process of subduction has driven mountain building around the globe.

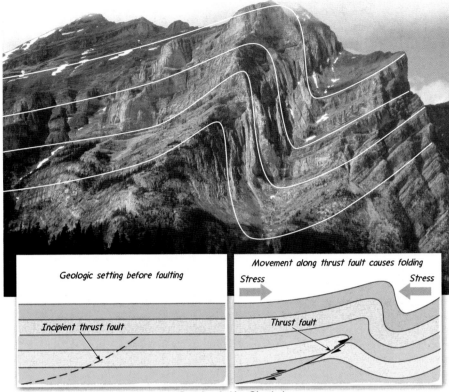

Geologist's Sketch

▲ **SmartFigure 14.2**
Mount Kidd, Alberta, Canada Highly deformed sedimentary strata exposed on Alberta's Mount Kidd. These sedimentary rocks are continental shelf deposits that were folded and displaced toward the interior of Canada by thrust faulting.

Tutorial
https://goo.gl/ KtGmFP

CONCEPT CHECKS 14.1

1. Define *orogenesis*.

2. Which type of plate boundary is most directly associated with Earth's major mountain belts?

 **Concept Checker**
https://goo.gl/kYYaaV

14.2 Subduction Zones

List and describe the four major features associated with subduction zones.

In their ongoing quest to unravel the events that produce mountains, researchers examine ancient mountain belts as well as sites where orogenesis is currently active. Of particular interest are convergent plate boundaries where lithospheric plates subduct. The subduction of oceanic lithosphere generates Earth's strongest earthquakes and most explosive volcanic eruptions, and it plays a pivotal role in generating most of Earth's mountain belts.

Features of Subduction Zones

Subduction zones can be roughly divided into four regions: (1) a *volcanic arc*, which is built on the overlying plate; (2) a *deep-ocean trench*, which forms where subducting slabs of oceanic lithosphere bend and descend into the asthenosphere; (3) a *forearc region*, which is located between a trench and a volcanic arc; and (4) a *back-arc region*, which is located on the side of the volcanic arc opposite the trench.

Volcanic Arcs Perhaps the most obvious structure generated by subduction is a *volcanic arc*. In settings

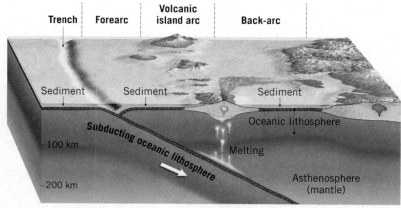

A. Convergent plate boundary involving two slabs of oceanic lithosphere.

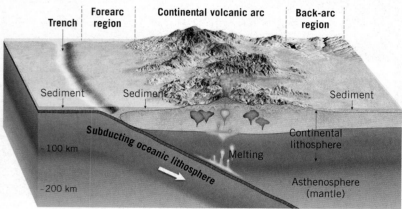

B. Convergent plate boundary where oceanic lithosphere is subducting beneath continental lithosphere.

▲ Figure 14.3
Development of two types of volcanic arcs

where two oceanic slabs converge, one plate gets pushed beneath the other, initiating partial melting of the mantle wedge located above the subducting plate (**Figure 14.3A**). This molten rock rises, eventually leading to the growth of a **volcanic island arc**, or simply an **island arc**, on the ocean floor. Examples of active island arcs include the Mariana, Tonga, and Aleutian arcs in the Pacific (see Figure 2.17, page 50).

By contrast, when oceanic lithosphere is subducted beneath a continental block, a **continental volcanic arc** results (**Figure 14.3B**). Continental volcanic arcs build on the topography of older, thicker continental blocks, resulting in volcanic peaks that may reach 6000 meters (nearly 20,000 feet) above sea level. The Cascade Range of the Pacific Northwest is a classic example.

Deep-Ocean Trenches   Where oceanic lithosphere bends as it descends into the mantle, a **deep-ocean trench** results. Trench depth relates strongly to the age—and therefore the temperature and density—of the subducting oceanic slab. In the western Pacific, where oceanic lithosphere is cold and dense, oceanic

slabs descend into the mantle at steep angles, producing trenches with average depths of about 8 kilometers (5 miles) below sea level. A well-known example is the Mariana trench, where the deepest area is an amazing 10,994 meters (36,069 feet) below sea level.

By contrast, the Cascadia subduction zone off the coasts of Washington and Oregon lacks a well-defined trench, partly because the warm, buoyant Juan de Fuca plate subducts at a very low angle. Trench depth is also related to the availability of sediments. A massive amount of sediment from the Columbia River basin fills most of what would otherwise be a shallow trench in this subduction zone—about 3 kilometers (2 miles) deep.

Forearc   The **forearc** region of a subduction zone is located between a deep-ocean trench and the associated volcanic arc (see Figure 14.3). Here pyroclastic material from the volcanic arc, as well as sediments eroded from the adjacent landmass, accumulate. Ocean-floor sediments are also carried to forearc regions by subducting plates.

The amount of sediment carried to a forearc region varies. The forearc region adjacent to the Mariana trench, for example, contains minimal sediment, partially because of its distance from a significant source of sediment. By contrast, the forearc region adjacent to the Cascadia subduction zone is choked with sediment derived from the nearby outlet of the Columbia River.

In addition, forearc width can vary significantly. Where an oceanic slab subducts at a steep angle, the forearc region is quite narrow, but when the angle of subduction is low, the forearc tends to be broad.

Back-Arc   Another site where sediments and volcanic debris accumulate is the **back-arc**, which is located on the backside of a volcanic arc when viewed from the trench (see Figure 14.3). In these regions, tensional forces tend to prevail, causing Earth's crust to stretch and thin and resulting in the formation of a down-faulted basin. (The reason for this development is considered in the next section.)

Back-arc regions associated with volcanic island arcs tend to be long, linear seas, such as the Sea of Japan and the Java Sea. In continental settings, the back-arc regions are located landward of the continental volcanic arc. Here stretching of the crust usually results in subsidence, forming basins that quickly fill with volcanic ash and sediments derived from the growing volcanic structures.

Extension and Back-Arc Spreading

Because subduction zones form where two plates converge, it is logical to assume that large compressional forces deform the plate margins. However, convergent margins are *not necessarily* regions dominated by compressional forces. As mentioned above, tensional stresses act on the overlying plates along some convergent plate margins and cause extension—stretching and thinning—of the crust. But how do extensional processes operate where two plates are moving together?

The age of the subducting oceanic slab is thought to play a significant role in determining the dominant forces acting on the overriding plate. When a relatively cold, dense slab subducts, it does *not* follow a fixed path into the asthenosphere (**Figure 14.4A**). Rather, it sinks vertically as it descends along an angled path. This causes the trench to retreat, or "roll back," as shown in **Figure 14.4B**. As the subducting plate sinks, it creates a flow in the asthenosphere called *slab suction* that "pulls" the upper plate toward the retreating trench. (Visualize what would have happened if you were in a lifeboat, unable to move away from the *Titanic* as it sank!)

Slab suction, in turn, produces tensional stress that elongates and thins the overriding plate, most often creating a basin in the region behind the volcanic arc (**Figure 14.4C**). Thinning of the crust results in upwelling of hot mantle rock and accompanying decompression melting. Continued extension may initiate seafloor spreading, which increases the size of the newly formed basin. Basins of this type within a back-arc region are termed **back-arc basins** (see Figure 14.4C). Seafloor spreading is currently enlarging the back-arc basins found landward of the Mariana and Tonga volcanic island arcs.

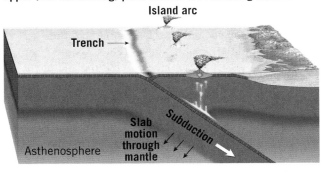

A. Subduction and "roll back" of an oceanic slab creates flow in the mantle, called slab suction, that "pulls" the upper (non-subducting) plate toward the retreating trench.

◀ Figure 14.4
Formation of a back-arc basin

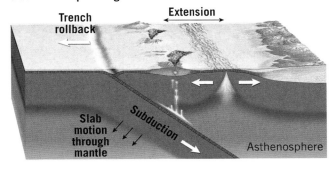

B. Slab suction causes the overlying plate to become elongated and thinned, often resulting in the formation of a basin and spreading center behind the volcanic arc.

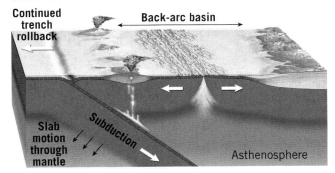

C. Some back-arc basins develop into well-developed spreading centers that generate a deep ocean basin behind the volcanic arc.

CONCEPT CHECKS 14.2

1. List the four major features of subduction zones.

 **Concept Checker**
https://goo.gl/422XhS

2. Briefly describe how back-arc basins form.

14.3 Subduction and Mountain Building

Sketch a cross-section of an Andean-type mountain belt and describe how its major features are generated.

Where oceanic lithosphere subducts beneath oceanic lithosphere, a *volcanic island arc* and related tectonic features develop. Subduction of oceanic lithosphere beneath continental lithosphere, on the other hand, results in the formation of a *continental volcanic arc* and mountainous topography along the margin of a continent. Acting like a conveyor belt, oceanic lithosphere may also bring volcanic island arcs and other crustal fragments to a subduction zone. These crustal elements are generally too buoyant to subduct to any great depth, and they become welded to the overriding plate, which may be another small crustal fragment or a continent. If subduction continues long enough, it can ultimately lead to the closure of an ocean basin and the ensuing collision of two continents.

Island Arc–Type Mountain Building

Island arcs result from the steady subduction of oceanic lithosphere under oceanic lithosphere, which may continue for 200 million years or more. Periodic volcanic activity, the emplacement of igneous plutons at depth, and the accumulation of sediment that is scraped from the subducting plate gradually increase the volume of crustal material capping the upper plate (see Figure 14.3A). Some large volcanic island arcs, such as Japan, owe their size to having been built on fragments of continental crust that have rifted from a large landmass or to the joining of multiple island arcs over time.

The continued growth of a volcanic island arc can generate mountainous topography consisting of nearly parallel belts of igneous and metamorphic rocks. This activity, however, is viewed as just one phase in the development of Earth's major mountain belts. As you will see later, a volcanic arc may be carried by a subducting plate to the margin of a large continental block, where it becomes involved in large-scale mountain-building episodes.

Andean-Type Mountain Building

Andean-type mountain building is characterized by subduction beneath a continent rather than oceanic lithosphere, as in the Andes Mountains of South America. Subduction along these active continental margins is associated with long-lasting magmatic activity that builds continental volcanic arcs. The result is crustal thickening: The crust may become more than 70 kilometers (45 miles) thick.

The first stage in the development of Andean-type mountain belts occurs along *passive continental margins*. The East Coast of the United States provides a modern example of a passive continental margin where sedimentation has produced a thick platform of shallow-water sandstones, limestones, and shales (**Figure 14.5A**). At some point, the forces that drive plate motions change, and a subduction zone develops along the margin of the continent. This subduction zone may form because the oceanic lithosphere has become so old and dense that it begins to sink of its own accord. Alternatively, strong compressional forces may help initiate subduction.

Building Volcanic Arcs

Recall that as oceanic lithosphere descends into the mantle, increasing temperatures and pressures drive volatiles (mostly water and carbon dioxide) from the crustal rocks. These mobile fluids migrate upward into the wedge-shaped region of mantle between the subducting slab and the upper plate. At a depth of about 100 kilometers (60 miles), these fluids reduce the melting point of hot mantle rock sufficiently to trigger partial melting (**Figure 14.5B**).

Partial melting of the ultramafic mantle rock peridotite generates magmas with mafic (basaltic) compositions. Because these newly formed mafic magmas are less dense than the rocks from which they originated, they will rise buoyantly. In continental settings, mafic magma often ponds beneath the less dense rocks of the crust. The hot magma may heat these overlying crustal rocks sufficiently to generate a silica-rich magma of intermediate and/or felsic (granitic) composition that can rise to form the continental volcanic arcs characteristic of an Andean-type subduction zone.

Emplacement of Batholiths

Because of its low density and great thickness, continental crust significantly impedes the ascent of molten rock. Consequently, much of the magma that intrudes Earth's crust never reaches the surface; instead, it crystallizes at depth to form

A. Passive continental margin.

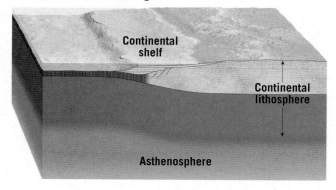

B. Partial melting produces a continental volcanic arc.

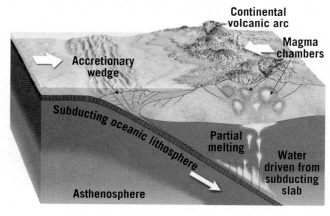

C. Subduction ends and is followed by a period of uplift.

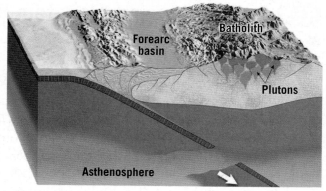

▲ Figure 14.5
Andean-type mountain building

▲ Figure 14.6
Torres del Paine National Park, Chile This area east of the high southern Andes consists mainly of large granite plutons (light color) between layers of sedimentary and metamorphic rocks (dark color). Magma intruded and metamorphosed the adjacent sedimentary host rock to form these rock bodies that were uplifted and exhumed by erosional forces.

massive collections of igneous plutons called *batholiths* (see Section 4.8). The result of this activity is thickening of Earth's crust.

Eventually, uplift and erosion exhume the batholiths. The American Cordillera contains several large batholiths, including the Sierra Nevada batholith of California, the Coast Range batholith of western Canada, and several large igneous bodies in the Andes (**Figure 14.6**). Most batholiths consist of intrusive igneous rocks that range in composition from granite to diorite.

Development of an Accretionary Wedge During the development of volcanic arcs, unconsolidated sediments that are carried on the subducting plate, as well as fragments of oceanic crust, may be scraped off and plastered against the edge of the overriding plate, like a wedge of soil scraped up by an advancing bulldozer. The resulting chaotic accumulation of deformed and thrust-faulted sediments and scraps of ocean crust is called an **accretionary wedge** (see Figure 14.5B).

Some of the sediments in an accretionary wedge are muds that accumulated on the ocean floor and were carried to the subduction zone by plate motion. Additional materials are derived from an adjacent continental

volcanic arc and consist of volcanic debris and products of weathering and erosion.

Where sediment is plentiful, prolonged subduction may thicken a developing accretionary wedge so that it protrudes above sea level. This has occurred along the southern end of the Puerto Rico trench, where the Orinoco River basin of Venezuela is a major source of sediments. The resulting wedge emerges to form the island of Barbados.

Forearc Basins As an accretionary wedge thickens, it acts as a barrier to the movement of sediment from the volcanic arc to the trench. As a result, sediments begin to collect between the accretionary wedge and the volcanic arc. This region, which is composed of relatively undeformed layers of sediment and sedimentary rocks, is called a **forearc basin** (see Figure 14.5C). Subsidence and continued sedimentation in forearc basins can generate a sequence of nearly horizontal sedimentary strata that can attain thicknesses of several kilometers.

Sierra Nevada, Coast Ranges, and Great Valley

California's Sierra Nevada, Coast Ranges, and Great Valley are excellent examples of the tectonic structures that are typically generated along an Andean-type subduction zone (**Figure 14.7**). These structures were produced by the subduction of a portion of the Pacific

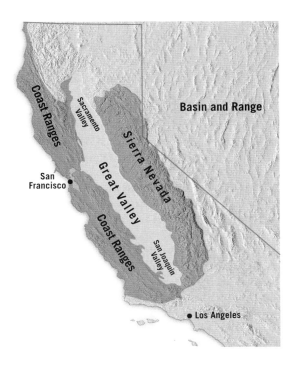

◄ Figure 14.7
Tectonic structures of California generated by an Andean-type subduction zone

basin (the Farallon plate) under the western margin of California (see Figure 14.5B). The Sierra Nevada batholith is a remnant of the continental volcanic arc that was produced by many intrusions of magma over a span of more than 100 million years. The Coast Ranges were built from the vast accumulation of sediments (accretionary wedge) that collected along the continental margin.

Beginning about 30 million years ago, subduction gradually ceased along much of the margin of North America, as the spreading center that produced the Farallon plate entered the California trench. The uplifting and erosion that followed removed most of the evidence of past volcanic activity and exposed the core of crystalline igneous and associated metamorphic rocks that make up the Sierra Nevada (see Figure 14.5C). The Coast Ranges were uplifted only recently, as evidenced by the young, unconsolidated sediments that currently blanket portions of these highlands.

California's Great Valley is a remnant of the forearc basin that formed between the Sierra Nevada and the accretionary wedge and trench that lay offshore. Throughout much of its history, portions of the Great Valley lay below sea level. This sediment-laden basin contains thick marine deposits and debris eroded from the adjacent continental volcanic arc.

CONCEPT CHECKS 14.3

1. In what ways are the Sierra Nevada and the Andes ranges similar?

2. What is an accretionary wedge? Briefly describe its formation.
Concept Checker
https://goo.gl/JLYZES

3. What is a batholith? In what tectonic setting are batholiths generated?

14.4 Collisional Mountain Belts

Compare and contrast the formation of an Alpine-type mountain belt with that of a Cordilleran-type mountain belt.

Most major mountain belts are generated when one or more buoyant crustal fragments collide with a continental margin as a result of subduction. Whereas the relatively dense oceanic lithosphere readily subducts, continental lithosphere contains significant amounts of low-density crustal rocks and is therefore too buoyant to be subducted deeply or permanently. Consequently, the arrival of a crustal fragment at a trench results in a collision between two relatively less dense continental blocks.

Alpine-Type Mountain Building: Continental Collisions

Alpine-type orogenies are episodes of mountain building that occur where two continental masses collide. They are named after the Alps, which have been intensively studied for more than 200 years. Mountain belts formed by the closure of major ocean basins include the Himalayas, Appalachians, Urals, and Alps. Continental collisions result in the development of mountains characterized by laterally shortened and vertically thickened crust, achieved through deformation such as folding and large-scale thrust faulting. Prior to the collision of the two large landmasses, this type of orogeny may also involve the accretion (joining) of smaller continental fragments or island arcs occupying the ocean basin that had once separated the two continental blocks.

The zone where two continents collide and are "welded" together is called a **suture**. The same term can be used to describe the boundary between two adjacent crustal blocks. This portion of a mountain belt often preserves slivers of oceanic lithosphere that were trapped between the colliding plates. The unique structure of these pieces of oceanic lithosphere, called *ophiolites*, helps researchers identify the collision boundary.

Noteworthy features of most collisional mountain ranges are **fold-and-thrust belts**. These mountainous zones result from the deformation of thick sequences of shallow marine sedimentary rocks, like those currently found along passive continental margins of the Atlantic. During continental collisions, sedimentary rocks are pushed inland, away from the core of the developing mountain belt and over the stable continental interior. In essence, crustal shortening is achieved by displacement along thrust faults where once relatively flat-lying strata are "sliced" into thick layers that are eventually stacked one upon another. During this displacement, material caught between the thrust faults is often folded, thereby forming the other major structure of a fold-and-thrust belt. Excellent examples of fold-and-thrust belts are found in the Appalachian Valley and Ridge Province, the Canadian Rockies, the Lesser (southern) Himalayas, and the northern Alps.

Next, we will take a closer look at two examples of collisional mountains: the Himalayas and the Appalachians. The Himalayas, Earth's youngest collisional mountains, are still rising. By contrast, the Appalachians are a much older mountain belt, in which active mountain building ceased about 250 million years ago.

The Himalayas

The mountain-building episode that created the Himalayas began between 50 and 30 million years ago, when India began to collide with Asia. Prior to the breakup of Pangaea, India was located between Africa and Antarctica in the Southern Hemisphere. As Pangaea fragmented, India moved rapidly, geologically speaking, a few thousand kilometers in a northward direction.

The Himalayas and Appalachians were formed by collisions between continents when the intervening ocean basin subducted completely.

The subduction zone that facilitated India's northward migration was near the southern margin of Asia (**Figure 14.8A**). Continued subduction along Asia's margin created an Andean-type plate margin that contained a well-developed continental volcanic arc and an accretionary wedge. India's northern margin, on the other hand, was a passive continental margin consisting of a thick platform of shallow-water sediments and sedimentary rocks.

Geologists have determined that two, or perhaps more, small crustal fragments were positioned on the subducting plate somewhere between India and Asia. During the closing of the intervening ocean basin, a small crustal fragment, which now forms southern Tibet, reached the trench and was accreted to Asia. This event was followed by the docking of India itself.

As the intervening ocean basin was closing up, the more deformable materials on the continental margins of these landmasses became highly folded and faulted (**Figure 14.8B**). Two major thrust faults and many smaller ones sliced through the Indian crust. Subsequent motion along these thrust faults caused slices of the Indian crust to be stacked one upon the other. Today, these slices make up the bulk of the highest peaks in the Himalayas—many of which are capped by tropical marine limestones that formed along what was once the continental shelf.

The formation of the Himalayas was followed by a period of uplift that raised the Tibetan Plateau. Seismic evidence suggests that a portion of the Indian

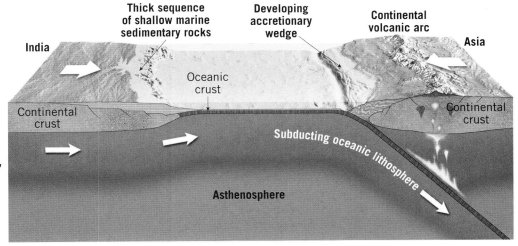

A. Prior to the collision of India and Asia, India's northern margin consisted of a thick platform of continental shelf sediments, whereas Asia's was an active continental margin with a well-developed accretionary wedge and volcanic arc.

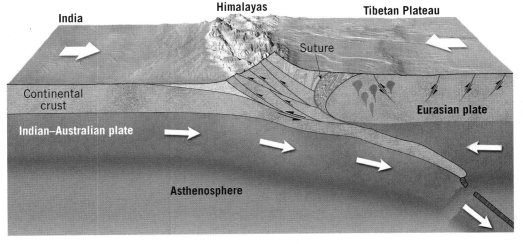

B. The continental collision folded and faulted crustal rocks along the margins of these continents to form the Himalayas. This event was followed by the gradual uplift of the Tibetan Plateau as the subcontinent of India was shoved under Asia.

◀ SmartFigure 14.8
Continental collision formed the Himalayas These diagrams illustrate the collision of India with the Eurasian plate that produced the spectacular Himalayas.

Animation
https://goo.gl/4BaBrn

▶ Figure 14.9
India's continued northward migration severely deformed much of China and Southeast Asia

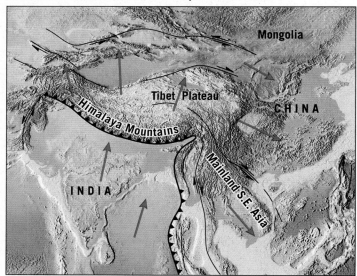

Map view showing the southeastward displacement of China and the mainland of Southeast Asia as India plowed into Asia.

subcontinent was thrust beneath Tibet—a distance of perhaps 400 kilometers (250 miles). If this occurred, the added crustal thickness would account for the lofty landscape of southern Tibet, which has an average elevation of more than 4500 meters (14,800 feet), higher than the tallest mountain in the contiguous United States.

The collision with Asia slowed but did not stop the northward movement of India, which has since penetrated at least 2000 kilometers (1200 miles) into the mainland of Asia. Crustal shortening and thickening accommodated some of this motion. Much of the remaining penetration into Asia caused lateral displacement of large blocks of the Asian crust by a mechanism described as *escape tectonics*. As shown in **Figure 14.9**, when India continued its northward trek, parts of Asia were "squeezed" eastward, out of the collision zone. These displaced crustal blocks include much of Southeast Asia (the region between India and China) and sections of China.

Why was the interior of Asia deformed to such a large extent, while India has remained essentially intact? The answer lies in the nature of these diverse crustal blocks. Much of India is a continental shield composed mainly of old Precambrian rocks (see Figure 14.1). This thick, cold slab of crustal material has been intact for more than 2 billion years and is mechanically strong as a result. By contrast, Southeast Asia was assembled more recently, from the collision of several smaller crustal fragments. Consequently, it is still relatively "warm and weak" from recent periods of mountain building (see Figure 14.1).

The Appalachians

The Appalachian Mountains provide great scenic beauty near the eastern margin of North America, from Alabama to Newfoundland. Mountain belts of similar origin that formed during the same period and

were once contiguous are found in the British Isles, Scandinavia, northwestern Africa, and Greenland (see Figure 2.6, page 41). The orogenies that generated this extensive mountain system lasted a few hundred million years and resulted in the assembly of the supercontinent Pangaea. Detailed studies of the Appalachians indicate that this mountain belt was the result of three distinct episodes of mountain building.

Our simplified overview begins roughly 750 million years ago, with the breakup of a supercontinent called Rodinia that predates Pangaea. Much like the breakup of Pangaea, this episode of continental rifting and seafloor spreading generated a new ocean between the rifted continental blocks. Located within this widening ocean basin was a microcontinent near the edge of ancestral Africa.

About 600 million years ago, for reasons geologists do not completely understand, plate motion changed dramatically, and this ancient ocean basin began to close. This led to the development of multiple subduction zones, and the stage was set for the three orogenic events that would lead to the collision of North America and Africa (**Figure 14.10A**).

Taconic Orogeny Around 450 million years ago, the marginal sea between the volcanic island arc and ancestral North America began to close. The collision that ensued, called the *Taconic Orogeny*, caused the volcanic arc along with ocean sediments located on the upper plate to be accreted to the edge of the larger continental block. The remnants of this volcanic arc and oceanic sediments are recognized today as the metamorphic rocks through much of the Appalachian mountain belt (**Figure 14.10B**). For example, schists beneath New York City and Washington, DC, formed at this time. In addition to this pervasive regional metamorphism, numerous magma bodies intruded the crustal rocks along the entire continental margin.

Acadian Orogeny A second episode of mountain building, called the *Acadian Orogeny*, occurred about 350 million years ago. The continued closing of this ancient ocean basin resulted in the collision of a microcontinent with North America (**Figure 14.10C**). This orogeny involved thrust faulting, metamorphism, and the intrusion of many large granite bodies. This event also added substantially to the width of North America, particularly in eastern New England.

Alleghanian Orogeny The final orogeny, called the *Alleghanian Orogeny*, occurred between 250 and 300 million years ago, when Africa collided with North America. This collision displaced material that was accreted earlier by as much as 250 kilometers (155 miles)

◀ SmartFigure 14.10
Formation of the Appalachian Mountains

Tutorial
https://goo.gl/
NAtW4E

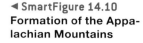

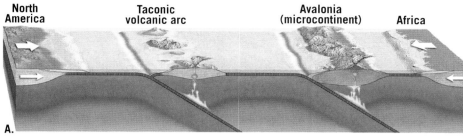

Closing of an Ocean Basin About 600 million years ago, the precursor to the North Atlantic began to close. Located within this ocean basin was an active volcanic island arc off the coast of North America and a microcontinent situated closer to Africa.

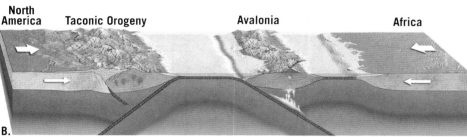

Taconic Orogeny Around 450 million years ago, the marginal sea between the volcanic island arc and North America closed. The collision, called the Taconic Orogeny, thrust the island arc over the eastern margin of North America.

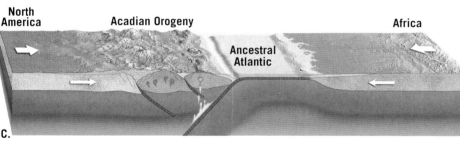

Acadian Orogeny A second episode of mountain building, called the Acadian Orogeny, occurred about 350 million years ago and involved the collision of a microcontinent with North America.

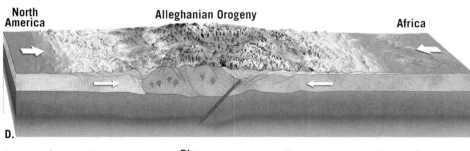

Alleghanian Orogeny The final event, the Alleghanian Orogeny, occurred between 250 and 300 million years ago, when Africa collided with North America. The result was the formation of the Appalachian Mountains.

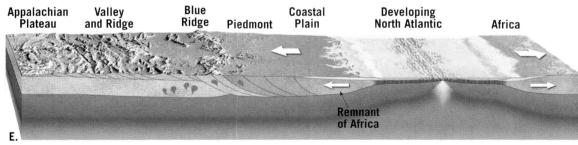

Rifting of Pangaea About 180 million years ago, Pangaea began to break into smaller fragments, a process that ultimately created the modern Atlantic Ocean. Because this new zone of rifting occurred east of the suture that formed when Africa and North America collided, remnants of African crust remain "welded" to the North American plate.

toward the interior of North America. This event also displaced and further deformed the continental shelf sediments and sedimentary rocks that had once flanked the eastern margin of North America (**Figure 14.10D**). Today these folded and thrust-faulted sandstones, limestones, and shales make up the largely unmetamorphosed rocks of the Valley and Ridge Province (**Figure 14.11**). This structural signature of mountain building can be found as far inland as central Pennsylvania and West Virginia.

With the collision of Africa and North America, the young Appalachians, perhaps as majestic as the Himalayas, lay along the suture, in the interior of Pangaea. The tectonic forces that built the mountains ceased to drive them upward. Then, about 180 million years ago, the new supercontinent began to break into smaller fragments, a process that ultimately created the modern Atlantic Ocean. Because this new zone of rifting occurred east of the suture that formed when

Appalachian Plateau   Valley and Ridge   Blue Ridge   Piedmont   Coastal Plain

many of the mountainous regions that rim the Pacific. These accreted blocks of crust are called **terranes**. Geologists use this term to describe any crustal fragment that consists of a distinct and recognizable series of rock formations and has been transported and accreted by plate tectonic processes. Notice that *terrane* is a different word from *terrain*; the two are pronounced the same, but *terrain* refers to the shape of the surface topography, or "lay of the land."

The Nature of Terranes

Some crustal fragments that have become terranes were once **microcontinents** similar to the modern-day island of Madagascar, located east of Africa in the Indian Ocean. Many others were island arcs similar to Japan, the Philippines, and the Aleutian Islands. Still others may have been submerged oceanic plateaus created by massive submarine outpourings of basaltic lavas (see Figure 13.5, page 366). More than 100 of these relatively small crustal fragments exist in the modern world.

Africa and North America collided, remnants of Africa remain stuck to the North American plate (**Figure 14.10E**). The crust underlying Florida is an example.

Other mountain ranges built from continental collisions include the Alps and the Urals. The Alps formed as Africa and several smaller crustal fragments collided with Europe during the closing of the Tethys Sea. Similarly, the Urals were deformed and uplifted during the assembly of Pangaea, when northern Europe and northern Asia collided, forming a major portion of Eurasia. Unlike the Appalachian belt, however, the Urals did not break apart again after their orogenesis.

Cordilleran-Type Mountain Building

A Cordilleran-type orogeny, named after the North American Cordillera, is associated with an ocean that, like the Pacific, may never close. The rapid rate of seafloor spreading in the Pacific basin is balanced by a high rate of subduction. In this setting, island arcs and small crustal fragments are often carried along until they collide with an active continental margin and accrete onto it. This process of collision and accretion has generated

A terrane is a relatively small crustal fragment (microcontinent, volcanic island arc, or oceanic plateau) that has been carried by an oceanic plate to a continental subduction zone and then accreted onto the continental margin.

Accretion and Orogenesis   Small structures such as seamounts are generally subducted along with a descending oceanic slab. However, thick sections of oceanic crust, such as the Ontong Java Plateau (which is almost as big as Alaska) or an island arc dominated by low-density andesitic igneous rocks, are too buoyant to subduct. In these situations, a collision between the crustal fragment and the continental margin occurs.

The sequence of events that occur when small crustal fragments reach a Cordilleran-type margin is shown in **Figure 14.12**. The upper crustal layers are "peeled" from the descending plate and thrust in relatively thin sheets onto the adjacent continental block. Convergence does not

A. A microcontinent and a volcanic island arc are being carried toward a subduction zone.

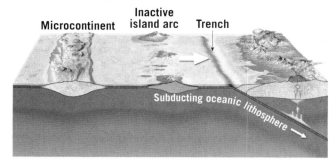

B. The volcanic island arc is sliced off the subducting plate and thrust onto the continent.

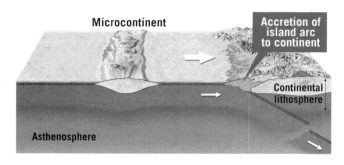

C. A new subduction zone forms seaward of the old subduction zone.

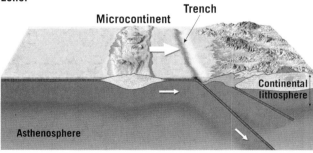

D. The accretion of the microcontinent to the continental margin shoves the remnant island arc further inland and grows the continental margin seaward.

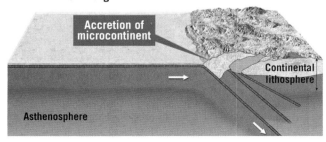

▲ SmartFigure 14.12
Collision and accretion of small crustal fragments to a continental margin

Tutorial
https://goo.gl/
DvC3ju

generally end with the accretion of a crustal fragment. Rather, new subduction zones typically form seaward of the accreted terrane, and they can carry other island arcs or microcontinents toward a collision with the continental margin. Each collision displaces earlier accreted terranes further inland, adding to the zone of deformation as well as to the thickness and lateral extent of the continental margin.

The North American Cordillera The correlation between mountain building and the accretion of crustal fragments was first developed in studies of the North American Cordillera (**Figure 14.13**). Researchers determined that some of the rocks in the orogenic belts of Alaska and British Columbia contained fossil and paleomagnetic evidence indicating that these strata previously lay much closer to the equator.

It is now known that many of the terranes that make up the North American Cordillera were scattered

◄ SmartFigure 14.13
Terranes added to western North America during the past 200 million years Paleomagnetic studies and fossil evidence indicate that some of these terranes originated thousands of kilometers to the south of their present locations.

Animation
https://goo.gl/
LqxYsr

throughout the Pacific, like the island arcs and oceanic plateaus currently distributed in the western Pacific. During the breakup of Pangaea, the eastern portion of the Pacific basin (the Farallon plate) began to subduct under the western margin of North America. This activity resulted in many additions of crustal fragments along the entire Pacific margin of the continent—from Mexico's Baja Peninsula to northern Alaska (see Figure 14.13). Geologists expect that many modern microcontinents will likewise be accreted to active continental margins surrounding the Pacific, producing new orogenic belts.

14.5 Fault-Block Mountains

Summarize the stages in the formation of a fault-block mountain range.

Most mountain belts form in compressional environments, as evidenced by the predominance of large thrust faults and folded strata. However, other tectonic processes, such as continental rifting, can also produce mountainous terrain. Recall that continental rifting occurs when tensional forces stretch and thin the lithosphere, resulting in upwelling of hot mantle rock. Upwelling heats the thinned lithosphere, which becomes less dense (more buoyant) and rises. This reduced density accounts, in part, for the elevated topography associated with continental rifts. Simultaneously, stretching elongates the rigid upper crust, which breaks into large crustal blocks that are bounded by high-angle normal faults. Continued rifting causes the blocks to tilt, with one edge rising as the other drops (see Figure 10.16, page 300). Mountains that form in these tectonic settings are termed **fault-block mountains**.

The Teton Range in western Wyoming is an excellent example of fault-block mountains. This lofty structure was faulted and uplifted along its eastern flank as the block tilted downward to the west. Looking west from Jackson Hole, Wyoming, the eastern front of this mountain rises more than 2 kilometers (1.2 miles) above the valley, making it one of the most imposing mountain fronts in the United States (**Figure 14.14**).

The Basin and Range Province

Located directly east of the Sierra Nevada is the Basin and Range Province—one of Earth's largest regions of fault-block mountains. This region extends in a roughly north–south direction for nearly 3000 kilometers (2000 miles) and encompasses all of Nevada and portions of the surrounding states, as well as a large area of western Mexico (**Figure 14.15**). In the Basin and Range Province, Earth's brittle upper crust has been broken into hundreds of fault blocks. Uplifting and tilting of these faulted structures produced nearly parallel mountain ranges, averaging about 80 kilometers (50 miles) in length, which rise above adjacent sediment-filled basins (see Figure 10.16, page 300).

Geologists have proposed several hypotheses to explain the events that generated the Basin and Range region. The most widely accepted view is that a change in the nature of the plate boundary along California's western margin led to the formation of this region. About 30 million years ago, the dominant forces acting on the western margin of North America were compressional, caused by the buoyant subduction of a segment of the Farallon plate (**Figure 14.16A**). Starting about 25 million years ago, subduction gradually ceased along the California coast as the spreading center that separated the Farallon plate from the Pacific began to subduct beneath North America. This event spawned the San Andreas Fault, with its strike-slip motion, that currently separates the Pacific plate from the North American plate (**Figure 14.16B**). According to this hypothesis, the northwestward motion of the Pacific plate

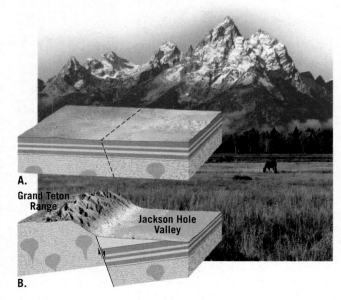

► Figure 14.14
Wyoming's Teton Range, an example of fault-block mountains

A.
Grand Teton Range

Jackson Hole Valley

B.

▲ Figure 14.15
Map of the Basin and Range Province in the United States

produced tensional forces that stretched and fractured the crust of the North American plate to produce the fault-block mountains of the Basin and Range Province (**Figure 14.16C**). Stretching and thinning of the lithosphere also led to mantle upwelling, which accounts for the higher-than-average elevation of this region.

Another model contends that about 20 million years ago, the cold, dense lithospheric mantle located beneath the Basin and Range region decoupled (separated) from the overlying crustal layer and slowly sank into the mantle. This process, called *delamination*, resulted in upwelling and lateral spreading of hot mantle rock, which produced tensional forces that stretched and thinned the overlying crust. According to this view, these elevated crustal blocks began to gravitationally slide from their lofty perches to generate the fault-block topography of the Basin and Range Province.

CONCEPT CHECKS 14.5

1. How does formation of fault-block mountains differ from the processes that generate most other major mountain belts?

2. Briefly describe the basic structure of the Basin and Range Province and identify its geographic extent.

🔊 **Concept Checker**
https://goo.gl/fjwuYs

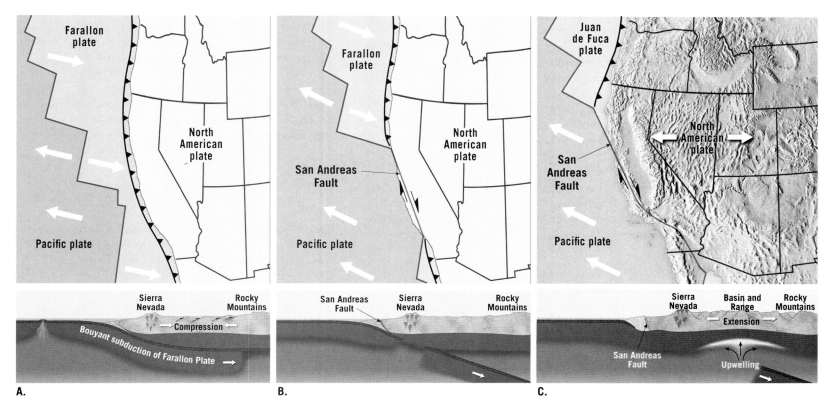

A.   B.   C.

▲ Figure 14.16
Proposed model for the formation of the Basin and Range Province The Basin and Range region consists of more than 100 fault-block mountains generated during the past 20 million years. Upwelling of hot mantle rock and gravitational collapse (crustal sliding) may have contributed to considerable stretching and thinning of the crust.

14.6 Vertical Motions of the Crust

Explain the principle of isostasy and how it contributes to the elevated topography of mountain belts.

The processes that produced Earth's varied topography are complex (see GEOgraphics 14.1). Beyond the tectonic forces that move rocks laterally and thicken them vertically to produce mountains, additional processes help to shape Earth's surface. As weathering and erosion work to lower mountains, a compensating process called *isostasy* causes them to rise, so they remain mountainous long after the tectonic processes that initially created them have ceased. Also, if tectonic processes raise a mountain belt "too high," the rock at its core will become too weak to support the load, and the mountain will spread.

The Principle of Isostasy

During the 1840s, researchers discovered that Earth's low-density crust "floats" on top of the high-density rocks of the mantle, much as a block of wood floats in water. We will use the floating wooden blocks in **Figure 14.17** to explore this idea. You can think of these blocks as a model mountain belt, floating in the mantle. Notice that only about one-quarter of each block projects above the water, and three-quarters is submerged. This is because wood is about three-quarters as dense as water. Similarly, most of the vertical thickness of a mountain belt forms a buoyant root that is "submerged" in the mantle; the remainder projects above the surrounding crust. This concept—that the crust floats in gravitational balance in the mantle—is called **isostasy**.

Notice that the tallest of the floating blocks in Figure 14.17 stands the highest above the water surface and also sits the deepest. Similarly, the greater the crustal thickness of a mountain belt, the higher it stands above sea level and the deeper its roots. Thus, the Himalayas, as the tallest range on Earth, also have the deepest roots.

Now, visualize what would happen if you placed a second small block on top of one of the blocks in Figure 14.17. The combined block would sink until it reached a new isostatic (gravitational) balance, at which point its top would be higher than before, and its bottom would be lower. This process of establishing a new gravitational balance in response to loading or unloading is called **isostatic adjustment**. Notice also that as a block rises or sinks, the surrounding water flows to accommodate

it. Similarly, the highly viscous mantle will flow, albeit at an excruciatingly slow rate, when weight is added to or subtracted from the overlying crustal blocks.

Applying the concept of isostatic adjustment, we should expect the crust to subside when weight is added, causing the underlying mantle rocks to flow away from the sinking crust. When the weight is removed, the crust rebounds, and the mantle rock flows back underneath. (Visualize what happens to a ship when its cargo is loaded or unloaded.) Scientists have found evidence for crustal subsidence followed by isostatic rebound in areas formerly overlain by Ice Age glaciers. When continental ice sheets covered portions of North America during the Pleistocene epoch, ice masses that averaged about 3 kilometers (2 miles) thick weighed down the crust and caused downwarping by hundreds of meters. In the 8000 years since this ice sheet melted, gradual uplift of as much as 330 meters (1000 feet) has occurred in Canada's Hudson Bay region, where the thickest ice had accumulated.

One of the consequences of isostatic adjustment is that, as erosion cuts into a mountain range and removes mass, the range rises in response to the reduced load (**Figure 14.18**). In fact, because erosion removes material mainly by carving canyons and valleys rather than by uniformly wearing down mountain peaks, isostasy may actually "push" the peaks higher than their original height.

The processes of uplift and erosion continue until the mountain block reaches average crustal thickness. When this occurs, these once-elevated structures are near sea level, and the once-deeply buried interior of the mountain is exposed at the surface. In addition, as mountains are worn down, the eroded sediment is deposited on adjacent landscapes, causing these areas to subside (see Figure 14.18).

How High Is Too High?

Where compressional forces are great, such as those driving India into Asia, lofty mountains such as the Himalayas result. Is there a limit on how high a mountain can rise? As mountaintops are elevated, gravity-driven processes such as erosion and mass

► **SmartFigure 14.17**
The principle of isostasy
This drawing shows how wooden blocks of different thicknesses float in water. In a similar manner, thick sections of crustal material float higher than thinner crustal slabs.

Animation
https://goo.gl/a5Yfvw

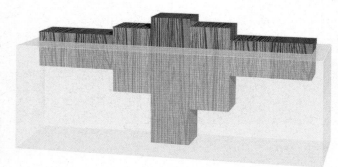

wasting accelerate, carving the deformed strata into rugged landscapes. Equally important, however, is the fact that gravity also acts on the rocks within the mountain belt. The higher the mountain, the greater the downward force on rocks near the base. Eventually, the rocks deep within the developing mountain, which are relatively warm and weak, begin to flow laterally, as shown in **Figure 14.19**. This process of **gravitational collapse** is analogous to what happens when a ladle of very thick pancake batter is poured onto a hot griddle. In addition to causing ductile spreading at depth, this process leads to normal faulting and subsidence in the upper, brittle portion of Earth's crust.

Considering these factors, what keeps the Himalayas standing? Simply, the horizontal compressional forces that are driving India into Asia are greater than the vertical force of gravity. However, when India's northward trek ends, the downward pull of gravity, weathering, and erosion will become the dominant forces acting on this mountainous region.

Mantle Convection: A Cause of Vertical Crustal Movement

Based on studies of Earth's gravitational field, it became apparent that up-and-down convective flow in the mantle also affects the elevation of Earth's major landforms. The buoyancy of hot rising material accounts for broad upwarping in the overlying lithosphere, whereas downward flow causes downwarping.

Uplifting Whole Continents Southern Africa is one region where large-scale vertical motion is evident. The topography of the region consists of an expansive plateau that has an average elevation of nearly 1500 meters (5000 feet)—much higher than what would be predicted for a stable continental platform.

Evidence from seismic tomography (see Figure 12.21, page 358) indicates that a large mass of hot mantle rock

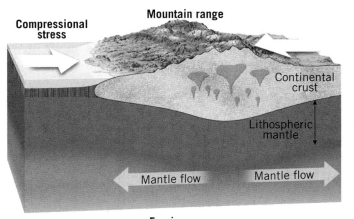

When compressional mountains are young, they are composed of thick, low-density crustal rocks that float on the denser mantle below.

Compressional stress — **Mountain range** — Continental crust — Lithospheric mantle — Mantle flow — Mantle flow

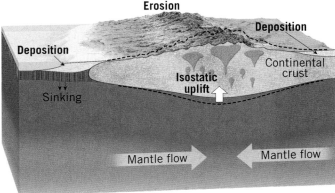

As erosion lowers the mountains, the crust rises in response to the reduced load in order to maintain isostatic balance.

Erosion — Deposition — Deposition — Sinking — Isostatic uplift — Continental crust — Mantle flow — Mantle flow

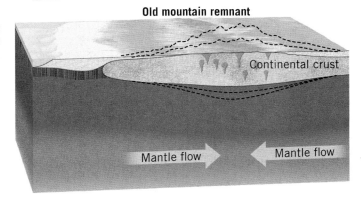

Erosion and uplift continue until the mountains reach "normal" crustal thickness.

Old mountain remnant — Continental crust — Mantle flow — Mantle flow

◄ **SmartFigure 14.18**
The effects of isostatic adjustment and erosion on mountainous topography This sequence illustrates how the combined effects of erosion and isostatic adjustment result in a thinning of the crust in mountainous regions.

Tutorial
https://goo.gl/oB66Rr

is centered below the southern tip of Africa. This structure, called a *superplume*, extends upward about 2900 kilometers (1800 miles) from the core–mantle boundary and has a lateral expanse of a few thousand kilometers. Researchers have determined that the upward flow associated with this huge mantle plume is sufficient to elevate southern Africa.

Crustal Subsidence Extensive areas of downwarping also occur on Earth's surface. In the United States, large, nearly circular basins are found in Michigan and Illinois. Similar structures are known on other continents as well.

The cause of the downwarping that created these basins may be linked to the subduction of slabs of oceanic lithosphere. One proposal suggests that when

The Laramide Rockies

Where are the Laramide Rockies?

Sometimes called the Central and Southern Rockies, this mountain belt lies to the east of the Colorado Plateau.

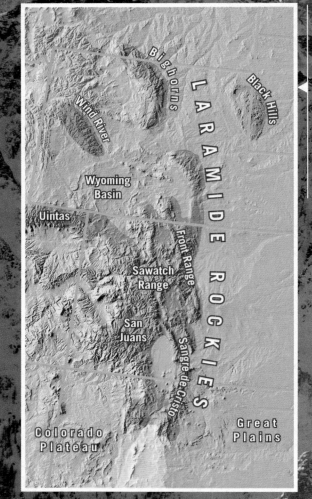

What is the geologic history of the Laramide Rockies?

These mountain ranges formed when Precambrian age basement rocks were uplifted nearly vertically along reverse and thrust faults, upwarping the overlying layers of younger sedimentary rocks. Uplifting accelerated the processes of weathering and erosion, which removed much of the younger sedimentary cover from the highest portions of the uplifted blocks.

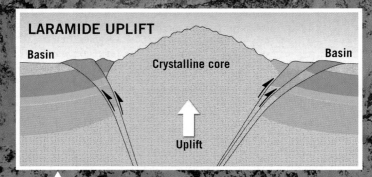

LARAMIDE UPLIFT

Basin — Crystalline core — Basin

Uplift

How did this mountain belt form?

Because the Laramide Rockies lie more than 1500 km from the plate boundary that produced the Cordilleran Orogeny, it has been difficult to identify the mechanism which produced them. The most widely accepted model proposes that a thickened slab of the Farallon Plate began to subduct under the west coast of California about 85 Ma. Because of its thickness, this buoyant slab resisted subduction as it was shoved beneath the continent. As the thick slab continued eastward under the Colorado Plateau it triggered uplift of at least 2 km. Upon reaching the Rockies compressional forces squeezed the crust, which responded by developing high angle faults along which igneous and metamorphic basement rock was uplifted. These crystalline blocks form the cores of many of the high peaks that comprise this mountain belt.

Sierra Nevada — Colorado Plateau — Laramide Rockies — Great Plains

Bouyant subduction of Farallon plate

Compressional stress

Partial melting

Over-thickened slab of oceanic crust

Asthenosphere

A. Horizontal compressional forces dominate: Compression causes shortening and thickening of the crust.

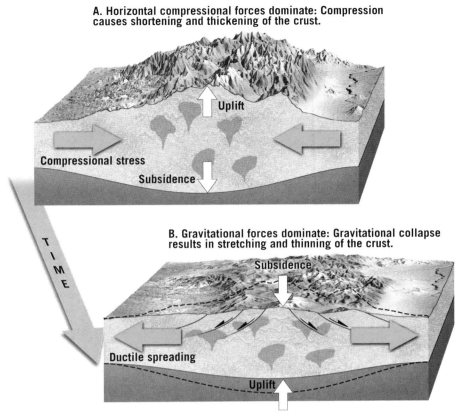

Uplift

Compressional stress

Subsidence

T
I
M
E

B. Gravitational forces dominate: Gravitational collapse results in stretching and thinning of the crust.

Subsidence

Ductile spreading

Uplift

▲ Figure 14.19
Gravitational collapse Without compressional forces to support them, mountains gradually collapse under their own weight. Gravitational collapse involves normal faulting in the upper, brittle portion of the crust and ductile spreading in the warm, weak rocks at depth.

subduction ceases, the descending slab detaches from the trailing lithosphere and continues its descent into the mantle. As this detached lithospheric slab sinks, it creates a downward flow in its wake that tugs at the base of the overriding continent. In some settings, the crust is apparently pulled down sufficiently to produce a large basin that eventually fills with sediments. As the oceanic slab sinks deeper into the mantle, the pull of the trailing wake weakens, and the continent "floats" back into isostatic balance.

CONCEPT CHECKS 14.6

1. Define *isostasy*. What happens to a floating object when weight is added? Removed?

2. Give one example of evidence that supports the concept of crustal uplift.

Concept Checker
https://goo.gl/EMfaA4

3. Explain the process whereby mountainous regions experience gravitational collapse.

14 CONCEPTS IN REVIEW

Mountain Building

14.1 Mountain Building

Name and locate Earth's major mountain belts on a world map.

Key Terms: orogeny
orogenesis   collisional mountain

- *Orogenesis* is the making of mountains. An episode of orogenesis is an *orogeny*. Most orogenesis occurs along convergent plate boundaries, where compressional forces cause folding and faulting, thickening the crust vertically and shortening it horizontally.

Q Look at the South American plate on the accompanying map. Explain why the Andes Mountains are located on the western margin of South America rather than the eastern margin.

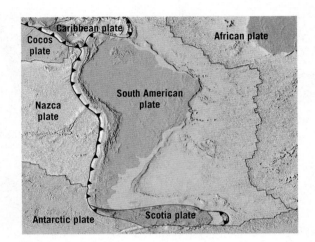

14.2 Subduction Zones

List and describe the four major features associated with subduction zones.

Key Terms: deep-ocean trench   forearc
volcanic island arc   continental volcanic   back-arc
                      arc   back-arc basin

- Sites of subduction are marked by *volcanic island arcs*, *deep-ocean trenches*, and *forearc* basins located between them. A subduction zone may also contain a *back-arc basin*, a site of tectonic extension that forms due to trench rollback and the sinking of old, cold, dense oceanic lithosphere.

Q Match the areas in the illustration represented by the letters A–D with the following labels: volcanic island arc, trench, forearc, and back-arc.

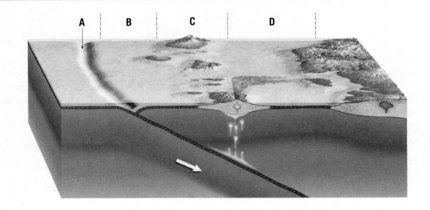

14.3 Subduction and Mountain Building

Sketch a cross-section of an Andean-type mountain belt and describe how its major features are generated.

Key Terms: accretionary wedge   forearc basin

- The type of convergent margin determines the type of mountains that form. Where one oceanic plate overrides another, a volcanic island arc forms. Where an oceanic plate subducts under a continent, Andean-type mountain building occurs.
- In either case, release of water from the subducted slab triggers melting in the overlying mantle wedge, generating mafic (basaltic)

magmas that rise to the base of the continental crust, where they often pond. The hot mafic magma may heat the overlying crustal rocks sufficiently to generate a silica-rich magma of intermediate or felsic (granitic) composition.
- Sediment scraped off the subducting plate builds an *accretionary wedge*. Between the accretionary wedge and the volcanic arc is a relatively calm site of sedimentary deposition, the *forearc basin*.
- The geography of central California preserves an accretionary wedge (Coast Ranges), a forearc basin (Great Valley), and the roots of an Andean-style mountain belt (Sierra Nevada).

14.4 Collisional Mountain Belts

Compare and contrast the formation of an Alpine-type mountain belt with that of a Cordilleran-type mountain belt.

Key Terms: fold-and-thrust belt   microcontinent
suture   terrane

- The Himalayas and Appalachians were formed by collisions between continents when the intervening ocean basin subducted completely. The Himalayas were formed by the collision of India and Eurasia starting around 50 million years ago, and they are still rising. The collision of ancestral North America with ancestral Africa more than 250 million years ago created the Appalachians.
- A *terrane* is a relatively small crustal fragment (a *microcontinent*, a volcanic island arc, or an oceanic plateau) that has been carried by an oceanic plate to a continental subduction zone and then accreted onto the continental margin. The North American Cordillera formed through the accretion of many successive terranes.

Q Which of the accompanying sketches best illustrates an Andean-type orogeny, which illustrates a Cordilleran-type orogeny, and which illustrates an Alpine-type orogeny?

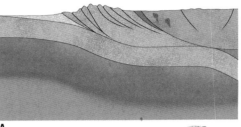
A.

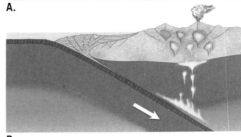
B.

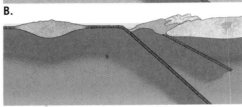
C.

14.5 Fault-Block Mountains

Summarize the stages in the formation of a fault-block mountain range.

Key Term: fault-block mountain

- Mountains can form in extensional tectonic settings. When the crust is thinned and stretched, normal faulting breaks the landscape into chunks, some of which slide down relative to their neighbors. *Fault-block mountains* are common in the Basin and Range Province in the western United States. The Teton Range is another example.

14.6 Vertical Motions of the Crust

Explain the principle of isostasy and how it contributes to the elevated topography of mountain belts.

Key Terms: isostatic adjustment
isostasy   gravitational collapse

- Earth's crust floats in the denser material of the mantle the way wood floats in water. This principle is termed *isostasy*. If additional weight is placed on the crust (an ice sheet, for example), the crust sinks, and if weight is removed (glacial melting), the crust rebounds. This process of maintaining gravitational equilibrium is called *isostatic adjustment*. For a mountain belt, isostasy partially offsets the effect of erosion, pushing the mountains up as erosion wears them down.
- When compressional forces raise a mountain belt too high, the rock at the belt's core becomes warm and weak, and the belt spreads, becoming broader and lower.
- Convection in the mantle can influence the vertical position of the crust. Areas of crust above sites of mantle upwelling may bulge upward, whereas the crust above sites of mantle downwelling may sink into broad basin-like structures.

GIVE IT SOME THOUGHT

1. Suppose that a sliver of oceanic crust were discovered in the interior of a continent. Would this refute the theory of plate tectonics? Explain.

2. Refer to the accompanying map, which shows the locations of the Galapagos Rise and the Rio Grande Rise, to answer the following questions:

   a. Compare the continental margin of the west coast of South America with the continental margin along the east coast.

   b. Based on your comparison in Question a, is the Galapagos Rise or the Rio Grande Rise more likely to end up accreted to a continent? Explain your choice.

   c. In the distant future, how might a geologist determine that this accreted landmass is distinct from the continental crust to which it accreted?

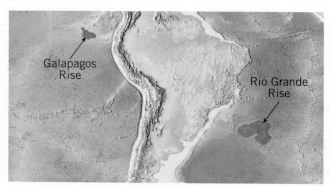

3. The Ural Mountains exhibit a north–south orientation through Eurasia. How does the theory of plate tectonics explain the existence of this mountain belt in the interior of an expansive landmass?

4. Mount Moran (3842 meters elevation) in the Teton Range rises 1.8 kilometers above Jackson Hole, Wyoming, and is capped by a layer of Flathead Sandstone. On the other side of the Teton Fault, the same Flathead Sandstone lies 7 kilometers below Jackson Hole. The Teton Fault dips to the east at 45 to 75 degrees, but for the sake of simplicity, use a value of 60 degrees to estimate displacement. Calculate the total offset on the Teton Fault. (Use the diagram as a guide.)

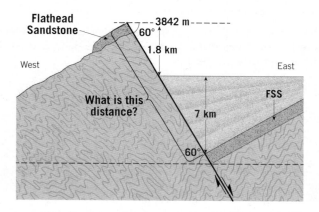

5. What processes (besides formation and melting of large ice sheets) could cause isostatic adjustments?

EYE ON EARTH

1. These interbedded layers of chert and shale were strongly folded during the growth of an accretionary wedge. A recent period of uplift has exposed these deformed strata near Marin Headlands, north of San Francisco, California.

   a. What is the nature of the stress that most likely generated these highly folded strata: compressional or tensional?

   b. Along what type of plate boundaries do accretionary wedges form?

   c. What type of plate boundary is found today in the San Francisco Bay area?

DATA ANALYSIS

Isostasy at Work

Despite the lowering forces of erosion, mountains persist in part because of isostasy, which determines how high a mountain grows and how deep its "root" extends below the surface. We can represent a mountain belt with a wooden block floating in a container of water and explore how the mountain belt will react under various scenarios.

https://goo.gl/eD226M

ACTIVITIES

Imagine that you add another block of wood of equal size on top of the one shown in the image.

1. **How would the height of the wood above the water change? How would the depth of wood below the water change?**

2. **What real-world process would result in more material being added to the top of a mountain?**

3. **The depth of the wood in the water represents the root of the mountain. How would the root change if more material were added to the top of a mountain?**

4. **List one mountain that is undergoing this process. How are the height and root of the mountain changing?**

Referring again to the original image, imagine that you cut the wooden block in half, so that it is only half as tall as it used to be.

5. **How would the block's height above water and depth below water change?**

6. **What real-world process would result in a reduction in the height of a mountain, and how would the root of the mountain change?**

7. **List one mountain currently undergoing this process. How are the height and root of the mountain changing?**

Referring again to the original image, imagine that you carve a hole (cavern) into the side of the block of wood.

8. **How would the block's height above water and depth below water change?**

9. **What real-world process would result in the creation of a cavern in a mountain? How would the root of the mountain change?**

Mastering Geology

Looking for additional review and test prep materials? Visit pearson.com/mastering/geology to enhance your understanding of this chapter's content by accessing a variety of resources, including **Self-Study Quizzes, Geoscience Animations, SmartFigure Tutorials,** *Project Condor* **Videos, Mobile Field Trips, Dynamic Study Modules,** and an optional **Pearson eText.**

Can Air Pollution Cause Killer Landslides?

From mountains to hills, the slopes on Earth's surface are involved in a balancing act between gravity and the strength of the material on which gravity pulls. A new study by scientists in China and New Zealand has found that severe acid rain, a product of coal-burning electricity generation, can weaken the rock masses that hold up mountain slopes. And that can, in turn, set the stage for mass movement.

Since 1978, China has experienced rapid industrial growth, fueled largely by major expansion of coal-burning power plants. The plants send large quantities of particulate sulfur dioxide and nitrogen dioxide into the atmosphere, and those particulates combine with water vapor to become acid rain. This type of rainfall has a pH as low as 2.8—about the same level as apple cider vinegar.

When such acidic water percolates into the ground, it dissolves carbonate minerals, such as calcite, in certain rock layers. The acidic water can also feed microbes that then eat organic material (such as coal) in some layers. The loss of these minerals weakens the layer, creating a slippery surface within the rock—think of sprinkling a layer of talcum powder within a tilted stack of books.

It is on these low-friction layers that landslides have developed in the Chongqing area of central China. Such pollution-related weakening may have played a role in the Jiweishan rock avalanche in 2009, which killed 74 people and destroyed many structures. Researchers studying this avalanche found that mining operations in the area allowed acid rain to penetrate cracks in shale rock, which lead to weakening of the mountain face.

▶ Rescuers climb over huge boulders after the Jiweishan landslide in Central China destroyed an iron ore plant and several homes, killing 74 people.

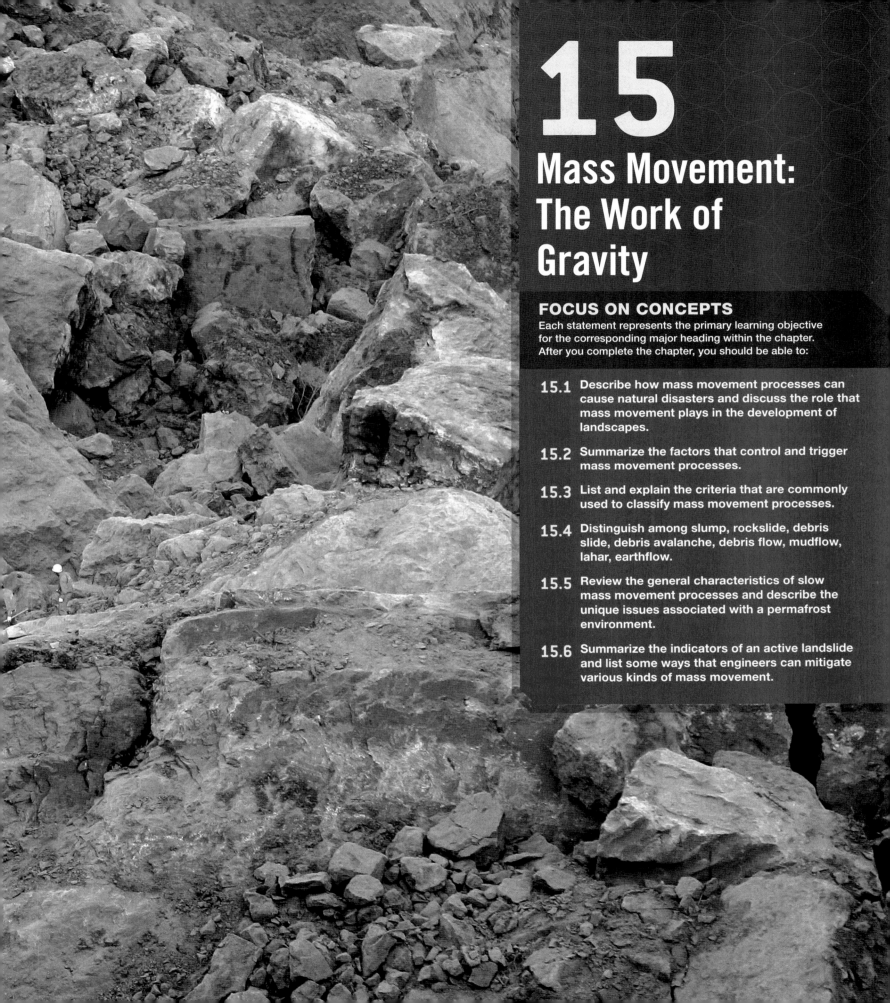

15

Mass Movement: The Work of Gravity

FOCUS ON CONCEPTS

Each statement represents the primary learning objective for the corresponding major heading within the chapter. After you complete the chapter, you should be able to:

15.1 Describe how mass movement processes can cause natural disasters and discuss the role that mass movement plays in the development of landscapes.

15.2 Summarize the factors that control and trigger mass movement processes.

15.3 List and explain the criteria that are commonly used to classify mass movement processes.

15.4 Distinguish among slump, rockslide, debris slide, debris avalanche, debris flow, mudflow, lahar, earthflow.

15.5 Review the general characteristics of slow mass movement processes and describe the unique issues associated with a permafrost environment.

15.6 Summarize the indicators of an active landslide and list some ways that engineers can mitigate various kinds of mass movement.

Many of the preceding chapters focus on *internal processes* that are powered by energy from Earth's interior and produce volcanoes, earthquakes, ocean basins, and mountains. Beginning with this chapter and extending through Chapter 20, we turn our attention to *external processes* that sculpt and erode Earth's surface. Driven by gravity and energy from the Sun, external processes are responsible for shaping a wide variety of landforms and creating many distinctive landscapes.

15.1 The Importance of Mass Movement

Describe how mass movement processes can cause natural disasters and discuss the role that mass movement plays in the development of landscapes.

In most places, Earth's surface is not perfectly flat but instead consists of slopes of many different varieties. Slopes are, indeed, a very common element in our physical landscape. They may be steep, moderate, or gentle. Some are long and gradual; others short and abrupt. Slopes can consist of barren rock and rubble, or they can be mantled with soil and covered by vegetation. Although most slopes may appear to be stable and unchanging, the force of gravity causes rock material to move downslope by way of processes we call **mass movement**. This term refers to the downslope movement of rock, regolith, and soil under the direct influence of gravity. Mass movement is distinct from the erosional processes that are examined in subsequent chapters in that mass movement does not require a transporting medium, such as water, wind, or glacial ice. At one extreme, the movement may consist of a roaring debris flow or a thundering rock avalanche. At the other extreme, it may be gradual and practically imperceptible.

Landslides as Geologic Hazards

The chapter-opening photo is a striking example of a type of mass movement, a phenomenon that many people would call a landslide. For most of us, *landslide* implies a sudden event in which large quantities of rock and soil plunge down steep slopes. But mass movement of rock and soil happens on many scales and at many speeds. When people and communities are in the way, a natural disaster may result. Though the two terms *hazard* and *risk* are often used in the same conversation, they have different meanings. A *geologic hazard* is a process that can cause harm to people or property if they are sufficiently exposed. *Risk* is the probability that exposure to a hazard will cause harm. Thus, we can think of Risk = Hazard × Exposure.

Landslides don't always occur in remote mountains and canyons. People frequently live where rapid but rare mass movement events occur—and usually they remain ignorant of the risks associated with their particular area of residence. However, media reports remind us that such events occur with some regularity around the world. Landslides constitute major geologic hazards that each year in the United States cause billions of dollars in damages and the loss of dozens of lives. As you will see, many landslides occur in connection with other major natural disasters, including earthquakes, volcanic eruptions, wildfires, and severe storms.

> *Although most slopes may appear to be stable and unchanging, the force of gravity causes rock material to move downslope by way of processes we call mass movement.*

The Role of Mass Movement in Landscape Development

In the evolution of most landscapes, mass movement is the step that follows weathering. Landscapes slowly change as rock and soil are removed from the places where they originate. Once weathering weakens and breaks apart rock, mass movement transfers the debris downslope, and a stream or glacier, acting as a conveyor belt, usually carries it away. Although there may be many intermediate stops along the way, the sediment is eventually transported to its ultimate destination: the sea.

The combined effects of mass movement and running water produce stream valleys, which are among the most common and conspicuous of Earth's landforms. If streams alone were responsible for creating the valleys in which they flow, the valleys would all be very narrow canyons. However, the fact that most river valleys are much wider than they are deep is a strong indication of the significance of mass movement processes in supplying material to streams. Consider the Grand Canyon, whose walls extend far from the Colorado River due to the transfer of weathered debris downslope to the river and its tributaries by mass movement processes (**Figure 15.1**). Streams and mass movement work together in this manner to modify and sculpt Earth's surface. Of course, glaciers, groundwater, waves, and wind are also important agents in shaping landforms and developing landscapes.

Slopes Change Through Time

It is clear that if mass movement is to occur, there must be slopes that rock, soil, and regolith can move down. Earth's mountain-building and volcanic processes, driven by plate tectonics, produce these slopes through sporadic changes in the elevations of landmasses and the ocean floor. If dynamic internal processes did not continually produce regions having higher elevations, the system that moves debris to lower elevations would gradually slow and eventually cease.

Most rapid and spectacular mass movement events occur in areas of rugged, geologically young mountains. Newly formed mountains are rapidly eroded by rivers and glaciers into

regions characterized by steep and unstable slopes. These areas often include valleys in thick deposits of unconsolidated material. It is in such settings that massive destructive landslides, such as those described at the beginning of the chapter, occur. As mountain building subsides, mass movement and erosional processes lower the land. Through time, steep and rugged mountain slopes give way to gentler, more subdued terrain. Thus, as a landscape ages, massive and rapid mass movement processes give way to smaller, less dramatic downslope movements that are often imperceptibly slow.

CONCEPT CHECKS 15.1

1. Define *mass movement.* How does it differ from erosional agents such as streams, glaciers, and wind?

2. In what sort of landscape are rapid mass movement processes most likely to occur? Describe how these geologic hazards might become geologic risks.

3. Sketch or describe how mass movement combines with stream erosion to expand valleys.

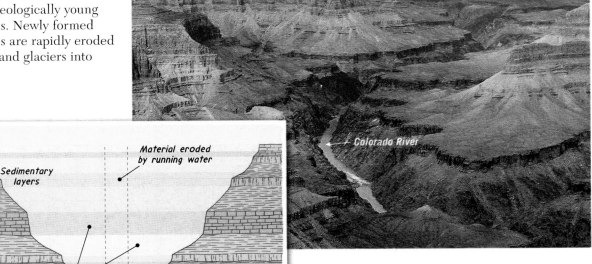

◀ SmartFigure 15.1
Excavating the Grand Canyon The walls of the canyon extend far from the channel of the Colorado River. This results primarily from the transfer of weathered debris downslope to the river and its tributaries by mass movement processes.

Tutorial
https://goo.gl/FubuFy

Geologist's Sketch

Sedimentary layers

Material eroded by running water

Weathered debris moved downslope by mass movement

Metamorphic and igneous rocks

Colorado River

15.2 Controls and Triggers of Mass Movement

Summarize the factors that control and trigger mass movement processes.

Gravity, the force pulling material downslope, is the driving force of mass movement. Whether a slope fails depends on both gravity and the resisting strength of the rock or soil. However, several other factors also play important roles in overcoming friction and material strength to create downslope movements.

Long before a landslide occurs, various processes work to weaken slope material, gradually making it more and more susceptible to the pull of gravity. During this span, the slope remains stable but gets closer and closer to being unstable. Eventually, the strength of the slope is weakened to the point that something, referred to as the **trigger**, causes the slope to cross the threshold from stability to instability. A trigger is not the sole cause of a mass movement event; it is just the last factor. Common factors that trigger mass movement processes are saturation of material with water, oversteepening of slopes, removal of anchoring vegetation, and ground vibrations from earthquakes.

However, many rapid mass movement events occur without discernible triggers. Slope materials gradually weaken over time under the influence of long-term weathering, infiltration of water, and other physical processes. Eventually, if the strength falls below what is necessary to maintain slope stability, a landslide will occur. The timing of such events is random, and thus accurate prediction is impossible.

Role of Water

Mass movement is sometimes triggered when heavy rains or periods of snowmelt saturate surface materials. In January 2018, deadly debris flows in Montecito, California, resulted from a sequence of deforestation by wildfire followed by heavy rains. In other situations, water does not transport the material. Rather, it allows gravity to more easily set the material in motion. This was the case in January 2005, when a massive debris flow (popularly called a mudslide) swept through La Conchita, California, a small coastal community northwest of Los Angeles (**Figure 15.2**). The La Conchita debris flow happened when older landslide deposits on the hillside became saturated during a winter of unusually high rainfall.

When the pores in sediment become filled with water, the cohesion among particles is destroyed, allowing them to move past one another with relative ease. For example, when sand is slightly moist, it sticks together quite well. However, if enough water is added to fill the openings between the grains, the sand will ooze in all directions (**Figure 15.3**). Thus, water saturation reduces the

► SmartFigure 15.2
Heavy rains trigger debris flows On January 10, 2005, a massive debris flow, originating from the saturated deposits of a 1995 landslide, swept through La Conchita, California, a small coastal town situated on a narrow coastal strip between the shoreline and a steep bluff. The event, which occurred after a span of near record rainfall, killed 10 people and destroyed many homes.

Tutorial https://goo.gl/7V64RH

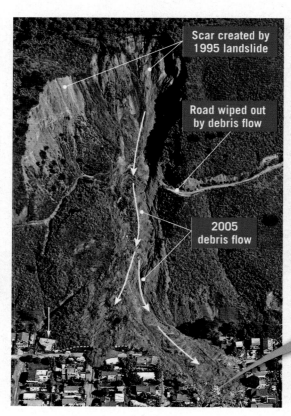
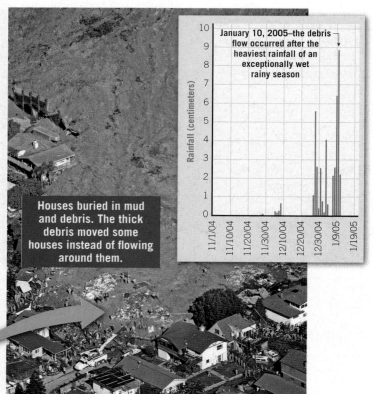

Dry sand

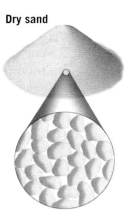

Dry sand grains are bound mainly by friction with one another

Damp sand

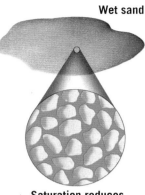

Small amounts of water increase the cohesion among sand grains

Wet sand

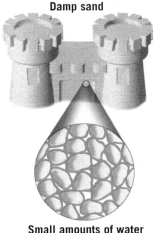

Saturation reduces friction and causes the sand to flow

◀ **Figure 15.3**
Saturation reduces friction When water saturates sediment, friction among particles is reduced, allowing material to move downslope.

it removes material from the base of the wall. This causes the slope to become too steep to remain stable, and material can then fall or slide into the stream. Furthermore, through their activities, people often create oversteepened and unstable slopes that become prime sites for mass movement (**Figure 15.4**).

Loose, granular particles (sand-size or coarser) assume a stable slope called the **angle of repose** (*reposen* = to be at rest). This is the steepest angle at which material remains in place (**Figure 15.5**). Depending on the size and shape of the particles, the angle of repose varies from 25 to 40 degrees. Larger, more angular particles maintain the steepest slopes. If the angle is increased, the rock debris will adjust by moving downslope.

Oversteepening is not just important because it triggers movements of unconsolidated granular materials. Oversteepening also produces unstable slopes and mass movements in cohesive soils, regolith, and bedrock because the steeper angle increases the effect of gravity. The response is not immediate, as with loose, granular material, but sooner or later, one or more mass movement processes will eliminate the oversteepening and restore stability to the slope.

internal resistance of materials, which are then easily set in motion by the force of gravity. When clay is wetted, it becomes very slick—another example of the "lubricating" effect of water. Water also adds considerable weight to a mass of material. The added weight in itself may cause the downslope forces to exceed the resisting forces, which leads to the mass of material sliding or flowing downslope.

Oversteepened Slopes

Oversteepening of slopes is another trigger of many mass movements. Many situations in nature result in oversteepening. For example, as a stream cuts into a valley wall,

Removal of Vegetation

Plants protect against erosion and contribute to the stability of slopes because their root systems bind together soil and regolith. In addition, plants shield the soil surface from the erosional effects of raindrop impact. Where plants are lacking, mass movement is enhanced, especially if slopes are steep and water is plentiful. When anchoring vegetation is removed by forest fires or by people (for timber, farming, or development), surface materials frequently move downslope.

◀ **Figure 15.4**
Unstable slopes Natural processes such as stream and wave erosion can oversteepen slopes. Changing the slope to accommodate a new house or road can also lead to instability and a destructive mass movement event.

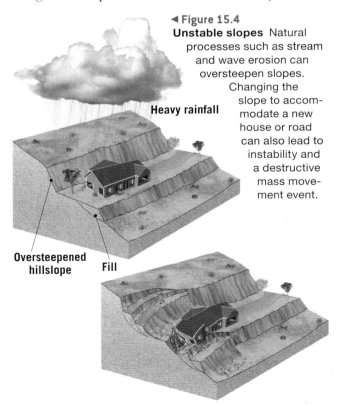

Heavy rainfall

Oversteepened hillslope   **Fill**

▼ **Figure 15.5**
Angle of repose The angle of repose is the steepest angle at which an accumulation of granular particles remains stable. Larger, more angular particles maintain the steepest slopes.

Angle of repose

▼ Figure 15.6
Wildfires contribute to mass movement In November 2018, the Woolsey wildfire consumed 100,000 near Los Angeles California. The fire destroyed 1,643 structures, killed three people, and prompted the evacuation of 295,000 more. Millions of acres are burned in the western U.S. each year, with the totals rising as the climate warms. The loss of anchoring vegetation sets the stage for accelerated mass movement such as debris flows.

An unusual example illustrating the anchoring effect of plants occurred several decades ago, on steep slopes near Menton, France. Farmers replaced olive trees, which have deep roots, with a more profitable but shallow-rooted crop: carnations. When the less stable slope failed, the landslide took 11 lives.

In July 1994 a severe wildfire swept Storm King Mountain, west of Glenwood Springs, Colorado, denuding the slopes of vegetation. Two months later, heavy rains resulted in numerous debris flows—rapid mass movement events involving water-saturated rock and soil. One debris flow blocked Interstate 70 and threatened to dam the Colorado River. A 5-kilometer (3-mile) length of the highway was inundated with tons of rock, mud, and burned trees. The closure of Interstate 70 imposed costly delays for travelers on this major highway.

Wildfires are inevitable in the western United States, and fast-moving, highly destructive debris flows triggered by intense rainfall are some of the most dangerous postfire hazards (**Figure 15.6**). Such events are particularly dangerous because they tend to occur with little warning. Their mass and speed make them particularly destructive. Postfire debris flows are most common in the 2 years after a fire. The very first intense rain event following a wildfire is likely to trigger a debris flow event. It takes much less rain to trigger debris flows in burned areas than in unburned areas. In southern California, as little as 7 millimeters (0.3 inch) of rain in 30 minutes has been enough to trigger debris flows.

In addition to eliminating plants that anchor the soil, fire can promote mass movement in other ways. Following a wildfire, the upper part of the soil may become dry and loose. As a result, even in dry weather, the soil tends to move down steep slopes. Fire can also "bake" the ground, creating a water-repellent layer at a shallow depth. This nearly impermeable barrier prevents or slows the infiltration of water, resulting in increased surface runoff during rains. The result can be dangerous torrents of viscous mud and rock debris.

Earthquakes as Triggers

Among the most important and dramatic triggers of mass movement are earthquakes. An earthquake and its aftershocks can dislodge enormous volumes of rock and unconsolidated material (see the chapter-opening photo).

Examples from Plate Boundaries: California and Nepal

A memorable U.S. example of an earthquake triggering mass movement occurred in January 1994, when a quake struck the Los Angeles region of southern California. Named for its epicenter in the town of Northridge, the magnitude 6.7 event produced estimated losses of $20 billion. Some of the losses resulted from more than 11,000 landslides in an area of about 10,000 square kilometers (3900 square miles) that were set in motion by the quake. Most were shallow rockfalls and slides, but some were much larger and filled canyon bottoms with jumbles of soil, rock, and plant debris. The debris in canyon bottoms created a secondary threat because it can mobilize during rainstorms, producing debris flows that can affect communities far downstream. Such flows are common and often disastrous in southern California.

On April 25, 2015, a magnitude 7.8 earthquake shook much of central Nepal. The earthquake and its aftershocks triggered thousands of landslides in the rugged Himalayas and caused nearly 8900 fatalities (**Figure 15.7**). Rock avalanches and debris slides thundered down steep mountain slopes, burying buildings and blocking roads and rail lines. The landslides also dammed rivers, creating many lakes. Earthquake-created lakes present a dual danger. Apart from the upstream floods that occur as a lake builds behind the natural dam, the piles of rubble that form the dam may be unstable. Another quake or simply the pressure of

◄ **Figure 15.7**
Earthquakes as triggers
A major earthquake in Nepal in April 2015 triggered hundreds of landslides, which destroyed roads and blocked rivers. The damage shown here occurred in the village of Singati in northeastern Nepal.

the water building up could burst the dam, sending a wall of water downstream. Such floods may also occur when water begins to cascade over the top of the dam.

Liquefaction   Intense ground shaking during earthquakes can cause water-saturated surface materials to lose their strength and behave as fluid-like masses that flow (Figures 11.22 and 11.23, page 324). This process, called **liquefaction**, was a major cause of property damage in Anchorage, Alaska, during the massive 1964 Good Friday earthquake—the largest magnitude quake to strike North America in the twentieth century.

Landslide Risk in the United States

GEOgraphics 15.1 shows the landslide potential for the contiguous United States. All states experience some damage from rapid mass movement processes, but it is obvious that not all areas of the country have the same landslide hazard potential. As you might expect, there are greater landslide risks in mountain areas. In the East, landslides are most common in the Appalachian Mountains. In the Pacific Northwest, water from winter rains and melting snow often triggers rapid forms of mass movement, including slumps in the glacial

sediments that underlie the hilly neighborhoods of Seattle and Portland. Coastal California's steep slopes have a high landslide potential. Here mass movement events may be triggered by winter storms or by the ground shaking associated with earthquakes. Landslides are also triggered when strong wave activity undercuts and oversteepens coastal cliffs.

A glance at the map shows that Florida and the adjacent Atlantic and Gulf coastal plains have some of the lowest landslide potentials because steep slopes are largely absent. In the center of the country, the Plains states are relatively flat, so landslide potential there is mostly low to moderate. High-potential areas are along the steep bluffs that flank river valleys.

CONCEPT CHECKS 15.2

1. How does water affect mass movement processes?

2. Describe the significance of the angle of repose.    **Concept Checker** https://goo.gl/PvuDWG

3. How might a wildfire influence mass movement?

4. Describe the relationship between earthquakes and landslides.

Landslide Risks: United States and Worldwide

According to the U. S. Geological Survey, each year in the United States, landslides cost nearly $4 billion (2010 dollars) in damage repair and cause between 25 and 50 deaths. All states experience rapid mass-wasting processes, but not all areas have the same landslide potential. What's the risk where you live?

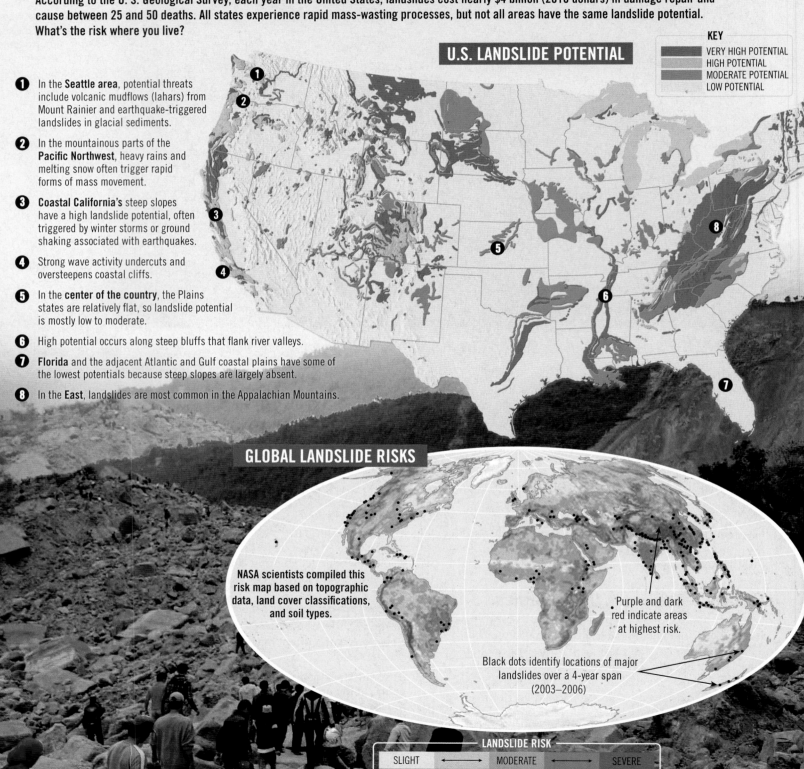

U.S. LANDSLIDE POTENTIAL

KEY
- VERY HIGH POTENTIAL
- HIGH POTENTIAL
- MODERATE POTENTIAL
- LOW POTENTIAL

❶ In the **Seattle area**, potential threats include volcanic mudflows (lahars) from Mount Rainier and earthquake-triggered landslides in glacial sediments.

❷ In the mountainous parts of the **Pacific Northwest**, heavy rains and melting snow often trigger rapid forms of mass movement.

❸ **Coastal California's** steep slopes have a high landslide potential, often triggered by winter storms or ground shaking associated with earthquakes.

❹ Strong wave activity undercuts and oversteepens coastal cliffs.

❺ In the **center of the country**, the Plains states are relatively flat, so landslide potential is mostly low to moderate.

❻ High potential occurs along steep bluffs that flank river valleys.

❼ **Florida** and the adjacent Atlantic and Gulf coastal plains have some of the lowest potentials because steep slopes are largely absent.

❽ In the **East**, landslides are most common in the Appalachian Mountains.

GLOBAL LANDSLIDE RISKS

NASA scientists compiled this risk map based on topographic data, land cover classifications, and soil types.

Purple and dark red indicate areas at highest risk.

Black dots identify locations of major landslides over a 4-year span (2003–2006)

LANDSLIDE RISK

| SLIGHT | ← → MODERATE ← → | SEVERE |

15.3 Classification of Mass Movement Processes

List and explain the criteria that are commonly used to classify mass movement processes.

Different types of mass movement are generally classified based on the type of material involved, the kind of motion displayed, and the velocity of the movement.

Type of Material

Classification of mass movement processes on the basis of the material involved in the movement depends on whether the descending mass began as unconsolidated material or as bedrock. If soil and regolith dominate, terms such as *debris*, *mud*, or *earth* are used in the description. In contrast, when a mass of bedrock breaks loose and moves downslope, the term *rock* may be part of the description.

Type of Motion

The way a material moves in a mass movement event may also be important. Generally, the kind of motion is described as either a fall, a slide, or a flow.

Fall When the movement in a mass movement event involves the free fall of detached individual pieces of any size, it is termed a **fall**. This is a common form of movement on slopes that are so steep that loose material cannot remain on the surface. The rock may fall directly to the base of the slope or may move in a series of leaps and bounds over other rocks along the way. Rockfalls commonly create and maintain **talus slopes** (**Figure 15.8A**). Many falls result when freeze–thaw cycles within rock cracks and/or the action of plant roots loosen rock to the point at which gravity overwhelms the resisting forces. Although signs along bedrock cuts on highways warn of falling rock, few of us have actually witnessed such an event in progress. However, as **Figure 15.8B** illustrates, they do indeed occur.

When large masses of rock plunge from great heights, they hit the ground with enormous force and often trigger additional mass movement events. One especially deadly example occurred in Peru. In May 1970, an earthquake caused a huge mass of rock and ice to break free from the precipitous north face of NevadoHuascarán, the loftiest peak in the Peruvian Andes. The material plunged nearly 1 kilometer (3000 feet) and was pulverized on impact. The rock avalanche that followed rushed down the mountainside, made fluid by trapped air and ice. Along the way, it ripped loose millions of tons of additional debris that ultimately and tragically buried more than 20,000 people in the towns of Yungay and Ranrahirca.

In the mountains, mechanical weathering produces angular rock fragments that fall to the base of a cliff. Over time, a talus slope forms.

Talus slopes

A.

B.

▲ **SmartFigure 15.8**
Talus slopes and rockfall A. These large talus slopes, sometimes called talus cones, are found in Canada's Banff National Park. **B.** Falling rocks can block highways, as shown here.

 Animation
https://goo.gl/
CPpWDk

A different effect triggered by a rockfall occurred in Yosemite National Park on July 10, 1996. When two large rock masses broke loose from steep cliffs and fell about 500 meters (1640 feet) to the floor of Yosemite Valley, the impacts were great enough to be recorded at seismic stations 200 kilometers (125 miles) from the site. As the dislodged rock masses struck the ground, they generated atmospheric pressure waves that were comparable in velocity to a tornado or hurricane. The force of the air blasts uprooted and snapped more than 1000 trees, including some that were 40 meters (130 feet) tall. Rockfalls continue to occur at Yosemite, including one in 2017 from the popular climbing wall known as El Capitan that killed one person.

Slide Many mass movement processes are described as **slides**. The term refers to mass movements in which there is a distinct zone of weakness separating the slide material from the more stable underlying material. A *rotational slide* is a slide in which the surface of a rupture is a concave-upward curve resembling the shape of a spoon, and the descending material exhibits a downward and outward rotation. By contrast, a *translational slide* is a slide in which a mass of material moves along relatively flat surfaces such as joints, faults, or bedding planes. Such slides exhibit little rotation or backward tilting.

Flow The third type of movement common to mass movement processes is termed **flow**. This is the term used to describe material that moves downslope as a viscous, often turbulent, fluid. Most flows are saturated with water and typically move as lobes or tongues. Frequently, when flows of saturated mud and debris occur, they are incorrectly described in the media as "mudslides."

Rate of Movement

Some of the events that have been described so far in this chapter involved very rapid rates of movement. For example, it is estimated that the debris that rushed down the slopes of Peru's Nevado Huascarán moved at speeds in excess of 200 kilometers (125 miles) per hour. This most rapid type of mass movement is termed a **rock avalanche**. Geologists used to think that rock avalanches, such as the one that produced the scene in **Figure 15.9A**, must literally "float on air" as they move downslope. That is, they thought that high velocities might result when air becomes trapped and compressed beneath the falling mass of debris, allowing it to move as a buoyant, flexible sheet across the surface. But more recent studies of landslide deposits on other planetary bodies have caused us to question this hypothesis. Similar deposits from long runout landslides on Mars seem to have been lubricated by interactions of rock and water (**Figure 15.9B**).

Most mass movements, however, do not occur at the speeds of rock avalanches. In fact, a great deal of

▼ **Figure 15.9**
Landslides on Earth and Mars A. The prehistoric Blackhawk rock avalanche is considered one of the largest non-volcanic landslides in North America. **B.** The Blackhawk event is just a fraction the size of the landslides in the Valles Marineris canyons of Mars. Water probably played an important role in both situations, allowing the deposits to extend so far from their sources.

San Bernardino Mountains

Between 9 and 30 meters thick

8 kilometers

50 kilometers

A.

B.

mass movement is imperceptibly slow. One process we will examine later, termed *creep*, results in particle movements that are usually measured in millimeters or centimeters per year. Thus, as you can see, rates of movement can be spectacularly sudden or exceptionally gradual. Although various types of mass movement are often classified as either rapid or slow, such a distinction is highly subjective because a wide range of rates exists between the two extremes.

CONCEPT CHECKS 15.3

1. List and sketch three ways material can move during mass movement events.

2. In what way has scientific thinking changed about how rock avalanches move at such great speeds?

Concept Checker
https://goo.gl/DBgxRp

15.4 Common Forms of Mass Movement

Distinguish among slump, rockslide, debris slide, debris avalanche, debris flow, mudflow, lahar, earthflow.

Classifying kinds of mass movement can be tricky because there are several variable factors: speed, three-dimensional geometry, and type of material. The common mass movement processes discussed in this section are slump, rockslide, debris flow, and earthflow. They are most common where slopes are steep. The speed of movement varies from barely perceptible to very rapid.

Slump

Slump refers to the downward sliding of a mass of rock or unconsolidated material that moves as a unit along a curved surface (**Figure 15.10**). Usually the slumped material does not travel spectacularly fast, and it does not travel very far. Slump is a common form of mass movement, especially in thick accumulations of cohesive materials such as clay. The ruptured surface is characteristically spoon shaped and concave upward or outward. As the movement occurs, a crescent-shaped scarp is created at the head, and the block's upper surface is sometimes tilted backward. Although slump may involve a single mass, it often consists of multiple blocks. Sometimes water collects between the base of the scarp and the top of the tilted block. As this water percolates downward along the surface of rupture, it may promote further instability and additional movement.

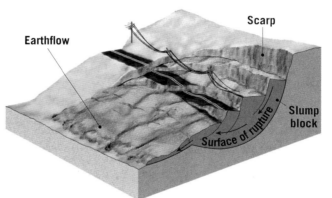

Earthflow   Scarp

Surface of rupture

Slump block

◀ **Figure 15.10**
Slump Slump occurs when material slips downslope en masse along a curved surface of rupture. It is an example of a rotational slide. Earthflows frequently form at the base of the slump.

Slump commonly occurs because a slope has been oversteepened. As the anchoring material at the base of the slope is removed, the material above becomes unstable and reacts to the pull of gravity. One relatively common example is a valley wall that becomes oversteepened by a meandering river. The photo in **Figure 15.11** provides

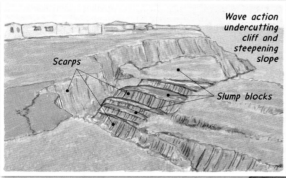

Wave action undercutting cliff and steepening slope

Scarps

Slump blocks

Geologist's Sketch

◀ **Figure 15.11**
Slump at Holderness peninsula in northern England Ocean waves undercut the base of the steep slope, making it unstable.

an example in which a coastal cliff has been undercut by wave action at its base. Slumping may also occur when a slope is overloaded, causing internal stress on the material below. This type of slump often occurs where weak, clay-rich material underlies layers of stronger, more resistant rock such as sandstone. The seepage of water through the upper layers reduces the strength of the clay below, resulting in slope failure.

Rockslide and Debris Avalanche

Rockslides occur when blocks of bedrock break loose and slide down a slope (**Figure 15.12**). If the material involved is largely unconsolidated, the term **debris slide** is used instead. After the initial sliding stage, the material can break up into a chaotic **debris avalanche**. Such events are among the fastest and most destructive mass movements. Usually rockslides take place in a geologic setting where the rock strata are inclined or where joints and fractures exist parallel to the slope. When such a rock unit is undercut at the base of the slope, it loses support, and the rock eventually gives way. Sometimes a rockslide is triggered when rain or melting snow lubricates the underlying surface to the point at which friction is no longer sufficient to hold the rock unit in place. As a result, rockslides tend to be most common during the spring, when heavy rains and melting snow are most prevalent.

As mentioned earlier, earthquakes can trigger rockslides and other mass movements. There are many well-known examples. On August 17, 1959, a severe earthquake west of Yellowstone National Park triggered a massive slide in the canyon of the Madison River in southwestern Montana. In a matter of moments, an estimated 27 million cubic meters (6 million dump truck loads) of rock, soil, and trees slid into the canyon. The debris dammed the river and buried a campground and highway. More than 20 unsuspecting campers perished.

Heavy rains and melting snow, rather than an earthquake, triggered another major rockslide—a debris avalanche in the Yellowstone region. The legendary Gros Ventre rockslide occurred not far from the site of the Madison Canyon slide. The Gros Ventre River flows west from the northernmost part of the Wind River Range in northwestern Wyoming, through Grand Teton National Park, and eventually empties into the Snake River. On June 23, 1925, a massive rockslide took place in its valley, just east of the small town of Kelly. In the span of only minutes, a great mass of sandstone, shale, and soil crashed down the south side of the valley, carrying with it a dense pine forest. The volume of debris, estimated at 38 million cubic meters (50 million cubic yards), created a dam on the Gros Ventre River 70 meters (230 feet) high. Because the river was completely blocked, a lake was formed. It filled so quickly that a house that had been 18 meters (60 feet) above the river was floated off its foundation 18 hours after the slide. In 1927, the lake overflowed the dam, partially draining the lake and resulting in a devastating flood downstream.

Why did the Gros Ventre rockslide take place? **Figure 15.13** shows a diagrammatic cross-sectional view of the geology of the valley. You will notice that (1) the sedimentary strata in this area dip (tilt) 15 to 21 degrees; (2) underlying the bed of sandstone is a relatively thin layer of clay; and (3) at the bottom of the valley, the river had cut through much of the sandstone layer. During the spring of 1925, water from heavy rains and melting snow seeped through the sandstone, saturating the clay below. Because much of the sandstone layer had been cut through by the Gros Ventre River, the layer had virtually no support at the bottom of the slope. Eventually the sandstone could no longer hold its position on the wetted clay, and gravity pulled the mass down the side of the valley. The circumstances at this location were such that the event was inevitable.

Debris Flow

Debris flow is a relatively rapid type of mass movement that involves a flow of soil and regolith containing a large amount of water. The La Conchita debris flow (see Figure 15.2) is one example. Debris flows are sometimes called **mudflows** when the material is primarily fine-grained. Although they can occur in many different climate settings, they tend to occur more frequently in semiarid mountainous regions. Because of their fluid properties, debris flows frequently follow canyons and stream channels. In populated areas, debris flows can pose significant hazards to life and property.

Debris Flows in Semiarid Regions When a cloudburst or rapidly melting mountain snows create a sudden flood in a semiarid region, large quantities of soil and regolith wash into nearby stream channels because there is usually little vegetation to anchor the surface material. The end product is a flowing tongue of well-mixed mud, soil,

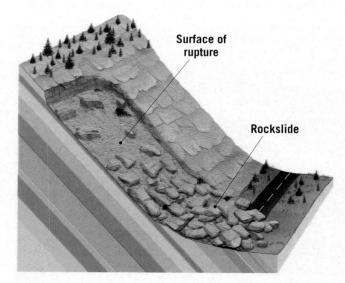

▶ **Figure 15.12**
Rockslide These rapid movements are classified as translational slides in which the material moves along a relatively flat surface with little or no rotation or backward tilting.

Surface of rupture

Rockslide

rock, and water. Its consistency may range from that of wet concrete to a soupy mixture not much thicker than muddy water. The rate of flow, therefore, depends not only on the steepness of the slope but also on the water content. Dense debris flows are capable of carrying or pushing large boulders, trees, and even houses with relative ease.

Debris flows pose a serious hazard to development in relatively dry mountainous areas such as southern California. The construction of homes on canyon hillsides and the removal of native vegetation by brush fires and other means have increased the frequency of these destructive events (**Figure 15.14**). Moreover, when a debris flow reaches the end of a steep, narrow canyon, it spreads out, covering the area beyond the mouth of the canyon with a mixture of wet debris. This material contributes to the buildup of fanlike deposits, called *alluvial fans*, at canyon mouths. The fans are relatively easy to build on, often have nice views, and are close to the mountains; in fact, like the nearby canyons, many have become preferred sites for development. Because debris flows occur only sporadically, the public is often unaware of the potential hazard of such sites.

Lahars   Debris flows composed mostly of volcanic materials on the flanks of volcanoes are called **lahars**. The word originated in Indonesia, a volcanic region that has experienced many of these often-destructive events. Historically, lahars have been some of the deadliest volcano-related hazards. They can occur either during an eruption or when a volcano is quiet. They take place when highly unstable layers of ash and debris become saturated with water and flow down steep volcanic slopes, generally following existing stream channels (**Figure 15.15**). Heavy rainfalls often trigger these flows. Others are initiated when large volumes of ice and snow are melted by heat flowing to the surface from within the volcano or by the hot gases and near-molten debris emitted during a violent eruption.

When Mount St. Helens erupted in May 1980, several massive lahars resulted. The flows and accompanying floods raced down the valleys of the north and south forks of the Toutle River at speeds that were often in excess of 30 kilometers (20 miles) per hour. Fortunately, the affected area was not densely settled. Nevertheless, more than 200

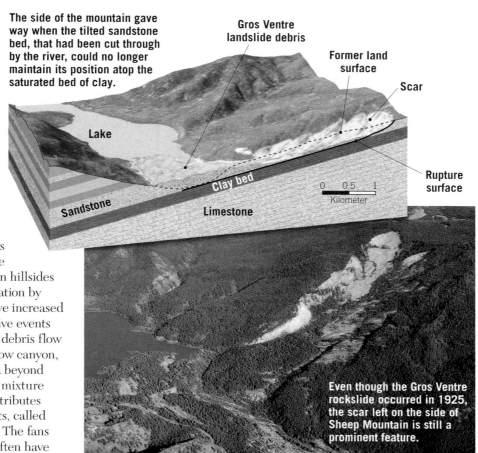

The side of the mountain gave way when the tilted sandstone bed, that had been cut through by the river, could no longer maintain its position atop the saturated bed of clay.

Gros Ventre landslide debris

Former land surface

Scar

Lake

Rupture surface

Clay bed

Sandstone

Limestone

0   0.5   1
Kilometer

◄ **SmartFigure 15.13**
Gros Ventre rockslide This massive slide occurred on June 23, 1925, just east of the small town of Kelly, Wyoming.

Tutorial
https://goo.gl/77XRQf

Even though the Gros Ventre rockslide occurred in 1925, the scar left on the side of Sheep Mountain is still a prominent feature.

▲ Figure 15.14

Debris flow at Montecito, California This debris flow, generated by heavy rain on burned slopes in southern California in January 2018, was a moving tongue of well-mixed mud, soil, rock, and water.

homes were destroyed or severely damaged (**Figure 15.16**). Most bridges in the vicinity met a similar fate.

In November 1985, lahars were produced during the eruption of Nevado del Ruiz, a 5300-meter (17,400-foot) volcano in the Andes Mountains of Colombia. The eruption melted much of the snow and ice that capped the uppermost 600 meters (2000 feet) of the peak, producing torrents of hot, viscous mud, ash, and debris. The lahars moved outward from the volcano, following the valleys of three rain-swollen rivers that radiate from the peak. The flow that moved down the valley of the Lagunilla River was the most destructive. It devastated the town of Armero, 48 kilometers (30 miles) from the mountain. Most of the more than 25,000 deaths caused by the event occurred in this once-thriving agricultural community.

Death and property damage due to the lahars also occurred in 13 other villages within the 180-square-kilometer (70-square-mile) disaster area. Although a great deal of pyroclastic material was explosively ejected from Nevado del Ruiz, it was the lahars triggered by this eruption that made this natural disaster so devastating. In fact, it was the worst volcanic disaster to occur since the 1902 eruption of Mount Pelée on the Caribbean island of Martinique, which killed 28,000 people.*

Earthflow

We have seen that debris flows are frequently confined to channels in semiarid regions. In contrast, **earthflows** most often form on hillsides in humid areas during times of heavy precipitation or snowmelt. When water saturates the soil and regolith on a hillside, the material may break away, leaving a scar on the slope and forming a tongue- or teardrop-shaped mass that flows downslope (**Figure 15.17**).

Earthflows, which contain only small proportions of sand and coarser particles, are most commonly composed of materials rich in clay and silt. These hazards range in size from bodies a few meters long, a few meters wide, and less than 1 meter deep to masses more than 1 kilometer long, several hundred meters wide, and more than 10 meters deep. Because earthflows are quite viscous, they generally move at slower rates than the more fluid debris flows described in the preceding section. They are characterized by gradual movement

---

*A discussion of the Mount Pelée eruption, as well as additional material on lahars, can be found in Chapter 5.

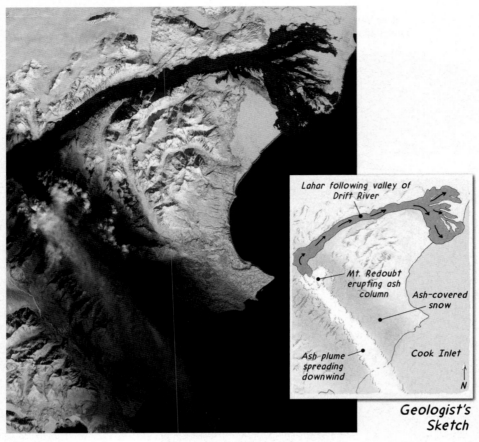

Geologist's Sketch

▲ **Figure 15.15**
Lahar at Redoubt Volcano, Alaska An eruption in 2009 sent lahars down the Drift River valley. The dark color of the lahar contrasts sharply with the surrounding snow-covered land.

◄ **Figure 15.16**
Lahar at Mount St. Helens The force of this viscous debris flow that followed the Toutle River tore the end section from a house and lodged it in the trees. Lahars generated by the Mount St. Helens eruption surged up valley walls as much as 110 meters (360 feet) and over hills as high as 76 meters (250 feet). Based on "bathtub-ring" mudlines, some lahars at their peak averaged 10 to 20 meters (33 to 66 feet) deep.

that may last for periods ranging from days to years. Depending on the steepness of the slope and the material's consistency, measured velocities range from less than a millimeter a day up to several meters a day. Over the time span during which an earthflow is active, movement is typically faster during wet periods than during drier times. In addition to occurring as isolated hillside phenomena, earthflows commonly take place in association with large slumps. In this situation, they may be seen as tongue-like flows at the base of the slump block.

CONCEPT CHECKS 15.4

1. Without looking back at the figures in this section, sketch and label a simple cross section (side view) of a slump.

2. What factors led to the massive rockslide at Gros Ventre, Wyoming?

  **Concept Checker** https://goo.gl/LzLArB

3. How is a lahar different from a debris flow that might occur in southern California?

4. Contrast earthflows and debris flows.

▲ Figure 15.17
Earthflow This small tongue-shaped earthflow occurred on a newly formed slope along a recently constructed highway in central Illinois. It formed in clay-rich material following a period of heavy rain. Notice the small slump at the head of the earthflow.

15.5 Very Slow Mass Movements

Review the general characteristics of slow mass movement processes and describe the unique issues associated with a permafrost environment.

Movements such as rockslides, rock avalanches, and lahars are certainly the most spectacular and catastrophic forms of mass movement. However, sudden movements are responsible for moving less material than the slower and far more subtle action of creep. Whereas rapid types of mass movement are characteristic of mountains and steep hillsides, creep takes place on both steep and gentle slopes and is thus much more widespread.

Creep

Creep is a type of mass movement that involves the gradual downhill movement of soil and regolith. One factor that contributes to creep is the alternating expansion and contraction of surface material caused by freezing and thawing or wetting and drying. As shown in **Figure 15.18**, freezing or wetting lifts particles at right angles to the slope, and thawing or drying allows the particles to fall back to a slightly lower level. Each cycle therefore moves the material a tiny distance downslope. Creep is aided by anything that disturbs the soil, such as raindrop impact or disruption by plant roots and burrowing animals. Creep is also promoted when the ground becomes saturated with water. Following a heavy rain or snowmelt, a water-logged soil may lose its internal cohesion, allowing gravity to pull the material downslope. Because creep is imperceptibly slow, the process cannot be observed in action. However, the effects

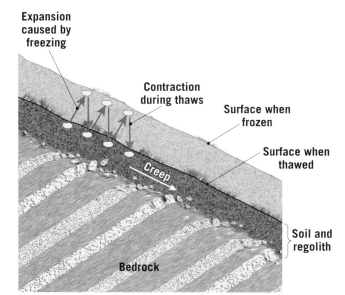

◄ **SmartFigure 15.18**
Creep The repeated expansion and contraction of the surface material causes a net downslope migration of soil and rock particles.

Tutorial https://goo.gl/bAS9Ao

Bent rock layers

Curved tree trunk

Tilted fence

Broken retaining wall

Tension cracks

Zone of creep

▲ **Figure 15.19**
Effects of creep Although creep involves imperceptibly slow movement, its effects are often visible.

of creep can be observed. Creep causes fences and utility poles to tilt and retaining walls to be displaced (**Figure 15.19**).

Solifluction

When soil is saturated with water, the soggy mass may flow downslope at a rate of a few millimeters or a few centimeters per day or per year. This process is

called **solifluction** (literally "soil flow"). It is a type of mass movement that is common wherever water cannot escape from the saturated surface layer by infiltrating to deeper levels. A dense clay hardpan in soil or an impermeable bedrock layer can promote solifluction.

Solifluction is also common in regions underlain by permafrost. *Permafrost* refers to the permanently frozen ground that occurs in association with Earth's harsh tundra and subarctic climates. (There is more about permafrost in the next section.) Solifluction occurs in a zone above the permafrost called the *active layer*, which thaws to a depth of about 1 meter (3 feet) during the brief high-latitude summer and then refreezes in winter. During the summer season, water is unable to percolate into the impervious permafrost layer below. As a result, the active layer becomes saturated and slowly flows. The process can occur on slopes as gentle as 2 to 3 degrees. Where there is a well-developed mat of vegetation, a solifluction sheet may move in a series of well-defined lobes or as a series of partially over-riding folds (**Figure 15.20**).

The Sensitive Permafrost Landscape

Many of the mass movement disasters described in this chapter had sudden and disastrous impacts on people. When the activities of people lead to the melting of ice contained in permanently frozen ground, the impact is more gradual and less deadly. Nevertheless, because permafrost regions are sensitive and fragile landscapes, the scars resulting from poorly planned actions can remain for generations.

Permanently frozen ground, known as **permafrost**, occurs where summers are too short and cool to melt

▶ **Figure 15.20**
Solifluction lobes near the Arctic Circle in Alaska
Solifluction occurs in permafrost regions when the active layer thaws in summer. Because summers are cool and very short, frozen soils generally thaw to depths of less than 1 meter (3 feet).

Active layer (thawed in summer)

Frozen layer (permafrost)

Geologist's Sketch

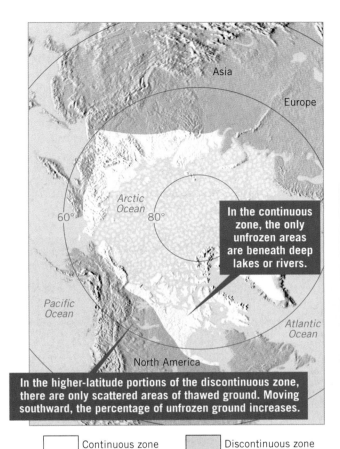

In the continuous zone, the only unfrozen areas are beneath deep lakes or rivers.

In the higher-latitude portions of the discontinuous zone, there are only scattered areas of thawed ground. Moving southward, the percentage of unfrozen ground increases.

☐ Continuous zone   ▨ Discontinuous zone

▲ **Figure 15.21**
Distribution of permafrost in the Northern Hemisphere More than 80 percent of Alaska and about 50 percent of Canada are underlain by permafrost.

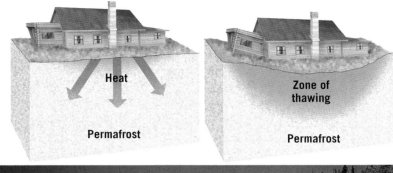

▲ **SmartFigure 15.22**
When permafrost thaws This building, located south of Fairbanks, Alaska, subsided because of thawing permafrost. Notice that the right side, which was heated, settled much more than the unheated porch on the left.

Tutorial
https://goo.gl/EaEiqR

more than a shallow surface layer. Deeper ground remains frozen year-round. Permafrost is extensive in the lands surrounding the Arctic Ocean (**Figure 15.21**). Strictly speaking, permafrost is defined only on the basis of temperature; that is, it is ground with temperatures that have remained below 0°C (32°F) continuously for 2 years or more. The degree to which ice is present in the ground strongly affects the behavior of the surface material. Knowing how much ice is present and where it is located is very important when it comes to constructing roads, buildings, and other projects in areas underlain by permafrost.

When people disturb the surface, such as by removing the insulating vegetation mat or by constructing roads and buildings, the delicate thermal balance is disturbed, and ice within the permafrost can thaw. When a heated structure is built directly on permafrost that contains a high proportion of ice, thawing creates soggy material into which a building can sink (**Figure 15.22**). One solution is to place buildings and other structures on piles, like stilts. The piles allow subfreezing air to

circulate between the floor of the building and the soil and thereby keep the ground frozen. Thawing related to human-induced rapid warming in the Arctic has produced unstable ground that may slide, slump, subside, and undergo severe frost heaving. Arctic communities are scrambling to deal with housing shortages caused by permafrost thawing.

CONCEPT CHECKS 15.5

1. Describe the basic mechanisms that contribute to creep. How might you recognize that creep is occurring?

2. During what season does solifluction in the Arctic occur? Explain why it occurs only during that season.

3. What is permafrost? How might disturbing permafrost lead to unstable ground that may slide, flow, or subside?

Concept Checker
https://goo.gl/HXPhz5

15.6 Detecting, Monitoring, and Mitigating Landslides

Summarize the indicators of an active landslide and list some ways that engineers can mitigate various kinds of mass movement.

Active landslides of all types present serious risks to human life and property worldwide. Detecting a developing case of mass movement can allow for evacuations that save lives, permit real-time monitoring of the situation, and trigger efforts to diminish the threat (mitigation). One method used to identify active landslides is field mapping of ground deformation. This involves locating cracks or bulges in the ground surface, cracked concrete foundations, tilted trees, and other disturbances. Remote mapping of ground deformation by aerial photography, lidar, and satellite imaging also helps identify active landslide sites.

Monitoring Active Landslides

When geologists detect an active landslide, various systems can be used to monitor the ongoing deformation of the soil or rock mass. Using a simple rotational landslide as an example, **Figure 15.23** illustrates several systems that geologists and engineers commonly use: displacement instruments (continuous sensors that extend like tape measures); wells that monitor groundwater levels, soil moisture, and tilting; geophones that detect sound energy generated by movement; and gauges that measure precipitation and soil moisture. In addition, repeat laser or radar surveys, photogrammetry from ground or air, and satellite measurements (InSAR) provide motion assessments of large areas (**Figure 15.24**).

Boundary of active landslide

Landslide monitoring site

A.

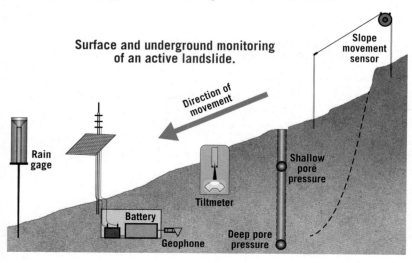

▼ Figure 15.23
Surface and underground monitoring of an active landslide

Surface and underground monitoring of an active landslide.

Slope movement sensor

Direction of movement

Rain gage

Shallow pore pressure

Tiltmeter

Battery

Geophone

Deep pore pressure

B.

▲ Figure 15.24
Two views of the same large earthflow in Colorado A. This photo shows the Slumgullion landslide in Colorado. **B.** The same earthflow shown in a false-color radar image. The yellow areas within the landslide area show the most rapid movement during a week-long period in 2011, when that section of the flow moved 9 cm (3.5 inches).

Mitigating Active Landslides

When active landslides threaten infrastructure such as roads or buildings, geologists and engineers employ a variety of stabilizing methods. As mass movements involve a balance of resisting versus driving forces, effective mitigation strategies usually involve increasing the strength of the failing mass. These include techniques to (1) "de-water" the failing mass (for example, rerouting stream channels or installing horizontal drain pipes); (2) buttress the slope with stronger material (broken rock or reinforced concrete); or (3) bind the sliding surfaces with rock bolts for fractures or steel (or timber) piles for unconsolidated masses.

CONCEPT CHECKS 15.6

1. List two observations which indicate that a slope may be failing.

2. Describe some monitoring methods used when an active landslide is detected.

Concept Checker
https://goo.gl/wLwqzd

3. Name two techniques that are used to strengthen failing slopes.

15 CONCEPTS IN REVIEW

Mass Movement: The Work of Gravity

15.1 The Importance of Mass Movement

Describe how mass movement processes can cause natural disasters and discuss the role that mass movement plays in the development of landscapes.

Key Term: mass movement

- After weathering breaks apart rock, gravity moves the debris downslope in a process called *mass movement*. Sometimes this occurs rapidly as a landslide, and at other times the movement is slower. Landslides are a significant geologic hazard, taking many lives and destroying property.

- Mass movement serves an important role in landscape development. It widens stream valleys and helps tear down the mountains thrust up by plate tectonics.

Q The deadly 2014 Oso landslide in Washington state surprised many people. In this detailed topographic map of the Stilliguamish River valley, the deposit of the 2014 Oso landslide is shown with the crosshatch pattern. Colored areas show older landslide deposits, distinguished by age. Study the map and explain whether you think the 2014 event should have been anticipated.

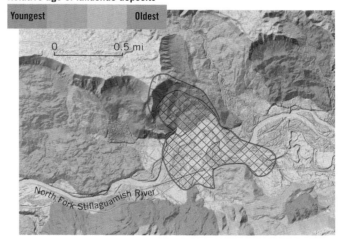

Relative age of landslide deposits

Youngest   Oldest

0   0.5 mi

North Fork Stillaguamish River

15.2 Controls and Triggers of Mass Movement

Summarize the factors that control and trigger mass movement processes.

Key Terms: angle of repose   liquefaction
trigger

- Most (but not all) mass movement events involve a *trigger* event. The addition of water, oversteepening of the slope, removal of vegetation, and shaking due to an earthquake are four important examples of triggers.

- Water added to a slope can expand the pores, separating grains and causing them to lose their cohesion. Water also lubricates the contacts between the grains and adds a significant amount of mass to a wetted slope.

- Granular materials can pile up to a certain angle of slope, but granular piles steeper than that critical angle will spontaneously collapse outward to form a gentler slope. For most geologic materials, this *angle of repose* varies between 25 and 40 degrees from horizontal. Oversteepened slopes are likely to fail, causing landslides.

- The roots of plants (especially deep ones) act as a three-dimensional "net" that holds soil and regolith particles in place. When the plants die, the soil loses an important support structure. Plants may be removed naturally (through wildfires, for instance) or by humans harvesting timber, planting crops, or building structures.

- Earthquakes deliver an energetic jolt to slopes poised on the brink of failure, triggering mass movement events.

Q Do all mass movement events have triggers? Explain.

15.3 Classification of Mass Movement Processes

List and explain the criteria that are commonly used to classify mass movement processes.

Key Terms: talus slope   flow
fall   slide   rock avalanche

- There are a variety of Earth materials (rock, soil, and regolith) and a variety of rates at which mass movements take place. The type of material and the nature of the motion are combined into classifying terms for the different types of mass movement.
- *Falls* occur when pieces of bedrock detach and fall freely through the air, slamming into the ground below with tremendous force. Repeated falls generate a *talus slope*, the characteristic "apron" of angular rock debris that accumulates below mountain cliffs. *Slides* occur when discrete blocks of rock or unconsolidated material slip downslope on a planar or curved surface. Unlike in a fall, the material in a slide does not drop through the air. *Flows* occur when individual grains or particles move randomly in a slurry, a viscous mixture of water-saturated materials.
- *Rock avalanches* and debris avalanches on Earth and Mars move incredibly rapidly over surprising distances (as much as tens of kilometers in a few minutes). Other forms of mass movement are much slower, moving perhaps only millimeters per year.

15.4 Common Forms of Mass Movement

Distinguish among slump, rockslide, debris slide, debris avalanche, debris flow, mudflow, lahar, earthflow.

Key Terms: debris slide   mudflow
slump   debris avalanche   lahar
rockslide   debris flow   earthflow

- A *slump* is a distinctive and common form of mass movement in which coherent blocks of material move downhill on a spoon-shaped slip surface. Slumps are often marked by curved scarps that open up at their tops. They are frequently triggered by oversteepening, such as that caused by stream erosion of a valley wall.
- *Rockslides* are rapid mass movement events in which a coherent block of rock slides downhill along a planar surface. Often this is a preexisting structure such as a joint or a bedding plane. Situations where these surfaces dip into a valley at an angle are especially dangerous.
- *Debris flows* occur when unconsolidated soil or regolith becomes saturated with water and moves downhill in a slurry, picking up other objects (trees, houses, livestock) along the way. Varieties of debris flow include *mudflows*, which are dominated by small particle sizes, and *lahars*, which involve volcanic materials. Debris flows can move quickly—up to 30 kilometers (nearly 20 miles) per hour.
- *Earthflow* is characterized by a similar loss of coherence between grains in unconsolidated material, but it is much slower than debris flow. Typically, sites of earthflow show an uphill scarp and a lobe of viscous soil on the downhill side.

Q This rockslide occurred in the rugged Himalayas of northern India. Identify a feature in the photo that may have been a factor that contributed to the slide. Speculate about what might have triggered the event.

15.5 Very Slow Mass Movements

Review the general characteristics of slow mass movement processes and describe the unique issues associated with a permafrost environment.

Key Terms: solifluction   permafrost
creep

- *Creep* is a very slow form of mass movement that is both important and widespread. It occurs when freezing (or wetting) causes soil particles to be pushed out away from the slope, only to drop down to a lower position following thawing (or drying). In contrast, *solifluction* is the gradual flow of a saturated surface layer that is underlain by an impermeable zone. In arctic regions, the impermeable zone is permafrost.
- *Permafrost*, permanently frozen ground, covers large portions of North America and Siberia. Constructing buildings and other infrastructure in such regions requires special planning. Leaking heat can melt permafrost, causing a loss of volume and triggering flow in the formerly frozen soil. The resulting subsidence can be devastating.

Q This pipeline carries heated oil from Alaska's North Slope to a port along the south coast. Notice that the pipeline is not buried but rather suspended above ground. Suggest a reason why the pipeline in this image is not buried.

Pipeline

15.6 **Detecting, Monitoring, and Mitigating Landslides**

Summarize the indicators of an active landslide and list some ways that engineers can mitigate various kinds of mass movement.

- Field and satellite mapping can be used to identify areas of active landslides.
- Monitoring unstable or actively deforming slopes allows for more accurate prediction of dangerous slope failures.
- A variety of techniques have been developed to stabilize failing slopes, mostly involving strengthening the rock or soil mass.

GIVE IT SOME THOUGHT

1. Describe a type of mass movement that might occur in your home area. Remember to consider characteristics such as climate, surface materials, and steepness of slopes. Does your example have a trigger?
2. Rivers, groundwater, glaciers, wind, and waves can all move and deposit sediment. Geologists refer to these phenomena as *agents of erosion*. Mass movement also involves the movement and deposition of sediment, yet it is *not* classified as an agent of erosion. How is mass movement different?
3. Describe at least one situation in which an internal process might cause or contribute to a mass movement event.
4. Do you think it is likely that landslides frequently occur on the Moon? Explain why or why not.
5. This image shows landslide debris atop Buckskin Glacier in Denali National Park in the rugged Alaska Range. The glacier feeds a river that flows into Cook Inlet, just west of the city of Anchorage. Cook Inlet is an arm of the North Pacific.

   a. Based on what you learned about sorting of sediment in Chapter 7 (see Figure 7.7, page 209), would you expect the material deposited by the landslide to be well sorted? Why or why not?

   b. Also referring to Chapter 7 and Figure 7.7, would you expect the particles in the landslide debris to be rounded or angular? Explain.

   c. These mountains are clearly being sculpted by glaciers. However, other processes also have important roles. Briefly describe some of these processes and the role that each plays in the evolution of this mountainous landscape.

6. Describe at least one situation in which an internal process might cause or contribute to a mass movement event.

7. GEOgraphics 15.1 on page 422 includes a world map showing global landslide risks. What criteria were used to construct the map? What additional data could be considered that would make this risk map even more useful?
8. When the rail line in the accompanying photo was built in rural Alaska in the 1930s, the terrain was relatively level. Not long after the railroad was completed, a great deal of subsidence and shifting of the ground occurred, turning the tracks into the "roller coaster" shown here. As a result, the rail line had to be abandoned. Suggest a reason the ground became unstable and shifted.

9. Mass movement is influenced by many processes associated with all four spheres of the Earth system. Select two items from the list below. For each one, outline a series of events that relate the item to various spheres and to a mass movement process. For example, if "frost wedging" were an item on the list, you might write the following: *Frost wedging involves rock (geosphere) being broken when water (hydrosphere) freezes. Freeze–thaw cycles (atmosphere) promote frost wedging. When frost wedging loosens a rock on a cliff, the fragment tumbles to the base of the cliff. This event, rockfall, is an example of mass movement.* Now you give it a try. Use your imagination.

   - Deforestation

   - Spring thaw/melting snow

   - Highway road cut

   - Crashing waves

   - Cavern formation (see Figure 17.35, page 495)

10. The Swift Creek landslide is a large, complex slump in northwest Washington State that transitions into an earthflow and also generates debris flows during heavy rain events. Explore the Swift Creek Landslide Observatory website (http://landslide.geol.wwu.edu). The landslide moves on average 3 to 4 meters per year—that's about 1 centimeter (one half inch) per day! To see this slow movement over an extended period, click "TimeLapse" (upper right) to try out the video generator, using these settings:

Start Date: 01/01/2007

End Date: 01/01/2014

Time of Day: Between 12pm-3pm

Camera: Upper

Framerate: Medium (20 frames/second)

Video Length: 10 seconds

Then click "Generate." (The site may take about 30 seconds to generate your custom time-lapse video.)

After you try this, use the video generator to determine whether the Swift Creek landslide moves more during the wet season (November–April) or during the dry season (May–October).

EYE ON EARTH

1. Heavy rains in late July 2010 triggered the mass movement that occurred in this mountain valley near Durango, Colorado. Heavy equipment is clearing away material that blocked railroad tracks and significantly narrowed the adjacent stream channel.

a. Was this event more likely a rockfall, creep, or a debris flow?

b. Most of us are familiar with the phrase "One thing leads to another." It certainly applies to the Earth system. Suppose the material from the mass movement event shown here had completely filled the stream. What other natural hazard might have developed?

DATA ANALYSIS

Landslides in Oregon

Many regions of Oregon are prone to landslides due to steep slopes and abundant rainfall, which works with gravity to destabilize the landscape.

https://goo.gl/eD226M

ACTIVITIES

Go to the *Statewide Landslide Information Database for Oregon (SLIDO):* https://www.oregongeology.org/slido/. Click on the "Help" link underneath the description. Skim the first few sections on how to navigate the map.

1. According to the "Mapped Landslide Data Inventory" bullet in the "Show/Hide Map Layer" section, what is talus-colluvium? What is a fan? What is a landslide?

Go back to the SLIDO home page and click on "Map." Click "OK" in the pop-up disclaimer box, and then a map of Oregon appears, with various menu options. In the box labeled "Layers Currently Shown," click on "Historic Landslide Data Inventory" and uncheck "Mapped Landslide Data Inventory." Directly above the layers menu are five square icons. Click on the second icon, which looks like four squares, to access the basemap gallery. Once there, select "Imagery with Labels."

2. In which portion of the state have most historic events occurred? (Make sure the map is zoomed out enough that you can see the entire state.)

3. Now zoom in and click on several events in the eastern half of Oregon. What type of landslide type (movement class) is common here? What do the number and type of events tell you about the land in eastern Oregon?

Zoom in until the scale on the lower left says 3 miles (or less). Look at northwestern Oregon. Follow the coastline down until Arch Cape is in the middle of your map.

4. Notice that events follow a path. Click on several events. What type of landslides (movement class) occur along this path?

5. "Zoom to" a few of these events. Along what feature do these land-slides occur? What is the most likely cause of these landslides?

Go to http://oregonstate.edu/ua/ncs/archives/2015/feb/map-outlines-western-oregon-landslide-risks-subduction-zone-earthquake.

6. What future event is likely to induce a significant number of mass movements in Oregon?

7. Why is the Coastal Range a special concern?

8. Why is it important to consider landslide and transportation issues in addition to damage to structures?

Mastering Geology

Looking for additional review and test prep materials? Visit pearson.com/mastering/geology to enhance your understanding of this chapter's content by accessing a variety of resources, including **Self-Study Quizzes, Geoscience Animations, SmartFigure Tutorials,** *Project Condor* **Videos, Mobile Field Trips, Dynamic Study Modules,** and an optional **Pearson eText.**

Using Social Media for More Accurate Flood Monitoring and Prediction

Traditionally, predictions of floods have been made by studying weather forecasts and past data on flooding in a given area. In areas where flooding is particularly common, sensors have sometimes been placed to measure water levels or drainage. However, most places where floods occur have no such sensing equipment or data available. This has for many centuries made communicating news of developing floods to nearby people difficult and chaotic. Social media is changing this.

Today most people carry a smartphone that enables them to quickly snap a photograph of a developing flood and then post the picture online for others to see. In the past few years, researchers have realized that posting patterns change in an area when a flood develops and that, based on key words or hashtags, it is often possible to get an almost real-time picture of a progressing flood. Some researchers have also looked into using Flickr photo tags and captions to monitor flooding.

▲ Researchers are experimenting with following social media posts to improve flood prediction and monitoring.

A research social enterprise project called FloodTags created an algorithm that monitors flood-related tweets and uses the data gleaned to create flood maps. The Philippines Red Cross has used FloodTags information to better anticipate floods, and to deploy resources to more specific areas. In addition, a study conducted by the University of Warwick in the United Kingdom created a flood-prediction model using geo-referenced social media messages from Twitter, along with data pulled from traditional flood-predicting sources (dubbed "authoritative data"). The researchers found that flood forecasting that combined both data sources had a 71 percent rate of accuracy—a significant improvement over the 39 percent accuracy rate researchers found existed when authoritative data alone were used for flood prediction. Crowd-sourced stream monitoring is also being piloted in Europe (CrowdWater) and the United States (StreamTracker). More research is needed, but the results so far are promising.

▶ During the Brisbane River floods of 2010 and 2011, community-based Facebook groups were a popular source for information.

16
Running Water

FOCUS ON CONCEPTS

Each statement represents the primary learning objective for the corresponding major heading within the chapter. After you complete the chapter, you should be able to:

16.1 List the hydrosphere's major reservoirs and describe the different paths that water takes through the hydrologic cycle.

16.2 Describe the nature of drainage basins and river systems. Sketch and briefly explain four basic drainage patterns.

16.3 Discuss streamflow and the factors that cause it to change.

16.4 Outline the ways in which streams erode, transport, and deposit sediment.

16.5 Compare and contrast bedrock and alluvial stream channels. Distinguish between two types of alluvial channels.

16.6 Describe alleys created by streams, including V-shaped, broad valleys with floodplains, and valleys that display incised meanders or stream terraces.

16.7 List the major depositional landforms associated with streams and describe the formation of these features.

16.8 Summarize various categories of floods and the common measures of flood control.

Consider the bittersweet relationship we have with rivers. They are vital economic tools—used as highways to move goods and as sources of water for drinking, sanitation, irrigation, and energy—as well as prime locations for recreation. When considered as part of the Earth system, rivers and streams represent a fundamental link in the constant cycling of our planet's water. Yet running water is the dominant agent of landscape alteration, eroding more terrain and transporting more sediment than any other natural process. (Humans, however, now move more material than rivers and glaciers combined.) Soils and other materials are constantly being washed away and then deposited elsewhere. Because human populations concentrate along rivers, flooding has enormous potential for destruction and loss of life.

This chapter provides an overview of the hydrologic cycle, the nature of river systems, the types of river channels and the factors that produce them, the influences of running water on our planet's landscapes, and the nature of floods and their impact on people.

16.1 Earth as a System: The Hydrologic Cycle

List the hydrosphere's major reservoirs and describe the different paths that water takes through the hydrologic cycle.

Water constantly moves among Earth's different spheres—the *hydrosphere*, the *atmosphere*, the *geosphere*, and the *biosphere*. This unending circulation of water is called the **hydrologic cycle**. Earth is the only planet in our solar system that has a global ocean and a hydrologic cycle.

Earth's Water

Water is almost everywhere on Earth—in the oceans, glaciers, rivers, lakes, air, soil, and living tissue. All these reservoirs constitute Earth's hydrosphere. A glance at **Figure 16.1** shows that the vast bulk of Earth's water, about 96.5 percent, is stored in the global ocean. Ice sheets, glaciers, and groundwater account for most of our planet's freshwater, leaving about 1 percent to be divided among lakes, streams, and the atmosphere. Although the percentage of Earth's total water found in each of the latter sources is just a small fraction of the total inventory, the absolute quantities are vast.

Water's Paths

The hydrologic cycle is a gigantic, worldwide system powered by energy from the Sun, in which the atmosphere provides a vital link between the oceans and continents (**Figure 16.2**). **Evaporation**, the process by which liquid water changes into water vapor (gas), is how water enters the atmosphere from the ocean and, to a lesser extent, from the land. Winds often transport moisture-laden air great distances. Complex processes of cloud formation eventually result in precipitation. The precipitation that falls into the ocean has completed its cycle and is ready to begin again. The water that falls on the continents, however, must make its way back to the ocean.

What happens to precipitation once it has fallen on land? A portion of the water soaks into the ground

▶ Figure 16.1
Distribution of Earth's water

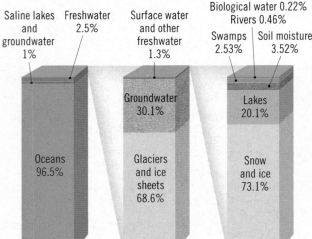

Total global water
Saline lakes and groundwater 1%
Freshwater 2.5%
Oceans 96.5%

Freshwater
Surface water and other freshwater 1.3%
Groundwater 30.1%
Glaciers and ice sheets 68.6%

Surface water and other freshwater
Atmosphere 0.22%
Biological water 0.22%
Rivers 0.46%
Swamps 2.53%
Soil moisture 3.52%
Lakes 20.1%
Snow and ice 73.1%

(called **infiltration**), slowly moving downward, then moving laterally, and finally seeping into lakes, streams, or directly into the ocean. When the rate of rainfall exceeds the ground's ability to absorb it, the surplus water flows over the surface into lakes and streams, a process called **runoff**. Much of the water that infiltrates or runs off eventually returns to the atmosphere because of evaporation from the soil, lakes, and streams. Also, some of the water that soaks into the ground is absorbed by plants, which then release it into the atmosphere. This process is called **transpiration**. Because both evaporation and transpiration involve the transfer of water from the surface directly to the atmosphere, they are often considered together as the combined process of **evapotranspiration**.

Hydrologic Cycle

Precipitation
96,000 km³

Precipitation
284,000 km³

Evaporation/Transpiration
60,000 km³

Evaporation
320,000

36,000 km³

Runoff

Infiltration

Oceans

◀ SmartFigure 16.2
The hydrologic cycle The primary movement of water through the cycle is shown by the large arrows. A number refers to the annual amount of water taking a particular path.

Tutorial
https://goo.gl/k3gvnd

Storage in Glaciers

When precipitation falls in very cold places—at high elevations or high latitudes—the water may not immediately soak in, run off, or evaporate. Instead, it may become part of a snowfield or glacier. Glaciers store extremely large quantities of water on land. If present-day glaciers were to melt and release all their water, sea level would rise by several dozen meters. Such a rise would submerge a number of coastal areas, many of them heavily populated coastal areas. As described in Chapter 18, over the past 2 million years, huge ice sheets have formed and melted on several occasions, each time affecting the balance of the hydrologic cycle.

Water Balance

Figure 16.2 shows Earth's overall *water balance*, or the volume that passes through each part of the cycle annually. The amount of water vapor in the air at any one time is just a tiny fraction of Earth's total water supply. But the *absolute* quantities that are cycled through the atmosphere over a 1-year period are immense—some 380,000 cubic kilometers (91,000 cubic miles)—enough to cover Earth's entire surface to a depth of about 1 meter (3 feet).

It is important to know that the hydrologic cycle is *balanced*. Because the total amount of water vapor in the atmosphere remains about the same, the average annual precipitation worldwide must be equal to the quantity of water evaporated. However, for all the continents taken together, precipitation exceeds evaporation. Conversely, over the oceans, evaporation exceeds precipitation. Because the level of the world ocean is not dropping, the system must be in balance. Balance is achieved because a vast amount of water—36,000 cubic kilometers (8600 cubic miles)—makes its way from the land back to the ocean annually.

About one-quarter of global precipitation falls on land and flows on and below the surface. This water is *the most important force sculpting Earth's land surface*. In the rest of this chapter, we will observe the work of water running over the surface, including floods, erosion, and the formation of valleys. In Chapter 17 we will look underground, at the slow labors of groundwater as it forms springs and caverns and provides water for people on its long migration to the sea.

16.2 River Systems

Describe the nature of drainage basins and river systems. Sketch and briefly explain four basic drainage patterns.

Much of the precipitation that falls on land either soaks into the ground (infiltration) or remains at the surface, moving downslope as *runoff*. The amount of runoff depends on several factors, including (1) the intensity and duration of rainfall, (2) the amount of water already in the soil, (3) the nature of the surface material, (4) the slope of the land, and (5) the extent and type of vegetation.

When the surface material is highly impermeable or when it becomes saturated, runoff is the dominant process. Runoff is also high in urban areas because many features are impermeable, including buildings, roads, and parking lots.

Runoff initially flows in broad, thin sheets across slopes in a process called *sheet flow*. This thin, unconfined flow eventually develops threads of current that form tiny channels called *rills*. Rills meet to form *gullies*, which join to form streams and, eventually, rivers. Although the terms *river* and *stream* are often used interchangeably, geologists define **stream** as water that flows in a channel, regardless of size. **River**, on the other hand, is typically used to describe streams that carry substantial amounts of water and have numerous tributaries.

In humid regions, the water to support streamflow comes from two sources: overland flow that sporadically enters the stream and groundwater that enters the channel. In areas where the bedrock is composed of soluble rocks such as limestone, large openings may exist that facilitate the transport of groundwater to streams. In 2018, the connection between rainfall, surface water, and groundwater was vividly displayed in Thailand as a youth soccer team was trapped by rising waters in a limestone cave system. (The team was eventually rescued by divers.) In arid regions, however, the *water table* may be below the level of the stream channel, in which case the stream loses water to the groundwater system by outflow percolating through the streambed.

Drainage Basins

Every stream drains an area of land called a **drainage basin**, or **watershed** (**Figure 16.3**). Each drainage basin is bounded by an imaginary line called a **divide**, which may be clearly visible as a sharp ridge in some mountainous areas or may be more difficult to determine when the topography is subdued. The outlet, where the stream exits the drainage basin, is at a lower elevation than the rest of the basin.

Drainage divides range in scale from a small ridge separating two gullies on a hillside to a *continental divide* that splits an entire continent into enormous drainage basins. The Mississippi River has the largest drainage basin in North America, collecting and

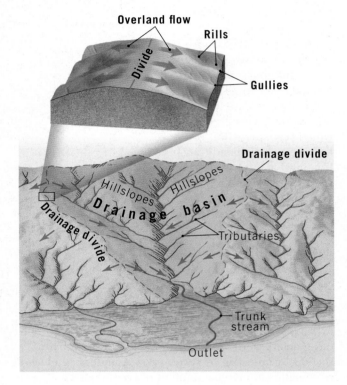

▲ **Figure 16.3**

Drainage basin and divide A drainage basin is the area drained by a stream and its tributaries. Boundaries between basins are called divides.

carrying 40 percent of the flow in the United States (**Figure 16.4**).

By looking at the drainage basin in Figure 16.3, it should be obvious that slopes cover most of the area. Water erosion on the hillsides is aided by the impact of raindrops and by sheet flow, moving downslope as sheets or in rills toward a stream channel. Hillslope erosion is the main source of fine particles (clays and fine sand) carried in stream channels.

If you could observe the streams in an area similar to that depicted in Figure 16.3 over several years, you would see many of them lengthen by **headward erosion**—that is, by extending the heads of their channels upslope. Headward erosion occurs when the surface flow converging at the head of a channel has enough power to cut the channel deeper (*downcut*): This

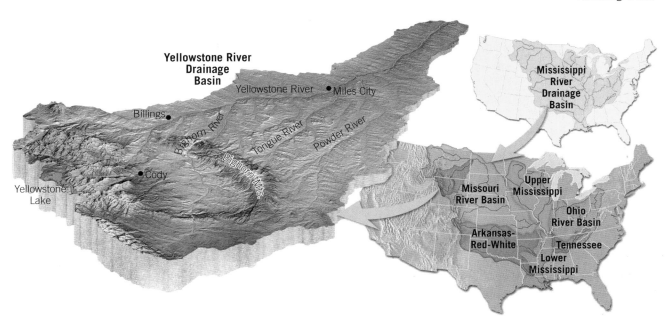

◀ **SmartFigure 16.4**
Mississippi River drainage basin The drainage basin of the Mississippi River forms a funnel that stretches from Montana and southern Canada in the west to New York State in the east and runs down to a spout in Louisiana. It consists of many smaller drainage basins. The drainage basin of the Yellowstone River is one of many that contribute water to the Missouri River, which, in turn, is one of many that make up the drainage basin of the Mississippi River.

Tutorial
https://goo.gl/
vTqboN

lowering of the stream leads to increased rates of erosion on the steeper slopes, which lie in the opposite direction of streamflow. Thus, through headward erosion, a valley extends into previously undissected terrain. In **Figure 16.5**, the tributaries of Utah's San Rafael River illustrate this process.

River Systems

Rivers drain much of the land area, with the exception of extremely arid regions and polar areas that are permanently frozen. To a large extent, the variety of rivers that exist is a reflection of the different environments in which they are found. For example, although the Paraná–La Plata River system in South America drains an area roughly the same size as the Nile in Egypt, it carries nearly 10 times more water to the ocean. Because its drainage basin is entirely in a rainy tropical climate, the Paraná–La Plata system has a huge discharge. By contrast, the Nile, which also originates in a humid region, flows through an expansive arid landscape, where significant amounts of water evaporate and are withdrawn to sustain agriculture. Thus, climatic differences and human intervention can significantly influence the character of a river. Later, we will examine other factors that contribute to stream variability.

A river system includes not only its network of stream channels but its entire drainage basin. It can be divided into three zones. In *zones of sediment production*, erosion dominates. The other two zones are the *zone of sediment transport* and the *zone of sediment deposition* (**Figure 16.6**). It is important to recognize that sediment is being eroded, transported, and deposited along the entire length of a stream, regardless of which process is dominant within each zone.

Tributaries to the San Rafael River are dissecting this flat upland area.

Headward erosion is occurring here.

◀ **SmartFigure 16.5**
Headward erosion A stream lengthens its course by extending the head of its valley upslope into previously undissected terrain.

Tutorial
https://goo.gl/
WnUUbt

▶ Figure 16.6
Zones of a river Each of the three zones is based on the dominant process that is operating in that part of the river system.

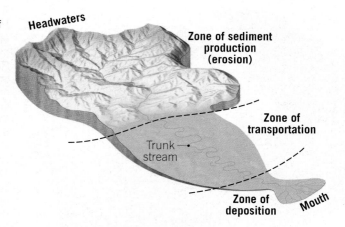

balance, the amount of sediment eroded from their banks equals the amount deposited elsewhere in the channel. Although trunk streams rework their channels over time, they are not a source of sediment, nor do they accumulate or store it.

Sediment Deposition   When a river reaches the ocean or another large body of water, it slows, and the energy to transport sediment is greatly reduced. Most of the sediment either accumulates at the mouth of the river to form a delta, is reconfigured by wave action to form a variety of coastal features, or is moved far offshore by ocean currents. Because coarse sediment tends to be deposited upstream, it is primarily the fine sediment (clay, silt, and fine sand) and dissolved ions that eventually reach the ocean. Taken together, erosion, transportation, and deposition are the processes by which rivers move Earth's surface materials and sculpt landscapes.

Sediment Production   The zone of *sediment production*, where most of the sediment is derived, is located in the headwaters region of the river system. Much of the sediment carried by streams begins as bedrock that is subsequently broken down by weathering and moved downslope by mass movement directly into the channel or by way of sheet flow and rills. Bank erosion can also contribute significant amounts of sediment. In addition, scouring of the channel bed deepens the channel and adds to the stream's sediment load.

Sediment Transport   Sediment acquired by a stream travels through the channel network along sections called *trunk streams*. When trunk streams are in

Drainage Patterns

Drainage systems, which are interconnected networks of streams, can exhibit a variety of patterns. The pattern that develops depends primarily on the kind of rock present and/or the structural pattern of joints, faults, and folds. **Figure 16.7** illustrates four drainage patterns.

The most common drainage pattern is the **dendritic pattern** (see Figure 16.7A). Its irregular tributary streams resemble the branching pattern of a deciduous tree (*dendritic* means "treelike"). The dendritic pattern

▶ Figure 16.7
Drainage patterns Networks of streams form a variety of patterns.

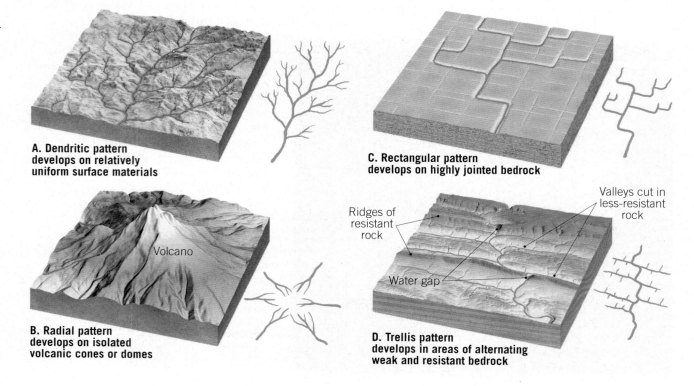

A. Dendritic pattern develops on relatively uniform surface materials

B. Radial pattern develops on isolated volcanic cones or domes

C. Rectangular pattern develops on highly jointed bedrock

D. Trellis pattern develops in areas of alternating weak and resistant bedrock

The river establishes its course on relatively uniform strata that cover a structural feature below.

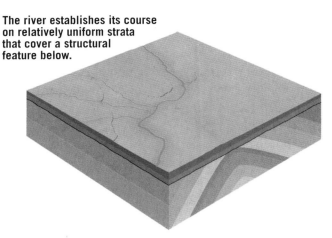

As the stream erodes downward, it encounters and cuts through the resistant rock to create a water gap. By this process, the river is superimposed on the ridge.

Water gap

Water gap

Bighorn River

The Bighorn River is a superposed stream that created a water gap through Wyoming's Sheep Mountain.

◄ **SmartFigure 16.8**
Development of a super-posed stream This is one way in which a water gap can form. The Bighorn River is a good example. The accompanying Mobile Field Trip shows some excellent views and additional explanation of the Bighorn River and how it created the water gap shown here.

Mobile Field Trip
https://goo.gl/cSXAGm

forms where the underlying material is relatively uniform. Because the surface material is essentially uniform in its resistance to erosion, it does not control the pattern of streamflow. Rather, the pattern is determined chiefly by the direction of slope of the land.

The **radial pattern** of drainage occurs where streams diverge from a central area like spokes from the hub of a wheel (see Figure 16.7B). This pattern typically develops on isolated volcanic cones and domal uplifts.

A **rectangular pattern** exhibits many right-angle bends (see Figure 16.7C). This pattern develops when the bedrock is crisscrossed by a series of joints and/or faults. Because these structures are eroded more easily than unbroken rock, their geometric pattern guides the directions of valleys.

A **trellis pattern** looks rectangular, with tributary streams running nearly parallel to one another, giving the appearance of a garden trellis (see Figure 16.7D). This pattern forms in areas underlain by alternating bands of resistant and less-resistant rock and is particularly well displayed in the folded Appalachian Mountains, where both weak and strong strata outcrop in nearly parallel belts.

To fully understand the drainage pattern displayed by a stream, it is often useful to consider the stream's entire history. For example, river valleys occasionally cut through a ridge or mountainous topography that lies across their path. This situation is illustrated by the trellis pattern in Figure 16.7D. The steep-walled notch followed by the river through a ridge of resistant rock is called a **water gap**.

Why do streams cut *across* such structures rather than flow *around* them? One possibility is that a stream existed before the ridge or mountain was uplifted. In this situation, the stream, called an **antecedent stream**, eroded its bed downward at a pace equal to the rate of uplift. That is, the stream maintained its course as folding or faulting gradually raised the structure across its path.

A second possibility is that a **superposed stream** eroded its channel into an existing structure (**Figure 16.8**). This can occur when folded beds or resistant rocks are buried beneath layers of relatively flat-lying sediments or sedimentary strata. Streams originating on the overlying strata establish their courses without regard to the structures below. Then, as the valley deepens, the river continues to cut its valley into the underlying structure. The folded Appalachians feature several superposed rivers, including the Potomac and the Susquehanna, which cut their channels through folded strata on their way to the Atlantic.

CONCEPT CHECKS 16.2

1. List several factors that cause infiltration and runoff to vary by place and time.

2. Draw a simple sketch of a drainage basin and a divide and label each.

3. What are the three main parts (zones) of a river system?

4. Briefly describe the four main stream patterns and what causes them.

 **Concept Checker**
https://goo.gl/wjPqb2

16.3 Streamflow Characteristics

Discuss streamflow and the factors that cause it to change.

The water in river channels moves under the influence of gravity. In very slowly flowing streams, water moves in nearly straight-line paths parallel to the stream channel; this is called **laminar flow** (Figure 16.9A). However, streams typically exhibit **turbulent flow** (Figure 16.9B). Strong turbulent behavior occurs in whirlpools and eddies, as well as in roiling whitewater rapids. Even streams that appear smooth on the surface often exhibit turbulent flow near the bottom and sides of the channel, where flow resistance is greatest. *Turbulence* contributes to a stream's ability to erode its channel because it acts to lift sediment from the streambed.

An important factor influencing stream turbulence is water's flow velocity. Flow becomes more turbulent as the velocity of a stream increases. Flow velocities can vary significantly from place to place along a stream channel, as well as over time, in response to variations in the amount and intensity of precipitation. If you have ever waded into a stream, you may have noticed that the strength of the current increased as you moved into deeper parts of the channel. This is related to the fact that frictional resistance is greatest near the banks and bed of a stream channel.

Flow velocities are determined at stream gaging stations by averaging measurements taken at various locations across the stream's channel. Some sluggish streams have flow velocities of less than 1 kilometer (0.6 mile) per hour, whereas stretches of some fast-flowing rivers may exceed 30 kilometers (19 miles) per hour.

Factors Affecting Flow Velocity

The ability of a stream to erode and transport material is directly related to its flow velocity. Even slight variations in flow rate can lead to significant changes in the

sediment load transported by a stream. Factors that influence flow velocities and, therefore, control a stream's potential to do its work include its gradient, the channel cross-sectional shape, channel size and roughness, and discharge, or the amount of water flowing in the channel.

Gradient The slope of a stream channel, expressed as the vertical drop of a stream over a specified distance, is referred to as its **gradient**. Portions of the Lower Mississippi River have very low gradients of 10 centimeters or less per kilometer (6 inches per mile). By contrast, some mountain streams have channels that drop at a rate of more than 40 meters per kilometer (200 feet per mile)—a gradient 400 times steeper than that of the Lower Mississippi. Gradient also varies along the length of a particular channel. When the gradient is steeper, more gravitational energy is available to drive channel flow.

Channel Shape As water in a stream channel moves downslope, it encounters a significant amount of frictional resistance. The *cross-sectional shape* (a slice taken across the channel) determines, to a large extent, the amount of flow that comes in contact with the banks and bed of the channel. This measure is referred to as the **wetted perimeter**. The most efficient channel is one with the least wetted perimeter for its cross-sectional area. **Figure 16.10** compares two channels that differ only in shape: One is wide and shallow, the other is narrow and deep. Although the two channels have identical cross-sectional areas, the narrow and deep one has less water in contact with the channel (a smaller wetted perimeter) and therefore less frictional drag. As a result, if all other factors are equal, the water will flow more efficiently and at a higher velocity in the deep and narrow channel than in the wide and shallow channel.

▶ Figure 16.9
Laminar and turbulent flow Most often streamflow is turbulent.

A.

B.

Running the rapids in the Grand Canyon— an extreme example of turbulent flow.

This water is not standing still. It is moving slowly toward the bottom of the image. The flow in the foreground is primarily laminar.

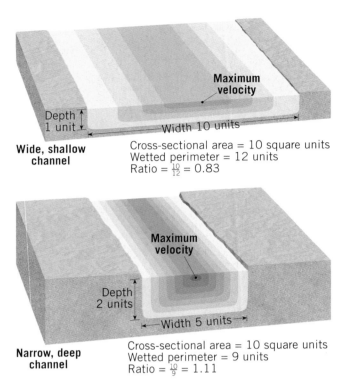

Cross-sectional area = 10 square units
Wetted perimeter = 12 units
Ratio = $\frac{10}{12}$ = 0.83

Wide, shallow channel

Cross-sectional area = 10 square units
Wetted perimeter = 9 units
Ratio = $\frac{10}{9}$ = 1.11

Narrow, deep channel

▲ **Figure 16.10**
Influence of channel shape on velocity The stream with the smaller wetted perimeter has less frictional drag and will flow more rapidly, all else being equal.

Channel Size and Roughness As just discussed, all other factors being equal, flow velocities are higher in large channels than in small channels. The depth of the water in a given channel also affects the frictional resistance exerted on flow. Maximum flow velocity occurs when a stream is *bankfull*—that is, before water starts to inundate the floodplain. At this stage, the channel's ratio of the cross-sectional area to wetted perimeter is highest, and streamflow is most efficient. A final factor is how *rough* a channel is. Elements such as boulders, irregularities in the channel bed, and woody debris create turbulence that significantly reduces flow velocity.

Discharge Streams vary in size from small headwater creeks less than a meter wide to large rivers with widths of several kilometers. The size of a stream channel is largely determined by the amount of water supplied from the drainage basin. The measure most often used to compare the size of streams is **discharge**—the volume of water flowing past a certain point in a given unit of time. Discharge, usually measured in cubic meters per second or cubic feet per second, is determined by multiplying a stream's cross-sectional area by its velocity.

The largest river in North America, the Mississippi, has an average discharge at its mouth of about 16,800 cubic meters (593,000 cubic feet) per second (**Figure 16.11**). That figure is dwarfed by South America's mighty Amazon River, which discharges nearly 12 times more water than the Mississippi (**Figure 16.12**).

◄ **SmartFigure 16.11**
Largest U.S. rivers The photo shows the Mississippi River near Helena, Arkansas.

Mobile Field Trip
https://goo.gl/kvBCZv

RANK	RIVER	AVERAGE DISCHARGE AT MOUTH (1000 ft³/sec)
1	Mississippi	593
2	St. Lawrence	348
3	Ohio	281
4	Columbia	265
5	Yukon	225
6	Missouri	76
7	Tennessee	68

▶ **Figure 16.12**
The world's great rivers

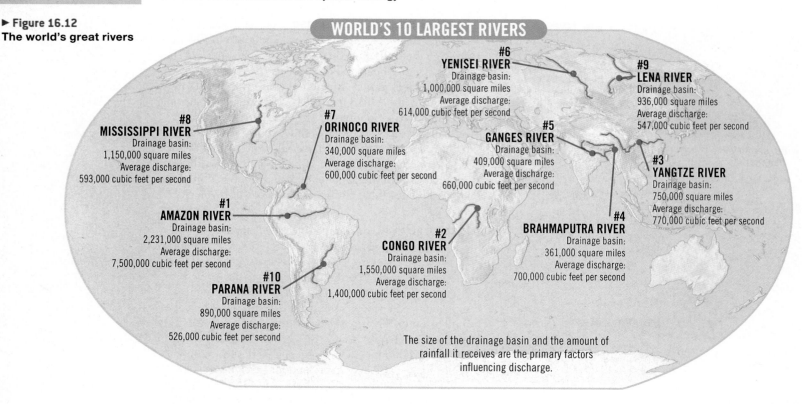

WORLD'S 10 LARGEST RIVERS

#6 YENISEI RIVER
Drainage basin:
1,000,000 square miles
Average discharge:
614,000 cubic feet per second

#9 LENA RIVER
Drainage basin:
936,000 square miles
Average discharge:
547,000 cubic feet per second

#8 MISSISSIPPI RIVER
Drainage basin:
1,150,000 square miles
Average discharge:
593,000 cubic feet per second

#7 ORINOCO RIVER
Drainage basin:
340,000 square miles
Average discharge:
600,000 cubic feet per second

#5 GANGES RIVER
Drainage basin:
409,000 square miles
Average discharge:
660,000 cubic feet per second

#3 YANGTZE RIVER
Drainage basin:
750,000 square miles
Average discharge:
770,000 cubic feet per second

#1 AMAZON RIVER
Drainage basin:
2,231,000 square miles
Average discharge:
7,500,000 cubic feet per second

#2 CONGO RIVER
Drainage basin:
1,550,000 square miles
Average discharge:
1,400,000 cubic feet per second

#4 BRAHMAPUTRA RIVER
Drainage basin:
361,000 square miles
Average discharge:
700,000 cubic feet per second

#10 PARANA RIVER
Drainage basin:
890,000 square miles
Average discharge:
526,000 cubic feet per second

The size of the drainage basin and the amount of rainfall it receives are the primary factors influencing discharge.

The discharge of a river system changes over time because of variations in the amount of precipitation received by the watershed. Studies show that when discharge increases, the width, depth, and flow velocity of the channel all increase predictably. As we saw earlier, when the size of the channel increases, proportionally less water is in contact with the bed and banks of the channel. This reduces friction, which acts to retard the flow, resulting in an increase in the rate of streamflow.

Monitoring Streamflow The U.S. Geological Survey maintains a network of about 7500 stream gaging stations that collect basic data about the country's surface-water resources (**Figure 16.13**). Data collected include flow velocity, discharge, and river stage. *Stage* is the height of the water surface relative to a fixed reference point. This measurement is frequently reported by the media, especially when a river approaches or surpasses *flood stage*. Streamflow measurements are essential components of river models used to make flood forecasts and issue

▶ **Figure 16.13**
Stream gaging stations The United States has a dense network of stream gaging stations. To determine a channel's discharge, it is necessary to survey the channel to determine its shape and calculate its area. A stilling well measures the stage, and a current meter determines flow velocity.

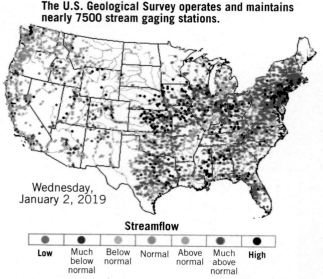

The U.S. Geological Survey operates and maintains nearly 7500 stream gaging stations.

Wednesday, January 2, 2019

Streamflow

Low · Much below normal · Below normal · Normal · Above normal · Much above normal · High

Colors indicate whether the current river stage is above or below normal.

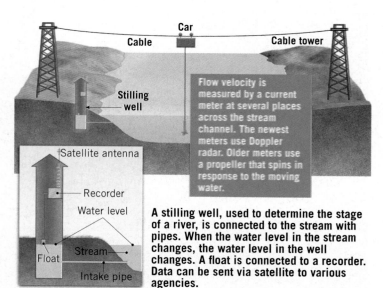

Car
Cable
Cable tower
Stilling well

Flow velocity is measured by a current meter at several places across the stream channel. The newest meters use Doppler radar. Older meters use a propeller that spins in response to the moving water.

Satellite antenna
Recorder
Water level
Stream
Float
Intake pipe

A stilling well, used to determine the stage of a river, is connected to the stream with pipes. When the water level in the stream changes, the water level in the well changes. A float is connected to a recorder. Data can be sent via satellite to various agencies.

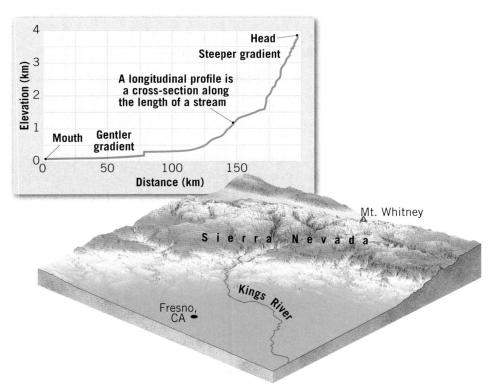

▲ Figure 16.14
Longitudinal profile California's Kings River originates high in the Sierra Nevada and flows into the San Joaquin Valley.

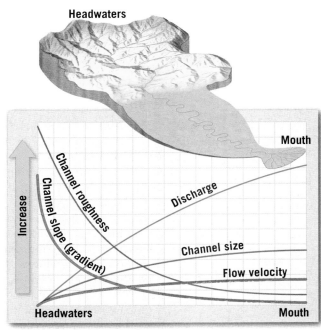

▲ SmartFigure 16.15
Channel changes from head to mouth Although the gradient decreases toward the mouth of a stream, increases in discharge and channel size and decreases in roughness more than offset the decrease in slope. Consequently, flow velocity usually increases toward the mouth.

Tutorial
https://goo.gl/Kcmmnd

warnings. There are other applications as well. The data are used to make decisions related to water supply allocations, operation of wastewater treatment plants, design of highway bridges, and recreation activities.

Changes from Upstream to Downstream

One useful way of studying a stream is to examine its **longitudinal profile**. Such a profile is simply a cross-sectional view of a stream from its source area (called the **head**, or **headwaters**) to its **mouth**, the point downstream where it empties into another water body— a river, a lake, or the ocean. As shown in **Figure 16.14**, the most obvious feature of a typical profile is a constantly decreasing gradient from the head to the mouth. Although many local irregularities may exist, the overall profile is a relatively smooth concave curve.

The change in slope observed on most stream profiles is usually accompanied by an increase in discharge and channel size, as well as a reduction in sediment particle size (**Figure 16.15**). Along most rivers in humid regions, discharge increases toward the mouth because farther downstream, more and more tributaries contribute water to the main channel. In the case of the Amazon, for example, about 1000 tributaries join the main river along its 6500-kilometer (4000-mile) course across South America. In order to accommodate the growing volume of water, channel size typically increases downstream as

well. Recall that flow velocities are higher in large channels than in small channels. Observations also show a general decline in sediment size downstream, making the channel smoother and more efficient (with less friction).

Although the channel slope decreases—the gradient becomes less steep—toward a stream's mouth, the flow velocity generally increases. This fact contradicts our intuitive assumptions of mountain rivers as swift, whereas we think of a broad river flowing across subtle topography as slow. Actually, the increase in channel size and discharge and decrease in channel roughness that occur downstream compensate for the decrease in slope, making the stream more efficient (see Figure 16.14). Thus, the average flow velocity is typically lower in headwater streams than in wide rivers that appear to be placid.

CONCEPT CHECKS 16.3

1. Contrast laminar flow and turbulent flow.

2. What is a longitudinal profile?

3. What typically happens to channel width, channel depth, flow velocity, and discharge between the headwaters and the mouth of a stream? Briefly explain why these changes occur.

 Concept Checker
https://goo.gl/ANxZcm

16.4 The Work of Running Water

Outline the ways in which streams erode, transport, and deposit sediment.

Streams are Earth's most important erosional agents. In addition to having the ability to deepen and widen their channels, streams also have the capacity to transport enormous quantities of sediment that are delivered by overland flow, mass wasting, and groundwater. Eventually, much of this material is deposited to create a variety of landforms.

Stream Erosion

A stream's ability to accumulate and transport soil and weathered rock is aided by the work of raindrops, which knock sediment particles loose (see Figure 6.25 on page 192). When the ground is saturated, rainwater cannot infiltrate, so it flows downslope, transporting some of the material it dislodges. On barren slopes the sheet flow often erodes small channels or rills, which in time may evolve into larger *gullies* (see Figure 6.26 on page 193).

Once flow is confined in a channel, the erosional power of a stream is related to its slope and discharge. The rate of erosion, however, also depends on the relative resistance of the bank and bed material. In general, channels composed of unconsolidated materials are more easily eroded than channels cut into bedrock.

When a channel is sandy, the particles are easily dislodged from the bed and banks and then lifted into the moving water. Moreover, sandy banks are often *undercut*, or eroded at their base, dumping even more loose debris into the water to be carried downstream. Banks that consist of coarse gravels or cohesive clay and silt particles tend to be relatively resistant to erosion. Thus, channels with cohesive silty banks are generally narrower and deeper than comparable ones with sandy banks.

Streams cut channels into bedrock through three main processes: *quarrying*, *abrasion*, and *corrosion*.

Quarrying **Quarrying** involves the removal of blocks from the bed of a stream channel. This process is aided by fracturing and weathering that loosen the blocks sufficiently so that they are movable during times of high flow rates. Quarrying is mainly a result of the impact forces exerted by flowing water.

Abrasion The process by which the bed and banks of a bedrock channel are bombarded by particles carried by the flow is termed **abrasion**. The individual sediment grains are also abraded by their many impacts with the channel and with one another. Thus, by scraping, rubbing, and bumping, abrasion erodes a bedrock channel and simultaneously smooths and rounds the abrading particles. That is why smooth, rounded cobbles and pebbles

Abrasion by long-lived whirlpools armed with sand and pebbles creates bowl-shaped bedrock depressions called potholes.

▲ Figure 16.16
Potholes The rotational motion of swirling pebbles acts like a drill to create potholes.

are found in streams. Abrasion also results in a reduction in the size of the sediments transported by streams.

Features common to some bedrock channels are circular depressions known as **potholes**, which are created by the abrasive action of particles swirling in fast-moving eddies (**Figure 16.16**). The rotational motion of sand and pebbles acts like a drill that bores the holes. As the particles wear down to nothing, they are replaced by new ones that continue to drill the streambed. Eventually, smooth depressions several meters across and deep may result.

Corrosion Bedrock channels formed in soluble rock such as limestone are susceptible to **corrosion**—a process in which rock is gradually dissolved by the flowing water. Corrosion is a type of chemical weathering between the solutions in the water and the mineral matter that makes up the bedrock.

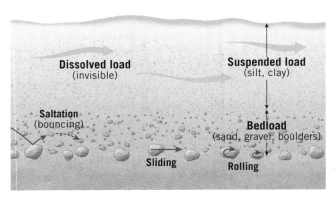

▲ SmartFigure 16.17
Transport of sediment Streams transport their load of sediment in three ways: in solution, in suspension, and by sliding, skipping, or rolling along the bottom. The dissolved and suspended loads are carried in the general flow. The bed load includes coarse sand, gravel, and boulders that move by rolling, sliding, and saltation.

Animation
https://goo.gl/SS3Kj1

◄ Figure 16.18
Suspended load An aerial view of the Colorado River in the Grand Canyon. Heavy rains washed sediment into the river.

Transport of Sediment

All streams, regardless of size, transport some weathered rock material (**Figure 16.17**). Streams also sort the solid sediment they transport because finer, lighter material is carried more readily than larger, heavier particles. Streams transport their load of sediment in three ways: in solution (**dissolved load**), in suspension (**suspended load**), and by sliding, skipping, or rolling along the bottom (**bed load**).

Dissolved Load Most of the dissolved load is brought to a stream by groundwater and is dispersed throughout the flow. When water percolates through the ground, it acquires soluble soil compounds. Then it seeps through cracks and pores in bedrock, dissolving additional mineral matter. Eventually much of this mineral-rich water finds its way into streams.

The velocity of streamflow has essentially no effect on a stream's ability to carry its dissolved load; material in the solution goes wherever the stream goes. Precipitation of the dissolved mineral matter occurs when the chemistry of the water changes, when organisms create hard parts, or when the water enters an inland "sea" located in an arid climate where the rate of evaporation is high.

Suspended Load Most streams carry the largest part of their load in *suspension*. Indeed, the muddy appearance created by suspended sediment is the most obvious portion of a stream's load (**Figure 16.18**).

Usually only very fine sand, silt, and clay particles are carried this way, but during flooding, larger particles are transported as well. During a flood, the total quantity of material carried in suspension also increases dramatically, as people whose homes have been sites for the deposition of this material can attest.

The type and amount of material carried in suspension are controlled by the flow velocity and the settling velocity of each sediment grain. **Settling velocity** is defined as the speed at which a particle falls through a still fluid. The larger the particle, the more rapidly it settles toward the streambed (**Figure 16.19**). In addition to size, the shape and specific gravity of particles also influence settling velocity. Flat grains sink through water more slowly than spherical grains, and dense particles fall toward the bottom more rapidly than less-dense particles. As long as flow velocity exceeds settling velocity, sediment remains suspended and is transported downstream. Deposition occurs when flow velocity falls below the settling velocity of a particle.

Bed Load A portion of a stream's solid-material load is sediment that is too large to be carried in suspension.

► **Figure 16.19**
Settling velocity The speed at which a particle falls through still water is its settling velocity. Deposition of suspended particles occurs when their settling velocities are greater than the stream's flow velocity.

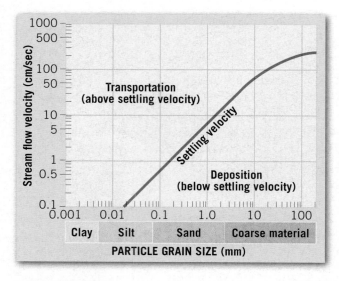

These coarser particles move along the bottom (bed) of the stream and constitute the bed load. Bed load is a key factor in the erosional work accomplished by a downcutting stream.

The particles that make up the bed load move by rolling, sliding, and **saltation**, which refers to irregular skipping or jumping motion along the streambed (see Figure 16.18). This occurs as particles are propelled upward by collisions or lifted by the current and then carried downstream a short distance until gravity pulls them back to the bed of the stream. Particles that are too large or heavy to move by saltation either roll or slide along the bottom, depending on their shapes.

Compared with the movement of suspended load, the movement of bed load through a stream network tends to be less rapid and more localized. A study conducted on a glacially fed river in Norway determined that suspended sediments took only a day to exit the drainage basin, while the bed load required several decades to travel the same distance. Depending on the discharge and slope of the channel, coarse gravels may only be moved during times of high flow, while boulders are moved only during exceptional floods. Once large particles are set in motion, they are usually carried short distances. Along some stretches of a stream, bed load cannot be carried at all until it is broken into smaller particles.

Capacity and Competence A stream's ability to carry solid particles is described using two criteria. **Capacity** is the maximum load of solid particles a stream can transport per unit time. The greater the discharge, the greater the stream's capacity for hauling sediment. Consequently, large rivers with high flow velocities have large capacities.

Competence is a measure of a stream's ability to transport particles based on size. Flow velocity is the key: A swift stream has greater competence than a slow stream, regardless of channel size. A stream's competence increases proportionately to the square of its velocity. Thus, if the velocity of a stream doubles, the impact force of the water increases four times; if the velocity triples, the force increases nine times, and so forth. Hence, large boulders that are often visible in low water and seem immovable can, in fact, be transported during exceptional floods because of the stream's increased competence.

By now it should be clear why the greatest erosion and transportation of sediment occur during floods. The increase in discharge results in greater capacity; the increased velocity produces greater competence. Rising velocity makes the water more turbulent, and larger particles are set in motion. In just a few days, or perhaps a few hours, a stream at flood stage can erode and transport more sediment than it does during several months of normal flow.

Deposition of Sediment

Deposition occurs whenever a stream slows, causing a reduction in competence. Put another way, particles are deposited when the flow velocity is less than the settling velocity. As a stream's flow velocity decreases, sediment begins to settle, largest particles first. In this manner, stream transport provides a mechanism by which solid particles of various sizes are separated. This process, called **sorting**, explains why particles of similar size are deposited together. For example, stretches of a riverbed may consist mainly of gravel or boulders, while sandbars may dominate another part of the stream.

The general term for sediment deposited by streams is **alluvium**. Many different depositional features are composed of alluvium. Some occur within stream channels, some occur on the valley floor adjacent to a channel, and some are found at the mouth of a stream. We will consider the nature of these features later in the chapter.

CONCEPT CHECKS 16.4

1. Describe two processes by which streams cut channels in bedrock.

2. In what three ways does a stream transport its load? Which part of the load moves most slowly?

3. What is settling velocity? What factors influence settling velocity? Does settling velocity affect the dissolved load?

🔊 **Concept Checker**
https://goo.gl/X7VrvF

16.5 Stream Channels

Compare and contrast bedrock and alluvial stream channels. Distinguish between two types of alluvial channels.

A basic characteristic of streamflow that distinguishes it from sheet flow is that streamflow is confined to a channel. A stream channel can be thought of as an open conduit that consists of the streambed and banks that act to confine the flow, except during floods. Although this is somewhat oversimplified, we can divide stream channels into two types: A bedrock channel is one in which the stream is actively cutting into solid rock. In contrast, when the bed and banks are composed mainly of unconsolidated sediment, the channel is called an alluvial channel.

Bedrock Channels

As the name suggests, **bedrock channels** are cut into the underlying strata and typically form in the headwaters of river systems where streams have steep slopes. The energetic flow tends to transport coarse particles that actively abrade the bedrock channel. Potholes are often visible evidence of the erosional forces at work.

Steep bedrock channels often develop a sequence of steps and pools. *Steps* are steep segments where bedrock is exposed. These steep areas contain rapids or, occasionally, waterfalls. *Pools* are relatively flat segments where alluvium tends to accumulate.

The channel pattern exhibited by streams cutting into bedrock is controlled by the underlying geologic structure. Even when flowing over rather uniform bedrock, streams tend to exhibit winding or irregular patterns rather than flow in straight channels. Anyone who has gone whitewater rafting has observed the steep, winding nature of a stream flowing in a bedrock channel.

Alluvial Channels

Many stream channels are composed of loosely consolidated sediment (alluvium) and therefore can undergo significant changes in shape because the sediments are continually being eroded, transported, and redeposited. The major factors affecting the shapes of these channels are the average size of the sediment being transported, the channel gradient, and the discharge.

Alluvial channel patterns reflect a stream's ability to transport its load at a uniform rate while expending the least amount of energy. Thus, the size and type of sediment being carried help determine the nature of the stream channel. Two common types of alluvial channels are *meandering channels* and *braided channels*.

Meandering Channels Streams that transport much of their load in suspension generally move in sweeping bends called **meanders**. These streams flow in relatively deep, smooth channels and primarily transport mud (silt and clay), sand, and occasionally fine gravel. The Lower Mississippi River provides an example of this type of channel.

Meandering channels evolve over time as individual bends migrate across the floodplain. Most of the erosion is focused at the outside of the meander, where velocity and turbulence are greatest. In time, the outside bank is undermined, especially during periods of high water. Because the outside of a meander is a zone of active erosion, it is often referred to as the **cut bank** (Figure 16.20). Debris acquired by the stream at the cut

▼ SmartFigure 16.20
Formation of cut banks and point bars By eroding its outer bank and depositing material on the inside of the bend, a stream is able to shift its channel.

Tutorial
https://goo.gl/htpmDS

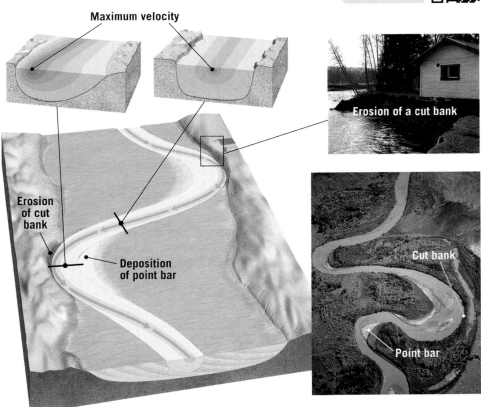

Maximum velocity

Erosion of a cut bank

Erosion of cut bank

Deposition of point bar

Cut bank

Point bar

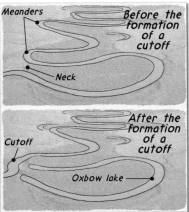

Geologist's Sketch

▲ SmartFigure 16.21
Formation of an oxbow lake Oxbow lakes occupy abandoned meanders. Aerial view of an oxbow lake created by the meandering Green River near Bronx, Wyoming.

Animation
https://goo.gl/ytN4qX

▲ Figure 16.22
Braided stream The Knik River is a classic braided stream with multiple channels separated by migrating gravel bars. The Knik is choked with sediment from four melting glaciers in the Chugach Mountains north of Anchorage, Alaska.

bank moves downstream, where the coarser material is generally deposited as **point bars** on the insides of bends. In this manner, meanders migrate laterally by eroding the outside of the bends and depositing sediment on the inside, moving sideways without appreciably changing their shape.

In addition to migrating laterally, the bends in a channel also migrate down the valley. This occurs because erosion is more effective on the downstream (downslope) side of the meander. Sometimes the downstream migration of a meander slows when it reaches a more resistant bank material. This allows the next meander upstream to gradually erode the material between the two meanders, as shown in **Figure 16.21**. Eventually, the river may erode through the narrow neck of land, forming a new, shorter channel segment called a **cutoff**. Because of its shape, the abandoned bend is called an **oxbow lake**.

Braided Channels   Some streams consist of a complex network of converging and diverging channels that thread their way among numerous islands or gravel bars (**Figure 16.22**). These interwoven channels, called **braided channels**, form where a large portion of a stream's sediment load consists of coarse material (sand and gravel) and the stream has a highly variable discharge. Because the bank material is readily erodible, braided channels are wide and shallow.

One setting in which braided streams form is at the end of glaciers, where there is large seasonal variation in discharge. During the summer, large amounts of ice-eroded sediment drop into the meltwater streams flowing away from the glacier. However, when flow is

sluggish, the stream deposits the coarsest material as elongated structures called *bars*. This process causes the flow to split into several paths around the bars. During the next period of high flow, the laterally shifting channels erode and redeposit much of this coarse sediment, thereby transforming the entire streambed. In some braided streams, the bars become semipermanent islands anchored by vegetation.

CONCEPT CHECKS 16.5

1. Are bedrock channels more likely to be found near the head or the mouth of a stream?

2. Describe or sketch the evolution of a meander, including how an oxbow lake forms.

3. Describe a situation that might cause a stream channel to become braided.

Concept Checker
https://goo.gl/ZJm5yr

16.6 Shaping Stream Valleys

Describe valleys created by streams, including V-shaped, broad valleys with floodplains, and valleys that display incised meanders or stream terraces.

A **stream valley** consists of a channel and the surrounding terrain that directs water into a stream. It includes the *valley floor*, the lower, flatter area partially or totally occupied by the stream channel, and the *valley walls* that rise above the valley floor on both sides. Alluvial channels often flow in valleys that have wide valley floors consisting of sand and gravel deposited in the channel and clay and silt deposited by floods. Bedrock channels, on the other hand, tend to be located in narrow V-shaped valleys. In some arid regions, where weathering is slow and rock is particularly resistant, narrow valleys develop with nearly vertical walls. Such features are called *slot canyons* (**Figure 16.23**). Stream valleys exist on a continuum from narrow, steep-sided valleys to valleys that are so flat and wide that the valley walls are not discernible. Streams, with the aid of weathering and mass wasting, shape the landscape through which they flow. As a result, streams continuously modify the valleys they occupy.

Base Level and Graded Streams

In 1875 John Wesley Powell, the pioneering geologist who first explored the Grand Canyon and later headed the U.S. Geological Survey, introduced the concept of a downward limit to stream erosion, which he called **base level**. A fundamental concept in the study of stream activity, base level is defined as the lowest elevation to which a stream can erode its channel. Essentially, it is the level at which the mouth of a stream enters the ocean, a lake, or a trunk stream. Powell determined that two types of base level exist: "We may consider the level of the sea to be a grand base level, below which the dry lands cannot be eroded; but we may also have, for local and temporary purposes, other base levels of erosion."°

Sea level, which Powell called "grand base level," is now referred to as **ultimate base level**. **Local (or temporary) base levels** include lakes, resistant layers of rock, and rivers that act as base levels for their tributaries. All limit a stream's ability to downcut its channel.

Changes in base level cause corresponding adjustments in the work that streams perform. When a dam is built along the course of a stream, the reservoir that forms behind it raises the base level of the stream (**Figure 16.24**). Upstream from the dam, the stream channel is flooded with still water, lowering the stream's velocity and, hence,

°Exploration of the Colorado River of the West (Washington, DC: Smithsonian Institution, 1875), p. 203.

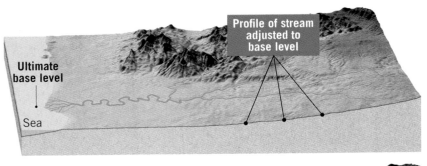

▼ **Figure 16.24**
Building a dam The base level upstream from the reservoir is raised, which reduces the stream's flow velocity and leads to deposition and a reduced gradient.

▲ **Figure 16.23**
Halls Creek Narrows This is a classic slot canyon in Utah's Capitol Reef National Park.

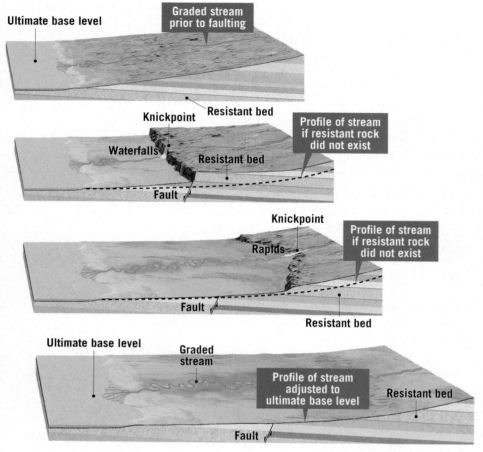

▲ Figure 16.25
Changes in base level A resistant layer of rock can act as a temporary base level. The stream concentrates its erosive energy on the resistant rock at the knickpoint. Eventually the river eliminates the knickpoint and reestablishes a smooth profile.

change in one characteristic causes a change in the others to counteract the effect.

Consider what happens if displacement along a fault raises a layer of resistant rock along the course of a graded stream. As shown in **Figure 16.25**, the resistant rock forms a waterfall and serves as a temporary base level for the stream. Because of the increased gradient, the stream concentrates its erosive energy on the resistant rock along an area called a *knickpoint*. Eventually, the river erases the knickpoint from its path and reestablishes a smooth profile.

Valley Deepening

When a stream's gradient is steep and the channel is well above base level, **downcutting** is the dominant activity. The streambed slowly deepens due to abrasion caused by bed load sliding and rolling along the bottom combined with the hydraulic power of fast-moving water. The result is usually a V-shaped valley with steep sides. A classic example of a V-shaped valley is the section of the Yellowstone River shown in **Figure 16.26**.

The most prominent features of V-shaped valleys are *rapids* and *waterfalls*. Both occur where the stream's gradient increases significantly, a situation usually caused by variations in the erodibility of the bedrock into which a stream channel is cutting. Resistant beds create rapids by acting as a temporary base level upstream while allowing downcutting to continue downstream. Recall that, over time, erosion usually eliminates the resistant rock.

its sediment-transporting ability. As a result, the stream deposits material, thereby building up its channel.

If, on the other hand, base level is lowered by a drop in sea level, the stream will downcut its channel to establish a balance with its new base level. Erosion first occurs near the mouth and then progresses upstream, creating a new stream profile.

Observing streams that adjust their profiles to changes in base level led to the concept of a graded stream. A **graded stream** has the necessary slope and other channel characteristics to maintain the minimum velocity required to transport the material supplied to it. On average, a graded system is neither eroding nor depositing material but simply transporting it. When a stream reaches equilibrium, it becomes a self-regulating system in which a

▲ Figure 16.26
Yellowstone River The V-shaped valley, rapids, and waterfalls indicate that the river is vigorously downcutting.

Geologist's Sketch

▶ **Figure 16.27**
The retreat of Niagara Falls The force of the plunging Niagara River is causing the falls to retreat.

As the river plunges over the falls, it erodes the weaker rocks beneath the more resistant Lockport Dolostone. As a section of dolostone is undercut, it loses support and breaks off.

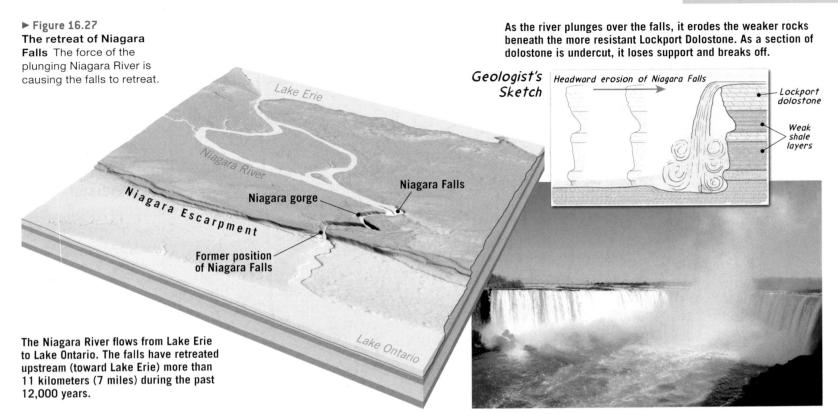

The Niagara River flows from Lake Erie to Lake Ontario. The falls have retreated upstream (toward Lake Erie) more than 11 kilometers (7 miles) during the past 12,000 years.

Waterfalls occur where streams make vertical drops. One type of waterfall is exemplified by Niagara Falls (**Figure 16.27**). These famous falls are supported by a resistant bed of dolostone that is underlain by less-resistant shale. As the water plunges over the lip of the falls, it erodes the less-resistant shale, undermining a section of overlying rock, which eventually breaks off. In this manner, the waterfall retains its vertical cliff while slowly, but continually, retreating upstream. Over the past 12,000 years, Niagara Falls has retreated upstream more than 11 kilometers (7 miles).

Valley Widening

As a stream approaches a graded condition—in which it primarily works to transport sediment—downcutting becomes less dominant. At this point, the stream's channel takes on a meandering pattern, and more of its energy is directed from side to side. As a result, the valley widens as the river cuts away at one bank and then the other. The continuous lateral erosion caused by shifting meanders gradually produces a broad, flat valley floor covered with alluvium (**Figure 16.28**). This feature is called a **floodplain** because it becomes inundated when a river overflows its banks during flood stage. Over time the floodplain will widen to a point where the stream is actively eroding the valley walls in

only a few places. In the case of the Lower Mississippi River, for example, the distance from one valley wall to another sometimes exceeds 160 kilometers (100 miles).

When a river erodes laterally and creates a floodplain as described, it is called an *erosional floodplain*. Floodplains can be depositional as well. *Depositional floodplains* are produced by major fluctuations in conditions, such as changes in base level or climate. The floodplain in California's Yosemite Valley is one such feature; it was produced when a glacier gouged the valley floor about 300 meters (1000 feet) deeper than its former level. After the

▶ **SmartFigure 16.28**
Development of an erosional floodplain Continuous side-to-side erosion by shifting meanders gradually produces a broad, flat valley floor. Alluvium deposited during floods covers the valley floor.

Condor Video
https://goo.gl/hKofCs

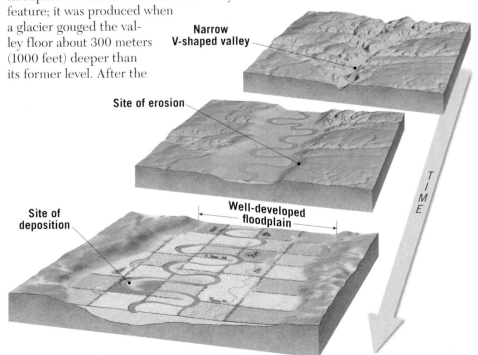

▲ SmartFigure 16.29
Incised meanders Aerial view of incised meanders of the Colorado River on the Colorado Plateau.

Before uplift of the Colorado Plateau, the river was meandering on a floodplain.

During uplift of the plateau, the meanders downcut because of the steepening gradient.

Tutorial
https://goo.gl/dv1tYs

glacial ice melted, running water refilled the valley with alluvium. The Merced River currently winds across a relatively flat floodplain that forms much of the floor of Yosemite Valley.

Changing Base Level: Incised Meanders and Stream Terraces

We usually find streams with highly meandering courses on floodplains in wide valleys. However, some rivers have meandering channels that flow in steep, narrow bedrock valleys. Such meanders, pictured in **Figure 16.29**, are called **incised meanders** (*incisum* = to cut into).

How do these features form? Originally the meanders probably developed on the floodplain of a stream that was in balance with

▼ SmartFigure 16.30
Stream terraces Terraces result when a stream adjusts to a relative drop in base level.

Condor Video
https://goo.gl/7soXyf

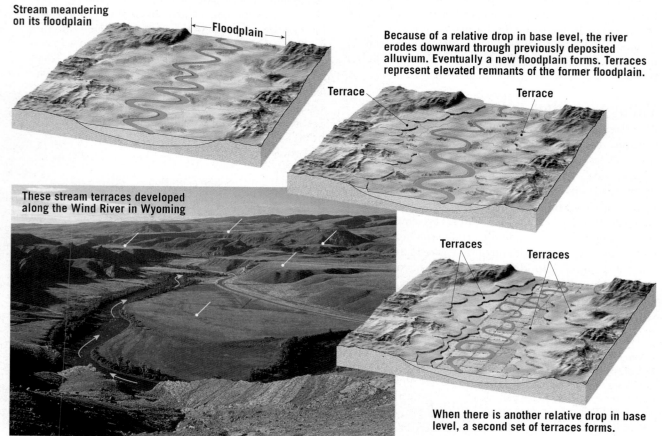

Stream meandering on its floodplain

←— Floodplain —→

Because of a relative drop in base level, the river erodes downward through previously deposited alluvium. Eventually a new floodplain forms. Terraces represent elevated remnants of the former floodplain.

Terrace

Terrace

These stream terraces developed along the Wind River in Wyoming

Terraces

Terraces

When there is another relative drop in base level, a second set of terraces forms.

its base level. Then a change in base level caused the stream to begin downcutting. Such a change can be caused either by a drop in a downstream base level or by uplift of the land on which the river is flowing. For example, regional uplifting of the Colorado Plateau in the southwestern United States generated incised meanders on several rivers. As the plateau gradually rose, meandering rivers began downcutting because of their steepening gradient.

Other features associated with a relative drop in base level are stream **terraces**. After a river that had been flowing on a floodplain has adjusted to a relative drop in base level, it may once again produce a floodplain at a level below the old one. As shown in **Figure 16.30**, the remnants of the former floodplain are present as relatively flat surfaces above the newly forming floodplain.

CONCEPT CHECKS 16.6

1. Define *base level* and distinguish between ultimate base level and local, or temporary, base level.

2. Explain why V-shaped valleys often contain rapids and/or waterfalls.

Concept Checker
https://goo.gl/UmRZSa

3. List two situations that would trigger the formation of incised meanders.

16.7 Depositional Landforms

List the major depositional landforms associated with streams and describe the formation of these features.

Recall that a stream continually picks up sediment in one part of its channel and deposits it downstream. These small-scale channel deposits, which are most often composed of sand and gravel, are called **bars**. Such features, however, are only temporary, as the material is picked up again and eventually carried to the ocean. In addition to sand and gravel bars, streams also create other depositional features that have somewhat longer life spans. These include *deltas*, *natural levees*, and *alluvial fans*.

Deltas

A **delta** forms where sediment-laden streams enters the relatively still waters of a lake, an inland sea, or the ocean (**Figure 16.31**). As the stream's forward motion slows, sediments are deposited by the dying current, producing three types of beds. *Foreset beds* are composed of coarse particles from the bed load deposited almost immediately upon entering the water to form layers that slope downcurrent from the delta front. The foreset beds are usually covered by thin, horizontal *topset beds* deposited during flood stage. The finer silts and clays in the stream's suspended load settle away from the mouth in nearly horizontal layers called *bottomset beds*.

As a delta grows outward from the shoreline, the stream's gradient continually decreases. This circumstance eventually causes the channel to become choked with sediment. As a consequence, the river seeks shorter, higher-gradient routes to base level. Figure 16.31 shows the main channel dividing into several smaller ones, called **distributaries**, that carry water away from the main channel in varying paths to base level. After numerous shifts in the main flow from one distributary to the next, a delta may grow into the triangular shape of the Greek letter delta (Δ), although several other shapes exist. Differences in the configurations of shorelines and variations in the nature and strength of wave activity are responsible for the shape and structure of each delta. Many of the world's great rivers have created massive deltas, each with its own peculiarities and typically more complex than the one illustrated in Figure 16.31.

Not all rivers have deltas. Even a river that transports a large sediment load may lack a delta if ocean waves and powerful currents quickly redistribute the material soon after it is deposited. The Columbia River in the Pacific Northwest is an example. In other cases, rivers do not carry sufficient quantities of sediment to build a delta. The St. Lawrence River, for example, has little opportunity to pick up much sediment between its head in Lake Ontario and its mouth in the Gulf of St. Lawrence.

▼ **Figure 16.31**
Formation of a simple delta Structure and growth of a simple delta that forms in relatively quiet waters.

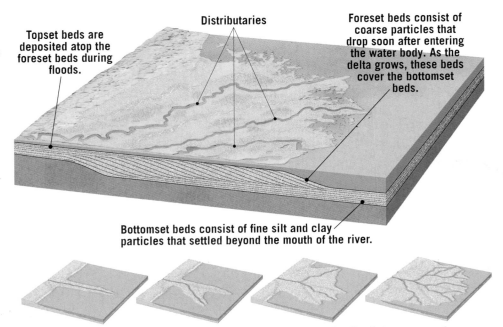

Topset beds are deposited atop the foreset beds during floods.

Distributaries

Foreset beds consist of coarse particles that drop soon after entering the water body. As the delta grows, these beds cover the bottomset beds.

Bottomset beds consist of fine silt and clay particles that settled beyond the mouth of the river.

As the stream extends its channel, the gradient is reduced. During flood stage some of the flow is diverted to a shorter, higher-gradient route forming a new distributary.

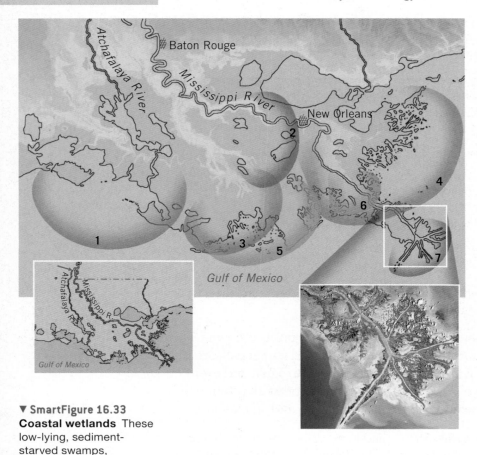

◀ **Figure 16.32**

Growth of the Mississippi River delta During the past 6000 years, the river has built a series of seven coalescing subdeltas. The numbers indicate the order in which the subdeltas were deposited. The present bird-foot delta (number 7) represents the activity of the past 500 years. The left inset shows the point where the Mississippi may sometime break through (arrow) and the shorter path it would take to the Gulf of Mexico.

▼ **SmartFigure 16.33**

Coastal wetlands These low-lying, sediment-starved swamps, marshes, and bayous are disappearing at an alarming rate.

Mobile Field Trip
https://goo.gl/tC8x5s

The Mississippi River Delta

The Mississippi River delta resulted from the accumulation of huge quantities of sediment derived from the vast region drained by the river and its tributaries. New Orleans currently rests where there was once ocean.

History and Structure The portion of the Mississippi delta that has formed during the past 6000 years is shown in **Figure 16.32**. As the figure illustrates, the delta is actually a series of seven coalescing subdeltas. Each subdelta formed when the main flow was diverted from one channel to a shorter, more direct path to the Gulf of Mexico. The individual subdeltas intertwine and partially cover each other, resulting in a complex structure. It is also apparent from Figure 16.32 that after each channel was abandoned, coastal erosion modified the newly formed subdelta. The present subdelta (number 7 in Figure 16.32), called a *bird-foot* delta because of the configuration of its distributaries, has been built by the Mississippi River over the past 500 years.

At present, this active bird-foot delta has extended seaward almost as far as natural forces will allow. In fact, for many years the river has been attempting to cut through a narrow neck of land and shift its course to the Atchafalaya River (see inset in Figure 16.32). If this were to happen, the Mississippi would abandon the lowermost 500 kilometers (300 miles) of its channel in favor of the Atchafalaya's much shorter 225-kilometer (140-mile) route to the Gulf.

In a concerted effort to keep the Mississippi on its present course, and out of adjacent farms and towns, a dam-like structure was erected at the site where the channel was trying to break through. A flood in 1973 weakened the control structure, and the river again threatened to shift. This event caused the U.S. Army Corps of Engineers to construct a massive auxiliary dam that was completed in the mid-1980s. For the time being, at least, the inevitable has been avoided, and the Mississippi River continues to flow past Baton Rouge and New Orleans on its way to the Gulf of Mexico.

Vanishing Wetlands The delta of the Mississippi River in Louisiana is a biologically significant region that includes about 12,000 square kilometers (3 million acres) of coastal wetlands—40 percent of all coastal wetlands in the contiguous United States (**Figure 16.33**).

These flat areas, only slightly above sea level, are sheltered from the wave action of hurricanes and winter storms by low-lying offshore barrier islands. Both the wetlands and offshore islands were formed and maintained by sediments carried to the Gulf of Mexico by the Mississippi River.

The coastal wetlands of Louisiana are disappearing at an alarming rate—accounting for 80 percent of the wetland loss in the lower 48 states. According to the U.S. Geological Survey, Louisiana has lost more than 5000 square kilometers (1900 square miles) of coastal land since the early 1930s. If wetland loss continues at this rate, another 3000 square kilometers (1170 square miles) will vanish by the year 2050.

Before Europeans settled the delta, the Mississippi River regularly overflowed its banks in seasonal floods. Huge quantities of sediment were deposited atop the delta, which kept the land elevated above sea level. However, with settlement came flood-control efforts and the desire to maintain and improve navigation on the river. Artificial levees (earthen mounds built parallel to the river) were constructed to contain the rising river during flood stage. Over time the levees were extended all the way to the mouth of the Mississippi to keep the channel open to navigation. The levees prevent sediment from being dispersed onto the wetlands. As a result, sediment is not added in sufficient quantities to offset compaction, subsidence, and wave erosion. Thus, the size of the delta and the extent of the wetlands are shrinking. The problem has been exacerbated by a decline in the sediment load of the Mississippi, which has decreased approximately 50 percent over the past 100 years. A substantial portion of the reduction is due to sediment being trapped upstream by dams on the river's many tributaries.

Natural Levees

Some rivers occupy valleys with broad floodplains and build **natural levees** that parallel their channels on both banks (**Figure 16.34**). Natural levees are built by years of successive floods. When a stream overflows onto the floodplain, the water flows over the surface as a broad sheet. Because the flow velocity drops significantly, the coarser portion of the suspended load is immediately deposited adjacent to the channel. As the water spreads across the floodplain, a thin layer of fine sediment is laid down over the valley floor. This uneven distribution of material produces the very gentle slope of the natural levee (see Figure 16.34).

The natural levees of the Lower Mississippi River rise 6 meters (20 feet) above the adjacent valley floor. The area behind a levee is characteristically poorly drained for the obvious reason that water cannot flow up the levee and into the river. Marshes called **back swamps** result. When a tributary stream enters a river valley that has a substantial natural levee, it often flows for many kilometers through the back swamp before finding an opening where it reenters the main river. Such streams are called **yazoo tributaries**, after the Yazoo River, which parallels the Lower Mississippi River for more than 300 kilometers (190 miles).

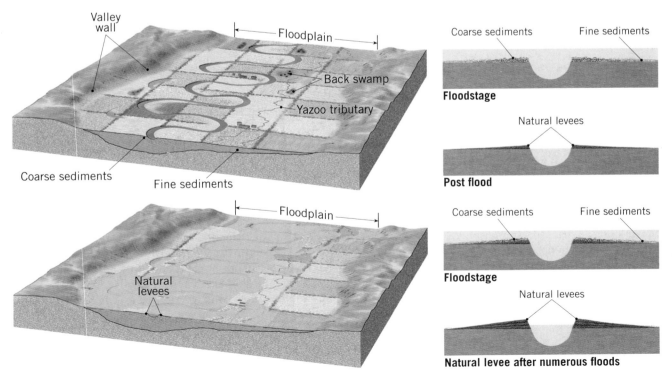

◄ SmartFigure 16.34
Formation of a natural levee These gently sloping structures that parallel a river channel are created by repeated floods. Because the ground next to the channel is higher than the adjacent floodplain, back swamps and yazoo tributaries may develop.

Animation
https://goo.gl/cNnWb3

▶ SmartFigure 16.35
Alluvial fan in Death Valley Fans are deposited at the mouth of a valley that emerges from a mountainous or upland area onto a relatively flat lowland. Usually coarse material is dropped near the apex of the fan, while finer materials are carried toward the base of the deposit. California's Death Valley has many large fans. As adjacent fans grow larger, they may coalesce to form a steep apron of sediment called a *bajada*.

Condor Video
https://goo.gl/Jun3Nd

Geologist's Sketch

Dry channels

Dry stream channel

Alluvial fan

Road

Death Valley

Alluvial Fans

Alluvial fans typically develop where a high-gradient stream leaves a narrow valley in mountainous terrain and comes out suddenly onto a broad, flat plain or valley floor (**Figure 16.35**). Alluvial fans form in response to the abrupt drop in gradient combined with the change from a narrow channel of a mountain stream to less confined channels at the base of the mountains. The sudden drop in velocity causes the stream to dump its load of sediment quickly in a distinctive cone- or fan-shaped accumulation. The surface of the fan slopes outward in a broad arc from an apex at the mouth of the steep valley. Usually, coarser material is dropped near the apex of the fan, while finer material is carried toward the base of the deposit.

Between rainy periods in deserts, little or no water flows across an alluvial fan, which is evident in the many dry channels that cross its surface. Thus, fans in dry regions grow intermittently, receiving considerable water and sediment only during wet periods. Because steep canyons in dry regions are prime locations for debris flows, many alluvial fans have debris flow deposits interbedded with the coarse alluvium.

CONCEPT CHECKS 16.7

1. Sketch a cross section of a simple delta and distinguish among the three types of beds that compose it.

2. Briefly describe the formation of a natural levee. How is this feature related to back swamps and yazoo tributaries?

3. Describe the formation of an alluvial fan.

Concept Checker
https://goo.gl/MvXqGW

16.8 Floods and Flood Control

Summarize the various categories of floods and the common measures of flood control.

A **flood** occur when the flow of a stream becomes so great that it exceeds the capacity of its channel and overflows its banks. Although floods are natural phenomena, the magnitude and frequency of flooding is often significantly influenced by human activities such as clearing forests, building cities, and constructing flood-control structures such as dams and levees. The occurrence of floods is often linked to other natural hazards, including severe storms and mass movement processes such as debris flows. For humans, floods are among the most disruptive, common, and costly natural hazards.

Types of Floods

Most floods are caused by atmospheric processes that can vary greatly in both time and space. An hour or less of intense thunderstorm rainfall can trigger flash floods in small valleys. By contrast, major floods in large river valleys often result from an extraordinary series of precipitation or snowmelt events over a broad region for many days or weeks.

Regional Floods   Most *regional floods* are seasonal. Rapid melting of snow in spring and/or heavy spring rain can overwhelm rivers. For example, the extensive 1997 flood along the Red River of the North, which forms the border between Minnesota and North Dakota, was preceded by an especially snowy winter and an early spring blizzard. Early April brought rapidly rising temperatures, melting the snow in a matter of days and causing a record-breaking 500-year flood. Roughly 4.5 million acres (18,000 square kilometers) were underwater, and the losses in the Grand Forks, North Dakota, region exceeded $3.5 billion.†

◄ **Figure 16.36**
The flooding Mississippi River, 2011 Extraordinary rains caused record floods from Illinois to Louisiana. This scene is from Vicksburg, Mississippi.

In April 2011, unrelenting storms brought record rains to the Mississippi watershed. The Ohio Valley, which makes up the eastern portion of the Mississippi's drainage basin, received nearly 300 percent of its normal springtime precipitation. When that rainfall combined with water from the rapidly melting large and extensive snowpack from the past winter, the Mississippi River and many of its tributaries began to swell to record levels by early May. The resulting floods were among the largest and most damaging in nearly a century (**Figure 16.36**). Like most other regional floods, these were associated with weather phenomena that could be forecast with a good deal of accuracy. Although economic losses approached $4 billion, loss of life was small because there was adequate time to warn and evacuate thousands of people who were in harm's way.

Flash Floods   Flash floods often occur with little warning and are potentially deadly because they produce rapid rises in water levels and can have devastating flow velocities (see **GEOgraphics 16.1**). Rainfall intensity and duration, surface conditions, and topography are among the factors that influence flash flooding. Mountainous areas are susceptible because steep slopes can quickly funnel runoff into narrow canyons.

A recent episode of flash flooding occurred in May 2018, when a storm dumped almost 10 inches of rain in 2 hours near Ellicott City, Maryland (**Figure 16.37**).

Graph showing river stages for Patapsco River near Ellicott City, Maryland.

The stream rose about 17 feet in less than a day and then subsided quickly.

▲ **Figure 16.37**
Flash flood in Ellicott City, Maryland An early summer storm in May 2018 brought torrential rains to this area. Ellicott City has experienced two historical floods in two years' time.

Flash floods

Flash floods are local floods of great volume and short duration. The rapidly rising surge of water usually occurs with little advance warning and can destroy roads, bridges, homes, and other substantial structures.

The power of a flash flood is illustrated by the Big Thompson River flood of July 31, 1976, in Colorado. During a 4-hour span, more than 30 centimeters (12 inches) of rain fell on portions of the river's small drainage basin. This amounted to nearly three-quarters of the average yearly total. The flash flood in the narrow canyon lasted only a few hours, but cost 139 people their lives.

Urban development increases runoff. As a result, peak discharge and flood frequency increase. A recent study indicated that the area of impervious surfaces in the 48 contiguous United States is roughly equal to the area of the state of Ohio (44,000 mi^2).

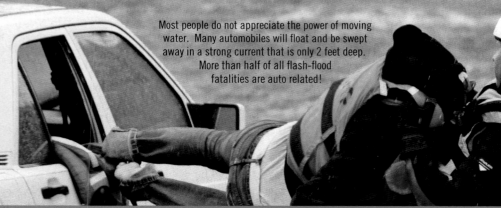

Most people do not appreciate the power of moving water. Many automobiles will float and be swept away in a strong current that is only 2 feet deep. More than half of all flash-flood fatalities are auto related!

Average Annual Storm-Related Deaths in the U.S. (1995–2017)

In most years floods are responsible for the greatest number of storm-related deaths. The average number of hurricane deaths was dramatically affected by Hurricane Katrina in 2005 (more than 1000). For all other years on this graph, hurricane fatalities numbered fewer than 20.

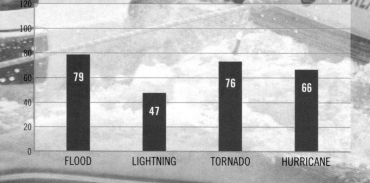

	Deaths
FLOOD	79
LIGHTNING	47
TORNADO	76
HURRICANE	66

Effect of Urban Development on Flooding

Streamflow in Mercer Creek, an urban stream in western Washington, increases more quickly, reaches a higher peak discharge, and has a larger volume during a 1-day storm on February 1, 2000, than streamflow in Newaukum Creek, a nearby rural stream. Streamflow during the following week, however, was greater in Newaukum Creek.

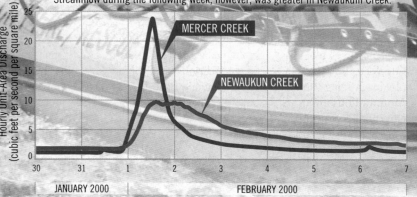

MERCER CREEK

NEWAUKUN CREEK

Hourly Unit-Area Discharge (cubic feet per second per square mile)

JANUARY 2000   FEBRUARY 2000

The resulting flash flood, which raised the level of the Patapsco River 17 feet (5 meters), washed out roads and bridges and severely damaged downtown buildings. This was the second major flooding disaster to hit Ellicott City in 2 years. These formerly rare intense precipitation events are now occurring more frequently. Climate models predict Earth will experience a more vigorous hydrologic cycle as the planet warms.

Urban areas are also susceptible to flash floods because a high percentage of the surface area is composed of impervious roofs, streets, and parking lots, where infiltration is minimal and runoff is rapid.

Ice-Jam Floods   Frozen rivers are especially susceptible to ice-jam floods. As the level of a stream rises, it breaks up ice and creates ice floes that can accumulate on channel obstructions. Jams of this nature create temporary ice dams across the channel. Water trapped upstream can rise rapidly and overflow the channel banks. When an ice dam fails, water behind the dam is often released with sufficient force to inflict considerable damage downstream.

These floods are often associated with northward-flowing rivers in the Northern Hemisphere. The Red River of the North, mentioned earlier in this chapter, is an example. The Siberian region of Russia has several northward-flowing rivers, such as the Ob, Lena, and Yenisei, that frequently experience these floods (**Figure 16.38**). When spring arrives, ice melts in the warmer southern portions of a river and its drainage basin, while farther north the river remains frozen. Water flowing from the south "backs up" behind the frozen northern portion of the river.

Dam-Failure Floods   Human interference with stream systems can cause floods. A prime example is the failure of a dam or an artificial levee designed to contain small or moderate floods. When larger floods occur, the dam

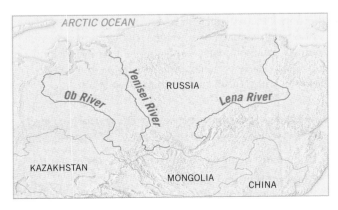

▲ **Figure 16.38**
Northward-flowing Siberian rivers Ice-jam floods are relatively common in late spring and early summer in Siberia because its large rivers flow north into the Arctic Ocean.

or levee may fail, resulting in the water behind it being released as a flash flood. The bursting of a dam in 1889 on the Little Conemaugh River caused the devastating Johnstown, Pennsylvania, flood that took more than 2200 lives.

Flood Recurrence Intervals

Land-use planning in river basins requires an understanding of the frequency and magnitude of floods. For every river, a relationship exists between the size of a flood and the frequency with which it occurs. The larger the flood, the less often it is expected to occur. You may have heard the term *100-year flood*. This describes the *recurrence interval*, which is an estimate of how often a flood of a given size can be expected to occur. A 25-year event would be much smaller than a 100-year flood, but it would be four times more likely to occur.

The term *100-year flood* does not mean that 100 years must pass between each flood of equal or greater magnitude. Rather, it means that there is a 1 percent (1 in 100) probability in a given year for a flood of that size. It is possible that 100-year floods could occur 2 or 3 years in a row. The 2016 and 2018 Ellicott City flood events described in Figure 16.37 were each 1000-year floods, according to historical flood records. It is also possible that such a flood might not occur over a span exceeding 100 years.

In order to make a reasonable calculation, stream gage data must be collected for at least 10 to 30 years. The longer the record, the better the prediction will likely be. Other factors also influence the accuracy of flood recurrence estimates. Climate cycles that involve extensive drought or rainy periods are an example. Land-use changes, such as when adjacent rural areas become urbanized, typically require reevaluation of flood recurrence intervals.

Flood Control

Several strategies have been devised to eliminate or reduce the catastrophic impact of floods on our lives and environment. Engineering efforts include the construction of artificial levees, the building of flood-control dams, and river channelization.

Artificial Levees   *Artificial levees* are earthen mounds built on the banks of a river to increase the volume of water the channel can hold. These most common of stream-containment structures have been used since ancient times. Artificial levees usually have steep slopes that make them easy to distinguish from more gently sloping natural levees. When exceptional floods threaten to overwhelm levees in densely populated areas, water is sometimes intentionally diverted from a river by creating openings in artificial levees. The purpose is to spare vulnerable urban areas by allowing water to flood sparsely populated rural areas. The areas that are intentionally flooded are called *floodways*. For example, to prevent the town of Cairo, Illinois, from being inundated during the 2011 floods along the Mississippi River, a 3-kilometer-wide (2-mile-wide) opening was blasted in a levee. This allowed water to spill into the 130,000-acre Birds Point–New Madrid Floodway. Similar steps were taken downstream in Louisiana to protect the cities of Baton Rouge and New Orleans.

Flood-Control Dams   *Flood-control dams* are built to store floodwater and then let it out slowly. This action lowers the flood crest by spreading it out over a longer time span. Since the 1920s, thousands of dams have been built on nearly every major river in the United States. Many dams have significant non-flood-related functions, such as providing water for irrigated agriculture and for hydroelectric power generation. Many reservoirs are also major regional recreational facilities.

Although dams may reduce flooding and provide other benefits, building these structures also has significant costs and consequences. For example, reservoirs created by dams may cover fertile farmland, forests, historic sites, and scenic valleys. Sedimentation behind a dam gradually diminishes the volume of its reservoir, reducing the long-term effectiveness of this flood-control measure (**Figure 16.39**). In addition, deltas and floodplains downstream erode because they are no longer replenished with silt during floods. Large dams also cause significant damage to river ecosystems that have developed over thousands of years (for example, blocking migrating salmon).

▼ Figure 16.39
Oroville, California, dam damage The Oroville dam is an earthen structure blocking the Feather River in northern California. Thousands of people downstream were evacuated in 2017 after heavy rainfall filled the reservoir and damaged the main and emergency spillways.

Damaged main spillway

Eroded emergency spillway

Channelization   *Channelization* involves altering a stream channel in order to make the flow more efficient. This may simply involve clearing a channel of obstructions or dredging a channel to make it wider and deeper.

Another alteration involves straightening a channel by creating artificial cutoffs. The idea is that by shortening the stream, the gradient and the flow velocity are both increased. By increasing velocity, the larger discharge associated with flooding can be dispersed more rapidly.

Since the early 1930s, the U.S. Army Corps of Engineers has created many artificial cutoffs on the Mississippi for the purpose of increasing the efficiency of the channel and reducing the threat of flooding. In all, the river has been shortened more than 240 kilometers (150 miles). These efforts have been somewhat successful in reducing the height of the river in flood stage. However, channel shortening led to steeper gradients and accelerated erosion of riverbank material, both of which necessitated further intervention. Following the creation of artificial cutoffs, massive riverbank protection to reduce erosion was installed along several stretches of the Lower Mississippi.

A similar case in which artificial cutoffs accelerated bank erosion occurred on the Blackwater River in Missouri, whose meandering course was shortened in 1910. Among the many effects of this project was a significant increase in the channel's width due to increased velocity. One particular bridge over the river collapsed in 1930 because of bank erosion. During the following 17 years, the bridge was replaced three times, each time requiring a longer span.

A Nonstructural Approach   All of the flood-control measures described so far have involved structural solutions aimed at "controlling" a river. These solutions are typically expensive and often give those who reside on the floodplain a false sense of security.

Today, many scientists and engineers advocate a nonstructural approach to flood control. They suggest that an alternative to artificial levees, dams, and channelization is floodplain management. By identifying high-risk areas, zoning regulations can promote more appropriate land use, reduce development, and allow rivers to maintain their dynamic equilibrium with the floodplain.

CONCEPT CHECKS 16.8

1. List and distinguish among four types of floods.

2. Describe three basic flood-control strategies. What are some drawbacks of each?
   https://goo.gl/FDfKTt

3. What is meant by a *nonstructural approach* to flood control?

16

CONCEPTS IN REVIEW

Running Water

16.1 Earth as a System: The Hydrologic Cycle

List the hydrosphere's major reservoirs and describe the different paths that water takes through the hydrologic cycle.

Key Terms:
hydrologic cycle
evaporation

infiltration
runoff
transpiration

evapotranspiration

- Water moves through the hydrosphere's many reservoirs by evaporating, condensing into clouds, and falling as precipitation. Once it reaches the ground surface, rain can either soak into the soil, evaporate, be returned to the atmosphere by plant *transpiration*, or run off the surface.
- Running water may be a small portion of the total water on Earth, but it is the most important agent in sculpting Earth's varied landscapes.

16.2 River Systems

Describe the nature of drainage basins and river systems. Sketch and briefly explain four basic drainage patterns.

Key Terms:
stream
river
drainage basin
divide

headward erosion
dendritic pattern
radial pattern
rectangular pattern
trellis pattern

water gap
antecedent stream
superposed stream

- The land area that contributes water to a *stream* is its *drainage basin*. Drainage basins are separated by imaginary lines called *divides*.
- As a generalization, the upstream portion of a drainage basin is a zone of sediment production, where most of a stream's sediment is derived. Sediment transport characterizes the middle section, and sediment deposition is associated with the downstream end.
- A stream erodes most effectively in a *headward* direction, thereby lengthening its course.
- A *water gap* is a steep-walled notch in a ridge through which a stream flows. Such streams may either be *antecedent* or *superposed*.

Q Identify each of the drainage patterns depicted in the accompanying sketch.

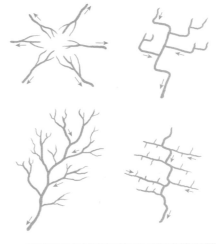

16.3 Streamflow Characteristics

Discuss streamflow and the factors that cause it to change.

Key Terms:
laminar flow
turbulent flow

gradient
wetted perimeter
discharge

longitudinal profile
head
mouth

- The flow of water in a stream may be *laminar* or *turbulent*. A stream's flow velocity is influenced by channel *gradient*; the size, shape, and roughness of the channel; and discharge.

- A cross-sectional view of a stream from *head* to *mouth* is a *longitudinal profile*. Usually the gradient and roughness of the stream channel decrease going downstream, whereas the size of the channel, stream discharge, and flow velocity increase in the downstream direction.

16.4 The Work of Running Water

Outline the ways in which streams erode, transport, and deposit sediment.

Key Terms:
quarrying
abrasion
pothole
corrosion

dissolved load
suspended load
bed load
settling velocity
saltation

capacity
competence
sorting
alluvium

- Streams are powerful agents of erosion that carve solid rock through *quarrying*, *abrasion*, and the focused "drilling" that results in *potholes*. Turbulent water also lifts loose particles from the streambed.

In areas of soluble bedrock such as limestone, stream water can also corrode the landscape by dissolving the rock.

- Sediment is transported in a stream either in solution, suspended in the water, or rolling or saltating along the bottom of the stream. Compared to slow-moving rivers, fast-moving rivers can carry a greater total amount of sediment (*capacity*) and larger individual particles (*competence*). Flooding increases both capacity and competence, which is why rivers do most of their work during short-lived times of peak flow.
- Sediment deposited by streams is called *alluvium*. Typically, streams are efficient agents of *sorting*, meaning that they deposit similarly sized grains in the same area.

16.5 Stream Channels

Compare and contrast bedrock and alluvial stream channels. Distinguish between two types of alluvial channels.

Key Terms:	meander	cutoff
bedrock channel	cut bank	oxbow lake
alluvial channel	point bar	braided channel

- *Bedrock channels* are cut into solid rock. They typically exhibit steps (highlighted by waterfalls or rapids) and pools (segments that are relatively flat).
- *Alluvial channels* are dominated by streamflow moving through sediment previously deposited by the river. A floodplain usually covers much of a valley floor, with the river itself moving through a channel that may meander or exhibit a braided pattern.
- *Meanders* are enhanced through erosion at the *cut bank* (outside edge of the meander) and deposition of sediment on the *point bar* (inside edge of the meander). The shape of the meander may grow more and more exaggerated until it loops back on itself, creating a *cutoff*. Once a cutoff forms, the main current abandons the old meander loop, which becomes an *oxbow lake*.
- *Braided channels* occur in streams that experience highly variable discharge. During times of low flow, the river moves in an interwoven network of channels between bars of coarse-grained alluvium.

Q The town of Carter Lake is the *only* portion of the state of Iowa that lies on the west side of the Missouri River. It is bounded on the north by its namesake, Carter Lake, on the south by the Missouri River, and on the east and west by Nebraska. After examining the map, prepare a hypothesis that explains how this unusual situation could have developed.

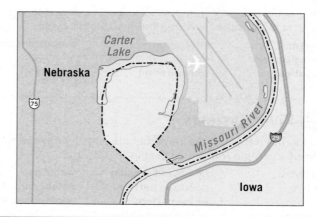

16.6 Shaping Stream Valleys

Describe valleys created by streams, including V-shaped, broad valleys with floodplains, and valleys that display incised meanders or stream terraces.

Key Terms:	local base level	incised meander
stream valley	graded stream	terrace
base level	downcutting	
ultimate base level	floodplain	

- A *stream valley* includes the channel itself, the adjacent floodplain, and relatively steep valley walls. Stream valley width and shape vary a lot from bedrock channels (which tend to be V-shaped, with narrow floodplains) to alluvial channels, which may have broad floodplains of alluvium and relatively subdued valley walls.
- Streams erode downward until they approach *base level*, which is usually the level at which the stream enters another stream, a lake, or the ocean. A river flowing toward the sea (*ultimate base level*) may encounter several *local base levels* along its route. These could be lakes or resistant rock units that retard the *downcutting* of the stream. A *graded stream* has reached equilibrium with its base level and primarily works to transport sediment.
- Stream valleys are widened through the meandering action of a stream, which erodes the valley walls and widens the *floodplain*. If the

base level drops, the stream *downcuts*. If it is underlain by bedrock, the stream may then develop *incised meanders*. Streams underlain by deep alluvium are likely to develop *terraces*.

Q Meanders are associated with a river that is eroding from side to side, whereas narrow canyons are associated with rivers that are vigorously downcutting. The river in this image is confined to a narrow canyon but is also meandering. Explain.

16.7 Depositional Landforms

List the major depositional landforms associated with streams and describe the formation of these features.

Key Terms:	distributary	yazoo tributary
bar	natural levee	alluvial fan
delta	back swamp	

- A *delta* may form where a river deposits sediment in another water body at its mouth. The partitioning of streamflow into multiple

distributaries spreads sediment in different directions. In the United States, the Mississippi River is an example of a major river with a dynamic delta system.

- *Natural levees* result from sediment deposited along the margins of a river channel by many years of successive floods. Because they slope gently away from the channel, the surrounding land is poorly drained, resulting in *back swamps*.
- *Alluvial fans* are fan-shaped deposits of alluvium that form where steep mountain fronts drop down into adjacent valleys.

16.8 Floods and Flood Control

Summarize various categories of floods and the common measures of flood control.

Key Term: flood

- When a stream receives more discharge than its channel can hold, a *flood* occurs. Four major factors that can cause a flood are large amounts of precipitation (or melting) across a region, sudden pulses of precipitation in areas of high runoff, temporary dams made of floating ice (which then break up, releasing impounded water), and failure of human-built dams, leading to the sudden escape of water from a reservoir.

- Three main structural strategies exist for coping with floods: construction of artificial levees to constrain streamflow to the channel, alterations to make a stream channel's flow more efficient, and building of dams on a river's tributaries so that a sudden influx of water can be temporarily stored and released slowly to the river system. A nonstructural approach is sound floodplain management. Here, a solid scientific understanding of flood dynamics informs policy and regulation of areas that are subject to flooding.

GIVE IT SOME THOUGHT

1. On the accompanying river system diagram, match each process with one of the three zones: sediment production (erosion), sediment deposition, sediment transportation.

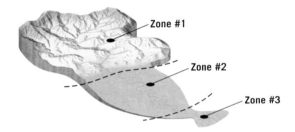

2. If you collect water from a stream in a jar, what part of its load will settle to the bottom, and what portion will remain in the water indefinitely? What part of the stream's load would probably not be represented in your sample?

3. Streamflow is affected by several variables, including channel roughness, size, and shape, as well as discharge and gradient. Develop a scenario in which a mass wasting event influences a stream's flow. Explain what led up to, or triggered, the event and describe how the mass wasting process influenced the stream's flow.

4. Examine the satellite image of central Pennsylvania. Identify the feature that occurs each time the Susquehanna River crosses one of the five mountain ridges and explain how these features likely formed.

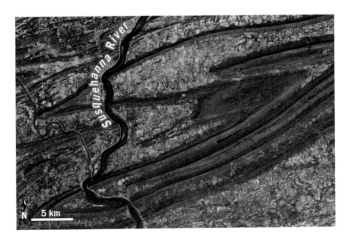

5. Several times during the past 2.5 million years, huge ice sheets (continental-size glaciers) formed and spread across large parts of Northern Hemisphere landmasses and then gradually melted away.

    a. How do you think the formation of ice sheets affected sea level?

    b. How would rivers flowing into the ocean have been affected as the ice sheets expanded?

    c. What kind of adjustments would these rivers make as the glacial ice melted?

6. Describe three ways that a stream can lengthen its course. How might a stream get shorter?

7. One day you and a friend are discussing the aftermath of a major (100-year) flood that occurred on a river in your area just a few months earlier. At the close of your conversation, your friend remarks, "At least we won't have to worry about another one of those in our lifetime." How would you respond?

8. This satellite image shows portions of the Ohio and Wabash Rivers in May 2011. What is the base level for the Wabash River? What is base level for the Ohio River? (*Hint:* Refer to Figure 16.4.) Is either of the base levels you just noted considered *ultimate base level*? Explain.

9. The accompanying diagrams show lag times between rainfall and peak flow (flooding) for an urban area and a rural area. Which graph (A or B) most likely represents the rural area? Explain your choice.

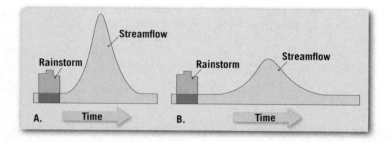

10. Building a dam is one method of regulating the flow of a river to control flooding. Dams and their reservoirs may also provide recreational opportunities and water for irrigation and hydroelectric power generation. This image, from near Page, Arizona, shows Glen Canyon Dam on the Colorado River upstream from the Grand Canyon and a portion of Lake Powell, the reservoir it created.

 a. How did the behavior of the river likely change upstream from Lake Powell?

 b. How might the behavior of the Colorado River downstream from the dam have been affected?

 c. Given enough time, how might the reservoir change?

 d. Speculate on the possible environmental impacts of building a dam such as this one.

EYE ON EARTH

1. The meandering White River in Arkansas is a tributary of the Mississippi River.

 a. In this aerial view, the color of the White River is brown. What part of the stream's load gives it this color?

 b. If a channel were created across the narrow neck of land shown by the arrow, how would the river's gradient change?

 c. How would the flow velocity be affected by the formation of such a channel?

2. The Middle Fork of the Salmon River flows for about 175 kilometers (110 miles) through a wilderness area in central Idaho.

 a. Is the river flowing in an alluvial channel or a bedrock channel? Explain.

 b. What process is dominant here: valley deepening or valley widening?

 c. Is the area shown in this image more likely near the mouth or the head of the river?

3. This satellite image shows the delta of the Yukon River. The river originates in northern British Columbia. It flows through the Yukon Territory and across the tundra of Alaska before entering the Bering Sea, a distance of nearly 3200 kilometers (about 2000 miles).

   a. Explain why the river breaks into numerous channels as it crosses the delta.

   b. What term is applied to the channels that radiate across the delta?

   c. Notice the cloud of sediment in the water surrounding the delta. Are these sediments more likely sand and gravel or silt and clay? Explain.

   d. After the cloud of material settles to the seafloor and the delta expands farther into the Bering Sea, which beds of the delta will these sediments become?

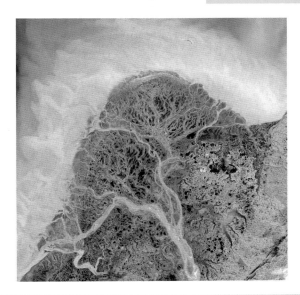

DATA ANALYSIS

Streamflow Rates Near You

Rivers are an important resource for communities, but they can cause dangerous flooding. River water levels are monitored across the United States to determine flow rates and flood stage (the level at which the water surface has risen enough to be a hazard).

https://goo.gl/eD226M

ACTIVITIES

Go to http://waterwatch.usgs.gov. Click on the "Current Streamflow" map.

1. Click on the word "Map" to bring up information about the map. What do the dots display? How long must data have been recorded at a stream gage for the gage to appear on this map?

2. Why might some states have few data points in the winter than during the rest of the year? What are the effects of ice?

3. Click anywhere in the "Explanation—Percentile Classes" table below the map to bring up information about the classes. What do the percentile classes mean?

4. Based on the Current Streamflow map, which areas of the United States are experiencing higher-than-normal streamflow? Which areas are experiencing lower-than-normal streamflow?

5. Explore the different water-resource regions by clicking on them in the "Water-Resources Regions" drop-down menu. Which water-resource region are you located in?

Click on a ranked (colored) stream gage near your current location that has Hydrograph, Peak, and Forecast graphs. Be sure the Peak graph also shows the National Weather Service flood stage level. You can click on each of these graphs to bring up a larger version.

6. Where is this streamflow site located?

7. What is the flood stage water height for this location? What is the current water height for this location?

8. What is the drainage area (in square miles)? How does this compare to nearby waterways?

9. What is the current discharge rate (in cubic feet per second)? How does this compare to nearby waterways?

10. Click on "Hydrograph." Is the current discharge rate higher or lower than the median daily discharge rate? Does this make sense, given what you know about the local conditions in your area and upstream? (You may need to do some investigating online to answer this.)

11. Click on "Peak." When was the most recent flood stage for this waterway? When was the largest flood stage?

Mastering Geology

Looking for additional review and test prep materials? Visit pearson.com/mastering/geology to enhance your understanding of this chapter's content by accessing a variety of resources, including **Self-Study Quizzes, Geoscience Animations, SmartFigure Tutorials,** *Project Condor* **Videos, Mobile Field Trips, Dynamic Study Modules,** and an optional **Pearson eText.**

A Sinkhole Swallows Florida Homes

On July 14, 2017, two homes were destroyed in Land O' Lakes, Florida, when a large sinkhole suddenly formed. Within a few days, five more homes were condemned as the depression grew larger. It was the largest sinkhole to form in the United States in 30 years.

▲ This sinkhole formed suddenly when a cavern roof collapsed.

Usually sinkholes form slowly, causing surface depressions to gradually develop over time. In fact, many of Florida's lakes occupy such depressions. But the type of sinkhole that occurred in Land O' Lakes is sudden, dramatic, and dangerous. It formed after acidic groundwater created a cave in the underlying limestone. As the cave grew larger, its roof was eventually unable to support its weight along with the soil and houses above. So it collapsed.

Land O' Lakes, about 20 miles north of Tampa, is in a region often called *Sinkhole Alley*. Florida is especially prone to sinkholes because much of the state is underlain by limestone. This is also true in areas of Alabama, Missouri, Texas, and Tennessee. According to the U.S. Geological Survey, over the preceding 15 years, sinkhole damages in the United States cost at least $300 million per year.

While sinkholes are natural phenomena, human activities can also promote their development. For example, pumping out too much groundwater reduces the support provided by the water, allowing the partially dissolved limestone and the overlying soil to collapse.

► Aerial view of sinkholes near Timaru, New Zealand. These depressions were created by the erosion of limestone by acidic groundwater,

17

Groundwater

Hidden from view, vast quantities of water exist in the cracks, crevices, and pore spaces of rock and soil. Groundwater can be found almost everywhere beneath Earth's surface and is a major source of water worldwide. Groundwater is a valuable natural resource that provides about half of our drinking water and is essential to the vitality of agriculture and industry. In addition to its utility, groundwater plays a crucial role in sustaining streamflow between precipitation events—especially during protracted dry periods. Many ecosystems depend on groundwater discharge into streams, lakes, and wetlands. In some regions, large-scale development has caused groundwater levels to decline, resulting in water shortages, streamflow depletion, land subsidence, and increased pumping costs. Groundwater pollution is also a serious issue in some places.

17.1 The Importance of Groundwater

Describe the importance of groundwater as a source of freshwater and the role of groundwater as a geologic agent.

Groundwater is one of our most important and widely available resources, yet people's perceptions of the subsurface environment from which it comes are often unclear and incorrect. This is because the groundwater environment is largely hidden from view except in caves and mines, and the impressions people gain from these subsurface openings are misleading. Observations on the land surface give an impression that Earth is "solid." This view remains when we enter a cave and see water flowing in a channel that appears to have been cut into solid rock.

Because of such observations, many people believe that groundwater occurs only in underground "rivers." In reality, most of the subsurface environment is not "solid" at all. It includes countless tiny *pore spaces* between grains of soil and sediment, plus narrow joints and fractures in bedrock. Together, these spaces add up to an immense volume. Where these subsurface pore spaces are saturated with water, that stored water is called **groundwater**. It is in these small openings that groundwater collects and moves.

Groundwater and the Hydrosphere

When we consider the entire hydrosphere, or all of Earth's water, only about six-tenths of 1 percent is located underground. Nevertheless, this small percentage, stored in the rocks and sediments beneath Earth's surface, is a vast quantity. When the oceans are excluded and only sources of freshwater are considered, the significance of groundwater becomes more apparent.

Figure 17.1 gives estimates of the distribution of freshwater in the hydrosphere. Clearly the largest volume occurs as glacial ice. Groundwater is ranked second, with slightly more than 30 percent of the total. However, when ice is excluded and just *liquid* freshwater is considered, 96 percent is groundwater. Without question, *groundwater represents the largest reservoir of freshwater that is readily available to humans.* Its value in terms of economics and human well-being is incalculable.

▶ Figure 17.1
Earth's freshwater
Groundwater is the major reservoir of liquid freshwater.

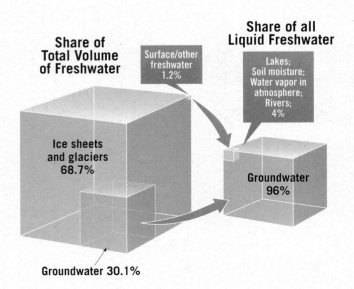

Share of Total Volume of Freshwater

Surface/other freshwater 1.2%

Ice sheets and glaciers 68.7%

Groundwater 30.1%

Share of all Liquid Freshwater

Lakes; Soil moisture; Water vapor in atmosphere; Rivers; 4%

Groundwater 96%

Geologic Importance of Groundwater

Geologically, groundwater is important as an erosional agent. The dissolving action of groundwater slowly removes soluble rock such as limestone, allowing surface depressions known as *sinkholes* to form (see the chapter-opening photo) as well as creating subterranean caverns (see Figure 17.33). The final section of this chapter describes the landforms associated with subsurface water. Groundwater is also an important equalizer of streamflow. Much of the water that flows in rivers is not direct runoff from rain and snowmelt. Rather, a large percentage of precipitation soaks into the ground and then moves slowly under the surface to stream channels. Groundwater is thus a form of storage that sustains streams during periods when rain does not fall. When we see water flowing in a river during a dry period, it is rain that fell at some earlier time and was stored underground.

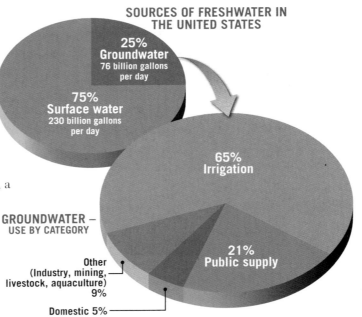

SOURCES OF FRESHWATER IN THE UNITED STATES

25% Groundwater
76 billion gallons per day

75% Surface water
230 billion gallons per day

GROUNDWATER – USE BY CATEGORY

65% Irrigation

21% Public supply

Other (Industry, mining, livestock, aquaculture) 9%

Domestic 5%

◀ Figure 17.2
Sources and uses of freshwater Each day in the United States, we use 306 billion gallons of freshwater. Groundwater is the source of nearly one-quarter of the total. More groundwater is used for irrigation than for all other uses combined.
(Data from U.S. Geological Survey)

Groundwater: A Basic Resource

Water is basic to life. It has been called the "bloodstream" of both the biosphere and society. Each day in the United States we use about 306 billion gallons of freshwater. According to the U.S. Geological Survey, about 75 percent of the water we use comes from surface sources. Groundwater provides the remaining 25 percent (**Figure 17.2**). One of the advantages of groundwater is that it exists almost everywhere across the country and thus is often available in places that lack reliable surface sources such as lakes and rivers. Water in a groundwater system is stored in subsurface pore spaces and fractures. As water is withdrawn from a well, the connected pore spaces and fractures act as a "pipeline" that allows water to gradually move from one part of the hydrologic system to where it is being withdrawn.

Primary Uses of Groundwater
The U.S. Geological Survey identifies several categories, shown in Figure 17.2. More groundwater is used for irrigation than for all other uses combined (**Figure 17.3**). There are nearly 60 million acres (nearly 243,000 square kilometers [about 93,700 square miles]) of irrigated land in the United States. That is an area nearly the size

of the state of Wyoming. The vast majority (75 percent) of the irrigated land is in the 17 conterminous western states, where annual precipitation is typically less than 20 inches. About 43 percent of the water used for irrigation is groundwater.

Public and domestic uses include water for indoor and outdoor household purposes as well as water used for commercial purposes. Common indoor uses include drinking, cooking, bathing, washing clothes and dishes, and flushing toilets. If you are curious about how much water an average American household uses each day for indoor domestic purposes, look

◀ Figure 17.3
Irrigation is number one Growing cotton in California's San Joaquin Valley would not be possible without irrigation. Contrast the irrigated field with the hills in the background. Irrigation is the number-one use of groundwater in the United States.

► Figure 17.4
How we use water The graph shows daily indoor use for an average U.S. household. By installing more efficient water fixtures and regularly checking for leaks, households could significantly reduce water consumption.

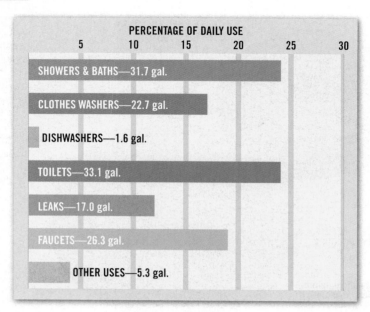

PERCENTAGE OF DAILY USE

SHOWERS & BATHS—31.7 gal.

CLOTHES WASHERS—22.7 gal.

DISHWASHERS—1.6 gal.

TOILETS—33.1 gal.

LEAKS—17.0 gal.

FAUCETS—26.3 gal.

OTHER USES—5.3 gal.

at **Figure 17.4**. Major outdoor uses are watering lawns and gardens. Water for domestic use may come from a public supply or may be self-supplied.° Practically all (98 percent) those whose water is self-supplied rely on wells that tap the local groundwater supply.

Another category, aquaculture, involves water used for fish hatcheries, fish farms, and shellfish farms. Many mining operations require significant quantities of water, as do industrial processes such as petroleum refining and the manufacture of chemicals, plastics, paper, steel, and concrete.

Trends in Water Use   How is water use in the United States changing? Demands for water are growing; population is increasing, bringing more industry and the need to grow more food (and hence more irrigation). Yet total water use in the United States is not rising. In 1980, when the U.S. population was 229.6 million, water withdrawals were 83 billion gallons per day. By 2010, the population had grown by 36 percent, to 313 million, but water use had dropped to 76 billion gallons per day. Water conservation efforts and greater efficiencies in using water clearly have had positive effects. Nevertheless, water use trends are not positive everywhere. As you will see later in the chapter, there are places where groundwater resources are declining because more water is being pumped from the ground than can be replenished.

CONCEPT CHECKS 17.1

1. What percentage of Earth's *total freshwater supply* is groundwater?

2. What share of Earth's *liquid freshwater* is groundwater?

   Concept Checker
   https://goo.gl/q4wkXe

3. List two geologic roles that groundwater plays.

4. What share of U.S. freshwater is provided by groundwater? What is most groundwater used for?

17.2 Groundwater and the Water Table

Prepare a sketch with labels that summarizes the distribution of water beneath Earth's surface. Discuss the factors that cause variations in the water table and describe the interactions between groundwater and streams.

When rain falls on Earth's land surface, some of the water runs off, some returns to the atmosphere by evaporation and transpiration, and the remainder soaks into the ground. This last path is the primary source of practically all subsurface water. The amount of water that takes each of these paths varies greatly both in time and space. Influential factors include steepness of slope, nature of the surface material, intensity of rainfall, and type and amount of vegetation. For example, when heavy rain falls on a steep slope underlain by impervious materials, the obvious result is a high percentage of the water running off. Conversely, when rain falls steadily and gently on more gradual slopes composed of materials that are easily penetrated by the water, a much larger percentage of water soaks into the ground.

Distribution of Groundwater

Some of the water that soaks in does not travel far because it is held by molecular attraction as a surface film on soil particles. This near-surface zone is called the **zone of soil moisture**. It is crisscrossed by roots, voids left by decayed roots, and animal and worm burrows that enhance the infiltration of rainwater into the soil. Soil water is used by plants in life functions and transpiration. Some water also evaporates directly back into the atmosphere.

Water that is not held as soil moisture percolates downward until it reaches a zone where all the open spaces in sediment and rock are completely filled with water (**Figure 17.5**). This is the **zone of saturation** (also called the **phreatic zone**). Water in the zone of saturation is called *groundwater*. The upper limit of this zone is known as the **water table**. Extending upward from the water table is the **capillary fringe** (*capillus* = hair). Here groundwater is held by surface tension in tiny passages between grains of soil or sediment. The area above the water table that includes the capillary fringe and the zone of soil moisture is called the **unsaturated zone** (also known as the **vadose zone**). The pore spaces in this zone contain both air and water. Although a considerable amount of water can be present in the unsaturated zone, this water cannot be

° According to the U.S. Geological Survey, *public supply* refers to water withdrawn by suppliers that furnish water to at least 25 people or have at least 15 connections.

pumped by wells because it clings too tightly to rock and soil particles. By contrast, below the water table, the water pressure is great enough to allow water to enter wells, thus permitting groundwater to be withdrawn for use. We will examine wells more closely later in the chapter.

Variations in the Water Table

The water table, the upper limit of the zone of saturation, is a very significant feature of the groundwater system. Knowing the water table level is important in predicting the productivity of wells, explaining the changes in the flow of springs and streams, and accounting for fluctuations in the levels of lakes. The water table level is highly variable and can range from zero, when the zone of saturation is at the surface, to hundreds of meters below the surface in some places. An important characteristic of the water table is that its configuration varies seasonally and from year to year because the addition of water to the groundwater system is closely related to the quantity, distribution, and timing of precipitation.

Monitoring and Mapping   Except where the water table is at the surface, we cannot observe it directly. However, the elevation of the water table is mapped and studied in detail by examining water levels in wells (**Figure 17.6**). The U.S. Geological Survey

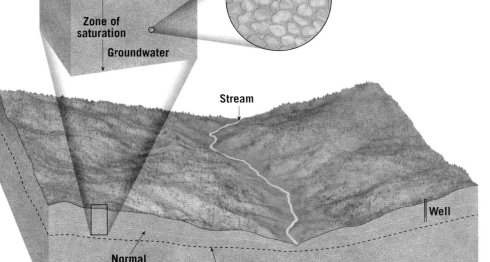

◀ **SmartFigure 17.5**
Water beneath Earth's surface The shape of the water table is usually a subdued replica of the surface topography. During periods of drought, the water table falls, reducing streamflow and drying up some wells.

Tutorial
https://goo.gl/iakFai

Water level data can be accessed remotely.

▼ **Figure 17.6**
Monitoring the water table Water level measurements from observation wells are a basic and important source of data. The graph shows data for the well highlighted on the map.

A groundwater network is an array of wells where water levels are routinely measured. The U.S. Geological Survey in conjunction with state agencies maintains an extensive network of about 20,000 observation wells. This map shows the network of wells in Missouri.

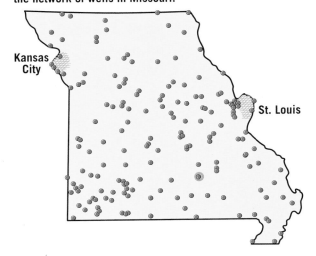

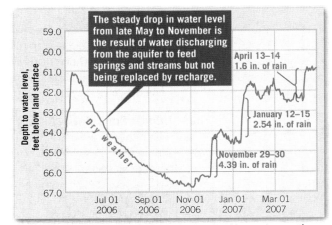

The steady drop in water level from late May to November is the result of water discharging from the aquifer to feed springs and streams but not being replaced by recharge.

April 13–14
1.6 in. of rain

January 12–15
2.54 in. of rain

November 29–30
4.39 in. of rain

Groundwater recharge and discharge at the Akers observation well in Shannon County, Missouri. The height of the water table in this well and many others in this region shows a steady decline between spring and fall. This occurs because recharge is low during these months. However, water continues to move through the aquifer to supply springs and streams in the area, maintaining their flow even in dry years.

▶ Figure 17.7
Mapping the water table
The water table coincides with the water level in observation wells.
(Based on data from U.S Geological Survey)

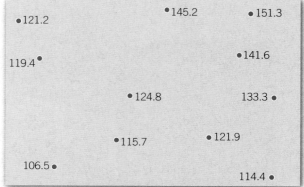

Step 1 The locations of observation wells and the elevations of the water table above sea level are plotted on the map.

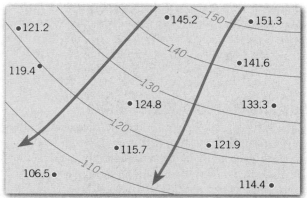

Step 2 Data points are used to guide the drawing of water-table contour lines. Groundwater flow lines can be added to show water movement in the upper portion of the zone of saturation. Groundwater moves perpendicular to the contours and down the slope of the water table.

and state agencies maintain and monitor an extensive network of observation wells to provide statistics about groundwater levels. Such data are the basis for maps that reveal that the water table—despite being called a "table"—is rarely level (**Figure 17.7**). Instead, its shape is usually a subdued replica of the surface topography, being highest beneath hills and descending into valleys (see Figure 17.5). Where a wetland (swamp) is encountered, the water table is right at the surface. Lakes and streams generally occupy areas low enough that the water table is above the land surface.

Several factors contribute to the irregular surface of the water table. One important influence is the fact that groundwater moves very slowly and at varying rates under different conditions. Because of this, water tends to "pile up" beneath high areas between stream valleys. If rainfall were to cease completely, these water table "hills" would slowly subside and gradually approach the level of the valleys. However, new supplies of rainwater are usually added frequently enough to prevent this. Nevertheless, in times of extended drought, the water table may drop enough to dry up shallow wells (see Figure 17.5). Other causes for the uneven water table are variations in precipitation and surface permeability from place to place.

Interactions Between Groundwater and Streams

The interaction between the groundwater system and streams is a basic link in the hydrologic cycle. This interaction can take place in one of three ways. A stream may gain water from the inflow of groundwater through the streambed. Such a stream is called a **gaining stream** (**Figure 17.8A**). For this to occur, the elevation of the water table must be higher than the level of the surface of the stream. Conversely, a **losing stream** loses water to the groundwater system by outflow through the streambed (**Figure 17.8B,C**). For a losing stream to form, the water table must sit lower than the surface of the stream.

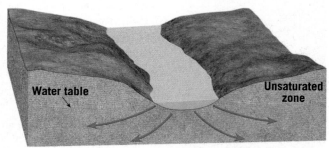

A. Gaining stream Gaining streams receive water from the groundwater system.

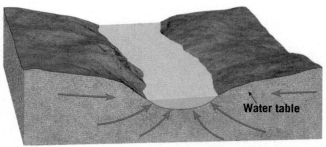

B. Losing stream (connected) Losing streams provide water to the groundwater system.

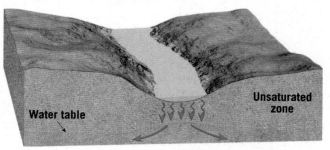

C. Losing stream (disconnected) When losing streams are separated from the groundwater system by the unsaturated zone, a bulge may form in the water table.

▲ Figure 17.8
Interactions between the groundwater system and streams

The third possibility is a combination of the first two: A stream may gain in some sections and lose in others. Moreover, the direction of flow can change over a short time span due to storms adding water near the stream bank or when temporary flood peaks move down the channel.

A losing stream can be connected to the groundwater system by a continuous saturated zone, or it can be disconnected from the groundwater system by an unsaturated zone (compare parts B and C in Figure 17.8). When the stream is disconnected, the water table may have a discernible bulge beneath the stream if the rate of water movement through the streambed and unsaturated zone is greater than the rate of groundwater movement away from the bulge.

Groundwater contributes to streams in most geologic and climatic settings. Even where streams are primarily losing water to the groundwater system, certain sections may receive groundwater inflow during some seasons. One study of 54 streams in all parts of the United States indicated that 52 percent of the stream-flow was contributed by groundwater. The groundwater contribution ranged from a low of 14 percent to a maximum of 90 percent. Groundwater is also a major source of water for lakes and wetlands.

CONCEPT CHECKS 17.2

1. When rain falls on land, what factors influence the amount of water that soaks in?

2. Define *groundwater* and relate it to the water table.

Concept Checker
https://goo.gl/3hcNSQ

3. A kitchen table is flat. Is this usually the case for a water table? Why?

4. Contrast a gaining stream and a losing stream.

17.3 Storage and Movement of Groundwater

Summarize the factors that influence the storage and movement of groundwater. Discuss how groundwater movement is measured and the different scales of movement.

The availability of groundwater depends not only on the amount of water stored in the saturated zone but also on the ability of groundwater to move through the subsurface environment. What factors influence the storage and movement of groundwater? What is the nature of groundwater movement? This section addresses these basic questions.

Influential Factors

The nature of subsurface materials strongly influences the rate of groundwater movement and the amount of groundwater that can be stored. Two factors are especially important: porosity and permeability.

Porosity Water soaks into the ground because bedrock, sediment, and soil contain countless voids or openings. These openings are similar to those of a sponge and are often called *pore spaces*. The quantity of groundwater that can be stored depends on the **porosity** of the material, which is the percentage of the total volume of rock or sediment that consists of pore spaces (**Figure 17.9**). Voids most often are spaces between sedimentary particles, but also common are joints, faults, cavities formed by the dissolving of soluble rock such as limestone, and vesicles (voids left by gases escaping from lava).

Variations in porosity can be great. Sediment is commonly quite porous, and open spaces may occupy 10 to 50 percent of the sediment's total volume. Pore space depends on the size and shape of the grains, how they are packed together, the degree of sorting, and, in sedimentary rocks, the amount of cementing material. For example, clay may have a porosity as high as 50 percent, whereas some gravels may have only 20 percent voids.

Where sediments are poorly sorted, the porosity is reduced because the finer particles tend to fill the openings among the larger grains. Most igneous and metamorphic rocks, as well as some sedimentary rocks, are composed of tightly interlocking crystals such that the voids between the grains may be negligible. In these rocks, fractures must provide the porosity.

▼ **Figure 17.9**
Porosity demonstration Porosity is the percentage of the total volume of rock or sediment that consists of pore spaces.

The beaker on the left is filled with 1000 ml of sediment. The beaker on the right is filled with 1000 ml of water.

The sediment-filled beaker now contains 500 ml of water. Pore spaces (porosity) must represent 50 percent of the volume of the sediment.

Permeability   Porosity alone cannot measure a material's capacity to yield groundwater. Rock or sediment may be very porous yet still not allow water to move through it. The pores must be *connected* to allow water flow, and they must be *large enough* to allow flow. Thus, the **permeability** (*permeare* = to penetrate) of a material—its ability to *transmit* a fluid—is also very important.

Groundwater moves by twisting and turning through small interconnected openings. The smaller the pore spaces, the more slowly the water moves. For example, the ability of a clay deposit to store water may be great, due to high porosity, but its pore spaces may be so small that water is unable to move through it. Thus, we say the clay is *impermeable*.

Aquitards and Aquifers   Impermeable layers that hinder or prevent water movement are termed **aquitards** (*aqua* = water, *tard* = slow). Clay is a good example. On the other hand, larger particles, such as sand or gravel, have larger pore spaces. Therefore, the water moves through with relative ease. Permeable rock strata or sediments that transmit groundwater freely are called **aquifers** (*aqua* = water, *fer* = carry). Sands and gravels are common examples.

In summary, porosity is not always a reliable guide to the amount of surface water that can be stored as groundwater, and permeability is significant in determining the rate of groundwater movement and the quantity of water that might be pumped from a well.

How Groundwater Moves

The movement of water in the atmosphere and on the land surface is relatively easy to visualize, but the movement of groundwater is not. Near the beginning of the chapter, we mentioned the common misconception that groundwater occurs in underground rivers that resemble surface streams. Although subsurface streams do exist, they are *not* common. Rather, as you learned in the preceding sections, groundwater exists in the pore spaces and fractures in rock and sediment. Thus, contrary to any impressions of rapid flow that an underground river might evoke, the movement of most groundwater is exceedingly slow, from pore to pore.

A Simple Groundwater Flow System   Figure 17.10 depicts a simple example of a *groundwater flow system*—a three-dimensional body of Earth material saturated with moving groundwater. It shows groundwater moving along flow paths from **recharge areas**, where groundwater is being replenished, to a **discharge area** along a stream where groundwater is flowing back to the surface. Discharge also occurs at springs, lakes, or wetlands, and in coastal areas, as groundwater seeps into bays or the ocean. Transpiration by plants whose roots extend to near the water table is

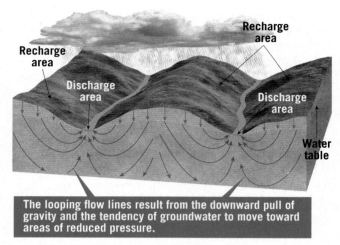

The looping flow lines result from the downward pull of gravity and the tendency of groundwater to move toward areas of reduced pressure.

▲ **Figure 17.10**
Groundwater movement Arrows show paths of groundwater movement through uniformly permeable material.

another form of groundwater discharge. Wells, where groundwater is being pumped to the surface, are artificial discharge areas.

The energy that makes groundwater move is provided by the force of gravity. In response to gravity, water moves from areas where the water table is high to zones where the water table is lower. Although some water takes the most direct path down the slope of the water table, much of the water follows long, curving paths.

Figure 17.10 shows water percolating into a stream from all possible directions. Some paths clearly turn upward, apparently against the force of gravity, and enter through the bottom of the channel. This is easily explained: The deeper you go into the zone of saturation, the greater the water pressure. Thus, the looping curves followed by water in the saturated zone may be thought of as a compromise between the downward pull of gravity and the tendency of water to move toward areas of reduced pressure. As a result, water at any given height is under greater pressure beneath a hill than beneath a stream channel, and the water tends to migrate toward points of lower pressure.

Measuring Groundwater Movement   Our modern understanding of groundwater movement began in the mid-nineteenth century with the work of the French scientist-engineer Henri Darcy. One of the experiments Darcy carried out showed that the velocity of groundwater flow is proportional to the slope of the water table: The steeper the slope, the faster the water moves (because the steeper the slope, the greater the pressure difference between two points). The water table slope, known as the **hydraulic gradient**, can be expressed as follows:

$$\text{Hydraulic gradient} = \frac{h_1 - h_2}{d}$$

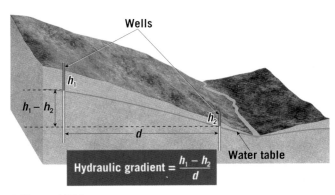

▲ **Figure 17.11**
Hydraulic gradient The hydraulic gradient is determined by measuring the difference in elevation between two points on the water table ($h_1 - h_2$) divided by the distance between them, d. Wells are used to determine the height of the water table.

where h_1 is the elevation of one point on the water table, h_2 is the elevation of a second point, and d is the horizontal distance between the two points (**Figure 17.11**).

Darcy also experimented with different materials, such as coarse sand and fine sand, measuring the rate of flow through sediment-filled tubes that were tilted at varying angles. He found that flow velocity varied with the permeability of the sediment: Groundwater flows more rapidly through sediments having greater permeability than through materials having lower permeability. He defined a coefficient known as **hydraulic conductivity** that takes into account the permeability of the aquifer and the viscosity of the fluid. To determine discharge (Q)—that is, the actual volume of water that flows through an aquifer in a specified time—the following equation is used:

$$Q = \frac{KA(h_1 - h_2)}{d}$$

where $\dfrac{h_1 - h_2}{d}$ is the hydraulic gradient, K is the coefficient that represents hydraulic conductivity, and A is the cross-sectional area of the aquifer. This expression has come to be called **Darcy's law**. Using this equation, if you know an aquifer's hydraulic gradient, conductivity, and cross-sectional area, you can calculate its discharge.

Different Scales of Movement The geographic extent of groundwater flow systems varies from a few square kilometers or less to tens of thousands of square kilometers. Flow paths range from a few meters to tens and sometimes hundreds of kilometers. **Figure 17.12** is a cross section of a hypothetical region in which a deep groundwater flow system is overlain by and connected to several shallower local flow systems. The subsurface geology exhibits a complicated arrangement of aquifer units with high hydraulic conductivity and aquitard units with low hydraulic conductivity. Starting near the top of Figure 17.12, blue arrows represent water movement in several local groundwater systems in the upper water table aquifer. These groundwater systems are separated by groundwater divides at the center of hills, and they discharge into the nearest surface water bodies. Beneath these most shallow systems, red arrows show water movement in a deeper system in which groundwater discharges into a more distant surface water body. Finally, black arrows show groundwater movement into a deep regional system. The horizontal scale of the figure could range from tens to hundreds of kilometers.

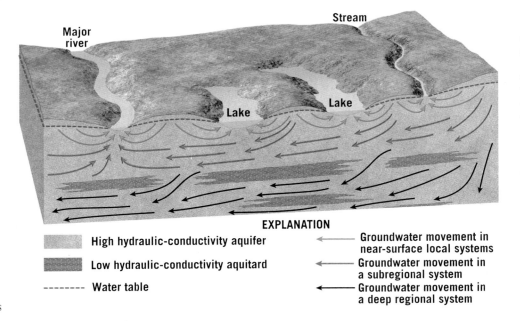

EXPLANATION

High hydraulic-conductivity aquifer	Groundwater movement in near-surface local systems
Low hydraulic-conductivity aquitard	Groundwater movement in a subregional system
-------- Water table	Groundwater movement in a deep regional system

◄ **SmartFigure 17.12**
Hypothetical groundwater flow system The diagram includes subsystems at three different scales. Variations in surface topography and subsurface geology can produce a complex situation. The horizontal scale of the figure could range from tens to hundreds of kilometers.

Tutorial
https://goo.gl/SdSHVb

CONCEPT CHECKS 17.3

1. Distinguish between porosity and permeability.

2. What is the difference between an aquifer and an aquitard?

3. What factors cause water to follow the paths shown in Figure 17.10?

 **Concept Checker** https://goo.gl/taeBVv

4. Relate groundwater movement to hydraulic gradient and hydraulic conductivity.

17.4 Wells and Artesian Systems

Discuss water wells and their relationship to the water table. Sketch and label a simple artesian system.

There are more than 20 million water wells for all purposes in the United States. Private household wells constitute the largest share—more than 80 percent. About 500,000 new residential wells are drilled each year.

Wells

The most common method for removing groundwater is a **well**, a hole bored into the zone of saturation (Figure 17.13). Wells serve as small reservoirs into which groundwater migrates so it can be pumped to the surface. Wells date back many centuries, and remain important today. Groundwater is the principal source of drinking water for about half the U.S. population and provides about 96 percent of water used in rural areas.

The water table level may fluctuate considerably during the course of a year, dropping during dry periods and rising following wet periods. Therefore, to ensure a continuous supply of water, a well must penetrate below the water table. Whenever substantial water is withdrawn from a well, the water table around the well may be lowered. This effect, termed **drawdown**, decreases with increasing distance from the well. The result is a depression in the water table, roughly conical in shape, known as a **cone of depression** (Figure 17.14). Because the cone of depression increases the hydraulic gradient near the well, groundwater will flow more rapidly toward the opening. For most small domestic wells, the cone of depression is negligible. However, when wells are heavily pumped for irrigation or for industrial purposes, the withdrawal of water can be great enough to create a very wide and steep cone of depression. This may substantially lower the water table in an area and cause nearby shallow wells to become dry. Figure 17.14 illustrates this situation.

Drilling a successful well is a familiar challenge for people in areas where groundwater is the primary source of supply. One well may be successful at a depth of 10 meters (33 feet), whereas a neighbor may have to go twice as deep to find an adequate supply. Still others may be forced to try a different site

▼ Figure 17.13
Wells Wells are the most common means by which people obtain groundwater.

▼ SmartFigure 17.14
Cone of depression For most small domestic wells, the cone of depression is negligible. When wells are heavily pumped, the cone of depression can be large and may lower the water table such that nearby shallower wells may be left dry.

Animation
https://goo.gl/EgC3Do

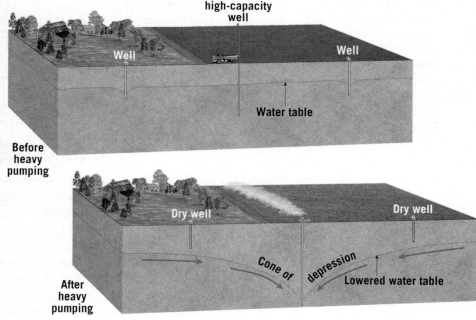

altogether. When subsurface materials are heterogeneous, the amount of water that a well can provide may vary a great deal over short distances. For example, when two nearby wells are drilled to the same level and only one is successful, it may be because there is a perched water table beneath one of them. As **Figure 17.15** illustrates, a **perched water table** forms where an aquitard is situated above the main water table. Massive igneous and metamorphic rocks provide a second example. These crystalline rocks are usually not very permeable, except where they are cut by many intersecting joints and fractures. Therefore, when a well drilled into such rock does not intersect an adequate network of fractures, it is likely to be unproductive.

Artesian Systems

In most wells, water cannot rise without the use of pumps. If water is first encountered at a depth of 30 meters (100 feet), it remains at that level, fluctuating perhaps 1 or 2 meters (3 to 6 feet) with seasonal wet and dry periods. However, in some wells, water rises, sometimes overflowing at the surface. Such wells are abundant in the Artois region of northern France, and so we call these self-rising wells *artesian*.

The term **artesian** is applied to *any* situation in which groundwater under pressure rises above

the level of the aquifer. For an artesian system to exist, two conditions usually are met (**Figure 17.16**): (1) Water is confined to an aquifer that is inclined so that one end can receive water, and (2) aquitards,

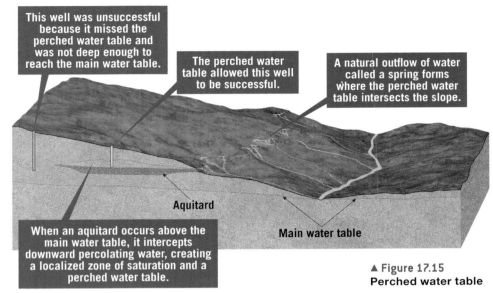

▲ Figure 17.15
Perched water table

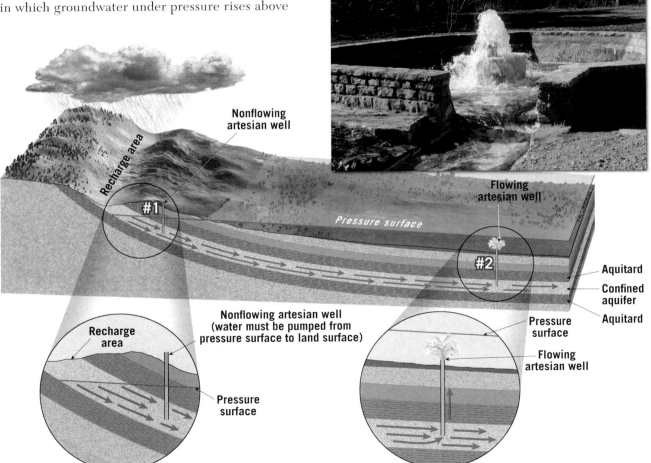

◀ SmartFigure 17.16
Artesian systems These groundwater systems occur where an inclined aquifer is surrounded by impermeable beds (aquitards). Such aquifers are called *confined aquifers*. The photo shows a flowing artesian well.

Tutorial
https://goo.gl/UHnjQR

► Figure 17.17
A classic artesian system This geologic cross section across South Dakota shows the major elements of the Dakota Sandstone artesian system.

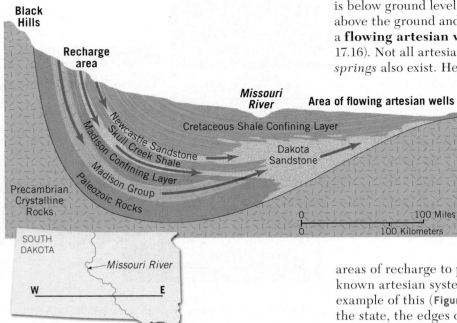

is below ground level. When the pressure surface is above the ground and a well is drilled into the aquifer, a **flowing artesian well** is created (well 2 in Figure 17.16). Not all artesian systems are wells. *Artesian springs* also exist. Here groundwater may reach the surface by rising along a natural fracture such as a fault rather than through an artificially produced hole. In deserts, artesian springs are sometimes responsible for creating oases.

Artesian systems act as conduits, often transmitting water great distances from remote areas of recharge to points of discharge. A well-known artesian system in South Dakota is a good example of this (**Figure 17.17**). In the western part of the state, the edges of a series of sedimentary layers have been bent up to the surface along the flanks of the Black Hills. One of these beds, the permeable Dakota Sandstone, is sandwiched between impermeable strata and gradually dips into the ground toward the east. When the aquifer was first tapped, water poured from the ground surface, creating fountains many meters high. In some places the force of the water was sufficient to power waterwheels. Such scenes no longer occur because thousands of additional wells now tap the same aquifer. This has depleted the reservoir and lowered the water table in the recharge area. As a consequence, the pressure has dropped to the point where many wells have stopped flowing altogether and now have to be pumped.

both above and below the aquifer, must be present to prevent the water from escaping. Such an aquifer is called a **confined aquifer**. When such a layer is tapped, the pressure created by the weight of the water above forces the water to rise. If there were no friction, the water in the well would rise to the level of the water at the top of the aquifer. However, friction reduces the height of the pressure surface—that is, the level to which the water in the aquifer would rise if not confined. The greater the distance from the recharge area (where water enters the inclined aquifer), the greater the friction and the less the water rises.

In Figure 17.16, well 1 is a **nonflowing artesian well** because at this location, the pressure surface

On a different scale, city water systems can be considered analogous to artificial artesian systems (**Figure 17.18**). The water tower, into which water is pumped, would represent the area of recharge, the pipes the confined aquifer, and the faucets in homes the flowing artesian wells.

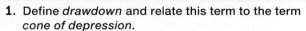

▲ Figure 17.18
City water systems City water systems can be considered artificial artesian systems.

17.5 Springs, Hot Springs, and Geysers

Distinguish among springs, hot springs, and geysers.

The phenomena described in this section often arouse people's curiosity and wonder. The fact that springs, hot springs, and geysers seem rather mysterious is not difficult to understand, for here is water (sometimes very hot water) flowing or spewing from the ground in all kinds of weather, in seemingly inexhaustible supply but with no obvious source.

Springs

Not until the middle of the seventeenth century did the French physicist Pierre Perrault invalidate the age-old assumption that precipitation could not adequately account for the amount of water emanating from springs and flowing in rivers. Over several years, Perrault computed the quantity of water that fell on France's Seine River basin. He then calculated the mean annual runoff by measuring the river's discharge. After allowing for the loss of water by evaporation, he showed that there was sufficient water remaining to feed the springs. Thanks to Perrault's pioneering efforts and the measurements by many afterward, we now know that the source of springs is water from the zone of saturation and that the ultimate source of this water is precipitation.

Whenever the water table intersects Earth's surface, a natural outflow of groundwater results, and we call this a **spring**. A spring may form when an aquitard blocks the downward movement of groundwater and causes the water to move laterally. Where the permeable bed crops out, a spring results. Another example is illustrated in Figure 17.15, which shows a perched water table intersecting a slope.

Springs, however, are not confined to places where a perched water table creates a flow at the surface. Many geologic situations lead to the formation of springs because subsurface conditions vary greatly from place to place. Even in areas underlain by impermeable crystalline rocks, permeable zones may exist in the form of fractures or solution channels. If these openings fill with water and intersect the ground surface along a slope, a spring results (**Figure 17.19**).

Hot Springs

There is no universally accepted definition of *hot spring*. One frequently used definition is that the water in a **hot spring** is 6°–9°C (10°–15°F) warmer than the mean annual air temperature for the locality where it occurs. In the United States alone, there are more than 1000 such springs.

Temperatures in deep mines and oil wells usually rise with increasing depth, an average of about 25°C (45°F) per kilometer. You learned in Chapter 4 that this is called the *geothermal gradient*. When groundwater circulates at great depths, it becomes heated. If the hot water rises rapidly to the surface, it may emerge as a hot spring. The water of some hot springs in the eastern United States is heated in this manner. The springs at Warm Springs, Georgia, the presidential retreat of Franklin Roosevelt, are one example.

The temperature of these hot springs is always near 32°C (90°F). Another example is Hot Springs National Park, Arkansas, where water temperatures average about 60°C (140°F).

The great majority (more than 95 percent) of the hot springs (and geysers) in the United States are in the West. A glance at **Figure 17.20** reinforces this fact. This is because the heat sources for most hot springs are magma bodies and hot igneous rocks, and igneous activity has occurred more recently in the West than in the rest of the country. The hot springs and geysers of the Yellowstone region are well-known examples.

Geysers

Geysers are intermittent hot springs or fountains in which columns of water are ejected with great force at various intervals, often rising 30 to 60 meters (100 to 200 feet) into the air. After the jet of water ceases, a column of steam rushes out, usually with a thunderous roar. Perhaps the most

▲ Figure 17.19
Vasey's Paradise A spring is a natural outflow of groundwater that occurs when the water table intersects the surface. This spring creates a waterfall after emerging from a steep rock wall of the Grand Canyon.

Yellowstone National Park

Hot Springs, Arkansas

Warm Springs, Georgia

0   400 mi
0   400 km

◄ Figure 17.20
Distribution of hot springs and geysers Note the concentration in the West, where igneous activity has occurred most recently.

▶ **Figure 17.21**
Old Faithful This geyser in Wyoming's Yellowstone National Park is one of the most famous in the world. Contrary to popular legend, it does not erupt every hour on the hour. Time spans between eruptions vary from about 65 minutes to more than 90 minutes and have generally increased over the years due to changes in the geyser's plumbing.

famous geyser in the world is Old Faithful in Yellowstone National Park (**Figure 17.21**). The great abundance, diversity, and spectacular nature of geysers and other thermal features in Yellowstone undoubtedly was the primary reason for its becoming the first national park in the United States. Geysers are also found in other parts of the world, notably New Zealand and Iceland. In fact, the Icelandic word *geysa*, meaning "to gush," gives us the name *geyser*.

How Geysers Work   Geysers occur where extensive underground chambers exist within hot igneous rocks. How they operate is shown in **Figure 17.22**. As relatively cool groundwater enters these chambers, it is heated by the surrounding rock. At the bottom of the chambers, the water is under great pressure because of the weight of the overlying water. This great pressure prevents the water from boiling at the normal surface temperature of 100°C (212°F). For example, water at the bottom of a 300-meter (1000-foot) water-filled chamber must attain nearly 230°C (450°F) before it will boil. The heating causes the water to expand, and as a result, some is forced out at the surface. This loss of water reduces the pressure on the remaining water in the chamber, which lowers the boiling point. A portion of the water deep within the chamber quickly turns to steam, and the geyser erupts. Following eruption, cool groundwater again seeps into the chamber, and the cycle begins anew.

Geyser Deposits   When groundwater from hot springs and geysers flows out at the surface, material in solution is often precipitated, producing an accumulation of chemical sedimentary rock. The material deposited at any given place commonly reflects the chemical makeup of the rock through which the water circulated. When the water contains dissolved silica, a material called *siliceous sinter*, or *geyserite*, is deposited around the spring. When the water contains dissolved calcium carbonate, a form of limestone called *travertine*, or

▼ **SmartFigure 17.22**
How a geyser works A geyser can form if the underground plumbing does not allow heat to be readily distributed by convection.

Tutorial
https://goo.gl/jYoqfU

A. Water near the bottom is heated to near its boiling point. The boiling point is higher at the bottom because pressure is high due to the weight of all the water above.

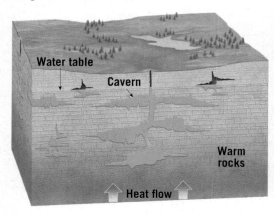

B. Higher up in the geyser the water is also heated and expands causing some to flow out at the top. This outflow reduces the pressure on the water at the bottom.

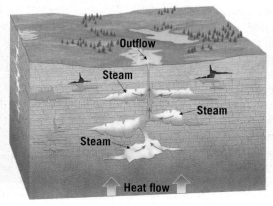

C. When pressure is reduced at the bottom, boiling occurs. Some of the bottom water flashes into steam. The expanding steam triggers an eruption. Then the water flows back in, and the whole process begins anew.

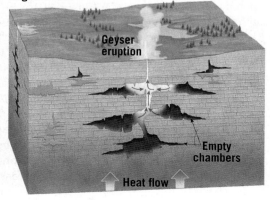

◄ **Figure 17.23**
Yellowstone's Mammoth Hot Springs Although most of the deposits associated with geysers and hot springs in Yellowstone National Park are silica-rich geyserite, the deposits here consist of a form of limestone called travertine.

calcareous tufa, is deposited. The latter term is used if the material is spongy and porous.

The deposits at Mammoth Hot Springs in Yellowstone National Park are more spectacular than most others (**Figure 17.23**). As the hot water flows upward through a series of channels and then out at the surface, the reduced pressure allows carbon dioxide to separate and escape from the water. The loss of carbon dioxide causes the water to become supersaturated with calcium carbonate, which then precipitates. In addition to containing dissolved silica and calcium carbonate, some hot springs contain sulfur, which gives water a poor taste and unpleasant odor. This is undoubtedly the case at Rotten Egg Spring, Nevada.

CONCEPT CHECKS 17.5

1. Describe some circumstances that lead to the formation of a spring.

2. What warms the waters that flow at Hot Springs National Park, Arkansas, and at Warm Springs, Georgia? **Concept Checker** https://goo.gl/oRjVui

3. What is the source of heat for most hot springs and geysers? How is this reflected in the distribution of these features?

4. Describe what occurs to cause a geyser to erupt.

17.6 Environmental Problems

List and discuss important environmental problems associated with groundwater.

Like many of our other valuable natural resources, groundwater is being exploited at an increasing rate. In some areas, overuse threatens the groundwater supply. In other places, groundwater withdrawal has caused the ground and everything resting on it to sink. Some localities are concerned with possible contamination of the groundwater supply.

Treating Groundwater as a Nonrenewable Resource

The height of the water table reflects a balance between the rate of recharge and the rate of discharge and withdrawal. Any imbalance will either raise or lower the water table. Long-term imbalances can lead to a significant drop in the water table if there is either an increase in groundwater withdrawal or a decrease in recharge due to prolonged drought (see **GEO-Graphics 17.1**).

In some regions, groundwater has been and continues to be treated as a *nonrenewable* resource because the amount of water available to recharge the aquifer is significantly less than the amount being withdrawn. In such situations, groundwater is essentially being mined.

Drought Impacts the Hydrologic System

Drought is a period of abnormally dry weather that persists long enough to produce a significant hydrologic imbalance such as crop damage or water supply shortages. Drought severity depends upon the degree of moisture deficiency, its duration, and the size of the affected area.

Drought status map for the western United States on April 7, 2015

Much of the West was experiencing drought at this time. California was suffering most. Entering its fourth year of drought, more than 93 percent of the state had at least *severe drought* conditions. *Exceptional drought* was affecting nearly 40 percent of the state. Mountain snowpack, the source that feeds California's rivers, lakes, and reservoirs, was just 19 percent of the late winter average.

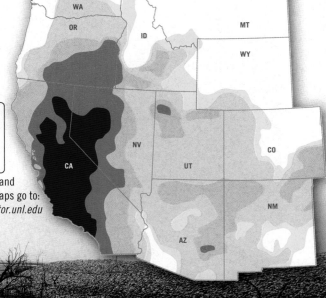

INTENSITY
- D0 Abnormally Dry
- D1 Drought - Moderate
- D2 Drought - Severe
- D3 Drought - Extreme
- D4 Drought - Exceptional

To examine current and archived drought maps go to: *http://droughtmonitor.unl.edu*

Impact on groundwater

Because drought depleted surface water sources, groundwater use soared to make up the shortfall with many areas experiencing a dramatic increase in well drilling—not only more wells, but *deeper* wells. Nearly 60 percent of the state's water needs were being met by groundwater, up from 40 percent in years when rain and snow were normal.

DROUGHT CATEGORIES

TIME: SEQUENCE OF DROUGHT IMPACTS →

METEOROLOGICAL DROUGHT	AGRICULTURAL DROUGHT	HYDROLOGICAL DROUGHT
PRECIPITATION DEFICIENCY (results in reduced runoff and infiltration)	**SOIL MOISTURE DEFICIENCY** (results in low crop yields)	**REDUCED STREAMFLOW, INFLOW TO RESERVOIRS, LAKES, AND PONDS; REDUCED WETLANDS** (results in reduced domestic water supply and wildlife habitat

After the onset of meteorological drought, agriculture is affected first, followed by reductions in streamflow and water levels in lakes, streams, and underground. When meteorological drought ends, agricultural drought ends as soil moisture is replenished. It can take much longer for hydrolgical drought to end.

COMPARING COSTS OF DROUGHTS, FLOODS, AND WILDFIRES (1980–2017)

WILDFIRE $3.7 billion per event

FLOOD $4.3 billion per event

DROUG $9.7 billion per eve

Although natural disasters such as wildfires and floods usually generate more attention, droughts can be just as devastating and often carry a bigger price tag. Unlike other hazards which are short-lived, drought occurs in a gradual "creeping" way, making its onset and end difficult to determine.

The High Plains aquifer provides one example (**Figure 17.24**). Underlying about 450,000 square kilometers (174,000 square miles) in parts of eight western states, it is one of the largest and most agriculturally significant aquifers in the United States. It accounts for about 30 percent of all groundwater withdrawn for irrigation in the country. Because evaporation rates are high and precipitation is modest, there is little rainwater to recharge the aquifer. Thus, in some parts of the region, where intense irrigation has been practiced for

The High Plains Aquifer provides 30% of the groundwater used for irrigation in the U.S.

an extended period, depletion of groundwater has been severe. Figure 17.24 bears this out.

Land Subsidence Caused by Groundwater Withdrawal

As you will see later in this chapter, surface subsidence can result from natural processes related to groundwater. However, the ground may also sink when water is pumped from wells faster than natural recharge processes can replace it. This effect is particularly pronounced in areas underlain by thick layers of unconsolidated sediments. As water is withdrawn, the water pressure drops, and the weight of the overburden is transferred to the sediment. As a result, sediment grains pack more tightly together, and the ground subsides.

Some Classic Examples   Many areas illustrate land subsidence caused by excessive pumping of groundwater. A classic U.S. example is in the San Joaquin Valley of California, where subsidence has approached 9 meters (30 feet) in some areas (**Figure 17.25**). Other

Because of its high porosity, excellent permeability, and great size, the High Plains aquifer, the largest in the United States, accumulated enough freshwater to fill Lake Huron.

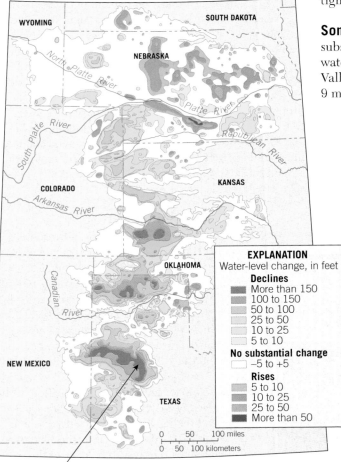

EXPLANATION
Water-level change, in feet
Declines
More than 150
100 to 150
50 to 100
25 to 50
10 to 25
5 to 10
No substantial change
−5 to +5
Rises
5 to 10
10 to 25
25 to 50
More than 50

0   50   100 miles
0   50   100 kilometers

The U.S. Geological Survey estimates that since 1950, water in storage in the High Plains aquifer declined by about 267 million acre feet (about 87 trillion gallons), with 60 percent of the total decline occurring in Texas.

▲ Figure 17.24
High Plains aquifer   The map shows changes in groundwater levels from predevelopment (about 1950) to 2013. Extensive pumping for irrigation has led to water level declines in excess of 45 meters (150 feet) in parts of four states. Water level rises have occurred where surface water is used for irrigation, such as along the Platte River in Nebraska.
(Based on U.S. Geological Survey)

Land in the San Joaquin Valley subsided by up to 10 m (30 ft) in the 50 years from 1925, when heavy pumping began, to 1977, when this photo was taken.

◄ Figure 17.25
That sinking feeling!   The San Joaquin Valley, an important agricultural area, relies heavily on irrigation. Between 1925 and 1977, withdrawal of groundwater caused sediments to compact and the ground to subside.

▶ Figure 17.26
Land subsidence in south-central Arizona In southern Arizona, heavy pumping has led to water table declines of up to 180 meters (600 feet). This has triggered extensive and uneven permanent compaction of sediments and the formation of large fissures (cracks) in the ground around the margins of subsiding basins. Some rural roads have signs that warn of the potential hazard.

well-known examples of land subsidence due to groundwater pumping include portions of southern Arizona (**Figure 17.26**); Las Vegas, Nevada; the New Orleans–Baton Rouge area of Louisiana; and the Houston–Galveston area of Texas. In the low-lying coastal area between Houston and Galveston, land subsidence ranges from 1.5 to 3 meters (5 to 10 feet). The result is that an area of about 78 square kilometers (30 square miles) is permanently flooded.

Outside the United States, one of the most spectacular examples of subsidence occurred in Mexico City, a portion of which is built on a former lakebed. In the first half of the twentieth century, thousands of wells were sunk into the water-saturated sediments beneath the city. As water was withdrawn, portions of the city subsided by as much as 6 to 7 meters (20 to 23 feet). In some places buildings sank so much that access to them from the street is located at what used to be the second-floor level!

Impact of Prolonged Drought   The map in **Figure 17.27A** shows the extent and severity of drought conditions in California in late January 2016, as the state entered its fifth year of drought. Nearly 64 percent of the state was experiencing extreme or exceptional drought. Because of the long span of abnormally dry weather, the amount of water stored in lakes and reservoirs was severely depleted. To make up for the shortfall, groundwater use soared. Many areas experienced a dramatic increase in well drilling—not only more wells but deeper wells to reach the dropping water table. Nearly 60 percent of the state's water needs were being met by groundwater, up from 40 percent in years when rain and snow were normal. As illustrated in **Figure 17.27B**, lowering the water table led to land subsidence of as much as 0.6 meter (2 feet) over a span of just a little over a year.

Saltwater Intrusion

In many coastal areas, groundwater resources are being threatened by the encroachment of saltwater. To understand this problem, let us examine the relationship between fresh and salty groundwater. **Figure 17.28** shows a cross section that illustrates this relationship in a coastal area underlain by permeable homogeneous materials. Freshwater is less dense than saltwater, so it floats on the saltwater and forms a large lens-shaped body that may extend to considerable depths below sea level. In such a situation, if the water table is 1 meter (3 feet) above sea level, the base of the freshwater body will extend to a depth of about 40 meters (130 feet) below sea level. Stated another way, the depth of the freshwater below sea level is about 40 times greater than the elevation of the water table above sea level. Thus, when excessive pumping lowers the water table by a certain amount, the bottom of the freshwater zone will rise by 40 times that

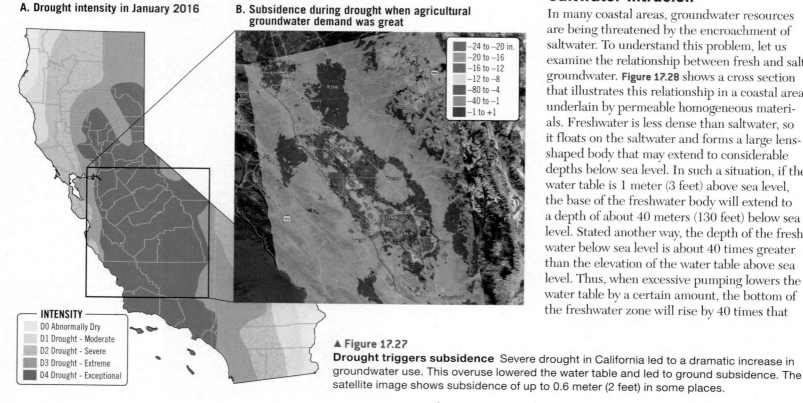

A. Drought intensity in January 2016

B. Subsidence during drought when agricultural groundwater demand was great

−24 to −20 in.	
−20 to −16	
−16 to −12	
−12 to −8	
−80 to −4	
−40 to −1	
−1 to +1	

INTENSITY
- D0 Abnormally Dry
- D1 Drought - Moderate
- D2 Drought - Severe
- D3 Drought - Extreme
- D4 Drought - Exceptional

▲ Figure 17.27
Drought triggers subsidence Severe drought in California led to a dramatic increase in groundwater use. This overuse lowered the water table and led to ground subsidence. The satellite image shows subsidence of up to 0.6 meter (2 feet) in some places.

Because freshwater is less dense than saltwater, it floats on the saltwater and forms a lens-shaped body that may extend to considerable depths below sea level.

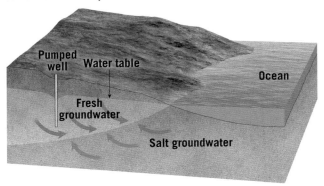

If excessive pumping lowers the water table, the base of the freshwater zone will rise 40 times that amount. The result may be saltwater contamination of wells.

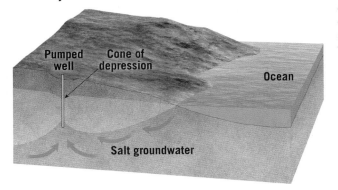

◄ **Figure 17.28**
Saltwater intrusion
Heavy pumping in coastal areas can cause encroachment of saltwater and threaten the supply of fresh groundwater.

amount. Therefore, if groundwater withdrawal continues to exceed recharge, at some point, the elevation of the saltwater will be sufficiently high that the saltwater will be drawn into wells, thus contaminating the freshwater supply. This is called *saltwater intrusion* or *saltwater contamination*. Deep wells and wells near the shore are usually the first to be affected.

In urbanized coastal areas, the problems created by excessive pumping are compounded by a decrease in the rate of natural recharge. As more and more of the surface is covered by streets, parking lots, and buildings, surface runoff increases, and infiltration into the soil is diminished.

One way to correct the problem of saltwater intrusion of groundwater resources is to use a network of recharge wells. These wells allow wastewater to be pumped back into the groundwater system. A second method of correction is accomplished by building large recharge basins. These basins collect surface drainage and allow it to seep into the ground (**Figure 17.29**). Recharge basins are not just used to help combat saltwater intrusion. They are used in any situation in which there is a need to enhance infiltration for the purpose of keeping the water table from declining.

Contamination of freshwater aquifers by saltwater is primarily a problem in coastal areas, but it can also threaten noncoastal locations. Many ancient sedimentary rocks of marine origin were deposited when the ocean covered places that are now far inland. In some instances, significant amounts of seawater were trapped and still remain in the rock. These strata sometimes contain quantities of freshwater that people may pump. However, if freshwater is removed more rapidly than it is replenished, saline water may encroach and render the wells unusable. Such a situation threatened users of a deep (Cambrian age) sandstone aquifer in the Chicago area. To counteract this, water from Lake Michigan was allocated to the affected communities to offset the rate of withdrawal from the aquifer.

Groundwater Contamination

The pollution of groundwater is a serious matter, particularly in areas where aquifers provide a large part of the

water supply. One common source of groundwater pollution is sewage. Its sources include an ever-increasing number of septic tanks, as well as inadequate or broken sewer systems and farm wastes.

If sewage water that is contaminated with bacteria enters the groundwater system, it may become purified through natural processes. The harmful bacteria may be mechanically filtered by the sediment through which the water percolates, destroyed by chemical oxidation, and/or assimilated by other organisms. For purification to occur, however, the aquifer must be of the correct composition. For example, extremely permeable aquifers (such as highly fractured crystalline rock, coarse gravel, or cavernous limestone) have such large openings that contaminated groundwater may travel long distances without being filtered and cleansed. In this case, the water flows too rapidly and is not in contact with the surrounding material

▼ **Figure 17.29**
Recharge basins
Recharge basins intercept surface runoff and allow the water to infiltrate. This helps to maintain the water table and prevent saltwater intrusion. Recharge basins are used in many places, not just coastal areas.

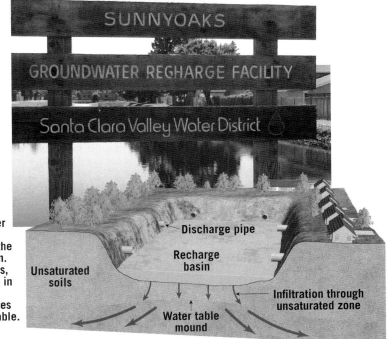

The stormwater system is connected to the recharge basin. Following rains, water collects in the basin and slowly infiltrates to the water table.

Although the contaminated water has traveled more than 100 meters before reaching Well 1, the water moves too rapidly through the cavernous limestone to be purified.

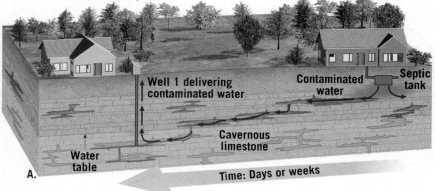

A.

As the discharge from the septic tank percolates through the permeable sandstone, it moves more slowly and is purified in a relatively short distance.

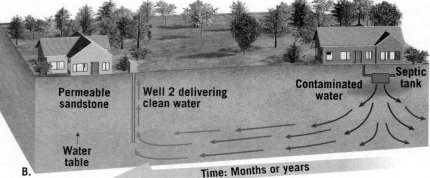

B.

▲ **Figure 17.30**
Comparing two aquifers
In this example, the limestone aquifer allowed the contamination to reach a well, but the sandstone aquifer did not.

▶ **SmartFigure 17.31**
Changing direction
Drawdown at the heavily pumped well changed the slope of the water table, which led to the contamination of the small well.

Tutorial
https://goo.gl/
8bZhjD

long enough for purification to occur. This is the problem at well 1 in **Figure 17.30A**.

On the other hand, when the aquifer is composed of sand or permeable sandstone, it can sometimes be purified after traveling only a few dozen meters through it. The openings

between sand grains are large enough to permit water movement, yet the movement of the water is slow enough to allow ample time for its purification (well 2, **Figure 17.30B**).

Sometimes sinking a well can lead to groundwater pollution problems. If a well pumps a sufficient quantity of water, the cone of depression will locally increase the slope of the water table. In some instances, the original slope may even be reversed. This could lead to the contamination of wells that yielded unpolluted water before heavy pumping began (**Figure 17.31**). Also recall that the rate of groundwater movement increases as the slope of the water table gets steeper. This could produce problems because a faster rate of movement allows less time for the water to be purified in the aquifer before it is pumped to the surface.

Other sources and types of contamination also threaten groundwater supplies. These include widely used substances such as highway salt, fertilizers that are spread across the land surface, and pesticides. In addition, a wide array of chemicals and industrial materials may leak from pipelines, storage tanks, landfills, and holding ponds. Some of these pollutants are classified as *hazardous*, meaning that they are either flammable, corrosive, explosive, or toxic. In landfills, potential contaminants are heaped onto mounds or spread directly over the ground. As rainwater oozes through the refuse, it may dissolve a variety of organic and inorganic materials. If the leached material reaches the water table, it will mix with the groundwater and contaminate the supply. Similar problems may result from leakage of shallow excavations called *holding ponds* into which a variety of liquid wastes are disposed.

Because groundwater movement is usually slow, polluted water can go undetected for a long time. In fact, contamination is sometimes discovered only after drinking water has been affected, causing people to become ill. By this time, the volume of polluted water may be very large, and even if the source of contamination is removed immediately, the problem is not solved.

Although the sources of groundwater contamination are numerous, there are relatively few solutions. The most effective solution is prevention.

Originally the outflow from the septic tank moved away from the small well.

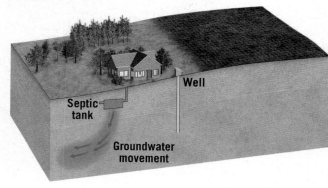

The heavily pumped well changed the slope of the water table, causing contaminated groundwater to flow toward the small well.

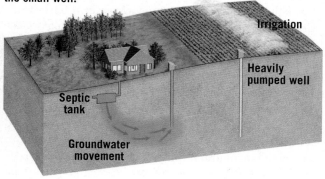

Whereas the sources of groundwater contamination are numerous, there are relatively few solutions. Once the source of the problem has been identified and eliminated, the most common practice is simply to abandon the water supply and allow the pollutants to be flushed away gradually. This is the least costly and easiest solution, but the aquifer must remain unused for many years. To accelerate this process, polluted water is sometimes pumped out and treated. Following removal of the tainted water, the aquifer is allowed to recharge naturally, or in some cases the treated water or other freshwater is pumped back in. This process is costly, time-consuming, and may be risky because there is no way to be certain that all of the contamination has been removed. Clearly, the most effective solution to groundwater contamination is prevention.

CONCEPT CHECKS 17.6

1. Describe the problem associated with pumping groundwater for irrigation in the southern High Plains.

2. Explain why the ground may subside after groundwater is pumped to the surface.

Concept Checker
https://goo.gl/RfCHig

3. Which aquifer would be most effective in purifying polluted groundwater: coarse gravel, sand, or cavernous limestone?

4. Describe a significant problem that may arise when groundwater is heavily pumped at a coastal site.

17.7 The Geologic Work of Groundwater

Explain the formation of caverns and the development of karst topography.

Groundwater dissolves rock. This fact is key to understanding how caverns and sinkholes form (**Figure 17.32**). Soluble rocks, especially limestone, underlie millions of square kilometers of Earth's surface, and it is in these rocks that groundwater carries on its important role as an erosional agent. Limestone is nearly insoluble in pure water but is quite easily dissolved by water containing small quantities of carbonic acid, and most groundwater contains this acid. It forms because rainwater readily dissolves carbon dioxide from the air and from decaying plants. When groundwater comes in contact with limestone, the carbonic acid reacts with the calcite (calcium carbonate) in the rocks to form calcium bicarbonate, a soluble material that is then carried away in solution.

Caverns

The most spectacular results of groundwater's erosional handiwork are limestone **caverns**. In the United States alone, about 17,000 caves have been discovered, and more are found every year. Although most are relatively small, some have spectacular dimensions. Mammoth Cave in Kentucky and Carlsbad Caverns in southeastern New Mexico are famous examples. The Mammoth Cave system is the most extensive in the world, with more than 540 kilometers (335 miles) of interconnected passages. The dimensions at Carlsbad Caverns are impressive in a different way. Here we find the largest and perhaps most spectacular single chamber. The Big Room at Carlsbad Caverns has an area equivalent to 14 football fields and enough height to accommodate the U.S. Capitol building.

Cavern Development Most caverns are created at or just below the water table, in the zone of saturation. Here acidic groundwater follows lines of weakness in the rock, such as joints and bedding planes. As time passes, the dissolving process slowly creates cavities and gradually enlarges them into caverns. Material that is dissolved by the groundwater is eventually discharged into streams and carried to the ocean.

In many cases, cavern development has occurred at several levels, with the current cavern-forming activity occurring at the lowest elevation. This situation reflects the close relationship between the formation of major subterranean passages and the river valleys into which they drain. As streams cut their valleys deeper and the elevation of the river drops, the water table also drops. Consequently, during periods when surface streams are rapidly

◀ SmartFigure 17.32
Kentucky's Mammoth Cave area Portions of Kentucky are underlain by limestone. Dissolution by groundwater has created a landscape characterized by caves and sinkholes.

Mobile Field Trip
https://goo.gl/Wb1Zhu

A.

B.

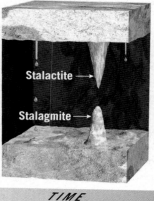

Soda straw →

Stalactite →

Stalagmite →

Column →

C.

TIME

▲ **Figure 17.33**
Cave decorations Speleothems are of many types, including stalactites, stalagmites, and columns. **A.** Close-up of a delicate live soda-straw stalactite in Chinn Springs Cave, Independence County, Arkansas. **B.** Stalagmites and stalactites in New Mexico's Carlsbad Caverns National Park. **C.** A column forms when a stalactite and a stalagmite join.

downcutting, surrounding groundwater levels drop rapidly, and cave passages are abandoned by the water while the passages are still relatively small in cross-sectional area. Conversely, when the entrenchment of streams is slow or negligible, there is time for large cave passages to form.

How Dripstone Forms   Certainly the features that arouse the greatest curiosity for most cavern visitors are the stone formations that give some caverns a wonderland

appearance. These are not erosional features, like the cavern itself, but depositional features created by the seemingly endless dripping of water over great spans of time. Recall from our discussion of hot springs that the calcium carbonate left behind produces the limestone we call *travertine*. These cave deposits, however, are also commonly called *dripstone*, an obvious reference to their mode of origin. Although the formation of caverns takes place in the zone of saturation, the deposition of dripstone is not possible until the caverns are above the water table, in the unsaturated zone. As soon as the chamber is filled with air, the stage is set for the decoration phase of cavern building to begin.

Dripstone Features—Speleothems   The various dripstone features found in caverns are collectively called **speleothems** (*spelaion* = cave, *thema* = put), and no two of them are exactly alike. Perhaps the most familiar speleothems are **stalactites** (*stalaktos* = trickling). These icicle-like pendants hang from the ceiling of a cavern and form where water seeps through cracks above. When the water reaches air in the cave, some of the dissolved carbon dioxide escapes from the drop, and calcite precipitates. Deposition occurs as a ring around the edge of the water drop. As drop after drop follows, each leaves an infinitesimal trace of calcite behind, and a hollow limestone tube is created. Water then moves through the tube, remains suspended momentarily at the end, contributes a tiny ring of calcite, and falls to the cavern floor. The stalactite just described is appropriately called a *soda straw* (**Figure 17.33A**). Often the hollow tube of the soda straw becomes plugged, or its supply of water increases. In either case, the water is forced to flow and hence deposit along the outside of the tube. As deposition continues, the stalactite takes on the more common conical shape.

Speleothems that form on the floor of a cavern and reach upward toward the ceiling are called **stalagmites** (*stalagmos* = dropping). The water supplying the calcite for stalagmite growth falls from the ceiling and splatters over the surface. As a result, stalagmites do not have a central tube and are usually more massive in appearance and rounded on their upper ends than stalactites (**Figure 17.33B**). Given enough time, a downward-growing stalactite and an upward-growing stalagmite may join to form a *column* (**Figure 17.33C**).

Thin films of groundwater moving down the walls or along the floor of a cavern may deposit layers of travertine called **flowstone** (**Figure 17.34**). Flowstone initially takes on the shape of the wall or floor beneath. Gradually deposits get thicker and more rounded as layers build on each other. Sometimes deposits are in the form of thin sheets called *draperies* or *curtains* where they appear to extend from overhanging portions of a cavern wall.

Karst Topography

Many areas of the world have landscapes that, to a large extent, have been shaped by the dissolving power of groundwater. Such areas are said to exhibit **karst topography**, named for the Kras Plateau in Slovenia, located along the northeastern shore of the Adriatic Sea, where such topography is strikingly developed. In the United States, karst landscapes occur in many areas that are underlain by limestone, including portions of Kentucky, Tennessee, Alabama, southern Indiana, and central and northern Florida (**Figure 17.35**). Generally, arid and semiarid areas are too dry to develop karst topography. When these features exist in such regions, they are

▲ **Figure 17.34**
Flowstone Thin films of water moving along the walls and/or floor of a cave deposit thin layers of travertine that gradually build up to create flowstone.

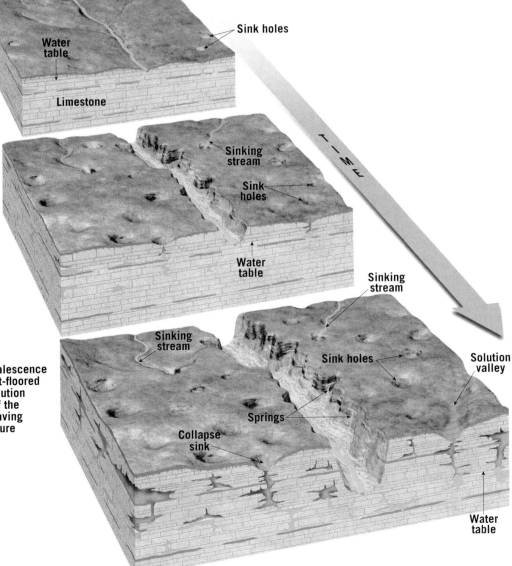

During early stages, groundwater percolates through limestone along joints and bedding planes. Solution activity creates and enlarges caverns at and below the water table.

With time, caverns grow larger and the number and size of sinkholes increase. Surface drainage is frequently funneled below ground.

Collapse of caverns and coalescence of sinkholes form larger, flat-floored depressions. Eventually solution activity may remove most of the limestone from the area, leaving isolated remnants, as in Figure 17.36.

Sink holes

Water table

Limestone

Sinking stream

Sink holes

Water table

TIME

Sinking stream

Sinking stream

Sink holes

Solution valley

Springs

Collapse sink

Water table

◀ **Figure 17.35**
Development of a karst landscape

► Figure 17.36
Tower karst in China One of the best-known and most distinctive regions of tower karst development is along the Li River in the Guilin District of southeastern China.

likely to be remnants of a time when rainier conditions prevailed.

Sinkholes   Karst areas typically have irregular terrain punctuated with many depressions, called **sinkholes**, or **sinks** (see chapter-opening photo). In the limestone areas of Florida, Kentucky, and southern Indiana, there are tens of thousands of these depressions, varying in depth from just 1 or 2 meters (3 or 7 feet) to more than 50 meters (165 feet).

Sinkholes commonly form in two ways. Some develop gradually over many years, without any physical disturbance to the rock. In these situations, the limestone immediately below the soil is dissolved by downward-sweeping rainwater that is freshly charged with carbon dioxide. With time, the bedrock surface is lowered, and the fractures into which the water seeps are enlarged. As the fractures grow in size, soil subsides into the widening voids; the soil is later removed by groundwater flowing in the passages below. These depressions are usually shallow and have gentle slopes.

In contrast, sinkholes can also form abruptly and without warning when the roof of a cavern collapses under its own weight. Typically, the depressions created in this manner are steep-sided and deep. When they form in populous areas, they may represent a serious geologic hazard. Such a situation is described in the "In the News" feature at the beginning of the chapter.

In addition to having surfaces pockmarked by sinkholes, karst regions characteristically show a striking

lack of surface drainage (streams). Following rainfall, the runoff is quickly funneled belowground through sinks. It then flows through caverns until it finally reaches the water table. Where streams do exist at the surface, their paths are usually short. The names of such streams often give a clue to their fate. In the Mammoth Cave area of Kentucky, for example, there is Sinking Creek, Little Sinking Creek, and Sinking Branch. Some sinkholes become plugged with clay and debris, creating small lakes or ponds.

Tower Karst   Some regions of karst development exhibit landscapes that look very different from the sinkhole-studded terrain depicted in the chapter-opening photo. One striking example is an extensive region in southern China that is described as exhibiting **tower karst**. As **Figure 17.36** shows, the term *tower* is appropriate because the landscape consists of a maze of isolated steep-sided hills that rise abruptly from the ground. Each is riddled with interconnected caves and passageways. This type of karst topography forms in wet tropical and subtropical regions having thick beds of highly jointed limestone. Here groundwater has dissolved large volumes of limestone, leaving only these residual towers. Karst development occurs more rapidly in tropical climates due to the abundant rainfall and the greater availability of carbon dioxide from the decay of lush tropical vegetation. The extra carbon dioxide in the soil means there is more carbonic acid for dissolving limestone. Other tropical areas of advanced karst development include portions of Puerto Rico, western Cuba, and northern Vietnam.

CONCEPT CHECKS 17.7

1. How does groundwater create caverns?

2. What causes cavern formation to stop at one level (depth) but continue or begin at a lower level?

Concept Checker
https://goo.gl/ZvS3MW

3. How do stalactites and stalagmites form?

4. Describe two ways in which sinkholes form.

17

CONCEPTS IN REVIEW

Groundwater

17.1 The Importance of Groundwater

Describe the importance of groundwater as a source of
freshwater and the role of groundwater as a geologic agent.

Key Term: groundwater

- *Groundwater* is water stored below Earth's surface, mainly in tiny
pore spaces between rock or sediment grains. Groundwater represents
the largest reservoir of freshwater that is readily available to humans
and is a critical resource for human civilization.

- Groundwater is important geologically because it dissolves rock to make
sinkholes and caverns and supplies surface streams with additional water.

- Each day in the United States we use about 306 billion gallons of
freshwater. Groundwater provides about 76 billion gallons, or
25 percent of the total. More groundwater is used for irrigation than
for all other uses combined.

Q Examine Figure 17.1 to answer these questions: How much of Earth's
total freshwater supply is groundwater? How much of Earth's *liquid*
freshwater is groundwater?

17.2 Groundwater and the Water Table

Prepare a sketch with labels that summarizes the distribution of
water beneath Earth's surface. Discuss the factors that cause
variations in the water table and describe the interactions
between groundwater and streams.

Key Terms:
zone of soil moisture
zone of saturation

water table
capillary fringe
unsaturated zone

gaining stream
losing stream

- Some of the rain that falls on land soaks into the ground. Typically, a
hole dug into the ground penetrates this *zone of soil moisture* and then
crosses the *unsaturated zone*, where pore spaces contain both water
and air. The soil grows moist again just above the water table, in the
capillary fringe. Crossing the *water table*, the boundary between the
groundwater below and the unsaturated zone above, the hole begins to
fill (to the height of the water table) with water that flows in from the
zone of saturation.

- Streams and groundwater interact in one of three ways: Streams gain
water from the inflow of groundwater (*gaining stream*); they lose water
through the streambed to the groundwater system (*losing stream*); or
they do both, gaining in some sections and losing in others.

Q Examine this cross section, which shows the distribution of water in
loose sediments. Provide the correct term for each lettered feature.

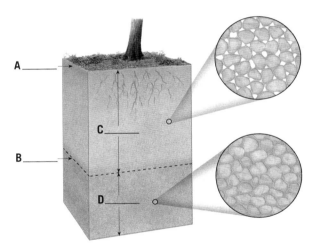

17.3 Storage and Movement of Groundwater

Summarize the factors that influence the storage and movement of groundwater. Discuss how groundwater movement is
measured and the different scales of movement.

Key Terms:
porosity
permeability
aquitard

aquifer
recharge area
discharge area
hydraulic gradient

hydraulic conductivity
Darcy's law

- The quantity of water that can be stored in a material depends on the
material's *porosity* (the volume of open spaces). The *permeability*
of a material (its ability to transmit a fluid through interconnected
pore spaces) is a very important factor controlling the movement of
groundwater.

- Materials with very small pore spaces (such as clay) hinder or prevent
groundwater movement and are called *aquitards*. *Aquifers* consist of

materials with larger pore spaces (such as sand) that are permeable
and transmit groundwater freely.

- Groundwater flows slowly through underground pore spaces, moving
on average only a few centimeters per day. Driven by gravity and
pressure, it moves as a three-dimensional mass from *recharge areas*
(where water is added) to *discharge areas* (where water leaves the
groundwater system), such as springs, gaining streams, or wells drilled
by people.

- French scientist-engineer Henri Darcy pioneered the quantification of
groundwater flow by measuring the slope of the water table (*hydraulic
gradient*) and the permeability of the sediment or rock (*hydraulic
conductivity*). *Darcy's law* combines these measurements in an
equation to estimate an aquifer's discharge.

- Groundwater flows both short and long distances at both shallow and
deep levels. Closer to the surface, the flow is more local in scale, while
at greater depths, the flow occurs over regional scales.

17.4 Wells and Artesian Systems

Discuss water wells and their relationship to the water table. Sketch and label a simple artesian system.

Key Terms:	cone of depression	confined aquifer
well	perched water table	nonflowing artesian well
drawdown	artesian	flowing artesian well

- For centuries, humans have been obtaining groundwater by drilling *wells*. As water is pumped out, the water table immediately adjacent to the well drops. This *drawdown* results in a "dimple" in the surface of the water table called the *cone of depression*. If there is sufficient drawdown, the cone of depression might encompass a large enough area to cause neighboring wells to go dry.
- A *perched water table* results from groundwater "piled up" atop an aquitard that is above the main body of groundwater. The shape of the water table is complex, which results in challenges for people trying to dig productive wells.
- *Artesian* wells tap into inclined aquifers bounded above and below by aquitards. For a system to qualify as artesian, the water in the well must be under sufficient pressure that it can rise above the top of the *confined aquifer*. Artesian wells may be *flowing* or *nonflowing*, depending on whether the pressure surface is above or below the ground surface.

Q In 1900, when this well was drilled near Woonsocket in eastern South Dakota, a "gusher" of water nearly 30 meters (100 feet) high resulted. Describe or sketch the subsurface geologic situation that was responsible for this fountain of water. What term is applied to a well such as this?

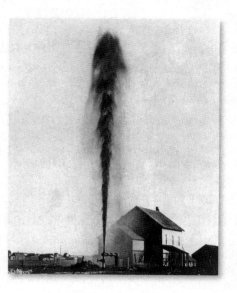

17.5 Springs, Hot Springs, and Geysers

Distinguish among springs, hot springs, and geysers.

Key Terms:	hot spring
spring	geyser

- *Springs* are naturally occurring spots where groundwater leaves the ground and flows out onto the surface. They may be due to the intersection of a perched water table and Earth's surface.
- *Hot springs* are like regular springs but hot. They transfer heat from the deeper crust to the surface. Most often, this heat comes from relatively shallow bodies of magma.
- *Geysers* are intermittent hot springs that "erupt" hot water periodically. They are fed by underground chambers that fill with water. Because this water is under pressure, it can be heated well above the normal surface boiling point. Heating causes the water to expand, forcing a little of it out of the spring and reducing the underground pressure. Once the pressure drops enough, the water in the lower chambers flashes into steam, propelling the eruption. Geysers can precipitate silica or calcium

carbonate around the geyser vent, producing the rocks siliceous sinter (geyserite) or travertine (tufa).

Q This photo from the 1930s shows Franklin Roosevelt enjoying the hot springs at the presidential retreat at Warm Springs, Georgia. The temperature of this water is always near 32°C (90°F). This area has no history of recent igneous activity. What is the likely reason these springs are so warm?

17.6 Environmental Problems

List and discuss important environmental problems associated with groundwater.

- Groundwater can be "mined" by being extracted at a rate that is greater than the rate of replenishment. When groundwater is treated as a nonrenewable resource, as it is in parts of the High Plains aquifer, the water table drops, in some cases by more than 45 meters (150 feet).
- The extraction of groundwater can cause pore spaces to decrease in volume and the grains of loose Earth materials to pack more closely together. This overall compaction of sediment volume results in subsidence of the land surface.

- Saltwater contamination is a common environmental problem near coastal areas. Fresh groundwater "floats" on salty groundwater due to its lower density. If sufficient freshwater is pumped out to lower the water table by some amount, the base of the freshwater lens will rise about 40 times that amount. Deep wells may begin to access the deeper, salty water instead of freshwater.
- Contamination of groundwater with sewage, highway salt, fertilizer, or industrial chemicals is another issue of critical concern. Once groundwater is contaminated, the problem is very difficult to solve, requiring expensive remediation or abandonment of the aquifer.

17.7 The Geologic Work of Groundwater

Explain the formation of caverns and the development of karst topography.

Key Terms:	stalactite	karst topography
cavern	stalagmite	sinkhole
speleothem	flowstone	tower karst

- Groundwater dissolves rock, in particular limestone, leaving behind void spaces in the rock. *Caverns* form in the zone of saturation, but later dropping of the water table may leave them open and dry—and available for people to explore.
- Dripstone is rock deposited by dripping of water that contains dissolved calcium carbonate inside caverns. *Speleothems*, features made of dripstone, include *stalactites*, *stalagmites*, columns, and *flowstone*.
- *Karst topography* is a distinctive type of landscape dominated by the dissolving of limestone near Earth's surface. Collapsing caverns show up as *sinkholes*. Streams flowing on the surface may "sink" into the subterranean cavern system, and in other places the same water may reemerge as a spring. If enough limestone is dissolved, only isolated pinnacles of limestone will remain, towering over the landscape as *tower karst*.

Q Identify the three speleothems shown in this photograph. Would these speleothems have formed in the saturated zone or the unsaturated zone? Why?

GIVE IT SOME THOUGHT

1. The cemetery in this photo is located in New Orleans, Louisiana. As in other cemeteries in the area, all the burial plots here are aboveground. Based on what you have learned in this chapter, suggest a reason for this rather unusual practice.

2. Imagine a water molecule that is part of a groundwater system in an area of gently rolling hills in the eastern United States. Describe some possible paths the molecule might take through the hydrologic cycle if:

   a. It is pumped from the ground to irrigate a farm field.

   b. There is a long period of heavy rainfall.

   c. The water table in the vicinity of the molecule develops a steep cone of depression due to heavy pumping from a nearby well.

   Combine your understanding of the hydrologic cycle with your imagination and include possible short-term and long-term destinations and information about how the molecule gets to these places via evaporation, transpiration, condensation, precipitation, infiltration, and runoff. Remember to consider possible interactions with streams, lakes, groundwater, the ocean, and the atmosphere.

3. The drainage basin of the Republican River occupies portions of Colorado, Nebraska, and Kansas. A significant part of the basin is considered semiarid. In 1943, the three states made a legal agreement regarding sharing the river's water. In 1998, Kansas went to court to force farmers in southern Nebraska to substantially reduce the amount of groundwater used for irrigation. Nebraska officials claimed that the farmers were not taking water from the Republican River and thus were not violating the 1943 agreement. The court ruled in favor of Kansas.

   a. Explain why the court ruled that groundwater in southern Nebraska should be considered part of the Republican River system.

   b. How might heavy irrigation in a drainage basin influence the flow of a river?

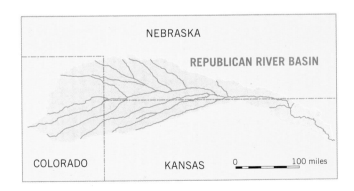

4. During a trip to a grocery store, your friend wants to buy some bottled water. Some brands promote the fact that their product is artesian. Other brands boast that their water comes from a spring. Your friend asks, "Is artesian water or spring water necessarily better than water from other sources?" How would you answer?

5. Imagine that you are an environmental scientist who has been hired to solve a groundwater contamination problem. Several homeowners have noticed that their well water has a funny smell and taste. Some think the contamination is coming from a landfill, but others think it might be due to a nearby cattle feedlot or chemical plant. Your first step is to gather data from wells in the area and prepare the map of the water table shown here.

   a. Based on your map, can any of the three potential sources of contamination be eliminated? If so, explain.

   b. What other steps would you take to determine the source of the contamination?

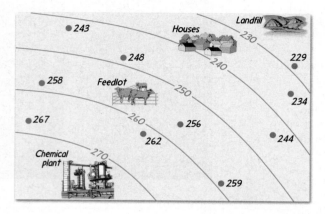

6. An acquaintance is considering purchasing a large tract of productive irrigated farmland in western Texas. His intention is to continue growing crops on the land for years to come. If he asked your opinion about the area he selected, what figure in this chapter would you consult before you responded? How would this figure help your friend evaluate his potential purchase?

EYE ON EARTH

1. This satellite image shows a portion of the desert in northern Saudi Arabia, a region known for its abundant sunshine, high temperatures, and meager rainfall. The green circles are agricultural fields that are about 1 kilometer (0.62 mile) in diameter. Water for irrigation is pumped from deep aquifers and distributed around a center point within each field—a technique known as center pivot irrigation. The deep aquifers contain water that dates to the Ice Age about 20,000 years ago, a time when the climate in this region was wetter and milder.

   a. Is it likely that agricultural activity in this region is sustainable indefinitely? Explain.

   b. A significant portion of the water placed on these fields is "lost" (not available to the crops). Suggest a reason for the loss of water.

   c. Relate what is likely occurring to the water table in the region pictured here to an example of a similar situation in the United States.

Each circle is 1 km in diameter

2. Sinkholes commonly form in one of two ways. Examine the accompanying photo, which shows a sinkhole in Winter Park, Florida, to answer the following questions.

   a. How did this sinkhole likely form?

   b. What aspects of the image support your answer?

DATA ANALYSIS

Sinkholes in Tennessee

Karst is common in regions that have carbonate bedrock, such as limestone, underneath the surface. Cracks allow rainwater to seep in and dissolve the bedrock, forming chambers underground. Sinkholes are formed when these chambers collapse.

https://goo.gl/eD226M

ACTIVITIES

Go to the Tennessee sinkhole count map at http://tnlandforms.us/landforms/sinks.php. Click on the word "map" in the sentence "Google map with 43 deep sinkholes" under the Tennessee map.

1. **Which regions of Tennessee are prone to sinkholes?**

2. **Identify three towns that have problems with sinkholes.**

3. **What overall pattern do you see in sinkhole distribution?**

Click on the "Map" drop-down menu (which is superimposed at the top-left the corner of the map) and choose "US topo."

4. **Do sinkholes tend to form along mountain ridges or in valleys? Why?**

Go to the Tennessee Geology map at http://mrdata.usgs.gov/sgmc/tn.html. Click on the hamburger button (the three horizontal lines) at the top left of the map and select the boxes next to "US States" and "OpenStreetMap." Click on the colored regions on the map to see geologic units.

5. **Click on several regions that show sinkholes on the Tennessee sinkhole count map you looked at earlier. Which rock types in Tennessee are prone to sinkholes?**

6. **Click on several regions that do not show sinkholes on the Tennessee sinkhole count map. Which rock types in Tennessee are not prone to sinkholes?**

7. **Based on the pattern that you see in Tennessee, in which portions of eastern states would you expect to find sinkholes?**

Mastering Geology

Looking for additional review and test prep materials? Visit pearson.com/mastering/geology to enhance your understanding of this chapter's content by accessing a variety of resources, including **Self-Study Quizzes, Geoscience Animations, SmartFigure Tutorials,** *Project Condor* **Videos, Mobile Field Trips, Dynamic Study Modules,** and an optional **Pearson eText.**

Will Glacier National Park Need a Name Change?

In Montana's Glacier National Park, some effects of climate change are strikingly clear. Glaciers are melting, and many glaciers have already disappeared. The rapid retreat of these small alpine glaciers reflects recent climate changes, which have brought altered temperature and precipitation. It has been estimated that approximately 150 glaciers were present in 1850. Most were still present in 1910, when the park was established and named. Recent measurements found only 26 glaciers that were large enough to merit the name—that is, larger than 0.1 square kilometer (25 acres), the size criterion used by U.S. Geological Survey researchers to define a glacier.

A U.S. Geological Survey study published in 2017 shows that these remaining glaciers lost an average of 39 percent of their area—and as much as 85 percent—in the approximately 50 years between 1966 and 2015. Most of the remaining glaciers are projected to disappear over the next few decades; by the end of the century, they will all be gone.

These changes affect more than the appropriateness of the park's name. Alpine glaciers act as natural reservoirs, storing water that falls as snow and releasing it gradually over the year. Without them, the amount, timing, and temperature of stream-flow change. Fish and other aquatic organisms are affected immediately, and the surrounding landscape responds as well. Similar changes are happening in other mountainous regions; with few exceptions, the world's alpine glaciers are in retreat.

▲ This hiker is standing above Grinnell Glacier, one of the shrinking alpine glaciers in Glacier National Park.

▶ Sunrise shines on the Garden Wall, a spine of rock shaped by glacial ice in Montana's Glacier National Park.

18

Glaciers and Glaciation

FOCUS ON CONCEPTS

Each statement represents the primary learning objective for the corresponding major heading within the chapter. After you complete the chapter, you should be able to:

18.1 Explain the role of glaciers in the hydrologic and rock cycles. Describe the different types of glaciers, their characteristics, and their present-day distribution.

18.2 Describe how glaciers move, the rates at which they move, and the significance of the glacial budget.

18.3 Discuss the processes of glacial erosion. Identify and describe major topographic features created by glacial erosion.

18.4 Distinguish between the two basic types of glacial drift. List and describe the major depositional features associated with glacial landscapes.

18.5 Describe and explain several important effects of Ice Age glaciers other than erosional and depositional landforms.

18.6 Briefly discuss the development of glacial theory and summarize current ideas on the causes of ice ages.

Climate has a strong influence on the nature and intensity of Earth's external processes. This fact is dramatically illustrated in this chapter because the existence and extent of glaciers are largely controlled by Earth's changing climate. Like the running water and groundwater that were the focus of the preceding two chapters, glaciers represent a significant erosional process. These moving masses of ice are responsible for creating many unique landforms and are part of an important link in the rock cycle in which the products of weathering are transported and deposited as sediment.

Today glaciers cover nearly 10 percent of Earth's land surface; however, in the recent geologic past, ice sheets were three times more extensive, covering vast areas with ice thousands of meters thick. Many regions still bear the marks of these glaciers. The landscapes of such diverse places as the Alps, Cape Cod, and Yosemite Valley were fashioned by now-vanished masses of glacial ice. Moreover, Long Island, the Great Lakes, and the fiords of Norway and Alaska all owe their existence to glaciers. Glaciers, of course, are not just a phenomenon of the geologic past. As you will see, they are still sculpting the landscape and depositing debris in many regions today.

18.1 Glaciers: A Part of Two Basic Cycles

Explain the role of glaciers in the hydrologic and rock cycles. Describe the different types of glaciers, their characteristics, and their present-day distribution.

Glaciers are a part of two fundamental cycles in the Earth system: the hydrologic cycle and the rock cycle. The water of the hydrosphere is constantly cycled through the atmosphere, biosphere, and geosphere. Time and time again, water evaporates from the oceans into the atmosphere, precipitates on the land, and flows in rivers and underground back to the sea. However, when precipitation falls at high elevations or high latitudes, the water may not immediately make its way toward the sea. Instead, it may become part of a glacier. Although the ice will eventually melt, allowing the water to continue its path to the sea, water can be stored as glacial ice for many tens, hundreds, or even thousands of years.

A **glacier** is a thick ice mass that forms over hundreds or thousands of years. It originates on land from the accumulation, compaction, and recrystallization of snow. A glacier appears to be motionless, but it is not; glaciers move very slowly. Like running water, groundwater, wind, and waves, glaciers are dynamic erosional agents that accumulate, transport, and deposit sediment. As such, glaciers are among the agents that perform a basic function in the rock cycle. Although glaciers are found in many parts of the world today, most are located in remote areas, either near Earth's poles or in high mountains.

Valley (Alpine) Glaciers

Literally thousands of relatively small glaciers exist in lofty mountain areas, where they usually follow valleys that were originally occupied by streams. Unlike the rivers that previously flowed in these valleys, glaciers advance slowly, perhaps only a few centimeters per day. Because of their setting, these moving ice masses are termed **valley glaciers**, or **alpine glaciers** (**Figure 18.1**). Each glacier actually is a stream of ice, bounded by precipitous rock walls, that flows downvalley from an accumulation center near its head. Like

▲ **Figure 18.1**
Valley glacier. This tongue of ice, also called an *alpine glacier*, is still eroding the Swiss Alps. Dark stripes of sediment within these glaciers are called medial moraines.

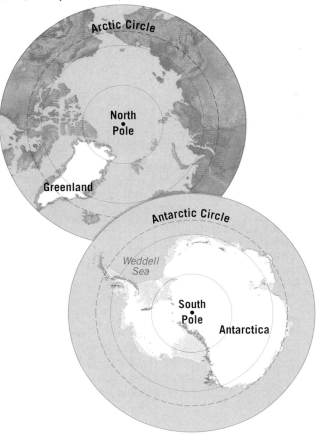

Greenland's ice sheet occupies 1.7 million square kilometers (663,000 square miles), about 80% of the island.

The area of the Antarctic Ice Sheet is almost 14 million square kilometers (5,460,000 square miles). Ice shelves occupy an additional 1.4 million square kilometers (546,000 square miles).

▲ **SmartFigure 18.2**
Ice sheets The only present-day ice sheets are those covering Greenland and Antarctica. Their combined areas represent almost 10 percent of Earth's land area.

Video
https://goo.gl/oCi42n

rivers, valley glaciers can be long or short, wide or narrow, single or with branching tributaries. Generally, alpine glaciers are longer than they are wide. Some extend for just a fraction of a kilometer, whereas others go on for many tens of kilometers. The west branch of the Hubbard Glacier, for example, runs through 112 kilometers (nearly 70 miles) of mountainous terrain in Alaska and the Yukon Territory.

Ice Sheets

In contrast to valley glaciers, **ice sheets** exist on a much larger scale. The low total annual solar radiation reaching the poles makes these regions hospitable to great ice accumulations. Presently both of Earth's polar regions support ice sheets: Greenland in the Northern Hemisphere and Antarctica in the Southern Hemisphere (**Figure 18.2**).

Ice Age Ice Sheets   About 18,000 years ago, glacial ice covered not only Greenland and Antarctica but also large portions of North America, Europe, and Siberia. That period in Earth history is appropriately known as the *Last Glacial Maximum*. The term implies that there were other glacial maximums, which is indeed the case. Throughout the Quaternary period, which began about 2.6 million years ago and extends to the present, ice sheets have formed, advanced over broad areas, and then melted away. These alternating glacial and interglacial periods have occurred over and over again.

Greenland and Antarctica   Some people mistakenly think that the North Pole is covered by glacial ice, but this is not the case. The ice that covers the Arctic Ocean is **sea ice**—frozen seawater. Sea ice floats because ice is less dense than liquid water. Although sea ice never completely disappears from the Arctic, the area covered with sea ice expands and contracts with the seasons (see Figure 21.28, page 612). The thickness of sea ice ranges from a few centimeters for new ice to 4 meters (13 feet) for sea ice that has survived for years. By contrast, glaciers can be hundreds or thousands of meters thick.

Glaciers form on land, and in the Northern Hemisphere, Greenland supports an ice sheet. Greenland extends between about 60 and 80 degrees north latitude. This largest island on Earth is covered by an imposing ice sheet that occupies 1.7 million square kilometers

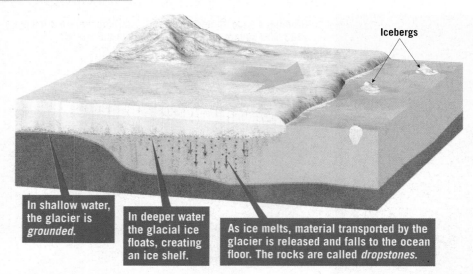

In shallow water, the glacier is *grounded*.

In deeper water the glacial ice floats, creating an ice shelf.

As ice melts, material transported by the glacier is released and falls to the ocean floor. The rocks are called *dropstones*.

Icebergs

▲ **Figure 18.3**
Ice shelves An ice shelf forms when a glacier or an ice sheet flows into the adjacent ocean.

beneath a glacier usually appear as relatively subdued undulations on the surface of the ice. Such topographic differences, however, affect the behavior of the ice sheets, especially near their margins, by guiding flow in certain directions and creating zones of faster and slower movement.

Ice Shelves   When a glacier or ice sheet flows into the ocean, it is called a **tidewater glacier**. In shallow water, the glacier remains grounded. Where the advancing glacier reaches deeper water, the ice floats, becoming an **ice shelf** (**Figure 18.3**). These large, relatively flat masses of floating glacial ice extend seaward from the coast but remain attached to the land along one or more sides. About 80 percent of the ice lies below the surface of the ocean. There are ice shelves along more than half of the Antarctic coast, but there are relatively few in Greenland.

(more than 660,000 square miles), or about 80 percent of the island. Averaging nearly 1500 meters (5000 feet) thick, the ice extends 3000 meters (10,000 feet) above the island's bedrock floor in some places.

In the Southern Hemisphere, practically all of Antarctica is covered by two huge ice sheets that extend over an area of more than 13.9 million square kilometers (5.4 million square miles). Because of the proportions of these huge features, they are often called *continental ice sheets*. The combined areas of present-day continental ice sheets represent almost 10 percent of Earth's land area.

These enormous masses flow out in all directions from one or more snow-accumulation centers and completely obscure all but the highest areas of underlying terrain. Even sharp variations in the topography

These shelves are thickest on their landward sides and become thinner seaward. They are sustained by ice from the adjacent ice sheet and may also be nourished by snowfall on their surfaces and freezing of seawater to their bases. Antarctica's ice shelves extend over approximately 1.4 million square kilometers (0.54 million square miles). The Ross and Ronne-Filchner Ice Shelves are the largest, with the Ross Ice Shelf alone covering an area approximately the size of Texas (**Figure 18.4A**). In recent years, satellite monitoring has shown that some ice

► **Figure 18.4**
Antarctica's ice shelves A. The map identifies the major ice shelves surrounding the continent. **B.** This satellite image, acquired in September 2017, shows an iceberg that had recently broken away from the Larsen C Ice Shelf. The iceberg is about the size of the state of Delaware. Warmer temperatures are likely increasing the frequency of calving events such as this.

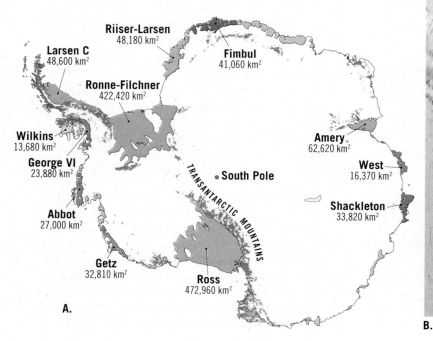

Riiser-Larsen
48,180 km²

Larsen C
48,600 km²

Fimbul
41,060 km²

Ronne-Filchner
422,420 km²

Wilkins
13,680 km²

Amery
62,620 km²

George VI
23,880 km²

West
16,370 km²

TRANSANTARCTIC MOUNTAINS

• South Pole

Abbot
27,000 km²

Shackleton
33,820 km²

Getz
32,810 km²

Ross
472,960 km²

A.

Larsen C
Ice Shelf

A-68A

B.

shelves are unstable and breaking apart. For example, in July 2017, satellite sensors determined that a massive iceberg was separating from the Larsen C Ice Shelf. By August, the first sunlit views became available (**Figure 18.4B**). The new iceberg, named A-68A by the U.S. National Ice Center, was about the size of the state of Delaware. This was not an isolated happening but part of a trend related to accelerated global warming.

Other Types of Glaciers

In addition to valley glaciers and ice sheets, scientists have identified other types of glaciers. Covering some uplands and plateaus are masses of glacial ice called **ice caps**. Like ice sheets, ice caps completely bury the underlying landscape, but they are much smaller than the continental-scale features. Ice caps occur in many places, including Iceland and several of the large islands in the Arctic Ocean (**Figure 18.5**).

Often ice caps and ice sheets feed **outlet glaciers**. These tongues of ice flow down valleys, extending outward from the margins of these larger ice masses. The tongues are essentially valley glaciers

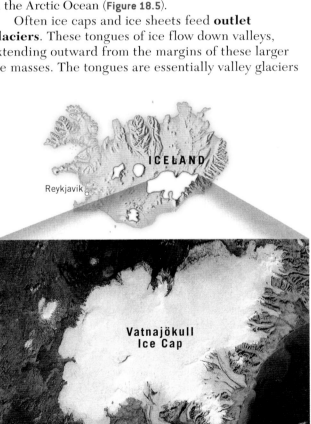

▲ SmartFigure 18.5
Iceland's Vatnajökull Ice Cap
In 1996 the Grímsvötn Volcano erupted beneath this ice cap, an event that triggered melting and floods.

Mobile Field Trip
https://goo.gl/ RsbHWM

Ice caps completely bury the underlying terrain but are much smaller than ice sheets.

◄ Figure 18.6
Piedmont glacier The ice of a piedmont glacier spills from a steep valley onto a relatively flat plain, where it spreads out. Malaspina Glacier, in southeastern Alaska, fills most of this image. Covering roughly 3880 square kilometers (1500 square miles), it extends nearly 45 kilometers (28 miles) from the mountain front nearly to the sea.

that are avenues for ice movement from an ice cap or ice sheet through mountainous terrain to the sea. Where they encounter the ocean, some outlet glaciers spread out as floating ice shelves. Often large numbers of icebergs are produced.

Piedmont glaciers occupy broad lowlands at the bases of steep mountains and form when one or more alpine glaciers emerge from the confining walls of mountain valleys. Here the advancing ice spreads out to form a broad lobe (**Figure 18.6**). The sizes of individual piedmont glaciers vary greatly. Among the largest is the broad Malaspina Glacier along the coast of southern Alaska. It covers thousands of square kilometers of the flat coastal plain at the foot of the lofty St. Elias Range.

CONCEPT CHECKS 18.1

1. Where are glaciers found today? What percentage of Earth's land surface do they cover?

2. Describe how glaciers fit into the hydrologic cycle. What role do they play in the rock cycle?

Concept Checker https://goo.gl/3EegWw

3. List and briefly distinguish among four types of glaciers.

4. Distinguish among the terms *ice sheet*, *sea ice*, and *ice shelf*.

18.2 Formation and Movement of Glacial Ice

Describe how glaciers move, the rates at which they move, and the significance of the glacial budget.

Snow is the raw material from which glacial ice originates; therefore, glaciers form in areas where more snow falls in winter than melts during the summer. Glaciers develop in the high-latitude polar realm because, even though annual snowfall totals are modest, temperatures are so low that little of the snow melts. Glaciers can form in mountains because temperatures drop with an increase in altitude. So even near the equator, glaciers may form at elevations above about 5000 meters (16,400 feet). For example, less than 10 degrees from the equator, the Cordillera Blanca, a part of the lofty Andes in Peru, is home to more than 700 alpine glaciers. The elevation above which snow remains throughout the year varies with latitude. Near the equator, this boundary occurs high in the mountains, whereas in the vicinity of 60 degrees north or south latitude, it is at or near sea level. Before a glacier is created, however, snow must be converted into glacial ice.

Glacial Ice Formation

When temperatures remain below freezing following a snowfall, the fluffy accumulation of delicate hexagonal crystals soon changes. As air infiltrates the spaces between the crystals, the extremities of the crystals evaporate, and the water vapor condenses near the centers of the crystals. This process of recrystallization makes the snowflakes smaller, thicker, and more spherical, giving them the consistency of coarse sand. The snow also compacts, reducing the pore spaces between grains. The resulting granular recrystallized snow, called **firn**, is the material that commonly makes up old snow banks near the end of winter. As more snow is added on top of firn, the pressure on the lower layers gradually increases, compacting the ice grains at depth. Once the thickness of ice and snow exceeds about 50 meters (165 feet), the weight is sufficient to fuse firn into a solid mass of interlocking ice crystals. *Glacial ice* has now been formed.

The rate at which this transformation occurs varies. In regions where the annual snow accumulation is great, burial is relatively rapid, and snow may turn to glacial ice in a matter of a decade or less. Where the yearly addition of snow is less abundant, burial is slow, and the transformation of snow to glacial ice may take hundreds of years.

Less than 10 degrees from the equator . . . a part of the lofty Andes in Peru is home to more than 700 alpine glaciers.

How Glaciers Move

The way in which ice moves is complex and is of two basic types. The first of these, *plastic flow*, involves movement *within* the ice. Ice behaves as a brittle solid until the pressure upon it is equivalent to the weight of about 50 meters (165 feet) of ice. Once that load is surpassed, ice behaves as a plastic material, and flow begins. Such flow occurs because of the molecular structure of ice. Glacial ice consists of layers of molecules stacked one upon the other. The bonds between layers are weaker than those within each layer.

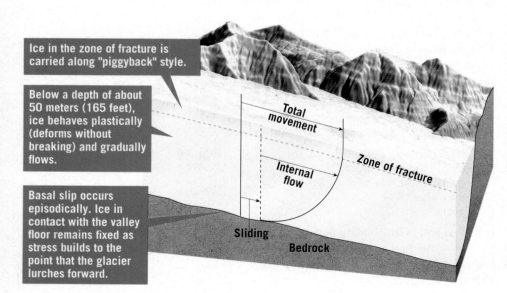

Ice in the zone of fracture is carried along "piggyback" style.

Below a depth of about 50 meters (165 feet), ice behaves plastically (deforms without breaking) and gradually flows.

Basal slip occurs episodically. Ice in contact with the valley floor remains fixed as stress builds to the point that the glacier lurches forward.

Total movement

Internal flow

Zone of fracture

Sliding

Bedrock

◄ SmartFigure 18.7
Movement of a glacier This vertical cross section through a glacier shows that movement is divided into two components. Also notice that the rate of movement is slowest at the base of the glacier, where frictional drag is greatest.

Tutorial
https://goo.gl/7LzafR

Therefore, when a stress exceeds the strength of the bonds between the layers, the layers remain intact and slide over one another.

A second and often equally important mechanism of glacial movement consists of an entire ice mass slipping along the ground. With the exception of some glaciers located in polar regions where the ice is probably frozen to the solid bedrock floor, most glaciers are thought to move by this sliding process, called *basal slip*. In this process, meltwater probably acts as a hydraulic jack and perhaps as a lubricant that helps the ice move over the rock. The source of the liquid water is related in part to the fact that the melting point of ice decreases as pressure increases. Therefore, deep within a glacier, the ice may be at the melting point even though its temperature is below 0°C (32°F).

Other factors may also contribute to the presence of meltwater deep within a glacier. Temperatures may be increased by plastic flow (an effect similar to heating due to friction), by heat added from Earth below, and by the refreezing of meltwater that has seeped down from above. This last process relies on the fact that as water changes state from liquid to solid, heat (termed *latent heat of fusion*) is released.

Figure 18.7 illustrates the effects of these two basic types of glacial motion. This vertical profile through a glacier also shows that not all the ice flows forward at the same rate. Frictional drag with the bedrock floor causes the lower portions of the glacier to move more slowly.

In contrast to the lower portion of the glacier, the upper 50 meters (165 feet) or so are not under sufficient pressure to exhibit plastic flow. Rather, the ice in this uppermost zone is brittle, and this zone is appropriately referred to as the **zone of fracture**. The ice in the zone of fracture is carried along "piggyback" style by the ice below. When the glacier moves over irregular terrain or down steep slopes, the zone of fracture is subjected to tension, resulting in cracks called **crevasses** (**Figure 18.8**). Sometimes these gaping cracks can be the only perceptible indication that the glacier is moving. This is the case in the aerial view of Antarctica in Figure 18.8B.

Observing and Measuring Movement

Unlike the movement of water in streams, the movement of glacial ice is not obvious. If we could watch a valley glacier move, we would see that, as with the water in a river, the ice moves downstream at different rates. Flow is greatest in the center of the glacier because the drag created by the walls and floor of the valley slows the base and sides.

The first experiments on glacier movement were designed and carried out in the Alps (**Figure 18.9**). Markers were placed in a straight line

▲ Figure 18.8
Crevasses A. As a glacier moves through narrow canyons and down steep slopes, internal stresses cause large cracks to develop in the brittle upper portion of the glacier, called the zone of fracture. Crevasses can extend to depths of 50 meters (165 feet) and can make travel across glaciers dangerous. **B.** In this aerial image from Antarctica, crevasses provide the only obvious indication that the ice is moving.

across an alpine glacier, and the line's position was marked on the valley walls so that if the ice moved, the change in position could be detected. Periodically positions of the markers were recorded, showing the movement. Although most glaciers move too slowly for direct visual detection, the experiments successfully demonstrated movement nevertheless occurs. While some glaciers move so slowly that trees and other vegetation may become well established in the debris on the surface, others advance up to several meters each day. Recent satellite imaging provided insights into

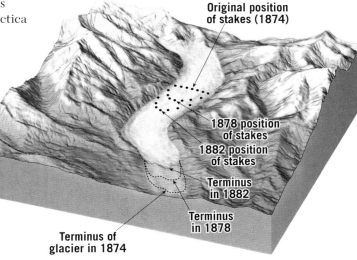

◀ Figure 18.9
Measuring the movement of a glacier Ice movement and changes in the terminus of Rhône Glacier, Switzerland. In this classic study of a valley glacier, the movement of stakes clearly shows that glacial ice moves and that movement along the sides of the glacier is slower than movement in the center. Also notice that even though the ice front was retreating, the ice within the glacier was advancing.

Original position of stakes (1874)

1878 position of stakes

1882 position of stakes

Terminus in 1882

Terminus in 1878

Terminus of glacier in 1874

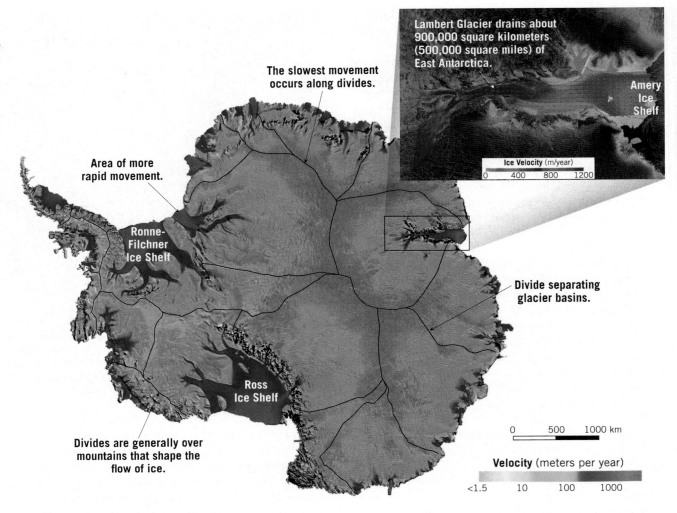

▶ **Figure 18.10**
Movement of Antarctic ice These maps were created based on thousands of satellite measurements taken over several years. Outflow from the interior is organized into a series of drainage basins separated by ice divides. Flow is concentrated into narrow, mountain-bound glaciers or into relatively fast-moving ice streams surrounded by slower-moving ice.
(NASA)

Lambert Glacier drains about 900,000 square kilometers (500,000 square miles) of East Antarctica.

Amery Ice Shelf

Ice Velocity (m/year)
0   400   800   1200

The slowest movement occurs along divides.

Area of more rapid movement.

Ronne-Filchner Ice Shelf

Divide separating glacier basins.

Ross Ice Shelf

Divides are generally over mountains that shape the flow of ice.

0   500   1000 km

Velocity (meters per year)
<1.5   10   100   1000

movements within the Antarctic Ice Sheet. An examination of **Figure 18.10** shows that portions of some outlet glaciers move at rates greater than 800 meters (2600 feet) per year; on the other hand, ice in some interior regions creeps along at less than 2 meters (6.5 feet) per year. Movement of some glaciers is characterized by occasional periods of extremely rapid advance called *surges*, followed by periods of much slower movement.

Budget of a Glacier: Accumulation Versus Wastage

Snow is the raw material from which glacial ice originates; therefore, glaciers form in areas where more snow falls in winter than melts during the summer. Glaciers are constantly gaining and losing ice.

Glacial Zones   Snow accumulation and ice formation occur in the **zone of accumulation** (**Figure 18.11**). Its outer limits are defined by the

ZONE OF ACCUMULATION
More snow falls each winter than melts each summer

ZONE OF WASTAGE
All the snow from the previous winter melts, along with some glacial ice

Snowline (Equilibrium line)

Crevasses

Braided streams

◀ **SmartFigure 18.11**
Zones of a glacier The snowline separates the zone of accumulation and the zone of wastage. Whether the ice front advances, retreats, or remains stationary depends on the balance or lack of balance between accumulation and wastage (ablation).

Tutorial
https://goo.gl/pTcWbd

◀ **Figure 18.12**
Examples of ablation
A. Melting at Alaska's Root Glacier created this river atop the glacier. Notice the large rocks exposed as the ice wastes away.
B. Ice loss by calving at Perito Moreno Glacier in the Andes of Argentina. When valley glaciers or outlet glaciers terminate in the ocean, they may also be called *tidewater glaciers*. It is common for large blocks to break off the front of such a glacier and form icebergs.

snowline, or **equilibrium line**—the elevation at which the accumulation and wasting of glacial ice are equal. As noted earlier, the elevation of this boundary varies greatly, from sea level in polar regions to altitudes approaching 5000 meters (16,000 feet) near the equator. Above the snowline, in the zone of accumulation, the addition of snow thickens the glacier and promotes movement. Below the snowline is the **zone of wastage**, where there is a net loss to the glacier as all of the snow from the previous winter melts, as does some of the glacial ice.

The loss of ice by a glacier is termed **ablation**. In addition to melting, glaciers waste away as large pieces of ice break off the front of the glacier in a process called **calving** (Figure 18.12). Calving creates **icebergs** in places where the glacier has reached the sea or a lake (Figure 18.13). Because icebergs are just slightly less dense than seawater, they float very low in the water, with more than 80 percent of their mass submerged. Along the margins of Antarctica's ice shelves, calving is the primary means by which these masses lose ice. The relatively flat icebergs produced here are often several kilometers across and up to about 600 meters (2000 feet) thick (see the icebergs in Figure

18.3 and 18.4B). By comparison, thousands of irregularly shaped icebergs are produced by outlet glaciers flowing from the margins of the Greenland Ice Sheet. Many drift southward and find their way into the North Atlantic, where they can be hazardous to navigation.

Glacial Budget   Whether the margin of a glacier is advancing, retreating, or remaining stationary depends

Geologist's Sketch

Only about 20 percent or less of an iceberg protrudes above the waterline.

◀ **Figure 18.13**
Icebergs Icebergs form when large masses of ice break off from the front of a glacier after it reaches a water body, in a process known as calving. Other examples of icebergs appear in Figures 18.3 and 18.4.

► **SmartFigure 18.14**
Retreating glacier
These false-color satellite images of Alaska's Columbia Glacier were taken about 28 years apart. During that span, the terminus retreated about 16 kilometers (10 miles). In addition, the glacier thinned substantially, as you can see by comparing the area of exposed bedrock (brown). Since the 1980s, in fact, Columbia Glacier has lost about half of its total thickness and volume.

Mobile Field Trip
https://goo.gl/pJvu7C

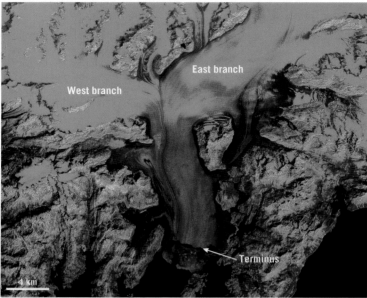
Columbia Glacier, July 1986

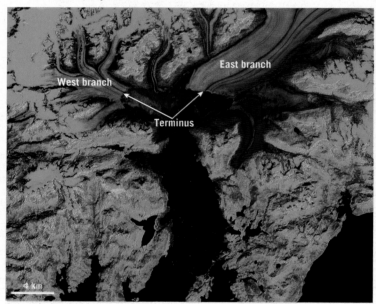
Columbia Glacier, July 2014

► **SmartFigure 18.15**
Declining ice mass in Greenland This graph shows monthly changes in Greenland's total ice mass between April 2002 and April 2016. The ice mass amounts measured are relative to the ice mass as of April 2002. The ups and downs in the graph track the accumulation of snow in winter and melting in summer.

Animation
https://goo.gl/Rr9WzJ

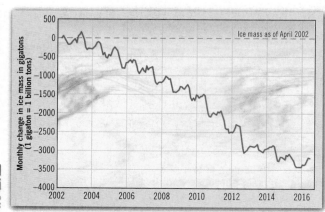

on the budget of the glacier. The **glacial budget** is the balance, or lack of balance, between accumulation at the upper end of the glacier and ablation at the lower end. If ice accumulation exceeds ablation, the glacial front advances until the two factors balance. When this happens, the terminus of the glacier is stationary.

If a warming trend increases ablation and/or if reduced snowfall decreases accumulation, the ice front will retreat. As the terminus of the glacier retreats, the extent of the zone of wastage diminishes. Therefore, in time a new balance will be reached between accumulation and wastage, and the ice front will again become stationary.

Whether the margin of a glacier is advancing, retreating, or stationary, the ice within the glacier continues to flow forward. In the case of a receding glacier, the ice still flows forward but not rapidly enough to offset ablation. This point is illustrated well in Figure 18.9. As the line of stakes within the Rhône Glacier continued to move downvalley, the terminus of the glacier slowly retreated upvalley.

Glaciers in Retreat: Unbalanced Glacial Budgets   Because glaciers are sensitive to changes in temperature and precipitation, they provide important clues about changes in climate. With few exceptions, glaciers around the world have been retreating at unprecedented rates over the past century. The alpine glaciers in Glacier National Park, highlighted in the chapter opening, provide a well-documented example. The satellite images of Alaska's Columbia Glacier in **Figure 18.14** provide another. The Greenland Ice Sheet and portions of Antarctica's ice are also shrinking. The graph in **Figure 18.15** documents the continuous loss of mass for the Greenland Ice Sheet between 2002 and 2016. The water added to the oceans by shrinking glaciers worldwide is a major contributor to global sea-level rise, which has been accelerating in recent decades. There is more about this in Chapter 21.

CONCEPT CHECKS 18.2

1. Describe two components of glacial movement.

2. What are crevasses, and where do they form?

Concept Checker
https://goo.gl/19Aaq1

3. Under what circumstances will the front of a glacier advance? Retreat? Remain stationary?

18.3 Glacial Erosion

Discuss the processes of glacial erosion. Identify and describe major topographic features created by glacial erosion.

Glaciers are capable of great erosion. To anyone who has observed the terminus of an alpine glacier, the evidence of its erosive force is clear. The release of rock material of various sizes from the ice as it melts leads to the conclusion that the ice has scraped, scoured, and torn rock from the floor and walls of the valley and carried it downvalley. It should be pointed out, however, that in mountainous regions, mass-movement processes also make substantial contributions to the sediment load of a glacier (see Figure 1.1A).

Once rock debris is acquired by a glacier, the enormous competence of ice will not allow the debris to settle out like the load carried by a stream or by the wind. For example, notice in Figure 18.12A the masses of rock exposed within the glacier as the ice melts. Indeed, as a medium of sediment transport, ice has no equal. Consequently, glaciers can carry large blocks that no other erosional agent could possibly budge. Although today's glaciers are of limited importance as erosional agents, many landscapes that were modified by the widespread glaciers of the Ice Age still reflect, to a high degree, the work of ice.

How Glaciers Erode

Glaciers erode the land primarily in two ways: plucking and abrasion. First, as a glacier flows over a fractured bedrock surface, it loosens and lifts blocks of rock and incorporates them into the ice. This process, known as **plucking**, occurs when meltwater penetrates the cracks and joints of bedrock beneath a glacier and freezes. As the water expands, it exerts tremendous leverage that pries the rock loose. In this manner, sediment of all sizes becomes part of the glacier's load.

The second major erosional process is **abrasion** (**Figure 18.16**). As the ice and its load of rock fragments advance over bedrock, they function like sandpaper, smoothing and polishing the surface below. The pulverized rock produced by the glacial "grist mill" is appropriately called **rock flour**. So much rock flour may be produced that meltwater streams flowing out of a glacier often have the cloudy appearance of skim milk and offer visible evidence of the grinding power of ice. Lakes fed by such streams frequently have a distinctive turquoise color (**Figure 18.17**).

When the ice at the bottom of a glacier contains large rock fragments, long scratches and grooves called

Glacial abrasion created the scratches and grooves in this bedrock.

A.

Glacially polished granite in California's Yosemite National Park.

B.

◄ Figure 18.16
Glacial abrasion Moving glacial ice, armed with sediment, acts like sandpaper, scratching and polishing rock.

▶ **Figure 18.17**
Distinctive color caused by rock flour Many lakes that are fed by glaciers have a distinctive turquoise color. This color occurs because the rock flour suspended in the lake reflects different parts of the visible spectrum more strongly than others.

the mountain landscape by creating steeper canyon walls and making bold peaks even more jagged. By contrast, continental ice sheets generally override the terrain and hence subdue rather than accentuate the irregularities they encounter. Although the erosional potential of ice sheets is enormous, landforms carved by these huge ice masses usually do not inspire the same wonderment and awe as do the erosional features created by valley glaciers. Much of the rugged mountain scenery so celebrated for its majestic beauty is produced by erosion by alpine glaciers. **Figure 18.18** shows a hypothetical mountain area before, during, and after glaciation. You will refer to this figure often in the following discussion.

Glaciated Valleys A hike up a glaciated valley reveals a number of striking ice-created features. The valley itself is often a dramatic sight. Unlike streams, which create their own valleys, glaciers take the path of least resistance and follow existing stream valleys. Prior to glaciation, mountain valleys are

glacial striations may even be gouged into the bedrock (see Figure 18.16A). These linear grooves provide clues to the direction of ice flow. By mapping the striations over large areas, patterns of glacial flow can often be reconstructed. On the other hand, not all abrasive action produces striations. The rock surfaces over which the glacier moves may also become highly polished by the ice and its load of finer particles. The broad expanses of smoothly polished granite in Yosemite National Park provide an excellent example (see Figure 18.16B).

As is the case with other agents of erosion, the rate of glacial erosion is highly variable. This differential erosion by ice is largely controlled by four factors: (1) rate of glacial movement; (2) thickness of the ice; (3) shape, abundance, and hardness of the rock fragments contained in the ice at the base of the glacier; and (4) erodibility of the surface beneath the glacier. Variations in any or all of these factors from time to time and/or from place to place mean that the features, effects, and degree of landscape modification in glaciated regions can vary greatly.

Landforms Created by Glacial Erosion

The erosional effects of valley glaciers and ice sheets are quite different. A visitor to a glaciated mountain region is likely to see a sharp and angular topography. This is because as the more confined alpine glaciers move downvalley, they tend to accentuate the irregularities of

characteristically narrow and V-shaped because streams are well above base level and are therefore downcutting (see Chapter 16). However, during glaciation, these narrow valleys are transformed as the glacier widens and deepens them, creating a U-shaped **glacial trough** (see Figure 18.18 and **Figure 18.19**). In addition to producing a broader and deeper valley, the glacier also straightens the valley. As ice flows around sharp curves, its great erosional force removes the spurs of land that extend into the valley. The results of this activity are triangular-shaped cliffs called **truncated spurs**.

The amount of glacial erosion that takes place varies in different valleys in a mountainous area. Prior to glaciation, the mouths of tributary streams join the main valley (or *trunk* valley) at the elevation of the stream in that valley. During glaciation, the amount of ice flowing through the main valley can be much greater than the amount advancing down each tributary. Consequently, the valley containing the main glacier (or *trunk* glacier) is eroded deeper than the smaller valleys that feed it. Thus, after the ice has receded, the valleys of tributary glaciers are left standing above the main glacial trough and are termed **hanging valleys** (see Figure 18.18C). Rivers flowing through hanging valleys can produce spectacular waterfalls, such as those in Yosemite National Park, California.

As hikers walk up a glacial trough, they may pass a series of bedrock depressions on the valley floor,

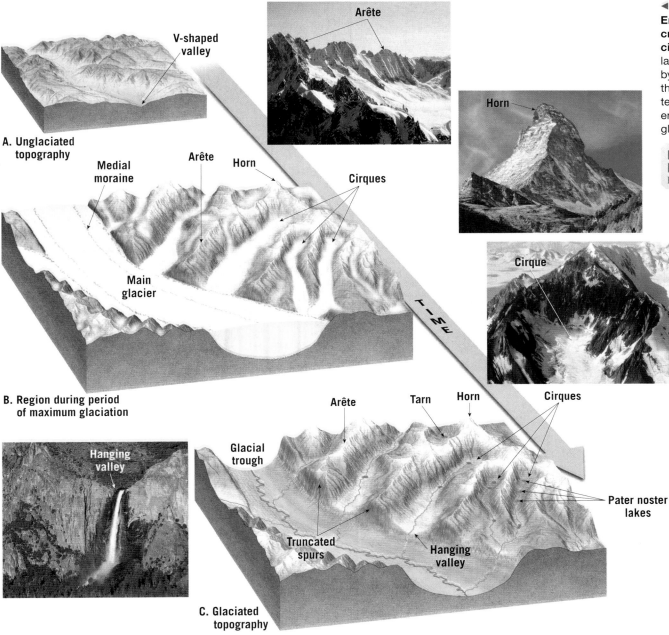

A. Unglaciated topography

B. Region during period of maximum glaciation

C. Glaciated topography

◄ **SmartFigure 18.18**
Erosional landforms created by alpine glaciers The unglaciated landscape (**A**) is modified by valley glaciers (**B**). After the ice recedes (**C**), the terrain looks very different than it looked before glaciation.

Mobile
Field Trip
https://goo.gl/QqtiRd

probably created by plucking and then scouring by the abrasive force of the ice. If these depressions are filled with water, they are called **pater noster lakes** (see Figure 18.18C). The Latin name means "our Father" and is a reference to a string of rosary beads.

Cirques At the head of a glacial valley is a characteristic and often imposing feature associated with an alpine glacier, called a **cirque**. As the photo in Figure 18.18 illustrates, these bowl-shaped depressions have precipitous walls on three sides but are open on the downvalley side. The cirque is the focal point of the glacier's growth because it is the area of snow accumulation and ice formation. Cirques begin as irregularities in the mountainside that are subsequently enlarged by frost wedging and

plucking along the sides and bottom of the glacier. The glacier in turn acts as a conveyor belt that carries away the debris. After the glacier has melted away, the cirque basin is sometimes occupied by a small lake called a **tarn** (see Figure 18.18C).

Sometimes, when two glaciers exist on opposite sides of a divide, each flowing away from the other, the dividing ridge between their cirques is largely eliminated as plucking and frost action enlarge each one. When this occurs, the two glacial troughs come to intersect, creating a gap or pass from one valley into the other. Such a feature is termed a **col**. Some important and well-known mountain passes that are cols include St. Gotthard Pass in the Swiss Alps, Tioga Pass in California's Sierra Nevada, and Berthoud Pass in the Colorado Rockies.

▶ SmartFigure 18.19
▶ SmartFigure 18.19
A U-shaped glacial trough Prior to glaciation, a mountain valley is typically narrow and V-shaped. During glaciation, an alpine glacier widens, deepens, and straightens the valley, creating the classic U-shape shown here. This glacial trough is in the Sierra Nevada, west of Bishop, California.

Animation
https://goo.gl/
p4znGf

Arêtes and Horns

The Alps, Northern Rockies, and many other mountain landscapes sculpted by valley glaciers reveal more than glacial troughs and cirques. In addition, sinuous, sharp-edged ridges called **arêtes** (French for "knife edge") and sharp, pyramid-like peaks termed **horns** project above the surroundings (see Figure 18.18C). Both features can originate from the same basic process: the enlargement of cirques produced by plucking and frost action. Several cirques around a single high mountain create the spires of rock called horns. As the cirques enlarge and converge, an isolated horn is produced. A famous example is the Matterhorn in the Swiss Alps.

Arêtes can form in a similar manner, except that the cirques are not clustered around a point but rather exist on opposite sides of a divide. As the cirques grow, the divide separating them is reduced to a very narrow, knifelike partition. An arête can also be created in another way. When two glaciers occupy parallel valleys, an arête can form when the land separating the moving tongues of ice is progressively narrowed as the glaciers scour and widen their valleys.

Roches Moutonnées

In many glaciated landscapes, but most frequently where continental ice sheets have modified the terrain, the ice carves small streamlined hills from protruding bedrock knobs. Such an asymmetrical knob of bedrock is called a **roche moutonnée** (French for "sheep rock"). These features are formed when glacial abrasion smooths the gentle slope facing the oncoming ice sheet and plucking steepens the opposite side as the ice rides over the knob (**Figure 18.20**). Roches moutonnées indicate the direction of glacial flow because the gentler slope is generally on the side from which the ice advanced.

Fiords

Sometimes spectacular steep-sided inlets of the sea called **fiords** are present at high latitudes where

▲ Figure 18.20
Roche moutonnée This classic example is in Yosemite National Park, California. The gentle slope was abraded, and the steep slope was plucked. The glacier moved from the right in this photo to the left.

Ice flow

Glacial plucking

Glacial abrasion

Bedrock

Geologist's Sketch

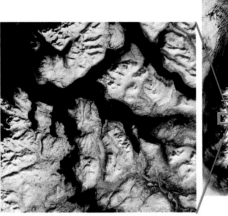

◄ **Figure 18.21**

Fiords The coast of Norway is known for its many fiords. Frequently these ice-sculpted inlets of the sea are hundreds of meters deep.

mountains are adjacent to the ocean (**Figure 18.21**). They are "drowned" glacial troughs that became submerged as the ice left the valleys and sea level rose following the Ice Age. The depths of fiords may exceed 1000 meters (3300 feet). However, the great depths of these flooded troughs are only partly explained by the post–Ice Age rise in sea level. Unlike the situation governing the downward erosional work of rivers, sea level does not act as base level for glaciers. As a consequence, glaciers are capable of eroding their beds far below the surface of the sea. For example, a 300-meter-thick (1000-foot-thick) glacier can carve its valley floor more than 250 meters (820 feet) below sea level before downward erosion ceases and the ice begins to float. Norway, British Columbia, Greenland, New Zealand, Chile, and Alaska all have coastlines characterized by fiords.

CONCEPT CHECKS 18.3

1. How do glaciers acquire their load of sediment?

2. What are some visible effects of glacial erosion?

3. How does a glaciated mountain valley differ in appearance from a mountain valley that was not glaciated? Describe the erosional features created by valley glaciers.

◄)) Concept Checker
https://goo.gl/5YGc6P

4. Relate fiords to glacial troughs.

18.4 Glacial Deposits

Distinguish between the two basic types of glacial drift. List and describe the major depositional features associated with glacial landscapes.

A glacier picks up and transports a huge load of rock debris as it slowly advances across the land. Ultimately when the ice melts, these materials are deposited. In regions where glacial sediment is deposited, it can play a significant role in forming the physical landscape. For example, in many areas once covered by the continental ice sheets of the recent Ice Age, the bedrock is rarely exposed because glacial deposits that are tens or even hundreds of meters thick completely mantle the terrain. The general effect of these deposits is to reduce the local relief and thus level the topography. Indeed, rural country scenes that are familiar to many of us—rocky pastures in New England, wheat fields in the Dakotas, rolling farmland in the Midwest—result directly from glacial deposition.

Glacial Drift

Long before the theory of an extensive Ice Age was ever proposed, much of the soil and rock debris covering portions of Europe was recognized as having come from somewhere else. At the time, these "foreign" materials were believed to have been "drifted" into their present positions by floating ice during an ancient flood. As a consequence, the term *drift* was applied to this

▶ **Figure 18.22**
Glacial till Unlike sediment deposited by running water and wind, material deposited directly by a glacier is not sorted.

Glacial till is an unsorted mixture of many different sediment sizes.

A close examination of glacial till often reveals cobbles that have been scratched as they were dragged along by the ice.

▼ **Figure 18.23**
Glacial erratic This large glacially transported boulder is a prominent feature along a trail in Squamish, British Columbia, Canada. Such boulders are called glacial erratics. Note the person (lower right) for scale.

Glacial erratic

glacial origin, no matter how, where, or in what shape they were deposited.

Geologists divide glacial drift into two distinct types: (1) materials deposited directly by the glacier, which are known as *till*, and (2) sediments laid down by glacial meltwater, called *stratified drift*.

Glacial Till   As glacial ice melts and drops its load of rock fragments, **till** is deposited. Unlike moving water and wind, ice cannot sort the sediment it carries; therefore, deposits of till are characteristically unsorted mixtures of many particle sizes (**Figure 18.22**). A close examination of this sediment shows that many of the pieces are scratched and polished as a result of being dragged along by the glacier. Such pieces help distinguish till from other deposits that are a mixture of different sediment sizes, such as material from a debris flow or a rockslide.

Boulders found in the till or lying free on the surface are called **glacial erratics** if they are different from the bedrock below (**Figure 18.23**). Of course, this means they must have been derived from a source outside the area where they are found. Although for most erratics the source is unknown, the origin of some can be determined. In many cases, boulders were transported as far as 500 kilometers (300 miles) from their source area and, in a few instances, more than 1000 kilometers (600 miles). Therefore, by studying glacial erratics as well as the mineral composition of the remaining till, geologists are sometimes able to trace the path of a lobe of ice.

In portions of New England and other areas, erratics dot pastures and farm fields. In fact, in some places, these large rocks were cleared from fields and piled to make fences and walls. Keeping the fields clear, however, is an ongoing chore because each spring, newly exposed erratics appear after wintertime frost heaving lifts them to the surface.

sediment. Although rooted in an incorrect concept, this term was so well established by the time the true glacial origin of the debris became widely recognized that it remained part of the basic glacial vocabulary. Today **glacial drift** is an all-embracing term for sediments of

Stratified Drift   As the name implies, **stratified drift** is sorted according to the size and weight of the particles. Ice is not capable of sorting sediment the way running water can. Therefore, stratified drift is not deposited directly by the glacier the way till is but instead reflects the sorting action of glacial meltwater.

Some deposits of stratified drift are made by streams issuing directly from the glacier. Other stratified deposits involve sediment that was originally laid down as till and later picked up, transported, and redeposited by meltwater beyond the margin of the ice. Accumulations of stratified drift often consist largely of sand and gravel because the meltwater is not capable of moving larger material and because the finer rock flour remains suspended and is commonly carried far from the glacier. Evidence that stratified drift consists primarily of sand and gravel can be seen in many areas where these deposits are actively mined as aggregate for road work and other construction projects.

Landforms Made of Till

Perhaps the most widespread features created by glacial deposition are *moraines*, which are simply layers or ridges of till. Some types of moraines are common in mountain valleys, and other types are associated with areas affected by either ice sheets or valley glaciers. Lateral and medial moraines fall in the first category, whereas end moraines and ground moraines are in the second.

Lateral and Medial Moraines   Alpine glaciers produce two types of moraines that occur exclusively in mountain valleys. The first of these is called a **lateral moraine**. As discussed earlier, when an alpine glacier moves down a valley, the ice erodes the sides of the valley with great efficiency. In addition, large quantities of debris are added to the glacier's surface as rubble falls or slides from higher up on the valley walls and collects on the margins of the moving ice. When the ice eventually melts, this accumulation of debris is dropped next to the valley walls. These ridges of till paralleling the sides of the valley constitute the lateral moraines.

The second type of moraine that is unique to alpine glaciers is the **medial moraine** (Figure 18.24). Medial moraines are created when two alpine glaciers coalesce to form a single ice stream. The till that was once carried along the sides of each glacier joins to form a single dark stripe of debris within the newly enlarged glacier. The sketch in Figure 18.24 illustrates this nicely. The creation of these dark stripes within the ice stream is one obvious proof that glacial ice moves because the moraine could not form if the ice did not flow downvalley. It is common to see several medial moraines within a single large alpine glacier because a streak forms whenever a tributary glacier joins the main valley glacier.

End and Ground Moraines   Sometimes a glacier is compared to a conveyor belt. No matter whether the front of a glacier or ice sheet is advancing, retreating, or stationary, it is constantly moving sediment forward and dropping it at its terminus. This is a useful analogy when considering end and ground moraines.

An **end moraine** is a ridge of till that forms at the terminus of a glacier. End moraines are characteristic of ice sheets and valley glaciers alike. These relatively common landforms are deposited when a state of equilibrium is attained between ablation and ice accumulation. That is, the end moraine forms when the ice is melting and calving near the end of the glacier at a rate equal to the forward advance of the glacier from its region of nourishment. Although the terminus of the glacier is stationary, the ice continues to flow forward, delivering a continuous supply of sediment in the same way a conveyor belt delivers goods to the end of a production line. As the ice melts, the till is dropped, and the end moraine grows. The longer the ice front remains stable (with ablation and accumulation in balance), the larger the ridge of till will become.

Eventually, ablation exceeds nourishment. At this point, the front of the glacier begins to recede in the direction from which it originally advanced. However, as the ice front retreats, the conveyor-belt action of the glacier continues to provide fresh supplies of sediment to the terminus. In this manner, a large quantity of till is deposited as the ice melts away, creating a rock-strewn, undulating plain. This gently rolling layer of till deposited as the ice front recedes is termed **ground moraine**. It has a leveling effect, filling in low spots and clogging old stream channels, often leading to a derangement of the existing drainage system. In areas where this layer of till is still relatively fresh, such as the northern Great Lakes region, poorly drained swampy lands are quite common.

Periodically, a glacier will retreat to a point where ablation and nourishment once again balance. When this happens, the ice front stabilizes, and a new end

▼ **SmartFigure 18.24 Formation of a medial moraine** Kennicott Glacier is a 43-kilometer-long (27-mile-long) valley glacier that is sculpting the mountains in Alaska's Wrangell–St. Elias National Park. The dark stripes of sediment are medial moraines.

Mobile Field Trip
https://goo.gl/xGKqUW

Geologist's Sketch

Medial moraines

Lateral moraine

Lateral moraines join to form medial moraine

Lateral moraine

Medial moraine

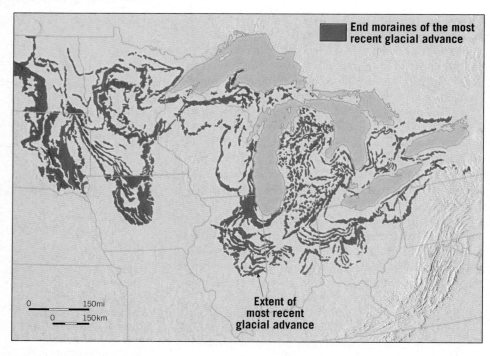

End moraines of the most recent glacial advance

Extent of most recent glacial advance

▲ Figure 18.25
End moraines of the Great Lakes region This map shows end moraines deposited during the most recent (Wisconsinan) glaciation.

moraine forms. The pattern of end moraine formation and ground moraine deposition may be repeated many times before the glacier has completely vanished. Such a pattern is illustrated in **Figure 18.25**. The very first end moraine to form signifies the farthest advance of the glacier and is called the *terminal end moraine*. End moraines that form as the ice front occasionally stabilizes during retreat are termed *recessional end moraines*. Terminal and recessional moraines are essentially alike; the only difference between them is their relative positions.

End moraines deposited by the most recent stage of Ice Age glaciation are prominent features in many parts of the U.S. Midwest and Northeast. In Wisconsin, the wooded, hilly terrain of the Kettle Moraine near Milwaukee is a particularly picturesque example. A well-known example in the Northeast is Long Island. This linear strip of glacial sediment extending northeastward from New York City is part of an end moraine complex

that stretches from eastern Pennsylvania to Cape Cod, Massachusetts (**Figure 18.26**).

Figure 18.27 represents a hypothetical area during glaciation and after the retreat of an ice sheet. This figure depicts landscape features such as the end moraines just described as well as depositional landforms, similar to what might be encountered if you were traveling in the upper Midwest or New England. You will refer to this figure several times as you read the following paragraphs on glacial deposits.

Drumlins   Moraines are not the only landforms composed of till. In some areas that were once covered by continental ice sheets, a special variety of glacial landscape exists—one characterized by smooth, elongate, parallel hills called **drumlins** (see Figure 18.27). Certainly one of the best-known drumlins is Bunker Hill in Boston, the site of the famous Revolutionary War battle in 1775.

An examination of Bunker Hill or other less famous drumlins would show that drumlins are streamlined, asymmetrical hills composed largely of till. They range in height from about 15 to 50 meters and may be up to 1 kilometer long. The steep side of the hill faces the direction from which the ice advanced, whereas the gentler, longer slope points in the direction the ice moved. Drumlins are not found as isolated landforms but rather occur in clusters called *drumlin fields* (**Figure 18.28**). One such cluster, east of Rochester, New York, is estimated to contain about 10,000 drumlins. Although drumlin formation is not fully understood, the streamlined shape of drumlins indicates that they were molded in the zone of plastic flow within an active glacier. It is believed that many drumlins originate when glaciers advance over previously deposited drift and reshape the material.

Landforms Made of Stratified Drift

Much of the material acquired and transported by a glacier is ultimately deposited by streams of glacial meltwater flowing on, within, beneath, and beyond a glacier. This sediment is termed *stratified drift*. Unlike glacial till, stratified drift shows some degree of sorting. There

▶ SmartFigure 18.26
Two significant end moraines in the Northeast The Ronkonkoma moraine, deposited about 20,000 years ago, extends through central Long Island, Martha's Vineyard, and Nantucket. The Harbor Hill moraine, formed about 14,000 years ago, extends along the north shore of Long Island, through southern Rhode Island and Cape Cod. A portion of the associated Mobile Field Trip "Cape Cod" explores the glacial origins of Cape Cod.

Mobile Field Trip
https://goo.gl/PEXEVr

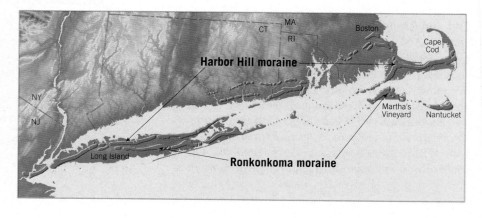

MA
CT   Boston
RI
Cape Cod
Harbor Hill moraine
NY
NJ
Martha's Vineyard   Nantucket
Long Island   **Ronkonkoma moraine**

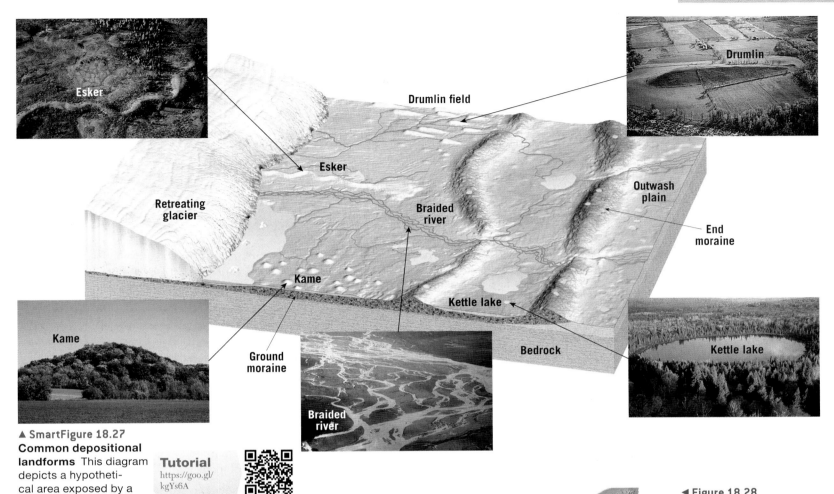

▲ SmartFigure 18.27
Common depositional landforms This diagram depicts a hypothetical area exposed by a retreating ice sheet.

Tutorial
https://goo.gl/kgYs6A

are two basic categories of features composed of stratified drift: *Ice-contact deposits* accumulate on, within, or immediately adjacent to a glacier. *Outwash sediment*, or simply *outwash*, is material deposited by meltwater streams beyond the terminus of a glacier.

Outwash Plains and Valley Trains

At the same time that an end moraine is forming, water from the melting glacier cascades over the till, sweeping some of it out in front of the growing ridge of unsorted debris. Meltwater generally emerges from the ice in rapidly moving streams that are often choked with suspended material and carry a substantial bed load as well. Water leaving the glacier moves onto the relatively flat surface beyond and rapidly loses velocity. As a consequence, much of its bed load is dropped, and the meltwater begins weaving a complex pattern of braided channels (see Figure 18.27). In this way, a broad, ramplike surface composed of stratified drift is built adjacent to the downstream edge of most end moraines. When the feature is formed in association with an ice sheet, it is termed an **outwash plain**, and when largely confined to a mountain valley, it is usually called a **valley train**.

◄ Figure 18.28
Drumlin field A portion of the drumlin field shown on the Palmyra, New York, 7.5-minute topographic map. North is at the top. The drumlins are steepest on the north side, indicating that the ice advanced from this direction.

Outwash plains and valley trains often are pock-marked with basins or depressions known as **kettles** (see Figure 18.27). Kettles also occur in deposits of till. Kettles are formed when blocks of stagnant ice become wholly or partly buried in drift and eventually melt, leaving pits in the glacial sediment. Although most kettles do not exceed 2 kilometers in diameter, some with diameters exceeding 10 kilometers occur in Minnesota. Likewise, the typical depth of most kettles is less than 10 meters, although the vertical dimensions of some approach 50 meters. In many cases water eventually fills the depression and forms a pond or lake. One well-known example is Walden Pond near Concord, Massachusetts. It is here that Henry David Thoreau lived alone for 2 years in the 1840s and about which he wrote his famous book *Walden; or, Life in the Woods*.

Ice-Contact Deposits

When the melting terminus of a glacier shrinks to a critical point, flow virtually stops, and the ice becomes stagnant. Meltwater that flows over, within, and at the base of the motionless ice lays down deposits of stratified drift. Then, as the supporting ice melts away, the stratified sediment is left behind in the form of hills, terraces, and ridges. Such accumulations are collectively termed **ice-contact deposits** and are classified according to their shapes.

When the ice-contact stratified drift is in the form of a mound or steep-sided hill, it is called a **kame** (see Figure 18.27). Some kames represent bodies of sediment deposited by meltwater in openings within or depressions on top of the ice. Others originate as deltas or fans built outward from the ice by meltwater streams. Later, when the stagnant ice melts away, these various accumulations of sediment collapse to form isolated, irregular mounds.

When glacial ice occupies a valley, **kame terraces** may be built along the sides of the valley. These features commonly are narrow masses of stratified drift laid down between the glacier and the side of the valley by meltwater streams that drop debris along the margins of the shrinking ice mass.

A third type of ice-contact deposit is a long, narrow, sinuous ridge composed largely of sand and gravel. Some are more than 100 meters high, with lengths in excess of 100 kilometers. The dimensions of many others are far less spectacular. Known as **eskers**, these ridges are deposited by meltwater rivers flowing within, on top of, and beneath a mass of motionless, stagnant glacial ice (see Figure 18.27). Many sediment sizes are carried by the torrents of meltwater in the ice-banked channels, but only the coarser material can settle out of the turbulent stream.

CONCEPT CHECKS 18.4

1. What term can be applied to any glacial deposit? Distinguish between till and stratified drift.

2. How are medial moraines and lateral moraines related? In what kind of setting do these features form?

3. Contrast end moraine and ground moraine. Relate these features to the budget of a glacier. **Concept Checker** https://goo.gl/FcYj4Q

4. Distinguish between outwash deposits and ice-contact deposits.

18.5 Other Effects of Ice Age Glaciers

Describe and explain several important effects of Ice Age glaciers other than erosional and depositional landforms.

In addition to the massive erosional and depositional work carried out by Ice Age glaciers, ice sheets had other effects, sometimes profound, on the landscape. For example, as the ice advanced and retreated, animals and plants were forced to migrate. This led to stresses that some organisms could not tolerate. Hence, a number of plants and animals became extinct. Other effects of Ice Age glaciers that are described in this section involve sea-level changes associated with the formation and wastage of ice sheets and adjustments in Earth's crust due to the addition and removal of ice. The advance and retreat of ice sheets also led to significant changes in the routes taken by rivers. In some regions, glaciers acted as dams that created large lakes. When these ice dams failed, the effects on the landscape were dramatic. In areas that today are deserts, lakes of another type, called pluvial lakes, formed.

Sea-Level Changes

One of the most interesting and perhaps dramatic effects of the Ice Age was the fall and rise of sea level that accompanied the advance and retreat of the glaciers. Although the total volume of glacial ice today is great, exceeding 25 million cubic kilometers, during the Last Glacial Maximum the volume of glacial ice amounted to about 70 million cubic kilometers, or 45 million cubic kilometers more than at present. Because we know that the snow from which glaciers form ultimately comes from the evaporation of ocean water, the growth of ice sheets must have caused a significant worldwide drop in sea level (**Figure 18.29**). Indeed, estimates suggest that sea level was as much as 100 meters (330 feet) lower than it is today. Thus, land that is presently flooded by the oceans was dry. The Atlantic coast of the United States lay more than 100 kilometers (60 miles) to the east of New York City, France and Britain were joined where the famous English Channel is today, Alaska and Siberia were connected across the Bering Strait, and Southeast Asia was tied by dry land to the islands of Indonesia. Conversely, if the water currently locked up in the Antarctic Ice Sheet were to melt completely, sea level would rise by an estimated 60 or 70 meters. Such an occurrence would flood many densely populated coastal areas.

Crustal Subsidence and Rebound

In areas that were major centers of ice accumulation, such as Scandinavia and the Canadian Shield, the land has been slowly rising over the past several thousand years. Uplifting of almost 300 meters (1000 feet) has occurred in the Hudson Bay region. This, too, is the result of the continental ice sheets. But how can glacial ice cause such vertical crustal movement? We now understand that the land is rising because the added weight of the 3-kilometer-thick (2-mile-thick) mass of ice caused downwarping of Earth's crust. For example, scientists have determined that Antarctica's ice sheets depress Earth's crust by an estimated 900 meters (3000 feet) or more in some places. Following the removal of such an immense load, the crust adjusts by gradually rebounding upward (**Figure 18.30**).°

Sea level was as much as 100 meters lower than now during the last Ice Age.

Today we know that sea level is rising worldwide because of global warming (see the section "Sea-Level Rise" in Chapter 21). Yet when measurements are made in the coastal areas of Finland and Sweden adjacent to the Gulf of Bothnia, sea level appears to be falling. What is the cause? The answer is post-Ice Age crustal rebound. Globally, sea level has been rising by about 3 millimeters per year over the past decade. During this same span, crustal rebound in this region has been as great as 9 millimeters per year. Thus, because the land is rising faster than the rise in sea level, *relative sea level* is dropping.

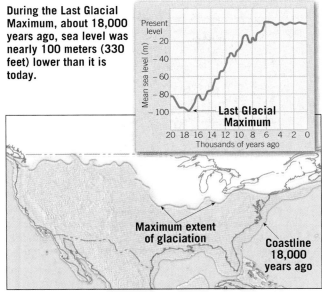

During the Last Glacial Maximum, about 18,000 years ago, sea level was nearly 100 meters (330 feet) lower than it is today.

Present level

Mean sea level (m)
−20
−40
−60
−80
−100

Last Glacial Maximum

20 18 16 14 12 10 8 6 4 2 0
Thousands of years ago

Maximum extent of glaciation

Coastline 18,000 years ago

During the Last Glacial Maximum, the shoreline extended out onto the present-day continental shelf.

◀ **SmartFigure 18.29**
Changing sea level As ice sheets form and then melt away, sea level falls and rises, causing the shoreline to shift.

Animation
https://goo.gl/Y58Mpp

One effect is that Scandinavia's land area is growing. Land that was once submerged is now dry. Not all of the effects are positive. For example, since 1980, relative sea level in the Swedish port city of Lulea has dropped about 50 centimeters (1.6 feet). Because the land is rising, the harbor is getting too shallow for some ships. In consequence, a large and costly dredging operation is being planned.

Changes to Rivers and Valleys

Among the effects associated with the advance and retreat of North American ice sheets were changes in the routes of many rivers and the modification in the size and shape of many valleys. If we are to understand the present pattern of rivers and lakes in the central and northeastern United States (and many other places as well), we need to be aware of glacial history. Two examples illustrate these effects.

° For a detailed explanation of this concept, termed *isostatic adjustment*, see the discussion of isostasy in the section "Vertical Motions of the Crust" in Chapter 14.

◀ **Figure 18.30**
Crustal subsidence and rebound These simplified diagrams illustrate subsidence and rebound resulting from the addition and removal of continental ice sheets.

In northern Canada and Scandinavia, where the greatest accumulation of glacial ice occurred, the added weight caused downwarping of the crust.

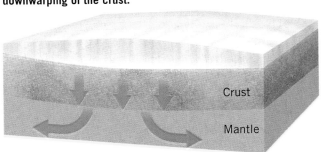

Crust

Mantle

Crust

Mantle

Ever since the ice melted, there has been gradual uplift, or rebound, of the crust.

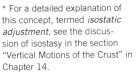

▶ **Figure 18.31**
Changing rivers The advance and retreat of ice sheets caused major changes in the routes followed by rivers in the central United States.

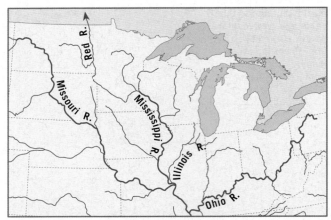

A. This map shows the Great Lakes and the familiar present-day pattern of rivers. Quaternary ice sheets played a major role in creating this pattern.

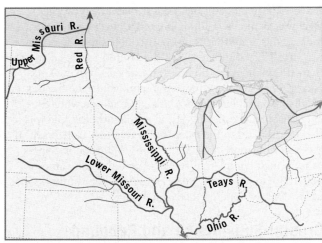

B. Reconstruction of drainage systems prior to the Ice Age. The pattern was very different from today, and the Great Lakes did not exist.

Upper Mississippi Drainage Basin   **Figure 18.31A** shows the familiar present-day pattern of rivers in the central United States, with the Missouri, Ohio, and Illinois Rivers as major tributaries to the Mississippi. **Figure 18.31B** depicts drainage systems in this region prior to the Ice Age. The pattern is *very* different from the present pattern. This remarkable transformation of river systems resulted from the advance and retreat of the ice sheets.

Notice that prior to the Ice Age, a significant part of the Missouri River drained north toward Hudson Bay. Moreover, the Mississippi River did not follow the present Iowa–Illinois boundary but rather flowed across west-central Illinois, where the lower Illinois River flows today. The preglacial Ohio River barely reached to the present-day state of Ohio, and the rivers that today feed the Ohio in western Pennsylvania flowed north and drained into the North Atlantic. The Great Lakes were created by glacial erosion during the Ice Age. Prior to the Pleistocene epoch, the basins occupied by these huge lakes were lowlands with rivers that ran eastward to the Gulf of St. Lawrence.

The large Teays River was a significant feature prior to the Ice Age (see Figure 18.31B). The Teays flowed from West Virginia across Ohio, Indiana, and Illinois, and it discharged into the Mississippi River not far from present-day Peoria. This river valley, which would have rivaled the Mississippi in size, was completely obliterated during the Pleistocene, buried by glacial deposits hundreds of feet thick. Today the sands and gravels in the buried Teays valley make it an important aquifer.

Lake Canandaigua

▲ **Figure 18.32**
New York's Finger Lakes The long, narrow basins occupied by these lakes were created when ice sheets scoured these river valleys into deep troughs.

New York's Finger Lakes   The recent geologic history of west-central New York State south of Lake Ontario was dominated by ice sheets. We have already noted the drumlins in the vicinity of Rochester (see Figure 18.28). Many other depositional features and erosional effects are found in the region. Perhaps the best known are the Finger Lakes, 11 long, narrow, roughly parallel water bodies oriented north–south like fingers on a pair of outstretched hands (**Figure 18.32**). Prior to the Ice Age, the Finger Lakes area consisted of a series of river valleys that were oriented parallel to the direction of ice movement. Multiple episodes of glacial erosion transformed these valleys into deep, steep-walled lakes. Two of the lakes are very deep—Seneca Lake is more than 180 meters

(600 feet) deep at its lowest point, and Cayuga Lake is nearly 135 meters (450 feet) deep—and the beds of both lakes lie below sea level. The depth to which the glaciers carved these basins is much greater. There are hundreds of feet of glacial sediment in the deep rock troughs below the lake beds.

Ice Dams Create Proglacial Lakes

Ice sheets and alpine glaciers can act as dams to create lakes by trapping glacial meltwater and blocking the flow of rivers. Some of these lakes are relatively small, short-lived impoundments. Others can be large and exist for hundreds or thousands of years.

Figure 18.33 is a map of Lake Agassiz—the largest lake to form during the Ice Age in North America. It came into existence about 12,000 years ago and lasted about 4500 years. With the retreat of the ice sheet came enormous volumes of meltwater. The Great Plains generally slope upward to the west. As the terminus of the ice sheet receded northeastward, meltwater was trapped between the ice on one side and the sloping land on the other, causing Lake Agassiz to deepen and spread across the landscape. Such water bodies are termed **proglacial lakes**, referring to their position just beyond the outer limits of a glacier or an ice sheet. The history of Lake Agassiz is complicated by the dynamics of the ice sheet, which, at various times, readvanced and affected lake levels and drainage systems. Where drainage occurred depended on the water level of the lake and the position of the ice sheet.

Lake Agassiz left marks over a broad region. Former beaches, now many kilometers from any water, mark former shorelines. Several modern river valleys, including the Red River and the Minnesota River, were originally cut by water entering or leaving the lake. Present-day remnants of Lake Agassiz include Lake Winnipeg, Lake Manitoba, Lake Winnipegosis, and Lake of the Woods. The sediments of the former lake basin are now fertile agricultural land.

Research shows that the shifting of glaciers and the failure of ice dams can cause the rapid release of huge volumes of water. Such events occurred during the history of Lake Agassiz. A dramatic example of such glacial outbursts occurred in the Pacific Northwest between about 15,000 and 13,000 years ago and is briefly described in Figure 18.34.

Pluvial Lakes

While the formation and growth of ice sheets was an obvious response to significant changes in climate, the existence of the glaciers themselves triggered important climatic changes in the regions beyond their margins. In arid and semiarid areas on all the continents, temperatures were lower, and thus evaporation rates were lower, but at the same time, precipitation totals were moderate. This cooler, wetter climate formed many **pluvial lakes** (*pluvia* = rain). In North America the greatest concentration of pluvial lakes occurred in the vast Basin and Range region of Nevada

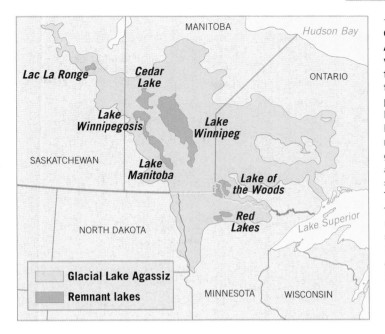

◄ **Figure 18.33 Glacial Lake Agassiz** This lake was an immense feature—bigger than all of the present-day Great Lakes combined. Modern-day remnants of this proglacial water body are still major landscape features. The lake is named for Louis Agassiz, a nineteenth-century scientist who helped develop the glacial theory of an ice age.

▼ **Figure 18.34**
Lake Missoula and the Channeled Scablands During a span of 1500 years, more than 40 megafloods from Lake Missoula carved the Channeled Scablands.

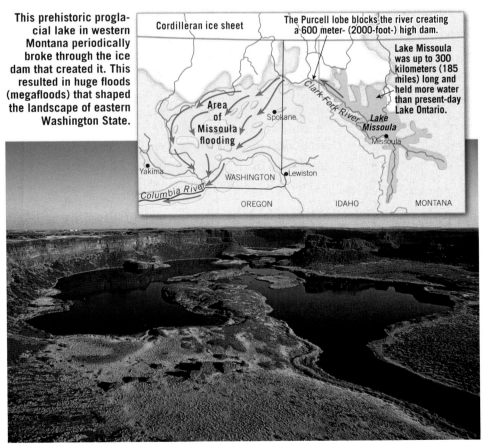

This prehistoric proglacial lake in western Montana periodically broke through the ice dam that created it. This resulted in huge floods (megafloods) that shaped the landscape of eastern Washington State.

Cordilleran ice sheet

The Purcell lobe blocks the river creating a 600 meter- (2000-foot-) high dam.

Lake Missoula was up to 300 kilometers (185 miles) long and held more water than present-day Lake Ontario.

The towering mass of rushing water from each megaflood stripped away layers of sediment and soil and cut deep canyons (coulees) into the underlying layers of basalt to create the Channeled Scablands.

▶ Figure 18.35
Pluvial lakes During the Ice Age, the Basin and Range region experienced a wetter climate than it has today. Many basins turned into large lakes.

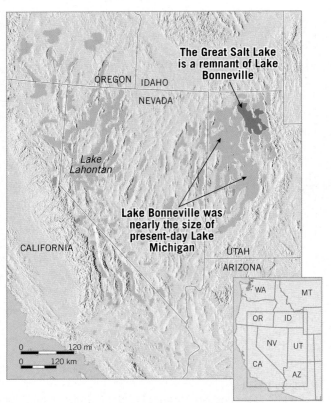

and Utah (**Figure 18.35**). By far the largest of the lakes in this region was Lake Bonneville. With maximum depths exceeding 300 meters (1000 feet) and an area of 50,000 square kilometers (20,000 square miles), Lake Bonneville was nearly the same size as present-day Lake Michigan. As the ice sheets waned, the climate again grew more arid, and the lake levels lowered in response. Although most of the lakes completely disappeared, a few small remnants of Lake Bonneville remain, the Great Salt Lake being the largest and best known.

CONCEPT CHECKS 18.5

1. List and briefly describe five effects of Ice Age glaciers aside from the formation of major erosional and depositional features.

2. Examine Figure 18.29 and determine how much sea level has changed since the Last Glacial Maximum.
   Concept Checker
   https://goo.gl/r6WNqE

3. Compare the two parts of Figure 18.31 and identify three major changes to the flow of rivers in the central United States during the Ice Age.

4. Contrast proglacial lakes and pluvial lakes. Give an example of each.

18.6 The Ice Age

Briefly discuss the development of glacial theory and summarize current ideas on the causes of ice ages.

In the preceding pages, we mentioned the Ice Age, a time when ice sheets and alpine glaciers were far more extensive than they are today. As noted, there was a time when the most popular explanation for what we now know to be glacial deposits was that the materials had been drifted in by means of icebergs or perhaps simply swept across the landscape by a catastrophic flood. What convinced geologists that an extensive ice age was responsible for these deposits and many other glacial features?

Historical Development of the Glacial Theory

In 1821 a Swiss engineer, Ignaz Venetz, presented a paper suggesting that glacial landscape features occurred at considerable distances from the existing glaciers in the Alps. This implied that the glaciers had once been larger and occupied positions farther down-valley. Another Swiss scientist, Louis Agassiz, doubted the proposal of widespread glacial activity put forth by Venetz. He set out to prove that the idea was not valid. However, his 1836 fieldwork in the Alps convinced him of the merits of his colleague's hypothesis. In fact, a year later Agassiz hypothesized a great ice age that had extensive and far-reaching effects—an idea that was to give Agassiz widespread fame.

The proof of the glacial theory proposed by Agassiz and others constitutes a classic example of applying the principle of uniformitarianism (see "The Birth of Modern Geology" in Chapter 1). Realizing that certain features are produced by no other known process but glacial action, the scientists were able to begin reconstructing the extent of now-vanished ice sheets based on the presence of features and deposits found far beyond the margins of present-day glaciers and ice sheets. In this manner, the development and verification of the glacial theory continued during the nineteenth century, and through the efforts of many scientists, a knowledge of the nature and extent of former ice sheets became clear.

Multiple Glaciations By the beginning of the twentieth century, geologists had largely determined the extent of the Ice Age glaciation. Further, during the course of their investigations, they discovered that many glaciated regions had not one but several layers of drift. Moreover, close examination of these older deposits showed well-developed zones of chemical weathering and soil formation, as well as the remains of plants that require warm temperatures. The evidence was clear: There had been not just one glacial advance but many, each separated by extended periods when climates were as warm as or even warmer than the present. The Ice Age had not simply been a time when the ice advanced over the land, lingered for a while, and then receded. Rather, the period was a very complex event, characterized by a number of advances and withdrawals of glacial ice.

By the early twentieth century, a fourfold division of the Ice Age had been established for both North America and Europe. These divisions were based largely on studies of glacial deposits. In North America each of the four major stages was named for the midwestern state where deposits of that stage were well exposed and/or were first studied. These are, in order of occurrence, the Nebraskan, Kansan, Illinoian, and Wisconsinan. These traditional divisions remained in place for many years, until it was learned that sediment cores from the ocean floor contain a much more complete record of climate change during the Ice Age (**Figure 18.36**). Unlike the glacial record on land, which is punctuated by many unconformities, seafloor sediments provide an uninterrupted record of climatic cycles for this period. Studies of these seafloor sediments showed that glacial–interglacial cycles had occurred about every 100,000 years. About 20 such cycles of cooling and warming have been identified for the span we call the Ice Age.

Extent of Glaciation
During the Ice Age, ice left its imprint on almost 30 percent of Earth's land area, including about 10 million square kilometers of North America, 5 million square kilometers of Europe, and 4 million square kilometers of Siberia (**Figure 18.37**). The amount of glacial ice in the Northern Hemisphere was roughly twice that in the Southern Hemisphere. The primary reason is that the southern polar ice could not spread far beyond the margins of Antarctica. By contrast, North America and Eurasia provided great expanses of land for the spread of ice sheets.

Today we know that the Ice Age began between 2 million and 3 million years ago. This means that most of the major glacial stages occurred during a division of the geologic time scale called the **Quaternary period**. However, this period does not encompass all of the last glacial period. The Antarctic Ice Sheet, for example, probably formed at least 30 million years ago.

Causes of Ice Ages
A great deal is known about glaciers and glaciation. Much has been learned about glacier formation and movement, the extent of glaciers past and present, and the features created by glaciers, both erosional and depositional. However, the causes of ice ages are not completely understood.

Although widespread glaciation has been rare in Earth's history, the Ice Age that encompassed most of the Quaternary period is not the only glacial period for which records exist. Earlier glaciations are indicated by deposits called **tillite**, a sedimentary rock formed when glacial till becomes lithified. Such deposits, found in strata of several different ages, usually contain striated rock fragments, and some overlie grooved and polished bedrock surfaces or are associated with sandstones and conglomerates that show features of outwash deposits. For example, our Chapter 2 discussion of evidence supporting the continental drift hypothesis mentioned a glacial period that occurred in late Paleozoic time (see Figure 2.7, page 42). Two Precambrian glacial episodes have been identified in the geologic record, the first approximately 2 billion years ago and the second about 600 million years ago.

◀ Figure 18.36
Evidence from the seafloor Cores of seafloor sediment provided data that led to a more complete understanding of the complexity of Ice Age climates.

▼ Figure 18.37
Where was the ice? This map shows the maximum extent of ice sheets in the Northern Hemisphere during the Ice Age.

RUSSIA

North Pole

Arctic Ocean

GREENLAND

Alps

Pacific Ocean

Atlantic Ocean

CANADA

UNITED STATES

Glacial ice

Sea ice

Any theory that attempts to explain the causes of ice ages must successfully answer two basic questions:

- *What causes the onset of glacial conditions?* For continental ice sheets to have formed, average temperature must have been somewhat lower than at present—and perhaps substantially lower than throughout much of geologic time. Thus, a successful theory would have to account for the cooling that finally leads to glacial conditions.
- *What caused the alternating glacial and interglacial stages that have been documented for the Quaternary period?* Whereas the first question deals with long-term trends in temperature on a scale of millions of years, this question relates to much shorter-term changes.

Although the scientific literature contains many hypotheses related to possible causes of glacial periods, we will discuss only a few major ideas to summarize current thought.

Plate Tectonics   Probably the most attractive proposal for explaining the fact that extensive glaciations have occurred only a few times in the geologic past comes from the theory of plate tectonics. Because glaciers can form only on land, we know that landmasses must exist somewhere in the higher latitudes before an ice age can commence. Many scientists suggest that ice ages have occurred only when Earth's shifting crustal plates have carried the continents from tropical latitudes to more poleward positions.

Glacial features in present-day Africa, Australia, South America, and India indicate that these regions, which are now tropical or subtropical, experienced an ice age near the end of the Paleozoic era, about 250 million years ago. However, there is no evidence that ice sheets existed during this same period in what are today the higher latitudes of North America and Eurasia. For many years this puzzled scientists. Was the climate in these relatively tropical latitudes once like it is today in Greenland and Antarctica? Why did glaciers not form in North America and Eurasia? Until the plate tectonics theory was formulated, there had been no reasonable explanation.

Today scientists understand that the areas containing these ancient glacial features were joined together as a single supercontinent called Pangaea, located at latitudes far to the south of their present positions. Later this landmass broke apart, and its pieces, each moving on a different plate, migrated toward their present locations (**Figure 18.38**). Now we know that during the geologic past, plate movements accounted for many dramatic climate changes, as landmasses shifted in relationship to one another and moved to different latitudinal positions. Changes in oceanic circulation also must have occurred, altering the transport of heat and moisture and, consequently, the climate as well. Because the rate of plate movement is very slow—a few centimeters annually—appreciable changes in the positions of the continents

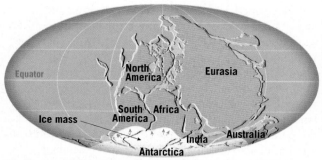

The supercontinent Pangaea showing the area covered by glacial ice near the end of the Paleozoic era.

The continents as they appear today. The white areas indicate where evidence of the late Paleozoic ice sheets exists.

▲ **Figure 18.38**
A late Paleozoic ice age Shifting tectonic plates sometimes move landmasses to high latitudes, where the formation of ice sheets is possible.

occur only over great spans of geologic time. Thus, climate changes triggered by shifting plates are extremely gradual and happen on a scale of millions of years.

Variations in Earth's Orbit   Because climatic changes brought about by moving plates are extremely gradual, the plate tectonics theory cannot be used to explain the alternating glacial and interglacial climates that occurred during the Pleistocene epoch. Therefore, we must look to some other triggering mechanism that might cause climate change on a scale of thousands, rather than millions, of years. Today many scientists strongly suspect that the climate oscillations that characterized the Quaternary period are linked to variations in Earth's orbit. This hypothesis, first developed and strongly advocated by the Serbian astrophysicist Milutin Milankovitch, is based on the premise that variations in incoming solar radiation are a principal factor in controlling Earth's climate.

Milankovitch formulated a comprehensive mathematical model based on the following elements (**Figure 18.39**):

- Variations in the shape (*eccentricity*) of Earth's orbit about the Sun
- Changes in *obliquity*—that is, changes in the angle that Earth's axis makes with the plane of our planet's orbit
- The wobbling of Earth's axis, called *precession*

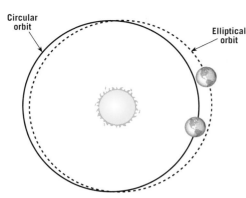

A. The shape of Earth's orbit changes during a cycle that spans about 100,000 years. It gradually changes from nearly circular to more elliptical and then back again.

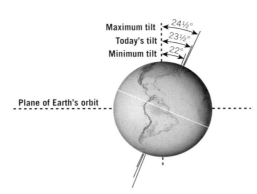

B. Today the axis of rotation is tilted about 23.5 degrees to the plane of Earth's orbit. During a cycle of 41,000 years, this angle varies from 22 to 24.5 degrees.

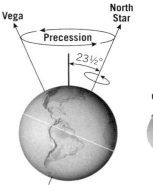

C. Earth's axis wobbles like a spinning top. Consequently, the axis points to different spots in the sky during a cycle of about 26,000 years.

▲ **SmartFigure 18.39**
Orbital variations Periodic variations in Earth's orbit are linked to alternating glacial and interglacial conditions during the Ice Age.

Animation
https://goo.gl/
7LRk5y

Using these factors, Milankovitch calculated variations in the receipt of solar energy and the corresponding surface temperature of Earth back into time, in an attempt to correlate these changes with the climate fluctuations of the Quaternary. In explaining climate changes that result from these three variables, note that they cause little or no variation in the *total* solar energy reaching the ground over the course of a year. Instead, their impact is felt because they change the degree of contrast between the seasons. Somewhat milder winters in the middle to high latitudes mean greater snowfall totals, whereas cooler summers bring a reduction in snowmelt.

Among the studies that have added credibility to the astronomical hypothesis of Milankovitch is one in which deep-sea sediments containing certain climatically sensitive microorganisms were analyzed to establish a chronology of temperature changes going back nearly a half-million years. This time scale of climate change was then compared to astronomical calculations of eccentricity, obliquity, and precession to determine whether a correlation existed. Although the study was very involved and mathematically complex, the conclusions were straightforward: The researchers found that major variations in climate over the past several hundred thousand years were closely associated with changes in the geometry of Earth's orbit; that is, cycles of climate change were shown to correspond closely with the periods of obliquity, precession, and orbital eccentricity. More specifically, the authors stated: "It is concluded that changes in the earth's orbital geometry are the fundamental cause of the succession of Quaternary ice ages."[†]

To briefly summarize the ideas just described, plate tectonics theory provides an explanation for the widely spaced and nonperiodic onset of glacial conditions at various times in the geologic past, whereas the theory proposed by Milankovitch and supported by the work of J. D. Hays and his colleagues furnishes an explanation for the alternating glacial and interglacial episodes of the Quaternary.

Other Factors Variations in Earth's orbit correlate closely with the timing of glacial–interglacial cycles. However, the variations in solar energy reaching Earth's surface caused by these orbital changes do not adequately explain the magnitude of the temperature changes that occurred during the most recent Ice Age. Other factors must also have contributed. One factor involves variations in the composition of the atmosphere. Other influences are related to changes in the reflectivity of Earth's surface and in ocean circulation. Let's take a brief look at these factors.

Chemical analyses of air bubbles trapped in glacial ice at the time of ice formation indicate that the Ice Age atmosphere contained less carbon dioxide and methane than the post–Ice Age atmosphere (**Figure 18.40**). Carbon dioxide and methane are important "greenhouse" gases, which means they trap radiation emitted by Earth and contribute to atmospheric heating. When the amount of carbon dioxide and methane in the atmosphere increases, global temperatures rise, and when there is a reduction in these gases, as occurred during the Ice Age, temperatures fall. Therefore, reductions in the concentrations of greenhouse gases help explain the magnitude

[†] J. D. Hays, John Imbrie, and N. J. Shackelton, "Variations in the Earth's Orbit: Pacemaker of the Ice Ages," *Science* 194 (1976): 1121–1132.

◄ **Figure 18.40**
Ice cores contain clues to shifts in climate This scientist is slicing an ice core from Antarctica for analysis. He is wearing protective clothing and a mask to minimize contamination of the sample. Chemical analyses of ice cores can provide important data about past climates.

of the temperature drop that occurred during glacial times. Although scientists know that concentrations of carbon dioxide and methane dropped, they do not fully understand what caused the drop. As often occurs in science, observations gathered during one investigation yield information and raise questions that require further analysis and explanation.

Obviously, whenever Earth enters an ice age, extensive areas of land that were once ice free are covered with ice and snow. In addition, a colder climate causes the area covered by sea ice (frozen surface seawater) to expand. Ice and snow reflect a large portion of incoming solar energy back to space. Thus, energy that would have warmed Earth's surface and the air above is lost, and global cooling is reinforced.

Yet another factor that influences climate during glacial times relates to ocean currents. Research has shown that ocean circulation changes during ice ages. For example, studies suggest that the warm current that transports large amounts of heat from the tropics toward higher latitudes in the North Atlantic was significantly weaker during the Ice Age. This would have led to a colder climate in Europe, amplifying the cooling attributable to orbital variations.

In conclusion, we emphasize that the ideas just discussed do not represent the only possible explanations for ice ages. Although interesting and attractive, these proposals are certainly not without critics, nor are they the only possibilities currently under study. Other factors may be, and probably are, involved.

CONCEPT CHECKS 18.6

1. What is the best source of data showing Ice Age climate cycles?

2. About what percentage of Earth's land surface has been affected by glaciers during the Quaternary period? During the Ice Age, were ice sheets more extensive in the Northern Hemisphere or in the Southern Hemisphere? Why?

   Concept Checker
   https://goo.gl/sD4ras

3. How does the theory of plate tectonics help us understand the causes of ice ages? Does this theory explain alternating glacial–interglacial climates during the Ice Age?

4. Briefly summarize the climate change hypothesis that involves variations in Earth's orbit.

18 CONCEPTS IN REVIEW

Glaciers and Glaciation

18.1 Glaciers: A Part of Two Basic Cycles

Explain the role of glaciers in the hydrologic and rock cycles. Describe the different types of glaciers, their characteristics, and their present-day distribution.

Key Terms:	sea ice	outlet glacier
glacier	tidewater glacier	piedmont glacier
valley glacier	ice shelf	
ice sheet	ice cap	

- A *glacier* is a thick mass of ice originating on land from the compaction and recrystallization of snow that shows evidence of past or present flow. Glaciers are part of both the hydrologic cycle and the rock cycle because they store and release freshwater, and they erode and transport rock material.

- *Valley glaciers* flow down mountain valleys, while *ice sheets* are very large masses of ice, such as those that cover Greenland and Antarctica. During the Last Glacial Maximum, Earth was in an Ice Age that covered large areas of the land surface with glacial ice.

- When valley glaciers leave the confining mountains, they may spread out into broad lobes called *piedmont glaciers*. Similarly, *ice shelves* form when glaciers flow into the ocean and spread out to form a wide layer of floating ice.

- *Ice caps* are like smaller ice sheets. Both ice sheets and ice caps may be drained by *outlet glaciers*.

Q What term is applied to the ice at the North Pole? What term best describes Greenland's ice? Are both considered glaciers? Explain.

18.2 Formation and Movement of Glacial Ice

Describe how glaciers move, the rates at which they move, and the significance of the glacial budget.

Key Terms:

firn	zone of accumulation	calving
zone of fracture	snowline	iceberg
crevasse	zone of wastage	glacial budget
	ablation	

- When snow piles up sufficiently, it recrystallizes to dense granules of *firn*, which can then pack together even more tightly to make glacial ice.
- When ice is put under pressure, it flows very slowly. In the uppermost 50 meters (165 feet) of a glacier, there is not sufficient pressure to flow, and dangerous cracks called *crevasses* open up in the *zone of fracture*. In addition, most glaciers also move by a sliding process called basal slip.
- Glaciers move at a slow but measurable rate. Fast glaciers may move 800 meters (2600 feet) per year, while slow glaciers may move only 2 meters (6.5 feet) per year. Some glaciers experience periodic surges of sudden movement.
- A glacier's rate of flow does not necessarily correlate to the position of its terminus. Instead, if the glacier has a positive *budget*, in which accumulation exceeds wastage, the terminus will advance. If *calving* of *icebergs*, melting, or other forms of ablation exceed the input of new ice, the glacier's terminus will retreat. Even a retreating glacier is still experiencing downstream flow.

Q This image shows that melting is one way that glacial ice wastes away. What is another way that ice is lost from a glacier? What is the general term for ice loss from a glacier?

Glacial meltwater

18.3 Glacial Erosion

Discuss the processes of glacial erosion. Identify and describe major topographic features created by glacial erosion.

Key Terms:

plucking	truncated spurs	col
abrasion	hanging valley	arête
rock flour	pater noster lakes	horn
glacial striations	cirque	roche moutonnée
glacial trough	tarn	fiord

- Glaciers are powerful agents of erosion and acquire sediment through *plucking* from the bedrock beneath the glacier, by *abrasion* of the bedrock using sediment already in the ice, and when mass-wasting processes drop debris on top of the glacier. Grinding of the bedrock produces grooves and scratches called *glacial striations*.
- *Glacial troughs* have a distinctive U-shaped profile, very different from a stream-carved mountain valley, with its typical V-shaped profile. Lining the edge of the valley may be triangle-shaped cliffs called *truncated spurs*. Higher up, *hanging valleys* mark the spots where tributary glaciers once flowed into the main glacier.
- At the head of a valley glacier is an amphitheater-shaped *cirque*, which may or may not hold a small lake called a *tarn*. Other glacial lakes in the bottom of the valley resemble beads on a string and are called *pater noster lakes*. The intersection of two cirques produced by glaciers flowing in opposite directions forms a *col*.
- If two glaciers flow parallel to each other, their troughs may intersect in a knife-edge ridge called an *arête*. A high point that had multiple glaciers flowing away from a point in a radial array may leave behind a pyramidal *horn*. Protrusions of bedrock may be glacially abraded on their upstream side and plucked on their downstream side. This produces asymmetric knobs of bedrock called *roches moutonnées*.
- In coastal mountain settings, a glacial trough may be eroded below sea level and subsequently flooded by the ocean to become a *fiord*, a narrow steep-sided inlet.

Q Examine the illustration of a mountainous landscape after glaciation. Identify the landforms that resulted from glacial erosion.

18.4 Glacial Deposits

Distinguish between the two basic types of glacial drift. List and describe the major depositional features associated with glacial landscapes.

Key Terms:	medial moraine	kettle
glacial drift	end moraine	ice-contact deposit
till	ground moraine	kame
glacial erratic	drumlin	kame terrace
stratified drift	outwash plain	esker
lateral moraine	valley train	

- Any sediment of glacial origin is called *glacial drift*. The two distinct types of glacial drift are *till*, which is unsorted material deposited directly by the ice, and *stratified drift*, which is sediment sorted and deposited by meltwater from a glacier.

- The most widespread features created by glacial deposition are layers or ridges of till, called moraines. Associated with valley glaciers are *lateral moraines*, formed along the sides of a valley, and *medial moraines*, formed between two valley glaciers that have merged. *End moraines* mark the former position of the front of a glacier, and *ground moraines*, layers of till deposited as the ice front retreats, are common to both valley glaciers and ice sheets.

- Stratified drift can be deposited immediately adjacent to a glacier or carried some distance away and laid down as outwash. Streams draining an ice sheet produce a broad *outwash plain* beyond the end moraine. A similar feature hemmed in by the walls of a mountain valley is called a *valley train*. Blocks of ice buried by the sediment may melt to produce depressions called *kettles*.

- *Kames* are mounds or steep-sided hills of stratified drift that represent the former positions of sediment-filled lakes on top of (or within) the glacier. *Kame terraces* are narrow masses of stratified drift deposited adjacent to a stagnant glacier. Meltwater streams flowing through tunnels in the ice may leave behind sinuous ridges of stratified drift called *eskers*.

Q Examine the illustration of depositional features formed in the wake of a retreating ice sheet. Name the features and indicate which landforms are composed of till and which are composed of stratified drift.

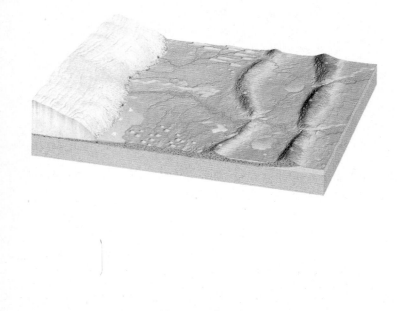

18.5 Other Effects of Ice Age Glaciers

Describe and explain several important effects of Ice Age glaciers other than erosional and depositional landforms.

Key Terms:	proglacial lake	pluvial lake

- Glaciers are heavy—so heavy that they can cause Earth's crust to flex downward under their tremendous load. After the glaciers melt off, that weight is released, and the crust slowly rebounds vertically upward.

- Ice sheets are nourished by water that ultimately comes from the ocean, so when ice sheets grow, sea level falls, and when they melt, sea level rises. At the Last Glacial Maximum, global sea level was about 100 meters (330 feet) lower than it is today. At that time, the coastlines of the modern continents were vastly different.

- The advance and retreat of ice sheets caused significant changes to the paths that rivers follow. In addition, glaciers deepened and widened stream valleys and lowlands to create features such as the Great Lakes.

- An ice sheet can act as a dam by trapping meltwater or blocking the flow of rivers to create *proglacial lakes*. Glacial Lakes Agassiz and Missoula both impounded tremendous quantities of water. In the case of Lake Missoula, the water drained out in huge torrents when the ice dam periodically broke.

- *Pluvial lakes* existed during the height of the Ice Age but occurred far from the actual glaciers, in a climate that was cooler and wetter than today's. Lake Bonneville is a classic example that existed in the area that is now Utah and Nevada.

18.6 The Ice Age

Briefly discuss the development of glacial theory and summarize current ideas on the causes of ice ages.

Key Terms:	Quaternary period	tillite

- The idea of a geologically recent Ice Age was born in the early 1800s in Switzerland. Louis Agassiz and others established that only the former presence of tremendous quantities of glacial ice could explain the landscape of Europe (and later North America and Siberia). As additional research accumulated, especially data from the study of seafloor sediments, it was revealed that the *Quaternary period* was marked by numerous advances and retreats of glacial ice.

- While rare, glacial episodes have occurred in Earth history prior to the recent glaciations we call the Ice Age. Lithified till, called *tillite*, is a major line of evidence for these ancient ice ages. Several factors explain why glacial ice might accumulate globally, including the position of the continents, which is driven by plate tectonics.

- The Quaternary period is marked by glacial advances alternating with episodes of glacial retreat. One way to explain these oscillations is through variations in Earth's orbit, which lead to seasonal variations in the distribution of solar radiation. The orbit's shape varies (eccentricity), the tilt of the planet's rotational axis varies (obliquity), and the axis slowly "wobbles" over time (precession). These three effects, all of which occur on different time scales, collectively account for alternating colder and warmer periods during the Quaternary.

- Additional factors that may be important for initiating or ending glaciations include falling or rising levels of greenhouse gases, changes in the reflectivity of Earth's surface, and variations in the ocean currents that redistribute heat energy from warmer to colder regions.

Q About 250 million years ago, parts of India, Africa, and Australia were covered by ice sheets, while Greenland, Siberia, and Canada were ice free. Explain why this occurred.

GIVE IT SOME THOUGHT

1. The accompanying diagram shows the results of a classic experiment used to determine how glacial ice moves in a mountain valley. The experiment occurred over an 8-year span. Refer to the diagram and answer the following:

   a. What was the average yearly rate at which ice in the center of the glacier advanced?

   b. About how fast was the center of the glacier advancing *per day*?

   c. Calculate the average rate at which ice along the sides of the glacier moved forward.

   d. Why was the rate at the center different than the rate along the sides?

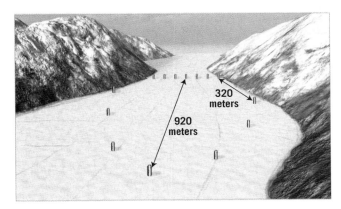

320 meters

920 meters

2. Studies have shown that during the Ice Age, the margins of some ice sheets advanced southward from the Hudson Bay region at rates ranging from about 50 to 320 meters per year.

   a. Determine the maximum amount of time required for an ice sheet to move from the southern end of Hudson Bay to the south shore of present day Lake Erie, a distance of 1600 kilometers.

   b. Calculate the minimum number of years required for an ice sheet to move this distance.

3. If Earth were to experience another Ice Age within the next few hundred thousand years, one hemisphere would have substantially more expansive ice sheets than the other. Would it be the Northern Hemisphere or the Southern Hemisphere? What is the reason for the large disparity?

4. While taking a break from a hike in the Northern Rockies with a fellow geology enthusiast, you notice that the boulder you are sitting on is part of a deposit consisting of a jumbled mixture of unsorted sediment. Since you are in an area that once had extensive valley glaciers, your colleague suggests that the deposit must be glacial till. Although you know this is certainly a good possibility, you remind your companion that other processes in mountain areas also produce unsorted deposits. What might such a process be? How might you and your friend determine whether this deposit is actually glacial till?

5. If the budget of a valley glacier were balanced for an extended time span, what feature would you expect to find at the terminus of the glacier? Is it composed of till or stratified drift? Now assume that the glacier's budget changes so that ablation exceeds accumulation. How would the terminus of the glacier change? Describe the deposit you would expect to form under these conditions.

6. Is the glacial deposit shown here an example of till or stratified drift? Is it more likely part of an end moraine or an esker?

7. Watch the Mobile Field Trip titled "Yosemite: Granite and Glaciers" (see SmartFigure 4.14). List and briefly describe the erosional and depositional features associated with glaciers depicted in this video.

8. This wall, located in New England, is built of diverse stones and boulders cleared from nearby fields. In 1914, Robert Frost wrote a now-famous poem titled "Mending Wall" about a feature like this one. It begins with these lines:

   > Something there is that doesn't love a wall,
   > That sends the frozen-ground-swell under it,
   > And spills the upper boulders in the sun;
   > And makes gaps even two can pass abreast.

   a. What is the likely weathering process causing the wall to swell and "spill" its boulders? (Think back to Chapter 6.)

   b. Is it likely that the source of all the rocks in the wall is bedrock in the immediate vicinity? Explain.

   c. What term applies to the rocks composing the wall?

9. Assume that you and a nongeologist friend are visiting the valley glacier pictured in Figure 18.1. After studying the glacier for quite a long time, your friend asks, "Do these things really move?" How would you convince your friend that this glacier does indeed move, using evidence that is clearly visible in this image?

EYE ON EARTH

1. The central focus of this satellite image is Byrd Glacier in Antarctica. In this region, the glacier advances at a rate of about 0.8 kilometer (0.5 mile) per year. Take note of its position in relationship to other labeled features.

   a. Is Byrd Glacier flowing toward the top of the image or toward the bottom of the image? How did you figure this out?

   b. What term describes this type of glacier?

2. This photo shows an iceberg floating in the ocean near the coast of Greenland.

   a. How do icebergs form? What term applies to this process?

   b. Using the knowledge you have gained about these features, explain the common phrase "only the tip of the iceberg."

   c. Is an iceberg the same as sea ice? Explain.

   d. If this iceberg were to melt, how would sea level be affected?

DATA ANALYSIS

Glacial Flow Patterns

Glacial ice flows much like water, only very slowly. Flow patterns are determined by snow accumulation and the local terrain.

https://goo.gl/eD226M

ACTIVITIES

Go to Earth Observatory's Columbia Glacier, at http://earthobservatory.nasa.gov/Features/WorldOfChange/, and select the "Columbia Glacier, Alaska" article. Note that in the time lapse, snow and ice appear as bright blue, vegetation is green, and open water is dark blue. Rocky debris on the glacier's surface appears gray, while exposed rock is brown.

Click the "Play" symbol. As you step forward in time, notice that sometimes the ice has flow streaks and sometimes it does not. The flowing part is the glacier, and the non-flowing part is chunks of icebergs that are rafted together so you can add information to the map.

1. **In 1986, what was the distance between the medial moraine and the underwater terminal moraine? Between the medial moraine and the terminus?**

2. **What role did the underwater terminal moraine play in ice accumulation downstream of the terminus?**

3. **Why was there no ice accumulation downstream of the terminus in some years?**

4. **How far did the terminus retreat in 1996, 2006, 2014, and 2017?**

Go to the BEDMAP2 map: http://cdn.antarcticglaciers.org/wp-content/uploads/2013/08/bedmap2_preview.png. This map shows a compilation of bedrock topography data for Antarctica, where red represents higher elevation.

In a separate window, go to the Ice Streams of Antarctica map: http://cdn.antarcticglaciers.org/wp-content/uploads/2012/09/Antarctic_icestreams.jpg.

5. **Compare the topography and ice streams maps. Which direction do ice streams flow? What controls the direction in which ice streams flow?**

6. **Where are the fastest ice velocities located? Why?**

In a third window, go to the South Pole Sterographic map: http://cdn.antarcticglaciers.org/wp-content/uploads/2012/06/Antarctica_LIMA.jpg.

7. **Which ice sheets have the fastest ice velocities?**

Mastering Geology

Looking for additional review and test prep materials? Visit pearson.com/mastering/geology to enhance your understanding of this chapter's content by accessing a variety of resources, including **Self-Study Quizzes, Geoscience Animations, SmartFigure Tutorials,** *Project Condor* **Videos, Mobile Field Trips, Dynamic Study Modules,** and an optional **Pearson eText.**

In The NEWS

Dust Storms Menace Desert Landscapes

In early May 2018, the dry lands of northern India experienced several unusually intense dust storms known as *haboobs*. During each of the storms, visibility dropped to nearly zero in a matter of seconds as strong winds downed trees and power lines and damaged poorly constructed dwellings. Some of the storms were deadly.

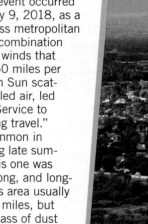

▲ The leading edge of a haboob moving through Phoenix, Arizona.

A haboob is created when strong, relatively cold downdrafts from powerful thunderstorms reach the surface in a phenomenon called a *gust front* and swiftly spread across the relatively barren desert landscape, lifting huge quantities of dust hundreds of meters high.

Most often associated with parts of North Africa, haboobs occasionally menace the American Southwest as well. An extraordinary event occurred on the afternoon of July 9, 2018, as a wall of dust rolled across metropolitan Phoenix, Arizona. The combination of intense dust, strong winds that sometimes exceeded 50 miles per hour, and the afternoon Sun scattered by the particle-filled air, led the National Weather Service to warn of "life-threatening travel." Haboobs are not uncommon in southern Arizona during late summer afternoons, but this one was exceptionally large, strong, and long-lasting. Haboobs in this area usually subside after 25 to 50 miles, but this nearly mile-high mass of dust traveled about 200 miles across the Sonoran Desert landscape; local meteorologists described it as "epic" and "historic." As a geologic force, these turbulent dust storms are capable of transporting and depositing enormous quantities of sediment.

► This huge dust storm, known as a haboob, rolled across Phoenix, Arizona, on July 9, 2018.

19

Deserts and Wind

FOCUS ON CONCEPTS

Each statement represents the primary learning objective for the corresponding major heading within the chapter. After you complete the chapter, you should be able to:

19.1 Describe the general distribution of Earth's dry lands and explain why deserts form in the subtropics and middle latitudes.

19.2 Summarize the geologic roles of weathering, running water, and wind in arid and semiarid climates.

19.3 Contrast the landscapes of the Basin and Range and the Colorado Plateau in the western United States.

19.4 Describe the ways that wind transports sediment and the processes and features associated with wind erosion.

19.5 Discuss dune formation and movement and distinguish among different dune types. Explain how loess deposits differ from deposits of sand.

Climate has a strong influence on the nature and intensity of Earth's external processes. This was clearly demonstrated in the preceding chapter, on glaciers. Desert landscapes and their development provide an other excellent example of the strong link between climate and geology. The word *desert* literally means "deserted" or "unoccupied." For many dry regions, this is an appropriate description, although where water is available in deserts, plants and animals thrive. Nevertheless, the world's dry regions are probably the least familiar land areas on Earth outside the polar realm.

As you will see, arid regions are not dominated by a single geologic process. Rather, the effects of tectonic forces, running water, and wind are all apparent. Because these processes combine in different ways from place to place, the appearance of desert landscapes varies a great deal.

19.1 Distribution and Causes of Dry Lands

Describe the general distribution of Earth's dry lands and explain why deserts form in the subtropics and middle latitudes.

Desert landscapes frequently appear stark. Their profiles are not softened by a continuous carpet of soil and abundant plant life. Instead, barren rocky outcrops with steep, angular slopes are common. At some places the rocks are tinted orange and red. At others they are gray and brown and streaked with black. For many visitors, desert scenery exhibits a striking beauty; to others, the terrain seems bleak. No matter which feeling is elicited, it is clear that deserts are very different from the more humid places where most people live.

What Is Meant by *Dry*

We all recognize that deserts are dry places, but just what is meant by the term *dry*? That is, how much rain defines the boundary between humid and dry regions? Sometimes dryness is arbitrarily defined by a single rainfall figure, such as 25 centimeters (10 inches) per year of precipitation. However, the concept of dryness is a relative one that refers to any situation in which an ongoing water deficiency exists.

Dry Climates Climatologists define **dry climate** as a climate in which yearly precipitation is not as great as the potential loss of water by evaporation. Therefore, dryness is not only related to annual rainfall totals but is also a function of evaporation, which in turn is closely dependent on temperature. As temperatures climb, potential evaporation also increases. As little as 15 to 25 centimeters (6 to 10 inches) of precipitation per year may be sufficient to support coniferous forests in northern Scandinavia or Siberia, where evaporation into the cool, humid air is slight and a surplus of water remains in the soil. However, the same amount of rain falling on Nevada or Iran supports only a sparse vegetative cover because evaporation into the hot, dry air is great. So, clearly, no specific amount of precipitation can serve as a universal boundary for dry climates.

The dry regions of the world encompass about 42 million square kilometers (16 million square miles), a surprising 30 percent of Earth's land surface. No other climate category covers so large a land area. In water-deficient regions, two climatic types are commonly recognized: **desert**, or arid, and **steppe**, or semiarid. The two share many features; their differences are primarily a matter of degree. The steppe is a marginal and more humid variant of the desert and is a transition zone that surrounds the desert and separates it from bordering humid climates. The world map of the distribution of desert and steppe regions in **Figure 19.1** shows that dry lands are concentrated in the subtropics and in the middle latitudes.

Are Deserts Expanding?

Desertlike conditions are expanding worldwide. This important environmental problem, called **desertification**, refers to the persistent degradation of dry-land ecosystems primarily due to human activities. It most often, but not always, takes place on the margins of desert and steppe regions and involves a continuum of change, from slight to severe alteration of plant and soil

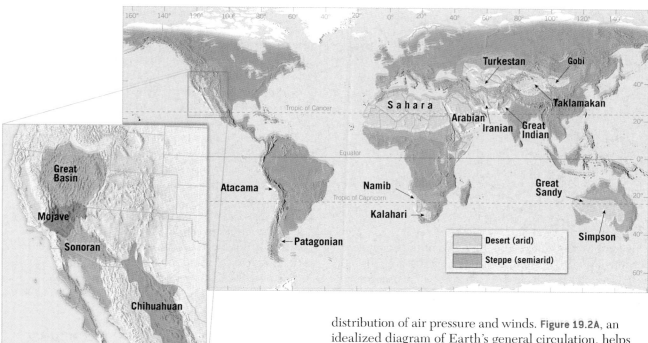

◄ SmartFigure 19.1
Dry climates Arid and semiarid climates cover about 30 percent of Earth's land surface. The dry region of the American West is commonly divided into four deserts, two of which extend into Mexico.

Tutorial
https://goo.gl/kmWns9

▼ SmartFigure 19.2
Subtropical deserts The distribution of subtropical deserts and steppes is closely related to the global distribution of air pressure.

Animation
https://goo.gl/9nQKvT

resources. Desertification occurs when deforestation and overgrazing reduce or completely remove the tree and plant cover that anchors the soil. In some regions, intensive and unsustainable farming practices destroy the natural vegetation and deplete soil nutrients. Wind and water erosion aggravate the damage by carrying away topsoil. When drought occurs on these lands without sufficient vegetation to hold the soil against erosion, the destruction can be irreversible. Desertification is occurring in many places but is particularly serious in the region south of the Sahara Desert known as the Sahel.

Subtropical Deserts and Steppes

The heart of the subtropical dry climates lies in the vicinities of the Tropics of Cancer and Capricorn. Figure 19.1 shows a virtually unbroken desert environment stretching for more than 9300 kilometers (5800 miles) from the Atlantic coast of North Africa to the dry lands of northwestern India. In addition to this single great expanse, the Northern Hemisphere contains another, much smaller area of subtropical desert and steppe in northern Mexico and the southwestern United States.

In the Southern Hemisphere, dry climates dominate Australia. Almost 40 percent of the continent is desert, and much of the remainder is steppe. In addition, arid and semiarid areas occur in southern Africa and make a limited appearance in coastal Chile and Peru.

Subsiding Air Masses
What causes these bands of low-latitude desert? The primary control is the global

distribution of air pressure and winds. **Figure 19.2A**, an idealized diagram of Earth's general circulation, helps visualize the relationship. Heated air in the pressure belt known as the *equatorial low* rises to great heights (usually between 15 and 20 kilometers) and then spreads out. As the upper-level flow reaches 20 to 30 degrees latitude, north or south, it sinks toward the surface. Air that rises through the atmosphere expands and cools, a process that leads to the development of clouds and precipitation. For

A. Subtropical deserts and steppes are centered between 20° and 30° north and south latitude in association with the subtropical high-pressure belts. Dry subsiding air inhibits cloud formation and precipitation.

B. In this view from space, the Sahara Desert, the adjacent Arabian Desert, and the Kalahari and Namib Deserts are clearly visible as tan-colored, cloud-free zones. The band of clouds across central Africa and the adjacent oceans coincides with the equatorial low-pressure belt.

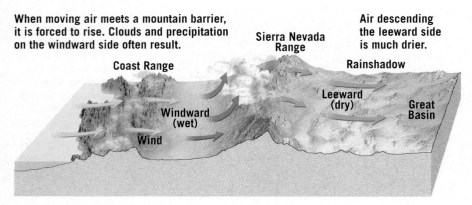

When moving air meets a mountain barrier, it is forced to rise. Clouds and precipitation on the windward side often result.

Sierra Nevada Range

Air descending the leeward side is much drier.

Coast Range

Rainshadow

Windward (wet)

Leeward (dry)

Great Basin

Wind

▲ SmartFigure 19.3
Rainshadow deserts Mountains frequently contribute to the aridity of middle-latitude deserts and steppes by creating a rainshadow.

Animation
https://goo.gl/4E4mFi

this reason, areas under the influence of the equatorial low are among the rainiest on Earth. Just the opposite is true for the regions in the vicinity of 30 degrees north and south latitude, where high pressure predominates. Here, in the zones known as the *subtropical highs*, air is subsiding. When air sinks, it is compressed and warmed. Such conditions are just the opposite of what is needed to produce clouds and precipitation. Consequently, these regions are known for their clear skies, sunshine, and ongoing dryness (**Figure 19.2B**).

West Coast Subtropical Deserts   Where subtropical deserts are found along the west coasts of continents, cold ocean currents have a dramatic influence on the climate. The principal examples are the Atacama Desert in South America, which is adjacent to the cold Peru Current, and the Namib Desert in southwestern Africa, which parallels the cold Benguela Current (see Figure 19.1). These dry lands deviate considerably from the general image we have of subtropical deserts.

► Figure 19.4
Distribution of precipitation in western Washington State
The Olympic and Cascade Mountains receive abundant rainfall. The semiarid eastern portion of the state is in a rainshadow.

ANNUAL PRECIPITATION

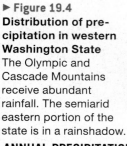

cm	in.
>500	>200
405–499	160–199
250–404	100–159
150–249	60–99
50–149	20–59
25–49	10–19
<25	<10

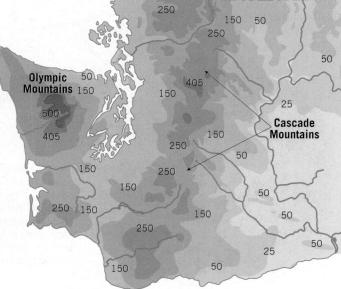

Olympic Mountains

Cascade Mountains

The most obvious effect of the cold current is reduced temperatures. Furthermore, although these deserts are adjacent to the ocean, their yearly rainfall totals are among the lowest in the world. The aridity of these coastal areas is intensified because the lower air is chilled by the cold offshore waters. When air is cooled from below, it resists the upward movement required for cloud formation and precipitation. In addition, the cold current often chills the air enough to cause fog to form. Thus, not all subtropical deserts are sunny and hot. Cold offshore currents cause west coast subtropical deserts to be relatively cool places that sometimes are foggy.

The Atacama along the west coast of South America has the distinction of being the world's driest desert. The average rainfall at the Atacama's wettest locations is not more than 3 millimeters (0.12 inch) per year. At Arica, a coastal town near Chile's border with Peru, the average annual rainfall is a mere 0.5 millimeter (0.02 inch). Further inland, some stations have *never* recorded rainfall.

Middle-Latitude Deserts and Steppes

Unlike their low-latitude counterparts, middle-latitude deserts and steppes are not controlled by the subsiding air masses associated with zones of high pressure. Instead, many of these dry lands exist because they are sheltered in the deep interiors of large landmasses. They are far removed from the ocean, which is the ultimate source of moisture for cloud formation and precipitation. One well-known example is the Gobi Desert of central Asia (see Figure 19.1).

The presence of high mountains across the paths of prevailing winds is another factor that separates middle-latitude arid and semiarid areas from water-bearing, maritime air masses. Mountains force the air masses to lose much of their water. The mechanism is straightforward: As prevailing winds meet mountain barriers, the air is forced to ascend. When air rises, it expands and cools, a process that can produce clouds and precipitation. The windward sides of mountains, therefore, often have high precipitation. By contrast, the leeward sides of mountains are usually much drier (**Figure 19.3**). This situation exists because air reaching the leeward side has lost much of its moisture, and if the air descends, it is compressed and warmed, making cloud formation even less likely. The dry region that results on this leeward side is often referred to as a **rainshadow**. **Figure 19.4**, which shows the distribution of precipitation in western Washington State, is a good example. When prevailing winds from the Pacific Ocean to the west (left) meet the mountains, rainfall totals are high. By comparison, precipitation on the leeward (eastern) side of the mountains is relatively meager. In North America, the foremost mountain barriers to moisture from the Pacific are the Coast Ranges, Sierra Nevada, and Cascades. In Asia, the great Himalaya Mountains prevent the summertime monsoon flow of moist Indian Ocean air from reaching the interior of the continent.

Because the Southern Hemisphere lacks extensive land areas in the middle latitudes, only small areas of desert and steppe occur in this latitude range, existing primarily near the southern tip of South America, in the rainshadow of the towering Andes.

The middle-latitude deserts provide an example of how tectonic processes affect climate. Rainshadow deserts exist in part because of the mountains produced when plates collide. Without such mountain-building episodes, wetter climates would prevail where many dry regions exist today.

CONCEPT CHECKS 19.1

1. Explain why the boundary between humid and dry climates cannot be defined by a single rainfall amount.

2. How extensive are the desert and steppe regions of Earth?

Concept Checker
https://goo.gl/ZSPdkd

3. What is the primary cause of subtropical deserts and steppes? How do cold ocean currents influence some subtropical deserts?

4. Why do middle-latitude dry regions exist?

19.2 Geologic Processes in Arid Climates

Summarize the geologic roles of weathering, running water, and wind in arid and semiarid climates.

The angular hills, the sheer canyon walls, and the desert surface of pebbles or sand contrast sharply with the rounded hills and curving slopes of more humid places. To a visitor from a humid region, a desert may seem to have been shaped by forces different from those operating in well-watered areas. However, although the contrasts may be striking, arid and humid landscapes do not reflect different processes. They merely disclose the differing effects of the same processes that operate under contrasting climatic conditions.

Dry-Region Weathering

Recall from Chapter 6 that water plays an important role in chemical weathering. Consequently, chemical weathering processes are not as prominent in regions with dry climates as in wetter regions. In humid regions, relatively fine-textured soils support an almost continuous cover of vegetation that mantles the surface. The slopes and rock edges are rounded, reflecting the strong influence of chemical weathering in a humid climate. By contrast, much of the weathered debris in deserts consists of unaltered rock and mineral fragments—the result of mechanical weathering processes. In dry lands, rock weathering of any type is greatly reduced because of the lack of moisture and the scarcity of organic acids from decaying plants. However, chemical weathering is not completely lacking in deserts. Over long spans of time, clays and thin soils do form, and many iron-bearing silicate minerals oxidize, producing the rust-colored stain that tints some desert landscapes.

Desert Varnish

The rocks in some deserts exhibit a dark reddish-brown or black coating called **desert varnish**. This microscopically thin layer is composed primarily of clay minerals along with manganese and iron oxides. The color of the varnish depends on the relative amounts of manganese and iron. Manganese-rich varnishes are black; iron-rich coatings are red to orange, and varnishes with similar amounts of manganese and iron oxide are some shade of brown.

Originally scientists thought that desert varnish was made from substances drawn from the rock it covers.

However, recent studies indicate that the coating likely comes from dust that settles on the surface of the rock. Desert varnish forms very slowly, over thousands of years, as microorganisms extract iron and manganese from the dust and transform it into iron and manganese oxides. This coating does not form in humid places because rain washes the dust from rock surfaces. Measuring the thickness of desert varnish allows scientists to approximate how long a rock has been exposed at the ground surface.

Desert varnish can play an important role in some archeological studies. In past centuries Native Americans created images and symbols by removing the varnish to reveal the underlying lighter-colored rock. These drawings, called *petroglyphs*, provide a valuable record of cultural expression (**Figure 19.5**).

▼ **Figure 19.5 Petroglyphs** Native Americans removed desert varnish to create images called petroglyphs. These petroglyphs are in Dinosaur National Monument, along the Colorado-Utah border.

An ephemeral stream shortly after a heavy shower. Although such floods are short-lived, they cause large amounts of erosion.

Most of the time desert stream channels are dry.

A familiar sign in desert areas. Roads dip into washes which can rapidly fill with water following a heavy rain.

POTENTIAL FLASH FLOOD AREAS

NEXT 21 MILES

▲ Figure 19.6
Ephemeral stream This example is near Arches National Park in southern Utah.

° See Chapter 16 for more information on flash flooding.

The wadi in its usual dry state.

Following a rainy period, freshly sprouted vegetation turns the wadi green.

▲ Figure 19.7
Wadi in North Africa These two satellite images show how rain transformed a wadi in Niger. (NASA)

The Role of Water

Deserts have scant precipitation and few major rivers. Nevertheless, water plays an important role in shaping landscapes in dry regions. Permanent streams are normal in humid regions, but practically all desert streambeds are dry most of the time. Deserts have intermittent steams, or **ephemeral streams** (*ephemero* = short-lived), which carry water only in response to specific episodes of rainfall (**Figure 19.6**).

A typical ephemeral stream might flow only a few days or perhaps just a few hours during the year, following sporadic rains. In some years the channel might carry no water at all. This fact is obvious even to a casual traveler who notices numerous bridges with no streams beneath them or numerous dips in the road where dry channels cross. When the rare heavy showers do come, however, so much rain falls in such a short time that all of it cannot soak in. Because desert vegetative cover is sparse, runoff is largely unhindered and consequently rapid, often creating

Running water does most of the erosional work in deserts.

flash floods along valley floors. These floods are quite unlike floods in humid regions. Whereas a flood on a river such as the Mississippi may take several days to crest and then subside, a desert flood arrives suddenly and subsides quickly. Because much surface material in a desert is not anchored by vegetation, the amount of erosional work that occurs during a single short-lived rain event is impressive.°

In the dry western United States, different names are used for ephemeral streams, including *wash* and *arroyo*. In other parts of the world, a dry desert stream may be a *wadi* (Arabian Peninsula and North Africa), a *donga* (South America), or a *nullah* (India). The satellite images in **Figure 19.7** show a wadi in the Sahara Desert.

Humid regions are notable for their integrated drainage systems. But in arid regions, streams usually lack an extensive system of tributaries. In fact, a basic characteristic of desert streams is that they are small and die out before reaching the sea. Because the water table is usually far below the surface, few desert streams can draw upon it as streams do in humid regions (see Figure 17.8). Without a steady supply of water, the combination of evaporation and infiltration soon depletes the stream.

The few permanent streams that do cross arid regions, such as the Colorado and Nile Rivers, originate *outside* the desert, often in well-watered mountains. Here the water supply must be great, or the stream will lose all its water as it crosses the desert. For example, after the Nile leaves its headwaters in the lakes and mountains of central Africa, it traverses almost 3000 kilometers (1900 miles) of the Sahara without a single tributary. By contrast, in humid regions the discharge of a river grows as it flows downstream because tributaries and groundwater contribute additional water along the way.

It should be emphasized that *running water, although infrequent, nevertheless does most of the erosional work in deserts*. This is contrary to the common belief that wind is the most important erosional agent sculpturing desert landscapes. Although wind erosion is indeed more significant in dry areas than elsewhere, most desert landforms are carved by running water. As you will see later in the chapter, the main role of wind is the transport and deposition of sediment, which creates and shapes the ridges and mounds we call dunes. Other misconceptions about deserts are presented in **GEOgraphics 19.1**.

CONCEPT CHECKS 19.2

1. How does the rate of rock weathering in dry climates compare to the rate in humid regions?

2. What is an ephemeral stream?

3. What is the most important erosional agent in deserts?

 Concept Checker
https://goo.gl/N35psN

Common Misconceptions About Deserts

Deserts are hot, lifeless, sand-covered landscapes shaped largely by the forces of wind. The preceding statement summarizes the image of arid regions that many people hold, especially those living in more humid places. Is it an accurate view? The answer is no. Although there are clearly elements of reality in such an impression, it is a generalization that contains several misconceptions.

As this image from the American Southwest illustrates, deserts are not necessarily hot and lifeless. Although reduced in amount and different in character, both plant and animal life are usually present. In addition to record-setting heat, cold temperatures are also experienced. The average daily minimum in January at Phoenix, Arizona, is 1.7°C (35°F), just barely above freezing.

A mistaken assumption is that deserts consist of mile after mile of drifting sand such as these giant dunes along the southwest coast of Africa in the Namib Desert. These huge dunes reach heights of 300 to 350 meters (1000 to 1167 feet). Although sand accumulations may be striking features in some areas, they represent only a small percentage of the total desert area. For example, in the Sahara accumulations of sand cover only one-tenth of its area. The sandiest of all deserts is the Arabian, one-third of which is sand covered.

Another mistaken assumption is the seemingly logical idea that wind is the most important agent of erosion in deserts. However, the greatest erosional work in deserts is done by running water. The infrequent rains often take the form of thunderstorms. Because the heavy rains cannot all soak in, rapid runoff results. Without a thick vegetative cover to protect the ground, erosion is great.

19.3 Arid Landscapes of the American West

Contrast the landscapes of the Basin and Range and the Colorado Plateau in the western United States.

This section examines the assemblage of landforms that characterize the two regions that occupy much of the arid interior of the western United States—the Basin and Range and the Colorado Plateau.

Basin and Range

As discussed earlier, arid regions typically lack permanent streams and often have **interior drainage**. This means they exhibit a discontinuous pattern of intermittent streams that do not flow out of the desert to the ocean. In the United States, the dry Basin and Range region provides an excellent example. The region includes southern Oregon, all of Nevada, western Utah, southeastern California, southern Arizona, and southern New Mexico. The name *Basin and Range* is an apt description for this almost 800,000-square-kilometer (300,000-square-mile) region because it is characterized by more than 200 relatively small mountain ranges that rise 900 to 1500 meters (3000 to 5000 feet) above the basins that separate them. The origin of these fault-block mountains is examined in Chapter 14. The Mobile Field Trip associated with Figure 10.16 also provides worthwhile insight into Basin and Range geology. In this discussion we look at how surface processes change the landscape.

In the Basin and Range region, as in other areas like it around the world, erosion mostly occurs without reference to the ocean (ultimate base level) because the interior drainage never reaches the sea. The block models in **Figure 19.8** show how the landscape has evolved in the Basin and Range region and illustrate the landforms described in the following paragraphs. During and following the uplift of the mountains, running water begins carving the elevated mass and depositing large quantities of debris in the basin. The relief is greatest during this early stage because as erosion lowers the mountains and sediment fills the basins, elevation differences gradually diminish.

When the occasional torrents of water produced by sporadic rains move down the mountain canyons, they are heavily loaded with sediment. Emerging from the confines of the canyon, the runoff spreads over the gentler slopes at the base of the mountains and quickly loses velocity. Consequently, most of its load is dumped within a short distance. The result is a cone of debris at the mouth of a canyon known as an **alluvial fan**. Because the coarsest (heaviest) material is dropped first, the head of the fan is steepest, having a slope of perhaps 10 to 15 degrees. Moving down the fan, the size of the sediment and the steepness of the slope decrease, and the fan merges imperceptibly with the basin floor. An examination of the fan's surface would likely reveal a braided channel pattern because of the water shifting its course as successive channels became choked with sediment. Over the years, a fan enlarges, eventually coalescing with fans from adjacent canyons to produce an apron of sediment called a **bajada** along the mountain front.

On the rare occasions of abundant rainfall, streams may flow across the bajada to the center of the basin, converting the basin floor into a shallow **playa lake**. Playa lakes are temporary features that last only a few days or at best a few weeks before evaporation and infiltration remove the water. The dry, flat lake bed

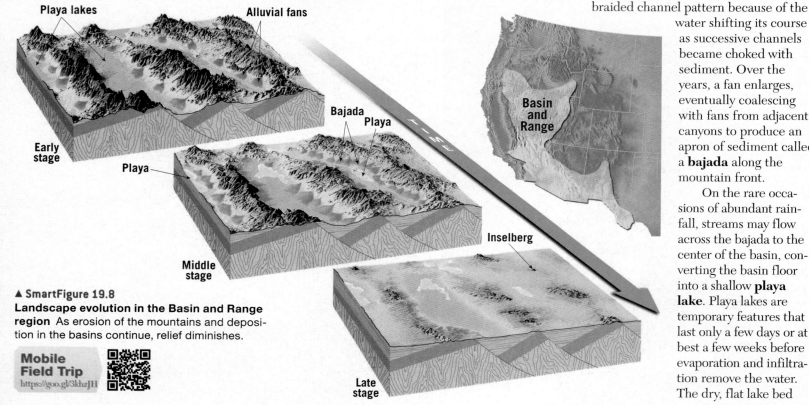

▲ SmartFigure 19.8
Landscape evolution in the Basin and Range region As erosion of the mountains and deposition in the basins continue, relief diminishes.

Playa lakes   Alluvial fans

Early stage

Playa

Bajada
Playa

Middle stage

Basin and Range

Inselberg

Late stage

that remains is called a **playa**. Playas are typically composed of fine silts and clays and are occasionally encrusted with salts precipitated during evaporation. These precipitated salts may be unusual. A case in point is the sodium borate (better known as borax) mined from ancient playa lake deposits in Death Valley, California.

With the ongoing erosion of the mountain mass and the accompanying sedimentation, the local relief continues to diminish. Eventually, nearly the entire mountain mass is gone. Thus, by the late stages of erosion, the mountain areas are reduced to a few large bedrock knobs projecting above the surrounding sediment-filled basin. These isolated erosional remnants on a late-stage desert landscape are called **inselbergs**, a German word meaning "island mountains."

Each of the stages of landscape evolution in an arid climate depicted in Figure 19.8 can be observed in the Basin and Range region. Recently uplifted mountains in an early stage of erosion are found in southern Oregon and northern Nevada. Death Valley, California, and southern Nevada fit into the more advanced middle stage, whereas the late stage, with its inselbergs, can be seen in southern Arizona.

The satellite images of Death Valley in **Figure 19.9** show these features as they appeared after a heavy rain. In the main image, a wide, shallow playa lake occupies the valley's lowest spot. This lake was gone after 3 months, and the playa was again an expanse of crusted salt.

Colorado Plateau

The Colorado Plateau got its name because most of the region is drained by the Colorado River and its tributaries. The plateau occupies about 337,000 square kilometers (130,000 square miles) centered on the Four Corners area where Utah, Arizona, New Mexico, and Colorado meet (**Figure 19.10**). The Colorado Plateau includes several

Geologist's Sketch

▲ SmartFigure 19.9
Death Valley: A classic Basin and Range landscape Shortly before this satellite image was taken in February 2005, heavy rains led to the formation of a small playa lake—the pool of greenish water on the basin floor. By May 2005, the lake had reverted to a salt-covered playa. The small photo is a closer view of one of Death Valley's many alluvial fans. The Condor Video associated with this figure takes a closer look at alluvial fans.

Condor Video
https://goo.gl/zptz86

◀ Figure 19.10
Colorado Plateau Along with the Basin and Range, this area, centered on the Four Corners area, occupies much of the interior of the American West.

▲ Figure 19.11
Buttes and mesas These rocky remnants give Monument Valley its name.

different degrees of resistance to weathering and erosion, so that abrupt changes in topography are the rule. Cliffs of resistant sandstone or limestone rise almost vertically from one plateau level to a higher one. In contrast, easily eroded shale generally forms gradual slopes that extend out from the base of the cliffs (see Figure 7.5).

Often the cliff-forming strata are jointed. By breaking along successive joints, cliffs retreat back into the land, maintaining their perpendicular faces. Cliff retreat along the margins of a plateau produce steep-sided, flat-topped hills called **mesas** (Spanish for "table"). Related but smaller features are called **buttes** (Figure 19.11). With further erosion, still smaller residual features called **pinnacles** may be all that remains—a final spire of resistant caprock protecting weaker beds below. Mesas, buttes, and pinnacles are remnants of a formerly more extensive surface, much of which has been eroded away (Figure 19.12).

prominent national parks, including Grand Canyon, Zion, Bryce, Capitol Reef, Canyonlands, and Arches.

A distinguishing feature is the region's elevation. Aside from canyon bottoms, no large area is lower than 1500 meters (5000 feet). The plateau has hundreds of remarkable canyons that largely result from the elevation of the land well above base level. In contrast to the fault-block mountains and sediment-filled basins of the Basin and Range region, sedimentary rocks that are mostly flat lying characterize most of the plateau (see Figure 9.16). The layers exhibit

CONCEPT CHECKS 19.3

1. Describe the features and characteristics associated with each stage in the evolution of a mountainous desert. Where in the United States can each stage of desert landscape evolution be observed?

2. How does the landscape of the Colorado Plateau differ from that of the Basin and Range?

▶ Figure 19.12
Monument Valley This iconic landscape on the Colorado Plateau in northern Arizona has many mesas, buttes, and pinnacles.

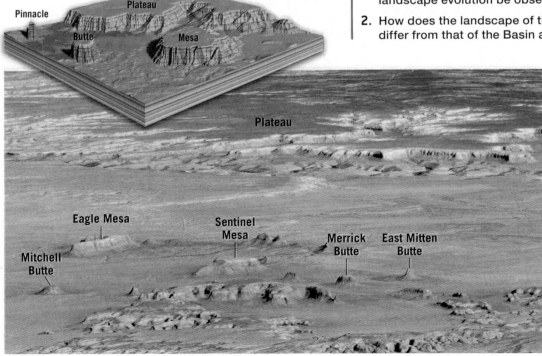

19.4 Wind Erosion

Describe the ways that wind transports sediment and the processes and features associated with wind erosion.

Wind erosion is more effective in arid lands than in humid areas because in humid places, moisture binds particles together and vegetation anchors the soil. When the ground is dry and vegetation is scant, however, wind may pick up, transport, and deposit great quantities of fine sediment. The Dust Bowl of the 1930s, in which vast dust storms devastated portions of the Great Plains, occurred when prolonged drought desiccated land that had been stripped of its natural vegetation by plowing.

Transportation of Sediment by Wind

Moving air, like moving water, is turbulent and able to pick up and transport loose debris. Just as in a stream, the velocity of wind increases with height above the surface. Also as with a stream, wind carries fine particles in suspension and moves heavier ones as bed load. However, the transport of sediment by wind differs from that of running water in two significant ways. First, wind's lower density compared to water renders it less capable of picking up and transporting coarse materials. Second, because wind is not confined to channels, it can spread sediment over large areas, as well as high into the atmosphere.

Bed Load   The **bed load** that wind carries consists of sand grains. Observations in the field and experiments using wind tunnels indicate that windblown sand moves mainly by skipping and bouncing along the surface—a process termed **saltation** (Latin for "to jump").

The movement of sand grains begins when wind reaches a velocity sufficient to overcome the inertia of the resting particles. At first the sand rolls along the surface. When a moving sand grain strikes another grain, one or both of them may jump into the air. Once in the air, the grains are carried forward by the wind until gravity pulls them back toward the surface. When the sand hits the surface, it either bounces back into the air or dislodges other grains, which then jump upward. In this manner, a chain reaction is established, filling the air near the ground with saltating sand grains in a short period of time (**Figure 19.13**).

Bouncing sand grains never travel far from the surface. Even when winds are very strong, the height of the saltating sand seldom exceeds 1 meter (3 feet) and usually is no greater than 0.5 meter (1.5 feet). Some sand grains are too large to be thrown into the air by impact from other particles. When this is the case, the energy provided by the impact of the smaller saltating grains drives the larger grains forward.

Suspended Load   Unlike sand, finer particles of dust can be swept high into the atmosphere by the wind. Because dust is often composed of rather flat particles that have large surface areas given their weight, it is relatively easy for turbulent air to counterbalance the pull of gravity and keep these fine particles airborne for hours or even days. Although both silt and clay can be carried in suspension, silt commonly makes up the bulk of the **suspended load** because the reduced level of chemical weathering in deserts provides only small amounts of clay.

Fine particles are easily carried by the wind, but they are not so easily picked up to begin with. This is because wind velocity is practically zero within a very thin layer close to the ground. Thus, the wind cannot lift the sediment by itself. Instead, the dust must be ejected or spattered into the moving air by bouncing sand grains or other disturbances. This idea is illustrated nicely by a dry, unpaved country road on a windy day. When the road is undisturbed, the wind raises little dust. However, as a car or truck moves over the road, the layer of silt is kicked up, creating a thick cloud of dust.

Although the suspended load is usually deposited relatively near its source, high winds are capable of carrying large quantities of dust great

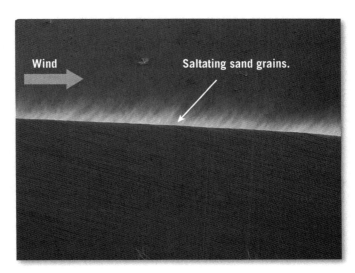

Wind   Saltating sand grains.

◀ **SmartFigure 19.13 Transporting sand** The bed load that wind carries consists of sand grains that move by bouncing along the surface. Sand never travels far from the surface, even when winds are very strong.

Animation
https://goo.gl/
rQBnMc

A dust storm blackens the Colorado sky in this historic image from the Dust Bowl of the 1930's.

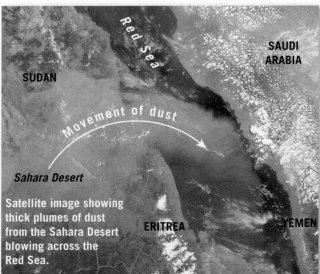

Satellite image showing thick plumes of dust from the Sahara Desert blowing across the Red Sea.

distances (**Figure 19.14**). In the 1930s, silt that was picked up in Kansas was transported to New England and beyond, into the North Atlantic. Similarly, dust blown from the Sahara has been traced as far as the Caribbean.

Erosional Features

Compared to running water and glaciers, wind is a relatively modest force in sculpting landforms. Recall that even in deserts, most erosion is performed by intermittent running water, not by wind. Nevertheless, features created by wind erosion are significant elements of some landscapes.

Blowout

In this example, deflation has removed about 1.2 meters (4 feet) of soil—the distance from the man's outstretched arm to his feet.

Deflation and Blowouts One way that wind erodes is by **deflation**—the lifting and removal of loose material. Deflation sometimes is difficult to notice because the entire surface is being lowered at the same time, but it can be significant. In portions of the 1930s Dust Bowl, vast areas of land were lowered by as much as 1 meter in only a few years.

The most noticeable results of deflation in some places are shallow depressions appropriately called **blowouts** (**Figure 19.15**). In the Great Plains region, from Texas north to Montana, thousands of blowouts are visible on the landscape. They range from small dimples less than 1 meter deep and 3 meters wide to depressions that approach 50 meters deep and several kilometers across (see Figure 6.27). The factor that controls the depths of these basins (that is, acts as base level) is the local water table. When blowouts are lowered to the water table, damp ground and vegetation prevent further deflation.

Ventifacts and Yardangs Like glaciers and streams, wind also erodes by **abrasion**. In dry regions as well as along some beaches, windblown sand cuts and polishes exposed rock surfaces. Abrasion sometimes creates interestingly shaped stones called **ventifacts** (**Figure 19.16A**). The side of such a stone that is exposed to the prevailing wind is abraded, leaving it polished, pitted, and with sharp edges. If the wind is not consistently from one direction, or if the pebble becomes reoriented, it may have several faceted surfaces.

Unfortunately, abrasion is often given credit for accomplishments beyond its capabilities. Such features as balanced rocks that stand high atop narrow pedestals and intricate detailing on tall pinnacles are not the results of abrasion. Sand seldom travels more than 1 meter above the surface, so the wind's sandblasting effect is obviously limited in vertical extent.

Ventifacts are stones that are shaped and polished by sandblasting.

A.

Yardangs are usually small, wind-sculpted landforms that are aligned parallel with the wind.

B.

◄ Figure 19.16
Shaped by the wind The sandblasting effect of the wind creates (**A**) ventifacts and (**B**) yardangs.

In addition to creating ventifacts, wind erosion is responsible for creating much larger features, called yardangs (from the Turkistani word *yar*, meaning "steep bank"). A **yardang** is a streamlined, wind-sculpted landform that is oriented parallel to the prevailing wind (**Figure 19.16B**). Individual yardangs are generally small features that stand less than 5 meters (16 feet) high and no more than about 10 meters (33 feet) long. Because the sandblasting effect of wind is greatest near the ground, these abraded bedrock remnants are usually narrower at their base. Sometimes yardangs are large features. Peru's Ica Valley contains yardangs that approach 100 meters (330 feet) in height and several kilometers in length. Some yardangs in the desert of Iran reach 150 meters (nearly 500 feet) in height.

Armoring the Desert Surface

As GEOgraphics 19.1 points out, desert surfaces are not always sand covered; far from it. In portions of many deserts, the surface consists of a closely packed layer of coarse particles. This veneer of pebbles and cobbles is only one or two stones thick (**Figure 19.17A**). When this layer is present, it sets an important control on wind erosion because the closely packed stones are too large for deflation to remove. If this stony armor is disturbed, wind can easily erode the underlying finer sediment (**Figure 19.17B**). There are two processes that lead to the formation of this stony layer: wind erosion and deposition.

Lag Deposits   In some cases, a stony veneer is left behind when wind removes sand and silt from *poorly sorted* surface deposits, so that the concentration of larger particles at the surface gradually increases as the finer particles are blown away.

Eventually the surface is completely covered with pebbles and cobbles too large to be moved by the wind. When this process dominates, the resulting layer of coarse particles is called a **lag deposit**. This term, although widely used, is somewhat misleading because the process that produced the feature resulted from erosion, not deposition.

A.

B.

◄ Figure 19.17
Stony armor A. Some desert surfaces consist of a veneer of closely packed pebbles that protects the underlying surface from wind erosion. **B.** If the protective stony layer is disturbed, the underlying surface becomes susceptible to wind erosion.

▶ SmartFigure 19.18
Lag deposits and desert pavement A. A lag deposit forms when an area with poorly sorted surface deposits undergoes deflation, which removes the fine particles and concentrates the coarse particles. **B.** Desert pavement forms when the surface is initially covered with cobbles and pebbles, and windblown dust accumulates at the surface and gradually sifts downward.

Tutorial
https://goo.gl/aTg4YY

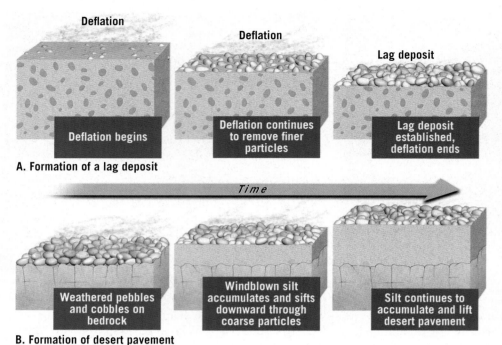

A. Formation of a lag deposit

Time

B. Formation of desert pavement

Desert Pavement

For many years, it was thought that all armored desert surfaces represented lag deposits. However, studies have shown that this cannot be the case. In many places, for example, the veneer is underlain by a relatively thick layer of silt that contains few if any pebbles and cobbles. In such a setting, deflation of fine sediment *could not* leave behind a layer of coarse particles. Geologists also determined that in these areas, the pebbles and cobbles composing the stony layer have all been exposed at the surface for about the same length of time, based on the thickness of desert varnish. This would not be the case for a lag deposit (**Figure 19.18A**), in which the coarse particles that make up the pavement reach the surface at different times over

an extended span, as deflation gradually removes the fine material.

For such deposits, which are called **desert pavement**, the mechanism shown in **Figure 19.18B** was proposed. According to this model, desert pavement develops on a surface that initially consists of coarse particles. Over time, protruding cobbles trap fine, windblown grains that settle and sift downward through the spaces between the larger surface stones. The process is aided by infiltrating rainwater. In this model, the cobbles composing the pavement were *never* buried. This mechanism successfully explains the lack of coarse particles beneath the desert pavement.

CONCEPT CHECKS 19.4

1. How does wind's bed load differ from its suspended load?

2. Why is wind erosion relatively more important in dry regions than in humid areas?

Concept Checker
https://goo.gl/EG8jNu

3. What are yardangs and ventifacts?

4. Contrast the formation of lag deposits and desert pavement.

19.5 Wind Deposits

Discuss dune formation and movement and distinguish among different dune types. Explain how loess deposits differ from deposits of sand.

Although wind is relatively unimportant in producing *erosional* landforms, significant *depositional* landforms are created by the wind in some regions. Accumulations of windblown sediment are particularly conspicuous in the world's dry lands and along many sandy coasts. Wind deposits are of two distinctive types: (1) mounds and ridges of sand from the wind's bed load, which we call *dunes*, and (2) extensive blankets of silt, called *loess*, that once were carried in suspension.

Sand Deposits

As is the case with running water, wind drops its load of sediment when velocity falls and the energy available for transport diminishes. Thus, sand begins to accumulate wherever an obstruction across the path of the wind slows its movement. Unlike many deposits of silt, which form blanketlike layers over large areas, winds commonly deposit sand in mounds or ridges called **dunes** (**Figure 19.19**).

Moving air encountering an object, such as a clump of vegetation or a rock, sweeps around and over the object, leaving a "shadow" of slower-moving air

Wind

Strong winds move sand up the relatively gentle windward slope.

As sand accumulates at the dune crest, the slope steepens, and some of the sand slides down the steep *slip face.*

▲ SmartFigure 19.19
White Sands National Monument The dunes at this landmark in southeastern New Mexico are composed of gypsum. The dunes slowly migrate with the wind.

Mobile Field Trip https://goo.gl/BYk7bt

behind the obstacle and a smaller zone of quieter air just in front of the obstacle. Some of the saltating sand grains moving with the wind come to rest in these wind shadows. As the accumulation of sand continues, it becomes a more imposing barrier to the wind and thus a more efficient trap for even more sand. If there is a sufficient supply of sand and the wind blows steadily for a long enough time, the mound of sand grows into a dune.

Many dunes have an asymmetrical profile, with the leeward (sheltered) slope being steep and the windward slope more gently inclined. The dunes in Figure 19.19 are a good example. Sand moves up the gentler slope on the windward side by saltation. Just beyond the crest of the dune, where the wind velocity is reduced, the sand accumulates. As more sand collects, the slope steepens, and eventually some of it slides under the pull of gravity. In this way, the leeward slope of the dune, called the **slip face**, maintains an angle of about 34 degrees, the angle of repose for loose dry sand. (Recall from Chapter 15 that the angle of repose is the

steepest angle at which loose material remains stable.) Continued sand accumulation, coupled with periodic slides down the slip face, results in the slow migration of the dune in the direction of air movement.

As sand is deposited on the slip face, layers form that are inclined in the direction the wind is blowing. These sloping layers are called **cross-beds** (Figure 19.20). When the dunes are eventually buried under other layers of sediment and become part of the sedimentary rock record, their asymmetrical shape is destroyed, but the cross-beds remain as testimony to their origin. Nowhere is cross-bedding more prominent than in the sandstone walls of Zion Canyon in southern Utah (see Figure 19.20).

▼ SmartFigure 19.20
Cross-bedding As sand is deposited on the slip face, layers form that are inclined in the direction the wind is blowing. With time, complex patterns develop in response to changes in wind direction.

Tutorial https://goo.gl/ FNG7Ha

Dunes commonly have an asymmetrical shape and migrate with the wind.

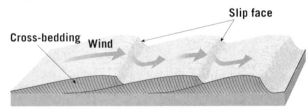

Sand grains deposited on the slip face at the angle of repose create the cross-bedding of dunes.

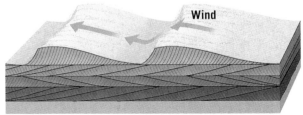

When dunes are buried and become part of the sedimentary rock record, the cross-bedding is preserved.

Cross-bedding is an obvious characteristic of the Navajo Sandstone in Zion National Park, Utah.

Types of Sand Dunes

Dunes are not just random heaps of windblown sediment. Rather, they usually assume patterns that are surprisingly consistent. A leading early investigator of dunes, the British engineer R. A. Bagnold, observed: "Instead of finding chaos and disorder, the observer never fails to be amazed at a simplicity of form, an exactitude of repetition, and a geometric order."

The main factors that influence the shape and size of dunes are the direction and velocity of wind, the availability of sand, and the amount of vegetation present. The broad range of dune forms are generally simplified into the six basic types shown in **Figure 19.21**. Bear in mind, however, that these forms intergrade, and that some irregular dunes do not fit easily into any category.

Barchan Dunes   Solitary sand dunes that are shaped like crescents with their tips pointing downwind are called **barchan dunes** (**Figure 19.21A**). These dunes form where supplies of sand are limited and the surface is relatively flat, hard, and lacking in vegetation. They migrate slowly with the wind at a rate of up to 15 meters (50 feet) per year. Their size is usually modest; the largest barchans reach heights of about 30 meters (100 feet), with a maximum spread between their tips approaching 300 meters (1000 feet). When the wind direction is nearly constant, the crescent form of these dunes is nearly symmetrical. However, when the wind direction is not perfectly fixed, one tip becomes larger than the other.

▼ **SmartFigure 19.21**
Types of sand dunes Factors that influence the form and size of dunes include wind direction and velocity, the availability of sand, and the amount of vegetation.

Tutorial
https://goo.gl/4mt1HB

Transverse Dunes   In regions where the prevailing winds are steady, sand is plentiful, and vegetation is sparse or absent, the dunes form a series of long ridges that are separated by troughs and oriented at right angles to the prevailing wind. Because of this orientation, they are termed **transverse dunes** (**Figure 19.21B**). Many coastal dunes are of this type. When transverse dunes form in arid regions, the extensive surface of wavy sand is sometimes called a *sand sea*. In some parts of the Sahara and Arabian Deserts, transverse dunes reach heights of 200 meters (660 feet), are 1 to 3 kilometers (0.6 to 2 miles) across, and can extend for distances of 100 kilometers (60 miles) or more.

Intermediate between isolated barchans and straight transverse dunes are **barchanoid dunes**, which form scalloped rows of sand oriented at right angles to the wind (**Figure 19.21C**). This is a common dune form.

Longitudinal Dunes   Long ridges of sand that form more or less *parallel* to the prevailing wind where sand supplies are moderate are called **longitudinal dunes** (see **Figure 19.21D**). Apparently, to form these dunes, the prevailing wind direction must vary somewhat but still remain in the same quadrant of the compass. Although the smaller types are only 3 or 4 meters high and several dozen meters long, in some large deserts, longitudinal dunes can reach great size. For example, in portions of North Africa, Arabia, and central Australia, these dunes may approach a height of 100 meters (330 feet) and extend for distances of more than 100 kilometers (62 miles).

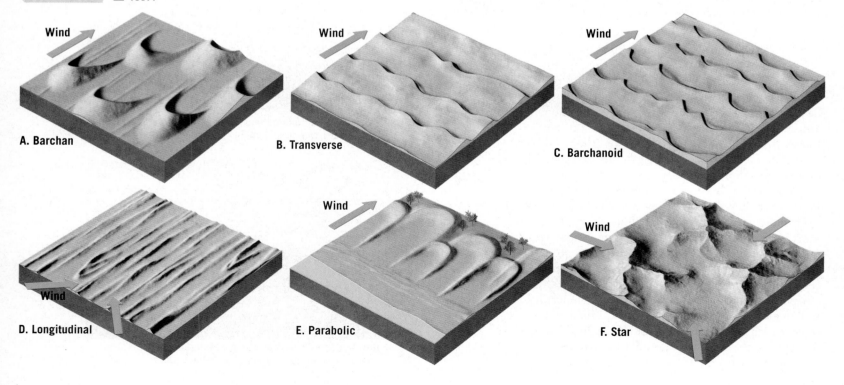

A. Barchan

B. Transverse

C. Barchanoid

D. Longitudinal

E. Parabolic

F. Star

Parabolic Dunes   Unlike the other dune types described thus far, **parabolic dunes** form where vegetation partially covers the sand. The shape of these dunes resembles the shape of barchans except that their tips point into the wind, rather than downwind (**Figure 19.21E**). Parabolic dunes often form along coasts where there are strong onshore winds and abundant sand. If the sand's sparse vegetative cover is disturbed at some spot, deflation creates a blowout. Sand is then transported out of the depression and deposited as a curved rim, which grows higher as deflation enlarges the blowout.

Star Dunes   Confined largely to parts of the Sahara and Arabian Deserts, **star dunes** are isolated hills of sand that exhibit a complex form (**Figure 19.21F**). Their name is derived from the fact that the bases of these dunes resemble multipointed stars. Usually three or four sharp-crested ridges diverge from a central high point that in some cases may approach a height of 90 meters (300 feet). As their form suggests, star dunes develop where wind directions are variable.

Loess (Silt) Deposits

In some parts of the world, the surface topography is mantled with deposits of windblown silt called **loess**. Over periods of perhaps thousands of years, dust storms deposited this material. When loess is breached by streams or road cuts, it tends to maintain vertical cliffs and lacks any visible layers, as you can see in **Figure 19.22**.

The distribution of loess worldwide indicates that there are two primary sources for this sediment: deserts and glacial outwash deposits. The thickest and most extensive deposits of loess on Earth occur in western and northern China. They were blown there from the extensive desert basins of central Asia. Accumulations of 30 meters (100 feet) are common, and thicknesses of more than 100 meters (330 feet) have been measured. It is this fine, buff-colored sediment that gives the Yellow River (Huang Ho) its name.

In the United States, deposits of loess are significant in many areas, including South Dakota, Nebraska, Iowa, Missouri, and Illinois, as well as portions of the Columbia Plateau in the Pacific Northwest. The correlation between the distribution of loess and

This vertical bluff near the Mississippi River in southern Illinois is about 3 meters (10 feet) high.

◄ **Figure 19.22 Loess** In some regions, the surface is mantled with deposits of windblown silt.

important farming regions in the Midwest and eastern Washington State is not just a coincidence: Soils derived from this wind-deposited sediment are among the most fertile in the world.

Unlike the deposits in China, which originated in deserts, the loess in the United States and Europe is an indirect product of glaciation. Its source is deposits of stratified drift. During the retreat of the ice sheets, many river valleys were choked with sediment deposited by meltwater. Strong westerly winds sweeping across the barren floodplains picked up the finer sediment and dropped it as a blanket on the eastern sides of the valleys. Such an origin is confirmed by the fact that loess deposits are thickest and coarsest on the leeward sides of such major glacial drainage outlets as the Mississippi and Illinois Rivers and rapidly thin with increasing distance from the valleys. Furthermore, the angular, mechanically weathered particles composing the loess are essentially the same as the rock flour produced by the grinding action of glaciers.

CONCEPT CHECKS 19.5

1. How do sand dunes migrate?

2. List and briefly distinguish among basic dune types.

3. How do loess deposits differ from sand deposits?   **Concept Checker** https://goo.gl/u4o43S

4. How are some loess deposits related to glaciers?

19

CONCEPTS IN REVIEW

Deserts and Wind

19.1 Distribution and Causes of Dry Lands

Describe the general distribution of Earth's dry lands and explain why deserts form in the subtropics and middle latitudes.

Key Terms:	desert	desertification
dry climate	steppe	rainshadow

- *Dry climates* cover about 30 percent of Earth's land area. These regions have yearly precipitation totals that are less than the potential loss of water through evaporation. Evaporation depends on temperature, and deserts may occur in hot or cold climates. *Deserts* are drier than *steppes*, but both climatic types are considered water deficient.
- Dry climates in subtropical latitudes are associated with the global distribution of air pressure and winds. Near the equator, warm, moist air rises (causing lots of rain) and then moves to 20 or 30 degrees latitude before sinking back toward Earth's surface. The subsiding air brings clear skies, copious sunshine, and dry conditions to these zones of subtropical high pressure.
- Deserts also occur in continental interiors of the middle latitudes. Most are due to the *rainshadow* effect, in which moist air moving inland from oceans is intercepted by mountainous obstacles. As the air is forced to rise, it cools, producing clouds and precipitation on the windward slopes. By contrast, the leeward side, called the rainshadow, tends to be quite dry.

Q This map shows annual precipitation across the continent of Africa. At which latitudes is the atmosphere rising? At which latitudes is it sinking? How does this atmospheric circulation influence the continent's climates?

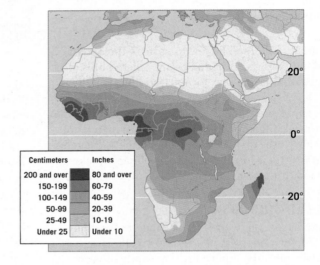

Centimeters	Inches
200 and over	80 and over
150-199	60-79
100-149	40-59
50-99	20-39
25-49	10-19
Under 25	Under 10

19.2 Geologic Processes in Arid Climates

Summarize the geologic roles of weathering, running water, and wind in arid and semiarid climates.

Key Terms:	desert varnish	ephemeral stream

- In dry lands, rock weathering of any type is greatly reduced because of the lack of water and the scarcity of organic acids from decaying plants.
- Practically all desert streams are dry most of the time and are said to be *ephemeral*. Their channels are carved out largely by flash floods that occur during sporadic storm events.
- Permanent desert streams originate in wetter climates. They must carry a great volume of water to keep from losing all their water as they cross the desert.
- Running water is responsible for most of the erosional work in a desert. Although wind erosion is more significant in dry areas than in other environments, it still cannot match running water as an agent of erosion in deserts.

Q This is a typical desert stream shortly after a heavy rain. What term is applied to such a stream? How will this scene likely change in the near future?

19.3 Arid Landscapes of the American West

Contrast the landscapes of the Basin and Range and the Colorado Plateau in the western United States.

Key Terms:	playa lake	mesa
interior drainage	playa	butte
alluvial fan	inselberg	pinnacle
bajada		

- The Basin and Range region is characterized by *interior drainage*, with streams eroding uplifted mountain blocks and depositing sediment in interior basins. Over time, relief diminishes as the mountains are lowered and the basins are filled. *Alluvial fans*, *bajadas*, *playas*, *playa lakes*, *salt flats*, and *inselbergs* are features often associated with these landscapes.

- The Colorado Plateau is an elevated region of relatively flat-lying strata that has hundreds of deep canyons. When plateau cliffs composed of resistant sandstone and limestone retreat, *mesas*, *buttes*, and *pinnacles* gradually form.

Q Identify the lettered features in this photo. How did they form?

19.4 Wind Erosion

Describe the ways that wind transports sediment and the processes and features associated with wind erosion.

Key Terms: deflation   yardang
bed load   blowout   lag deposit
saltation   abrasion   desert pavement
suspended load   ventifact

- A current of air can pick up and transport sediment, though with a lower competence than a current of water or glacial ice. Wind cannot pick up the coarser particles that water can but is capable of transporting sediment over vast areas and even high into the atmosphere.
- Sand can be transported by wind as *bed load* that bounces along Earth's surface. Generally, these *saltating* particles never get more than 0.5 meter above the ground.

- Clay and silt are fine-grained enough to be carried by wind as *suspended load*. Once aloft, they can be transported very far—across entire continents and ocean basins.
- Wind is capable of erosion, though it should be emphasized that water is the most important agent of erosion in desert regions. The Dust Bowl is a classic example of a massive episode of soil erosion by wind during the 1930s. Localized *deflation* may produce shallow depressions known as *blowouts*.
- *Ventifacts* are produced when individual rocks become abraded by windblown sediment, giving the rocks a polished, pitted surface. Similar "sand-blasting" of the land surface may sculpt rock outcrops into stream-lined *yardangs* that are longest parallel to the prevailing wind direction.
- A thin layer of coarse pebbles and cobbles covers some desert surfaces. *Lag deposits* form when deflation strips the finer grains from a deposit of poorly sorted sediments. *Desert pavement* forms when a pebble- and cobble-covered surface traps windblown dust, which sifts downward and gradually lifts the stony surface.

19.5 Wind Deposits

Discuss dune formation and movement and distinguish among different dune types. Explain how loess deposits differ from deposits of sand.

Key Terms: barchan dune   parabolic dune
dune   transverse dunes   star dune
slip face   barchanoid dune   loess
cross-bed   longitudinal dunes

- Wind deposits are of two distinct types: (1) mounds and ridges of sand, called *dunes*, which are formed from sediment that is carried as part of the wind's bed load, and (2) extensive blankets of silt, called *loess*, carried by wind in suspension.
- Dunes accumulate due to the difference in wind energy on the upwind and downwind sides of some obstacle. Wind moves sand up the more gently sloping upwind side and across the crest of the dune. The sand settles out in the calmer air on the downwind side, forming the steeply sloped *slip face*. When the slip face exceeds the angle of repose, it collapses in small "avalanches" of sand, causing the dune to move slowly in the direction of the prevailing wind. Inside the dune, the buried slip faces may be preserved as *cross-beds*.

- Dunes are categorized into six major types. Their shapes result from the pattern of prevailing winds, the amount of available sand, and the presence of vegetation.
- Loess is windblown silt, deposited over large areas and sometimes in thick blankets. Most loess is derived either from deserts or from the stratified drift in areas that have recently been glaciated.

Q This is an aerial view of dunes in northern Arizona. Which one of the basic dune types does it show? Sketch a simple profile (side view) of this dune. Include an arrow to show the prevailing wind direction and label the dune's slip face.

GIVE IT SOME THOUGHT

1. Albuquerque, New Mexico, receives an average of 20.7 centimeters (8.07 inches) of rainfall annually and is considered a desert. The Russian city of Verkhoyansk, near the Arctic Circle in Siberia, receives about 5 centimeters (2 inches) less annual precipitation than Albuquerque and yet is classified as having a humid climate. Explain why this is the case.

2. Are middle-latitude deserts most common in the Northern Hemisphere or the Southern Hemisphere? Explain why this is the case.

3. When a permanent stream such as the Nile River crosses a desert, does the river's discharge increase or decrease further downriver? How does this compare to a river in a humid area?

4. Compare and contrast the sediment deposited by a stream, the wind, and a glacier. Which deposit should have the most uniform grain size? Which one would exhibit the poorest sorting? Explain your choices.

5. Examine the precipitation map for the state of Nevada. Notice that the areas receiving the most precipitation resemble long, slender "islands" scattered across the state. Provide an explanation for this pattern.

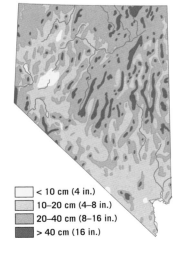

☐ < 10 cm (4 in.)
☐ 10–20 cm (4–8 in.)
☐ 20–40 cm (8–16 in.)
■ > 40 cm (16 in.)

6. **Is either of the following statements true? Are they both true? Explain your answer.**

   a. Wind is more effective as an agent of erosion in dry places than in humid places.

   b. Wind is the most important agent of erosion in deserts.

7. Bryce Canyon National Park, shown in the photo, is in dry southern Utah. It is carved into the eastern edge of the Paunsaugunt Plateau. Erosion has sculpted the colorful limestone into bizarre shapes, including spires called "hoodoos." As you and a companion (who has not studied geology) view Bryce Canyon, your friend says, "It's amazing how wind has created this incredible scenery!" Now that you have studied arid landscapes, how would you respond to your companion's statement?

EYE ON EARTH

1. This satellite image shows a small portion of the Zagros Mountains in dry southern Iran. Streams in this region flow only occasionally. The green tones on the image identify productive agricultural areas.

   a. Identify the large feature that is labeled with a question mark.

   b. Explain how that feature formed.

   c. What term is used to describe streams like the one in this image?

   d. Speculate on the likely source of water for the agricultural areas in this image.

2. This satellite image shows a large plume of windblown sediment covering large portions of Iran, Afghanistan, and Pakistan in March 2012. The airborne material is thick enough to completely hide the area beneath it. On either side of the plume, skies are mostly clear.

   a. What term is applied to the erosional process responsible for producing this plume?

   b. Is the wind-transported material in the image more likely bed load or suspended load?

   c. People sometimes refer to events like the one pictured here as "sandstorms." Is that an appropriate description? Why or why not?

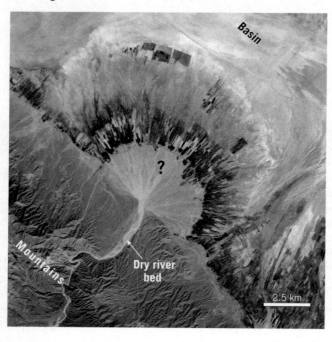

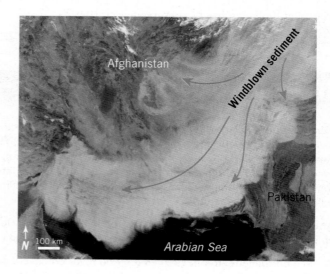

DATA ANALYSIS

The Aral Sea

The Aral Sea was once the fourth-largest lake in the world. This lake has now decreased in size by more than 80 percent, and the southern Aral Sea has disappeared altogether. This has had devastating effects on the communities around the lake.

https://goo.gl/eD226M

ACTIVITIES

Go to NASA's Earth Observatory site at http://earthobservatory.nasa.gov. Click on the Search icon and enter "Aral Sea" in the search box. Select "World of Change: Shrinking Aral Sea." Read the article and use the graphic to observe the lake as you step forward in time. (Green regions are lake; white regions are salt deposits.)

1. **When did the Aral Sea begin to shrink? Why?**

2. **How has the shrinking lake affected the quality of the water and farmland in the area?**

3. **What was the maximum east–west width of the lake in 1960? How far had the easternmost shore receded to the west by 2000? By 2005? By 2015?**

4. **How has the lake's reduction affected summer and winter temperatures?**

5. **What is the average rate at which the easternmost shore receded westward between 2000 and 2015? (Remember that rate of change is the distance change divided by the number of years.)**

6. **Why was there a significant decline in the overall size of the southern Aral Sea after 2005?**

Return to the Search icon at the top of the page and use it to search for "April 29 2008 Aral Sea dust storm." Choose the image showing a dust storm blowing west from the Aral Sea.

7. **Based on what you saw earlier, do you think dust storms like this one occurred in 1960? Why or why not?**

8. **Approximately how far does this dust storm extend?**

Mastering Geology

Looking for additional review and test prep materials? Visit pearson.com/mastering/geology to enhance your understanding of this chapter's content by accessing a variety of resources, including **Self-Study Quizzes, Geoscience Animations, SmartFigure Tutorials,** *Project Condor* **Videos, Mobile Field Trips, Dynamic Study Modules,** and an optional **Pearson eText.**

In The NEWS

Erosion Threatens the Alaska Coast

Erosion is putting communities along Alaska's north and west coasts in jeopardy. Even waves from modest storms are eating away at the shore, so that scenes of houses collapsing into the sea have become familiar. Why are Alaskans losing the ground beneath their feet? One important factor is that global warming has led to a significant decline in the formation of winter sea ice. In the past, sea ice reliably armored this low-lying coast against the storms of fall and winter. Not only did sea ice absorb the energy of storm waves, it also reduced the amount of open water over which such waves could form.

▲ Erosion by storm waves destroyed this house on the Alaska coast.

However, coastal sea ice is now forming about a month later than in 1979, when satellites began monitoring the region, and the ice that does form is thinner and less extensive. As a result, much of the coast is unprotected during the fall, the prime season for storms. Storm waves that once would have spent their energy crashing over sea ice now surge ashore. The resulting erosion can be especially severe where the ground is underlain by permafrost. In such places, once storm waves expose the layer of permafrost, they can literally melt the ground, causing coastal bluffs to collapse.

As the shield of coastal sea ice continues to decline, erosion along Alaska's shoreline is expected to grow worse. Some threatened coastal communities are seeking ways to relocate.

► Sea ice hugs Alaska's north coast near Barrow.

20
Shorelines

FOCUS ON CONCEPTS

Each statement represents the primary learning objective for the corresponding major heading within the chapter. After you complete the chapter, you should be able to:

20.1 Explain why the shoreline is considered a dynamic interface. List the factors that influence the height, length, and period of a wave and describe the motion of water within a wave.

20.2 Explain how waves erode and move sediment along the shore.

20.3 Describe the features typically created by wave erosion and those resulting from sediment deposited by longshore transport processes.

20.4 Distinguish between emergent and submergent coasts. Contrast the erosion problems faced on the Atlantic and Gulf coasts with those along the Pacific coast.

20.5 Describe the basic structure and characteristics of a hurricane and the three broad categories of hurricane destruction.

20.6 Summarize the ways in which people deal with shoreline erosion problems.

20.7 Explain the cause of tides, their monthly cycles, and tidal patterns. Describe the horizontal flow of water that accompanies the rise and fall of tides.

The restless waters of the ocean are constantly in motion. Winds generate surface currents, the gravity of the Moon and Sun produces tides, and density differences create deep-ocean circulation. Further, waves carry the energy from storms to distant shores, where their impact erodes the land.

Shorelines are dynamic environments. Their topography, geologic makeup, and climate vary greatly from place to place. Continental and oceanic processes converge along the shore to create landscapes that frequently undergo rapid change. When it comes to the deposition of sediment, shore areas are transition zones between marine and continental environments.

20.1 The Shoreline and Ocean Waves

Explain why the shoreline is considered a dynamic interface. List the factors that influence the height, length, and period of a wave and describe the motion of water within a wave.

The **shoreline** is the line that marks the contact between land and sea. Each day, as tides rise and fall, the position of the shoreline migrates. Over longer time spans, as sea level rises or falls, the average position of the shoreline gradually shifts.

A Dynamic Interface

Nowhere is the restless nature of the ocean's water more noticeable than along the shore—the dynamic interface where air, land, and sea come together. An **interface** is a common boundary where different parts of a system interact. This is certainly an appropriate designation for the coastal zone. Here we can see the rhythmic rise and fall of tides and observe waves constantly rolling in and breaking.

Although it may not be obvious, the shoreline is constantly being modified by waves. Crashing surf erodes the land. Wave activity also moves sediment toward and away from the shore, as well as along it. Such activity sometimes produces narrow sandbars and fragile offshore islands that frequently change size and shape as storm waves come and go.

Present-Day Shorelines The nature of present-day shorelines is not just the result of the sea's relentless attack on the land. Rather, the shore has a complex character that results from multiple geologic processes. For example, practically all coastal areas were affected by the worldwide rise in sea level that accompanied the melting of glaciers following the Last Glacial Maximum (see Figure 18.29, page 523). As the sea encroached landward, the shoreline retreated, becoming superimposed upon existing landscapes that had resulted from such diverse processes as stream erosion, glaciation, volcanic activity, and the forces of mountain building.

▲ Figure 20.1
Teetering on the edge Bluff failure caused by storm waves in March 2016 resulted in these apartments in Pacifica, California, being condemned. When these buildings were erected in the 1970s, they were safely away from the cliffs. Over the years, although several measures were attempted to reduce erosion of the sandstone cliffs, they all proved to be inadequate.

Human Activity Today the coastal zone is experiencing intensive human activity (**Figure 20.1**). About half of the world's human population lives on or within about 100 kilometers (60 miles) of a coast. Such large numbers of people so near the shore means that hurricanes and tsunamis place millions at risk. Unfortunately, people often treat the shoreline as if it were a stable platform on which structures can safely be built. This attitude inevitably leads to conflicts between people and nature. As you will see, many coastal land-forms, especially beaches and barrier islands, are relatively fragile, short-lived features that are inappropriate sites for development. The image of the New Jersey shoreline in **Figure 20.2** is a good example. In the years to come, coastal areas will be even more vulnerable because sea level is rising due to human-induced global warming (see Chapter 21).

Ocean Waves

Ocean waves travel along the interface between ocean and atmosphere. They can carry energy from a storm far out at sea over distances of several thousand kilometers. That's why even on calm days, the ocean still has waves that travel across its surface. When observing waves, you are watching *energy* travel through a medium (water). If you make waves by tossing a pebble into a pond, or by splashing in a pool, or by blowing across the surface of a cup of coffee, you are imparting *energy* to the liquid, and the waves you see are visible evidence of the energy passing through.

Wind-generated waves provide most of the energy that shapes and modifies shorelines. Where the land and sea meet, waves that may have traveled unimpeded for hundreds or thousands of kilometers suddenly encounter a barrier that will not allow them to advance farther and must absorb much of their energy. Stated another way, the shore is the location where a practically irresistible force confronts an almost immovable object. The conflict that results is never-ending and sometimes dramatic.

Wave Characteristics

Most ocean waves derive their energy and motion from the wind. When the velocity of a breeze is less than 3 kilometers (2 miles) per hour, only small wavelets appear. At greater wind speeds, more stable waves gradually form and advance with the wind.

Characteristics of ocean waves are illustrated in **Figure 20.3**, which shows a simple, nonbreaking wave form. The tops of the waves are the *crests*, which are separated by *troughs*. Halfway between the crests and troughs is the *still water level*, which is the level the water would occupy if there

▲ Figure 20.2

Hurricane Sandy A portion of the New Jersey shoreline just south of New York City shortly after this huge storm, also called Superstorm Sandy, struck in late October 2012. The extraordinary storm surge caused the damage pictured here. The fact that it struck the most populated metropolitan region in the United States clearly contributed to the storm's great financial impact. Many shoreline areas are intensively developed. Often the shifting shoreline sands and the desire of people to occupy these areas are in conflict.

were no waves. The vertical distance between trough and crest is called the **wave height**, and the horizontal distance between successive crests (or troughs) is the **wavelength**. The time it takes one full wave—one wavelength—to pass a fixed position is the **wave period**.

The height, length, and period that are eventually achieved by a wave depend on three factors: (1) the wind speed, (2) the length of time the wind has blown, and (3) the **fetch**, or distance that the wind has traveled across open water. As the quantity of energy transferred from the wind to the water increases, the height and steepness of the waves increase as well. Eventually a critical point is reached where waves grow so tall that they topple over, forming ocean breakers called *whitecaps*.

◄ SmartFigure 20.3
Wave basics The basic parts of an idealized nonbreaking wave and the movement of water with increasing depth.

Animation
https://goo.gl/zDRGFE

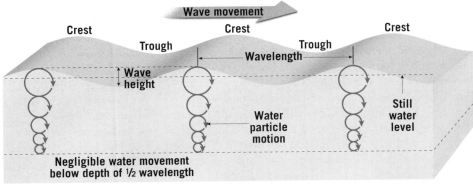

Wave movement

Crest   Crest   Crest

Trough   Trough

Wavelength

Wave height

Water particle motion

Still water level

Negligible water movement below depth of ½ wavelength

For a particular wind speed, there is a maximum fetch and duration of wind beyond which waves will no longer increase in size. When the maximum fetch and duration are reached for a given wind velocity, the waves are said to be "fully developed." At this point, the waves are losing as much energy through the breaking of whitecaps as they are receiving from the wind.

When wind stops or changes direction, or if waves leave the stormy area where they were created, the waves continue on without a relationship to local winds. The waves also undergo a gradual change to *swells*, which are lower and longer and may carry a storm's energy to distant shores. Because many independent wave systems exist at the same time, the sea surface acquires a complex, irregular pattern. Hence, the sea waves we watch from the shore are often a mixture of swells from faraway storms and waves created by local winds.

Circular Orbital Motion

Waves can travel great distances across ocean basins. One study tracked waves generated near Antarctica as they traveled through the Pacific Ocean basin. After traveling more than 10,000 kilometers (more than 6000 miles), the waves finally expended their energy a week later along the shoreline of Alaska's Aleutian Islands. The water itself doesn't travel this distance, but the wave form does. As a wave travels, the water passes the energy along by moving in a circle. This movement is called *circular orbital motion*.

Observation of an object floating in waves reveals that it moves not only up and down but also slightly forward and backward with each successive wave. **Figure 20.4** shows that a floating object moves up and backward as the crest approaches, up and forward as the crest passes, down and forward after the crest, and down and backward as the trough approaches; it rises and moves backward again as the next crest advances. When we trace the movement of the toy boat shown in Figure 20.4 as a wave passes, we see that the boat moves in a circle and returns to essentially the same place. Circular orbital motion allows a wave form (the wave's shape) to move forward *through the water* while the individual water particles that transmit the wave move in a circle. Wind moving across a field of wheat causes a similar phenomenon: The wheat itself doesn't travel across the field, but the waves do.

The wind energy given to the water is transmitted not only along the surface of the sea but also downward. However, beneath the surface, the circular motion rapidly diminishes until, at a depth equal to one-half the wavelength measured from the still water level, the movement of water particles becomes negligible. This depth is known as the *wave base*. The dramatic decrease of wave energy with depth is shown by the rapidly diminishing diameters of water-particle orbits in Figure 20.3.

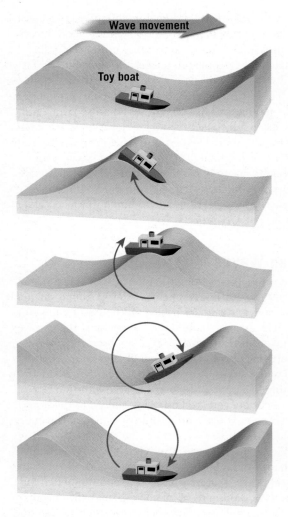

▲ **SmartFigure 20.4**
Passage of a wave The movements of the toy boat show that the wave form advances, but the water does not advance appreciably from the original position. In this sequence, the wave moves from left to right as the boat (and the water in which it is floating) rotates in an imaginary circle.

 Tutorial
https://goo.gl/
Eu9E7R

Waves in the Surf Zone

As long as a wave is in deep water, it is unaffected by water depth (**Figure 20.5, left**). However, when a wave approaches the shore, the water becomes shallower and influences wave behavior. The wave begins to "feel bottom" at a water depth equal to its wave base. Such depths interfere with water movement at the base of the wave and slow its advance (see **Figure 20.5, center**).

As a wave advances toward the shore, the slightly faster waves farther out to sea catch up, decreasing the wavelength. As the speed and length of the wave diminish, the wave steadily grows higher.

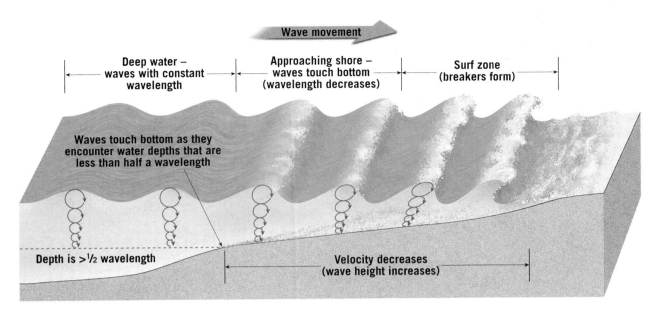

Wave movement

Deep water – waves with constant wavelength

Approaching shore – waves touch bottom (wavelength decreases)

Surf zone (breakers form)

Waves touch bottom as they encounter water depths that are less than half a wavelength

Depth is >½ wavelength

Velocity decreases (wave height increases)

◄ SmartFigure 20.5
Waves approaching the shore When water depth is less than half the wavelength, wave speed decreases, and the faster-moving waves farther from shore begin to catch up, causing the distance between waves (the wavelength) to decrease. This causes an increase in wave height, to the point where the waves finally pitch forward and break in the surf zone. The first portion of the animation illustrates the ideas presented in this figure.

Animation
https://goo.gl/eBJH6d

Finally, a critical point is reached when the wave is too steep to support itself, and the wave front collapses, or *breaks* (see **Figure 20.5, right**), causing water to advance up the shore.

The turbulent water created by breaking waves is called **surf**. On the landward margin of the surf zone, the *swash*—the turbulent sheet of water from collapsing breakers—moves up the slope of the beach. When the energy of the swash has been expended, the water flows back down the beach toward the surf zone as *backwash*.

CONCEPT CHECKS 20.1

1. Why is the shoreline described as being an interface?

2. List three factors that determine the height, length, and period of a wave.

3. Describe the motion of a floating object as a wave passes.

Concept Checker
https://goo.gl/zVU6E1

4. How do a wave's speed, wavelength, and height change as the wave moves into shallow water and breaks?

20.2 Beaches and Shoreline Processes

Explain how waves erode and move sediment along the shore.

For many, a beach is the sandy area where people lie in the Sun and walk along the water's edge. Technically, a **beach** is an accumulation of sediment found along the landward margin of a water body. Along straight coasts, beaches may extend for tens or hundreds of kilometers. Where coasts are irregular, beach formation may be confined to the relatively quiet waters of bays.

Beaches are composed of whatever material is locally abundant. The sediment for some beaches is derived from the erosion of adjacent cliffs or nearby coastal mountains. Other beaches are built from sediment delivered to the coast by rivers. Although the mineral makeup of many beaches is dominated by durable quartz grains, other minerals may be dominant. For example, in areas such as southern Florida, where there are no mountains or other sources of rock-forming minerals nearby, most beaches are composed of shell fragments and the remains of organisms that live in coastal waters (**Figure 20.6A**). Some beaches on volcanic islands in the open ocean are composed of weathered

grains of basaltic lava or of coarse debris eroded from coral reefs that develop around islands in low latitudes (**Figure 20.6B**).

Regardless of the composition, the material that comprises a beach does not stay in one place. Instead, crashing waves are constantly moving it. Thus, beaches can be thought of as material in transit along the shore.

Wave Erosion

During calm weather, wave action is minimal. However, just as streams do most of their work during floods, so too do waves accomplish most of their work during

▶ **Figure 20.6**
Beaches A beach is an accumulation of sediment on the landward margin of an ocean or a lake and can be thought of as material in transit along the shore. A beach is composed of whatever material is locally available.

This beach on Florida's Sanibel Island consists of shells and shell fragments.

A.

The black sands on this beach in Hawaii were derived from the weathering of nearby basaltic lava flows.

B.

storms. The impact of storm-induced waves against the shore can be awesome in its violence (**Figure 20.7**). Each breaking wave may hurl thousands of tons of water against the land, sometimes causing the ground to literally tremble. It is no wonder that cracks and crevices are quickly opened in cliffs, seawalls, breakwaters, and anything else that is subjected to these enormous shocks. Water is forced into every opening, causing air in the cracks to become highly compressed by the thrust of crashing waves. When the wave subsides, the air expands rapidly, dislodging rock fragments and enlarging and extending fractures.

In addition to the erosion caused by wave impact and pressure, **abrasion**—the sawing and grinding action of the water armed with rock fragments—is also important. In fact, abrasion is probably more intense in

▶ **Figure 20.7**
Storm waves along the Portugal coast When large waves break against the shore, the force of the water can be powerful, and the erosional work that is accomplished can be great.

▲ **Figure 20.8**
Rocky beach Smooth, rounded rocks along the shore are an obvious reminder that abrasion can be intense in the surf zone.

the surf zone than in any other environment. Smooth, rounded stones and pebbles along the shore are obvious reminders of the relentless grinding action of rock against rock in the surf zone (**Figure 20.8**). Further, where a cliff occurs, waves use rock fragments as "tools" as they cut horizontally into the land, creating a notch in the base of the cliff (**Figure 20.9A**). The cliff then retreats as the rock above the notch collapses (**Figure 20.9B**).

Sand Movement on the Beach

Energy from breaking waves can move large quantities of sand along the beach face and in the surf zone roughly parallel to the shoreline. Wave energy also causes sand to move perpendicular to (toward and away from) the shoreline.

Movement Perpendicular to the Shoreline

If you stand ankle deep in water at the beach, you will see that swash and backwash move sand toward and away from the shoreline. Whether there is a net loss or addition of sand depends on the level of wave activity. When wave activity is relatively light (less energetic waves), much of the swash soaks into the beach, which reduces the backwash. Consequently, the swash dominates and causes a net movement of sand up the beach face.

When high-energy waves prevail, the beach is saturated from previous waves, and much less of the swash soaks in. As a result, erosion occurs because backwash is strong and causes a net movement of sand down the beach face.

Along many beaches, light wave activity is the rule during summer. Therefore, a wide sandy beach gradually develops. During winter, when storms are frequent and more powerful, strong wave activity erodes and narrows the beach. A wide beach that may have taken months to build can be dramatically narrowed in just a few hours by the high-energy waves created by a strong winter storm.

Wave Refraction   The bending of waves, called **wave refraction**, plays an important part in shoreline

◄ **Figure 20.9**
Cliff retreat Breaking waves armed with rock debris can do a great deal of erosional work.

Waves undercut cliff

Original position of cliff

This sandstone cliff at Gabriola Island, British Columbia, was undercut by wave action.

A.   **B.**

As these waves approach nearly straight on, refraction causes the wave energy to be concentrated at headlands (resulting in erosion) and dispersed in bays (resulting in deposition).

Beach deposits

Headland

▲ SmartFigure 20.10
Wave refraction As waves first touch bottom in the shallows along an irregular coast, they slow down and bend (refract) so that they align nearly parallel to the shoreline.

Tutorial
https://goo.gl/
Y7c5Sp

Waves travel at original speed in deep water

Waves "feel bottom" and slow down in surf zone

Shoreline

Result: waves bend so that they strike the shore more directly

Wave refraction at Rincon Point, California

processes (**Figure 20.10**). It affects the distribution of energy along the shore and thus strongly influences where and to what degree erosion, sediment transport, and deposition will take place.

The shore is seldom oriented exactly parallel to approaching ocean waves. Rather, most waves move toward the shore at an angle. However, when they reach the shallow water of a smoothly sloping bottom, they are bent and tend to become parallel to the shore. Such bending occurs because the part of the wave nearest the shore reaches shallow water and slows first, whereas the end that is still in deep water continues forward at its full speed. The net result is a wave front that may approach nearly parallel to the shore, regardless of the original direction of the wave.

Because of refraction, wave impact is concentrated against the sides and ends of headlands that project into the water, whereas wave attack is weakened in bays. This differential wave attack along irregular coastlines is illustrated in Figure 20.10. As the waves reach the shallow water in front of the headland sooner than they do in adjacent bays, they are bent more nearly parallel to the protruding land and strike it from all three sides. By contrast, refraction in the bays causes waves to diverge and expend less energy. In these zones of weakened wave activity, sediments can accumulate and form sandy beaches. Over a long period, erosion of the headlands and deposition in the bays will straighten an irregular shoreline.

Path of sand particles

Beach drift

Net movement of sand grains

Longshore current

Beach drift occurs as incoming waves carry sand at an angle up the beach, while the water from spent waves carries it directly down the slope of the beach. Similar movements occur offshore in the surf zone to create the longshore current.

Longshore current

These waves approaching the beach at a slight angle along the Oregon coast produce a longshore current moving parallel to the shore.

▲ SmartFigure 20.11
The longshore transport system The two components of the transport system, beach drift and longshore currents, are created by breaking waves that approach the beach at an angle. These processes transport large quantities of material along the beach and in the surf zone.

Tutorial
https://goo.gl/
LBphUX

Longshore Transport   Although waves are refracted, most still reach the shore at some angle, however slight. Consequently, the uprush of water from each breaking wave (the swash) is at an oblique angle to the shoreline. However, the backwash is straight down the slope of the beach. The effect of this pattern of water movement is to transport sediment in a zigzag pattern along the beach face (**Figure 20.11**). This movement, called **beach drift**, can transport sand and pebbles hundreds or even thousands of meters each day. However, a more typical rate is 5 to 10 meters per day.

Waves that approach the shore at an angle also produce currents within the surf zone that flow parallel to the shore and move substantially more sediment than beach drift. Because the water here is turbulent, these **longshore currents** easily move the fine suspended sand and roll larger sand and gravel along the bottom. When the sediment transported by longshore currents is added to the quantity moved by beach drift, the total amount can be very large. At Sandy Hook, New Jersey, for example, the quantity of sand transported along the shore over a 48-year period averaged almost 750,000 tons annually. For a 10-year period in Oxnard, California, more than 1.5 million tons of sediment moved along the shore each year.

Both rivers and coastal zones move water and sediment from one area (*upstream*) to another (*downstream*). As a result, the beach is often characterized as a "river of sand." Beach drift and longshore currents, however, move in a zigzag pattern, whereas rivers flow mostly in a turbulent, swirling fashion. In addition, the direction of flow of longshore currents along a shoreline can change, whereas rivers always flow in the same direction (downhill). Longshore currents change direction because the direction that waves approach the beach changes seasonally. On balance, however, longshore currents flow southward along both the Atlantic and Pacific shores of the United States.

Rip Currents   **Rip currents** are concentrated movements of water that flow in the *opposite* direction of breaking waves. (Sometimes rip currents are incorrectly called *rip tides*, although they are unrelated to tidal phenomena.) Most of the backwash from spent waves finds its

Rip current extends outward from shore and interferes with incoming waves.

◀ **Figure 20.12**
Rip current These concentrated movements of water flow opposite the direction of breaking waves.

way back to the open ocean as an unconfined flow across the ocean bottom called *sheet flow*. However, sometimes a portion of the returning water moves seaward in the form of surface rip currents. Rip currents do not travel far beyond the surf zone before breaking up, and they can be recognized by the way they interfere with incoming waves or by the sediment that is often suspended within them (**Figure 20.12**). They can be hazardous to swimmers, who, if caught in them, can be carried out away from shore. The best strategy for exiting a rip current is to swim *parallel* to the shore for a few tens of meters.

CONCEPT CHECKS 20.2

1. Why do waves approaching the shoreline often bend?

2. What is the effect of wave refraction along an irregular coastline?

   🔊 **Concept Checker**
   https://goo.gl/eeFUcy

3. Describe the two processes that contribute to longshore transport.

20.3 Shoreline Features

Describe the features typically created by wave erosion and those resulting from sediment deposited by longshore transport processes.

A fascinating assortment of shoreline features can be observed along the world's coastal regions. Although the same processes cause change along every coast, not all coasts respond in the same way. Interactions among different processes and the relative importance of each process depend on local factors. The factors include (1) the proximity of a coast to sediment-laden rivers, (2) the degree of tectonic activity, (3) the topography and composition of the land, (4) prevailing winds and weather patterns, and (5) the configuration of the coastline and near-shore areas. Features that originate primarily because of erosion are called *erosional features*, whereas accumulations of sediment produce *depositional features*.

► **Figure 20.13**
Wave-cut platform and marine terrace This wave-cut platform is exposed at low tide along the coast near Kaikoura, New Zealand. A wave-cut platform was uplifted to create the marine terrace.

Marine terrace
(elevated wave-cut platform)

Wave-cut platform

Erosional Features

Many coastal landforms owe their origin to erosional processes. Such erosional features are common along the rugged and irregular New England coast and along the steep shorelines of the West Coast of the United States.

► **Figure 20.14**
Sea stack and sea arch These features along the coast of Portugal resulted from vigorous wave attack on a headland.

Sea arches

Sea stack

Wave-Cut Cliffs, Wave-Cut Platforms, and Marine Terraces   As the name implies, **wave-cut cliffs** originate in the cutting action of the surf against the base of coastal land (see Figure 20.9B). As erosion progresses, rocks overhanging the notch at the base of the cliff crumble into the surf, and the cliff retreats. A relatively flat, benchlike surface, called a **wave-cut platform**, is left behind by the receding cliff (**Figure 20.13**). The platform broadens as wave attack continues. Some debris produced by the breaking waves remains along the water's edge as sediment on the beach, and the remainder is transported farther seaward. If a wave-cut platform is uplifted above sea level by tectonic forces, it becomes a **marine terrace** (see Figure 20.13). Marine terraces are easily recognized by their gentle seaward-sloping shape and are often desirable sites for coastal roads, buildings, or agriculture.

Sea Arches and Sea Stacks   Because of refraction, waves vigorously attack headlands that extend into the sea. The surf erodes the rock selectively, wearing away the softer or more highly fractured rock at the fastest rate. At first, sea caves may form. When two caves on opposite sides of a headland unite, a **sea arch** results (**Figure 20.14**). Eventually the arch falls in, leaving an isolated remnant, or

A.

▲ SmartFigure 20.15
Coastal Massachusetts
A. High-altitude image of a well-developed spit and baymouth bar along the coast of Martha's Vineyard. **B.** This photograph, taken from the International Space Station, shows Provincetown Spit at the tip of Cape Cod.

Mobile Field Trip
https://goo.gl/tRBXSq

sea stack, on the wave-cut platform. In time, it too will be consumed by the action of the waves.

Depositional Features

Sediment eroded from the beach is transported along the shore and deposited in areas where wave energy is low. Such processes produce a variety of depositional features.

Spits, Bars, and Tombolos
Where beach drift and longshore currents are active, several features related to the movement of sediment along the shore may develop. A **spit** (*spit* = spine) is an elongated ridge of sand that projects from the land into the mouth of an adjacent bay. Often the end of a spit that is in the water hooks landward in response to the dominant direction of the longshore current (**Figure 20.15**). The term **baymouth bar** is applied to a sandbar that completely crosses a

bay, sealing it off from the open ocean. Such a feature tends to form across a bay where currents are weak, allowing a spit to extend to the other side. A **tombolo** (*tombolo* = mound), a ridge of sand that connects an island to the mainland or to another island, forms in much the same manner as a spit.

Barrier Islands The Atlantic and Gulf coastal plains are relatively flat and slope gently seaward. The shore zone is characterized by **barrier islands**. These low ridges of land parallel the coast at distances from 3 to 30 kilometers offshore. From Cape Cod, Massachusetts, to Padre Island, Texas, nearly 300 barrier islands rim the coast (**Figure 20.16**).

Most barrier islands are 1 to 5 kilometers wide and 15 to 30 kilometers long. The tallest features are sand dunes, which usually reach heights of 5 to 10 meters; in a few areas, unvegetated dunes are more than 30 meters high. The lagoons separating these narrow islands from the shore are zones of

► Figure 20.16
Barrier islands Nearly 300 barrier islands rim the Gulf and Atlantic coasts. The islands along the coast of North Carolina are excellent examples. In this view, south is at the top of the photo.

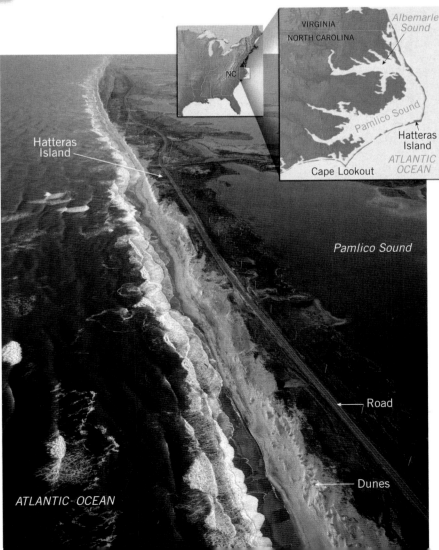

relatively quiet water that allow small craft traveling between New York and northern Florida to avoid the rough waters of the North Atlantic.

Over time, many lagoons gradually fill with sediment contributed by mainland streams, wind-deposited sand from adjacent barrier islands, and tidal deposits if the lagoon has an opening to the sea. Unless tidal inlets permit strong tidal currents to move lagoon sediment seaward, many lagoons are slowly transformed into coastal marshes.

Barrier islands form in several ways. Some originated as spits that were severed from the mainland by wave erosion or by the general rise in sea level after the last episode of glaciation. Others are created when turbulent waters in the line of breakers heap up sand scoured from the ocean bottom. Because these sand barriers rise above normal sea level, the sand likely piles up as a result of the work of storm waves at high tide. Finally, some barrier islands may be former sand

dune ridges that originated along the shore during the last glacial period, when sea level was lower. When the ice sheets melted, sea level rose and flooded the area behind the beach–dune complex.

The Evolving Shore

A shoreline continually undergoes modification, regardless of its initial configuration. At first, most coastlines are irregular, although the degree of and reason for the irregularity may vary considerably from place to place. Along a coastline that is characterized by varied geology, the pounding surf may at first increase its irregularity because the waves will erode the weaker rocks more easily than the stronger ones. However, if a shoreline remains tectonically stable, marine erosion and deposition will eventually produce a straighter, more regular coast. **Figure 20.17** illustrates the evolution of an initially

▶ **Figure 20.17**
The evolving shore These diagrams illustrate changes that can take place through time along an initially irregular coastline that remains relatively stable. The diagrams also illustrate many of the features described in Section 20.3.

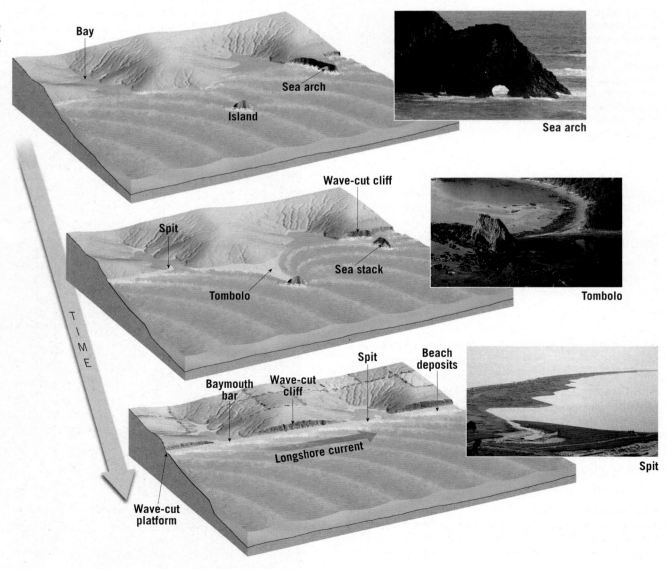

irregular coast. As waves erode the headlands, creating cliffs and a wave-cut platform, sediment is carried along the shore. Some material is deposited in the bays, while other debris is formed into spits and baymouth bars. At the same time, rivers fill the bays with sediment. Ultimately, a generally straight, smooth coast results.

20.4 Contrasting America's Coasts

Distinguish between emergent and submergent coasts. Contrast the erosion problems faced on the Atlantic and Gulf coasts with those along the Pacific coast.

The shoreline along the Pacific coast of the United States is strikingly different from that of the Atlantic and Gulf coast regions. Some of the differences are related to plate tectonics. As the leading edge of the North American plate, the west coast experiences active uplift and deformation. By contrast, the east coast is far from any active plate boundary and relatively quiet tectonically. Because of this basic geologic difference, the nature of shoreline erosion problems along America's opposite coasts is different.

Coastal Classification

The great variety of shorelines demonstrates their complexity. Indeed, to understand any particular coastal area, many factors must be considered, including rock types, size and direction of waves, frequency of storms, tidal range, and offshore topography. In addition, recall from Chapter 18 that practically all coastal areas were affected by the worldwide rise in sea level that accompanied the melting of Ice Age glaciers at the close of the Pleistocene epoch. Finally, tectonic events that elevate or drop the land or change the volume of ocean basins must be taken into account. The myriad factors that influence coastal areas make shoreline classification difficult.

Many geologists classify coasts based on the changes that have occurred with respect to sea level. This commonly used classification divides coasts into two general categories: emergent and submergent. **Emergent coasts** develop because an area experiences either uplift or a drop in sea level. Conversely, **submergent coasts** are created when sea level rises or the land adjacent to the sea subsides.

Emergent Coasts   In some areas, the coast is clearly emergent because rising land or a falling water level exposes wave-cut cliffs and platforms above sea level (see Figure 20.13). In the United States, excellent examples include portions of coastal California, where uplift has occurred in the recent geologic past. In the case of the Palos Verdes Hills, south of Los Angeles, seven different terrace levels exist, indicating seven episodes of uplift. The ever-persistent sea is now cutting a new platform at the base of the cliff. If uplift follows, it too will become an elevated marine terrace.

Other examples of emergent coasts include regions that were once buried beneath great ice sheets. When glaciers were present, their weight depressed the crust, and when the ice melted, the crust began gradually to spring back. Consequently, prehistoric shoreline features may now be found high above sea level. The Hudson Bay region of Canada is such an area; portions of it are still rising at a rate of more than 1 centimeter per year.

Submergent Coasts   In contrast to the preceding examples, other coastal areas show definite signs of submergence. Shorelines that have been submerged in the relatively recent past are often highly irregular because the sea typically floods the lower reaches of river valleys flowing into the ocean. The ridges separating the valleys, however, remain above sea level and project into the sea as headlands. These drowned river mouths, which are called **estuaries**, characterize many coasts today. Along the Atlantic coastline, the Chesapeake and Delaware Bays are examples of large estuaries created by submergence (**Figure 20.18**). The picturesque coast of Maine, particularly in the vicinity of Acadia National Park, is another excellent example of an area that was flooded by the postglacial rise in sea level and transformed into a highly irregular coastline.

Keep in mind that most coasts have complicated geologic histories. With respect to sea level, at various

East coast estuaries The lower portions of many river valleys were flooded by the rise in sea level that followed the end of the Quaternary Ice Age, creating large estuaries such as Chesapeake and Delaware Bays.

Tutorial
https://goo.gl/zZSWou

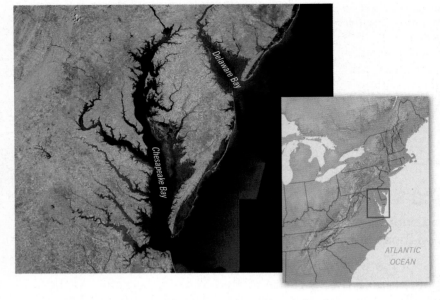

times many coasts have emerged and then submerged. Each time, they may retain some of the features created during the previous situation.

▼ **Figure 20.19**
Relocating the Cape Hatteras lighthouse After the failure of a number of efforts to protect this 21-story lighthouse, the nation's tallest, from being destroyed due to a receding shoreline, the structure finally had to be moved.

Atlantic and Gulf Coasts

Much of the coastal development along the Atlantic and Gulf coasts has occurred on barrier islands. Typically, a barrier island, also termed a *barrier beach*, or *coastal barrier*, consists of a wide beach that is backed by dunes and separated from the mainland by marshy lagoons.

The broad expanses of sand and exposure to the ocean have made barrier islands exceedingly attractive sites for development. Unfortunately, development has taken place more rapidly than our understanding of barrier island dynamics has increased.

Because barrier islands face the open ocean, they receive the full force of major storms that strike the coast. When a storm occurs, the narrow islands absorb the energy of the waves primarily through the movement of sand. **Figure 20.19**, which shows changes at Cape Hatteras National Seashore, reinforces this point. The process and problems that result were recognized years ago and accurately described as follows:

Waves may move sand from the beach to offshore areas or, conversely, into the dunes; they may erode the dunes, depositing sand onto the beach or carrying it out to sea; or they may carry sand from the beach and the dunes into the marshes behind the barrier, a process known as overwash. The common factor is movement. Just as a flexible reed may survive a wind that destroys an oak tree, so the barriers survive hurricanes and nor'easters not through unyielding strength but by giving before the storm.

This picture changes when a barrier is developed for homes or as a resort. Storm waves that previously rushed harmlessly through gaps between the dunes now encounter buildings and roadways. Moreover, since the dynamic nature of the barriers is readily perceived only during storms, homeowners tend to attribute damage to a particular storm, rather than to the basic mobility of coastal barriers. With their homes or investments at stake, local residents are more likely to seek to hold the sand in place and the waves at bay than to admit that development was improperly placed to begin with.*

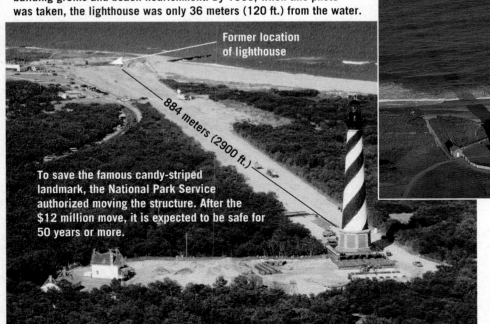

Various attempts to protect the lighthouse failed. They included building groins and beach nourishment. By 1999, when this photo was taken, the lighthouse was only 36 meters (120 ft.) from the water.

Former location of lighthouse

884 meters (2900 ft.)

To save the famous candy-striped landmark, the National Park Service authorized moving the structure. After the $12 million move, it is expected to be safe for 50 years or more.

*Frank Lowenstein, "Beaches or Bedrooms—The Choice as Sea Level Rises," *Oceanus* 28 (No. 3, Fall 1985): p. 22 © Woods Hole Oceanographic Institute.

◄ Figure 20.20

Pacoima Dam and Reservoir Dams such as this one in the San Gabriel Mountains near Los Angeles trap sediment that otherwise would have nourished beaches along the nearby coast.

Dam

irrigation and flood control. The reservoirs effectively trap the sand that would otherwise nourish the beach environment (**Figure 20.20**). When the beaches were wider, they protected the cliffs behind them from the force of storm waves. Now, however, the waves move across the narrowed beaches without losing much energy and cause more rapid erosion of the sea cliffs.

Although the retreat of the cliffs provides material to replace some of the sand impounded behind dams, it also endangers homes and roads built on the bluffs. In addition, development atop the cliffs aggravates the problem. Urbanization increases runoff, which, if not carefully controlled, can result in serious bluff erosion. Watering lawns and gardens high on the cliffs adds significant quantities of water to the slope. This water percolates downward toward the base of the cliff, where it may emerge in small seeps. This action reduces the slope's stability and facilitates mass wasting.

Shoreline erosion along the Pacific coast varies considerably from one year to the next, largely because of the sporadic occurrence of storms. As a consequence, when the infrequent but serious episodes of erosion occur, the damage is often blamed on the unusual storms and not on coastal development or the sediment-trapping dams that may be great distances away. As sea level rises at an increasing rate in the years to come, increased shoreline erosion and sea-cliff retreat should be expected along many parts of the Pacific coast. Coastal vulnerability to sea-level rise is examined in more detail as part of a discussion of the possible consequences of global warming in Chapter 21.

Pacific Coast

In contrast to the broad, gently sloping coastal plains of the Atlantic and Gulf coasts, much of the Pacific coast is characterized by relatively narrow beaches that are backed by steep cliffs and mountain ranges. Recall that America's western margin is a more rugged and tectonically active region than the eastern margin. Because uplift continues, a rise in sea level in the west is not so readily apparent. Nevertheless, like the shoreline erosion problems facing the Atlantic coast's barrier islands, west coast difficulties also stem largely from the alteration of a natural system by people.

A major problem facing the Pacific shoreline, and especially portions of southern California, is a significant narrowing of many beaches. The bulk of the sand on many of these beaches is supplied by rivers that transport it from the mountains to the coast. Over the years, this natural flow of material to the coast has been interrupted by dams built for

Destructive coastal erosion is often blamed on unusual storms and not on coastal development or sediment-trapping dams.

CONCEPT CHECKS 20.4

1. Are estuaries associated with submergent or emergent coasts? Explain.

2. What observable features would lead you to classify a coastal area as emergent?

 Concept Checker
https://goo.gl/SH4kvY

3. How might building a dam on a river that flows to the sea affect a coastal beach?

20.5 Hurricanes: The Ultimate Coastal Hazard

Describe the basic structure and characteristics of a hurricane and the three broad categories of hurricane destruction.

Many view the weather in the tropics with favor—and rightfully so. Places such as islands in the South Pacific and the Caribbean are known for their lack of significant day-to-day variations. Warm breezes, steady temperatures, and rains that occur as heavy but brief tropical showers are expected. It is ironic that these relatively tranquil regions occasionally produce some of the world's most violent storms. Once formed, these storms carry severe conditions far from the tropics.

▶ SmartFigure 20.21
Hurricane Maria Maria was the strongest storm of the 2017 Atlantic hurricane season. **A.** The well-developed eye of the storm is visible in this satellite image. On September 20, while southeast of Puerto Rico, peak winds reached 280 kilometers (180 miles) per hour. **B.** When Hurricane Maria struck, Puerto Rico was still recovering from Hurricane Irma, which had occurred 2 weeks earlier. Damages from the deadly storm were devastating. In addition to damage from strong winds and coastal storm surge, many landslides were triggered by the heavy rains.

Video
https://goo.gl/jWrZvJ

A.

B.

Whirling tropical cyclones—the greatest storms on Earth—occasionally have wind speeds exceeding 300 kilometers (185 miles) per hour. In the United States they are known as **hurricanes**, in the western Pacific they are called *typhoons*, and in the Indian Ocean they are simply called *cyclones*. No matter which name is used, these storms are among the most destructive of natural disasters.

The vast majority of hurricane-related deaths and damage are caused by relatively infrequent yet powerful storms. However, during August and September 2017, the Caribbean and Gulf of Mexico experienced three very strong and deadly hurricanes—Harvey, Irma, and Maria (**Figure 20.21**). The Gulf Coast of Texas and Puerto Rico and the Virgin Islands in the Caribbean were especially hard hit. Estimated destruction approached $300 billion.

Our coasts are vulnerable. People are flocking to live near the ocean. Over half of the U.S. population resides within 75 kilometers (45 miles) of a coast, placing millions at risk. Moreover, the potential costs of property damage are incredible. As sea level continues to rise in coming decades, low-lying, densely populated coastal areas will become even more vulnerable to the destructive effects of major storms.

Profile of a Hurricane

A hurricane is a heat engine fueled by the energy liberated when huge quantities of water vapor condense. The amount of energy produced by a typical hurricane in just a single day is truly immense. To get the engine started, a large quantity of warm, moist air is required, and a continuous supply is needed to keep it going.

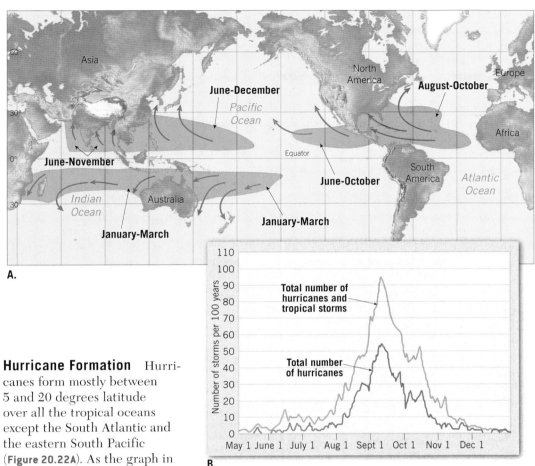

A.

◄ SmartFigure 20.22
The where and when of hurricanes A. The world map shows the regions where most hurricanes form as well as their principal months of occurrence and the tracks they most commonly follow. Hurricanes do not develop within about 5 degrees of the equator because the Coriolis effect (a force related to Earth's rotation that gives storms their "spin") there is too weak. Because warm ocean-surface temperatures are necessary for hurricane formation, hurricanes seldom form poleward of 20 degrees latitude or over the cool waters of the South Atlantic and the eastern South Pacific. **B.** The graph shows the frequency of tropical storms and hurricanes from May 1 through December 31 in the Atlantic basin. It shows the number of storms to be expected over a span of 100 years. The period from late August through October is clearly the most active.

Tutorial
https://goo.gl/
Swq5jJ

Hurricane Formation

Hurricanes form mostly between 5 and 20 degrees latitude over all the tropical oceans except the South Atlantic and the eastern South Pacific (**Figure 20.22A**). As the graph in **Figure 20.22B** illustrates, hurricanes most often form in late summer and early fall. It is in the tropics during this time period that sea-surface temperatures reach 27°C (80°F) or higher and are thus able to provide the necessary heat and moisture to the air (**Figure 20.23**). This ocean-water temperature requirement explains why hurricane formation over the relatively cool waters of the South Atlantic and eastern South Pacific is extremely rare. For the same reason, few hurricanes form poleward of 20 degrees latitude. Although water temperatures are sufficiently high, hurricanes do not develop within about 5 degrees of the equator because the Coriolis effect (a force related to Earth's rotation that gives storms their "spin") is too weak there (see Figure 20.22A).

Pressure Gradient

Hurricanes are intense low-pressure centers, which means that air pressure gets lower and lower closer to the center of the storm. Such storms are said to have a very steep *pressure gradient*, which refers to how rapidly the pressure changes per unit distance. A steep pressure

B.

▼ SmartFigure 20.23
Sea-surface temperatures Among the necessary ingredients for a hurricane is warm ocean temperatures above 27°C (80°F). The map shows sea-surface temperatures in the Atlantic Ocean, Caribbean Sea, and Gulf of Mexico on September 5, 2017. The yellow-to-red line represents Irma's track from September 3 to 6.

Video
https://goo.gl/
8Acgj1

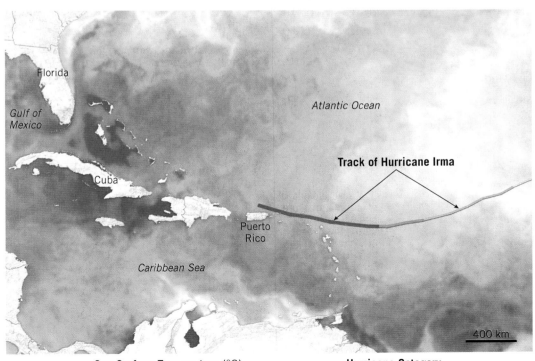

Sea Surface Temperature (°C)			Hurricane Category				
≤24.8	27.8	≥30.8	2	3	4	5	

gradient generates the rapid, inward-spiraling winds of a hurricane. As the air rushes toward the center of the storm, its velocity increases. This is similar to skaters with their arms extended spinning faster as they pull their arms in close to their bodies.

Storm Structure   As the inward rush of warm, moist surface air approaches the core of the storm, it turns upward and ascends in a ring of cumulonimbus cloud towers (**Figure 20.24A**). This doughnut-shaped wall of intense thunderstorm activity surrounding the center of the hurricane is called the **eye wall**. It is here that the greatest wind speeds and heaviest rainfall occur (**Figure 20.24B**). Surrounding the eye wall are curved bands of clouds that trail away in a spiral fashion.

Near the top of the hurricane, the airflow is outward, carrying the rising air away from the storm center, thereby providing room for more inward flow at the surface.

At the very center of the storm is the **eye** of the hurricane (see Figure 20.24A). This well-known feature is a zone about 20 kilometers (12.5 miles) in diameter where precipitation ceases and winds subside. It offers a brief but deceptive break from the extreme weather in the enormous curving wall clouds that surround it. The air within the eye gradually descends and heats by compression, making it the warmest part of the storm. Although many people believe that the eye is characterized by clear blue skies, this is usually not the case because the subsidence in the eye is seldom strong enough to produce cloudless conditions. Although the sky appears much brighter in this region, scattered clouds at various levels are common.

Hurricane Destruction

Although hurricanes are tropical or subtropical in origin, their destructive effects can be experienced far from where they originate. For example, in 2012 Hurricane Sandy (also called Superstorm Sandy) originated in the Caribbean Sea and affected the entire eastern seaboard from Florida to Maine. Destruction was especially great in New Jersey and New York, even though Sandy was downgraded from hurricane status by that time.

The amount of damage caused by a hurricane depends on several factors, including the size and population density of the area affected and the shape of the ocean bottom near the shore. The most significant factor, of course, is the strength of the storm. By studying past storms, a scale called the *Saffir–Simpson*

▶ **SmartFigure 20.24**
Conditions inside a hurricane
(Data from World Meteorological Organization)

Video
https://goo.gl/cptEbP

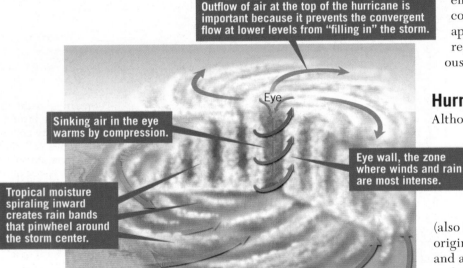

Outflow of air at the top of the hurricane is important because it prevents the convergent flow at lower levels from "filling in" the storm.

Eye

Sinking air in the eye warms by compression.

Eye wall, the zone where winds and rain are most intense.

Tropical moisture spiraling inward creates rain bands that pinwheel around the storm center.

A. Cross section of a hurricane. Note that the vertical dimension is greatly exaggerated. (After NOAA)

B. Measurements of surface pressure and wind speed during the passage of Cyclone Monty at Mardie Station, Western Australia, between February 29 and March 2, 2004. (Hurricanes are called "cyclones" in this part of the world.)

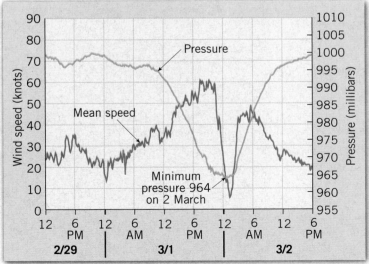

Saffir–Simpson Hurricane Wind Scale*				
Category (scale number)	Central pressure (millibars)	Winds (km/hr)	Storm surge (meters)	Damage at landfall
1	980	119–153	1.2–1.5	Minimal
2	965–979	154–177	1.6–2.4	Moderate
3	945–964	178–209	2.5–3.6	Extensive
4	920–944	210–250	3.7–5.4	Extreme
5	<920	>250	>5.4	Catastrophic

*The Saffir–Simpson Hurricane Scale became the Saffir-Simpson Hurricane Wind Scale in May 2010.

▲ **Figure 20.25**
Hurricane categories

hurricane wind scale was established to rank the relative intensities of hurricanes. As **Figure 20.25** indicates, a category 5 storm is the worst possible, and a category 1 hurricane is least severe.

During hurricane season, it is common to hear scientists and reporters alike use the numbers from the Saffir–Simpson scale. As Hurricane Maria approached Puerto Rico in September 2017, it had sustained winds of 250 kilometers (155 miles) per hour, making it a rare category 5 storm. Although category 5 storms are rare, Hurricane Irma also achieved this status less than 2 weeks earlier. When Hurricane Harvey approached the Texas coast in late August 2017 with maximum winds of 215 kilometers (130 miles) per hour, it was a category 4 storm.

Once a hurricane makes landfall, it loses energy because it is cut off from its energy source—warm ocean water—and is usually downgraded to a lower category. However, these storms are so large and violent that their effects are felt far inland. Damage caused by hurricanes can be divided into three categories: storm surge, wind damage, and inland flooding.

Storm Surge   Without question, the most devastating damage in the coastal zone is caused by storm surge. It not only accounts for a large share of coastal property losses but is also responsible for a high percentage of all hurricane-caused deaths. A **storm surge** is a dome of water 65 to 80 kilometers (40 to 50 miles) wide that sweeps across the coast near the point where the eye makes landfall. If all wave activity were smoothed out, the storm surge would be the height of the water above normal tide level. In addition, tremendous wave activity is superimposed on the surge. This surge of water can inflict immense damage on low-lying coastal areas (**Figure 20.26**). The worst surges occur in places like the Gulf of Mexico, where the continental shelf is very shallow and gently sloping. In addition, local features such as bays and rivers can cause the surge to double in height and increase in speed.

As a hurricane advances toward the coast in the Northern Hemisphere, storm surge is always most intense on the right side of the eye, where winds are blowing *toward* the shore. On this side of the storm, the forward movement of the hurricane also contributes to the storm surge. In **Figure 20.27**, assume that a hurricane with peak winds of 175 kilometers (109 miles) per hour is moving toward the shore at 25 kilometers (15.5 miles) per hour. In this case, the net wind speed on the right side of the advancing storm is 200 kilometers (124 miles) per hour. On the left side, the hurricane's winds are blowing opposite the direction of storm movement, so the net winds are *away* from the coast at 150 kilometers (93 miles) per hour. Along the shore facing the left side of the oncoming hurricane, the water level may actually decrease as the storm makes landfall.

◀ **Figure 20.26**
Storm surge destruction This is Crystal Beach, Texas, on September 16, 2008, 3 days after Hurricane Ike came ashore. At landfall the storm had sustained winds of 165 kilometers (105 miles) per hour. The extraordinary storm surge caused most of the damage shown here.

► **Figure 20.27**
Wind speeds of an approaching hurricane Winds associated with a hypothetical Northern Hemisphere hurricane that is advancing toward the coast. Storm surge will be greatest along the part of the coast hit by the right side of the advancing hurricane when looking in the direction the storm is moving.

North Carolina

South Carolina

Georgia

Florida

Speed and direction of hurricane movement 25 kph

Hurricane movement 25 kph

Net wind 200 kph from the southeast

Wind on right side 175 kph

Hurricane movement 25 kph

Net wind 150 kph from the northwest

Wind on left side 175 kph

that people should stay below the 10th floor of a building but remain above any floors at risk for flooding. In regions with good building codes, wind damage is usually not as catastrophic as storm-surge damage. However, hurricane-force winds affect a much larger area than storm surge and can cause huge economic losses. For example, in 1992 it was largely the winds associated with Hurricane Andrew that produced more than $25 billion of damage in southern Florida and Louisiana.

A hurricane may produce tornadoes that contribute to the storm's destructive power. Studies have shown that more than half of the hurricanes that make landfall produce at least one tornado. When Hurricane Harvey struck the coast of Texas in 2017, it was a prolific tornado producer, spawning 52 tornadoes in and near the Houston metro area. Additional tornadoes were reported as the remnants of the hurricane moved northeastward.

Fortunately, in this case, almost all the tornadoes were relatively weak.

Wind Damage   Destruction caused by wind is perhaps the most obvious of the classes of hurricane damage. Debris such as signs, roofing materials, and small items left outside become dangerous flying missiles in hurricanes. For some structures, the force of the wind is sufficient to cause total ruin. Mobile homes are particularly vulnerable. High-rise buildings are also susceptible to hurricane-force winds. Upper floors are most vulnerable because wind speeds usually increase with height. Recent research suggests

Harvey was the most significant tropical rainfall event in recorded U.S. history.

Heavy Rains and Inland Flooding   The torrential rains that accompany most hurricanes bring a third significant threat: flooding. Whereas the effects of storm surge and strong winds are concentrated in coastal areas, heavy rains may affect places hundreds of kilometers from the coast for up to several days after the storm has lost its hurricane-force winds. Often elevated terrains enhance rainfall amounts as storm remnants move inland.

Rainfall amounts are not just related to the intensity of a hurricane but also to the storm's size and rate of movement. For example, after Hurricane Harvey made landfall along the Texas coast in late August 2017, it stalled for 4 days, dropping historic amounts of rain. Much of southeastern Texas received more than 100 centimeters (40 inches) of rain, and some areas recorded more than 150 centimeters (60 inches) of rain. The result was catastrophic flooding (**Figure 20.28**). According to the National

► **Figure 20.28**
Hurricane Harvey When this storm struck the Texas coast in August 2017, record-breaking rains led to catastrophic flooding.

Hurricane Center, "Harvey was the most significant tropical rainfall event in United States history . . . since reliable rainfall records began in the 1880s." Heavy rains continued as the remnants of Harvey eventually moved northward, with parts of western Kentucky receiving more than 20 centimeters (8 inches).

Monitoring Hurricanes

Today we have the benefit of numerous observational tools for monitoring hurricanes and other tropical storms. Using input from satellites, aircraft reconnaissance, coastal radar, and remote data buoys in conjunction with sophisticated computer models, meteorologists monitor and forecast storm movements and intensity. The goal is to issue timely watches and warnings.

The Role of Satellites   The greatest single advancement in tools used for observing hurricanes has been the development of meteorological satellites. Vast areas of open ocean must be observed in order to detect a hurricane. Before satellites, this was an impossible task (**Figure 20.29**). Today instruments aboard satellites can detect a potential storm even before it develops its characteristic circular cloud pattern.

In recent years two methods of using satellite-acquired data to monitor hurricane intensity have been developed. One technique involves using instruments aboard a satellite to estimate wind speeds within a storm. A second method involves using satellites to identify areas of extraordinary cloud development, called *hot towers*, in the eye wall of an approaching hurricane (**Figure 20.30**).

Track Forecasts   The predicted path of a hurricane is called the *track forecast*. The track forecast is probably the most basic information because accurate prediction of other storm characteristics (winds and rainfall) is of little value if there is significant uncertainty about where the storm is going. Accurate track forecasts are important because they can lead

◄ **Figure 20.29**
Aftermath of the historic Galveston hurricane The storm struck an unsuspecting and unprepared city on September 8, 1900. The strength of the storm, together with the lack of adequate warning, caught the population by surprise and took the lives of 8000 people. It was the worst natural disaster in U.S. history. Entire blocks were swept clean, and mountains of debris accumulated around the few remaining buildings.

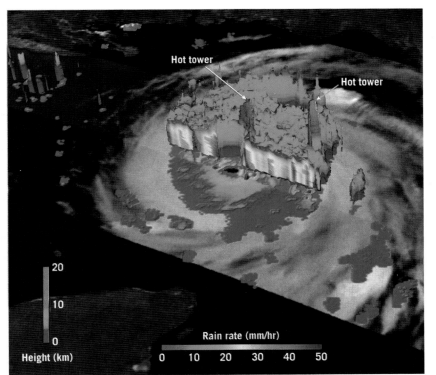

Rain rate (mm/hr)

Height (km)

◄ **SmartFigure 20.30**
Hot towers This *Tropical Rainfall Measuring Mission* (*TRMM*) satellite image of Hurricane Katrina was acquired early on August 28, 2005. The cutaway view of the inner portion of the storm shows cloud height on one side and rainfall rates on the other. Two hot towers (in red) are visible: one in an outer rain band and the other in the eye wall. The eye wall tower rises 16 kilometers (10 miles) above the ocean surface and is associated with an area of intense rainfall. Towers this tall near the core often indicate that a storm is intensifying. Katrina grew from a category 3 to a category 4 storm soon after this image was received.

Video
https://goo.gl/BzjUSM

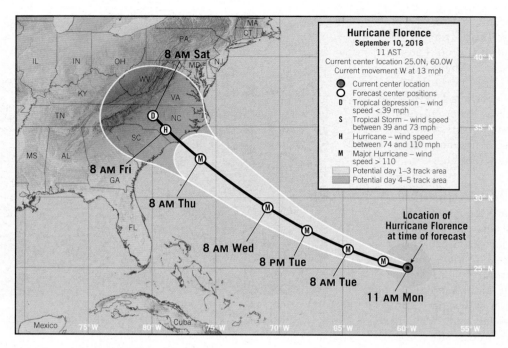

▶ Figure 20.31
Five-day track forecast for Hurricane Florence, issued at 11A.M. AST, Monday, September, 2018 When a hurricane track forecast is issued by the National Hurricane Center, it is termed a *forecast cone*. The cone represents the probable track of the center of the storm and is formed by enclosing the area swept out by a set of circles along the forecast track (at 12 hours, 24 hours, 36 hours, etc.). The circles get larger as they extend further into the future. The entire track of an Atlantic tropical cyclone can be expected to remain entirely within the cone roughly 60 to 70 percent of the time.

to timely evacuations from the surge zone, where the greatest number of deaths usually occur. Track forecasts have been steadily improving. Current 5-day track forecasts are now as accurate as the 3-day forecasts of 20 years ago (**Figure 20.31**). Despite improvements in accuracy, forecast uncertainty still requires that hurricane warnings be issued for relatively large coastal areas. Only about one-quarter of an average warning area experiences hurricane conditions.

CONCEPT CHECKS 20.5

1. What factors influence where and when hurricane formation takes place?

2. Distinguish between the eye and eye wall of a hurricane.

Concept Checker
https://goo.gl/NSmwoM

3. What are the three broad categories of hurricane damage?

20.6 Stabilizing the Shore

Summarize the ways in which people deal with shoreline erosion problems.

Compared with natural hazards such as earthquakes, volcanic eruptions, and landslides, shoreline erosion is often perceived to be a more continuous and predictable process that may cause relatively modest damage to limited areas. In reality, the shoreline is a dynamic place that can change rapidly in response to natural forces. Exceptional storms are capable of eroding beaches and cliffs at rates that greatly exceed the long-term average. Such bursts of accelerated erosion not only significantly affect the natural evolution of a coast but also can have a profound impact on people who reside in the coastal zone. Erosion along our coasts causes significant property damage. Huge sums are spent annually not only to repair damage but also to prevent or control erosion. Already a problem at many sites, shoreline erosion is certain to become an increasingly serious problem as extensive coastal development continues.

During the past 100 years, growing affluence and increasing demands for recreation have brought unprecedented development to many coastal areas. As both the number and the value of buildings have increased, so too have efforts to protect property from storm waves by stabilizing the shore. Also, controlling the natural migration of sand is an ongoing struggle in many coastal areas.

Such interference can result in unwanted changes that are difficult and expensive to correct.

Hard Stabilization

Structures built to protect a coast from erosion or to prevent the movement of sand along a beach are collectively known as **hard stabilization**. Hard stabilization can take many forms and often results in predictable yet unwanted outcomes. Hard stabilization includes jetties, groins, breakwaters, and seawalls.

Jetties Since relatively early in America's history, a principal goal in coastal areas has been the development and maintenance of harbors. In many cases, this has involved the construction of jetty systems. **Jetties** are usually built in pairs and extend into the ocean at the entrances to rivers and harbors. With the flow of water confined to a narrow zone, the ebb and flow caused by the rise and fall of the tides keep the sand in motion and prevent deposition in the channel. However, as illustrated in **Figure 20.32**, a jetty may act as a dam against which the longshore current and beach drift deposit sand. Because the other side is not receiving any new sand, there is soon no beach at all.

Groins To maintain or widen beaches that are losing sand, groins are sometimes constructed. A **groin** is a barrier built at a right angle to the beach to trap sand that is moving parallel to the shore. Groins are usually constructed of large rocks but may also be composed of wood. These structures often do their job so effectively that the longshore current beyond the groin becomes sand starved. As a result, the current erodes sand from the beach on the downstream side of the groin.

To offset this effect, property owners downstream from the structure may erect groins on their property. In this manner, the number of groins multiplies, resulting in a *groin field* (**Figure 20.33**). The New Jersey shoreline is a good example of groin proliferation, as hundreds of these structures have been built there. Because it has been shown that groins often do not provide a satisfactory solution, using them is no longer the preferred method of keeping beach erosion in check.

Breakwaters and Seawalls Hard stabilization can be built parallel to the shoreline. One such structure is a **breakwater**, which protects boats from the force of large breaking waves by creating a quiet water zone near the shoreline. However, when a breakwater is constructed, the reduced wave activity along the shore behind the structure may allow sand to accumulate. If this happens, the marina will eventually fill with sand, while the downstream beach erodes

and retreats. At Santa Monica, California, where the building of a breakwater has created such a problem, the city uses a dredge to remove sand from the protected quiet water zone and deposit it downstream, where longshore currents and beach drift continue to move the sand down the coast (**Figure 20.34**).

Another type of hard stabilization built parallel to the shoreline is a **seawall**, which is designed to armor the coast and defend property from the force of breaking waves. Waves expend much of their energy as they move across an open beach. Seawalls cut this process short by reflecting the force of unspent waves seaward.

◄ Figure 20.32 Jetties These structures are built at the entrances to rivers and harbors and are intended to prevent deposition in the navigation channel. The photo is an aerial view at Santa Cruz Harbor, California.

◄ SmartFigure 20.33 Groins These wall-like structures trap sand that is moving parallel to the shore. This series of groins is along the shoreline near Chichester, Sussex, England.

Mobile Field Trip https://goo.gl/LynnwD

▶ Figure 20.34
Breakwater These two aerial photos show the beach and pier at Santa Monica **A.** before the breakwater was built and **B.** after construction. The breakwater disrupted longshore transport and caused the seaward growth of the beach. After the breakwater was destroyed by storm waves in 1983, the bulge disappeared, and the shoreline returned to its pre-breakwater appearance.

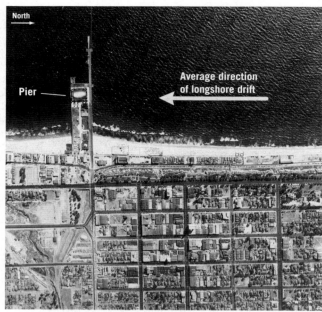

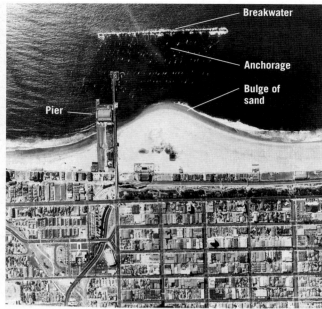

A. The shoreline and pier at Santa Monica as it appeared in September 1931, before the breakwater was constructed in 1933. Note that the pier is on stilts and thus does not affect longshore transport.

B. The same area in 1949. Construction of the breakwater to create a boat anchorage disrupted the longshore transport of sand and caused a bulge of sand on the beach in the wave shadow behind the breakwater.

As a consequence, the beach to the seaward side of the seawall experiences significant erosion and may in some instances be eliminated entirely (**Figure 20.35**). Once the width of the beach is reduced, the seawall is subjected to even greater pounding by the waves. Eventually this battering will cause the wall to fail, and a larger, more expensive wall must be built to take its place.

The wisdom of building temporary protective structures along shorelines is increasingly questioned.

Many coastal scientists and engineers are of the opinion that halting an eroding shoreline with protective structures benefits only a few and seriously degrades or destroys the natural beach and the value it holds for the majority. Protective structures divert the ocean's energy temporarily from private properties but usually refocus that energy on the adjacent beaches. Many structures interrupt the natural sand flow in coastal currents, robbing affected beaches of vital sand replacement.

▶ Figure 20.35
Seawall Sea Bright in northern New Jersey once had a broad, sandy beach. A seawall 5 to 6 meters (16 to 18 feet) high and 8 kilometers (5 miles) long was built to protect the town and the railroad that brought tourists to the beach. After the wall was built, the beach narrowed dramatically.

Alternatives to Hard Stabilization

Armoring the coast with hard stabilization has several potential drawbacks, including the cost of the structures and the loss of sand on the beach. Alternatives to hard stabilization include beach nourishment and changing land use.

Beach Nourishment One approach to stabilizing shoreline sands without hard stabilization is **beach nourishment**. As the term implies, this practice involves adding large quantities of sand to the beach system (**Figure 20.36**). Extending beaches seaward makes buildings along the shoreline less vulnerable to destruction by storm waves and enhances recreational uses. Without sandy beaches, tourism suffers.

The process of beach nourishment is straightforward. Sand is pumped by dredges from offshore or trucked from inland locations. The "new" beach, however, will not be the same as the former beach. Because replenishment sand is from somewhere else, typically not another beach, it is new to the beach environment. The new sand is often different in size, shape, sorting, and composition. Such differences pose problems in terms of erodibility and the kinds of life the new sand will support.

Beach nourishment is not a permanent solution to the problem of shrinking beaches. The same processes that removed the sand in the first place will eventually remove the replacement sand as well. Nevertheless, the number of nourishment projects has increased in recent years, and many beaches, especially along the Atlantic coast, have had their sand replenished many times. Virginia Beach, Virginia, has been nourished more than 50 times.

Beach nourishment is costly. For example, a modest project might involve 38,000 cubic meters (50,000 cubic yards) of sand distributed across about 1 kilometer (0.6 mile) of shoreline. A good-sized dump truck holds about 7.6 cubic meters (10 cubic yards) of sand. So this small project would require about 5000 dump-truck loads. Many projects extend for many miles. Nourishing beaches typically costs millions of dollars per mile.

Changing Land Use Instead of building structures such as groins and seawalls to hold the beach in place or adding sand to replenish eroding beaches, another option is available. Many coastal scientists and planners are calling for a policy shift from defending and rebuilding beaches and coastal property in high-hazard areas to relocating or abandoning storm-damaged buildings and

◄ **Figure 20.36**
Beach nourishment If you visit a beach along the Atlantic coast, it is more and more likely that you will walk into the surf zone atop a beach composed of sand from somewhere else. In this image, an offshore dredge is pumping sand to a beach.

letting nature reclaim the beach. This option is similar to an approach the federal government adopted for river floodplains following the devastating 1993 Mississippi River floods, in which vulnerable structures are either abandoned or relocated on higher, safer ground.

A recent example of changing land use occurred on New York's Staten Island following Hurricane Sandy in 2012. The state turned some vulnerable shoreline areas of the island into waterfront parks. The parks act as buffers to protect inland homes and businesses from strong storms while providing the community with needed open space and access to recreational opportunities.

Land-use changes can be controversial. People with significant near-shore investments want to rebuild and defend coastal developments from the erosional wrath of the sea. Others, however, argue that with sea level rising, the impact of coastal storms will get worse in the decades to come, and vulnerable or oft-damaged structures should be abandoned or relocated to improve personal safety and reduce costs. Such ideas will no doubt be the focus of much study and debate as states and communities evaluate and revise coastal land-use policies.

CONCEPT CHECKS 20.6

1. List at least three examples of hard stabilization and describe what each is intended to do. How does each affect distribution of sand on a beach?

   Concept Checker
   https://goo.gl/oTymA3

2. What are two alternatives to hard stabilization, and what are the potential problems associated with each?

20.7 Tides

Explain the cause of tides, their monthly cycles, and tidal patterns. Describe the horizontal flow of water that accompanies the rise and fall of tides.

Tides are daily changes in the elevation of the ocean surface caused by gravitational interactions of Earth with the Moon and Sun. Their rhythmic rise and fall along coastlines have been known since antiquity. After waves, they are the easiest ocean movements to observe (**Figure 20.37**).

▶ **Figure 20.37**
Bay of Fundy tides High tide and low tide at Hall's Harbor along the Bay of Fundy.

In contrast, the world ocean, which is mobile, is deformed quite dramatically by this effect, producing the two opposing tidal bulges.

Because the position of the Moon changes only moderately in a single day, the tidal bulges remain in place while Earth rotates "through" them. For this reason, if you stand on the seashore for 24 hours, Earth will rotate you through alternating areas of deeper and shallower water. As you are carried into each tidal bulge, the tide rises, and as you are carried into the intervening troughs between the tidal bulges, the tide falls. Therefore, most places on Earth experience two high tides and two low tides each day.

Further, the tidal bulges migrate as the Moon revolves around Earth about every 29 days. As a result, the tides, like the time of moonrise, shift about 50 minutes later each day. After 29 days the cycle is complete, and a new one begins.

In many locations, there may be an inequality between the high tides during a given day. Depending on the position of the Moon, the tidal bulges may be inclined to the equator, as in Figure 20.38. This figure illustrates

Although known for centuries, tides were not explained satisfactorily until Sir Isaac Newton applied the law of gravitation to them. Newton showed that there is a mutual attractive force between two bodies and that because oceans are free to move, they are deformed by this force. Hence, ocean tides result from the gravitational attraction exerted upon Earth by the Moon and, to a lesser extent, by the Sun.

Causes of Tides

It is easy to see how the Moon's gravitational force can cause the water to bulge on the side of Earth nearest the Moon. In addition, however, an equally large tidal bulge is produced on the side of Earth directly opposite the Moon (**Figure 20.38**).

Both tidal bulges are caused, as Newton discovered, by the pull of gravity. Gravity is inversely proportional to the square of the distance between two objects, meaning simply that it quickly weakens with distance. In this case, the two objects are the Moon and Earth. Because the force of gravity decreases with distance, the Moon's gravitational pull on Earth is slightly greater on the near side of Earth than on the far side. The result of this differential pulling is to stretch (elongate) the "solid" Earth very slightly.

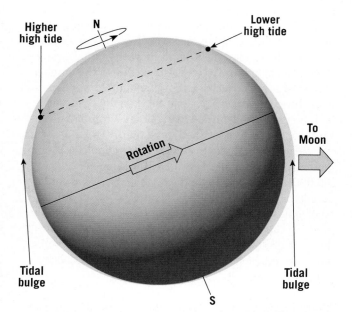

▲ **Figure 20.38**
Idealized tidal bulges caused by the Moon If Earth were covered to a uniform depth with water, there would be two tidal bulges: one on the side of Earth facing the Moon (right) and the other on the opposite side of Earth (left). Depending on the Moon's position, tidal bulges may be inclined relative to Earth's equator. In this situation, Earth's rotation causes an observer to experience two unequal high tides during a day.

that one high tide experienced by an observer in the Northern Hemisphere is considerably higher than the high tide half a day later. In contrast, a Southern Hemisphere observer would experience the opposite effect.

Monthly Tidal Cycle

The primary body that influences the tides is the Moon, which makes one complete revolution around Earth every 29.5 days. The Sun, however, also influences the tides. It is far larger than the Moon, but because it is much farther away, its effect is considerably less. In fact, the Sun's tide-generating effect is only about 46 percent that of the Moon.

Near the times of new and full moons, the Sun and Moon are aligned, and their forces on tides are added together (**Figure 20.39A**). The combined gravity of these two tide-producing bodies causes larger tidal bulges (higher high tides) and deeper tidal troughs (lower low tides), producing a large tidal range. These are called the **spring tides** (*springen* = to rise up), which have no connection with the spring season but occur twice a month, during the time when the Earth–Moon–Sun system is aligned. Conversely, at about the time of the first and third quarters of the Moon, the gravitational forces of the Moon and Sun act on Earth at right angles, and each partially offsets the influence of the other (**Figure 20.39B**). As a result, the daily tidal range is smaller. These are called **neap tides** (*nep* = scarcely, or barely, touching), and they also occur twice each month. Each month, then, there are two spring tides and two neap tides, each about 1 week apart.

Tidal Patterns

Although the basic causes and types of tides have been explained, these theoretical considerations cannot be used to predict either the height or the time of actual tides at a particular place. This is because many factors—including the shape of the coastline, the configuration of ocean basins, and water depth—greatly influence the tides. Consequently, tides at various locations respond differently to tide-producing forces. This being the case, the nature of the tide at any coastal location can be determined most accurately by actual observation. The predictions in tidal tables and tidal data on nautical charts are based on such observations.

Three main tidal patterns exist worldwide (**Figure 20.40**). A *diurnal tidal pattern* (*diurnal* = daily) is characterized by a single high tide and a single low tide each tidal day (24-hour period). Tides of this type occur along the northern shore of the Gulf of Mexico, among other locations. A *semidiurnal tidal pattern* exhibits two

▶ **SmartFigure 20.39**
Spring and neap tides Earth–Moon–Sun positions influence the tides.

Animation
https://goo.gl/qDBG7y

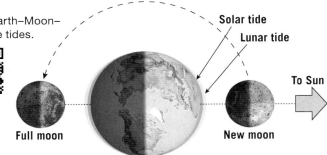

A. Spring Tide When the Moon is in the full or new position, the tidal bulges created by the Sun and Moon are aligned, and there is a large tidal range.

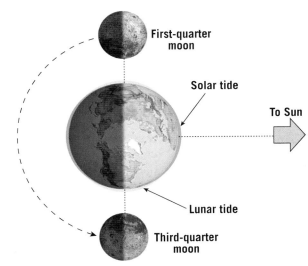

B. Neap Tide When the Moon is in the first- or third-quarter position, the tidal bulges produced by the Moon are at right angles to the bulges created by the Sun, and the tidal range is smaller.

Two high tides and two low tides of approximately equal heights during each tidal day

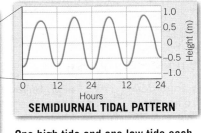

SEMIDIURNAL TIDAL PATTERN

One high tide and one low tide each tidal day

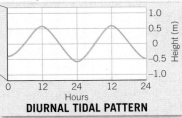

DIURNAL TIDAL PATTERN

Two high tides and two low tides of unequal heights during each tidal day

MIXED TIDAL PATTERN

◀ **SmartFigure 20.40**
Tidal patterns A diurnal tidal pattern (lower right) shows one high tide and one low tide each tidal day. A semidiurnal pattern (upper right) shows two high tides and two low tides of approximately equal heights during each tidal day. A mixed tidal pattern (left) shows two high tides and two low tides of unequal heights during each tidal day.

Tutorial
https://goo.gl/JrfTym

▶ Figure 20.41
Tidal deltas As a rapidly moving tidal current (flood current) moves through a barrier island's inlet into the quiet waters of the lagoon, the current slows and deposits sediment, creating a tidal delta. Because this tidal delta has developed on the landward side of the inlet, it is called a *flood delta*. Such a tidal delta is shown in Figure 20.15A.

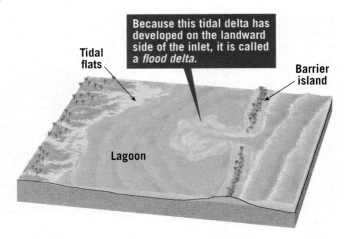

Tidal flats

Because this tidal delta has developed on the landward side of the inlet, it is called a *flood delta*.

Barrier island

Lagoon

high tides and two low tides each tidal day, with the two highs about the same height and the two lows about the same height. This type of tidal pattern is common along the Atlantic coast of the United States. A *mixed tidal pattern* is similar to a semidiurnal pattern except that it is characterized by a large inequality in high water heights, low water heights, or both. In this case, there are usually two high tides and two low tides each day, with high tides of different heights and low tides of different heights. Such tides are prevalent along the Pacific coast of the United States and in many other parts of the world.

Tidal Currents

Tidal current is the term used to describe the *horizontal* flow of water that accompanies the rise and fall of the tide. These water movements induced by tidal forces can be important in some coastal areas. Tidal currents flow in one direction during a portion of the tidal cycle and reverse their flow during the remainder. Tidal currents that advance into the coastal zone as the tide rises are called **flood currents**. As the tide falls, seaward-moving

water generates **ebb currents**. Periods of little or no current, called *slack water*, separate flood and ebb. The areas affected by these alternating tidal currents are **tidal flats**. Depending on the nature of the coastal zone, tidal flats vary from narrow strips seaward of the beach to extensive zones that may extend for several kilometers.

Although tidal currents are generally not important in the open sea, they can be rapid in bays, river estuaries, straits, and other narrow places. Off the coast of Brittany in France, for example, tidal currents that accompany a high tide of 12 meters (40 feet) may attain a speed of 20 kilometers (12 miles) per hour. While tidal currents are not generally major agents of erosion and sediment transport, notable exceptions occur where tides move through narrow inlets. Here they constantly scour the small entrances to many good harbors that would otherwise be blocked.

Sometimes deposits called **tidal deltas** are created by tidal currents (**Figure 20.41**). They may develop either as *flood deltas* landward of an inlet or as *ebb deltas* on the seaward side of an inlet. Because wave activity and longshore currents are reduced on the sheltered landward side, flood deltas are more common and more prominent (see Figure 20.15A). They form after the tidal current moves rapidly through an inlet. As the current emerges from the narrow passage into more open waters, it slows and deposits its load of sediment.

CONCEPT CHECKS 20.7

1. Explain why an observer can experience two unequal high tides during a single day.

2. Distinguish between *neap tides* and *ebb tides*.

 Concept Checker https://goo.gl/VTaJSJ

3. Contrast *flood current* and *ebb current*.

20 CONCEPTS IN REVIEW

Shorelines

20.1 The Shoreline and Ocean Waves

Explain why the shoreline is considered a dynamic interface. List the factors that influence the height, length, and period of a wave and describe the motion of water within a wave.

Key Terms:

wave height	fetch	
shoreline	wavelength	surf
interface	wave period	

- The *shoreline* is a transition zone between marine and continental environments. It is a dynamic *interface*, a boundary where land, sea, and air meet and interact.

- Energy from waves plays an important role in shaping the shoreline, but many factors contribute to the character of particular shorelines.

- Waves are moving energy, and most ocean waves are initiated by wind. The three factors that influence the *height*, *wavelength*, and *period* of a wave are (1) wind speed, (2) length of time the wind has blown, and (3) *fetch*, the distance that the wind has traveled across open water. Once waves leave a storm area, they are termed swells, which are symmetrical, longer-wavelength waves.

- As waves travel, water particles transmit energy by circular orbital motion, which extends to a depth equal to one-half the wavelength (the wave base). When a wave enters water that is shallower than the wave base, it slows, allowing waves farther from shore to catch up. As a result, wavelength decreases and wave height increases. Eventually the wave breaks, creating turbulent *surf* in which water rushes toward the shore.

20.2 Beaches and Shoreline Processes

Explain how waves erode and move sediment along the shore.

Key Terms:

beach	wave refraction	rip current
abrasion	beach drift	
	longshore current	

- A *beach* is composed of any locally derived material that is in transit along the shore.
- Waves provide most of the energy that modifies shorelines. Wave erosion is caused by wave impact pressure and *abrasion* (the sawing and grinding action of water armed with rock fragments).
- As they approach the shore, waves refract (bend) to align nearly parallel to the shore. Refraction occurs because a wave travels more slowly in shallower water, allowing the part that is still in deeper water to catch up. *Wave refraction* causes wave erosion to be concentrated against the sides and ends of headlands and dispersed in bays.
- Waves that approach the shore at an angle transport sediment parallel to the shoreline. On the beach face, this longshore transport is called *beach drift*, and it is due to the fact that the incoming swash pushes sediment obliquely upward, whereas the backwash pulls it directly downhill. *Longshore currents* are a similar phenomenon in the surf zone, capable of transporting very large quantities of sediment parallel to a shoreline.

Q What process is causing wave energy to be concentrated on the headland? Predict how this area will appear in the future.

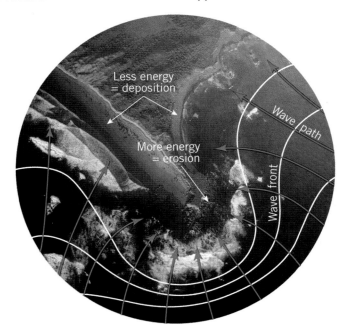

Less energy = deposition

More energy = erosion

Wave path

Wave front

20.3 Shoreline Features

Describe the features typically created by wave erosion and those resulting from sediment deposited by longshore transport processes.

Key Terms:

wave-cut cliff	sea arch	tombolo
wave-cut platform	sea stack	barrier island
marine terrace	spit	
	baymouth bar	

- Erosional features include *wave-cut cliffs* (created by the cutting action of the surf against the base of coastal land), *wave-cut platforms* (relatively flat surfaces left behind by receding cliffs), and *marine terraces* (uplifted wave-cut platforms). Erosional features also include *sea arches* (formed when a headland is eroded and two sea caves from opposite sides unite) and *sea stacks* (formed when the roof of a sea arch collapses).
- Depositional features that form when sediment is moved by beach drift and longshore currents include *spits* (elongated ridges of sand that project from the land into the mouth of a bay), *baymouth bars* (sandbars that completely cross a bay), and *tombolos* (ridges of sand connecting an island to the mainland or another island). The Atlantic and Gulf coastal region is characterized by offshore *barrier islands*, which are low ridges of sand that parallel the coast.
- Over time, irregular, rocky shorelines are modified by erosion and deposition to become smoother and straighter.

Q Identify the lettered features in this diagram.

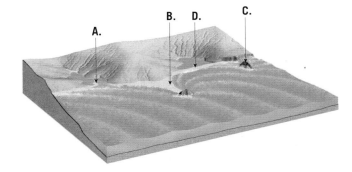

A.   B.   D.   C.

20.4 Contrasting America's Coasts

Distinguish between emergent and submergent coasts. Contrast the erosion problems faced on the Atlantic and Gulf coasts with those along the Pacific coast.

Key Terms: emergent coast   submergent coast   estuary

- Coasts may be classified by their changes relative to sea level. *Emergent coasts* are sites of either land uplift or sea-level fall. Marine terraces are features of emergent coasts. *Submergent coasts* are sites of land subsidence or sea-level rise. One characteristic of submergent coasts is drowned river valleys called *estuaries*.
- The Atlantic and Gulf coasts are lined in many places by barrier islands—dynamic expanses of sand that see a lot of change during storm events. Many of these low and narrow islands have also been prime sites for real estate development.
- The Pacific coast's big issue is the narrowing of beaches due to sediment starvation. Rivers that drain to the coast (bringing it sand) have been dammed, resulting in reservoirs trapping sand before it can make it to the coast. Narrower beaches offer less resistance to incoming waves, often leading to erosion of bluffs behind the beach.

Q Is this an emergent coast or a submergent coast? Provide an easily seen line of evidence to support your answer. Is the location more likely along the coast of North Carolina or California? Explain.

20.5 Hurricanes: The Ultimate Coastal Hazard

Describe the basic structure and characteristics of a hurricane and the three broad categories of hurricane destruction.

Key Terms: hurricane   eye wall   eye   storm surge

- *Hurricanes* are fueled by warm, moist air and usually form in the late summer and early fall, when sea-surface temperatures are highest. Water vapor in rising warm air condenses, releasing heat and triggering the formation of dense clouds and heavy rain. Because of a steep pressure gradient, air rushes into the center of the storm. The Coriolis effect and ocean-water temperatures strongly influence where hurricanes form.
- The *eye* at the center of a hurricane has the lowest pressure, is relatively calm, and lacks rain. The surrounding *eye wall* has the strongest winds and most intense rainfall. The Saffir–Simpson hurricane wind scale classifies storms based on their air pressure and wind speed.
- Most hurricane damage results from one or a combination of three causes: storm surge, wind damage, or inland freshwater flooding due to heavy rains. *Storm surge* occurs when ocean water gets pushed up above the normal water level by the strong winds. In the Northern Hemisphere hurricane winds rotate counterclockwise, and storm surge is greatest on the right side of an advancing hurricane. This is due to the combination of the storm's forward movement and strong winds blowing toward the shore.

Q This coastal scene shows hurricane destruction. Which one of the three basic classes of damage was most likely responsible for this destruction? What is your reasoning?

20.6 Stabilizing the Shore

Summarize the ways in which people deal with shoreline erosion problems.

Key Terms:
hard stabilization
jetty
groin
breakwater
seawall
beach nourishment

- *Hard stabilization* refers to any structures built along the coastline to prevent movement of sand. *Jetties* project out from the coast with the goal of keeping inlets open. *Groins* are also oriented perpendicular to the coast, but their goal is to slow beach erosion by longshore currents. Offshore *breakwaters* are constructed parallel to the coast to blunt the force of incoming ocean waves, often to protect boats. Like breakwaters, *seawalls* are oriented parallel to the coast, but they are built on the shoreline itself. Hard stabilization measures often result in increased erosion elsewhere.
- Beach *nourishment* is an expensive alternative to hard stabilization. Sand is pumped onto a beach from some other area to temporarily replenish the sediment supply. Another option is relocating buildings away from high-risk areas and leaving the beach to be shaped by natural processes.

Q Based on their position and orientation, identify the four kinds of hard stabilization illustrated in this diagram.

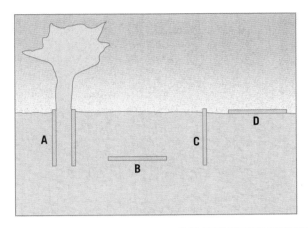

20.7 Tides

Explain the cause of tides, their monthly cycles, and tidal patterns. Describe the horizontal flow of water that accompanies the rise and fall of tides.

Key Terms:
tide
spring tide
neap tide
tidal current
flood current
ebb current
tidal flat
tidal delta

- *Tides* are daily changes in ocean-surface elevation. They are caused by gravitational pull on ocean water by the Moon and, to a lesser extent, the Sun. When the Sun, Earth, and Moon all line up about every 2 weeks (full moon or new moon), the tides are most exaggerated. When a quarter Moon is in the sky, the Moon is pulling on Earth's water at a right angle relative to the Sun, and the daily tidal range is minimized as the two forces partially counteract one another.
- A *flood current* is the landward movement of water during the shift between low tide and high tide. When high tide transitions to low tide again, the movement of water away from the land is an ebb current. *Ebb currents* may expose tidal flats to the air. If a tide passes through an inlet, the current may carry sediment that gets deposited as a *tidal delta*.

GIVE IT SOME THOUGHT

1. **This surfer is enjoying a ride on a large wave along the coast of Maui.**

   a. What was the source of energy that created this wave?

   b. How was the wavelength changing just prior to the time when this photo was taken?

   c. Why was the wavelength changing?

   d. Many ocean waves exhibit circular orbital motion. Is that true of the wave in this photo? Explain.

2. **You and a friend set up an umbrella and chairs at a beach. Your friend then goes into the surf zone to play Frisbee with another person. Several minutes later, your friend looks back toward the beach and is surprised to see that she is no longer near the umbrella and chairs. Although she is still in the surf zone, she is 30 yards away from where she started. How would you explain to your friend why she moved along the shore?**

3. What term is applied to the masses of rock protruding from the water in this photo? How did they form? Is the location more likely along the U.S. Gulf Coast or the Pacific Coast? Explain.

4. A friend wants to purchase a vacation home on a barrier island. If consulted, what advice would you give your friend?

5. Hurricane Rita was a major storm that struck the Gulf coast in late September 2005, less than a month after Hurricane Katrina. The accompanying graph shows changes in air pressure and wind speed from the storm's beginning as an unnamed tropical disturbance north of the Dominican Republic on September 18 until its last remnants faded away in Illinois on September 26. Use the graph to answer these questions:

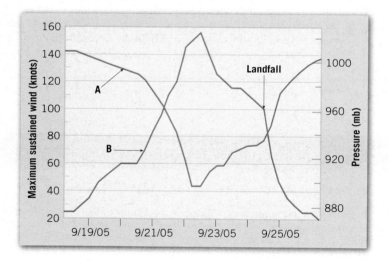

a. Which line represents air pressure, and which line represents wind speed? How did you figure this out?

b. What was the storm's maximum wind speed, in knots? Convert this answer to kilometers per hour by multiplying by 1.85.

c. What was the lowest pressure attained by Hurricane Rita?

d. Using wind speed as your guide, what was the highest category reached on the Saffir–Simpson hurricane wind scale? On what day was this status reached?

e. When landfall occurred, what was the category of Hurricane Rita?

6. Examine the photo, which shows a portion of the coast near Cape May, New Jersey. What term is applied to the wall-like structures that extend into the water? What is their purpose? In what direction are beach drift and longshore currents moving sand: toward the right or toward the left of the photo?

7. The force of gravity plays a critical role in creating ocean tides. The more massive an object, the stronger its pull of gravity. Explain why the Sun's influence is only about half that of the Moon, even though the Sun is much more massive than the Moon.

8. This photo shows a portion of the Maine coast. The brown muddy area in the foreground is influenced by tidal currents. What term is applied to this muddy area? Name the type of tidal current this area will experience in the hours to come.

EYE ON EARTH

1. This satellite image shows Cyclone Favio as it came ashore along the coast of Mozambique, Africa, on February 22, 2007. This powerful storm was moving from east (right) to west (left). Since it is a Southern Hemisphere storm, the cloud pattern shows that the winds circulate in a clockwise spiral instead of the counterclockwise pattern typical of storms in the Northern Hemisphere. Portions of the cyclone had sustained winds of 203 kilometers (126 miles) per hour as it made landfall.

   a. Identify the eye and eye wall of the storm.

   b. Based on wind speed, classify the storm using the Saffir–Simpson scale.

   c. Which one of the lettered sites should experience the strongest storm surge? Explain.

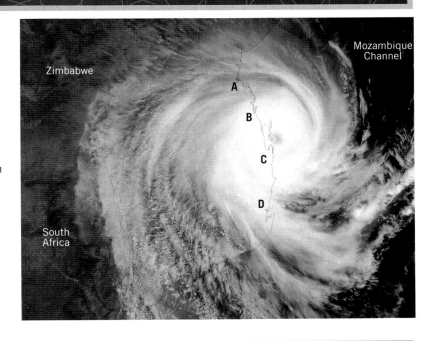

DATA ANALYSIS

Tides and Sea-Level Rise

Tides account for short-term changes in coastal water levels. Long-term changes in water levels are also observed as the level of the ocean itself changes.

https://goo.gl/eD226M

ACTIVITIES

Go to the Tides and Currents page, at http://tidesandcurrents.noaa.gov. Click on "Tides and Currents Map" and zoom in to view Miami, Florida. Click on the "Virginia Key" station next to Miami. (You can click on "Help" in the top-right corner to learn how to use this map.)

1. What time is the next tide? Will it be a high tide or low tide?

2. What is the predicted height (in feet) of this tide?

3. How long does it take to go from high tide to low tide? High tide to high tide?

4. Is this a diurnal, semidiurnal, or mixed tidal pattern?

5. Zoom in on the Seattle area in Washington State. Note the time and height of the next high tide at the La Push station on the west coast of the Olympic Peninsula and at the Seattle station. Do the time and/or height differ? If so, what is your explanation?

Set the "Require Type" drop-down menu to "Sea Level Trends." Click on a station (such as Boston) and select "Sea Level Trends" from the pop-up.

6. How many years are shown on the graph? How long is the data record for this station?

7. Use the Linear Mean Sea Level Trend line to determine the total sea-level change over the data record.

Go to the Sea Level Rise and Coastal Flooding Impacts site, at https://coast.noaa.gov/slr/. Zoom in to a state along the East Coast. Use the "Sea Level Rise" slider to raise the sea level.

8. Which locations did you visualize, and what effects did you see?

Mastering Geology

Looking for additional review and test prep materials? Visit pearson.com/mastering/geology to enhance your understanding of this chapter's content by accessing a variety of resources, including **Self-Study Quizzes, Geoscience Animations, SmartFigure Tutorials,** *Project Condor* **Videos, Mobile Field Trips, Dynamic Study Modules,** and an optional **Pearson eText.**

The Arctic's Permafrost Is Thawing

The loss of frozen ground in the Arctic is a striking result of global warming and a contributor to more warming to come. Permafrost covers about 24 percent of the exposed landmass of the Northern Hemisphere—nearly 14.4 million square kilometers (9 million square miles). It has characterized the Arctic for tens of thousands of years. However, we are learning that the Arctic's permafrost is not as permanent as its name implies. It is starting to thaw. By 2050, much of this frozen ground could be gone. Should we be alarmed?

▲ This road in Alaska was built on permafrost that subsequently thawed.

Permafrost stores the carbon-rich remains of plants and animals that froze before they could decompose. Scientists estimate that the world's permafrost holds 1500 billion tons of carbon, almost double the amount of carbon that is currently in the atmosphere. Permafrost soils are warming even faster than Arctic air temperatures—as much as 1.5° to 2.5°C (2.7° to 4.5°F) in just the past 3 decades. Heat from the warming surface penetrates and thaws the permafrost. Once the ancient organic material thaws, microbes convert some of it to carbon dioxide and methane. These important greenhouse gases escape into the atmosphere, where they contribute to additional global warming. Thus, thawing permafrost could be a tipping point that triggers an irreversible cycle: When permafrost releases carbon dioxide and methane, it accelerates warming, which then leads to more permafrost thaw, and so on.

In addition to greenhouse gas emissions, thawing permafrost causes the once rock-solid ground to become unstable. The shifting ground damages buildings, roads, power lines, and other infrastructure.

► The treeless expanse of the Arctic tundra. Much of Alaska and Canada are underlain by permanently frozen ground called permafrost.

21

Global Climate Change

Climate has a significant impact on people, and people have a strong influence on climate. Today global climate change caused by humans is a major environmental issue. Unlike changes in the geologic past, which were natural variations, modern climate change is dominated by human influences that are sufficiently large that they exceed the bounds of natural variability. Moreover, these changes are likely to continue for many centuries. The effects of this venture into the unknown with climate could be very disruptive not only to humans but to many other life-forms as well. The latter portion of this chapter examines the ways in which humans may be changing global climate.

21.1 Climate and Geology

List the major parts of the climate system and some connections between climate and geology.

The term **weather** refers to the state of the atmosphere at a given time and place. Changes in the weather are frequent and sometimes seemingly erratic. In contrast, **climate** is a description of aggregate weather conditions, based on observations over many decades. Climate is often defined simply as "average weather," but this definition is inadequate because variations and extremes are also important parts of a climate description.

The Climate System

Throughout this text, you have frequently seen that Earth is a complex system that consists of many interacting parts. A change in any one part can produce changes in any or all of the other parts—often in ways that are neither obvious nor immediately apparent. Key to understanding climate change and its causes is the fact that climate is related to all parts of the Earth system.

Earth's **climate system** derives its energy from the Sun and includes the atmosphere, hydrosphere, geosphere, biosphere, and cryosphere. The first four were discussed in Chapter 1; the **cryosphere** refers to the portion of Earth's surface where water is in solid form. This includes snow, glaciers, sea ice, freshwater ice, and frozen ground (termed *permafrost*). The climate system, illustrated in **Figure 21.1**, involves the exchanges of energy and moisture that occur among the five spheres. These exchanges link the atmosphere to the other spheres so that the whole functions as an extremely complex interactive unit.

Climate–Geology Connections

Climate has a profound impact on many geologic processes. When climate changes, these processes respond. A glance back at the rock cycle in Chapter 1 reminds us about many of the connections. Of course, rock weathering has an obvious climate connection, as do processes associated with arid, tropical, and glacial landscapes.

Phenomena such as debris flows and river flooding are often triggered by atmospheric events such as periods of extraordinary rainfall. Clearly, the atmosphere is a basic link in the hydrologic cycle. Other climate–geology connections involve the impact of internal processes on the atmosphere. For example, the particles and gases emitted by volcanoes can change the composition of the atmosphere, and mountain building can have a significant impact on regional temperature, precipitation, and wind patterns.

The study of sediments, sedimentary rocks, and fossils clearly demonstrates that, during Earth's long and complex history, practically every place on our planet has experienced wide swings in climate, from ice ages to conditions associated with subtropical coal swamps or desert dunes. Timescales for climate change vary from decades to millions of years. Chapter 22, "Earth's Evolution Through Geologic Time," documents many of these shifts in climate.

Climates Change

Using fossils and many other geologic clues, scientists have reconstructed Earth's climate going back hundreds of millions of years. Over long timescales (tens to hundreds of millions of years), Earth's climate can be broadly characterized as being a warm "greenhouse" or a cold "icehouse."

During greenhouse times, there is little, if any, permanent ice at either pole, and relatively warm

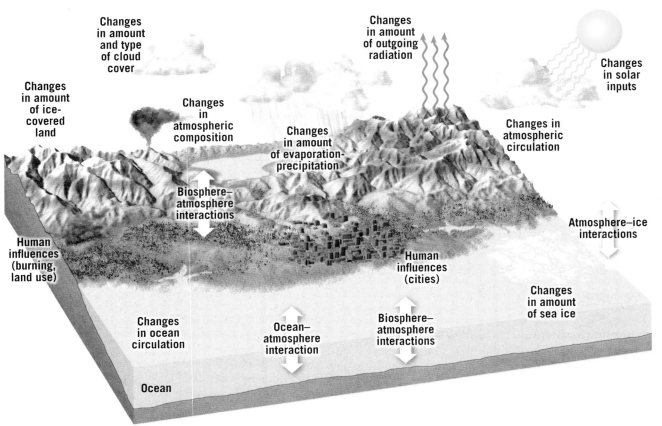

temperate climates are found even at high latitudes. During icehouse conditions, global climate is cool enough to support ice sheets at one or both poles. Earth's climate has gradually transitioned between these two categories only a few times in the past 541 million years, the span known as the Phanerozoic ("visible life") eon. The rocks and deposits of the Phanerozoic eon contain abundant fossils that document major environmental and evolutionary trends. The most recent transition occurred during the Cenozoic era.

The early Cenozoic was a time of greenhouse climates like those the dinosaurs experienced during the preceding Mesozoic era. Peak temperatures occurred about 50 million years ago and were followed by a period of gradual cooling. By about 34 million years ago, permanent ice sheets were present at the South Pole, ushering in icehouse conditions (**Figure 21.2**). In North America, the lush "greenhouse" forests, marked by palm trees in Wyoming and banana plants in Oregon, were replaced by open grasslands. Grassland ecosystems are better suited for a cooler, drier "icehouse" climate. By about 2.6 million years ago (the start of the Quaternary epoch), Earth's climate was cold enough to

support vast ice sheets at both poles. In the Northern Hemisphere ice advanced nearly as far south as the present-day Ohio River, then subsequently retreated to Greenland. For the past 800,000 years, this cycle

▼ **Figure 21.2**
Relative climate change during the Cenozoic era During the past 65 million years, Earth's climate shifted from being a warm "greenhouse" to being a cool "icehouse." Climate is not stable when viewed over long time spans. Earth has experienced several back-and-forth shifts between warm and cold.

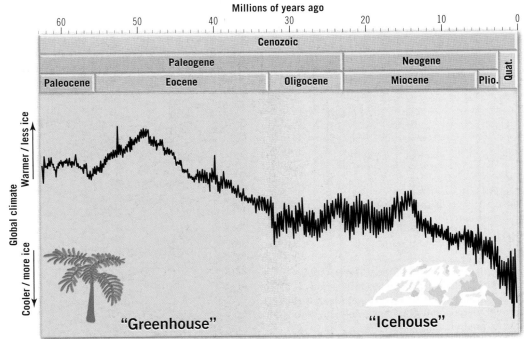

of ice advance and retreat has occurred about every 100,000 years. The last major ice sheet advance reached a maximum about 18,000 years ago.

The next section examines how scientists decipher Earth's climate history. We then explore some natural causes of climate change before returning to the question of how humans have changed climate.

CONCEPT CHECKS 21.1

1. Distinguish between *weather* and *climate*.

2. What are the five major parts of the climate system?

Concept Checker
https://goo.gl/y2yz7a

3. List at least five connections between climate and geology.

21.2 Detecting Climate Change

Discuss several ways in which past climate changes are detected.

Climate not only varies from place to place but is also naturally variable over time. During the great expanse of Earth history, and long before humans were roaming the planet, there were many shifts—from warm to cold and from wet to dry and back again. Nearly every place on our planet has experienced wide swings in climate.

Proxy Data

The high-tech digital and precision instruments currently used to study the composition and dynamics of the atmosphere are recent inventions and therefore have been providing data for only a short time. To understand fully the behavior of the atmosphere and to anticipate future climate change, we must somehow discover how climate has changed over broad expanses of time.

Instrumental records of climate go back only a few centuries, at best, and the further back we go, the less complete and more unreliable the data become. To overcome this lack of direct measurements, scientists must decipher and reconstruct past climates by using indirect evidence called **proxy data**. The main goal of

such work is to understand climates of the past, a study termed **paleoclimatology**, in order to assess potential future climate changes in the context of natural climate variability.

Proxy data come from natural recorders of climate variability, such as glacial ice, seafloor sediments, oxygen isotopes, corals, tree-growth rings, and fossil pollen (**Figure 21.3**).

Glacial Ice

Cores of glacial ice are an indispensable source of data for reconstructing past climates. Scientists collect ice cores mainly from the Greenland and Antarctic Ice Sheets using a drilling rig—a small version of an oil-well

▶ Figure 21.3
Ancient bristlecone pines Some of these trees in California's White Mountains are more than 4000 years old. The study of tree-growth rings is one way that scientists reconstruct past climates.

A. National Ice Core Laboratory

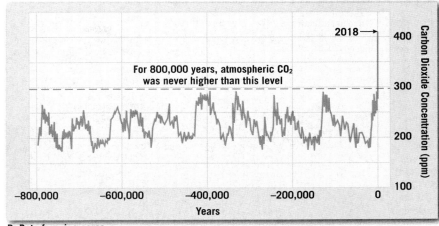

For 800,000 years, atmospheric CO_2 was never higher than this level

B. Data from ice cores

▲ **SmartFigure 21.4**
Ice cores: Important sources of climate data
A. The National Ice Core Laboratory is a physical plant for storing and studying cores of ice taken from glaciers around the world. These cores represent a long-term record of the composition of the atmosphere in the distant past.
B. This graph, showing carbon dioxide variations over the past 800,000 years, is derived from oxygen-isotope analysis of ice cores recovered from the Antarctica and Greenland Ice Sheets.

Tutorial
https://goo.gl/
GjwGPz

drill. A hollow shaft follows the drill head into the ice, and a cylindrical ice core is extracted (**Figure 21.4A**). Air bubbles trapped in ice cores contain samples of the atmosphere at the time snow was converted to ice. From these trapped air samples, researchers measure the concentration of greenhouse gases such as carbon dioxide and methane.* Whereas modern continuous measurements of the atmosphere, taken at Mauna Loa Observatory in Hawaii, extend back only to 1958, ice core data allow us to "see" up to hundreds of thousands of years into the past.

The longest ice core comes from East Antarctica, at 75 degrees south latitude, where the ice sheet is the thickest. A multinational team extracted a 10-centimeter (4-inch) ice core to a depth that exceeded 3 kilometers (2 miles). The analysis has provided climate and atmospheric data back 800,000 years—encompassing eight glacial–interglacial cycles (**Figure 21.4B**).

Antarctic ice cores show that the concentration of carbon dioxide (CO_2) during the past 800,000 years never exceeded 300 parts per million (ppm) until the early nineteenth century, when it started to rise (see Figure 21.4B). In 2018, the concentration of CO_2 exceeded 410 ppm, which is about 40 percent higher than it was before the Industrial Revolution.

Seafloor Sediments

Most seafloor sediments contain the tiny shells of microorganisms, including foraminifera, that once lived in the ocean. When these organisms die, their shells slowly settle to the floor of the ocean, where they become part of the sedimentary record. When recovered in sediment cores, such as the one in **Figure 21.5**, they carry information about past ocean conditions and, thus, about climate.

Different types of marine microorganisms thrive in different conditions; for instance, some are characteristic of tropical oceans and some of temperate or polar conditions. Thus, one way to gain information from sediment cores is to count the numbers and types of microfossils along the length of a core and in cores from different parts of the oceans.

Oxygen–Isotope Analysis   The fossilized calcium carbonate ($CaCO_3$) shells of some marine microorganisms, particularly foraminifera, also provide climate information by recording the composition of the seawater in which they formed. In particular, these shells record the ratio in seawater of two isotopes of oxygen: ^{16}O, which is the most common, and the heavier ^{18}O. A molecule of water (H_2O) can contain either ^{16}O or ^{18}O. However, water molecules containing the lighter ^{16}O isotope evaporate more readily than those with the heavier

*Greenhouse gases absorb infrared (longwave) radiation emitted by Earth, resulting in warming of the atmosphere.

◄ **Figure 21.5**
Ocean floor sediment core The study of seafloor sediments is an important source of proxy data.

▶ **Figure 21.6**
The concentration of oxygen isotopes in seawater varies with climate
A. Water molecules that contain an atom of ^{16}O are lighter than those that contain ^{18}O, and hence they evaporate more readily. During a glacial period, these atoms fall as snow and become locked up in ice.
B. During warmer periods, this water is returned to the oceans, changing the oxygen-isotope ratio.

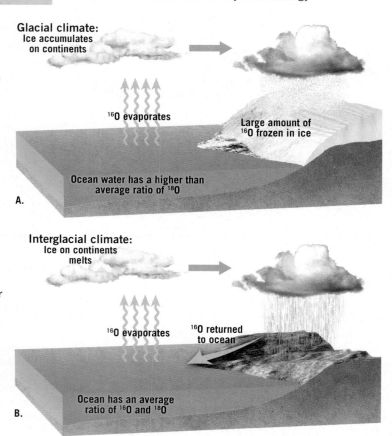

Glacial climate:
Ice accumulates on continents

^{16}O evaporates

Large amount of ^{16}O frozen in ice

Ocean water has a higher than average ratio of ^{18}O

A.

Interglacial climate:
Ice on continents melts

^{16}O evaporates

^{16}O returned to ocean

Ocean has an average ratio of ^{16}O and ^{18}O

B.

measuring the changing ratio of ^{18}O to ^{16}O along the length of a core, they obtain a record of climate over the time interval represented by the sediment core. Oxygen-isotope analysis of foraminifera shells found in seafloor sediments also confirms the accuracy and reliability of climate data extracted from the ice cores discussed in the previous section.

Corals

Coral reefs are built by colonies of corals, invertebrates that live in warm, shallow waters, with each generation growing on top of earlier generations. Because corals, like foraminifera, have hard skeletons built from the $CaCO_3$ extracted from the surrounding seawater, the ratio of ^{16}O to ^{18}O in corals can be used to determine the temperature of the water at the time the coral reef was forming. Think of coral as another *paleothermometer* that reveals important data about climate variability in the world's oceans.

Oxygen-isotope analysis of coral can also serve as a proxy measurement for precipitation, particularly in areas where large variations in annual rainfall occur. High concentrations of ^{16}O (light oxygen) indicate that the area draining into the sea experienced heavy rains when the corals were forming. (Recall that rainwater contains a higher-than-average amount of light oxygen.)

^{18}O isotope (**Figure 21.6A**). Thus, water that evaporates from the sea and falls as rain or snow is enriched in ^{16}O compared to ocean water.

During a warm climate interval, water that falls on land returns quickly to the ocean in rivers, bringing back its ^{16}O (see **Figure 21.6B**). However, consider what happens as ice sheets form during a glacial interval. As they grow, they lock away huge amounts of ^{16}O-enriched water, leaving the oceans progressively richer in ^{18}O (see Figure 21.6A). In other words, the $^{18}O/^{16}O$ ratio in seawater serves as a proxy for global temperature: The *higher* the $^{18}O/^{16}O$ ratio, the *cooler* the climate, and the *lower* the ratio, the *warmer* the climate.

To obtain this information, climate researchers carry out **oxygen-isotope analysis** on the fossilized shells of foraminifera in sediment cores. By

Wider rings indicate good growing conditions

Narrower rings indicate harsh growing conditions

A.

▶ **Figure 21.7**
Tree rings
A. These rings are useful records of past climate because the amount of growth (the thickness of a ring) depends on precipitation and temperature.
B. Scientists are not limited to trees that have been cut down. Small, non-destructive core samples can be taken from living trees.

B.

Tree Rings

Tree trunks grow in thickness by laying down new wood under the bark. In temperate regions, where trees grow faster in summer than in winter, each year's new wood forms a visually distinct layer called a *tree ring* (**Figure 21.7A**). Tree rings are also the rule in places that have wet and dry seasons. Because one ring is laid down each year, you can tell the age of a tree by counting its rings. Scientists can do this by taking a small, nondestructive core sample from the trunk, as shown in **Figure 21.7B**.

The width, density, and other characteristics of each tree ring reflect the environmental conditions (especially temperature and precipitation) that prevailed when it formed. Favorable growth conditions produce a wide ring; unfavorable ones produce a narrow ring. Trees that grow at the same time in the same region show similar tree-ring patterns. Therefore, tree rings constitute a sensitive proxy for local climatic conditions. The dating and study of tree rings is called **dendrochronology**.

To make the most effective use of tree rings, dendrochronologists establish extended *ring chronologies* by comparing the patterns of rings among trees of various ages in an area. For instance, a particular pattern of broad and narrow rings that can be dated to, say, the 1830s in living trees from a given mountain range might be identified in much older logs from log cabins built in the 1850s, allowing the rings in those logs to be dated precisely. Tree-ring chronologies extending back for thousands of years have been established for some regions. Thus, tree rings can be used to reconstruct climate variations within a region for spans of thousands of years prior to human historical records.

Fossil Pollen

Climate is a major factor influencing the distribution of vegetation, so the nature of the plant community occupying an area at some past time serves as its climate proxy. Pollen and plant spores have very resistant cell walls, so these are often the most abundant, easily identifiable, and best-preserved plant fossils in sediments. Analyzing pollen from accurately dated sediments makes it possible to obtain high-resolution records of vegetation changes in an area. If the pollen recovered is indicative of an arid ecosystem, we can be fairly certain that when this vegetation was alive, the area was dry. Past climates can be reconstructed from such information.

CONCEPT CHECKS 21.2

1. What are proxy data, and why are they necessary in the study of climate change?

2. Explain how past temperature and precipitation characteristics are determined by using oxygen-isotope analysis.

Concept Checker
https://goo.gl/Y8Xu3w

3. Describe how each of the following are useful in the study of past climates: seafloor sediment, glacial ice, coral, tree rings, fossil pollen, and oxygen-isotope analysis.

21.3 Some Atmospheric Basics

Describe the composition of the atmosphere and the atmosphere's vertical changes in pressure and temperature.

To better understand climate change, it is helpful to possess some basic knowledge about the composition and structure of the atmosphere.

Composition of the Atmosphere

Air is *not* a unique element or compound. Rather, air is a *mixture* of many discrete gases, each with its own physical properties, in which varying quantities of tiny solid and liquid particles are suspended.

Clean, Dry Air   As you can see in **Figure 21.8**, clean, dry air is composed almost entirely of two gases—about 78 percent nitrogen and 21 percent oxygen. Although these gases are the most plentiful components of air and are of great significance to life on Earth, they are of little or no importance in affecting weather phenomena. The remaining 1 percent of dry air is mostly the inert gas argon (0.93 percent) plus tiny quantities of a number of other gases. Carbon dioxide, although present in only minute amounts (0.0405 percent, or 405 parts per

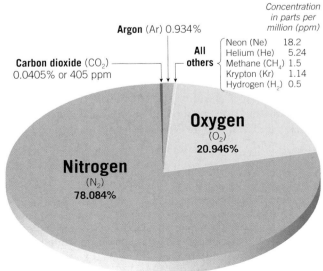

Argon (Ar) 0.934%

All others
Neon (Ne) 18.2
Helium (He) 5.24
Methane (CH_4) 1.5
Krypton (Kr) 1.14
Hydrogen (H_2) 0.5

Concentration in parts per million (ppm)

Carbon dioxide (CO_2) 0.0405% or 405 ppm

Oxygen (O_2) 20.946%

Nitrogen (N_2) 78.084%

◄ SmartFigure 21.8 **Composition of the atmosphere** Proportional volume of gases composing dry air. Nitrogen and oxygen obviously dominate.

Tutorial
https://goo.gl/vUW9ht

million), is nevertheless an important constituent of air because it has the ability to absorb heat energy radiated by Earth and thus influences the heating of the atmosphere.

Air includes many gases and particles that vary significantly from time to time and from place to place. Important examples of these variable gases include water vapor, ozone, and tiny solid and liquid particles.

Water Vapor   The amount of *water vapor* in the air varies considerably, from practically none at all up to about 4 percent by volume. Why is such a small fraction of the atmosphere so significant? Certainly the fact that water vapor is the source of all clouds and precipitation makes it important. However, water vapor has other roles. Like carbon dioxide, it has the ability to absorb heat energy given off by Earth, as well as some incoming solar energy. It is, therefore, important when we examine the heating of the atmosphere.

Ozone   Another important component of the atmosphere is *ozone*. It is a form of oxygen that combines three oxygen atoms into each molecule (O_3). Ozone is not the same as the oxygen we breathe, which has two atoms per molecule (O_2). There is very little ozone in the atmosphere, and its distribution is not uniform. It is concentrated well above Earth's surface in a layer called the *stratosphere*, at an altitude of between 10 and 50 kilometers (6 and 31 miles). The presence of the ozone layer in our atmosphere is crucial to those who dwell on Earth. The reason is that ozone absorbs the potentially harmful ultraviolet (UV) radiation from the Sun. If ozone did not filter out a great deal of the ultraviolet radiation—allowing the Sun's UV rays to reach the surface of Earth undiminished—our planet would be uninhabitable for most life as we know it.

Aerosols   The movements of the atmosphere are sufficient to keep a large quantity of solid and liquid particles suspended within it. Although visible dust sometimes clouds the sky, these relatively large particles are too heavy to stay in the air for very long. Still, many particles are microscopic and remain suspended for considerable periods of time. They may originate from many sources, both natural and human made, and include sea salts from breaking waves, fine soil blown into the air, smoke and soot from fires, pollen and microorganisms lifted by the wind, ash and dust from volcanic eruptions, and

more. Collectively, these tiny solid and liquid particles are called **aerosols**.

From a meteorological standpoint, these tiny, often invisible particles can be significant. First, many act as surfaces on which water vapor can condense, an important function in the formation of clouds and fog. Second, aerosols can absorb or reflect incoming solar radiation. Thus, when an air pollution episode is occurring, or when ash fills the sky following a volcanic eruption, the amount of sunlight reaching Earth's surface can be measurably reduced (**Figure 21.9**).

Extent and Structure of the Atmosphere

To say that the atmosphere begins at Earth's surface and extends upward is obvious. But where does the atmosphere end, and where does outer space begin? There is no sharp boundary; the atmosphere rapidly thins as you travel away from Earth, until there are too few gas molecules to detect.

Pressure Changes with Height   To understand the vertical extent of the atmosphere, let us examine changes in atmospheric pressure with height. Atmospheric pressure is simply the weight of the air above. At sea level, the average pressure is slightly more than 1000 millibars. This corresponds to a weight of slightly more than 1 kilogram per square centimeter (14.7 pounds per square inch). Obviously, the pressure at higher altitudes is lower (**Figure 21.10**).

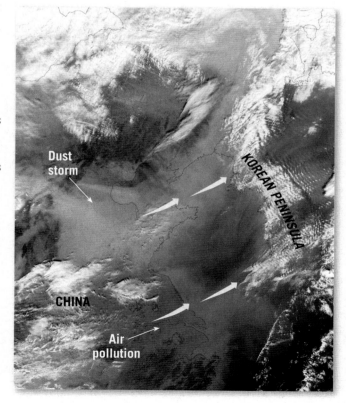

▶ **SmartFigure 21.9**
Aerosols This satellite image shows two examples of aerosols. First, a large dust storm is moving across northeastern China toward the Korean Peninsula. Second, a dense haze toward the south (bottom center) is human-generated air pollution.

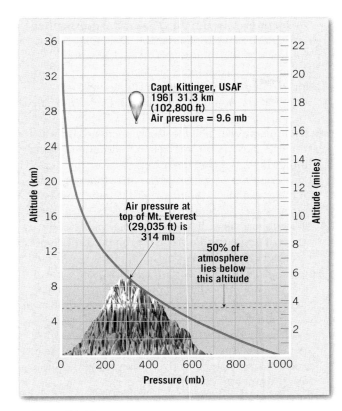

▲ **Figure 21.10**
Vertical changes in air pressure Pressure decreases rapidly near Earth's surface and more gradually at greater heights. Put another way, the graph shows that the vast bulk of the gases making up the atmosphere is near Earth's surface, and the gases gradually merge with the emptiness of space.

One-half of the atmosphere lies below an altitude of 5.6 kilometers (3.5 miles). At about 16 kilometers (10 miles), 90 percent of the atmosphere has been traversed, and above 100 kilometers (62 miles), only 0.00003 percent of all the gases making up the atmosphere remains. Even so, traces of our atmosphere extend far beyond this altitude, gradually merging with the emptiness of space.

Temperature Changes   In addition to vertical changes in air pressure, there are also changes in air temperature as we ascend through the atmosphere. Earth's atmosphere is divided vertically into four layers, on the basis of temperature (**Figure 21.11**):

- **Troposphere.** We live in the bottom layer, which is characterized by a decrease in temperature with increasing altitude and is called the **troposphere**. The term literally means the region where air "turns over," a reference to the appreciable vertical mixing of air in this lowermost zone. The troposphere is the chief focus of meteorologists because it is in this layer that essentially all important weather phenomena occur.
  The temperature decrease in the troposphere is called the *environmental lapse rate*. Its average

value is 6.5°C per kilometer (3.5°F per 1000 feet), a figure known as the *normal lapse rate*. It should be emphasized, however, that the environmental lapse rate is not a constant; it changes over time and from one place to another, and it must be measured regularly. To determine the environmental lapse rate at a particular time and place, as well as to gather information about vertical changes in pressure, wind, and humidity, radiosondes are used. A **radiosonde** is an instrument package that is attached to a weather balloon and transmits data by radio as it ascends through the atmosphere (**Figure 21.12**).
  The thickness of the troposphere is not the same everywhere; it varies with latitude and season. On the average, the temperature drop continues to a height of about 12 kilometers (7.4 miles). The outer boundary of the troposphere is the *tropopause*.

- **Stratosphere.** Beyond the tropopause is the **stratosphere**. In the stratosphere, the temperature remains constant to a height of about 20 kilometers (12 miles) and then begins a gradual increase that continues until the *stratopause*, at a height of nearly 50 kilometers (30 miles) above Earth's surface. Below the tropopause, atmospheric properties such as temperature

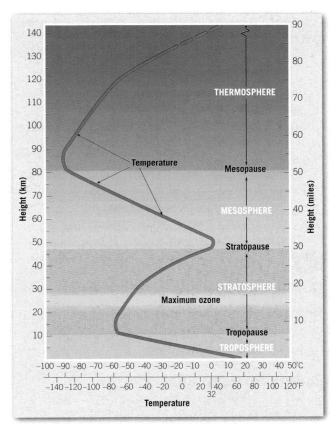

▲ **Figure 21.11**
Thermal structure of the atmosphere Earth's atmosphere is traditionally divided into four layers, based on temperature.

▶ Figure 21.12
Radiosonde A radiosonde is a lightweight package of instruments that is carried aloft by a small weather balloon. It transmits data on vertical changes in temperature, pressure, and humidity in the troposphere. These measurements are very important because the troposphere is where practically all weather phenomena occur; therefore, it is very important to have frequent measurements.

Radar unit for tracking radiosonde

Weather balloon

Radiosonde (instrument package)

• **Mesosphere.** In the third layer, the **mesosphere**, temperatures again decrease with height until, at the *mesopause*, about 80 kilometers (50 miles) above Earth's surface, the temperature approaches −90°C (−130°F). The coldest temperatures anywhere in the atmosphere occur at the mesopause.

• **Thermosphere.** The fourth layer extends outward from the mesopause and has no well-defined upper limit. This is the **thermosphere**, a layer that contains only a *tiny fraction* of the atmosphere's mass. In the extremely rarefied air of this outermost layer, temperatures again increase, due to the absorption of very short-wave, high-energy solar radiation by atoms of oxygen and nitrogen. Temperatures rise to extremely high values of more than 1000°C (1800°F) in the thermosphere. But such temperatures are not comparable to those experienced near Earth's surface. Temperature is defined in terms of the average speed at which molecules move. Because the gases of the thermosphere are moving at very high speeds, the temperature is very high. But the gases are so sparse that, collectively, they process only an insignificant amount of heat.

CONCEPT CHECKS 21.3

1. What are the major components of clean, dry air? List two significant components of the atmosphere that vary from place to place and time to time.

2. Describe how air pressure changes with an increase in altitude. Does it change at a constant rate?

    Concept Checker https://goo.gl/k8CH1F

3. The atmosphere is divided vertically into four layers, on the basis of temperature. Name the layers from bottom to top and indicate how temperatures change in each.

---

and humidity are readily transferred by large-scale turbulence and mixing. Above the tropopause, in the stratosphere, they are not. Temperatures increase in the stratosphere because it is in this layer that the atmosphere's ozone is concentrated. Recall that ozone absorbs ultraviolet radiation from the Sun. As a consequence, the stratosphere is heated.

21.4 Heating the Atmosphere

Outline the basic processes involved in heating the atmosphere.

Before examining various causes of climate change, both those that are natural and those related to human activities, it is useful to have a basic understanding of the processes responsible for heating Earth's atmosphere.

Energy from the Sun

Nearly all the energy that drives Earth's variable weather and climate comes from the Sun. From our everyday experience, we know that the Sun emits light and heat as well as the ultraviolet rays that cause suntan. Although these forms of energy comprise a major portion of the total energy that radiates from the Sun, they are only part of a large spectrum of energy called *radiation*, or *electromagnetic radiation*. The spectrum of electromagnetic energy is shown in **Figure 21.13**. All radiation—whether X-rays, microwaves, or radio waves—transmits energy through the vacuum of space at 300,000 kilometers (186,000 miles) per second and only slightly more slowly through our atmosphere. When an object absorbs any form of radiant energy, the result is an increase in molecular motion, which causes a corresponding increase in temperature.

To better understand how the atmosphere is heated, it is useful to have a general understanding of the basic laws governing radiation:

- **All objects, at any temperature, emit radiant energy.** All objects—not only hot objects like the Sun but also Earth, including its polar ice caps—continually emit energy.
- **Hotter objects radiate more total energy per unit area than do colder objects.**
- **The hotter the radiating object, the shorter the wavelength of maximum radiation.** The Sun, with a surface temperature of about 5700°C, radiates maximum energy at 0.5 micrometer, which is in the visible range. Earth's radiation peaks at a wavelength of 10 micrometers, well within the infrared (heat) range. Because the peak wavelength for Earth radiation is roughly 20 times longer than that for the Sun, Earth radiation is often called *long-wave radiation*, and solar radiation is called *short-wave radiation*.
- **Objects that are good absorbers of radiation are good emitters as well.** Earth's surface and the Sun approach being perfect radiators because they absorb and radiate with nearly 100 percent efficiency for their respective temperatures. On the other hand, *gases are selective absorbers and emitters of radiation*. For some wavelengths, the atmosphere is nearly transparent (that is, little radiation is absorbed). For other wavelengths, however, the atmosphere is nearly opaque (that is, it is a good absorber). Experience tells us that the atmosphere is transparent to visible light; hence, these wavelengths readily reach Earth's surface. This is not the case for the longer-wavelength radiation emitted by Earth.

The Paths of Incoming Solar Energy

Figure 21.14 shows the paths taken by incoming solar radiation averaged for the entire globe. Notice that the atmosphere is quite transparent to incoming solar radiation. On average, about 50 percent of the solar energy reaching the top of the atmosphere passes through the atmosphere and is absorbed at Earth's surface. Another 20 percent is absorbed directly by clouds and certain atmospheric gases (including oxygen and ozone) before reaching the surface. The remaining 30 percent is reflected back to space by the atmosphere, clouds, and reflective surfaces such as snow and ice. The fraction of the total radiation that is reflected by an object is called

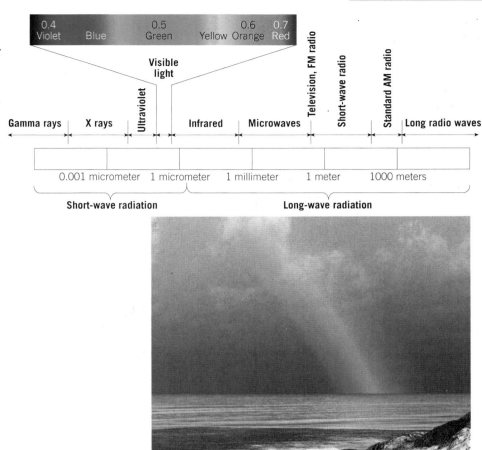

▲ SmartFigure 21.13
The electromagnetic spectrum This diagram illustrates the wavelengths and names of various types of radiation. Visible light consists of the array we commonly call the "colors of the rainbow."

Video
https://goo.gl/aBy7Yu

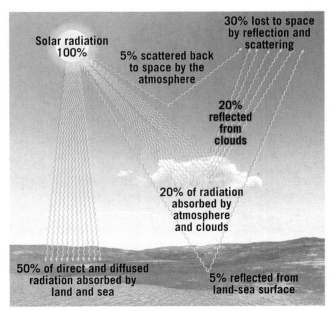

◀ SmartFigure 21.14
Paths taken by solar radiation This diagram shows the average fate of solar radiation that reaches Earth. More solar radiation is absorbed by Earth's surface than by the atmosphere.

Tutorial
https://goo.gl/VngXxB

energy being transmitted, as well as on the nature of the intervening material.

The numbers shown in Figure 21.14 represent global averages. The actual percentages can vary greatly, primarily due to changes in the percentage of light reflected and scattered back to space. For example, if the sky is overcast, a higher percentage of light is reflected back to space than when the sky is clear.

The Greenhouse Effect

If Earth had no atmosphere, it would experience an average surface temperature far below freezing. But the atmosphere warms the planet and makes Earth livable. The extremely important role the atmosphere plays in heating Earth's surface has been named the **greenhouse effect**.

As discussed earlier, cloudless air is largely transparent to incoming short-wave solar radiation and, hence, transmits it to Earth's surface. By contrast, a significant fraction of the long-wave radiation emitted by Earth's land–sea surface is absorbed by water vapor, carbon dioxide, and other trace gases in the atmosphere. This energy heats the air and increases the rate at which it radiates energy, both out to space and back toward Earth's surface. Without this complicated game of "pass the hot potato," Earth's average temperature would be $-18°C$ ($-0.4°F$) rather than the current average temperature of $15°C$ ($59°F$) (**Figure 21.16**). These absorptive

its **albedo** (**Figure 21.15**). Thus, the albedo for Earth as a whole (the *planetary albedo*) is 30 percent. The atmosphere also *scatters* some of the incoming radiation— that is, reflects it in random directions.

What determines whether solar radiation will be transmitted to the surface, scattered, or reflected outward? It depends greatly on the wavelength of the

Airless bodies like the Moon All incoming solar radiation reaches the surface. Some is reflected back to space. The rest is absorbed by the surface and radiated directly back to space. As a result the lunar surface has a much lower average surface temperature than Earth.

Bodies with modest amounts of greenhouse gases like Earth The atmosphere absorbs some of the longwave radiation emitted by the surface. A portion of this energy is radiated back to the surface and is responsible for keeping Earth's surface 33°C (59°F) warmer than it would otherwise be.

Bodies with abundant greenhouse gases like Venus Venus experiences extraordinary greenhouse warming, which is estimated to raise its surface temperature by 523°C (941°F).

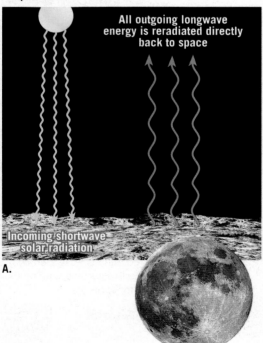

A.

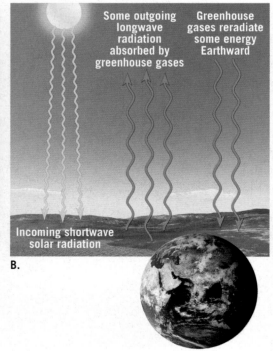

B.

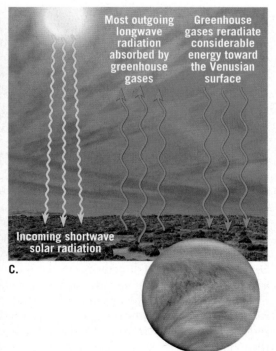

C.

gases in our atmosphere make Earth habitable for humans and other life-forms. As you will see in the sections that follow, changes in the air's composition, both natural and human caused, impact the greenhouse effect in ways that can cause the atmosphere to become either warmer or cooler.

This natural phenomenon was named the greenhouse effect because it was once thought that greenhouses are heated in a similar manner. The glass in a greenhouse allows short-wave solar radiation to enter and be absorbed by the objects inside. These objects, in turn, radiate energy but at longer wavelengths, to which glass is nearly opaque. The heat, therefore, is "trapped" in the greenhouse. It has been shown, however, that air inside greenhouses becomes warmer than outside air mainly because greenhouses restrict the exchange of air between the inside and outside. Nevertheless, the term *greenhouse effect* remains.

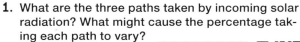

CONCEPT CHECKS 21.4

1. What are the three paths taken by incoming solar radiation? What might cause the percentage taking each path to vary?

2. Explain why the atmosphere is heated chiefly by radiation from Earth's surface.

 Concept Checker
https://goo.gl/3Y5kP6

3. Prepare a sketch with labels that explains the greenhouse effect.

21.5 Natural Causes of Climate Change

Discuss hypotheses that relate to natural causes of climate change.

In this section we examine phenomena that can drive climate change but are *unrelated* to human activities. These phenomena include plate tectonics, variations in Earth's orbit, volcanic activity, and solar variability. More than one of these mechanisms may interact to change Earth's climate. Also, no single mechanism explains climate change on all timescales. A mechanism that changes climate over millions of years generally does not explain fluctuations that take place over hundreds of years and vice versa.

Plate Movements and Orbital Variations

In Chapter 18, the section "Causes of Ice Ages" describes two natural mechanisms of climate change. Recall that the movement of lithospheric plates gradually shifts Earth's continents closer to or farther from the equator. Although these plate movements are very slow, they have a dramatic impact on climate over spans of millions of years. In addition to the effects discussed in Chapter 18, plate tectonics can lead to significant shifts in ocean circulation, influencing the transport of heat around the globe.[†]

A second natural mechanism of climate change related to the causes of ice ages involves variations in Earth's orbit. Changes in the shape of the orbit (*eccentricity*), variations in the angle that Earth's axis makes with the plane of its orbit (*obliquity*), and the wobbling of the axis (*precession*) cause fluctuations in the seasonal and latitudinal distribution of solar radiation (see Figure 18.39, page 529). These variations contributed to the alternating glacial–interglacial episodes of the Ice Age.

Volcanic Activity and Climate Change

Volcanic eruptions also account for some aspects of climate variability. Explosive eruptions emit huge quantities of gases and fine-grained debris into the atmosphere (**Figure 21.17**). The largest eruptions are sufficiently powerful to inject material high into the atmosphere, where it spreads around the globe and remains suspended for many months or even years.

◄ Figure 21.17
Eruption of volcanic ash and gases from Mount Bromo, 2011, on the island of Java in Indonesia

[†] For more on this, see the section "Supercontinents, Mountain Building, and Climate" in Chapter 22.

► Figure 21.18
Sulfur dioxide from Hawaii's Kilauea Volcano
The volcano's ongoing eruption intensified in spring 2018. When SO$_2$ occurs above about 16 kilometers (10 miles), it is transformed into tiny sulfate particles that create a bright haze that reflects sunlight back to space.

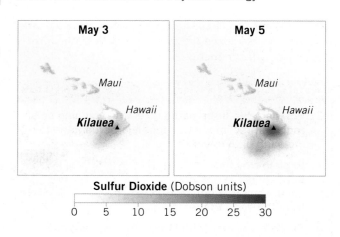

Sulfur Dioxide (Dobson units)

0   5   10   15   20   25   30

This suspended volcanic material is able to reflect or scatter a portion of the incoming solar radiation, which in turn lowers temperatures in the troposphere. Perhaps the most notable cool period linked to a volcanic event is the "year without a summer" that followed the 1815 eruption of Mount Tambora, Indonesia, the largest of modern times. During April 7–12, 1815, this volcano violently expelled more than 100 cubic kilometers (24 cubic miles) of volcanic debris. These tiny particles are believed to have been responsible for widespread cooling in the Northern Hemisphere over the subsequent months. From May through September 1816, for instance, an unprecedented series of cold spells affected the northeastern United States and adjacent portions of Canada. The region experienced heavy snow in June and frost in July and August. Much of Western Europe also experienced abnormal cold.

El Chichón and Mount Pinatubo

Two more recent volcanic events have provided considerable data and insight regarding the impact of volcanoes on global temperatures. The eruptions of El Chichón in Mexico in 1982 and Mount Pinatubo in the Philippines in 1991 gave scientists opportunities to study the atmospheric effects of volcanic eruptions with the aid of sophisticated technology. Satellite images and remote-sensing instruments allowed scientists to closely monitor the effects of the clouds of gases and ash that these volcanoes emitted.

Two years of monitoring and studies following the 1982 El Chichón eruption indicated that it had a cooling effect on global mean temperature, on the order of about 0.5°C (0.9°F). The historic eruption, although not particularly explosive, emitted huge quantities of sulfur dioxide high into the atmosphere. This gas combines with water vapor in the atmosphere to produce a dense cloud of tiny sulfuric acid particles, as shown for Kilauea in **Figure 21.18**. These particles take several years to settle out completely. Like volcanic ash, tiny sulfuric acid droplets lower the mean surface temperature because they reflect solar radiation back to space (**Figure 21.19**).

We now understand that volcanic debris suspended in the atmosphere for a year or more is composed largely of sulfuric acid droplets and not of volcanic ash. Thus, the volume of volcanic ash emitted during an explosive event is not an accurate criterion for predicting the global atmospheric effects of an eruption.

The Impact of Volcanic Eruptions

The impact on climate of a single volcanic eruption, as just described for El Chichón, is relatively small and short-lived. The graph in Figure 21.19 reinforces this point. If volcanism is to have a pronounced impact over an extended period, either a much larger volcanic eruption or many eruptions rich in sulfur dioxide and closely spaced in time must occur. Events like these could load the atmosphere with enough gases and volcanic ash to significantly diminish the amount of solar radiation reaching the surface.

Although no such period of explosive volcanism has occurred in historic times, we know that eruptions on this scale have occurred at various times in the geologic past. The eruption that produced the

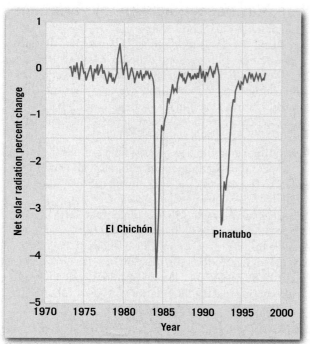

Net solar radiation at Hawaii's Mauna Loa Observatory relative to 1970 (zero on the graph).

▲ Figure 21.19
Volcanic debris reducing sunlight at Earth's surface
The eruptions of El Chichón and Mount Pinatubo clearly caused temporary drops in solar radiation reaching Earth's surface.

landscape of Yellowstone National Park 630,000 years ago emitted more than 100 times as much volcanic ash and gas as the eruption of Mount Pinatubo in 1991. This catastrophic eruption sent showers of debris as far as the Gulf of Mexico. Eruptions on the scale of the Yellowstone volcanism and other massive events would have clearly had a marked impact on past climate.

Volcanism and Global Warming

The Cretaceous period is the last period of the Mesozoic era, the era of *middle life* that is often called the "age of dinosaurs." It began about 145 million years ago and ended about 65 million years ago, with the extinction of the dinosaurs and many other life-forms.[‡]

The Cretaceous climate was among the warmest in Earth's long history. Dinosaur fossils of that period have been discovered north of the Arctic Circle. Tropical forests existed in Greenland and Antarctica, and deposits of peat that would eventually form widespread coal beds accumulated at high latitudes. Sea level was as much as 200 meters (650 feet) higher than it is today, consistent with a lack of polar ice sheets.

What was the cause of the unusually warm climates of the Cretaceous period? Among the factors thought to have contributed was an enhanced greenhouse effect due to an increase in the amount of carbon dioxide in the atmosphere. Geologists have concluded that the probable source of the CO_2 was volcanic activity. Carbon dioxide is one of the gases emitted during volcanism, and there is now considerable geologic evidence that the Middle Cretaceous was a time when there was an unusually high rate of volcanic activity. Several huge oceanic lava plateaus were produced on the floor of the western Pacific during this span. These vast features were associated with hot spots that may have been produced by large mantle plumes. Massive outpourings of lava over millions of years would have been accompanied by the release of huge quantities of CO_2, which in turn would have enhanced the atmospheric greenhouse effect. *Thus, the warmth that characterized the Cretaceous may have had its origins deep in Earth's mantle.*

This example illustrates the interrelationships among parts of the Earth system. Seemingly unrelated materials and processes turn out to be linked. Here you have seen how processes originating deep in Earth's interior are connected directly or indirectly to the atmosphere, the oceans, and the biosphere.

Solar Variability and Climate

Researchers have asked whether any changes in Earth's temperature can be ascribed to changes in the Sun's output of energy. An increase in solar output would be expected to warm Earth, and a decrease in

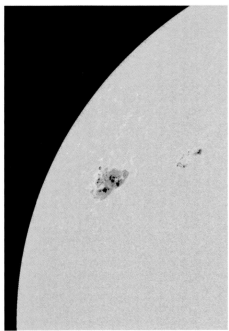

A. The black spots surrounded by deep orange is a sunspot region where magnetic activity is extremely intense.

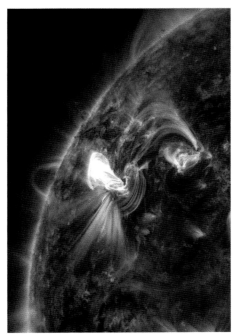

B. The looping lines connect sunspots, a type of solar storm that enhances solar output.

▲ SmartFigure 21.20 **Sunspots** Both images show sunspot activity at the same location on the solar disk at the same time on March 5, 2012, using two different instruments from NASA's Solar Dynamics Observatory.

Video
https://goo.gl/
6HuYuu

output would be expected to cool Earth. Actual measurement of solar radiation outside the atmosphere has only been possible since satellite technology became available. Within this era, the only detectable changes in solar output have been related to the sunspot cycle.

Sunspots are huge magnetic storms that extend from the Sun's surface deep into its interior (**Figure 21.20**). Sunspots occur in cycles, and the number of sunspots reaches a maximum about every 11 years. The Sun emits slightly more energy during periods of maximum sunspot activity than during sunspot minimums. (This may seem counterintuitive, as sunspots are dark. However, each sunspot is surrounded by a brighter area that more than makes up for it.) Sunspots can also be associated with the ejection of huge clouds of charged particles; when these clouds encounter Earth, their interaction with the upper atmosphere produces intense auroral displays.

Based on measurements from space that began in 1978, the variation in solar output over an 11-year sunspot cycle is about 0.1 percent. Scientists have concluded that this change is too small, and the cycles are too short, to have any appreciable effect on global temperatures.

However, it remains possible that longer-term variations in solar output occur and could affect Earth's climate. One candidate is a period between 1645 and 1715 during which sunspots were largely

[‡] For more about the end of the Cretaceous, see Chapter 22.

absent. This interval of missing sunspots, called the *Maunder minimum*, corresponds closely with an especially cold period in Europe known as the *Little Ice Age*. For some scientists, this correlation suggests that a reduction in the Sun's output was responsible at least in part for this cold episode. Other scientists seriously question this notion. Their hesitation stems in part from subsequent investigations using different climate records from around the world that failed to find a significant correlation between sunspot activity and climate.

21.6 Human Impact on Global Climate

Summarize the nature and cause of the atmosphere's changing composition since about 1750. Describe the climate's response.

Having examined *natural causes* of climate change, we now shift our focus to how *humans* contribute to global climate change. Human influence on regional and global climate did not begin with the onset of the modern industrial age. There is evidence that people have been modifying the environment over extensive areas for thousands of years. The use of fire and the overgrazing of marginal lands by domesticated animals have both reduced the abundance and distribution of vegetation. By altering ground cover, humans have modified such important climate factors as surface albedo and evaporation rates.

However, the most significant impact humans have had on climate is the recent addition of carbon dioxide and other greenhouse gases into the atmosphere. A secondary impact is the addition of human-generated aerosols to the atmosphere.

Rising Carbon Dioxide Levels

Earlier you learned that carbon dioxide (CO_2) represents about 0.0410 percent (410 parts per million) of clean, dry air. Nevertheless, carbon dioxide is influential because it is transparent to incoming short-wavelength solar radiation but not to some of the longer-wavelength, outgoing Earth radiation. A portion of the energy leaving Earth's surface is absorbed by atmospheric CO_2. This energy is subsequently re-emitted, part of it back toward the surface, thereby keeping the air near the ground warmer than it would be without CO_2. Thus, along with water

vapor, carbon dioxide is largely responsible for the atmosphere's greenhouse effect.

The world's industrialization over the past two centuries has been fueled mainly by burning fossil fuels: coal, natural gas, and petroleum (**Figure 21.21**). Combustion of these fuels has added great quantities of carbon dioxide to the atmosphere. **Figure 21.22** shows changes in CO_2 concentrations at Hawaii's Mauna Loa Observatory, where measurements have been made since 1958. The graph shows an annual seasonal cycle and a steady upward trend over the years.

The use of fossil fuels is the most prominent means by which humans add CO_2 to the atmosphere, but it is not the only way. The clearing of forests also contributes substantially because CO_2 is released as vegetation is burned or decays (**Figure 21.23**). Deforestation is particularly pronounced in the tropics, where vast tracts are cleared for ranching and agriculture or subjected to inefficient commercial logging operations. All major tropical forests—including those in South America, Africa, Southeast Asia, and Indonesia—are disappearing. According to United Nations estimates, more than 10 million hectares (25 million acres) of tropical forest were permanently destroyed *each year* during the decades of the 1990s and 2000s.

Some of the CO_2 that has been released by humans is taken up by plants or is dissolved in the ocean, but an estimated 45 percent remains in the atmosphere. A graphic record of changes in atmospheric CO_2 extending back 800,000 years was

▶ **Figure 21.21**
U.S. energy consumption The graph shows energy consumption in 2017. The total was 97.7 quadrillion Btu. A quadrillion is 10 raised to the 15th power, or a billion million. The burning of fossil fuels represents about 80 percent of the total. (Data from U.S. Energy Information Administration)

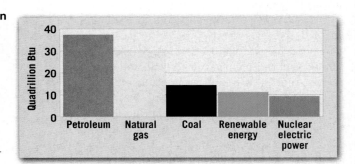

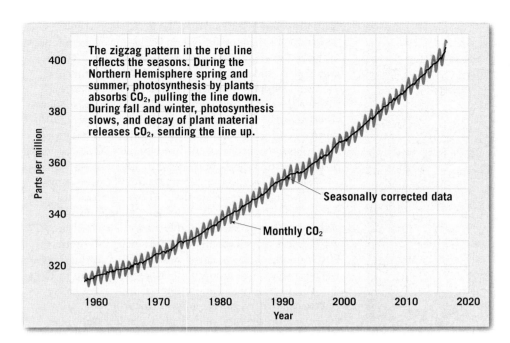

The zigzag pattern in the red line reflects the seasons. During the Northern Hemisphere spring and summer, photosynthesis by plants absorbs CO₂, pulling the line down. During fall and winter, photosynthesis slows, and decay of plant material releases CO₂, sending the line up.

Seasonally corrected data

Monthly CO₂

◄ **SmartFigure 21.22**
Monthly CO₂ concentrations Atmospheric CO_2 has been measured at Mauna Loa Observatory, Hawaii, since 1958. There has been a consistent increase since monitoring began. This graphic portrayal is known as the *Keeling Curve*, in honor of the scientist who originated the measurements.

Tutorial
https://goo.gl/fXDPZL

provided earlier, in Figure 21.4B. Over this long span, atmospheric CO_2 has fluctuated naturally from about 180 to 300 ppm. However, as of 2017 the level was about 40 percent higher than the levels prior to the Industrial Revolution. Even more alarming is that over the past several decades, there has been an acceleration of the annual rate at which atmospheric carbon dioxide concentrations have increased.

The Role of Trace Gases

Carbon dioxide is clearly the most important contributor to human-accentuated greenhouse warming. However, atmospheric scientists have come to realize that human industrial and agricultural activities have increased the concentration of several trace gases that also play significant roles. The substances are called **trace gases** because their concentrations are much lower than the concentration of carbon dioxide. The most important trace gases are methane (CH_4) and nitrous oxide (N_2O). Like carbon dioxide and water vapor, these are "greenhouse gases" that absorb wavelengths of outgoing radiation from Earth that would otherwise escape into space. Although individually their impact is modest, taken together they play a significant role in warming the atmosphere.

Methane   Although methane is present in much smaller amounts than CO_2, it matters because it is about 20 times more effective than CO_2 at absorbing infrared radiation emitted by Earth.

Methane is produced by anaerobic bacteria in wet places where oxygen is scarce. (*Anaerobic* means "without air," referring specifically to oxygen.) Such places include swamps, bogs, wetlands, and the guts of termites

and grazing animals such as cattle and sheep. Methane is also generated in flooded paddy fields used for growing rice (**Figure 21.24**).

The concentration of methane in the atmosphere has increased in sync with the growth in human population. This relationship reflects the close link between methane production and agriculture. As population increases, so do the numbers of cattle and rice paddies. Mining for coal and drilling for oil and natural gas are other sources because methane is released during these activities.

▲ **Figure 21.23**
Tropical deforestation Clearing the tropical rain forest is a serious environmental issue. In addition to causing a loss of biodiversity, deforestation is a significant source of carbon dioxide. Fires are frequently used to clear the land. This scene is in Brazil's Amazon basin.

▶ **Figure 21.24**
Methane The flooded paddy fields used to grow rice represent a source of methane—one that has been growing with the human population for thousands of years.

Intergovernmental Panel on Climate Change (IPCC), "Warming of the climate system is unequivocal, and many of the observed changes are unprecedented. . . . The atmosphere and ocean have warmed, the amounts of snow and ice have diminished, and sea level has risen."[§] Most of the observed increase in global average temperatures since the mid-twentieth century is *extremely likely* due to the observed increase in human-generated greenhouse gas concentrations. As used by the IPCC, *extremely likely* indicates a probability of 95 to 100 percent.

Globally, average temperatures in 2017 were nearly 1.0°C (1.8°F) warmer than the mid-twentieth century mean. This upward trend in surface temperatures is shown in **Figure 21.25**. The graph shows that 17 of the warmest 18 years have all occurred since 2001, with 2016 ranking as the warmest year and 2017 as the second warmest year since modern recordkeeping began in 1880.

Weather patterns and other natural cycles cause fluctuations in average temperatures from year to year, especially on regional and local levels. For example, while the globe experienced record high temperatures in 2016, the continental United States, Iceland, and parts of China experienced unusually cold January weather. Yet it was a record warm year for North America and the third-warmest year on record for Asia. Regardless of regional and seasonal differences in any year, increases in greenhouse gas levels are causing a long-term rise in global temperatures (**Figure 21.26A**). Although each calendar year will not necessarily be warmer than the one before, scientists expect each decade to be warmer than the previous one. An examination of the decade-by-decade temperature trend in **Figure 21.26B** bears this out.

Nitrous Oxide   Nitrous oxide, sometimes called "laughing gas," is also building up in the atmosphere, although not as rapidly as methane. The increase is primarily a result of agricultural activity. When farmers use nitrogen fertilizers to boost crop yield, some of the nitrogen enters the air as nitrous oxide. This gas is also produced by high-temperature combustion of fossil fuels. Although the annual release into the atmosphere is small, the lifetime of a nitrous oxide molecule in the atmosphere is about 150 years! If nitrogen fertilizer and fossil fuel use grow at projected rates, nitrous oxide's contribution to greenhouse warming may approach half that of methane.

The Atmosphere's Response

Given the increase in the atmosphere's greenhouse gas content, have global temperatures actually increased? The answer is "yes." According to a 2014 report by the

[§]IPCC, "Observed Changes and Their Causes," in *Climate Change 2014 Synthesis Report*. The IPCC is an authoritative group of scientists that provides advice to the world community through periodic reports that assess the state of knowledge of the causes and effects of climate change.

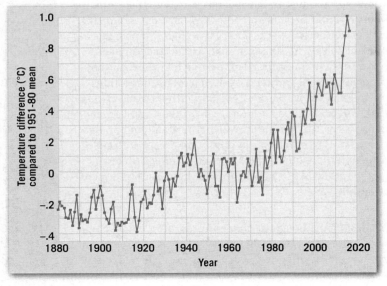

▲ **Figure 21.25**
Global temperatures, 1880–2017 Over the 137-year span from 1880 to 2017, 17 of the 18 warmest years occurred after 2001.

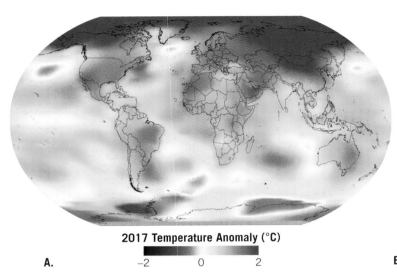

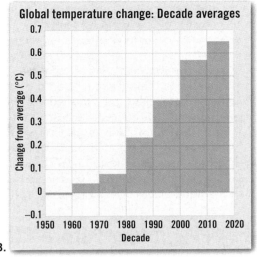

2017 Temperature Anomaly (°C)

−2   0   2

A.

B.

▲ **SmartFigure 21.26**
Global temperature trend A. The world map shows how the average temperatures in 2017 deviated from the mean for the 1951–1980 base period. **B.** The decade-by-decade temperature trend is increasing: Since 1950, each decade has been warmer than the previous one. Continued increases in the atmosphere's greenhouse gas levels are driving a long-term increase in global temperatures.

Video
https://goo.gl/
FgFafg

How Aerosols Influence Climate

Global climate is also affected by human activities that contribute to the atmosphere's aerosol content. Recall that *aerosols* are the tiny, often microscopic, liquid and solid particles that are suspended in the air. Unlike cloud droplets, aerosols are present even in relatively dry air. Natural sources are numerous and include such phenomena as wildfires, dust storms, breaking waves, and volcanoes. Most human-generated aerosols come from burning of vegetation to clear agricultural land and from the sulfur dioxide emitted during the combustion of fossil fuels. Chemical reactions in the atmosphere convert the sulfur dioxide into sulfate aerosols, the same material that produces acid precipitation. The satellite images in **Figure 21.27** provide an example.

How do aerosols affect climate? Most aerosols act directly by reflecting sunlight back to space and indirectly by making clouds "brighter" reflectors. The latter effect relates to the fact that many aerosols (such as those composed of salt or sulfuric acid) attract water and thus are especially effective as cloud condensation nuclei. The large quantity of aerosols produced by human activities (especially industrial emissions) triggers an increase in the number of cloud droplets that form within a cloud. A greater number of small droplets increases the cloud's brightness, causing more sunlight to be reflected back to space.

One category of aerosols, called **black carbon**, is soot generated by combustion processes and fires. Unlike most other aerosols, black carbon warms the atmosphere because it is an effective absorber of

incoming solar radiation. In addition, when deposited on snow and ice, black carbon reduces surface albedo, thus increasing the amount of radiation absorbed. Nevertheless, despite the warming effect of black carbon, the overall effect of atmospheric aerosols is to cool Earth.

Studies indicate that the cooling effect of human-generated aerosols offsets a portion of the global warming caused by the growing quantities of greenhouse gases in the atmosphere. The magnitude and extent of the cooling effect of aerosols are uncertain.

▼ **Figure 21.27**
Human-generated aerosols These satellite images show a serious air pollution episode in China on October 8, 2010.

This satellite image shows the extremely high levels of aerosols associated with this air pollution episode. At an index value of 4, aerosols are so dense that you would have difficulty seeing the midday Sun.

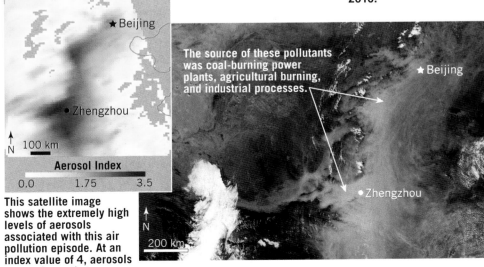

It is important to point out some significant differences between global warming caused by greenhouse gases and aerosol cooling. After being emitted, carbon dioxide and trace gases remain in the atmosphere for many years. By contrast, aerosols released into the troposphere remain there for only a few days or, at most, a few weeks before they are "washed out" by precipitation, limiting their effects. Because of their short lifetime in the troposphere, aerosols are distributed unevenly over the globe. As expected, human-generated aerosols are concentrated near the areas that produce them—namely industrialized regions that burn fossil fuels and places where vegetation is burned.

CONCEPT CHECKS 21.6

1. Why has the **CO₂** level of the atmosphere been increasing over the past 200 years? Aside from **CO₂**, what trace gases are contributing to global temperature change?

2. How have temperatures in the lower atmosphere changed as greenhouse gas levels have increased?

3. List the main sources of human-generated aerosols and describe their net effect on atmospheric temperatures.

🔊 **Concept Checker**
https://goo.gl/fY6CtZ

21.7 Predicting Future Climate Change

Contrast positive- and negative-feedback mechanisms and provide examples of each.

Earth's climate is complex and difficult to predict in part because it contains many *feedback mechanisms*. In a **feedback mechanism**, a change in one part of a system leads to changes in another part of the system, which then feed back on the initial effect in a way that either amplifies or diminishes it. Feedback mechanisms add to the complexity of the climate system and the difficulty of forecasting its behavior.

Types of Feedback Mechanisms

One significant feedback mechanism associated with global warming is that warmer ocean surface temperatures increase the rate of evaporation. This, in turn, increases the amount of water vapor in the atmosphere. Remember that water vapor is a more powerful absorber of radiation emitted by Earth than carbon dioxide. Therefore, with more water vapor in the air, the temperature increase caused by greenhouse gases is reinforced. Because this effect enhances the initial change, it is called a **positive-feedback mechanism**.

Another well-understood positive-feedback mechanism relates to the extent of sea ice and continental ice sheets. Because ice reflects a much larger percentage of incoming solar radiation than open water or land (which are relatively dark), the melting of ice substantially increases the solar energy absorbed at the surface, which leads to higher surface temperatures and therefore even more melting (**Figure 21.28**).

The climate system also contains **negative-feedback mechanisms**, so called because they produce results that oppose the initial change and tend to offset it. For example, the increase in evaporation that results from a rise in temperature can produce an increase in cloud cover because the atmosphere contains more water vapor. Most clouds are good reflectors of incoming solar radiation. Thus, by providing shade, clouds can cool the underlying surface, opposing the increase in temperature that produced them.

The climate system is complex because it contains many interacting feedback mechanisms, some positive and some negative. For instance, while clouds can reflect sunlight, thus tending to cool the surface, they are also good at absorbing and re-emitting the long-wavelength radiation emitted by Earth, thus preventing heat from escaping to space. In this role, they provide positive

▶ SmartFigure 21.28
Sea ice as a feedback mechanism The image shows the springtime breakup of sea ice near Antarctica. The diagram shows a likely feedback loop. A reduction in sea ice acts as a positive-feedback mechanism because surface albedo decreases, and the amount of energy absorbed at the surface increases.

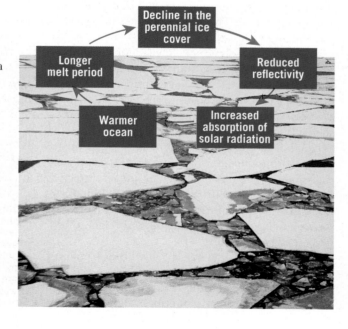

feedback, tending to reinforce rather than reduce surface warming.

Which effect, if either, is stronger? Recent studies suggest that this question is more complex than first thought. Whether clouds exert positive or negative feedback on surface warming depends mainly on the type of clouds. For example, a thin veil of stratus clouds increases albedo more than it absorbs outgoing Earth radiation, so thin stratus clouds tend to have a cooling effect. By comparison, a thick blanket of cumulus cloud absorbs more outgoing Earth radiation than it reflects incoming solar radiation, thus leading to warming. The combined effect that clouds have on climate change isn't fully understood, making it a significant obstacle to modeling future climate change.

Computer Models of Climate: Important yet Imperfect Tools

In many fields of study, hypotheses can be tested with direct experimentation in a laboratory or by field observations and measurements. However, such testing is often not possible in the study of climate. Rather, scientists must construct computer models of how our planet's climate system works. If we understand the climate system correctly and construct the model appropriately, then the behavior of the model climate system should mimic the behavior of Earth's climate system (**Figure 21.29**), allowing scientists to explore possible climate-change scenarios.

Earth's climate system is amazingly complex. Today's computer models incorporate the fundamental laws of physics and chemistry as well as human and biological interactions. Called *General circulation models (GCMs)*, they are used to simulate many variables, including temperature, rainfall, snow cover, soil moisture, winds, clouds, sea ice, and ocean circulation over the entire globe through the seasons and over spans of decades.

What factors influence the accuracy of climate models? Clearly, mathematical models are *simplified* versions of the real Earth and cannot capture its full complexity, especially at smaller geographic scales. Moreover, computer models used to simulate future climate change must make assumptions about important human variables, such as population, economic growth, fossil fuel consumption, technological development, and improvements in energy efficiency. Typically, these models are run for a range of possible scenarios.

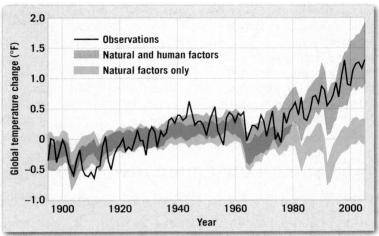

◄ **Figure 21.29**
Separating human and natural influences on climate The blue band shows how global average temperatures would have changed due to natural forces only, as simulated by climate models. The pink band shows model projections of the effects of human and natural forces combined. The black line shows actual observed global average temperatures. As the blue band indicates, without human influences, temperatures over the past century would actually have first warmed and then cooled slightly over recent decades. Bands of color are used to express the range of uncertainty.

Despite many obstacles, our ability to use supercomputers to simulate climate is very good and continues to improve. Although today's models are far from infallible, they are powerful tools for understanding what Earth's future climate might be like.

What do computer models tell us about the future? Projections for the years ahead depend, in part, on the quantities of emitted greenhouse gases. **Figure 21.30** shows the IPCC report's best estimates of global warming for two different emission scenarios. The highest-emission scenario estimates that if the level of carbon dioxide in the atmosphere were to become twice the pre-industrial level—rising from 280 ppm to 560 ppm—the *likely* temperature increase would be about 4.5°C (8.1°F) by the end of the century. Values higher than 4.5°C (8.1°F) are also possible. By contrast, under the lowest-emission scenario, surface air temperatures would rise only about 1.5°C (2.7°F) above current levels. The report goes on to say that a temperature increase of less than 1.5°C (2.7°F) is *very unlikely*.

Sophisticated computer models also show that the warming of the lower atmosphere caused by CO_2 and trace gases will not be the same everywhere. Rather, the

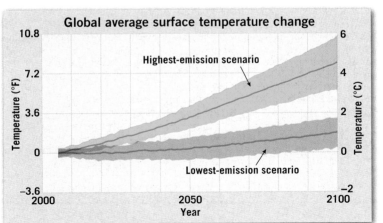

◄ **SmartFigure 21.30**
Projected temperature changes based on two emission scenarios

Animation
https://goo.gl/vD3ePS

temperature response in polar regions could be two to three times greater than the global average. One reason is the fact that the polar atmosphere is very stable, suppressing vertical mixing and thus limiting the amount of surface heat transferred upward. In addition, as described in the preceding section, an expected reduction in sea ice would contribute to the greater temperature increase.

CONCEPT CHECKS 21.7

1. Distinguish between positive- and negative-feedback mechanisms and provide an example of each.

🔊 **Concept Checker**
https://goo.gl/c7yPh3

2. What factors influence the accuracy of computer models of climate?

21.8 Some Consequences of Global Warming

Discuss several likely consequences of global warming.

At present, we can predict the likely broad-scale consequences for various future scenarios. Because the climate system is complex, we cannot yet accurately predict what will happen on a local scale.

As we've said, the magnitude of the temperature increase will not be the same everywhere. The temperature rise will be smallest in the tropics and increase toward the poles. As for precipitation, models indicate that some regions will experience significantly more precipitation and runoff than currently, whereas others will experience a decrease in runoff due to reduced precipitation and/or greater evaporation caused by higher temperatures.

Figure 21.31 lists possible effects of global warming based on the IPCC's projections for the late twenty-first century, ranked in decreasing order of certainty. Probabilities are based on the quality, volume, and consistency of the evidence and the extent of agreement among scientists. Notice that likelihood does not equal risk: Even a low-probability outcome represents a serious risk if its consequences would be severe.

▶ **Figure 21.31 Summary of the likely climate changes and their predicted impacts**

IPCC Projections for the Late Twenty-First Century

Projection	Probability
• Cold days and nights will be warmer and less frequent over most land areas • Hot days and nights will be warmer and more frequent over most land areas • The extent of permafrost will decline • Ocean acidification will increase as the atmosphere accumulates CO_2 • Northern Hemisphere glaciation will not initiate before the year 3000 • Global mean sea level will rise and continue to do so for many centuries	**Virtually certain (99–100%)**
• Arctic sea ice cover will continue to shrink and thin, and Northern Hemisphere spring snow cover will decrease • The dissolved oxygen content of the ocean will decrease by a few percent • The rate of increase in atmospheric CO_2, methane, and nitrous oxide will reach levels unprecedented in the past 10,000 years • The frequency of warm spells and heat waves will increase • The frequency of heavy precipitation events will increase • Precipitation amounts will increase in high latitudes • The ocean's conveyor-belt circulation will weaken • The rate of sea-level rise will exceed that of the late twentieth century • Extreme high sea-level events will increase, as will ocean wave heights of midlatitude storms	**Very likely (90–100%)**
• If the atmospheric CO_2 level stabilizes at double the present level, global temperatures will rise by between 1.5°C (2.7°F) and 4.5°C (8.1°F) • Areas affected by drought will increase • Precipitation amounts will decline in the subtropics • The loss of glaciers will accelerate in the next few decades	**Likely (66–100%)**
• Intense tropical cyclone activity will increase • The West Antarctic Ice Sheet will pass the melting point if global warming exceeds 5°C (9°F)—this is relative, not absolute	**About as likely as not (33–66%)**
• Antarctic and Greenland Ice Sheets will collapse due to surface warming	← **Not likely (0–33%)**

0 10 20 30 40 50 60 70 80 90
Probability (%)

Sea-Level Rise

One important impact of human-induced global warming is a rise in sea level. As this occurs, coastal cities, wetlands, and low-lying islands will be threatened with more frequent flooding. In addition, increased shoreline erosion and saltwater encroachment into coastal rivers and aquifers that supply freshwater to nearby cities are expected.

How is a warmer atmosphere related to a rise in sea level? One significant factor is thermal expansion of the uppermost layer of the global ocean. Higher air temperatures warm the adjacent upper portion of the ocean, which in turn causes the water to expand and sea level to rise.

A second factor contributing to global sea-level rise is melting glaciers (**Figure 21.32**). With few exceptions, glaciers around the world have been retreating at unprecedented rates over the past century. A satellite study spanning 16 years showed that the combined mass of the Greenland and Antarctic Ice Sheets dropped an average of 413 gigatons per year. (A gigaton is 1 billion metric tons.) That is enough water to raise sea level about 1.5 millimeters (0.05 inch) *per year*. The rate of ice loss accelerated over the study period. Mountain glaciers are also shrinking at alarming rates and adding substantial quantities of water to the ocean. (For more information, review the section "Glaciers in Retreat: Unbalanced Glacial Budgets" in Chapter 18, especially Figures 18.14 and 18.15 on page 512.)

Research indicates that sea level has risen about 25 centimeters (9.75 inches) since 1870. The pace has accelerated in recent decades; global sea-level rise averaged 1.7 millimeters per year throughout most of the twentieth century, but since 1993, the rate has doubled to 3.4 millimeters per year. As **Figure 21.33** indicates, the estimated future sea-level rise depends on the degree of ocean warming and ice sheet loss, ranging from 0.2 meter (8 inches) to 2 meters (6.6 feet). The lowest curve simply extrapolates the annual rate of sea-level rise that occurred between 1870 and 2000 (1.7 millimeters per year). However, as just mentioned, the *actual* rate of sea-level rise between 1993 and the present is about twice that. Such data show that there is a reasonable chance that sea level will rise considerably more than the lowest scenario indicates.

Even modest rises in sea level along a *gently* sloping shoreline, such as the Atlantic and Gulf coasts of the United States, will lead to significant erosion and severe permanent inland flooding. If this happens, many beaches and wetlands will be eliminated,

and heavily populated coastal areas will be inundated. Low-lying and densely populated places such as Bangladesh and the small island nation of the Maldives are especially vulnerable. The average elevation in the Maldives is 1.5 meters (less than 5 feet), and its highest point is just 2.4 meters (less than 8 feet) above sea level.

The Changing Arctic

The effects of global warming are most pronounced in the high latitudes of the Northern Hemisphere. For more than 30 years, the extent and thickness of sea ice have been rapidly declining. In addition, permafrost temperatures have been rapidly rising, and the area affected by permafrost has been decreasing. Meanwhile, alpine glaciers and the Greenland ice sheet have been shrinking. Another sign that the Arctic is rapidly

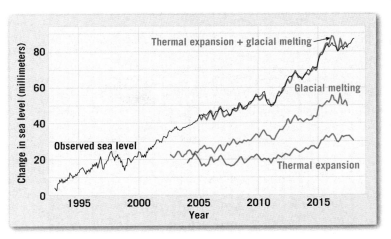

◄ **Figure 21.32**
Sea-level rise 1993–2017 The black line on this graph shows observed sea-level rise between 1993 and 2017. The red and blue lines show model estimates of contributions to sea-level rise due to thermal expansion and melting glaciers. When these two sources are added together (purple line), they closely match the observed sea level.

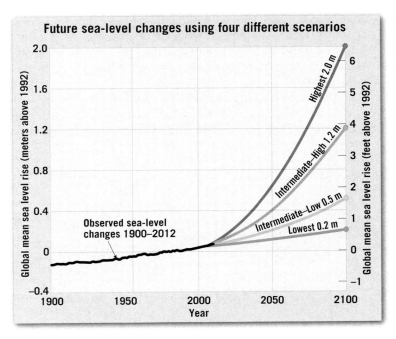

◄ **Figure 21.33**
Projecting sea level to 2100 This graph shows changes in sea level between 1900 and 2012 and projections to 2100 using four different scenarios. Currently the highest and lowest projections are considered to be extremely unlikely. The greatest uncertainty surrounding estimates is the rate and magnitude of ice sheet loss from Greenland and Antarctica. Zero on the graph represents mean sea level in 1992.

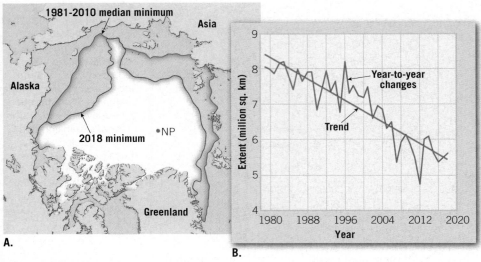

A.

B.

▲ **SmartFigure 21.34**
Tracking sea ice changes Sea ice is frozen seawater. In winter the Arctic Ocean is completely ice covered. In summer, a portion of the ice melts.
A. This map shows the extent of sea ice in early September 2018 compared to the average extent for the period 1981 to 2010. The sea ice that does not melt in summer is getting thinner.
B. The graph clearly depicts the trend in the area covered by sea ice at the end of the summer melt period.
(Data from National Snow and Ice Data Center)

Animation
https://goo.gl/
UsCgWk

warming is related to plant growth. A 2013 study showed that vegetation growth at northern latitudes had grown to resemble that which had characterized areas 4 to 6 degrees of latitude farther south as recently as 1982—a distance of 400 to 700 kilometers (250 to 430 miles). One researcher characterized the finding this way: "It's like Winnipeg, Manitoba, moving to Minneapolis-St. Paul in only 30 years."

Arctic Sea Ice   Climate models generally assert that one of the strongest signals of global warming should be a loss of sea ice in the Arctic. Sea ice grows in extent and thickness over the Arctic Ocean when temperatures drop to below freezing (about $-1°C$ because the water is salty) during the fall and winter, reaching a maximum extent in March. The ice can be as thick as 6 meters (20 feet) and is usually thickest in areas where it makes contact with land and stays there year-round. In the summer, as temperatures climb, Arctic sea ice decreases in extent and thickness, often breaking into large slabs near the ice margins (see Figure 21.23).

The image in **Figure 21.34** compares the average Arctic sea ice extent for September 2017 to the long-term average for the period 1981–2010. Observations

indicated that on September 13, 2017, ice extent shrunk to the eighth lowest minimum extent since 1979, the beginning of the satellite record. (September represents the end of the melt period, when the area covered by sea ice is at a minimum.) Not only is the area covered by sea ice declining, but the remaining sea ice has become thinner, making it more vulnerable to further melting.

The Arctic's sea ice *maximum* extent has also dropped by an average of 2.8 percent per decade since 1979. In fact, the sea ice maximum in 2017 was the lowest on record. Models that best match historical trends project that Arctic waters may be virtually ice free in the late summer by the 2030s. As noted earlier, a reduction in sea ice is a positive-feedback mechanism that reinforces global warming.

Permafrost   Chapter 15 includes a brief discussion of permafrost landscapes. The map in Figure 15.21 (page 431) shows that permafrost occupies much of the high latitudes of the Northern Hemisphere. During the past decade, however, mounting evidence indicates that the extent of permafrost in the Northern Hemisphere has decreased, as would be expected under long-term warming conditions.

One way to observe changes in the extent of permafrost is to survey the thousands of pools that dot the Arctic summer landscape. Where permafrost is healthy, only the top layer of the frozen ground thaws in summer. The permafrost beneath this *active layer* acts like the cement bottom of a swimming pool, forcing meltwater to accumulate in surface ponds. Where permafrost thaws, however, the water is able to percolate downward, draining the pools. Satellite imagery shows that a significant number of lakes have shrunk or disappeared altogether.

Studies in Alaska show that thawing is occurring in interior and southern parts of the state where permafrost temperatures are near the thaw point. As Arctic temperatures continue to rise, some models project that by the end of the century, near-surface permafrost may be lost entirely from large parts of Alaska.

Thawing permafrost represents a potentially significant positive-feedback mechanism that reinforces global warming. When vegetation dies in the Arctic, cold

▶ **Figure 21.35**
The pH scale This is the common measure of the degree of acidity or alkalinity of a solution. The scale ranges from 0 to 14, with a value of 7 indicating a solution that is neutral. Values below 7 indicate greater acidity, whereas numbers above 7 indicate greater alkalinity. It is important to note that the pH scale is logarithmic; that is, each whole number increment indicates a tenfold difference. Thus, pH 4 is 10 times more acidic than pH 5 and 100 times (10 × 10) more acidic than pH 6.

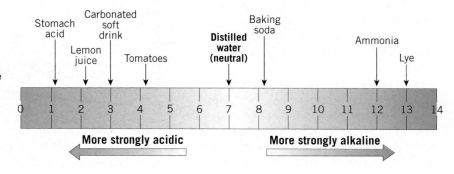

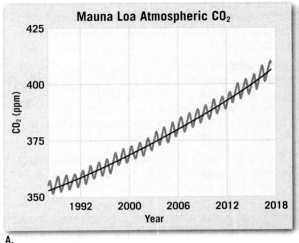

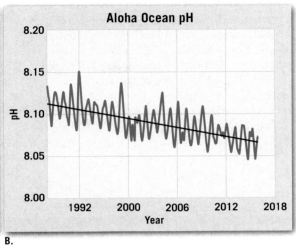

◀ **Figure 21.36**
Oceans becoming more acidic
The graphs show the correlation between rising levels of CO_2 in the atmosphere as measured at Mauna Loa Observatory (**A**) and falling pH in the nearby ocean (**B**). As CO_2 accumulates in the ocean, the water becomes more acidic (pH declines).

temperatures inhibit its decomposition. As a consequence, over thousands of years, a great deal of organic matter has become stored in the permafrost. When the permafrost thaws, organic matter that may have been frozen for millennia comes out of "cold storage" and decomposes. The result is the release of carbon dioxide and methane—greenhouse gases that contribute to global warming.

Increasing Ocean Acidity

The human-induced increase in the amount of carbon dioxide in the atmosphere has some serious implications for ocean chemistry and for marine life. Nearly half of the human-generated CO_2 currently ends up dissolved in the oceans. When atmospheric CO_2 dissolves in seawater (H_2O), it forms carbonic acid (H_2CO_3). This lowers the ocean's pH (**Figure 21.35**). In fact, the oceans have already absorbed enough carbon dioxide for surface waters to have experienced a decrease of 0.1 pH units since preindustrial times—which means that seawater is more acidic (**Figure 21.36**). Moreover, if the current trend in carbon dioxide emissions continues, the ocean will experience a pH decrease of at least 0.3 pH units by 2100, which represents a change in ocean chemistry that has not occurred for millions of years.

This shift toward acidity and the resulting changes in ocean chemistry make it more difficult for certain marine creatures to build hard parts (shells) out of calcium carbonate. This effect threatens shell-secreting organisms as diverse as microbes and corals, and it concerns marine scientists because of the potential consequences for other sea life that depend on the health and availability of these organisms.

The Potential for "Surprises"

You have seen that climate in the twenty-first century, unlike during the preceding 1000 years, is not expected to be stable. The amount and rate of future climate shifts depends primarily on current and future

human-caused emissions of heat-trapping gases and airborne particles. Many of the changes will probably be gradual, imperceptible from year to year. Nevertheless, the effects accumulated over decades will have powerful economic, social, and political consequences.

Despite our best efforts to understand future climate shifts, there is also the potential for "surprises." This simply means that the complexity of Earth's climate system might lead to relatively sudden, unexpected shifts in some aspects of climate. Many projections predict steadily changing conditions, giving the impression that humanity will have time to adapt. However, the scientific community has been paying attention to the possibility that at least some changes will be abrupt, perhaps crossing a threshold, or "tipping point," so quickly that there will be little time to react.

This is a reasonable concern because abrupt changes occurring over periods as short as decades or even years have been a natural part of the climate system throughout Earth history. The paleoclimate record described earlier in the chapter contains ample evidence of such abrupt changes. One such abrupt change occurred at the end of a time span known as the *Younger Dryas*, a time of abnormal cold and drought in the Northern Hemisphere that occurred about 12,000 years ago. Following a 1000-year-long cold period, the Younger Dryas abruptly ended in a few decades or less. This change occurred at the same time as the extinction of more than 70 percent of large-bodied mammals in North America.

There are many examples of potential surprises, each of which would have large consequences. We simply do not know how far the climate system, or other systems it affects, can be pushed before responding in unexpected ways. Even if the chance of any particular surprise happening is small, the chance that at least one such surprise will occur is much greater. In other words, although we may not know which of these events will occur, it is likely that one or more will eventually occur.

Among scientists worldwide, the impact of an increase in atmospheric CO_2 and trace gases is *not* in question: Global warming is real and *anthropogenic* (human caused). Policy makers are confronted with the fact that climate-induced environmental changes cannot be reversed quickly, if at all, due to the lengthy timescales associated with the climate system. The solution is clear, there must be a significant and rapid shift away from reliance on fossil fuels if disastrous consequences are to be averted or substantially reduced.

CONCEPT CHECKS 21.8

1. Describe the factors that are causing sea level to rise.

2. How is Arctic sea ice changing, and what impact might this have on future climate change? Does this change reflect positive- or negative-feedback mechanisms?

   Concept Checker
   https://goo.gl/1zmxPC

3. Based on Figure 21.31, what projected changes relate to something other than temperature?

21 CONCEPTS IN REVIEW

Global Climate Change

21.1 Climate and Geology

List the major parts of the climate system and some connections between climate and geology.

Key Terms: climate cryosphere
weather climate system

- *Climate* is the aggregate *weather* conditions for a place or region over a long period of time. If those conditions shift toward a new situation over time, with hotter or cooler temperatures and/or more or less precipitation, the climate is said to have changed.

- Earth's *climate system* is a complex interchange of energy and moisture among the atmosphere, hydrosphere, geosphere, biosphere, and *cryosphere* (ice and snow). When the climate changes, geologic processes such as weathering, mass movement, and erosion may change as well.

Q Which sphere of the climate system dominates this image? What other sphere or spheres are present?

21.2 Detecting Climate Change

Discuss several ways in which past climate changes are detected.

Key Terms: paleoclimatology dendrochronology
proxy data oxygen-isotope analysis

- The geologic record yields multiple kinds of indirect evidence about past climate. These *proxy data* are the focus of *paleoclimatology* and can be found in glacial ice cores, seafloor sediment, oxygen isotopes, corals, tree rings, and fossil pollen.

- *Oxygen-isotope analysis* is based on the difference between heavier ^{18}O and lighter ^{16}O and their relative amounts in water molecules (H_2O). This ratio can be used to gauge how warm or cold temperatures are. Oxygen isotopes can also be measured in the shells of fossil marine organisms, in coral structures, or in the water molecules that make up glacial ice.

- Trees grow thicker rings in warmer, wetter years and thinner rings in colder, drier years. The pattern of ring thickness can be matched up between trees of overlapping ages to form a long-term record of a region's climate.

21.3 Some Atmospheric Basics

Describe the composition of the atmosphere and the atmosphere's vertical changes in pressure and temperature.

Key Terms: radiosonde   mesosphere
aerosols   stratosphere   thermosphere
troposphere

- Air is a mixture of many discrete gases, and its composition varies from time to time and place to place. Two gases, nitrogen and oxygen, make up about 99 percent of the volume of clean, dry air. Carbon dioxide, although present in only minute amounts (0.405 percent, or 405 parts per million), is an efficient absorber of energy emitted by Earth and thus influences the heating of the atmosphere.
- Two important variable components of air are water vapor and *aerosols*. Like carbon dioxide, water vapor can absorb heat given off by Earth. Aerosols are important because these often invisible particles act as surfaces on which water vapor can condense and are also good absorbers and reflectors (depending on the particles) of incoming solar radiation.
- The atmosphere is densest closest to the surface of Earth. It thins rapidly with increasing altitude and gradually fades off into space. Temperature varies through a vertical section of the atmosphere. Generally, temperatures drop in the *troposphere*, warm in the *stratosphere*, cool in the *mesosphere*, and increase in the *thermosphere*.

Q This graph shows changes in an atmospheric element from Earth's surface to a height of about 140 kilometers (90 miles). Which element is being depicted: air pressure, humidity, or temperature? Show how this graph is used to divide the atmosphere into layers.

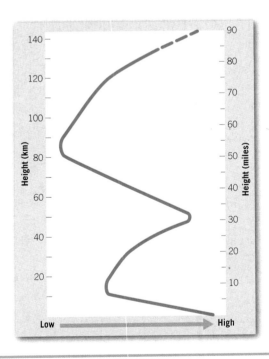

21.4 Heating the Atmosphere

Outline the basic processes involved in heating the atmosphere.

Key Terms: albedo   greenhouse effect

- Radiation (electromagnetic radiation) is the only mechanism to transfer heat through the vacuum of space. It consists of a large array of wavelengths that includes X-rays, visible light, infrared (heat), microwaves, and radio waves. Shorter wavelengths have greater energy.
- These are four basic laws that of radiation: (1) All objects emit radiant energy; (2) hotter objects radiate more total energy per unit area than colder objects; (3) the hotter the radiating body, the shorter the wavelengths of maximum radiation; and (4) objects that are good absorbers of radiation are good emitters. Gases are selective absorbers, meaning that they absorb and emit certain wavelengths but not others.
- Approximately 50 percent of the solar energy striking the top of the atmosphere reaches Earth's surface. About 30 percent is reflected back to space. Clouds and the atmosphere's gases absorb the remaining 20 percent.
- Radiant energy absorbed at Earth's surface is eventually reradiated skyward. Because Earth has a much lower surface temperature than the Sun, its radiation is in the form of long-wave infrared radiation.
- A significant fraction of the long-wave radiation emitted by Earth is absorbed by certain atmospheric gases, primarily water vapor and carbon dioxide. This absorbed energy heats the air and increases the rate at which it radiates energy, both out to space and back toward Earth's surface. The fraction that does not radiate to space keeps Earth warmer than it otherwise would be. This is the *greenhouse effect*.

21.5 Natural Causes of Climate Change

Discuss hypotheses that relate to natural causes of climate change.

Key Term: sunspot

- The natural functions of the Earth system produce climate change. The position of lithospheric plates can influence the climate of the continents as well as oceanic circulation. Variations in the shape of Earth's orbit, angle of axial tilt, and orientation of the axis all cause changes in the distribution of solar energy.
- Volcanic aerosols act like a Sun shade, screening out a portion of incoming solar radiation. Volcanic sulfur dioxide emissions that reach high into the atmosphere are particularly important. Combined with water to form tiny droplets of sulfuric acid, these aerosols can remain aloft for several years.
- Volcanoes also emit carbon dioxide. During times of especially large eruptions, such as those that produced oceanic lava plateaus during the Cretaceous period, volcanic CO_2 emissions may contribute to the greenhouse effect sufficiently to cause global warming.
- Since Earth's climate is fueled by solar energy, variations in the Sun's energy output affect Earth temperatures. *Sunspots* are dark features on the surface of the Sun associated with periods of slightly increased solar energy output. The number of sunspots rises and drops in an 11-year cycle.

21.6 Human Impact on Global Climate

Summarize the nature and cause of the atmosphere's changing composition since about 1750. Describe the climate's response.

Key Terms: trace gases black carbon

- By altering ground cover with the use of fire and the overgrazing of land, people have modified climatic factors such as surface albedo and evaporation rates for thousands of years.

- Human activities also produce climate change by releasing carbon dioxide (CO_2) and *trace gases*, particularly methane and nitrous oxide. Humans release CO_2 when they cut down forests or burn fossil fuels such as coal, oil, and natural gas. A steady rise in atmospheric CO_2 levels has been documented at Mauna Loa, Hawaii, and other locations around the world.

- Air bubbles trapped in glacial ice reveal that the atmosphere currently contains about 40 percent more CO_2 than at any time in the past 800,000 years.

- As a result of extra heat retained by added greenhouse gases, Earth's atmosphere has warmed by about 0.8°C (1.4°F) in the past 100 years, most of it since the 1970s. Temperatures are projected to increase by another 2° to 4.5°C (3.5° to 8.1°F) in the future.

- Overall, aerosols reflect a portion of incoming solar radiation back to space and therefore have a cooling effect.

Q Do aerosols spend more or less time in the atmosphere than greenhouse gases such as carbon dioxide? What is the significance of this difference in residence time? Explain.

21.7 Predicting Future Climate Change

Contrast positive- and negative-feedback mechanisms and provide examples of each.

Key Terms: positive-feedback negative-feedback
feedback mechanism mechanism mechanism

- A change in one part of the climate system may trigger changes in other parts of the climate system that amplify or diminish the initial effect. These *feedback mechanisms* are called *positive-feedback mechanisms* if they reinforce the initial change and *negative-feedback mechanisms* if they counteract the initial effect.

- The melting of sea ice due to global warming (decreasing albedo and increasing the initial effect of warming) is one example of a positive-feedback mechanism. The production of more clouds (blotting out incoming solar radiation, leading to cooling) is an example of a negative-feedback mechanism.

- Computer models of climate give scientists tools for testing hypotheses about climate change. Although these models are simpler than the real climate system, they are useful tools for predicting the future climate.

Q Changes in precipitation and temperature due to climate change can increase the risk of forest fires. Describe two ways that the event shown in this photo could contribute to global warming.

21.8 Some Consequences of Global Warming

Discuss several likely consequences of global warming.

- In the future, temperature increases will likely be greatest in the polar regions and least in the tropics. Some areas will get drier, and other areas will get wetter.

- Sea level is predicted to rise because of melting of glacial ice and thermal expansion of seawater. Low-lying, gently sloped, highly populated coastal areas are most at risk.

- Sea ice cover and thickness in the Arctic have been declining since satellite observations began in 1979.

- Because of the warming of the Arctic, permafrost is melting, releasing CO_2 and methane to the atmosphere in a positive-feedback mechanism.

- Because the climate system is imperfectly understood, it could produce sudden, unexpected changes with little warning.

Q This ice breaker is plowing through sea ice in the Arctic Ocean. What spheres of the climate system are represented in this photo? How has the area covered by summer sea ice been changing since satellite monitoring began in 1979? How does this change influence temperatures in the Arctic?

GIVE IT SOME THOUGHT

1. Refer to Figure 21.1, which illustrates various components of Earth's climate system. The labels represent interactions or changes that occur in the climate system. Select three labels and provide an example of an interaction or change associated with each. Explain how these interactions may influence temperature.

2. When this weather balloon was launched, the surface air temperature was 17°C. The balloon is now at an altitude of 1 kilometer. What term is applied to the instrument package being carried aloft by the balloon? In what layer of the atmosphere is the balloon? If average conditions prevail, what is the air temperature at this altitude? How did you figure this out?

3. Explain how oxygen-isotope analysis of the fossil shells of marine animals can tell us how much water is trapped in glacial ice and, hence, the status of global climate at the time those animals lived.

4. Figure 21.14 shows that about 30 percent of the Sun's energy intercepted by Earth is reflected or scattered back to space. If Earth's albedo were to increase to 50 percent, how would you expect Earth's average surface temperature to change? Explain.

5. Volcanic events, such as the eruptions of El Chichón and Mount Pinatubo, have been associated with drops in global temperatures. During the Cretaceous period, volcanic activity was associated with global warming. Explain the apparent paradox.

6. The accompanying photo is a recent view of Athabasca Glacier in the Canadian Rockies. The marker indicates the location of the outer limit of this glacier in 1925. Is the behavior of Athabasca Glacier shown in this image typical of other glaciers around the world? Describe a significant impact of such behavior.

7. Motor vehicles are a significant source of CO_2. Using electric cars, such as the one pictured here, is one way to reduce emissions from this source. Even though these vehicles emit little or no CO_2 or other pollutants directly into the air, can they still be connected to such emissions? Explain.

8. During a conversation, an acquaintance indicates that he is skeptical about global warming. When you ask him why, he says, "The past couple of years in this area have been among the coolest I can remember." While you assure this person that it is useful to question scientific findings, you suggest to him that his reasoning in this case may be flawed. Use your understanding of the definition of *climate* along with one or more graphs in the chapter to persuade this person to reevaluate his reasoning.

9. This large cattle feedlot is in the Texas panhandle. How might consuming less beef influence global climate change?

EYE ON EARTH

1. This satellite image shows an extensive plume of ash from an explosive volcanic eruption of Indonesia's Sinabung Volcano on January 16, 2014.

   a. How might the volcanic ash from this eruption influence air temperatures?

   b. Would this effect likely be long-lasting—perhaps extending for years? Explain.

   c. What "invisible" volcanic emission might have a greater effect than the volcanic ash?

2. This satellite image from August 2007 shows the effects of tropical deforestation in a portion of the Amazon basin in western Brazil. Intact forest is dark green, whereas cleared are as are tan (bare ground) or light green (crops and pasture). Notice the relatively dense smoke in the left center of the image.

   a. How does the destruction of tropical forests change the composition of the atmosphere?

   b. Describe the effect that tropical deforestation has on global warming.

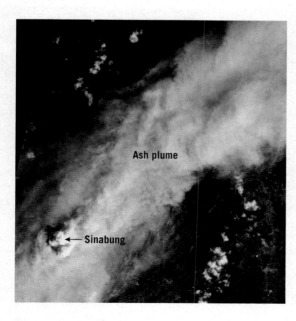

Ash plume

← Sinabung

DATA ANALYSIS

Arctic Sea Ice

Climate models are in general agreement that one of the strongest signals of climate change should be a loss of sea ice in the Arctic.

https://goo.gl/eD226M

ACTIVITIES

Go to the Earth Observatory page at http://earthobservatory.nasa.gov/. Click on the Search icon in the upper left and enter "Arctic sea ice" in the search field. From the list of search results, select "World of Change: Arctic Sea Ice." Watch the animation and read the information below the images.

1. **In what month does Arctic sea ice reach its maximum extent and concentration?**

2. **In what month does the sea ice reach its minimum extent and concentration?**

3. **Explain what the yellow-orange lines on these images illustrate.**

4. **How is the area with the highest ice concentration shown on these images? How is open water shown?**

5. **In which year did the Arctic sea ice seem to reach the smallest extent and concentration?**

6. **Compare and contrast the September 1999 image with the September 2016 image.**

7. **Compare and contrast the March 2000 image with the March 2017 image.**

8. **Arctic sea ice floats on the Arctic Ocean, so when it melts, it does not cause a rise in sea level. With that in mind, explain the role that Arctic sea ice plays in climate change.**

Mastering Geology

Looking for additional review and test prep materials? Visit pearson.com/mastering/geology to enhance your understanding of this chapter's content by accessing a variety of resources, including **Self-Study Quizzes, Geoscience Animations, SmartFigure Tutorials,** *Project Condor* **Videos, Mobile Field Trips, Dynamic Study Modules,** and an optional **Pearson eText.**

Earth's First Known Animal

For about 3 billion years, life on Earth was simple and largely microbial. Complex, mobile animals with hard parts such as shells and jaws, arose in a burst of innovation only 451 million years ago.

But between the ancient microbial world and our familiar world, the oceans were home to a biosphere unlike anything that survives. During this time, called the Ediacaran, the seafloor was carpeted with expanses of microbial mat bearing a garden of large, soft-bodied, mostly static lifeforms, including upright fronds, coiled blobs, and flat-lying organisms.

▲ A *Dickinsonia* fossil, one of the oldest multicellular animals identified to date.

A number of localities around the world preserve these creatures as detailed impressions. Yet despite years of study, scientists did not know what they were. Animals, algae, fungi, lichens?

For one of these creatures, we now have an answer. *Dickinsonia*, which had no mouth or eyes and resembled a quilted placemat, was an animal. Scientists recently made this determination by examining molecules from the organic film left in well-preserved *Dickinsonia* fossils. When an organism dies, some of its molecules may form chemical substances that can survive in rock for hundreds of millions of years.

Examining a *Dickinsonia* fossil, they found biomarkers derived from cholesterol. Only animals produce cholesterol—so *Dickinsonia* was an animal. As a result we are confident that the appearance of *Dickinsonia* and its relatives was the prelude to the Cambrian explosion of animal life that followed.

▶ Mistaken Point, Newfoundland, a site where many fossil impressions of Ediacaran life forms have been discovered.

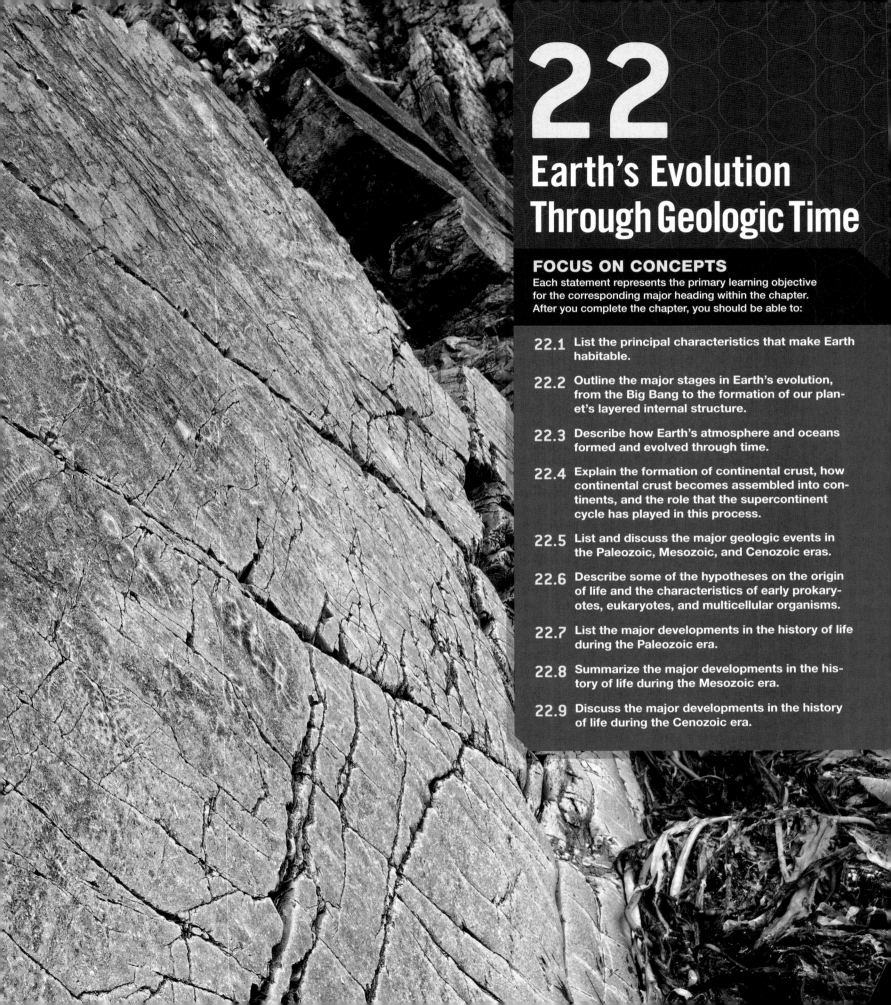

22

Earth's Evolution Through Geologic Time

FOCUS ON CONCEPTS

Each statement represents the primary learning objective for the corresponding major heading within the chapter. After you complete the chapter, you should be able to:

22.1 List the principal characteristics that make Earth habitable.

22.2 Outline the major stages in Earth's evolution, from the Big Bang to the formation of our planet's layered internal structure.

22.3 Describe how Earth's atmosphere and oceans formed and evolved through time.

22.4 Explain the formation of continental crust, how continental crust becomes assembled into continents, and the role that the supercontinent cycle has played in this process.

22.5 List and discuss the major geologic events in the Paleozoic, Mesozoic, and Cenozoic eras.

22.6 Describe some of the hypotheses on the origin of life and the characteristics of early prokaryotes, eukaryotes, and multicellular organisms.

22.7 List the major developments in the history of life during the Paleozoic era.

22.8 Summarize the major developments in the history of life during the Mesozoic era.

22.9 Discuss the major developments in the history of life during the Cenozoic era.

Earth has a long and complex history. Time and again, the splitting and coming of continents has led to the formation of new ocean basins, the creation of great mountain ranges, and the evolution of Earth's continents. Furthermore, the degree to which life on our planet has evolved cannot be overstated. In this chapter, we will first explore the characteristics that make Earth habitable and then briefly examine Earth's physical and biological evolution though geologic time.

22.1 What Makes Earth Habitable?

List the principal characteristics that make Earth habitable.

Not since Galileo first pointed his telescope toward the sky has there been as much excitement and popular interest as there is today in astronomy. People from all walks of life the world over are interested in the discovery of planets that orbit other stars, called **exoplanets**. Much of this interest undoubtedly stems from human curiosity—more specifically, our desire to know if we are alone in the universe.

The discoveries of most of the exoplanets in recent years have been made using NASA's Kepler telescope (**Figure 22.1**). As of 2016, this space telescope had identified more than 2000 exoplanets and numerous planetary candidates. One goal of the Kepler mission is to survey nearby sections of the Milky Way in search of Earth-size planets orbiting in or near the habitable zones of planetary systems. The **habitable zone** can be broadly defined as the region around a host star where a planet with sufficient atmospheric pressure can maintain liquid water on its surface.

> The habitable zone can be broadly defined as the region around a host star where a planet with sufficient atmospheric pressure can maintain liquid water on its surface.

Figure 22.1 shows some of the small exoplanets recently discovered using the Kepler space telescope. All these planets are located in the habitable zones of one of three types of stars. G-type stars are Sun-like, yellow stars having masses about 0.8 to 1.2 times the mass of the Sun, with surface temperatures of about 5800°C (about 10,000°F). The K-type stars, which are somewhat smaller, are orange stars with masses between 0.45 and 0.8 that of the Sun and surface temperatures of about 4000°C (about 7000°F). These stars are considered likely candidates for hosting habitable planets because they have longer life spans than Sun-like stars and emit less ultraviolet radiation, which is harmful to an organism's DNA. M-type stars are reddish in color and less than half the size of the Sun. For various reasons, they were once thought to be unlikely candidates to host habitable planets. However, because they are by far the most common type of star, they might be a likely place to find planets capable of supporting life.

Astronomers generally consider habitable planets to be *terrestrial* (Earth-like) planets that orbit within the habitable zone, with conditions roughly comparable to those of Earth and, thus, favorable for supporting Earth-like life. The planet Venus offers a sobering view of the importance of the habitable zone. In terms of size, mass, and composition, Venus is nearly Earth's twin. However, it orbits a little closer to the Sun than does Earth—and its atmosphere has evolved into a thick blanket that keeps temperatures at the surface hot enough to melt lead—most certainly too hot to sustain life.

What fortuitous events produced Earth, a planet so hospitable to life? Earth was not always as we find it today. During its formative years, our planet supported a magma ocean. It also survived frequent bombardments by large asteroids. The oxygen-rich atmosphere that makes

▼ Figure 22.1
Small exoplanets found in the habitable zones of their host stars A sampling of the exoplanets discovered by the Kepler space telescope that are small enough to be Earth-like and are located within their host stars' habitable zones.

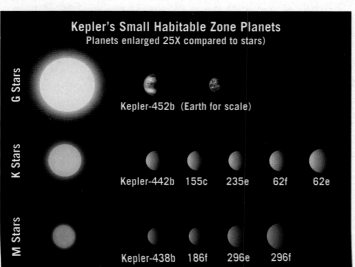

Kepler's Small Habitable Zone Planets
Planets enlarged 25X compared to stars)

G Stars
Kepler-452b (Earth for scale)

K Stars
Kepler-442b   155c   235e   62f   62e

M Stars
Kepler-438b   186f   296e   296f

many modern life-forms possible developed long after the solar system was born. Serendipitously, Earth seems to be the right planet, in the right location, at the right time.

The Right Planet

What are some of the characteristics that make Earth unique among the planets of our solar system? Consider the following:

- If Earth were considerably larger (more massive), its force of gravity would be proportionately greater. Like the giant planets, Earth might have retained a thick, hostile atmosphere consisting of ammonia and methane, and possibly hydrogen and helium.
- If Earth were significantly smaller, oxygen, water vapor, and other volatiles would escape to space and be lost forever. Thus, like the Moon and Mercury, both of which lack appreciable atmospheres, Earth would be devoid of life.
- If Earth did not have a rigid lithosphere overlaying a weak asthenosphere, plate tectonics would not operate. The continental crust (Earth's "highlands") would not have formed without the recycling of plates. Consequently, the entire planet would likely be covered by an ocean a few kilometers deep. As author Bill Bryson so aptly stated, "There might be life in that lonesome ocean, but there certainly wouldn't be baseball."*
- Most surprising, perhaps, is the fact that if our planet did not have a molten metallic outer core, most of the life-forms on Earth would not exist. Fundamentally, without the flow of iron in the core, Earth could not support a magnetic field. It is the magnetic field that prevents lethal cosmic rays from showering Earth's surface and stripping away our atmosphere.

The Right Location

A primary factor that determines whether a planet is habitable is its location with respect to its host star. The following scenarios substantiate Earth's advantageous position:

- If Earth were about 10 percent closer to the Sun, our atmosphere would be more like that of Venus and consist mainly of the greenhouse gas carbon dioxide. Earth's surface temperature would then be too hot to support higher life-forms.
- If Earth were about 10 percent farther from the Sun, the problem would be reversed: It would be too cold. The oceans would freeze over, and Earth's active water cycle would not exist. Without liquid water, most life would perish.
- Earth is near a star of modest size. Stars like the Sun have a life span of roughly 10 billion years and emit radiant energy at a fairly constant level during most

of this time. Giant stars, on the other hand, consume their nuclear fuel at very high rates and "burn out" in a few hundred million years. Earth's proximity to a modest-sized star allowed enough time for the evolution of humans, who first appeared on this planet only a few million years ago.

The Right Time

The last, but certainly not the least, fortuitous factor for Earth is timing. The first organisms to inhabit Earth came into existence roughly 3.8 billion years ago. From that point in Earth's history, innumerable changes occurred: Life-forms came and went, and the physical environment of our planet was transformed in many ways. Consider two of the many timely Earth-altering events:

- Earth's atmosphere has developed over time. Earth's primitive atmosphere is thought to have been composed mostly of nitrogen, water vapor, methane, and carbon dioxide—but no free oxygen (that is, oxygen not combined with other elements). Fortunately, microorganisms evolved that released oxygen into the atmosphere through the process of *photosynthesis*. About 2.5 billion years ago, an atmosphere with free oxygen came into existence. The result was the evolution of the ancestors of the vast array of multicellular organisms that we find on Earth today.
- About 66 million years ago, our planet was struck by an asteroid 10 kilometers (6 miles) in diameter. This impact contributed to a mass extinction that obliterated nearly three-quarters of all plant and animal species—including dinosaurs other than birds (**Figure 22.2**).[†] Although it may not seem lucky, the extinction of dinosaurs opened new habitats for small mammals that

▼ Figure 22.2
Reconstruction of dinosaurs during the Late Cretaceous An asteroid impact about 66 million years ago contributed to the extinction of the dinosaurs, which opened new habitats for mammals that survived the event.

---

* *A Short History of Nearly Everything* (Broadway Books, 2003).

[†] We use the term dinosaurs here to refer to all members of this group except birds.

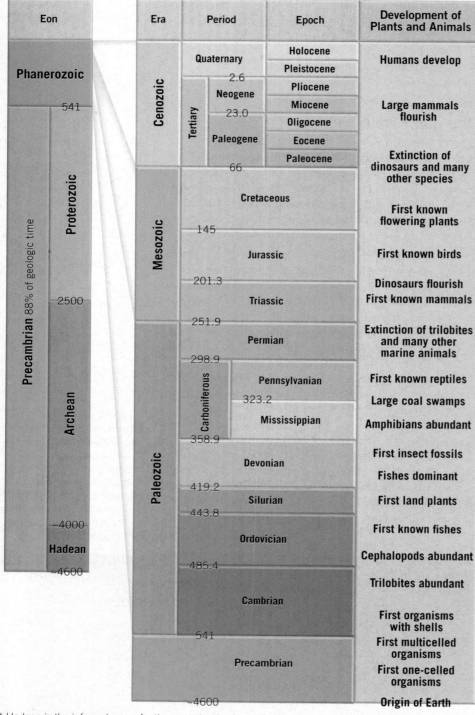

Eon	Era	Period		Epoch	Development of Plants and Animals
Phanerozoic	**Cenozoic**	Quaternary		Holocene	**Humans develop**
				Pleistocene	
		2.6			
		Neogene		Pliocene	
				Miocene	**Large mammals flourish**
		23.0		Oligocene	
		Paleogene		Eocene	
				Paleocene	**Extinction of dinosaurs and many other species**
		66			
	Mesozoic	Cretaceous			**First known flowering plants**
		145			
		Jurassic			**First known birds**
		201.3			**Dinosaurs flourish**
		Triassic			**First known mammals**
		251.9			
	Paleozoic	Permian			**Extinction of trilobites and many other marine animals**
		298.9			
		Carboniferous	Pennsylvanian		**First known reptiles**
			323.2		**Large coal swamps**
			Mississippian		**Amphibians abundant**
		358.9			
		Devonian			**First insect fossils**
					Fishes dominant
		419.2			
		Silurian			**First land plants**
		443.8			
		Ordovician			**First known fishes**
					Cephalopods abundant
		485.4			
		Cambrian			**Trilobites abundant**
					First organisms with shells
		541			
Precambrian 88% of geologic time — **Proterozoic** 541 / 2500, **Archean** ~4000, **Hadean** ~4600		Precambrian			**First multicelled organisms**
					First one-celled organisms
		~4600			**Origin of Earth**

* Hadean is the informal name for the span that begins at Earth's formation and ends with Earth's earliest-known rocks.

◀ **Figure 22.3**
The geologic timescale Numbers represent time in millions of years before the present. The Precambrian accounts for about 88 percent of geologic time.

survived the impact. These habitats, along with evolutionary forces, led to the development of the many large mammals that occupy our modern world. Without this event, mammals might have remained mostly small and inconspicuous.

As various observers have noted, Earth developed under "just right" conditions to support higher life-forms. Astronomers refer to this as the *Goldilocks scenario.* As in the classic "Goldilocks and the Three Bears" fable, Venus is too hot (Papa Bear's porridge), and Mars is too cold (Mama Bear's porridge), but Earth is just right (Baby Bear's porridge).

Viewing Earth's History

The remainder of this chapter focuses on the origin and evolution of planet Earth—the one place in the universe that we know fosters life. As you learned in Chapter 9, researchers utilize many tools to interpret clues about Earth's past. Using these tools, as well as clues contained in the rock record, scientists continue to unravel many complex events of the geologic past. This chapter provides a brief overview of the history of our planet and its life-forms—a journey that takes us back about 4.6 billion years, to the formation of Earth. Later, we will consider how our physical world assumed its present state and how Earth's inhabitants changed through time. As you read this chapter, refer to the *geologic timescale* presented in **Figure 22.3**.

CONCEPT CHECKS 22.1

1. Explain why Earth is just the right size to support life.

2. In what way does Earth's molten, metallic core help protect Earth's life-forms?

 Concept Checker https://goo.gl/gKzB4Y

3. Why is Earth's location in the solar system ideal for the development of complex life-forms such as humans?

22.2 Birth of a Planet

Outline the major stages in Earth's evolution, from the Big Bang to the formation of our plant's layered internal structure.

The universe began about 13.8 billion years ago with the *Big Bang*, when all matter and space came into existence. Shortly thereafter, the two simplest elements, hydrogen and helium, formed. These basic elements were the ingredients for the first star systems. Several billion years later, our home galaxy, the Milky Way, came into existence. It was within a band of stars and nebular debris in an arm of this spiral galaxy that the Sun and planets took form nearly 4.6 billion years ago.

From the Big Bang to Heavy Elements

One of the products of the Big Bang was an array of subatomic particles, including protons, neutrons, and electrons (**Figure 22.4**). Later, as the debris cooled, these subatomic particles combined to generate atoms of hydrogen and helium, the two lightest elements. A few hundred million years after the Big Bang, the primordial hydrogen and helium had condensed and coalesced to form the first stars and galaxies (see Figure 22.4C). Within the cores of these early stars, heating triggered the process of *nuclear fusion*, in which hydrogen nuclei combine to form helium nuclei, releasing enormous amounts of radiant energy (heat, light, and cosmic rays). As these stars aged and died (in some cases via cataclysmic **supernova** explosions), other nuclear reactions generated all the elements on the periodic table, sending them into interstellar space. It is from this material, as well as primordial hydrogen and helium, that our Sun and the rest of the solar system formed. Based on this scenario, all the atoms in your body except for hydrogen and helium were produced billions of years ago in now-defunct stars, and the gold in your jewelry was produced mainly during supernova explosions and neutron star mergers.

From Planetesimals to Protoplanets

Recall that the solar system, including Earth, formed about 4.6 billion years ago from the **solar nebula**, a large rotating cloud of interstellar dust and gas (see Figure 22.4E). As the solar nebula contracted, most of the matter collected in the center to create the hot *protosun*. The remaining materials formed a thick, flattened, rotating disk, within which matter gradually cooled and condensed into grains and clumps of icy, rocky, and metallic material. Repeated collisions resulted in most of this solid material eventually collecting into asteroid-sized objects called **planetesimals**.

The composition of planetesimals was largely determined by their proximity to the protosun. As you might expect, temperatures were highest in the inner solar system and decreased toward the outer edge of the disk. Therefore, between the present orbits of Mercury and Mars, the planetesimals were composed mainly of materials with high melting temperatures—metals and rocky substances. The planetesimals that formed beyond the orbit of Mars, where temperatures are low, contained high percentages of ices—water ice and the frozen forms of carbon dioxide, ammonia, and methane—as well as smaller amounts of rocky and metallic debris.

Through repeated collisions and accretion (sticking together), these planetesimals grew into eight **protoplanets**, as well as dwarf planets and some larger moons (see Figure 22.4G). During this process, matter was concentrated into fewer and fewer bodies, each having greater and greater mass.

At some point in Earth's early evolution, a giant impact occurred between a Mars-sized object and a young, semimolten Earth. This collision ejected huge amounts of debris into space, some of which coalesced to form the Moon (see Figure 22.4J,K,L).

Earth's Early Evolution

As material continued to collide and accumulate, the high-velocity impacts of interplanetary debris (planetesimals) and the decay of radioactive elements caused Earth's temperature to steadily increase. This early period of heating resulted in a magma ocean that was perhaps several hundred kilometers deep. Within the magma ocean, buoyant masses of molten rock rose toward the surface and eventually solidified to produce thin rafts of crustal rocks. Geologists call this early period of Earth's history the **Hadean**, and it began with Earth's formation about 4.6 billion years ago and ended roughly 4 billion years ago (**Figure 22.5**). The name *Hadean* is derived from the Greek word *Hades*, meaning "the underworld," referring to the "hellish" conditions on Earth at the time.

During this period of intense heating, Earth became so hot that iron and nickel began to melt. Melting produced liquid blobs of heavy metal that gravitationally sank toward the center of Earth. This process occurred rapidly on the scale of geologic time and produced Earth's dense iron-rich core. As you learned

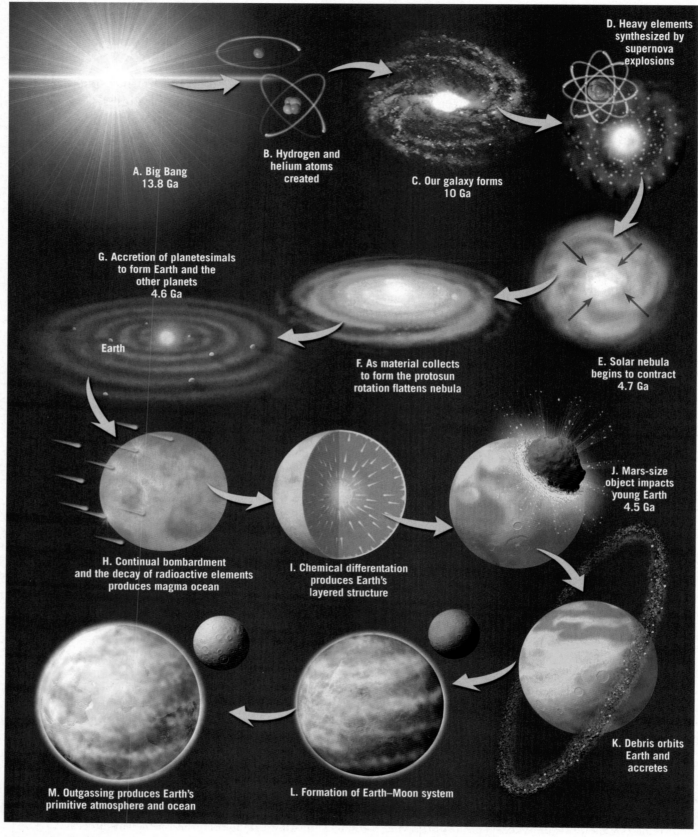

▲ SmartFigure 22.4

Major events that led to the formation of early Earth Ages are in billions of years (Ga).

▲ **Figure 22.5**
Artistic depiction of Earth during the Hadean The Hadean is an unofficial eon of geologic time that occurred before the Archean. Its name refers to the "hellish" conditions on Earth. During the early Hadean, Earth had a magma ocean and experienced intense bombardment by nebular debris.

in Chapter 12, the formation of a molten iron core was one of many stages of chemical differentiation in which Earth converted from a more homogeneous body, with roughly the same matter at all depths, to a layered planet with material sorted by density and composition (see Figure 22.4I).

This period of chemical differentiation established the three major divisions of Earth's interior—the iron-rich *core*; the thin *primitive crust*; and Earth's thickest layer, the *mantle*, located between the core and the crust. In addition, the lightest materials—including water vapor, carbon dioxide, and other gases—escaped to form a primitive atmosphere and, shortly thereafter, the oceans.

CONCEPT CHECKS 22.2

1. What two elements made up most of the observable matter in the very early universe?

2. What is the name for a cataclysmic event in which an exploding massive star produces heavy elements?

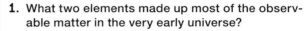 
Concept Checker
https://goo.gl/EJwxRp

3. Briefly describe the formation of the planets from the solar nebula.

22.3 Origin and Evolution of the Atmosphere and Oceans

Describe how Earth's atmosphere and oceans formed and evolved through time.

We can be thankful for our atmosphere; without it, there would be no greenhouse effect, and Earth would be nearly 60°F colder. Earth's water bodies would be frozen nearly solid, making the hydrologic cycle nonexistent.

The air we breathe is a relatively stable mixture of 78 percent nitrogen, 21 percent oxygen, about 1 percent argon (an inert gas), and small amounts of other gases such as carbon dioxide and water vapor. However, our planet's original atmosphere was substantially different.

Earth's Primitive Atmosphere

Late in Earth's formative period, its atmosphere probably consisted of gases most common in the early solar system: hydrogen, helium, methane, ammonia, carbon dioxide, and water vapor. The lightest of these—hydrogen and helium—likely escaped into space because Earth's gravity was too weak to hold them. The gases that remained—methane, ammonia, carbon dioxide, and water vapor—contain the basic ingredients of life: carbon, hydrogen, oxygen, and nitrogen.

Over time, Earth's atmosphere was enhanced by a process called **outgassing**, by which gases trapped in the planet's interior are released. Outgassing from hundreds of active volcanoes continues to be an important

planetary process worldwide (**Figure 22.6**). These eruptions release mainly water vapor, carbon dioxide, and sulfur dioxide, with minor amounts of other gases. As a result, Earth's early atmosphere gradually became enriched in carbon dioxide (most of the water vapor condensed to form liquid water) and was probably similar in this respect to the atmospheres of Venus and Mars.

Equally important, molecular oxygen (O_2) was not present in Earth's atmosphere in appreciable amounts for at least the first 2 billion years of Earth history. Molecular oxygen is often called "free oxygen" because it consists of oxygen atoms that are not bound to other elements, such as hydrogen (in water molecules, H_2O) or carbon (in carbon dioxide, CO_2).

Oxygen in the Atmosphere

As Earth's surface cooled, water vapor condensed to form clouds, and torrential rains formed the oceans. Based on evidence from trace fossils from Western Australia, ancient microorganisms began to flourish in

▲ Figure 22.6
Outgassing produced Earth's first enduring atmosphere Outgassing continues today from hundreds of active volcanoes worldwide.

these oceans at least 3.5 billion years ago. Descendants of these primitive life-forms known as *cyanobacteria* (once called blue-green algae) developed the ability to carry out the type of photosynthesis performed by plants (which we will call *plant photosynthesis*) and began to release oxygen into the water. **Plant photosynthesis** is the production of energy-rich molecules of sugar from molecules of carbon dioxide (CO_2) and water (H_2O), using sunlight as the energy source. The sugars (glucose and other sugars) generated by photosynthesis are used in metabolic processes by living things, and the by-product of photosynthesis is molecular oxygen.

Initially, the newly released molecular oxygen likely combined with other elements through processes that included the chemical weathering of rocks. Scientists have also found evidence that appreciable quantities of oxygen were "soaked up" by iron that was dissolved in the young ocean. Apparently, large quantities of iron were released into the young ocean by hydrothermal vents that spewed hot water solutions containing iron and other metals.

Iron has tremendous affinity for oxygen. When these two elements join, they become iron oxide (rust). These early iron oxide accumulations on the seafloor created alternating layers of iron-rich rocks and chert, called **banded iron formations**. Most banded iron deposits accumulated in the Precambrian eon, between 3.5 and 2 billion years ago, and these deposits represent the world's most important reservoirs of iron ore. As a result, the percentage of oxygen in the early oceans remained low.

As photosynthesizing organisms proliferated, oxygen began to build in the oceans as well as the atmosphere. Chemical analysis of rock suggests that molecular oxygen began to appear in significant amounts in the atmosphere around 2.3 billion years ago, a phenomenon termed the **Great Oxygenation Event**. One positive benefit of the Great Oxygenation Event is that, when struck by sunlight, oxygen molecules form a compound called *ozone* (O_3), a type of oxygen molecule composed of three oxygen atoms. Ozone, which absorbs much of the Sun's harmful ultraviolet radiation before it reaches Earth's surface, is concentrated between 10 and 50 kilometers (6 and 30 miles) above Earth's surface, in a layer called the *stratosphere*. Thus, as a result of the Great Oxygenation Event, Earth's landmasses were protected from ultraviolet radiation, which is particularly harmful to DNA—the genetic blueprints for living organisms. Marine organisms had always been shielded from harmful ultraviolet radiation by seawater, and the development of the atmosphere's protective ozone layer made the continents more hospitable as well.

During the billion years following the Great Oxygenation Event, oxygen levels in the atmosphere probably fluctuated but remained below current levels. Then, just prior to the start of the Cambrian period 541 million years ago, the level of free oxygen in the atmosphere began to increase. The availability of abundant oxygen in the atmosphere contributed to the proliferation of aerobic life-forms (oxygen-consuming organisms). On the other hand, it likely wiped out huge portions of Earth's anaerobic organisms (organisms that do not require oxygen for respiration), for which oxygen is poisonous.

One apparent spike in oxygen levels occurred during the Pennsylvanian period (about 300 million years ago), when oxygen made up as much as 35 percent of the atmosphere, compared to today's level of 21 percent. One possible effect of this increase in oxygen is the occurrence of unusually large insects from that time period. (Studies have shown that, in some insects at least, higher oxygen levels promote larger size.) A fossil dragonfly found in 1979 had a wingspan of 50 centimeters (20 inches). An even larger specimen, which had a 75-centimeter (30-inch) wingspan, was named *Meganeura* (*mega* = large). One hypothesis proposes that the climate during the Pennsylvanian period was ideal for plant growth, both in the extensive swampy areas on land and in the oceans (see Figure 22.26). With all the trees and plankton in the sea producing oxygen via photosynthesis, the environment was favorable for the development of large insects.

Evolution of Earth's Oceans

When Earth cooled sufficiently to allow water vapor to condense, rainwater fell and collected in low-lying areas. By 4 billion years ago, scientists estimate that as much as

These prominent chalk cliffs are composed largely of tiny shells of marine organisms, such as foraminifera.

◀ **Figure 22.7**
White Cliffs of Dover, England Similar chalk deposits are also found in northern France.

90 percent of the current volume of seawater was contained in the developing ocean basins. Because volcanic eruptions released into the atmosphere large quantities of sulfur dioxide, which readily combines with water to form sulfuric acid, the earliest rainwater was highly acidic. The level of acidity was even greater than the acid rain that damaged lakes and streams in eastern North America during the latter part of the twentieth century. Consequently, Earth's rocky surface weathered at an accelerated rate. The products released by chemical weathering included atoms and molecules of various substances—including sodium, calcium, potassium, and silica—that were carried by running water into the newly formed oceans. Some of these dissolved substances precipitated to become chemical sediment that mantled the ocean floor. Other substances formed soluble salts, which increased the salinity of seawater. Research suggests that the salinity of the oceans initially increased rapidly, but it has remained relatively constant over the past 2 billion years.

Earth's oceans also serve as a repository for tremendous volumes of carbon dioxide, a major constituent of the primitive atmosphere. This is significant because carbon dioxide is a greenhouse gas that strongly influences the heating of the atmosphere. Venus, once thought to be very similar to Earth, has an atmosphere composed of 97 percent carbon dioxide, which produces an extreme greenhouse effect. As a result, Venus's surface temperature is 475°C (nearly 900°F).

Carbon dioxide is readily soluble in seawater, where it often combines with other atoms or molecules to produce various chemical precipitates. One of the most

common compounds generated by mineral precipitation is calcium carbonate ($CaCO_3$). Crystalline calcium carbonate is the mineral calcite, the main component of the sedimentary rock limestone. About 541 million years ago, marine organisms began to extract large quantities of calcium carbonate from seawater to make their shells and other hard parts. Trillions of tiny marine organisms, such as foraminifera, deposited their shells on the seafloor at the end of their life cycle. Some of these deposits can be observed today in the chalk beds exposed along the White Cliffs of Dover, England (**Figure 22.7**). By "locking up" carbon dioxide, these limestone deposits store massive amounts of this greenhouse gas so that it cannot easily reenter the atmosphere. Thus, the evolution of lifeforms that secrete calcium carbonate shells aided in the removal of this greenhouse gas from the atmosphere.

CONCEPT CHECKS 22.3

1. List the major gases that were added to Earth's early atmosphere through the process of outgassing.

2. Why was the evolution of photosynthesizing cyanobacteria important for the evolution of large, oxygen-consuming organisms such as humans?

Concept Checker
https://goo.gl/mCdNNu

3. How does the ocean remove carbon dioxide from Earth's atmosphere? What role do tiny marine organisms, such as foraminifera, play in the removal of carbon dioxide?

22.4 Precambrian History: The Formation of Earth's Continents

Explain the formation of continental crust, how continental crust becomes assembled into continents, and the role that the supercontinent cycle has played in this process.

Earth's first 4 billion years are encompassed in the time span called the *Precambrian*. Representing nearly 90 percent of Earth's history, the Precambrian is divided into the *Archean eon* ("ancient age"), the *Proterozoic eon* ("early life age"), and the informal time span referred to as the Hadean. Our knowledge of this ancient time is limited because much of the early rock record has been obscured by the very Earth processes you have been studying, especially plate tectonics, erosion, and deposition. Most Precambrian rocks lack fossils, which hinders correlation of rock units (see Chapter 9). In addition, rocks this old are often metamorphosed and deformed, extensively eroded, and frequently concealed by younger strata. Indeed, Precambrian history is written in scattered, speculative episodes, like a long book with many missing chapters.

Earth's First Continents

Geologists have discovered tiny crystals of the mineral zircon in continental rocks that formed 4.4 billion years ago—evidence that the continents began to form early in Earth's history. By contrast, the oldest rocks in the ocean basins are generally less than 200 million years old.

What differentiates continental crust from oceanic crust? Recall that oceanic crust is a relatively dense (3.0 g/cm^3) homogeneous layer of basaltic rocks derived from partial melting of the rocky upper mantle. In addition, oceanic crust is thin, averaging only 7 kilometers (4 miles) thick. Continental crust, on the other hand, is composed of a variety of rock types, has an average thickness of nearly 40 kilometers (25 miles), and contains a large percentage of low-density (2.7 g/cm^3), silica-rich rocks such as granite.

The significance of the differences between continental crust and oceanic crust cannot be overstated in a review of Earth's geologic evolution. The relatively thin and dense oceanic crust is found several kilometers below sea level—unless of course it has been pushed onto a landmass by tectonic forces. Continental crust, because of its great thickness and lower density, may extend well above sea level. Also, recall that dense oceanic crust of normal thickness readily subducts, whereas thick, buoyant blocks of continental crust resist being recycled into the mantle.

Making Continental Crust   The formation of continental crust is a continuation of the gravitational segregation of Earth materials that began during the final stage of our planet's formation. Dense metallic material, mainly iron and nickel, sank to form Earth's core, leaving behind the less dense rocky material that forms the mantle. It is from Earth's rocky mantle that low-density, silica-rich minerals were gradually distilled to form continental crust. In a similar manner, partial melting of mantle rocks generates low-density, silica-rich melts that buoyantly rise to the surface to form Earth's crust, leaving behind the dense mantle rocks (see Chapter 5). However, little is known about the details of the mechanisms that generated these silica-rich melts during the Archean eon.

Earth's first crust was probably ultramafic in composition, but because physical evidence no longer exists, we are not certain. The hot, turbulent mantle that most likely existed during the Archean eon recycled most of this crustal material back into the mantle. In fact, it may have been continuously recycled, in much the same way that the "crust" that forms on a lava lake is repeatedly replaced with fresh lava from below (**Figure 22.8**).

The oldest preserved continental rocks occur as small, highly deformed *terranes*, which are incorporated within somewhat younger blocks of continental crust (**Figure 22.9**). One of these is a 3.8-billion-year-old terrane located near Isua, Greenland. Slightly older crustal rocks, called the Acasta Gneiss, have been discovered in Canada's Northwest Territories.

Some geologists think that a type of plate-like motion operated early in Earth's history. In addition, hot-spot volcanism was likely active during this time.

▼ Figure 22.8
Earth's early crust was continually recycled

The crust covering this lava lake is continually being replaced with fresh lava from below, much like the way Earth's crust was recycled early in its history.

However, because the mantle was hotter in the Archean than it is today, both of these phenomena would have progressed at faster rates than their modern counterparts. Hot-spot volcanism, due to mantle plumes, is thought to have created immense shield volcanoes as well as oceanic plateaus. Simultaneously, subduction of oceanic crust generated volcanic island arcs. Collectively, these relatively small crustal fragments represent the first phase in creating stable, continent-size landmasses.

From Continental Crust to Continents The growth of larger continental masses was accomplished through collision and accretion of many thin, highly mobile crustal fragments, as illustrated in **Figure 22.10**. This type of collisional tectonics deformed and metamorphosed sediments caught between converging crustal fragments, thereby shortening and thickening the developing crust. In the deepest regions of these collision zones, partial melting of the thickened crust generated silica-rich magmas that ascended and intruded the rocks above. This led to the formation of large crustal provinces that, in turn, accreted with others to form even larger crustal blocks called **cratons**.

The assembly of a large craton involves the accretion of several crustal blocks that cause major mountain-building episodes similar to India's collision with Asia. **Figure 22.11** shows the extent of crustal material that was produced during the Archean and Proterozoic eons and remains today. The regions within a modern continent

These rocks at Isua, Greenland, some of the world's oldest, have been dated at 3.8 billion years.

▲ Figure 22.9
Earth's oldest preserved continental rocks are more than 3.8 billion years old

where these ancient cratons are exposed at Earth's surface are called **shields** (see Figure 14.1, page 392).

Although the Precambrian was a time when much of Earth's continental crust was generated, a substantial amount of crustal material was destroyed as well. Some was lost via weathering and erosion. In addition, during much of the Archean, it appears that thin slabs of continental crust were subducted into the mantle. However, by about 3 billion years ago, cratons grew sufficiently

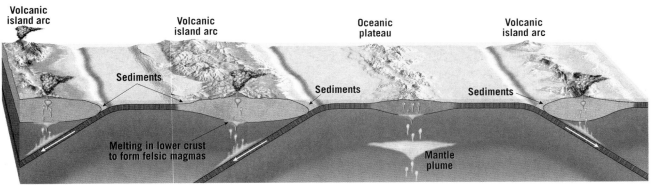

Volcanic island arc   Volcanic island arc   Oceanic plateau   Volcanic island arc

Sediments   Sediments   Sediments

Melting in lower crust to form felsic magmas   Mantle plume

A. Scattered crustal fragments separated by ocean basins

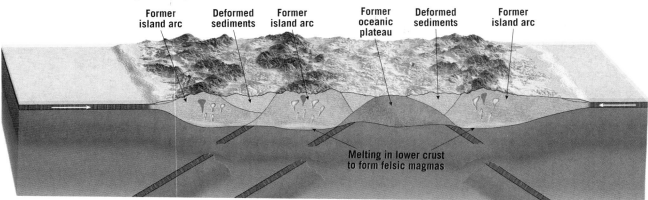

Former island arc   Deformed sediments   Former island arc   Former oceanic plateau   Deformed sediments   Former island arc

Melting in lower crust to form felsic magmas

B. Collision of volcanic island arcs and oceanic plateau to form a larger crustal block

◄ SmartFigure 22.10
The formation of continents The growth of large continental masses occurs through the collision and accretion of smaller crustal fragments.

Tutorial
https://goo.gl/
M9bie6

▶ Figure 22.11
Distribution of crustal material remaining from the Archean and Proterozoic eons Ages are in billions of years (Ga).

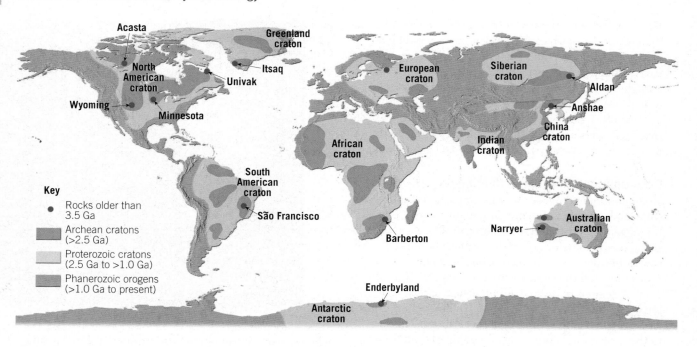

Key

- Rocks older than 3.5 Ga
- Archean cratons (>2.5 Ga)
- Proterozoic cratons (2.5 Ga to >1.0 Ga)
- Phanerozoic orogens (>1.0 Ga to present)

large and thick to resist subduction. After that time, weathering and erosion became the primary processes of crustal destruction. By the close of the Precambrian, an estimated 85 percent of the modern continental crust had formed.

The Making of North America

North America provides an excellent example of the development of continental crust and its piecemeal assembly into a continent. With the exception of a few locations noted in Figure 22.11, very little continental crust older than 3.5 billion years remains. In the late Archean, between 3 and 2.5 billion years ago, there was a period of major continental growth, which is shown in purple in **Figure 22.12**. During this span, the accretion of numerous island arcs and other fragments generated several large crustal provinces. North America contains some of these crustal units, including the Superior and Hearne-Rae cratons shown in Figure 22.12, but just where these ancient continental blocks formed is unknown.

About 1.9 billion years ago, these crustal provinces collided to produce the Trans-Hudson mountain belt (see Figure 22.12). (Such mountain-building episodes were not restricted to North America; ancient deformed strata of similar age are also found on other continents.) This event built the North American craton, around which several large and numerous small crustal fragments were later added. One of these late arrivals is the Appalachian province. In addition, several terranes were added to the western margin of North America during the Mesozoic and Cenozoic eras to

North America was assembled from crustal blocks that were joined by processes very similar to modern plate tectonics. Ancient collisions produced mountain belts that include remnant volcanic island arcs, trapped by colliding continental fragments.

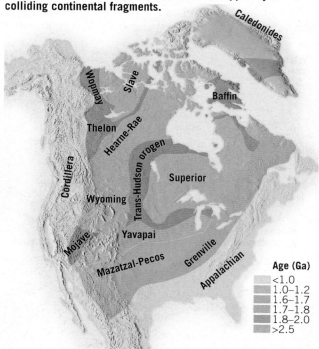

Age (Ga)
- <1.0
- 1.0–1.2
- 1.6–1.7
- 1.7–1.8
- 1.8–2.0
- >2.5

▲ SmartFigure 22.12
The major geologic provinces of North America The age of each province is in billions of years (Ga).

Tutorial
https://goo.gl/qTzVX7

generate the mountainous North American Cordillera (see Figure 14.13, page 403).

Supercontinents of the Precambrian

At different times, parts of what is now North America combined with other continental landmasses to form a supercontinent. **Supercontinents** are large landmasses that contain all, or nearly all, the existing continents. Pangaea was the most recent, but certainly not the only, supercontinent to exist in the geologic past. The earliest well-documented supercontinent, *Rodinia*, formed during the Proterozoic eon, about 1.1 billion years ago (**Figure 22.13**). Although geologists are still studying its construction, it is clear that Rodinia's configuration was quite different from Pangaea's. One obvious distinction is North America's position near the center of this ancient landmass.

Between 800 and 600 million years ago, Rodinia gradually split apart. By the end of the Precambrian many of the fragments had reassembled, producing a large landmass in the Southern Hemisphere called *Gondwana*, composed mainly of present-day South America, Africa, India, Australia, and Antarctica (**Figure 22.14**). Some continental fragments—including portions of North America, Siberia, and Northern Europe—remained separate landmasses. We consider the fate of these Precambrian landmasses in the next section.

Supercontinent Cycle

The **supercontinent cycle** involves rifting and dispersal of one supercontinent followed by a long period during which the fragments gradually reassemble into a new supercontinent with a different configuration. The assembly and dispersal of supercontinents had a profound impact on the evolution of Earth's continents. In addition, this phenomenon greatly influenced global climates and contributed to periodic episodes of rising and falling sea level.

Supercontinents and Climate

The movement of continents changes the patterns of ocean currents and global winds, which influences the global distribution of temperature and precipitation. The formation of the Antarctic's vast ice sheet is one example of how the movement of a continent is thought to have contributed to climate change. Although eastern Antarctica remained over the South Pole for more than 100 million years, Antarctica was not covered by a stable continental-scale ice sheet until about 34 million years ago. Prior to this period of glaciation, South America and Antarctica were connected. As shown in **Figure 22.15A**, this arrangement of landmasses helped maintain a circulation pattern in which warm ocean currents reached the coast of Antarctica and helped to keep Antarctica mostly ice free. This is similar to the way in which the modern Gulf Stream helps keep Iceland mostly ice free, despite the island country's name.

As South America separated from Antarctica and moved northward, a pattern of ocean circulation

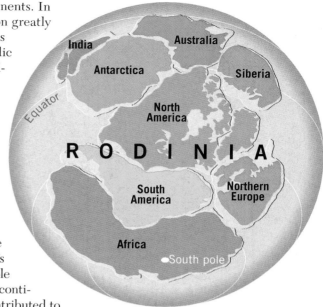

▲ **Figure 22.13
Possible configuration of the supercontinent Rodinia** For clarity, the continents are drawn with somewhat modern shapes, not their actual shapes from 1 billion years ago.

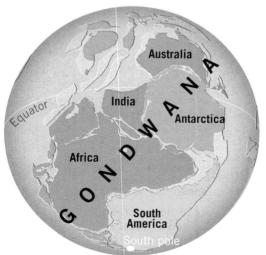

A. Continent of Gondwana

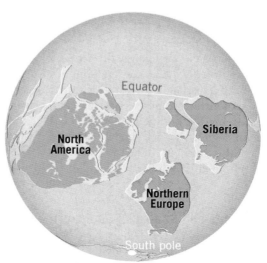

B. Continents not a part of Gondwana

◀ **Figure 22.14
Reconstruction of Earth as it may have appeared in late Precambrian time** The southern continents were joined into a single landmass called Gondwana. Other landmasses that were not part of Gondwana include North America, northwestern Europe, and northern Asia.

50 million years ago warm ocean currents kept Antarctica nearly ice free.

As South America separated from Antarctica, the West Wind Drift developed. This newly formed ocean current effectively cut Antarctica off from warm currents and contributed to the formation of its vast ice sheets.

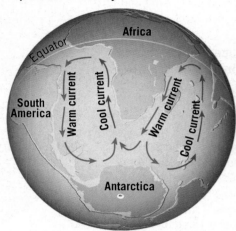
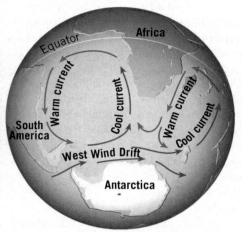

A. Antarctica not extensively glaciated   **B. Antarctica covered by continental-size ice sheets**

developed that flowed from west to east around the entire continent of Antarctica (Figure 22.15B). This cold current, called the West Wind Drift, effectively isolated the entire Antarctic coast from the warm, poleward-directed currents in the southern oceans. This change in circulation, along with a period of global cooling, probably resulted in the growth of Antarctica's massive ice sheet.

Local and regional climates are also influenced by large mountain systems created by the collision of large cratons. One example is the collision of the Indian subcontinent with southern Asia that generated the Himalayas. Because of their high elevations, mountains exhibit markedly lower average temperatures than surrounding lowlands. In addition, air rising over these lofty structures promotes condensation and precipitation, leaving the region downwind relatively dry. A modern analogy is the wet, heavily forested western slopes of the Sierra Nevada compared to the dry climate of the Great Basin Desert that lies directly to the east.

The supercontinent cycle involves rifting and dispersal of one supercontinent followed by a long period during which the fragments gradually reassemble into a new supercontinent with a different configuration.

Supercontinents and Sea-Level Changes

Significant and numerous sea-level changes have been documented in geologic history, and many of them appear to have been related to the assembly and dispersal of supercontinents. If sea level rises, shallow seas advance onto the continents. Evidence for periods when the seas advanced onto the continents include thick sequences of ancient marine sedimentary rocks that blanket large areas of modern landmasses—including much of the eastern two-thirds of the United States.

The supercontinent cycle and sea-level changes are directly related to rates of *seafloor spreading*. Recall from Chapter 13 that when the rate of spreading is rapid, as it is along the East Pacific Rise today, the production of warm oceanic crust is also high. Because new, warm oceanic crust is less dense (takes up more space) than cold crust, fast-spreading ridges occupy more volume in the ocean basins than do slow-spreading centers. (Think of getting into a tub filled with water.) As a result, when rates of seafloor spreading increase, more seawater is displaced, which results in the sea level rising. This, in turn, causes shallow seas to advance onto the low-lying portions of the continents.

CONCEPT CHECKS 22.4

1. Describe how cratons came into being.

2. What is the supercontinent cycle? What supercontinent preceded Pangaea?

Concept Checker
https://goo.gl/8ar2hm

3. Explain how the rate of seafloor spreading is related to changes in sea level.

22.5 Geologic History of the Phanerozoic: The Formation of Earth's Modern Continents

List and discuss the major geologic events in the Paleozoic, Mesozoic, and Cenozoic eras.

The time span since the close of the Precambrian, called the *Phanerozoic eon*, encompasses 541 million years and is divided into three eras: *Paleozoic*, *Mesozoic*, and *Cenozoic*. The beginning of the Phanerozoic is marked by the appearance of the first life-forms with hard parts such as shells, scales, bones, or teeth—all of which greatly enhance the chances for an organism to be preserved in the fossil record. Thus, the study of Phanerozoic crustal history was aided by the availability of fossils, which improved our ability to date and correlate geologic events. Moreover, because every organism is associated with its own particular environmental niche, the greatly improved fossil record provided invaluable information for deciphering ancient environments.

Paleozoic History

As the Paleozoic era opened, what is now North America hosted no land plants or animals large enough to be seen—just tiny microorganisms such as bacteria. There were no Appalachian or Rocky Mountains; the continent was largely a barren lowland. Several times during the early Paleozoic, shallow seas moved inland and then receded from the continental interior, leaving behind the thick deposits of limestone, shale, and clean sandstone that mark the shorelines of these previously midcontinent shallow seas.

Formation of Pangaea One of the major events of the Paleozoic era was the formation of the supercontinent **Pangaea**. This event began with a series of collisions that over millions of years joined North America, Europe, Siberia, and other smaller crustal fragments to form a large continent called **Laurasia** (Figure 22.16). Located south of Laurasia was the vast southern continent called **Gondwana**, which encompassed five modern landmasses—South America, Africa, Australia, Antarctica, and India—and perhaps portions of China. Evidence of extensive continental glaciation places this landmass near the South Pole. By the late Paleozoic, Gondwana had migrated northward and collided with Laurasia to begin the final stage in the assembly of Pangaea.

The accretion of all of Earth's major landmasses to form Pangaea spans more than 300 million years and resulted in the formation of several mountain belts. The collision of northern Europe (mainly Norway) with Greenland produced the Caledonian Mountains, and the joining of northern Asia (Siberia) and Europe created the Ural Mountains. Northern China is also thought to have accreted to Asia by the end of the Paleozoic, whereas southern China may not have become part of Asia until after Pangaea had begun to rift. (Recall that India did not begin to accrete to Asia until about 50 million years ago.)

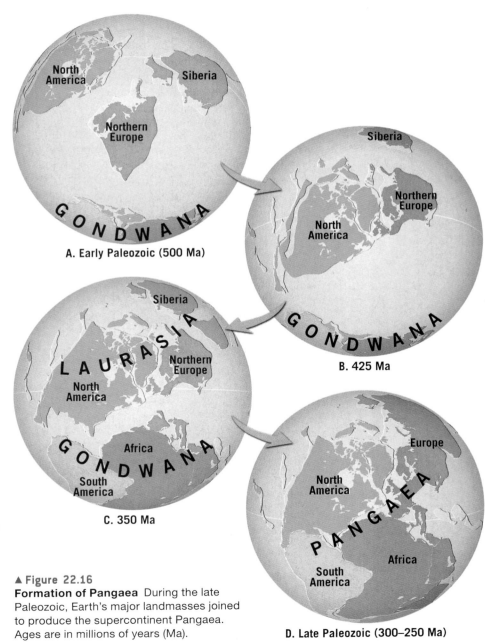

A. Early Paleozoic (500 Ma)

B. 425 Ma

C. 350 Ma

D. Late Paleozoic (300–250 Ma)

▲ Figure 22.16
Formation of Pangaea During the late Paleozoic, Earth's major landmasses joined to produce the supercontinent Pangaea. Ages are in millions of years (Ma).

Pangaea reached its maximum size between 300 and 250 million years ago, as Africa collided with North America (see Figure 22.16D). This event marked the final and most intense period of mountain building in the long history of the Appalachian Mountains (see Figure 14.10, page 401). This mountain-building event produced the central Appalachians of the Atlantic states, as well as New England's northern Appalachians and mountainous structures that extend into Canada (**Figure 22.17**).

Mesozoic History

Spanning about 186 million years, the Mesozoic era is divided into three periods: the *Triassic, Jurassic,* and *Cretaceous.* Major geologic events of the Mesozoic include the breakup of Pangaea and the evolution of our modern ocean basins.

Changes in Sea Levels   The Mesozoic era began with much of the world's continents above sea level. The exposed Triassic strata are primarily red sandstones and mudstones that lack marine fossils, features that indicate a terrestrial environment. (The red color in sandstone comes from the oxidation of iron.)

As the Jurassic period opened, the sea invaded western North America. Adjacent to this shallow sea, extensive continental sediments were deposited on what is now the Colorado Plateau. The most prominent is the Navajo Sandstone, a cross-bedded, quartz-rich layer that in some places approaches 300 meters (1000 feet) thick. These remnants of massive dunes indicate that an enormous desert occupied much of the American Southwest during early Jurassic times. Another well-known Jurassic deposit is the Morrison Formation—one of the world's richest storehouses of dinosaur fossils. Included are the fossilized bones of massive dinosaurs such as *Apatosaurus, Brachiosaurus,* and *Stegosaurus.*

The Breakup of Pangaea   Another major event of the Mesozoic era was the breakup of Pangaea. About 185 million years ago, a rift developed between what is now North America and western Africa, marking the birth of the Atlantic Ocean. As Pangaea gradually broke apart, the westward-moving North American plate began to override the Pacific basin. This tectonic event triggered a continuous wave of deformation that moved inland along the entire western margin of North America.

Formation of the North American Cordillera   By Jurassic times, subduction of the Pacific basin under the North American plate began to produce the chaotic mixture of rocks that exist today in the Coast Ranges of California (see Figure 14.7, page 397). Further inland, igneous activity was widespread, and for more than 100 million years, volcanism was rampant as huge masses of magma rose to within a few kilometers of Earth's surface. The remnants of this activity include the granitic plutons of the Sierra Nevada, as well as the Idaho batholith and British Columbia's Coast Range batholith.

The subduction of the Pacific basin under the western margin of North America also resulted in the piecemeal addition of crustal fragments to the entire Pacific margin of the continent—from Mexico's Baja Peninsula to northern Alaska (see Figure 14.13, page 403). Each collision displaced crustal fragments (terranes) accreted earlier farther inland, adding to the zone of deformation as well as to the thickness and lateral extent of the continental margin.

Northern Appalachians

Central Appalachians

Appalachian Plateau

Valley and Ridge

Blue Ridge Mountains

Piedmont

Coastal Plain

A.

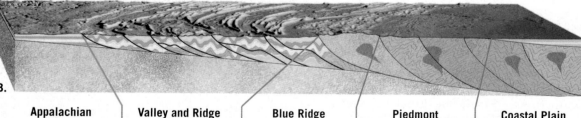

B.

▶ SmartFigure 22.17
Major provinces of the Appalachian Mountains

Tutorial
https://goo.gl/75f4xN

Appalachian Plateau (Underlain by nearly flat-lying sedimentary strata of Paleozoic age.)	Valley and Ridge (Highly folded and thrust-faulted sedimentary rocks of Paleozoic age.)	Blue Ridge (Hilly to mountainous terrain consisting of slices of basement rock of Precambrian age.)	Piedmont (Crustal fragments of metamorphosed sedimentary and igneous rocks that were added to North America.)	Coastal Plain (Area of low relief underlain by gradually sloping sedimentary strata and unlithified sediments.)

Compressional forces moved huge rock units in a shingle-like fashion toward the east. Across much of North America's western margin, older rocks were thrust eastward over younger strata, for distances exceeding 150 kilometers (90 miles). Ultimately, this activity was responsible for generating a vast portion of the North American Cordillera that extends from Wyoming to Alaska.

Toward the end of the Mesozoic, the southern portions of the Rocky Mountains developed (**Figure 22.18**). This mountain-building event, called the *Laramide Orogeny*, occurred when large blocks of deeply buried Precambrian rocks were lifted nearly vertically along steeply dipping faults, upwarping the overlying younger sedimentary strata. The mountain ranges produced by the Laramide Orogeny include Colorado's Front Range, the Sangre de Cristo of New Mexico and Colorado, and the Bighorns of Wyoming (see GEOgraphics 14.1).

▲ Figure 22.18
Autumn at Maroon Bells These snow-covered peaks are a part of the Laramide Rockies.

Cenozoic History

The Cenozoic era, or "era of recent life," encompasses the past 66 million years of Earth history. It was during this span that the physical landscapes of our modern world came into existence, along with many familiar life-forms. The Cenozoic era represents a considerably smaller fraction of geologic time than either the Paleozoic or the Mesozoic, but we know much more about this time span because the rock formations are more widespread and less disturbed than those of any preceding era.

Most of North America was above sea level during the Cenozoic era. However, the eastern and western margins of the continent experienced markedly dissimilar events because of their different plate boundary relationships. The Atlantic and Gulf coastal regions, far removed from active plate boundaries, were tectonically stable. By contrast, western North America was the leading edge of the North American plate, and the plate interactions during the Cenozoic account for many events of mountain building, volcanism, and earthquakes.

Eastern North America   The stable continental margin of eastern North America was the site of abundant marine sedimentation. The most extensive deposits surrounded the Gulf of Mexico, from the Yucatan Peninsula to Florida, where a massive buildup of sediment caused the crust to downwarp. In many instances, faulting created structures in which oil and natural gas accumulated. Today, these and other petroleum traps (see Figure 23.6, page 666) are the Gulf coast's most economically important resource, as evidenced by numerous offshore drilling platforms.

Early in the Cenozoic, the Appalachians had eroded to create a low plain. Later, isostatic adjustments again raised the region and rejuvenated its rivers. Streams eroded with renewed vigor, gradually sculpting the surface into its present-day topography. Sediments from this erosion were deposited along the eastern continental margin, where they accumulated to a thickness of many kilometers. Today, portions of the strata deposited during the Cenozoic are exposed as the gently sloping Atlantic and Gulf coastal plains, where a large percentage of the eastern and southeastern United States population resides.

Western North America   In the West, the Laramide Orogeny, responsible for building the southern Rocky Mountains, was coming to an end. As erosional forces lowered the mountains, the basins between uplifted ranges began to fill with sediment. East of the Rockies, a large wedge of sediment from the eroding mountains created the gently sloping Great Plains.

Beginning in the Miocene epoch, about 20 million years ago, a broad region from northern Nevada into Mexico experienced crustal extension that created more than 100 fault-block mountain ranges. Today, they rise abruptly above the adjacent basins, forming the Basin and Range Province (see Figure 14.15, page 405).

During the development of the Basin and Range Province, the entire western interior of the continent gradually uplifted. This event elevated the Rockies and rejuvenated many of the West's major rivers. As the rivers became incised, many spectacular gorges were created, including the Grand Canyon of the Colorado River, the Grand Canyon of the Snake River, and the Black Canyon of the Gunnison River.

Volcanic activity was also common in the West during much of the Cenozoic. Beginning in the Miocene epoch, great volumes of fluid basaltic lava flowed from fissures in portions of present-day Washington, Oregon, and Idaho. These eruptions built the 3.4-million-square-kilometer (1.3-million-square-mile) Columbia Plateau. Immediately west of the vast Columbia Plateau, volcanic activity was different in character. Here, more viscous magmas with higher silica content erupted explosively to create the Cascades, a chain of stratovolcanoes extending from northern California into Canada. Some of these volcanoes are still active (see Figure 5.30, page 162).

As the Cenozoic was drawing to a close, the effects of mountain building, volcanic activity, isostatic adjustments, and extensive erosion and sedimentation created the physical landscape we know today. All that remained of the Cenozoic era was the final 2.6-million-year episode called the Quaternary period. During this most recent, and ongoing, phase of Earth's history, humans evolved and the action of glacial ice, wind, and running water added to our planet's long, complex geologic history.

CONCEPT CHECKS 22.5

1. During which period of geologic history did the supercontinent Pangaea come into existence?

2. Compare and contrast eastern and western North America's geology during the Cenozoic era.

Concept Checker
https://goo.gl/6Fdhhr

22.6 Earth's First Life

Describe some of the hypotheses on the origin of life and the characteristics of early prokaryotes, eukaryotes, and multicellular organisms.

The oldest fossils provide evidence that life on Earth was established at least 3.5 billion years ago (**Figure 22.19**). Microscopic fossils similar to modern cyanobacteria have been found in silica-rich chert deposits worldwide. Notable examples include southern Africa, where rocks date to more than 3.1 billion years ago, and the Lake Superior region of western Ontario and northern Minnesota, where the Gunflint Chert contains some fossils older than 2 billion years. Chemical traces of organic matter in even older rocks have led paleontologists to conclude that life may have existed much earlier.

Origin of Life

How did life begin? This question sparks considerable debate, and hypotheses abound. Requirements for life, in addition to a hospitable environment, include the chemical raw materials—principally complex organic compounds—that are essential for life. The organic molecules that provide the primary structural material for life and contribute to its functioning are **proteins** (from the Greek *proteiso*, which means "primary"—hence "primary substance"). Proteins consist of long chains made up of comparatively small molecular units called *amino acids*. Because proteins cannot make copies of themselves—a necessary condition for the proliferation of life—the production of proteins from simpler amino acids requires a template. These information-carrying, self-replicating (or *genetic*) components of life are types of *nucleic acids* with which you are likely familiar—DNA and RNA.

What was the source of the basic organic molecules that became Earth's first life? The first organic molecules may have been synthesized from carbon dioxide and nitrogen, both of which were plentiful in Earth's primitive atmosphere. Some scientists suggest that these gases could have been easily reorganized into amino acids by ultraviolet light. Others believe lightning was the impetus, as the groundbreaking experiments conducted by biochemists Stanley Miller and Harold Urey attempted to demonstrate.

Still other researchers suggest that amino acids arrived "ready-made," delivered by asteroids or comets that collided with a young Earth. Evidence for this hypothesis comes from a group of meteorites, called *carbonaceous chondrites*, which contain amino acid–like organic compounds.

Yet another hypothesis proposes that the organic material needed for life came from the methane and hydrogen sulfide that spews from deep-sea hydrothermal vents (black smokers). It is also possible that life originated in hot springs similar to those in Yellowstone National Park.

Earth's First Life: Prokaryotes

Regardless of where or how life originated, it is clear that the journey from "then" to "now" involved change (see Figure 22.19). The first known organisms were simple, single-cell **prokaryotes** (bacteria and similar microbes), in which DNA is not segregated from the rest of the cell in a nucleus.

A major triumph in the history of prokaryote life was the evolution of a primitive type of photosynthesis, which provided the energy that allowed life to proliferate. Recall that photosynthesis by *cyanobacteria* contributed to the

Evolution of Life Through Geologic Time

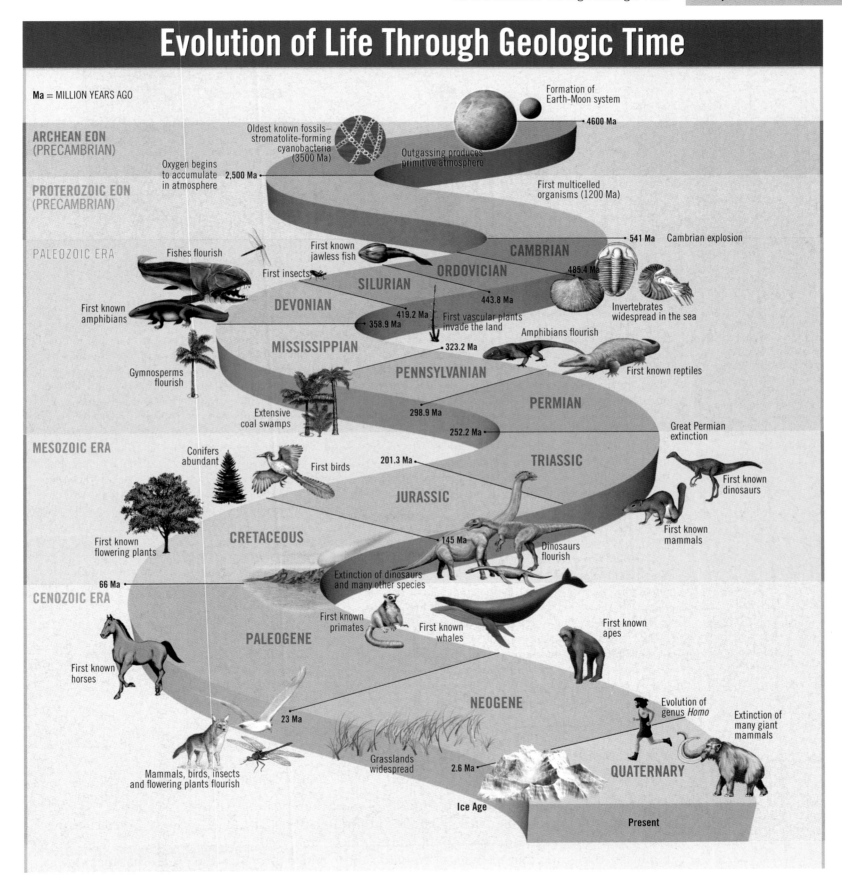

▲ Figure 22.19
Evolution of life though geologic time Ages are in millions of years (Ma).

A.

B.

▲ Figure 22.20
Stromatolites are among the most common Precambrian fossils
A. Cross-section though fossil stromatolites deposited by cyanobacteria.
B. Modern stromatolites exposed at low tide in western Australia.

gradual rise in the level of oxygen in the atmosphere. Fossil evidence for the existence of these microscopic bacteria are distinctively layered mats, called **stromatolites**, which are composed of slimy material secreted by these organisms, along with trapped sediments (**Figure 22.20A**). What is known about these ancient fossils comes mainly from the study of modern stromatolite structures found in Shark Bay, Australia (**Figure 22.20B**). Today's stromatolites look like stubby pillars built as these microbes slowly move upward to avoid being buried by the sediment that is continually deposited on them.

Evolution of Eukaryotes

The oldest fossils of larger and more complex organisms, called **eukaryotes**, are about 2.1 billion years old. Eukaryotic cells have their genetic material segregated into a nucleus, and they are more complex in other ways than their prokaryotic precursors. While the first eukaryotes were single celled, all the complex multicellular organisms that now inhabit our planet—trees, birds, fish, reptiles, and humans—are eukaryotes.

During much of the Precambrian, life consisted largely of single-celled organisms. It wasn't until about 1.2 billion years ago that multicellular eukaryotes

evolved. Green algae, one of the first types of multicellular eukaryote, contained chloroplasts (used in photosynthesis) and were the likely ancestors of modern plants. The first marine multicellular animals did not appear until somewhat later, perhaps 600 million years ago.

Fossil evidence suggests that life underwent evolutionary change at an excruciatingly slow pace until nearly the end of the Precambrian. At that time, Earth's continents were largely barren, and the oceans were populated mainly with tiny organisms, many too small to be seen with the naked eye. Nevertheless, the stage was set for the evolution of much more diverse plants and animals.

CONCEPT CHECKS 22.6

1. What group of organic compounds combine to form proteins and is therefore necessary for life as we know it?

2. What are stromatolites?

3. Compare prokaryotes with eukaryotes. To which group do most modern multicellular organisms belong?

 Concept Checker
https://goo.gl/4QEmi4

22.7 Paleozoic Era: Life Explodes

List the major developments in the history of life during the Paleozoic era.

The Cambrian period marks the beginning of the Paleozoic era, a time span that saw the emergence of a spectacular variety of new life-forms. All major **invertebrate** (animals lacking backbones) groups became widespread during the Cambrian, including jellyfish, sponges, worms, mollusks (the group that includes clams and snails), and arthropods (the group that includes insects and crabs). This expansion in biodiversity, which began about 541 million years ago, is known as the **Cambrian explosion**.

Early Paleozoic Life-Forms

The Cambrian explosion, which lasted about 20 to 30 million years, led to an immense variety of invertebrate animals inhabiting Earth's oceans. The development of hard shells and skeletons resulted in predators evolving to have sharp claws and modified mouth parts to capture and break apart their prey. Other animals developed defense mechanisms, including spikes or armor.

The Cambrian period was the golden age of *trilobites* (**Figure 22.21**). Like modern crabs and lobsters, trilobites had a jointed external skeleton, which enabled them to be mobile and obtain food. More than 600 genera of these mud-burrowing scavengers and grazers flourished worldwide.

The Ordovician period marked the appearance of abundant cephalopods—mobile, highly developed mollusks that became the major predators of their time (**Figure 22.22**). Descendants of these cephalopods include the squid, octopus, and chambered nautilus that inhabit our modern oceans. Cephalopods were the first truly large organisms on Earth, including one species that reached a length of nearly 10 meters (more than 30 feet).

The early diversification of animals was driven, in part, by the emergence of predatory lifestyles. The larger mobile cephalopods preyed on trilobites that were typically smaller than a child's hand. The evolution of efficient movement was often associated with the development of greater sensory capabilities and more complex

◄ Figure 22.21
Fossil of a trilobite Trilobites, which were abundant in the early Paleozoic, found food on the ocean bottom.

◄ Figure 22.22
Artistic depiction of a shallow Ordovician sea During the Ordovician period (488–444 million years ago), the shallow waters of an inland sea over central North America contained an abundance of marine invertebrates. Shown in this reconstruction are (1) corals, (2) a trilobite, (3) a snail, (4) brachiopods, and (5) a straight-shelled cephalopod.

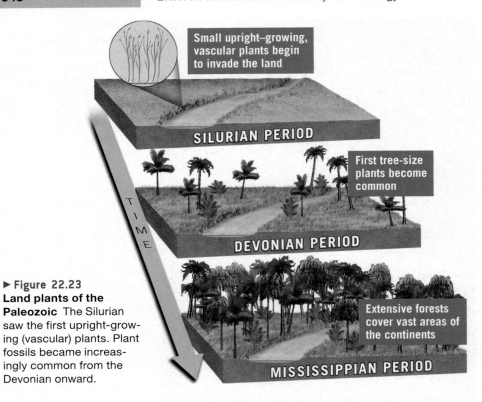

Small upright–growing, vascular plants begin to invade the land

SILURIAN PERIOD

First tree-size plants become common

DEVONIAN PERIOD

Extensive forests cover vast areas of the continents

MISSISSIPPIAN PERIOD

T I M E

▶ Figure 22.23
Land plants of the Paleozoic The Silurian saw the first upright-growing (vascular) plants. Plant fossils became increasingly common from the Devonian onward.

nervous systems. These early animals elaborated sensory devices for detecting light, odor, and touch.

Mid-Paleozoic Life

Approximately 450 million years ago, green algae that had adapted to survive at the water's edge gave rise to the first multicellular land plants. Some of the adaptations needed for sustaining plant life on land were the ability to transport water and minerals internally as well as stay upright, despite gravity and winds. By 420 million years ago, plants stood upright and possessed a primitive *vascular system* to transport ion-rich water. These leafless vertical spikes were about the size of a human index finger, about 10 centimeters (4 inches) tall. By the beginning of the Mississippian period, there were forests with treelike plants tens of meters tall (**Figure 22.23**).

In the ocean, the **vertebrates** (animals with backbones) began to thrive—primarily the ancestors of modern fish. An important derived characteristic of these early vertebrates was jaws, which enabled them to grab, pry, or bite off a chunk of flesh. One example was the armor-plated fishes called *placoderms* (meaning "plate-skinned") that evolved during the Ordovician (**Figure 22.24**).

Other fish evolved during the Devonian, including sharks with cartilage skeletons and bony fish with hard internal skeletons—the two groups to which most modern fish belong. The first large vertebrates, fish proved to be faster swimmers than most invertebrates and possessed acute senses and large brains. They became dominant predators of the sea, which is why the Devonian period is sometimes referred to as the "Age of the Fishes."

Vertebrates Move to Land

At the beginning of the Devonian period, all vertebrates shared the same fish-like anatomy. Shortly thereafter, a group of fishes called *lobe-finned fishes* or simply *lobe-fins* began to adapt to terrestrial environments (**Figure 22.25A**). The Devonian landscape included abundant coastal wetlands, which were home to a wide range

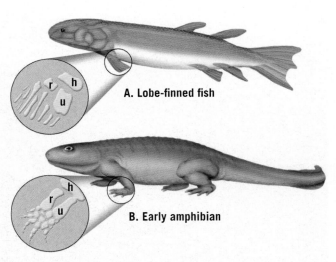

A. Lobe-finned fish

B. Early amphibian

▲ Figure 22.25
Comparison of the anatomical features of a lobe-finned fish and an early amphibian A. The fins on the lobe-finned fish contained the same basic elements (*h*, humerus, or upper arm; *r*, radius, and *u*, ulna, which correspond to the lower arm) as those of the amphibians. **B.** This amphibian is shown with the standard five toes, but early amphibians had as many as eight toes, as well as other characteristics that made them quite different from modern amphibians. Eventually the amphibians evolved to have a standard toe count of five.

▶ Figure 22.24
Fossil of an armored fish belonging to a group called the placoderms This particular placoderm was a formidable predator that grew up to 10 meters (30 feet) in length, although most were much smaller.

1 meter

of lobe-fins. Like some modern fishes, such as the African lungfish, when lobe-fins occupied oxygen-poor water, they would use their primitive lungs for breathing. Some lobe-finned fishes also no doubt used their stout bony fins to move from one pond to another. By the late Devonian, the fins of one group of lobe-finned fishes had evolved into four limbs and feet with digits (**Figure 22.25B**). The limbs were used to support the weight of the animal on land, and the feet aided in walking. By about 365 million years ago, this group of lobe-fins had given rise to the four-legged ancestors of the first amphibians and eventually to other land-dwelling vertebrate groups, including reptiles and mammals (**Figure 22.26**).

Modern amphibians, such as frogs, toads, and salamanders, are small and occupy mainly ponds or damp habitats such as swamps and rain forests. Conditions during the late Paleozoic were ideal for these newcomers to land. Large tropical swamps teeming with insects and millipedes extended across North America, Europe, and Siberia. These tropical swamps produced the coal deposits found in much of the eastern third of the United States (**Figure 22.27**). With an abundance of ideal habitats,

amphibians diversified rapidly. Some even took on lifestyles and forms similar to those of modern reptiles such as crocodiles.

Despite their success on land, amphibians are not fully adapted to drier habitats. Amphibians must lay their eggs in water or moist environments on land because their eggs lack a shell and dehydrate in dry air. In fact, **amphibian** means "both ways of life." Frogs, for instance, lay their eggs in water and develop as aquatic tadpoles with gills and tails (fish characteristics) before maturing into air-breathing adults with four legs.

Reptiles: The First True Terrestrial Vertebrates

Fossil evidence indicates that about 310 million years ago, the first reptiles evolved from an ancestor common to both reptiles and amphibians. The earliest known reptiles resembled lizards with small, sharp teeth. The **reptile** group includes snakes, turtles, lizards, and crocodiles, as well as extinct groups such as

▼ SmartFigure 22.26
Relationships of major land-dwelling vertebrate groups and their divergence from lobe-finned fish

Tutorial
https://goo.gl/
gLc65G

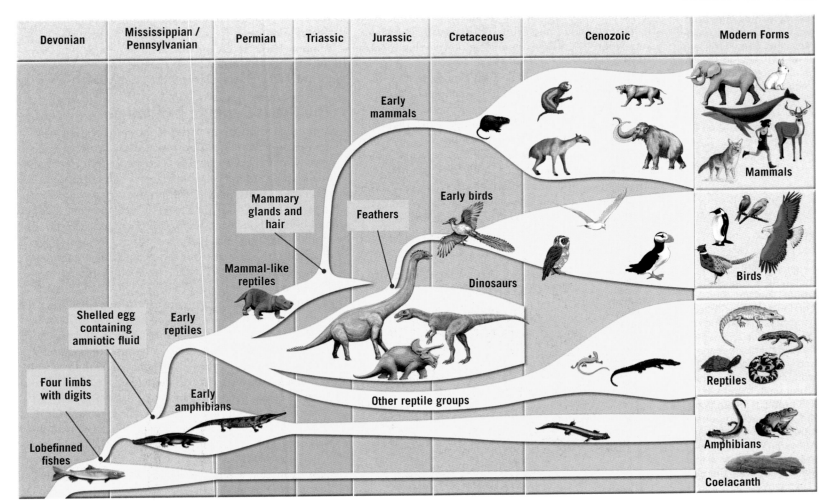

▲ Figure 22.27
Artistic depiction of a Pennsylvanian-age coal swamp Shown are scale trees (left), seed ferns (lower left), and horsetails (right). Also note the large dragonfly.

dinosaurs, ichthyosaurs, and plesiosaurs. (Birds, which are descendants of one group of dinosaurs, can also be considered reptiles, depending on the classification system used.)

Reptiles share several derived characteristics that are beneficial for life on land. For example, unlike amphibians, most reptiles have scales made of the protein keratin (also found in human fingernails) that help prevent the loss of body fluids and resist abrasion. More importantly, most reptiles lay shell-covered eggs, called **amniotic eggs** that contain a fluid that bathes the embryo—a significant evolutionary step (**Figure 22.28**). Because the reptile embryo matures in this watery environment, the shelled egg has been characterized as a "private aquarium" in which the embryos of these land

vertebrates spend their water-dwelling stage of life. With this amniotic egg, the remaining ties to a watery environment were finally broken. Thus, unlike amphibians, the first reptiles were able to occupy a wider range of terrestrial habitats—most notably arid regions.

The Great Permian Extinction

A **mass extinction** occurred at the close of the Permian period, and in it a large number of Earth's species became extinct. During this mass extinction, 70 percent of all land-dwelling vertebrate species and perhaps 90 percent of all marine organisms were obliterated; it was the most severe of five mass extinctions to occur over the past 500 million years. Each extinction wreaked havoc with the existing biosphere, wiping out large numbers of species. In each case, however, survivors created new biological communities, which were often more diverse. Therefore, mass extinctions can actually invigorate life on Earth, as the few hardy survivors eventually filled the environmental niches left behind by the victims.

Several mechanisms have been proposed to explain these ancient mass extinctions. Initially, paleontologists believed these were gradual events caused by a combination of climate change and biological forces, such as predation and competition. Other research groups have attempted to link certain mass extinctions to the explosive impact of a large asteroid striking Earth's surface.

The most widely held view is that the Permian mass extinction was driven mainly by volcanic activity because it coincided with a period of voluminous eruptions of flood basalts that blanketed about 1.6 million square kilometers (624,000 square miles), an area nearly

▶ Figure 22.28
The shelled egg of a reptile The amniotic egg contains specialized membranes, including the *amnion*, which encloses the fluid that bathes the embryo and acts as a hydraulic shock absorber to protect it. Mammals also produce an amniotic egg, but in most groups of mammals the outer membranes form a placenta and an umbilical cord that is attached to the uterus of a pregnant adult.

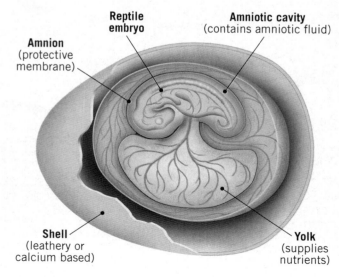

Amnion
(protective membrane)

Reptile embryo

Amniotic cavity
(contains amniotic fluid)

Shell
(leathery or calcium based)

Yolk
(supplies nutrients)

the size of Alaska. This series of eruptions, which lasted roughly 1 million years, occurred in northern Russia, in an area called the Siberian Traps. It was the largest volcanic eruption in the past 500 million years. The release of huge amounts of carbon dioxide likely generated a period of accelerated greenhouse warming, while the emission of sulfur dioxide is credited with producing copious amounts of acid rain and low-oxygen conditions in marine environments. These drastic environmental changes likely put excessive stress on many of Earth's life-forms.

22.8 Mesozoic Era: Dinosaurs Dominate the Land

Summarize the major developments in the history of life during the Mesozoic era.

The life-forms that existed at the dawn of the Mesozoic era were the survivors of the great Permian extinction. These organisms diversified in many ways to fill the biological voids created at the close of the Paleozoic. While life on land underwent a radical transformation with the rise of the dinosaurs, life in the sea also entered a dramatic phase of transformation that produced many of the animal groups that prevail in the oceans today, including modern groups of fish, crustaceans, mollusks, and starfish and their relatives.

Gymnosperms: The Dominant Mesozoic Trees

Conditions on land favored organisms that could adapt to drier climates. One useful evolutionary adaptation was the **seed**, an embryo packaged with a supply of nutrients inside a protective coating. Two groups of seed plants exist: *gymnosperms* and *angiosperms*. **Gymnosperms** produce "naked seeds" that develop on modified leaves, usually scale-like structures that form a cone. Think of the rather flat structures that make up a pinecone. Gymnosperm seeds are referred to as naked because they are not enclosed in a structure, such as those found in flowers that eventually become fruits—apple seeds, for example. Once gymnosperm seeds mature, the scales of the cone separate, and the seeds are released. Unlike the first plants to invade the land, the more primitive ferns, seed-bearing gymnosperms no longer had to depend on a water body for fertilization. Consequently, they readily adapted to drier habitats.

The gymnosperms dominated terrestrial ecosystems throughout much of the Mesozoic era, which lasted from 252 to 66 million years ago. Examples of this group include cycads, which resemble large pineapple plants; ginkgo trees, which have fan-shaped leaves; and the largest plants of the time, the *conifers*, whose modern descendants include pines, junipers, and redwoods (**Figure 22.29**). The best-known fossil occurrence of these ancient trees is in northern Arizona's Petrified Forest National Park. Here, huge petrified logs lie exposed at

◄ Figure 22.29
The giant sequoia is a cone-bearing conifer The giant sequoia is perhaps the largest living organism, weighing as much as 2,500 metric tons, equivalent to about 24 blue whales or 40,000 humans. The natural distribution of giant sequoias is the western Sierra Nevada, California. A close relative is the coast redwood, the tallest living trees, which can grow to about 115 meters (380 feet) tall.

▲ **Figure 22.30**
Petrified logs of Triassic age, Arizona's Petrified Forest National Park

▲ **Figure 22.31**
Reptiles returned to the sea Ichthyosaurs are one of the groups of reptiles that returned to the sea during the Mesozoic.

the surface, exhumed by the weathering of rocks of the Triassic Chinle Formation (**Figure 22.30**).

Reptiles Take Over the Land, Sea, and Sky

Among the animals, reptiles readily adapted to the drier Permian and Triassic environment. On land, the dinosaurs ranged in size from small bird-sized bipeds (animals that move on two feet) to 40-meter-long (130-foot-long) quadrupeds (animals that move on four feet) with necks long enough to allow them to graze high up in tall trees. Other large dinosaurs were carnivorous, such as the infamous bipedal *Tyrannosaurus*.

Some reptiles evolved with specialized characteristics that allowed them to inhabit drastically different habitats. One group, the pterosaurs, became airborne. How the largest pterosaurs—some of which had wing spans greater than 11 meters (35 feet) and weighed more than 90 kilograms (200 pounds)—took flight is still unknown. Other reptiles returned to the sea, including fish-eating *plesiosaurs* and *ichthyosaurs* (**Figure 22.31**). These reptiles became proficient swimmers, breathing by means of lungs rather than gills, much like modern sea-going mammals such as whales and dolphins.

By about 160 million years ago, a group of feathered dinosaurs gave rise to modern birds (**Figure 22.32**). Many researches consider *Archaeopteryx* to be the first known bird. *Archaeopteryx* had feathered wings but

retained teeth, clawed digits in its wings, and a long tail with many vertebrae. A recent study concluded that *Archaeopteryx* flew well at high speeds, but unlike most modern birds, it could not take off from a standing position. Rather,

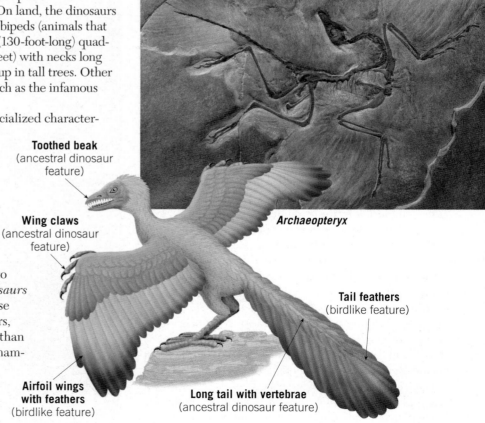

Archaeopteryx

Toothed beak
(ancestral dinosaur feature)

Wing claws
(ancestral dinosaur feature)

Tail feathers
(birdlike feature)

Airfoil wings with feathers
(birdlike feature)

Long tail with vertebrae
(ancestral dinosaur feature)

▲ **Figure 22.32**
***Archaeopteryx*, a transitional form related to modern birds** *Archaeopteryx* had feathered wings and a feathered tail that appear intended for flight, but in many ways it resembled ancestral dinosaurs. The sketch shows an artist's reconstruction of *Archaeopteryx*.

these descendants of bird-like dinosaurs took flight by running and leaping into the air. Other researchers disagree and envision them as climbing animals that glided down to the ground, following the idea that birds evolved from tree-dwelling gliders. Scientists continue to debate whether the first birds took to the air from the ground *up* or from the trees *down* or by both mechanisms.

Dinosaurs flourished for nearly 160 million years. However, by the close of the Mesozoic, all dinosaurs (except for the ancestors of modern birds) and many related reptile groups became extinct. The huge land-dwelling dinosaurs, the marine plesiosaurs, and the flying pterosaurs are known only through the fossil record. What caused this great extinction?

Demise of the Dinosaurs

The boundaries between divisions on the geologic timescale represent times of significant geologic and/or biological change. Of special interest is the boundary between the Mesozoic era ("middle life") and the Cenozoic era ("recent life"), about 66 million years ago. During this transition, roughly three-quarters of all plant and animal species died out in another mass extinction. This boundary marks the end of the era in which dinosaurs and other large reptiles dominated the landscape (**Figure 22.33**) and the beginning of the era when mammals assumed that role.

What could have triggered the extinction of one of the most successful groups of land animals? An increasing number of researchers support the view that the dinosaurs fell victim to a "one–two punch." The first blow occurred during the last few million years of the Mesozoic era; climate data indicate that during this time, average temperatures over the land may have increased by more than 20°C (about 40°F) in a few tens of thousands of years—a blink of the eye in geologic time. This episode of global warming is thought to have coincided with massive basaltic eruptions, which produced the Deccan Plateau, located in what is now India. The Deccan eruptions released massive amounts of carbon dioxide, which caused a period of greenhouse warming that resulted in a dramatic rise in temperatures. Presumably, this period of global warming snuffed out some species and hobbled others.

The final blow came about 66 million years ago, when our planet was struck by a stony meteorite, a relic from the formation of the solar system. The errant mass of rock was approximately 10 kilometers (6 miles) in diameter and was traveling at about 90,000 kilometers per hour at the time of impact. It collided with the southern portion of North America in a shallow tropical sea—now Mexico's Yucatan Peninsula (**Figure 22.34**).

Following the impact, suspended dust greatly reduced the amount of sunlight reaching Earth's surface, which resulted in global cooling ("impact winter") and inhibited photosynthesis, disrupting food production. Long after the dust settled, the sulfur compounds

▲ Figure 22.33
Artist's rendering of *Allosaurus* *Allosaurus* was a large carnivorous dinosaur that lived in the late Jurassic period (155–145 million years ago).

added to the atmosphere by the blast remained suspended and reflected solar radiation back to space, perpetuating the unusually cold temperatures.

One piece of evidence that points to a catastrophic collision 66 million years ago is a thin layer of sediment, less than 1 centimeter thick, discovered in several places around the globe. This sediment contains a high level

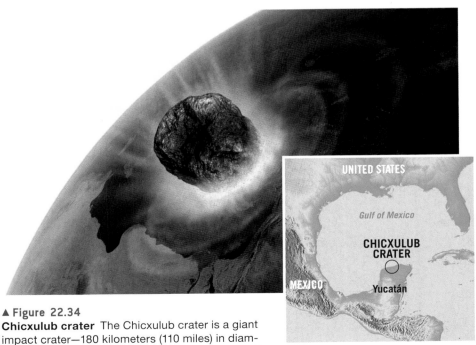

▲ Figure 22.34
Chicxulub crater The Chicxulub crater is a giant impact crater—180 kilometers (110 miles) in diameter—that formed about 66 million years ago and has since filled with sediment. The impact that created it likely contributed to the demise of the dinosaurs.

► **Figure 22.35**
Iridium layer: Evidence for a catastrophic impact 66 million years ago A thin layer of sediment has been discovered worldwide at Earth's physical boundary separating the Mesozoic and Cenozoic eras. This sediment contains a high level of iridium, an element that is rare in Earth's crust but found in much higher concentrations in stony meteorites.

Cenozoic rocks

IRIDIUM LAYER

Mesozoic rocks

of the element *iridium*, rare in Earth's crust but found in high proportions in stony meteorites (**Figure 22.35**). This layer presumably contains scattered remains of the meteorite responsible for the environmental changes that provided the second and final blow that led to the demise of many reptile groups.

Regardless of what caused this massive extinction, its outcome provides a valuable lesson in understanding the role that catastrophic events play in shaping our planet's physical landscape and biosphere. The extinction of the dinosaurs opened habitats for the mammals that survived. These new habitats, along with evolutionary forces, led to the development of the great diversity of mammals that occupy our modern world.

CONCEPT CHECKS 22.8

1. What group of plants became the dominant trees during the Mesozoic era? Name a modern descendant of this group.

2. What was the dominant reptile group on land during the Mesozoic?

3. What group of reptiles gave rise to modern birds?

 **Concept Checker**
https://goo.gl/dh8wKR

22.9 Cenozoic Era: Mammals Diversify

Discuss the major developments in the history of life during the Cenozoic era.

During the Cenozoic era, from about 65.5 million years ago to the present, mammals replaced dinosaurs as the dominant vertebrates on land. Among plants, **angiosperms** (flowering plants), which appeared about 140 million years ago, came to occupy most terrestrial environments. Today, angiosperms make up roughly 90 percent of all plant species.

The development of flowering plants strongly influenced the evolution of both birds and mammals that feed on seeds and fruits, as well as many insect groups. During the middle of the Cenozoic, one type of angiosperm, grasses, spread rapidly to form grasslands—a completely novel type of landscape. At the same time, herbivorous mammal groups gave rise to a great variety of grazers, and carnivores evolved to prey on them (**Figure 22.36**).

The Cenozoic era ocean teemed with modern fish such as tuna, swordfish, and barracuda. In addition, some mammals, including seals, whales, and walruses, took up life in the sea.

From Dinosaurs to Mammals

Mammals coexisted with dinosaurs for nearly 100 million years but remained mostly small and probably mostly nocturnal. Then, about 66 million years ago, fate intervened when the large meteorite collided with Earth and dealt the final crashing blow to the reign of the dinosaurs. This transition from one prominent group to another is clearly visible in the fossil record.

Mammals are named for their *mammary glands* that produce milk for their offspring. Another distinctive characteristic of mammals is hair. Also, like birds, mammals are **endothermic**, which means they are capable of maintaining a constant body temperature through metabolic activity, and they have large brains for their size compared to other vertebrate groups.

With the demise of the large Mesozoic dinosaurs, Cenozoic mammals diversified rapidly. The many forms that exist today evolved from small primitive mammals that were characterized by short legs; flat, five-toed feet; and small brains. Their development and specialization took four principal directions: increase in size, increase in brain capacity, specialization of teeth to accommodate more diverse diets, and specialization of limbs to be better equipped for a particular lifestyle or habitat.

Mammal Groups

The three major lineages of living mammals, the *monotremes* (egg-laying mammals), *marsupials* (mammals with a pouch), and *placentals* (the group to which humans belong), emerged during the Mesozoic. The groups differ principally in their modes of reproduction. The monotremes lay hard-shelled eggs, an ancestral characteristic retained in this group as well as by most modern reptiles.

◀ **Figure 22.36**
Angiosperms became the dominant plants during the Cenozoic Angiosperms, commonly known as flowering plants, consist of a group of seed plants that have reproductive structures called flowers and fruits. **A.** The most diverse and widespread of modern plants, many angiosperms display easily recognizable flowers. **B.** Some angiosperms, including grasses, have very tiny flowers. The expansion of the grasslands during the Cenozoic era greatly increased the diversity of grazing mammals and the predators that feed on them.

The platypus, which is native to Australia and Tasmania, is one of only a handful of monotreme species remaining.

Young marsupials, on the other hand, are born live at a very early stage of development. At birth, the tiny and immature young enter the mother's pouch to suckle as they complete their development. Today, marsupials are found primarily in Australia, where they took a separate evolutionary path largely isolated from placental mammals. Modern marsupials include kangaroos, opossums, and koalas (**Figure 22.37**).

Placental mammals, also called *eutherians*, complete their embryonic development in their mother's uterus and are nourished via a placenta, to which they are connected by an umbilical cord. Most mammals, including humans, are placental. Other members of this group include wolves, elephants, bats, manatees, and monkeys.

Humans: Mammals with Large Brains and Bipedal Locomotion

Both fossil and genetic evidence suggest that by 6.5 million years ago, our own group, the hominins, had branched off from the line leading to modern chimpanzees. Scientists have a good record of this evolution in fossils found in several sedimentary basins in Africa, including the Rift Valley system in East Africa. To date, anthropologists have unearthed roughly 20 extinct species of primates that are more closely related to humans than to chimpanzees.

The genus *Australopithecus*, which came into existence about 4.2 million years ago, showed skeletal characteristics that were intermediate between our apelike ancestors and modern humans. In particular, *Australopithecus* walked upright. Evidence for this bipedal stride includes footprints preserved in 3.2-million-year-old ash

deposits at Laetoli, Tanzania (**Figure 22.38**). This new way of moving around made it possible for our human ancestors to leave forested habitats and to travel long distances for hunting and gathering food.

The earliest fossils of our genus, *Homo*, include the remains of *Homo habilis*, nicknamed "handy man" because these early humanoids were often found with sharp stone tools in sedimentary deposits from 2.4 to 1.6 million years ago. *Homo habilis* had a shorter jaw and a larger brain than its predecessor.

During the next 1.3 million years, our ancestors developed substantially larger brains and long slender legs with hip joints adapted for long-distance walking. These species (including *Homo erectus*) ultimately gave rise to our species, *Homo sapiens*, as well as to some extinct related species, including the Neanderthals (*Homo neanderthalensis*). Despite having larger brains than present-day humans and being able to fashion hunting tools from wood and stone, Neanderthals became extinct about 28,000 years ago. At one time, Neanderthals were considered a stage in the evolution of *Homo sapiens*, but that view has been abandoned.

Based on our current understanding, *Homo sapiens* originated in Africa about 200,000 years ago and began to spread around the globe. The oldest-known *Homo sapiens* fossils outside Africa were found in the Middle East and date to 115,000 years ago. Humans are known to have coexisted with Neanderthals and other prehistoric populations, with remains found in Siberia, China, and Indonesia.

▼ **Figure 22.37**
Kangaroos, examples of marsupial mammals After the breakup of Pangaea, the Australian marsupials evolved independently.

▲ Figure 22.38
Footprints of *Australo-pithecus* in ash deposits at Laetoli, Tanzania

▲ Figure 22.40
Artist's rendition of mammoths These relatives of modern elephants were among the large mammals that became extinct at the close of the Ice Age.

Further, there is genetic evidence that our species interbred with Neanderthals that lived in Eurasia.

By 36,000 years ago, humans were producing spectacular cave paintings in Europe (**Figure 22.39**). About 28,000 years ago, all prehistoric hominin populations except for *Homo sapiens* died out.

Large Mammals and Extinction

During the rapid mammal diversification of the Cenozoic era, some species became very large. For example, by the Oligocene epoch (about 23 million years ago), a hornless

▼ Figure 22.39
Cave painting of animals by early humans

rhinoceros evolved that stood nearly 5 meters (16 feet) high. It is the largest land mammal known to have existed. As time passed, many other mammals evolved to large forms—more, in fact, than exist now. Many of these large forms were common as recently as 11,000 years ago, but a wave of late Pleistocene extinctions rapidly eliminated many of these animals from the landscape.

North America experienced the extinction of mastodons and mammoths, both large relatives of the modern elephants (**Figure 22.40**). In addition, saber-toothed cats, giant beavers, large ground sloths, horses, giant bison, and others died out. In Europe, late Pleistocene extinctions included woolly rhinos, large cave bears, and Irish elk. Scientists remain puzzled about the reasons for these more recent extinctions of large mammals. Because these large animals survived several major glacial advances and interglacial periods, it is difficult to ascribe their extinction to climate change. Some scientists hypothesize that early humans hastened the decline of these mammals by selectively hunting large forms.

CONCEPT CHECKS 22.9

1. Explain how the demise of the dinosaurs impacted the development of mammals.

2. Where did researchers discover most of the evidence for the early evolution of our hominin ancestors?

3. What two characteristics best separate humans from other mammals?

Concept Checker
https://goo.gl/qp5WDo

22

CONCEPTS IN REVIEW

Earth's Evolution Through Geologic Time

22.1 What Makes Earth Habitable?

List the principal characteristics that make Earth habitable.

Key Terms: exoplanet   habitable zone

- As far as we know, Earth is unique among planets in hosting life. The planet's size, composition, and distance from the Sun all contribute to conditions that support life.

22.2 Birth of a Planet

Outline the major stages in Earth's evolution, from the Big Bang to the formation of our planet's layered internal structure.

Key Terms: solar nebula   protoplanet
supernova   planetesimal   Hadean

- The universe is thought to have formed about 13.8 billion years ago, with the Big Bang, which generated space, time, energy, and matter, including the elements hydrogen and helium. Elements heavier than hydrogen and helium were synthesized by nuclear reactions that occur in stars.
- Earth and the solar system formed around 4.6 billion years ago, with the contraction of a *solar nebula*. Collisions between clumps of matter in this spinning disk resulted in the growth of *planetesimals* and then *protoplanets*. Over time, the matter of the solar nebula was concentrated into a smaller number of larger bodies: the Sun, the rocky inner planets, the icy outer planets, moons, comets, and asteroids.
- The early Earth was hot enough for rock and iron to melt, thanks to the kinetic energy of impacting asteroids and planetesimals as well as the decay of radioactive isotopes. This allowed iron to sink to form Earth's core and rocky material to rise to form the mantle and crust.

Q The accompanying image shows the impact of Comet Shoemaker-Levy 9 with Jupiter in 1994. After this event, what happened to Jupiter's total mass? How was the number of objects in the solar system affected?

Comet Shoemaker-Levy 9 impacted Jupiter in 1994.

22.3 Origin and Evolution of the Atmosphere and Oceans

Describe how Earth's atmosphere and oceans formed and evolved through time.

Key Terms: plant photosynthesis   Great Oxygenation
outgassing   banded iron formation   Event

- Earth's atmosphere formed as volcanic *outgassing* added mainly water vapor and carbon dioxide to the primordial atmosphere of gases common in the early solar system: methane and ammonia.
- Free oxygen began to accumulate through *plant photosynthesis* by cyanobacteria, which released oxygen as a waste product. Much of this early oxygen immediately reacted with iron dissolved in seawater and settled to the ocean floor as chemical sediments called *banded iron formations*. The *Great Oxygenation Event* of 2.3 billion years ago marks the first evidence of significant amounts of free oxygen in the atmosphere.
- Earth's oceans formed after the planet's surface had cooled. Soluble ions weathered from the crust were carried to the ocean, making it salty. The oceans also absorbed tremendous amounts of carbon dioxide from the atmosphere.

22.4 Precambrian History: The Formation of Earth's Continents

Explain the formation of continental crust, how continental crust becomes assembled into continents, and the role that the supercontinent cycle has played in this process.

Key Terms: shield   supercontinent cycle
craton   supercontinent

- The Precambrian includes the Archean and Proterozoic eons. Our knowledge of these eons is limited because erosion has destroyed much of the rock record.
- Continental crust was produced over time through the recycling of ultramafic and mafic crust in an early version of plate tectonics. Small crustal fragments formed and amalgamated into large crustal provinces called *cratons*. Over time, North America and other continents grew through the accretion of new terranes around the edges of this central "nucleus" of crust.
- Early cratons not only merged but sometimes rifted apart. The *supercontinent* Rodinia formed around 1.1 billion years ago and then rifted apart, opening new ocean basins. Over time, these ocean basins closed to form a new supercontinent called Pangaea around 250 million years ago. Like Rodinia before it, Pangaea broke up as part of the ongoing *supercontinent cycle*.
- The formation of elevated oceanic ridges following the breakup of a supercontinent displaced enough water that sea level rose, and shallow seas flooded the continents. The breakup of continents can also influence the direction of ocean currents, with important consequences for climate.

22.5 Geologic History of the Phanerozoic: The Formation of Earth's Modern Continents

List and discuss the major geologic events in the Paleozoic, Mesozoic, and Cenozoic eras.

Key Terms: Pangaea   Laurasia   Gondwana

- The Phanerozoic eon began 545 million years ago and is divided into the Paleozoic, Mesozoic, and Cenozoic eras.
- In the Paleozoic era, North America experienced a series of collisions that resulted in the rise of the young Appalachian mountain belt, as part of the assembly of *Pangaea*. High sea levels caused the ocean to cover vast areas of the continent and resulted in a thick sequence of sedimentary strata.
- During the Mesozoic, Pangaea broke up, and the Atlantic Ocean began to form. As the North American continent moved westward, the Cordillera began to rise due to subduction as well as the accretion of terranes along the West Coast. In the Southwest, vast deserts accumulated thick layers of dune sand.
- In the Cenozoic era, a thick sequence of sediments was deposited along North America's Atlantic margin and the Gulf of Mexico. Meanwhile, western North America experienced an extraordinary episode of crustal extension that formed the Basin and Range Province.

22.6 Earth's First Life

Describe some of the hypotheses on the origin of life and the characteristics of early prokaryotes, eukaryotes, and multicellular organisms.

Key Terms: prokaryote   eukaryote   protein   stromatolite

- Life began from nonlife. Amino acids, a necessary building block for *proteins*, may have been assembled with energy from ultraviolet light or lightning, or in a hot spring, or may have been delivered later to Earth via meteorites.
- The first organisms were relatively simple single-celled *prokaryotes* that thrived in the absence of oxygen. They may have formed by 3.8 billion years ago. The advent of photosynthesis allowed microbial mats to build up and form *stromatolites*.
- *Eukaryotes* have larger, more complex cells than prokaryotes. The oldest-known eukaryotic cells date from around 2.1 billion years ago. Eukaryotic cells gave rise to the great diversity of multicellular organisms.

22.7 Paleozoic Era: Life Explodes

List the major developments in the history of life during the Paleozoic era.

Key Terms: vertebrate   amniotic egg   invertebrate   amphibian   mass extinction   Cambrian explosion   reptile

- Abundant fossil hard parts appear in sedimentary rocks at the beginning of the Cambrian period. These shells and other skeletal material came from a profusion of new animals, including trilobites and cephalopods.
- Plants colonized the land around 400 million years ago and soon diversified into forests.
- In the Devonian, some lobe-finned fishes gradually evolved into the first *amphibians*. A subset of the amphibian population evolved waterproof skin and shelled eggs and split off to become the *reptile* line.
- The Paleozoic era ended with the largest *mass extinction* in the geologic record. This deadly event may have been related to the eruption of the Siberian Traps flood basalts.

Q Briefly describe the advantages that the amniotic egg gave reptiles for life on dry land.

Amniotic egg

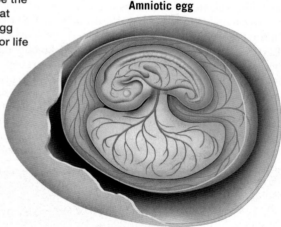

22.8 Mesozoic Era: Dinosaurs Dominate the Land

Summarize the major developments in the history of life during the Mesozoic era.

Key Terms: seed   gymnosperm

- Plants diversified during the Mesozoic. The flora of that time was dominated by *gymnosperms*, the first plants with *seeds*.
- The dinosaurs came to dominate the land, pterosaurs took to the air, and a suite of marine reptiles swam the seas. The first birds evolved during the Mesozoic, as evidenced by *Archaeopteryx*, a transitional fossil.
- Like the Paleozoic, the Mesozoic ended with a mass extinction. This extinction was due to a massive meteorite impact and a period of extensive volcanism, both of which released particulate matter into the atmosphere and dramatically altered Earth's climate and disrupted its food chain.

22.9 Cenozoic Era: Mammals Diversify

Discuss the major developments in the history of life during the Cenozoic era.

Key Terms: angiosperm   mammal   endothermic

- Flowering plants, called *angiosperms*, diversified and spread around the world through the Cenozoic era.
- Once the giant Mesozoic reptiles were extinct, *mammals* were able to diversify on the land, in the air, and in the oceans. Mammals are *endothermic* (control their body temperature metabolically), have hair, and nurse their young with milk. Marsupial mammals are born very immature and then move to a pouch on the mother, while placental mammals spend a longer time *in utero* and are born in a relatively mature state compared to marsupials.
- Humans evolved from primate ancestors in Africa over a period of about 6.5 million years. They are distinguished from their ape ancestors by an upright, bipedal posture and large brains, as well as the use of elaborate tools. The oldest anatomically modern human fossils are 200,000 years old. Some of these humans migrated out of Africa and coexisted with Neanderthals and other human-like populations.

GIVE IT SOME THOUGHT

1. Referring to Figure 22.4, write a brief summary of the events that led to the formation of Earth.

2. The accompanying photograph shows layered iron-rich rocks called banded iron formations. What does the existence of these 2.3-billion-year-old rocks tell us about the evolution of Earth's atmosphere?

3. Describe how the sudden appearance of oxygen in the atmosphere about 2.3 billion years ago influenced the development of modern life-forms.

4. Five mass extinctions, during each of which 50 percent or more of Earth's marine species became extinct, are documented in the fossil record. The accompanying graph depicts the time and extent of each mass extinction.

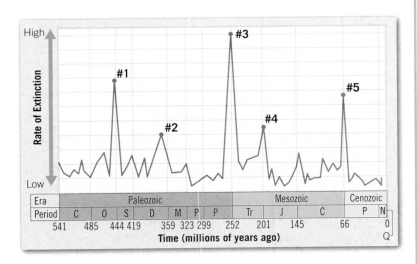

   a. Which of the five mass extinctions was the *most extreme*? Identify this extinction by name and when it occurred.

   b. When did the *most recent* mass extinction occur?

   c. During the most recent mass extinction, what prominent group of land animals was eliminated?

   d. What group of terrestrial animals experienced a major period of diversification following the most recent mass extinction?

5. Currently, oceans cover about 71 percent of Earth's surface. However, early in Earth history the oceans covered a greater percentage of the surface. Explain.

6. Contrast the eastern and western margins of North America during the Cenozoic era in terms of their relationships to plate boundaries.

7. Between 300 and 250 million years ago, plate movement assembled all the previously separated landmasses together to form the supercontinent Pangaea. Pangaea's formation resulted in deeper ocean basins and a drop in sea level, causing shallow coastal areas to dry up. Thus, in addition to rearranging the geography of our planet, continental drift had a major impact on life on Earth. Use the accompanying diagram showing the movement of hypothetical landmasses and the information above to answer the following:

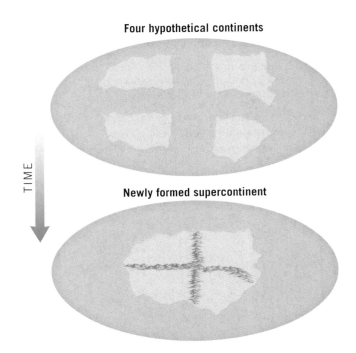

Four hypothetical continents

TIME

Newly formed supercontinent

   a. Which of the following types of habitats would likely diminish in size during the formation of a supercontinent: deep-ocean habitats, wetlands, shallow marine environments, or terrestrial (land) habitats? Explain.

   b. During the breakup of a supercontinent, would sea level remain the same, rise, or fall?

   c. Explain how and why the development of an extensive oceanic ridge system that forms during the breakup of a supercontinent affects sea level.

EYE ON EARTH

1. The oldest known sample of Earth is a 4.4-billion-year-old zircon crystal found in a metaconglomerate in the Jack Hills area of western Australia. Zircon is a silicate mineral that occurs in trace amounts in most granitic rocks.

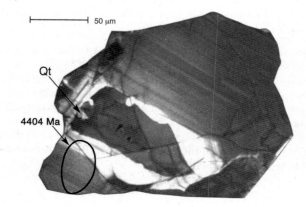

a. What is the parent rock of a metaconglomerate?

b. Assuming that this zircon crystal originated as part of a granite intrusion, briefly describe its journey from the time of its formation until it was discovered in the Jack Hills.

c. Is this zircon crystal younger or older than the metaconglomerate in which it was found? Explain.

2. The rocks shown here are Cambrian-age stromatolites of the Hoyt Limestone, exposed at Lester Park, near Saratoga Springs, New York.

a. Using Figure 22.3, determine approximately how many years ago these rocks were deposited.

b. What is the name of the group of organisms that likely produced these limestone deposits?

c. What was the environment like in this part of New York when these rocks were deposited?

DATA ANALYSIS

Fossils Used to Decipher Earth History

Fossils have been found around the world and help us to decipher the history of our planet.

https://goo.gl/eD226M

ACTIVITIES

Go to the Paleobiology Database at https://paleobiodb.org and click the blue "Explore" button. Use the examples or the "Help" link to learn how to use the application, and click + to go to the application. Zoom in to your state. Hovering over a circle will highlight the relevant time period on the geologic timeline. Hovering over the colored blocks on the timeline will display the name of the period.

1. Which geologic time periods are represented? Which time period has the most sites?

2. Zoom to the fossil site closest your location. Which eon does this fossil find belong to? Which era? Period? Epoch? Stage (time period shown at the bottom of the timeline), if available?

3. Click on the fossil site. If the site has multiple fossil collections, select one of them.

4. What is the lithology or environment of this site (if reported)? Click on the "Occurrences" tab. What group or groups of organisms (at the class or family level) are listed? In common language, what kinds of creatures are these? (Use the Internet, as needed.)

5. Zoom out until you can see several fossil sites in your region. What other eras and periods are shown for your area, if any?

6. Investigate the fossil sites by clicking on them. What other types of fossils have been found in your region?

7. Based on the fossils in the area, describe the likely environmental conditions at that time. Information in the chapter will help you answer this question.

Zoom in to a region were fossils come from a different era (shown by different colors) than the ones in your region.

8. What geologic time periods are shown?

9. What types of fossils have been found in this region?

Mastering Geology

Looking for additional review and test prep materials? Visit pearson.com/mastering/geology to enhance your understanding of this chapter's content by accessing a variety of resources, including **Self-Study Quizzes, Geoscience Animations, SmartFigure Tutorials,** *Project Condor* **Videos, Mobile Field Trips, Dynamic Study Modules,** and an optional **Pearson eText.**

Sand Pirates Get Rich Stealing Beaches, Riverbeds, and Islands

Sand pirates sound like fanciful enemies dreamed up for a movie. But a global construction boom has led to a great demand for sand, a key ingredient in concrete and other building materials. In 2017 the United States produced 890 million tons of construction sand and gravel, according to the U.S. Geological Survey. Worldwide, the United Nations estimates that between 47 and 59 billion tons of sand are mined annually. Sand and gravel are among the most commonly extracted materials in the world. Demand for this nonrenewable resource far outstrips supply, especially in Asia. In some areas, this sets the stage for "sand mafias" to take advantage of the shortage and make a profit.

▲ Illegal sand dredging from rivers in India.

For the past decade, officials in Indonesia have battled sand pirates, who dredge up entire small islands and float them away on boats for use in construction. Crime organizations in Morocco and India also illegally mine sand from beaches and rivers. Unlawful operations don't follow safety practices and frequently harm the ecosystems of shorelines and river basins. In India, illegal sand harvesting has even undermined bridges.

You might think that places like the Sahara Desert, with its towering dunes, should have more than enough sand for our needs. Unfortunately, sand that weathers in the desert has rounded grains that don't blend well with a concrete mix. The angular, gritty sands that make excellent building materials usually form only in riverbeds or along coasts.

► This large aggregate mine is near Caledon, Ontario, Canada. The demand for sand and gravel is on the rise.

23
Energy and Mineral Resources

FOCUS ON CONCEPTS

Each statement represents the primary learning objective for the corresponding major heading within the chapter. After you complete the chapter, you should be able to:

23.1 Distinguish between renewable and nonrenewable resources.

23.2 Compare and contrast fossil-fuel types and describe how each of them contributes to U.S. energy consumption.

23.3 Describe the importance of nuclear energy and discuss its pros and cons.

23.4 List and discuss the major sources of renewable energy. Describe the contribution of renewable energy to the overall U.S. energy supply.

23.5 Distinguish among resources, reserves, and ores.

23.6 Explain how different igneous and metamorphic processes produce economically significant mineral deposits.

23.7 Discuss ways in which surface processes produce ore deposits.

23.8 Distinguish between two broad categories of nonmetallic mineral resources and list examples of each.

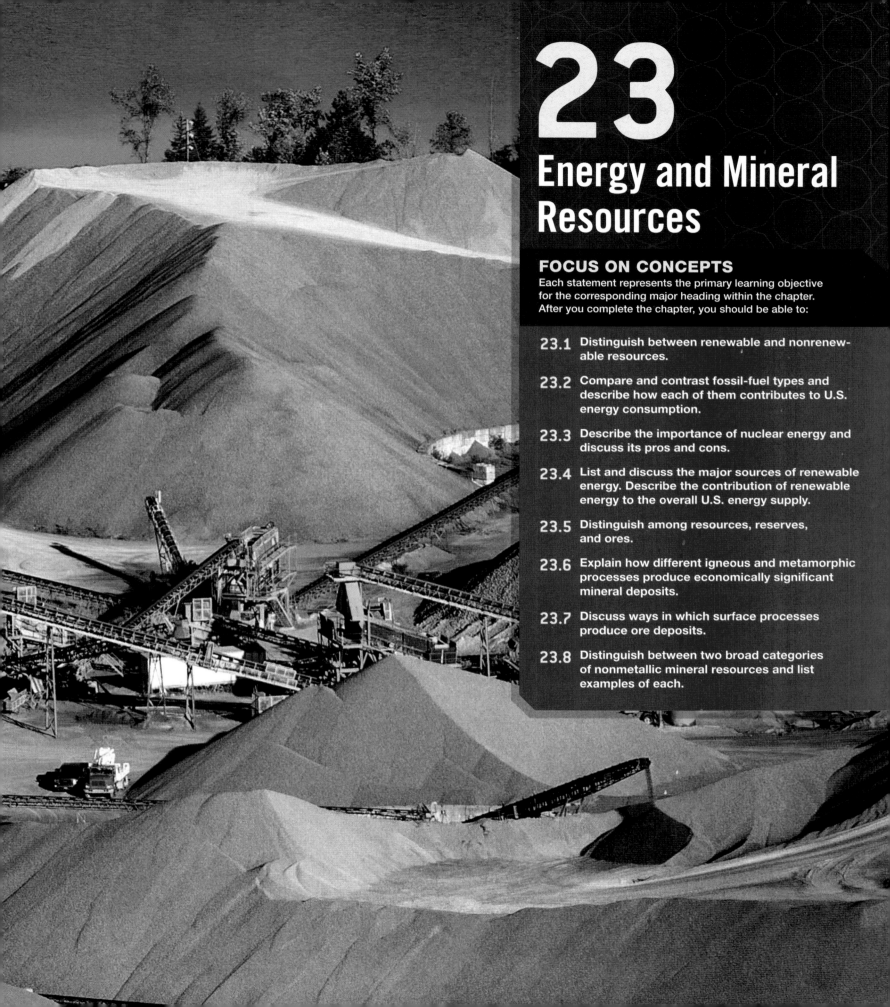

Materials that we extract from Earth are the basis of modern civilization. A large percentage of the products and substances we use every day were mined from Earth's crust and/or manufactured using energy extracted from the crust. Humans consume huge quantities of energy and mineral resources, and our demand continues to grow. In 1960 the world population was 3 billion, and by 2018 it had grown to 7.6 billion. By 2025, the planet's population is expected to exceed 8 billion. Increased population and improving living standards in many parts of the world are contributing to the demand for resources.

How long can our remaining resources sustain the rising standard of living in today's industrialized countries and still provide for the growing needs of developing regions? How much environmental deterioration are we willing to accept in pursuit of resources? Can alternatives be found? If we are to satisfy an increasing per capita demand and a growing world population, we must understand our resources and their limits.

23.1 Renewable and Nonrenewable Resources

Distinguish between renewable and nonrenewable resources.

We commonly divide resources into two broad categories, according to their ability to be regenerated: renewable and nonrenewable. **Renewable resources** can be replenished over relatively short time spans. Common examples are corn used for food and ethanol, natural fibers such as cotton for clothing, and forest products for lumber and paper. Energy from flowing water, wind, and the Sun are also considered renewable.

By contrast, many important metals, such as iron, aluminum, and copper, as well as many widely used fuels, such as oil, natural gas, and coal, are **nonrenewable resources** (Figure 23.1). While this category of resources may form continuously, the processes that create them are so slow that significant deposits take millions of years to accumulate. Thus, for all practical purposes, Earth contains fixed quantities of these substances. The present supplies will be depleted as they are mined or pumped from the ground. Although some nonrenewable resources, such as the aluminum we use for containers, can be recycled, others, such as the oil burned for fuel, cannot.

Most of the energy and mineral resources we use are nonrenewable. How much of these does the average person in the United States use in a lifetime?

▶ Figure 23.1
Utah's Bingham Canyon copper mine This excavation, one of the largest open-pit mines in the world, is nearly 4 kilometers (2.5 miles) across and 900 meters (3000 feet) deep. Although the amount of copper in the rock is less than 0.5 percent, the huge volume of material removed each day yields enough metal to be profitable. In addition to copper, the mine produces gold, silver, and molybdenum—all nonrenewable resources.

TABLE 23.1   How Much Do We Use?

Mineral Commodity	Amount Used in a Lifetime*	
	in Pounds	in Kilograms
Aluminum	5677	2555
Cement	65,480	29,466
Clays	19,245	8660
Copper	1309	589
Gold	99	45
Iron ore	29,608	13,324
Lead	928	417
Phosphate rock	19,815	8917
Stone, sand, and gravel	1,610,000	724,500
Zinc	671	302

Source: Data from U.S. Geological Survey and U.S. Energy Information Administration.

*Every American born in 2008 is estimated to use these amounts in his or her lifetime.

Table 23.1 provides an estimate of the amounts of several nonfuel mineral commodities each of us will use. The amounts shown in Table 23.1 may seem extraordinary until you learn that an average American automobile contains more than a ton of iron and steel, 240 pounds of aluminum, 42 pounds of copper, 41 pounds of silicon, 22 pounds of zinc, and more than 30 other mineral commodities, including titanium, platinum, and gold.

CONCEPT CHECKS 23.1

1. Distinguish between renewable and nonrenewable resources.

2. Are resources such as aluminum, that can be recycled, considered renewable?

 Concept Checker
https://goo.gl/vo4isx

23.2 Energy Resources: Fossil Fuels

Compare and contrast fossil-fuel types and describe how each of them contributes to U.S. energy consumption.

Earth's tremendous industrialization over the past 2 centuries has been largely powered by burning coal, petroleum, and natural gas—all commonly known as **fossil fuels**, since they are the remains of organisms that lived millions of years ago. About 80 percent of the energy consumed in the United States today comes from these sources. Figure 23.2 shows where our energy comes from and what it is used for. Our reliance on fossil fuels is obvious.

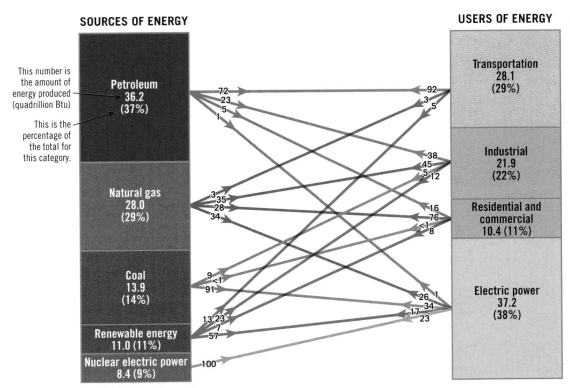

SOURCES OF ENERGY

This number is the amount of energy produced (quadrillion Btu)

This is the percentage of the total for this category.

Petroleum
36.2
(37%)

Natural gas
28.0
(29%)

Coal
13.9
(14%)

Renewable energy
11.0 (11%)

Nuclear electric power
8.4 (9%)

USERS OF ENERGY

Transportation
28.1
(29%)

Industrial
21.9
(22%)

Residential and commercial
10.4 (11%)

Electric power
37.2
(38%)

◀ SmartFigure 23.2

U.S. energy consumption, 2017 The total was 97.7 quadrillion Btu (British thermal units). A quadrillion is 10^{15}, or a billion million. A quadrillion Btu is a convenient unit for referring to U.S. energy use as a whole. (Data from the U.S. Department of Energy/Energy Information Agency)

Tutorial
https://goo.gl/76jKAm

Reading this double graph:
The left side indicates what energy sources we use. The right side shows where we use the energy.
The lines with numbers that connect the graphs provide more details. Use the top line as an example. It shows that 72% of the petroleum is used by the transportation sector. It also indicates that 92% of the energy used by the transportation sector is petroleum.

▶ **Figure 23.3**
Coal fields of the United States Production is greatest in the western region, with about 40 percent of U.S. production coming from Wyoming. West Virginia (11 percent) and Pennsylvania (6 percent), in the Appalachian region, rank second and third among coal-producing states. Bituminous and subbituminous each account for about 45 percent of production, and lignite accounts for about 10 percent. Anthracite production is negligible.

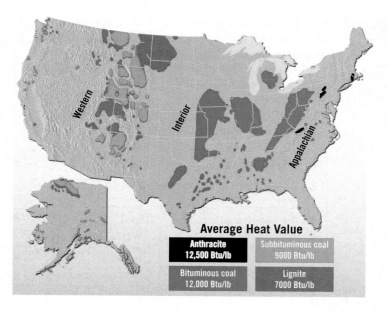

Average Heat Value

Anthracite 12,500 Btu/lb	**Subbituminous coal** 9000 Btu/lb
Bituminous coal 12,000 Btu/lb	**Lignite** 7000 Btu/lb

Coal

Long cheap and plentiful, coal has been an important fuel for centuries. In the nineteenth and early twentieth centuries, it powered the Industrial Revolution. In 1900 coal provided 90 percent of the energy used in the United States. In 2017 coal accounted for a much smaller proportion of the nation's energy usage: about 14 percent. In the United States, coal fields are widespread and contain supplies that should last hundreds of years (**Figure 23.3**). (A discussion of coal formation can be found in Chapter 7.)

As of 2017, about 91 percent of U.S. coal usage occurred in power plants to generate electricity. As shown in Figure 23.2, coal is used to generate about one-third of all U.S. electricity today. As mentioned previously, coal usage in all sectors has steadily dropped for decades, and an especially steep decline has occurred recently. In 2014 coal represented 17 percent of the nation's total energy consumption. Just 3 years later, it was 14 percent.

The recovery and use of coal present a number of challenges. Surface mining can turn the countryside into a scarred wasteland if careful (and costly) reclamation is not carried out to restore the land. Today, all U.S. surface mines must undertake land reclamation efforts. Although underground mining does not scar the landscape to the same degree, it has been costly in terms of human life and health. Strong federal safety regulations have made U.S. mining quite safe. However, collapsing roofs, gas explosions, and the required heavy equipment remain hazards. Over the years, the share of coal produced from surface mines has increased significantly, from 51 percent in 1949 to about 65 percent in 2017 (**Figure 23.4**).

Burning coal produces many harmful emissions, including these:

- Sulfur dioxide (SO_2), which contributes to acid rain and respiratory illnesses
- Nitrogen oxides (NO_x), which contribute to smog and respiratory illnesses
- Particulate matter, which contributes to smog, haze, respiratory illnesses, and lung disease
- Carbon dioxide (CO_2), the primary greenhouse gas produced from the burning of fossil fuels, which plays a significant role in the heating of our atmosphere (Chapter 21 examines this issue in some detail.)

The coal industry now utilizes several methods to reduce sulfur, nitrogen oxides, and other impurities from coal, and it has developed more effective ways of cleaning coal after it is mined. Coal consumers have shifted toward greater use of less-polluting low-sulfur coal. However, significant challenges remain.

Oil and Natural Gas

Together, oil and natural gas provided 66 percent of the energy consumed in the United States in 2017. By examining Figure 23.2, you can see that the transportation sector of the U.S. economy relies almost totally on petroleum as an energy source. In 2011 natural gas surpassed coal for the first time in more than 30 years as a source of energy in the United States. An important reason for this is the development of technologies that have increased production from shale formations. Users of natural gas are almost evenly divided among the three categories other than transportation in Figure 23.2.

▼ **Figure 23.4**
Surface coal mine About 65 percent of U.S. coal production is from surface mines. This mine is in Campbell County, Wyoming.

Petroleum Formation   Petroleum and natural gas are found in similar environments, frequently together. Both consist of hydrocarbons (compounds consisting of hydrogen and carbon). They may also contain smaller quantities of elements such as sulfur, nitrogen, and oxygen. Like coal, petroleum and natural gas are biological products derived from remains of organisms. However, environments in which they form are very different, as are the organisms. Coal forms mostly from plant material that accumulated in a swampy environment above sea level (see Figure 7.20, page 217), while oil and gas derive from the remains of marine plants and animals.

Formation of oil and natural gas begins with the accumulation of marine sediment rich in plant and animal remains. Accumulations must occur where biological activity is high, such as near-shore areas. However, most marine environments are oxygen rich, which leads to decay of organic remains before they can be buried by other sediments. Therefore, accumulations of oil and gas are not as widespread as are the marine environments that support abundant biological activity. Nevertheless, large quantities of organic matter are buried and protected from oxidation in many offshore sedimentary basins. With ever-deeper burial over millions of years, chemical reactions gradually transform some of the original organic matter into the liquid and gaseous hydrocarbons we call petroleum and natural gas.

The newly created petroleum and natural gas are mobile. A portion of these fluids gradually squeezes from the compacting, mud-rich layers where they originate into adjacent permeable beds such as sandstone, where openings between sediment grains are larger. This occurs underwater, so the rock layers containing the oil and gas are saturated with water. Because oil and gas are less dense than water, they migrate upward through the water-filled pore spaces of the enclosing rocks. Unless something halts this upward migration, the fluids will eventually reach the surface, at which point the volatile components will evaporate.

Traps for Oil and Gas   Sometimes the upward migration of oil and natural gas is halted. A geologic environment that allows for economically significant amounts of oil and gas to accumulate underground is termed an **oil trap**. Several geologic structures may act as oil traps, but all have two basic conditions in common: a porous, permeable **reservoir rock** that will yield petroleum and natural gas in sufficient quantities to make drilling worthwhile, and a **cap rock**, such as shale, that is virtually impermeable to oil and gas. The cap rock halts the upwardly mobile oil and gas and keeps both from escaping at the surface. **Figure 23.5** illustrates some common oil and natural gas traps, described in the following list:

- **Anticline.** One of the simplest traps is an *anticline*, an uparched series of sedimentary strata (see Figure 23.5A). As the strata are bent, the rising oil and gas collect at the apex (top) of the fold. Because of its lower

▼ SmartFigure 23.5
Common oil traps

Tutorial
https://goo.gl/
DzBBAZ

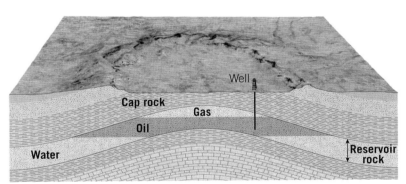

A. Anticline

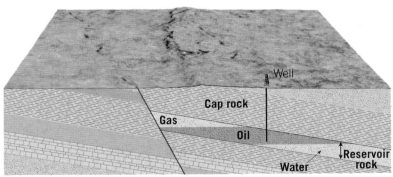

B. Fault trap

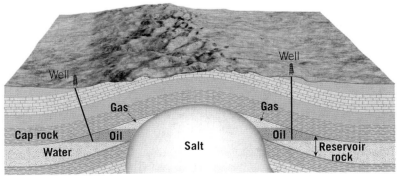

C. Salt dome

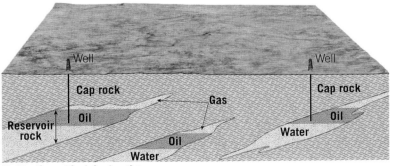

D. Stratigraphic (pinchout) trap

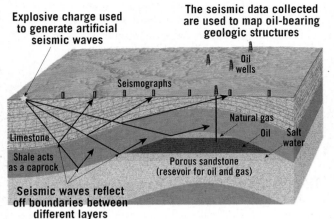

density, the natural gas collects above the oil. Both rest upon the denser water that saturates the reservoir rock.

- **Fault trap.** When strata are displaced in such a manner as to bring a dipping reservoir rock into position opposite an impermeable bed, a *fault trap* forms, as shown in Figure 23.5B. In this case, the upward migration of the oil and gas stops where it encounters the fault.

- **Salt dome.** In the Gulf coastal plain region of the United States, important accumulations of oil occur in association with *salt domes*. Such areas have thick accumulations of sedimentary strata, including layers of rock salt. Salt occurring at great depths was forced to rise in columns by the pressure of overlying beds. These rising salt columns gradually deform

the overlying strata. Because oil and gas migrate to the highest level possible, they accumulate in the upturned sandstone beds adjacent to the salt column (see Figure 23.5C).

- **Stratigraphic (pinchout) trap.** Another important geologic circumstance that may lead to significant accumulations of oil and gas is termed a *stratigraphic trap*. These oil-bearing structures result primarily from the original pattern of sedimentation rather than structural deformation. The stratigraphic trap illustrated in Figure 23.5D exists because a sloping bed of sandstone thins to the point of disappearance.

How do geologists locate oil traps? Recall that changes in the composition or structure of rock cause seismic waves to reflect off boundaries between layers. This characteristic of waves is especially useful in exploration for oil and natural gas, as artificially generated seismic waves can be used to probe the crust (**Figure 23.6**). Oil and natural gas would be much more difficult and expensive to find without seismic imaging because a huge number of wells would have to be randomly drilled to locate oil traps.

When the lid created by the cap rock is punctured by drilling, the oil and natural gas, which are under pressure, migrate from the pore spaces of the reservoir rock to the drill hole. On rare occasions, when fluid pressure is great, it may force oil up the drill hole to the surface, causing a "gusher," or an oil fountain, at the surface. Usually, however, a pump is required to lift out the oil (**Figure 23.7**).

Modern offshore oil production platform in the North Sea.

The first successful oil well was completed by Edwin Drake (right) on August 27, 1859, near Titusville, PA. The oil-bearing reservoir rock was encountered at a depth of 21 meters (69 feet).

A drill hole is not the only means by which oil and gas can escape from a trap. Traps can be broken by natural forces. For example, Earth movements may create fractures that allow the hydrocarbon fluids to escape. Surface erosion may breach a trap—with similar results.

Oil Sands　Oil sands are a somewhat unconventional yet significant source of oil that is likely to become increasingly important in decades to come. Unlike the oil that accumulates in traps, **oil sands** are usually mixtures of clay and sand combined with water and varying amounts of a black, highly viscous tar-like material known as *bitumen*. The use of the term *sand* can be misleading because not all deposits are associated with sands and sandstones. Some occur in other materials, including shales and limestones. The oil in these deposits is very similar to heavy crude oils pumped from wells. The major difference between conventional oil reservoirs and oil sand deposits is in the viscosity (resistance to flow) of the oil they contain. In oil sands, the oil is much more viscous and cannot simply be pumped out (**Figure 23.8**).

Substantial oil sand deposits occur in several locations around the world. By far the largest are in the Canadian province Alberta (**Figure 23.9**). Some oil sands are removed at the surface in a manner similar to the strip mining of coal. The excavated material is heated with pressurized steam until the bitumen softens and rises. Once collected, the oily material is treated to remove impurities, and then hydrogen is added. This last step upgrades the material to a synthetic crude, which can then be refined. Extracting and refining oil sands requires a great deal of energy—almost half as much as the end product yields! In 2017 nearly 2.8 million barrels of crude

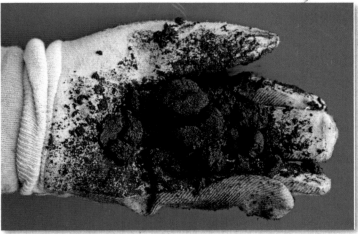

◀ **Figure 23.8**
A close-up of oil sand Notice how solid the material is. Sometimes the term *tar sand* is used to describe bitumen deposits, but this is inaccurate because tar is a human-made substance produced by distilling organic matter. Bitumen looks like tar but is a naturally occurring substance.

▼ **SmartFigure 23.9**
Alberta's oil sands Huge oil reserves occur across more than 140,000 square kilometers (54,000 square miles) of northern Alberta. Some bitumen-rich material can be mined at the surface, as in the photo, but most will be pumped to the surface after injecting the material with steam. The Mobile Field Trip explores this unconventional source of oil.

Mobile Field Trip
https://goo.gl/MR9PZb

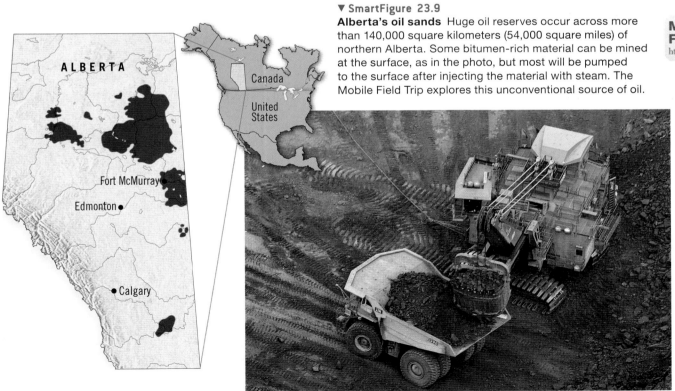

bitumen were produced per day. Obtaining oil from oil sand has environmental drawbacks. Substantial land disturbance is associated with mining huge quantities of rock and sediment. Moreover, large quantities of water are required for processing, and when processing is completed, contaminated water and sediments accumulate in toxic disposal ponds.

About 80 percent of the oil sands in Alberta are buried too deep for surface mining. Oil from these deep deposits must be recovered by using *in situ* (Latin for "in place") techniques. Using drilling technology, steam is injected into the deposit to heat the oil sand, which reduces the viscosity of the bitumen. The hot, mobile bitumen migrates toward producing wells, which pump it to the surface, while the sand is left in place. Production using *in situ* techniques already rivals open-pit mining and in the future will replace mining as the main source of bitumen production from oil sands.

> *Hydraulic fracturing (often called fracking) shatters shale, opening up cracks through which natural gas—and oil, if present— can flow into wells and then be brought to the surface.*

Challenges facing *in situ* processes include increasing the efficiency of oil recovery, managing the water used to make steam, and reducing the costs of energy required for the process.

Hydraulic Fracturing   Not all natural gas and oil migrates into source rocks and accumulates in oil traps. Huge quantities remain confined in impermeable source rocks, usually shale. Extracting trapped gas and oil was not economical until advances in drilling and methods of opening up fractures in the shale were developed. **Hydraulic fracturing** (often called *fracking*) shatters the shale, opening up cracks through which natural gas—and oil, if present—can flow into wells and then be brought to the surface. The recovery of *shale gas* and *shale oil* created an energy boom in the United States, and with it have come some environmental concerns.

Figure 23.10 illustrates the fracking process. Pumping fluids into the rock at very high pressures fractures the shale. The fluid used is mostly water but also includes other chemicals that aid the process. Some of

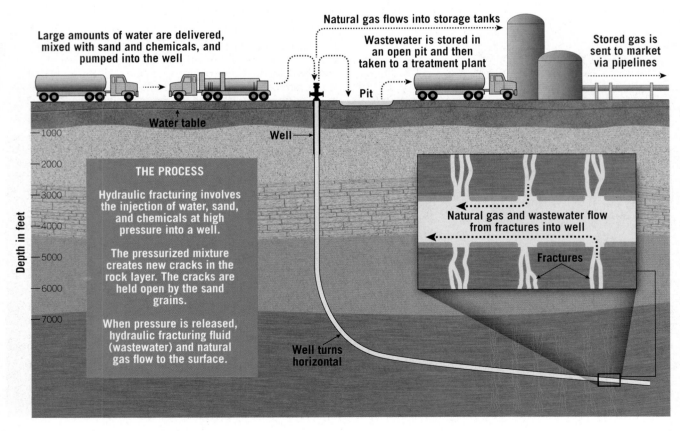

▶ **Figure 23.10**
Hydraulic fracturing ("fracking") This well-stimulation process is commonly used in low-permeability rocks such as shale to increase oil and/or natural gas flow.

Large amounts of water are delivered, mixed with sand and chemicals, and pumped into the well

Natural gas flows into storage tanks

Wastewater is stored in an open pit and then taken to a treatment plant

Stored gas is sent to market via pipelines

Pit

Water table

Well

Depth in feet

—1000
—2000
—3000
—4000
—5000
—6000
—7000

THE PROCESS

Hydraulic fracturing involves the injection of water, sand, and chemicals at high pressure into a well.

The pressurized mixture creates new cracks in the rock layer. The cracks are held open by the sand grains.

When pressure is released, hydraulic fracturing fluid (wastewater) and natural gas flow to the surface.

Natural gas and wastewater flow from fractures into well

Fractures

Well turns horizontal

these chemicals may be toxic, and there are concerns about fracking fluids leaking into freshwater aquifers that supply drinking water. The injected fluid also includes sand, and once fractures open up in the shale, the sand grains can keep them propped open and permit the gas to continue to flow. Next, the fracking fluid is brought back to the surface. The retrieved wastewater is then injected into deep disposal wells. In some locations, these injections appear to trigger numerous minor earthquakes. Due to concerns about potential groundwater contamination and induced seismicity, hydraulic fracturing remains a controversial practice. Its environmental effects are a focus of continuing research.

CONCEPT CHECKS 23.2

1. Contrast the formation of coal with the formation of oil and natural gas. What part of U.S. energy consumption does each of these fossil fuels represent?

Concept Checker
https://goo.gl/J1AEPU

2. What is an oil trap? What do all oil traps have in common?

3. What are oil sands, and where are they most plentiful?

4. Describe the circumstances in which hydraulic fracturing is used.

23.3 Nuclear Energy

Describe the importance of nuclear energy and discuss its pros and cons.

Nuclear energy meets an important part of U.S. energy needs. A glance back at Figure 23.2 shows that nuclear power was the source of about 9 percent of U.S. energy consumption in 2017. Figure 23.2 also shows that 100 percent of that energy was used to produce electricity. The right side of Figure 23.2 also shows that 23 percent of U.S. electricity is from nuclear energy. Worldwide, 11 percent of electricity is generated by nuclear power plants.

The fuel for these nuclear power plants comes from radioactive materials that release energy through **nuclear fission**. Fission occurs when the nucleus of a heavy atom, such as uranium-235, becomes unstable and splits into fragments, emitting neutrons and heat in the process. In the case of uranium-235, the ejected neutrons can trigger fission in other uranium-235 nuclei, producing a *chain reaction*. If the supply of fissionable material is sufficient and if the reaction is allowed to proceed in an uncontrolled manner, an enormous amount of energy is released, in the form of an atomic explosion.

In a nuclear power plant, the fission reaction is controlled by moving neutron-absorbing rods into or out of the nuclear reactor. The result is a controlled nuclear chain reaction that releases great amounts of heat. The heat is transported from the reactor and used to drive steam turbines that turn electrical generators, which is similar to what occurs in most conventional power plants.

Uranium

Uranium-235 is the only naturally occurring isotope that is readily fissionable, and it is therefore the primary fuel used in nuclear power plants.* Although large quantities of uranium ore have been discovered, most contain less than 0.05 percent uranium. Of this small amount, 99.3 percent is the nonfissionable isotope uranium-238, and just 0.7 percent consists of the fissionable isotope uranium-235. Because most nuclear reactors operate with fuels that are at least 3 percent uranium-235, the two isotopes must be separated in order to concentrate the fissionable uranium-235. The isotope separation process is difficult and substantially increases the cost of nuclear power generation.

Although uranium is a rare element in Earth's crust, it does occur in enriched deposits. Some of the most important occurrences are associated with what are believed to be ancient placer deposits in streambeds.† For example, in Witwatersrand, South Africa, grains of uranium ore (as well as rich gold deposits) were found concentrated in rocks made largely of quartz pebbles. In the United States, the richest uranium deposits are found in Jurassic and Triassic sandstones in the Colorado Plateau and in younger rocks in Wyoming. Most of these deposits have formed through the precipitation of uranium compounds from groundwater. Here, precipitation of uranium occurs as a result of a chemical reaction with organic matter, as evidenced by the concentration of uranium in fossil logs and organic-rich black shales.

It should be emphasized that the concentrations of fissionable uranium-235 and the design of reactors are such that nuclear power plants cannot explode like atomic bombs. Nonetheless, spent fuel from a nuclear reactor includes plutonium, which can be reprocessed (if one has the proper equipment) into weapons.

*Thorium, although not capable by itself of sustaining a chain reaction, can be used with uranium-235 as a nuclear fuel.

†Placer deposits are discussed later in the chapter.

Nuclear Power Challenges

At one time, proponents heralded nuclear power as the clean, cheap source of energy that would replace fossil fuels. Among the benefits of nuclear energy is the fact that nuclear power plants do not emit carbon dioxide—the greenhouse gas that contributes most significantly to global warming (see Chapter 21). By contrast, the generation of electricity from fossil fuels produces large quantities of carbon dioxide. Thus, substituting nuclear power for power generated by fossil fuels represents one option for reducing carbon emissions.

However, despite its attractiveness as a clean power source, nuclear power has not developed into a major energy source throughout the world. Lack of a permanent storage facility for nuclear power plant waste in the United States has been a problem. The skyrocketing cost of building nuclear facilities that must contain numerous safety features held off construction of many plants in the United States. The need for expensive preventive safety measures hinges on the fact that cleaning up a nuclear power plant accident is difficult; contaminated topsoil and other materials must be physically removed in order to make a site safe again, and radioactivity that drifts on air currents from the accident site can impact people far and wide.

Rare but Serious Accidents While accidents have been extremely rare, a few serious incidents have occurred. In 2011 a powerful earthquake near Japan's Fukushima nuclear power plant generated a tsunami that devastated the coastal zone where the plant was situated. A series of equipment failures, nuclear meltdowns, and releases of radioactive materials followed. The plant and immediate surrounding area were evacuated, and the cleanup is expected to take decades. In 1986 an accident at the Chernobyl nuclear plant in Ukraine killed about three dozen people initially, and radioactive fallout has been blamed for later causing more than 6000 cases of thyroid cancer in Ukraine, Russia, and Belarus. The International Atomic Energy Agency estimates that the region around the Chernobyl plant probably won't be fit for long-term human habitation for 20,000 years!

CONCEPT CHECKS 23.3

1. What portion of U.S. energy consumption is provided by nuclear power?

2. What is the primary fuel used in nuclear power plants?

Concept Checker
https://goo.gl/Annax4

3. List some pros and cons of nuclear energy.

23.4 Renewable Energy

List and discuss the major sources of renewable energy. Describe the contribution of renewable energy to the overall U.S. energy supply.

The use of renewable energy is not new. More than 150 years ago, wood supplied a high percentage of our energy needs. Today nonrenewable coal, oil, and natural gas dominate, but the use of renewable forms of energy is on the rise. In 2017 renewable energy represented about 11 percent of all energy used in the United States (**Figure 23.11**). About 17 percent of U.S. electricity was generated from renewable sources in 2017, up from 13 percent just 3 years earlier. The use of renewable fuels is expected to continue growing in the decades to come. Nevertheless, the U.S. Department of Energy projects that we will still rely on nonrenewable fuels to meet a significant portion of our needs.

Solar-Generated Electricity

The Sun is a powerful energy source that can be harnessed to produce electricity. Energy from the Sun has significant advantages: It is renewable, pollution free, and creates no greenhouse gases. Between 2008 and 2017, use of solar energy to generate electricity grew from 1.2 gigawatts (abbreviated GW) to 53.3 gigawatts. In 2017 nearly 2 percent of U.S. electrical generation originated with solar energy methods—enough electricity to supply more than 10 million homes. The U.S. Department of Energy expects this figure to double by 2023. This extraordinary growth reflects the fact that solar electricity is now economically competitive with conventional sources in many places. Solar electricity is created in two different ways: *photovoltaic* and *concentrating solar power*.

Photovoltaic (PV) Solar photovoltaic devices, or solar cells, change sunlight directly into electricity. A single PV device is known as a cell. An individual PV cell is usually small, producing about 1 or 2 watts of power. PV cells of this type can power calculators, watches, and other electronic devices. Arrangements of many solar cells into PV panels and arrangements of multiple PV panels into PV arrays can produce electricity for an entire house. PV power plants have

TOTAL: 11.0 quadrillion Btu

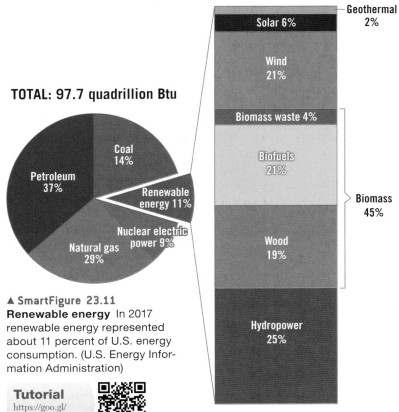

TOTAL: 97.7 quadrillion Btu

- Solar 6%
- Geothermal 2%
- Wind 21%
- Biomass waste 4%
- Biofuels 21%
- Biomass 45%
- Wood 19%
- Hydropower 25%

Coal 14%
Petroleum 37%
Renewable energy 11%
Nuclear electric power 9%
Natural gas 29%

▲ SmartFigure 23.11
Renewable energy In 2017 renewable energy represented about 11 percent of U.S. energy consumption. (U.S. Energy Information Administration)

Tutorial
https://goo.gl/hcUC6U

generator. One advantage of trough systems is that the heated fluid can be stored and used later to keep making electricity when the Sun isn't shining. Sunny skies and hot temperatures make the Southwest an ideal place for these kinds of power plants.

Wind Energy

Air has mass, and when it moves (that is, when wind blows), it contains the energy of that motion—kinetic energy. A portion of that energy can be converted into other forms—mechanical force or electricity—that we can use to perform work.

Mechanical energy from wind is commonly used to pump water in rural or remote places. The "farm windmill," still a familiar sight in many rural areas, is an example. Mechanical energy converted from wind can also be used to saw logs, grind grain, and propel sailboats. More importantly, modern wind-powered electric turbines contribute ever-growing amounts of electricity in

large arrays that cover many acres and produce electricity for thousands of homes (**Figure 23.12**). In 2017 PV electricity accounted for more than 10 percent of the total electricity generation in four states: California (15 percent), Hawaii (11.8 percent), Nevada (10.8 percent), and Vermont (11.5 percent).

Concentrating Solar Power (CSP)   Most power plants use steam to spin large turbines that drive generators and produce electricity. Typically fossil fuels are tapped to boil water and generate the steam. However, a new generation of power plants with CSP systems uses the Sun as a heat source. One type of CSP system uses parabolic troughs to concentrate sunlight (**Figure 23.13**). A parabolic trough is a large mirror shaped like a giant U. Parabolic troughs are connected in long lines and track the Sun throughout the day. When the Sun's heat is reflected off the mirror, the curved shape sends most of that reflected heat onto a receiver. The receiver tube is filled a fluid substance that holds the heat well, such as oil or molten salt. The superhot liquid (200°C) heats water in a heat exchanger, causing the water to turn to steam, which is then used to drive an electrical

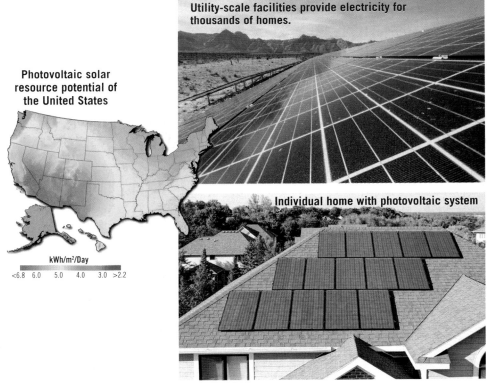

Utility-scale facilities provide electricity for thousands of homes.

Photovoltaic solar resource potential of the United States

kWh/m²/Day
<6.8   6.0   5.0   4.0   3.0   >2.2

Individual home with photovoltaic system

▲ Figure 23.12
Photovoltaic (solar) cells *Photovoltaic (PV) cells* turn solar radiation directly into electricity. The desert Southwest, with its generally cloud-free skies, has the greatest potential for solar energy development. (Map data from National Renewable Energy Laboratory)

► **Figure 23.13**
Parabolic troughs concentrate sunlight These solar collectors, an example of *concentrating solar power* (*CSP*), focus sunlight onto collection pipes filled with a fluid. The heat is used to make steam that drives turbines used to generate electricity.

United States (about 89,000 megawatts in 2017). According to the World Wind Energy Association, wind turbines were capable of supplying about 5 percent of worldwide electricity demand at the beginning of 2018.

Wind speed is crucial in determining suitable sites for installing a wind energy facility. Generally a minimum average wind speed of 21 kilometers (13 miles) per hour (or 6 meters [20 feet] per second) is necessary for a large-scale wind-power plant to be profitable. A small difference in wind speed results in a large difference in energy production and, therefore, a large difference in the cost of the electricity generated. For example, a turbine operating on a site with an average wind speed of 19.3 kilometers (12 miles) per hour would generate about 33 percent more electricity than one operating at 17.7 kilometers (11 miles) per hour. Also, there is little energy to be harvested at low wind speeds; air moving at 9.6 kilometers (6 miles) per hour contains less than one-eighth the energy of air moving at twice that speed.

the United States and worldwide. Like solar energy, wind power is not only renewable but pollution free, and it creates no greenhouse gases.

Worldwide, the installed wind power capacity was nearly 540,000 megawatts (million watts; abbreviated MW) in 2017, an increase of 270 percent over 2010 (**Figure 23.14**).‡ China has the greatest installed capacity (nearly 188,000 megawatts in 2017), followed by the

‡One megawatt is enough electricity to supply 250-300 average American households.

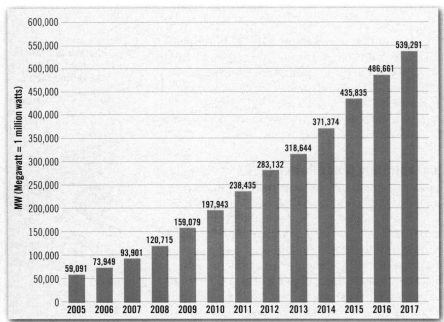

▲ **Figure 23.14**
Global cumulative installed wind capacity, 2000–2017 For many countries, wind power has become an important part of the strategy for phasing out fossil and nuclear energy. In 2017 Denmark set a new world record, with 43 percent of its power coming from wind. An increasing number of countries have reached a double-digit wind power share, including Germany, Ireland, Portugal, Spain, Sweden, and Uruguay. (Data from the World Wind Energy Association)

◀ **Figure 23.15**
Wind energy potential for the United States
At the end of 2017, Texas (23,262 megawatts) had the most installed wind capacity, followed by Oklahoma (7495 megawatts), Iowa (7312 megawatts), and California (5686 megawatts). Many coastal areas have significant potential because offshore winds are stronger and blow more consistently than winds over land. (Data from National Renewable Energy Laboratory/DOE)

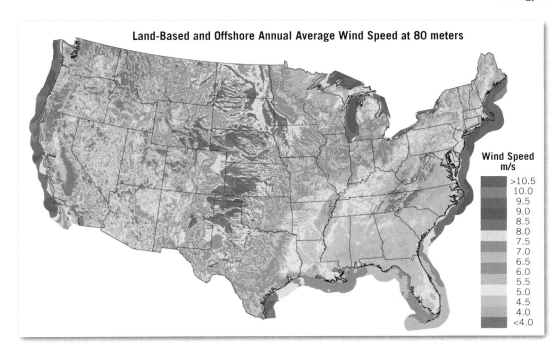

Land-Based and Offshore Annual Average Wind Speed at 80 meters

Wind Speed
m/s

>10.5
10.0
9.5
9.0
8.5
8.0
7.5
7.0
6.5
6.0
5.5
5.0
4.5
4.0
<4.0

Figure 23.15 shows wind energy potential for land areas of the 48 contiguous states. Places with average wind speeds greater than about 6 meters per second (13 miles per hour) are considered to have potential for development. Figure 23.15 also shows the potential for offshore areas. Offshore winds are stronger and blow more consistently than winds over land. Although wind energy development of offshore areas is in an early stage, the potential is significant. U.S. offshore wind has a technical resource potential of more than 2000 gigawatts of generation per year. For context, this is nearly double the nation's current electricity use. This means that if only 1 percent of the technical potential is recovered, nearly 6.5 million homes could be powered by offshore wind energy.

Hydroelectric Power

Through much of human history, waterwheels built on rivers or streams harnessed the energy of running water to power mills and other machinery. Today, the power generated by falling water is used to drive turbines that produce electricity—hence the term **hydroelectric power**. In the United States, hydroelectric power plants satisfied about 7.5 percent of the country's demand in 2017. Most of this energy is produced at large dams, which allow for a controlled flow of water (**Figure 23.16**). The water impounded in a reservoir is a form of stored energy that can be released at any time to produce electricity.

Although water power is considered a renewable resource, the dams built to provide hydroelectricity have finite lifetimes. Recall from Chapter 16 that rivers carry suspended sediment, which begins being deposited behind the dam as soon as it is built. Eventually

sediment may completely fill the reservoir. This may take just decades or hundreds of years, depending on the quantity of suspended material transported by the river. An example is Egypt's huge Aswan High Dam. Completed in the 1960s, it is estimated that half of the reservoir will be filled with sediment from the Nile River by 2025.

The availability of appropriate sites is an important limiting factor in the development of large-scale hydroelectric power plants. A good site provides significant height for the water to fall and a high rate of flow. Hydroelectric dams exist in many parts of the United States, with the greatest concentrations occurring in the Southeast and the Pacific Northwest. Most of the

▼ **Figure 23.16**
Grand Coulee Dam
More than one-half of U.S. hydroelectric capacity is concentrated in Washington, Oregon, and California. In 2017 about 29 percent of the total hydropower was generated in Washington, the location of the nation's largest hydroelectric facility—Grand Coulee Dam.

best U.S. sites have already been developed, limiting the future expansion of hydroelectric power. The total power produced by hydroelectric sources might still increase, but the relative share provided by this source will likely decline because other alternate energy sources will increase at a faster rate.

Geothermal Energy

Geothermal energy is heat within Earth. People can use this energy to heat buildings or to generate electricity. Most geothermal energy is harnessed by tapping deep natural underground reservoirs of steam and hot water. These occur where subsurface temperatures are high due to relatively recent volcanic activity or nearby magma chambers. The Italians began generating electricity geothermally in 1904, so the idea is not new.

Tapping Deep Reservoirs Iceland is a large volcanic island with ongoing volcanic activity (**Figure 23.17**). In Iceland's capital, Reykjavik, underground steam and hot water are pumped into buildings throughout the city for space heating. Steam and hot water also warm greenhouses, where fruits and vegetables are grown all year. In the United States, localities in several western states use hot water from geothermal sources for space heating.

Modern geothermal power plants tap high-temperature (150° to 370°C) hydrothermal reservoirs by drilling wells that can be up to 3 kilometers (1.8 miles) deep. The steam or hot water is piped to the surface, where it powers turbines that generate electricity. In 2017 two dozen countries produced more than 13,270 megawatts of power this way. The United States was the leading producer. Nevertheless, geothermal power represents just 0.4 percent of U.S. utility-scale electric power.

The first commercial geothermal power plant in the United States was built in 1960 at The Geysers, north of San Francisco (**Figure 23.18**). The Geysers remains the world's largest geothermal power plant, generating nearly 1000 megawatts annually. In addition to The Geysers, geothermal development is occurring in Nevada, Utah, and the Imperial Valley in southern California. The 2017 U.S. geothermal generating capacity of 3567 megawatts was enough to supply more than 3 million homes. This is comparable to burning about 70 million barrels of oil each year.

Certain geologic factors make a geothermal reservoir favorable for power plant development:

- *A potent source of heat*, such as a large magma chamber deep enough to ensure adequate pressure and slow cooling but not so deep that the natural water circulation is inhibited. Such magma chambers are most likely in regions of recent volcanic activity.
- *Large and porous reservoirs with channels connected to the heat source*, near which water can circulate and then be stored in the reservoir.
- *A cap of low-permeability rocks* that inhibits the flow of water and heat to the surface. A deep, well-insulated reservoir contains much more stored energy than a similar but uninsulated reservoir.

▶ Figure 23.17
Geothermal development in Iceland Geothermal sources account for 66 percent of Iceland's primary energy use. Much is used directly for space heating: 9 of 10 homes are heated this way. About 25 percent of Iceland's electricity is generated by geothermal sources.

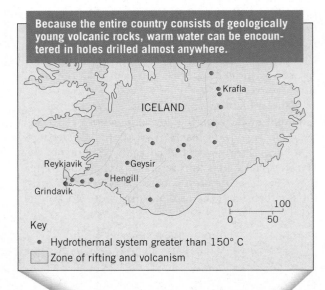

Because the entire country consists of geologically young volcanic rocks, warm water can be encountered in holes drilled almost anywhere.

Key
- • Hydrothermal system greater than 150° C
- ☐ Zone of rifting and volcanism

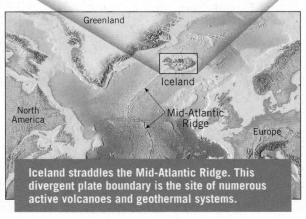

Iceland straddles the Mid-Atlantic Ridge. This divergent plate boundary is the site of numerous active volcanoes and geothermal systems.

The steam at this power station in southwestern Iceland is used to generate electricity. Hot (83°C) water from the plant is sent via an insulated pipeline to Reykjavik for space heating.

◀ **Figure 23.18**
The Geysers This facility, near the city of Santa Rosa in northern California, is the world's largest electricity-generating geothermal development. Most of the steam wells are about 3000 meters (10,000 feet) deep. California's 35 geothermal power plants were responsible for about 80 percent of U.S. geothermal power production in 2017.

Geothermal sources are not expected to provide a high percentage of the world's growing energy needs. Nevertheless, in regions where its potential can be developed, usage continues to grow due to the clean and renewable characteristics of this type of energy.

Geothermal Heat Pumps Although air temperatures above ground change throughout the day and with the seasons, temperatures 3 meters (10 feet) below Earth's surface frequently remain between 10° and 16°C (50° and 60°F). For many areas of the United States, this means near-surface temperatures are usually warmer than the air in winter and cooler than the air in summer. Geothermal heat pumps use Earth's relatively constant temperature to heat and cool buildings. They transfer heat from the ground into buildings during the winter and reverse the process in the summer. Geothermal heat pumps are energy efficient and cost-effective, and they have almost no negative environmental effects. In fact, they can have a positive effect by reducing the use of energy sources that have more or greater negative effects on the environment.

Biomass: Renewable Energy from Plants and Animals

Biomass is organic material made from plants and animals. It is a renewable energy source because we can always grow more trees and crops, and waste will always exist. Some examples of biomass fuels are wood, crops, manure, and some garbage. When burned, the chemical energy in biomass is released as heat. A common example is burning a log in a woodstove or fireplace. Wood and garbage can be burned to produce steam for generating electricity or to provide heat to industries and homes.

Burning biomass is not the only way to release its energy. Biomass can be converted to other usable forms of energy, such as methane gas, or transportation fuels, such as ethanol and biodiesel. In 2017 biomass fuels provided more than 4.9 percent of the energy used in the United States. Primary sources of biomass energy are as follows:

- *Wood biomass* includes wood chips from forestry operations, residues from lumber, pulp/paper, and furniture mills, and fuel wood for space heating. The largest single source of wood energy is "black liquor," a residue of pulp, paper, and paperboard production.
- *Biofuels* include alcohol fuels such as ethanol, and "biodiesel," a fuel from grain oils and animal fats. Most biofuel used in the United States is ethanol produced from corn.
- *Municipal waste* contains biomass such as paper, cardboard, food scraps, grass clippings, and leaves. Waste can be recycled, composted, sent to landfills, or used in waste-to-energy plants. Hundreds of landfills in the United States recover *biogas*, which is methane that forms when waste matter decomposes in low-oxygen (anaerobic) conditions. They then burn the methane to produce electricity and heat.

Tidal Power

Several methods of generating electrical energy from the oceans have been proposed, but the ocean's energy potential remains largely untapped. The development of tidal power is the principal example of energy production from the ocean.

Tides have been used as a source of power for centuries. Beginning in the twelfth century, waterwheels driven by the tides were used to power gristmills and sawmills. During the seventeenth and eighteenth centuries, much of Boston's flour was produced at a tidal mill. Today far greater energy demands must be satisfied, and more sophisticated ways of using the force created by the perpetual rise and fall of the ocean must be employed.

Tidal power is harnessed by constructing a dam across the mouth of a bay or an estuary in a coastal area

▶ Figure 23.19
Tidal power A. Simplified diagram showing the principle of the tidal dam. Electricity is generated only when a sufficient water-height difference exists between the bay and the ocean. **B.** The Sihwa Lake tidal power plant along the northwest coast of South Korea is the largest facility in the world.

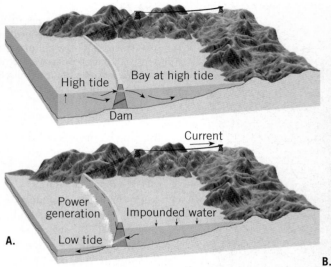

having a large tidal range (**Figure 23.19A**). The narrow opening between the bay and the open ocean magnifies the variations in water level that occur as the tides rise and fall. The strong in-and-out flow that results at such a site is then used to drive turbines and electrical generators.

For decades commercial tidal energy production was exemplified by a tidal power plant at the mouth of the Rance River in France. This plant went into operation in 1966 and continues to produce enough power to satisfy the needs of Brittany. In 2011 this French installation, with a capacity of 240 megawatts, was surpassed in capacity by the Sihwa Lake project along the northwestern coast of South Korea. With a capacity of 254 megawatts, the electricity generated by this power plant each year is equivalent to the energy created by about 860,000 barrels of oil (**Figure 23.19B**).

Unfortunately, it is not possible to harness tidal energy along most coasts. If the tidal range is less than 8 meters (25 feet), or if there are no narrow, enclosed bays, tidal power development is not economically feasible.

For this reason, tides will never provide a very high portion of our ever-increasing electrical energy requirements. Nevertheless, the development of tidal power is worth pursuing at feasible sites because electricity produced by the tides consumes no exhaustible fuels and creates no noxious wastes.

CONCEPT CHECKS 23.4

1. How important is renewable energy to the overall U.S. energy supply?

2. Describe two ways in which solar energy is used to produce electricity. How does solar energy consumption compare with wind energy consumption in the United States?

 **Concept Checker** https://goo.gl/Vyooh4

3. What is biomass? List four examples.

4. Where in the United States is hydroelectric power development most concentrated? What about geothermal power development?

23.5 Mineral Resources

Distinguish among resources, reserves, and ores.

Earth's crust is the source of a wide variety of useful and essential substances. In fact, practically every manufactured product contains substances derived from minerals. The "Economic Use" column in Figure 3.39 (page 94) provides some important examples.

Mineral resources are the endowment of useful minerals ultimately available commercially. Resources include already identified deposits from which minerals can be extracted profitably, called **reserves**, as well as known deposits that are not yet economically or technologically recoverable.

Deposits inferred to exist but not yet discovered are also considered mineral resources. Materials such as building stone, road aggregate, abrasives, ceramics, and fertilizers are not usually called mineral resources; rather, they are classified as *industrial rocks and minerals*.

An **ore** deposit is a naturally occurring concentration of one or more metallic minerals that can be extracted for use in economic goods. In common usage, the term *ore* is also applied to some nonmetallic minerals, such as fluorite and sulfur. Recall from Chapter 3 that more than 98 percent of the continental crust is composed of only eight elements. Except for oxygen and silicon, all other elements make up a relatively small fraction of common crustal rocks (**Figure 23.20**). Indeed, the natural concentrations of many elements are exceedingly small. A deposit containing the average percentage of a valuable element is worthless if the cost of extracting it exceeds the value of the material recovered.

In order to have economic value, an ore deposit must be highly concentrated. For example, copper makes up about 0.0068 percent of the crust. For a deposit to be considered a copper ore, it must contain a concentration of copper that is about 100 times this amount, or about 0.68 percent. Aluminum, on the other hand, represents about 8.1 percent of the crust and can be extracted profitably when it is found in concentrations 3 or 4 times that amount.

It is important to understand that due to economic or technological changes, a deposit may either become profitable to extract or lose its profitability. If the demand for a metal increases and its value rises sufficiently, a previously unprofitable deposit can be upgraded in status from mineral to ore. Technological advances that allow a resource to be extracted more efficiently and, thus, more profitably than before may also trigger a change of status. This occurred at the copper-mining operation located at Bingham Canyon, Utah, one of the largest open-pit mines on Earth (see Figure 23.1). Mining halted there in 1985 because outmoded equipment had driven the cost of extracting the copper beyond the selling price at the time. In 1989 new owners responded by replacing an antiquated 1000-car railroad with conveyor belts and pipelines for transporting the ore and waste. The advanced equipment reduced extraction costs by

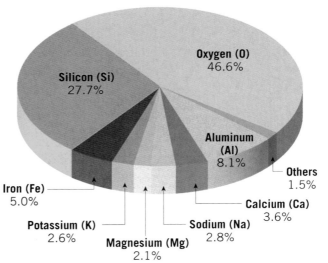

◀ Figure 23.20
Most abundant elements in the continental crust More than 98 percent of the continental crust is composed of just eight elements. The concentrations of many elements that are important mineral resources are part of the "others" category in the chart.

Oxygen (O) 46.6%
Silicon (Si) 27.7%
Aluminum (Al) 8.1%
Others 1.5%
Calcium (Ca) 3.6%
Sodium (Na) 2.8%
Magnesium (Mg) 2.1%
Potassium (K) 2.6%
Iron (Fe) 5.0%

nearly 30 percent, ultimately returning the copper mine operation to profitability.

Over the years, geologists have been keenly interested in learning how natural processes produce localized concentrations of essential metallic minerals. One well-established fact is that occurrences of valuable mineral resources are closely related to the rock cycle. That is, the mechanisms that generate igneous, sedimentary, and metamorphic rocks, including the processes of weathering and erosion, play a major role in producing concentrated accumulations of useful elements. Moreover, with the development of plate tectonics theory, geologists added yet another tool for understanding the processes by which one rock is transformed into another.

CONCEPT CHECKS 23.5

1. Contrast *resource* and *reserve*.

2. What is an ore?

 Concept Checker
https://goo.gl/z2VqTC

23.6 Igneous and Metamorphic Processes

Explain how different igneous and metamorphic processes produce economically significant mineral deposits.

Some of the most important accumulations of metals, such as gold, silver, copper, mercury, lead, platinum, and nickel, are produced by igneous and metamorphic processes (**Table 23.2**). These mineral resources, like most others, result from processes that concentrate desirable materials to such an extent that extraction is economically feasible.

Magmatic Differentiation and Ore Deposits

The igneous processes that generate some metal deposits are quite straightforward. For example, as a large basaltic magma body cools, the heavy minerals that crystallize early tend to settle to the lower portion of the magma chamber (*crystal settling*).

▶ **Figure 23.21**
Pegmatite This granite pegmatite, exposed in the inner gorge of the Grand Canyon, is composed mostly of quartz and potassium feldspar.

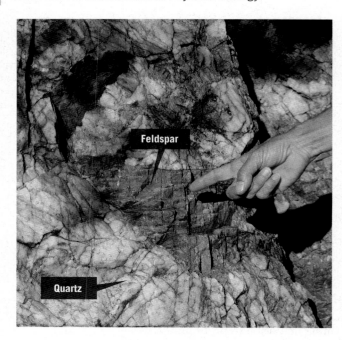

Feldspar

Quartz

This type of magmatic differentiation (see Figure 4.22, page 121) serves to concentrate some metals, producing major deposits of chromite (ore of chromium), magnetite, and platinum. Layers of chromite, interbedded with other heavy minerals, are mined from such deposits in Montana's Stillwater Complex. Another example is the Bushveld Complex in South Africa, which contains more than 70 percent of the world's known reserves of platinum.

Pegmatite Deposits   Magmatic differentiation is important in the late stages of the magmatic process. This is particularly true of granitic magmas, in which the residual melt can become enriched in rare elements, including some heavy metals. Further, because water and other volatile substances do not crystallize along with the bulk of the magma body, these fluids make up a high percentage of the melt during the final phase of solidification. Crystallization in a fluid-rich environment, where ion migration is enhanced, results in the formation of crystals several centimeters, or even a few meters, in length. The resulting rocks, called **pegmatites**, are composed of these unusually large crystals (**Figure 23.21**).

Most pegmatites are granitic in composition and consist of unusually large crystals of quartz, feldspar, and muscovite. Feldspar is used in the production of ceramics, and muscovite is used for electrical insulation and glitter. Further, pegmatites often contain some of the least abundant elements. Thus, in addition to the common silicates, some pegmatites include semiprecious gems such as beryl, topaz, and tourmaline. Moreover, minerals containing the elements lithium, cesium, uranium, and the rare earths[§] are occasionally found. Most pegmatites are located within large igneous masses or occur as dikes or veins that cut into the host rock surrounding the magma chamber (**Figure 23.22**).

Not all late-stage magmas produce pegmatites, nor do all have a granitic composition. Rather, some magmas become enriched in iron or, occasionally, copper. For example, at Kiruna, Sweden, magma composed of more than 60 percent magnetite solidified to produce one of the largest iron deposits in the world.

Hydrothermal Deposits

Among the best-known and most important ore deposits are those generated from hot, ion-rich fluids

TABLE 23.2	Occurrence of Metallic Minerals	
Metal	**Principal Ores**	**Geologic Occurrences**
Aluminum	Bauxite	Residual product of weathering
Chromium	Chromite	Magmatic differentiation
Copper	Chalcopyrite Bornite Chalcocite	Hydrothermal deposits; contact metamorphism; enrichment by weathering processes
Gold	Native gold	Hydrothermal deposits; placers
Iron	Hematite Magnetite Limonite	Banded sedimentary formations; magmatic differentiation
Lead	Galena	Hydrothermal deposits
Magnesium	Magnesite Dolomite	Hydrothermal deposits
Manganese	Pyrolusite	Residual product of weathering
Mercury	Cinnabar	Hydrothermal deposits
Molybdenum	Molybdenite	Hydrothermal deposits
Nickel	Pentlandite	Magmatic differentiation
Platinum	Native platinum	Magmatic differentiation; placers
Silver	Native silver Argentite	Hydrothermal deposits; enrichment by weathering processes
Tin	Cassiterite	Hydrothermal deposits; placers
Titanium	Ilmenite Rutile	Magmatic differentiation; placers
Tungsten	Wolframite Scheelite	Pegmatites; contact metamorphic deposits; placers
Uranium	Uraninite (pitchblende)	Pegmatites; sedimentary deposits
Zinc	Sphalerite	Hydrothermal deposits

[§]The rare earths are a group of 15 elements (atomic numbers 57 through 71) that possess similar properties. They are useful catalysts in petroleum refining and are used to manufacture strong magnets for turbines and rechargeable batteries for cell phones and computers.

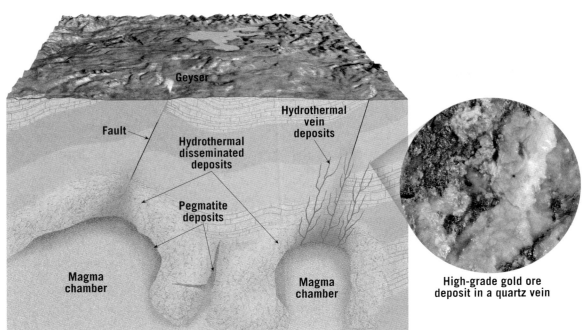

◀ **SmartFigure 23.22**
Pegmatites and hydrothermal deposits Illustration of the relationship between an igneous body and associated pegmatites and hydrothermal mineral deposits.

Tutorial
https://goo.gl/
k7qQaS

High-grade gold ore deposit in a quartz vein

called **hydrothermal solutions** (*hydro* = water, *thermal* = hot).°° Included in this group are the gold deposits of the Homestake Mine in South Dakota; the lead, zinc, and silver ores near Coeur d'Alene, Idaho; the silver deposits of the Comstock Lode in Nevada; and the copper ores of the Keweenaw Peninsula in Michigan (**Figure 23.23**).

Hydrothermal Vein Deposits   The majority of hydrothermal deposits originate from hot, metal-rich fluids that are remnants of late-stage magmatic processes. During solidification, liquids plus various metallic ions accumulate near the top of the magma chamber. These ion-rich solutions can migrate great distances through the surrounding rock before they are eventually deposited, usually as sulfides of various metals. Some of this fluid moves along openings such as fractures or bedding planes, where it cools and precipitates the metallic ions to produce **vein deposits** (see the photo in Figure 23.22). Many of the most productive deposits of gold, silver, and mercury occur as hydrothermal vein deposits.

Disseminated Deposits   Another important type of accumulation generated by hydrothermal activity is called a **disseminated deposit**. Rather than being concentrated in narrow veins and dikes, these ores exist as minute masses throughout the entire rock mass. Much of the world's copper is extracted from disseminated

deposits, including those at Chuquicamata, Chile, and the huge Bingham Canyon copper mine in Utah (see Figure 23.1). Because these accumulations contain only 0.4 to 0.8 percent copper, between 125 and 250 kilograms of ore must be mined to recover 1 kilogram of metal. The environmental impact of these large excavations, including the problem of waste disposal, is significant.

Some hydrothermal deposits are generated from the circulation of ordinary groundwater in regions where magma was emplaced near the surface. The Yellowstone National Park area is a modern example of such a situation. When groundwater invades a zone of recent

◀ **Figure 23.23**
Native copper This nearly pure metal from northern Michigan's Keweenaw Peninsula is an excellent example of a hydrothermal deposit. At one time this area was an important source of copper, but it is now largely depleted.

°° Because these hot, ion-rich fluids tend to chemically alter the host rock, this process is called *hydrothermal metamorphism*; it is discussed in Chapter 8.

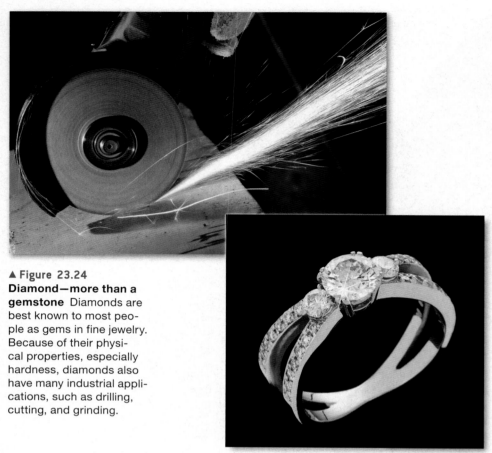

▲ **Figure 23.24**
Diamond—more than a gemstone Diamonds are best known to most people as gems in fine jewelry. Because of their physical properties, especially hardness, diamonds also have many industrial applications, such as drilling, cutting, and grinding.

the dissolved mineral matter precipitates to form massive metallic sulfide deposits. GEOgraphics 13.1 (page 378) illustrates the process.

Origin of Diamonds

Another economically important mineral with an igneous origin is diamond. Although best known as gems, diamonds are used extensively as abrasives in industry and manufacturing. Diamond is one of the hardest natural materials, much harder than corundum, the mineral adjacent to it on the Mohs scale (see Figure 3.17, page 81). Diamonds are used in many tools, such as saws, drills, and grinders (**Figure 23.24**).

Diamonds originate at depths of nearly 200 kilometers (125 miles), where the confining pressure is great enough to generate this high-pressure form of carbon. Once crystallized, the diamonds are carried upward through pipe-shaped conduits that increase in diameter toward the surface. In diamond-bearing pipes, nearly the entire pipe contains diamond crystals that are disseminated throughout an ultramafic rock called *kimberlite*. The most productive kimberlite pipes are found in Russia. The only equivalent source of diamonds in the United States is located near Murfreesboro, Arkansas, but this deposit is exhausted and today serves merely as a tourist attraction.

Metamorphic Processes

The role of metamorphism in producing mineral deposits is frequently linked to igneous processes. For example, contact metamorphism produces many of the most important metamorphic ore deposits. In this process, the host rock recrystallizes and chemically changes due to heat, pressure, and hydrothermal solutions emanating from an intruding igneous body. The extent to which the host rock is altered depends on the nature of the intruding igneous mass as well as the nature of the host rock.

Some resistant materials, such as quartz sandstone, may show very little alteration, whereas others, including limestone, might exhibit the effects of metamorphism for several kilometers from the igneous pluton. As hot, ion-rich fluids move through limestone, chemical reactions take place, producing useful minerals such as garnet and corundum. Further, these reactions release carbon dioxide, which greatly facilitates the outward migration of metallic ions. Thus, extensive zones of metal-rich deposits may surround igneous plutons that have invaded limestone strata.

The most common metallic minerals associated with contact metamorphism are sphalerite (zinc), galena (lead), chalcopyrite (copper), magnetite (iron), and bornite (copper). The hydrothermal ore deposits may

igneous activity, its temperature rises, greatly enhancing its ability to dissolve minerals. These migrating hot waters remove metallic ions from intrusive igneous rocks and carry them upward, where they may be deposited as an ore body. Depending on the conditions, the resulting accumulations may occur as vein deposits, as disseminated deposits, or, where hydrothermal solutions reach the surface in the form of hot springs or geysers, as surface deposits.

Hydrothermal Activity at Oceanic Ridges
Along oceanic ridges, interactions between newly formed oceanic crust alter both the seawater and the crust. The permeable and highly fractured basalt of the upper crust allows seawater to penetrate to depths of 2 to 3 kilometers (1 to 2 miles). As the water circulates, it is heated and reacts with the basaltic crust, stripping metals such as iron, copper, and (occasionally) silver and gold, as well as other elements, such as sulfur, from the surrounding rock. Once the seawater reaches a few hundred degrees Celsius, these mineral-rich fluids buoyantly rise along fractures and eventually reach the ocean floor. As the hot liquid spews from vents and mixes with the cold seawater,

be disseminated throughout the altered zone or may exist as concentrated masses that are located either next to the intrusive body or along the margins of the metamorphic zone.

Regional metamorphism can also generate useful mineral deposits. Recall that at convergent plate boundaries, the oceanic crust and sediments that have accumulated at the continental margins are carried to great depths. In these high-temperature, high-pressure environments, the mineralogy and texture of the subducted materials are altered, producing deposits of nonmetallic minerals such as talc and graphite.

CONCEPT CHECKS 23.6

1. Describe two examples of mineral resources associated with magmatic differentiation.

2. What are hydrothermal deposits? List two general types.

3. Describe the process at oceanic ridges that produces sulfide ore deposits.

Concept Checker
https://goo.gl/GsefUf

4. Relate ore deposits to contact metamorphism.

23.7 Mineral Resources Related to Surface Processes

Discuss ways in which surface processes produce ore deposits.

The preceding discussion focused on mineral resources closely linked to igneous activity and metamorphism. In this section, we look at examples of metallic mineral resources that accumulate as a result of surface processes.

Weathering and Ore Deposits

Weathering creates many important mineral deposits by concentrating minor amounts of metals that are scattered through unweathered rock into economically valuable concentrations. Such a transformation is often termed **secondary enrichment**, and it takes place in one of two ways. In one situation, chemical weathering coupled with downward-percolating water removes undesirable materials from decomposing rock, leaving the desirable elements enriched in the upper zones of the soil. The second way is basically the reverse of the first. That is, the desirable elements that are found in low concentrations near the surface are removed and carried to lower zones, where they are redeposited and become more concentrated.

Bauxite *Bauxite*, the principal ore of aluminum, is one important example of an ore created as a result of enrichment by weathering processes (**Figure 23.25**). Although aluminum is the third-most-abundant element in Earth's crust, economically valuable concentrations of this important metal are not common because most aluminum is tied up in silicate minerals, from which it is extremely difficult to extract.

Bauxite forms in rainy tropical climates, in association with deeply weathered soils known as *laterites.* (In fact, bauxite is sometimes called *aluminum laterite.*) When aluminum-rich source rocks are subjected to the intense and prolonged chemical weathering of the tropics, most of the common elements, including calcium, sodium, and silicon, are removed by leaching.

Because aluminum is extremely insoluble, it becomes concentrated in the soil as bauxite, a hydrated aluminum oxide. Thus, the formation of bauxite depends both on climatic conditions in which chemical weathering and leaching are pronounced and on the presence of an aluminum-rich source rock. Important deposits of nickel and cobalt are also found in laterite soils that develop from igneous rocks rich in ferromagnesian silicate minerals.

Other Deposits Many copper and silver deposits result when weathering processes concentrate metals that are deposited through a low-grade primary ore. Usually such enrichment occurs in deposits

◄ Figure 23.25
Bauxite This ore of aluminum forms as a result of weathering processes under tropical conditions. Its color varies from red or brown to nearly white.

containing pyrite (FeS_2), the most common and widespread sulfide mineral. Pyrite is important because sulfuric acid forms when it chemically weathers, which enables percolating waters to dissolve the ore metals. Once dissolved, the metals gradually migrate downward through the primary ore body until they are precipitated. Deposition takes place because of changes that occur in the chemistry of the solution when it reaches the saturated zone (the zone beneath the surface where all pore spaces are filled with groundwater). In this manner, the small percentage of dispersed metal can be removed from a large volume of rock and redeposited as a higher-grade ore in a smaller volume of rock.

This enrichment process is responsible for the economic success of many copper deposits, including one located in Miami, Arizona. Here the ore was upgraded from less than 1 percent copper in the primary deposit to as much as 5 percent copper in some localized zones of enrichment. When pyrite weathers (oxidizes) near the surface, residues of iron oxide remain. The presence of these rusty-colored caps at the surface indicates the possibility of an enriched copper ore below and is a visible sign for prospectors.

Placer Deposits

The sorting of sediments by wind or water usually results in like-sized grains being deposited together. However, sorting according to the specific gravity of particles also occurs. This latter type of sorting is responsible for the creation of **placers**, which are deposits formed when heavy minerals are mechanically concentrated by currents. Placers associated with streams are among the most common and best known,

but the sorting action of waves can also create placers along the shore. Placer deposits usually involve minerals that are not only heavy but also durable (to withstand physical destruction during transportation) and chemically resistant (to endure weathering processes). Placers form because heavy minerals settle quickly from a current, whereas less dense particles remain suspended and are carried onward. Common sites of accumulation include point bars on the insides of meanders as well as cracks, depressions, and other irregularities on streambeds.

Many economically important placer deposits exist; accumulations of gold are the best known. Indeed, it was the placer deposits discovered in 1848 that led to the famous California Gold Rush. Years later, similar deposits created a gold rush to Alaska (**Figure 23.26**). Panning for gold by washing sand and gravel from a flat pan to concentrate the fine "dust" at the bottom was a common method early prospectors used to recover the precious metal, and it is a process similar to that which created the placer in the first place.

In addition to gold, other heavy and durable minerals form placers. These include platinum, diamonds, and tin. The Ural Mountains contain placers rich in platinum, and placers are important sources of diamonds in southern Africa. Significant portions of the world's supply of cassiterite, the principal ore of tin, have come from placer deposits in Malaysia and Indonesia. Cassiterite is often widely disseminated through granitic igneous rocks. In this state, the mineral is not sufficiently concentrated to be extracted profitably. However, as the enclosing rock dissolves and disintegrates, the heavy and durable cassiterite grains are set free. Eventually the freed particles are washed to a stream, where they are deposited in placers that are significantly more concentrated than the original deposit. Similar

▶ **Figure 23.26**
Placers These deposits form when heavy minerals are mechanically concentrated by currents.

circumstances and events are common to many minerals mined from placers.

In some cases, if the source rock for a placer deposit can be located, it too may become an important ore body. By following placer deposits upstream, one can sometimes locate the original ore body. This is how the gold-bearing veins of the Mother Lode in California's Sierra Nevada batholith were found, as well as the famous Kimberley diamond mine of South Africa. The placers were discovered first, and their source was found at a later time.

CONCEPT CHECKS 23.7

1. What is secondary enrichment?

2. Name the primary ore of aluminum and describe its formation.

 **Concept Checker** https://goo.gl/eHHf6q

3. How might the mineral pyrite play a role in creating an ore deposit?

4. Describe the way in which minerals accumulate in placers. List four minerals that are mined from such deposits.

23.8 Nonmetallic Mineral Resources

Distinguish between two broad categories of nonmetallic mineral resources and list examples of each.

Earth materials that are not used as fuels or processed for the metals they contain are referred to as **nonmetallic mineral resources**. Realize that use of the word *mineral* is very broad in this economic context and is quite different from the geologist's strict definition of *mineral* found in Chapter 3. Nonmetallic mineral resources are extracted and processed either for the elements they contain or for the physical and chemical properties they possess.

People often do not realize the importance of nonmetallic minerals because they see only the products that result from their use and not the minerals themselves. That is, many nonmetallics are used up in the process of creating other products. Examples include the fluorite and limestone that are part of the steelmaking process, the abrasives required to make a piece of machinery, and the fertilizers needed to grow a food crop (**Table 23.3**).

The quantities of nonmetallic minerals used each year are enormous. A glance back at Table 23.1 reminds us that our consumption of nonfuel resources in the United States is great and that a very high percentage of these resources are nonmetallics. Nonmetallic mineral resources are commonly divided into two broad groups: *building materials* and *industrial minerals*. Because some substances have many different uses, they fit in both categories. Limestone, perhaps the most versatile and widely used rock of all, is the best example. As a building material, it is used not only as crushed rock and building stone but also in the making of cement. Moreover, as an industrial mineral, limestone is an ingredient in the manufacture of steel and is used in agriculture

---

TABLE 23.3 Occurrences and Uses of Nonmetallic Minerals

Mineral	Uses	Geologic Occurrences
Apatite	Phosphorus fertilizers	Sedimentary deposits
Asbestos	Incombustible fibers	Metamorphic alteration (chrysotile)
Calcite	Aggregate; steelmaking; soil conditioning; chemicals; cement; building stone	Sedimentary deposits
Clay minerals	Ceramics; china	Residual product of weathering (kaolinite)
Corundum	Gemstones; abrasives	Metamorphic deposits
Diamond	Gemstones; abrasives	Kimberlite pipes; placers
Fluorite	Steelmaking; aluminum refining; glass; chemicals	Hydrothermal deposits
Garnet	Abrasives; gemstones	Metamorphic deposits
Graphite	Pencil lead; lubricant; refractories	Metamorphic deposits
Gypsum	Plaster of Paris	Evaporite deposits
Halite	Table salt; chemicals; ice control	Evaporite deposits; salt domes
Muscovite	Insulator in electrical applications	Pegmatites
Quartz	Primary ingredient in glass	Igneous intrusions; sedimentary deposits
Sulfur	Chemicals; fertilizer manufacture	Sedimentary deposits; hydrothermal deposits
Sylvite	Potassium fertilizers	Evaporite deposits
Talc	Powder used in paints, cosmetics, etc.	Metamorphic deposits

▶ **Figure 23.27**
Limestone quarry Limestone is considered both a building material and an industrial mineral. This quarry is near Amsterdam, Indiana.

to reduce the acidity of soils (see GEOgraphics 7.1, page 213). **Figure 23.27** shows a limestone quarry.

Building Materials

Natural aggregate consists of crushed stone, sand, and gravel. From the standpoint of quantity and value, aggregate is a very important building material. The United States produces nearly 2 billion tons of aggregate per year, which represents about one-half of the country's entire nonenergy mining volume. Aggregate is produced commercially in every state and is used in nearly all building construction and in most public works projects.

Other important building materials include gypsum for plaster and wallboard, clay for tile and bricks, and limestone and shale, which combine to form cement. Cement and aggregate go into the making of concrete, a material that is essential to practically all construction. Aggregate gives concrete its strength and volume, and cement binds the mixture into a rock-hard substance. Building just 2 kilometers of four-lane highway requires more than 85 thousand tons of aggregate. Building an average six-room house requires 90 tons of aggregate.

Because most building materials are widely distributed and present in almost unlimited quantities, they have little intrinsic value. Their economic worth comes only after the materials are removed from the ground and processed. Since the per-ton value compared with the values of metals and industrial minerals is low, mining and quarrying operations are usually undertaken to satisfy local needs. Except for special types of cut stone used for building and monuments, transportation costs greatly limit the distance most building materials can be profitably moved.

Industrial Minerals

Many nonmetallic resources are classified as industrial minerals. In some instances, these materials are important because they are sources of specific chemical elements or compounds. Such minerals are used to manufacture chemicals and produce fertilizers. In other cases, their importance is related to the physical properties they exhibit. Examples include minerals such as corundum and garnet, which are used as abrasives. Although supplies of industrial minerals are generally plentiful, most of these minerals are not nearly as abundant as building materials. Moreover, deposits are far more restricted in distribution and extent. As a result, many of these nonmetallic resources must be transported considerable distances, which of course adds to their cost. Unlike most building materials, which need a minimum of processing before they are ready to use, many industrial minerals require considerable processing to extract the desired substance at the proper degree of purity for its ultimate use.

Fertilizers   To meet the needs of the growing world population, food production must continue expanding. Therefore, fertilizers—primarily nitrate, phosphate, and potassium compounds—are extremely important to agriculture. The synthetic nitrate industry, which derives nitrogen from the atmosphere, is the source of practically all the world's nitrogen fertilizers. The primary source of phosphorus and potassium, however, remains Earth's crust. Apatite, a group of calcium phosphate minerals, is the primary source of phosphate. In the United States, most production comes from marine sedimentary deposits in Florida and North Carolina. Although potassium is an abundant element in many minerals, the primary commercial sources are evaporite deposits containing the mineral sylvite. In the United States, deposits near Carlsbad, New Mexico, have been especially important. **Figure 23.28** shows a mining operation in southern Utah that produces sylvite (commonly called *potash*).

Sulfur   Sulfur is used in so many ways that the quantity of sulfur used by a nation is considered an indicator of its level of development. More than 80 percent of sulfur worldwide is used to produce sulfuric acid, which is necessary for the manufacture of phosphate fertilizer. Sources include deposits of native sulfur associated with salt domes and volcanic areas, as well as common iron sulfides such as pyrite. In recent years an increasingly important source has been the sulfur removed from coal, oil, and natural gas in order to make these fuels less polluting.

Salt   Common salt (*halite*) is another important and versatile resource. It is among the most prominent

◄ **Figure 23.28**
Potash mine evaporation ponds west of Moab, Utah Potash (potassium chloride) at this site is obtained using a system that combines solution mining and solar evaporation. Water from the nearby Colorado River is injected into deposits that are 900 meters (3000 feet) below the surface. The mineral-rich water is brought to the surface and piped to 400 acres of shallow ponds. The water evaporates, leaving behind deposits of potash and common salt (sodium chloride). A blue dye is added to assist the evaporation process. The crystals are harvested and sent to a mill, where the potash is separated from the salt by using a flotation process.

nonmetallic minerals used as a raw material in the chemical industry. In addition, large quantities are used to "soften" water and to keep streets and highways free of ice. Of course, most people are aware that it is also a basic nutrient and a part of many food products.

Salt is a common evaporite, and thick deposits are exploited using conventional underground mining techniques (see Figure 3.40, page 96). Subsurface deposits are also tapped, using brine wells in which a pipe is introduced into a salt deposit and water is pumped down the pipe. The salt dissolved by the water is brought to the surface through a second pipe. This is also the process used to obtain the potassium chloride, or potash (see Figure 23.28). In addition, seawater

continues to serve as a source of salt, as it has for centuries. The salt is harvested after the Sun evaporates the water (see Figure 7.18, page 216).

CONCEPT CHECKS 23.8

1. Which is greater, the per capita consumption of metallic resources or of nonmetallic mineral resources?

2. Distinguish between *building materials* and *industrial minerals*.

Concept Checker
https://goo.gl/BRpyVb

3. List two examples of building materials. What are three examples of industrial minerals?

23

CONCEPTS IN REVIEW

Energy and Mineral Resources

23.1 Renewable and Nonrenewable Resources

Distinguish between renewable and nonrenewable resources.

Key Terms: renewable resource   nonrenewable resource

• *Renewable resources* can be replenished over relatively short time spans (for example, trees used for lumber and paper pulp).

Nonrenewable resources form so slowly that, from a human standpoint, Earth contains fixed supplies (for example, fuels such as oil and coal and metals such as copper and gold).

Q Some resources can be recycled. Does this mean they can be classified as renewable? Explain.

23.2 Energy Resources: Fossil Fuels

Compare and contrast fossil-fuel types and describe how each of them contributes to U.S. energy consumption.

Key Terms:
reservoir rock
oil sands
fossil fuel
cap rock
hydraulic fracturing
oil trap

- Coal, oil, and natural gas are all *fossil fuels* that store energy from hydrocarbons of plants or other organisms that were buried by sediments for millions of years.
- Coal forms from compressed plant fragments that were originally deposited in ancient swamps. Burning it supplies about 14 percent of U.S. energy use. Coal mining can be risky and environmentally damaging, and burning coal generates several kinds of pollution.
- Oil and natural gas are formed from the heated remains of marine plankton. Together, they account for about 66 percent of U.S. energy use. Both oil and natural gas leave their source rock (typically shale) and migrate to an *oil trap* made up of other, more porous rocks, called *reservoir rocks*, covered by a suitable impermeable *cap rock*.

- *Oil sands* are sedimentary deposits that contain bitumen in their pore spaces. Because of its high viscosity, bitumen cannot be pumped out of the rock and must be processed with steam (a significant energy input).
- *Hydraulic fracturing* (fracking) is a method of opening up pore space in otherwise impermeable rocks, permitting natural gas to flow out into wells.

Q As you can see here, the Sinclair Oil Company's logo speaks directly to the "fossil" nature of the fuel it sells. However, is it likely that any *dinosaur* carbon ended up in Sinclair's oil? Explain.

23.3 Nuclear Energy

Describe the importance of nuclear energy and discuss its pros and cons.

Key Terms:
nuclear fission

- Nuclear power plants use a controlled *nuclear fission* chain reaction, in which heavy atoms such as uranium are bombarded with neutrons to cause atoms to split. This releases heat that boils water and drives steam turbines that generate electricity.
- Uranium ore is mined from old placer deposits or sites of groundwater precipitation. The proportion of uranium-235 must be concentrated before it is capable of supporting a fission reaction.
- Nuclear power plants carry significant risks, and building a safe plant is very expensive.

23.4 Renewable Energy

List and discuss the major sources of renewable energy. Describe the contribution of renewable energy to the overall U.S. energy supply.

Key Terms:
geothermal energy
biomass
hydroelectric power

- Renewable energy resources are replenished rapidly, and their use can be sustained for an indefinite period of time.
- Solar energy is used to produce electricity in two ways. Photovoltaic cells change sunlight directly into electricity. Another process involves concentrating solar power that uses sunlight as a fuel source to create steam to drive electrical generators.
- Wind energy is harvested with windmills or turbines. Wind speed is an important variable when considering a site to erect a turbine.

- When running water is dammed, its energy may be harvested as *hydroelectric power*.
- *Geothermal energy* uses heat in Earth's subsurface to produce hot water or electricity. A commercial geothermal site requires (1) a potent source of heat, (2) large subsurface reservoirs, and (3) a cap of low-permeability rock.
- *Biomass* is animal or plant matter that is burned directly or converted into a fuel. Campfires, corn ethanol gasoline additives, and methane capture at landfills are all examples of biomass energy.
- Tidal energy works by impounding the waters of high tide and then releasing them at low tide through a dam, much as with hydroelectric power. It requires a suitable coastline and tidal range, so it is not applicable in most areas.

23.5 Mineral Resources

Distinguish among resources, reserves, and ores.

Key Terms:
reserve
ore
mineral resource

- *Ores* are metallic mineral resources that can be economically mined. We refer to as-yet-unmined but economically recoverable *mineral*

resources as *reserves*. To be considered valuable, a mineral resource must have the element of interest concentrated above the level of its average crustal abundance.

Q Give two examples of events that would cause estimates of U.S. platinum reserves to be revised.

23.6 Igneous and Metamorphic Processes

Explain how different igneous and metamorphic processes produce economically significant mineral deposits.

Key Terms: hydrothermal solution disseminated deposit
pegmatite vein deposit

- Igneous processes concentrate some elements through magmatic differentiation, emplacement of *pegmatites*, and intrusion of kimberlites.
- Magmas may give off *hydrothermal solutions* that penetrate surrounding rock, carrying dissolved metals in them. The metal ores may be precipitated as fracture-filling deposits (*veins*) or may penetrate surrounding strata, producing a great many tiny *disseminated deposits* throughout the host rock. Deep-sea vents along ocean ridges are another example of metal-rich hydrothermal deposits.
- Contact and regional metamorphism may both produce concentrations of nonmetallic and metallic mineral resources, particularly where igneous plutons intrude limestone.

Q This sample consists mainly of quartz, potassium feldspar, and tourmaline. The large tourmaline crystal at the top right is more than 4 centimeters (1.5 inches) long. What term is applied to rocks such as this that are composed of unusually large crystals? With what process is this rock associated?

23.7 Mineral Resources Related to Surface Processes

Discuss ways in which surface processes produce ore deposits.

Key Terms: secondary enrichment placer

- Weathering creates ore deposits by concentrating minor amounts of metals into economically valuable deposits. The process, called *secondary enrichment*, is accomplished either by (1) removing undesirable materials and leaving the desired elements enriched in the upper zones of the soil or (2) removing and carrying the desirable elements to lower zones, where they are redeposited and become more concentrated. Bauxite, the principal ore of aluminum, is associated with the first of these processes.
- *Placer* deposits form when tough, dense minerals such as gold are sorted by water currents and separated from lower-density sediments.

Q Would Jamaica or Sweden be more likely to have bauxite deposits forming at the present time? Explain.

23.8 Nonmetallic Mineral Resources

Distinguish between two broad categories of nonmetallic mineral resources and list examples of each.

Key Term: nonmetallic mineral resource

- Earth materials that are not used as fuels or processed for the metals they contain are referred to as *nonmetallic mineral resources*. Many are sediments or sedimentary rocks. The two broad groups of nonmetallic resources are building materials and industrial minerals.

Q Is limestone considered an industrial mineral or a building material? Include examples as part of your explanation.

GIVE IT SOME THOUGHT

1. At one time most of the energy used in the United States was renewable. What was the renewable energy source that was once dominant? What replaced it?

2. While you're in the car with a friend, a radio news story mentions a coal-mine accident in which some miners have been injured. After hearing this, your friend says, "I thought coal was sort of an old-fashioned fuel that wasn't used much anymore." How would you convince your friend that he's mistaken?

3. The scene in this photo seems an unlikely source of renewable energy, but it is. Describe how this could be the case.

4. As you and a companion pass a refuse container with a sign that says "Aluminum cans only," your companion says, "They sure recycle lots of aluminum these days. That makes it a renewable resource, right?" How would you reply?

5. Are industrial minerals actually minerals? That is, do they meet the definition of *mineral* outlined in Chapter 3? Explain.

6. Assume that you just read a magazine article about a copper mine that is closing. In the article, a mining geologist states that there is plenty of copper-bearing rock remaining and that the concentration of copper is uniform and of equal quality to what has been mined in the past. Later in the article, a spokesperson for the mine owner states, "No ore remains." Assume that both the geologist and the spokesperson are correct and suggest an explanation for the apparently contradicting statements.

EYE ON EARTH

1. The image shows an active landfill where tons of trash and garbage are dumped every day. Eventually this site will be reclaimed to resemble the area shown on the right and will become a source of energy.

   a. Explain how an area filled with trash and waste could become a source of energy. What form of energy will it be? How might it be used?

   b. Will this energy be considered renewable or nonrenewable? Explain.

DATA ANALYSIS

Fossil Fuels Near You

Fossil fuels are the primary energy source for the United States. Coal, natural gas, and oil are mined throughout the United States and shipped across the country, sometimes over very long distances, before they make their way to power plants.

https://goo.gl/eD226M

ACTIVITIES

Go to the U.S. Energy Information Administration site at www.eia.gov/state/maps.php, which shows a map of U.S. energy sources. Click on "Layers/Legend" and select "Remove All" to remove all symbols from the map. Add the appropriate map layer for each question. Click next to a category to see subcategories when needed.

1. Where are the majority of all coal mines located? Are the coal mines in Pennsylvania predominately surface coal mines or underground coal mines?

2. What regions contain the largest concentrations of oil wells? Gas wells?

3. Where are the majority of active offshore oil and gas platforms located?

Ensure that "All Power Plants" is checked and that all the power plant types under this heading are checked. Deselect all other map layers. Click "Find Address," enter your city and state, and click "Locate." Click on a power plant location to see information about the plant. Add map layers as needed.

4. Where (in what city, county, and state) is the power plant nearest your location? What is the primary fuel type used by this plant? Approximately how far away is this power plant from your location?

5. Find the nearest natural gas or petroleum power plant. Where (in what city, county, and state) is this power plant located? What is the primary fuel type used by this plant? Approximately how far away is the nearest refining and processing plant for this type of fuel?

6. Find the nearest coal power plant. Where (in what city, county, and state) is this power plant located? Approximately how far away is the nearest coal mine?

Mastering Geology

Looking for additional review and test prep materials? Visit pearson.com/mastering/geology to enhance your understanding of this chapter's content by accessing a variety of resources, including **Self-Study Quizzes, Geoscience Animations, SmartFigure Tutorials,** *Project Condor* **Videos, Mobile Field Trips, Dynamic Study Modules,** and an optional **Pearson eText.**

Ancient Lunar Lava Tube Could House a Moon Base

In 2009 researchers at the Japan Aerospace Exploration Agency (JAXA) were collecting radar data with their Moon orbiter to see if lunar lava tubes, long surmised to be hidden among the Moon's craters, really existed. The scientists spotted evidence of a hole near the Marius Hills, a lunar region noted for its volcanic domes. But it wasn't until

▲ The Marius Hills Skylight.

Japanese and American scientists teamed up to share JAXA and NASA data several years later that the hole's size could be estimated. In 2017 scientists announced that the hole, now named the Marius Hills Skylight, tops a massive cavern. How big is it? This ancient lava tube could easily house a city larger than Philadelphia. Some scientists immediately proclaimed the Marius Hills Skylight to be a perfect place to house a lunar base.

The lack of oxygen isn't the only reason living on the Moon would be problematic for humans. Without an Earth-like atmosphere and magnetic field, temperatures on the Moon's surface vary wildly, radiation levels are dangerously high, and there is a greater risk of spacecraft or habitats being destroyed by micrometeor hits. Therefore, a cavern-sheltered spot is an important finding for a potential Moon base.

The discovery of the Marius Hills Skylight was especially timely. In December 2017, NASA had a new focus: Concentrate its exploration efforts on returning astronauts to the Moon, with the ultimate goal of exploring Mars—and beyond. Other recent discoveries, such as ice found in caverns near the Moon's north pole, also improve the odds that a lunar base could someday become reality.

► A Moon base on the surface would face more hazards than one built below the surface.

24

Touring Our Solar System

FOCUS ON CONCEPTS

Each statement represents the primary learning objective for the corresponding major heading within the chapter. After you complete the chapter, you should be able to:

24.1 Describe the formation of the solar system according to the nebular theory. Compare and contrast the terrestrial and Jovian planets.

24.2 List and describe the major features of Earth's Moon and explain how maria basins were formed.

24.3 Outline the principal characteristics of Mercury, Venus, and Mars. Describe their similarities to and differences from Earth.

24.4 Summarize and compare the features of Jupiter, Saturn, Uranus, and Neptune, including their ring systems.

24.5 List and describe the principal characteristics of the solar system's small bodies.

P lanetary scientists study the formation and evolution of the bodies in our solar system and beyond—including the eight planets and myriad smaller objects, such as moons, dwarf planets, asteroids, comets, and meteoroids. Studying these objects provides valuable insights into the dynamic processes that operate on Earth. Understanding how other atmospheres evolve helps scientists build better models for predicting climate change on Earth. In addition, seeing how erosional forces work on other bodies allows us to observe the many ways landscapes are created. Finally, the uniqueness of Earth, a body that harbors life, is revealed through the exploration of other planetary bodies.

24.1 Our Solar System: An Overview

Describe the formation of the solar system according to the nebular theory. Compare and contrast the terrestrial and Jovian planets.

The Sun is the center of a revolving system, trillions of miles wide, consisting of eight planets and their satellites, along with numerous smaller bodies—dwarf planets, asteroids, comets, and meteoroids. An estimated 99.85 percent of the mass of our solar system is contained within the Sun. Collectively, the planets account for most of the remaining 0.15 percent. Starting from the Sun, the planets are Mercury, Venus, Earth, Mars, Jupiter, Saturn, Uranus, and Neptune (**Figure 24.1**).

Tethered to the Sun by gravity, the planets travel in the same direction, on slightly elliptical orbits; those nearest the Sun travel fastest (**Table 24.1**). Mercury has the highest orbital velocity, 48 kilometers (30 miles) per second, and the shortest period of revolution around the Sun, 88 Earth days. By contrast, Neptune has an orbital speed of just 5.3 kilometers (3.3 miles) per second and requires 165 Earth-years to complete one revolution. Most large bodies orbit the Sun approximately in the same plane. The planets' inclination with respect to the Earth–Sun orbital plane, known as the *ecliptic*, is shown in Table 24.1.

Nebular Theory: Formation of the Solar System

The **nebular theory**, which describes the formation of the solar system, proposes that the Sun and the planets formed from the same rotating cloud of interstellar gases, called the *solar nebula* (see Figure 1.18, page 19). Much of what we know about the formation of the Sun and other stars comes from observing regions in which stars are currently forming. In all cases, star formation occurs within expansive interstellar clouds. Based on this observation, we conclude that the **solar nebula** began as a slowly rotating cloud of cold, mostly gaseous material composed of over 98 percent hydrogen and helium; all other elements combined made up less than 2 percent of the total. The solid material

consisted of tiny grains often referred to as *interstellar dust*. As this enormous cloud of dust and gases collapsed, gravitational energy was converted into energy of motion and into thermal energy, causing the contracting gases to greatly increase in temperature. Most of the material in the solar nebula collected in the center, where temperatures were hottest, forming the *protosun*.

As the solar nebula collapsed, it also spun faster and faster, as happens when an ice skater pulls her arms in as she spins. The rotational motion caused some of the material to form a flattened, rotating disk that surrounded the hot protosun. Within this rotating disk, matter gradually cooled and condensed into tiny metallic grains and clumps of icy and rocky material. Repeated collisions resulted in most of that material clumping together into increasingly larger chunks that eventually became boulder-sized objects called **planetesimals**, which means "pieces of planets." The stronger gravitational attraction of the largest planetesimals caused them to rapidly grow to hundreds of kilometers in diameter.

The compositions of the planetesimals were determined largely by their proximity to the protosun. As you might expect, temperatures were highest in the inner solar system and decreased toward the outer edge of the rotating disk. Between the 88-day orbit of Mercury and the 687-day orbit of Mars, the planetesimals were composed mainly of materials with high melting

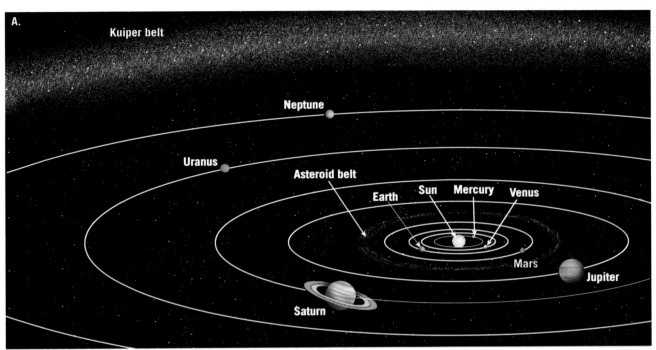

A.

Kuiper belt

Neptune

Uranus

Asteroid belt

Earth Sun Mercury Venus

Mars

Jupiter

Saturn

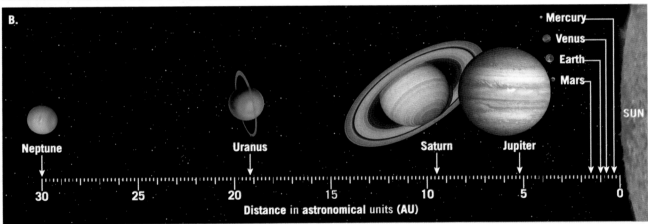

B.

Mercury
Venus
Earth
Mars

SUN

Neptune   Uranus   Saturn   Jupiter

30   25   20   15   10   5   0

Distance in astronomical units (AU)

◄ **SmartFigure 24.1**
Planetary orbits
A. Artistic view of the solar system (not drawn to scale).
B. Distances of planets from the Sun and relative sizes of planets shown using two different scales. Distances are given in astronomical units (AU), where 1 AU equals the mean distance from Earth to the Sun— 150 million kilometers (93 million miles). If the Sun and planets were shown at the scale used for the distances, they would be about 5000 times smaller than illustrated here.

Tutorial
https://goo.gl/
GfPteG

temperatures—metals and rocky substances. In this region, through repeated collisions and accretion, these asteroid-sized rocky bodies combined to form four **protoplanets** that eventually became the terrestrial planets Mercury, Venus, Earth, and Mars.

By contrast, Jupiter, Saturn, Uranus, and Neptune—collectively known as the *gas giants*—have orbital periods that range from about 12 to 165 years and contain more than 150 times the mass of the terrestrial planets. These planets accreted from planetesimals that originated beyond the orbit of Mars (beyond the *frost line*), where temperatures were low enough so

Jupiter, Saturn, Uranus, and Neptune—collectively known as the gas giants—have orbital periods that range from about 12 to 165 years and contain more than 150 times the mass of the terrestrial planets.

that some of the compounds that existed as gases in the inner solar system condensed to form ices. As a result, these planetesimals contained high percentages of ices—mainly ices of water, carbon dioxide, ammonia, and methane—as well as smaller amounts of rocky and metallic debris. The fact that the outer reaches of the solar system contained much larger quantities of ices than metallic and rocky material accounts in part for the large sizes and low densities of the outer planets. The two most massive planets, Jupiter and Saturn, also had surface gravities sufficient to attract and retain even

TABLE 24.1 Planetary Data

Planet	Symbol	Mean Distance from Sun			Orbital Period	Inclination of Orbit	Orbital Velocity	
		AU*	Millions of Miles	Millions of Kilometers			mi/s	km/s
Mercury	☿	0.39	36	58	88 days	7°00'	29.5	47.5
Venus	♀	0.72	67	108	245 days	3°24'	21.8	35.0
Earth	⊕	1.00	93	150	365.25 days	0°00'	18.5	29.8
Mars	♂	1.52	142	248	687 days	1°51'	14.9	24.1
Jupiter	♃	5.20	483	778	12 years	1°18'	8.1	13.1
Saturn	♄	9.54	886	1427	30 years	2°29'	6.0	9.6
Uranus	♦	19.18	1783	2870	84 years	0°46'	4.2	6.8
Neptune	Ψ	30.06	2794	4497	165 years	1°46'	3.3	5.3

Planet	Period of Rotation Around Axis	Diameter		Relative Mass (Earth = 1)	Average Density (g/cm³)	Polar Flattening (%)	Eccentricity**	Number of Known Satellites†
		Miles	Kilometers					
Mercury	59 days	3015	4878	0.06	5.4	0.0	0.206	0
Venus	243 days	7526	12,104	0.82	5.2	0.0	0.007	0
Earth	23^{h}56^{m}04^s	7920	12,756	1.00	5.5	0.3	0.017	1
Mars	24^{h}37^{m}23^s	4216	6794	0.11	3.9	0.5	0.093	2
Jupiter	9^{h}56^m	88,700	143,884	317.87	1.3	6.7	0.048	79
Saturn	10^{h}30^m	75,000	120,536	95.14	0.7	10.4	0.056	62
Uranus	17^{h}14^m	29,000	51,118	14.56	1.2	2.3	0.047	27
Neptune	16^{h}07^m	28,900	50,530	17.21	1.7	1.8	0.009	14

*AU = astronomical unit, Earth's mean distance from the Sun
**Eccentricity is a measure of the amount an orbit deviates from a circular shape. The larger the number, the less circular the orbit.
†Includes all satellites discovered as of August 2018. Satellites are celestial bodies that orbit a planet rather than orbiting a star like the Sun.

large quantities of hydrogen and helium gas, the lightest elements.

The first few hundred million years of the solar system's existence were dynamic and violent. The planets "cleared" their orbits by accreting (combining) much of the remaining leftover material in a period of intense bombardment. The scars of this period are still evident on the Moon's cratered surface. Because of the gravitational pull of the planets, particularly Jupiter, small bodies were flung into planet-crossing orbits or interstellar space. The small fraction of interplanetary matter that survived this violent period became either asteroids or comets, the latter residing mainly in the outer reaches of the solar system.

Lingering Questions About Solar System Formation

While the nebular theory successfully accounts for nearly all of the major features of the solar system, recent discoveries have led some planetary scientists to conclude that the model may need some tweaking. In particular, based on thousands of *exoplanets* (worlds that orbit other stars; see Section 22.1, page 626) discovered by surveys such as NASA's *Kepler* mission, planetary scientists began to realize that solar systems resembling ours are relatively rare. Many of the first-known

exoplanets were "hot Jupiters," gas giants speeding around their stars with orbital periods of a few days. Furthermore, of the planetary systems discovered thus far, most contain one or more "super Earths" (planets a few times larger than Earth) with orbital periods of less than 100 days. Recall that Mercury, the runt of the solar system and the innermost planet, has an orbital period of 88 days.

How do giant planets end up in scorching proximity to a star, where ices can't possibly exist? Why does our inner solar system lack super Earths that orbit close to the Sun? Finding answers to these and other questions related to the development of our solar system continues to give planetary scientists much to contemplate.

The Planets: Internal Structures and Atmospheres

The planets fall into two groups, based on location, size, and density: the **terrestrial (Earth-like) planets** (Mercury, Venus, Earth, and Mars) and the **Jovian (Jupiter-like) planets** (Jupiter, Saturn, Uranus, and Neptune). Because of their locations relative to the Sun, the four terrestrial planets are also known as *inner planets*, and the four Jovian planets are known as *outer planets*. A correlation exists between planetary locations and sizes: The inner planets are substantially smaller than the outer planets. For example, Neptune (the smallest Jovian planet) has nearly 4 times the diameter and 17 times the mass of Earth or Venus.

Other properties that differ among the planets include density, chemical composition, orbital period, and number of satellites (see Table 24.1). Variations in the chemical compositions of planets are largely responsible for their density differences. The average density of the terrestrial planets, which consist mainly of rock and metal, is about 5 times the density of water, whereas the average density of the Jovian planets, with their high proportions of hydrogen, helium, and low-density compounds, is only 1.5 times that of water. In fact, Saturn's density is only 0.7 times that of water, which means it would float in a sufficiently large tank of water. The outer planets are also characterized by long orbital periods and numerous satellites.

Recall from Chapter 12 that shortly after Earth formed, segregation of material formed three major layers, defined by their chemical composition: the crust, mantle, and core. This type of chemical differentiation also occurred in the other planets, but because the terrestrial planets are compositionally different from the Jovian planets, the nature of these layers differs as well (**Figure 24.2**).

Interiors of Terrestrial Planets

The terrestrial planets are dense, having relatively large cores composed mainly of iron and nickel. Silicate minerals and other lighter compounds make up the mantles of the terrestrial planets. The silicate crusts of terrestrial planets are relatively thin compared to their mantles.

Based on seismological evidence, we know that Earth's outer core is molten. We also know that Earth's strong magnetic field is generated by convection within the molten outer core, aided by moderately rapid planetary rotation.

Mars's core is thought to be partially molten but not hot enough to support convection; as a result, Mars lacks a magnetic field. Venus does not have a magnetic field either, even though its core is thought to have a molten metallic layer much like Earth's. Presumably, it lacks a magnetic field either because the core is not hot enough to drive convection or because the planet's 243-day rotation period is too slow to generate a magnetic field. Researchers were surprised to discover that Mercury, despite its small size and slow 59-day period of rotation, possesses a measurable magnetic field, albeit only 1 percent the strength of Earth's. This may be the result of Mercury's partially molten metallic core, which is unusually large compared to the size of the planet.

Interiors of the Jovian Planets

The two largest Jovian planets, Jupiter and Saturn, likely have small, solid cores consisting of iron compounds, like the cores of the terrestrial planets, and rocky material similar to Earth's mantle. Progressing outward, the layer above the core consists of liquid hydrogen that is under extremely high temperatures and pressures. Substantial evidence indicates that under these conditions, hydrogen behaves like a metal in that its electrons move freely about and efficiently conduct both heat and electricity. Jupiter's intense magnetic field is thought to be the result of electric currents flowing within a spinning layer of liquid metallic hydrogen. Saturn's magnetic field is much weaker than Jupiter's due to its smaller shell of liquid metallic hydrogen. Above this metallic layer, scientists believe both Jupiter and Saturn are composed of molecular liquid hydrogen intermixed with helium. The outermost layer is a gaseous atmosphere consisting mainly of hydrogen and helium, along with small amounts of water, ammonia, methane, and other substances.

Uranus and Neptune also have small iron-rich rocky cores, but their mantles are likely hot, dense water, methane, and ammonia. Like Jupiter and Saturn, they have atmospheres dominated by hydrogen and helium.

The Atmospheres of the Planets

The Jovian planets have very thick atmospheres composed mainly of hydrogen and helium, with lesser amounts of water, methane, ammonia, and other hydrogen compounds. By contrast, the

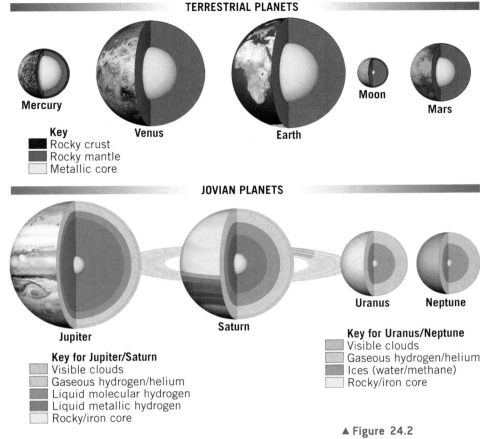

TERRESTRIAL PLANETS

Mercury

Venus

Earth

Moon

Mars

Key
- Rocky crust
- Rocky mantle
- Metallic core

JOVIAN PLANETS

Jupiter

Saturn

Uranus

Neptune

Key for Jupiter/Saturn
- Visible clouds
- Gaseous hydrogen/helium
- Liquid molecular hydrogen
- Liquid metallic hydrogen
- Rocky/iron core

Key for Uranus/Neptune
- Visible clouds
- Gaseous hydrogen/helium
- Ices (water/methane)
- Rocky/iron core

▲ Figure 24.2
Comparing internal structures of the planets

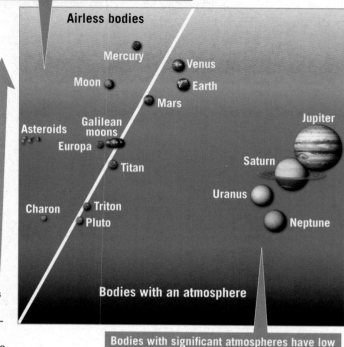

Airless worlds have relatively warm surface temperatures and/or weak gravities.

Airless bodies

Solar heating (temperature)

Mercury

Venus

Moon

Earth

Mars

Asteroids

Galilean moons

Jupiter

Europa

Saturn

Titan

Uranus

Charon

Triton

Neptune

Pluto

Bodies with an atmosphere

Bodies with significant atmospheres have low surface temperatures and/or strong gravities.

Gravity

► SmartFigure 24.3
Bodies with atmospheres versus airless bodies Airless worlds have relatively warm surface temperatures and/or weak gravities. Bodies with significant atmospheres have low surface temperatures and/or strong gravities.

Tutorial
https://goo.gl/
aARBNx

terrestrial planets, including Earth, have relatively meager atmospheres that are typically dominated by carbon dioxide or nitrogen, and most small solar system bodies are airless. Two factors explain these significant differences: solar heating (temperature) and gravity (**Figure 24.3**). These variables determine what planetary gases, if any, were captured by the planets during the formation of the solar system and which were ultimately retained.

During planetary formation, the inner regions of the developing solar system were too hot for ices and gases to condense, but the outer planets formed where temperatures were low and solar heating of planetesimals was minimal. This allowed water vapor, ammonia, and methane to condense into ices. Hence, the Jovian planets contain large amounts of these volatiles. As the planets grew, the strong gravitational fields of the largest Jovian planets, Jupiter and Saturn, also attracted large quantities of the lightest gases, hydrogen and helium. This explains why the Jovian planets have thick atmospheres.

How did Earth acquire water and other volatile gases? It seems that early in the history of the solar system, gravitational tugs by the gas giants, mainly Jupiter and Saturn, sent planetesimals into very eccentric orbits. As a result, Earth was bombarded with icy objects (mainly comet-like objects) that brought in water and other elements. This was a fortuitous event for organisms that currently inhabit our planet. Mercury, our Moon, and numerous other small bodies lack significant

atmospheres, although they certainly would have been bombarded by icy bodies early in their development.

Airless bodies develop where solar heating is strong and/or gravities are weak. Simply stated, *small warm bodies* have a better chance of losing their atmospheres: Gas molecules are more energetic (and hence faster-moving) on a warm body, and they need less speed to escape the weak gravity of a small body. Mercury, the smallest and least massive of the eight planets, has a low surface gravity, which makes holding on to an atmosphere challenging. In addition, because it is the planet closest to the Sun, it is constantly bombarded by solar wind. Mercury therefore has the thinnest atmosphere of all the planets.

The somewhat larger terrestrial planets—Earth, Venus, and Mars—retain some heavy gases, including nitrogen, carbon dioxide, oxygen, and even water vapor. However, their atmospheres are miniscule compared to their total masses. Early in their development, the terrestrial planets may have had thicker atmospheres. Over time, however, these primitive atmospheres gradually changed as light gases trickled away into space. Earth's atmosphere continues to leak hydrogen and helium (the two lightest gases) into space near the top of Earth's atmosphere, where air is so tenuous that nothing stops the fastest-moving particles from flying off into space. The speed required to escape a planet's gravity is called **escape velocity**. Because hydrogen is the lightest gas, it most easily reaches the speed needed to overcome Earth's gravity. Billions of years in the future, the loss of hydrogen (one of the components of water) will eventually "dry out" Earth's oceans, ending its hydrologic cycle.

The massive Jovian planets have strong gravitational fields, which partially explains their thick atmospheres. Furthermore, because of their great distances from the Sun, solar heating is minimal. Because the molecular motion of a gas is temperature dependent, even hydrogen and helium move too slowly to escape the gravitational pull of the Jovian planets. Cold temperatures also explain why Saturn's moon Titan, which is small compared to Earth but much farther from the Sun and therefore colder, retains an atmosphere.

Planetary Impacts

Impacts between solar system bodies have occurred throughout the history of the solar system. On bodies that have little or no atmosphere (like the Moon) and, therefore, no air resistance, even the smallest pieces of interplanetary debris (meteorites) can reach the surface. At high enough velocities, this debris can produce microscopic cavities on individual mineral grains. By contrast, large **impact craters** result from collisions with massive bodies, such as asteroids and comets.

Planetary impacts were considerably more common in the early history of the solar system than they are today. Following that early period of intense bombardment, the rate of cratering diminished dramatically, and it now remains essentially constant. Because weathering and erosion are almost nonexistent on the Moon and Mercury, evidence of past impacts remain clearly evident.

On larger bodies, the presence of an atmosphere may cause the impacting objects to break up and/or decelerate. For example, Earth's atmosphere causes meteoroids with masses of less than 10 kilograms (22 pounds) to lose up to 90 percent of their speed as they penetrate the atmosphere. Therefore, impacts of low-mass bodies produce only small craters on Earth. Our atmosphere is much less effective at slowing large bodies, but fortunately they very rarely make appearances.

The formation of a large impact crater is illustrated in **Figure 24.4**. The meteoroid's high-speed

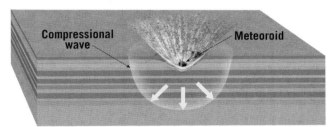

A. The energy of a rapidly moving body is transformed into heat and shock waves.

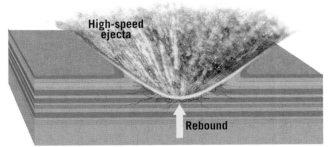

B. The rebound of over-compressed rock causes debris to be explosively ejected from the crater. Some of this material may melt and be deposited as glass beads.

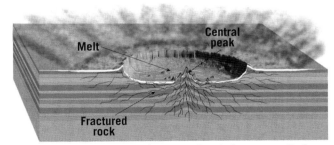

C. Large craters may contain areas of rock that was melted by the impact and a rebounded central peak.

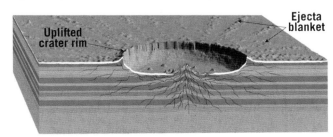

D. Ejected material forms a "blanket" around the crater.

▲ Figure 24.4
Formation of an impact crater

▲ **Figure 24.5**
Lunar crater Euler This 20-kilometer-wide (12-mile-wide) crater is located in the southwestern part of Mare Imbrium. Clearly visible are the bright rays, central peak, secondary craters, and large accumulation of ejecta near the crater rim.

impact compresses the material it strikes, causing an almost instantaneous rebound, which ejects material from the surface. On Earth, impacts can occur at speeds that exceed 50 kilometers (30 miles) per second. Impacts at such high speeds produce shock waves that compress both the impactor and the material being impacted. Almost instantaneously, the over-compressed material rebounds and explosively ejects material out of the newly formed crater. In addition, large craters often exhibit a central peak, such as the one shown in **Figure 24.5**. These central peaks also result from crustal rebound.

Much of the material expelled, called *ejecta*, lands in or near the crater, where it accumulates to form a rim. The remaining material forms a blanket around the crater. Upon impact, large meteoroids may eject large projectiles that strike the surrounding landscape to generate smaller structures called *secondary craters*. Large meteoroids also generate sufficient heat to melt and then eject some of the impacted rock as glass beads. Specimens of glass beads produced in this manner, as well as melt breccia consisting of broken fragments welded by the heat of impact, have been collected on Earth and on the Moon.

CONCEPT CHECKS 24.1

1. Briefly outline the steps in the formation of our solar system, according to the nebular theory.

2. By what criteria are planets considered either terrestrial or Jovian?

3. What accounts for the large density differences between the terrestrial and Jovian planets?

24.2 Earth's Moon: A Chip Off the Old Block

List and describe the major features of Earth's Moon and explain how maria basins were formed.

The Earth–Moon system is unique partly because of the Moon's large size relative to other bodies in the inner solar system. The Moon's diameter is 3475 kilometers (2160 miles), about one-fourth of Earth's 12,756 kilometers (7926 miles), and its surface temperature averages about 107°C (225°F) during daylight hours and −153°C (−243°F) at night. Because its period of rotation on its axis equals its period of revolution around Earth, the same lunar hemisphere always faces Earth. All of the landings of staffed *Apollo* missions were confined to the side of the Moon that faces Earth.

The Moon's density is 3.3 times that of water—comparable to the density of mantle rocks on Earth but considerably less than Earth's average density (5.5 times of water). The Moon's low mass relative to Earth results in a lunar gravitational attraction that is one-sixth that of Earth. The Moon's small mass (and low gravity) is the primary reason it was not able to retain an atmosphere.

How Did the Moon Form?

Current models show that Earth is too small to have formed with a moon, particularly one so large. Furthermore, a captured moon would likely have an eccentric orbit similar to the captured moons that orbit the Jovian planets.

Astronomers generally agree that the Moon formed as a result of a collision between a Mars-sized body and a youthful, semi-molten Earth about 4.5 billion years ago. During this collision, some of the ejected debris was thrown into orbit around Earth and gradually coalesced to form the Moon. Computer simulations show that most of the ejected material would have come from the rocky mantle of the impactor, while its core was assimilated into the growing Earth. This *impact model* is consistent with the Moon having a proportionately smaller core than Earth's and, hence, a lower density.

> *Astronomers generally agree that the Moon formed as a result of a collision between a Mars-sized body and a youthful, semi-molten Earth about 4.5 billion years ago.*

The Lunar Surface

When Galileo first pointed his telescope toward the Moon, he observed two different types of terrain: dark lowlands and brighter, highly cratered highlands (**Figure 24.6**). Because the dark regions appeared to be smooth, resembling seas on Earth, they were called **maria** (*mar* = sea, singular *mare*). The *Apollo 11* mission showed conclusively that the maria are exceedingly smooth plains composed of basaltic lavas. These vast plains are strongly concentrated on the side of the Moon facing Earth and cover about 16 percent of the lunar surface. The lack of large volcanic cones on these surfaces is evidence of high eruption rates of very fluid basaltic lavas similar to the Columbia Plateau flood basalts on Earth.

By contrast, the Moon's light-colored areas resemble Earth's continents, so the first observers dubbed them *terrae* (Latin for "lands"). These areas are now called **lunar highlands** because they are elevated several kilometers above the maria. Rocks retrieved from the highlands are mainly breccias, pulverized by massive bombardment early in the Moon's history. The arrangement of terrae and maria has resulted in the legendary "face" of the "man in the Moon."

Some of the most obvious lunar features are impact craters. A meteoroid 3 meters (10 feet) in diameter can blast out a crater 50 times larger, or about 150 meters

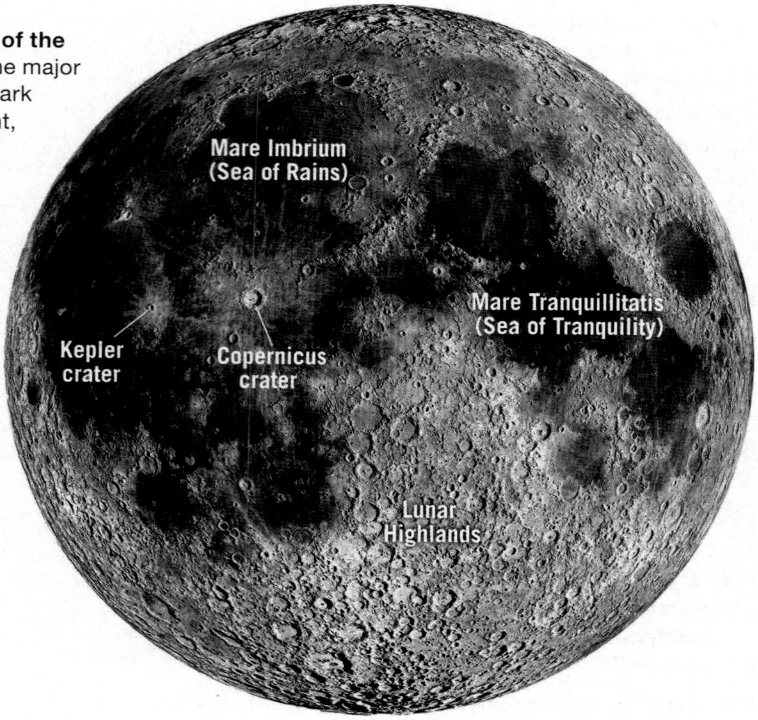

► **Figure 24.6**
Telescopic view of the lunar surface The major features are the dark maria and the light, highly cratered highlands.

Mare Imbrium (Sea of Rains)

Mare Tranquillitatis (Sea of Tranquility)

Kepler crater

Copernicus crater

Lunar Highlands

(500 feet) in diameter. The larger craters shown in Figure 24.6, such as Kepler and Copernicus (32 and 93 kilometers [20 and 58 miles] in diameter, respectively), were created from bombardment by bodies 1 kilometer (0.62 mile) or more in diameter.

History of the Lunar Surface   The evidence used to unravel the history of the lunar surface comes primarily from radiometric dating of rocks returned from *Apollo* missions and studies of crater densities—counting the number of craters per unit area. Because planets and moons have continually been struck by meteoroids throughout their history, the greater the density of craters on a surface feature, the older the feature is inferred to be. Such evidence suggests that, after the Moon coalesced, it passed through four phases: (1) formation of the original crust, (2) excavation of the large impact basins, (3) filling of maria basins, and (4) formation of rayed craters.

During the late stages of its accretion, the Moon's outer shell was most likely completely melted—literally a magma ocean. Then, about 4.4 billion years ago, the magma ocean began to cool and underwent magmatic differentiation (see Chapter 4). Most of the dense minerals, olivine and pyroxene, sank, while less-dense silicate minerals floated to form the Moon's crust. The highlands are made of these igneous rocks, which rose buoyantly like "scum" from the crystallizing magma. The most common highland rock type is *anorthosite*, composed mainly of calcium-rich plagioclase feldspar.

Once formed, the lunar crust was continually impacted as the Moon swept up debris from the solar nebula. During this time, several large impact basins were created (**Figure 24.7**). Then, about 3.8 billion years ago, the Moon, as well as the rest of the solar system, experienced a sudden drop in the rate of meteoritic bombardment.

The Moon's next major event was the filling of the large impact basins that had been created at least 300 million years earlier (see Figure 24.7). Radiometric dating of the maria basalts puts their age between 3.0 billion and 3.5 billion years—considerably younger than the initial lunar crust.

The maria basalts are thought to have originated at depths between 200 and 400 kilometers (125 and 250 miles). They were likely generated by a slow rise in temperature attributed to the decay of radioactive elements. Partial melting probably occurred in several isolated pockets, as indicated by the diverse chemical makeup of the rocks retrieved during the *Apollo* missions. Recent evidence suggests that some mare-forming eruptions may have occurred as recently as 1 billion years ago.

Other lunar surface features related to this period of volcanism include small shield volcanoes (8 to 12 kilometers [5 to 7.5 miles] in diameter), evidence of pyroclastic

Impact of an asteroid-size body produced a huge crater hundreds of kilometers in diameter and disturbed the lunar crust far beyond the crater.

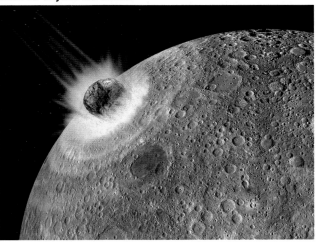

◀ **SmartFigure 24.7**
Formation and filling of large impact basins

Tutorial
https://goo.gl/PtkG7y

Filling of the impact crater with fluid basalts, perhaps derived from partial melting deep within the lunar mantle.

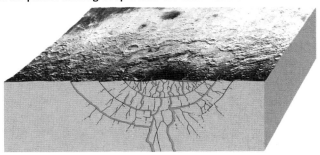

Today these lava-filled basins make up the lunar maria and similar large structures on Mercury.

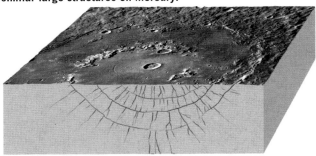

eruptions, narrow winding valleys (*rilles*) thought to be collapsed lava tubes, and long linear depressions similar to down-faulted valleys (*grabens*) on Earth.

The last prominent features to form were rayed craters, as exemplified by the roughly 93-kilometer-wide (58-mile-wide) Copernicus crater shown in Figure 24.6. Light-colored material ejected from these craters, called *rays* because they radiate outward, blankets the maria surfaces and many older, rayless craters. The relatively young Copernicus crater is thought to be about 1 billion years old. Had it formed on Earth, weathering and erosion would have long since obliterated it.

from space (*micrometeorites*) that continually bombard its surface and gradually smooth the landscape. This activity has crushed and repeatedly mixed the upper portions of the lunar crust.

Both the maria and lunar highlands are mantled with a layer of gray, unconsolidated debris derived from a few billion years of meteoric bombardment (**Figure 24.8**). This soil-like layer, properly called **lunar regolith** (*rhegos* = blanket, *lithos* = stone), is anywhere from 2 to 20 meters (6.6 to 66 feet) thick, composed of igneous rocks, breccia, glass beads, and fine *lunar dust*.

▲ **Figure 24.8**
Astronaut Harrison Schmitt, sampling the lunar surface Notice the footprint (inset) in the lunar "soil," called regolith, which lacks organic material and is therefore not a true soil.

Today's Lunar Surface   Weathering and erosion as seen on Earth are nearly absent on the Moon due to low gravity, lack of atmosphere, and absence of water. In addition, tectonic forces are no longer active on the Moon, so quakes and volcanic eruptions have ceased. Erosion is dominated by the impact of tiny particles

---

CONCEPT CHECKS 24.2

1. Briefly describe the origin of the Moon.

2. Compare and contrast the Moon's maria and highlands.   **Concept Checker** https://goo.gl/s6niNd

3. How is crater density used in the relative dating of the Moon's surface features?

24.3 Terrestrial Planets

Outline the principal characteristics of Mercury, Venus, and Mars. Describe their similarities to and differences from Earth.

The terrestrial planets in order from the Sun are Mercury, Venus, Earth, and Mars. Here we consider Mercury, Venus, and Mars and compare their features to those of Earth.

Mercury: The Innermost Planet

▼ **Figure 24.9**
Mercury This high-resolution, color-enhanced image is a mosaic constructed from thousands of images obtained by the Mercury-orbiting spacecraft *Messenger*.

Mercury, the innermost and smallest planet, revolves around the Sun quickly (88 days) but rotates so slowly on its axis that a day–night cycle lasts 176 Earth days. Thus, 1 "night" on Mercury is roughly equivalent to 3 months on Earth and is followed by the same duration of daylight. Mercury has the greatest extremes of surface temperature among the planets, from as low as −173°C (−280°F) at night to noontime temperatures exceeding 427°C (800°F), hot enough to melt tin and lead—and making life as we know it impossible.

Mercury absorbs most sunlight that strikes it, reflecting only 6 percent into space, a characteristic of terrestrial bodies with little or no atmosphere. The minuscule amount of gas that makes up Mercury's atmosphere may have originated from several sources, including ionized gas from the Sun, ices that vaporized during a relatively recent comet impact, and outgassing of the planet's interior.

Although Mercury is small and scientists initially expected the planet's interior to have already cooled, NASA's *Messenger* spacecraft found in 2012 that Mercury has a magnetic field, although it is about 100 times weaker than Earth's. This suggests that an outer layer of Mercury's large core remains molten and capable of convection—a requirement for generating a magnetic field.

Mercury resembles Earth's Moon in that it has very low reflectivity, no sustained atmosphere, numerous volcanic features, and a heavily cratered terrain (**Figure 24.9**). The planet's largest-known impact crater, Caloris Basin (1300 kilometers [800 miles] in diameter), is shown in Figure 24.9 as a large circular tan feature in the upper right. Like our Moon, Mercury has extensive smooth plains. Most of these smooth areas are associated with large impact basins, including Caloris Basin, where lava partially filled the basins and surrounding lowlands. Consequently, they appear to be similar in origin to lunar maria. Recently, *Messenger* found evidence of other types of volcanism on Mercury, including a huge flood basalt province reminiscent of, but much larger than, Earth's Columbia Plateau. Researchers also confirmed the presence of substantial deposits of ice within perpetually shadowed polar craters.

Unique to Mercury are hundreds of *lobate scarps* (**Figure 24.10**), which appear as scalloped-edged cliffs when viewed from space. These cliffs, which cut across numerous craters, are thousands of kilometers long, and some are elevated as much as 3 kilometers above the surrounding landscape. They may have resulted from crustal shortening as the interior of the planet cooled and shrank. As the planet contracted, compressional forces displaced large slabs of crustal rocks over one another along large thrust faults (see Figure 24.10).

Venus: The Veiled Planet

Venus, second only to the Moon in brilliance in the night sky, is named for the Roman goddess of love and beauty. It orbits the Sun in a nearly perfect circle once every 225 Earth days. However, Venus rotates in the opposite direction of the other planets (*retrograde motion*) at an agonizingly slow pace: a single Venus day is equivalent to about 243 Earth days. Venus has the densest atmosphere of the terrestrial planets, consisting mostly of carbon dioxide (97 percent)—and it is the prototype for an extreme *greenhouse effect*. As a consequence, the surface temperature of Venus averages more than 450°C (900°F) both day and night. Temperature variations at the surface are generally minimal because of the intense mixing within the planet's dense atmosphere. Investigations of the planet's extreme and uniform surface temperature led scientists to more fully understand how the greenhouse effect operates on Earth.

The composition of the Venusian interior is probably similar to Earth's. Although Venus, like Earth, probably has a molten outer core, Venus essentially lacks a magnetic field, which indicates that the core does not convect. That could be due to some combination of the planet's slow rotation and an insufficient thermal gradient. However, scientists believe that mantle convection does operate on Venus, but the processes of plate tectonics do not appear to have contributed to the present Venusian topography.

The surface of Venus is completely hidden from view by a thick cloud layer composed mainly of tiny sulfuric acid droplets. Between 1961 and 1984, despite extreme temperatures and pressures, 10 Russian spacecraft landed successfully and transmitted data, including surface images. As expected, however, all the probes were crushed by the planet's immense atmospheric pressure, approximately 90 times that on Earth, within an hour of landing. Using radar imaging, the unstaffed spacecraft *Magellan* mapped Venus's surface in stunning detail (**Figure 24.11**).

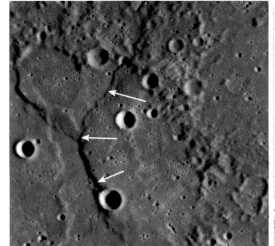
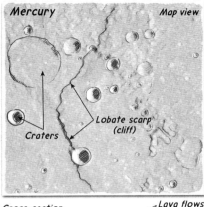

Cross-section · *Lava flows* · *Thrust fault*

Geologist's Sketch

▲ **Figure 24.10**
Lobate scarps on Mercury These clifflike structures, which are often more than 1.6 kilometers (1 mile) high, formed when Mercury's crust was contracting as the planet cooled. This image is from the *Messenger* orbiter.

A few thousand impact craters have been identified on Venus—far fewer than on Mercury and Mars but more than on Earth. Researchers expected that Venus would show evidence of extensive cratering from the heavy bombardment period but found instead that a period of extensive volcanism was responsible for resurfacing Venus. The planet's thick atmosphere also limits the number of impacts by breaking up large incoming meteoroids and incinerating most of the small debris.

About 80 percent of the Venusian surface consists of low-lying plains covered by lava flows, some of which traveled along lava channels that extend hundreds of kilometers. Venus's Baltis Vallis, the longest-known lava channel in the solar system, meanders 6800 kilometers (4255 miles) across the planet. More than 1000 volcanoes

► **Figure 24.11**
Global view of the surface of Venus This computer-generated, false-color image of Venus was constructed from years of investigations, culminating with the *Magellan* mission. The twisting bright features that cross the globe are highly fractured ridges and canyons of the eastern Aphrodite highland.

Elevation of surface
Low ———————→ High

▶ Figure 24.12
Venus's Maat Mons The largest volcano on Venus, Maat Mons, is about 8.5 kilometers (5 miles) high and 400 kilometers (250 miles) wide. The bright areas in the foreground are lava flows. Because the vertical scale in this image has been exaggerated about 22 times, the flanks of this volcano appear steeper than they actually are.

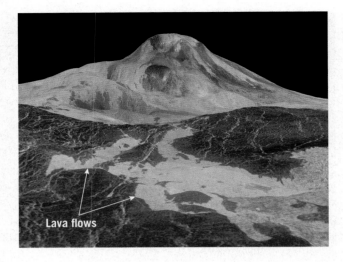

with diameters greater than 20 kilometers (12 miles) have been identified on Venus. However, high surface pressures keep the gaseous components in lava from escaping and limit production of pyroclastic material and lava fountaining, phenomena that tend to steepen volcanic cones. In addition, Venus's high temperatures allow lava to remain mobile longer and, thus, flow far from the vent. Both of these factors result in volcanoes that tend to be flatter and wider than those on Earth or Mars. Maat Mons, the largest volcano on Venus, is about 8.5 kilometers (5 miles) high and 400 kilometers (250 miles) wide (**Figure 24.12**). By comparison, Mauna Loa, Earth's largest volcano, is about 9 kilometers high (5.5 miles) and only 120 kilometers (75 miles) wide.

Venus also has major highlands consisting of plateaus, ridges, and topographic rises that stand above the plains. The rises are thought to have formed where hot mantle plumes encountered the base of the planet's crust, causing uplift. Much as with mantle plumes on Earth, abundant volcanism is associated with mantle upwelling on Venus. Recent data collected by the European Space Agency's (ESA's) *Venus Express* suggest that Venus's highlands contain silica-rich granitic rock. These elevated landmasses resemble Earth's continents, albeit on a smaller scale.

Mars: The Red Planet

Mars, approximately one-half the diameter of Earth, revolves around the Sun in 687 Earth days. Mean surface temperatures range from lows of −140°C (−220°F) at the poles in the winter to highs of 20°C (68°F) at the equator in the summer. Although seasonal temperature variations are similar to Earth's, daily temperature variations are greater due to the very thin atmosphere of Mars (only 1 percent as dense as Earth's). The tenuous Martian atmosphere consists primarily of carbon dioxide (95 percent), with small amounts of nitrogen, oxygen, and water vapor.

Martian Topography   Mars, like the Moon, is pitted with impact craters. The smaller craters are usually filled with windblown dust—confirming that Mars is a dry, desert world. The reddish color of the Martian landscape is due to iron oxide (rust). Large impact craters provide information about the nature of the Martian surface. For example, where the surface is composed of dry rocky debris, ejecta similar in size and shape to that surrounding craters on Earth's Moon is found. However, some Martian craters feature ejecta that looks like muddy slurry was splashed from the crater. Planetary geologists infer that a layer of permafrost (frozen, icy soil) lies below portions of the Martian surface and that the heat of impact melted the ice to produce the fluidlike appearance of these ejecta. About two-thirds of the surface of Mars consists of heavily cratered highlands, concentrated mostly in its southern hemisphere (**Figure 24.13**). The period of extreme cratering occurred

◀ Figure 24.13
Two hemispheres of Mars Color represents height above (or below) the mean planetary radius: White is about 12 kilometers above average, and dark blue is 8 kilometers below average.

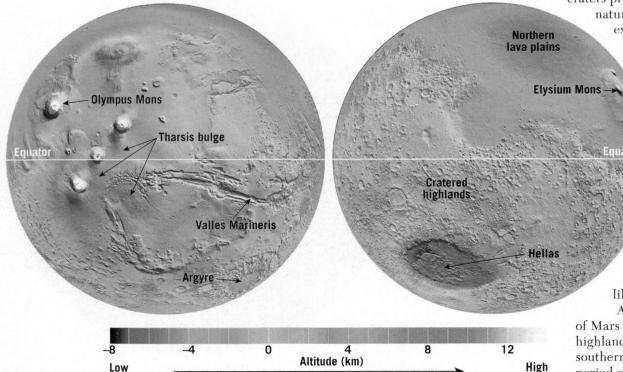

early in the planet's history and ended about 3.8 billion years ago, as it did in the rest of the solar system. Thus, Martian highlands are similar in age to the lunar highlands.

The remaining one-third of the planet, located in the northern hemisphere, is covered by low plains. Based on their relatively low crater counts, these northern plains are younger than the highlands. Their flat topography, possibly the smoothest surface in the solar system, is consistent with vast outpourings of fluid basaltic lavas. Visible on the plains are volcanic cones, some with summit pits (craters) and lava flows with wrinkled edges. If Mars once had abundant water, it would have flowed to the north, which is lower in elevation, possibly forming an expansive ocean.

Located along the Martian equator is an enormous elevated region about the size of North America, called the *Tharsis bulge*. This feature, about 10 kilometers (6 miles) high, appears to have been uplifted and capped with a massive accumulation of volcanic rock that includes the solar system's largest volcanoes.

The tectonic forces that created the Tharsis region also produced fractures that radiate from its center, like spokes on a bicycle wheel. Along the eastern flanks of the bulge, a series of vast canyons called *Valles Marineris* (Mariner Valleys) developed (see Figure 24.13). This canyon network was largely created by down-faulting rather than the stream erosion that carved Arizona's Grand Canyon. Thus, it consists of graben-like valleys similar to the East African Rift Valley. Once formed, Valles Marineris grew thanks to water erosion and the collapse of the rift walls. The main canyon is more than 5000 kilometers (3000 miles) long, 7 kilometers (4 miles) deep, and 100 kilometers (60 miles) wide.

Large impact basins are another prominent type of feature on the Martian landscape. Hellas, the largest visible impact basin, is about 2300 kilometers (1400 miles) in diameter and has the planet's lowest elevation. Debris ejected from this basin contributed to the elevation of the adjacent highlands. Other buried crater basins even larger than Hellas exist, including Utopia Basin, where *Viking 2* landed.

Volcanoes on Mars

Volcanism has been prevalent on Mars during most of its history. The scarcity of impact craters on some volcanic surfaces suggests that the planet is still active. Mars has several of the solar system's largest known volcanoes, including the massive, Olympus Mons, which is about the size of Arizona and stands nearly three times higher than Mount Everest. This enormous volcano was active as recently as a few million years ago and resembles Earth's Hawaiian shield volcanoes (**Figure 24.14**).

How did the volcanoes on Mars grow so much larger than similar structures on Earth? The largest

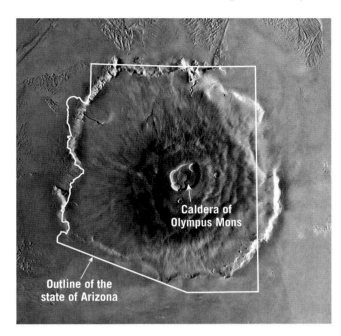

◄ SmartFigure 24.14 Olympus Mons This massive inactive shield volcano on Mars covers an area about the size of the state of Arizona.

Tutorial https://goo.gl/ xzW6Uk

volcanoes on the terrestrial planets tend to form where plumes of hot rock rise from deep within their interiors. On Earth, moving plates keep the crust in constant motion. Consequently, mantle plumes tend to produce a chain of volcanic structures, like the Hawaiian Islands. By contrast, Mars experiences no plate movement, so successive eruptions accumulate in the same location, creating enormous volcanoes rather than a string of smaller ones.

Wind Erosion on Mars

The dominant force currently shaping the Martian surface is wind erosion. Extensive dust storms with winds up to 270 kilometers (170 miles) per hour can persist for weeks. Planetary scientists have also recorded dust devils like the one shown in **Figure 24.15**. Most of the Martian landscape resembles Earth's rocky deserts, with abundant dunes and low areas partially filled with dust.

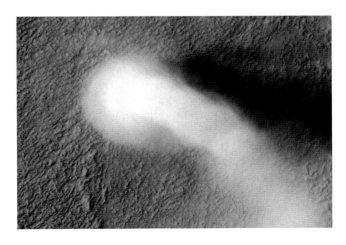

◄ Figure 24.15 A Martian dust devil roughly 20 kilometers (12 miles) high

► **Figure 24.16**
Earth-like stream channels These streamlike channels are strong evidence that Mars once had flowing water. Inset shows a close-up of a streamlined island where running water encountered resistant material along its channel.

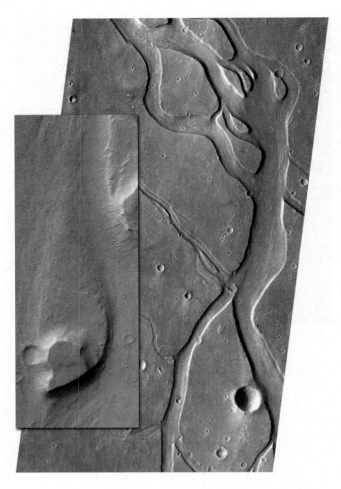

sediment called Mount Sharp. As of October 2018, *Curiosity* had traveled about 20 kilometers (13 miles), examining the lower slopes of this layered mountain (see **GEOgraphics 24.1**). At a target zone NASA calls "Big Arm," the Mars Hand Lens Imager (MAHLI) took images of sediment containing rounded grains, which must have traveled long distances before being deposited. Analysis of these sediments indicates that Gale Crater periodically filled with water, forming a lake that lasted hundreds or possibly even thousands of years. This is strong evidence that Mars must have had a much thicker atmosphere that supported a hydrologic cycle similar to Earth's. It also means that other craters most likely supported long-lived lakes that may have provided suitable habitat for microbial life.

Does Liquid Water Exist on Present-Day Mars?

Without liquid water, life as we know it could not exist. Therefore, there is great interest in detecting liquid water on other bodies in our solar system.

Recent images from a high-resolution camera aboard NASA's *Mars Reconnaissance Orbiter* show dark streaks on Mars, called *recurring slope lineae* (**Figure 24.17**). Researchers think these streaks, which appear seasonally on steep, relatively warm Martian slopes, are caused by the flow of briny (salty) liquid water. Although these dark streaks are just 0.5 to 5 meters (1.6 to 16 feet) wide, they can extend for hundreds of meters downslope. In addition, these features

Water on Mars in the Past Considerable evidence indicates that in the first 1 billion years of the planet's history, liquid water flowed on Mars's surface, creating stream valleys and related features. One location where running water was involved in carving valleys can be seen in the *Mars Reconnaissance Orbiter* image in **Figure 24.16**. Notice the streamlike banks that contain numerous teardrop-shaped islands. These valleys appear to have been cut by catastrophic floods with discharge rates more than 1000 times greater than those of the Mississippi River. Most of these large flood channels emerge from areas of chaotic topography that appear to have formed when the surface collapsed. The most likely source of water for these flood-created valleys was the melting of subsurface ice. However, not all Martian valleys were generated from water released in this manner. Some exhibit branching, tree-like patterns resembling dendritic stream drainage networks on Earth.

On August 6, 2012, the Mars rover *Curiosity* landed in Gale Crater, an impact crater containing a 5-kilometer-high (3-mile-high) accumulation of

Dark streaks
(Recurring slope lineae)

▲ **Figure 24.17**
Dark streaks on Mars, thought to be caused by the flow of briny (salty) water These streaks, called *recurring slope lineae*, are found on steep, warm Martian slopes and disappear during the cold season.

Mars Exploration

Since the first close-up picture of Mars was obtained in 1965, spacecraft voyages to the fourth planet from the Sun have revealed a world that is strangely familiar. Mars has a thin atmosphere, polar ice caps, volcanoes, lava plains, sand dunes, and seasons. Unlike Earth, Mars appears to lack any appreciable amount of liquid water on its surface. However many Martian landscapes suggest that, in the past, running water was an effective erosional agent. The defining question for Mars exploration is "Has Mars ever harbored life?"

NASA's *Phoenix* lander dug into the Martian surface to uncover water ice in a northern region of the planet. Whether ice becomes available as liquid water to support microbial life remains unanswered.

PHOENIX

VIKING 2

VIKING 1

PATHFINDER

OPPORTUNITY

CURIOSITY

SPIRIT

MARS LANDING SITES

The U.S. has successfully landed seven spacecrafts on the surface Mars. The most recent was NASA's *Curiosity*, which landed in Gale Crater in August, 2012.

	Spacecraft*	Type	Landed or entered orbit	Years Active***
1	*Curiosity*	Rover	August 2012	Remains in operation
2	*Phoenix*	Lander	May 2008	Ran out of power during its first Martian winter
3	*Mars Reconnaissance*	Orbiter	March 2006	Planned 2-year mission, remains in operation
4	*Spirit*	Rover	January 2004	Planned 4-year mission, operated for more than 6 years
5	*Opportunity*	Rover	January 2004	Planned 90-day mission, remains in operation
6	*Odyssey*	Orbiter	October 2001	Remains active, longest active spacecraft in orbit around another planet
7	*Viking I*	Lander**	July 1976	Operational for more than 6 years
8	*Viking II*	Lander**	September 1976	Operational for more than 3 years
9	*Maven*	Orbiter	September 2014	Remains in operation

* Selected NASA missions; ** Also had an orbiter; *** As of 2019

NASA's *Curiosity* rover, the size of a car, gracefully landed on Mars after decelerating from 13,000 miles per hour to a complete stop. The landing, described as "seven minutes of terror," began a two-year mission in and around Gale Crater to discover signs of past or present microbial life.

appear during warm weather but fade away when temperatures drop, providing further evidence that liquid water is involved in their formation.

The discovery of these dark streaks has important implications for future missions. In the late 2030s, when NASA plans to send astronauts to the "red planet," the presence of liquid water—even very salty water—would provide a much-needed water source for human explorers on Mars.

NASA's *Phoenix Mars Lander*, which dug into the Martian surface, showed us that ice lies within 1 meter (3 feet) of the surface at latitudes poleward of about 30 degrees. Furthermore, Mars's permanent polar ice caps are composed of mainly water ice, blanketed by a thin layer of carbon dioxide ice during the cold season. Current estimates place the maximum amount of water ice held by the Martian polar ice caps at about 1.5 times the amount covering Greenland.

More than 30 years ago, scientists hypothesized that liquid water might be present below the Martian polar ice caps. It is well known from studies on Earth that the melting point of water decreases under the pressure of an overlying ice mass. In addition, salts, which are widespread on the Martian surface, would lower the freezing point of water from 0°C (32°F) to −70°C (−94°F).

In 2018 researchers found strong evidence that a relatively large body of liquid water may exist below the polar ice cap close to the Martian south pole—a discovery based on data obtained from ground-penetrating radar aboard ESA's *Mars Express*

> *In 2018 researchers found strong evidence that a relatively large body of liquid water may exist below the polar ice cap close to the Martian south pole.*

spacecraft. This instrument sends radar pulses toward the icy surface and records the amount of time it takes for them to be reflected back to the spacecraft. The radar data obtained indicate that the Martian south polar region is covered by many layers of ice and debris to a depth of about 1.5 kilometers (1 mile). A particularly strong radar reflection at the base of the icy layers was also detected. Analysis of the properties of these strong radar signals indicates that they were likely reflected from a boundary between the polar ice cap and a body of liquid water below. The finding is reminiscent of the discovery of Lake Vostok, which lies about 4 kilometers (2.5 miles) below the Antarctic Ice Sheet on Earth. Some microbial life is known to thrive in Earth's subglacial environments, but whether the salty liquid water on Mars could provide a suitable habitat for similar life-forms remains a question.

CONCEPT CHECKS 24.3

1. What other body in our solar system is most like Mercury?

2. Venus was once referred to as "Earth's twin." How are these two planets similar? How do they differ from one another?
   Concept Checker
   https://goo.gl/r5X3NU

3. What surface features do Mars and Earth have in common?

24.4 Jovian Planets

Summarize and compare the features of Jupiter, Saturn, Uranus, and Neptune, including their ring systems.

The four Jovian planets, in order from the Sun, are Jupiter, Saturn, Uranus, and Neptune. They are also commonly called the *outer planets* and the *gas giants*, due to their location, size, and composition.

Jupiter: Lord of the Heavens

The giant among planets, Jupiter has a mass 2.5 times greater than the combined masses of all other planets, satellites, and asteroids in the solar system. However, it pales in comparison to the Sun, with only 1/800 of the Sun's mass.

Jupiter orbits the Sun once every 12 Earth years, and it rotates more rapidly than any other planet, completing one rotation in slightly less than 10 hours. When viewed telescopically, the effect of this fast spin is evident in the bulge of the equatorial region and the slight flattening at the poles (see the "Polar Flattening" column in Table 24.1).

Jupiter's appearance is mainly attributable to the colors of light reflected from its three main cloud layers (**Figure 24.18**). The warmest, and lowest, layer is composed mainly of water ice and appears blue-gray, while the middle layer, where temperatures are lower, consists of brown to orange-brown clouds of ammonium hydrosulfide droplets. These colors are thought to be by-products of chemical reactions occurring in Jupiter's atmosphere. Near the top of its atmosphere lie wispy white clouds of ammonia ice.

Because of its immense gravity, Jupiter shrinks a few centimeters each year. This contraction generates most of the heat that drives Jupiter's atmospheric circulation. Thus, unlike winds on Earth, which are driven by solar energy, the heat emanating from Jupiter's interior produces the huge convection currents observed in its atmosphere.

Jupiter's convective flow produces alternating dark-colored *belts* and light-colored *zones*. The light clouds (zones) are regions where warm material is ascending and cooling, whereas the dark belts represent cool material that is sinking and warming. This convective circulation, along with Jupiter's rapid rotation, generates the high-speed, east–west flow observed between the belts and zones.

The largest storm on the planet is the Great Red Spot. This enormous anticyclonic storm, twice the size of Earth, has been observed for 300 years. In addition to the Great Red Spot, there are various white and brown oval-shaped storms. The white ovals are the cold cloud tops of huge storms many times larger than hurricanes on Earth. The brown storm clouds reside at lower levels in the atmosphere. Lightning in the white oval storms has been photographed by the *Cassini* spacecraft, but the strikes appear to be less frequent than those on Earth.

In 2016 *Juno* spacecraft sent back the first images of Jupiter's north polar region. This region of Jupiter is bluer than the rest of the planet and exhibits numerous storm systems, unlike anything previously seen on any of our solar system's gas giants.

Jupiter's magnetic field, the strongest in the solar system, is probably generated by a rapidly rotating, liquid metallic hydrogen layer surrounding its core. Bright auroras associated with the magnetic field have been photographed over Jupiter's poles. Unlike Earth's auroras, which occur only in conjunction with heightened solar activity, Jupiter's auroras are continuous.

Jupiter's Moons

Jupiter's satellite system, consisting of 79 moons discovered thus far, resembles a miniature solar system. Galileo discovered the 4 largest satellites, called Galilean satellites, in 1610 (**Figure 24.19**). The two largest, Ganymede and Callisto, are roughly the size of Mercury, whereas the two smaller ones, Europa and Io, are about the size of Earth's Moon. The eight largest moons appear to have formed around Jupiter as the solar system condensed.

Jupiter also has many very small satellites, most of which revolve in the opposite direction (retrograde motion) of the largest moons and have eccentric (elongated) orbits steeply inclined toward the Jovian equator.

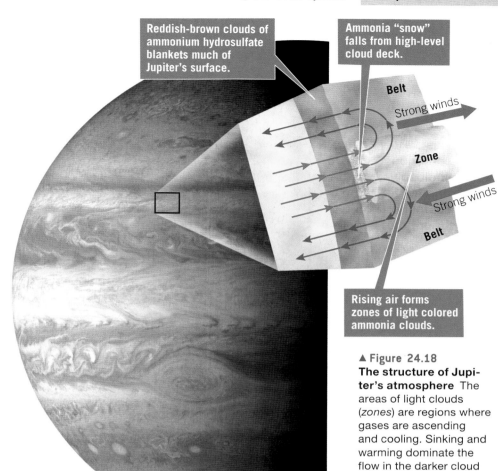

Reddish-brown clouds of ammonium hydrosulfate blankets much of Jupiter's surface.

Ammonia "snow" falls from high-level cloud deck.

Belt

Strong winds

Zone

Strong winds

Belt

Rising air forms zones of light colored ammonia clouds.

▲ **Figure 24.18**
The structure of Jupiter's atmosphere The areas of light clouds (*zones*) are regions where gases are ascending and cooling. Sinking and warming dominate the flow in the darker cloud layers (*belts*). This convective circulation, along with the planet's rapid rotation, generates the high-speed winds observed between the belts and zones.

These satellites appear to be asteroids or comets that passed near enough that either they are remnants of the collisions of larger bodies or they were gravitationally captured by Jupiter.

The Galilean moons can be observed with binoculars or a small telescope and are interesting in their own right. Images from *Voyagers 1* and 2 revealed, to the surprise of most geoscientists, that each of the four Galilean satellites is a unique world (see Figure 24.19). The *Galileo* mission also unexpectedly revealed that the composition of each satellite is strikingly different, implying a different evolution for each. For example, Ganymede has a dynamic core that generates a strong magnetic field not observed in other satellites.

The innermost Galilean moon, Io, is thought to be the most volcanically active body in our solar system. More than 80 active, sulfurous volcanic centers have been discovered. Umbrella-shaped plumes have been observed rising from Io's surface to heights exceeding

► **Figure 24.19**
Jupiter's four largest moons These moons are often referred to as the Galilean moons because Galileo discovered them.

A. Io, perhaps the most volcanically active body in our solar system, has more than 80 active, sulfurous volcanic structures.

B. Europa's icy surface is quite flat and thought to cover a vast ocean composed of briny water.

C. Ganymede, the largest of the Jovian satellites, contains both smooth as well as cratered regions, which suggest this body is still active.

D. Callisto, the outermost of the Galilean satellites, is densely cratered, much like Earth's Moon.

100 kilometers (60 miles) (**Figure 24.20A**). The heat source for volcanic activity is tidal energy generated by a relentless "tug of war" between Jupiter and the other Galilean satellites—with Io as the rope. The gravitational field of Jupiter and the other nearby satellites pull and push on Io's tidal bulge as its slightly eccentric orbit takes it alternately closer to and farther from Jupiter. This gravitational flexing of Io is transformed into heat (similar to the back-and-forth bending of a

piece of sheet metal) and results in Io's spectacular sulfurous volcanic eruptions. Lava, thought to be mainly composed of silicate minerals, regularly erupts on its surface (**Figure 24.20B**).

Of all the bodies in our solar system beyond Earth, Jupiter's icy moon Europa may hold the most promise for harboring life. Most researchers agree that Europa's icy crust, which may be only 25 kilometers (15 miles) thick, conceals a vast underground liquid ocean (**Figure 24.21**). Because the surface of Europa exhibits few impact craters despite its age, it has been hypothesized that the icy crust moves in a manner similar to plate tectonics on Earth. As the crust moves over the ocean below, fragments collide, buckle, and slide underneath one another. Some fragments may even split apart, producing younger slabs of icy crust. These processes may have the potential to transport ocean water and organisms, if they exist, to the surface. Liquid water is a necessity for life as we know it on Earth. Thus, the potential for liquid water existing on Europa means there is considerable interest in sending an orbiter to Europa—and, eventually, a lander capable of launching a robotic submarine for exploration.

Jupiter's Rings   Both the *Voyager 1* and *Galileo* missions studied Jupiter's ring system. By analyzing how these rings scatter light, researchers determined that they are composed of fine, dark particles similar in size to smoke particles. Furthermore, the faint nature of the rings indicates that these minute particles are widely dispersed. The main ring is composed of particles believed to be fragments blasted from the surfaces of Metis and Adrastea, two small moons of Jupiter. Impacts on Jupiter's moons Amalthea and Thebe are believed to be the source of the debris from which the outer gossamer ring formed.

Saturn: The Elegant Planet

Requiring more than 29 Earth years to make one revolution, Saturn is almost twice as far from the Sun as Jupiter, yet their atmospheres, compositions, and internal structures are remarkably similar. The most striking feature of Saturn is its system of rings, first observed by astronomer Galileo Galilei (namesake of the twentieth-century *Galileo* spacecraft) in 1610 (**Figure 24.22**). Through his primitive telescope, the rings appeared as two small bodies adjacent to the planet. Their ring nature was determined 50 years later by Dutch astronomer Christiaan Huygens.

Saturn's atmosphere, like Jupiter's, is dynamic. Although

▼ **Figure 24.20**
A volcanic eruption on Jupiter's moon Io

A.

This plume of volcanic gases and debris is rising more than 100 kilometers (60 miles) above Io's surface.

B.

The bright red area on the left side of the image (see arrow) is newly erupted lava.

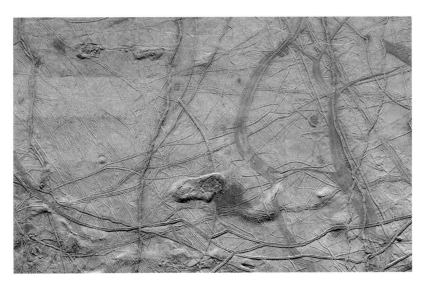

▲ **Figure 24.21**
Europa's icy crust It is believed that the ice conceals a vast liquid ocean.

Saturn's Moons The Saturnian satellite system consists of 62 known moons, of which 53 have been named. The moons vary significantly in size, shape, surface age, and origin. Some of the moons are "original" satellites that formed in tandem with their parent planet. Most of Saturn's smallest moons have irregular shapes and are only a few tens of kilometers in diameter.

Saturn's largest moon, Titan, is larger than Mercury and is the second-largest satellite in the solar system. Titan and Neptune's Triton are the only satellites in the solar system known to have substantial atmospheres. Titan was visited and photographed by the *Cassini–Huygens* probe in 2005. The atmospheric pressure at Titan's surface is about 1.5 times that at Earth's surface, and the atmospheric composition is about 98 percent nitrogen and 2 percent methane, with trace organic compounds. Titan has Earth-like landforms and geologic processes, such as dune formation and streamlike erosion caused by methane "rain." In addition, the northern latitudes appear to have lakes of liquid methane.

Enceladus is another unique satellite of Saturn—one of a few icy moons that erupt "fluid" ice containing minor amounts of other debris. This amazing manifestation of volcanism, called **cryovolcanism** (from the Greek *kryos*, meaning "frost") describes the eruption of magmas derived from the partial melting of ice instead

the bands of clouds are fainter and wider near the equator, rotating "storms" similar to Jupiter's Great Red Spot occur in Saturn's atmosphere, as does intense lightning. Although the atmosphere is about 90 percent hydrogen and nearly 10 percent helium, the clouds (or condensed gases) are composed mainly of ammonia, ammonium hydrosulfide, and water, each segregated by temperature. Also, much like Jupiter, Saturn emits roughly twice as much energy as it receives from the Sun. This implies that it must have an internal heat source, which may come from chemical differentiation in its interior.

The innermost Galilean moon, Io, is thought to be the most volcanically active body in our solar system.

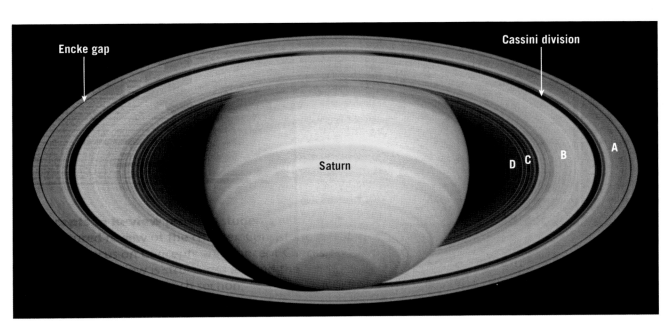

◄ **Figure 24.22**
Saturn's major rings The two bright rings, called A ring (outer) and B ring (inner), are separated by the Cassini division. A second small gap (Encke gap) is also visible as a thin line in the outer portion of the A ring.

▶ **Figure 24.23**
Enceladus, one of Saturn's tectonically active, icy satellites Enceladus has active linear features, called tiger stripes, which are a source region for cryovolcanic activity. Inset image shows jets spurting ice particles, water, and organic compounds from the area of the tiger stripes.

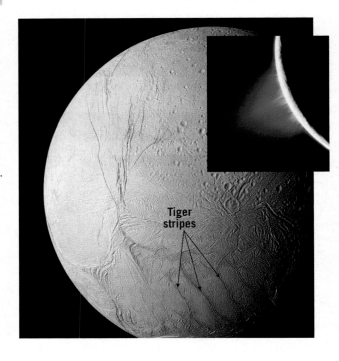

Tiger stripes

of silicate rocks (**Figure 24.23**). In the south polar region, areas called "tiger stripes," consisting of large fractures with ridges on either side, erupt geyser-like plumes. The material ejected by these eruptions is thought to be the source of material that replenishes Saturn's E ring.

Saturn's Ring System
In the early 1980s, *Voyagers 1* and 2 explored Saturn to within 160,000 kilometers (100,000 miles) of its cloud tops. More information was collected about Saturn in that short time than had been acquired since astronomer Galileo first viewed this "elegant planet" in the early 1600s. More

recently, observations from ground-based telescopes, the Hubble Space Telescope, and the *Cassini–Huygens* spacecraft, have added to our knowledge of Saturn's ring system. In 1995 and 1996, when the positions of Earth and Saturn allowed the rings to be viewed edge-on, Saturn's faintest rings and satellites first became visible to us. These rings were once again visible edge-on in 2009.

Saturn's ring system is more like a large rotating disk of varying densities and brightnesses than a series of independent ringlets. Each ring is composed of individual particles—mainly water ice, with lesser amounts of rocky debris—that circle the planet while regularly impacting one another. There are only a few gaps; most of the areas that look like empty space contain either fine dust particles or coated ice particles that are inefficient light reflectors.

Most of Saturn's rings fall into one of two categories, based on density. Saturn's main (bright) rings, designated A and B, are tightly packed and contain particles ranging in size from a few centimeters (pebble-size) to tens of meters (house-size), with most of the particles being roughly the size of a large snowball (see Figure 24.22). In these dense rings, particles collide frequently as they orbit the planet. Although Saturn's main rings (A and B) are 40,000 kilometers (25,000 miles) wide, they are very thin, only 10 to 30 meters (30 to 100 feet) from top to bottom.

At the other extreme are Saturn's faint rings. Saturn's outermost ring (E ring), not visible in Figure 24.22, is composed of widely dispersed, tiny particles. Recall that cryovolcanism (eruption of a water/ice mixture) on Saturn's satellite Enceladus is thought to be the source of material for the E ring.

Studies have shown that the gravitational tugs of nearby moons tend to "shepherd" the ring particles by gravitationally altering their orbits (**Figure 24.24**). For example, the F ring, which is very narrow, appears to be the work of satellites located on either side that confine the ring by pulling back particles that try to escape. On the other hand, the Cassini division, a clearly visible gap in Figure 24.22, arises from the gravitational pull of Mimas, one of Saturn's moons.

Some of the ring particles are believed to be debris ejected from the moons. It is also possible that material is continually recycled between the rings and the ring moons. The ring moons gradually sweep up particles, which are subsequently ejected by collisions with large chunks of ring material or perhaps by energetic collisions with other moons. It seems, then, that planetary rings are continually recycled and changing.

The origin of planetary ring systems is still being debated. Saturn's rings probably formed when objects such as comets, asteroids, or perhaps even moons were pulled apart by Saturn's strong gravity. Pieces of these objects would have collided with each other, breaking into even smaller pieces.

▼ **Figure 24.24**
Two of Saturn's ring moons

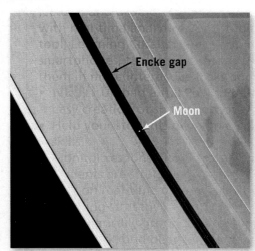

Encke gap

Moon

A. Pan is a small moon about 30 kilometers in diameter that orbits in the Encke gap, located in the A ring. It is responsible for keeping the Encke gap open by sweeping up any stray material that may enter.

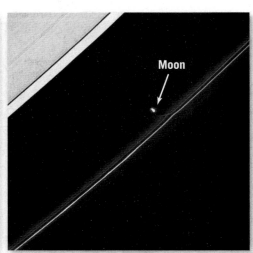

Moon

B. Prometheus, a potato-shaped moon, acts as a ring shepherd. Its gravity helps confine the particles that make up Saturn's thin F ring.

Collisions among these fragments would tend to jostle one another and cause them to spread out to form the flat, thin ring system we observe today. Saturn's rings, as well as those of the other planets, are thought to be short-lived compared to the age of our solar system. This means that Saturn probably lacked rings early in its history, and its existing ring system will likely dissipate in the distant future.

Uranus and Neptune: Twins

Although Earth and Venus have many similar traits, Uranus and Neptune are perhaps more deserving of being called "twins." Nearly equal in diameter (both about four times the size of Earth), they are both bluish in appearance, as a result of methane in their atmospheres. Their days are nearly the same length (about 17 and 16 hours respectively), and their cores are made of rocky silicates and iron—similar to the other Jovian planets. However, their mantles, made mainly of water, ammonia, and methane, are thought to be very different from those of Jupiter and Saturn. One of the most pronounced differences between Uranus and Neptune is the time they take to complete one revolution around the Sun—84 and 165 Earth years, respectively.

Uranus: The Sideways Planet
Unique to Uranus is the orientation of its axis of rotation. Whereas the other planets resemble spinning toy tops as they circle the Sun, Uranus is like a top that has been knocked on its side but remains spinning (**Figure 24.25**). This unusual characteristic of Uranus is likely due to one or more impacts that essentially knocked the planet sideways from its original orientation early in its evolution.

Uranus shows evidence of huge storm systems larger than Earth's continents. Recent photographs from the Hubble Space Telescope also reveal banded clouds composed mainly of ammonia and methane ice—similar to the cloud systems of the other Jovian planets.

Uranus's Moons
Spectacular views from *Voyager 2* showed that Uranus's five largest moons have varied terrains. Some have long, deep canyons and linear scars, whereas others possess large, smooth areas on otherwise crater-riddled surfaces. Studies conducted at NASA's Jet Propulsion Laboratory suggest that Miranda, the innermost of the five largest moons, was recently geologically active—most likely driven by gravitational heating, as occurs on Io.

Uranus's Rings
A surprise discovery in 1977 showed that Uranus has a ring system. The discovery was made as Uranus passed in front of a distant star and blocked its view, a process called *occultation* (*occult* = hidden). Observers saw the star "wink" briefly five times (indicating five rings) before the primary occultation and again five times afterward. More recent ground- and space-based observations indicate that Uranus has at least 10 sharp-edged, distinct rings orbiting its equatorial region. Interspersed among these distinct structures are broad sheets of dust.

◀ Figure 24.25
Uranus, surrounded by its major rings and a few of its known moons Also visible in this image are cloud patterns and several oval storm systems. This false-color image was generated from data obtained by Hubble's Near Infrared Camera.

Neptune: The Windy Planet
Because of Neptune's great distance from Earth, astronomers knew very little about it until 1989 when, 12 years and nearly 3 billion miles into *Voyager 2*'s mission, we obtained a more detailed view of the outermost planet. Neptune has a dynamic atmosphere, much like that of the other Jovian planets (**Figure 24.26**). Record wind speeds approaching 2400 kilometers (1500 miles) per hour encircle the planet, making Neptune the windiest place in the solar system.

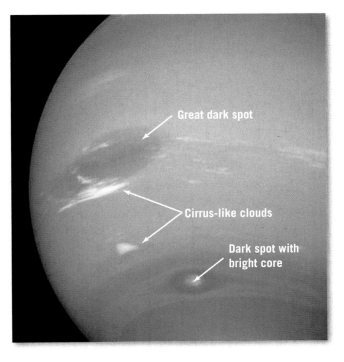

Great dark spot

Cirrus-like clouds

Dark spot with bright core

◀ Figure 24.26
Neptune's dynamic atmosphere

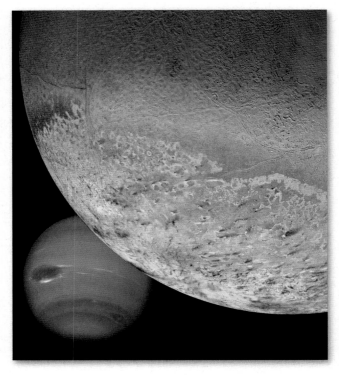

► Figure 24.27
Triton, Neptune's largest moon The bottom of the image shows Triton's wind- and sublimation-eroded south polar cap. Sublimation is the process whereby a solid (ice) changes directly to a gas.

The planet also exhibits large dark spots, thought to be rotating storms similar to Jupiter's Great Red Spot. However, Neptune's storms appear to have comparatively short life spans—usually only a few years. Another feature that Neptune has in common with Uranus is layers of white, cirrus-like clouds—probably frozen methane—about 50 kilometers (30 miles) above the main cloud deck.

Neptune's Moons   Neptune has 14 known satellites, the largest of which is Triton; the remaining 13 are

small, irregularly shaped bodies (**Figure 24.27**). Like Saturn's Enceladus, Triton exhibits cryovolcanism. Triton's icy magma is most likely a mixture of water ice, methane, and ammonia. When partially melted, this mixture behaves as molten rock does on Earth. In fact, upon reaching the surface, these magmas can generate quiet outpourings of ice lavas that can flow great distances from their source—similar to the fluid basaltic flows on Hawaii. They also occasionally produce explosive eruptions that can generate the ice equivalent of volcanic ash. In 1989, *Voyager 2* detected active plumes on Triton that rose 8 kilometers (5 miles) above the surface and were blown downwind for more than 100 kilometers (60 miles).

Neptune's Rings   Neptune has five named rings; two of them are broad, and three are narrow, perhaps no more than 100 kilometers (60 miles) wide. The outermost ring appears to be partially confined by the satellite Galatea. Neptune's rings, like Jupiter's, appear faint, which suggests that they are composed mostly of dust-size particles. Neptune's rings also display red colors, indicating that the dust is composed of organic compounds.

CONCEPT CHECKS 24.4

1. What is the nature of Jupiter's Great Red Spot?

2. What is distinctive about Jupiter's satellite Io?   Concept Checker   https://goo.gl/dqCiJU

3. What two roles do ring moons play in the nature of planetary ring systems?

24.5 Small Solar System Bodies

List and describe the principal characteristics of the solar system's small bodies.

Countless chunks of debris occupy the vast spaces separating the eight planets and the outer reaches of the solar system. In 2006 the International Astronomical Union organized solar system objects not classified as planets or moons into two broad categories: (1) **small solar system bodies**, including *asteroids*, *comets*, and *meteoroids*, and (2) **dwarf planets**. The newest grouping, dwarf planets, includes Ceres, a body about 1000 kilometers (600 miles) in diameter and the largest-known object in the asteroid belt, and Pluto, formerly considered a planet.

Asteroids and meteoroids are compositionally quite similar, both being composed of rocky and/or metallic materials similar to those making up the terrestrial planets. They are usually distinguished by size, with asteroids being much larger than meteoroids, although the exact size difference is not well defined. Comets, on the other hand, are collections of ices, with lesser amounts of dust and small rocky particles. Comets mainly inhabit the outer reaches of the solar system.

Asteroids: Leftover Planetesimals

Asteroids are small bodies (planetesimals) that remain from the formation of the solar system, which means they are about 4.6 billion years old. The orbits of more than 100,000 asteroids have been accurately measured, and thousands more have orbits that are incompletely known, keeping them off the "official" list.

Most asteroids orbit the Sun between Mars and Jupiter, in the region known as the **asteroid belt**

(**Figure 24.28**). Only about 2 dozen asteroids are more than 200 kilometers (125 miles) across. However, our solar system hosts an estimated 1 to 2 million asteroids larger than 1 kilometer (0.6 mile) across and many millions that are smaller.

A smaller number of asteroids travel along eccentric orbits that take them near the Sun, and about 1300 of these, called *Earth-crossing asteroids*, will eventually collide with Earth. Many of the recent large impact craters discovered on Earth resulted from collisions with asteroids. Although these events are rare, they merit our attention because of their potential destruction. As a result, observational initiatives that aim to measure asteroid orbits with great accuracy are ongoing.

Asteroid Structure and Composition

The largest asteroids are roughly spherical because, as is the case for planets and large moons, gravity determines their shape. Indirect evidence from meteorites suggests that the largest asteroids were heated by impact events early in their history, which caused them to melt. This resulted in an early period of chemical differentiation that produced their dense iron-rich cores and rocky mantles.

Most asteroids, however, are small and have irregular shapes, which led planetary geologists to conclude that they are leftover debris from the solar nebula. In addition, these small asteroids have densities lower than scientists originally predicted, indicating that they are relatively porous bodies, like "piles of rubble," loosely bound together by their weak gravitational fields (**Figure 24.29**).

In February 2001, an American spacecraft became the first visitor to an asteroid. Although it was not designed for landing, *NEAR Shoemaker* landed successfully on Eros and collected information that has planetary geologists both intrigued and perplexed. Images obtained as the spacecraft drifted toward Eros revealed a barren, rocky surface composed of particles ranging in size from fine dust to boulders up to 10 meters (30 feet) across. Researchers unexpectedly discovered that fine debris tends to concentrate in the low areas, where it forms flat deposits resembling ponds. Surrounding the low areas, the landscape is marked by an abundance of large boulders. One of several hypotheses to explain the boulder-strewn topography is seismic shaking, which would cause the boulders to move upward as the finer materials sink. This is analogous to what happens when a jar of sand and various-sized pebbles is shaken: The larger pebbles rise to the top, while the smaller sand grains settle to the bottom (sometimes referred to as the "Brazil-nut effect").

Exploring Asteroids

In November 2005, the Japanese probe *Hayabusa* made a soft landing on the small near-Earth asteroid Itokawa and picked up some rocky debris before returning to Earth in June 2010 (see Figure 24.29). Chemical analyses of samples from this mission show that

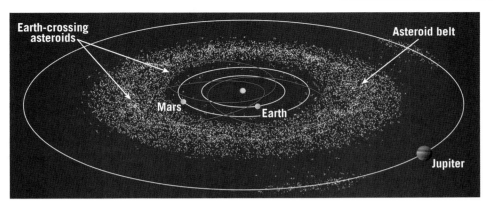

▲ **Figure 24.28**
The asteroid belt The orbits of most asteroids lie between Mars and Jupiter. Shown in red are the orbits of a few known near-Earth asteroids.

the surface of this asteroid is nearly identical in composition to rocky meteorites. This finding strongly supports the idea that asteroids are the source of most meteoroids large enough to reach Earth's surface.

In 2014 Japan launched *Hayabusa 2*, which arrived at asteroid named Ryugu in June 2018. Ryugu, which contains a high percentage of carbon, is called a *carbonaceous asteroid*. For 18 months, the spacecraft is expected to probe and impact the asteroid to create an artificial crater on its surface. It will also deploy a small lander and three rovers that will bounce along the surface, make

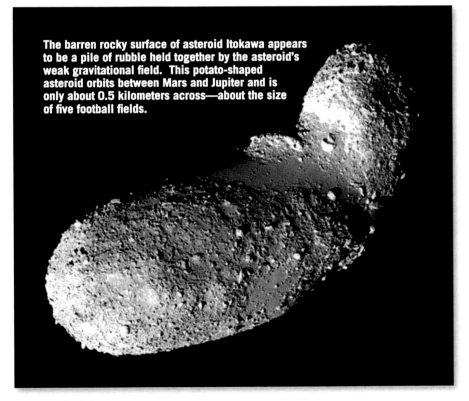

▲ **Figure 24.29**
Asteroid Itokawa The barren rocky surface of asteroid Itokawa appears to be a pile of rubble held together by the asteroid's weak gravitational field. This potato-shaped asteroid orbits between Mars and Jupiter and is only about 0.5 kilometers across—about the size of five football fields.

Gas tail swept back by solar wind

Dust tail is pushed by sunlight

Tails point away from the Sun

Sun

Tails form when comet is about 1 AU from Sun

Coma begins to form when comet is about 5 AU from Sun

Orbit

▲ **Figure 24.30**
Changing orientation of a comet's tail as it orbits the Sun

Bright comets have two visible tails: a dark blue tail that points straight away from the Sun and a brighter tail that points away from the Sun but also curves slightly in the direction from which the comet has traveled. Scientists have determined the mechanisms that account for the formation of these tails. The faint, straight *gas tail* consists of ionized cometary gases that are pushed away from the coma by the pressure of the *solar wind* of charged particles emitted by the Sun. The brighter, curved *dust tail* consists of dust particles that are pushed away from the coma by the much weaker pressure of sunlight (*radiation pressure*). The dust tail is curved because its bright, relatively slow-moving particles record how the direction toward the Sun changes as the comet moves along its orbit.

As a comet's orbit carries it away from the Sun, the gases forming the coma dissipate, the tails disappear, and the comet returns to cold storage. Material that was blown from the coma to form the tails is lost forever. When all the gases are expelled, the inactive comet continues its orbit without a coma or tail. Sometimes the comet literally disintegrates into small fragments that continue to orbit the Sun. Scientists believe that few comets remain active for more than a few hundred close orbits of the Sun.

In 2015 the ESA spacecraft *Rosetta* obtained a close-up image of the nucleus of Comet 67P/

close-up observations, and collect data. The spacecraft will then head back to Earth and is scheduled to return in late 2020.

Comets: Dirty Snowballs

Comets, like asteroids, are leftover material from the formation of the solar system. They are loose collections of rocky material, dust, water ice, and frozen gases (ammonia, methane, and carbon dioxide), thus the nickname "dirty snowballs." Recent space missions to comets have shown their surfaces to be dry and dusty, so that their ices are hidden beneath a layer of rocky debris.

Most comets reside in the outer reaches of the solar system and take hundreds of thousands of years to complete a single orbit around the Sun. However, a smaller number of *short-period comets* (those having orbital periods of less than 200 years), such as the famous Halley's Comet, make regular encounters with the inner solar system (**Figure 24.30**). The shortest-period comet (Encke's Comet) orbits the Sun once every 3 years.

Comet Structure and Composition The phenomena associated with comets come from a small central body called the **nucleus**. These structures are typically 1 to 10 kilometers in diameter, but comet nuclei 40 kilometers across have been observed. When a comet comes within about 5 astronomical units (AU) of the Sun, solar energy heats its surface sufficiently to cause its icy components to begin vaporizing into gas. The escaping gases carry dust from the comet's surface, producing a huge dusty atmosphere called a **coma** (**Figure 24.31**). As a comet approaches the inner solar system, the coma grows, and some of the dust and gas are pushed away from the Sun to form the comet's tails, which can grow to hundreds of millions of kilometers in length.

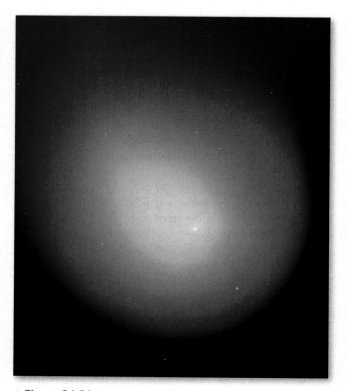

▲ **Figure 24.31**
Coma and nucleus of Comet Holmes The nucleus of Comet Holmes is the bright yellow spot within the reddish-orange coma.

▲ Figure 24.32
Jets of gas and dust erupting from the nucleus of Comet 67P This comet's full name is Comet 67P/Churyumov–Gerasimenko; like most other comets, it is named after its discoverers. It is a regular visitor to the inner solar system, orbiting the Sun every 6.5 years.

Churyumov–Gerasimenko that offered a new perspective on the extent of the comet's activity. **Figure 24.32** shows jets of gas and dust emanating from the central region of the nucleus and extending toward the upper right of the image. This image also shows the nebulous glow of material escaping the bright sunlit surface of the comet.

The Realm of Comets: The Kuiper Belt and Oort Cloud
Most comets originate in one of two regions: the *Kuiper belt* or the *Oort cloud*. Named in honor of astronomer Gerald Kuiper, who predicted its existence,

the **Kuiper belt** hosts a large group of icy objects that reside in the outer solar system, beyond the orbit of Neptune. Pluto's orbit lies within the Kuiper belt, and in 2005 scientists discovered Eris, a Kuiper belt body more massive than Pluto.

Like most other bodies located in the inner solar system, Kuiper belt comets orbit the Sun in the same direction and along roughly the same plane as the planets (**Figure 24.33A**). This disc-shaped structure is thought to contain about 100,000 bodies more than 100 kilometers (60 miles) across, as well as many smaller objects. The largest Kuiper belt objects are much larger than the comets we observe in the inner solar system, probably because relatively small objects are more likely to have their orbits altered than are larger bodies.

Kuiper belt objects are thought to be leftover planetesimals that formed and remain in the frigid outer reaches of the solar system. Interactions with other nearby bodies, or the gravitational influence of one of the Jovian planets, occasionally alter their orbits sufficiently to send them into our view.

Named for Dutch astronomer Jan Oort, the **Oort cloud** consists of icy planetesimals that form a roughly spherical shell around the outer reaches of the solar system (**Figure 24.33B**). Oort cloud objects have random orbits at distances often greater than 50,000 times the Earth–Sun distance. This places them at nearly half the distance to Proxima Centauri, the star nearest the Sun. These icy objects are so loosely bound to the solar system that the gravitational effect of a distant passing star may send an occasional Oort cloud body into a highly eccentric orbit that carries it toward the Sun. Because most comets that enter the inner solar system have random orbits, they are assumed to have come from the Oort cloud.

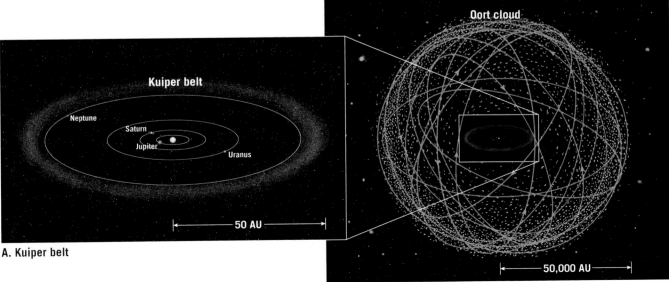

◀ Figure 24.33
The realm of the comets: The Oort cloud and Kuiper belt Most comets reside in one of two places. **A.** Located beyond the orbit of Neptune is the Kuiper belt, a disk of icy objects with roughly circular orbits that travel in the same direction as the planets. **B.** The Oort cloud is a spherical cloud containing icy planetesimals located roughly 50,000 times farther from the Sun than Earth.

A. Kuiper belt

B. Oort cloud

The Oort cloud is estimated to contain at least a trillion comets. How did so many comets end up in the outer reaches of the solar system? The most widely accepted hypothesis is that these bodies formed early in our solar system's history, in the region occupied by the still-developing Jovian planets. Rather than being captured by one of these growing planets, these comets were gravitationally flung in all directions.

Meteors, Meteoroids, and Meteorites

Nearly everyone has seen **meteors**, commonly (but inaccurately) called "shooting stars." These streaks of light can be observed in as little as the blink of an eye or can last as "long" as a few seconds. They occur when a small solid particle, a **meteoroid**, enters Earth's atmosphere from interplanetary space. Heat, created by friction between the meteoroid and Earth's atmosphere, produces the streak of light we see trailing across the sky. Most meteoroids originate from one of three sources: (1) interplanetary debris missed by the gravitational sweep of the planets during solar system formation, (2) material ejected from the asteroid belt, or (3) rocky remains of comets that once passed through Earth's orbit.

Meteoroids smaller than about 1 meter (3 feet) in diameter generally vaporize before reaching Earth's surface. Some, called *micrometeorites*, are so tiny and their rate of fall so slow that they drift to Earth continually as space dust. Researchers estimate that thousands of meteoroids enter Earth's atmosphere every day. After sunset on a clear, dark night, many are bright enough to be seen with the naked eye.

Meteor Showers Occasionally, meteor sightings increase dramatically to 60 or more per hour. Such displays, called **meteor showers**, result when Earth encounters a swarm of meteoroids traveling in the same direction at nearly the same speed as Earth. The close association of these swarms to the orbits of some short-term comets strongly suggests that they represent material lost by these comets (**Table 24.2**). Some swarms, not associated with the orbits of known comets, are probably the scattered remains of the nucleus of a long-defunct comet. The notable *Perseid meteor shower* that occurs each year around August 12 is likely material ejected from the comet Swift–Tuttle on previous approaches to the Sun.

Meteorites: Visitors to Earth The remains of meteoroids that impact Earth's surface are called **meteorites** (**Figure 24.34**).

TABLE 24.2 Major Meteor Showers

Shower	Approximate Dates	Associated Comet
Quadrantids	January 4–6	Unknown*
Lyrids	April 20–23	Thatcher
Eta Aquarids	May 3–5	Halley's
Delta Aquarids	July 30	96P Machholz**
Perseids	August 12	Swift–Tuttle
Draconids	October 7–10	Giacobini–Zinner
Orionids	October 20	Halley's
Taurids	November 3–13	Encke
Andromedids	November 14	Biela
Leonids	November 18	Tempel–Tuttle
Geminids	December 4–16	Unknown*

*The Quadrantids and Geminids are believed to be associated with asteroids that are the remains of extinct comets.

**Astronomers don't all agree on the comet associated with the Delta Aquarids, but 96P Machholz is one candidate.

Most meteoroids large enough to survive passage through the atmosphere likely started out as asteroids; a chance collision or gravitational interaction with Jupiter modifies the orbit and sends the asteroid toward Earth, and then Earth's gravity pulls it in. A few meteorites are fragments of the Moon, Mars, or possibly even Mercury, ejected by a violent asteroid impact. Before *Apollo* astronauts brought Moon rocks back to Earth, meteorites were the only extraterrestrial materials that could be studied in laboratories.

More than 40 terrestrial craters with diameters larger than 20 kilometers (12 miles) exist from large meteorite strikes. These craters exhibit features that could only have been produced by an explosive impact

▼ **SmartFigure 24.34**
Iron meteorite found near Meteor Crater, Arizona

 Tutorial https://goo.gl/aq7aT7

Meteor Crater (cross-section)

Geologist's Sketch

Overturned rock units → Ejecta
Kaibab ls →
Moenkopi
Coconino ss

Original rim profile
Moenkopi
Kaibab ls
Coconino ss
Recent sediment
Fallback breccia
Highly fractured bedrock

◄ **SmartFigure 24.35**
Meteor Crater, near Winslow, Arizona
This cavity is about 1.2 kilometers (0.75 mile) across and 170 meters (560 feet) deep. The solar system is cluttered with asteroids and comets that can strike Earth with explosive force.

Tutorial
https://goo.gl/S43wUh

of an asteroid or perhaps even a comet nucleus. More than 250 smaller craters are also thought to have impact origins. Notable among them is Arizona's Meteor Crater, a cavity 1.2 kilometers (0.75 mile) wide and 170 meters (560 feet) deep, with an upturned rim that rises above the surrounding countryside (**Figure 24.35**). More than 30 tons of iron fragments have been found in the immediate area, but attempts to locate the main body have been unsuccessful. Based on the amount of erosion observed on the crater rim, the impact likely occurred within the past 50,000 years.

Types of Meteorites   Meteorites are classified by their composition into three basic groups: (1) *irons*, mostly aggregates of iron with 5–20 percent nickel and trace elements; (2) *stony*, silicate minerals with inclusions of other minerals; or (3) *stony–irons*, mixtures of the two that account for less than 2 percent of all known meteorites. Although stony meteorites are the most common, irons are found in large numbers because metallic meteorites withstand impacts better, weather more slowly, and are easily distinguished from terrestrial rocks. Iron meteorites are probably fragments of once-molten cores of large asteroids or small planets.

One subcategory of stony meteorite, called a *carbonaceous chondrite*, contain a high percentage of water as well as organic compounds and occasionally simple amino acids, which are some of the basic building blocks of life. This discovery confirms similar findings in observational astronomy, which indicate that numerous organic compounds exist in interstellar space. The presence of water and volatile organic materials also indicates that this type of carbonaceous chondrite has not undergone significant heating since it formed. Therefore, scientists consider these meteorites to be compositionally similar to the solar nebula from which the solar system formed.

Data from meteorites have been used to ascertain the internal structure of Earth and the age of the solar system. If meteorites represent the composition of the terrestrial planets, as some planetary geologists suggest, our planet must contain a much larger percentage of iron than is indicated by surface rocks. This is one reason that geologists think Earth's core is mostly iron and nickel. In addition, radiometric dating of meteorites indicates that the age of our solar system is about 4.6 billion years. This "old age" has been confirmed by data obtained from lunar samples.

Dwarf Planets

Pluto, once considered the ninth planet, was discovered in 1930 by Clyde Tombaugh, who was searching for a yet-undiscovered planet in order to explain irregularities in the orbit of Uranus. Astronomers quickly realized that Pluto was too tiny and too distant to account for such irregularities. Pluto has a diameter of about 2370 kilometers (1470 miles), or about one-fifth that of Earth and less than half that of Mercury, which has long been considered the solar system's "runt."

A.

B.

▲ **Figure 24.36**
Images of Pluto obtained from Nasa's *New Horizons* spacecraft A. This color-enhanced image was generated to detect differences in the composition and texture of Pluto's surface. The bright area in the lower central region, informally named Sputnik Planitia, consists mainly of soft, nitrogen-rich ices that, in some locations, appear to flow like glaciers on Earth. **B.** This close-up view shows Sputnik Planitia in the lower right, which in this region forms a nearly level plane broken into cellular units. The area in the upper left consists of blocks of water ice, some standing almost 2.5 kilometers (about 1.5 miles) above the surrounding plains.

More attention was given to Pluto's status as a planet when astronomers began to discover other large Kuiper belt bodies. Clearly, Pluto and these other large Kuiper belt bodies were completely different from either the terrestrial planets or the Jovian planets. In 2006 the International Astronomical Union, the group responsible for naming and classifying celestial objects, voted to designate a new class of solar system objects called *dwarf planets*. To be classified as a dwarf planet, a celestial body must orbit the Sun and must be essentially spherical due to its own gravity but must not be large enough to sweep its orbits clear of other debris. By this definition, Pluto is recognized as a dwarf planet, and it is the prototype for this new category of planetary objects. Other dwarf planets include Eris, Makemake, and Haumea; many Kuiper belt objects; and Ceres, the largest-known asteroid.

Exploration of Pluto   In July 2015, NASA's *New Horizons* spacecraft shot past Pluto after a 9-year journey that brought it within 12,500 kilometers (7800 miles) of Pluto's surface. Images transmitted from *New Horizons* show Pluto to be an active body with several distinct terrains, including mountainous areas consisting of blocks of water ice, flat ice plains composed of nitrogen ice, and rugged areas indicating a long history of impacts.

One of the most interesting regions, named Sputnik Planitia (*planitia* = plain, or flat area), is the left part of Pluto's iconic "heart" feature (**Figure 24.36A**). Sputnik Planitia is a large ice field about 1050 kilometers (650 miles) in diameter, with a maximum depth of about 4 kilometers (2.5 miles). Found in Pluto's northern hemisphere, it contains lobes of ice flowing around its edges, similar to glaciers on Earth. With surface temperatures of about −235°C (just shy of −400°F),

Pluto is too cold for Sputnik Planitia's ice field to be made of water. Instead, it is composed mainly of frozen nitrogen, with lesser amounts of carbon monoxide and methane ices, which flow at these frigid temperatures.

As shown in **Figure 24.36B**, much of the surface of Sputnik Planitia consists of irregular polygons separated by linear troughs. This structure led researchers to conclude that heat from the planet's interior has generated a type of convective flow, whereby upwelling of nitrogen ice occurs in the center of the cells and spreads out along the surface, eventually sinking at the linear troughs. This convective model is supported by the nearly 100-meter (300-foot) height difference between the centers of these cells and their lower margins. Because no craters were discovered on the surface of Sputnik Planitia, researchers conclude that its surface is relatively young—less than 100 million years old.

One hypothesis proposes that Sputnik Planitia began as a large impact basin, much like the maria on Earth's Moon (see Figure 24.7). Once formed, the impact basin on Pluto began to fill with nitrogen ice rather than the basaltic magma that fills the maria. Pluto's atmosphere, a tenuous layer of gases consisting mainly of nitrogen, is thought to be the source of nitrogen ice that blankets Sputnik Planitia. Two processes are involved: *sublimation* (the change of a solid directly to a gas) of nitrogen ice in regions where the planet receives the most direct sunlight, and *deposition* of nitrogen gas to form ice in the much colder impact basin.

What's Next   Having completed its flyby of Pluto, *New Horizons* changed course and took an image of another Kuiper Belt object nicknamed Ultima Thule. This distant object consists of two small spherical bodies that likely collided when the solar system was first forming. Higher resolution images are expected to be provided by NASA in 2019.

CONCEPT CHECKS 24.5

1. Compare and contrast asteroids and comets. Where are most asteroids found?

2. Where are most comets thought to reside? What eventually becomes of comets that orbit close to the Sun?

Concept Checker
https://goo.gl/wdSj5k

3. What characteristics does a celestial body possess to be classified as a dwarf planet?

24 CONCEPTS IN REVIEW

Touring Our Solar System

24.1 Our Solar System: An Overview

Describe the formation of the solar system according to the nebular theory. Compare and contrast the terrestrial and Jovian planets.

Key Terms:

nebular theory	planetesimal	Jovian planet
solar nebula	protoplanet	escape velocity
	terrestrial planet	impact crater

- Our Sun is the most massive body in our solar system, which includes planets, dwarf planets, moons, and other small bodies. The planets orbit in the same direction and at speeds proportional to their distance from the Sun, with inner planets moving faster and outer planets moving more slowly.
- The solar system began as a *solar nebula* before condensing due to gravity. While most of the matter ended up in the Sun, some material formed a thick disk around the early Sun and later clumped together into larger and larger bodies. *Planetesimals* collided to form *protoplanets*, and protoplanets grew into planets.
- The four *terrestrial planets* are enriched in rocky and metallic materials, whereas the *Jovian planets* have a higher proportion of ice and gas. The terrestrial planets are relatively dense, with thin atmospheres, while the Jovian planets are less dense and have thick atmospheres.
- Smaller planets have less gravity to retain gases in their atmosphere. Lightweight gases such as hydrogen and helium more easily reach escape velocity, so the atmospheres of the terrestrial planets tend to be enriched in heavier gases, such as water vapor, carbon dioxide, and nitrogen.

24.2 Earth's Moon: A Chip Off the Old Block

List and describe the major features of Earth's Moon and explain how maria basins were formed.

Key Terms: maria   lunar highlands   lunar regolith

- The Moon has a composition that is approximately the same as that of Earth's mantle. The Moon likely formed from a collision between a Mars-sized protoplanet and the early Earth.
- The lunar surface is dominated by light-colored *lunar highlands* (or terrae) and darker lowlands called *maria*, the latter formed primarily from flood basalts. Both terrae and maria are partially covered by *lunar regolith* produced by micrometeorite bombardment.

24.3 Terrestrial Planets

Outline the principal characteristics of Mercury, Venus, and Mars. Describe their similarities to and differences from Earth.

- Mercury has a very thin atmosphere and a weak magnetic field. Like Earth's Moon, Mercury has both heavily cratered areas and smooth plains; the smooth plains are similar to lunar maria.
- Venus has a very dense atmosphere, dominated by carbon dioxide. The resulting extreme greenhouse effect produces surface temperatures around 450°C (900°F). The topography of Venus has been resurfaced by active volcanism.
- Mars has about 1 percent as much atmosphere as Earth, so it is relatively cold ($-140°$C to 20°C [$-220°$F to 68°F]). Mars appears to be the closest planetary analog to Earth, showing surface evidence of rifting, volcanism, and modification by flowing water. Volcanoes on Mars are much bigger than volcanoes on Earth because of the lack of plate motion on Mars.

Q As you can see from this graph, Mercury's temperature varies significantly from "day" to "night," but Venus's temperature is relatively constant "around the clock." Suggest a reason for this difference.

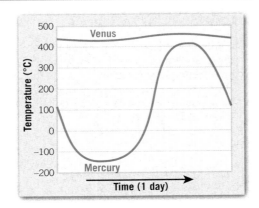

24.4 Jovian Planets

Summarize and compare the features of Jupiter, Saturn, Uranus, and Neptune, including their ring systems.

Key Term: cryovolcanism

- Jupiter's mass is several times larger than the combined masses of everything else in the solar system except for the Sun. Convective flow, combined with its three cloud layers, produces its banded appearance. Persistent, giant rotating storms exist between bands. Many moons orbit Jupiter, including Io, which shows active volcanism, and Europa, which is believed to have a liquid ocean under its icy shell.
- Saturn, like Jupiter, is big and gaseous, and it is endowed with dozens of moons. Some moons show evidence of tectonics, and Titan has its own atmosphere. Saturn's well-developed rings are made of many particles of water ice and rocky debris.
- Uranus, like its "twin" Neptune, has a blue atmosphere dominated by methane, and its diameter is about four times greater than Earth's. Uranus rotates sideways relative to the plane of the solar system. It has a relatively thin ring system and at least five moons.
- Neptune has an active atmosphere, with fierce wind speeds and giant storms. It has 1 large moon, Triton, which shows evidence of *cryovolcanism*, as well as 13 smaller moons and a ring system.

24.5 Small Solar System Bodies

List and describe the principal characteristics of the solar system's small bodies.

Key Terms:

small solar system body	asteroid belt	Oort cloud
dwarf planet	comet	meteor
asteroid	nucleus	meteoroid
	coma	meteor shower
	Kuiper belt	meteorite

- *Small solar system bodies* include rocky asteroids and icy comets. Both are basically scraps left over from the formation of the solar system or fragments from later impacts.
- Most *asteroids* are concentrated in a wide belt between the orbits of Mars and Jupiter. Some are rocky, some are metallic, and some are basically "piles of rubble," loosely held together by their own weak gravity.
- *Comets* are dominated by ices, "dirtied" by rocky material and dust. Most originate in either the *Kuiper belt* beyond Neptune or the *Oort cloud*. When a comet's orbit brings it through the inner solar system, solar radiation causes its ices to vaporize, generating the coma and its characteristic "tail."
- A *meteoroid* is a small rocky or metallic body traveling through space. When it enters Earth's atmosphere, it flares briefly as a *meteor* before either burning up or striking Earth's surface to become a *meteorite*. Asteroids and material lost from comets as they travel through the inner solar system are the most common sources of meteoroids.
- Bodies massive enough to have a spherical shape but not so massive as to have cleared their orbits of debris are classified as *dwarf planets*.

They include the rocky asteroid Ceres as well as the icy worlds Pluto and Eris, which are located in the Kuiper belt.

Q Shown here are four small solar system bodies. Identify each and explain the differences among them.

GIVE IT SOME THOUGHT

1. Assume that a solar system was discovered in a nearby region of the Milky Way Galaxy. The accompanying table shows data that have been gathered about three of the planets orbiting the central star of this newly discovered solar system. Using Table 24.1 as a guide, classify each planet as Jovian, terrestrial, or neither. Explain your reasoning.

	Planet 1	Planet 2	Planet 3
Relative Mass (Earth = 1)	1.2	15	0.1
Diameter (km)	11,000	52,000	2200
Mean Distance from Star (AU)	1.4	17	35
Density (g/cm³)	4.8	1.22	1.8
Orbital Eccentricity	0.01	0.05	0.23

2. In order to conceptualize the size and scale of Earth and Moon as they relate to the rest of the solar system, complete the following:

   a. Approximately how many Moons (diameter 3475 kilometers [2160 miles]) would fit side by side across the diameter of Earth (diameter 12,756 kilometers [7926 miles])?

   b. Given that the Moon's orbital radius is 384,798 kilometers, approximately how many Earths would fit side by side between Earth and the Moon?

   c. Approximately how many Earths would fit side by side across the Sun, whose diameter is about 1,390,000 kilometers?

   d. Approximately how many Suns would fit side by side between Earth and the Sun, a distance of about 150,000,000 kilometers?

3. The accompanying graph shows the temperatures at various distances from the Sun during the formation of our solar system. (You can assume that the planets formed at roughly their current distances from the Sun, although that may not be strictly true.) Use it to answer the following questions:

   a. Which planet or planets formed at locations in the solar system where the temperature was hotter than the boiling point of water?

   b. Which planet or planets formed at locations in the solar system where the temperature was cooler than the freezing point of water?

Condition	Temperature (Fahrenheit)	Temperature (kelvin)
Water freezes	32	199
Room temp.	72	296
Human body	98.6	310
Water boils	212	373

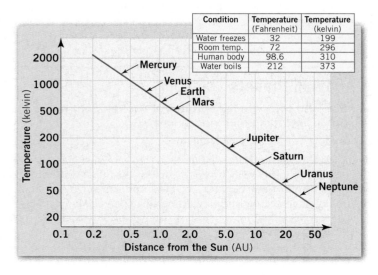

4. **This sketch shows four primary craters (A, B, C, and D). The impact that produced Crater A produced two secondary craters (labeled a) and three rays. Crater D has one secondary crater (labeled d). Rank the four primary craters from oldest to youngest and explain your ranking.**

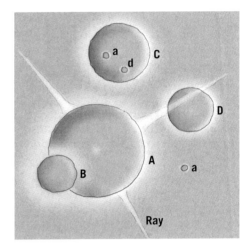

Ray

5. **Halley's Comet has a mass estimated at 100 billion tons. Furthermore, it is estimated to lose about 100 million tons of material when its orbit brings it close to the Sun. With an orbital period of 76 years, calculate the maximum remaining life span of Halley's Comet.**

6. **The accompanying diagram shows two of Uranus's moons, Ophelia and Cordelia, which act as shepherd moons for the Epsilon ring. Explain what would happen to the Epsilon ring if a large asteroid struck Ophelia, knocking it out of the Uranian system.**

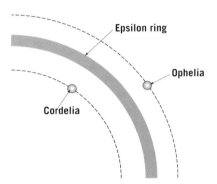

Epsilon ring

Ophelia

Cordelia

7. **Assume that three irregularly shaped planet-like objects, each smaller than our Moon, have just been discovered orbiting the Sun at a distance of 35 AU. One of your friends argues that the objects should be classified as planets because they are large and orbit the Sun. Another friend argues that the objects should be classified as dwarf planets, like Pluto. State whether you agree or disagree with either or both of your friends. Explain your reasoning.**

EYE ON THE UNIVERSE

1. *Mariner 9* **obtained this image of Phobos, one of two tiny moons of Mars. Phobos has a diameter of only 24 kilometers (15 miles). The two moons of Mars were not discovered until the late 1800s because their size made them nearly impossible to see with earlier telescopes.**

   a. In what way is Phobos similar to Earth's Moon?

   b. List characteristics of Phobos that make it different from Earth's Moon.

   c. Do an Internet search to learn how Phobos and its companion moon, Deimos, got their names.

DATA ANALYSIS

Topography of Venus

Rocky planets and natural satellites exhibit geology only recently photographed by passing spacecraft. Impact craters are present on every rocky planet, but these features are not preserved in the same way on all planets.

https://goo.gl/eD226M

ACTIVITIES

Go to the USGS Astrogeology Science Center at https://astrogeology.usgs.gov and click on "Planetary Nomenclature" at the bottom of the page. Click on "Nomenclature" and select "Venus." To navigate these data, hover over a feature type in the list to see a short description of the feature. Click on the feature type to see a table of named features of this type. You can reorder column data by clicking on the up and down arrows at the top of a selected column, such as name or diameter. (Note that reordering the table may take 1–2 minutes.)

1. Click on "Crater, craters" to examine the craters of Venus. How many of the planet's craters have been named? What is the name of the largest approved crater on Venus? What is its diameter?

2. Note that the largest crater in the list has been disallowed. In fact, it has been reclassified as a corona. The exact origin of coronae is unknown, but they appear to be unique to Venus. Go back to the "Search by Feature Types" page for Venus and click on "Corona, coronae." How many coronae on Venus have been named?

3. What is the name of the largest corona? What is its diameter? Return to the "Search by Feature Types" page for Venus and click on "Venus Map" under "Interactive Images & Maps." Click on the "Venus Global GIS Viewer." Check the box next to "Venus Global GIS" and add only the "Nomenclature" layer. To display the legend, click the + next to the "Nomenclature" box and then click the + next to the "Venus Nomenclature" label.

4. Based on the colors on the map, what is the most commonly named feature on Venus?

5. Check the box next to "Volcanic Catalogs" and click the + next to that box. Then check only the box next to the "USGS Venus Volcano Catalog 1999" layer. How common are volcanoes on Venus, compared to other features?

Now uncheck the "Venus Global GIS" box. Without closing the "Venus" map tab, go back to the "Planetary Names" tab, and in a second window, open the "Lunar Global GIS" map by choosing "The Moon" from the "Nomenclature" list and clicking "Lunar Map" under "Interactive Images & Maps." Then open the "Moon Global GIS Viewer." Do the same for Mars: In a third window, open the "Mars Global GIS" map by choosing "Mars" from the "Nomenclature" list and then select "Mars" to display a list of its features. Click on "Map of Mars" under "Interactive Images & Maps" and open the "Mars Global GIS Viewer." Compare the Venus, Mars, and lunar maps. You will need to zoom in and pan around to see surface features clearly.

6. Rank these bodies from most densely packed craters to least densely packed craters. On which body are craters most easily distinguished?

7. Based on what you have learned from this chapter, suggest reasons craters on Venus differ in size, shape, and density compared to craters on Mars and Earth's Moon.

Mastering Geology

Looking for additional review and test prep materials? Visit pearson.com/mastering/geology to enhance your understanding of this chapter's content by accessing a variety of resources, including **Self-Study Quizzes, Geoscience Animations, SmartFigure Tutorials,** *Project Condor* **Videos, Mobile Field Trips, Dynamic Study Modules,** and an optional **Pearson eText.**

APPENDIX

Metric and English Units Compared

Units

1 kilometer (km) = 1000 meters (m)
1 meter (m) = 100 centimeters (cm)
1 centimeter (cm) = 0.39 inch (in.)
1 mile (mi) = 5280 feet (ft)
1 foot (ft) = 12 inches (in.)
1 inch (in.) = 2.54 centimeters (cm)
1 square mile (mi²) = 640 acres (a)
1 kilogram (kg) = 1000 grams (g)
1 pound (lb) = 16 ounces (oz)
1 fathom = 6 feet (ft)

Conversions

When you want
to convert: multiply by: to find:

Length

inches	2.54	centimeters
centimeters	0.39	inches
feet	0.30	meters
meters	3.28	feet
yards	0.91	meters
meters	1.09	yards
miles	1.61	kilometers
kilometers	0.62	miles

Area

square inches	6.45	square centimeters
square centimeters	0.15	square inches
square feet	0.09	square meters
square meters	10.76	square feet
square miles	2.59	square kilometers
square kilometers	0.39	square miles

Volume

cubic inches	16.38	cubic centimeters
cubic centimeters	0.06	cubic inches
cubic feet	0.028	cubic meters
cubic meters	35.3	cubic feet
cubic miles	4.17	cubic kilometers
cubic kilometers	0.24	cubic miles
liters	1.06	quarts
liters	0.26	gallons
gallons	3.78	liters

Masses and Weights

ounces	28.35	grams
grams	0.035	ounces
pounds	0.45	kilograms
kilograms	2.205	pounds

Temperature

When you want to convert degrees Fahrenheit (°F) to degrees Celsius (°C), subtract 32 degrees and divide by 1.8.

When you want to convert degrees Celsius (°C) to degrees Fahrenheit (°F), multiply by 1.8 and add 32 degrees.

When you want to convert degrees Celsius (°C) to Kelvins (K), delete the degree symbol and add 273. When you want to convert Kelvins (K) to degrees Celsius (°C), add the degree symbol and subtract 273.

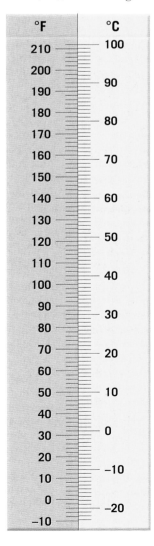

Figure A.1
A comparison of Fahrenheit and Celsius temperature scales.

GLOSSARY

A

Aa flow A type of lava flow that has a jagged, blocky surface.

Ablation A general term for the loss of ice and snow from a glacier.

Abrasion The grinding and scraping of a rock surface by the friction and impact of rock particles carried by water, wind, and ice.

Abyssal plain A very level area of the deep-ocean floor, usually lying at the foot of the continental rise.

Accretionary wedge A large wedge-shaped mass of sediment that accumulates in subduction zones. Here sediment is scraped from the subducting oceanic plate and accreted to the overriding crustal block.

Active continental margin A margin that is usually narrow and consists of highly deformed sediments. Such margins occur where oceanic lithosphere is being subducted beneath the margin of a continent.

Active layer The zone above the permafrost that thaws in summer and refreezes in winter.

Aerosols Tiny solid and liquid particles suspended in the atmosphere.

Aftershock A smaller earthquake that follows the main earthquake.

Albedo The reflectivity of a substance, usually expressed as a percentage of the incident radiation reflected.

Alluvial channel A stream channel in which the bed and banks are composed largely of unconsolidated sediment (alluvium) that was previously deposited in the valley.

Alluvial fan A fan-shaped deposit of sediment formed when a stream's slope is abruptly reduced.

Alluvium Unconsolidated sediment deposited by a stream.

Alpine glacier See *Valley glacier*.

Ambiguous properties Properties of minerals that may vary among different samples of the same mineral, such as color.

Amniotic egg A fluid-filled, shelled egg that can be laid on land. Amniotic eggs are among the adaptations that enabled early reptiles to occupy environments not open to amphibians.

Amphibian A vertebrate such as a frog or salamander that has legs but must lay its eggs in water.

Andesite A gray, fine-grained igneous rock, primarily of volcanic origin and commonly exhibiting a porphyritic texture.

Andesitic composition See *Intermediate composition*.

Angiosperm A flowering plant in which fruits contain the seeds.

Angle of repose The steepest angle at which loose material remains stationary without sliding downslope.

Angular unconformity An unconformity in which the older strata dip at an angle different from that of the younger beds.

Antecedent stream A stream that continued to downcut and maintain its original course as an area along its course was uplifted by faulting or folding.

Anthracite A hard metamorphic form of coal that burns cleanly and hot.

Anticline A fold in sedimentary strata that resembles an arch.

Aphanitic A texture of igneous rocks in which the crystals are too small for individual minerals to be distinguished without the aid of a microscope.

Aquifer Rock or sediment through which groundwater moves easily.

Aquitard An impermeable bed that hinders or prevents groundwater movement.

Archean eon The first eon of Precambrian time. The eon preceding the Proterozoic. It extends between 4.5 and 2.5 billion years ago.

Arête A narrow, knifelike ridge separating two adjacent glaciated valleys.

Arkose A feldspar-rich sandstone.

Artesian Describing a well in which the water rises above the level where it was initially encountered.

Assimilation In igneous activity, the process of incorporating country rock into a magma body.

Asteroid One of thousands of small planetlike bodies, ranging in size from a few hundred kilometers to less than 1 kilometer across. The orbits of most asteroids lie between those of Mars and Jupiter.

Asteroid belt The region between Mars and Jupiter in which most asteroids orbit the Sun.

Asthenosphere A subdivision of the mantle situated below the lithosphere. This zone of weak material exists below a depth of about 100 kilometers (60 miles) and in some regions extends as deep as 700 kilometers (430 miles). The rock within this zone is easily deformed.

Atmosphere The gaseous portion of a planet, the planet's envelope of air. One of the traditional subdivisions of Earth's physical environment.

Atoll A coral island that consists of a nearly continuous ring of coral reef surrounding a central lagoon.

Atom The smallest particle that exists as an element.

Atomic mass unit A mass unit equal to exactly one-twelfth the mass of a carbon-12 atom.

Atomic number The number of protons in the nucleus of an atom.

Atomic weight The average of the atomic masses of isotopes for a given element.

Augite A black, opaque silicate mineral of the pyroxene group that is a dominant component of basalt.

Aureole A zone or halo of contact metamorphism found in the country rock surrounding an igneous intrusion.

B

Back-arc The backside of a volcanic arc when viewed from the trench.

Back-arc basin A basin that forms on the side of a volcanic arc away from the trench.

Backshore The inner portion of the shore, lying landward of the high-tide shoreline. It is usually dry, being affected by waves only during storms.

Back swamp A poorly drained area on a floodplain that results when natural levees are present.

Bajada An apron of sediment along a mountain front created by the coalescence of alluvial fans.

Banded iron formation A finely layered iron and silica-rich (chert) layer deposited mainly during the Precambrian.

Bar Common term for sand and gravel deposits in a stream channel.

Barchan dune A solitary sand dune shaped like a crescent, with its tips pointing downwind.

Barchanoid dune A type of dune that forms scalloped rows of sand oriented at right angles to the wind. This form is intermediate between isolated barchans and extensive waves of transverse dunes.

Barrier island A low, elongate ridge of sand that parallels the coast.

Basal slip A mechanism of glacial movement in which the ice mass slides over the surface below.

Basalt A fine-grained igneous rock of mafic composition.

Basaltic composition A compositional group of igneous rocks indicating that the rock contains substantial dark silicate minerals and calcium-rich plagioclase feldspar.

Basalt plateau The broad and extensive accumulation of lava from a succession of flows emanating from fissure eruptions.

Base level The level below which a stream cannot erode.

Basin A circular downfolded structure.

Batholith A large mass of igneous rock that formed when magma was emplaced at depth, crystallized, and subsequently exposed by erosion.

Bathymetry The measurement of ocean depths and the charting of the topography of the ocean floor.

Baymouth bar A sandbar that completely crosses a bay, sealing it off from the main body of water.

Beach An accumulation of sediment found along the landward margin of the ocean or a lake.

Beach drift The transport of sediment in a zigzag pattern along a beach, caused by the uprush of water from obliquely breaking waves.

Beach face The wet, sloping surface that extends from the berm to the shoreline.

Beach nourishment A process in which large quantities of sand are added to the beach system to offset losses caused by wave erosion. Building beaches seaward improves beach quality and storm protection.

Bed load Sediment moved along the bottom of a stream by moving water, or particles moved along the ground surface by wind.

Bedding plane A nearly flat surface that separates two beds of sedimentary rock. Each bedding plane marks the end of one deposit and the beginning of another one that has different characteristics.

Bedrock channel A channel in which a stream is cutting into solid rock. Such channels typically form in the headwaters or river systems where gradients are high.

Beds See *Strata*.

Benioff zone See *Wadati–Benioff zone*.

Berm The dry, gently sloping zone on the backshore of a beach at the foot of coastal cliffs or dunes.

Biochemical A type of chemical sediment that forms when material dissolved in water is precipitated by water-dwelling organisms. Shells are common examples.

Biogenous sediment Seafloor sediments consisting of material of marine-organic origin.

Biomass Organic material that is renewable energy derived from trees, crops, and waste. Examples include biofuels such as ethanol and biodiesel, as well as biogas, which is methane recovered from landfills.

Biosphere The totality of life-forms on Earth.

Biotite A dark, iron-rich mineral and a member of the mica family that displays excellent cleavage.

Bituminous coal The most common form of coal, often called soft, black coal.

Black carbon Soot generated by combustion processes and fires.

Black smoker A hydrothermal vent on the ocean floor that emits a black cloud of hot, metal-rich water.

Block diagram A three-dimensional view of a portion of Earth's crust that makes it possible to visualize rock layers at the surface as well as underground.

Block lava Lava that has a surface of angular blocks associated with material having andesitic and rhyolitic compositions.

Blowout A depression excavated by wind in easily eroded materials.

Body wave A seismic wave that travels through Earth's interior.

Bottomset bed A layer of fine sediment deposited beyond the advancing edge of a delta and then buried by continued delta growth.

Bowen's reaction series A concept proposed by N. L. Bowen that illustrates the relationships between magma and the minerals crystallizing from it during the formation of igneous rocks.

Braided channel A stream channel that is interwoven with other stream channels. Such channels form where a large portion of a stream's sediment load consists of coarse material (sand and gravel) and the stream has a highly variable discharge.

Braided stream A stream that consists of numerous intertwining channels.

Breakwater A structure that protects a near-shore area from breaking waves.

Breccia A sedimentary rock composed of angular fragments that were lithified.

Brittle deformation Deformation that involves the fracturing of rock. Associated with rocks near the surface.

Burial metamorphism Low-grade metamorphism that occurs in the lowest layers of very thick accumulations of sedimentary strata.

Butte An isolated hill with steep sides and a flat top.

C

Calcite Calcium carbonate ($CaCO_3$), one of the two most common carbonate minerals.

Caldera A large depression typically caused by collapse or ejection of the summit area of a volcano.

Caliche A hard layer, rich in calcium carbonate, that forms beneath the *B* horizon in soils of arid regions.

Calving Wastage of a glacier that occurs when large pieces of ice break into the water.

Cambrian explosion The huge expansion in biodiversity that occurred at the beginning of the Paleozoic era.

Cap rock A necessary part of an oil trap. The cap rock is impermeable and hence keeps upwardly mobile oil and gas from escaping at the surface.

Capacity The total amount of sediment that a stream is able to transport.

Capillary fringe A relatively narrow zone at the base of the zone of aeration were water rises from the water table in tiny, threadlike openings between grains of soil or sediment.

Carbon cycle An Earth system in which carbon moves through the atmosphere, hydrosphere, biosphere, and geosphere, in different directions.

Cassini gap A wide gap in the ring system of Saturn between the A ring and the B ring.

Catastrophism The concept that Earth was shaped by catastrophic events of a short-term nature.

Cavern A naturally formed underground chamber or series of chambers most commonly produced by solution activity in limestone.

Cementation One way in which sedimentary rocks are lithified. As material precipitates from water that percolates through the sediment, open spaces are filled, and particles are joined into a solid mass.

Cenozoic era A time span on the geologic time scale beginning about 65.5 million years ago, following the Mesozoic era.

Chemical bond A strong attractive force that exists between atoms in a substance. It involves the transfer or sharing of electrons that allows each atom to attain a full valence shell.

Chemical compound A substance formed by the chemical combination of two or more elements in definite proportions and usually having properties different from those of its constituent elements.

Chemical sedimentary rock Sedimentary rock consisting of material that was precipitated from water by either inorganic or organic means.

Chemical weathering The processes by which the internal structure of a mineral is altered by the removal and/or addition of elements.

Chert A durable sedimentary rock formed of microcrystalline quartz.

Cinder cone A rather small volcano built primarily of ejected lava fragments that consist mostly of pea- to walnut-size lapilli.

Circum-Pacific belt An area approximately 40,000 kilometers (24,000 miles) in length surrounding the basin of the Pacific Ocean where oceanic lithosphere is continually subducted beneath the surrounding continental plates, causing most of Earth's largest earthquakes.

Cirque An amphitheater-shaped basin at the head of a glaciated valley produced by frost wedging and plucking.

Clastic A sedimentary rock texture consisting of broken fragments of preexisting rock.

Clastic sedimentary rock Transported accumulations of rocks weathered by both mechanical and chemical processes.

Clay A group of light-colored silicates that typically form as products of chemical weathering of igneous rocks. It is a major component of soil and sedimentary rocks. Kaolinite is a common clay mineral derived from the weathering of feldspar.

Cleavage The tendency of a mineral to break along planes of weak bonding.

Climate A description of aggregate weather conditions; the sum of all statistical weather information that helps describe a place or region.

Climate system Exchanges of energy and moisture occurring among the atmosphere, hydrosphere, lithosphere, biosphere, and cryosphere.

Closed system A system that is self-contained with regard to matter—that is, no matter enters or leaves.

Coal A sedimentary rock consisting primarily of organic matter, formed in stages from accumulations of large quantities of undecayed plant material. It is used as a fossil fuel.

Coarse-grained texture See *Phaneritic texture*.

Coast A strip of land that extends inland from the coastline as far as ocean-related features can be found.

Coastline The coast's seaward edge. The landward limit of the effect of the highest storm waves on the shore.

Col A pass between mountain valleys where the headwalls of two cirques intersect.

Collisional mountain A mountain in which compressive horizontal forces have shortened and thickened the crust. Most major mountain belts are of this type.

Color A phenomenon of light by which otherwise identical objects may be differentiated.

Column A feature found in caves that is formed when a stalactite and stalagmite join.

Columnar jointing A pattern of cracks that forms during cooling of molten rock to generate columns.

Coma The fuzzy, gaseous component of a comet's head.

Comet A small body that generally revolves about the Sun in an elongated orbit.

Compaction A type of lithification in which the weight of overlying material compresses more deeply buried sediment. It is most important in the fine-grained sedimentary rocks such as shale.

Competence A measure of the largest particle a stream can transport; a factor dependent on velocity.

Composite volcano A volcano composed of both lava flows and pyroclastic material.

Compound A substance formed by the chemical combination of two or more elements in definite proportions and usually having properties different from those of its constituent elements.

Compressional mountains Mountains in which great horizontal forces have shortened and thickened the crust. Most major mountain belts are of this type.

Compressional stress Differential stress that shortens a rock body.

Concordant A term used to describe intrusive igneous masses that form parallel to the bedding of the surrounding rock.

Conduction The transfer of heat through matter by molecular activity.

Conduit A pipelike opening through which magma moves toward Earth's surface. It terminates at a surface opening called a vent.

Cone of depression A cone-shaped depression in the water table immediately surrounding a well.

Confined aquifer An aquifer that has impermeable layers (aquitards) both above and below.

Confining pressure Stress that is applied uniformly in all directions.

Conformable Referring to rock layers that were deposited without interruption.

Conglomerate A sedimentary rock composed of rounded, gravel-size particles.

Contact metamorphism Changes in rock caused by the heat from a nearby magma body.

Continent A large, continuous area of land that includes the adjacent continental shelf and islands that are structurally connected to the mainland.

Continental drift hypothesis A hypothesis, credited largely to Alfred Wegener, which suggests that all present continents once existed as a single supercontinent. Further, beginning about 200 million years ago, the supercontinent began breaking into smaller continents, which then "drifted" to their present positions.

Continental margin The portion of the seafloor that is adjacent to the continents. It may include the continental shelf, continental slope, and continental rise.

Continental rift A linear zone along which continental lithosphere stretches and pulls apart. Its creation may mark the beginning of a new ocean basin.

Continental rise The gently sloping surface at the base of the continental slope.

Continental shelf The gently sloping submerged portion of the continental margin, extending from the shoreline to the continental slope.

Continental slope The steep gradient that leads to the deep-ocean floor and marks the seaward edge of the continental shelf.

Continental volcanic arc Mountains formed in part by igneous activity associated with the subduction of oceanic lithosphere beneath a continent. Examples include the Andes and the Cascades.

Convection The transfer of heat by the mass movement or circulation of a substance.

Convergent plate boundary A boundary in which two plates move together, resulting in oceanic lithosphere being thrust beneath an overriding plate, eventually to be reabsorbed into the mantle. It can also involve the collision of two continental plates to create a mountain system.

Coral reef A structure formed in a warm, shallow, sunlit ocean environment that consists primarily of the calcite-rich remains of corals as well as the limy secretions of algae and the hard parts of many other small organisms.

Core The innermost layer of Earth. It is thought to be largely an iron–nickel alloy, with minor amounts of oxygen, silicon, and sulfur.

Correlation The process of establishing the equivalence of rocks of similar age in different areas.

Corrosion The process by which soluble rock is gradually dissolved by flowing water.

Country rock See *Host rock*.

Covalent bond A chemical bond produced by the sharing of electrons.

Crater The depression at the summit of a volcano or a depression that is produced by a meteorite impact.

Craton The part of the continental crust that has attained stability; that is, it has not been affected by significant tectonic activity during the Phanerozoic eon. It consists of the shield and the stable platform.

Creep The slow downhill movement of soil and regolith.

Crevasse A deep crack in the brittle surface of a glacier.

Cross-bed A structure in which relatively thin layers are inclined at an angle to the main bedding. It is formed by currents of wind or water.

Cross-bedding A structure in which relatively thin layers are inclined at an angle to the main bedding. Cross-bedding is formed by currents of wind or water.

Cross-cutting A principle of relative dating which says that a rock or fault is younger than any rock (or fault) through which it cuts.

Crust The very thin, outermost layer of Earth.

Cryosphere The portion of Earth's surface where water is in solid form, including snow, glaciers, sea ice, freshwater ice, and frozen ground. It is one of the spheres of the climate system.

Cryovolcanism A type of volcanism that results from the eruption of magmas derived from the partial melting of ice.

Crystal Any natural solid with an ordered, repetitive atomic structure.

Crystal settling A process that occurs during the crystallization of magma, in which the earlier-formed minerals are denser than the liquid portion and settle to the bottom of the magma chamber.

Crystal shape See *Habit*.

Crystalline See *Nonclastic*.

Crystallization The formation and growth of a crystalline solid from a liquid or gas.

Curie point The temperature above which a material loses its magnetization.

Cut bank The area of active erosion on the outside of a meander.

Cutoff A short channel segment created when a river erodes through the narrow neck of land between meanders.

D

D" layer (also known as D double prime) A region in roughly the lowermost 200 kilometers (125 miles) of the mantle where P waves experience a sharp decrease in velocity.

Darcy's law An equation which states that groundwater discharge depends on the hydraulic gradient, hydraulic conductivity, and cross-sectional area of an aquifer.

Dark silicate A silicate mineral that contains ions of iron and/or magnesium in its structure. Dark silicates are dark in color and have a higher specific gravity than nonferromagnesian silicates.

Daughter product An isotope that results from radioactive decay.

Debris avalanche One of the fastest and most destructive types of rockslides consisting of unconsolidated rock fragments.

Debris flow A flow of soil and regolith that contains a large amount of water. Most common in semiarid mountainous regions and on the slopes of some volcanoes.

Debris slide A type of rockslide consisting of largely unconsolidated rock bebris.

Decompression melting Melting that occurs as rock ascends due to a drop in confining pressure.

Deep-ocean basin The portion of seafloor that lies between the continental margin and the oceanic ridge system. This region comprises almost 30 percent of Earth's surface.

Deep-ocean trench A narrow, elongated depression of the seafloor.

Deep-sea fan A cone-shaped deposit at the base of the continental slope. The sediment is transported to the fan by turbidity currents that follow submarine canyons.

Deflation The lifting and removal of loose material by wind.

Deformation General term for the processes of folding, faulting, shearing, compression, or extension of rocks as the result of various natural forces.

Delta An accumulation of sediment formed where a stream enters a lake or an ocean.

Dendritic pattern A stream system that resembles the pattern of a branching tree.

Dendrochronology The dating and study of tree rings.

Density A property of matter defined as mass per unit volume.

Desalination The removal of salts and other chemicals from seawater.

Desert One of the two types of dry climate; the driest of the dry climates.

Desert pavement A layer of closely spaced coarse pebbles and gravel that cover barren, rocky deserts to form a relatively smooth surface.

Desert varnish A reddish-brown or black coating that presents in some desert rocks.

Desertification The degradation of dryland ecosystems on desert margins, primarily due to human activities such as deforestation and overgrazing that remove tree and plant cover anchoring the soil.

Detachment fault A nearly horizontal fault that may extend for hundreds of kilometers below the surface. Such a fault represents a boundary between rocks that exhibit ductile deformation and rocks that exhibit brittle deformation.

Detrital sedimentary rocks Rocks that form from the accumulation of materials that originate and are transported as solid particles derived from both mechanical and chemical weathering.

Diagenesis A collective term for all the chemical, physical, and biological changes that take place after sediments are deposited and during and after lithification.

Diagnostic properties Properties of minerals that aid in mineral identification. Taste or feel, crystal shape, and streak are examples of diagnostic properties.

Differential stress Forces that are unequal in different directions.

Differential weathering The variation in the rate and degree of weathering caused by such factors as mineral makeup, degree of jointing, and climate.

Diffraction The bending of waves as they pass by a curved surface between two compositionally different layers. For example, seismic waves diffract at the boundary between Earth's mantle and outer core.

Dike A tabular-shaped intrusive igneous feature that cuts through the surrounding rock.

Diorite A coarse-grained, intrusive igneous rock primarily composed of plagioclase feldspar and amphibole minerals.

Dip The angle at which a rock layer or fault is inclined from the horizontal. The direction of dip is at a right angle to the strike.

Dip-slip fault A fault in which the movement is parallel to the dip of the fault.

Discharge The quantity of water in a stream that passes a given point in a period of time.

Discharge area A location, such as a spring or a stream, where groundwater flows back to the surface.

Disconformity A type of unconformity in which the beds above and below are parallel.

Discontinuity A sudden change of depth in one or more of the physical properties of the material making up Earth's interior. The boundary between two dissimilar materials in Earth's interior as determined by the behavior of seismic waves.

Discordant A term used to describe plutons that cut across existing rock structures, such as bedding planes.

Disseminated deposit Any economic mineral deposit in which the desired mineral occurs as scattered particles in the rock but in sufficient quantity to make the deposit an ore.

Dissolution The process of dissolving into a homogeneous solution, as when an acidic solution dissolves limestone; a common form of chemical weathering.

Dissolved load The portion of a stream's load that is carried in solution.

Distributary A section of a stream that leaves the main flow.

Divergent plate boundary A boundary in which two plates move apart, resulting in upwelling of material from the mantle to create new seafloor.

Divide An imaginary line that separates the drainage of two streams, often found along a ridge.

Dolomite Calcium/magnesium carbonate, $CaMg(CO_3)_2$, one of the two most common carbonate minerals.

Dolostone A chemical sedimentary rock formed from dolomite, a calcium-magnesium carbonate mineral.

Dome A roughly circular upfolded structure.

Downcutting The lowering of a streambed toward base level as turbulent water lifts unconsolidated material or when bedrock channels are lowered by means of quarrying, abrasion, and corrosion.

Drainage basin The land area that contributes water to a stream. Also called a *watershed*.

Drawdown The difference in height between the bottom of a cone of depression and the original height of the water table.

Drift See *Glacial drift*.

Drumlin A streamlined symmetrical hill composed of glacial till. The steep side of the hill faces the direction from which the ice advanced.

Dry climate A climate in which yearly precipitation is less than the potential loss of water by evaporation.

Ductile deformation A type of solid-state flow that produces a change in the size and shape of a rock body without fracturing. Occurs at depths where temperatures and confining pressures are high.

Dune A hill or ridge of wind-deposited sand.

Dwarf planet Celestial bodies that orbit stars and are massive enough to be spherical but have not cleared their neighboring regions of planetesimals.

E

Earth system science An interdisciplinary study that seeks to examine Earth as a system composed of numerous interacting parts or subsystems.

Earthflow The downslope movement of water-saturated, clay-rich sediment. Most characteristic of humid regions.

Earthquake Vibration of Earth produced by the rapid release of energy.

Ebb current The movement of tidal current away from the shore.

Echo sounder An instrument used to determine the depth of water by measuring the time interval between emission of a sound signal and the return of its echo from the bottom.

Economic mineral A concentration of a mineral resource or reserve that can be profitably extracted from Earth.

Effusive eruption A quiescent eruption that produces mainly outpourings of fluid lava.

Elastic deformation Rock deformation in which the rock returns to nearly its original size and shape when the stress is removed.

Elastic rebound The sudden release of stored strain in rocks that results in movement along a fault.

Electron A negatively charged subatomic particle that has a negligible mass and is found outside an atom's nucleus.

Element A substance that cannot be decomposed into simpler substances by ordinary chemical or physical means.

Eluviation The washing out of fine soil components from the *A* horizon by downward-percolating water.

Emergent coast A coast where land formerly below sea level has been exposed by crustal uplift or a drop in sea level or both.

End moraine A ridge of till marking a former position of the front of a glacier.

Endothermic Capable of maintaining a constant body temperature through metabolic activity.

Energy levels, or shells Spherically shaped, negatively charged zones that surround the nucleus of an atom.

Environment of deposition A geographic setting where sediment accumulates. Each site is characterized by a particular combination of geologic processes and environmental conditions.

Eon The largest time unit on the geologic time scale, next in order of magnitude above era.

Ephemeral stream A stream that is usually dry because it carries water only in response to specific episodes of rainfall. Most desert streams are of this type.

Epicenter The location on Earth's surface that lies directly above the focus of an earthquake.

Epoch A unit of the geologic time scale that is a subdivision of a period.

Equilibrium line See *Snowline*.

Era A major division on the geologic time scale; eras are divided into shorter units called periods.

Erosion The incorporation and transportation of material by a mobile agent, such as water, wind, or ice.

Eruption column Buoyant plumes of hot, ash-laden gases that can extend thousands of meters into the atmosphere.

Escape velocity The initial velocity that an object needs to escape from the surface of a celestial body.

Esker A sinuous ridge composed largely of sand and gravel deposited by a stream flowing in a tunnel beneath a glacier near its terminus.

Estuary A funnel-shaped inlet of the sea that formed when a rise in sea level or subsidence of land caused the mouth of a river to be flooded.

Eukaryote An organism whose genetic material is enclosed in a nucleus; plants, animals, and fungi are eukaryotes.

Evaporite A sedimentary rock formed of material deposited from solution by evaporation of the water.

Evapotranspiration The combined effect of evaporation and transpiration.

Exfoliation dome A large, dome-shaped structure, usually composed of granite, that is formed by sheeting.

Exoplanet A planet that orbits a star other than the Sun.

Exotic stream A permanent stream that traverses a desert and has its source in well-watered areas outside the desert.

External process A process such as weathering, mass wasting, or erosion that is powered by the Sun and contributes to the transformation of solid rock into sediment.

Extrusive igneous rock Igneous rock formed when magma solidifies at Earth's surface.

Eye A zone of scattered clouds and calm averaging about 20 kilometers (12 miles) in diameter at the center of a hurricane.

Eye wall The doughnut-shaped area of intense cumulonimbus development and very strong winds that surrounds the eye of a hurricane.

F

Facies A portion of a rock unit that possesses a distinctive set of characteristics that distinguishes it from other parts of the same unit.

Fall A type of movement that is common to mass-wasting processes that refers to the free falling of detached individual pieces of any size.

Fault A break in a rock mass along which movement has occurred.

Fault breccia Loosely coherent rock composed of broken and crushed rock fragments.

Fault creep Gradual displacement along a fault. Such activity occurs relatively smoothly and with little noticeable seismic activity.

Fault scarp A cliff created by movement along a fault. It represents the exposed surface of the fault prior to modification by weathering and erosion.

Fault-block mountain A mountain that is formed by the displacement of rock along a fault.

Feedback mechanism Various outcomes that may result when one of a complex interactive physical system's elements is altered.

Felsic composition See *Granitic composition*.

Ferromagnesian silicate See *Dark silicate*.

Fetch The distance that the wind has traveled across open water.

Fine-grained texture See *Aphanitic texture*.

Fiord A steep-sided inlet of the sea formed when a glacial trough is partially submerged.

Firn Granular, recrystallized snow. A transitional stage between snow and glacial ice.

Fissility The property of splitting easily into thin layers along closely spaced, parallel surfaces, such as bedding planes in shale.

Fission (nuclear) The splitting of a heavy nucleus into two or more lighter nuclei, caused by the collision with a neutron. During this process, a large amount of energy is released.

Fissure A crack in rock along which there is a distinct separation.

Fissure eruption An eruption in which lava is extruded from narrow fractures or cracks in the crust.

Flood The overflow of a stream channel that occurs when discharge exceeds the channel's capacity. The most common and destructive geologic hazard.

Flood basalts Flows of basaltic lava that issue from numerous cracks or fissures and commonly cover extensive areas to thicknesses of hundreds of meters.

Flood current The tidal current associated with the increase in the height of the tide.

Floodplain The flat, low-lying portion of a stream valley subject to periodic inundation.

Flow A type of movement common to mass-wasting processes in which water-saturated material moves downslope as a viscous fluid.

Flowing artesian well An artesian well in which water flows freely at Earth's surface because the pressure surface is above ground level.

Flowstone Layers of travertine deposited by films of groundwater moving down the walls or along the floor of a cavern.

Fluorescence The absorption of ultraviolet light, which is reemitted as visible light.

Focus (earthquake) See *Hypocenter*.

Fold A bent layer or series of layers that were originally horizontal and subsequently deformed.

Fold-and-thrust belt A region within a compressional mountain system where large areas have been shortened and thickened by the processes of folding and thrust faulting, as exemplified by the Valley and Ridge province of the Appalachians.

Foliated texture A texture of metamorphic rocks that gives the rock a layered appearance.

Foliation A term for a linear arrangement of textural features often exhibited by metamorphic rocks.

Footwall block The rock surface below a fault.

Forced subduction A process that occurs at Peru–Chile–type subduction zones in which lithosphere is too buoyant to subduct spontaneously but is forced beneath the overriding plate.

Forearc A subduction zone located between a deep-ocean trench and an associated volcanic arc.

Forearc basin The region located between a volcanic arc and an accretionary wedge where shallow-water marine sediments typically accumulate.

Foreset bed An inclined bed deposited along the front of a delta.

Foreshock A small earthquake that may precede a major earthquake.

Foreshore The portion of the shore that lies between the normal high and low water marks; the intertidal zone.

Fossil The remains or traces of organisms preserved from the geologic past.

Fossil assemblage The overlapping ranges of a group of fossils (assemblage) collected from a layer. By examining such an assemblage, the age of the sedimentary layer can be established.

Fossil fuel General term for any hydrocarbon that may be used as a fuel, including coal, oil, natural gas, bitumen from tar sands, and shale oil.

Fossil magnetism See *Paleomagnetism*.

Fossil succession The definite and determinable order in which fossil organisms occur. Fossil succession enables us to identify many time periods by their fossil content.

Fracture Any break or rupture in rock along which no appreciable movement has taken place.

Fracture zone A linear zone of irregular topography on the deep-ocean floor that follows transform faults and their inactive extensions.

Fragmental texture See *Pyroclastic texture*.

Frost wedging The mechanical breakup of rock caused by the expansion of freezing water in cracks and crevices.

Fumarole A vent in a volcanic area from which fumes or gases escape.

G

Gabbro A dark-green to black intrusive igneous rock composed of dark silicate minerals. Gabbro makes up a significant percentage of oceanic crust.

Gaining stream A stream that gains water from the inflow of groundwater through the streambed.

Garnet A silicate mineral composed of individual silica tetrahedra. Garnet is most often brown to deep red and has a glassy luster, lacks cleavage, and exhibits conchoidal fracture.

Gas hydrates Compact chemical structures made of water and natural gas (usually methane) that occur in permafrost and under the ocean floor at depths greater than 525 meters (1720 feet).

Geodynamo The generation and maintenance of Earth's magnetic field by the rising iron-rich fluid in the outer core.

Geologic map Graphic depiction of an area of geologic study, with labels and annotations.

Geologic structure See *Rock structure*.

Geologic time scale The division of Earth history into blocks of time—eons, eras, periods, and epochs. The time scale was created using relative dating principles.

Geologic time The span of time since the formation of Earth, about 4.6 billion years.

Geology The science that examines Earth, its form and composition, and the changes that it has undergone and is undergoing.

Geosphere The solid Earth; one of Earth's four basic spheres.

Geotherm See *Geothermal gradient*.

Geothermal energy Natural steam used for power generation.

Geothermal gradient The gradual increase in temperature with depth in the crust. The average is 30°C per kilometer in the upper crust.

Geyser A fountain of hot water ejected periodically from the ground.

Glacial budget The balance, or lack of balance, between ice formation at the upper end of a glacier and ice loss in the zone of wastage.

Glacial drift An all-embracing term for sediments of glacial origin, no matter how, where, or in what shape they were deposited.

Glacial erratic An ice-transported boulder that was not derived from the bedrock near its present site.

Glacial striations Scratches and grooves on bedrock caused by glacial abrasion.

Glacial trough A mountain valley that has been widened, deepened, and straightened by a glacier.

Glacier A thick mass of ice originating on land from the compaction and recrystallization of snow that shows evidence of past or present flow.

Glass (volcanic) Natural glass that is produced when molten lava cools too rapidly to permit recrystallization. Volcanic glass is a solid composed of unordered atoms.

Glassy texture A term used to describe the texture of certain igneous rocks, such as obsidian, that contain no crystals.

Gneiss Medium- to coarse-grained banded metamorphic rocks in which granular and elongated minerals dominate.

Gneissic banding See *Gneissic texture*.

Gneissic texture A texture of metamorphic rocks in which dark and light silicate minerals are separated, giving the rock a banded appearance.

Gondwana The southern portion of Pangaea, consisting of South America, Africa, Australia, India, and Antarctica.

Graben A valley formed by the downward displacement of a fault-bounded block.

Graded bed A sediment layer characterized by a decrease in sediment size from bottom to top.

Graded stream A stream that has the correct channel characteristics to maintain exactly the velocity required to transport the material supplied to it.

Gradient The slope of a stream, generally expressed as the vertical drop over a fixed distance.

Granite An abundant, coarse-grained igneous rock composed of about 10–20 percent quartz and 50 percent potassium feldspar. Granite is used as a building material.

Granitic composition A compositional group of igneous rocks indicating the rock is composed almost entirely of light-colored silicates.

Gravitational collapse The gradual subsidence of mountains caused by lateral spreading of weak material located deep within the mountains.

Gravity anomaly A gravitational force that deviates from the average.

Great Oxygenation Event A time about 2.5 billion years ago, when a significant amount of oxygen appeared in the atmosphere.

Greenhouse effect The transmission of short-wave solar radiation by the atmosphere coupled with the selective absorption of longer-wavelength terrestrial radiation, especially by water vapor and carbon dioxide, resulting in warming of the atmosphere.

Groin A short wall built at a right angle to the seashore to trap moving sand.

Ground moraine An undulating layer of till deposited as an ice front retreats.

Groundmass The matrix of smaller crystals within an igneous rock that has porphyritic texture.

Groundwater Water in the zone of saturation.

Guyot A submerged, flat-topped seamount.

Gymnosperm A group of seed-bearing plants that includes conifers and Ginkgo. The term means "naked seed," a reference to the unenclosed condition of the seeds.

Gypsum A hydrated calcium sulfate mineral. It is the mineral of which plaster, drywall, and other similar building materials are composed.

H

Habit Refers to the common or characteristic shape of a crystal or an aggregate of crystals.

Habitable zone The region around a host star where a planet with sufficient atmospheric pressure can maintain liquid water on its surface.

Hadean The earliest time interval (eon) of Earth history. The time before the planet's first rocks.

Halite The mineral name for common table salt (NaCl); a nonsilicate mineral commonly found in sedimentary rocks.

Half graben A tilted fault block in which the higher side is associated with mountainous topography and the lower side is a basin that fills with sediment.

Half-life The time required for one-half of the atoms of a radioactive substance to decay.

Hanging valley A tributary valley that enters a glacial trough at a considerable height above the floor of the trough.

Hanging wall block The rock surface immediately above a fault.

Hard stabilization Any form of artificial structure built to protect a coast or to prevent the movement of sand along a beach. Examples include groins, jetties, breakwaters, and seawalls.

Hardness A mineral's resistance to scratching and abrasion.

Head (stream) The beginning or source area for a stream. Also called the *headwaters*.

Headward erosion The extension upslope of the head of a valley due to erosion.

Headwaters See *Head*.

Historical geology A major division of geology that deals with the origin of Earth and its development through time. Usually involves the study of fossils and their sequence in rock beds.

Hogback A narrow, sharp-crested ridge formed by the upturned edge of a steeply dipping bed of resistant rock.

Horizon A layer in a soil profile.

Horn A pyramid-like peak formed by glacial action in three or more cirques surrounding a mountain summit.

Hornblende A dark green to black mineral of the amphibole group, often found in igneous rocks.

Hornfels A fine-grained nonfoliated metamorphic rock formed from various minerals.

Horst An elongate, uplifted block of crust bounded by faults.

Host rock Preexisting crustal rocks intruded by magma. Host rock may be displaced or assimilated by magmas.

Hot spot A concentration of heat in the mantle, capable of producing magma that, in turn, extrudes onto Earth's surface. The intraplate volcanism that produced the Hawaiian Islands is one example.

Hot-spot track A chain of volcanic structures produced as a lithospheric plate moves over a mantle plume.

Hot spring A spring in which the water is 6–9°C (10–15°F) warmer than the mean annual air temperature of its locality.

Humus Organic matter in soil that is produced by the decomposition of plants and animals.

Hurricane A tropical cyclonic storm having winds in excess of 119 kilometers (74 miles) per hour.

Hydraulic conductivity A factor relating to groundwater flow; it is a coefficient that takes into account the permeability of the aquifer and the viscosity of the fluid.

Hydraulic fracturing A method of opening up pore space in otherwise impermeable rocks, permitting natural gas to flow out into wells.

Hydraulic gradient The slope of the water table. It is determined by finding the height difference between two points on the water table and dividing by the horizontal distance between the two points.

Hydroelectric power Electricity generated by falling water that is used to drive turbines.

Hydrogenous sediment Seafloor sediment consisting of minerals that crystallize from seawater. An important example is manganese nodules.

Hydrologic cycle The unending circulation of Earth's water supply. The cycle is powered by energy from the Sun and is characterized by continuous exchanges of water among the oceans, the atmosphere, and the continents.

Hydrolysis A chemical weathering process in which minerals are altered by chemically reacting with water and acids.

Hydrosphere The water portion of our planet; one of the traditional subdivisions of Earth's physical environment.

Hydrothermal metamorphism Chemical alterations that occur as hot, ion-rich water circulates through fractures in rock.

Hydrothermal solution The hot, watery solution that escapes from a mass of magma during the latter stages of crystallization. Such solutions may alter the surrounding country rock and are frequently the source of significant ore deposits.

Hypocenter The zone within Earth where rock displacement produces an earthquake.

Hypothesis A tentative explanation that is then tested to determine if it is valid.

I

Ice cap A mass of glacial ice covering a high upland or plateau and spreading out radially.

Ice sheet A very large, thick mass of glacial ice flowing outward in all directions from one or more accumulation centers.

Ice shelf A large, relatively flat mass of floating ice that forms where glacial ice flows into bays and that extends seaward from the coast but remains attached to the land along one or more sides.

Iceberg A mass of floating ice produced by a calving glacier. Usually 20 percent or less of the iceberg protrudes above the waterline.

Ice-contact deposit An accumulation of stratified drift deposited in contact with a supporting mass of ice.

Igneous rock Rock formed from the crystallization of magma.

Immature soil A soil that lacks horizons.

Impact crater A depression that results from collisions with bodies such as asteroids and comets.

Impact metamorphism Metamorphism that occurs when meteorites strike Earth's surface.

Incised meander A meandering channel that flows in a steep, narrow valley. It forms either when an area is uplifted or when the base level drops.

Inclusion A piece of one rock unit that is contained within another. Inclusions are used in relative dating. The rock mass adjacent to the one containing the inclusion must have been there first in order to provide the fragment.

Index fossil A fossil that is associated with a particular span of geologic time.

Index mineral A mineral that is a good indicator of the metamorphic environment in which it formed. Used to distinguish different zones of regional metamorphism.

Inertia A property by which objects at rest tend to remain at rest, and objects in motion tend to stay in motion unless either is acted upon by an outside force.

Infiltration The movement of surface water into rock or soil through cracks and pore spaces.

Infiltration capacity The maximum rate at which soil can absorb water.

Inner core The solid innermost layer of Earth, about 1216 kilometers (754 miles) in radius.

Inner planets The innermost planets of our solar system, which include Mercury, Venus, Earth, and Mars. Also known as the *terrestrial planets* because of their Earth-like internal structure and composition.

Inselberg An isolated mountain remnant characteristic of the late stage of erosion in a mountainous arid region.

Intensity (earthquake) A measure of the degree of earthquake shaking at a given locale, based on the amount of damage.

Interface A common boundary where different parts of a system interact.

Interior drainage A discontinuous pattern of intermittent streams that do not flow to the ocean.

Intermediate composition A compositional group of igneous rocks that contains at least 25 percent dark silicate minerals. The other dominant mineral is plagioclase feldspar.

Internal process A process such as mountain building or volcanism that derives its energy from Earth's interior and elevates Earth's surface.

Intraplate volcanism Igneous activity that occurs within a tectonic plate, away from plate boundaries.

Intrusion See *Pluton*.

Intrusive igneous rock Igneous rock that formed below Earth's surface.

Invertebrate An animal that does not have a vertebral column (backbone); in other words, any animal that is not a vertebrate.

Ion An atom or a molecule that possesses an electrical charge.

Ionic bond A chemical bond between two oppositely charged ions that is formed by the transfer of valence electrons from one atom to the other.

Iron meteorite One of the three main categories of meteorites. This group is composed largely of iron, with varying amounts of nickel (5–20 percent). Most meteorite finds are irons.

Island arc See *Volcanic island arc*.

Isostasy The concept that Earth's crust is "floating" in gravitational balance upon the material of the mantle.

Isostatic adjustment Compensation of the lithosphere when weight is added or removed. When weight is added, the lithosphere responds by subsiding, and when weight is removed, there is uplift.

Isotopes Varieties of the same element that have different mass numbers; their nuclei contain the same number of protons but different numbers of neutrons.

J

Jetty A structure extending into the ocean at the entrance to a harbor or river that is built for the purpose of protecting against storm waves and sediment deposition.

Joint A fracture in rock along which there has been no movement.

Jovian (Jupiter-like) planet One of the Jupiter-like planets, Jupiter, Saturn, Uranus, and Neptune. These planets have relatively low densities.

K

Kame A steep-sided hill composed of sand and gravel that originates when sediment collects in openings in stagnant glacial ice.

Kame terrace A narrow, terracelike mass of stratified drift deposited between a glacier and an adjacent valley wall.

Karst A type of topography formed on soluble rock (especially limestone) primarily by dissolution. It is characterized by sinkholes, caves, and underground drainage.

Kettle A depression created when a block of ice that became lodged in glacial deposits subsequently melts.

Klippe A remnant or an outlier of a thrust sheet that was isolated by erosion.

Kuiper belt A region outside the orbit of Neptune where most short-period comets are thought to originate.

L

Laccolith A massive igneous body intruded between preexisting strata.

Lag deposit The layer of coarse particles that is left behind when wind removes sand and silt from *poorly sorted* surface deposits, so that the concentration of larger particles at the surface gradually increases as the finer particles are blown away.

Lag time The amount of time between a rainstorm and the occurrence of flooding.

Lahar A debris flow on the slopes of a volcano that results when unstable layers of ash and debris become saturated and flow downslope, usually following stream channels.

Laminar flow The movement of water particles in straight-line paths that are parallel to the channel. The water particles move downstream without mixing.

Large igneous province Voluminous accumulations of lava extruded along fissures that produce broad, flat features that are also referred to as basalt plateaus.

Lateral moraine A ridge of till along the sides of a valley glacier composed primarily of debris that fell to the glacier from the valley walls.

Laterite A red, highly leached soil type found in the tropics that is rich in oxides of iron and aluminum.

Laurasia The northern portion of Pangaea, consisting of North America and Eurasia.

Lava Magma that reaches Earth's surface.

Lava dome A bulbous mass associated with an old-age volcano, produced when thick lava is slowly squeezed from the vent. Lava domes may act as plugs to deflect subsequent gaseous eruptions.

Lava tube A tunnel in hardened lava that acts as a horizontal conduit for lava flowing from a volcanic vent. Lava tubes allow fluid lavas to advance great distances.

Law A formal statement of the regular manner in which a natural phenomenon occurs under given conditions.

Law of constancy of interfacial angles A law which states that the angle between equivalent faces of the same mineral is always the same.

Leaching The depletion of soluble materials from the upper soil by downward-percolating water.

Light silicate A silicate mineral that lacks iron and/or magnesium. Light silicates are generally lighter in color and have lower specific gravities than dark silicates.

Limestone A chemical sedimentary rock composed chiefly of calcite. Limestone can form by inorganic means or from biochemical processes.

Liquefaction The transformation of a stable soil into a fluid that is often unable to support buildings or other structures.

Lithification The process, generally involving cementation and/or compaction, of converting sediments to solid rock.

Lithosphere The rigid outer layer of Earth, including the crust and upper mantle.

Lithospheric mantle The uppermost portion of the mantle, below Earth's crust, ranging from a few kilometers thick to about 200 kilometers thick under continental crust.

Lithospheric plate A coherent unit of Earth's rigid outer layer that includes the crust and upper unit.

Local base level The level of a lake, resistant rock layer, or any other base level that stands above sea level. Also called *temporary base level*.

Loess Deposits of windblown silt, lacking visible layers, generally buff-colored, and capable of maintaining a nearly vertical cliff.

Long (L) waves Earthquake-generated waves that travel along the outer layer of Earth and are responsible for most of the surface damage. L waves have longer periods than other seismic waves.

Longitudinal dunes Long ridges of sand oriented parallel to the prevailing wind; these dunes form where sand supplies are limited.

Longitudinal profile A cross section of a stream channel along its descending course from the head to the mouth.

Longshore current A near-shore current that flows parallel to the shore.

Losing stream A stream that loses water to the groundwater system by outflow through the streambed.

Lower mantle The part of the mantle that extends from the core–mantle boundary to a depth of 660 kilometers (410 miles).

Low-velocity zone A subdivision of the mantle located between 100 and 250 kilometers (60 and 150 miles) and discernible by a marked decrease in the velocity of seismic waves. This zone does not encircle Earth.

Lunar breccia A lunar rock formed when angular fragments and dust are welded together by the heat generated by the impact of a meteoroid.

Lunar highlands The extensively cratered highland areas of the Moon. Also known as *terrae*.

Lunar regolith A thin, gray layer on the surface of the Moon, consisting of loosely compacted, fragmented material believed to have been formed by repeated meteoritic impacts.

Luster The appearance or quality of light reflected from the surface of a mineral.

M

Mafic composition See *Basaltic composition*.

Magma A molten rock found at depth, including any dissolved gases and crystals.

Magma mixing The process of altering the composition of a magma through the mixing of material from another magma body.

Magmatic differentiation The process of generating more than one rock type from a single magma.

Magnetic field A phenomenon occurring around a magnet or an electric charge, characterized by a magnetic force at every point in the region. Earth's magnetic field is dipolar and extends from the core out to the solar wind.

Magnetic reversal A change in Earth's magnetic field from normal to reverse or vice versa.

Magnetic time scale A scale that shows the ages of magnetic reversals and is based on the polarity of lava flows of various ages.

Magnetometer A sensitive instrument used to measure the intensity of Earth's magnetic field at various points.

Magnitude (earthquake) An estimate of the total amount of energy released during an earthquake, based on seismic records.

Mammal Vertebrates that possess a neocortex (region of the brain) hair, three middle ear bones, and mammary glands.

Manganese nodules A type of hydrogenous sediment scattered on the ocean floor, consisting mainly of manganese and iron and usually containing small amounts of copper, nickel, and cobalt.

Mantle One of Earth's compositional layers. The solid rocky shell that extends from the base of the crust to a depth of 2900 kilometers (1800 miles).

Mantle plume A mass of hotter-than-typical mantle material that ascends toward the surface, where it may lead to igneous activity. These plumes of solid yet mobile material may originate as deep as the core–mantle boundary.

Marble A soft metamorphic rock formed from limestone or dolostone. Marble of various colors is used for building stones and monuments.

Maria The smooth areas on the Moon's surface that were incorrectly thought to be seas.

Marine terrace A wave-cut platform that has been exposed above sea level.

Mass extinction An event in which a large percentage of species become extinct.

Mass number The sum of the number of neutrons and protons in the nucleus of an atom.

Mass movement The downslope movement of rock, regolith, and soil under the direct influence of gravity.

Massive An igneous pluton that is not tabular in shape.

Meander A looplike bend in the course of a stream.

Mechanical weathering The physical disintegration of rock, resulting in smaller fragments.

Medial moraine A ridge of till formed when lateral moraines from two coalescing alpine glaciers join.

Megathrust fault The plate boundary separating a subducting slab of oceanic lithosphere and the overlying plate.

Melt The liquid portion of magma excluding the solid crystals.

Mercalli intensity scale See *Modified Mercalli Intensity scale.*

Mesa A term for a tableland, an elevated area of land with a flat top and sides that usually form steep clifts.

Mesosphere The part of the mantle that extends from the core–mantle boundary to a depth of 660 kilometers (410 miles). Also known as the *lower mantle.*

Mesozoic era A time span on the geologic time scale between the Paleozoic and Cenozoic eras—from about 248 to 65.5 million years ago.

Metallic bond A chemical bond that is present in all metals that may be characterized as an extreme type of electron sharing in which the electrons move freely from atom to atom.

Metamorphic facies A group of associated minerals that are used to establish the pressures and temperatures at which rocks undergo metamorphism.

Metamorphic grade The degree to which a parent rock changes during metamorphism. It varies from low grade (low temperatures and pressures) to high grade (high temperatures and pressures).

Metamorphic rock Rock formed by the alteration of preexisting rock deep within Earth (but still in the solid state) by heat, pressure, and/or chemically active fluids.

Metamorphism The changes in mineral composition and texture of a rock subjected to high temperatures and pressures within Earth.

Meteor The luminous phenomenon observed when a meteoroid enters Earth's atmosphere and burns up; popularly called a "shooting star."

Meteor shower Numerous meteoroids traveling in the same direction and at nearly the same speed. They are thought to be material lost by comets.

Meteorite Any portion of a meteoroid that survives its traverse through Earth's atmosphere and strikes the surface.

Meteoroid Any small, solid particle that has an orbit in the solar system.

Microcontinent A relatively small fragment of continental crust that may lie above sea level, such as the island of Madagascar, or that may be submerged, as exemplified by the Campbell Plateau near New Zealand.

Micrometeorite A very small meteorite that does not create sufficient friction to burn up in the atmosphere but slowly drifts down to Earth.

Mid-ocean ridge A continuous mountainous ridge on the floor of all the major ocean basins and varying in width from 500 to 5000 kilometers (300 to 3000 miles). The rifts at the crests of these ridges represent divergent plate boundaries.

Migmatite A rock exhibiting both igneous and metamorphic rock characteristics. Such rocks may form when light-colored silicate minerals melt and then crystallize, while the dark silicate minerals remain solid.

Mineral A naturally occurring, inorganic crystalline material with a unique chemical structure.

Mineral phase change A change that occurs when a mineral is subjected to intense pressure; in this change, the structure of a mineral may become unstable, causing its atoms to rearrange into a denser, more stable structure.

Mineral resource All discovered and undiscovered deposits of a useful mineral that can be extracted now or at some time in the future.

Mineralogy The study of minerals.

Modified Mercalli Intensity scale A 12-point scale developed to evaluate earthquake intensity, based on the amount of damage to various structures.

Moho The boundary separating the crust and the mantle, discernible by an increase in seismic velocity. Also known as the *Mohorovičić discontinuity.*

Mohs scale A series of 10 minerals used as a standard in determining hardness.

Moment magnitude A more precise measure of earthquake magnitude than the Richter scale that is derived from the amount of displacement that occurs along a fault zone.

Monocline A one-limbed flexure in strata. The strata are usually flat-lying or very gently dipping on both sides of the monocline.

Mountain belt A geographic area of roughly parallel and geologically connected mountain ranges developed as a result of plate tectonics.

Mouth The point downstream where a river empties into another stream or water body.

Mud crack A feature in some sedimentary rocks that forms when wet mud dries out, shrinks, and cracks.

Mudflow See *Debris flow.*

Muscovite A common member of the mica family of minerals, with excellent cleavage.

N

Natural levee An elevated landform composed of alluvium that parallels some streams and acts to confine their waters, except during floodstage.

Neap tide The lowest tidal range, occurring near the times of the first and third quarters of the Moon.

Near-shore The zone of a beach that extends from the low-tide shoreline seaward to where waves break at low tide.

Nebular theory A model for the origin of the solar system that supposes a rotating nebula of dust and gases that contracted to form the Sun and planets.

Negative-feedback mechanism As used in climatic change, any effect that is opposite the initial change and tends to offset it.

Neutron A subatomic particle in the nucleus of an atom. The neutron is electrically neutral, with a mass approximately equal to that of a proton.

Nonclastic A term for the texture of sedimentary rocks in which the minerals form a pattern of interlocking crystals.

Nonconformity An unconformity in which older metamorphic or intrusive igneous rocks are overlain by younger sedimentary strata.

Nonferromagnesian silicate See *Light silicate.*

Nonflowing artesian well An artesian well in which water does not rise to the surface because the pressure surface is below ground level.

Nonfoliated Describes metamorphic rocks that do not exhibit foliation.

Nonmetallic mineral resource A mineral resource that is not a fuel or processed for the metals it contains.

Nonrenewable resource A resource that forms or accumulates over such long time spans that it must be considered as fixed in total quantity.

Nonsilicates Mineral groups that lack silicas in their structures and account for less than 10 percent of Earth's crust.

Normal fault A fault in which the rock above the fault plane has moved down relative to the rock below.

Normal polarity A magnetic field the same as that which presently exists.

Nuclear decay (radioactive decay) The spontaneous decay of certain unstable atomic nuclei. Also known as *radioactive decay.*

Nuclear fission The splitting of atomic nuclei into smaller nuclei, causing neutrons to be emitted and heat energy to be released.

Nucleus The small, heavy core of an atom that contains all of its positive charge and most of its mass.

Nuée ardente Incandescent volcanic debris buoyed up by hot gases that moves downslope in an avalanche fashion.

Numerical date The number of years that have passed since an event occurred.

O

Oblique-slip fault A fault that exhibits both dip-slip and strike-slip movement.

Obsidian A volcanic glass of felsic composition.

Occultation The disappearance of light that results when one object passes behind an apparently larger one (for example, the passage of Uranus in front of a distant star).

Ocean basin A deep submarine region that lies beyond the continental margins.

Oceanic plateau An extensive region on the ocean floor that is composed of thick accumulations of pillow basalts and other mafic rocks that, in some cases, exceed 30 kilometers (20 miles) in thickness.

Oceanic ridge (or rise) See *Mid-ocean ridge.*

Oceanic ridge system A continuous elevated zone on the floor of all the major ocean basins and varying in width from 500 to 5000 kilometers (300 to 3000 miles). The rifts at the crests of ridges represent divergent plate boundaries.

Octet rule A rule which states that atoms combine in order that each may have the electron arrangement of a noble gas (that is, the outer energy level contains eight neutrons).

Offshore The relatively flat submerged zone that extends from the breaker line to the edge of the continental shelf.

Oil sands Mixtures of clay, sand, water, and a black viscous form of petroleum known as bitumen.

Oil shale A fine-grained sedimentary rock that contains a solid mixture of organic compounds from which liquid hydrocarbons called shale oil can be produced.

Oil trap A geologic structure that allows for significant amounts of oil and gas to accumulate.

Olivine A high-temperature, dark silicate mineral typically found in basalt.

Oort cloud A spherical shell composed of comets that orbit the Sun at distances generally greater than 10,000 times the Earth–Sun distance.

Open system A system in which both matter and energy flow into and out of the system. Most natural systems are of this type.

Ophiolite complex The sequence of rocks that make up the oceanic crust. The three-layer sequence includes an upper layer of pillow basalts, a middle zone of sheeted dikes, and a lower layer of gabbro.

Ore Usually a useful metallic mineral that can be mined at a profit. The term is also applied to certain nonmetallic minerals such as fluorite and sulfur.

Organic sedimentary rock Sedimentary rock composed of organic carbon from the remains of plants that died and accumulated on the floor of a swamp. Coal is the primary example.

Original horizontality Layers of sediment that are generally deposited in a horizontal or nearly horizontal position.

Orogenesis The processes that collectively result in the formation of mountains.

Orogeny A specific episode of orogenesis (mountain building).

Outcrop Sites where bedrock is exposed at the surface.

Outer core A layer beneath the mantle about 2270 kilometers (1410 miles) thick, which has the properties of a liquid.

Outer planets The outermost planets of our solar system, which include Jupiter, Saturn, Uranus, and Neptune. They are also known as the Jovian planets.

Outgassing The escape of dissolved gases from molten rocks.

Outlet glacier A tongue of ice normally flowing rapidly outward from an ice cap or ice sheet, usually through mountainous terrain to the sea.

Outwash plain A relatively flat, gently sloping plain consisting of materials deposited by melt-water streams in front of the margin of an ice sheet.

Oxbow lake A curved lake that is created when a stream cuts off a meander.

Oxidation The removal of one or more electrons from an atom or ion. So named because elements commonly combine with oxygen.

Oxygen-isotope analysis A method of deciphering past temperatures based on the precise measurement of the ratio between two isotopes of oxygen, ^{16}O and ^{18}O. Analysis is commonly made of seafloor sediments and cores from ice sheets.

P

P wave The fastest earthquake wave, which travels by compression and expansion of the medium.

Pahoehoe flow A lava flow with a smooth to ropy surface.

Paleoclimatology The study of ancient climates; the study of climate and climate change prior to the period of instrumental records using proxy data.

Paleomagnetism The natural remnant magnetism in rock bodies. The permanent magnetization acquired by rock that can be used to determine the location of the magnetic poles and the latitude of the rock at the time it became magnetized.

Paleontology The systematic study of fossils and the history of life on Earth.

Paleoseismology The study of the timing, location, and size of prehistoric earthquakes.

Paleozoic era A time span on the geologic time scale between the Precambrian and Mesozoic eras—from about 542 million to 251 million years ago.

Pangaea The proposed supercontinent that 200 million years ago began to break apart and form the present landmasses.

Parabolic dune A sand dune that is similar in shape to a barchan dune except that its tips point into the wind. These dunes often form along coasts that have strong onshore winds, abundant sand, and vegetation that partly covers the sand.

Parasitic cone A volcanic cone that forms on the flank of a larger volcano.

Parent material The material on which a soil develops.

Parent rock The rock from which a metamorphic rock formed.

Partial melting The process by which most igneous rocks melt. Since individual minerals have different melting points, most igneous rocks melt over a temperature range of a few hundred degrees. If the liquid is squeezed out after some melting has occurred, a melt with a higher silica content results.

Passive continental margin A margin that consists of a continental shelf, continental slope, and continental rise. They are not associated with plate boundaries and therefore experience little volcanism and few earthquakes.

Pater noster lakes A chain of small lakes in a glacial trough that occupies basins created by glacial erosion.

Pegmatite A very coarse-grained igneous rock (typically granite) commonly found as a dike associated with a large mass of plutonic rock that has smaller crystals. Crystallization in a water-rich environment is believed to be responsible for the very large crystals.

Pegmatitic texture A texture of igneous rocks in which the interlocking crystals are all larger than one centimeter in diameter.

Perched water table A localized zone of saturation above the main water table, created by an impermeable layer (aquiclude).

Peridotite An igneous rock of ultramafic composition thought to be abundant in the upper mantle.

Period A basic unit of the geologic time scale that is a subdivision of an era. Periods may be divided into smaller units called epochs.

Periodic table An arrangement of the elements in which atomic number increases from the left to right and elements with similar properties appear in columns called families or groups.

Permafrost Any permanently frozen subsoil. Usually found in the subarctic and arctic regions.

Permeability A measure of a material's ability to transmit water.

Phaneritic An igneous rock texture in which the crystals are roughly equal in size and large enough so the individual minerals can be identified without the aid of a microscope.

Phanerozoic eon The part of geologic time that is represented by rocks containing abundant fossil evidence. The eon extending from the end of the Proterozoic eon (540 million years ago) to the present.

Phase change In geology, the process by which the atomic structure of a mineral changes although its composition remains the same.

Phenocryst A conspicuously large crystal embedded in a matrix of finer-grained crystals.

Phreatic zone See *Zone of saturation*.

Phyllite A metamorphic rock composed mainly of fine crystals of muscovite, chlorite, or both.

Physical geology A major division of geology that examines the materials of Earth and seeks to understand the processes and forces acting beneath and upon Earth's surface.

Piedmont glacier A glacier that forms when one or more alpine glaciers emerge from the confining walls of mountain valleys and spread out to create a broad sheet in the lowlands at the base of the mountains.

Pillow lava Basaltic lava that solidifies in an underwater environment and develops a structure that resembles a pile of pillows.

Pinnacle An isolated erosional remnant found in association with buttes and mesas.

Pipe A vertical conduit through which magmatic materials have passed.

Placer A deposit formed when heavy minerals are mechanically concentrated by currents, most commonly streams and waves. Placers are sources of gold, tin, platinum, diamonds, and other valuable minerals.

Plagioclase feldspar A relatively hard light silicate mineral containing both sodium and calcium ions that freely substitute for one another depending on the crystallization environment.

Planetesimal A solid celestial body that accumulated during the first stages of planetary formation. Planetesimals aggregated into increasingly larger bodies, ultimately forming the planets.

Plant photosynthesis The production of energy-rich molecules of sugar from molecules of carbon dioxide (CO_2) and water (H_2O) using sunlight as the energy source.

Plastic flow A type of glacial movement that occurs within a glacier, below a depth of approximately 50 meters (165 feet), in which the ice is not fractured.

Plate See *Lithospheric plate*.

Plate resistance A force that counteracts plate motion as a subducting plate scrapes against an overriding plate.

Plate tectonics A theory which proposes that Earth's outer shell consists of individual plates that interact in various ways and thereby produce earthquakes, volcanoes, mountains, and the crust itself.

Playa The flat central area of an undrained desert basin.

Playa lake A temporary lake in a playa.

Pleistocene epoch An epoch of the Quaternary period that began about 2.6 million years ago and ended about 10,000 years ago. Best known as a time of extensive continental glaciation.

Plucking A process by which pieces of bedrock are lifted out of place by a glacier.

Plug See *Volcanic neck.*

Pluton A structure that results from the emplacement and crystallization of magma beneath the surface of Earth.

Plutonic rock Igneous rocks that form at depth. Named after Pluto, the god of the lower world in classical mythology.

Pluvial lake A lake formed during a period of increased rainfall. For example, this occurred in many nonglaciated areas during periods of ice advance elsewhere.

Point bar A crescent-shaped accumulation of sand and gravel deposited on the inside of a meander.

Polymerization The ability of silicate tetrahedra to link to one another in a variety of configurations, including chains, sheets, and three-dimensional structures.

Polymorphs Two or more minerals that have the same chemical composition but different crystalline structures. Exemplified by the diamond and graphite forms of carbon.

Porosity The volume of open spaces in rock or soil.

Porphyritic texture An igneous rock texture characterized by two distinctively different crystal sizes. The larger crystals are called phenocrysts, whereas the matrix of smaller crystals is termed the groundmass.

Porphyroblastic texture A texture of metamorphic rocks in which particularly large grains (porphyroblasts) are surrounded by a fine-grained matrix of other minerals.

Porphyry An igneous rock that has a porphyritic texture.

Positive feedback mechanism As used in climatic change, any effect that acts to reinforce the initial change.

Potassium feldspar An abundant, relatively hard light silicate mineral containing potassium ions in its structure.

Pothole A depression formed in a stream channel by the abrasive action of the water's sediment load.

Precambrian All geologic time prior to the Phanerozoic eon. A term encompassing both the Archean and Proterozoic eons.

Precursor Events or changes that precede an earthquake and may provide a warning.

Preserved magnetism See *Paleomagnetism.*

Pressure solution A processes involving hot water solutions where mineral matter moves from areas of high pressure to areas of low pressure.

Primary (P) waves Seismic waves that involve alternating compression and expansion of the material through which they pass.

Principal shell The shell or energy level an electron occupies.

Principle of cross-cutting relationships The geologic principle which states that geologic features that cut across rocks must form after the rocks they cut through.

Principle of fossil succession A principle by which fossil organisms succeed one another in a definite and determinable order, and any time period can be recognized by its fossil content.

Principle of inclusions The principle which states that a rock mass adjacent to one containing inclusions must have been there first in order to provide the rock fragments and is therefore the older rock mass.

Principle of lateral continuity A principle which states that sedimentary beds originate as continuous layers that extend in all directions until they grade into a different type of sediment or thin out at the edge of a sedimentary basin.

Principle of original horizontality A principle by which layers of sediment are generally deposited in a horizontal or nearly horizontal position.

Principle of superposition A principle which states that in any undeformed sequence of sedimentary rocks, each bed is older than the one above and younger than the one below.

Proglacial lake A lake created when a glacier acts as a dam blocking the flow of a river or trapping glacial meltwater. The term refers to the position of such lakes just beyond the outer limits of a glacier.

Prokaryote A cell or an organism such as bacteria whose genetic material is not enclosed in a nucleus.

Protein A class of organic molecules that provide the primary structural material for life and contribute to the functioning of cells.

Proterozoic eon The eon following the Archean and preceding the Phanerozoic. It extends between 2500 and 542 million years ago.

Proton A positively charged subatomic particle found in the nucleus of an atom.

Protoplanet A developing planetary body that grows by the accumulation of planetesimals.

Proxy data Data gathered from natural recorders of climate variability such as tree rings, ice cores, and ocean-floor sediments.

Pumice A light-colored, glassy vesicular rock commonly having a granitic composition.

Pyroclastic flow A highly heated mixture, largely of ash and pumice fragments, that travels down the flanks of a volcano or along the surface of the ground.

Pyroclastic material The volcanic rock ejected during an eruption. Pyroclastics include ash, bombs, and blocks.

Pyroclastic rocks Rocks composed of consolidated rock fragments ejected during a volcanic eruption.

Pyroclastic texture An igneous rock texture resulting from the consolidation of individual rock fragments that are ejected during a violent volcanic eruption.

Q

Quarrying Removing loosened blocks from the bed of a channel during times of high flow rates.

Quartz A common silicate mineral consisting entirely of silicon and oxygen that resists weathering.

Quartzite A hard metamorphic rock formed from quartz sandstone.

Quaternary period The most recent period on the geologic time scale. It began about 2.6 million years ago and extends to the present.

R

Radial pattern A system of streams running in all directions, away from a central elevated structure, such as a volcano.

Radioactive decay See *Nuclear decay.*

Radioactivity See *Nuclear decay.*

Radiocarbon (carbon-14) dating Dating of events from the very recent geologic past (the past few tens of thousands of years) based on the fact that the radioactive isotope of carbon is produced continuously in the atmosphere.

Radiometric dating The procedure of calculating the absolute ages of rocks and minerals that contain certain radioactive isotopes.

Rainshadow A dry area on the lee side of a mountain range. Many middle-latitude deserts are produced by this effect.

Radiosonde A lightweight package of weather instruments fitted with a radio transmitter and carried aloft by a balloon.

Rapids A part of a stream channel in which the water suddenly begins flowing more swiftly and turbulently because of an abrupt steepening of the gradient.

Rays Bright streaks that appear to radiate from certain craters on the lunar surface. The rays consist of fine debris ejected from the primary crater.

Recessional moraine An end moraine formed as the ice front stagnated during glacial retreat.

Recharge area An area where groundwater is replenished.

Recrystallization The formation of new mineral crystals in a rock that tend to be larger than the original crystals.

Rectangular pattern A drainage pattern characterized by numerous right angle bends that develops on jointed or fractured bedrock.

Recurrence interval The average time interval between occurrences of hydrologic events such as floods of a given or greater magnitude.

Reflection (seismic) The redirection of some waves back to the surface when seismic waves hit a boundary between different Earth materials.

Refraction A change in direction of waves as they enter shallow water. The portion of the wave in shallow water is slowed, which causes the waves to bend and align with the underwater contours.

Regional metamorphism Metamorphism associated with large-scale mountain building.

Regolith The layer of rock and mineral fragments that nearly everywhere covers Earth's land surface.

Rejuvenation A change in relation to base level, often caused by regional uplift, which causes the forces of erosion to intensify.

Relative date The chronological order of events, determined by placing rocks and structures in their proper sequence or order.

Renewable resource A resource that is virtually inexhaustible or that can be replenished over relatively short time spans.

Reptile A group of animals that includes turtles, snakes, lizards, and crocodiles, and that traditionally also includes extinct groups such as dinosaurs, ichthyosaurs, and plesiosaurs. Modern representatives are generally scaled and ectothermic.

Reserve Already identified deposits from which minerals can be extracted profitably.

Reservoir rock The porous, permeable portion of an oil trap that yields oil and gas.

Residual soil Soil developed directly from the weathering of the bedrock below.

Return period See *Recurrence interval.*

Reverse fault A fault in which the material above the fault plane moves up in relation to the material below.

Reverse polarity A magnetic field opposite that which presently exists.

Rhyolite The fine-grained equivalent of the igneous rock granite, composed primarily of the light-colored silicates.

Richter scale A scale of earthquake magnitude based on the amplitude of the largest seismic wave.

Ridge push A mechanism that may contribute to plate motion. It involves the oceanic lithosphere sliding down the oceanic ridge under the pull of gravity.

Rift valley A long, narrow trough bounded by normal faults. It represents a region where divergence is taking place.

Rills Tiny channels that develop as unconfined flow begins producing threads of current.

Ring of Fire The zone of active volcanoes surrounding the Pacific Ocean.

Rip current A strong, narrow surface or near-surface current of short duration and high speed that moves seaward through the breaker zone at nearly a right angle to the shore.

Ripple marks Small waves of sand that develop on the surface of a sediment layer by the action of moving water or air.

River A general term for a stream that carries a substantial amount of water and has numerous tributaries.

Roche moutonnée An asymmetrical knob of bedrock that is formed when glacial abrasion smooths the gentle slope facing the advancing ice sheet and plucking steepens the opposite side as the ice overrides the knob.

Rock A consolidated mixture of minerals.

Rock avalanche Very rapid downslope movement of rock and debris. These rapid movements may be aided by a layer of air trapped beneath the debris, and they have been known to reach speeds of over 200 kilometers (125 miles) per hour.

Rock cleavage The tendency of rocks to split along parallel, closely spaced surfaces. These surfaces are often highly inclined to the bedding planes in the rock.

Rock cycle A model that illustrates the origin of the three basic rock types and the interrelatedness of Earth materials and processes.

Rock flour Ground-up rock produced by the grinding effect of a glacier.

Rock structure All features created by the processes of deformation from minor fractures in bedrock to a major mountain chain.

Rock-forming minerals The relatively few minerals that make up most of the rocks in Earth's crust.

Rockslide The rapid slide of a mass of rock downslope, along planes of weakness.

Rock texture A term used to describe the appearance of a rock based on the size, shape, and arrangement of its mineral grains.

Runoff Water that flows over land rather than infiltrating into the ground.

S

S wave An earthquake wave, slower than a P wave, that travels only in solids.

Salinity The proportion of dissolved salts to pure water, usually expressed in parts per thousand (‰).

Salt flat A white crust on the ground that is produced when water evaporates and leaves behind its dissolved materials.

Saltation Transportation of sediment through a series of leaps or bounces.

Sandstone An abundant, durable sedimentary rock primarily composed of sand-size grains.

Schist Medium- to coarse-grained metamorphic rocks having a foliated texture, in which platy minerals dominate.

Schistosity A type of foliation that is characteristic of coarser-grained metamorphic rocks. Such rocks have a parallel arrangement of platy minerals (for example, the micas).

Scientific method The process by which researchers raise questions, gather data, and formulate and test scientific hypotheses.

Scoria Vesicular ejecta that is the product of basaltic magma.

Scoria cone See *Cinder cone*.

Sea arch An arch formed by wave erosion when caves on opposite sides of a headland unite.

Sea ice Frozen seawater that is associated with polar regions. The area covered by sea ice expands in winter and shrinks in summer.

Sea stack An isolated mass of rock standing just offshore, produced by wave erosion of a headland.

Seafloor spreading A hypothesis, first proposed in the 1960s by Harry Hess, which suggests that new oceanic crust is produced at the crests of mid-ocean ridges, which are the sites of divergence.

Seamount An isolated volcanic peak that rises at least 1000 meters (3300 feet) above the deep-ocean floor.

Seawall A barrier constructed to prevent waves from reaching the area behind the wall. Its purpose is to defend property from the force of breaking waves.

Secondary (S) waves Seismic waves that involve oscillation perpendicular to the direction of propagation.

Secondary enrichment The concentration of minor amounts of metals that are scattered through unweathered rock into economically valuable concentrations by weathering processes.

Sediment Unconsolidated particles created by the weathering and erosion of rock by chemical precipitation from solution in water, or from the secretions of organisms, and transported by water, wind, or glaciers.

Sedimentary environment See *Environment of deposition*.

Sedimentary rock Rock formed from the weathered products of preexisting rocks that have been transported, deposited, and lithified.

Seed An embryo packaged with a supply of nutrients inside a protective coating.

Seiche The sloshing of water in an enclosed basin, generated by seismic waves.

Seismic gap A segment of an active fault zone that has not experienced a major earthquake over a span when most other segments have. Such segments are probable sites for future major earthquakes.

Seismic reflection profiler An instrument for viewing the rock structure beneath a blanket of sediment that uses strong, low-frequency sound waves that penetrate the sediments and reflect off the contacts between rock layers and fault zones.

Seismic tomography A technique in which seismic signals collected from many earthquakes are used to make images that map locations within Earth where seismic waves travel slower or faster.

Seismic wave A rapidly moving ocean wave generated by earthquake activity capable of inflicting heavy damage in coastal regions.

Seismogram A record made by a seismograph.

Seismograph An instrument that records earthquake waves. Also known as a *seismometer*.

Seismology The study of earthquakes and seismic waves.

Seismometer See *Seismograph*.

Settling velocity The speed at which a particle falls through a still fluid. The size, shape, and specific gravity of particles influence settling velocity.

Shadow zone The zone between 105 and 140 degrees from an earthquake epicenter. Direct waves do not penetrate the shadow zone because of refraction by Earth's core.

Shale The most common sedimentary rock, consisting of silt- and clay-size particles.

Shear Stress that causes two adjacent parts of a body to slide past one another.

Sheet flow Runoff moving in unconfined thin sheets.

Sheeted dike complex A large group of nearly parallel dikes.

Sheeting A mechanical weathering process that is characterized by the splitting off of slablike sheets of rock.

Shelf break The point at which a rapid steepening of the gradient occurs, marking the outer edge of the continental shelf and the beginning of the continental slope.

Shield A large, relatively flat expanse of ancient igneous and metamorphic rocks within the craton.

Shield volcano A broad, gently sloping volcano built from fluid basaltic lavas.

Shock metamorphism See *Impact metamorphism*.

Shore Seaward of the coast, a zone that extends from the highest level of wave action during storms to the lowest tide level.

Shoreline The line that marks the contact between land and sea. It migrates up and down as the tide rises and falls.

Silicate Any one of numerous minerals that have the silicon-oxygen tetrahedron as their basic structure.

Silicon-oxygen tetrahedron A structure composed of four oxygen atoms surrounding a silicon atom that constitutes the basic building block of silicate minerals.

Sill A tabular igneous body that was intruded parallel to the layering of preexisting rock.

Sink See *Sinkhole*.

Sinkhole A depression produced in a region where soluble rock has been removed by groundwater. Also known as a *sink*.

Slab pull A mechanism that contributes to plate motion in which cool, dense oceanic crust sinks into the mantle and "pulls" the trailing lithosphere along.

Slab suction One of the driving forces of plate motion, which arises from the drag of the subducting plate on the adjacent mantle. It is an induced mantle circulation that pulls both the subducting and overriding plates toward the trench.

Slate A very fine-grained metamorphic rock containing platy minerals and having excellent rock cleavage.

Slaty cleavage A type of foliation that is characteristic of slates, in which there is a parallel arrangement of fine-grained metamorphic minerals.

Slickenslide Polished and grooved rock surfaces etched as crustal rocks slide past one another.

Slide A movement common to mass-wasting processes in which the material moving downslope remains fairly coherent and moves along a well-defined surface.

Slip face The steep, leeward surface of a sand dune that maintains a slope of about 34 degrees.

Slump The downward slipping of a mass of rock or unconsolidated material moving as a unit along a curved surface.

Small solar system body Solar system objects such as asteroids, comets, and meteoroids.

Snowball Earth A hypothesis that relates a period of global glaciation to the Great Oxygenation Event.

Snowfield An area where snow persists throughout the year.

Snowline The lower limit of perennial snow.

Soil A combination of mineral and organic matter, water, and air; the portion of the regolith that supports plant growth.

Soil horizon A layer of soil that has identifiable characteristics produced by chemical weathering and other soil-forming processes.

Soil profile A vertical section through a soil, showing its succession of horizons and the underlying parent material.

Soil Taxonomy A soil classification system that consists of six hierarchical categories, based on observable soil characteristics. The system recognizes 12 soil orders.

Soil texture The relative proportions of clay, silt, and sand in a soil. A soil's texture strongly influences its ability to retain and transmit water and air.

Solar nebula The cloud of interstellar gas and/or dust from which the bodies of our solar system formed.

Solid-state flow The solid-state flow of rock that occurs when rocks are exposed to extreme conditions of heat and pressure.

Solifluction The slow, downslope flow of water-saturated materials common to permafrost areas.

Solum The *O*, *A*, and *B* horizons in a soil profile. Living roots and other plant and animal life are largely confined to this zone.

Sonar An instrument that uses acoustic signals (sound energy) to measure water depths. Sonar is an acronym for *so*und *na*vigation and *r*anging.

Sorting The degree of similarity in particle size in sediment or sedimentary rock.

Specific gravity The ratio of a substance's weight to the weight of an equal volume of water.

Speleothem A collective term for the dripstone features found in caverns.

Spheroidal weathering Any weathering process that tends to produce a spherical shape from an initially blocky shape.

Spit An elongate ridge of sand that projects from the land into the mouth of an adjacent bay.

Spontaneous subduction A process that occurs at Mariana-type subduction zones in which old, dense lithosphere sinks into the mantle at a steep angle by its own weight creating a deep trench.

Spreading center See *Divergent plate boundary*.

Spring A flow of groundwater that emerges naturally at the ground surface.

Spring tide The highest tidal range. Occurs near the times of the new and full moons.

Stable platform That part of a carton that is mantled by relatively undeformed sedimentary rocks and underlain by a basement complex of igneous and metamorphic rocks.

Stalactite An icicle-like structure that hangs from the ceiling of a cavern.

Stalagmite A columnlike form that grows upward from the floor of a cavern.

Star dune An isolated hill of sand that exhibits a complex form and develops where wind directions are variable.

Steno's law See *Law of constancy of interfacial angles*.

Steppe One of the two types of dry climate. A marginal and more humid variant of the desert that separates it from bordering humid climates.

Stock A pluton similar to but smaller than a batholith.

Stony meteorite One of the three main categories of meteorites. Such meteorites are composed largely of silicate minerals with inclusions of other minerals.

Stony-iron meteorite One of the three main categories of meteorites. This group, as the name implies, is a mixture of iron and silicate minerals.

Storm surge The abnormal rise of the sea along a shore as a result of strong winds.

Strain An irreversible change in the shape and size of a rock body caused by stress.

Strata Parallel layers of sedimentary rock.

Stratified drift Sediments deposited by glacial meltwater.

Stratosphere The layer of the atmosphere immediately above the troposphere, characterized by increasing temperatures with height, due to the concentration of ozone.

Stratovolcano See *Composite cone*.

Streak The color of a mineral in powdered form.

Stream A general term to denote the flow of water within any natural channel. Thus, a small creek and a large river are both streams.

Stream piracy Diversion of the drainage of one stream that results from the headward erosion of another stream.

Stream valley The channel, valley floor, and sloping valley walls of a stream.

Stress The force per unit area acting on any surface within a solid.

Striations (glacial) Scratches or grooves in a bedrock surface caused by the grinding action of a glacier and its load of sediment.

Strike The compass direction of the line of intersection created by a dipping bed or fault and a horizontal surface. A strike is always perpendicular to the direction of dip.

Strike-slip fault A fault along which movement occurs horizontally.

Stromatolite A distinctively layered mound of calcium carbonate, which is fossil evidence for the existence of ancient microscopic bacteria.

Subduction The process by which oceanic lithosphere plunges into the mantle along a convergent zone.

Subduction erosion A process in subduction zones in which sediment and rock are scraped off the bottom of the overriding plate and transported into the mantle.

Subduction zone A long, narrow zone where one lithospheric plate descends beneath another.

Subduction zone metamorphism High-pressure, low-temperature metamorphism that occurs where sediments are carried to great depths by a subducting plate.

Submarine canyon A seaward extension of a valley that was cut on the continental shelf during a time when sea level was lower, or a canyon carved into the outer continental shelf, slope, and rise by turbidity currents.

Submergent coast A coast whose form is largely a result of the partial drowning of a former land surface due to a rise of sea level or subsidence of the crust, or both.

Subsoil A term applied to the *B* horizon of a soil profile.

Sulfur dioxide A gas with the chemical formula SO_2, that is associated naturally with volcanic activity, and as a waste gas (air pollutant) with the burning of fossil fuels and various industrial processes.

Sunspot A dark area on the Sun that is associated with powerful magnetic storms that extend from the Sun's surface deep into the interior.

Supercontinent A large landmass that contains all, or nearly all, of the existing continents.

Supercontinent cycle The idea that the rifting and dispersal of one supercontinent is followed by a long period during which the fragments gradually reassemble into a new supercontinent.

Supernova An exploding star that increases its brightness many thousands of times.

Superplume A large area of the mantle that is dominated by the upwelling of hot mantle rock, which originates near the mantle-core boundary.

Superposed stream A stream that cuts through a ridge lying across its path. The stream established its course on uniform layers at a higher level without regard to underlying structures and subsequently downcut.

Surf A collective term for breakers; also the wave activity in the area between the shoreline and the outer limit of breakers.

Surface waves Seismic waves that travel along the outer layer of Earth.

Surge A period of rapid glacial advance. Surges are typically sporadic and short-lived.

Suspended load Fine sediment carried within the body of flowing water or air.

Suture A zone along which two crustal fragments are jointed together. For example, following a continental collision, the two continental blocks are sutured together.

Swells Wind-generated waves that have moved into an area of weaker winds or calm.

Syncline A linear downfold in sedimentary strata; the opposite of anticline.

System A group of interacting or interdependent parts that form a complex whole.

T

Tablemount See *Guyot*.

Tabular Describing a feature such as an igneous pluton that has two dimensions that are much longer than the third.

Talus slope The characteristic "apron" of angular rock debris that accumulates below mountain cliffs.

Tarn A small lake in a cirque.

Tectonic plate See *Lithospheric plate*.

Tectonic structure A basic geologic feature, such as a fold, fault, or rock foliation, that results from forces associated with the interaction of tectonic plates.

Tectonics The study of the large-scale processes that collectively deform Earth's crust.

Temporary base level See *Local base level*.

Tenacity Describes a mineral's toughness or resistance to breaking or deforming.

Tensional stress The type of stress that tends to pull a body apart.

Tephra See *Pyroclastic materials*.

Terminal moraine The end moraine that marks the farthest advance of a glacier.

Terrace A flat, benchlike structure produced by a stream, which was left elevated as the stream cut downward.

Terrane A crustal block bounded by faults, whose geologic history is distinct from the histories of adjoining crustal blocks.

Terrestrial (Earth-like) planet One of the Earth-like planets: Mercury, Venus, Earth, and Mars. These planets have similar densities.

Terrigenous sediment Seafloor sediments derived from terrestrial weathering and erosion.

Texture The size, shape, and distribution of the particles that collectively constitute a rock.

Theory A well-tested and widely accepted view that explains certain observable facts.

Theory of plate tectonics A theory which proposes that Earth's outer shell consists of individual plates that interact in various ways and thereby produce earthquakes, volcanoes, mountains, and the crust itself.

Thermal metamorphism See *Contact metamorphism*.

Thermosphere The region of the atmosphere immediately above the mesosphere and characterized by increasing temperatures due to absorption of very shortwave solar energy by oxygen.

Thrust fault A low-angle reverse fault.

Tidal current The alternating horizontal movement of water associated with the rise and fall of the tide.

Tidal delta A deltalike feature created when a rapidly moving tidal current emerges from a narrow inlet and slows, depositing its load of sediment.

Tidal flat A marshy or muddy area that is alternately covered and uncovered by the rise and fall of the tide.

Tide The periodic change in the elevation of the ocean surface.

Tidewater glacier Any glacier that flows into the ocean.

Till Unsorted sediment deposited directly by a glacier.

Tillite A rock formed when glacial till is lithified.

Tombolo A ridge of sand that connects an island to the mainland or to another island.

Topset bed An essentially horizontal sedimentary layer deposited on top of a delta during flood-stage.

Tower karst Steep-sided hills formed in wet tropical and subtropical regions with thick beds of highly jointed limestone. The limestone is dissolved by groundwater, leaving residual towers.

Trace gases Gases present in Earth's atmosphere at concentrations much lower than that of carbon dioxide. Methane and nitrous oxide are important trace gases that absorb outgoing radiation and help warm the atmosphere.

Transform fault A major strike-slip fault that cuts through the lithosphere and accommodates motion between two plates.

Transform plate boundary A boundary in which two plates slide past one another without creating or destroying lithosphere.

Transition zone The lowest portion of the upper mantle.

Transpiration The release of water vapor to the atmosphere by plants.

Transported soil Soil that forms on unconsolidated deposits.

Transverse dunes A series of long ridges oriented at right angles to the prevailing wind; these dunes form where vegetation is sparse and sand is very plentiful.

Travertine A form of limestone ($CaCO_3$) that is deposited by hot springs or as a cave deposit.

Trellis pattern A system of streams in which nearly parallel tributaries occupy valleys cut in folded strata.

Trench See *Deep-ocean trench*.

Trigger A factor or event, such as soil saturation, oversteepened slopes, removal of vegetation, or ground shaking, that initiates downslope movement of rock material.

Triple junction A point where three lithospheric plates meet.

Troposphere The lowermost layer of the atmosphere. It is generally characterized by a decrease in temperature with height.

Truncated spurs Triangular-shaped cliffs produced when spurs of land that extend into a valley are removed by the great erosional force of a valley glacier.

Tsunami The Japanese word for a seismic sea wave.

Tuff A rock composed mainly of tiny, ash-size fragments cemented together.

Turbidite A turbidity current deposit characterized by graded bedding.

Turbidity current A downslope movement of dense, sediment-laden water created when sand and mud on the continental shelf and slope are dislodged and thrown into suspension.

Turbulent flow Erratic movement of water often characterized by swirling, whirlpool-like eddies. Most streamflow is of this type.

U

Ultimate base level Sea level; the lowest level to which stream erosion could lower the land.

Ultramafic composition A compositional group of igneous rocks containing mostly olivine and pyroxene.

Unconformity A surface that represents a break in the rock record, caused by erosion and nondeposition.

Uniformitarianism The concept that the processes that have shaped Earth in the geologic past are essentially the same as those operating today.

Unit cell The smallest group of atoms, ions, or molecules that form the building block of a crystal.

Unsaturated zone The area above the water table where openings in soil, sediment, and rock are not saturated but filled mainly with air. Also known as the *vadose zone*.

Upper mantle The top portion of the mantle extending from the Moho to a depth of about 660 km and comprising the lithospheric mantle, asthenosphere, and transition zone.

V

Vadose zone See *Unsaturated zone*.

Valence electron The electrons involved in the bonding process; the electrons occupying the highest principal energy level of an atom.

Valley glacier A glacier confined to a mountain valley, which in most instances had previously been a stream valley.

Valley train A relatively narrow body of stratified drift deposited on a valley floor by meltwater streams that issue from the terminus of an alpine glacier.

Vein deposit A mineral that fills a fracture or fault in a host rock. Such deposits have a sheetlike, or tabular, form.

Vent The surface opening of a conduit or pipe.

Ventifact A cobble or pebble polished and shaped by the sandblasting effect of wind.

Vertebrate A group of animals that are characterized by the possession of a vertebral column (backbone).

Vesicles Spherical or elongated openings on the outer portion of a lava flow that were created by escaping gases.

Vesicular texture A term applied to aphanitic igneous rocks that contain many small cavities called vesicles.

Viscosity A measure of a fluid's resistance to flow.

Volatiles Gaseous components of magma dissolved in the melt. Volatiles will readily vaporize (form a gas) at surface pressures.

Volcanic Pertaining to the activities, structures, or rock types of a volcano.

Volcanic bomb A streamlined pyroclastic fragment ejected from a volcano while still semimolten.

Volcanic cone A cone-shaped structure built by successive eruptions of lava and/or pyroclastic materials.

Volcanic island A seamount that has grown large enough to rise above sea level.

Volcanic island arc A chain of volcanic islands generally located a few hundred kilometers from a trench where there is active subduction of one oceanic plate beneath another.

Volcanic neck An isolated, steep-sided, erosional remnant consisting of lava that once occupied the vent of a volcano.

Volcanic rock See *Extrusive igneous rock*.

Volcano A mountain formed from lava and/or pyroclastics.

W

Wadati–Benioff zone The narrow zone of inclined seismic activity that extends from a trench downward into the asthenosphere.

Water gap A pass through a ridge or mountain in which a stream flows.

Water table The upper level of the saturated zone of groundwater.

Waterfall A precipitous drop in a stream channel that causes water to fall to a lower level.

Watershed See *Drainage basin*.

Wave height The vertical distance between the trough and crest of a wave.

Wave of oscillation A water wave in which the wave form advances as the water particles move in circular orbits.

Wave of translation The turbulent advance of water created by breaking waves.

Wave period The time interval between the passage of successive crests at a stationary point.

Wave refraction See *Refraction*.

Wave-cut cliff A seaward-facing cliff along a steep shoreline formed by wave erosion at its base and mass wasting.

Wave-cut platform A bench or shelf along a shore at sea level, cut by wave erosion.

Wavelength The horizontal distance separating successive crests or troughs.

Weather The state of the atmosphere at any given time.

Weathering The disintegration and decomposition of rock at or near the surface of Earth.

Welded tuff A pyroclastic deposit composed of particles fused together by the combination of heat still contained in the deposit after it has come to rest and the weight of overlying material.

Well An opening bored into the zone of saturation.

Wetted perimeter The total distance in a linear cross-section of a stream that is in contact with water.

Wilson Cycle See *Supercontinent cycle*.

Wind gap An abandoned water gap. These gorges typically result from stream piracy.

X

Xenolith An inclusion of unmelted country rock in an igneous pluton.

Xerophyte A plant that is highly tolerant of drought.

Y

Yardang A streamlined, wind-sculpted ridge that has the appearance of an inverted ship's hull that is oriented parallel to the prevailing wind.

Yazoo tributary A tributary that flows parallel to the main stream because a natural levee is present.

Z

Zone of accumulation The part of a glacier that is characterized by snow accumulation and ice formation. The outer limit of this zone is the snowline.

Zone of fracture The upper portion of a glacier consisting of brittle ice.

Zone of saturation The zone where all open spaces in sediment and rock are completely filled with water. Also known as the *phreatic zone*.

Zone of soil moisture A zone in which water is held as a film on the surface of soil particles and may be used by plants or withdrawn by evaporation. The uppermost subdivision of the unsaturated zone.

Zone of wastage The part of a glacier beyond the snowline, where annually there is a net loss of ice.

CREDITS

Cover: Climbers rappelling from the Getu Arch, China. Photo by Jimmy Chin

Design Elements: Concept Checker icon: Nurlan Aliyev/Shutterstock.

Chapter 1 page 2 Chapter Opener: Planetpix/Alamy Stock Photo, inset: George Rose/Getty Images; page 4 Figure 1.1a: Michael Collier, b: Terray Sylvester/Thomson Reuters; page 5 Figure 1.2a: USGS, b: Philippe Psaila/Science Source; page 6 Figure 1.4: Bloomberg/Getty Images; Figure 1.5: Ball Miwako/Alamy Stock Photo; page 8 Figure 1.6: Michael Collier; page 11 Figure 1.9: David Parker/Science Source; page 12 Figure 1.11: NASA Johnson Space Center; page 13 Figure 1.12: MedioTuerto/Getty Images; page 14 Figure 1.13 (left): Bernhard Edmaier/Science Source, (center, right): Michael Collier; page 15 Figure 1.15 (left): Darryl Leniuk/age Fotstock, (right): age Fotostock/Superstock; page 16 Figure 1.16: Mike Eliason/SBC Fire Department/ZUMA Press/Newscom; page 17 Figure 1.17a, b: USGS, c: Terry Donnelly/Alamy Stock Photo; page 20 Figure 1.19: NASA; page 21 GeoGraphics 1.1: NASA; page 24 Figure 1.21a: geoz/Alamy Stock Photo, b: Tyler Boyes/Shutterstock; page 25 Figure 1.22 (top): Michael Collier, (center, bottom): Dennis Tasa; page 26 Figure 1.23 (clockwise from top left): U.S. Geological Survey Library, E. J. Tarbuck, Dennis Tasa; page 30 Figure 1.25 (left): Radius Images/SuperStock, (center): Michael Collier, (right): Image Source/Alamy Stock Photo; page 32 Concepts in Review 1.4: Michael Collier; page 33 Concepts in Review 1.7a: U.S. Geological Survey Library, b, e: Dennis Tasa, c: Bogdan Ionescu/Shutterstock, d: E. J. Tarbuck; page 34 Give It Some Thought 1.3: Michael Collier, 1.5: Bill Swindaman/Getty Images; page 35 Eye on Earth 1.1: William Perry/123RF.

Chapter 2 page 36 Chapter Opener: NASA, inset: John Cancalosi/age Fotostock; page 38 Figure 2.1: Daniel Prudek/123RF; page 41 Figure 2.5: Used by permission of John C. Holden; page 43 Figure 2.8: Archive of Alfred Wegener Institute; page 46 Figure 2.12: Arctic Images/Alamy Stock Photo; page 50 Figure 2.16: Wallace Garrison/Getty Images; Figure 2.17: USGS; page 53 Figure 2.21: Michael Collier; page 56 Figure 2.25: Katsumi Kasahara/AP Images; page 57 Figure 2.26: NASA; page 64 Concepts in Review 2.2: John Cancalosi/age Fotostock; page 68 Eye on Earth 2.1, 2.2: NASA.

Chapter 3 page 70 Chapter Opener: Carsten Peter/Speleoresearch & Films/Getty Images; page 72 Figure 3.1: Jeffrey A. Scovil; page 73 Figure 3.3: E. J. Tarbuck; page 75 Figure 3.6: Dennis Tasa; page 78 Figure 3.11: Jeffrey A. Scovil; page 79 Figure 3.12: E. J. Tarbuck; Figure 3.13: E. J. Tarbuck; page 80

Figure 3.14, 3.15: Dennis Tasa; page 81 Figure 3.16a: E. J. Tarbuck, b–d: Dennis Tasa; Figure 3.17: Dennis Tasa; page 82 Figure 3.19: Dennis Tasa; Figure 3.18: Chip Clark/Fundamental Photographs; page 83 Figure 3.20a: Dennis Tasa, b: E. J. Tarbuck; Figures 3.21, 3.22: Chip Clark/Fundamental Photographs; page 84 Figure 3.24d: Dennis Tasa; page 85 Figure 3.25: Dennis Tasa; page 86 Figure 3.27a: USGS, b: E. J. Tarbuck; page 90 Figure 3.31a–c, e: Dennis Tasa, d, f, g: E. J. Tarbuck; page 91 Figure 3.33a, c: Dennis Tasa, b, d: E. J. Tarbuck; Figure 3.34: Dennis Tasa; page 92 Figure 3.35: E. J. Tarbuck; Figure 3.36: Dennis Tasa; page 93 Figure 3.38: E. J. Tarbuck; page 94 Figure 3.39a–d, g–l: Dennis Tasa, e, f: E. J. Tarbuck; page 95 GeoGraphics 3.1 (top right): Hal Beral Visual&Written/Newscom, (bottom left): Charles D. Winters/Science Source, (bottom center): Dorling Kindersley/Getty Images, (bottom right): Greg C Grace/Alamy Stock Photo; page 96 Figure 3.40: Pearson Education; page 97 Concepts in Review 3.4: Dennis Tasa; page 98 Concepts in Review 3.8: Dennis Tasa; page 99 Give It Some Thought 3.4, 3.6: Dennis Tasa; page 100 Eye on Earth 3.1: blickwinkel/Alamy Stock Photo, 3.2: chrisbrignell/Shutterstock.

Chapter 4 page 102 Chapter Opener: Philippe Bourseiller/Getty Images; page 104 Figure 4.1: Zoonar GmbH/Alamy Stock Photo; page 105 Figure 4.2: Dennis Tasa; page 106 Figure 4.4: Barbara A. Harvey/Shutterstock; page 108 Figure 4.6: E. J. Tarbuck; page 109 Figure 4.7a, b, d–f: E. J. Tarbuck, c: Dennis Tasa; page 110 Figure 4.8: Dennis Tasa; Figure 4.9: USGS, inset: E. J. Tarbuck; page 111 Figure 4.10: Jeffrey A. Scovil; Figure 4.11: Stocktrek Images/Richard Roscoe; page 112 Figure 4.12: E. J. Tarbuck; page 113 Figure 4.13: Dennis Tasa, E. J. Tarbuck; page 114 GeoGraphics 4.1(1): Chris Joseph/Glow Images, (2): Andrew McLachlan/All Canada Photos/SuperStock, (3): Lisa Seaman/Aurora Open/SuperStock, (4): Erin Paul Donovan/age Fotostock, (5): JRC, Inc./Alamy Stock Photo, (6): William Helsel/age Fotostock, (7): Alan Majchrowicz/age Fotostock, (8): Nick Rothman; page 115 Figure 4.14: Michael Collier, (left inset): Jimmy Chin/National Geographic Creative/Alamy Stock Photo, (right inset): Dennis Tasa; page 116 Figure 4.15a: E. J. Tarbuck, b: Chip Clark/Fundamental Photographs, NYC; Figure 4.16: David Reggie/Getty Images; page 117 Figure 4.17: Marli Miller, inset: E. J. Tarbuck; page 121 Figure 4.23: Terese Loeb Kreuzer/Alamy Stock Photo; page 124 Figure 4.28: intoit/Shutterstock; page 125 Figure 4.29: Michael Collier; Figure 4.30: Michael Collier; page 126 Figure 4.31: E. J. Tarbuck; page 127 Figure 4.33: Mark A. Wilson (Department of Geology, The College of Wooster); page 128 Figure 4.34: Michael DeFreitas North America/Alamy Stock Photo; page 129 Concepts in Review 4.2:

E. J. Tarbuck; page 131 Give It Some Thought 4.2: E. J. Tarbuck, 4.4: Dennis Tasa; page 132 Give It Some Thought 4.9: John Greim/Photolibrary/Getty Images; Eye on Earth 4.1: Michael Collier.

Chapter 5 page 134 Chapter Opener: David Jon Ogmundsson/Getty Images, inset: USGS; page 136 Figure 5.1: U.S. Geological Survey, Denver; page 137 Figure 5.2: Lyn Topinka/AP Photo/U.S. Geological Survey, inset: John M. Burnley/Science Source; page 139 Figure 5.4: Rosario Patanè/Alamy Stock Photo; Figure 5.5: Ness Kerton/Getty Images; page 140 Figure 5.6: USGS; page 141 Figure 5.7a: Dave Bunnell/Under Earth Images, b: USGS; Figure 5.8d: USGS; page 142 Figure 5.9 (top to bottom): USGS, Dennis Tasa, USGS; page 143 Figure 5.10: E. J. Tarbuck; page 145 Figure 5.12 (top): Greg Vaughn/Alamy Stock Photo, (bottom): NASA; page 147 GeoGraphics 5.1 (top left): Greg Vaughn/Alamy Stock Photo, (bottom left): David Reggie/Getty Images, (top right): Michael Collier, (bottom right): Terray Sylvester/Thomson Reuters; page 148 Figure 5.14, 5.15: Michael Collier; page 150 Figure 5.16: Koji Nakano/Getty Images; page 151 Figure 5.17a: USGS, b: Chaideer Mahyuddin/Getty Images; page 152 Figure 5.18a: Library of Congress Prints and Photographs Division, b: Archive Farms/Getty Images; Figure 5.19a: Olivier Goujon/Robert Harding World Imagery, b: Leonard Von Matt/Science Source; page 153 Figure 5.20: USGS; page 154 Figure 5.21 (top): Aman Rochman/Getty Images, (bottom): USGS; page 155 Figure 5.22: Michael Collier, inset: USGS; page 156 Figure 5.24a: USGS, b: NASA; page 157 Figure 5.25b: Williamborg; Figure 5.26a: USGS; page 158 Figure 5.27: Dennis Tasa; page 160 Figure 5.29a, c: USGS; page 161 Figure 5.29b: Bettmann/Getty Images, f: Fuse/Corbis/Getty Images; page 164 Figure 5.33: ESA/NASA/JPL-Caltech; Figure 5.34: NASA/MODIS Rapid Response Team; page 166 Concepts in Review 5.2: USGS, 5.3: Photo by Dr. Erik W. Klemetti; page 169 Give It Some Thought 5.1, 5.6: USGS; page 170 Eye on Earth 5.1: Ulet Ifansasti/Getty Images, 5.2: taolmor/123RF.

Chapter 6 page 172 Chapter Opener: Science History Images/Alamy Stock Photo, inset: Tim McCabe/USDA; page 174 Figure 6.1: Dennis Tasa; page 176 Figure 6.3: Renate Frost/EyeEm/Getty Images; Figure 6.4: Martyn F. Chillmaid/Science Source; Figure 6.5: Marli Miller; page 177 Figure 6.6: Gary Moon/age Fotostock; Figure 6.7: Michael Collier; page 178 Figure 6.8: Scott Linneman; page 179 Figure 6.10: Günter Lenz/imageBROKER/age Fotostock; page 180 Figure 6.11a: Vladimir Melnik/Shutterstock, b: Cedric Weber/Shutterstock; page 181 Figure 6.12: Whit Richardson/Alamy Stock Photo; page 182 Figure 6.13: E. J. Tarbuck; page 183 Figure 6.14: E. J. Tarbuck; Figure 6.15: jose garcia/Alamy Stock Photo; page 184 Figure 6.16: Michael Collier; page 185 Figure 6.17: Juice Images/ILI/Alamy Stock Photo; page 186 Figure 6.19 (left, center): E. J. Tarbuck, (right): Lucarellli Temistocle/Shutterstock; page 187 Figure 6.20 (left): Bill Brooks/Alamy Stock Photo, (center): Nickolay Stanev/Shutterstock, (right): USDA; page 189 Figure 6.22 (left): USDA, (right): E. J. Tarbuck; page 191 Figure 6.24: Wesley Bocxe/Science Source; page 192 Figure 6.25: Photo courtesy U.S. Department of the Navy/Soil Conservation Service/USDA; page 193 Figure 6.26a: Lynn Betts/NRCS, b: D.P. Burnside/Science Source; Figure 6.27: USDA/NRCS/Natural Resources Conservation Service; page 194 GeoGraphics 6.1: Library of Congress Prints and Photographs Division; page 195 Figure 6.28: Erwin C. Cole/USDA/NRCS/Natural Resources Conservation Service; Figure 6.29: USDA; Figure 6.30: Bob Nichols/NRCS; page 196 Concepts in Review 6.1: Bare Essence Photograph/Alamy Stock Photo; page 197 Concepts in Review 6.4 (top): Kevin Schafer/Alamy Stock Photo, (bottom): Kushch Dmitry/Shutterstock, inset: E. J. Tarbuck; page 198 Concepts in Review 6.8: Erwin C. Cole/NRCS; Give It Some Thought 6.1, 6.4: Michael Collier; page 199 Give It Some Thought 6.9: Lynn Betts/NRCS; Eye on Earth 6.1: Marli Miller, 6.2: E. J. Tarbuck; page 200 Eye on Earth 6.3: Mark Brine/Alamy Stock Image, 6.4: Sandra A. Dunlap/Shutterstock.

Chapter 7 page 202 Chapter Opener: Martin Rügner/age Fotostock, inset: Stefan Bengtson; page 204 Figure 7.1: S Sailer/age Fotostock; Figure 7.3: Dennis Tasa, E. J. Tarbuck; Figure 7.4: E. J. Tarbuck; page 208 Figure 7.5: Dennis Tasa; Figure 7.6: E. J. Tarbuck; page 209 Figure 7.8a: Dennis Tasa, b: George H.H. Huey/Alamy Stock Photo; page 210 Figure 7.9: E. J. Tarbuck; Figure 7.10: Michael Collier; page 211 Figure 7.11: E. J. Tarbuck; Figure 7.12: Amar and Isabelle Guillen–Guillen Photo LLC/Alamy Stock Photo, inset: Dante Fenolio/Science Source; page 212 Figure 7.13: David R. Frazier Photolibrary, Inc./Alamy Stock Photo, inset: E. J. Tarbuck; Figure 7.14 (left): JC Photo/Shutterstock, (right): Michael Collier; page 213 GeoGraphics 7.1 (top): USGS, (bottom left): Martin Bond/Science Source, (bottom right): B Christopher/Alamy Stock Photo; page 214 Figure 7.15: David Wall/Alamy Stock Photo, inset: Science Photo Library/Alamy Stock Photo; Figure 7.16: Marli Miller; page 215 Figure 7.17 (left to right): E. J. Tarbuck, Daniel Sambraus/Science Source, gracious_tiger/Shutterstock; page 216 Figure 7.18: Geof Kirby/Alamy Stock Photo; Figure 7.19: Jupiterimages/Stockbyte/Getty Images, inset: NASA; page 217 Figure 7.20: E. J. Tarbuck; page 220 Figure 7.23: E. J. Tarbuck; page 221 Figure 7.24: Michael Collier; page 222 Figure 7.25 (clockwise from top left): Dennis Tasa, Michael Collier, Marli Miller, USGS, Michael Collier, Alan Majchrowicz/age Fotostock, Masonjar/Shutterstock, Dave Wald/USGS; page 223 Figure 7.26 (clockwise from top left): Michael Collier, Pixachi/Shutterstock,

Barrett/MacKay/Glow Images, Radius/SuperStock, E. J. Tarbuck, Marli Miller; page 225 Figure 7.28: Colin D Young/Shutterstock; page 226 Figure 7.29 (left): Dan Ross/Alamy Stock Photo, (right): Dennis Tasa; Figure 7.30: Marli Miller; page 227 Figure 7.31a: Tim Graham/Alamy Stock Photo, b: Marli Miller; page 231 Give It Some Thought 7.8: Michael Collier; page 232 Give It Some Thought 7.10: E. J. Tarbuck; Eye on Earth 7.1: Dennis Tasa, 7.2: Jesse Kraft/Alamy Stock Photo, 7.3: Michael Collier.

Chapter 8 page 234 Chapter Opener: Peter Langer/Robert Harding, inset: Nika Lerman/Alamy Stock Photo; page 237 Figure 8.2: Dennis Tasa; page 239 Figure 8.5: E. J. Tarbuck; page 240 Figure 8.6: Alan P. Trujillo; page 241 Figure 8.7: E. J. Tarbuck; page 242 Figure 8.9: Fred Bruemmer/Getty Images, inset: E. J. Tarbuck; page 243 Figure 8.11c: Dennis Tasa; page 244 Figure 8.12: E. J. Tarbuck; page 245 Figures 8.13, 8.14: E. J. Tarbuck; page 246 Figure 8.15: Michael Collier; Figure 8.16: Sam Dcruz/Alamy Stock Photo; page 247 Figure 8.17: Dennis Tasa; page 248 Figure 8.19: Michael Collier; Figure 8.21: Pavel Svoboda/Fotolia, inset: E. J. Tarbuck; page 249 Figure 8.22b: Fisheries and Oceans Canada/Uvic–Verena Tunnicliffe/Newscom; Figure 8.23: Dennis Tasa; page 250 Figure 8.25a: Alan P. Trujillo, b: Ann Bykerk-Kauffman; page 251 GeoGraphics 8.1 (top left): Courtesy of Aerolite Meteorites Inc., (center): Brian Mason/Smithsonian Institution/Office of Imaging, Printing, and Photographic Services, (bottom): Michael Collier; page 252 Figure 8.26: E. J. Tarbuck; page 253 Figure 8.29: Michael Collier; page 254 Figure 8.30: Dirk Wiersma/Science Source; page 255 Figure 8.33, 8.34: Dennis Tasa; page 257 Concepts in Review 8.3, 8.4: Dennis Tasa; page 258 Concepts in Review 8.6: Marli Miller; Give It Some Thought 8.2, 8.3: Dennis Tasa; page 259 Give It Some Thought 8.4: E. J. Tarbuck, 8.7: Dennis Tasa, 8.8: Callan Bentley; page 260 Eye on Earth 8.1: Dmitry Pichugin/Shutterstock, inset: E. J. Tarbuck, 8.2: E. J. Tarbuck.

Chapter 9 page 262 Chapter Opener: Jon Arnold Images Ltd/Alamy Stock Photo, inset: Xinhua/Alamy Stock Photo; page 264 Figure 9.1: Michael Collier; page 265 Figure 9.2: Dennis Tasa; page 266 Figure 9.3: Marco Simoni/Robert Harding World Imagery; Figure 9.4c: bcampbell65/Shutterstock; page 267 Figure 9.5: Morley Read/Alamy Stock Photo; Figure 9.6: Jonathan. S Kt; page 268 Figure 9.9: Marli Miller; page 269 Figure 9.12 (top, bottom): E. J. Tarbuck, (center): Marli Miller; page 271 GeoGraphics 9.1: NASA/Goddard Space Flight Center Scientific Visualization Studio; page 272 Figure 9.14: Reed Saxon/AP Images, inset: Martin Shields/Alamy Stock Photo; page 273 Figure 9.15a: Bernhard Edmaier/Science Source, b, d, f: E. J. Tarbuck, c: Florissant Fossil Beds National Monument, e: Colin Keates/Getty Images; page 274 Figure 9.16: E. J. Tarbuck; page 275 Figure 9.17: Biophoto Associates/Science Source; page 276 GeoGraphics 9.2 (top left): Gianni Tortoli/Science

Source, (top right, bottom): Richard T. Nowitz/Science Source; page 281 Figure 9.23: Javier Trueba/MSF/Science Source; page 285 Concepts in Review 9.2: E. J. Tarbuck; page 286 Give It Some Thought 9.1: Marli Miller, 9.3: Michael Collier; page 287 Give It Some Thought 9.4: Callan Bentley, 9.5: Francois Gohier/Science Source, 9.8: Marli Miller; page 288 Eye on Earth 9.1: Michael Collier, 9.2: Marli Miller.

Chapter 10 page 290 Chapter Opener: ArthurBullock/Getty Images, inset: USGS; page 292 Figure 10.1a: E. J. Tarbuck; page 294 Figure 10.5: Callan Bentley; page 296 Figure 10.7 (left, center): E. J. Tarbuck, (right): Michael Collier; Figure 10.8c: Javier Trueba/MSF/Science Source; page 298 Figure 10.12a: Michael Collier; page 299 Figure 10.13: E. J. Tarbuck; Figure 10.14a: Marli Miller; page 300 Figure 10.16: Michael Collier; page 301 Figure 10.19: Thomas Fikar/123RF; Figure 10.20b: David Parker/Science Source; page 302 Figure 10.22: Katherine Pedley; Figure 10.23: E. J. Tarbuck; page 303 GeoGraphics 10.1(top right): ZUMA Press Inc/Alamy Stock Photo, (left): USGS, (bottom right): Chad Ehlers/Alamy Stock Photo; page 304 Figure 10.24, 10.25: Michael Collier; page 307 Concepts in Review 10.1: Anthony Pleva/Alamy Stock Photo, 10.2, 10.3: E. J. Tarbuck; page 308 Give It Some Thought 10.3: USGS, 10.5: Marli Miller; page 309 Eye on Earth 10.1: Michael Collier, 10.2: NASA.

Chapter 11 page 310 Chapter Opener: adamkaz/Getty Images, inset: GeoPic/Alamy Stock Photo; page 313 Figure 11.2 (left): Library of Congress Prints and Photographs Division, (right): Hal Garb/Getty Images; Figure 11.3: G. K. Gilbert/USGS; page 316 Figure 11.8: Photo from the James E. Patterson Collection, courtesy of F. K. Lutgens; Figure 11.9: Zephyr/Science Source; page 323 Figure 11.21: Ronaldo Schemidt/AFP/Getty Images; page 324 Figure 11.22: USGS; Figure 11.23: Diarmuid/Alamy Stock Photo; page 325 Figure 11.24: Paul Sakuma/AP Images; page 326 Figure 11.25: USGS; page 327 Figure 11.27: Afp/Getty Images; Figure 11.28: Jiji Press/Afp/Getty Images; page 329 Figure 11.32: USGS; page 331 GeoGraphics 11.1 (top): USGS, (bottom left): AFP/Getty Images, (bottom right): Tom McHugh/Science Source; page 332 Figure 11.33b: USGS; page 333 Figure 11.35: Mel Melcon/Los Angeles Times/Getty Images; page 334 Figure 11.36: Peter Turnley/Corbis Vgc/Getty Images; Figure 11.37: Xiaolei Wu/Alamy Stock Photo; page 339 Eye on Earth 11.1: Michael Collier.

Chapter 12 page 340 Chapter Opener: Christian Handl/imageBROKER/Alamy Stock Photo, inset: Pictorial Press Ltd/Alamy Stock Photo; page 343 Figure 12.2: Siim Sepp/Alamy Stock Photo; page 344 GeoGraphics 12.1 (left): Lawrence Livermore National Laboratory, (right): Douglas L. Peck Photography; page 345 Figure 12.3: The Natural History Museum/Alamy Stock Photo.

Chapter 13 page 362 Chapter Opener: Fotosearch/ Getty Images, inset: NOAA; page 364 Figure 13.1: Library of Congress Prints and Photographs Division; page 371 Figure 13.10a: NOAA, b: Mark Thiessen/ AFP/Getty Images/Newscom; Figure 13.11: Image courtesy Woods Hole Oceanographic Institution, Charles Hollister, translator. Used with permission of the Woods Hole Oceanographic Institution; page 372 Figure 13.12 (left): ImageBROKER/FB-Fischer/ Alamy Stock Photo, (center): Matteo Colombo/The Image Bank/Getty Images, (right): David Hiser/ Getty Images; page 374 Figure 13.14: Rod Catanach/ Krt/Newscom; page 377 Figure 13.20: G.R.Roberts/ Science Source; page 378 GeoGraphics 13.1: NOAA; page 383 Figure 13.26: PaKos Photography/Alamy Stock Photo; page 385 Figure 13.29: SeaWiFS Project/ORBIGMAGE/Goddard Space Flight Center/NASA; page 388 Give It Some Thought 13.5: Ken C. Macdonald; 13.8: NOAA.

Chapter 14 page 390 Chapter Opener: Alice Cahill/ Getty Images, inset: Scott London/Alamy Stock Photo; page 393 Figure 14.2: Callan Bentley; page 397 Figure 14.6: Michael Collier; page 402 Figure 14.11: GSFC/JPL, MISR Science Team/NASA; page 404 Figure 14.14: Rob Marmion/Shutterstock; page 408 GeoGraphics 14.1: Michael Collier; page 412 Eye on Earth 14.1: Michael Collier.

Chapter 15 page 414 Chapter Opener: AP Images; page 417 Figure 15.1: Bryan Brazil/Shutterstock; page 418 Figure 15.2: Kevork Djansezian/AP Images; page 419 Figure 15.5: George Leavens/Science Source; page 420 Figure 15.6: Kevork Djansezian/ Stringer/Getty Images; page 421 Figure 15.7: Prakash Mathema/AFP/Getty Images; page 422 GeoGraphics 15.1: Chine Nouvelle/SIPA/Newscom; page 423 Figure 15.8a: Marli Miller, b: Europics/Newscom; page 424 Figure 15.9a: Michael Collier, b: NASA; page 425 Figure 15.11: robertharding/Alamy Stock Photo; page 427 Figure 15.13b: Michael Collier; Figure 15.14: Mike Eliason/ZUMA Press/Newscom; page 428 Figure 15.15: NASA; Figure 15.16: Crandell/USGS; page 429 Figure 15.17: E. J. Tarbuck; page 430 Figure 15.19 (left): D. Bradley/NOAA, (center): Marli Miller, (right): Tim McGuire Images, Inc./Alamy Stock photo; Figure 15.20: Photo from the James E. Patterson Collection, courtesy of F. K. Lutgens; page 431 Figure 15.22: Steve McCutcheon/ USGS; page 432 Figure 15.24a: NASA, b: William Schultz/USGS; page 434 Concepts in Review 15.4: Sajjad Hussain/AFP/Getty Images, 15.5: Michael Collier; page 435 Give It Some Thought 15.5: Michael Collier, 15.9: L.A. Yehle/O. J. Ferrians, Jr./ Denver/USGS; Eye on Earth 15.1: Soaring Tree Adventures.

Chapter 16 page 438 Chapter Opener: Janie Barrett/ ZUMA Press/Newscom, inset: comzeal/123RF; page 443 Figure 16.5: Michael Collier; page 445 Figure 16.8: Michael Collier; page 446 Figure 16.9: Michael Collier;

page 447 Figure 16.11a: Michael Collier; page 450 Figure 16.16: StormStudio/Alamy Stock Photo; page 451 Figure 16.18: Michael Collier; page 453 Figure 16.20 (top): Michael Collier, (bottom): P A Glancy/USGS; page 454 Figure 16.21a: Michael Collier; Figure 16.22: Colin Monteath/age Fotostock/ Alamy Stock Photo; page 455 Figure 16.23: Michael Collier; page 456 Figure 16.26a: Charles A. Blakeslee/ age Fotostock; page 457 Figure 16.27b: Michael Collier; page 458 Figure 16.29a: Michael Collier; Figure 16.30: Gregory S. Hancock; page 460 Figure 16.33: Michael Collier; page 462 Figure 16.35: Michael Collier; page 463 Figure 16.36: Scott Olson/Getty Images; Figure 16.37: Katherine Frey/The Washington Post/Getty Images; page 464 GeoGraphics 16.1 (top left): USGS, (top right): Michael Collier, (bottom): Sue Ogrocki/AP Images; page 466 Figure 16.39: Dale Kolke/California DWR via Planetpix/Alamy Stock Photo; page 468 Concepts in Review 16.6: Michael Collier; page 469 Give It Some Thought 16.4, 16.8: NASA; page 470 Give It Some Thought 16.10: Michael Collier; Eye on Earth 16.1, 16.2: Michael Collier; page 471 Eye on Earth 16.3: NASA.

Chapter 17 page 472 Chapter Opener: David Wall Photo/Getty Images, inset: Luis Santana Tampa Bay Times/AP Photo; page 475 Figure 17.3: Debra Ferguson/Design Pics Inc/Alamy; page 477 Figure 17.6: Missouri Department of Natural Resources; page 482 Figure 17.13: ASP/YPP/age Fotostock, inset: Mikael Damkier/Shutterstock; page 483 Figure 17.16: Photo from the James E. Patterson Collection, Courtesy of F. K. Lutgens; page 485 Figure 17.19: Michael Collier; page 486 Figure 17.21: Gary J Weathers/Getty Images; page 487 Figure 17.23: Danita Delimon/ Alamy Stock Photo; page 488 GeoGraphics 17.1 (top): Albert Cesare/The Odessa American/AP Images, (bottom left): EOS Earth Observing System/NASA, (bottom center): Jeff Roberson/AP Images, (bottom right): Daniel Acker/Bloomberg/Getty Images; page 489 Figure 17.25: U.S. Geological Survey Library; page 490 Figure 17.26: Todd Shipman/Courtesy of the Arizona Geological Survey; Figure 17.27b: Data SIO, NOAA, US Navy, NGA GECO; page 491 Figure 17.29: Mary Pope-Handy; page 493 Figure 17.32: Michael Collier; page 494 Figure 17.33a: Dante Fenolio/Science Source, b: Fritz Poelking/age Fotostock; page 495 Figure 17.34: Tiffany Nardico Photography/Alamy Stock Photo; page 496 Figure 17.36: Philippe Michel/age Fotostock; page 498 Concepts in Review 17.4: N.H. Darton/USGS, 17.5: New York Daily News/Getty Images; page 499 Concepts in Review 17.7: Miroslav Krob/age Fotostock; Give It Some Thought 17.1: Zack Frank/Shutterstock; page 500 Eye on Earth 17.1: NASA, 17.2: AP Images.

Chapter 18 page 502 Chapter Opener: Keith Ladzinski/National Geographic/Getty Images, inset: Feng Wei Photography/Getty Images; page 505 Figure 18.1: Brännhage Bo/age Fotostock; page 506 Figure 18.4b: NASA; page 507 Figure 18.5: NASA;

Figure 18.6: NASA Earth Observatory image by Jesse Allen, using Landsat data from the U.S. Geological Survey; page 509 Figure 18.8a: Wave/Glow Images, b: NASA; page 511 Figure 18.12a: Michael Collier, b: Durk Talsma/123RF; Figure 18.13: Radius Images/Alamy Stock Photo; page 512 Figure 18.14: NASA Earth Observatory image by Jesse Allen, using Landsat data from the U.S. Geological Survey; Figure 18.15: NOAA; page 513 Figure 18.16: Michael Collier; page 514 Figure 18.17: Bradley L. Grant/ Shutterstock; page 515 Figure 18.18 (clockwise from top left): Photo from the James E. Patterson Collection, courtesy of F. K. Lutgens, Andy Selinger/age Fotostock, Marli Miller, Danita Delimont/Alamy Stock Photo, page 516 Figure 18.19: Michael Collier; Figure 18.20: E. J. Tarbuck; page 517 Figure 18.21 (top): NASA, (bottom): Yoshio Tomii/SuperStock; page 518 Figure 18.22: Michael Collier, inset: E. J. Tarbuck; Figure 18.23: Kristin Piljay; page 519 Figure 18.24: Michael Collier; page 521 Figure 18.27 (clockwise from top left): Grambo Photography/ Alamy Stock Photo, Ward's Natural Science Establishment, Carlyn Iverson/Science Source, Michael Collier, Drake Fleege/Alamy Stock Photo; page 524 Figure 18.32 (left): NASA, (right): James Schwabel/ Alamy Stock Photo; page 525 Figure 18.34b: Greg Vaughn/VW PICS/age Fotostock; page 527 Figure 18.36: Science History Images/Alamy Stock Photo; page 529 Figure 18.40: Simon Fraser/Science Source; page 530 Concepts in Review 18.1: NASA; page 531 Concepts in Review 18.2: Robbie Shone/Science Source; page 533 Give It Some Thought 18.6: E. J. Tarbuck, 18.8: Kenneth Wiedermann/Getty Images; page 534 Eye on Earth 18.1: NASA, 18.2: Andrzej Gibasiewicz/Shutterstock.

Chapter 19 page 536 Chapter Opener: Andrew Pielage/ZUMA Press/Newscom, inset: Barcroft Media/Getty Images; page 539 Figure 19.2b: NASA; page 541 Figure 19.5: Heeb Christian/Alamy Stock Photo; page 542 Figure 19.6: E. J. Tarbuck, inset: Dorling Kindersley Limited; Figure 19.7: NASA; page 543 GeoGraphics 19.1 (top left): alantobey/ Getty Images, (top right): Universal Images Group/ SuperStock, (bottom left): Francois van der Merwe/ Shutterstock, (bottom right): Michael Collier; page 545 Figure 19.9: NASA; page 546 Figure 19.11: Mimi Ditchie Photography/Getty Images; Figure 19.12: NASA; page 547 Figure 19.13: Bernd Zoller/ Photolibrary/Getty Images; page 548 Figure 19.14 (left): Natural Resources Conservation Service/ USDA, (right): NASA; Figure 19.15: USDA/NRCS/ Natural Resources Conservation Service; page 549 Figure 19.16a: Dirk Wiersma/Science Source, b: Mike P Shepherd/Alamy Stock Photo; Figure 19.17a: Michael Collier, b: Maria Nelasova/Shutterstock; page 551 Figure 19.19: Michael Collier; Figure 19.20: Dennis Tasa; page 553 Figure 19.22: Photo from the James E. Patterson Collection, courtesy of F. K. Lutgens; page 554 Concepts in Review 19.2: Leon Werdinger/Alamy Stock Photo, 19.3: Michael

Collier; page 555 Concepts in Review 19.5: Michael Collier; page 556 Give It Some Thought 19.7: Ozoptimist/Shutterstock; Eye on Earth 19.1, 19.2: NASA.

Chapter 20 page 558 Chapter Opener: Michael Collier, inset: Gabriel Bouys/AFP/Getty Images; page 560 Figure 20.1: Terry Chea/AP Images; page 561 Figure 20.2: Mario Tama/Getty Images; page 564 Figure 20.6a: David R. Frazier/Photolibrary/Alamy Stock Photo, b: E. J. Tarbuck; Figure 20.7: Zacarias da Mata/Fotolia; page 565 Figure 20.8: Michael Collier; Figure 20.9a: Fletcher/Baylis/Science Source; page 566 Figure 20.10: Rich Reid/National Geographic/Getty Images; Figure 20.11b: Marli Miller; page 567 Figure 20.12: Alan P. Trujillo; page 568 Figure 20.13: Marli Miller; Figure 20.14: Mikehoward 1/Alamy Stock Photo; page 569 Figure 20.15a: ASCS/USDA, b: NASA; Figure 20.16: Michael Collier; page 570 Figure 20.17 (top, bottom): E. J. Tarbuck, (center): Michael Collier; page 572 Figure 20.18: NASA; Figure 20.19 (left): Don Smetzer/PhotoEdit Inc., (right): Virginian-Pilot, Drew C. Wilson/AP Images; Figure 20.20: Michael Collier; page 574 Figure 20.21a: NOAA, b: Andrea Booher/Alamy Stock Photo; page 575 Figure 20.23: NASA Earth Observatory images by Joshua Stevens and Jesse Allen, using VIIRS day-night band data from the Suomi National Polar-orbiting Partnership, MODIS data from the Land Atmosphere Near real-time Capability for EOS (LANCE), sea surface temperature data from Coral Reef Watch, and storm track information from Unisys; page 577 Figure 20.26: Smiley N. Pool/Newscom; page 578 Figure 20.27: Jeff Schmaltz/MODIS Land Rapid Response Team/NASA GSFC; Figure 20.28: CBP Photo/Alamy Stock Photo; page 579 Figure 20.29: AP Images; Figure 20.30: NASA; page 581 Figure 20.32: U.S. Army Corps of Engineers; page 582 Figure 20.34: University of California, Santa Barbara Library Special Collections; Figure 20.35: Rafael Macia/Science Source; page 583 Figure 20.36: Michael Weber/imagebroker/Alamy Stock Photo; page 584 Figure 20.37: Eric Carr/Alamy Stock Photo; page 587 Concepts in Review 20.2: Michael Collier; page 588 Concepts in Review 20.4: Michael Collier, 20.5: Julie Dermanksy/Getty Images; page 589 Give It Some Thought 20.1: Ron Dahlquist/Perspectives/Getty Images; page 590 Give It Some Thought 20.3: Michael Collier, 20.6: Edwin Remsberg/Getty Images, 20.8: Marli Miller; page 591 Eye on Earth 20.1: NASA.

Chapter 21 page 592 Chapter Opener: John E Marriott/Getty Images, inset: The Asahi Shimbun/Getty Images; page 596 Figure 21.3: Bill Stevenson/Stock Connection Blue/Alamy Stock Photo; page 597 Figure 21.4a: National Ice Core Laboratory; Figure 21.5: Ingo Wagner/Newscom; page 598 Figure 21.7a: Victor Zastol'skiy/Fotolia, b: Gregory K. Scott/Science Source; page 600 Figure 21.9: NASA; page 602 Figure 21.12: David R. Frazier Photolibrary, Inc./Alamy Stock Photo; page 603 Figure 21.13: Dennis

Tasa; page 604 Figure 21.16a: David Cole/Alamy Stock Photo, b, c: NASA; page 605 Figure 21.17: Richard Roscoe/Stocktrek Images, Inc./Alamy Stock Photo; page 606 Figure 21.18: NASA Earth Observatory images by Joshua Stevens, using OMPS data from the Goddard Earth Sciences Data and Information Services Center (GES DISC); page 607 Figure 21.20: NASA; page 609 Figure 21.23: Nigel Dickinson/Alamy Stock Photo; page 610 Figure 21.24: Robert Harding/Alamy Stock Photo; page 611 Figure 21.27: NASA; page 612 Figure 21.28: Radius Images/Alamy Stock Photo; page 616 Figure 21.34: National Snow and Ice Data Center; page 618 Concepts in Review 21.1: Photo Researchers, Inc./Science Source; page 620 Concepts in Review 21.7: Michael Collier, 21.8: Wolfgang Bechtold/ImageBroker/Alamy Stock Photo; page 621 Give It Some Thought 21.2: Photo Researchers, Inc./Science Source, 21.6: Alan Marsh/Getty Images, 21.7: Marc Zakian/Alamy Stock Photo; page 622 Give It Some Thought 21.9: Glow Images; Eye on Earth 21.1, 21.2: NASA.

Chapter 22 page 624 Chapter Opener: Barrett & MacKay/Getty Images, inset: Verisimilus; page 627 Figure 22.2: Esteban De Armas/Shutterstock; page 632 Figure 22.6: Tom Pfeiffer/VolcanoDiscovery/Getty Images; page 633 Figure 22.7: Image Source/Glow Images; page 634 Figure 22.8: USGS; page 635 Figure 22.9: James L. Amos/Corbis Documentary/Getty Images; page 641 Figure 22.18: Sean Xu/Alamy Stock Photo; page 644 Figure 22.20a: Sinclair Stammers/Science Source, b: William D. Bachman/Science Source; page 645 Figure 22.21: sarkao/Shutterstock; Figure 22.22: Ron Testa/Field Museum Library/Getty Images; page 646 Figure 22.24: Field Museum Library/Getty Images; page 648 Figure 22.27: John Weinstein/Field Museum Library/Premium Archive/Getty Images; page 649 Figure 22.29: Marc Jankow/Shutterstock; page 650 Figure 22.30: Bernd Siering/Premium/age Fotostock; Figure 22.31: Blickwinkel/Koenig/Alamy Stock Photo; Figure 22.32: Michael Collier; page 651 Figure 22.33: Photo Researchers, Inc./Science Source; page 652 Figure 22.35: National Parks; page 653 Figure 22.36a: WDG Photo/Shutterstock, b: Torleif Svensson/Corbis/Getty Images; Figure 22.37: Martin Harvey/Photolibrary/Getty Images; page 654 Figure 22.38: Photo Researchers, Inc./Science Source; Figure 22.39: Sisse Brimberg/National Geographic/Getty Images; Figure 22.40: Interfoto/Alamy Stock Photo; page 655 Concepts in Review 22.2: NASA; page 657 Give It Some Thought 22.2: Graeme Churchard (Bristol, UK); page 658 Eye on Earth 22.1: John W. Valley/NSF, 22.2: Michael C. Rygel.

Chapter 23 page 660 Chapter Opener: James MacDonald/Bloomberg/Getty Images, inset: Rajesh Kumar Singh/AP Images; page 662 Figure 23.1: Michael Collier; page 664 Figure 23.4: David R. Frazier Photolibrary, Inc./Alamy Stock Photo; page 666 Figure 23.7 (left): Peter Bowater/Science Source,

(right): Hulton Archive/Getty Images; page 667 Figure 23.8: Peter Essick/Aurora Photos/Alamy Stock Photo; Figure 23.9: Michael Collier; page 671 Figure 23.12 (top): Andrei Orlov/Shutterstock, (bottom): BanksPhotos/Getty Images; page 672 Figure 23.13: Jim West/Alamy Stock Photo; Figure 23.14: Axel Heimken/dpa/Alamy Stock Photo; page 673 Figure 23.16: Photo Researchers, Inc./Science Source; page 674 Figure 23.17: Simon Fraser/Science Source; page 675 Figure 23.18: Theodore Clutter/Science Source; page 676 Figure 23.19b: Topic Images Inc./Getty Images; page 678 Figure 23.21: E. J. Tarbuck; page 679 Figure 23.22: GS International/Greenshoots Communications/Alamy Stock Photo; Figure 23.23: E. J. Tarbuck; page 680 Figure 23.24 (left): tuja66/Alamy Stock Photo, (right): Lubos Chlubny/Alamy Stock Photo; page 681 Figure 23.25: E. J. Tarbuck; page 682 Figure 23.26 (left): Underwood Archives/Getty Images, (right): Library of Congress; page 684 Figure 23.27: Daniel Dempster Photography/Alamy Stock Photo; page 685 Figure 23.28: Michael Collier; page 686 Concepts in Review 23.2: Vespasian/Alamy Stock Photo; page 687 Concepts in Review 23.6: Harry Taylor/Getty Images; page 688 Give It Some Thought 23.3: Radharc Images/Alamy Stock Photo, 23.4: Andre Babiak/Alamy Stock Photo; Eye on Earth 23.1 (left): NHPA/Superstock, (right): Jim West/Alamy Stock Photo.

Chapter 24 page 690 Chapter Opener: Gene Cernan/NASA, inset: NASA/Goddard/Arizona State University; page 697 Figure 24.5: NASA; page 698 Figure 24.6: NASA/ASU/GSFC; page 699 Figure 24.7: NASA; page 700 Figure 24.8: NASA; Figure 24.9: NASA/Johns Hopkins University Applied Physics Laboratory/Carnegie Institution of Washington; page 701 Figures 24.10, 24.11: NASA; page 702 Figures 24.12, 24.13: JPL/NASA; page 703 Figure 24.14: NASA; Figure 24.15: JPL-Caltech/Univ. of Arizona/NASA; page 704 Figures 24.16, 24.17: NASA; page 705 GeoGraphics 24.1: NASA; page 707 Figure 24.18a: NASA; page 708 Figure 24.19: NASA; Figure 24.20a: NASA, b: University of Arizona/JPL/NASA; page 709 Figure 24.21: JPL-Caltech/SETI Institute/NASA; Figure 24.22: NASA; page 710 Figures 24.23, 24.24: NASA; page 711 Figure 24.25: Erich Karkoschka (Univ. Arizona)/NASA; Figure 24.26: NASA; page 712 Figure 24.27: NASA; page 713 Figure 24.29: Japan Aerospace Exploration Agency; page 714 Figure 24.30: Dan Schechter/Science Source; Figure 24.31: NASA; page 715 Figure 24.32: European Space Agency; page 716 Figure 24.34: M2 Photography/Alamy Stock Photo; page 717 Figure 24.35a: Michael Collier; page 718 Figure 24.36a: Johns Hopkins University Applied Physics Laboratory/Southwest Research Institute/NASA, b: NASA; page 720 Concepts in Review 24.5 (clockwise from top left): Muellek Josef/Shutterstock, National Science Foundation, NASA/JPL-Caltech/UCAL/MPS/DLR/IDA, Jerry Schad/Science Source; page 721 Eye on the Universe 24.1: NASA.

INDEX